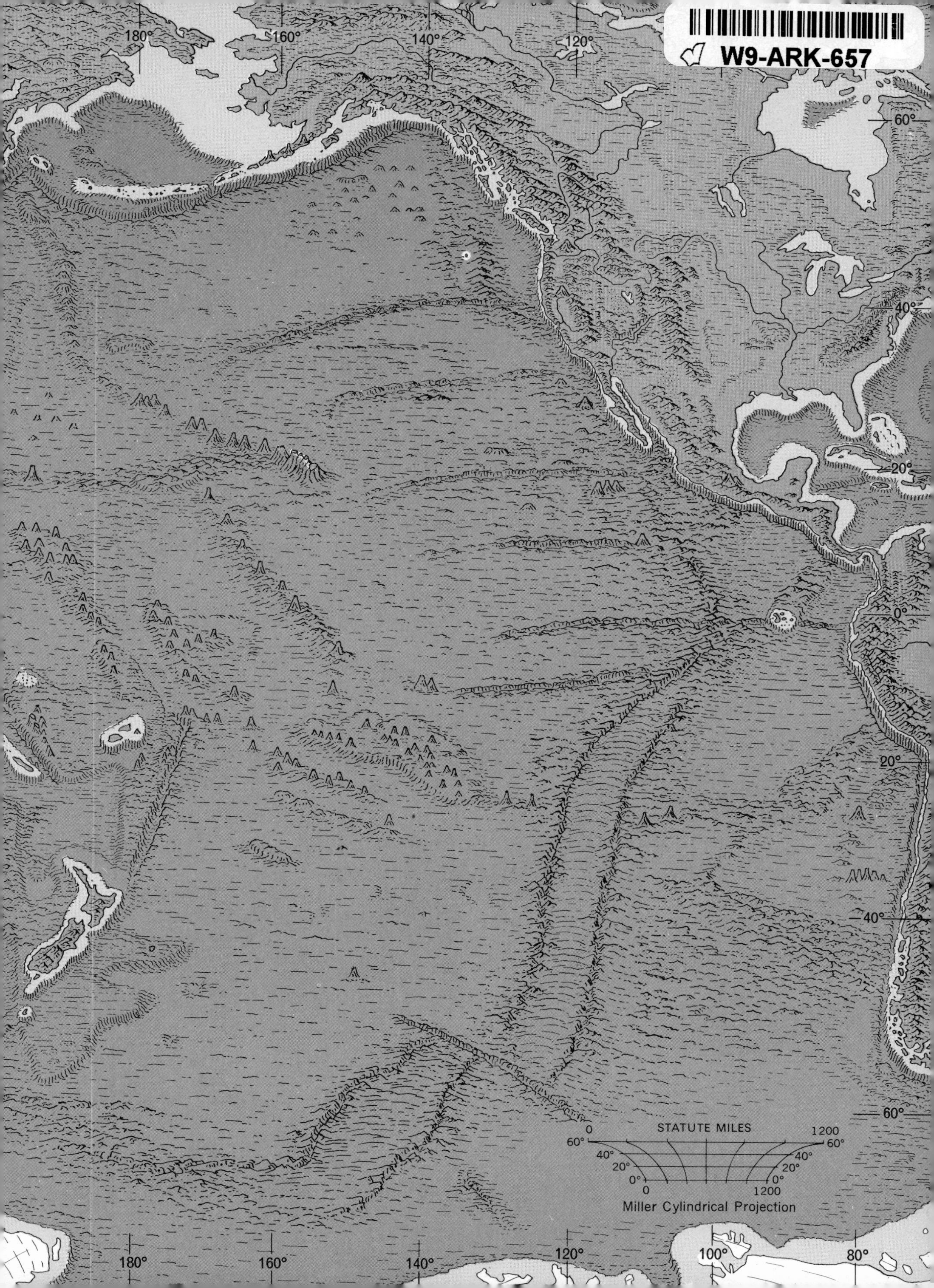

W9-ARK-657
180°
160°
140°
120°
60°
40°
20°
0°
20°
40°
60°
STATUTE MILES
0
1200
60°
40°
20°
0°
0
1200
Miller Cylindrical Projection
180°
160°
140°
120°
100°
80°

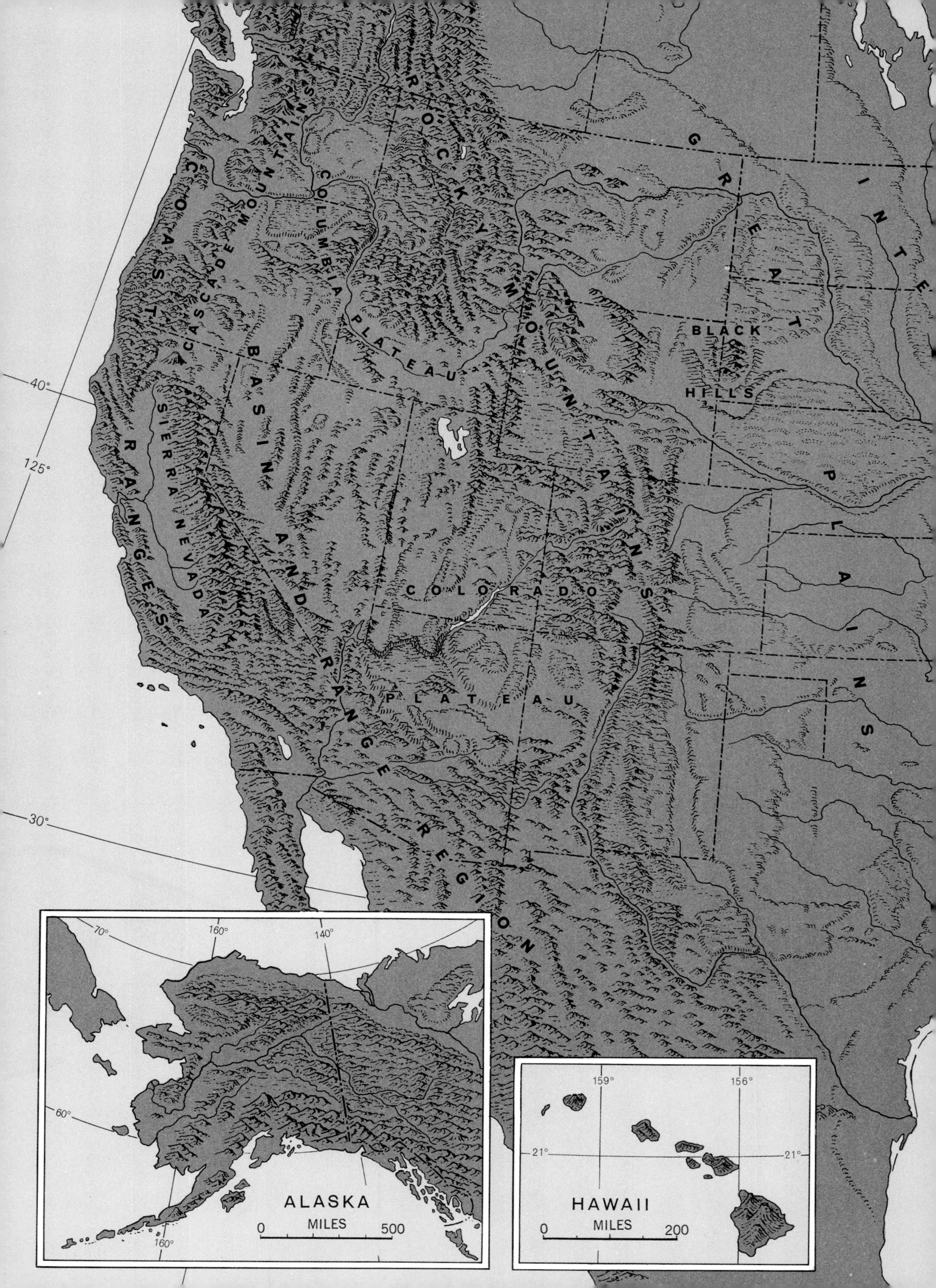
COAST RANGES
CASCADE MOUNTAINS
SIERRA NEVADA
COLUMBIA PLATEAU
ROCKY MOUNTAINS
BASIN AND RANGE REGION
COLORADO PLATEAU
GREAT PLAINS
BLACK HILLS
INTE
40°
125°
30°
ALASKA
MILES
0
500
70°
160°
140°
60°
160°
HAWAII
MILES
0
200
159°
156°
21°
21°

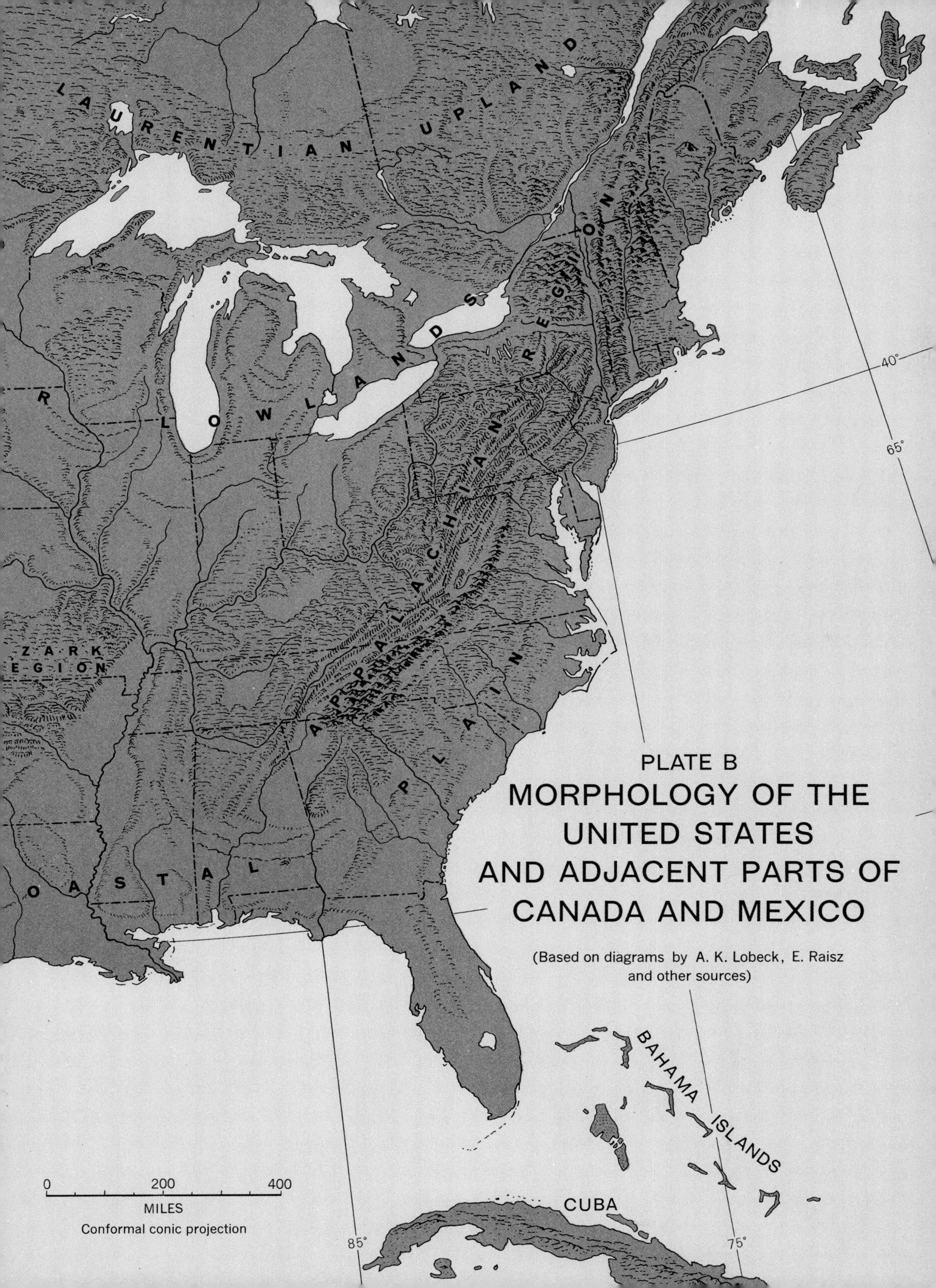

LAURENTIAN UPLAND

LOWLANDS

APPALACHIAN REGION

ZARK
EGION

OASTAL PLAIN

PLATE B

MORPHOLOGY OF THE UNITED STATES AND ADJACENT PARTS OF CANADA AND MEXICO

(Based on diagrams by A. K. Lobeck, E. Raisz and other sources)

BAHAMA ISLANDS

CUBA

40°

65°

85°

75°

0 200 400
MILES
Conformal conic projection

Physical Geology

Physical Geology

CHESTER R. LONGWELL

Emeritus Professor of Geology
Yale University

RICHARD FOSTER FLINT

Professor of Geology
Yale University

JOHN E. SANDERS

Visiting Professor of Geology
Barnard College, and
Senior Research Associate
Columbia University
Adjunct Professor of Geology
Rensselaer Polytechnic Institute

JOHN WILEY AND SONS, INC.

NEW YORK LONDON SYDNEY

10 9 8 7 6 5 4 3 2

Library of Congress Catalog Card Number: 68-22893
SBN 471 54475 2
Printed in the United States of America

Preface

This volume is not just a revision of the Longwell-Flint *Introduction to Physical Geology* (1962); most of it is quite new. Certain rewritten chapters follow the pattern of those in the earlier book, but all the rest have been written *de novo*, with the treatment of topics reorganized along a central corridor that represents the geologic cycle as first set forth by James Hutton. The concept of the geologic cycle is introduced in Chapter 1, is extended in Chapter 5, and is repeated in subsequent chapters in connection with each process (external or internal) to which it is relevant. The cycle concept is reinforced by repetition of a small version of the rock-cycle diagram on the title pages of most chapters.

In analyzing the various aspects of the geologic cycle, we emphasize again and again the concept of the open system in nature and of the steady-state condition. Also we repeatedly stress the influence of environment on the internal and external dynamic processes and the transformations that occur as the environment changes.

The general arrangement of topics represented by the sequence of chapters is broadly conventional, but cycles are discussed in a chapter by themselves. Chapters 1 to 6 constitute a first synthesis of the dynamics of the Earth with discussions of materials and geologic time, in order to enable the reader to grasp with more insight the specifics of the external processes in Chapters 7 to 15. Chapter 16 is concerned with strata, which are treated as a bridge between the external processes that create them and the internal processes (Chapters 17 to 21) that deform and transform them. This sequence leads to Chapter 22, a synthesis of crustal evolution. On a different trend, Chapter 23 discusses several of the principles that have emerged in foregoing chapters, in their applications to economic geology.

We introduce no physical or chemical relationship that is not subsequently made use of in a geologic context in one or several places. In several chapters we expand certain topics beyond their apparent ideal proportions, because we believe comprehension of these processes is basic to geologic understanding.

The discussion as a whole has been tightened by the interrelation of topics. Examples are the close relation of mass-wasting to weathering, of volcanic to plutonic features, of strata to sediments, and of structures to strata. The interrelations of topics are emphasized further with transition statements at the beginnings or ends of chapters.

Much of the mechanics of the Earth now appears to be related in one way or another to processes most apparent on the floors of the ocean basins, and this is leading to exciting new hypotheses. We therefore include hypotheses (clearly labeled as such) about which discussion is wide and current. We think this will help the reader to understand how geologic science grows, and will illustrate the scientific method by example. We carry this aim further by describing interesting hypotheses that have been abandoned, and by showing why they are no longer tenable.

The discussion of processes has been made somewhat more quantitative. We emphasize the speeds of processes and other measurements. Likewise we introduce additional, simple chemical and mathematical expressions. These, however, are explained in the text in such a way that instructors who wish to skip over them can do this with little or no loss of continuity.

We also analyze geologic processes more fully than in the earlier book, and relate them more closely to the physical or chemical concepts that underlie them. We especially emphasize the force of gravity, analyzing its components in Chapter 2 and thereafter referring to them when discussing each external process, generally with a diagram. We believe the text will be readily comprehensible by students with no more than a secondary-school background in physics, chemistry, and mathematics.

Wave motion, an expression of energy being transferred, is a topic vital to numerous aspects of geology. The kinds of waves reflect the source of the energy or the properties of the material through which the waves pass, or both. Electromagnetic waves and the concept of the wave spectrum begin in Chapter 2 and continue in Chapter 5 in their relation to solar energy. Chapter 14 elaborates the spectrum concept with surface waves in water, because these are responsible for many geologic phenomena. In Chapter 15 we explain the use of sound waves in exploring the sea floor, introducing the principles required for appreciation of waves traveling within the solid Earth. These principles lead to the analysis of earthquakes and the interpretation of the Earth's interior.

Apart from these various new approaches, we continue features of the Longwell-Flint *Introduction to Physical Geology* that have proved popular. Among these are the review summaries at the ends of chapters and the unique scheme of defining technical terms within the body of the text. A term defined in the text appears in ***boldface italic*** type, and the defining phrase or clause appears in *lightface italics*. The page on which the term is defined is indicated by boldface italic type in the Index. The definition thus occurs in context, in the presence of additional information, and perhaps also in one or more clarifying illustrations. Thus, in some instances, the reader can see how a definition is built and even why some definitions are difficult to frame. In addition, a glossary is included for those who may wish it.

The selected bibliographies at the ends of the chapters include both source material and suggested reading for students. In the captions accompanying illustrations, references to these bibliographies cite only the last name of the author and the date of publication. References to sources not listed in the bibliographies cite full names and dates to lead the reader to the *Bibliography of North American Geology* published by the U.S. Geological Survey. The few references to titles not listed in that publication are cited in full.

Most of the illustrations are new, and most of the line drawings employ color effectively. Diagrams retained from the older book have been redrawn.

As to units of measurement, we have compromised between the metric and

English systems. We abbreviate metric units (10cm, 10km) but spell out English units (50 feet, 15 miles). In general we express in metric units very small or very large dimensions, nearly all measurements of depths and distances at sea, and nearly all values involved in calculations. We use English units in general statements about familiar distances or rates; Figure 8-3 serves as a graphic conversion scale of rates. Maps and sections follow the standard usage in their country of origin. For example, a map of an area in the Rocky Mountains is scaled in English units, whereas a map of the Alps is in metric units. In some cases we have scaled a single measurement in units of both systems. For temperatures the Centigrade scale is used throughout the book.

Our sincere thanks go to colleagues and others who gave us willing help. Those who critically read parts of the text include Mrs. H. J. Ardrey, R. L. Armstrong, J. S. Creager, G. H. Crowl, R. F. Dill, D. B. Doan, Margaret Flint, G. M. Friedman, B. C. Heezen, Vernon Hughes, Earl Ingerson, Bates McKee, M. F. Meier, Barbara Sanders, E. C. Stoever, Jr., C. P. Walters, A. L. Washburn, Peter Weyl, and H. H. Woodard. Those who supplied unpublished information include H. W. Borns, E. J. Bowen, Brian Daily, W. O. Field, M. L. Jensen, R. W. Legget, J. A. Mabbutt, R. L. Nace, P. M. Orville, B. J. Skinner, and Kemble Widmer. Naresh Kumar assisted in checking references. Among coworkers at John Wiley and Sons, Inc., we thank D. H. Deneck for help of many kinds, A. W. Hepner for editorial help on parts of the text, and members of the Illustration Department for aid well beyond the call of duty in the preparation of line illustrations. We are grateful to the many persons who freely supplied photographs, each of which carries its own credit line.

C. R. L.

R. F. F.

J. E. S.

Contents

Part One: The Earth as a Whole

CHAPTER 1 Science of the Earth 1
2 General View of Earth 23
3 Matter and Minerals 45
4 Rocks and Regolith 73
5 Major Geologic Cycles 95
6 The Geologic Column and Geologic Time 119

Part Two: External Processes and Configurations

CHAPTER 7 Weathering and Soils 135
8 Mass-Wasting 159
9 Running Water 183
10 Sculpture of the Land by Running Water and Mass-Wasting 213
11 Ground Water 231
12 Glaciers and Glaciation 253
13 Deserts and Wind Action 285
14 Coasts and Continental Shelves 315
15 The Deep-Sea Floor 343
16 Sedimentary Strata 375

Part Three: Internal Processes and Configurations

CHAPTER 17 Deformation of Sedimentary Strata 395
18 Earthquakes: The Earth's Interior 425
19 Volcanoes 453
20 Plutonism and Metamorphism 485
21 Mountains 513

Part Four: Evolution of the Lithosphere

CHAPTER 22 Evolution of the Lithosphere 541

Part Five: Geology in Industry

CHAPTER 23 Geology in Industry 567

Appendices 593

A Some Physical and Chemical Concepts 595
B Minerals 603
C Rocks 613
D Maps, Cross-Sections, and Field Measurements 631
E Metric—English Conversion Tables 642

Glossary 644

Index 663

CHAPTER 1

Science of the Earth

Man and the Earth
Geology and geologic record
Earth's morphology, regolith, and bedrock
History derived from the geologic record
James Hutton's "great geological cycle"
Where does Hutton's cycle stand today?
Geology and the citizen

Plate 1

Surf at Point Firmin, California is eroding cliff composed of gently inclined layers of sedimentary rock that were deposited in an ancient ocean. (George Cox from Philip Gendreau.)

CHAPTER 1 *Science of the Earth*

MAN AND THE EARTH

Earth is responsible for man, but to what extent is man responsible to the Earth? Life-giving energy from the Sun, air, water, and soil have made peaceful, pleasant environments on Earth, in which plant growth is luxuriant and animals in great variety (among them man) have proliferated. Man, the creature with prehensile hands and large brain, has reacted to his surroundings in various ways. He has used natural materials for his own purposes. To satisfy his curiosity and assuage his fears of violent natural phenomena he has studied the Earth and has formulated various ideas about it. Violent activities of his environment have damaged him, and his activities have changed and damaged his environment.

Primitive men simply picked up pieces of stone and fashioned crude tools from them. By trying out various kinds, men probably learned that they could make better tools from some stones than from others. They even may have made a special note of the localities from which the better stones could be obtained. When the human population was small and human activity limited, man's consumption of natural materials and his effects on his environment were negligible.

Later, man began to use natural materials on a greater scale and to exert larger influences on his environment. With more elaborate tools, such as axes made from metals melted out of rocks, he cleared forests and in the clearings planted crops (Fig. 1-1). He learned to make dams and to channel water through long distances, to quarry and transport great blocks of stone for construction, and to build cities. Today civilization depends on the Earth and its materials. Where would we be without the soil, which produces food and other necessities; the land itself; water; coal, oil, and natural gas; and metals and other natural products? Toward the beginning of the nineteenth century mankind consciously realized that the study of the Earth is a valuable aid in locating and exploiting natural materials. Since that time, ever-growing requirements for natural resources have stimulated studies of the Earth on an ever-increasing scale.

But the natural environment, which spawned man, is not always peaceful and serene. Violent waves at the shore (Fig. 1-2), floods (Fig. 1-3), earthquakes (Fig. 1-4), avalanches, storms, and eruptions of volcanoes (Fig. 1-5) have killed millions of people. These acts of natural violence, the uncertainties of everyday weather, and the wide variety of climates on Earth have stimulated men to think about the Earth and to study it. Primitive peoples were terrified and frightened by natural phenomena. They attributed violent events to evil spirits and tried to placate these spirits and assuage their own fears of them by making up myths and enacting elaborate rituals. More sophisticated men have been curious about their environment rather than just frightened by it. They wanted to know how everything works. What makes volcanoes, and why are most of them aligned in narrow belts? How do mountains originate, and why are no mountains present in some vast regions? Why are some hills steep, even jagged, whereas others are smooth and gently rounded? The questions *what? how? why?* arose the moment any landscape came before their eyes. Patient study gave answers to some of these questions; a few brilliant men even realized the astonishing fact that the history of the Earth is written in the rocks. Some men set about learning the language of the rocks in order to read this history. They did so purely as an intellectual pursuit to satisfy their burning desire to know.

In our present stage of industrial development, we are not only voracious consumers of natural resources but also great changers of the natural scene. Without second thoughts we have altered the face of the Earth to accommodate a burgeoning

Fig. 1-1. Flat valley floor cultivated by man contrasts with rolling desert in the distance. Nearly flat-lying sedimentary strata are exposed in side of valley. Indian farms near Tuba City, Arizona. (Ansel Adams from Magnum.)

Fig. 1-2. Two great waves wash over Pier 1, Kuhio Wharf, Hilo, Hawaii, April 1, 1946. In *A*, retouched, arrow points to a man swept to his death seconds later. In *B* a second wave, which arrived a little more than 10 minutes after the first, approaches moored ship *Brigham Victory*, whose deck, ventilators, and railing show in foreground. (Wide World Photos.)

population. We have built great cities in places vulnerable to natural disasters. We have realized that we must study our environment not only because we need improved warning systems to ease its destructive effects on us but also because we can no longer ignore our destructive influences on it. We have scarred the countryside with careless mining (Fig. 1-6) and have damaged the soil with bad farming. We have polluted rivers with industrial wastes, sewage, and mining refuse and have contaminated

Fig. 1-3. Missouri River in flood, April, 1952. (W. J. Forsythe, U. S. Dept. Agriculture.)

the atmosphere with the products of combustion and other chemical reactions. Study of the Earth is no longer simply a matter of finding more natural resources or of satisfying our curiosity, important as these are; it looms as a necessity for survival. Our rising standard of living and an exploding population are creating environmental problems of the utmost urgency. If we expect to carry on, we must change our previous practices and monitor our future activities. If we continue to act irresponsibly toward the Earth, our environment (which gave us birth and permitted us to grow) could be responsible for our extinction. Systematic study of the Earth has resulted in geology and other natural sciences, which have expanded and must continue to expand in the future.

GEOLOGY AND THE GEOLOGIC RECORD

Geology (Gr. *Ge,* "earth," and *logos,* "discourse"), *the science of the Earth,* is the systematic study of the materials, processes, environments, and history of our planet. The purpose of this book is to explain the Earth as a dynamic body by use of examples and problems drawn from the observation of the Earth.

Geologic lessons and illustrations exist at every turn. They can be derived directly from visual study of three features, (1) the varied form of the Earth's surface, (2) loose surficial material, and (3) firm bedrock (Fig. 1-7). These three features, wherever seen, compose the *geologic record,* so named because a study of it makes possible the reconstruction of Earth history. The central principles of geology can be derived from the record with an inquiring mind and a discerning eye; no tricks or gadgets are needed. Therefore a substantial appreciation of geology is available to anyone who will look and will reason about what he sees. This fundamental simplicity is responsible for the basic appeal of geology, a science built on the systematic study of common natural features. Nature is a willing partner; she does not hide her secrets from those who look and think.

EARTH'S MORPHOLOGY, REGOLITH, AND BEDROCK

Morphology. The ***Earth's morphology*** is *the shape of its surface.* This surface is a dynamic product; the only constant is continual change. Many things happen on the Earth's surface as it adjusts itself to the internal and external forces that operate on it. Internal forces elevate some parts of the surface and cause other parts to subside. External forces, resulting from the activities of air, water, and ice powered by gravity, concentrate on high places and strive to destroy them (Fig. 1-8). Mechanically and chemi-

Fig. 1-4. Houses damaged in landslide triggered by earthquake of March 27, 1964 at Turnagain-by-the-Sea, Anchorage, Alaska. (Steve McCutcheon, Alaska Pictorial Service.)

cally, air and water attack solid rock and break it down into loose particles. These move downslope, responding to the pull of gravity, and are delivered to streams. While carrying the particles away, streams dissect the land into vast networks of gullies, valleys, and canyons (Pl. 10).

Water, seeping underground, dissolves rock material, carries it away in solution, and eventually delivers it to the sea. When water in the ground freezes, the ice can wrench apart huge masses of rock. Thick masses of ice that flow over the Earth's surface as glaciers gouge and polish the solid rock. Glaciers quarry out great chunks, grind up the pieces, and spread them out as distinctive carpets of deposits, heaping some into ridges. The wind blows loose sand grains across the surface. It lifts dust particles high into the atmosphere and transports them through long distances. Blown sand abrades solid rock, but generally it passes the rocky areas quickly and accumulates in dunes. At the seashore, waves attack and wear away coastal cliffs.

These varied activities contribute to the process of ***erosion***, a general term that describes *the physical breaking down, chemical solution, and movement of broken-down and dissolved rock materials from place to place on the Earth's surface.* The agents of erosion leave their imprints on the loose material they transport and on the surfaces over which they move.

Subsiding parts of the Earth's surface act as catchment basins for the products of erosion removed from high places. Eventually most products of erosion reach the sea. The surface of the sea is a fundamental physical boundary; areas above it tend to be eroded, whereas areas below it accumulate products of erosion. Because it sets a lower limit to the leveling effects of the gravitational processes active on the land, sea level is a *base level* for streams.

Regolith. The loose surficial materials, *the non-cemented rock fragments, and mineral grains derived from rocks, which overlie the bedrock in most places,*

Fig. 1-5. Devastation created by volcanic activities. *A.* Rock formed from molten material erupted by Parícutin Volcano (*right*) has completely covered the village of San Juan Parangaricutiro, Mexico, except for towers of the church. (Tad Nichols.)

***B.* In 1902, all the buildings of St. Pierre, Martinique, and the vegetation of the surrounding hills, were reduced to utter ruin by a fast-moving sheet of hot, solid particles and a glowing cloud of hot gases erupted from Mont Pelée Volcano. (Underwood and Underwood.)**

are known as ***regolith*** (rĕǵ-ō-lĭth; Gr. *regos,* "blanket," and *lithos,* "rock"). Bedrock and regolith are closely connected. Regolith originates in the destruction of bedrock, yet by cementation regolith can be converted back into bedrock, although possibly into rock of a kind unlike that from which the regolith was derived. Regolith, whether on land or on the sea floor, is of two kinds, residual and transported. *Regolith that has been transported* is ***sediment*** (Lat. *sedimentum,* "a settling"). The term was first used in the days when transported rock materials were supposed to have been deposited from suspension in the fluids, water and air, the most obvious agents of transport. Although much sediment does settle out of water and air, much is transported and deposited by other mechanisms. Regolith can slide, creep, or flow down slopes without ever being in a state of suspension; a pebble can be pushed along a stream bed without being lifted into suspension. Accordingly the definition of sediment has been broadened to take account of particles moved in *any* way. It may be difficult to distinguish between regolith formed essentially in place and sediment that has been moved only a short distance down a slope. The problem arises because gradational series are common in nature.

Bedrock. *Continuous solid rock, **bedrock**,* underlies regolith everywhere. In places where no regolith is present, exposed bedrock forms the surface itself (Fig. 1-9). Bedrock consists of many varieties, most of which can be fitted into three rock groups: *igneous, sedimentary,* and *metamorphic.* The groups are based on the two major environmental realms where rocks originate and on the many rock-making processes that operate within these realms. The two realms are: (1) at or near the Earth's surface, and (2) within the Earth. The three most common rock-making processes can be compared with the making of three familiar building materials: steel, concrete, and firebrick.

Most igneous rocks originate by only one process, the solidification by cooling of materials that became molten in the interior realm. Although molten material ancestral to igneous rocks comes from the interior, it can solidify either at or beneath the Earth's surface. We can see igneous rocks being created at the surface from molten material discharged by some volcanoes. Commonly, liquid and

Fig. 1-6. Ravaged landscape in a combination gravel pit and coal strip mine near Du Quoin, Illinois. April 30, 1963. (A. P. Wirephoto from Wide World Photos.)

solid material from volcanoes build characteristic conical hills or mountains. *Basalt,* a black, heavy fine-grained igneous rock with particles so small that a magnifying lens is required to see them, is the most common igneous rock formed by the cooling of molten material discharged from volcanoes. *Granite,*

Fig. 1-7. Three components of geologic record, Earth's surface, bedrock, and regolith, seen in a borrow pit, Sixmile Canyon, Utah. Nearly horizontal sedimentary strata (*left*) are covered by thin regolith. Strata terminate (*center*) at steep side of former valley. Thick regolith at right lacks strata. (J. E. Sanders.)

Fig. 1-8. In Monument Valley, Arizona, isolated remnants are all that remain of a thick sandstone layer, formerly continuous across the view. Erosion has removed all the rest. Compare Fig. 1-11, *B*. (Frank Jensen).

an igneous rock lighter colored, less dense, and coarser grained than basalt, does not resemble any material discharged from volcanoes and, as we shall see later, indicates by the large size of its grains that it solidified slowly beneath the surface. The particles of a coarse-grained igneous rock are angular and interlock (Fig. 1-10, *A*).

Sedimentary rocks originate by many processes, but they all operate in the surface or near-surface realm. All sedimentary rocks share one common feature: the material composing them has come from the destruction, physical and chemical, of older rocks. The most typical process in their origin is cementation of surficial sediments. We are able to watch most surficial sediments being accumulated but generally are unable to see them being cemented. Cementation of sediments makes sedimentary rocks that are analogous to concrete; the particles, many of which have become rounded (Fig. 1-10, *B*), correspond to the aggregate. Particle size varies from *gravel* through *sand* to *silt* and *clay* (Fig. 4-11). When converted into sedimentary rock the particles form *conglomerates, sandstones, siltstones,* and *claystones.* Other kinds of sedimentary rocks do not originate by cementation of sediments. *Limestones* contain calcium carbonate secreted by organisms or precipitated from the supersaturated waters of seas, lakes, and springs. *Rock salt* and related sedimentary rocks originate by direct chemical precipitation from saline waters.

A characteristic feature of sediments (Fig. 1-11, *A*), hence of sedimentary rocks, is their arrangement in *definite layers* called ***strata*** (sing., *stratum;* Lat. *sternere,* "to spread"). Strata are commonly called *beds,* and sedimentary rocks, *stratified* or *bedded rocks* (Fig. 1-11, *B*). Because molten rock material from volcanoes spreads out on the Earth's surface as layers, the term strata has been extended to include them, albeit the rocks are igneous. Thickness of strata varies from fractions of a centimeter to many meters.

Metamorphic rocks are created in only one realm, the interior. They come into being through various processes that change preexisting rocks belonging to any of the three groups, with or without the addition of new material. A typical process of origin of metamorphic rocks is analogous to the firing of clay to make bricks; the preexisting material (clay) is heated sufficiently to recrystallize the constituents but not intensely enough to melt them. Other metamorphic rocks originate by mechanical squeezing resulting from static pressure or differential pressure; this aligns the particles (Fig. 1-10, *C*) and may stretch or flatten them. Still others are created by various combinations of these processes. Whatever the process, the final result is to transform preexisting rocks into new kinds of rocks. When the temperature becomes high enough, metamorphic rocks melt and ultimately become igneous rocks.

In summary, most igneous rocks result from one process that can act in both rock-making realms: the interior and the surface. Sedimentary rocks result from many processes, but all acting at or near the surface. Metamorphic rocks result from the transformation of preexisting rocks by processes that operate only in the interior.

Fig. 1-9. Bedrock exposed at the Earth's surface. Prescott County, Ontario. (Geol. Survey of Canada.)

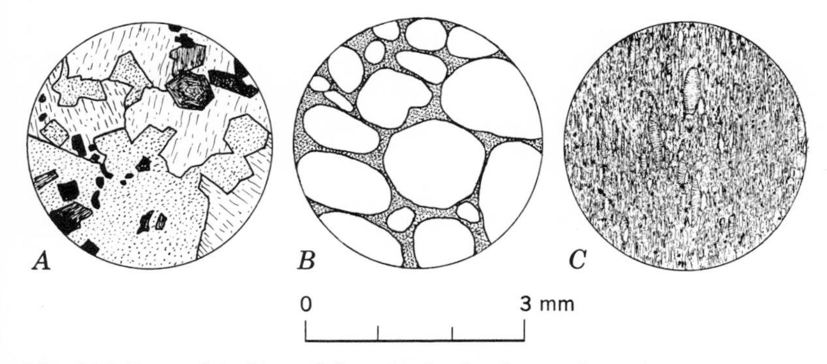

Fig. 1-10. Paper-thin slices of three kinds of rock seen through a microscope. (*A*) In igneous rock particles are angular and interlock. (*B*) Particles (white) of sedimentary rock have been rounded and the spaces between particles filled with cement. (*C*) In metamorphic rock particles are elongate and their long dimensions are parallel.

Fig. 1-11. Original horizontal position of strata in sediments and sedimentary rocks. *A.* Sediments exposed at time of low water in a reservoir near Pound Ridge, New York. Flat-lying strata are exposed in section on far shore; shrinkage cracks and faint animal footprints on horizontal surface in foreground. (Nancy Palmer.) *B.* Nearly horizontal sedimentary strata underlying Colorado Plateau exposed in walls of Colorado River. East end of Grand Canyon, near mouth of Little Colorado River. (Spence Air Photos.)

HISTORY DERIVED FROM THE GEOLOGIC RECORD

By taking a geologic excursion we can gather the evidence we need to demonstrate the use of the three components of the geologic record for reconstructing geologic history. Let us visit part of the Colorado Plateau and Grand Canyon in northern Arizona (Fig. 1-12, *A*), where regolith is scarce and where the features are so large that we can see them easily, even from an airplane.

On the north side of the Grand Canyon stands the Uinkaret Plateau. Near its southern end, its otherwise flat top gives way to the Pine Mountains, a range of conical hills the largest of which is Mount Trumbull. The plateau is bounded on its west, south, and east sides by an escarpment almost 1,500 feet high. Black rock caps the southern end of the Uinkaret Plateau. It forms the conical hills and projects as distinct tongues from the top of the Plateau to the floor of Whitmore Canyon on the west and to the floor of Tuweep Valley on the east. Another great tongue of black rock extends southeastward from Mount Emma and completely covers the bench lying north of the Grand Canyon and stretching for about 8 miles between Whitmore Canyon and Tuweep Valley (Fig. 1-12, *A*). The black rock fills recessed areas in the edge of this bench, drapes over the edge, and extends into the Grand Canyon. At the south end of Tuweep Valley, along the north edge of the Grand Canyon lies Vulcan's Throne, a small double conical hill. A tongue of black rock extends into the Grand Canyon from near Vulcan's Throne (Fig. 1-12, *B*). Locally in Whitmore Canyon and Tuweep Valley surficial sediments cover the edges of the tongues of black rock. A closer look indicates that the black rock is basalt, a common volcanic material. With only basalt and conical shapes for clues we can feel reasonably confident that the hills are cones built by volcanoes now inactive. We infer that the basalt solidified from formerly molten fluid discharged from the cones. Horizontal strata of sedimentary rocks visible in the escarpment between tongues of basalt extend laterally underneath the volcanic cones.

The four essential parts of the local scene are: (1) plateau north of Grand Canyon underlain by nearly horizontal sedimentary strata; (2) Whitmore Canyon, Tuweep Valley, and Grand Canyon; (3) volcanic basalt and conical hills; and (4) valley-floor sediments. From the relationships among them we reconstruct the following sequence. (a) The strata underlying the plateau were deposited as sediments. (b) These sediments were converted into rock, uplifted and eroded. During the vast erosion Whitmore Canyon, Tuweep Valley, and the Grand Canyon were excavated and the escarpments and benches were carved. (c) Volcanoes became active on the surface of the Uinkaret Plateau. They discharged molten material that formed conical hills, covered the southern end of the plateau, flowed down its east and west sides into Whitmore Canyon, Tuweep Valley, and the Grand Canyon, spread out on the valley floors, and cooled to form basalt. (d) Surficial sediments spread across parts of the valley floors and covered the margins of the basalt locally. Other events took place in the area, but these four illustrate the geologic method of studying the Earth's present features to reconstruct its past history.

Think a moment about the sedimentary strata. Where was the source of all that sediment? In what direction was it transported to the place where it is seen now? What became of all the rock material excavated from Whitmore Canyon, Tuweep Valley, and the Grand Canyon? By what process was this material removed? In 1882 Clarence Dutton of the U. S. Geological Survey asked himself these same questions; his answer is worth remembering:

"Erosion viewed in one way is the supplement of the process by which strata are accumulated. The materials which constitute the stratified rocks were derived from the degradation of the land. This proposition is fundamental in geology—nay, it is the broadest and most comprehensive proposition with which that science deals. It is to geology what the law of gravitation is to astronomy. We can conceive of no other origin for the materials of the strata, and no other is needed, for this one is sufficient and its verity a thousand times proven. Erosion and 'sedimentation' are the two half phases of one cycle of causation—the debit and credit sides of one system of transactions."[1]

Dutton's "one cycle of causation" logically answers these questions about the Colorado Plateau. But Dutton was not the first person to conceive of such a "cycle." A Scot, James Hutton, had formulated the cycle concept nearly a century earlier. Hutton studied at the Universities of Edinburgh and Paris, took a medical degree at Leiden, and then returned to Scotland. In Edinburgh, he became an active member of a group devoted to "natural philosophy," the term used in the eighteenth century for natural history.

Dutton called it "one cycle of causation"; Hutton invented a more dramatic phrase, the "great geological cycle." By any name the idea embodies so much of geology that it deserves further analysis.

[1] Dutton, C. E., 1882, The Tertiary history of the Grand Canyon district: U. S. Geol. Survey, Mon. 2, p. 61–62.

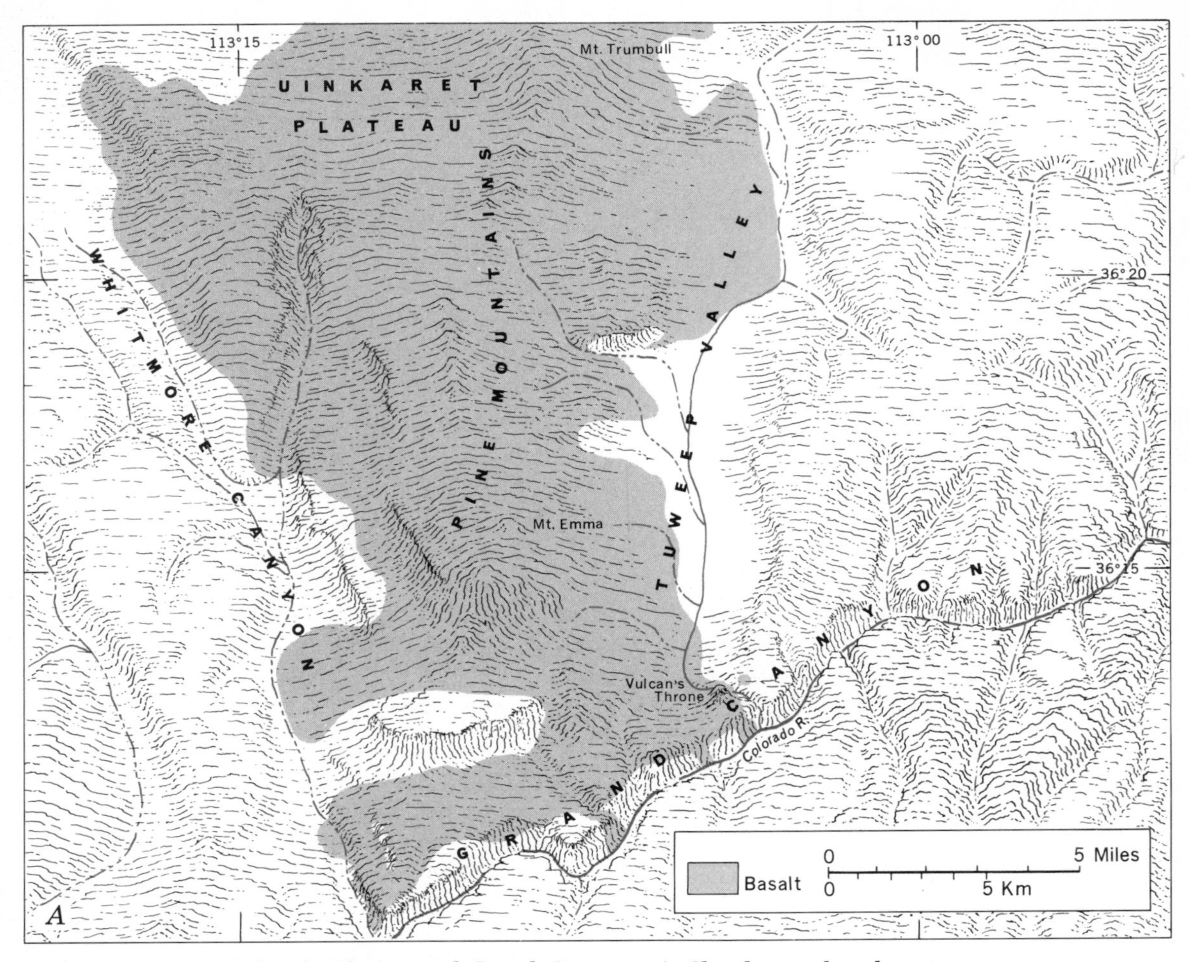

Fig. 1-12. Parts of Colorado Plateau and Grand Canyon. A. Sketch map based on topographic map by U. S. Geol. Survey.

JAMES HUTTON'S "GREAT GEOLOGICAL CYCLE"

Hutton's remarkable insights into geology are found in his two-volume book, *Theory of the Earth* (1795). This book contains ideas presented originally as two lectures to the Royal Society of Edinburgh on March 7 and April 4, 1785, under the title, "Theory of the Earth, or an Investigation of the Laws observable in the Composition, Dissolution and Restoration of Land upon the Globe." Hutton's genius remained obscure, however, until his theories were rewritten and extended by his friend and pupil, Professor John Playfair, whose book, *Illustrations of the Huttonian Theory* (1802), is available now in a reprinted edition. In the century and a half that has elapsed since first publication it has become a classic.

In Hutton's time all theories of the Earth agreed on the following points: (1) Solid bedrock exists everywhere beneath loose "soils" or "dirt," now called regolith. (2) Much bedrock is stratified and consists of layers in parallel series, whose thickness persists through long distances. (3) Strata are horizontal in many localities, but in others they are inclined at various angles. (4) The particles that compose strata were deposited under water as loose sediments and were cemented afterwards to form stone. But Hutton alone perceived the staggering lesson of the "system of universal decay and degradation" which he saw in action all around him. He realized that even the most ancient layers consist

***B.* Oblique air view northward showing Grand Canyon, Vulcan's Throne, tongue of basalt extending into Grand Canyon, and horizontal sedimentary strata underlying plateau in distance. (W. K. Hamblin.)**

of "materials furnished from the ruins of former continents." Everywhere at the Earth's surface Hutton saw evidence of rock decay. He speculated further that below the surface, at the bottom of the sea, loose materials are converted back into stone. This great partnership between wasting away and renovation of the land is Hutton's "great geological cycle." Its driving force is the circulation of moisture through the atmosphere. Hutton knew that this cycle had been completed innumerable times. As Playfair put it:

"How often these vicissitudes of decay and renovation have been repeated, is not for us to determine; they constitute a series, of which, as the author of this theory has remarked, we neither see the beginning nor the end . . ."[2]

Hutton fully understood that areas uplifted during one episode can be depressed later and receive sediments, still later can be reuplifted, and so on, and on, and on, and on. This back-and-forth relationship compares with the repeated tipping of an hourglass. Like the contents of an hourglass, the geologic "sands of time" record events of Earth history (Fig. 1-13).

[2]Playfair, John, 1802, Illustrations of the Huttonian theory of the Earth, repr. 1956; Urbana, Univ. of Illinois Press, par. 118, p. 119.

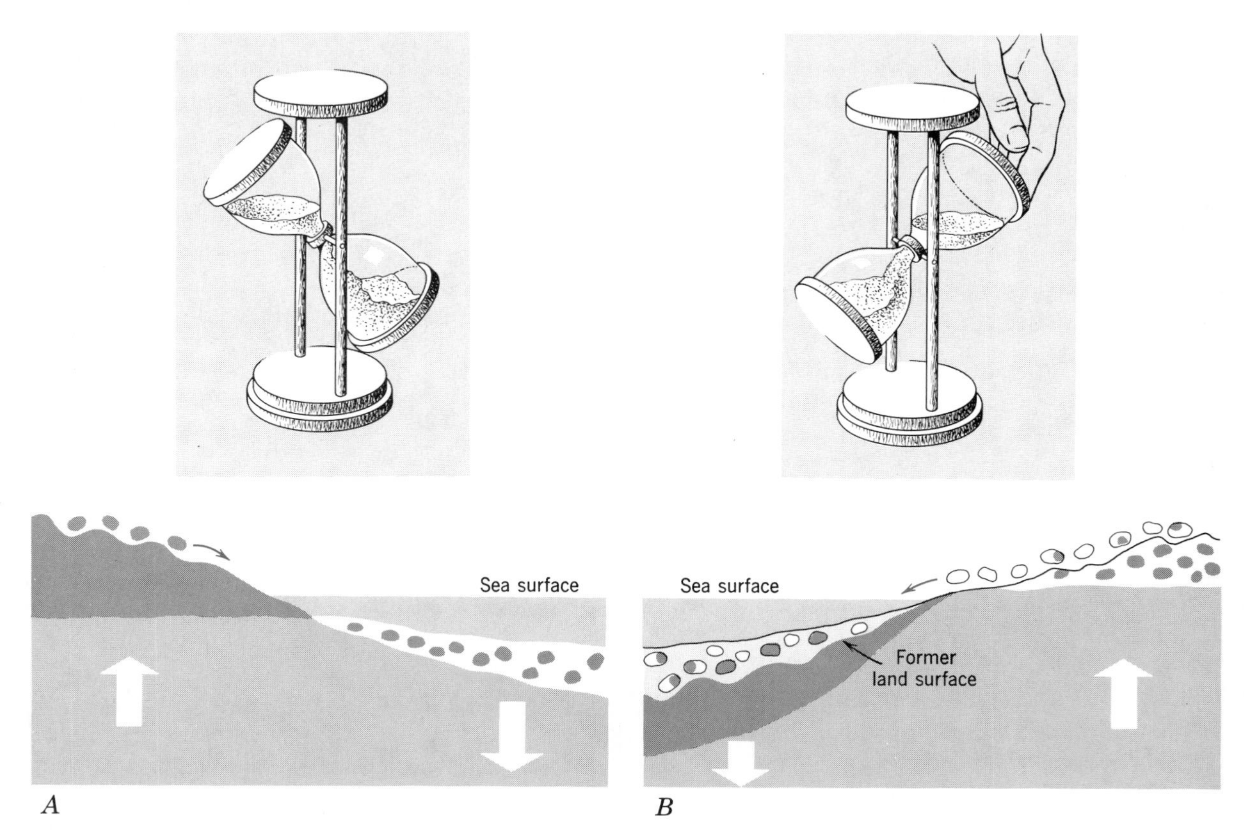

Fig. 1-13. Uplift and subsidence of the Earth's crust transposes sites of erosion and deposition, shifting sediment back and forth like sand in an hourglass that is turned up and down.
A. **Schematic section shows sediment eroded from uplifted area (arrow pointing up), transported to area of subsidence (arrow pointing down), and deposited on sea floor.** ***B.*** **Directions of uplift, subsidence, and transport are reversed: former sea floor is exposed (*right*) and new sea-floor sediments accumulate over former land surface (*left*).**

Hutton's tremendous concept was made use of by Sir Charles Lyell, an Englishman who started out to be a lawyer before devoting his life to geology. Lyell showed how it could be used to construct a logical basis for interpreting Earth history. His textbook, *Principles of Geology,* expressed the philosophic impact of Hutton's concepts in the following words:

"... The declaration was the more startling when coupled with the doctrine, that all past changes on the globe had been brought about by the slow agency of existing causes. The imagination was first fatigued and overpowered by endeavouring to conceive the immensity of time required for the annihilation of whole continents by so insensible a process. Yet when the thoughts had wandered through these interminable periods, no resting place was assigned in the remotest distance. The oldest rocks were represented to be of a derivative nature, the last of an antecedent series, and that perhaps one of the many pre-existing worlds. Such views of the immensity of past time, like those unfolded by the Newtonian philosophy in regard to space, were too vast to awaken ideas of sublimity unmixed with a painful sense of our incapacity to conceive a plan of such infinite extent."[3]

WHERE DOES HUTTON'S CYCLE STAND TODAY?

Modern scientific discoveries, however fascinating, have not altered the fundamental Hutton-Lyell proposition that the bedrock contains numerous an-

[3] Lyell, Sir Charles, 1830, Principles of geology, v. 1: London, John Murray, p. 63.

cient former surfaces of the Earth buried and preserved by the deposition on them of surficial sediments. These sediments later were "turned to stone," raised up as bedrock, and eroded, thus providing new sediments to cover other parts of the Earth's surface. In turn, these sediments were cemented into bedrock, uplifted, and eroded, and so it has gone, up and down, back and forth, apparently without end. The most ancient sedimentary rocks imply even more ancient erosion.

Although they have quantified many of its aspects, later scientists have not changed the concept of Hutton's "great geological cycle" significantly. We now call it simply the geologic cycle and shall examine it and its significance in detail in Chapter 5. Geology has advanced by applying modern concepts of matter and energy to Earth materials and processes, and has benefited from sophisticated tools for analyzing materials on which the cycle operates. Of particular importance to geology was the invention, in the middle of the nineteenth century, of the polarizing or petrographic microscope and of the technique for making paper-thin slices of rocks. These made possible the minute study of composition and texture of rocks, the basic materials of the geologic cycle. Geologic interpretations of the history of life were revised completely as a result of Darwin's theory of the origin of species. The discovery of radioactivity (Chap. 6) revolutionized geology by enabling scientists to measure geologic time. Radioactivity emits great quantities of heat, and this fact gave geologists a basis for new ideas on the thermal history of the Earth. The discovery of X-rays ushered in a new era in the study of crystals; these rays showed the diagnostic internal arrangement of particles in crystals and revealed the identity of very small particles whose composition formerly was unknown. Laboratory experiments at high temperatures and pressures can reproduce conditions and rock materials found within the outer parts of the Earth and thus give us a basis for understanding the environment in which rocks originate. Optical spectrometers, mass spectrometers (Fig. 1-14), and similar analytical tools enable chemists to determine the composition of Earth materials in their natural solid state even on the level of parts per billion, although some of these materials are so complex that in the days when chemical techniques dealt principally with material dissolved in liquid solutions chemists paid no attention to them. Precise recording of waves propagated through the ground and computer analysis of wave records give us a glimpse of the Earth's interior. Attempts to monitor underground nuclear explosions by recording waves propagated from them have stimulated many new developments. Computers have made possible calculations, storage of data, and rapid plotting of graphs and maps, all tasks that were too complex or too time-consuming in the days before computers.

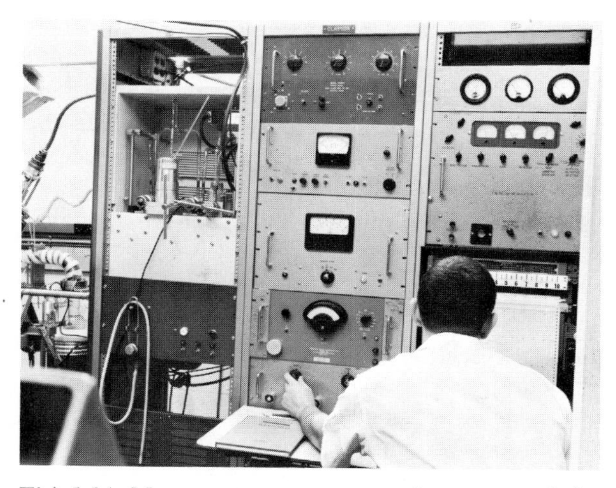

Fig. 1-14. Mass spectrometer measuring masses of the particles in a sample of argon. Particles pass through system of tubes (*left*) to be accelerated through a magnetic field, in which they are dispersed according to their masses. Control panels and chart recorder are at right. (E. F. Patterson, U. S. Geol. Survey.)

All these modern tools and techniques give us a quantitative understanding of various parts of the geologic cycle and of invisible parts of the Earth, but they do not invalidate Hutton's basic principles, which he derived with his eyes and his intelligence. Just because these tools are new, deal with things that are invisible, and give spectacular results, we are not justified in concluding that they have made direct observation obsolete. Instruments extend the geologists' vision, but they do not replace it. Although now very dependent on instruments and measurements of things that can not be seen, geology is still a science based on human eyes and brains. Much of what remains to be done requires almost no other equipment.

SCIENTIFIC METHOD IN GEOLOGY

As geology has grown more quantitative, more of its parts have become simplified by the reduction of verbal descriptions to numerical measurements and by the use of mathematics to express relationships among the numbers. However, large areas of geology have stubbornly resisted numerical treatment. The complexity of some problems and the incompleteness of the geologic record limit this approach, even in the age of computers. Nevertheless, geologists have made great progress despite those characteristics of their subject which deny them the use of elegant mathematical analyses. They have done so by using sound philosophic principles and the *scientific method.*

We begin by asking a question or formulating an hypothesis. We collect data and on the basis of them, make predictions. Next we test the predictions and modify our hypothesis in the light of the test results. We make further predictions, test these, and so on until we have established our hypothesis with greater probability. Because the laws of Nature express regular or systematic relationships, they possess powers of prediction: the unknown can be predicted from what is known.

We can never overemphasize the importance of converting new ideas into questions to be tested, of making predictions from the tests, and of further testing the predictions. Science is not blind data gathering but rather the establishment of predictable relationships among natural phenomena. Sometimes the first step, formulation of a question, is a more significant contribution than the last step, establishment of the answer. This is because systematic attention is not given to a subject until someone poses a significant question about it. After that many scientists respond with efforts to find the answer.

To many scientific questions, particularly in geology, there is no certain answer. The responses to questions are expressed as probabilities. The progress of science is marked by changing the status of theories from one of lower to higher probability.

IS THE PRESENT THE KEY TO THE PAST?

One of the main goals of geology is to reconstruct past events; therefore, geologists need sound principles as a basis for interpreting these events. Geology's working basis is expressed generally by the statement: "The present is the key to the past."

Early growth of the concept. This concept, first formulated by Hutton and reiterated by Playfair, did not make much headway at the outset. Our present appreciation of it derives largely from Lyell, whose greatest contribution to geology was to establish the key principle that "existing causes" are the basis for understanding the geologic record. The impact of Lyell's argument had the effect of an exploding bomb. Early in the nineteenth century most people believed that the correct history of the Earth was contained in the Book of Genesis, that few if any changes in the morphology of the Earth had taken place, and that the age of the Earth was only about 6,000 years. In the twentieth century it is difficult to imagine how the minds of men must have been staggered when Hutton and Lyell set forth proofs that enormous, apparently endless changes in the Earth had occurred. What is more, Hutton and Lyell argued that these changes had been brought about by processes, still operating, whose day-to-day effects are too small to be noticed by most people.

Once geologic study had made it clear that the Earth's morphology had indeed changed greatly, all geologists abandoned the static part of the Genesis creation story. But one group of geologists, though admitting that the Earth had changed, supposed that all changes had occurred within the time scale of the biblical chronology. This meant that the changes had to be catastrophic, much had to be done within a short time. As geologic knowledge expanded, the job of rationalizing a growing list of events with a small but fixed supply of time became hopeless. The catastrophes had to occur faster and faster. Eventually such catastrophism became absurd even to its most ardent scientific proponents and the whole concept was abandoned.

A second group of geologists, led by Lyell, argued that the geologic record can be interpreted without calling upon catastrophes. They based their claim on observations that even the most ordinary, commonplace natural activities of the sea, streams, gravity, and wind cause small geologic changes. Given enough time, they argued, such minute changes would gradually accumulate and produce large results. Today the views of this group are accepted generally but with an important qualification. Before radioactive measurements established the length of geologic time, a tendency had developed to emphasize the vast length of geologic time by concentrating too much on small natural activities. The importance of geologic work done by occasional vigorous natural events, such as floods, great storms, severe earthquakes, and landslides, tended to be ignored, because such events had been overemphasized in the past. Now that the great length of geologic time has been firmly established, the overemphasizing of small, slow activities at the expense of big, fast events gains nothing. Both occur, and we should learn to identify the effects of each.

Modern view. We have seen how the concept that the present is the key to the past became established. But what attributes of the present serve as reliable baselines for extrapolating backward into geologic time? Have present geographic conditions prevailed always more or less as we now see them? Does geology depend entirely on the constancy of natural laws? If so, where does geology stand in relation to contemporary science which keeps discovering that many natural laws are not so "constant" after all?

Despite the demonstrated variability of many supposed constants, geologists infer that observations and measurements made today can establish fundamental relationships between processes and materials that are valid independent of time and place. This *principle of the uniformity of processes and materials* does not mean that the present level of intensity of processes has prevailed always; it states that changes of intensity leave behind in the materials definite clues by which such changes can be inferred.

Let us give a concrete example of this abstract concept, which we will call, for short, the *principle of uniformity*. The present regimen of air temperatures and moisture is such that glaciers are confined to high mountains (Fig. 1-15) and to two continental ice sheets, on Greenland and Antarctica (Chap. 12). A rigid extension of present geographic conditions into the past would force the conclusion that former glaciers were no larger than existing glaciers. Yet by studying modern glaciers we see their distinctive effects on bedrock, morphology, and sediments. These same effects can be seen again in regions where no glaciers exist today. Using the principle of uniformity of processes and materials we infer that formerly glaciers were far more extensive than at present.

If we are careful how we use "the present" as a "key to the past" and concentrate on processes and relationships between materials and processes to establish fundamental generalizations, we can interpret Earth history correctly. Moreover, we can work out such variations as changes in the distribution of lands and seas through time, changes in the salinity of seawater, changes in the composition of the atmosphere, changes in the polarity of the Earth's magnetic field, changes in the positions of the magnetic poles, and changes in the relative positions of continents with respect to one another. We need not suppose that exact duplication of present-day geographic conditions is the only basis for extending the present into the past.

GEOLOGY AND THE CITIZEN

Another aspect of geology, not directly practical nor purely scientific, consists of its impact on the everyday lives of individuals and on organized human affairs. Even a casual acquaintance with geology can greatly enrich such an informal pastime as viewing landscapes. In a more philosophic vein knowledge of geology yields perspectives in time as familiarity with astronomy does in distance. Principles of geology are useful in aspects of the legal profession, in formulating enlightened public policies on land use, and in making decisions about control of environments. Political power in the modern world rests heavily on industrial capability, which, in turn, depends entirely on natural resources. Problems of developing and conserving the available supplies of these resources, as well as competition for their control, affect the policies of nations. Military operations depend upon knowledge of topography, environment, and the availability of certain natural materials. In these and in other ways geology reaches into the life of everyone.

So much for the preliminaries. We now begin our systematic study with a general view of the Earth.

Summary

1. Geology is the systematic study of the materials, processes, environments, and history of the Earth.
2. The essentials of geology can be comprehended by visual study of the three components of the geologic record: Earth's morphology, regolith, and bedrock.
3. The Earth's surface moves up and down and is eroded; it changes continuously.
4. Regolith and materials dissolved in solution result from erosion of bedrock. Regolith can be moved as sediment and/or converted back into bedrock. Materials in solution are transported to the oceans.

Fig. 1-15. Glacier at head of Johns Hopkins Inlet, Alaska, formerly extended downvalley and around corner at lower left. Widespread features, characteristic of erosion by glaciers, indicate that glacier ice not only filled the valley but covered nearly the entire field of view. (A. S. Post.)

5. Bedrock consists of three varieties: igneous, sedimentary, and metamorphic.

6. The concept of the geologic cycle was stated by James Hutton in the late eighteenth century, established as the basis of geology by Sir Charles Lyell in the nineteenth century, and is still valid today.

7. By operation of the geologic cycle, uplifted bedrock breaks down to regolith; regolith is transported to areas of subsidence, where it buries parts of the Earth's surface; and buried regolith can be transformed into bedrock and uplifted to start another round in the cycle.

8. The scientific method uses questions, observations, predictions, and repeated tests of predictions to establish the validity of hypotheses.

9. Generalizations drawn from the study of modern processes, and the effects of processes on materials, can be used to decipher the geologic record of past events.

Selected References

Albritton, C. C., Jr., 1963, The fabric of geology: Reading, Mass., Addison-Wesley. Article on James Hutton by D. B. McIntyre, p. 1–11, should be read in conjunction with article by G. L. Davies (1967) listed below.

Bailey, Sir E. B., 1963, Charles Lyell: Garden City, N. Y., Doubleday.

Bailey, Sir E. B., 1967, James Hutton—the founder of modern geology: New York, American Elsevier Pub.

Davies, G. L., 1967, George Hoggart Toulmin and the Huttonian theory of the Earth: Geol. Soc. America Bull., v. 78, p. 121–124.

Hutton, James, 1795, Theory of the Earth with proofs and illustrations: Edinburgh. (Facsimile repr., 1959, New York, Hafner, 2.)

Koons, E. D., 1945, Geology of the Uinkaret Plateau, northern Arizona: Geol. Soc. America Bull., v. 56, p. 151–180.

Krutch, J. W., 1962, Grand Canyon: New York, Doubleday and Co.

Leet, L. D., 1948, Causes of catastrophe: New York, McGraw-Hill.

Mather, K. F., and Mason, S. L., 1939, A source book in geology: New York, McGraw-Hill. (Esp. p. 92–100, Hutton; p. 131–135, Playfair; and p. 263–273, Lyell.)

Playfair, John, 1802. Illustrations of the Huttonian theory of the Earth: (Facsimile repr., 1956, Urbana, Univ. of Illinois Press.)

CHAPTER 2

General View of the Earth

What makes the geologic cycle go?
Sun-Earth-Moon system
Spectrum of solar radiation
Size and shape of the Earth
Gravity
Earth's rotation and its effect on movements of air and water
Lithosphere
Magnetism in rocks
What makes the crust go up and down?

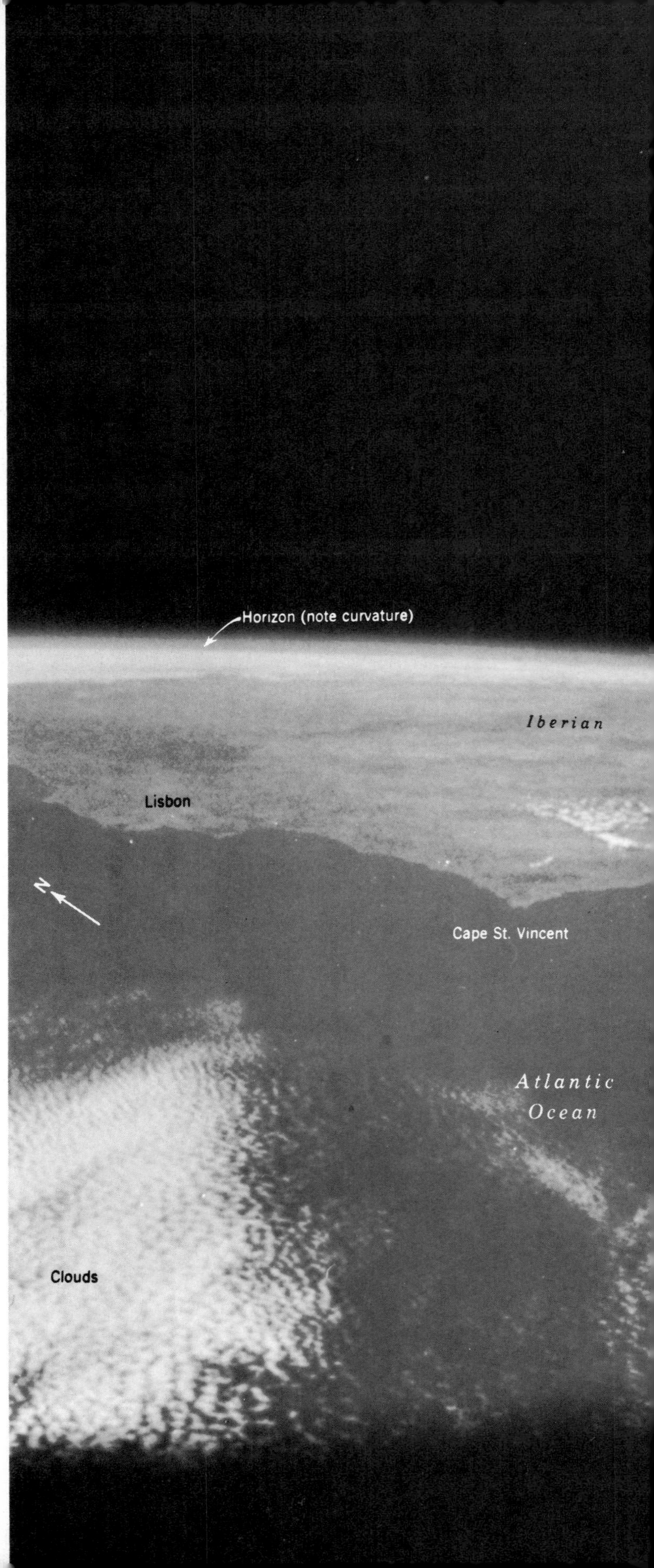

Plate 2

Atlantic Ocean, Strait of Gibraltar, and western end of Mediterranean Sea photographed from Gemini X, showing curvature of the Earth. As the distance from Cape St. Vincent to Casablanca is 400km, the photo is about 800km wide. (National Aeronautics and Space Administration.)

Peninsula
Mediterranean
Sea
Strait of
Gibraltar
Africa
Storm
Casablanca

CHAPTER 2 *General View of the Earth*

WHAT MAKES THE GEOLOGIC CYCLE GO?

The geologic cycle consists of processes so ordinary that most of us take them for granted. Our experience tells us that the events within the cycle are normal for the Earth; so we may think nothing more about them. Yet if we extend our view beyond the Earth we find that the cycle is really quite extraordinary and depends on delicate adjustments among several factors. What are some of these factors?

The cycle would grind to a halt if the supply of solar energy were shut off or if this energy became too intense. Consider the surface of the Moon (Fig. 2-1). The Moon resembles the Earth in possessing a lithosphere ("rock sphere") in which oxygen and silicon predominate and in receiving solar energy. Nevertheless, we are certain that nothing like the geologic cycle has ever operated on the Moon because its landscape lacks the characteristic features made by running water, wind, and glaciers. Why are these features not present? One reason is that the Moon lacks an atmosphere ("vapor sphere") and a hydrosphere ("water sphere"), both of which are present on Earth and which are essential to the geologic cycle. A second reason why the cycle does not operate on the Moon is that the distribution of solar energy on the Moon's surface is extremely uneven. The Moon always keeps the same face toward the Earth. This means that during each of its revolutions around the Earth, which requires 29½ days, it rotates once with respect to the Sun (Fig. 2-2). Accordingly the length of a lunar day, as well as of a lunar night, is equal to 14¾ Earth days. During the lunar day, solar energy sears the surface of the Moon, but during the lunar night solar energy is cut off for so long and heat radiates out into space so fast that the surface of the Moon becomes very cold. Such extreme conditions would obliterate all earthly organisms not equipped with special protective devices.

Does the geologic cycle require anything else beyond atmosphere and hydrosphere and a supply of some, but not too much, solar energy uniformly distributed? On Earth, one of the major activities of the cycle, erosion, consumes lands relentlessly yet lands still exist. Why did the geologic cycle not destroy all lands long ago? Is their existence guaranteed by some powerful ally within the Earth with restorative powers equal to the destructive power of erosion? Clearly the geologic cycle would cease to operate if no lands were available.

The geologic cycle, then, depends on the right combination of external energy from the Sun; on air, water, and land at the Earth's surface; and on conditions within the Earth. Before we turn our attention to the individual processes at work within the cycle, let us take a general view of the Earth to become acquainted with the major factors influencing the cycle. We need to include the influences on the Earth from outer space, enough information on nuclear physics to appreciate solar energy, and the Earth's broader features.

INFLUENCES ON THE EARTH FROM OUTER SPACE

Sun-Earth-Moon system. At the center of the Solar System is the Sun, around which nine planets, including the Earth, revolve in a common plane, *the plane of the ecliptic.* Six of the nine planets are orbited by one or more satellites or moons. The Earth's one natural satellite is the Moon. Only the Sun, the Earth, and the Moon relate intimately to our study of geology; so we center our attention on these three bodies. The Sun lies at one focus of the ellipse traced by the orbiting Earth-Moon pair (Fig. 2-2).

The dimensions of the Sun-Earth-Moon system (Table 2-1) can be visualized if we scale them down to more familiar objects. Tiny models of these three tiny celestial bodies fit on a collegiate football field

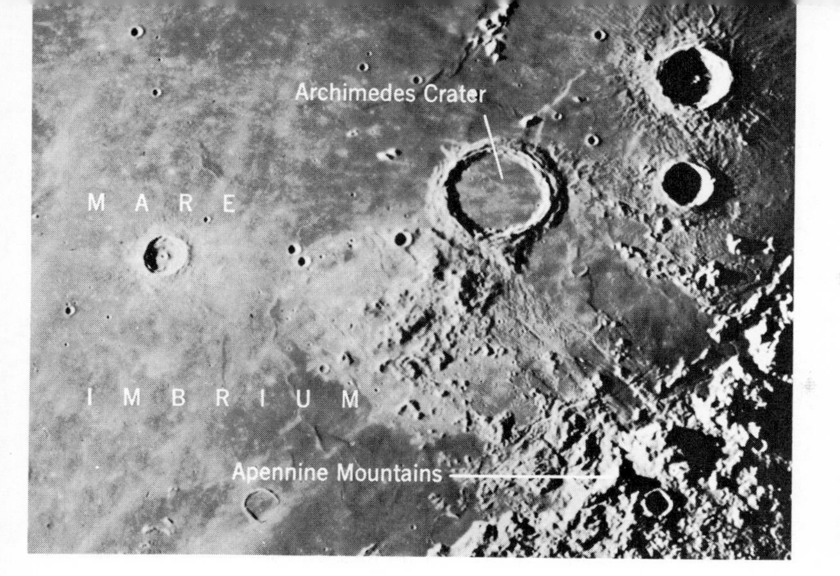

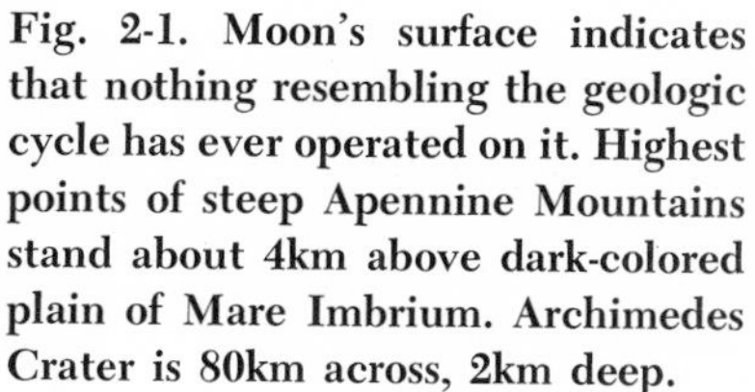
Fig. 2-1. Moon's surface indicates that nothing resembling the geologic cycle has ever operated on it. Highest points of steep Apennine Mountains stand about 4km above dark-colored plain of Mare Imbrium. Archimedes Crater is 80km across, 2km deep.

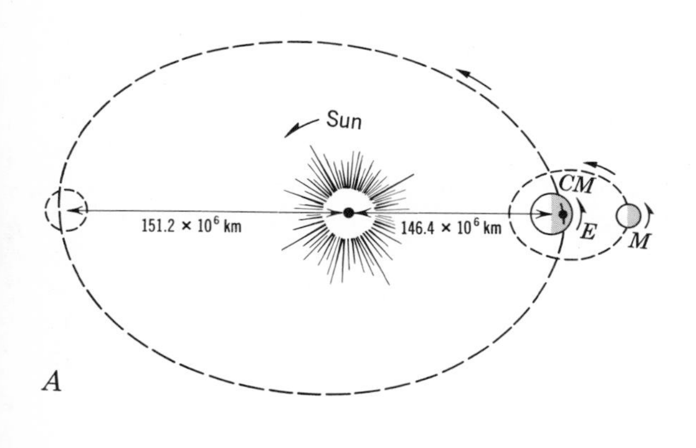

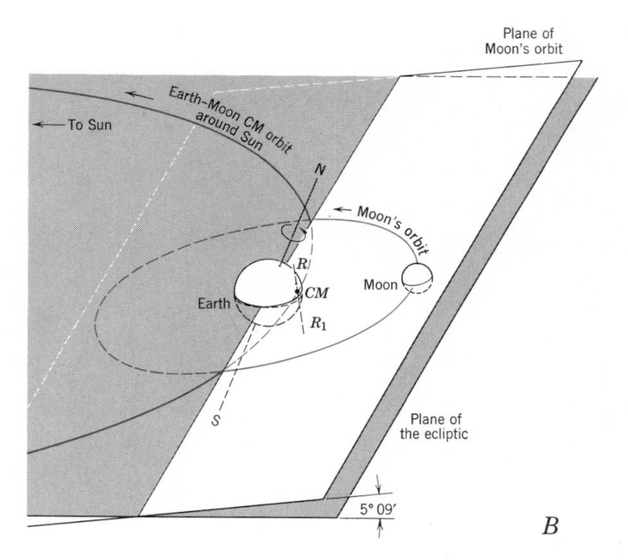

Fig. 2-2. Schematic relationships among orbits of Sun, Earth (*E*) and Moon (*M*). *A*. Orbit around Sun is generated not by Earth alone but by center of mass (*CM*) of Earth-Moon pair (RR_1 in *B*). Arrows show directions of revolution; half arrows, directions of rotation; numbers, center-to-center distances between Earth and Sun. *B*. Detail showing small angle between plane of Moon's orbit and plane of the ecliptic. Axis RR_1, passing through center of mass (*CM*) of Earth-Moon pair, is perpendicular to plane of Moon's orbit.

TABLE 2-1. DIMENSIONS OF SUN-EARTH-MOON SYSTEM (compiled from various sources)

Sun	Center-to-Center Separation	Earth	Center-to-Center Separation	Moon
Diameter		Equatorial diameter		Diameter
1.393×10^6km (0.865×10^6 miles)		12,756km (7,926 miles)		3,456km (2,160 miles)
		Polar diameter		
		12,714km (7,900 miles)		
	Mean: 148.6×10^6km (92.9×10^6 miles)		Mean: 384,411km (238,862 miles)	
	Minimum: 146.4×10^6km (91.342×10^6 miles)		Minimum: 360,000km (225,000 miles)	
	Maximum: 151.7×10^6km (94.452×10^6 miles)		Maximum: 404,800km (253,000 miles)	

as illustrated in Fig. 2-3.

Gravity. Sun, Earth, and Moon remain in their orbits because of the effects of the ***universal law of gravitation,*** formulated in 1687 by Sir Isaac Newton. The law states that *every particle in the universe attracts every other particle with a force directly proportional to the product of their masses and inversely as the square of the distance between their centers.* If we introduce the *gravitational constant, G,* the precise value of the attractive force between two masses of 1g each, exactly 1cm apart, this relationship can be written as the familiar equation:

$$F = \frac{G M_1 M_2}{d^2}$$

where F = the force of attraction (in dynes)

M_1 and M_2 = any two masses (in grams)

d = the distance between their centers (in centimeters)

Although we shall have much more to say about gravity, we simply mention it here because it is an external influence that keeps order in the movements of all bodies in the Universe.

Solar radiation. We have seen that solar energy is the driving force behind the geologic cycle, but what creates solar energy? How is this energy transmitted to the Earth? Full answers to these questions would require a treatise on nuclear physics. We shall deal only with the highlights, which include concepts of nuclear reactions, the continuous spectrum of electromagnetic radiation, and relationships of the spectrum to matter and energy.

Nuclear reactions. One of the great unifying principles of physics is the equivalence of matter and energy (App. A). Matter can be consumed and converted into energy in two ways: (1) heavy elements can break down into lighter elements, and (2) light elements can build up heavier elements. The first process includes fission and radioactivity; the second is fusion.

Fission reactions need not enter our study of geology, but radioactivity is vital to geology because it forms a basis for measuring time (Chap. 6) and for understanding heat within the Earth (Chap. 20).

Inside the Sun the main process taking place is thought to be fusion of hydrogen into helium. During every second of time, 4.7 million tons of solar hydrogen disappear and are converted into helium and energy. The energy radiates outward in all directions.

Spectrum of electromagnetic radiation. Our everyday experience tells us that light and heat come from the Sun. But what are light and heat, and are they the only things that come from the Sun? Physicists have demonstrated that light and heat are electromagnetic waves, which span only a small part of the ***spectrum*** (Lat. "an appearance"), *a collection of waves of different lengths* usually covering a considerable range. Visible light occupies only a small span in the middle of the spectrum; most electromagnetic waves are so long or so short that they are invisible (Fig. A-3). Just because we are unable to see them does not mean that they are not important. We use the longer waves for communications. Heat is radiated as infrared waves. Some of the longer ultraviolet waves are beneficial to life, but the shorter ultraviolet waves and other very short waves, such as X-rays and γ-rays, can be lethal in large doses.

The sizes of the vibrating objects that originate the waves determine the lengths of electromagnetic waves; small objects generate waves of short length. For example, molecules emit light waves, whereas the nuclei of atoms send out γ-rays. By

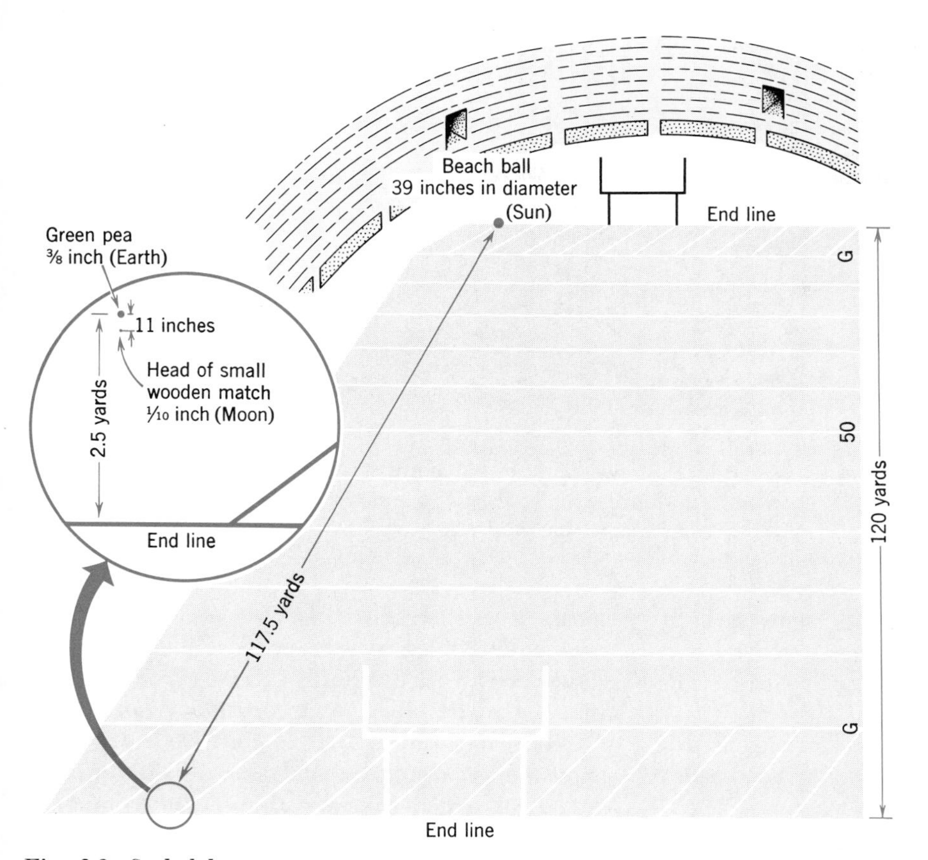

Fig. 2-3. Scaled-down sizes and spacings of Sun, Earth, and Moon shown by sketches of familiar objects on a collegiate football field. Natural dimensions are 1.393×10^9 times as large as those shown.

the kinds of waves it emits, an object tells much about itself. This principle has enabled physicists to use electromagnetic waves for making detailed projections of nuclear reactions inside the Sun, and chemists to analyze accessible materials of the Earth as well as inaccessible materials of the stars. Geophysicists have used other waves as a means of understanding the structure of the Earth's interior.

What are the important things for geologists to know about solar radiation? First of all we want to know if energy from the Sun is constant or variable. On this topic physicists disagree. Some say that solar radiation is essentially constant and refer to solar radiation received at the Earth's surface as the *solar constant.* The measured value has been found to be 3.2×10^{-2} cal*/cm^2/sec.

Constancy is what we would expect from the very nature of the nuclear reactions taking place in the Sun. But other physicists have found measurable variations in the output of the Sun's energy in parts of the electromagnetic spectrum. Outbursts of energy are associated with *sunspots,* features whose frequency has been found to occur in a cycle with a period of $11\frac{1}{6}$ years. Some scientists have observed variations in growth rings of trees and in certain sediments on Earth that they infer have been caused by temperature variations controlled by the sunspot cycle. Other scientists claim the opposite; they contend that temperatures recorded at various weather stations during the past 150 years lack variations related to the sunspot cycle. We shall have to be content to leave this subject in an unsettled condition and wait for future measurements to give us more definitive evidence about it.

The second question geologists ask about solar energy is: how much of it gets down to Earth? Many things happen to solar radiation as it reacts with the Earth's magnetic field and atmosphere. We need not

* Appendix A explains calorie.

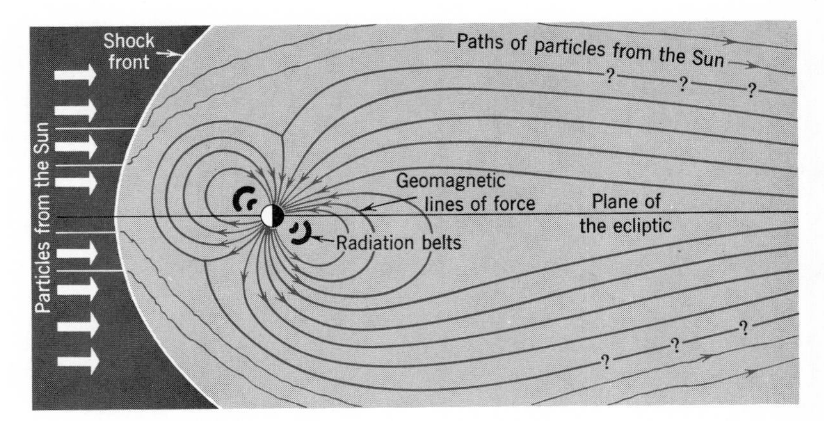

Fig. 2-4. Section through Earth and its magnetic field in plane perpendicular to plane of the ecliptic and normal to Earth's direction of travel. Stream of ionized particles from the Sun distorts magnetic field. Scale is given by diameter of Earth. (U. S. Navy.)

describe all the details, but we should know the highlights of the reactions. We can get a general idea of the major points by following solar energy from Sun to Earth.

All electromagnetic waves and associated particles travel with the speed of light, approximately 299,800 km/sec (186,000 miles per second). Therefore they reach the Earth only 8 minutes after they leave the Sun. While still traveling through space they encounter magnetic lines of force (Fig. 2-4), the Earth's most far-reaching influence on solar energy.

Reaction with Earth's magnetic field. The magnetic lines of force surrounding the Earth constitute a *magnetic field* extending far out into space. The shape of the field is such that we can picture the Earth as being a very large dipole bar magnet. The dipole shape of the field clearly indicates that the field owes its existence to conditions inside the Earth. The ionized particles streaming away from the Sun distort the Earth's magnetic field. On the solar side of the Earth they compress the lines of force into the shape of a shock wave and on the opposite side spread them out like a comet's tail (Fig. 2-4). Some of the arriving particles are reflected back into space by the magnetic field. Other particles are trapped by it and flow along the lines of force, where they are concentrated in two zones, inner and outer *radiation belts,* so called because of their high concentration of charged particles.

The magnetic field serves as a protective shield against charged particles streaming toward the Earth. If the magnetic field were to disappear the charged particles now trapped within it would shower down on the Earth's surface. Some organisms would be killed by them; others might survive but would undergo rapid genetic changes. We know these things would take place because we have seen them happen when organisms in the laboratory are subjected to radiation by charged particles.

Reaction with Earth's atmosphere. Further reactions occur when solar radiation encounters *the gaseous envelope that surrounds the Earth,* the ***atmosphere.*** The top of the atmosphere reflects long waves back into space. Reactions between the upper part of the atmosphere and the rest of the spectrum of solar radiation consume approximately 20 per cent of the total supplied. In the absence of clouds the remaining 80 per cent reaches the ground. Clouds can reflect as much as 60 per cent of the solar energy incident upon them and can absorb as much as 20 per cent of the total passing through them. A cover of clouds can reduce the amount of solar energy received at the ground by one half or more. The effect of dust particles on incoming energy is nearly the same as that of clouds. Dense concentrations of dust can prevent a large proportion of the solar energy from reaching the Earth's surface.

Despite these barriers some solar energy does reach the ground. Many geologic processes depend on the effects of the energy at the surface.

Climatic regions. The ancient Greeks noted that in warm lands, such as Egypt, the Sun at noon is located high in the sky. Northward, in cooler regions,

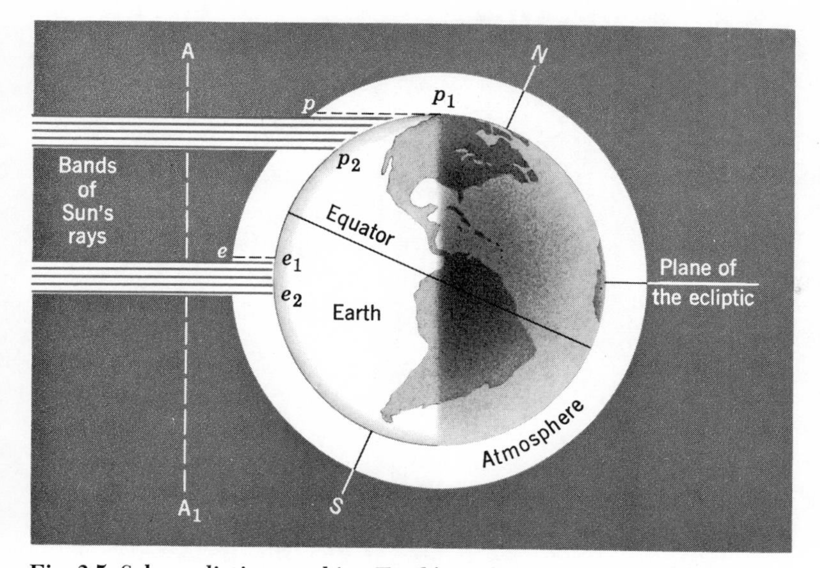

Fig. 2-5. Solar radiation reaching Earth's surface varies with latitude. Two bands of Sun's rays, of equal intensity in plane A A_1, are of unequal intensity at Earth's surface. Rays arriving near plane of the ecliptic traverse minimum thickness of atmosphere (e e_1) and strike ground perpendicularly through distance e_1 e_2. Rays arriving near poles traverse maximum thickness of atmosphere (p p_1) and strike ground obliquely through distance p_1 p_2.

they observed that the Sun's track became progressively lower toward the southern horizon. They reasoned, correctly, that solar rays strike the Earth's surface more and more obliquely toward the north; accordingly, the ancient Greeks divided the Earth into climatic regions bounded by parallels of latitude (Fig. D-1). Their explanation of the variation in the Sun's position was that the Earth's surface slopes away from the Equator. Our word *climate* is based on the Greek word *klima,* which means slope or incline.

Solar energy in polar regions is less than that near the Equator for another reason. The path of energy through the atmosphere is longest near the poles and more of it is absorbed by the atmosphere. In the equatorial belt the Sun's rays traverse a smaller thickness of atmosphere and so lose less energy to the atmosphere by absorption (Fig. 2-5).

Because of the effects of the Earth's curvature and of differences in thickness of atmosphere solar radiation must traverse, the equatorial region receives solar heat in excess of the average amount, whereas the polar regions receive less than the average amount. This perennial imbalance of energy creates systems of air circulation at the Earth's surface that act to distribute heat more uniformly.

Meteorites. *Particles of solid matter from outer space that fall to the ground through the atmosphere* are ***meteorites.*** They reach the ground in a wide range of sizes and in quantities estimated to be as large as 2.4×10^9 tons annually.

Meteorites consist of four major groups (Fig. 2-6): (1) *iron meteorites,* composed of iron and nickel; (2) *stony meteorites,* which lack metallic iron and nickel; (3) *stony-iron meteorites,* an intermediate group containing both metallic iron and stony material; and (4) *tektites* ("molten bodies"), glassy objects formed by the rapid cooling of particles that are thought to have fused during impact.

All meteorites, with the possible exception of tektites, are thought to have originated within the Solar System, possibly from the broken bits of a former planet. Whatever their origin, meteorites indicate that solid particles in outer space consist of substances similar to those found on Earth.

SHAPE AND SIZE OF THE EARTH

As a first approximation we can consider the Earth, apart from surface irregularities, to be a sphere with a diameter about 12,600km, circum-

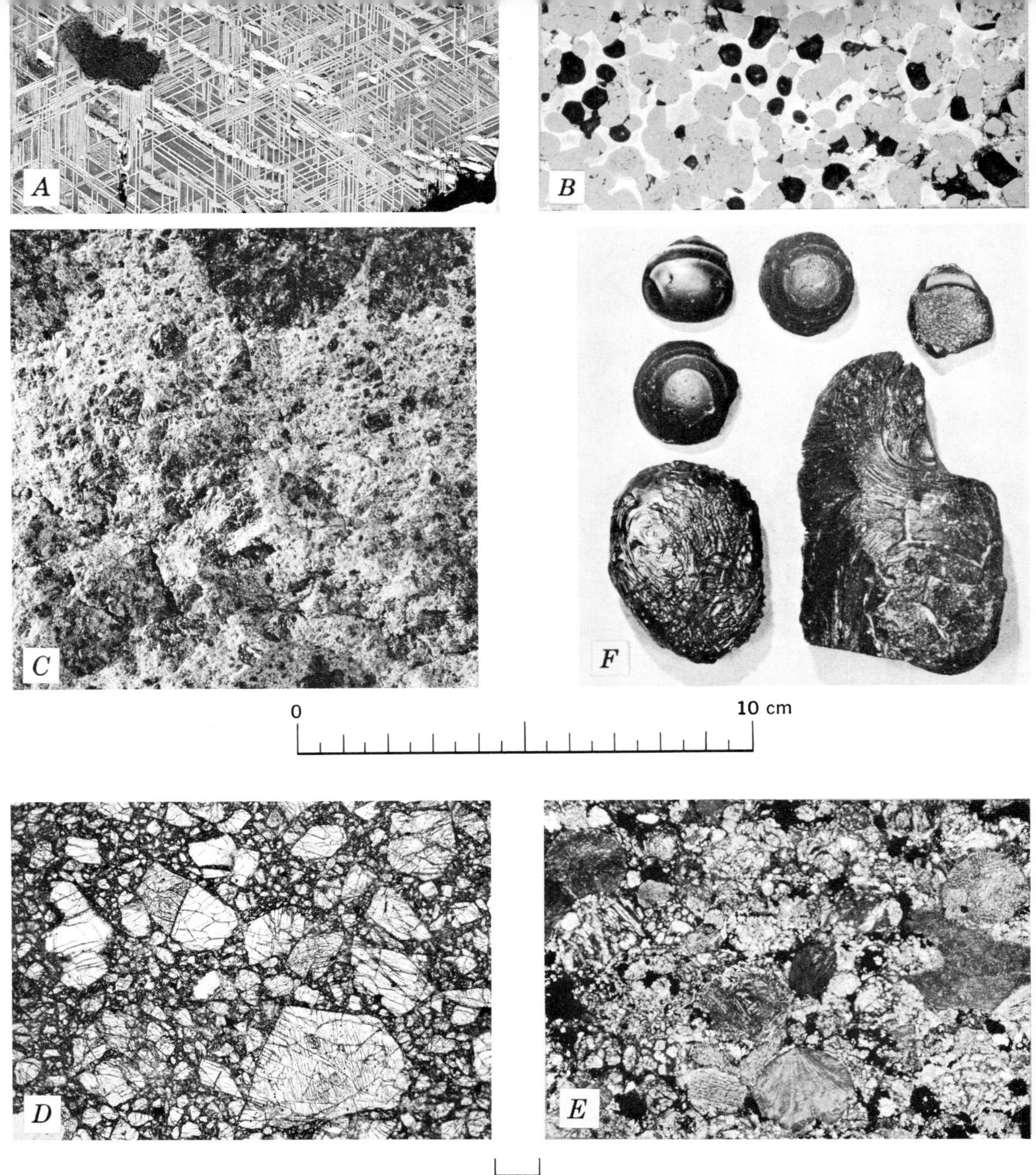

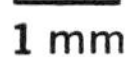

Fig. 2-6. Various meteorites and tektites. A. Polished surface of nickel-iron meteorite, Edmonton, Kentucky. Black area (*upper left*) is a cavity; black at lower right is shadow. (Smithsonian Institution.) *B*. Polished surface of stony-iron meteorite, Brenham, Kansas. Rounded black areas are stony material; gray areas, nickel-iron. *C*. Interior of stony meteorite seen on rough broken surface. *D* and *E*. Microscopic views of paper-thin slices of stony meteorites: (*C* and *D*: Johnstown, Colorado; *E*: Knyahinya, U.S.S.R.) *F*. Various tektites from Australia. (*B–F*, American Museum of Natural History.)

ference about 39,800km and surface area of 504×10^6km^2. Any doubts about the spherical shape of the Earth have been removed many times over since we entered the Space Age in 1957. Using rockets and artificial satellites, we can now get a camera lens so far away from the Earth that photographs show the curvature of its surface (P1. 2). Closer measurement indicates that the Earth is not a perfect sphere but an *oblate spheroid* with its polar diameter slightly shorter than its equatorial diameter. If we consider gravity and centrifugal forces as they act on the Earth, we can understand why this is so.

Gravity, the sphere-maker and leveler. The ***Earth's gravity,*** also called the ***Earth's gravitational acceleration,*** and represented by the lower-case letter *g*, is *the inward-acting force with which the Earth tends to pull all objects toward its center;* it tends to make the Earth a sphere (Fig. 2-7). Actually, however, the Earth is not quite a sphere because, on parts of its surface, forces of rotation exceed those of gravity, as we shall see presently.

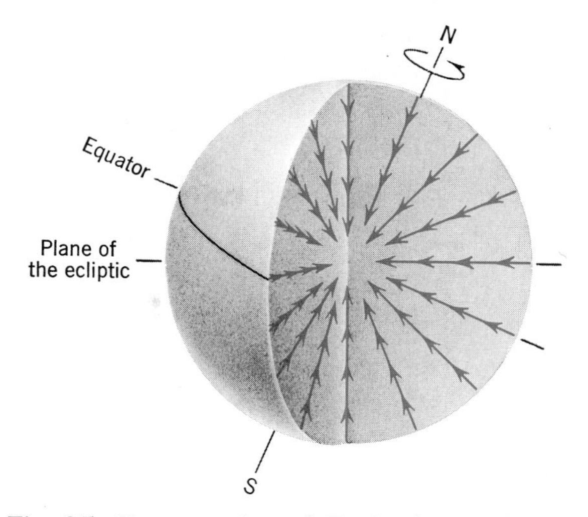

Fig. 2-7. Cut-away view of Earth, showing how at every point along each radius the force of gravity pulls constantly toward the center of mass, tending to create a spherical form.

Throughout our lives we strive against or take advantage of the Earth's gravity. This force is strong enough to prevent the escape of the gases composing our atmosphere and to compress the greater part of them into a zone a few kilometers thick directly above the Earth. The Earth's gravity shapes the surface of the sea to a nearly spherical form, disturbed only by tides and winds; it makes water flow down slopes and so provides energy for stream erosion; it pulls persistently on rock material on slopes and does enormous work in moving such material to lower places.

On a horizontal surface the Earth's gravity holds objects in place by pulling in a direction normal to the surface. On slopes, gravity can be resolved into two components. The *normal component of gravity* (g_n in Fig. 2-8) acts at right angles to the slope and tends to hold the particles in place as on a horizontal surface. The magnitude of g_n is determined by the cosine of the angle of slope. The *tangential component of gravity* (g_t in Fig. 2-8) acts along the slope and tends to pull the particles downslope. The magnitude of g_t is determined by the sine of the angle of slope.

On gentle slopes the normal component of gravity holds loose particles in place. As slope increases, g_t increases and g_n decreases. On a slope of 45° the two components are equal. On steeper slopes g_t predominates. Arrows showing the relationships of g_n, g_t, and g to slope are included in diagrams that illustrate natural activities, such as movements of regolith, water, and ice, which occur on slopes.

Effects of Earth's rotation. Rotation creates the

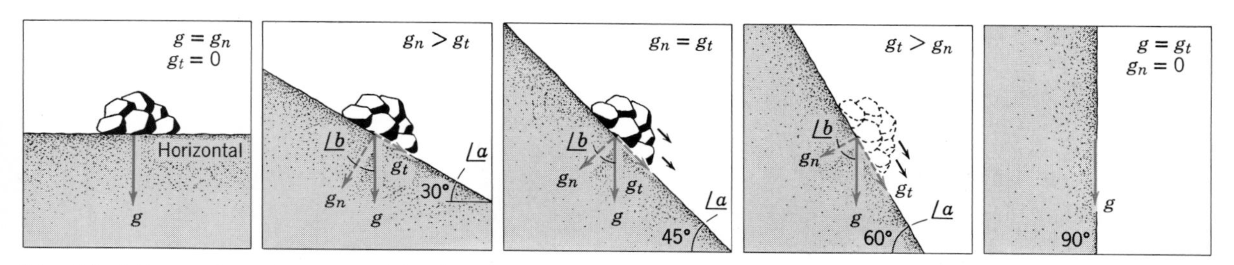

Fig. 2-8. Effects of gravity on particles lying on slopes of various inclinations. Angle *b* between g_n and g equals slope angle *a*. Therefore $g_n = g \cos a$ and $g_t = g \sin a$. Further explanation in text.

oblate shape of the Earth and deflects systematically all currents of air and water. Rotation creates forces that are equal and opposite and that act at right angles to the axis of rotation. Centrifugal force acts *away* from the axis of rotation and centripetal force *toward* this axis. We can calculate the magnitude of either one of these forces, F, from a single formula.

$$F = \frac{m\ v^2}{r}$$

where m is the mass of the moving body, v its velocity, and r the length of the radius of rotation.

The interaction between gravity, which acts toward the center of the Earth along all radii (Fig. 2-7), and centrifugal force resulting from rotation, which acts at right angles to the axis of rotation, determines the shape of the Earth. Centrifugal force reaches its maximum value at the Equator (C_{max} in Fig. 2-9) because the instantaneous linear velocity of rotation is greatest here. The instantaneous linear velocity of rotation at any point depends upon the circumference of the Earth in a plane perpendicular to the axis of rotation and passing through the point. This velocity can be calculated by dividing the appropriate circumference by the time required to make 1 revolution (24 hours).

Centrifugal force becomes progressively less at points between Equator and poles, where it equals zero.

Because both the instantaneous linear velocity of rotation and the length of the radius of rotation decrease progressively from Equator to poles, centrifugal force decreases poleward and at the poles it vanishes.

At the Equator all the centrifugal force acts along the same radius as gravity (r_E in Fig. 2-9) but in the opposite direction. At any intermediate point, the centrifugal force (C in Fig. 2-9) acting outward along a radius r perpendicular to the axis of rotation (NS in Fig. 2-9) no longer exerts its entire effort along the same radius as gravity does. Only a component of centrifugal force (C_1, dashed white arrow in Fig. 2-9) acts opposite to gravity along a radius from the intermediate point to the center of the Earth.

Because centrifugal force bulges the Equator outward, the Earth is not a perfect sphere but an oblate spheroid, whose polar axis is 42km shorter than its equatorial axis. A plane section through the Earth at the Equator is a circle, but such a section through the Earth including the poles is an ellipse.

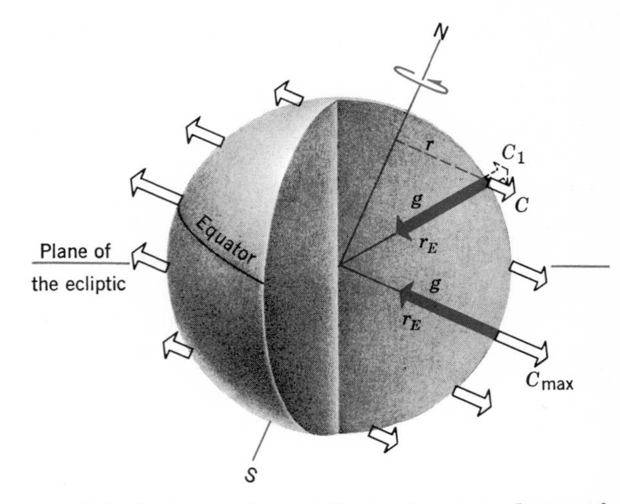

Fig. 2-9. Cut-away view of Earth, showing relation of centrifugal force (white arrows) set up by rotation about polar axis, NS, to Earth's gravitational acceleration, g (blue arrows, shown for 2 radii only; Fig. 2-7). Further explanation in text.

By pushing out the Earth's surface at the Equator, centrifugal force is responsible for variation in the Earth's gravitational acceleration. Measured values of g are about 978 gals (gal for Galileo; 1 gal = 1 cm/sec^2 of acceleration) at the Equator and about 983 gals at the poles, a difference of a little more than 5 parts in 1,000. This means that a man whose weight registers 200 pounds on a spring balance at the Equator would register about 201 pounds on the same balance at the poles.

Coriolis force. The solid Earth is our frame of reference for measuring motions. The motion of an object, such as an automobile, that stays in contact with the solid Earth, can be analyzed by assuming that the frame of reference is fixed. Rotation of the Earth must be taken into account in calculations involving the motion of any object shot into the air (except one shot vertically upward at one of the poles) and of any current of wind or water that can move with respect to the solid Earth. In all such motions the frame of reference is not fixed but rotates. In 1835 Coriolis, a French engineer and mathematician, showed that where the frame of reference rotates an inertial force must be added to all motions. This inertial force acts to the right where the frame of reference rotates counterclockwise and to the left where the frame of reference rotates clockwise. If we view the Earth from a fixed point above the North Pole we see only the Northern Hemisphere and it appears to be rotating counterclockwise. Therefore all motions in the Northern Hemisphere obey the Coriolis right-deflection rule. If we select a fixed point above the South Pole, however, we see only the Southern Hemisphere. Although the direction of rotation has not changed, from our new fixed point of reference we see only clockwise rotation. Therefore the left-deflection rule applies in the Southern Hemisphere. The inertia term in the calculations has been named the *Coriolis force.*

ATMOSPHERE AND HYDROSPHERE

Because they bear separate names, the atmosphere and hydrosphere tend to be considered as distinct entities, but they interact so completely that we had best describe them together. Water vapor is distributed widely in the atmosphere, and much air and most of the Earth's free carbon dioxide are dissolved in the hydrosphere. Fine particles from the lithosphere are suspended in both atmosphere and hydrosphere.

The atmosphere consists of about 4 parts nitrogen and 1 part oxygen; a few other gases compose less than 2 per cent (Fig. 5-10). Of particular importance in the circulation of moisture in the atmosphere is ***relative humidity,*** *the ratio of amount of water vapor present to the maximum amount that the air mass can contain without condensation or precipitation.* Depending on the temperature the atmosphere can retain up to 4 weight per cent of water. At 30°C, $1m^3$ of air can hold 30g of water; but at −20°C, the same volume of air can hold only 1g of water.

The water of the hydrosphere exists as a liquid in oceans, seas, lakes, rivers, and in openings underground; as a vapor in the atmosphere; and as a solid in ice, most of which is in glaciers. If we increase the size of our Sun-Earth-Moon model (Fig. 2-3) 109 times, then the Earth is represented by the beach ball 1m in diameter which formerly served as the Sun. If we dip the ball in water and take it out again, the thin film of water adhering to its surface represents an ocean about 0.8km deep. This is about one-third as deep as the layer of water that would cover the Earth if its surface were made smooth by planing off the continental masses and filling in the ocean basins with the planed-off material. Water that flows through the lithosphere is chemically active; it dissolves soluble rock materials and transports them to the oceans. This is part of the geochemical cycle (Chap. 5) whose operation through long ages has made seawater salty and unfit for most human uses. Water from the oceans is being evaporated continually, however, and circulates up into the atmosphere and back down again as part of the hydrologic cycle. Water that falls out of the atmosphere has been distilled naturally and so is fit for human consumption. As the water requirements of an enlarging population increase, we are actively seeking economical ways to increase the quantity of water available for human use. Two possibilities are (1) conversion of saltwater to freshwater, and (2) interception and storage of more of the water that falls on the lands to be used before it flows back to the oceans.

Movements of air. As we have seen, circulation of the atmosphere results chiefly from the unequal distribution of solar energy. In addition, the Earth's rotation greatly influences all movements of air.

Perennial differences in temperature between equatorial and polar regions divide the atmosphere into northern and southern cells that are mirror images of each other (Fig. 2-10). When heated at the Equator, air becomes light and rises. At higher levels it spreads in both directions and flows toward the poles. Close to the Earth's surface, air flowing toward the Equator creates the ***trade winds,*** *the prevailing winds in tropical regions.* Because of the Coriolis effect the trade winds in the Northern Hemisphere are deflected toward their right and blow generally from the northeast. In the Southern Hemisphere they are deflected toward their left and blow from the southeast. Some of the air flowing poleward from the Equator is cooled at high altitudes and descends. Part of the air that descends flows back toward the Equator as the trade winds, and part moves toward the poles as the ***westerlies,*** *the prevailing winds in the middle latitudes.* The Coriolis effect causes the westerlies to blow generally from the southwest in the Northern Hemisphere and from the northwest in the Southern.

Air cooled at the poles becomes denser and tends to flow along the Earth's surface toward the equator, forming the ***polar easterlies,*** *the common winds in polar regions.* The Coriolis effect influences the polar winds just as it does the trade winds; hence, polar winds blow generally from the northeast in the Arctic zone and from the southeast in the Antarctic zone.

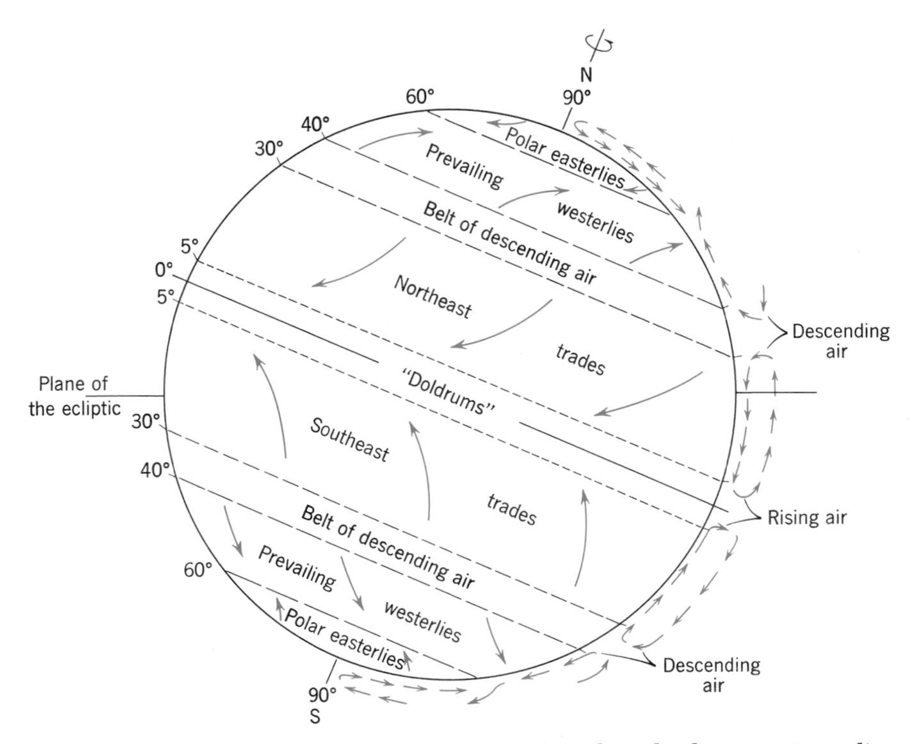

Fig. 2-10. Idealized wind circulation on an Earth without land masses. In reality the wind belts shown prevail over oceans but are distorted over land areas.

The simplicity of this ideal circulation is disturbed by many influences, especially irregular land masses. Lands are heated and cooled more rapidly than are adjoining seas, and this sets up special circulation cells, which tend to diverge from the ideal main system. Violent tornadoes that affect some areas are but one example of disturbances within a large circulation. High mountain ranges force air currents to turn sharply upward, even dissipating them or perhaps steering them into conflicting winds.

Movements of ocean water. The uneven distribution of solar heat over the Earth is responsible not only for wind but also (through wind) for major ocean currents. Frictional drag of prevailing winds on water creates most of the surface currents in the oceans. The trade winds generate broad equatorial currents that flow westward in the Atlantic and Pacific Oceans. At the west margin of each ocean, each of these currents divides and completes its circulation in both belts of westerlies (Fig. 2-11). The major circulation in the Northern Hemisphere is clockwise and in the Southern Hemisphere counterclockwise. Of course, continental masses are more effective in deflecting and controlling circulation in the oceans than in the atmosphere. A conspicuous example of deflected flow is the North Atlantic current, a continuation of the Gulf Stream, which is responsible for the warmth in the British Isles and western Norway that is exceptional for latitudes 50° to 70°N. At these latitudes in northeastern North America, the climate not only is much colder than in England and Norway but also is made even bleaker by a cold current from the Arctic Ocean flowing southward past Greenland. Ocean currents affect the climates of many lands.

The effects of uneven distribution of solar heat over the Earth are not restricted to the surfaces of the oceans but extend as well to deeper zones. Surface water that flows poleward is cooled and becomes denser, whereupon it gradually sinks to the bottom. Flowing in deep, complex paths it returns toward the Equator. The cold, dense water brings oxygen to bottom-living organisms and thus makes life possible at great depths. Gradually mixing with warmer water, it rises to the surface, ready to make another trip toward the polar regions.

World precipitation. Movements of air govern the *precipitation* of moisture as rain, snow, sleet, or hail. In general, precipitation is abundant wherever air

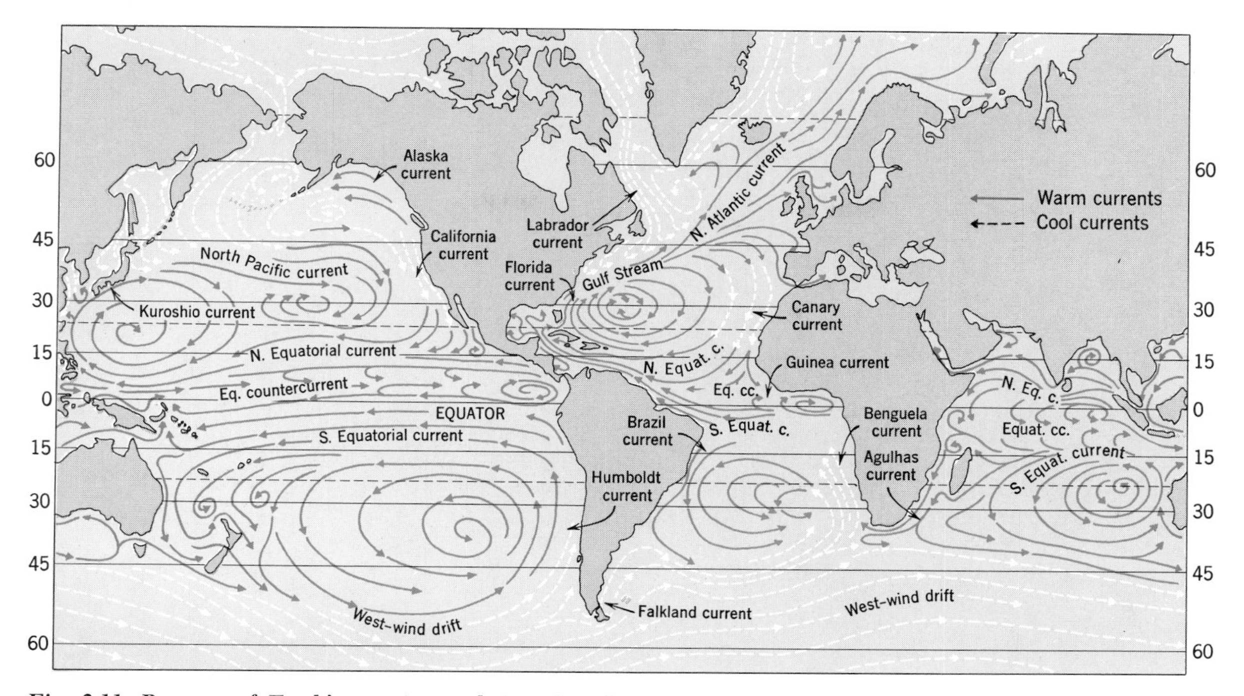

Fig. 2-11. Because of Earth's rotation and Coriolis effect surface currents in the oceans curve right in the Northern Hemisphere and left in the Southern. (After U. S. Navy Oceanographic Office.)

flows predominantly upward. At higher altitudes the rising air is cooled; it expands and loses its capacity to retain moisture. The excess moisture, which can be held no longer, is precipitated. Descending air, on the other hand, is heated, is compressed, and its capacity to retain moisture is increased. As a result, descending air absorbs moisture from the Earth by evaporation. These principles, together with our map of idealized air circulation, lead us to expect that the greatest precipitation should fall where air rises, near the Equator and around latitude 60° N and S, and that the most arid regions should occur where air descends in the latitudes around 30° and the polar regions (Fig. 2-12).

The arrangement of land masses and seas is an additional influence on precipitation. Masses of cold, heavy air flow far southward from the Arctic zone over the low interior of North America. Their leading edges force up warm, moist air that has moved northeastward from the Gulf of Mexico and the nearby Atlantic. The rising air precipitates widespread rain or snow. Because of this precipitation an arid belt is not present near 30°N in the midwestern and eastern United States, a region where we would otherwise expect to find one. In Asia the high Himalaya Mountain chain displaces the belt of deserts northward into central Asia by causing large seasonal rainfall near latitude 30°N. High mountain chains that extend generally north-south, as in North and South America, interrupt the normal east-west trends of climatic belts. In the western United States, air currents moving eastward are forced upward by the Sierra Nevada and the Cascade Mountains and drop most of their moisture on westward slopes and summits. As the air flows down the eastward slopes, it is warmed and takes up moisture from the land, thus creating an arid belt extending nearly north-south in this region. In many areas, ocean currents, such as the Gulf Stream, conspicuously influence precipitation by carrying heat and moisture beyond the normal limits of the climatic zones.

LITHOSPHERE

The outer zone of the solid Earth is known as the ***lithosphere.*** *The outer part of the lithosphere* is the ***crust,*** so called on the basis of an old and now-discarded concept that apart from a thin outer shell the Earth is molten. Despite its dubious origin, "crust" is a popular and convenient term, and by employing it we do not imply anything about the

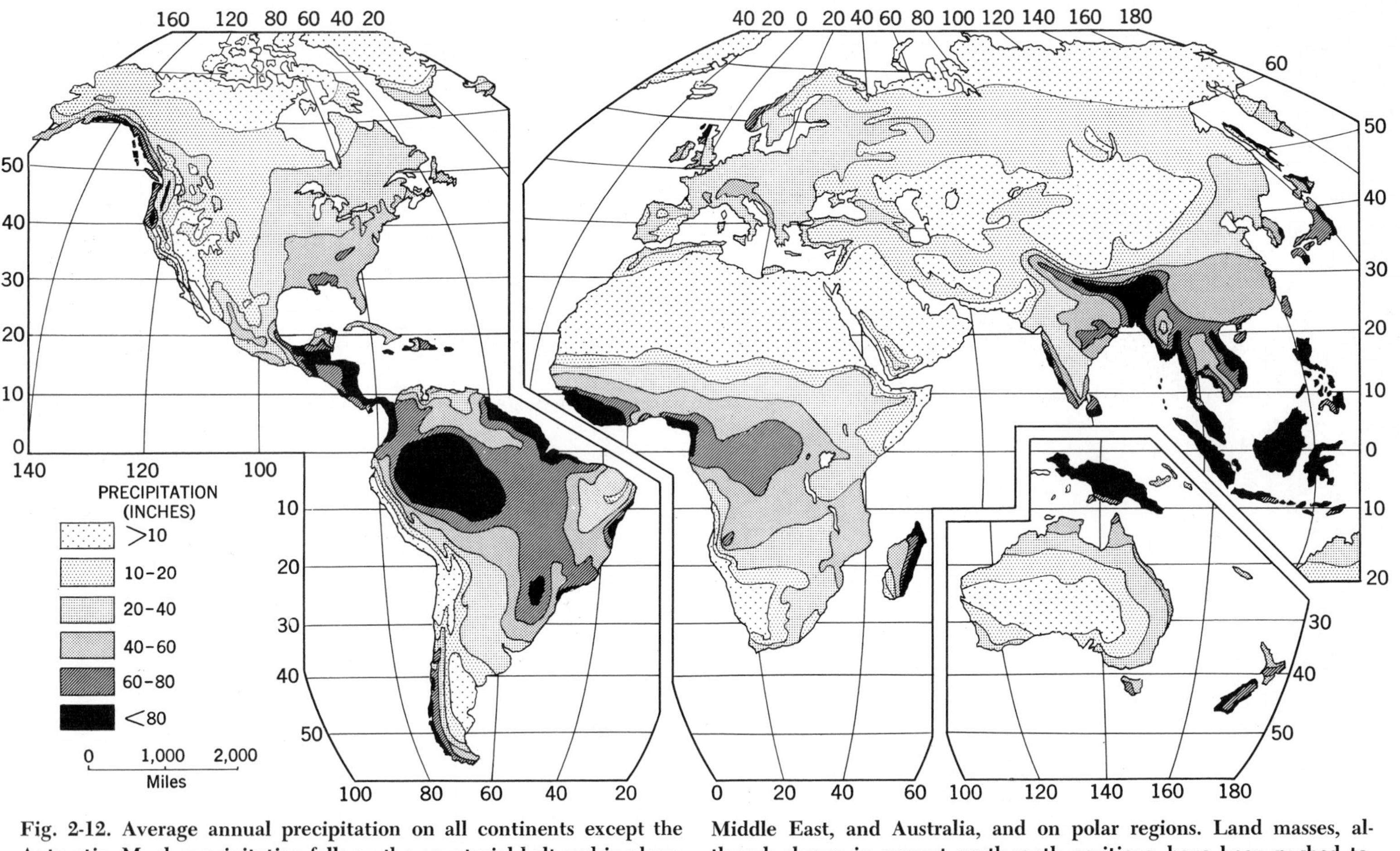

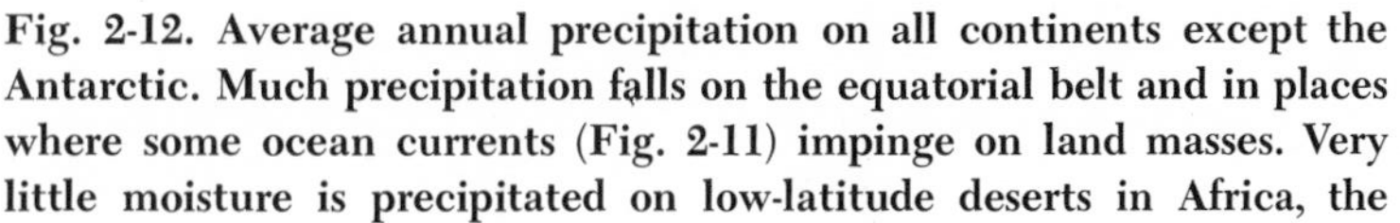

Fig. 2-12. Average annual precipitation on all continents except the Antarctic. Much precipitation falls on the equatorial belt and in places where some ocean currents (Fig. 2-11) impinge on land masses. Very little moisture is precipitated on low-latitude deserts in Africa, the Middle East, and Australia, and on polar regions. Land masses, although shown in correct north-south positions, have been pushed together laterally, greatly reducing the oceans. (V. C. Finch, G. T. Trewartha, and M. H. Shearer, 1959.)

condition and origin of the Earth's interior.

The lithosphere is divided into two groups of relief features of the first order: (1) ***continental masses,*** *the major high-standing parts of the lithosphere,* and (2) **ocean basins,** *the lower parts of the lithosphere that lie between the continental masses and are covered by seawater.* Thanks to modern exploration of the sea floor, we can now describe the major features of the entire outer part of the lithosphere (Pls. A and C).

Continental masses. The aggregate area of continental masses, divided from the ocean basins by the 2,000m depth contour, is about 40 per cent of the Earth's surface area. We can subdivide this area into two parts: (1) ***continents,*** *major land areas that stand above sea level,* constituting 29.2 per cent, and (2) ***continental shelves,*** *submerged marginal zones of continental masses,* 10.8 per cent.

Continents on the average stand slightly more than 800m above sea level. On continents we recognize relief features of the second order, such as mountains, continental shields, plateaus, and plains.

A ***mountain*** is *any land mass that stands conspicuously higher than its surroundings.* Mountains are of many varieties. Some are *resistant remnants standing high as a result of long-continued erosion within a wide area.* Included among such ***residual mountains*** are Stone Mountain, Georgia and the high points of the Catskills. The high peaks of the Andes, the Cascades, and Hawaiian Islands are examples of ***volcanic mountains,*** *conical accumulations of volcanic materials.* The great mountain belts of the Earth are known as *cordilleras* (Span., "strings"; pronounced either kôr-dĭl-ȳa′ras or kŏr-dil′ēr-ăs). In these great tracts the strata have been crumpled and broken; many such belts include in addition metamorphic rocks, granites, and volcanic rocks. The individual parts of mountain belts are designated by the terms *range, system,* and *chain.*

A single large complex ridge or a series of clearly related ridges that constitute a fairly continuous and compact unit is a ***mountain range.*** The Sierra Nevada in eastern California and the Front Range in Colorado exemplify mountain ranges.

A ***mountain system*** consists of *a group of ranges similar in general form, alignment, and structure, which presumably originated from the same general causes.* The Rocky Mountain system, which extends from Mexico northward into Canada, is a great assemblage of ranges, formed at approximately the

Fig. 2-13. Deformed metamorphic rocks of the Canadian Shield, north of Yellowknife and Great Slave Lake, Northwest Territories, Canada, seen from the air. Dark areas are lakes. (Royal Canadian Air Force, courtesy A. W. Jolliffe.)

Fig. 2-14. Strata, originally horizontal, have been tilted, bent, and broken by great pressures within the Earth. Such deformed strata characterize mountain chains. Northern Rocky Mountains south of Heavens Peak, Glacier National Park, Montana. (R. H. Chapman, U. S. Geol. Survey.)

same time by great crustal disturbances.

Mountain chain, a less exact term than range or system, designates *an elongate unit consisting of numerous ranges or groups, regardless of similarity in form or of equivalence in age.*

The foundation rocks of continental masses, with strata bent, broken, and interrupted by great bodies of granite, are exposed in *continental shields* (Fig. 2-13). The rocks of shields resemble those of many mountain chains (Fig. 2-14), but the surfaces of shields lack mountainous relief. Because of the similarity of rocks in shields to those in mountain chains, we infer that shields are the roots of former mountains, whose relief has been largely destroyed by erosion.

Rocks of continental foundations are blanketed by nearly horizontal strata of bedrock or sediments. The strata underlying plateaus consist of bedrock (Fig. 1-11, *B*) whereas those underlying plains consist of sediments, still loose or only partly converted into sedimentary rocks.

With the aid of the sketches in Pls. A, B, and C, let us scan the continents rapidly to mark their major features. The continent having the greatest relief is Eurasia. Large low parts of this great continent lie in Europe; despite its Alps and other commanding mountain chains, the average altitude in the European sector is little more than 300m. The average altitude of the Asiatic part is about 1,000m. The Tibetan Plateau and the great Himalayan chain, with Mount Everest reaching above 8,840m, have earned for this part of Asia the title "top of the world." At the eastern end of the Himalayas the mountains change their trend abruptly from east-west to north-south. Despite this abrupt change the mountains continue southward across southeast Asia, down the Malay Peninsula, and into Indonesia, where they gradually resume their east-west trend.

Large parts of Africa are composed of several continental shields and plateaus; prominent mountain chains occur only at the extreme northern and southern edges. Australia consists largely of monotonous plains and plateaus, in the eastern part relieved by a mountain chain of moderate height.

North America, with average altitude of about 800m, contains two mountain tracts of unequal size that trend generally north-south; the higher one, on the west, known collectively as the ***North American Cordillera,*** embraces *all the mountain units in western North America, from the eastern border of the Rocky Mountains to the Pacific coast.* Similarly, *the entire broad mountain belt that extends continuously from Alaska to Cape Horn* is the ***American Cordillera.*** On the east is the Appalachian chain with moderate heights. Between these two chains is a vast region of interior plains. Extending from east of the Appalachians around their southern termination and continuing on toward the southwest is the Atlantic and Gulf Coastal Plain.

In South America the great Cordilleran belt includes the Andes and the high plateau of Bolivia; to the east, much lower ground extends to the continental shield in eastern Brazil.

A thick ice sheet conceals much of the surface of the Antarctic Continent. If the ice sheet were removed we would find mountainous relief.

Ocean basins. Although seawater covers about 71 per cent of the Earth, the aggregate area of the ocean basins totals only 60 per cent of the Earth's surface. The greatest depths occur, not in the centers of the basins as we might suppose, but near their margins. This is illustrated strikingly by the Pacific Ocean basin, which occupies one-third of the Earth's area and in which the average depth is about 4km. In many places, deep, narrow trenches, in which depths range from 7.6km to about 11km, lie along the boundary between this basin and adjoining continental masses (Fig. 22-3). The vertical distance from the top of Mount Everest to the bottom of the deepest trench in the Pacific Ocean is 19.75km. Great though this may seem by human standards, it is only 1.5 thousandths of the diameter of the Earth and on a global scale is negligible. Another large feature of the Pacific basin is the East Pacific Rise, a low rocky ridge that extends across the southeastern part from Central America to the Antarctic Ocean.

The Atlantic Ocean basin contrasts with the Pacific basin in several important respects. Its middle third is occupied by the vast rocky Mid-Atlantic Ridge, whose summit is marked by a narrow, steep-sided valley paralleling the trend of the ridge. Along most parts of the borders of the Atlantic basin are great bodies of sediment instead of trenches as seen in the Pacific. Only two small arcuate trenches are present, one between the Caribbean and North Atlantic basins, and the other along the southwestern part of the South Atlantic basin.

The smaller Indian Ocean basin bears a closer resemblance to the Atlantic than to the Pacific basin. Except for the gently curving trench along its northeastern part, the margins of the Indian basin are occupied by great bodies of sediment. Several rocky ridges with median valleys occur on the basin floor.

Features common to both continental masses and ocean basins. Other prominent features, found in all oceans, also occur on continental masses. Lined

Fig. 2-15. Atitlán and Tolimán Volcanoes, Guatemala, with Lake Atitlán in foreground. On lower slopes of cones tonguelike masses of igneous rock have formed by solidification of what was viscous molten material. At right three tongues have been superposed; another (*lower center*) is bordered by conspicuous ridges (blue arrows). (U. S. Air Force.)

up in great rows and locally scattered at random on continental masses and in ocean basins are volcanic cones (Fig. 2-15). Many such cones form majestic mountain peaks on land and islands within the oceans. Others, equally large, lie beneath the surface of the sea. Some of these were never built up to sea level; others were formerly islands, but their tops were truncated by wave erosion, and afterward they sank down out of sight.

Great linear systems of breaks in the Earth's crust are ***fracture zones*** (Fig. 2-16). Conspicuous fractures with nearly east-west orientation are located in the eastern Pacific, tropical Atlantic, and central Indian Oceans. Other fracture zones, diversely oriented, occur on lands and on other parts of the ocean basins.

We shall meet these features again on later pages. We leave them now to examine the Earth's internal heat and magnetism in rocks, and to ponder the cause of crustal movements.

INTERNAL HEAT

Molten rock material from volcanoes and hot water from thermal springs disclose that the interior of the Earth is hotter than its surface. Volcanoes and hot springs are extremes, but the conclusion that the Earth's interior is hotter than its surface is supported by an increase of temperature downward in deep mine shafts and borings. Temperatures have been measured likewise by lowering sensitive heat probes into sediments of the ocean floor. From many temperature measurements the average ***geothermal gradient****, the rate of increase of temperature downward in the Earth,* has been determined to be about 1°C/30m.

To calculate the rate of terrestrial heat flow from

Fig. 2-16. Small segment of an ancient linear fracture, whose overall length (not shown) is 300 miles. Scarp, 900 feet high, was formed not by movement along the fracture but by erosion of weaker rocks to the right. McDonald Lake, near Great Slave Lake, northwestern Canada. (Royal Canadian Air Force, courtesy A. W. Jolliffe.)

these measurements we need to know the thermal conductivities of rock materials, which can be measured in the laboratory. Calculations made from temperature measurements in many areas and from the laboratory study of rocks show that the average rate of heat flow is 1.62 microcalories/cm^2/sec (1 microcalorie equals 10^{-6} calories). This amounts to only one twenty thousandth (0.5×10^{-4}) of the average influx of solar heat. Except near volcanoes and thermal springs, therefore, temperatures of rocks and regolith near the Earth's surface are determined largely by mean annual air temperature rather than by terrestrial heat flow. Near the poles air temperatures are so low throughout the year that the ground is frozen perenially (Fig. 2-17). In cool-temperate belts water in the ground freezes and thaws seasonally.

MAGNETISM IN ROCKS

We mentioned previously that the Earth's magnetic field serves as a protective shield against charged particles arriving from outer space. The magnetic field serves men in other ways. Since at least A.D. 1000, mariners have employed compasses oriented by magnetic lines of force to guide their ships across the sea. Compasses permit geologists and other mapmakers to orient their maps and make various field measurements.

Careful magnetic measurements made through many years and at many localities have shown that the intensity and orientation of the Earth's magnetic field vary from time to time at any point and from place to place. Many of the variations with time at individual localities are caused by changes in the Earth's interior, but some result from atmospheric phenomena. Magnetic variations from place to place result from differences in the abundance of magnetic-iron compounds near the Earth's surface. Close to rocks containing abundant magnetic-iron compounds the orientation and intensity of the magnetic lines of force deviate from their regional pattern. Magnetic surveys, therefore, can be used to locate some kinds of iron deposits (Fig. 23-9) and to follow distinctive magnetic patterns of rocks from areas of exposure to areas where the bedrock is covered. Magnetic surveys are now made with airborne instruments that record continuously.

Even where particles of magnetic-iron compounds are not abundant enough in rocks to distort the regional magnetic field, they are still of geologic interest. The reason is that each particle, like the Earth as a whole, is surrounded by its own magnetic field. The orientations of the fields have been determined by sensitive instruments for many rock samples, with surprising results. In any rock sample, the fields surrounding all particles show the same orientation. In samples from many rock bodies, the fields are oriented the same as the Earth's present magnetic field, whereas in others the fields are not so oriented but diverge from the Earth's field at various angles. Clearly, those that are unlike the Earth's existing magnetic field could not have originated under its influence. In some samples, although the tiny fields parallel the Earth's field, their sense of polarity is reversed. That is, the south poles of the tiny fields coincide with the existing north pole of the Earth's field.

From these measurements we can draw the following inferences. (1) The tiny fields were created at the time the rock originated. (2) The tiny fields were aligned parallel with the Earth's magnetic field prevailing at that time. (3) Subsequently either the orientation of the specimen changed because of crustal movement, or the Earth's field changed its position and/or its polarity, or both possibilities occurred. The changes of orientation made by crustal movement can be corrected for by studying rocks in the field. When this has been done we find that some samples still show differences between their tiny fields and the Earth's present field. The differences are confirmed by measurement of rock samples that have not been subjected to crustal movement. The relict magnetism retained in the particles, which makes the tiny fields persist, is *remanent magnetism.*

Particles of magnetic iron become aligned with the Earth's magnetic field, and then become "frozen" in position as a result of two chief rock-making processes: (1) solidification of particles in molten rock material, and (2) settling of particles of suspended sediment onto the bottom of a lake or of the sea.

Molten rock material is hotter than the *Curie point,* the temperature above which a substance loses its magnetism. Hence particles that solidify from molten liquid, having lost all traces of any older magnetic fields because of the Curie effect, acquire a new field after they cool below their Curie point. The magnetism acquired is that of the Earth at the time of cooling. Particles settling through water possess remanent magnetism because they are cool. Accordingly, they behave as tiny compass needles and align themselves in the Earth's magnetic field then prevailing.

Studies of remanent magnetism raise many profound questions about the history of the Earth. What is the significance of ancient pole positions that are at variance with the present Magnetic Poles? What have been the effects of reversals of polarity of the

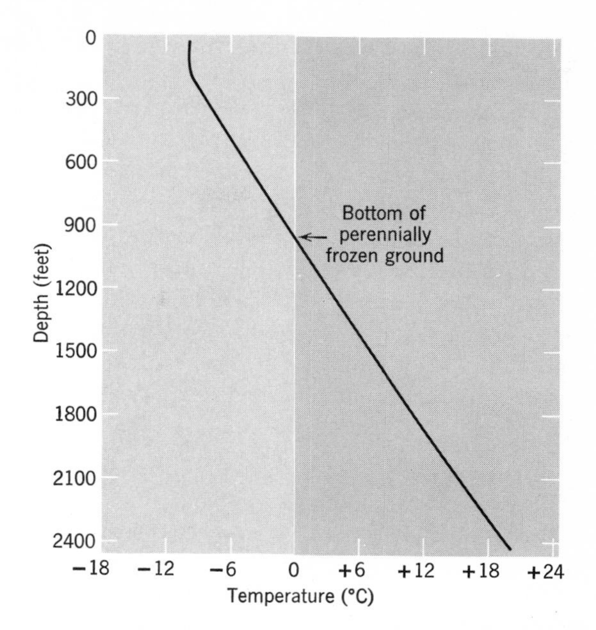

Fig. 2-17. Relation of temperature to depth in Cape Simpson well 28, southeast of Point Barrow, Alaska, measured May 22, 1953. (Max Brewer, 1958.)

present field? We need to know more about geologic materials and processes before we can come to grips with these questions. Although we set them aside now, we shall return to them again in Chapters 6 and 22.

WHAT MAKES THE CRUST GO UP AND DOWN?

At the beginning of this chapter we remarked that lands have persisted despite the tendency of erosion to eliminate them. The obvious implication is that internal forces keep elevating continental masses, thus replacing land destroyed by erosion. Strata of plateaus, consisting of sediments deposited in shallow seas, reveal more of the story. Sediments do not form piles of very large size; instead they spread out in horizontal layers. Strata composed of sediments that were deposited in shallow seas, but that are thick as well, record subsidence of the crust. Here is evidence that crustal blocks move downward. By their present positions high above the sea, the strata of plateaus indicate they have been raised without being much damaged. Marine strata, both thick and deformed, which today form parts of mountains, tell us not only that they subsided, but that later they were crumpled and lifted above the sea. What is responsible for these movements downward and upward and for the crumpling, in many places intense?

The explanation of these and related phenomena is so complex that we shall not try to deal with it further until after we have examined more evidence. From what we have seen already we realize that the Earth is not an inert body, bearing only the record of past activity. On the contrary it contains mighty forces, hidden yet still active, which loom large in the geologic cycle and whose effects are profound. Let us now take a closer look at the materials on which the cycle operates.

Summary

1. The geologic cycle requires air, water, land, solar energy, and internal forces within the Earth.

2. Nuclear reactions within the Sun generate energy that is transmitted outward, at the speed of light, by electromagnetic waves having a broad spectrum.

3. The Earth's magnetic field and atmosphere shield the Earth from charged particles and from energy transmitted by very short waves. Clouds and dust can reduce greatly the amount of energy that reaches the Earth's surface.

4. Gravity, acting inward, tends to shape the Earth into a sphere; centrifugal force, generated by rapid rotation, bulges out the Equator, shaping the Earth into an oblate spheroid.

5. Rotation deflects currents of air and water toward their right in the Northern Hemisphere and toward their left in the Southern Hemisphere.

6. Unequal distribution of solar energy over the Earth makes air and water circulate. Circulating moisture is a prime mover in the geologic cycle.

7. The solid part of the Earth consists of high-standing continental masses and low-standing ocean basins.

8. Plateaus, plains, and mountain chains are the larger features of continents; trenches, mountainous rocky ridges, and great bodies of sediment characterize ocean basins.

9. Volcanic cones and linear fracture zones are distributed widely, both on continental masses and in ocean basins.

10. Some rocks preserve a record of the orientations and polarities of ancient magnetic fields. Some ancient fields coincide with the Earth's present field; others differ from the Earth's present field in orientation, in polarity, or in both orientation and polarity.

Selected References

Bates, D. R., ed., 1957, The Earth and its atmosphere: New York, John Wiley.
Galley, D. P., ed., 1963, Space science: New York, John Wiley.
Gamow, George, 1963, A planet called Earth: New York, Viking Press.
Kuiper, G. P., ed., 1954, The Earth as a planet: Chicago, Univ. of Chicago Press.
Larousse Encyclopedia of the Earth, 1961, New York, Prometheus Press.
Lee, W. H. K., 1963, Heat flow data analysis: Rev. Geophysics, v. 1, p. 449–479.
Mason, Brian, 1962, Meteorites: New York, John Wiley.
Newell, R. E., 1965, The upper atmosphere [wall chart]: U. S. Navy Fleet Weather Research Facility 26-0665-106.
Shaw, D. M., 1965, Sunspots and temperatures: Jour. Geophys. Research, v. 70, p. 4997–4999.
Strahler, A. N., 1963, The Earth sciences: New York, Harper and Row.
Whipple, F. L., 1963, Earth, Moon, and planets: Cambridge, Mass., Harvard Univ. Press.

Matter and Minerals

The states of matter
Changes of state
Minerals as letters of the geologic alphabet
Definition of minerals
Rock-making silicates
Shapes of mineral particles
Minerals as indicators of environment

Plate 3

Quartz crystals, up to 1m long, at entrance to University Museum, Geneva, Switzerland. (B. M. Shaub.)

0
50 cm

CHAPTER 3 *Matter and Minerals*

The basis for recognizing the lithosphere, the hydrosphere, and the atmosphere depends on distinctions among the physical properties of matter in its three states of aggregation. The lithosphere implies the solid state, hydrosphere the liquid state, and atmosphere the gaseous state. In Chapter 2 we stated that solids, liquids, and gases of the three "spheres" tend to intermingle. Hence, even where the physical properties of matter in each of the three states contrast sharply, distinct boundaries do not always exist in nature among the "spheres" based on these states.

Many natural processes depend on the properties of matter in one of its three states of aggregation or upon a change from one state to another. A review of some of the properties of matter will facilitate our study of these processes. Because the lithosphere is of special concern to geology our discussion emphasizes the solid state. However, rocks and regolith consist not only of a framework of solid mineral particles but also of spaces occupied by fluids. Although we emphasize the solids we do not intend to impugn the importance of the fluids. The significance of fluids is threefold: (1) fluids can react with solids, (2) fluids can exert pressures and thus modify the behavior of solids; and (3) locally some fluids are themselves valuable substances (Chaps. 11 and 23).

MATTER

States of aggregation. The states of aggregation of matter are solids, liquids, and gases. Familiar physical properties of matter in each state are expressions of the spacing of particles (atoms or molecules) and of the degree of order in the arrangement of particles. Particles respond to three factors: (1) interatomic or intermolecular forces, which are strong at close range but act through only very small distances; (2) temperature, and (3) pressure. Interatomic and intermolecular forces tend to keep particles together and create order among them. Temperature governs thermal motions. At high temperatures thermal motions are intense and tend to drive particles apart and create disorder among them. At low temperatures thermal motions are less intense. At absolute zero thermal motions cease. Pressure tends to force particles together. The state of aggregation of a given body of matter, therefore, represents an adjustment of the particles to the conditions of temperature and pressure. We can satisfy ourselves on this point by the simple experiment of heating a substance that is solid at ordinary temperature (15 to 20°C). At higher temperatures the solid melts to become a liquid.

The boundaries between bodies of matter having different states of aggregation (or between bodies of matter having the same state of aggregation but different physical properties) are ***interfaces.*** The surface of the sea is a liquid/gas interface. The contact between a body of oil and a body of water is a liquid/liquid interface. Boundaries are called interfaces only where, as in these examples, they are well defined or where the transition zone between them is thin. The characteristics of each state can be shown by examples of various solids, liquids, and gases.

Solids. In solids, particles are packed so closely together that we can regard them as touching one another; they are arranged in what for most purposes can be considered as fixed frameworks. Solids are of two kinds. In *crystalline solids,* the patterns of the particles (molecules, atoms, or ions) are repeated and symmetrical (Fig. 3-1, *A*); in *amorphous solids,* patterns are irregular (Fig. 3-1, *B*). We shall have more to say presently about the internal arrangement of particles in solids.

Because particles are closely packed and arranged in fixed frameworks, solids tend to maintain their shapes and volumes. *Resistance to change of shape* is ***rigidity;*** *resistance to change of volume* is ***compressibility.*** The ability of a solid to resist changes

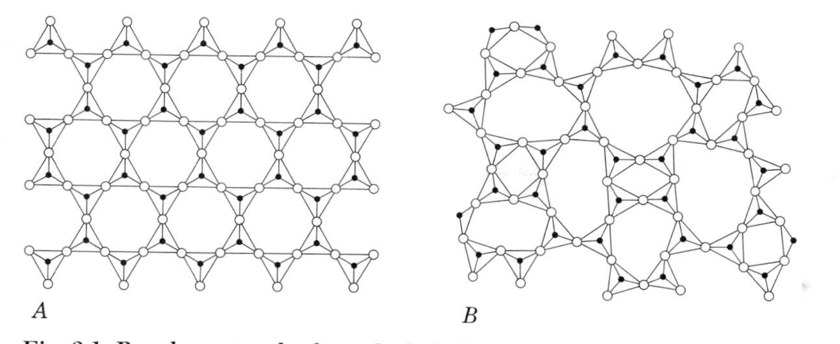

Fig. 3-1. Regular network of tetrahedral clusters of particles of crystalline solid and irregular network in amorphous solid. (Vanders and Kerr, 1967.)

of shape or volume is a measure of its strength. Under many conditions solids can be made to fracture (rupture) along well-defined surfaces; solids that fracture readily are said to be *brittle.* Under other conditions solids do not fracture but undergo ***plastic flow,*** *a continuous and permanent change of shape in any direction without breakage;* solids that flow readily are said to be *plastic.* The same solid can be brittle under some conditions and plastic under others. For example, the novelty substance called Silly Putty is brittle when struck a sharp blow, yet will flow plastically under its own weight if left undisturbed. Likewise natural solids, such as rocks, regolith, and ice, fracture under some conditions yet flow plastically under other conditions. As we shall see in later chapters, the outcome of many processes depends on the behavior of solid materials.

Solids react to the passage of waves in characteristic ways. In fact one of the most fruitful methods of studying solids is to transmit through them waves of light or sound, X-rays, and other waves and to record the results. Analysis of wave behavior in solids is a fundamental research method.

Another physical phenomenon related to the state of aggregation of matter is ***diffusion,*** *the transport of particles in the absence of bulk flow.* Diffusion can occur in solids but the rate is very slow.

Liquids. In contrast to solids the particles in fluids are not arranged in fixed frameworks. Fluids, which include liquids and gases, lack rigidity and tend to flow under almost all conditions. *The tendency within a body to resist flow is* ***viscosity.*** Because the viscosity of most fluids is low they must be placed in a container if they are to be prevented from flowing. At ordinary temperatures liquids, like most solids, are essentially incompressible and occupy a definite volume; but unlike solids, liquids require a container in order to retain their shape, which follows that of the container.

Although not set in a fixed framework the molecules of liquids are so closely packed that the free volume among them is estimated to be only 3 per cent of total volume. These molecules move readily through short distances, as we can see easily by suspending in water solids whose diameters are less than 2 microns (Fig. 4-11). The movement of the tiny suspended particles, called *Brownian movement,* results from impacts upon them of moving molecules of water. The relatively greater ease of movement of particles in liquids as compared with solids is also expressed by rates of diffusion, which are greater in liquids than in solids.

Within a body of liquid the resultant forces of attraction among molecules are equal in all directions because the number of molecules is the same in all directions. Along liquid/solid or liquid/gas interfaces, however, the number of liquid molecules is much smaller on the nonliquid side than on the liquid side. Although along the interface the molecules of the liquid continue to attract one another in all directions as they do within the main body of the liquid, their unequal distribution along the interface causes unequal resultant forces of attraction. These forces reach their maximum both perpendicular and parallel to the interface (Fig. 3-2). *Interfacial tension* is a consequence of this unequal attraction.

Surface tension is a special case of interfacial tension along a liquid/gas interface. Surface tension tends to shape a body of liquid, in contact with gas, into a sphere, the geometric shape with the least surface area. This tension is the force per centimeter

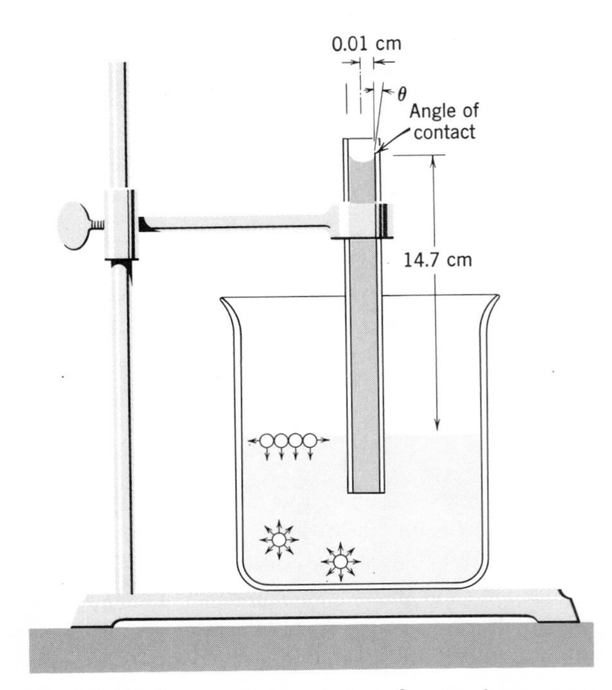

Fig. 3-2. Surface tension and rise of water in an open capillary tube. Resultant forces on water molecules (small arrows) are equal in all directions within body of liquid. Resultant forces are at maximum parallel with and perpendicularly away from a liquid/air interface. Further explanation in text.

that opposes increase of the surface area of a body of liquid; surface tension varies inversely with temperature.

When liquid is placed along a solid/air interface it forms a characteristic angle of contact. For water the angle of contact with most solids is nearly 0°; hence water tends to wet the surface of the solid. When a glass tube is inserted into a body of water, surface tension pulls the water up the inside of the tube. For mercury the angle of contact with most solids is 180°; hence mercury tends not to spread along a solid surface. When a glass tube is inserted into a body of mercury the mercury does not rise up the tube. Instead the mercury surface is depressed by the glass.

As a result of its tiny angle of contact and its surface tension, water rises through very small openings, called *capillaries,* whose diameters are smaller than that of a human hair. At ordinary temperature water will rise to a height of 14.7cm inside a vertical glass tube having an inside radius of 0.01cm (Fig. 3-2). Because many openings between the mineral particles of regolith and rocks are small enough to behave as capillaries, water rises within these openings just as it does inside the vertical glass tube.

Gases. In gases the particles are neither set in a fixed framework as in solids nor packed closely as in liquids. Instead, they are in constant random motion. Because of these motions, gases expand to fill the container into which they are placed and diffuse readily. Not only do gas molecules constantly move through long distances compared with the diameters of the molecules but they are spaced widely. Under ordinary temperature and 1 atmosphere of pressure molecules occupy only about 1/1,000 of the space in a given volume of gas.

At some temperatures and pressures two fluid phases, liquids and gases, exist. At other temperatures and pressures only one fluid phase exists. The point at which the distinction between gas and liquid disappears is known as the *critical temperature* (Fig. 3-3). When the temperature is lower than the critical temperature, pressure determines which fluid phase exists. The pressure at which both fluid phases exist is the *vapor pressure.* Here the denser fluid phase is a liquid, and the less dense phase a vapor or gas. The liquid is stable at pressures greater than the vapor pressure, and the gas is stable at pressures less than the vapor pressure. On a pressure-temperature graph (Fig. 3-3) the point at which solid, liquid, and gaseous states coexist is called the *triple point.*

Changes of state. Many geologic reactions involve changes of matter from one state to another. Of particular importance to geology are changes to and within the solid state. Solids can form from gases by condensation; from liquids by two processes, precipitation from solutions and solidification by cooling from melts; and from other solids by recrystallization or change of phase (Fig. 3-4).

Changes of state can occur not only without changes of chemical composition but also while chemical changes are in progress. The relationships between changes of state and the presence or absence of changes of chemical composition can be illustrated by two simple experiments. (1) Melt 10g of ice to water, vaporize the water, and then cool the vapor to liquid and freeze it into ice again. Because no chemical change has occurred we get back the same material, ice, and if none of the water vapor escaped, we recover the original amount, 10g. (2) Burn 10g of coal and convert most of it into a gas. In this case a chemical change accompanies the change of state: carbon has been oxidized to carbon

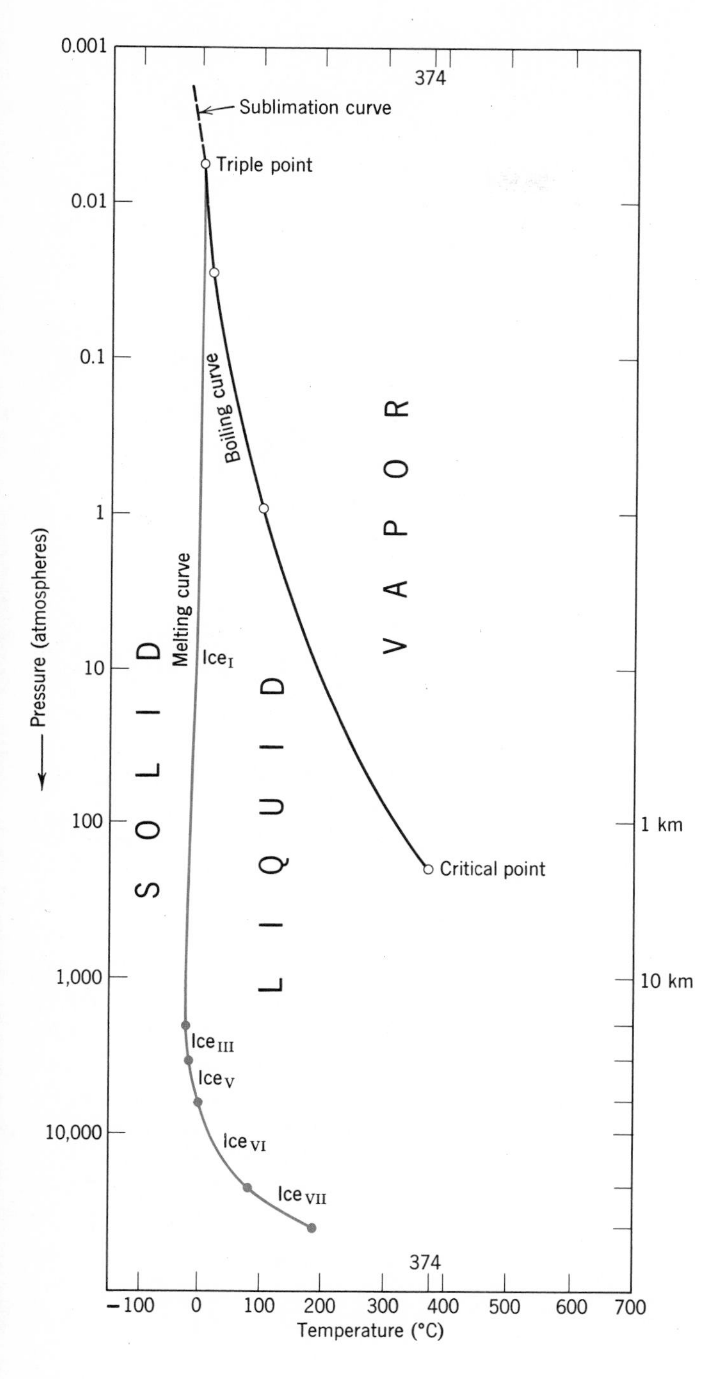

Fig. 3-3. The physical states of water and boundaries between them, shown on a graph of pressure versus temperature. Further explanation in text. (After Fyfe, 1964.)

dioxide gas. The carbon dioxide can be cooled and converted into a solid, but the result is not the original coal but a chunk of "dry ice." Because oxygen has been added to the carbon the "dry ice" weighs more than 10g.

Let us look more closely at the changes of state in water by graphing temperature versus energy added or released at atmospheric pressure (Fig. 3-5). Consider first the liquid/gas interface at the right side of the graph. *The amount of energy added or released when particles move across the liquid/gas interface* is the ***latent heat of vaporization.*** This amount of energy must be added before liquid particles cross the interface and become gas particles.

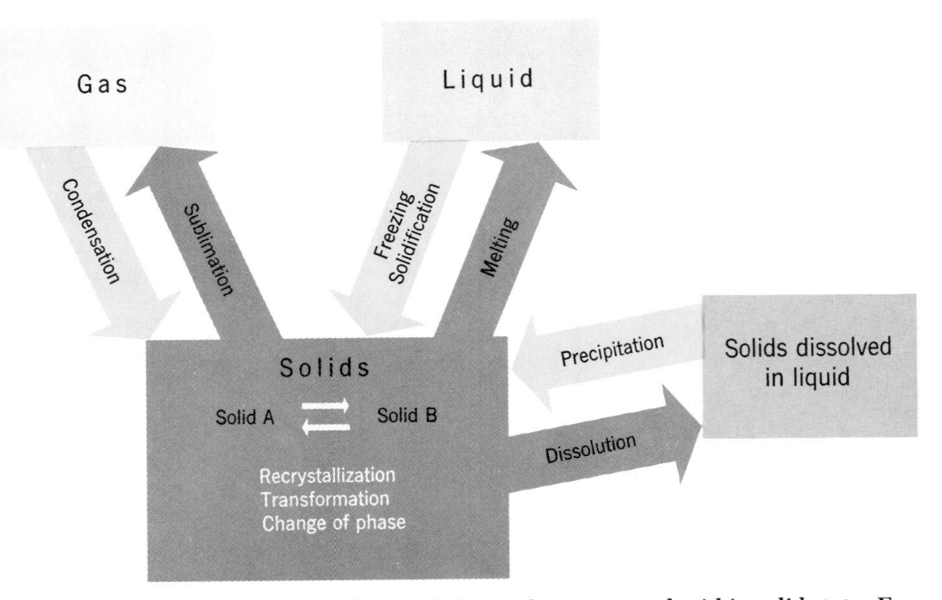

Fig. 3-4. Sketch of changes of solids to and from other states and within solid state. Formation of solids shown by processes and arrows colored blue; destruction of solids by processes and arrows colored black.

Energy is released when gas particles cross the interface in the opposite direction and become liquid particles. Because the latent heat of vaporization of water is so large (539.0 cal/gm at 100°C), tremendous quantities of heat are exchanged in the atmosphere when water evaporates or precipitates.

At the solid/liquid interface on the left side of the graph particles can change from solid to liquid or *vice versa.* Although in vaporization and condensation we can think of particles as crossing the liquid/gas interface, in melting and freezing we must visualize the solid/liquid interface as moving through the aggregate of particles. *The amount of energy added or released in changing state from a solid to a liquid* is the ***latent heat of fusion.*** An amount of energy equal to the latent heat of fusion is required to melt a given quantity of solid into a liquid without raising its temperature. The same amount of energy is released when the liquid freezes or solidifies. At 0°C the latent heat of fusion of water is 79.7 cal/gm.

The temperature at which changes of state occur is influenced by pressure (Fig. 3-3). The three change-of-state curves meet at the triple point. Conditions at the triple point of water, a pressure of 0.00603 atmosphere and a temperature of +0.0098°C, never vary. Accordingly the triple point of water serves as a fundamental calibration point; it is 273.16° above absolute zero on the Kelvin (or Absolute) temperature scale.

Increasing the pressure raises the boiling temperature of water, but decreases the melting temperature of ordinary ice. Such ice, which is less dense than water, is one of the few substances whose melting temperatures are *lowered* by increased pressure. The pressure of a skater's weight concentrated in a narrow steel blade lowers the melting temperature of ordinary ice. Therefore, the ice beneath the skate blade melts and a thin film of water lubricates the steel on the ice, thus making ice skating possible. Anyone who doubts the importance of this phenomenon should try to skate on "dry ice" or extremely cold ice. Denser forms of ice, indicated by Roman-numeral subscripts in Fig. 3-3, originate as pressure is increased. At pressures greater than 3,400 atmospheres the denser forms of ice behave like most other substances—their melting temperatures are *raised* when pressure increases. We mention these denser forms of ice to illustrate changes of phase in solids; such ice does not occur in nature.

At the *critical point,* defined by the *critical temperature* and *critical pressure,* the liquid/gas interface disappears. The critical temperature (374.0°C for water) is the temperature above which the liquid state can not exist. The critical pressure (218.4

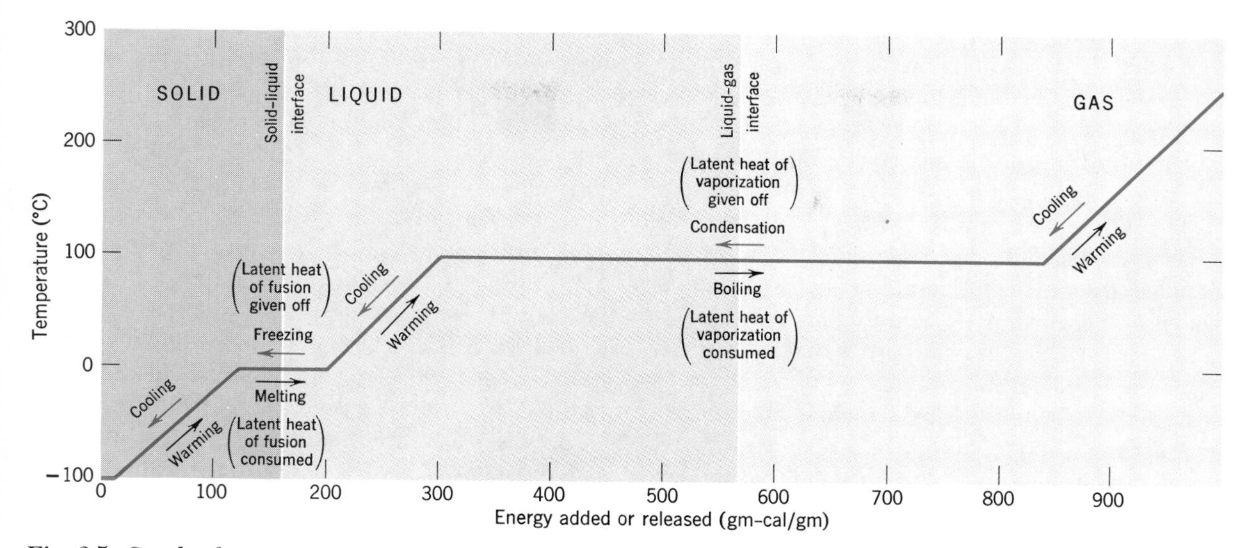

Fig. 3-5. Graph of temperature versus energy added or released for water at atmospheric pressure. Diagonal segments of curve, where temperature and energy are directly proportional, represent changes of temperature within the same physical state. Horizontal segments, where changes of energy do not change temperature, are levels of temperature at which changes of state occur. Lengths of horizontal segments indicate amount of energy involved in changes of physical state. (After Slabaugh and Parsons, 1966.)

atmospheres for water) is the pressure at which the liquid state reappears after the critical temperature has been reached. Pressure cannot be increased beyond this level until all vaporized water present has condensed into the liquid state.

At temperatures and pressures less than those at the triple point, solid and vapor states are in direct contact. *Sublimation* is the process by which particles of a solid pass directly into the gaseous state. *The amount of energy involved in moving particles across the solid/gas interface* is the ***latent heat of sublimation;*** it is equal to the latent heat of fusion plus the latent heat of vaporization even though the liquid state is not involved. Sublimation causes the gradual disappearance of snow at temperatures below 0°C.

Summary. The states of aggregation of matter can be characterized by the degree of order among particles, by the spacing between them, and by the ease of particle movement. In gases particle arrangement is chaotic or completely disordered, particles are widely spaced, and move easily and constantly. In liquids particle arrangement is more ordered than in gases, particles are closely spaced, and move readily through short distances. In solids particles are arranged in fixed networks, are closely spaced, and can be assumed to be fixed. Perfect order and lack of particle movement exist only in a crystalline solid at a temperature of absolute zero.

Changes in energy can cause matter to change its state of aggregation. The relationships between energy and the three states of matter are summarized graphically in Fig. 3-6.

MINERALS

As we have seen, in the lithosphere matter in the solid state predominates. The lithosphere consists of bedrock and regolith, in both of which the chief solids are minerals. In Chapter 1 we learned that much of geologic history is written in bedrock and regolith. If we regard bedrock and regolith as the principal components of the "language" of geologic history we should treat minerals as the "letters" of the "alphabet" in which this "language" is written. Minerals are not only the keys to rock and regolith but also the source of many raw materials required to sustain industrial civilization. Here we shall concentrate on minerals as letters of the geologic alphabet; we take up industrial aspects in Chapter 23.

Minerals can be studied with the naked eye, with eye plus low-power magnification provided by a hand lens or a binocular microscope, with eye plus high-power microscope, with various instruments that employ waves lying outside the range of human vision, and with chemical reagents. For our introductory study we can rely on observations with the naked eye, with eye plus low-power magnification, and on a few simple chemical tests. Although this choice introduces many ambiguities in the identification of complex minerals, these ambiguities are readily clarified with the aid of the instruments we have here excluded. Our discussion of minerals is intended to supplement laboratory study of mineral models and specimens. Appendix B contains reference material on mineral properties and descriptions of individual minerals.

Definition. Beyond being the chief solid constituents of rocks and regolith, what are minerals? Any attempt to formulate a more exact definition of them is complicated because many kinds of material have been classified as minerals. It will be useful to review some of these materials before we try to define minerals.

Most minerals are *crystalline solids.* As we mentioned previously the patterns of particles in crystalline solids are repeated and symmetrical. In other words, the particles are arranged in ***crystal lattices,*** *systematic, regular networks whose geometric configurations are symmetrical.* Crystallinity is an internal characteristic. It is always present in crystalline solids even though not expressed externally as ***crystals,*** *solids bounded by natural, regular plane surfaces formed by growth of crystal lattices* (Pl. 3). The building blocks of crystals are ***unit cells,*** *the smallest parts of crystal lattices that display the systematic pattern of the particles.*

Much of our present understanding of minerals stems from studies employing X-rays, which were discovered in 1895. X-ray studies of crystals began after 1912 and have been increasing ever since. The use of X-rays has established *crystal chemistry,* an entirely new discipline within ***mineralogy,*** *the study of minerals.*

Two important discoveries resulted from X-ray analysis of crystals: (1) the sizes of various ions within crystal lattices were determined, and (2) the arrangements of the particles were disclosed and particle spacing was measured. In crystals, ions can be regarded as rigid spheres having various diameters (Fig. 3-7). Not only ionic charge but also ionic diameter influences whether an ion can fit into the structure of a given crystal lattice. Ions of identical charge and about the same diameter, such as Fe^{2+} and Mg^{2+}, readily substitute for each other. The lattice changes only little when iron substitutes for magnesium, or vice versa, but the chemical composition changes considerably. When such substitution is made no other adjustment is required elsewhere in the lattice.

If ions are of about the same size their charges do not have to be identical in order for them to substitute for each other within a crystal lattice. Substitution of Ca^{2+} for Na^{+} is possible if the imbalance of the resulting electrical charge can be equalized by a compensatory substitution, such as an exchange of Al^{3+} and Si^{4+}, somewhere else in the lattice. The result is a coupled substitution.

Because of substitution, the major constituents of some minerals can vary. Despite such variation the basic structure of the unit cell remains the same even though cell dimensions change slightly with changes of composition. Some of the major parts are interchangeable but the structural framework remains about the same. Most minerals contain minor constituents that do not form integral parts of the unit cell. These fit into spaces between the major ions. Variation of minor constituents does not greatly influence the essential properties of the mineral, which are controlled by proportions of major constituents. Minor constituents can be regarded as impurities within the mineral. Other impurities result from growth of a large lattice around smaller lattices of other minerals or of small quantities of fluids. Such impurities acquired by growth are *inclusions.*

Although the crystal lattices of many minerals are complex structures whose composition is not easily described by chemical formulas devised for molecules, the lattices of some minerals are simple and their compositions can be described by molecular formulas. Because the particles of some lattices consist entirely of atoms of one kind, the mineral is classed as an element. *Diamond,* for example, is one of the mineral forms of the element carbon. In other minerals, the particles consist of more than one element and the lattice is built of molecules; hence, the mineral is a compound. *Ice* is an example of a mineral consisting of the compound water.

Minerals include some hydrocarbon compounds. Hydrocarbons are termed *organic* because they are synthesized by organisms, but organisms synthesize not only hydrocarbons but other substances as well. For example, the calcite in some limestone is derived principally from shells secreted by marine organisms

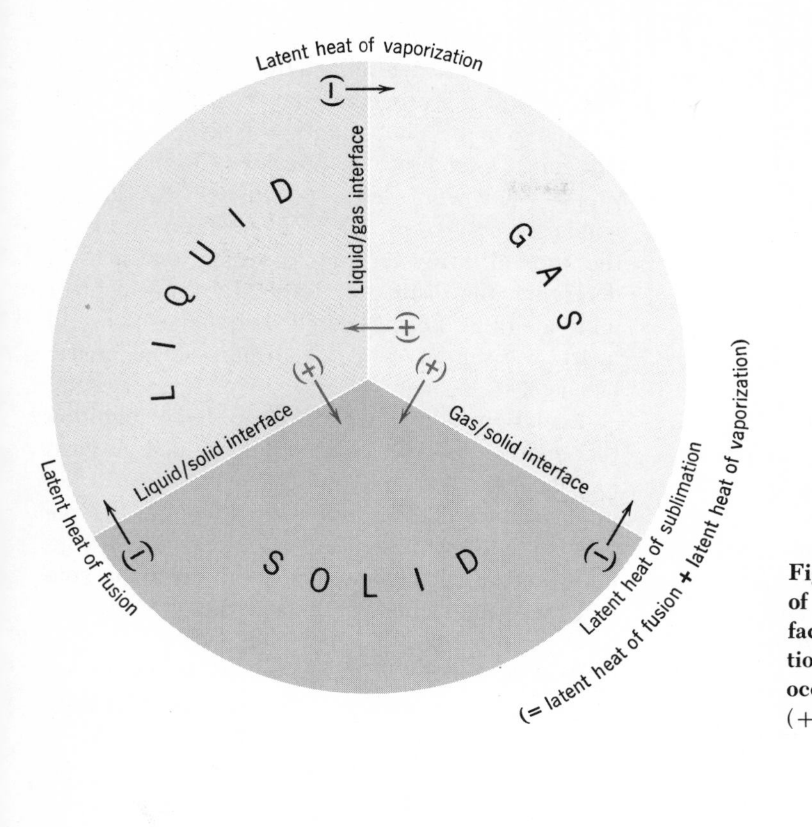

Fig. 3-6. Relationships of the three states of matter. When a particle and an interface change relative positions in the directions shown by arrows, a change of energy occurs. Release of energy is indicated by (+); consumption of energy by (−).

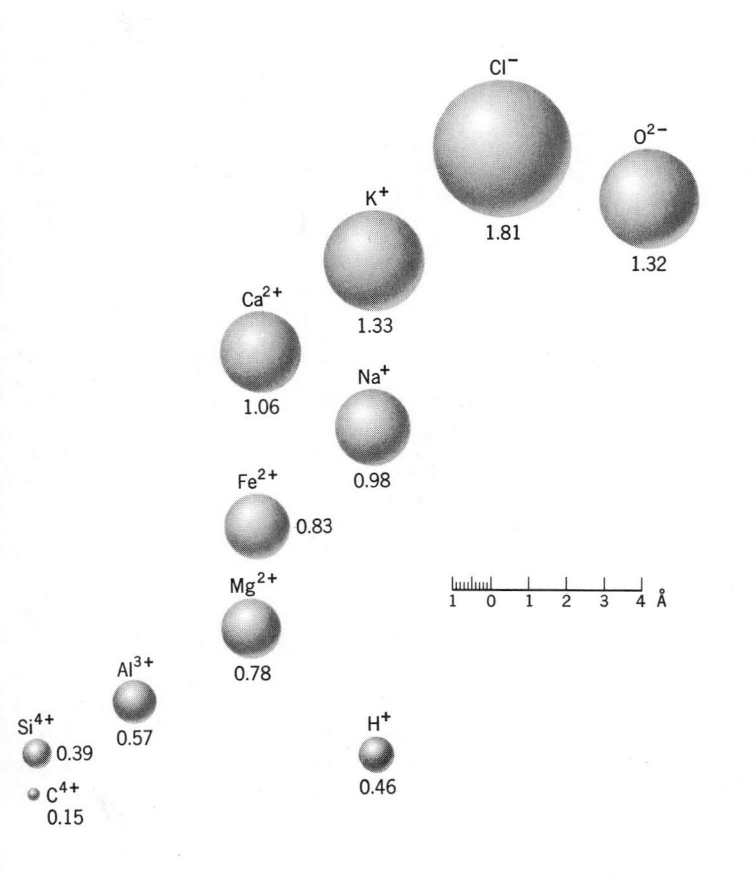

Fig. 3-7. Sizes of ions of 11 common elements, based on X-ray studies of crystals, range from C^{4+} at lower left to O^{2-} at upper right. Ions are arranged in vertical groups based on valence, from 4^+ at left to 2^- at right. Ions in each of the pairs Si^{4+} and Al^{3+}, Mg^{2+} and Fe^{2+}, and Na^+ and Ca^{2+} are about the same size and commonly substitute for each other in crystal lattices. Numbers show ionic radius (angstroms). (Data from Kalervo Rankama and T. G. Sahama, 1950.)

yet is not a hydrocarbon compound. Because *organic* has come to be associated with hydrocarbon compounds we shall use *biologic* as a general term for material created by organisms.

Most minerals are solids, although mercury is generally recognized as an exception and is considered to be a mineral despite being liquid under conditions prevailing at the Earth's surface. A final, and probably the only absolute qualification, is that a mineral must occur in nature. This excludes artificial substances that are produced only in the laboratory.

Taking into account these complications, we can define ***minerals*** as *naturally occurring substances, most but not all of which are crystalline solids, whose exteriors may or may not consist of crystal faces, whose atoms or ions of one or more elements or molecules of compounds are arranged regularly in a definite lattice, and whose chemical composition, though constant or variable within limits, bears fixed relationships to certain physical properties.*

Chemical composition. The chemical composition of a mineral is an important characteristic but by itself is not a reliable index to the physical properties of the mineral, as with many elements and compounds. For example, carbon occurs in two mineral forms: (1) as diamond, one of the hardest substances known, and (2) as *graphite,* one of the softest. Diamond and graphite are identical in chemical composition but because the arrangement of the atoms in the lattice is not the same in each, their physical properties differ. Likewise, calcium carbonate ($CaCO_3$) occurs in several mineral forms, each having the same chemical composition but distinct physical properties.

Nevertheless, chemical composition constitutes a basis on which minerals can be grouped. Some of the common groups (with examples of each) are: elements (*diamond, graphite, sulfur*) and compounds: carbonates (*calcite, aragonite, dolomite*); halides (*halite*); sulfides (*pyrite, galena*); sulfates (*gypsum, anhydrite*); oxides (*hematite, magnetite*); and silicates (*olivine, pyroxenes, amphiboles, micas, feldspars, quartz, clay minerals*).

Prior to development of X-ray techniques, grouping minerals by chemical composition alone was the only means of classification available. *Quartz* provides an example of the limitations of such classification. In the old days minerals were analyzed by grinding the specimen to powder and dissolving the powder. Relative amounts of elements in the resulting solution were determined by wet-chemical analysis. Because this procedure showed the composition of quartz to be SiO_2 quartz was assigned to the category of oxides. The powder-and-solution approach, however, destroyed the crystal lattice, which we now know is a fundamental property of the mineral. X-ray techniques, which permit us to determine the distinctive internal arrangement of minerals (Fig. 3-8), disclosed that the structure of quartz was similar to that of silicates; hence, quartz has been reclassified.

Analysis by X-rays not only permitted recognition that quartz was a silicate but also provided the basis for understanding and classifying all silicates. So important are silicates in rocks and regolith that we must study these minerals closely. Appendix B contains further details to augment the following general background information on rock-making silicates.

Rock-forming silicate minerals. The few silicate minerals that occur again and again in rocks are known as rock-forming silicates. The most significant discovery made by the X-ray analysis of silicates was the realization that the basic unit in their architecture was a tetrahedron having the chemical formula SiO_4. Structurally each tetrahedron consists of a small silicon ion surrounded by four larger oxygen ions (Fig. 3-9). In common rock-forming silicates the SiO_4 groups form combinations resembling those made by the CH_4 groups in hydrocarbon compounds. The structural arrangement of the tetrahedra exerts profound influences on the properties of the minerals, governing not only their physical properties but also their chemical properties. Because it determines which ions can be built into the crystal lattice, structural arrangement influences chemical composition.

On the basis of the relations of the silica tetrahedra to one another and to other ions, rock-forming silicate minerals can be divided into five groups. The ratio of silicon to oxygen varies from group to group. Mineral groups form the rows in Table 3-1. Individual minerals are arranged in columns by their occurrence in the three major rock groups.

Many of the silicate minerals of igneous rocks contain chiefly ions of ferrous iron and magnesium in addition to those of silicon and oxygen. Other ions, such as calcium and aluminum, if present, are much less abundant. Silicate minerals containing

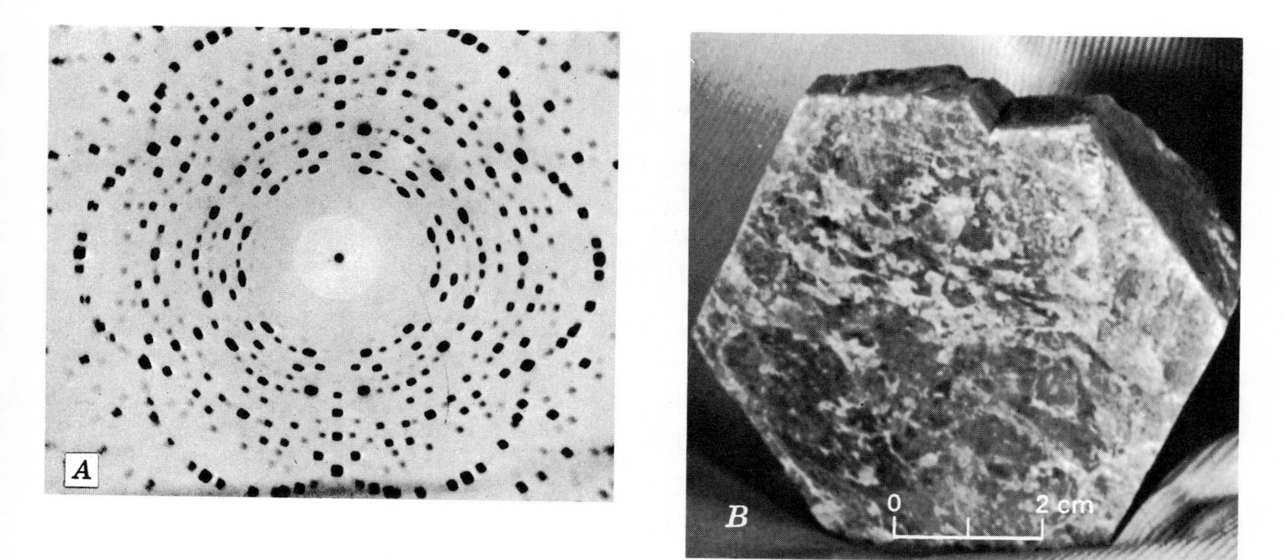

Fig. 3-8. A. X-ray diffraction photograph shows regular internal structure with six-sided symmetry of *beryl* (beryllium aluminum silicate). *B.* Compare the symmetry of beryl crystal. (A, C. S. Hurlbut; *B*, B. M. Shaub.)

abundant iron and magnesium are called *ferromagnesian silicate minerals.* Olivine and certain pyroxenes occur only in igneous rocks; other pyroxenes, biotite, and hornblende appear in both igneous and metamorphic rocks.

Most silicate minerals of metamorphic rocks contain only minor amounts of ferrous iron and magnesium, or none at all. Aluminum and calcium are the dominant ions in many minerals of metamorphic rocks. "Excess" oxygen occurs outside the silica tetrahedra in some metamorphic minerals; in others, hydroxyl and other ions are present.

Silicates of group 1 contain isolated individual silica tetrahedra (group 1a) or isolated groups of tetrahedra (group 1b) bound together by various metallic ions. *Olivine* is an example of a mineral in group 1a, and *beryl* of a mineral in group 1b.

In group-2 silicates the tetrahedra form single chains bound together by metallic ions. *Pyroxenes* are the principal minerals of group 2.

The tetrahedra form double chains in the *amphiboles,* characteristic minerals of group 3. The double-chain units are wider than the single-chain units of the pyroxenes but, like them, are bound together by metallic ions.

The minerals of silicate-group 4, the *micas, chlorites,* and *clay minerals,* are characterized by

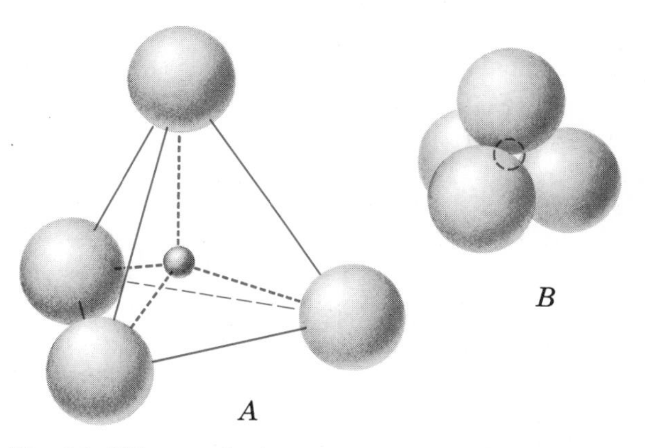

Fig. 3-9. Silica tetrahedron. A. Expanded view showing large oxygen ions at the four corners, equidistant from a small silicon ion. Dotted lines show bonds between silicon and oxygen ions: solid lines outline the tetrahedron. *B.* Tetrahedron with oxygen ions touching each other in natural positions. Silicon ion (dashed circle) occupies central space.

sheets of silica tetrahedra. Successive sheets are bound together by single ions or ion groups.

In group-5 silicates the tetrahedra form complex, three-dimensional networks. Each tetrahedron shares one of its oxygen ions with four other tetrahedra. In *quartz* no other metallic ions are present as major

TABLE 3-1. ROCK-FORMING SILICATE MINERALS SUBDIVIDED ACCORDING TO ARRANGEMENT

Group No.	Arrangement of SiO_4 Tetrahedra	Sketch of Tetrahedra and Corresponding Proportions of Si and O
1	Independent tetrahedra or groups of tetrahedra a. Independent tetrahedra; no shared oxygens.	$(SiO_4)^{4-}$
	b. Groups of tetrahedra; 2 shared oxygens	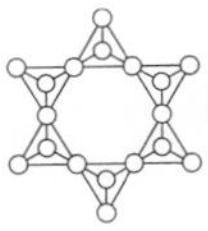 $(Si_6O_{18})^{12-}$ 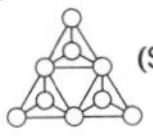 $(Si_3O_9)^{6-}$
2	Single chains; each tetrahedron shares 2 oxygens	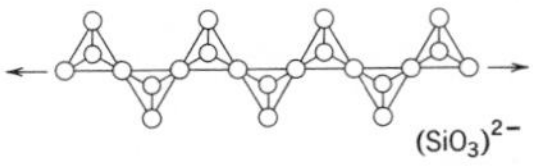 $(SiO_3)^{2-}$
3	Double chains; all tetrahedra share at least 2 oxygens and every other one shares 3 oxygens. Al^{3+} can substitute for Si^{4+} in the tetrahedra	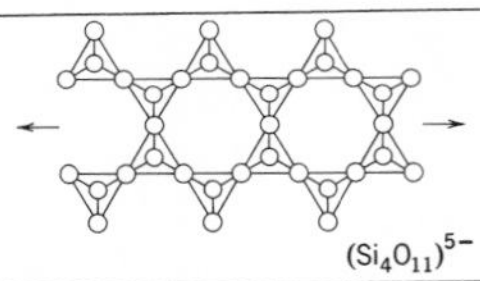 $(Si_4O_{11})^{5-}$
4	Layer-type silicates or sheets; each tetrahedron shares 3 oxygens. Al^{3+} can substitute for Si^{4+} in the tetrahedra	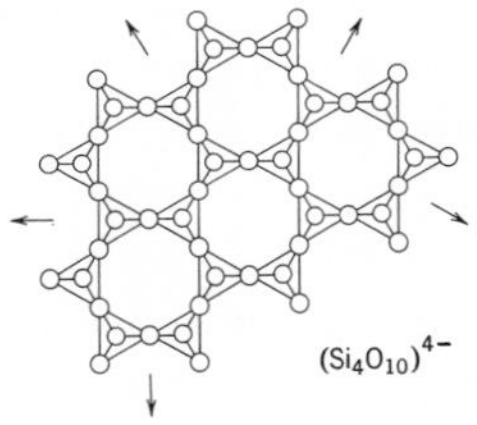 $(Si_4O_{10})^{4-}$
5	Three-dimensional networks; each tetrahedron shares 4 oxygens, one each with 4 other tetrahedra a. Al^{3+} can substitute for Si^{4+} in the tetrahedra	(Three-dimensional networks are too complex to be shown by a simple two-dimensional sketch) Si_4O_8 $(Al, Si_3O_8)^-$ $(Al_2Si_2O_8)^{2-}$
		$(SiO_3)^{2-}$ $(Al_3Si_3O_{12})^{3-}$ $(Al, Si_2O_6)^-$
		$(SiO_3)^{2-}$
	b. No substitutions of Al^{3+} for Si^{4+}; only Si and O are present	SiO_2

° Denotes heavy minerals; if present, these generally occur only in small amounts

OF SILICA TETRAHEDRA, CHEMICAL COMPOSITION, AND OCCURRENCE IN THE THREE MAJOR GROUPS OF ROCKS

Minerals Arranged According to Occurrence in the Three Major Rock Groups							
Igneous Rocks				Metamorphic Rocks		Sedimentary Rocks	
		Mineral	Chemical Composition	Mineral	Chemical Composition	Mineral	Chemical Composition
Ferromagnesian Silicate Minerals		Olivine	$(Mg, Fe^{2+})_2SiO_4$	Andalusite Kyanite Sillimanite	$Al_2O(SiO_4)$	Olivine* Andalusite* Kyanite* Sillimanite*	[See columns at left under Igneous and Metamorphic rocks.]
Ferromagnesian Silicate Minerals				Staurolite	$Fe^{2+}Al_4O_2(SiO_4)_2(OH)_2$	Staurolite*	
Ferromagnesian Silicate Minerals				Garnet[1]	$3X^{2+}2Y^{3+}(SiO_4)_3$ $X = Ca, Mg, Fe^{2+}, Mn$ $Y = Al, Fe^{3+}, Cr$	Garnet*	
Ferromagnesian Silicate Minerals				Epidote	$Ca_2(Al, Fe^{3+})_3(SiO_4)_3(OH)$	Epidote*	
Ferromagnesian Silicate Minerals				Wollastonite Beryl	$Ca_3(Si_3O_9)$ $Be_3Al_2(Si_6O_{18})$	Wollastanite* Beryl*	
Ferromagnesian Silicate Minerals	Pyroxenes	Orthopyroxenes: Hypersthene Clinopyroxenes: Augite	$(Mg, Fe^{2+})SiO_3$ $Ca(Mg, Fe^{2+})(SiO_3)_2$	Diopside	$CaMg(SiO_3)_2$	Hypersthene* Augite* Diopside*	
Ferromagnesian Silicate Minerals	Amphiboles	Hornblende	$NaCa_2(Mg, Fe, Al)_5[(Al, Si)_4O_{11}]_2(OH)_2$	Tremolite Actinolite	$Ca_2Mg_5(Si_4O_{11})_2(OH)_2$ $Ca_2(Mg, Fe^{2+})_5(Si_4O_{11})_2(OH)_2$	Hornblende* Tremolite* Actinolite*	
	Micas	Biotite	$K(Mg, Fe^{2+})_3(Al, Si_3)O_{10}(OH, F)_2$			Biotite*	
	Micas	Muscovite	$KAl_2(Al, Si_3)O_{10}(OH)$			Muscovite*	
				Chlorite	$(Mg, Fe^{2+})_5Al(Al, Si_3)O_{10}(OH)_8$	Chlorite*	
						Clay Minerals: Kaolinite	$Al_4Si_4O_{10}(OH)_8$
						Clay Minerals: Montmorillonite	$(Al_{1.67}, Mg_{0.33})_2(Si_4O_{10})_2(OH)_4$
						Clay Minerals: Illite	$Mg_2, Fe^{2+}, Al(Al_y, Si_{4-y}O_{10})_2(OH)_4K_y$
Feldspars		Plagioclase	Calcic plagioclase (Anorthite) Sodic plagioclase (Albite)		$Ca(Al_2Si_2)O_8$ $Na(Al, Si_3)O_8$	Rare in sediments	
Feldspars		Orthoclase	$K(Al, Si_3)O_8$				
Feldspathoids		Nepheline	$Na_3(Al_3, Si_3)O_{12}$			Rare in sediments	
Feldspathoids		Sodalite	$Na_4Cl(Al_3, Si_3)O_{12}$				
Feldspathoids		Leucite	$K(Al, Si_2)O_6$				
				Zeolites	Hydrous silicates of Al, alkalis, and alkali-earth metals		
		Quartz	Si_6O_{12}				

[1] Garnet occurs in a few varieties of igneous rock.

Fig. 3-10. Crystal structure of *quartz* (silicon dioxide) viewed parallel with long axis of prismatic crystal, with ions in normal positions as in Fig. 3-9, *B*. (R. W. G. Wyckoff, 1963.)

constituents (Fig. 3-10); in *feldspars* they are. Sodium, potassium, and calcium are located in larger spaces outlined by the network of tetrahedra, and within the tetrahedra aluminum substitutes for some of the silicon. Coupled substitutions occur commonly in plagioclase feldspars: each exchange of Ca^{2+} and Na^{+} in the spaces between the tetrahedra is accompanied by a compensating exchange of Al^{3+} and Si^{4+} within the tetrahedra. Neither iron nor magnesium occurs in the minerals of this group.

Shapes of mineral particles. The surfaces of mineral particles result from crystal growth or breakage and can be plane or irregular. Growth surfaces are governed by two factors: (1) the regular internal arrangement of the ions in the crystal lattice, and (2) the external conditions under which the lattice grew. Breakage surfaces may have originated naturally or may have been made as a deliberate test for mineral identification.

The shapes of mineral particles disclose so much about the internal structure of the particles and their geologic history that we can profitably explore some of the factors influencing shape. The value of particle shape as an aid to mineral identification will become obvious during the laboratory study of specimens. The aim of the following discussion, which emphasizes what can be learned from studying particle shapes, is to make laboratory study more meaningful through sharpening the reader's eyes.

Surfaces formed by growth. Depending on the conditions under which growth occurs, growing lattices can construct smooth planes or irregular surfaces. Where the lattice itself is the sole governing factor the result is ***crystal faces,*** *smooth plane surfaces of crystals.* Some crystal faces are so regular as to seem to have been shaped artifically. Mineral specimens with crystal faces on all sides display the characteristic *crystal form* of that mineral. Where present, crystal form is a useful guide to mineral identification. For example, halite (Fig. 3-11, *C*) and pyrite (Fig. 3-12, *A*) form cubes, each of the twelve sides of a garnet crystal is a parallelogram (Fig. 3-12, *B*), and quartz occurs in six-sided shapes whose sides are called prisms and whose slanting terminations are termed rhombohedra (Fig. 3-13).

Not many specimens occur as complete crystals. Some lack crystal faces altogether and are bounded entirely by irregular surfaces. Others display a few crystal faces and are bounded elsewhere by irregular surfaces. The degree of completion of crystal form, or lack of it, permits us to infer something about the conditions under which the lattice grew. The orderly internal arrangement and other physical properties are the same whether or not external conditions have permitted growth of crystal faces. Complete crystals (Fig. 3-13) imply that growth was equally free in all directions. From complete crystals we conclude that the internal lattice acted as the sole determinant of particle shape.

Almost all minerals that grow while surrounded by fluids or by soft mud develop complete crystals. The power of some minerals, such as garnet, to crystallize is so great that complete crystals grow even in the solid state during metamorphism. Where many lattices, growing simultaneously, interfere with one another, crystal faces do not develop and the particles are bounded by irregular surfaces. Particles with irregular surfaces result from a dominance of external conditions over the individual lattice. Crystal faces occur only on the ends of many specimens, and the sides and opposite ends of these specimens consist of irregular surfaces. Many specimens of this kind come from the walls of fractures and the linings of open cavities (Fig. 11-12). They are the obvious result of freedom to grow in only a few directions and the lack of such freedom in almost all others.

Surfaces formed by breaking. Minerals break in characteristic ways. Although the origin of surfaces formed by breaking contrasts markedly with that of surfaces formed by growth, nevertheless both are controlled by the crystal lattice. Breakage is determined by the strength of bonds between particles of the lattice, which varies according to the kind of bond. Four kinds of bonds between particles of

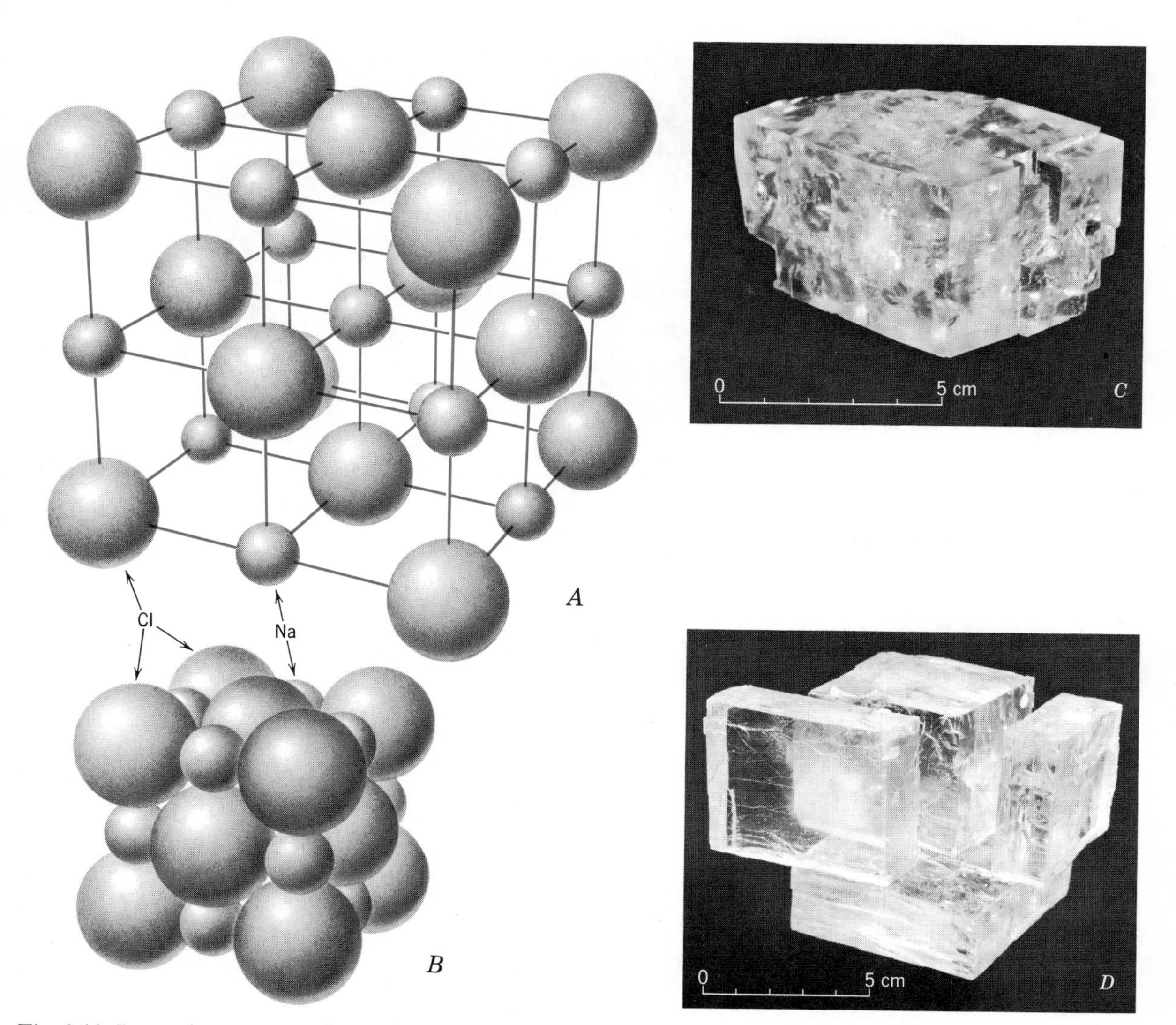

Fig. 3-11. Internal structure and morphology of *halite* (sodium chloride). *A.* Schematic diagram of internal structure, with small ions of sodium and larger ions of chlorine pulled apart. Blue lines show bonds between ions. *B.* Ions in natural positions. *C.* Photograph of halite crystal. *D.* Photograph of cleavage fragments of halite. Internal cubic pattern is reflected not only in shapes of crystals but also in fragments bounded by cleavage planes. (*C* and *D*, B. M. Shaub.)

crystals are: (1) molecular, (2) ionic, (3) covalent, and (4) metallic (Table 3-2).

In crystals with *molecular bonds,* molecules having electrically balanced internal atomic charges are held together weakly by forces of attraction between unlike partial charges distributed unevenly over the surfaces of the molecules. An example is *ice.*

In crystals with *ionic bonds,* strong nondirectional electrostatic forces of attraction between oppositely charged particles hold the ions together. In crystals having ionic bonding each negative ion is surrounded by positive ions, and each positive ion by negative ions. An example is *halite* (Fig. 3-11, *A* and *B*).

In crystals with *covalent bonds,* particles are held together firmly by forces of attraction between atoms that share electrons. An example is *quartz* (Fig. 3-10).

In crystals with *metallic bonds,* the positive ions form a fixed framework upon which is superimposed

Fig. 3-12. Distinctive crystal forms of pyrite and garnet. A. Cubic crystals of *pyrite* (iron sulfide) commonly are striated and intergrown. (B. M. Shaub.) *B. Garnet* (complex silicate of aluminum and various metallic ions) crystal showing six of the twelve characteristic faces, each of which is an identical parallelogram. (Ward's Natural Science Establishment.)

TABLE 3-2. KINDS OF BONDS AND PHYSICAL PROPERTIES OF CRYSTALLINE SOLIDS

	General Characteristics of Crystalline Substances Having Type of Bonding Indicated			
Type of Bond	Melting Point	Electrical Conductivity	Other Properties	Example
Molecular	Low	Low	Brittle; sublimes easily	Ice
Ionic	Low	Low; conducts electricity only in molten state	Brittle	Halite
Covalent	High	Low	—	Quartz
Metallic	Moderately high	Excellent	Malleable; ductile	All metals

another network consisting of freely moving electrons. Examples include all metals.

In many minerals more than one kind of bond is present within the lattice. As a result not all bonds are of equal strength—some are weaker than others. Minerals tend to break readily parallel to the planes of weak bonds. Each direction of weak bonding defines a direction of ***cleavage,*** *the capacity of a mineral to break in preferred directions along surfaces parallel to lattice planes.* Cleavage can be expressed as a series of nearly invisible cracks in the mineral. Where present, the cracks are seen easily with a microscope; at high magnifications they are prominent (Fig. 3-14, B). If bonds are especially weak, as in *micas* (Fig. 3-15) and *clay minerals* (Fig. 3-14), cleavage occurs with ease. A few minerals, such as halite (Fig. 3-11, *D*), calcite (Fig. 3-16), and galena (Fig. B-1), cleave along three directions.

Pyroxenes can be distinguished from amphiboles by the angles between their cleavage directions. In pyroxenes the cleavage directions are nearly at right angles (Fig. 3-17, A), whereas in amphiboles the

Fig. 3-13. Small hexagonal quartz crystal showing three of the six prismatic sides and rhombohedra at both ends (compare Pl. 3). The characteristic crystal shape of quartz persists even below the limits of visibility with unaided eye. (B. M. Shaub.)

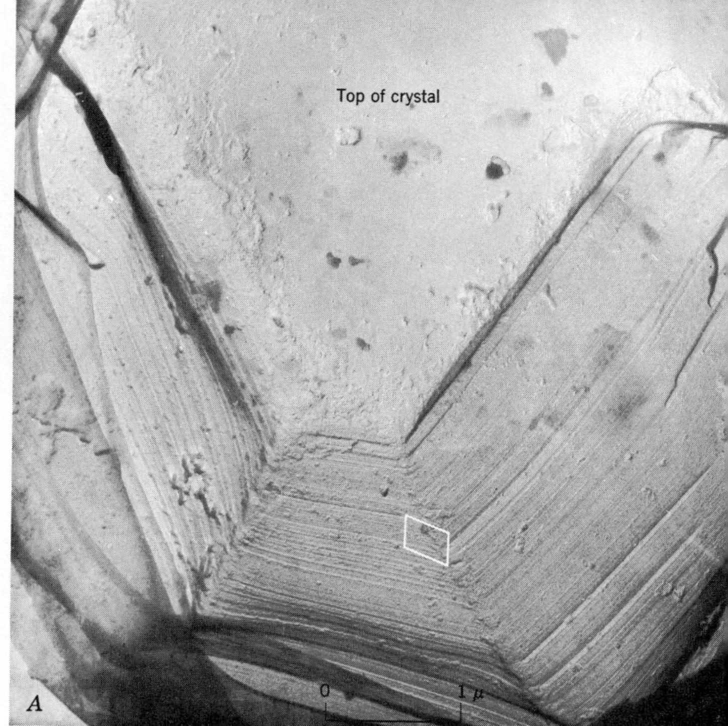

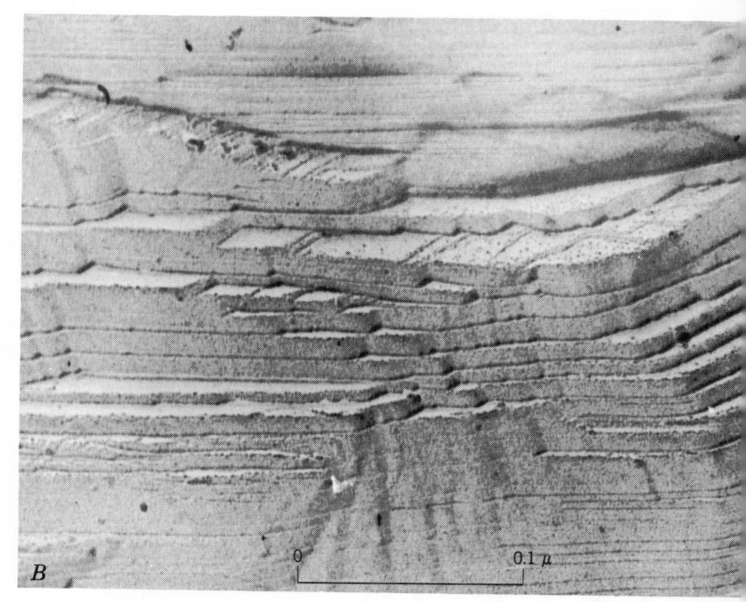

Fig. 3-14. Electron micrographs (pictures made with an electron microscope) of the clay mineral *dickite* (hydrous aluminum silicate). *A.* Crystal enlarged 11,100 times showing edges of closely spaced cleavage planes in direction parallel with face at top of crystal. With this enlargement 1 inch would be more than a sixth of a mile long. *B.* Part of face of same crystal enlarged 37,500 times, to show prominent horizontal cleavage planes and less-perfect cleavage in vertical direction at approximately 45° to plane of photograph. Crystal faces and cleavage defined in text. (T. F. Bates.)

cleavage directions form an angle of 56° (Fig. 3-17, *B*). In both groups the cleavage directions are controlled by directions of weak bonds between adjacent chains of silica tetrahedra; the angle between cleavage directions is controlled by the width of the chains.

The smooth, plane surfaces of crystal faces are easily confused with the smooth, plane surfaces of cleavage. Both are controlled by the regular internal arrangement of particles in the lattice. The difference between crystal faces and cleavage surfaces is that the former are initial growth features whereas the latter are the result of differential bonding. Both kinds of surfaces are parallel to lattice planes. Most cleavage directions are parallel to directions along which crystal faces have grown or can grow, but many crystal faces exist in directions along which no cleavage occurs (Fig. 3-14, *A*).

Minerals in which bonds are strong in all directions lack distinct cleavage and break along irregular surfaces. As applied to minerals, the word ***fracture*** refers to *the capacity of a mineral to break along irregular surfaces.* Quartz and garnet are examples of common minerals that lack good cleavage and that break with irregular fracture (Fig. 3-18). The fracture of quartz typically occurs along smooth curved surfaces; such fracturing is called *conchoidal.*

Minerals as indicators of environment. We have remarked that the state of aggregation of matter

Fig. 3-15. Perfect cleavage of *muscovite* (hydrous potassium-aluminum silicate) shown by very thin, plane flakes into which this six-sided crystal has been split. The cleavage flakes suggest leaves of a book, a resemblance embodied in the name "books of mica" for crystals elongated in a direction perpendicular to the cleavage flakes. Compare Fig. 3-14. (Ward's Natural Science Establishment.)

Fig. 3-16. Perfect rhombs of *calcite* (calcium carbonate) formed by cleavage planes in three directions. The dark areas in parts of the rhombs are caused by double refraction of light, a phenomenon especially well displayed by calcite. (Ward's Natural Science Establishment.)

represents an adjustment of particles to conditions of temperature and pressure. The mere fact that most minerals occur as solids tells us that one particular adjustment has taken place in response to environment. But minerals can tell us much more about environment. They originate in response to the three main factors that characterize environment: (1) temperature, (2) pressure, and (3) chemical characteristics of solutions.

Ever since late in the eighteenth century, attempts have been made in the laboratory to measure the conditions under which various minerals form. Early in the twentieth century experimental programs became systematic. Since 1960 it has been possible to reproduce controlled temperatures up to 2,000°C and pressures as great as those prevailing beneath a column of rock 400km thick. The general objective is to establish a firm basis for interpreting the environments in which individual minerals and mineral assemblages are stable.

Laboratory results are displayed on graphs of temperature versus pressure. On such graphs boundaries between fields of stability of various substances appear as lines. In order to convey the concept of increase of pressure with depth within the Earth, all temperature-pressure graphs in this book employ the same convention: pressure is represented on the ordinate, with values increasing toward the base of the graph; and temperature is represented on the abscissa, with values increasing toward the right.

The principle underlying the use of minerals as bases for inferring environments is that minerals originate only within a particular range of conditions. Within this range a given mineral is *stable;* outside this range it is *unstable.* Where a mineral is within its unstable range it generally does not form, or if it already exists it will break down into a stable form. The fact that under the temperatures and pressures prevailing at the Earth's surface silicates react slowly introduces a complication. Accordingly many silicate minerals are *metastable;* that is, even under environmental conditions in which they are not the most stable form, they can persist. Ordinarily a mineral that is metastable at low temperatures and/or pressures is stable at high temperatures and/or pressures.

To employ a mineral to infer environment we must assume that a condition of equilibrium existed between the environment and the growing mineral. Two kinds of equilibrium are possible depending on

A. Pyroxene

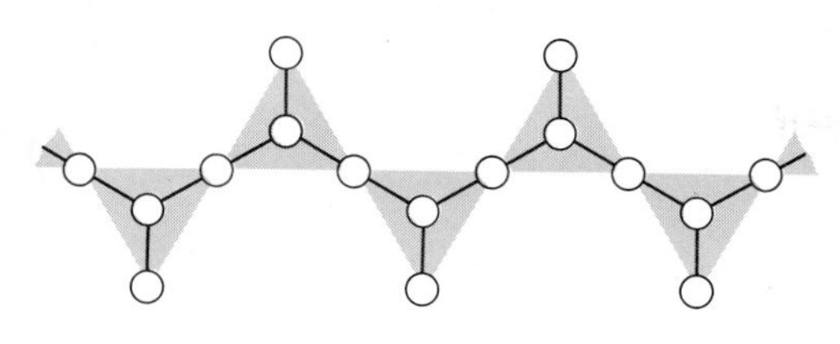
Top view of single chain of tetrahedra

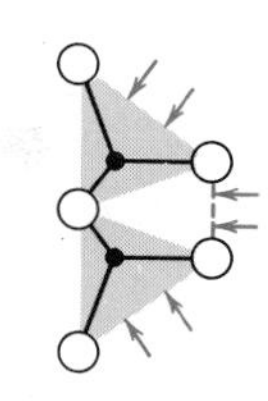
End view of single chain

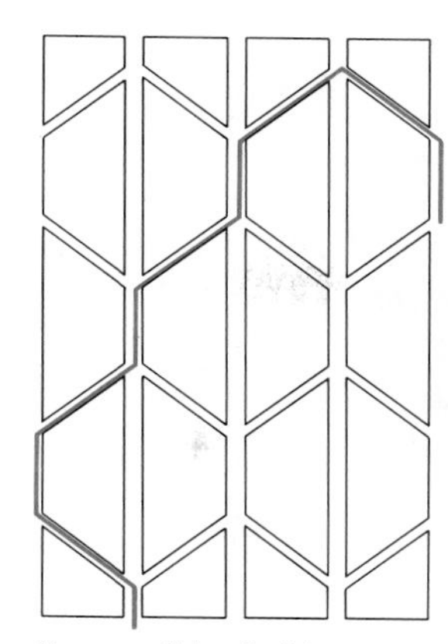
Cleavage (blue line) in relation to end views of chains

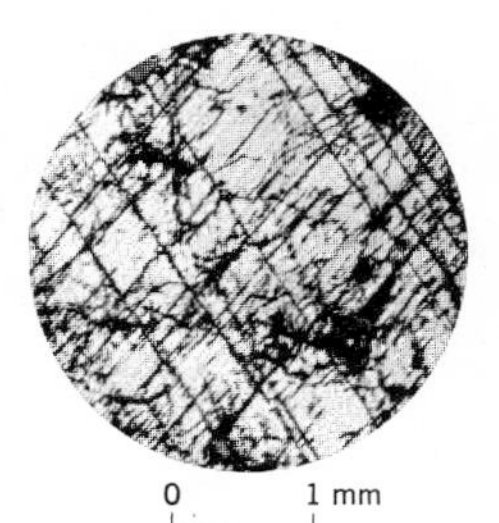

Photograph of pyroxene showing cleavage

B. Amphibole

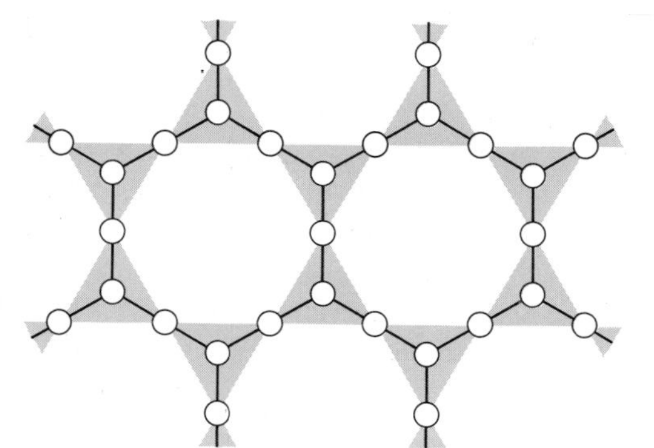
Top view of double chain of tetrahedra

End view of double chain

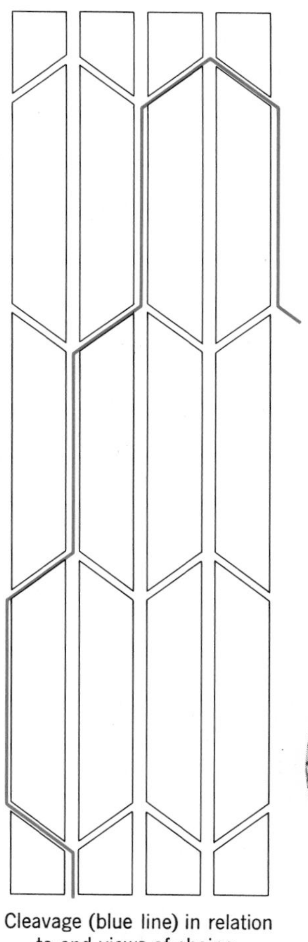
Cleavage (blue line) in relation to end views of chains

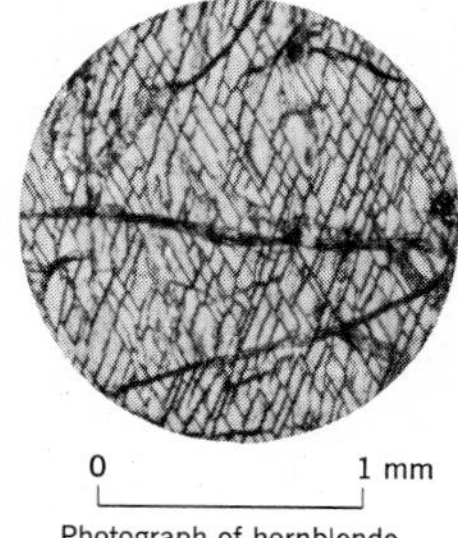

Photograph of hornblende showing cleavage

Fig. 3-17. Sketches and photomicrographs showing cleavage in pyroxene and amphibole. A. *Pyroxene* (complex silicate of iron, magnesium, and/or calcium). *B. Amphibole* (complex hydrous silicate of iron, magnesium, and/or calcium and other ions). Small arrows on sketches of end views of pyroxene single chain and amphibole double chain show surfaces of parting along which bond between adjacent tetrahedra is weakest. (Cleavage sketches: Fyfe, 1964; photomicrographs: B. M. Shaub.)

the degree of confinement of the materials. In a closed system, *static equilibrium* is attained; in an open system, either static or *dynamic equilibrium* can result.

Closed systems. When all materials involved in a given reaction remain on the scene they form a closed system. Under a given set of conditions the materials attain a static equilibrium and maintain this status as long as the conditions remain unchanged.

Minerals express static equilibrium with their environment in several ways. Some substances, such as carbon or SiO_2, are stable under a wide range of conditions but within various parts of this range they respond by building new lattice configurations with the same materials. Each distinct lattice configuration defines a distinct mineral. Other substances do not build many lattice types but, depending on conditions, either form or break down.

The slope of a stability boundary on a temperature-pressure graph expresses the relative influence of temperature and pressure. A vertical line reflects the dominance of temperature; a horizontal line results from the influence of pressure alone; an inclined line indicates the influence of both temperature and pressure. The boundary line between graphite and diamond (Fig. 3-19) displays the dominant influence of pressure in the origin of diamonds. The graph indicates that even at the rather modest temperature of 400°C the pressure required to convert graphite to diamond is slightly more than 10 kilobars, equivalent to the weight of a column of rock about 35km thick. The temperature of the transition is higher at increased pressures. If the laboratory experiments tell the whole story, we must conclude that rocks at the Earth's surface containing diamonds must have originated at depths greater than 35km.

A similar graph displays the stability fields of many mineral forms of SiO_2 (Fig. 3-20). Most of these SiO_2-minerals are exceedingly rare. Nevertheless, they are worthy of mention because of their great theoretical significance. Because coesite is not present in ordinary rocks we can place a lower limit on the depth to which the rocks have been buried within the lithosphere. The graph shows that at pressures less than 5 kilobars silica melts at temperatures greater than 1,730°C (in the absence of water). The vertical boundary between *cristobalite* and *tridymite* lies at 1,470°C. The strongly curved lines indicate the large influence of pressure on the stability temperatures of the boundaries between tridymite and *high-quartz* and between high-quartz and *low-quartz*. Pressure becomes the more dominant factor in the transition to the two denser forms *coesite* and *stishovite*. Because low-quartz and high-quartz lie within the temperature range of the environment in which most ordinary rocks originate, because ideal specimens are easily recognized, and because their limiting temperatures are only slightly influenced by pressures up to 5 kilobars, they are convenient examples of "geologic thermometers."

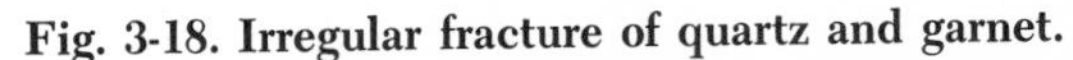

Fig. 3-18. Irregular fracture of quartz and garnet.
A. **Curved fracture surfaces at end of quartz crystal from Arkansas.** ***B.*** **Irregular fracture surfaces at corner of large garnet crystal from Australia. (B. M. Shaub.)**

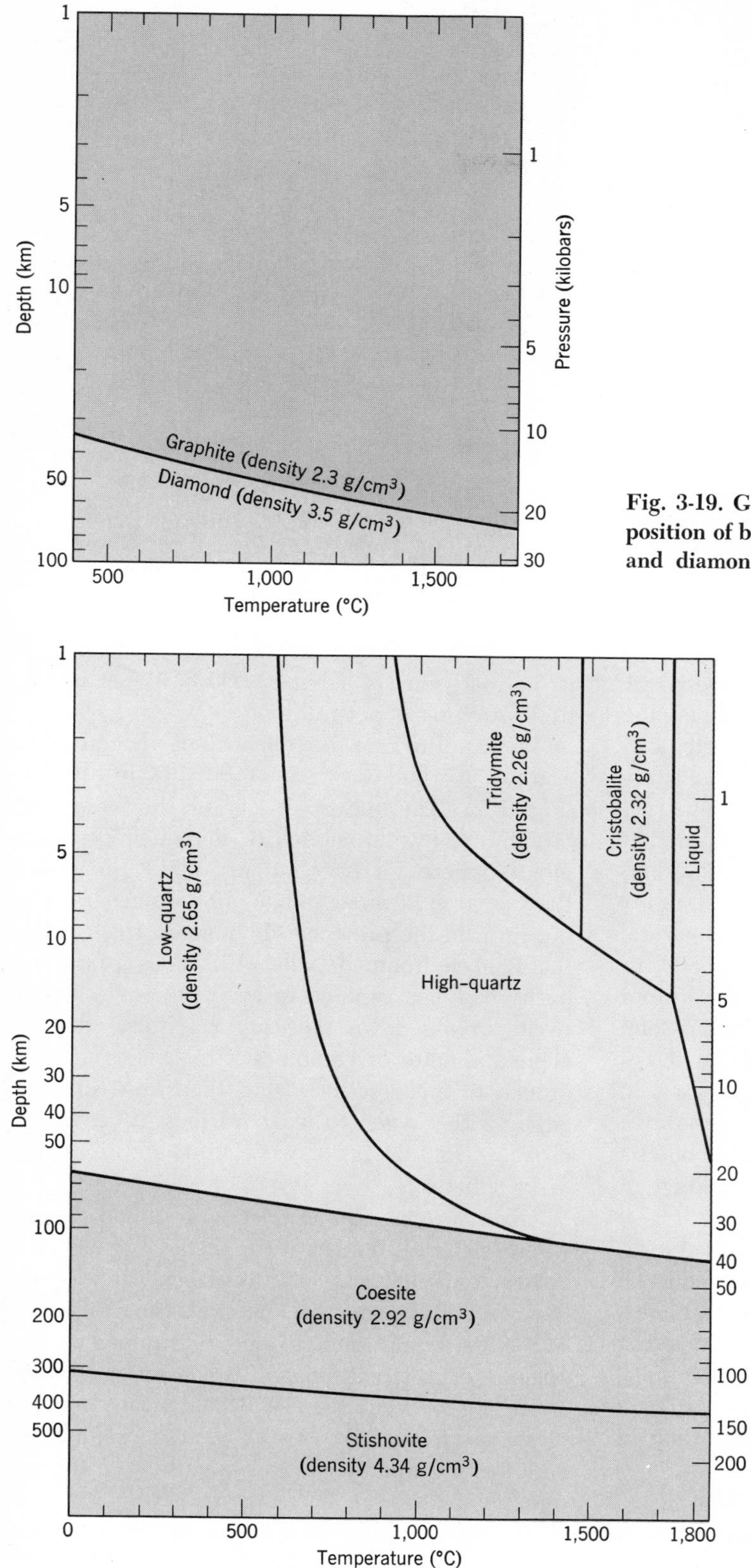

Fig. 3-19. Graph of temperature versus pressure shows position of boundary between stability fields of graphite and diamond. (After F. P. Bundy and others, 1961.)

Fig. 3-20. Graph of temperature versus pressure shows stability fields of various forms of SiO_2. (After F. R. Boyd and J. L. England, 1959.)

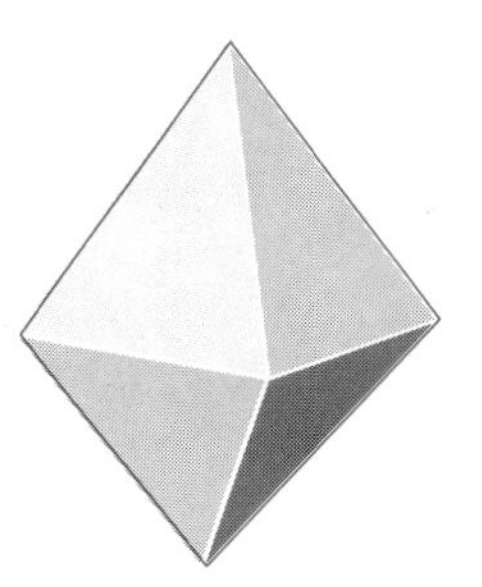

Fig. 3-21. Bipyramidal crystal of high-quartz. Compare Fig. 3-13. (Pirsson and Knopf, 1947, cited at end of Chap. 4.)

Typical low-quartz is shown in Fig. 3-13: the hexagonal crystals consist of median prisms and terminating rhombohedra. High-quartz (Fig. 3-21) lacks median prisms and rhombohedra; its terminations are pyramids. Because the pyramid faces touch one another in the absence of prisms the crystals are *bipyramids.* At temperatures and pressures on the right side of the high-quartz/low-quartz line of the graph high-quartz crystallizes. The characteristic bipyramids grow if conditions permit. When the temperature descends across the high-quartz/low-quartz line, high-quartz structure immediately changes to low-quartz structure. Despite this structural change, if bipyramidal crystals are formed at all, their shape persists. Unfortunately, conditions under which high-quartz generally crystallizes are such that crystal faces rarely grow. The structural change is so complete that if the bipyramidal crystals could not grow originally, no clues to the former existence of high-quartz remain. The prisms and rhombohedra of low-quartz originate only if the quartz initially crystallized under conditions within the low-quartz field on the graph. For all we know, every bit of quartz lacking prisms and rhombohedra could have passed through the high-quartz stage at one time.

The chemical characteristics of solutions at ordinary temperature and atmospheric pressure influence the origin of some minerals. Three such characteristics are (1) ***salinity,*** *proportion of dissolved solids;* (2) ***pH,*** *a measure of acidity or alkalinity of solutions;* and (3) ***Eh,*** *the oxidation-reduction potential,* which like ordinary electricity, is measured in volts (or millivolts, 1/1,000 of a volt).

Salinity is a factor in mineral formation during the evaporation of seawater. For example, halite is not precipitated until the salinity has increased from

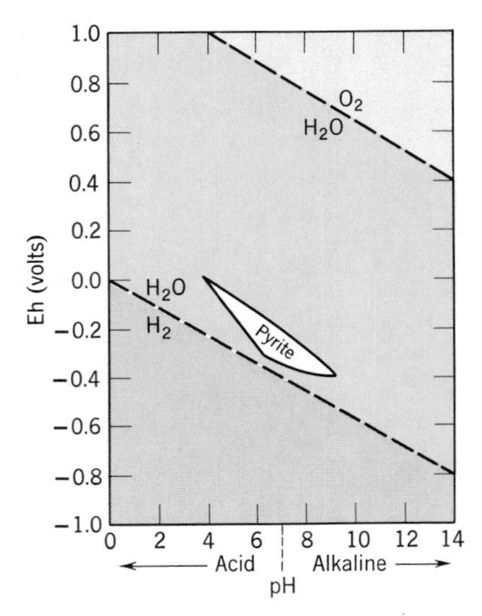

Fig. 3-22. Graph of Eh versus pH for water solution at 1 atmosphere total pressure and 25°C shows stability field of pyrite and water. (R. M. Garrels, 1960.)

its normal value of about 35,000 parts per million to 337,000 parts per million.

We can illustrate the influence of pH and Eh by a graph in which these characteristics are the axes (Fig. 3-22). The lines on the graph show the fields of stability of water and pyrite. In reducing environments, where Eh is negative, sulfides are stable. Pyrite and most other sulfide minerals generally do not form in the presence of abundant free oxygen; that is, in environments where Eh values are strongly positive. When exposed to oxidizing environments pyrite breaks down. The sulfur is released in its elemental state or combines with oxygen and hydrogen to make sulfuric acid. The iron combines either with oxygen to make various oxides or hydroxides (Fig. 23-18) or with sulfur and oxygen to make sulfates.

Open systems. In contrast with closed systems where materials remain at the scene of reaction, in open systems materials initially present can leave the scene of reaction and new materials can arrive from elsewhere. Equilibrium may not be attained in open systems, or if it is attained, it is complex and varies according to circumstances. Because many geologic reactions are open systems we need to examine how such systems operate. Let us begin with a mineral example.

Within certain environmental ranges some minerals can coexist with each other in the same rock.

At higher temperatures and pressures, however, these minerals react to make new ones. For example, calcite ($CaCO_3$) and quartz (SiO_2) can coexist at low temperatures but react to form *wollastonite* ($CaSiO_3$) and carbon dioxide (CO_2) at high temperatures. If the carbon dioxide does not escape, the system is closed and the transition to wollastonite is defined by the curve at left in Fig. 3-23. If carbon dioxide does escape, however, the system is open and the transition to wollastonite is defined by the curve at the right.

In an open system, equilibrium can be reached if the rate of arrival of one material equals the rate of escape of another. Under such circumstances a *steady state* prevails, which is somewhat comparable to the static equilibrium of a closed system. In many natural reactions, however, a much more complicated situation prevails: the rates of arrival and departure of material not only vary but as they vary they control the kinds of reactions that occur. In such circumstances a condition of dynamic equilibrium can be reached among all variables, but such equilibrium is ephemeral and shifts with the variables. Such dynamic equilibrium can be visualized as a series of changing steady states.

Much of geology deals with the analysis of complex, natural open systems. In a stream, for example, the flowing water may appear about the same from day to day, yet changes are constantly taking place. The molecules of water that were present yesterday are not present today. They have flowed downstream and new molecules of water have taken their place. The width, height, and depth of the stream, speed of flow, and amount of sediment carried are some of the properties that change as the amount of water varies (Fig. 9-14).

The three components of the geologic record—bedrock, regolith, and the Earth's morphology—form a complex open system that reaches a steady state under many conditions. Bedrock breaks down to form regolith, regolith moves away downslope, and the surface is lowered. Like the water in a stream the regolith may appear about the same from age to age, but the particles are constantly in a state of flux. Older particles move downslope and their places are taken by new particles released from the bedrock beneath. These are only two of the many examples of natural open systems. Numerous others will be discussed in later chapters.

Mineral particles and aggregates. Minerals occur

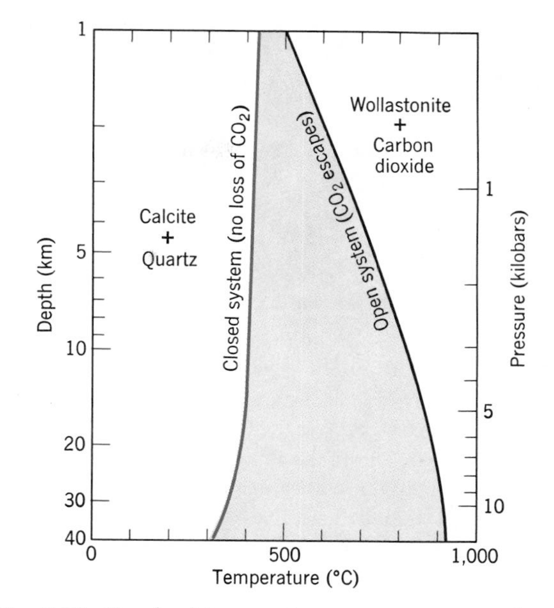

Fig. 3-23. Graph of temperature versus pressure shows differences between closed and open systems in transformation of calcite and quartz to wollastonite and carbon dioxide. (After T. F. W. Barth, 1962.)

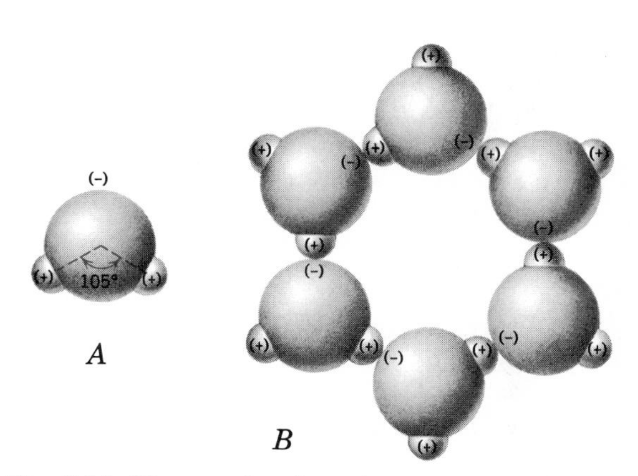

Fig. 3-24. Water molecule and ice crystal. *A*. Water molecule with large oxygen ion and smaller hydrogen ions. Weak, partial surface charges are shown: (+) at each hydrogen ion and (−) on the oxygen ion on the side opposite the hydrogen ions. *B*. Hexagonal pattern of water molecules in an ice crystal. Bonds between molecules result from attraction of partial surface charges having opposite signs. Other oxygen ions of water molecules can attach themselves to the hydrogen ions on the outside of the ring structure to build elaborate branching crystals, as in snowflakes (compare Fig. 3-25, *A*).

Fig. 3-25. Water in the solid state as a mineral only, and as a mineral aggregate in the form of a sediment and a metamorphic rock. *A.* Water as a mineral: a snowflake, showing elaborate branching hexagonal crystal structure. Many times natural size. Compare Fig. 3-24, *B.* (W. J. Pentley, U. S. Weather Bureau.) *B.* Water as a mineral aggregate in the form of a sediment: a pile of granular snow about 8 months old, collected just below the surface of a glacier in Yukon Territory, Canada. Grain diameters are about 1.5mm. (R. P. Sharp.) *C.* Water as a mineral aggregate in the form of a metamorphic rock: glacier ice from the Swiss Alps. A sample has been sliced thin and is seen through a microscope. The broad areas are individual grains. The small, dark, circular objects are bubbles of air that did not escape during compaction and recrystallization. (M. P. Perutz, courtesy Gerald Seligman.)

both as particles and as aggregates of particles. The distinction is that mineral particles include only one crystal lattice. Mineral aggregates, which contain more than one crystal lattice, can include one or more minerals. We can illustrate the differences between particles and aggregates by the solid states of water. To do so we must start with the structure of water.

A molecule of water consists of a large oxygen ion, O^{2+}, and two much smaller hydrogen ions, H^- (Fig. 3-24,*A*). Internally, water molecules are electrically balanced (internal charges are not shown in Fig. 3-24). Because of the way they are shared, the electrons tend to cluster about the nucleus of the oxygen ion. This leaves an excess partial positive charge near the proton of each hydrogen nucleus. The uneven distribution of partial charges causes each water molecule to behave as a tiny, weak dipolar magnet. Water molecules are attracted by other water molecules or by the surfaces of crystals having unevenly distributed partial surface charges.

When water freezes into snowflakes or ice, the molecules arrange themselves in a hexagonal crystal lattice (Fig. 3-24, *B;* Fig. 3-25, *A*). In a snowflake, the size of the crystal lattice can be less than a millimeter in diameter; in a large homogeneous mass of ice covering the surface of a lake or pond, the lattice may extend through long distances. In either case, if only one lattice is present the mass of ice is one mineral particle.

Snowflakes, hailstones, and grains of sleet or granular snow (Fig. 3-25, *B*) are transported mineral aggregates consisting of more than one crystal lattice. Solid water in each particle is a mineral, but the particles in the aggregates constitute a sediment or sedimentary rock, depending on the degree of coherence among the particles. Aggregates of loose particles are sediment; aggregates of particles bound together are sedimentary rock (Fig. 4-12). Where granular snow has become glacier ice by flow and recrystallization (Fig. 3-25, *C*), the particles of the aggregate are tightly bound because they interlock complexly. The interlocked arrangement of particles in glacier ice resembles that in many metamorphic rocks.

Many mineral aggregates, then, are rocks. After we have become familiar with common minerals the next order of business is to learn how minerals occur in rocks. This we shall try to do in Chapter 4.

Summary

Matter

1. Matter exists as solids, liquids, and gases.

2. The physical properties of matter in each of its states of aggregation are distinct under ordinary conditions but become less distinct as temperature and pressure are varied.

3. Particles of crystalline solids are arranged in crystal lattices having regular geometric configura-

tions. Particles in amorphous solids are arranged in irregular networks.

4. Rise of liquids in capillary openings is an important phenomenon resulting from surface tension and the interfacial behavior of liquids.

5. Pressure and temperature influence the stability of the states of matter.

Minerals

6. Minerals are naturally occurring substances, many of which are crystalline solids whose ions, elements, or molecules are arranged in unit cells. The chemical compositions of minerals, though variable, bear fixed relationships to certain physical properties.

7. Most rock-forming minerals are silicates whose frameworks are built of tetrahedra composed of one small silicon ion surrounded by four larger oxygen ions. Five main groups of silicate minerals are based on varying degrees of sharing of oxygen ions among tetrahedra.

8. The shapes of mineral particles result from opposite kinds of processes: crystal growth and crystal breaking. The crystal lattice exerts strong controls on both kinds of processes.

9. Because they are stable under specific environmental conditions and unstable under other conditions, minerals indicate the environments under which they formed.

Selected References

Fyfe, W. S., 1964, Geochemistry of solids, an introduction: New York, McGraw-Hill.

Hurlbut, C. S., Jr., 1959, Dana's manual of mineralogy, 17th ed.: New York, John Wiley.

Hurlbut, C. S., Jr., and Wenden, H. E., 1964, Changing science of mineralogy: New York, D. C. Heath and Co.

Ingerson, Earl, 1955, Methods and problems of geologic thermometry, *in* Econ. Geology, 50th Anniversary Volume, p. 341–410.

Mason, Brian, 1966, Principles of geochemistry, 3rd ed.: New York, John Wiley.

Pauling, Linus, 1960, The nature of the chemical bond, 3rd ed.: Ithaca, New York, Cornell Univ. Press.

Ransom, J. E., 1965, A range guide to mines and minerals: New York, Harper and Row.

Slabaugh, W. H., and Parsons, T. D., 1966, General chemistry: New York, John Wiley.

Vanders, Iris, and Kerr, P. F., 1967, Mineral recognition: New York, John Wiley.

Wade, F. A., and Mattox, R. B., 1960, Elements of crystallography and mineralogy: New York, Harper and Brothers.

Rocks and Regolith

The "words" of geologic language
An exposure of bedrock
Partings
Layers
Major classification of rocks
Relationships among particles
What holds particles together?
Go and see

Plate 4

Contrast between metamorphic rock (dark gray) with particles stretched, bent, and aligned, and extremely coarse-grained igneous rock (light gray), with particles not stretched or bent but aligned by growth perpendicular to the surface of contact with the metamorphic rock. Length of small sledge hammer showing 43cm. North of Parry Sound, Ontario, Canada. (J. E. Sanders.)

CHAPTER 4 *Rocks and Regolith*

THE "WORDS" OF GEOLOGIC LANGUAGE

The geologic record, consisting of bedrock, regolith, and the Earth's morphology, is the archive of Earth history. To read this history we must learn the language of geology, a language whose "words" include rocks "spelled" in an "alphabet" of minerals.

Once familiar with some of the letters of the mineral alphabet, we can try reading some of the words of geologic language. We begin by examining the ways in which mineral letters join together.

The best way to master ***petrology***, *the study of rocks*, is to scrutinize specimens; books are inadequate substitutes for the rocks themselves. But if we are to do more than memorize the appearances of many rocks, we must be able to direct our attention to the critical points when we examine rocks. To help us do this we need to understand the processes by which rocks originate. Such understanding poses a dilemma: in order to comprehend rocks we need to know about geologic processes, yet analysis of these processes is facilitated by familiarity with rocks. At some point we must break into this closed circle of rocks and processes.

Recall what we read in Chapter 1 about James Hutton's great insights into the geologic cycle. In modern terms we would paraphrase one of Hutton's conclusions by remarking that the sedimentary group of rocks originates by the conversion of regolith into bedrock. At the Earth's surface, bedrock becomes regolith; yet by subsidence and burial regolith may be converted back into bedrock.

We can solve the rocks-processes dilemma by taking a cue from Hutton. By summarizing the salient points of all rocks we are equipped to make a journey around the geologic cycle the easy way. We start with familiar processes, many of which can be seen operating at ordinary temperatures nearly everywhere at the Earth's surface. When we have mastered the processes by which regolith originates we shall have acquired a firm basis for a comprehensive understanding of sedimentary strata. We can then take advantage of strata to analyze the results of processes that operate below the Earth's surface, an inaccessible region outside our vision and beyond the range of our direct experience.

Most processes forming igneous and metamorphic rocks operate at high temperatures within the Earth. Analysis of these requires an indirect geologic approach based on data and kinds of reasoning not generally familiar except to geologists. Our purpose is to build toward the subjects requiring the less familiar geologic methods of analysis by first expanding our experience with common observations of readily accessible processes. Therefore we defer detailed treatment of igneous and metamorphic rocks until we are ready to cope with the problems of the Earth's deeper zones.

Within the framework we have outlined, our general plan for studying rocks is patterned after the procedure employed by many geologists in the field. It consists of three parts: (1) form an impression of some of the gross features of an exposure from a distant view; (2) move up close and examine a small specimen with a hand lens; and (3) enlarge the investigation from a small specimen to larger bodies of rock, based on the study of many exposures of rocks and boundaries between bodies of rock. In this chapter we shall deal with part 1 and lay the groundwork for study of specimens in the laboratory, which is the only way to become skilled in part 2. We shall study the subjects involved in part 3 in Chapters 16 through 23.

Our encounter with rocks in the present chapter has two objectives: (1) to become acquainted with some of the gross aspects of rock bodies, in particular with the partings and layers in rocks; and (2) to master the basis for classifying rocks and recognizing rocks of the three great groups. We begin with some of the larger features visible at an exposure of bedrock.

Fig. 4-1. Prominent partings, parallel and nearly vertical, transecting coarse-grained metamorphic rock. Less well-defined parting, dipping at a small angle to the left, permits fragments to break loose from bedrock. Shore of Long Island Sound, Branford, Connecticut. (J. E. Sanders.)

AN EXPOSURE OF BEDROCK

Once we acknowledge that rocks contain valuable information about the history of the Earth, we inevitably look at exposures of bedrock with a new interest and ask questions that never occurred to us before. In due course we shall be asking many questions and suggesting where and how to look for answers. At the moment, however, let us concentrate on the gross appearance of an exposure as seen from a distance of a hundred meters or so. Many exposures display two prominent features: partings and layers.

Partings. Characteristic of nearly all exposures of bedrock is a network of partings cutting through the rock (Fig. 4-1). Without inquiring further into their origin, we realize that partings exert at least three profound influences on the body of rock: (1) fragments of rock break loose along partings; (2) fluids can travel over and through the rock along partings; and (3) the total surface area of the rock body is greatly increased by the formation of partings.

Sizes of broken fragments. Obviously the sizes of fragments broken from bedrock vary according to the spacing of the partings. Large blocks originate where partings are widely spaced (Fig. 7-13) and small blocks where partings are closely spaced (Fig. 4-2).

Clearly many of man's activities involving bedrock, such as excavations, tunnels, and quarrying operations, must take account of the spacing of partings. Small blocks can be quarried nearly everywhere. Because partings are so ubiquitous, the number of possible sites for quarrying blocks more than a meter or so in dimension is small.

Partings are significant for their influence not only on man's activities but also on the origin of regolith. The range in the sizes of fragments broken from bedrock along partings lends great variety to the initial distribution of particle sizes in regolith.

Passage of fluids over and through rock bodies. Partings in bedrock offer ready-made avenues for movement of fluids both over the ground and through bodies. Water flowing over a bedrock surface tends to travel along the partings, with the result that surface streams become aligned along

Fig. 4-2. Small fragments (most being 30 to 50cm long and 10 to 20cm across) broken along numerous partings that cut fine-grained igneous rock. Compare Figs. 7-13, 17-13. Devil's Post Pile National Monument, California. (H. L. Mackay, from Design Photographers International, Inc.)

them (Fig. 4-3). Likewise partings direct water moving underground (Fig. 11-13).

Increased surface area created by partings. Where a new parting forms in a body of unbroken bedrock, the area of rock surface exposed to moisture increases by twice the actual area of the parting. This great increase results from the creation of two distinct rock faces, one along each side of the parting (Fig. 7-4). For every square meter of its own extent, then, a parting creates two square meters of new rock surface which becomes subject to the effects of fluids (Chap. 7).

Kinds of partings. Looking more closely we discover that there are at least three kinds of partings. (1) Some coincide with the surfaces separating adjacent strata and are known by the general term *bedding-plane partings.* Two other kinds cut across strata at varying angles and originate from large-scale fracturing of the rock. These fractures reveal that the rock behaved as a brittle solid when it broke. Fractures resulting from the breaking of rock are subdivided on the basis of evidence of displacement or lack of displacement of the blocks on opposite sides of the fracture. (2) ***Joints*** are *fractures on which no appreciable movement parallel to the fracture has occurred.* (3) ***Faults*** are *fractures along which the opposite sides have been relatively displaced* (Chap. 17).

Layers. Another large-scale aspect of an exposure of bedrock visible at a distance is the presence or absence of layers. In Chapter 1 we read that sedimentary rocks are arranged in strata, whose initial positions generally are horizontal. Strata can be seen in nearly all exposures of sedimentary rocks. Even where their thickness is great and the rock is composed of particles of nearly uniform size, subtle features generally disclose the presence of strata (Fig. 4-4, *A*).

Although layers are so characteristic of sedimentary rocks that such rocks have been called *stratified rocks,* layers are not the exclusive property of sedimentary rocks. Many igneous rocks display layers some of which are strata (Fig. 4-4, *B*); that is to say they have been spread out, one at a time, the oldest at the bottom. Although the materials of some igneous rocks are arranged in layers those of many such rocks are not so arranged. In fact, in what we may regard as a typical igneous rock the materials appear uniform in all directions and layers are not present.

Many metamorphic rocks are layered (Fig. 4-4, *C*) but some are not, depending on the condition of the parent rocks and the processes of metamorphism. In general, metamorphic rocks derived from layered rocks likewise display layers. By contrast, metamorphic rocks derived from nonlayered rocks may lack layers. The intense deforming movements accompanying some metamorphic processes are capable of creating new layers where none formerly existed, or of modifying preexisting layers. In appropriate chapters dealing with processes, we shall examine the origins and the significance of the positions of layers. Some thin layers may be visible even in laboratory specimens (Fig. C-1).

MAJOR CLASSIFICATION OF ROCKS BY PROCESSES AND PLACES OF ORIGIN

A useful prelude to laboratory study of small specimens of rocks is an understanding of the principles upon which rocks are classified into the three great groups. Two major factors are involved: process, and place of origin. As we noted in Chapter 1, rocks originate by many processes. In contrast, the places where rocks originate can be generalized into two great realms: (1) deep within the lithosphere and (2) at or near its surface. The essence of rock classification depends on the constancy of one major factor and the variability of the other. For

Fig. 4-3. Prominent vertical partings at right angles in strata of flat-lying sandstone underlying Colorado Plateau. Two streams have developed parallel to north-south partings. View south from an airplane near junction of Colorado and Green Rivers, southern Utah. (J. S. Shelton.)

example, in igneous rocks process is constant whereas place of origin varies. In sedimentary and metamorphic rocks place of origin is constant but process varies.

A schematic diagram (Fig. 4-5) represents the relationships of the two major factors involved in the origin of the three great groups of rocks. It shows the Earth's surface and the outer part of the crust, with depth relationships highly generalized. A clear-cut boundary (the Earth's surface) exists between magma and intrusive igneous rocks on the one hand and lava and extrusive igneous rocks on the other. No such clear-cut boundary separates the lower limit of sedimentary rocks and the upper limit of metamorphic rocks. The single dashed line separating these two great groups of rocks is diagrammatic only. Actually no single depth limit can be used as a boundary. Rather the lower limit of sedimentary rocks overlaps the upper limit of metamorphic rocks through an ill-defined zone of variable thickness. Let us look more closely at the relationship of the three great rock groups to these two major factors.

The constant factor in ***igneous rocks*** (Lat., "pertaining to fire"), *rocks formed by solidification of molten silicate materials,* is process: solidification resulting from cooling. The variable factor is place of origin. Solidification occurs in two places: (1) within the Earth, and (2) at the Earth's surface. This duality of place of origin permeates all aspects of igneous rocks. *Molten silicate materials beneath the Earth's surface, including associated crystals derived from them and gases dissolved in them,* constitute ***magma*** (Gr., "to squeeze" or "to knead"). *Molten silicate materials reaching the Earth's surface* constitute ***lava*** (from Ital. *lavare,* "to flood"). Because magma and lava differ in place of origin they also differ in gas content. Within the Earth, the weight of overlying rocks creates high pressure which tends to keep gases dissolved in magma; at the Earth's surface, the weight of the atmosphere exerts low pressure which permits gases to escape from lava.

Not only the molten materials but also the rocks derived from them are classified according to place of origin. Hence *rocks that originated from solidification of magma* are ***intrusive igneous rocks*** and *rocks that originated from solidification of lava* are ***extrusive igneous rocks.*** These terms for the two major classes of igneous rocks derive from the days

Fig. 4-4. Layers in rocks of each of the three great groups. *A.* Boundaries between extremely thick, horizontal layers of uniform, fine-grained sandstone, are indicated by pits and subtle changes in surface characteristics of exposure (*middle*). Narrow gorge appears to be parallel with direction of prominent vertical partings, which are widely spaced. (Frank Jensen.) *B.* Horizontal sheets of basalt. Closely spaced partings, rather inconspicuous, cause rocks to break into small fragments. Columbia Plateau at Moses Coulee, Washington. (Howard Coombs.) *C.* Inclined layers in metamorphic rocks. Partings, rather inconspicuous, are parallel with layers and with smooth plane face parallel with plane of photograph. White Plains, New York. (J. E. Sanders.)

when geologists thought that the Earth's interior beneath a thin, rocky crust was molten. When this concept was prevalent, igneous rocks that formed within the crust were explained as products of magma which supposedly derived from a still-deeper molten zone and had been *intruded* upward into the crust. Igneous rocks that formed at the Earth's surface were explained as products of lava *extruded* from the molten interior onto the surface. Even though the concept of a molten interior beneath a thin rocky crust has been abandoned (Chap. 18), the names live on.

Some geologists employ the term *lava* for both molten rock material at the Earth's surface and for extrusive igneous rock. Because dual usage tends to confuse, we shall define ***lava flow*** as *a hot stream or sheet of molten silicate material that is flowing or has flowed over the ground.* We term formerly molten material that has cooled a ***former lava flow,*** *a body of extrusive igneous rock derived from solidification of a lava flow.*

The constancy of process and variability of place of origin in igneous rocks are easier to visualize as bases for classification than constancy of place of origin and variability of process in sedimentary and metamorphic rocks. The reasons will appear shortly.

Rocks formed by cementation of sediments or by other processes acting at ordinary temperatures at or close below the Earth's surface are ***sedimentary rocks.*** Numerous sedimentary rocks result from ***lithification*** (literally "rock making"), *a general term for the conversion of sediments into sedimentary rocks.* Two other groups of processes give rise to noteworthy varieties: (1) physical precipitation and biologic secretion, and (2) accumulation of plant matter and hydrocarbons.

Rocks formed within the Earth's crust by transformation of preexisting rocks in the solid state, without fusion and with or without addition of new material, as a result of high temperature, high pressure, or both, are ***metamorphic rocks.*** Such rocks are united in that all of them form deep within the Earth and by the fact that all metamorphic reactions take place in the solid state, without fusion. The variability of metamorphic rocks comes from many processes responsible for ***metamorphism*** (Gr. *meta,* "beyond," and *morphe,* "form"), *the changes in mineral composition, arrangement of minerals, or both, that take place in the solid state at high*

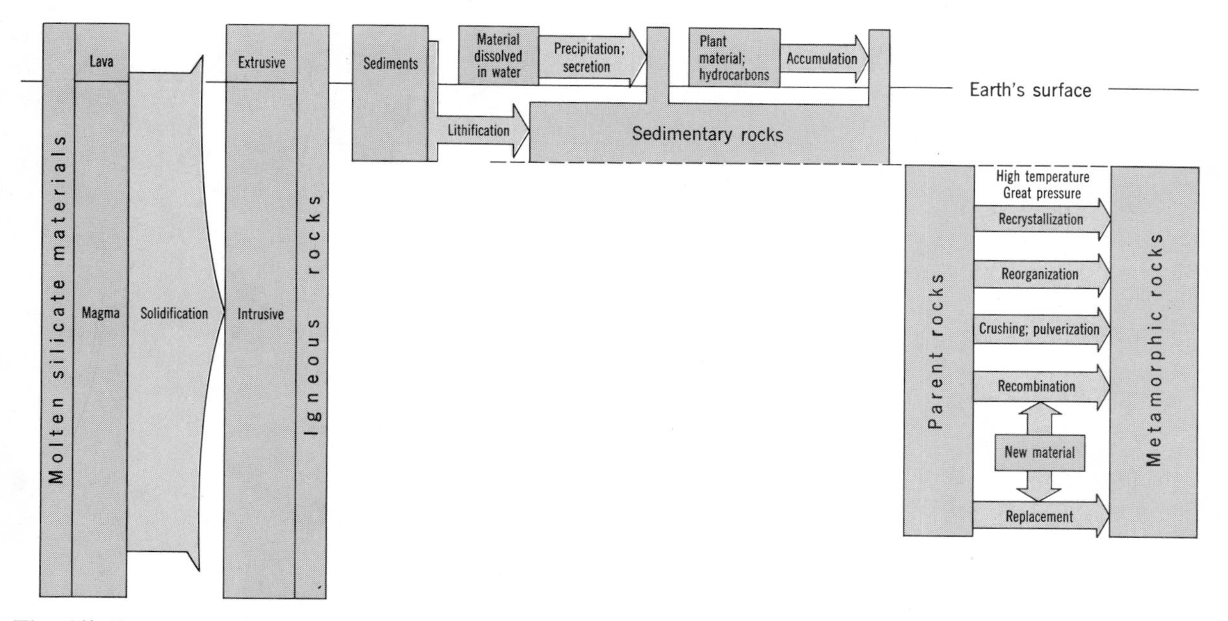

Fig. 4-5. Factors of process and location in classification of the three great groups of rocks shown on schematic profile through a segment of the Earth's crust. Boxes represent materials, arrows processes. Arrows showing metamorphic processes are arranged arbitrarily without depth significance. Further explanation in text.

temperatures, high pressures, or both, within the Earth's crust. The *processes* of metamorphism reflect the effects of the *agents* of metamorphism on the *materials* available. Agents of metamorphism include heat, pressure (including both pressures that are uniform in all directions and those that are not) and chemically active fluids. The materials available may come solely from the *parent rock* (the kind of rock that existed prior to metamorphism) or may include not only materials from the parent rock but also new materials added during metamorphism. The processes of metamorphism include: (1) *recrystallization,* simple enlargement of the particles in rocks containing only one mineral; (2) *reorganization* of two or more original constituents of the parent rock into new minerals; (3) *recombination* of the parent rock with material added to form new minerals; (4) *replacement* of minerals of the parent rock by new minerals whose material in whole or part were added during metamorphism; and (5) *crushing* and *pulverization* of mineral particles of the parent rock.

Eventually we shall come to examples resulting from these processes. Now let us explore the geologic method of analyzing rocks. We do this by discussing individual particles and relationships among particles in examples of rocks from each of the three groups. One example of each is sufficient to illustrate the principles involved; these principles can be extended to other examples as the need arises or according to the material available in various localities.

PARTICLES IN ROCKS

A rock's particles convey fundamental information concerning its history. Mineral composition forms the basis for recognizing and naming varieties within each of the great rock groups and also discloses chemical and environmental data. The kinds and origins, sizes, and shapes of particles give us clues to the sources of materials composing a rock and processes by which the rock was made.

Mineral composition. The major mineral constituents in rocks, which are the basis for classifying and naming many varieties within each great group (App. C), are termed *essential minerals.* The great variety of minor constituents are known as *accessory minerals* and do not enter into classification. Because of the varied origins of sedimentary rocks it is useful to recognize still other categories of particles within this rock group. All particles of some and the larger particles of other sedimentary rocks

Fig. 4-6. Specimens of rocks from each of the three great groups and minerals (shown by large crystals) composing them. (B. M. Shaub.) A. Granite. St. Cloud, Minnesota. *B.* Sandstone containing feldspar. East Haven, Connecticut. *C.* Coarse metamorphic rock. Uxbridge, Massachusetts. Further explanation in text.

0
5 cm
Quartz
Muscovite
Feldspar
B
0
5 cm
Feldspar
Muscovite
Quartz
Biotite
C

Fig. 4-7. Small rock particles in terrigenous sediment near Hartsdale, New York. (B. M. Shaub.)

constitute a *framework* within which particles support one another at points of contact. Between the particles of the framework may lie (1) spaces filled with fluids, (2) smaller particles, known collectively as the *matrix*, or (3) minerals precipitated from solution, the *cement*.

Although the number of known minerals reaches into thousands, most common varieties of rocks in each of the three great groups consist of only a few minerals. For instance in a granite (Fig. 4-6, *A*), a sandstone containing feldspar (Fig. 4-6, *B*), and a metamorphic rock derived from a granite (Fig. 4-6, *C*), only five minerals—feldspar, quartz, muscovite, biotite, and hornblende—are present as essential constituents, and no more than four are present in any one of the specimens figured. The photographs show the minerals by specimens of large crystals to emphasize that the same minerals occur within various rocks. Neither the sizes nor the shapes of the particles within the rock coincide with those of the crystals depicting the minerals, as can be seen by a glance at the rocks themselves.

The occurrence of common rock-forming silicate minerals in the three great groups of rocks is indicated in Table 3-1. The use of minerals in classifying and naming varieties of rocks is summarized in Appendix C. The chemical information contained in minerals is discussed in Chapter 5 and the environmental data in Chapters 7 and 20.

Kind and origin. Even after we have determined the composition of a particle, the whole story has not been told. For example, by simply identifying a particle as quartz we do not necessarily tell what kind of particle it is. It might be a crystal that grew in a magma or in mud on the sea floor. It might be a piece broken from an older igneous, sedimentary, or metamorphic rock. Here again are contrasting effects of growth and breakage.

All particles of igneous rocks result from cooling; hence, the dominant process in their origin is growth. Almost all particles of igneous rocks are crystalline; many display well-developed crystal faces. The only noncrystalline particles consist of natural glass, to be explained later.

Each major variety of sedimentary rocks is characterized by its own special kind of particles. The largest variety, rocks derived from sediments, consists of particles from four sources: (1) older rocks, (2) organic skeletons, (3) molten rock material, and (4) extra-terrestrial bodies.

(1) *Broken pieces of older rocks, minerals, or skeletal remains of organisms* are known collectively as ***detritus*** or by a synonym, particles of ***clastic*** (Gr. *klastos*, "broken") ***sediment*** (Fig. 4-7). *Solid particles of sediment derived from erosion of the lands* constitute ***terrigenous*** (Lat., "derived from the lands") ***sediment.*** Detritus and terrigenous

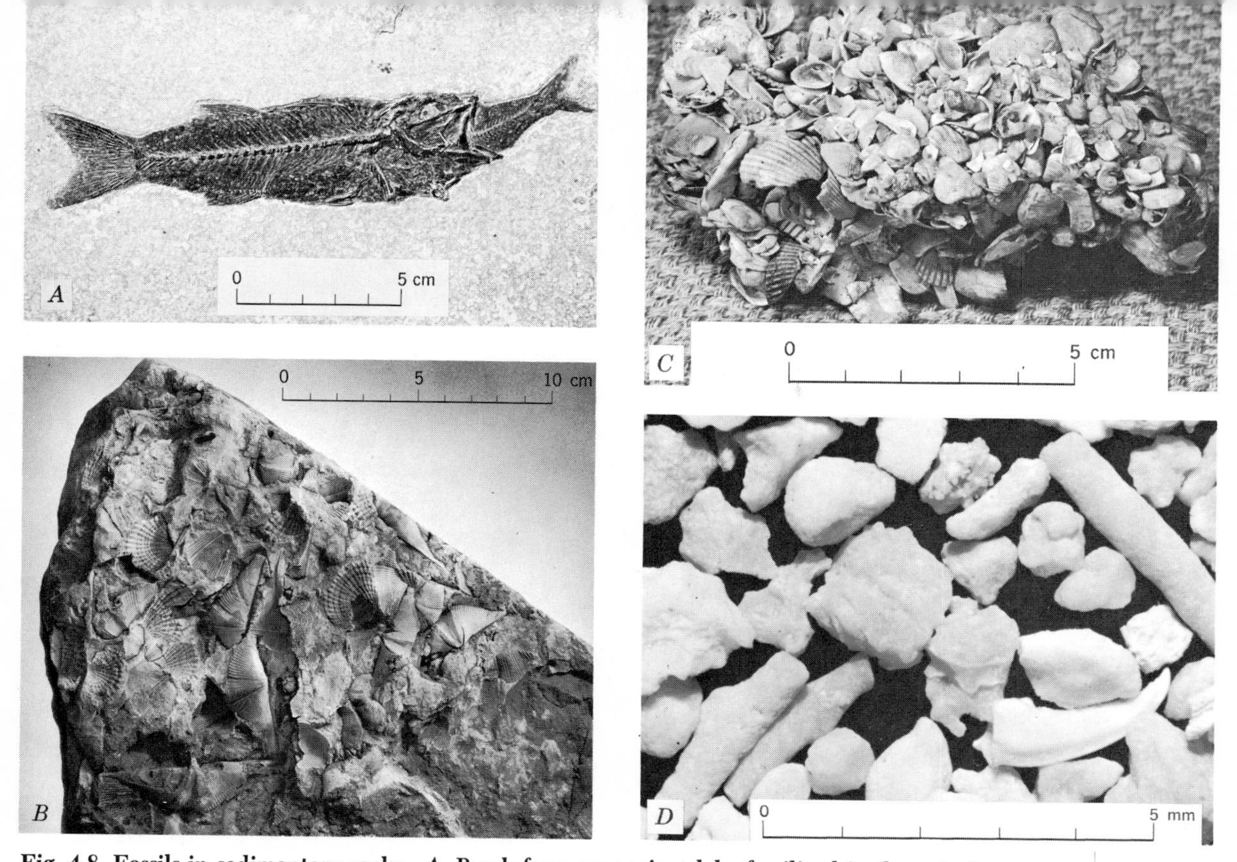

Fig. 4-8. Fossils in sedimentary rocks. ***A.*** **Perch from an ancient lake fossilized in the act of swallowing a small herring. Fossil Butte, Wyoming. (Princeton Univ. Nat. History, Photo 14705.)** ***B.*** **Impressions of hard parts of marine invertebrates in sandstone, eastern New York. (C. O. Dunbar.)** ***C.*** **Fragments of hard parts of marine invertebrates weakly cemented with calcite to make** ***coquina,*** **a rock. St. Augustine, Florida. (Yale Peabody Museum.)** ***D.*** **Organic skeletal remains, mostly of corals (cylindrical objects and irregular shapes). Curved object at right of center is part of a crab's claw. North of Discovery Bay, Jamaica, at depth of 22m. (B. M. Shaub.)**

sediment overlap in the sense that all terrigenous sediment is detritus. The converse is not necessarily true because much detritus consists of shells not eroded from the lands but secreted and accumulated on the sea floor. In the definition of terrigenous sediment the qualification of *solid particles* is added to exclude material carried in solution and that of *erosion* to exclude material derived from volcanic explosions. Material from both these other sources—solutions and volcanic explosions—notwithstanding their derivation from the lands, are assigned to their own distinctive categories.

(2) Particles of skeletal material secreted by organisms are abundant in many sedimentary rocks; indeed *calcareous*[1] skeletal material alone is present in many limestones. ***Fossils*** (Lat. *fossilis,* from *fodere,* "to dig"), *the naturally preserved remains or traces of animals and plants,* are criteria for recognizing sedimentary rocks. The definition of fossils includes not only the hard parts secreted by organisms but also impressions made by the hard parts, and tracks, trails, and burrow structures made by organisms. The hard parts can occur as complete skeletons of vertebrates or shells of invertebrates, or as broken pieces of skeletal material (Fig. 4-8), both ranging in size from the gigantic bones of whales and dinosaurs to bits of microorganisms invisible to the naked eye. Some tiny remains are extraordinarily abundant and tell fascinating stories. Minute unicellular plants and animals that abound in the sea contribute their hard parts to sediments deposited on the sea floor (Fig. 15-10,

[1] *Calcareous* means containing calcium carbonate.

B). Land plants generate countless billions of seeds in the form of pollen and spores. Although they are scarcely visible to the naked eye, pollen and spores travel widely; they are spread by winds and fall out in all places where sediments are accumulating. Because they are extremely resistant and can survive nearly all reactions to which sediments are subjected at low temperatures, these tiny particles can be extracted from most sediments and sedimentary rocks.

(3) Molten rock material can be converted into sediment directly by being blown out of the Earth during volcanic explosions. Collectively, *all particles ejected from volcanoes, irrespective of size, shape, or composition,* are known as ***tephra.*** Many tephra were blown into the atmosphere as clots of lava and cooled in flight to form glass particles (Fig. 4-9). Others came out as solids, hot or cold, and consist of crystals, crystalline particles, or rock particles of any of the three rock groups. Because of their mode of origin and composition, tephra form a special class of *pyroclastic* (Gr. *pyros,* "fire," and *klastos,* "broken") materials intermediate between igneous rocks and sediments. Strictly speaking, tephra that cooled from clots of lava are igneous rocks, but like other particles of sediment all tephra have been transported at the Earth's surface. In their distribution and final accumulation, tephra obey the same laws that govern cold mineral particles of whatever origin; hence, we treat them as sediment. Tephra can be deposited by one process, such as directly from the atmosphere or from water, or by several processes. They can be deposited from the atmosphere only to be eroded and redeposited later by water, or *vice versa.*

(4) Tiny particles from outside the Earth are *micrometeorites* or *cosmic dust.* Today these tiny particles of nickel-iron or silicate materials, generally spherical by the time they have passed through the Earth's atmosphere, constitute a minute proportion of the sediment being deposited; they are noticed only where particles of other kinds accumulate at a slow rate. However, early in the Earth's history the accumulation of cosmic particles may have been much greater.

Among the other varieties of sedimentary rocks we need mention only the features built *in situ* by organisms. Nontransported frameworks constructed by corals (Fig. 4-10) in limestones and remains of plants in coals (Fig. 23-2) can be recognized without further explanation.

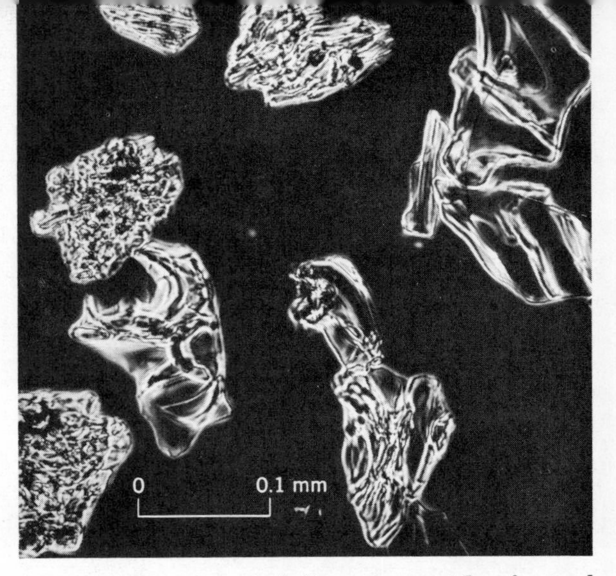

Fig. 4-9. Tephra composed of irregular particles of natural glass. Creston Bog, Washington. (R. B. Taylor, U. S. Geol. Survey.)

Like particles in sedimentary rocks, those in metamorphic rocks may result from either growth or breakage. In many metamorphic processes well-formed crystals or irregular particles will grow. In contrast, in other metamorphic processes particles can be stretched (Fig. 20-20) and broken (Fig. 20-19).

Size, shape, and arrangement. *The sizes and shapes of particles and mutual relationships among them* together constitute the ***texture*** of a rock. Many significant aspects of texture can be studied in small specimens with a hand lens or a binocular microscope. Even though we may not be able to identify some mineral particles with these simple tools, with them we can nevertheless observe important textural details. What we are unable to see with hand lens or binocular microscope we can readily discern with microscopes in which polarized light is either transmitted through paper-thin slices of rocks composed of transparent minerals or reflected from polished surfaces of opaque minerals.

The two contrasting processes, growth and breakage, are responsible for the two major textural categories. *Textures resulting from growth of associated crystalline particles* are ***crystalline textures.*** The clue to them is that irregular, angular particles interlock (Fig. 1-10, *A*). Crystalline textures are found in nearly all igneous rocks, in many metamorphic rocks, and in a few varieties of sedimentary rocks. *Textures resulting from break-*

Fig. 4-10. Coral limestone formed by organic secretion of massive skeletons composed of calcium carbonate. La Romana, Dominican Republic. (G. M. Friedman.)

age or abrasion of particles are ***clastic textures.*** Evidence that particles have been broken or abraded usually is easy to recognize (Fig. 1-10, *B*) and serves to identify clastic textures. Common in sedimentary rocks, clastic texture serves as a good clue for identifying such rocks. The importance of texture will become apparent through an example from each of the great rock groups.

Texture of an igneous rock. Refer again to the specimen of granite in Fig. 4-6, *A*. Notice that all particles are of about the same size (5 to 10mm), that they interlock complexly, and that none displays its characteristic crystal form. *The orientation of the particles* or the ***fabric*** is random; the arrangement is about the same in any direction. The interlocking of the particles and lack of crystal forms are the expected consequences of simultaneous growth of many crystal lattices. The random orientation suggests that growth occurred without any outside interference. No motion took place and the crystals must have been far from any fixed surface that would orient them. In contrast, observe the orienting effect of crystal growth from a fixed curved surface clearly displayed by the coarse particles in the upper part of Pl. 4.

The useful inference that the textures of igneous rocks reflect conditions during cooling was first stated late in the eighteenth century by Sir James Hall, a Scottish colleague of James Hutton. Hall melted rocks in a forge and allowed the liquids to cool at various rates. Natural glass resulted from extremely rapid cooling. When cooled slowly a melt made from this glass or from a fresh rock specimen yielded fine-grained crystalline aggregates. If cooled at a still slower rate the resulting crystalline aggregate became coarse grained.

Because Hall experimented only with dry solids and dry melts, his results can be interpreted simply in terms of one factor, rate of cooling. Subsequent experiments with both silicate melts and volatiles, especially water, indicate that particle size in igneous rocks depends not only on rate of cooling but also on a second factor, presence of volatiles. When volatiles are present particles can become extremely coarse.

Texture of a clastic sedimentary rock. At first sight the specimen of sandstone containing feldspar illustrated in Fig. 4-6 does not differ significantly from the specimen of granite. The range of size of the particles in the sandstone (1 to 10mm) is greater than that in the granite, but the particles appear to interlock; none displays its characteristic crystal form, and the fabric is random. A closer view is needed before we realize that the framework of the rock consists of broken particles of minerals from an older rock, in this case an older granite. By selecting this specimen on the basis of its mineral composition we have sacrificed a clearer display of the effects of wear on particles that results from many geologic processes. Both a matrix of smaller particles and a mineral cement occupy the spaces between particles of the framework, but neither is visible in the photograph. This specimen, therefore, is a rock formed by lithification of a clastic terrigenous sediment, which contains particles of minerals derived from a granite.

Earlier we mentioned that sedimentary rocks originate by lithification of regolith. This proposition is so fundamental to geology that we need to satisfy ourselves as to its validity beyond the slightest doubt. Therefore we set forth the proofs based on comparison of particles in regolith with particles in sedimentary rocks. Later in the book we shall add still other evidence based on features larger than particles.

Particles in regolith. As we have learned, the spacing of partings determines the sizes of fragments broken from bedrock. Once broken loose, rock fragments of whatever size can crumble into a heap of individual mineral particles (Fig. 7-11). In many kinds of bedrock the longest dimensions of individual particles do not exceed a few millimeters. Initially, therefore, particles in regolith can range from large fragments of rocks whose longest dimensions amount to several meters to microscopic particles of individual minerals.

Since the beginning of the twentieth century geologists have been working on the hypothesis that sizes of clastic particles deposited are related to the mechanics of the processes of transport and deposition of sediment. In attempting to prove this hypothesis they have made many measurements of the sizes of particles in sediments. To facilitate comparisons of samples they have devised a standard scale, whose size classes are shown in Fig. 4-11 alongside a $\log_{10}$ scale of centimeters.

Because the range of the aspects of so many natural features is extremely great, logarithmic scales are useful in many parts of geology. We shall employ logarithmic scales repeatedly in this book. A $\log_{10}$ scale is known also as an *order of magnitude scale.* Each of its divisions, marked off by a power of 10, covers a span that is 10 times larger than the span of the adjoining smaller division.

The basis for the scale of sediment particles is a geometric progression of particle diameters, in millimeters, built on powers of 2. In each class, diameters of particles are twice as large as those in the next smaller class. To avoid writing cumbersome fractions in the sand, silt, and clay classes, geologists have replaced the millimeter values of the class intervals with equivalent logarithms. Because the geometric progression is built on powers of 2, the logarithms employed in the scale have base 2. On any logarithmic scale, 1 is represented by 0, and the range can extend to both positive and negative values. The negative of the true logarithms has been adopted to make values in the finer-sized classes positive. Diameters larger than 1mm are represented by negative $\log_2$ values: 2mm by -1, 4mm by -2, and so forth. Diameters smaller than 1mm are represented by positive $\log_2$ values: $\frac{1}{2}$mm by $+1$, $\frac{1}{4}$mm by $+2$, and so forth. Perhaps from chemistry courses many readers are familiar with the scale of pH values, which employs negative values of logarithms of base 10.

Particles are named simply by measuring their diameters and then comparing the measurements with the size scale. Figure 4-12 displays particles separated into the classes of the standard scale by passing sediments through sieves whose screen openings correspond to the boundaries of the size classes. The particles in the figure, shown at natural size, range from fine sand ($\frac{1}{8}$mm or $+3$ on the $-\log_2$ scale) to very coarse pebbles (64mm or -6 on the $-\log_2$ scale).

So much for the sizes of individual particles. How do we deal with the aggregates of particles composing natural sediments? Observation tells us that natural transporting agents carry sediment particles of varying sizes, specific gravities, and shapes. Transporting processes are detailed in later chapters; here we are concerned only with the highlights of these processes as they relate to particle size in sediments.

Flowing fluids, water and air, carry fine particles (silt and clay) at nearly all speeds but transport coarse particles (sand, pebbles, and larger particles) only at high speeds. As speed of flow changes so do

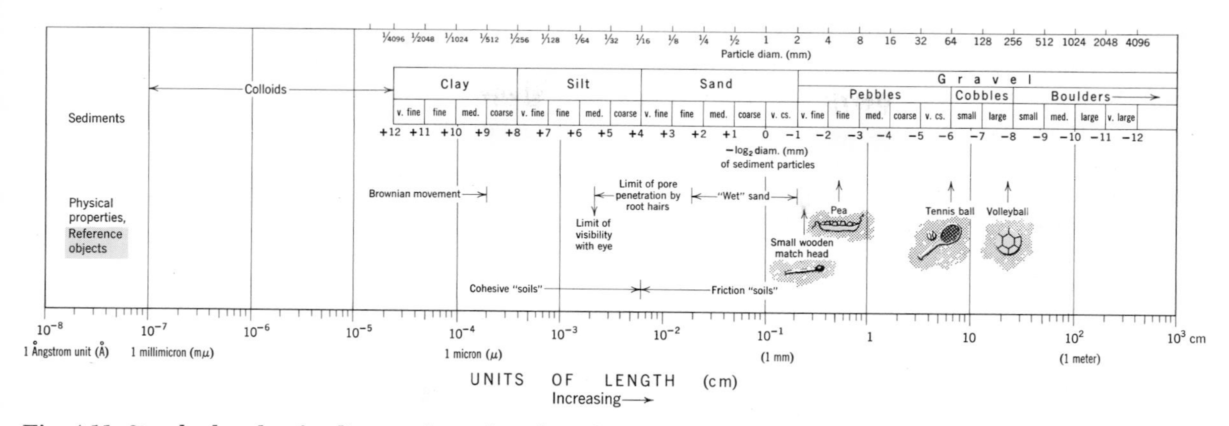

Fig. 4-11. Standard scale of sediment classes based on diameters of particles, shown on $\log_{10}$ scale of centimeters (*bottom*). Notes on physical behavior of sediments having various diameters and sketches of common objects for reference are below sediment scale. Text explains logarithmic scales.

the sizes of particles transported. Likewise particles are selectively transported and deposited according to their specific gravities and shapes. Transport in water or wind results in ***sorting,*** *the selection, by natural processes during transport, of rock particles or other particles according to size, specific gravity, shape, durability, or other characteristics.* In contrast with this sorting action of fluids, mass transport by gravity or by glaciers involves particles of all sizes, specific gravities, and shapes. Therefore, if we know that particles of matrix and framework were deposited together, we can rely on sorting as an indicator of the transporting agent.

One way to attack the problem of dealing with aggregates of particles of varying sizes in sediments is to measure particle sizes in the laboratory and to describe their distribution by statistical formulas. Such techniques, used by specialists, are beyond the scope of this book. Our purposes require only general categories for sediments whose particle sizes have been estimated just by inspection. The generally accepted definition of *well-sorted sediment* is a sediment consisting of particles all having about the same size. *Poorly sorted sediment* refers to a sediment consisting of particles of many sizes.

Such definitions of sorting generally coincide with standard geologic usage. But we must point out that geologists and some engineers differ in their ways of expressing sorting. In some engineering terminology *well-graded* ("well-sorted") means that the sediment includes a large assortment of sizes. This engineering "well-sorted" equals the *poorly sorted* of geologic usage.

Typically in regolith the shapes of most particles are angular to begin with. That is to say they are bounded by many plane surfaces which meet other plane surfaces at sharp edges and corners. Initially angular particles do not remain angular when subjected to the abrasion that results from repeated collisions among particles during certain kinds of transport. Edges and corners wear down as pieces break off; eventually flat surfaces disappear and the exteriors become smooth and rounded (Figs. 4-12; 16-7; 16-8, *C*). Although the results of abrasion are widespread among pebbles and larger particles, well-rounded quartz particles that are smaller than medium sand are rare. Does this mean that the smaller particles do not travel as much as the large ones? Very likely not. If anything, the small particles probably travel more than the large ones. Instead of differences in amounts of travel the cause of the angularity of small particles probably stems from insufficient momentum of sand grains. As particles become smaller their momentum is not large enough to cause chipping of edges and corners during a collision with another small particle. Furthermore, in water the viscosity tends to serve as a cushion to soften the effects of collisions. Some distinctive shapes resulting from abrasion of particles are described and illustrated in Chapters 12, 13, and 16.

Alongside many of the photographs of sediments in the standard classes shown in Fig. 4-12 we have placed photographs of sedimentary rocks. The

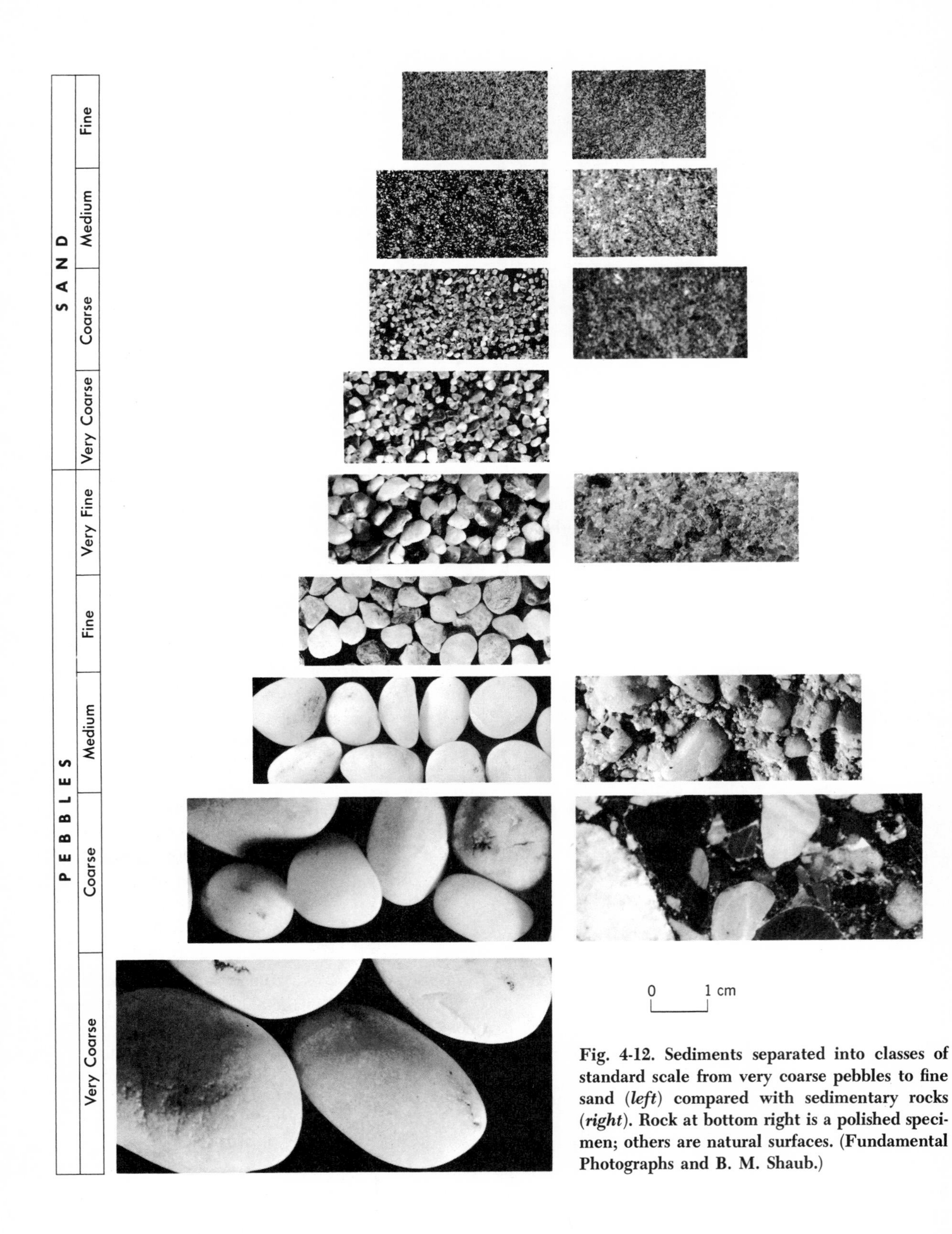

Fig. 4-12. Sediments separated into classes of standard scale from very coarse pebbles to fine sand (*left*) compared with sedimentary rocks (*right*). Rock at bottom right is a polished specimen; others are natural surfaces. (Fundamental Photographs and B. M. Shaub.)

reader can judge for himself how well this figure supports the general proposition that many sedimentary rocks are derived from lithification of regolith.

Texture of a metamorphic rock. The prominent characteristic of the metamorphic rock shown in Fig. 4-6, *C* is the decidedly streaked appearance presented by the arrangement of its light and dark minerals. Particle size and particle interlocking cannot be easily discerned, but fabric is obvious. Fabrics with particles aligned are *fabrics with preferred orientations.*

Study of the specimen itself discloses that the particles are intergrown in a crystalline texture. From study with a hand lens we are unable to determine whether the sizes of the particles were inherited from the parent rock or were changed during metamorphism. We infer that the fabric with preferred orientation resulted from the influence of deforming movements that were active during metamorphism. In this case, the parent rock was a granite but in other cases metamorphic rocks closely resembling this one have originated from the transformation of parent rocks of kinds other than granite.

WHAT HOLDS PARTICLES TOGETHER?

No very extensive experience is required to realize that the particles of some rocks hold together with great tenacity whereas those of others separate rather easily. The tenacity of particles of igneous and metamorphic rocks is easily visualized from their crystalline textures of interlocked particles. The particles grow close to one another and in some rocks are intricately interlocked. A similar tenacity based on interlocked particles is evident in steel.

The tenacity with which the particles of sedimentary rocks hold together displays a considerable range. The wonder is that particles of regolith evince any ability at all to stick together. Yet they do, after a fashion. Let us explore the reasons why particles of sedimentary rocks and regolith hold together, starting with regolith.

Togetherness in regolith. Particles of regolith remain together for two principal reasons, which vary with grain size: (1) friction and gravity hold coarse particles together; (2) cohesion operates among fine particles.

Friction and gravity. In sand and coarser particles, togetherness is achieved by the combined effects of friction at points of contact between grains and the normal component of gravity (g_n of Fig. 2-8). As Fig. 2-8 shows, the value of g_n varies with the cosine of the angle of slope on which the particle rests. As slope angle increases, g_n decreases and g_t increases. Eventually a slope is reached on which the value of g_t exceeds that of g_n plus friction, and the particles no longer remain together but begin to move down the slope (Chap. 8). As the particles begin to move they tend to rise upward as they pass over one another (Fig. 4-13). This creates a force, normal to the slope and opposite to g_n, which tends to disperse the particles and make all of them move with less internal friction at points of contact. When all the particles are in contact with one another, the body of solid particles displays many characteristics of one solid body. However, when the particles are no longer in contact with one another but are in a dilatant state, they behave in mass like a fluid in that they flow readily. A body of solid particles in a dilatant state is said to be *fluidized.*

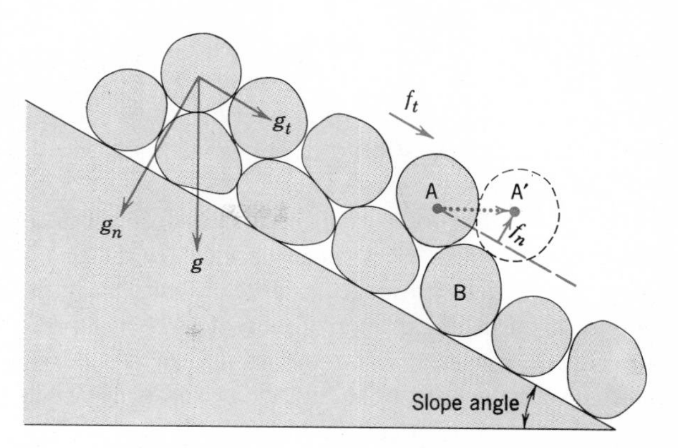

Fig. 4-13. Rock particles held on a slope by friction and gravity. If tangential force, f_t causes particle A to move over particle B, either by rotation (as shown) or by sliding, the center of gravity (large dot) of A is raised from A to A′. Therefore, motion of particles parallel to the slope creates an upward force, f_n acting normal to the slope in the direction opposite to g_n. The effect of f_n is to disperse the particles, enabling them to move more easily as a single mass in response to g_t.

Cohesion. In particles of silt size and smaller, the ratio of mass to surface area decreases progressively. Eventually this decrease causes the particles to respond to electrostatic forces of attraction. Electrostatic forces originate because atomic charges are unevenly distributed throughout crystal lattices. Where electrostatic forces operate between fine particles *cohesion* results. Figure 4-11 (*bottom*) labels the boundary between friction "soils" (engineers' term for regolith) and cohesive "soils."

Lithification. Previously we have mentioned that by lithification regolith is transformed into sedimentary rocks. Now let us investigate the processes by which lithification takes place. All the processes of lithification are influenced by the relationships between the framework and matrix particles and the spaces unoccupied by solid materials (*pore spaces, interstices,* or *interstitial spaces*). The processes include cementation, changes in packing, compaction, and recrystallization; all involve loss of original pore space and reduction in the sizes of the remaining pores.

Cementation. Coarse sediments are generally lithified by ***cementation,*** *the binding together of particles of framework and matrix by precipitation of mineral cement in former pore spaces.* Calcite, quartz, and iron oxides are the commonest cements. Because the cement is precipitated its minerals exhibit crystalline textures on a small scale.

Changes in packing. Many clastic sediments consist of elliptical particles whose initial arrangement (*packing*) varies. We can illustrate packing by groups of spheres (Fig. 4-14). Particles with open packing are separated by maximal pore space. Pressure of overlying material and earthquake vibrations in the ground can change open packing to close packing and thus decrease pore space. A change from open to close packing of sand grains having irregular shapes can convert them from loose grains into a coherent mass.

Compaction. *The reduction in pore space within a body of sediment, in response to the weight of overlying material or to pressures within the Earth's crust* is ***compaction,*** a process that forces particles closer together and drives out any water that initially separates them. Compaction can bring about great changes of thickness; in the process fragile shells within the sediment can be pressed flat. After clay has been compacted beneath the great weight of a thick column of overlying materials it no longer becomes sticky when wetted.

Recrystallization. Particles of sediments can interlock when they enlarge by recrystallizing. Recrystallization is a process that commonly affects soluble particles of framework or matrix. The difference between recrystallization and cementation is that framework particles, passive during cementation, are reorganized and generally enlarged during recrystallization. The effects of recrystallization can be so thorough that a crystalline texture resembling that of an igneous or metamorphic rock is formed.

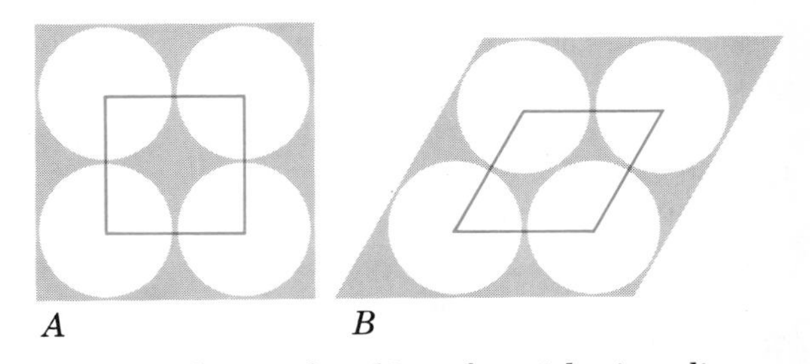

Fig. 4-14. Change of packing of particles in sediments shown by sections of four contiguous spheres of equal size. *A.* Open packing; pore space (shaded) 47.6 per cent of total volume. *B.* Closest packing; pore space 25.9 per cent of total volume. As the individual particles in natural sediments are neither perfect spheres nor of equal diameter, these sections illustrate only the principle involved. (After C. S. Slichter, 1899.)

GO AND SEE

Study of this chapter should sharpen the reader's vision in looking at rocks and regolith. The printed page can offer suggestions and show pictures but can never substitute for direct observation of the real thing. Rocks and regolith must be seen to be appreciated. Attentive study of them in the laboratory and in the field provides the great satisfaction that comes from being able to decipher the "words" of geologic language and thereby to read the fascinating stories that constitute geologic history.

As we move on to other subjects we shall never be far away from some aspects of rocks and regolith. Before getting to these other subjects let us survey the relationships among rock-forming processes by analyzing the major geologic cycles.

Summary

1. Partings and layers, two gross features of most exposures visible even from a distance, permit fragments of rock to break loose from the bedrock, facilitate the flow of fluids over and through the rock body, and increase the total surface area of the rock body.

2. Combinations of two major factors—process and place of origin—form the basis for classifying rocks into three great groups: igneous, sedimentary, and metamorphic.

3. The mineral compositions, kinds and origins, sizes, shapes, and arrangements of particles convey significant information about their history and that of the rock in which they occur.

4. Mineral composition is employed for classifying and naming varieties within each great rock group and also reflects chemical composition and environmental conditions under which particles originated.

5. The particles in igneous rocks are governed by the conditions under which the molten material cooled.

6. In regolith and sedimentary rocks particles come from older rocks, organic skeletons, explosive volcanoes, and extra-terrestrial bodies.

7. The particles in metamorphic rocks originate by recrystallization or breakage of particles of the parent rocks.

8. Texture—the sizes, shapes, and mutual relationships of particles—results from rock-forming processes. Crystalline textures predominate in igneous and metamorphic rocks; clastic texture is common only in sedimentary rocks.

9. The effects of friction and gravity hold coarse particles of regolith together; the effects of cohesion hold fine particles together.

10. By various processes particles of regolith can be lithified to become sedimentary rocks.

Selected References

Fisher, R. V., 1964, Maximum size, median diameter, and sorting of tephra: Jour. Geophys. Research, v. 69, p. 341–355.

Loomis, F. B., 1948, Field book of common rocks and minerals, rev. ed.: New York, G. P. Putnam's Sons.

Pearl, R. M., 1965, How to know the minerals and rocks: New York, New American Library of World Literature, Inc.

Pirsson, L. V., and Knopf, Adolph, 1947, Rocks and rock minerals, 3rd ed.: New York, John Wiley.

Shand, S. J., 1952, Rocks for chemists: London, Thomas Murby.

Spock, L. E., 1953, Guide to the study of rocks: New York, Harper and Bros.

Tyrrell, G. W., 1929, The principles of petrology, 2nd ed.: New York, E. P. Dutton.

Zim, H. S., and Shaffer, P. R., 1957, Rocks and minerals: New York, Golden Press. (Paperback.)

Major Geologic Cycles

The geologic cycle
Hydrologic cycle in various climates
Rock cycle
Geochemical cycle
Tectonic cycle
Other cycles

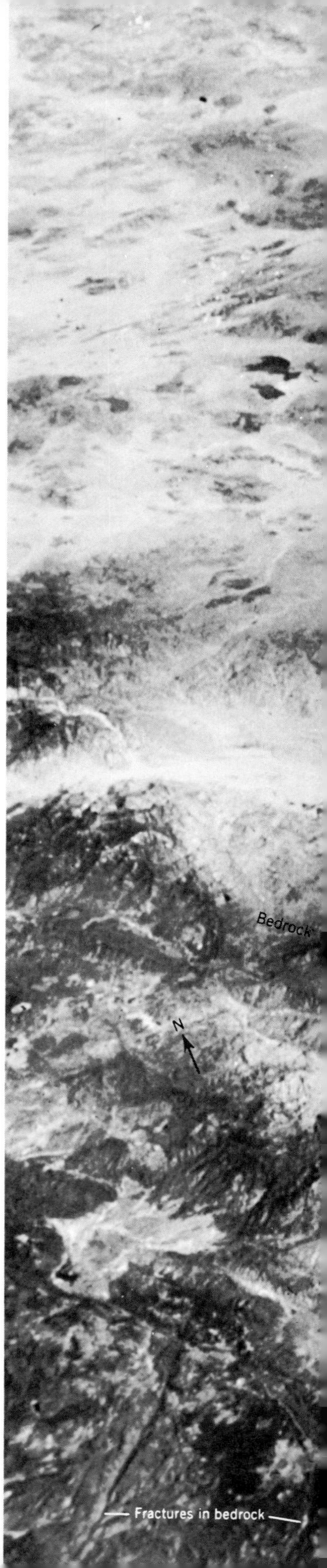

Plate 5

By operation of the geologic cycle nearly half this region in Southern California has been covered with sediment. Light-colored areas in dark water of Salton Sea show flow of currents carrying sediment in suspension. Large-scale irrigation employing water from the Colorado River sustains the cultivated areas of Imperial Valley. Salton Sea is about 30 miles long. (*Gemini V, from National Aeronautics and Space Administration.*)

Colorado River
San Bernardino Mountains
Sediment
SALTON SEA
Sediment
Imperial Valley

CHAPTER 5 *Major Geologic Cycles*

THE GEOLOGIC CYCLE

In Chapter 1 we explained that the geologic cycle proceeds unceasingly as moisture falls on the lands and, in flowing back to the oceans, carries with it products of erosion. The idea that this kind of activity has been going on throughout countless ages is a truly extraordinary and fundamental precept of geology. As Dutton wrote, the concept is to geology what the law of gravitation is to astronomy.

Now that we have taken a larger view of the Earth and a closer look at the materials composing its outer parts, we are in a position to be specific in our second encounter with the geologic cycle. How much water is available to travel from the oceans through the atmosphere and back to the oceans? Is seawater becoming saltier from all the dissolved substances continually being delivered to it by streams? How rapidly are continents being eroded and ocean basins filled with sediment? Even when we have considered some of these large questions we shall not have finished with the geologic cycle. Indeed we shall be looking into some aspects of the cycle in every succeeding chapter of this book. Our purpose in the present chapter is to build a framework of broader relationships to establish a perspective for our detailed studies of the individual processes operating within the cycle and of the products which these processes produce.

The geologic cycle consists of four interrelated cycles (Fig. 5-1). The whole operation depends on the *hydrologic cycle,* the circulation of moisture up into the atmosphere and back down to the Earth. Transport of products of erosion at the Earth's surface and processes operating within the Earth combine to create, to destroy, and to recreate rocks. These activities are grouped into the *rock cycle.* Two important aspects related to operation of the rock cycle are the movements of chemical elements and the motions of the Earth's crust. These are isolated for special study in the *geochemical cycle* and *tectonic cycle,* each of which we shall examine presently after a look at the *hydrologic cycle* and the *rock cycle.*

HYDROLOGIC CYCLE

The ***hydrologic cycle,*** *the system of water circulation from oceans to atmosphere and back to oceans, either directly or via the lands,* was perceived in part by King Solomon. According to Ecclesiastes 1:7 (King James version) Solomon remarked: "The rivers run into the sea, yet the sea *is* not full; unto the place from whence the rivers come, thither they return again." Ever since Hutton's time geologists have been particularly interested in understanding the hydrologic cycle, not only because of the geologic work done by the cycle but also because of the cycle's implications for man's water requirements.

The total quantity of water in the hydrosphere has been estimated to a first approximation from summation of areas and depths of oceans, lakes, and rivers; from observed vapor pressures in the atmosphere; and from measurements of areas and thickness of ice in major glaciers (Fig. 5-2). A quantitative statement of the amounts of water circulating through various paths in the hydrologic cycle is a *water balance* or *water economy.* The International Hydrologic Decade (1965 to 1974) has been organized and dedicated to massive, worldwide measurements designed to yield precise details on the world's water balance. The hydrologic cycle is nearly a closed system. Some water escapes from the cycle but more is added from the Earth's interior during volcanic eruptions (Chap. 19).

According to where the precipitation falls we recognize a short cycle and a long cycle. In the short cycle water evaporated from the oceans is precipitated directly back into the oceans without ever falling on land (Fig. 5-3, *center*). Apart from the expenditure of heat energy involved, water accomplishes little work in a short cycle. In the long cycle precipitation falls on the lands and returns to the oceans, either immediately through streams or later after storage on land for various lengths of time. In this cycle water does extensive geologic work; the details of work done vary according to climate.

Cycle in humid-temperate climate. In areas having humid-temperate climate, such as many parts of the conterminous United States, some rainfall seeps into the ground and some flows along the surface to streams, to lakes, and eventually back to the oceans. From measurements made at many localities during the twentieth century it is possible to estimate a water balance for the conterminous United States (Fig. 5-4). The values are shown as reported, in inches. From them we learn that only about a third of one per cent seeps back to the ocean through the ground, 30 per cent flows along the surface, and the remainder, about 70 per cent, is evaporated directly back into the atmosphere or moves from the ground through plant tissues and back into the atmosphere by ***transpiration,*** *the passing of water vapor into the atmosphere from pores of plant tissues.*

From these measurements we see that nearly three-quarters of the water falling on the land in a humid-temperate climate is delayed in its return to the oceans. The delayed water is stored in lakes, in pores or in combination with minerals of rocks and regolith underground, and in the tissues of plants and animals.

Flow of freshwater into the sea. Where precipitated back into the sea from a circuit around a short cycle freshwater mixes immediately with saltwater. But where introduced in large quantities from a circuit around a long cycle, freshwater and saltwater do not mix immediately. Mixing is delayed whether the freshwater enters the sea from streams or from pores in sediments and rocks underground. Instead of mixing immediately the freshwater flows over the saltwater along an interface that slopes landward (Fig. 5-5). This can happen because freshwater is less dense than seawater. The local circulation pattern established by the interaction of freshwater and seawater governs the availability of potable water in coastal areas and influences the discharge of sediment from streams into the sea.

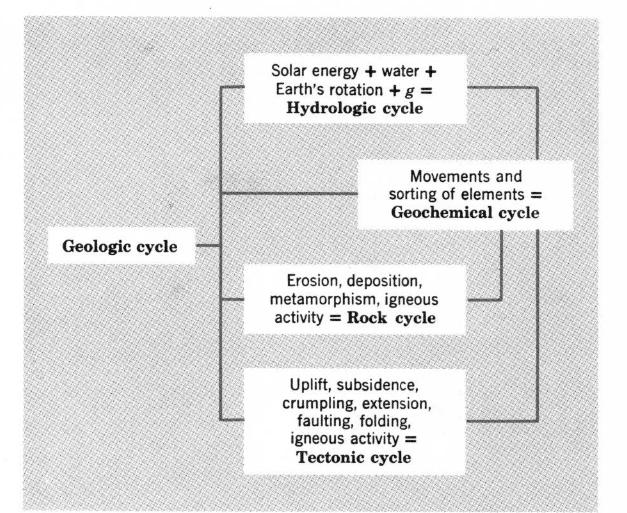

Fig. 5-1. Major component cycles of the geologic cycle.

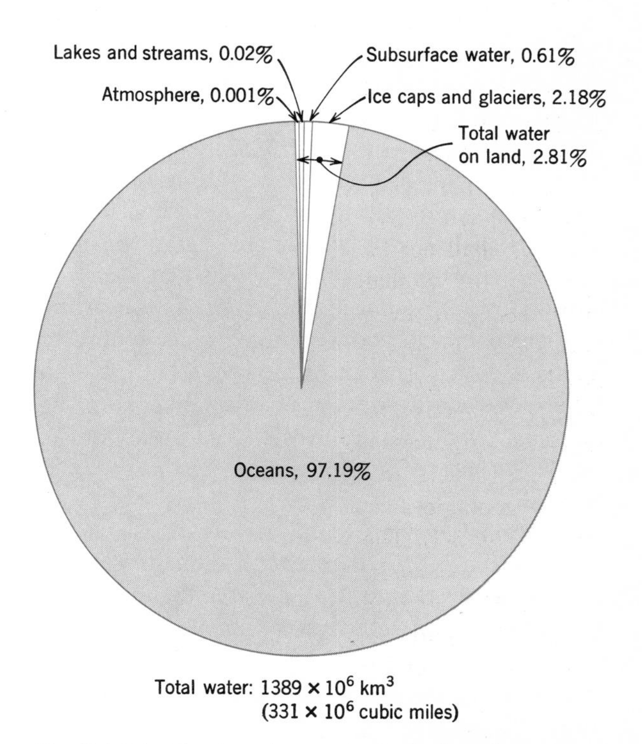

Fig. 5-2. Distribution of water in the hydrosphere. (Data from R. L. Nace, 1960 and H. W. Menard and S. M. Smith, 1966.)

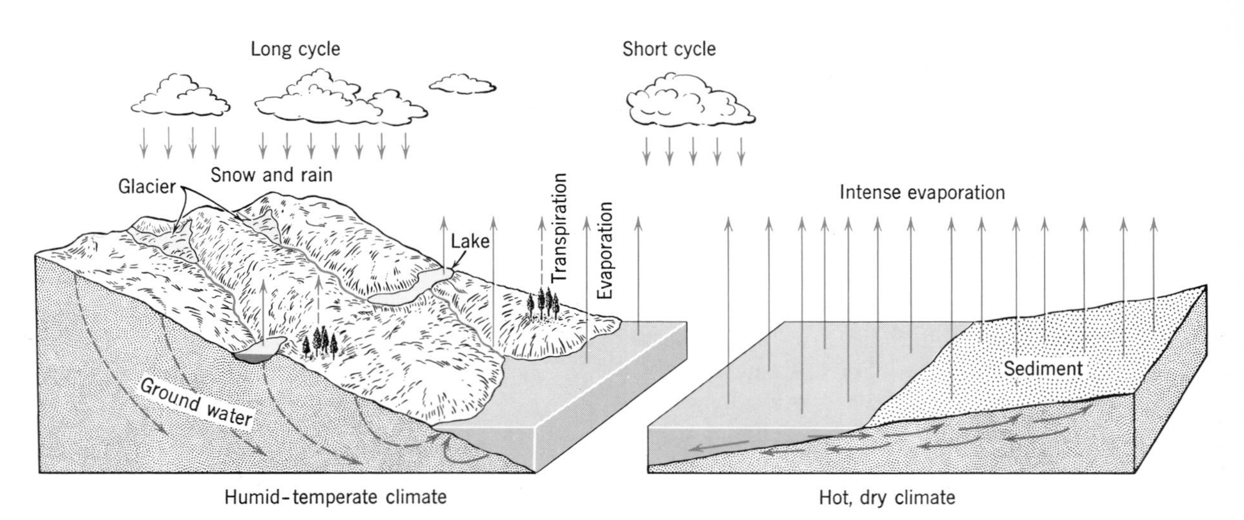

Fig. 5-3. The hydrologic cycle under contrasting climates. Cycle in humid-temperate climate (*left*) includes a short cycle and a long cycle. Intense evaporation in hot, dry climate (*right*) draws up water from ocean and from sediment, causing water to circulate landward. Blue curved arrows on front of block (dashed, freshwater; solid, saltwater) indicate flow of water underground. Further explanation in text.

Cycle in cold climate. In areas having a cold climate much or all precipitation falls as snow and thus, depending on time of melting, may be delayed in returning to the oceans. The delay may be only seasonal or as much as thousands of years when snow accumulates in thick piles and is converted to glacier ice (Fig. 3-25). Although most glacier ice remains on the land until the ice melts, some returns directly to the sea as icebergs.

Cycle in hot, dry climate. In regions having a hot, dry climate, evaporation greatly exceeds precipitation thus obscuring the cyclic flow of water. In such regions the only conspicuous flow of water is one-way, from ocean or land up into the atmosphere. Eventually, of course, atmospheric moisture becomes cyclic and returns to the ocean but its trip down may occur far away from the site of its trip up.

Continual evaporation of a body of seawater or of interstitial water in coastal sediments not only lowers the level but also increases the concentration of dissolved solids in the unevaporated solution. As concentration increases, the water becomes progressively denser. The changes of level create an imbalance which sets up a local circulation that tends to restore the level. Water from the sea is drawn landward at the surface. As it flows landward it evaporates, becomes denser, and sinks, returning seaward at lower levels (Fig. 5-4, *right*). In local circulation systems driven by evaporation, the directions of flow are just the reverse of those in systems driven by inflowing freshwater: at the surface water flows landward and lower down seaward.

If evaporation increases the concentration of dissolved solids sufficiently, minerals are precipitated. *Minerals precipitated as a result of evaporation* are ***evaporite minerals.*** A few examples include *aragonite, dolomite, gypsum, anhydrite,* and *halite.* Although they can be precipitated through evaporation some of these minerals form by other processes as well.

ROCK CYCLE

The basic precept of the rock cycle is that bedrock exposed at the Earth's surface breaks down into regolith, which subsides beneath the surface and is converted there back into bedrock. Furthermore, in deep zones within the Earth, rocks either are being transformed from one kind into another by metamorphism or are being melted. When the molten material crystallizes it begins a new career.

The rock cycle is not a completely closed system because from time to time new material is added from beneath the crust, some water escapes to the hydrosphere and some volatiles to the atmosphere, and a few soluble elements leave the rego-

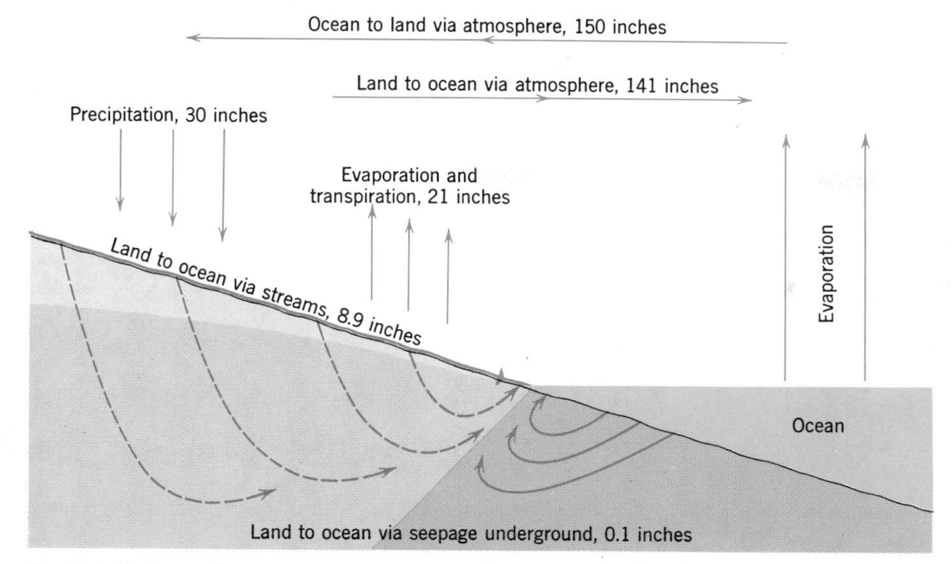

Fig. 5-4. Estimated average water economy for conterminous United States. Curved arrows (dashed, freshwater; solid, saltwater) indicate directions water flows underground. (Numbers based on publications of U. S. Dept. of Agriculture.)

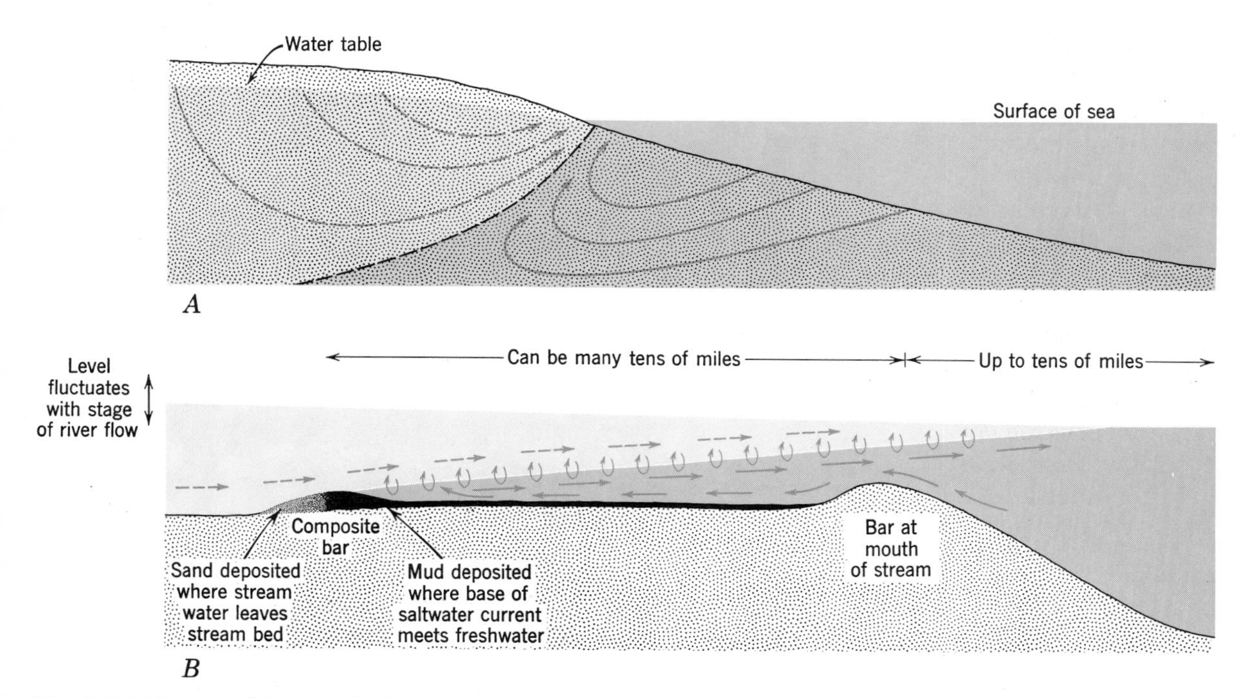

Fig. 5-5. Behavior of flowing freshwater at boundary with saltwater, shown by profiles at right angles to shore. A. Flow of water underground. (After F. A. Kohout, 1960.) *B.* Flow of freshwater (gray) of stream entering the sea (blue, right) shown by idealized profile down channel. Further explanation in text. (After E. A. Schultz and H. B. Simmons, 1957.)

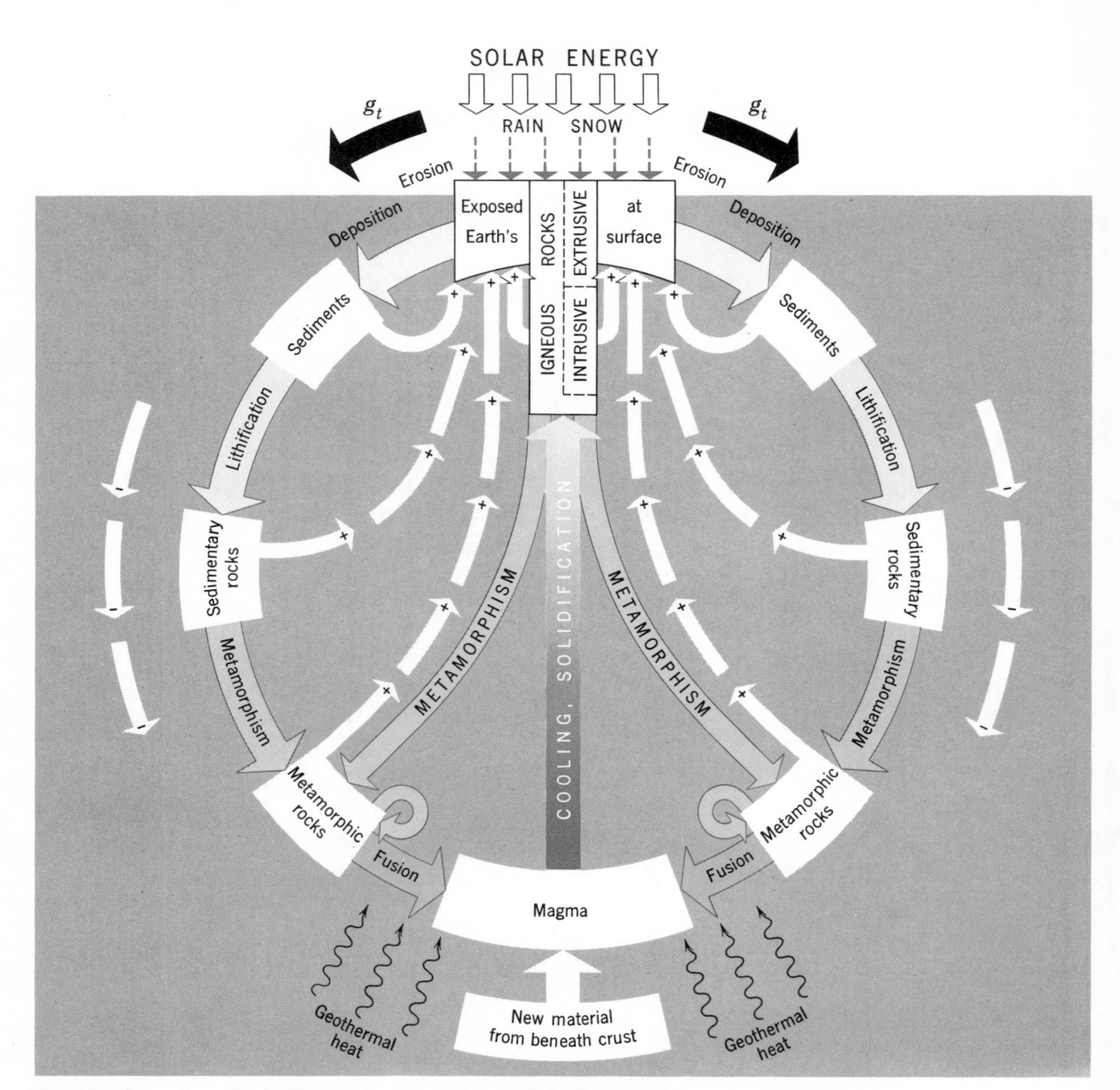

Fig. 5-6. The rock cycle. Full cycle is repeated in each half of the diagram to show relative position in Earth's crust (Earth's surface, above; subcrustal zones, below). Further explanation in text.

lith in solution and are delivered to the sea. On occasion, some of these soluble elements are precipitated out of seawater and return to the rock cycle as new rocks.

We can illustrate the rock cycle with a circular diagram (Fig. 5-6) representing a schematic section through the outer part of the crust, with the Earth's surface at the top and the deeper subcrustal zones at the bottom. White boxes designate rock-making materials. Arrows indicate rock-making processes (blue arrows, with darker shading for higher temperatures), energy added (open, black arrows for solar energy at top; black arrows with wavy shafts for geothermal heat at bottom), and generalized vertical crustal movements (open arrows with positive signs pointing up within the circle for uplifts; open arrows with minus signs pointing down along the outside of the circle for subsidence).

The materials and processes involved in a complete cycle are represented by a semicircle that includes the vertical center line through the middle of the circle and either the right or left half circumference. A convenient starting point is the large box at the top, which designates all kinds of materials (sediments, sedimentary rocks, metamorphic rocks, intrusive and extrusive igneous rocks) exposed at the Earth's surface. Such exposure typically results from crustal uplift and erosion. Extrusive igneous rocks are an exception; they originate at the surface by solidification of lava extruded from below.

Eroded sediment is deposited in strata. At this point (and at each of the following steps in the cycle), materials can be uplifted and returned to the surface, where they start around the cycle once again. The diagram shows return to the surface by open arrows with plus signs, leading from various boxes on the circumference of the circle back to the large box at the top. (Arrows also lead back to the surface from the lower part of this large box, signifying uplift of intrusive igneous rocks.) Ignore uplifts at intermediate points and follow the steps in the full cycle. Sediments become more deeply buried (shown on the diagrams by the arrows with minus signs pointing downward along the outside of the circle) and, by lithification, become sedimentary rocks. Further subsidence brings the materials into zones of still-higher temperatures and pressures, where they are likely to be metamorphosed and transformed into metamorphic rocks. Sedimentary rocks are not the only parents of metamorphic rocks; metamorphism can affect both igneous rocks (shown by long blue arrows leading downward through the center of the circle) and older metamorphic rocks (shown by small curved arrows starting and ending at the boxes labeled *metamorphic rocks*).

At still-higher temperatures rock materials are fused into magma. The composition of the material melted determines the kind of magma formed in this way. Some magma, most of it granitic, originates by fusion of materials that have passed through the rock cycle. Other magma, much of it basaltic but including notable amounts that are granitic, originates from deeper zones within the Earth. When this material enters the cycle from below, it mingles with crustal rocks for the first time. Whatever its origin or composition magma occupies a special position in the rock cycle, because fusion generally obliterates all record of the previous history of the material. Magma can be injected into the crust at various levels or can be extruded onto the Earth's surface as lava; whatever the location, cooling of molten material yields igneous rocks (shown in Fig. 5-6 by the elongated box with vertical labels *intrusive* and *extrusive*). When igneous rocks become exposed at the Earth's surface the full cycle can begin again.

It is not possible to show in the rock-cycle diagram two important by-products of this cycle: (1) the morphology of the land surface, which constantly changes while the cycle operates; and (2) the configuration of the layers of bedrock, which not only originates but also changes in response to parts of the cycle. We shall devote large parts of the remainder of this book to these important by-products of the rock cycle not discussed here.

We can see, then, that the rock cycle is not a simple process recurring regularly, like the stroke of a piston. It is a complex of interlocking activities; at any moment, all are in progress somewhere. The cycle does not progress in a regular, unvarying order; rather it is like a complex electrical system, full of possible short circuits. Nearly every aspect of physical geology belongs in the rock cycle. Accordingly, on the title pages of nearly all the following chapters, we include a simplified miniature reproduction of the rock-cycle diagram, without its labels but with blue color highlighting the particular segment of the cycle examined in that chapter.

GEOCHEMICAL CYCLE

Many natural chemical reactions take place at the Earth's surface during the unremitting operation of the hydrologic and rock cycles. Not only these reactions but also others involving organisms at the Earth's surface and processes operating within and beneath the crust create profound changes in the distribution of the chemical elements. *The study of chemistry of natural reactions* is ***geochemistry***, whose goal is to interpret natural processes in terms of the properties of the chemical elements involved. During natural reactions many elements follow cyclic paths; their collective movements are known as the *geochemical cycle.* In discussing the geochemical cycle here, we exclude reactions that have occurred in the Earth's deep interior and those that took place early in its history. Instead we restrict ourselves to chemical changes accompanying the hydrologic and rock cycles.

Chemical compositions of the four "spheres." In order to trace the operation of the geochemical cycle, we must begin with knowledge of the chemical compositions of the lithosphere, hydrosphere, atmosphere, and *biosphere,* the collective term for the somewhat spherical habitat of the swarm of living organisms. Let us begin with the lithosphere.

Chemical composition of the lithosphere. The first step in attempting to ascertain the chemical composition of the lithosphere is to perform chemical analyses of rock samples following standard wet-chemical procedures.

After we have determined the abundances of elements in many samples, however, we still do not know the chemical composition of the lithosphere. Obviously the lithosphere is too vast to be characterized by a single rock sample. We can arrive at a statement of abundance of elements in the lithosphere only by calculations. First we must compute its total volume. Next we decide what proportion of the total is occupied by the various kinds of rocks we have analyzed.

Our most detailed information about the lithosphere comes from the continents. Three-quarters or more of the land surface of the Earth is covered by sediments and sedimentary rocks. But these, like proverbial beauty, are only skin deep. When we consider its total volume we find that the vast bulk of the lithosphere consists of igneous rocks. Metamorphic rocks also contribute to the total.

In Fig. 5-7 the results of chemical analyses of various materials are shown graphically. The three groups of igneous rocks, distinguished by their contents of SiO_2, are known as silicic, intermediate, and mafic (Fig. 5-8).

The chemical compositions of sediments and sedimentary rocks vary widely. Some sedimentary materials differ but little from igneous rocks; others differ strikingly. In Fig. 5-7 we have included four analyses of sedimentary materials: average sandstone, silt from the Mississippi River, an average of 78 samples of clay from Norway, and average limestone. Similarity of composition with igneous rocks is exemplified by the Mississippi River silt and Norway clays, both of which contain most of the oxides reported in the analyses of granite and basalt. Neither the silt nor the clays contain any FeO and only the silt contains CO_2. Great differences with igneous rocks are exhibited by average sandstone and average limestone. Average sandstone contains nearly 80 per cent SiO_2, much more than the most silicic igneous rocks, and only small amounts of other oxides. The concentration of SiO_2 in sands and sandstones is an important result of the geochemical cycle. Average limestone contains much more CaO and CO_2 and slightly more MgO than any of the other materials shown.

The chemical compositions of metamorphic rocks reflect the compositions of varied parent rocks plus any change made during metamorphism. Because many metamorphic rocks are similar in composition to igneous and sedimentary rocks, our figure shows no metamorphic rocks. The results of a calculation of the abundances of elements in the crust beneath ocean basins and beneath continental shields round out the group of analyses shown.

Chemical composition of the hydrosphere. The sea constitutes such a large proportion of the total water that we are justified in representing the composition of the hydrosphere by an analysis of typical seawater. Many analyses of water samples collected from near the surface of the sea display a remarkable uniformity despite local variations of salinity caused by climate and streams. In tropical areas of great evaporation salinity is higher than normal; in humid climates or near the mouths of large streams, where fresh water is abundant, salinity is less than normal.

Although we can ignore differences in the composition of water in the sea and water draining off the continents when we calculate the composition of the hydrosphere, we dare not ignore these differences when we analyze the geochemical cycle. Figure 5-9 graphically compares the proportion of major ions dissolved in seawater and average river water determined from many rivers throughout the world.

Chemical composition of the atmosphere. Numerous analyses have shown that the chemical composition of samples of dry air collected in the lowest 100km of the atmosphere is remarkably uniform. The samples consist of "fixed" ingredients, whose concentrations are essentially constant, and "variable" components, whose concentrations vary from place to place. The fixed gases include approximately 78 per cent nitrogen, 21 per cent oxygen,

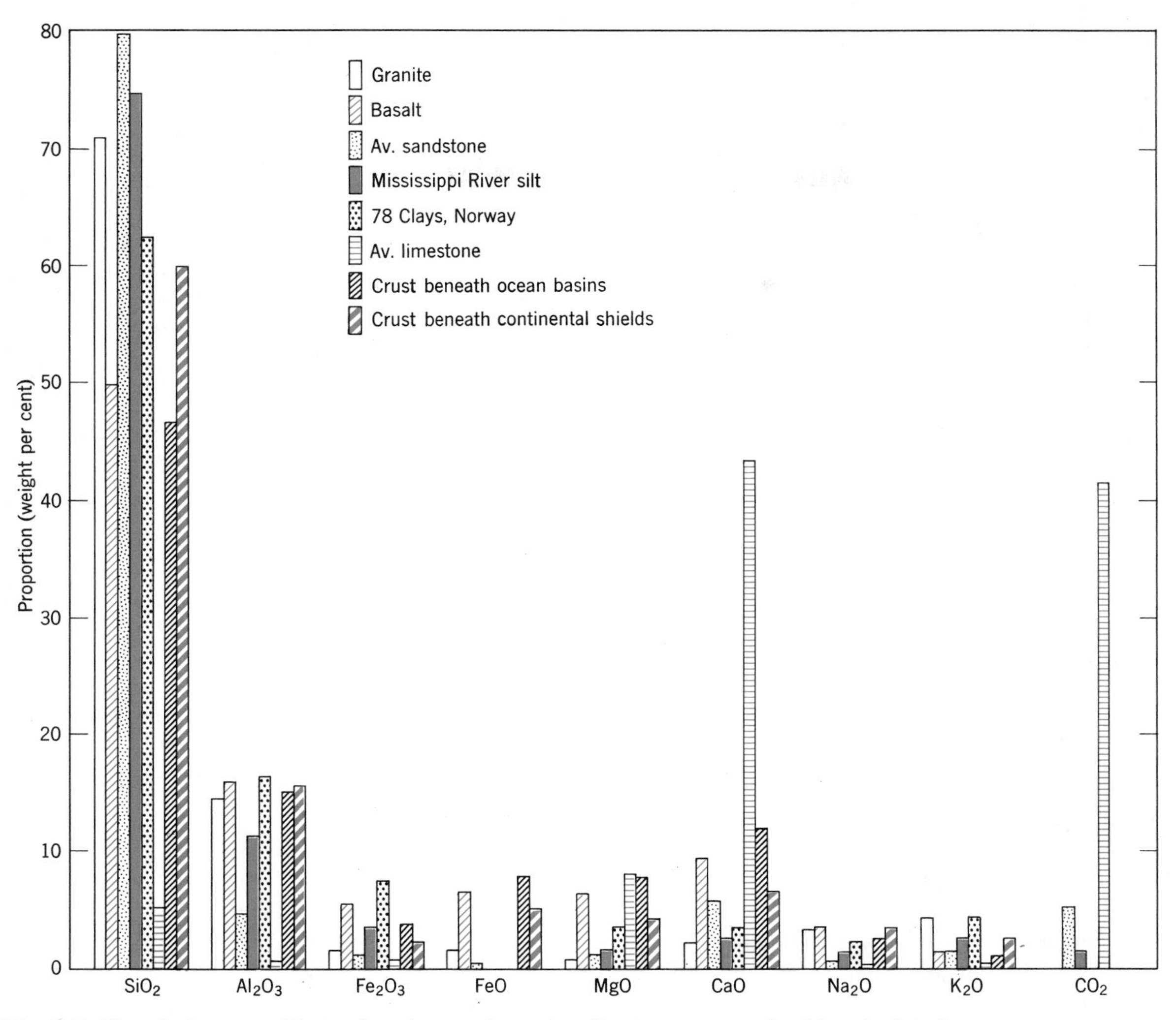

Fig. 5-7. Chemical compositions of various rocks and sediments compared with calculated compositions of the crust beneath ocean basins and beneath continental shields. (Data from Poldervaart, 1955.)

and nearly 1 percent argon (Fig. 5-10). The variables, constituting the remaining fraction of 1 per cent, include water vapor, carbon dioxide, products of combustion of fuels, and industrial wastes, such as other carbon dioxide, carbon monoxide, sulfur dioxide, nitrous oxide, and ammonia.

Chemical composition of the biosphere. We can acquire a general appreciation of the chemical composition of the biosphere by considering the elements in hydrocarbons, fats, and proteins, the major components of living organisms. Hydrocarbons include chiefly carbon, hydrogen, and oxygen; fats and proteins include not only these but other elements as well. The principal other elements in fats are phosphorus, nitrogen, and sulfur; in proteins they are phosphorus, sulfur, iron, and nitrogen.

Behavior of elements during melting and cooling. During the formation of igneous rocks, melting and cooling are the chief factors affecting their elements. But the elements of rocks occur in minerals. Therefore to know how elements behave during melting and cooling we must first know how minerals behave. Then we can reconstruct the movements of elements from our knowledge of the chemical compositions of minerals. Laboratory experiments and textural studies of rocks have provided the principal information on how minerals behave during igneous processes.

The first lesson taught by laboratory experiments is that at atmospheric pressure nearly all minerals melt at a definite temperature and, when the molten liquid cools, the same kind of crystal lattice with which we started reappears. Such *melting without change of composition* is **congruent melting.** In contrast, melting and cooling processes are much more complex in rocks than in minerals. Rocks containing more than one mineral do not melt at a given temperature; instead some minerals of the rock melt at one temperature and others at higher temperatures. Because the entire rock does not melt at a given temperature but requires a range of temperatures, such rocks exhibit *melting accompanied by a change of composition,* known as ***incongruent melting*** or *fractional melting.* Through incongruent melting, the molten liquid derived from partially melting rock A can solidify to form rock B, unlike A, and the unmelted residue composes still another variety, rock C, unlike either A or B.

The second lesson about the behavior of elements in igneous processes, derived both from studies of rocks and from laboratory experiments with minerals, is that minerals resulting from solidification of complex silicate melts vary according to circumstances, some of which are (1) chemical composition of the melt, (2) pressure, and (3) partial pressures of various volatiles.

Critical relationships of the composition of the melt depend upon the ratio of silica to other cations. Three possibilities exist. (1) In *oversaturated* magmas silica is present in excess with respect to other cations. (2) In *saturated* magmas silica balances other cations. (3) In *undersaturated* magmas silica is deficient with respect to other cations.

Perhaps the most far-reaching lesson from laboratory experiments is that as crystals of common rock-forming silicate minerals grow, they are not inert with respect to the melt, as we might imagine them to be. On the contrary, they react with the melt as cooling proceeds. This behavior prompted the name *reaction principle,* which has been found to apply to ferromagnesian silicates, plagioclase feldspars, and other minerals as well. The principle was derived from experiments with dry silicate melts at atmospheric pressure. We can understand the principle better if we review the results of two critical experiments.

A dry silicate mixture including magnesium, iron, and calcium was melted and allowed to cool slowly. The first mineral to crystallize was magnesian olivine, $MgSiO_4$. Olivine crystals continued to grow, but as cooling proceeded, they reacted with the melt: Mg^{2+} ions left the lattices and exchanged with Fe^{2+} ions from the melt. The Ca^{2+} ions remained in the melt. When the temperature decreased to about 1,550°C, a new reaction began abruptly: the olivine crystals started to dissolve. As the olivine dissolved, pyroxene began to form. As the temperature dropped, olivine continued to dissolve and pyroxene to crystallize. Like olivine at temperatures above 1,500°C, the pyroxene crystals at first reacted with the liquid without dissolving; Mg^{2+} ions from pyroxene lattices exchanged with Ca^{2+} ions in the melt. When the reaction had been completed, all the olivine had disappeared and all the pyroxene had become enriched in calcium.

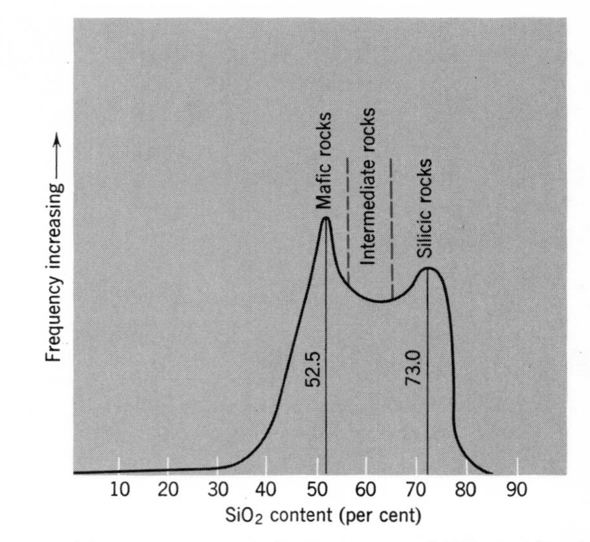

Fig. 5-8. Frequency graph of contents of SiO_2 in chemical analyses of igneous rocks displays two maxima corresponding to typical basaltic rocks (52.5 per cent SiO_2) and granitic rocks (73.0 per cent SiO_2). (After Mason, 1966.)

Other comparable experiments made with other materials showed that pyroxene crystals could dissolve and amphibole could crystallize. When one mineral, such as olivine, dissolves as another mineral, such as pyroxene, starts to form, the reaction is called *discontinuous.* When one mineral continues to crystallize but, during growth, exchanges some of its ions with those in the melt, as Fe^{2+} exchanged with Mg^{2+} in olivine, or Ca^{2+} with Mg^{2+} in pyroxene, the reaction is said to be *continuous.*

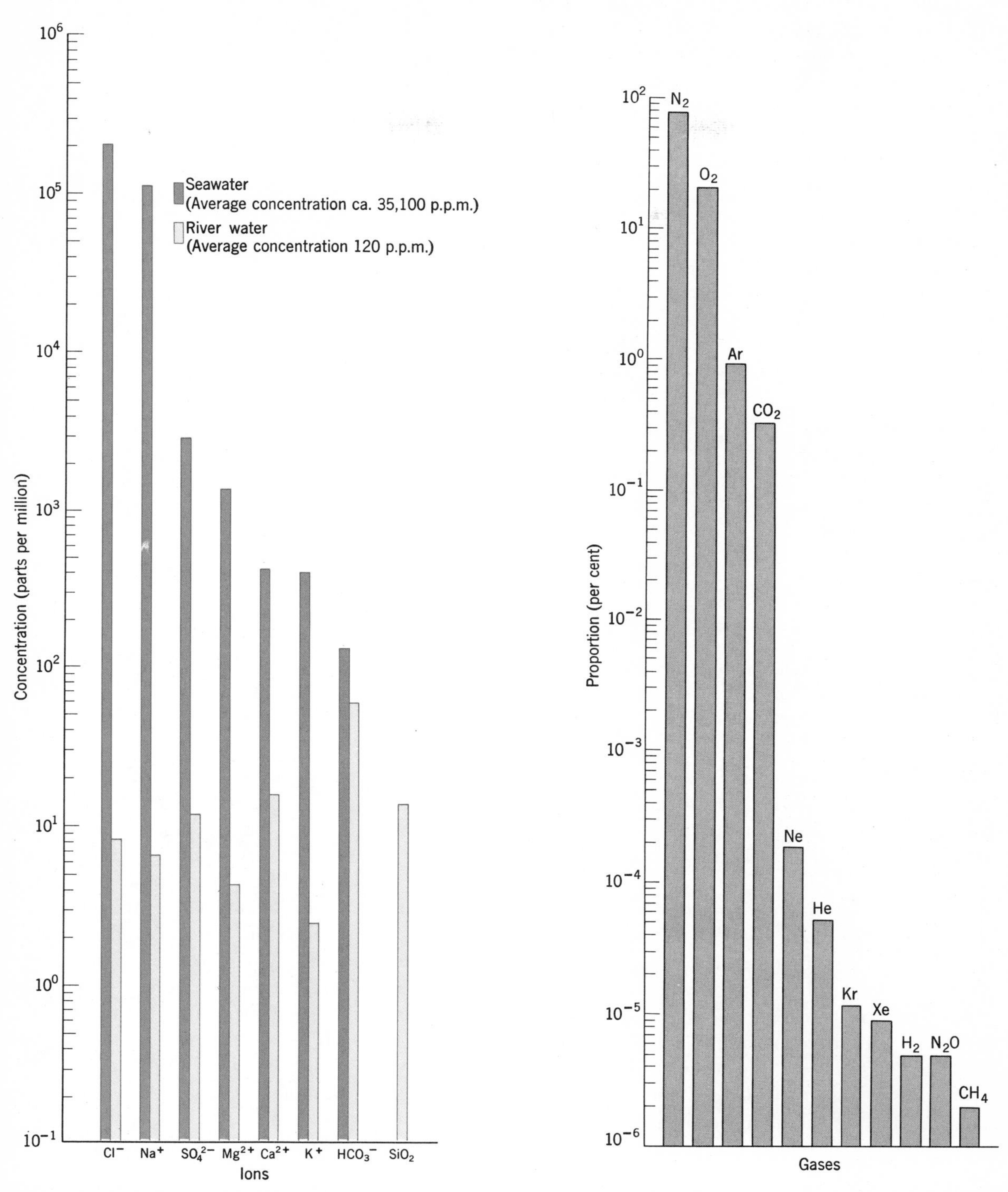

Fig. 5-9. Concentrations of ions in average seawater and world average river water shown by $\log_{10}$ scale. (Data for seawater from Poldervaart, 1955 and E. D. Goldberg, 1962; world average river water calculated by Livingstone, 1963.)

Fig. 5-10. Proportions of gases in the atmosphere, shown on $\log_{10}$ scale. (Data from American Meteorological Society.)

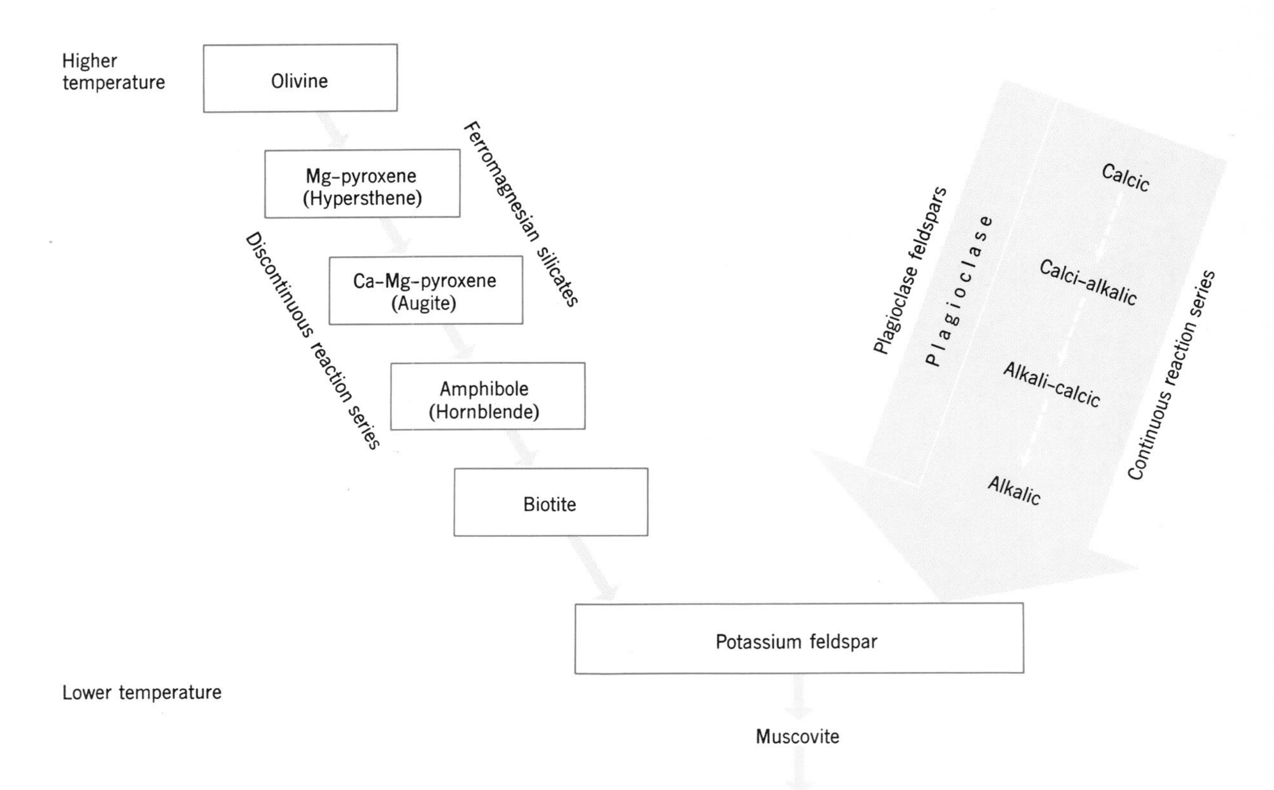

Fig. 5-11. Reaction relationships among silicate minerals of many igneous rocks. ***At top:*** **minerals that crystallize at high temperatures;** ***at bottom:*** **those that crystallize at lower temperatures. Further explanation in text. (After N. L. Bowen, 1922.)**

Plagioclase feldspars were discovered to undergo continuous reaction with silicate melts when an artificial mixture containing an initial composition of equal parts of albite ($NaAlSi_3O_8$) and anorthite ($CaAl_2Si_2O_8$), abbreviated as Ab_1An_1, was melted and allowed to cool. Such a mixture melts at temperatures higher than 1,450°C. When the liquid cooled to this temperature, crystals having a composition of Ab_1An_4 began to grow. As cooling progressed they reacted continuously with the melt: coupled substitutions occurred in which pairs of ions (Ca^{2+} and Al^{3+}) left the crystals and other pairs (Na^+ and Si^{4+}) took their places. After further cooling to 1,370°C and additional coupled substitutions, the composition of the crystals had changed to Ab_1An_2. By the time the melt had cooled to 1,100°C and the reaction had been completed, the composition of all crystals present was albite, Ab_1. All the Ca^{2+} and some of the Al^{3+} remained in the melt.

These reaction relationships between silicate melts and minerals, based on laboratory experiments with various simple mixtures and results of textural analysis of many igneous rocks, were combined by N. L. Bowen into a Y-shaped diagram showing the inferred relationships among the common silicate minerals crystallizing from oversaturated and nonmafic magmas (Fig. 5-11). The left branch of the Y shows ferromagnesian silicates and the right branch plagioclase feldspars. In the diagram ferromagnesian minerals forming *continuous reactions* with the liquid are shown within rectangles; the continuous reactions among plagioclase feldspars are shown by white arrows within the large rectangle. *Discontinuous-reaction relationships* are shown by small blue arrows.

We can trace the paths of ions during the formation of typical igneous rocks on another diagram like Bowen's (Fig. 5-12). In Fig. 5-12, ions substituting in the silica tetrahedra are shown on the outer sides of the branches of the Y, and ions occupying lattice positions between tetrahedra are shown inside the branches of the Y. The Si^{4+} and O^{2-} re-

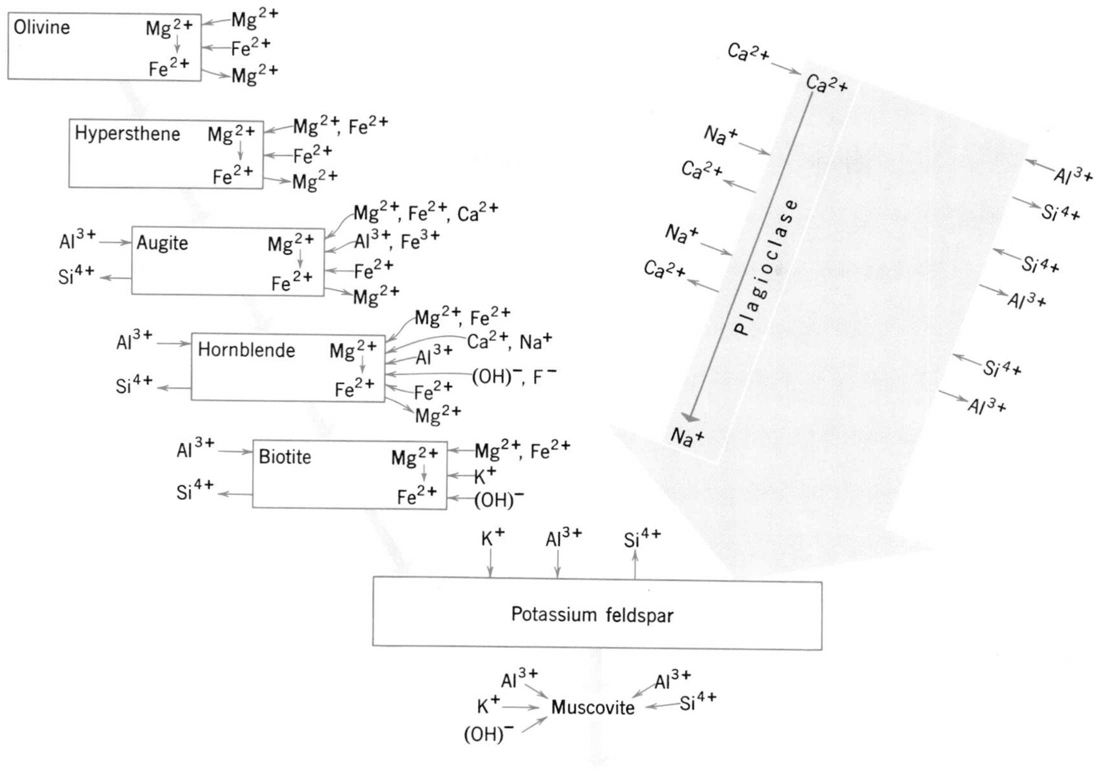

Fig. 5-12. Movements of ions between a silicate melt and lattices of growing crystals, shown on reaction-principle diagram. Further discussion in text.

quired for the silica tetrahedra are not shown. The first reactions, which take place at higher temperatures, are at the top of the diagram. Small arrows pointing toward the rectangles denote ions leaving the melt and entering crystal lattices of the mineral named within the rectangle; small arrows pointing away from the rectangles indicate ions leaving the lattices and reentering the melt.

So far we have not dealt with any of the effects of pressure, which have been shown by experiment to be considerable. The effects of the partial pressure of oxygen, pressure of water, and total pressure give us important examples.

The partial pressure of oxygen exerts a profound influence on the behavior of iron-oxide minerals, which are not shown on Fig. 5-11 but which are present in nearly all igneous rocks. Where the partial pressure of oxygen is high, magnetite (Fe_3O_4) crystallizes *before* most of the ferromagnesian silicates. Where oxygen is deficient, magnetite does not form. Accordingly much iron remains in the melt and iron oxides other than magnetite crystallize *after* most of the ferromagnesian silicates.

Water vapor lowers the melting temperatures of many minerals. The melting point of SiO_2 (Fig. 3-20) is lowered from 1,713°C under atmospheric pressure to 1,125°C under a water-vapor pressure of 1.4 kilobars. A water-vapor pressure of 2 kilobars reduces the crystallization temperatures of orthoclase and albite by about 300°C.

Large total pressures can change the melting behavior of both minerals and rocks. Increasing dry pressure creates an effect opposite to that of water pressure—melting temperatures are increased. For example, albite, pure sodium plagioclase, is entirely liquid at about 1,100°C under atmospheric pressure, but at a dry pressure of 20 kilobars, the liquid temperature is raised to about 1,325°C. At pressures greater than 32 kilobars the melting behavior of albite changes from congruent to incongruent. Pure calcium plagioclase begins to melt incongruently at about 10 kilobars. In magnesium-rich garnet, congruent melting takes place at dry pressures greater than 36 kilobars, but incongruent melting occurs at lower pressures. In the range of 20 to 36 kilobars the garnet melts incongruently to the gem mineral *spinel* ($MgAl_2O_4$) and a silica-rich liquid.

During igneous processes the external effects of temperature and pressure and the internal effects of bulk chemical composition exert major controls on the behavior of elements. Within the framework created by these factors the movement of elements is determined by the sizes and valences of their ions and by the kinds of crystal lattices growing.

Behavior of elements during erosion and deposition. When igneous rocks become exposed at the Earth's surface (our arbitrary starting point in a circuit around the geologic cycle), silicate minerals are subjected to low temperatures, to reactions with various solutions, and, generally, to reactions with abundant free oxygen. Under these conditions many silicate minerals are unstable and tend to be destroyed with the result that the rocks containing them break down into soluble ions and particles of insoluble materials, both of which are delivered to streams and eventually reach the sea. The soluble ions travel in solution as the dissolved load of streams, whereas the insoluble particles are transported mechanically by streams. For our present purposes we shall consider only the fine-grained fraction of the mechanical load, which constitutes the predominant proportion. Hereafter we designate these fine particles collectively by *suspended load.* Chapter 9 explains the processes by which streams transport their mechanical loads.

Samples of water from streams are now collected systematically from many points in the conterminous United States. From these samples the amounts of dissolved and suspended loads are determined. The measurements enable us to place limits on two important activities: (1) rates and kinds of erosion of the lands, and (2) rates of accretion of material in the oceans.

The rates and kinds of erosion in the conterminous United States can be calculated from selected examples of stream measurements averaged through various lengths of time and displayed graphically (Fig. 5-13). The proportions of the two loads vary according to local geologic conditions and to climate. In rivers draining the northern, eastern, and extreme western parts, where rainfall is abundant, dissolved loads exceed suspended loads. In streams draining the western interior parts, where rainfall is not abundant, suspended loads exceed dissolved loads. The striking difference in appearance of streams from these two contrasting climatic regions is apparent where the Missouri River joins the Mississippi River (Fig. 5-14).

No matter how the material is removed the net result is the same: land is eroded. An average of all the measurements from the United States gives a rate of lowering of the surface of the land by erosion that is approximately 6cm/1,000 years.

Naturally we view erosion of the lands as a serious problem. Erosion not only threatens agriculture, but in the long run also looms as a menace to the continued existence of continents. Geologists are concerned not only with erosion but also with the other end of the transaction—the ways in which the sea reacts to this tremendous influx of materials. We are certain that the material dissolved in seawater has come from erosion of the lands. When we consider the long-range implications of this disposition of the dissolved materials washed from the lands we are faced with one of two possibilities: (1) by the addition of dissolved loads of streams the sea has become saltier with time; or (2) after reaching its present concentration the salinity of seawater has remained more or less constant, either because new water is added at the same rate as new dissolved substances or because dissolved materials are removed from the sea about as fast as they are added, or both.

Early geologists thought that the sea had become saltier with time and that the annual increments of sodium would serve as a basis for calculating the age of the oceans. The results of these calculations indicated that oceans should be much younger than other evidence demonstrates (Chap. 6). The modern view is that the salinity of the world's oceans reached its present level at least 500 million years ago and has not changed substantially since.

Recent experiments suggest the possibility that the ocean possesses mechanisms for disposing of dissolved materials delivered to it by streams. We can illustrate the problem of disposition of dissolved materials by considering silica. The average concentration of silica in solution in the world's rivers is about 15 parts per million (p.p.m.), whereas in seawater the range is only 0.5 to 6.0p.p.m. Clearly silica is not accumulating in solution in seawater, because its concentration is greater in water from

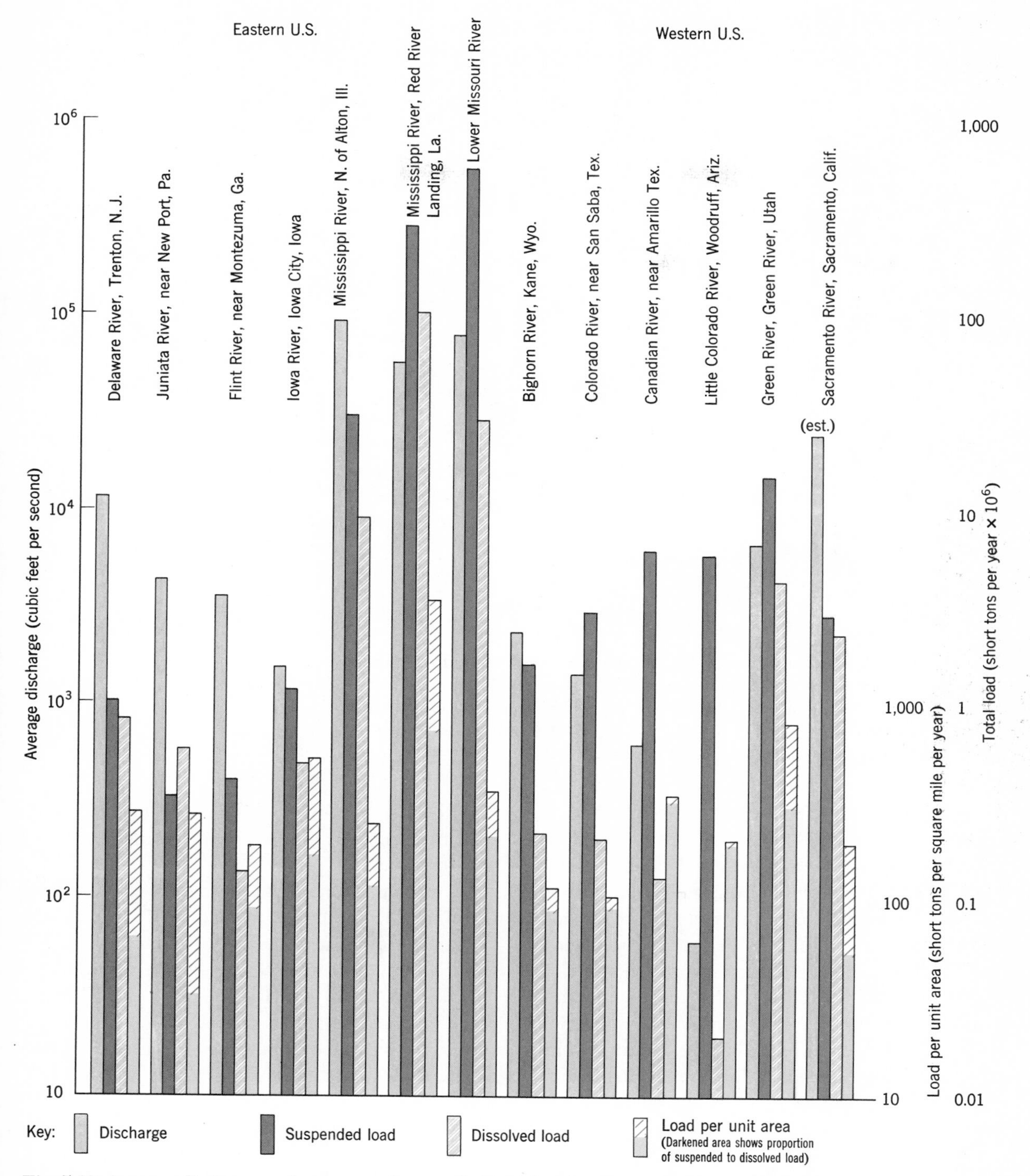

Fig. 5-13. Average discharge and amounts of suspended and dissolved loads of selected rivers. Left bar of each group shows average discharge (*scale at left*). Middle two bars show yearly totals for suspended load and dissolved load (*longer scale at extreme right*). Right bar indicates total load per unit area per year (*shorter scale at right*). Proportions of suspended and dissolved loads for each river are represented using the length of each right bar as 100 per cent. (No per cent scale shown because lengths of bars vary.) (Data from Leopold, Wolman, and Miller, 1964 and U. S. Geol. Survey.)

Fig. 5-14. Muddy water of Missouri River maintains its identity throughout a long distance downstream from junction with Mississippi River. Where width of clear water from upstream decreases after it has been joined by muddy water, the channel deepens. White areas on concave bank of Missouri River are sand bodies. View northwest toward Alton, Illinois, September 21, 1962. (U. S. Army Engineers.)

streams than in seawater. At least two mechanisms are at work to remove silica from seawater. (1) Diatoms, microscopic plants, are great users of silica. They take it from solution and build it into their tiny skeletons. Eventually the skeletons reach bottom and are incorporated into sea-floor sediments (Fig. 15-10). (2) Silicate minerals in suspension react with dissolved ions to re-create silicate minerals, which are deposited on the sea floor. The implication of the second mechanism is that the chemical composition of the suspended sediment reaching the sea floor is not the same as that of the suspended sediment entering the sea from rivers.

Other substances dissolved in seawater can likewise be removed from solution and returned to the lithosphere. Many kinds of organisms build skeletons of calcium carbonate. Tremendous quantities of such skeletons go into calcium-carbonate sediments and thence into limestones. Through intense evaporation of the water some materials become so concentrated that they precipitate as evaporite minerals. The chief elements of evaporites include calcium, carbon, oxygen, magnesium, sulfur, sodium, chlorine, and potassium.

As erosion and deposition progress, elements are distributed according to the solubilities of their compounds in liquids at ordinary temperatures and by the reactions of elements and compounds with oxygen. The dominant activity in erosion is the breakdown of silicate minerals and other constituents of rocks. During deposition some of these products of breakdown are rearranged to form new rock materials.

Behavior of elements during metamorphism. During metamorphism manufacturing new silicate minerals from sedimentary parent materials is a dominant process. For example, clay minerals, the products of destruction of feldspars of igneous rocks, are converted by metamorphism into distinctive alumino-silicate minerals. During the process, new elements can be added; these enter growing silicate lattices or make nonsilicate minerals.

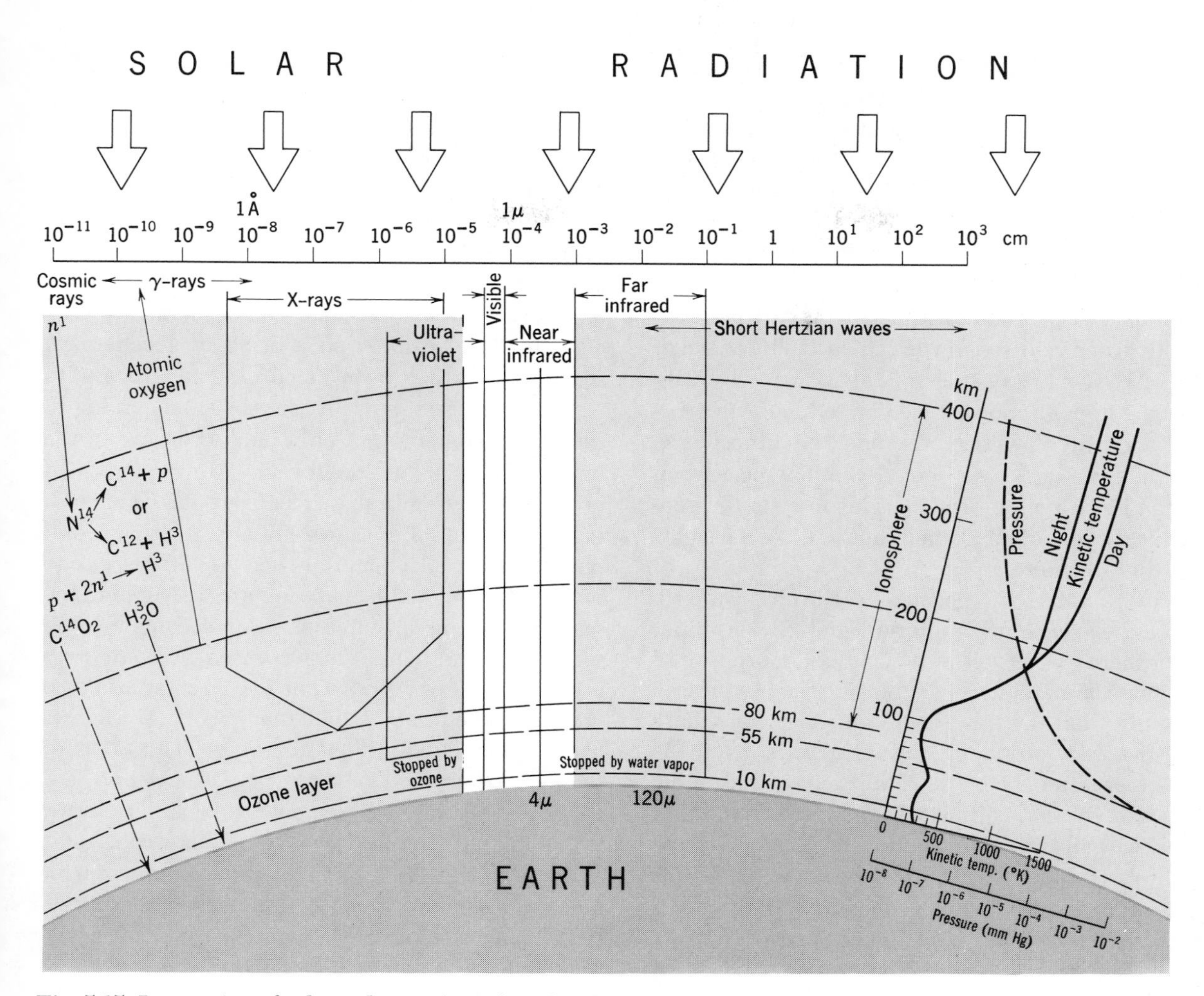

Fig. 5-15. Interaction of solar radiation (wavelengths shown by centimeter scale at top) with Earth's atmosphere shown by schematic profile. Narrow window (white) permits visible light and adjacent wavelengths to reach the Earth's surface essentially unmodified. Wavelengths of energy reflected upward from Earth's surface range from 4 to 120μ. Scale of height is not related to size of Earth, whose curvature is schematic. Heights where C^{14} and H^3 originate (*left*) are schematic also. Relationships of temperature and pressure with height shown at right. Further explanation in text.

Numerous chemical reactions accompany metamorphism but from the point of view of the geochemical cycle most of them can be considered as a reorganization of materials available. Accordingly we defer discussion of metamorphic reactions until Chapter 20.

Behavior of elements in the atmosphere. Many chemical reactions important to living things and likewise to the geochemical cycle take place in the upper atmosphere, a veritable laboratory of atomic activity. The reactions depend on the particles present and the wavelengths and energies of arriving solar radiation. Figure 5-15 summarizes many of the interactions between wavelengths of solar radiation and levels of the atmosphere. Two reactions, important not only to the geochemical cycle but also to geology because they create radioactive isotopes useful in the measurement of geologic time (Chap. 6), are: (1) creation of C^{14} from atmospheric nitrogen, and (2) creation of tritium, H^3, from nitrogen and from the neutrons of cosmic radiation.

Solar radiation includes cosmic rays consisting largely of neutrons (n^1 at upper left, Fig. 5-15). When a neutron collides with the nucleus of a common atom of atmospheric nitrogen, N^{14}, either of two reactions can occur. (1) The neutron remains in the nucleus and ejects a proton, p, and radioactive carbon, C^{14}, is formed. (2) The neutron splits the nitrogen nucleus, creating tritium, H^3, and a stable carbon isotope, C^{12}. Tritium can be created directly when two neutrons combine with a single proton, p.

Radioactive carbon forms less than 0.01 per cent of all carbon and tritium less than 0.001 per cent of all hydrogen. Nevertheless, both isotopes circulate from the upper atmosphere down to the Earth's surface and spread widely through the lithosphere, hydrosphere, and biosphere. By following carbon we can illustrate the principles involved in the geochemical cycle of an element and also gain insights into the biosphere.

Carbon enters the geochemical cycle from two sources: (1) the Earth's interior, and (2) the upper atmosphere. From volcanoes and metamorphism of carbonate minerals, carbon dioxide gas is released naturally from the lithosphere into the atmosphere (Fig. 5-16). Atmospheric carbon dioxide, from the lithosphere and the upper atmosphere, enters the hydrosphere as bicarbonate ion. Unlike all other gases, carbon dioxide is most abundant not in the atmosphere but in the hydrosphere. When plants take up atmospheric carbon dioxide during photosynthesis the carbon goes from atmosphere to biosphere and hydrosphere. Because plants are eaten by animals and animals likewise incorporate carbon in their tissues, the carbon is distributed throughout the rest of the biosphere. Some carbon reenters the atmosphere directly by respiration or indirectly by oxidation of dead organic matter.

Carbon enters the biosphere not only from the atmosphere but also from the hydrosphere, when marine organisms secrete skeletal material of calcium carbonate or incorporate carbon in their soft tissues.

Carbon enters the lithosphere directly from the hydrosphere as a constituent of calcite or aragonite in limestones and of dolomite in dolostones, and from the biosphere as calcium carbonate in shell material or as hydrocarbons and related compounds in coal and petroleum. Carbon from the lithosphere returns to the atmosphere, as carbon dioxide, when the calcite in carbonate rocks breaks down during metamorphism or during the manufacture of cement, and when coal and petroleum are burned as industrial fuels or are oxidized under natural conditions at the Earth's surface.

Sorting of elements. During the geochemical cycle many elements such as carbon can circulate freely among all four "spheres." Such elements may be removed from the active cycle and stored for indefinite intervals but their overall behavior is one of constant circulation. Many other elements, however, tend to be *sorted* as a result of geochemical reactions. Sorting varies according to process. In igneous processes elements are sorted according to their valence and ionic radii and according to the kinds of crystal lattices forming. In processes of erosion and deposition, elements are sorted according to the stabilities and solubilities of their compounds. Insoluble materials accumulate as clastic sediments, and soluble materials are delivered to the sea. Many soluble materials have accumulated in seawater; others are removed from seawater, by organisms or by evaporation and precipitation, to create sediments and sedimentary rocks. Most gases have accumulated in the atmosphere. This effect of the geochemical cycle, the *sorting of elements,* has continued through a tremendous length of time, bringing about colossal changes. By covering most of the Earth's surface with sediments and sedimentary rocks it has been responsible for the atmosphere, the oceans, and the geologic record.

TECTONIC CYCLE

The hydrologic cycle, rock cycle, and geochemical cycle are so intimately bound together that they form a nearly inseparable unit. Although these cycles operate everywhere, observation tells us that in mountain chains and on parts of the sea floor the products of the rock cycle are extraordinarily abundant and that the crust has been particularly active. In mountain chains we find not only abundant products of the rock cycle but also great structural features that have been built by powerful crustal movements. We relate the larger structural features of the Earth's crust to the kinds of rocks that form at various stages of development in the *tectonic* (Gr. *tekton,* "a builder") *cycle.* What we have read in previous chapters has been sufficient background for our rather detailed discussion in this chapter of the hydrologic and geochemical cycles. But at this point we do not have sufficient knowledge to understand a

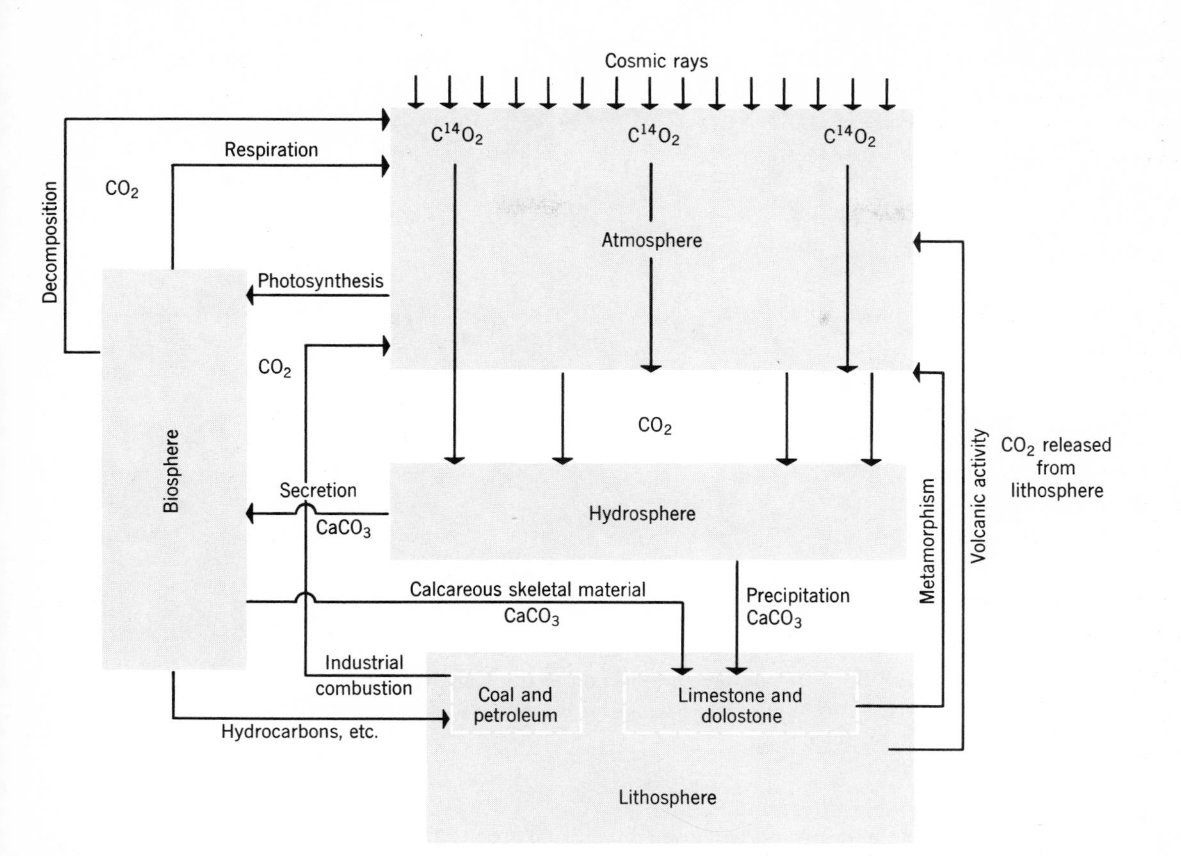

Fig. 5-16. Cycle of carbon in nature. (After Brian Mason, 1958.)

detailed discussion of the tectonic cycle. Accordingly we simply sketch the highlights of this cycle here and shall pick it up again in Chapter 21.

The tectonic cycle of a typical mountain chain begins when an elongate area of the Earth's surface subsides below sea level. Sediments accumulate on the sinking sea floor, gradually forming a body that is elongate and very thick. Eventually these sediments are converted into sedimentary rocks, are squeezed and deformed, and parts or all of them are metamorphosed. Large-scale deformation can occur one or more times within a single major cycle; the result is to thicken the crust locally. Eventually the thickened crust rises upward and a mountain chain is formed. Subsequent prolonged erosion may obliterate the relief of the mountains, but the characteristic structures created by previous deformation and metamorphism remain in the rocks. The cycle is completed when the forces responsible for the squeezing type of deformation become inactive. Igneous rocks accompany many parts of the tectonic cycle; the composition of such rocks changes as the cycle develops.

OTHER CYCLES

Other geologic activities can be visualized as cycles. The wearing down of lands results in a sequence of morphologies which has been called the *cycle of erosion* (Fig. 13-10). The activities of some volcanic eruptions are repeated in cyclic order; these constitute *cycles of volcanic eruption.*

In the cycles discussed so far the factor of time is involved but not in any regular way. In other words the cyclic events are not necessarily periodic. Many of Nature's cycles are periodic. Rotation of the Earth on its axis creates the daily cycle of daylight and darkness. Revolution of the Earth in an orbit around the Sun and the geometry of the Earth's axis of rotation (Fig. 5-17) create the yearly cycle of the seasons. Astronomers have described

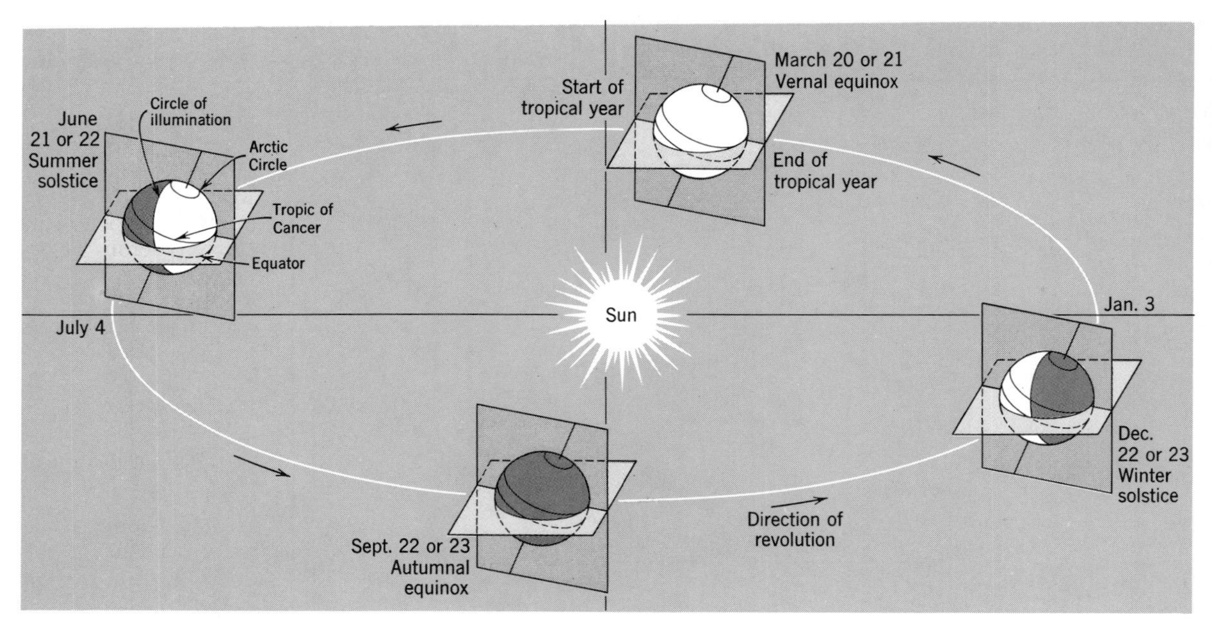

Fig. 5-17. Schematic sketches show how the constant orientation of the plane containing the Earth's polar axis controls the yearly cycle of seasons based on the Earth's orbit around the Sun in the plane of the ecliptic. No scale.

various other cycles having longer periods. These include the Moon's orbit around the Earth every $29\frac{1}{2}$ days; the cycle of sunspots on the surface of the Sun every $11\frac{1}{6}$ years; and the wobble of the Earth's polar axis, which is not fixed but describes a small cone, every 25,800 years. Although small and incomplete this list suffices to show that periodic cycles are inherent in Nature's operations and form integral parts of the larger geologic cycles.

As a prelude to detailed discussion of geologic processes, we turn next to the ways in which geologic time is measured.

Summary

1. The geologic cycle consists of four subcycles, each likewise entitled a cycle. These are hydrologic cycle, rock cycle, geochemical cycle, and tectonic cycle.

Hydrologic cycle

2. The hydrologic cycle, driven by solar energy, maintains a moisture circulation system in the atmosphere. Water evaporates, chiefly from the sea, and returns to the sea, directly or by various circuitous paths.

3. Water circulation depends on climate. Water falling on land can be stored and removed temporarily from the cycle. Glaciers provide the greatest storage capacity.

Rock cycle

4. The rock cycle results from activities within the Earth and at the Earth's surface.

5. Magma crystallizes to form igneous rocks. Rocks exposed at the Earth's surface are eroded, creating sediment that is deposited in strata and can be later transformed into sedimentary rocks. These and other rocks can be metamorphosed into metamorphic rocks or melted to create magma.

Geochemical cycle

6. The geochemical cycle results from the natural reactions of elements during the rock cycle and hydrologic cycle.

7. The most abundant elements in the lithosphere are oxygen (which also occurs in the atmosphere, hydrosphere, and biosphere); silicon, aluminum, iron, and potassium; and sodium, calcium, and magnesium (which are also abundant in the hydrosphere). Other important elements are chlorine (hydrosphere), carbon and sulfur (hydrosphere and

biosphere), and nitrogen (atmosphere and biosphere).

8. During the geochemical cycle many elements are sorted. Insoluble materials and precipitates of some soluble materials have accumulated as sediments and many soluble materials have become dissolved in seawater.

9. Carbon dioxide, the only gas that is more abundant in the hydrosphere than in the atmosphere, is an important compound by which carbon cycles through the lithosphere, atmosphere, hydrosphere, and biosphere.

Tectonic cycle

10. During the tectonic cycle, parts of the crust are raised and lowered, and thickened locally by forces that cause squeezing types of deformation.

Selected References

Ahrens, L. H., 1965, Distribution of the elements in our planet: New York, McGraw-Hill. (Paperback.)

Kuenen, Ph. H., 1955, Realms of water. Some aspects of its cycle in nature: Revised edition, 1963, translated by May Hollander. New York, John Wiley, Science Editions.

Leopold, L. B., Wolman, M. G., and Miller, J. P., 1964, Fluvial processes in geomorphology: San Francisco, W. H. Freeman. (Chapter 3 discusses dissolved and suspended loads of rivers.)

Livingstone, D. A., 1963, Data of geochemistry, 6th ed., Chap. G, Chemical composition of rivers and lakes: U. S. Geol. Survey Prof. Paper 440-G, p. Gl-G64.

Mason, Brian, 1966, Principles of Geochemistry, 3rd ed.: New York, John Wiley.

Nace, R. L., 1964, The international hydrologic decade: Am. Geophys. Union Trans., v. 45, p. 413–421.

Poldervaart, Arie, 1955, Chemistry of the Earth's crust, *in* Poldervaart, Arie, Crust of the Earth: Geol. Soc. America Spec. Paper 62, p. 119–144.

Strahler, A. N., 1965, Introduction to physical geography: New York, John Wiley.

U. S. Department of Agriculture, 1955, Water: Yearbook for 1955: Washington, D. C., U. S. Govt. Printing Office.

The Geologic Column and Geologic Time

Geologic column
Principle of superposition
Time significance of fossils
Isotopic dating
Scale of geologic time
Age of the Earth

Plate 6

Effect of radioactivity in rock. Photograph of a small part of the shell of a fossil clam collected from bedrock in California. Uranium (U^{238}) in the shell substance has undergone spontaneous fission, emitting high-energy particles. The linear black marks are tracks of these particles, which have damaged the surrounding mineral matter. By counting the number of tracks per unit area and by determining U^{238} content, we can calculate the age of the substance. Age of this clam has been determined tentatively at about 140,000 years. (D. S. Haglund.)

0 0.01 mm

CHAPTER 6 *The Geologic Column and Geologic Time*

THE GEOLOGIC COLUMN

A record of past events. The implications of time are all around us. No thoughtful person can look at a succession of rock layers like that shown in Fig. 1-11, *B* without at least beginning to speculate as to *how much time* was involved in their creation. Since the establishment of the principle of uniformity, probably the most significant contribution of geology has been the concept of the immense length of geologic time. The geologists' view of time, contrasted with the small quantities of time embraced in written human history, has expanded our historical horizon far beyond the limits that confined it a hundred years ago.

Today, looking at the charts showing the sequence of strata in the form of a column (frequently reproduced in encyclopedias and textbooks —including this one; see Table 6-1), we accept without question the sequence represented in them as a record of past events, the oldest unit at the bottom of the sequence and the youngest at the top. We accept also, without much difficulty, the approximate dates with which some of the units in the column are provided. Yet the column is the result of a century and a half of patient work, and the approximate dates are the fruit of a half century of thoughtful and painstaking measurement made with laboratory equipment of ever-increasing sophistication.

This column is a document of such basic importance for understanding geology that we need to look into the way it was constructed, as well as to note the ways in which it is being continually refined today. Despite the fact that its basic outline has withstood the test of more than a hundred years of critical examination, the column is a piece of perennially unfinished business, continually capable of further improvement.

The sequence of strata; principle of stratigraphic superposition. In the second part of the eighteenth century ***stratigraphy,*** *the systematic study of stratified rocks,* emerged. Scientists recognized that layers of rock had been deposited in sequence, one on top of another and therefore one after the other. From this characteristic arrangement of strata is derived the ***principle of stratigraphic superposition,*** which says that *in any sequence of strata, not later disturbed, the order in which they were deposited is from bottom to top.*

This principle implies the existence of a *relative time scale,* by which the age of a stratum could be fixed with reference to a second stratum according to whether the latter underlies or overlies it. Although such relative time determination is very useful, the principle says nothing about the absolute time we measure in years.

Although it was grasped by at least two or three people well before the end of the eighteenth century, the principle of stratigraphic superposition was first presented forcefully and first widely introduced to science by William Smith, an English civil engineer and land surveyor, shortly before the nineteenth century began. His profession gave him an ideal opportunity to observe not only terrain but the rocks that underlie it. While surveying for new canals in western England, he observed the sedimentary strata and soon realized that they lie, as he put it, "like slices of bread and butter" in a definite, unvarying sequence. It is likely that Smith rediscovered the principle of superposition, because we know of no evidence that he had read the works of any of his predecessors, which were written in German, Italian, and Latin. At any rate, using the principle of stratigraphic superposition, he became so familiar with the sequence that by looking at a specimen of sedimentary rock collected from anywhere within a wide region, he could name the layer from which it had come and, of course, the position of the layer in the sequence.

Time significance of fossils. In the region where

TABLE 6-1. THE GEOLOGIC COLUMN

Subdivisions Derived from Strata				Notable Events in Evolution of Organisms
	Systems	Series	Stages	
CENOZOIC	Quaternary	(Recent) Pleistocene	More than twenty widely recognized	Man appears
	Tertiary	Pliocene		Elephants, horses, large carnivores become dominant
		Miocene		Mammals diversify
		Oligocene		Grasses become abundant; grazing animals spread
		Eocene		Primitive horses appear
		Paleocene		Mammals develop rapidly
MESOZOIC	Cretaceous	Two or more series in each system	About thirty widely recognized	Dinosaurs become extinct; flowering plants appear
	Jurassic			Dinosaurs reach climax Birds appear
	Triassic			Primitive mammals appear; conifers and cycads become abundant Dinosaurs appear
PALEOZOIC	Permian		Many recognized	Reptiles spread; conifers develop
	Pennsylvanian (Upper Carboniferous)			Primitive reptiles appear; insects become abundant Coal-forming forests widespread
	Mississippian (Lower Carboniferous)			Fishes diversify
	Devonian			Amphibians, first known land vertebrates, appear Forests appear
	Silurian			Land plants and animals first recorded
	Ordovician			Primitive fishes, first known vertebrates, appear
	Cambrian			Marine invertebrate faunas become abundant
PRECAMBRIAN Complex assemblages of rocks, largely metamorphosed				Seas characterized by simple marine plants

Smith worked, the strata contained abundant fossils of marine invertebrate animals. Smith collected the fossils and soon realized that each layer contained distinctive kinds, enabling him to identify it by its fauna independently of its physical characteristics. In other words, he recognized that each assemblage of fossils was peculiar to the stratum in which it occurred and thus constituted an identification tag to the stratum. In so doing, Smith discovered what we now call the ***law of faunal succession,*** which says that *fossil faunas and floras succeed one another in a definite, recognizable order.*

As we now know, this relationship between a stratum and its fossils is the effect of the evolution of living things through time. As successive generations of living things gradually change their form, the changes are carried from one part of the world to another through the spreading or shifting of organisms as they expand or otherwise change their living areas. Rates of spreading and rates of evolution were about comparable, so that the major evolutionary changes affected large parts of the world within the generous time intervals represented by the major groups of strata. All this, however, was unknown in Smith's day and did not become clear until after 1859, when Charles Darwin put forth his famous theory of evolution.

Probably, then, William Smith not only rediscovered the principle of stratigraphic superposition but in addition discovered the law of faunal succession. Because of his dual insight, and because of the clarity with which he presented his results and the impact these made on further research, Smith is known as "the father of stratigraphy."

Smith's discovery that strata containing similar assemblages of fossils are broadly similar in age, no matter where they occur, was empirical. Nevertheless, it opened the door to the correlation of sedimentary rocks through increasingly wider areas. By ***correlation*** we mean *determination of equivalence, in geologic age and position in the sequence of strata, in different areas.* We shall have more to say about this in Chapter 16. Smith had been correlating strata on the dual basis of physical similarity and fossil content through distances measured in miles and then in tens of miles. But by means of fossils alone it became possible to correlate through hundreds and then thousands of miles.

Long-range correlation by means of fossils soon led to the establishment of a general ***geologic column*** (Table 6-1), *a composite diagram combining in a single column the succession of all known strata, fitted together on the basis of their fossils or of other evidence of relative age.* Because the groups of fossils in each layer of the column record the gradual progress of evolution through time, the column represents not only a succession of layers but also the concept (though of course not the measurement) of the passage of time. By the middle of the nineteenth century the column had attained a good degree of development. In elaborating it, geologists had made a major contribution to knowledge of the Earth's history, one that has stood the test of subsequent research. But at that early period no means were available for calibrating the column in terms of actual years.

The names of the twelve geologic periods shown in Table 6-1 are the work of nearly as many different scientists, of five nationalities. The oldest name is *Tertiary,* proposed in 1760; the youngest is *Pennsylvanian,* proposed in 1891. Most of the names are geographical, denoting regions in which the related strata occur. Of these *Devonian* (Devonshire), *Mississippian* (Mississippi valley), *Pennsylvanian* (State of Pennsylvania), *Permian* (region of Perm in Russia), and *Jurassic* (Jura Mountains) are self-evident. Less evident geographic names are *Cambrian* (the Roman name for Wales) and *Ordovician* and *Silurian* (Ordovices and Silures, two tribes of ancient Britons that lived in Wales). Of the others, *Triassic* was chosen because its strata consisted of three distinct layers, a triad, *Cretaceous* referred to the abundant presence of chalk (Lat. *creta*) in its strata, and *Tertiary* and *Quaternary* were the surviving members of a group of names that originally included *Primary* and *Secondary,* both long since abandoned.

EARLY ATTEMPTS TO CALIBRATE THE COLUMN

With the publication of Darwin's theory of evolution it became clear why successive strata could be identified by their fossils. At the same time the theory stimulated attempts to find means of calibrating the geologic column in units of absolute time, because such calibration would make it possible to determine the rates at which the evolution of various organisms had occurred. In the effort to calibrate the column, attempts were made to derive the actual times elapsed by measuring both the present rates of transfer of land-derived material to the sea and the total amounts of such material brought into the sea. For example the quantity of sodium in the oceans was divided by the annual increment from the lands, estimated from sampling rivers. Again, the aggregate thickness of the known marine sedimentary strata now exposed on the continents was divided by the thickness of the aggregate annual contribution of sediment by rivers. In all cases various assumptions had to be made, some of them unrealistic in the light of our present knowledge. Even so, the values that resulted from these computations ranged from around 100 million to more than 300 million years, and calculations from other measurements reached as high as 700 million years. Although such values were recognized as merely minimum ages of oceans and rivers and were small compared with the values we accept today, they were far greater than the age of the Earth had been thought to be only a few decades earlier. Indeed, throughout the second half of the nineteenth century, geologists were constrained by the estimate made by Lord Kelvin, one of the most famous physicists of his time, that the age of the Earth could not exceed 25 million years, the time necessary for the crust to have solidified from a supposedly formerly molten state. It was not then known that any other source of heat existed in the crust. From his long study of the evolutionary process, from Cambrian organisms to those living today, Darwin believed that evolution would have required at least 100 million years, and he, too, was very unhappy with the time limitations imposed by Lord Kelvin's estimate.

This was the existing state of uncertainty and conflict about the measurement of geologic time, when the discovery of radioactivity in 1896 opened the door to a new viewpoint and to wholly new kinds of measurement.

ISOTOPIC DATING

Natural radioactivity. Natural radioactivity is the spontaneous decay of the atoms of certain isotopes (Fig. A-1) into new isotopes, which may be stable or may undergo further decay until a stable isotope is finally created. Heat produced by radioactive decay destroyed the basis of Kelvin's time restriction as to the age of the Earth; radioactive decay also made possible the measurement of time and, therefore, determination of the ages of rocks. Under favorable conditions we can now calculate the ages of minerals and thus date rock bodies, even where fossils or other stratigraphic evidence of the positions of the bodies in the geologic column are not available.

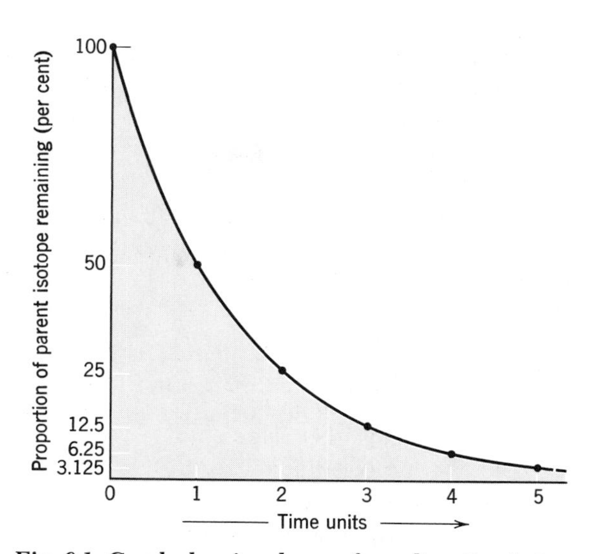

Fig. 6-1. Graph showing decay of a radioactive isotope. During each time unit half the atoms remaining in the parent isotope decay into daughter isotopes.

Basis of dating. We can determine the age of minerals because the rate at which radioactive isotopes decay is constant. Decay rates have not changed significantly through a length of time as great as the oldest material yet dated. A *decay constant*, λ, expresses the proportion of atoms of each isotope that decay in a unit of time. The decaying, or *parent isotope, P*, continually decreases in amount, whereas the isotope created by decay, the *daughter, D*, continually increases. The fraction of parent atoms that decay during a given time remains constant, but the actual *number* decaying decreases with time because the parent atoms are being used up. The activity of the parent decreases exponentially with time (Fig. 6-1). The decrease in the activity of a radioactive substance can be expressed in terms of the ***half-life*** of the substance, *the time required to reduce the number of parent atoms by one-half.*

At any moment the rate of change of the number of parent atoms is equal to -1 times the product of the decay constant times the number of parent atoms present. When the mathematical statement of this relationship is integrated, the result is:

$$P_t = P_0 e^{-\lambda t}$$

where P_t = number of atoms of the parent isotope at time t

P_0 = initial number of atoms of parent isotope

e = base of natural logarithm (ln). (This term appears as a result of the process of integration.)

This is the equation of the curve in Fig. 6-1. We can solve this expression for t. First we eliminate the unknown, P_0, by substituting for it

$$P_0 = P_t + D$$

where P_t = amount of parent present at time t
and D = amount of daughter present at time t

Both P_t and D can be measured; so we can write an expression for the unknown, t, as follows:

$$t = \frac{1}{\lambda} \ln\left(1 + \frac{D}{P_t}\right)$$

The decay constant and half-life, $t_{1/2}$, can be related by setting $P_t = P_0/2$ in the decay equation. This gives

$$t_{1/2} = \frac{0.693}{\lambda}$$

The isotopes commonly used in age determination, together with some of their properties and the substances in which they occur, are listed in Table 6-2. As has been said, radioactive decay leads ultimately to the creation of stable end products. However, in the complex radioactive series associated with uranium and thorium (Fig. 6-2), the decay takes place in steps in which are formed intermediate daughters that then decay into other daughters, each daughter having a different half-life.

In the uranium and thorium decay series, the decay occurs by ejection of an alpha-particle (α-particle) ($_2He^4$, a helium nucleus) or by activities involving beta-particles (β-particles) (Fig. A-1). The emission of particles is accompanied by the emission of γ-rays (high-energy photons). Loss of an α-particle decreases the mass of the parent nucleus by four units and the positive charge by two units. Hence the loss of one α-particle from $_{92}U^{238}$ results in the formation of a new isotope $_{90}Th^{234}$. This is shown on the upper right in Fig. 6-2.

In contrast, emission of a β-particle results in a unit increase of positive charge on the parent nucleus but does not change the mass number. Hence the $_{90}Th^{234}$ in Fig. 6-2 becomes, by loss of two β-particles, $_{92}U^{234}$. As shown in the zigzag path in the figure, decay continues through a succession of daughters until the final stable product $_{82}Pb^{206}$ is reached.

The half-lives of the primary uranium and thorium isotopes are much longer than the half-lives of any of the intermediate daughter products. Therefore, the actual amount of intermediate daughters present in a uranium or thorium mineral at any moment is small and can be disregarded in the computation of a uranium-lead or thorium-lead age that exceeds 1 million years. Below that point there are complications, because the amounts of intermediate daughters build up from zero to the quantities present in *secular equilibrium*. When such equilibrium exists, the rate of decay of the parent (the product of amount P and decay constant λ_p) is exactly matched by the rate of decay of every intermediate daughter ($\lambda_p P = \lambda_1 D_1 = \lambda_2 D_2$, etc.). Hence, statistically, each time an atom of parent decays, an atom of every intermediate daughter decays. The concentrations of intermediate daughters thus remain virtually constant while the amount of stable daughter steadily increases.

Age determination. Through the use of a variety of analytical techniques it is possible to determine, with a precision and accuracy close to 1 per cent, the decay rates of the isotopes important for radioactive dating and the concentrations of parents and daughters in rocks and minerals. The mass spectrometer (Fig. 1-14) is an important tool for the precise analysis of minute quantities of material by a method known as *isotope dilution*. By use of the simple equation derived above, we can calculate from the decay constant and analytical data a value for time, t. Before this calculated time, t (the *isotopic date*), can be regarded as an age determination, certain additional conditions must be met. The most important is that the mineral or rock dated must have remained a *closed system* since the time it was formed. No loss or gain of either parent or daughter can have occurred. Equally important, the amount of daughter present at the time of formation must be either zero or a quantity that can be determined accurately by the analysis, so that a correction for it can be applied before

TABLE 6-2. PRINCIPAL METHODS OF ISOTOPIC AGE DETERMINATION

Isotopes	Half-Life of Radioactive Isotope	Effective Dating Range	Some Materials That Can Be Dated
Uranium-238/Lead-206	4.50×10^9 years	10^7 years to T_0[a]	Zircon Uraninite Pitchblende
Uranium-235/Lead-207	0.71×10^9 years	10^7 years to T_0[a]	Zircon Uraninite Pitchblende
Potassium-40/Argon-40	1.30×10^9 years	10^4 years to T_0[a]	Muscovite Biotite Hornblende Whole volcanic rock Arkose[b] Sandstone[b] Siltstone[b]
Rubidium-87/Strontium-87	4.7×10^{10} years	10^7 years to T_0[a]	Muscovite Biotite Microcline Whole metamorphic rock
Carbon-14	5730 ± 30 years	0 to 50,000 years	Wood Charcoal Peat Grain Tissue Charred bone Cloth Shells Tufa Ground water Ocean water

[a] T_0 = Age of the Earth, about 4.6×10^9 years.

[b] For paleogeographic studies.

Source: After Isotopes, Inc.

the age calculation is made. If the assumptions and conditions have been satisfied, the calculated isotopic date is an *age determination*, that is, a determination of the age of the mineral or minerals measured, and therefore in many cases the age of the rock body of which the mineral forms a part. If not, it is only a number with perhaps no geologic significance.

Ages from uranium, thorium, and lead. The first attempts at isotopic dating were made by using the radioactive series ending in lead (Fig. 6-2), and, although techniques have changed, these series are still used. Invention of the mass spectrometer made it possible to calculate more than one age on the same sample. For instance, independent dates based on the isotopes U^{235}/Pb^{207}, U^{238}/Pb^{206}, and Pb^{207}/Pb^{206} can be calculated for uranium minerals. Where uranium and thorium occur together, as they do in zircon, an additional age can be calculated using Th^{232}/Pb^{208}. The age is considered to be reliable if the various results agree within a few per cent. Where the results do not agree, complications have occurred in the history of the sample. The Pb^{207}/Pb^{206} ages are commonly observed to be the most reliable.

The long half-lives (4.5 billion years for U^{238}, 0.71 billion years for U^{235}, and 13.9 billion years for Th^{232}) make the U-Th-Pb methods useful for dating rocks more than a few million years old. Before 1950, routine chemical analyses required such large amounts of uranium and thorium that the only minerals that could be dated by this

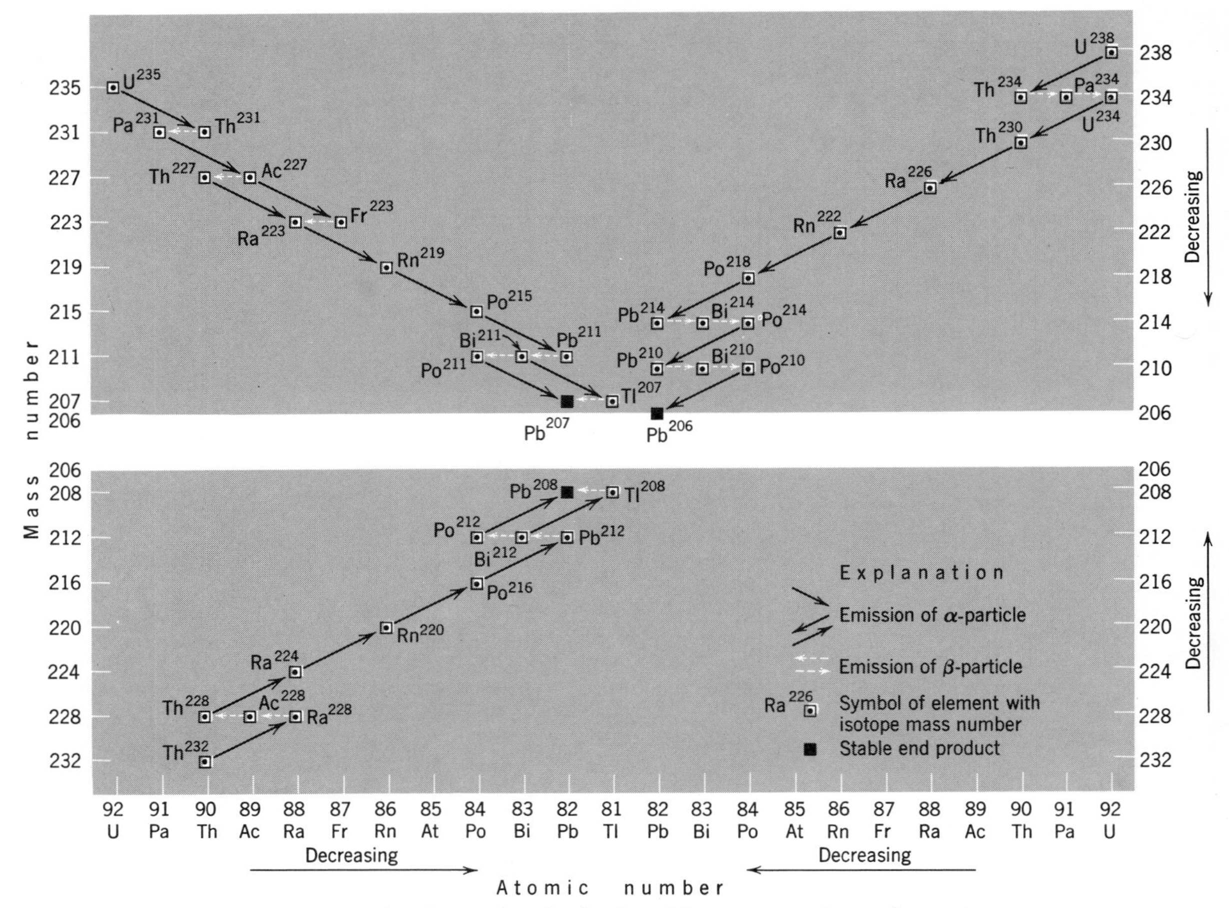

Fig. 6-2. The three radioactive series that end in lead, plotted by mass number and atomic number. Horizontal and vertical scales are symmetrical around center of diagram, with values decreasing toward center. When equilibrium has been established among all daughter isotopes intervening between the radioactive parent isotopes and the stable daughters, amounts of the intervening isotopes remain constant. The parents are gradually depleted and the stable daughters accumulate.

method were large-sized minerals from pegmatites. The more sensitive techniques now available, based on the mass spectrometer, require only a few micrograms of parent and daughter isotopes. Hence small-sized mineral grains can be used, and this freedom makes it possible to date a wide range of rocks. A mineral now commonly used is zircon, an accessory mineral in most granitic rocks.

Ages from rubidium and strontium. The decay of Rb^{87} to Sr^{87} by emission of a β-particle is the simplest decay scheme commonly used for isotopic dating. Because the half-life of Rb^{87} is approximately 47 billion years (Table 6-2), the method is suitable for rocks more than a few million years old.

Ages from potassium and argon. The potassium isotope K^{40} decays to two daughters simultaneously; 89 per cent of the decaying parent nuclei emit β-particles to become Ca^{40}, while 11 per cent capture an electron from the innermost shell and become Ar^{40}. The Ar^{40} from the specimen must be distinguished from the Ar^{40} of the atmosphere, but this can be done accurately. Atmospheric argon contains Ar^{36} in constant ratio to Ar^{40}, so that by measuring both the Ar^{36} and Ar^{40} obtained from a sample the correction for daughter contamination from the atmosphere can be made.

As the half-life of K^{40} is 1.3 billion years, the method is suitable for dating the entire range of geologic time beyond approximately 100,000 years.

Under particularly suitable circumstances, even younger ages may be measured. The only limitation is the correction for atmospheric Ar^{40} contamination. Quantities of argon in young samples as small as fractions of millionths of cubic centimeters per gram can be measured accurately.

Although it is an inert gas, Ar^{40} produced by radioactive decay is quantitatively retained at room temperature in micas, amphiboles, some feldspars, and a few volcanic rocks that solidified rapidly. Some minerals contain excess Ar^{40} and others lose Ar^{40} even at relatively low temperatures. These, of course, are not suitable for isotopic dating. Fortunately, usable minerals are common in all kinds of igneous and metamorphic rocks.

When subjected to high temperatures, all minerals and rocks lose argon. Thus K/Ar dates are very sensitive to the thermal history of the samples. The K^{40}/Ca^{40} method is rarely used for dating, because it is difficult to correct for Ca^{40} contamination.

Ages from radiocarbon. The foregoing methods are applicable to very old rocks. In 1947, a method using radioactive C^{14} for determining the ages of young samples was discovered. The parent isotope, C^{14}, known as *radiocarbon*, is continuously created in the atmosphere through bombardment of N^{14} by neutrons created by cosmic radiation (Fig. 5-15). Radiocarbon decays by β-radiation to N^{14}; the half-life is 5,730 years.[1]

The radioactive carbon mixes with ordinary carbon (C^{12}) and diffuses rapidly through the atmosphere, hydrosphere, and biosphere (Fig. 5-16). Because the rates of mixing and exchange are rapid compared with the half-life, the proportion of radiocarbon is nearly constant throughout the system. As long as the production rate remains constant, the radioactivity of natural carbon remains constant because rate of production is in equilibrium with rate of decay. It is true that changes in production rate and mixing rates and admixture of old, inactive carbon into the natural systems are genuine complications that upset radiocarbon dates. Nevertheless, the effects require, at worst, refinements in interpretation of the results and not rejection of the method.

[1] Dates are still quoted, in the interest of uniformity, on a basis of an earlier-determined half-life value of 5,568 years. Any published date can be adjusted to the newer half-life value by multiplying the date by 1.03.

As long as an organism is alive, it contains the equilibrium proportion of C^{14}. However, at death the equilibrium is destroyed because replenishment by life processes ceases and the C^{14} present continually decreases by radioactive decay. Thus the activity decreases with time, like the amounts of parent isotopes shown in Fig. 6-1. The analysis for a radiocarbon date requires only determination of the C^{14}-activity level. This, compared with the modern-activity level (excluding effects of nuclear-weapons tests and the burning of fossil fuels by man), gives sufficient information for calculating the age, using the equation of the curve, Fig. 6-1. It is not necessary (indeed it would be impossible) to determine the amount of daughter N^{14} in the analysis for age determination. Because the half-life is short, apparent C^{14} ages (by which we really mean apparent times elapsed since death) of as little as 100 years can be determined.

Because of its application to organisms and its short half-life, radiocarbon has proved to be enormously valuable in establishing dates for prehistoric races of man and for recently extinct animals and, in this way, is of extreme importance in archeology. Also it is of comparable value in the most recent part of geologic history, particularly the latest of the glacial ages. For example, the dates of many samples of wood taken from trees overrun by the advance of the latest of the great ice sheets and buried in the rock debris it deposited, show that that glacier reached its greatest extent in the Ohio-Indiana-Illinois region not more than 18,000 C^{14}-years[2] ago. Dates of other samples of wood and peat indicate that the edge of the glacier retreated through as much as 500 miles and then made a conspicuous readvance of diminished extent, which culminated less than 11,000 C^{14} years before our time. Similarly, radiocarbon dates afford the means for determining rates of movement, such as the rate of advance of the last ice sheet across Ohio, the rates of rise of the sea against the land as glaciers melted throughout the world, and the rates of local uplift of the crust that raised ancient beaches above the sea.

The dates just quoted are, of course, the apparent ages of samples. How closely do they approximate

[2] C^{14} years are years calculated from radiocarbon measurements. Hereafter we shall write C^{14} dates as C^{14} years, implying but without including the commonly used letters B.P. (before present), that is, before A.D. 1950.

real ages? The accuracy of the method has been checked against samples whose dates are known independently through historical information. Among these samples are grains of corn, wooden beams, furniture from ancient tombs, prehistoric garments, and samples of "dead" heartwood from huge *Sequoia* trees that are still living. In these samples, C^{14} dates and historical dates compare fairly well, particularly when the known modest fluctuations of C^{14} activity in the past are taken into account. However, none of the historically dated samples is as much as 5,000 years old, whereas many dates of 50,000 C^{14} years and more have been calculated. These greater dates are certainly less accurate than the smaller ones, as can be surmised from the curve in Fig. 6-1, but we do not yet know the extent of their inaccuracy. This is why it is wise to label them *apparent dates.*

Between the upper limit of C^{14} dating and the lower limit of K/Ar dating, we have very few dates so far. This is not the fault of technology, but results from the scarcity of suitable samples. For this reason no definitive comparisons of the two methods have yet been made. As such samples are discovered, the apparent gap in our sequence of dates will vanish, and a chronologic record of the glacial ages will be established.

GEOLOGIC TIME

Scale of geologic time. Through the methods described, chiefly the one based on K/Ar ratios, apparent ages have been obtained for igneous rocks that have identifiable positions in the geologic column. Here very careful field study is needed, for an obvious reason. The standard units of the geologic column consist of sedimentary rocks containing characteristic fossils, but the typical rocks from which apparent ages (apart from C^{14} ages) are measured are igneous rocks. It is necessary, therefore, to be sure of the time relations between an igneous body that is datable and a sedimentary stratum whose fossils closely indicate its position in the column.

Figure 6-3 shows in an idealized manner how apparent ages of sedimentary strata are approximated from the apparent ages of igneous bodies. The age of a stratum is bracketed between bodies of igneous rock, the apparent ages of which are known.

In the figure, four series of sedimentary strata, whose geologic ages are known from their fossils, are separated by surfaces of erosion. Related to the strata are two intrusive bodies of igneous rock (A, B) and two sheets of extrusive igneous rock (C, D). From the apparent dates of the igneous bodies and the geologic relations shown, we can draw these inferences as to the ages of the sedimentary strata:

Stratum	Age (millions of years)	that is,
4	$<34 <30 >20$	age lies between 20 and 30 million years
3	$<60 >34 >30$	age lies between 34 and 60 million years
2	$>60 >34$	age of both is more than 60 million years
1	$>60 >34$	

To separate 1 from 2, dates from other localities are needed. Dates from igneous rocks elsewhere could also narrow the possible ages of 3 and 4.

Through this combination of geologic relations and isotopic dating the geologic column has been calibrated (Table 6-3), and the calibration is being continually refined. As we have said, it is a great tribute to the work of geologists during the first half of the nineteenth century that isotopic dating has confirmed the geologic column they established. Comparison of Table 6-2 with Table 6-1 shows this. It shows also that the grouping of strata into the successively smaller subdivisions called systems, series, and stages is matched by the corresponding time units called periods, epochs, and ages. The time units, of course, can be referred to whether or not their apparent ages are known. That is, we could speak of events that occurred in the Devonian Period (or just Devonian time) even if we did not know that the apparent age of that period is between 400 and 340 million years ago.

We do not wish to give the impression, however, that the column is calibrated thoroughly. Some units are bracketed more closely than others, and there are long gaps without reliable dates. In time, however, the gaps will surely be filled in.

Dates of metamorphic events. As we learned in Chapter 4, when intense metamorphism affects rocks at great depth, recrystallization occurs, with new minerals formed from the constituents of older minerals that have become unstable in the high-pressure environment. The new minerals (like those in igneous rocks) include radioactive isotopes, which are therefore datable. Where crystallization has been complete, the apparent ages of such minerals,

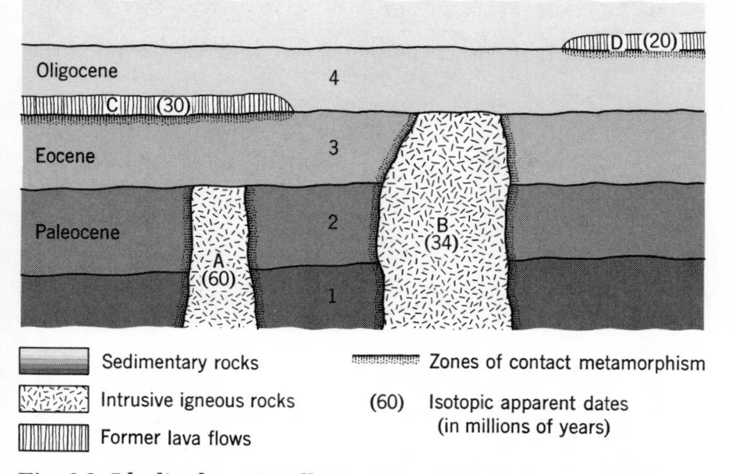

Fig. 6-3. Idealized section illustrating application of isotopic dating to the geologic column. For method see text.

however, are not the dates of the original rock; rather they date the end of a metamorphic event: the time when the rocks cooled below about 200°C as, through erosion of the strata overlying them, they approached the Earth's surface. Isotopic dating of metamorphic rocks enables us to assign upper limits to the ages of periods of metamorphism and thus provides an additional kind of information concerning the chronology of events in Earth history. We need to emphasize, however, that all such dates are minimum dates.

Dates of magnetic reversals. We can not leave the subject of geologic time without mentioning the fruitful result of comparing K/Ar ages with remanent magnetism in the same rocks. We remember from the discussion in Chapter 2 that in both former lava flows and sedimentary strata the remanent magnetism of tiny particles of magnetite and certain other minerals reflects both orientation and polarity of the Earth's magnetic field at time of solidification or deposition. When the K/Ar date and remanent magnetism of a sample of each of a large number of former lava flows, not more than a few million years old, are compared, a consistent pattern appears. The polarity of the remanent magnetism of some samples is normal (that is, similar to that of the Earth's field at present), whereas that of others is reversed. But the dates of nearly all the normal samples fall within two spans of geologic time, whereas the dates of nearly all the reversed samples fall within two different spans of time (Fig. 6-4).

As seen from the figure, these data tell us that at three times during the last few million years the polarity of the Earth's magnetic field has reversed. Even though we do not know what caused the reversals, we know approximately when they occurred, and we are sure their effects were felt simultaneously throughout the world. The reversals therefore are worldwide events, the dates of which are known within narrow limits of accuracy. These facts are useful in correlating strata that are not themselves capable of being dated isotopically but that have measurable polarity. Examples are certain deep-sea sediments (Chap. 15) represented by cores, within which various layers are characterized by opposite polarities. It should be possible to obtain isotopic dates for earlier magnetic reversals, at least somewhat farther down in the geologic column, and thus to extend the range of correlation by polarity in rocks for which no isotopic dates exist.

PRE-GEOLOGIC TIME; AGE OF THE EARTH

A glance at Table 6-1 shows that the oldest rocks consist of the great assemblage, mainly metamorphic, known as Precambrian. Of the large number of isotopic dates derived from Precambrian rocks, the youngest are around 600 million years and the oldest around 3.5 billion years, a number that sets a lower limit, or at least a temporary lower limit, to the apparent range of geologic time. One member of the group of very old rocks is a granite occurring in South Africa. Although it is an igneous

TABLE 6-3. SCALE OF GEOLOGIC TIME

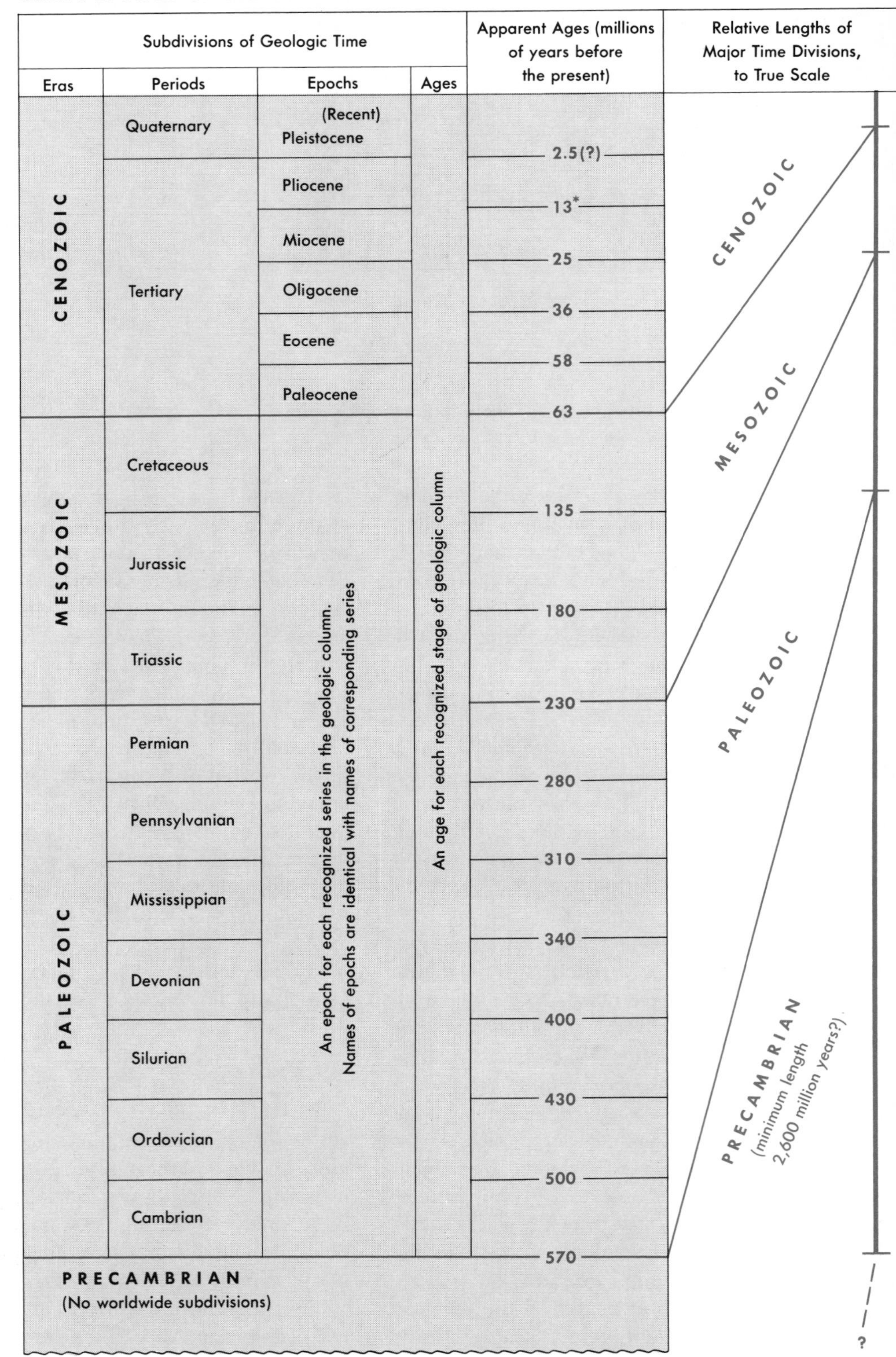

Subdivisions of Geologic Time: Eras	Periods	Epochs	Ages	Apparent Ages (millions of years before the present)
CENOZOIC	Quaternary	(Recent) Pleistocene		2.5(?)
	Tertiary	Pliocene		13*
		Miocene		25
		Oligocene		36
		Eocene		58
		Paleocene		63
MESOZOIC	Cretaceous	An epoch for each recognized series in the geologic column. Names of epochs are identical with names of corresponding series	An age for each recognized stage of geologic column	135
	Jurassic			180
	Triassic			230
PALEOZOIC	Permian			280
	Pennsylvanian			310
	Mississippian			340
	Devonian			400
	Silurian			430
	Ordovician			500
	Cambrian			570
PRECAMBRIAN (No worldwide subdivisions)				

* Dates based on stratigraphy in Europe differ, varying to as little as 7 million years.

body, this ancient granite surrounds detached masses of quartzite, which in turn must once have been sandstone. Hence the rock cycle must have been in operation well before the granite was formed. Analogous occurrences, implying very early operation of the rock cycle, are found in other continents. Geologic time, then, began still earlier. But how much earlier? We do not yet know.

The part of Earth history that antedates the oldest rocks falls within *pre-geologic time,* a span not measured by geologic methods. The age of the Earth is not precisely known, but information from astronomic data, from isotopic abundances in rocks and in meteorites, and from age determinations indicates that the Earth's age is 4.5 billion years, with an uncertainty not greater than a few hundred million years.

In its impact on our thinking about the length of geologic time and the age and earliest history of the Earth, isotopic age determination has been perhaps an even more important development for the twentieth century than the construction of the geologic column was for the nineteenth. These two great developments are complementary in a quite remarkable way.

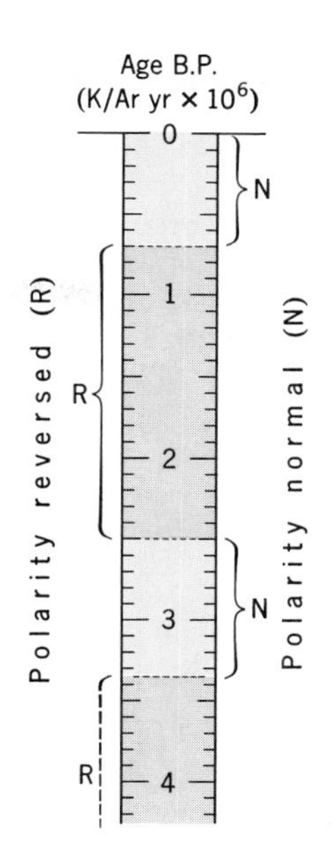

Fig. 6-4. Time scale of magnetic-polarity episodes through the last few million years. Braces mark time spans during which Earth's magnetic polarity was normal (N) and reversed (R). Dotted lines represent times of polarity reversal. (Adapted from Cox and others, 1966, Fig. 13.)

Summary

The geologic column

1. Sedimentary strata have been deposited in sequence, the oldest at the bottom and youngest on top.

2. Major groups of strata contain distinctive assemblages of fossils by which they can be identified.

3. The geologic column has been established through correlation of strata by means of their fossils.

4. Early attempts at time calibration of the column, by measuring rates of transport of substances from land to sea, were unsuccessful because the underlying assumptions were incorrect.

Isotopic dating

5. Decay of radioactive isotopes of Rb/Sr, K/Ar, and uranium and thorium series in minerals is the basis for determining the apparent ages of rocks.

6. Apparent ages exceeding 3 billion years have been determined from Precambrian rocks.

7. Fossil-bearing sedimentary rocks are assigned minimum and maximum dates on the basis of their positions between igneous bodies of known isotopic ages. In this way the geologic column has been calibrated to produce a tentative scale of geologic time.

8. Apparent ages of metamorphic rocks in most cases represent the end of an episode of metamorphism, not the age of the original rock material.

9. Measurement of radiocarbon activity, mainly in fossil organic matter, yields apparent ages for samples as old as 50,000 years.

10. Some strata of rather recent origin can be correlated by means of their magnetic polarities.

11. The age of the Earth appears to be about 4.5 billion years.

Selected References

Cox, Allan, and others, 1966, Geomagnetic polarity epochs: Pribilof Islands, Alaska: Geol. Soc. America Bull., v. 77, p. 883–910.

Darton, N. H., 1917, Story of the Grand Canyon: Grand Canyon, Arizona, Fred Harvey.

Faul, Henry, 1966, Ages of rocks, planets, and stars: New York, McGraw-Hill.

Hamilton, E. I., 1965, Applied geochronology: London and New York, Academic Press.

Harland, W. B., and others, eds., 1964, The Phanerozoic time-scale: Geol. Soc. London, Quart. Jour., v. 120s.

Holmes, Arthur, 1959, A revised geological time-scale: Edinburgh Geol. Soc. Trans., v. 17, p. 183–216.

Kulp, J. L., ed., 1961a, Geochronology of the rock systems: New York Acad. Sci. Annals, v. 91, p. 159–594.

——1961b, Geologic time scale: Science, v. 133, p. 1105–1114.

Libby, W. F., 1961, Radiocarbon dating: Science, v. 133, p. 621–629.

Tilton, G. R., and Hart, S. R., 1963, Geochronology: Science, v. 140, p. 357–366.

CHAPTER 7

Weathering and Soils

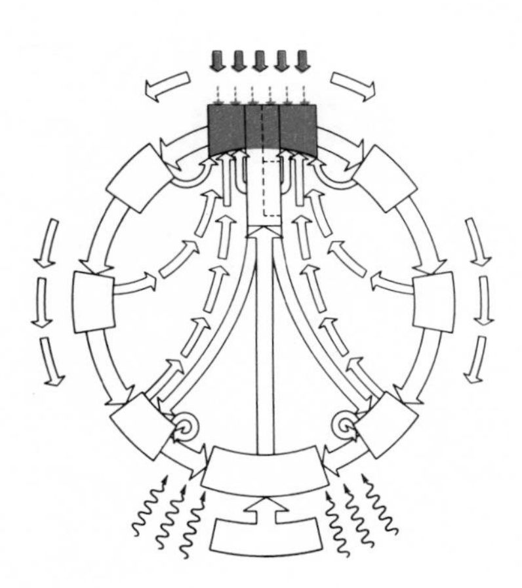

Environment of weathering
Weathering processes: chemical and mechanical
Rates of weathering
Influence of climate and other factors
Soils
Records of climate in soils
Role of weathering in the geologic cycle

Plate 7

Relics of an old civilization at Chichén-Itzá, Yucatán, México, have been etched by weathering through about eight centuries. Age is known from radiocarbon dates of related cultural material. (Sawders from Cushing.)

CHAPTER 7 *Weathering and Soils*

At this point we shall need to change the emphasis of our discussion. In the first six chapters our view has been a general one, embracing the Earth's three spheres, the minerals and rocks that constitute its crust, the movement of matter from one place to another through the geologic cycle, and finally the vital factor of geologic time. Now, in the next nine chapters, we shall look, one by one, at the external processes by which rock is broken up and moved to form sediments. This will enable us (in Chap. 16) to understand how sediments form new strata and how sequences of strata are analyzed in the reconstruction of Earth history.

The discussion of external processes begins logically with weathering, the step in the geologic cycle in which rocks are broken down and prepared for transport.

ENVIRONMENT OF WEATHERING

As far as we know, people since the beginning of civilization have tried to select, for buildings (Fig. 7-1), monuments, and tombs (Fig. 7-2), stone that would be durable. Their success has been mixed. Durability seems to have varied with climate and, in any one place, with characteristics of the stone used and degree and angle of exposure to the weather. From these differences we can extract useful information about rates of natural alteration of the kinds of rock from which the stone was cut. An inscription two or three centuries old, already partly obliterated, shows that in some kinds of rock and in some climates alteration can affect rock surfaces to a depth of several millimeters at least. Fast or slow, chemical and mechanical alteration occurs everywhere at the interface between the atmosphere and the Earth's solid surface. It has been said that we live at the bottom of a great sea of air which, although essential to our existence, is an agent of distinctive change.

The interface between lithosphere and atmosphere is not a sharply defined surface but is commonly a zone that extends into the ground to any depth to which air and water penetrate. In this critical zone the hydrosphere and biosphere are involved likewise; all four "spheres" are continually interacting. The activities are both physical and chemical, and whereas many of them are inorganic it is becoming increasingly apparent that organisms, especially plants, play a large part in the reactions and the movements of substances that take place at and near the surface.

The rock constitutes a framework, full of openings that are vulnerable to attack: joints; other cracks large and small, some of them minute; the tiny interfaces between the grains that make up the rock.

Fig. 7-1. Seventeenth-century façade built of limestone blocks, Trinity College, Oxford, photographed in 1961. Detail of buttress showing two or three shells peeling off the surface, most effectively where front and side of buttress meet. Stone has lost as much as 1cm of thickness during 300 years of weathering (R. F. Flint.)

Fig. 7-2. Three dated gravestones in a Connecticut cemetery show the strong influence of rock composition on rate of chemical weathering. (R. F. Flint). *Left* (1806): Marble, consisting of very soluble calcite, is greatly corroded. Urn (in medallion at top) is almost obliterated; inscription is hard to read. Entire surface is rough. *Center* (1779): Medium-grained quartz sandstone containing feldspar and micas, much less soluble, is slightly roughened overall and exfoliated in a patch just above center. *Right* (1769): Fine-grained sandstone consisting almost wholly of quartz, very insoluble, is virtually unaltered. Incised lettering is sharp and clear.

This framework is attacked by solutions that are chemically active and that can move in all directions, upward as well as downward and laterally.

The solutions in the ground are active both in the water-saturated zone and in the overlying zone that is not saturated (Fig. 11-1). In the latter zone much of the upward movement of solutions is the result of capillary tension in extremely narrow openings (Fig. 3-2); this is the force that causes ink to be drawn upward into blotting paper. Some of the upward movement is organic. Substances such as silica are pumped up through the roots and stems of plants; as the plants die these substances reach the ground in dead tissue and are released by decay.

The environment in which all this activity takes place, compared with environments deep within the Earth's crust, is one of low temperature and low pressure. It is one in which water substance is present as a liquid and also as vapor, and in which free oxygen is dissolved in the water in addition to the oxygen that constitutes part of the water molecule. In this environment the rock framework is continually attacked both chemically and physically and is conspicuously altered.

The results of such alteration are seen exposed rather commonly in cuts along highways and in other large excavations. Figure 7-3 shows fresh, unaltered granite-gneiss bedrock, grading upward imperceptibly through altered rock that still retains its organized appearance into loose, unorganized, earthy regolith in which the original texture of the rock has disappeared. Clearly alteration of the rock is developing from the surface downward. When exposed to the atmosphere no rock, whether bedrock or building stone, escapes the effects of ***weathering,*** *the chemical alteration and mechanical breakdown of rock materials during exposure to air, moisture, and organic matter.*

The regolith shown in Fig. 7-3 is residual, because it was formed in place by alteration of the bedrock beneath it. In many places, however, bedrock is overlain abruptly by regolith, the composition of which can not be explained by alteration of the underlying rock. This relationship implies that the residual material, formerly present, was removed and that sediment, transported from elsewhere, was deposited in its place. Both removal of the residual material and deposition of the sediment could have been performed by a single agency, such as a river, the surf along a coast, a glacier, or by several agencies combined.

Sediment is discussed in subsequent chapters, in connection with the various agencies that transported and deposited it. In this chapter attention is

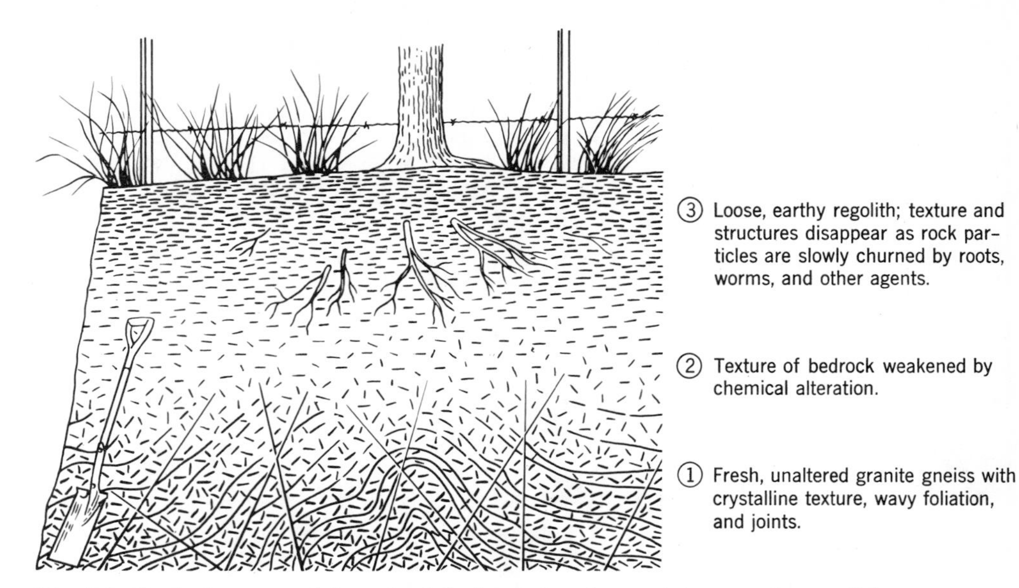

Fig. 7-3. Gradation upward from fresh bedrock (granite gneiss) to earthy regolith.

focused on residual material, which is of basic importance for these principal reasons:

1. First and foremost, it is the primary source of sediments and therefore of sedimentary rocks. It is one of the links in the chain that constitutes the rock cycle.

2. The character of some weathered regolith indicates the kind of rock from which the regolith developed. Where the rock itself is not exposed, it is an aid in mapping the local geology (App. D).

3. As some weathered regolith bears the stamp of the climate under which it was formed, we can infer in some cases that a change in the climate has taken place since the weathering occurred.

4. Some regolith is in itself an ore, in which useful minerals are concentrated. Bauxite (Chap. 23), a source of aluminum, is an example.

5. Regolith forms the basis of soil, the support and sustenance of rooted plants that directly or indirectly nourish most of the biosphere, including man.

WEATHERING PROCESSES

If we could look closely at the bedrock in Fig. 7-3, we could see that near the bottom of the exposure the cleavage surfaces of feldspar grains flash brightly between the rather dull quartz grains. Higher up, these surfaces are lusterless and stained, and near the top the grains of quartz, still distinguishable, are separated by soft, earthy material that in no way resembles the former feldspar, which has literally rotted away. Evidently the changes that have occurred are mainly chemical. On the other hand, some regolith consists of fragments identical with the adjacent bedrock. In them the minerals are quite fresh or have been only slightly altered. This relationship is seen commonly in the aprons of loose sliderock that mantle the lower parts of bedrock cliffs, from which the sliderock must have been derived (Fig. 8-18). As the coarse fragments of sliderock show little or no chemical change as compared with the bedrock, we conclude that bedrock can be broken down mechanically as well as decayed chemically. So we speak of *mechanical weathering* as distinct from *chemical weathering,* though the two processes work in close coordination, and commonly their effects are inseparably blended. ***Disintegration,*** *the mechanical breakup of rocks,* exposes additional fresh surfaces to air and water. Therefore disintegration aids ***decomposition,*** *the chemical alteration of rock materials.* On the other hand, some chemical changes are directly responsible for mechanical disruption on a large scale.

At this point we should note two significant relationships. (1) The effectiveness of chemical reactions increases with increased surface area available for reaction. (2) Increased surface area results simply from subdivision of large blocks of rock into smaller blocks. Take, for example, a cubic block, 1cm on a side. Its volume is 1cm^3, and its surface area is 6cm^2 or 600mm^2. If this block is cut along planes that bisect each of its edges, the result is 8 cubes each 5mm on a side (Fig. 7-4). The area of each side is 5 $\times$ 5mm or 25mm^2. The number of blocks has been multiplied by 8, the total number of sides is now 8 $\times$ 6, or 48, and the aggregate surface area (the sum of the areas of all the sides) has become 25mm^2 $\times$ 48, or 1,200mm^2 (12cm^2). In other words, by merely subdividing the cube, while adding nothing to its volume, we have doubled the surface area available for reaction. If our block formed part of a mass of bedrock subjected to weathering, we would have doubled its aggregate surface open to attack by both chemical and mechanical processes and so would have hastened its decomposition and disintegration.

The graph (Fig. 7-5) shows the rate at which the aggregate surface area of cubes increases as the length of cube sides decreases. From the graph we can calculate that our 1cm cube [sliced into eight smaller cubes each of which is sliced in a similar manner, and so on, each cube in each new generation through thirteen generations until the length of the cube sides becomes 1 micron (1μ)] yields about 1 trillion (10^{12}) cubes having an aggregate surface area of 5m^2, about ten thousand (10^4) times that of the original cube. No wonder the gradual subdivision of rock accelerates weathering in a very important way.

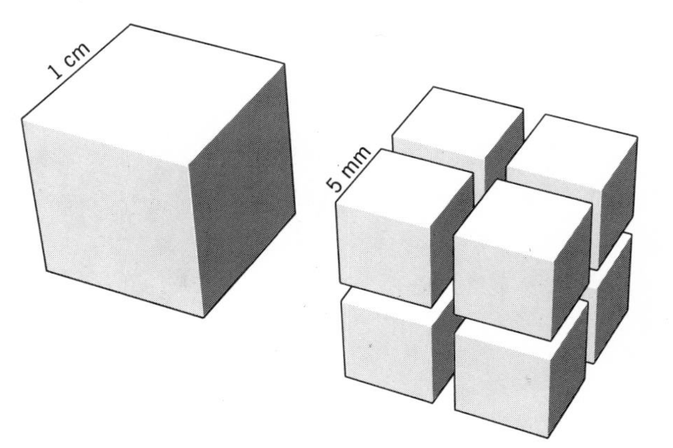

Fig. 7-4. Subdivision of a cube into eight smaller cubes.

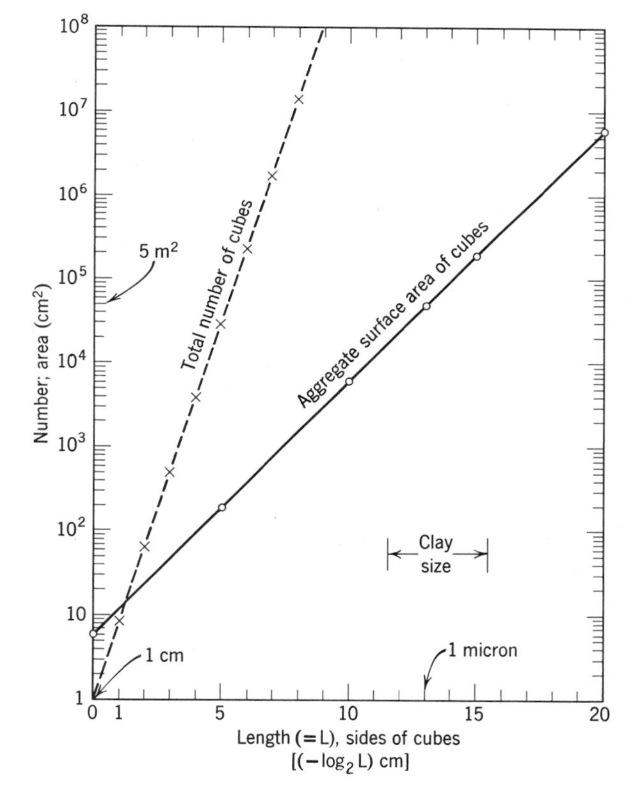

Fig. 7-5. Graph showing length of sides of cubes successively subdivided into 8 smaller cubes (abscissa) versus total number of cubes formed and aggregate surface area of the cubes, with volume of solid material remaining constant at 1cm^3. At each subdivision the number of cubes (dashed line) increases by a factor of 8 (2^3), and aggregate surface area (solid line) is doubled. Ordinate scale is logarithmic to base 10. Numbers on abscissa are negative logarithms (base 2) of length of cube sides (centimeters). These numbers indicate the number of times cube subdivisions have been made.

CHEMICAL WEATHERING: DECOMPOSITION

Agents of decomposition. We discuss decomposition first, because some disintegration is caused by it, and because the general effect of chemical weathering is further reaching than that of disintegration. The active agents of decomposition consist of chemically active aqueous solutions and water vapor. Rainwater brings to the ground with it small amounts of the carbon dioxide present in the air. This gas combines with water to make carbonic acid. As the water percolates down through the soil the strength of the acid solution is increased many times by addition of carbon dioxide created by the bacterial decay of vegetation.

The carbonic acid ionizes to form hydrogen ions and bicarbonate ions:

$$\underset{\text{(Water)}}{H_2O} + \underset{\text{(Carbon dioxide)}}{CO_2} \longrightarrow \underset{\text{(Carbonic acid)}}{H_2CO_3} \longrightarrow \underset{\text{(Hydrogen ion)}}{H^+} + \underset{\text{(Bicarbonate ion)}}{(HCO_3)^-}$$

Fig. 7-6. Crystals of the clay mineral kaolinite enlarged 26,900 times by an electron microscope, resemble plates of mica (compare Fig. 3-14). (P. F. Kerr.)

Hydrogen ions are extremely effective in decomposing minerals. They are very small (Fig. 3-7) and can squeeze in between the atoms of the crystal lattice and disrupt its structure.

Effects on potassium feldspar. The effectiveness of H^+ ions is illustrated by the way in which potassium feldspar (orthoclase) is decomposed by hydrogen ions and water:

$$\underset{\text{(Potassium feldspar)}}{2KAlSi_3O_8} + \underset{\text{(Hydrogen ions)}}{2H^+} + \underset{\text{(Water)}}{H_2O} \longrightarrow$$

$$\underset{\text{(Potassium ions)}}{2K^+} + \underset{\text{(Kaolinite)}}{Al_2Si_2O_5(OH)_4} + \underset{\text{(Silica)}}{4SiO_2}$$

Here the H^+ ions forcibly enter the potassium feldspar and displace potassium ions, which then leave the crystal lattice. Water combines with the remaining aluminum silicate radical to create the clay mineral kaolinite (Fig. 7-6). This combination of water with other molecules is *hydrolysis,* one of the chief processes in chemical weathering. The resulting kaolinite we call a *secondary mineral,* because it was not present in the original rock. Kaolinite is the most conspicuous of the three products of the reaction. It is a common member of the group of very insoluble minerals that constitute clay, and as clay it accumulates and forms a substantial part of the regolith. Many of the potassium ions released during the decomposition of orthoclase are taken up by plants; others enter into clay minerals other than kaolinite.

The silica, less insoluble than clay minerals, in part remains in the clay regolith and in part moves away in solution. Much of the rather soluble potassium carbonate likewise escapes in water solution; and some of it, together with some of the dissolved silica, eventually finds its way through streams to the sea. Some, however, is held in the clay regolith and can be utilized by plants as food.

We speak of the matter carried away in solution as having been *leached* out of the parent rock. ***Leaching*** *is the continued removal, by water, of soluble matter from bedrock or regolith.*

Effects on granite and on basalt. Our look at the chemical weathering of potassium feldspar has prepared us for a broader look at the weathering of two kinds of igneous rocks, granite and basalt. We can now trace the chemical and mineral transformations that occur in the minerals characteristic of each kind, while these rocks decompose. Table 7-1 shows that chemical weathering converts granite to clay in which some of the original micas, most or all of the original quartz, and some iron oxides, are present. As the feldspar decays, the quartz grains are loosened like bricks in a wall when the mortar between them crumbles. In contrast, basalt is converted into quartz-free clay, because little or no quartz was present in the original rock. Besides clay, the products of weathering of basalt include iron oxides, especially the hydrous iron oxides collectively called *limonite* (not a mineral but a group of minerals) and also hematite, Fe_2O_3. Much of the clay created during weathering of basaltic rocks is yellowish or brownish, the distinctive color of limonite. In clay derived from

TABLE 7-1. CHEMICAL WEATHERING OF TWO GREAT GROUPS OF IGNEOUS ROCKS, REPRESENTED BY GRANITE AND BASALT

	Primary Constituents		Weathering Products			
	Minerals	Metallic Ions	Colloids	Secondary Minerals That Form from Colloids and Ions	Primary Minerals That Persist	Soluble Ions Removed in Solution
GRANITE	ALKALI FELDSPARS	K^+ Na^+	Silica, alumina	Clay minerals		Na^+ K^+
	QUARTZ				Quartz	
	MICAS	K^+ Fe^{2+} Mg^{2+}	Silica, alumina	Clay minerals	Some mica	
	FERRO-MAGNESIAN MINERALS	Mg^{2+} Fe^{2+}	Silica, alumina	Clay minerals		Mg^{2+}
			Iron oxides	Hematite, "limonite"		
BASALT	PLAGIOCLASE FELDSPARS	Ca^{2+} Na^+	Silica, alumina	Clay minerals		Na^+ Ca^{2+}
	FERRO-MAGNESIAN MINERALS	Mg^{2+} Fe^{2+}	Silica, alumina	Clay minerals		Mg^{2+}
	MAGNETITE	Fe^{2+}	Iron oxides	Hematite, "limonite"		

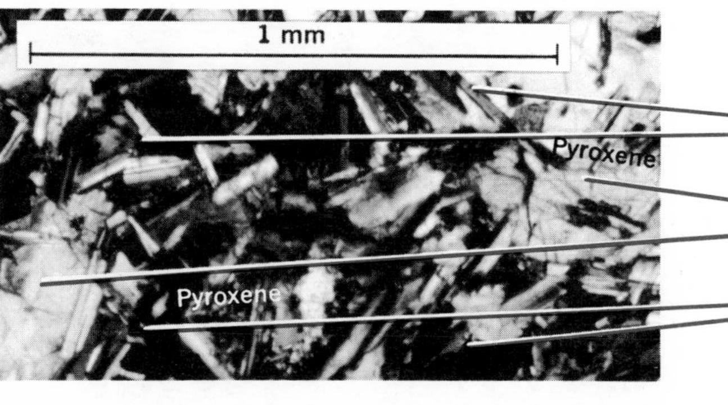

Primary minerals shown in capital letters; minerals constituting the ultimate product of complete weathering shown in shaded boxes. Photographs (*left*) are of paper-thin slices of granite and basalt, with key minerals identified. Uncommon primary minerals (gold, for instance) are not included. Like quartz they tend to persist and show up in the weathered product in very small quantities. (Upper photo G. M. Friedman; lower photo B. M. Shaub.)

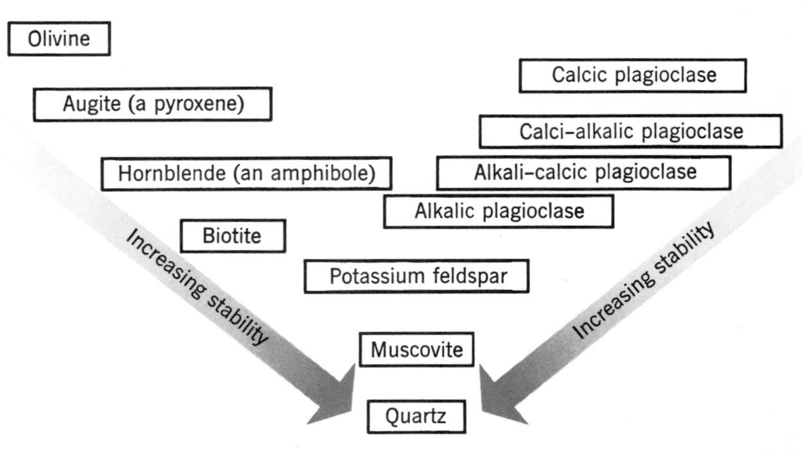

Fig. 7-7. Mineral-stability series in weathering (Goldich, 1938). Minerals common in igneous rocks are arranged in order of decreasing rate of chemical weathering—in other words, in order of increasing stability in the environment where weathering occurs. Compare the reaction series for the same minerals (Fig. 5-11).

granitic rocks, this color is much fainter, because in granite ferromagnesian minerals such as biotite and hornblende are much less abundant.

Effect on limestone. Because theoretically pure limestone consists solely of carbonate minerals ($CaCO_3$), which are highly soluble in carbonic acid, the chemical weathering of limestone is a simple process. Carbonic acid in the ground readily dissolves the $CaCO_3$, leaving behind only the nearly insoluble impurities that are always present, at least in small amounts. The most common impurities are clay particles and quartz particles that were deposited in the sea along with the $CaCO_3$; therefore, as a limestone becomes weathered, a residual regolith gradually develops over it, consisting of clay and particles of quartz. As most limestone contains pyrite, FeS_2 (Fig. 3-22), a mineral unstable in the presence of oxygen, weathering results in oxidation of the pyrite, with development of iron oxides. These oxides are what produce the yellowish and brownish tints commonly seen in limestone near the surface of the ground.

Rates of weathering of various minerals. It is evident from Table 7-1, *A* that quartz and some of the micas break down much less rapidly than feldspars, which are transformed into clay. After close examination of decomposed rock from hundreds of localities, geologists noted that the minerals fall into a kind of susceptibility-to-decomposition sequence, regardless of the kinds of rocks in which they occur. The reality of the sequence was confirmed by making chemical analyses of rocks that had been decomposed to various extents. Comparison of the analyses shows that chemical constituents are lost in about the same order, again regardless of rock type. Calcium, sodium, magnesium, and potassium tend to be lost first, followed by silica. Alumina and iron are affected least rapidly. This implies that the minerals from which these constituents were lost possess definite degrees of stability in relation to chemical weathering. In fact they have been arranged in a stability series (Fig. 7-7), a Y-shaped list with the plagioclase feldspars forming one arm and the ferromagnesian minerals the other arm, with quartz at the base.

This Y-shaped list is identical with the reaction-series diagram (Fig. 5-11) derived from the order in which minerals crystallize from a cooling magma. This means that the first minerals to be created from a magma show the fastest rate of decay when attacked by weathering, whereas the last mineral to crystallize from the magma shows the slowest rate of decay. The obvious inference is that minerals such as olivine and calcic plagioclase, which were most stable in the high-temperature, high-pressure environment early in the life of the magma, are least stable under the low-pressure, low-temperature environment of weathering, and hence decompose most readily; whereas quartz, stable under the much lower temperatures and pressures that prevailed as the magma finally solidified, is more "at home"—that is, stable—in the environment of weathering.

Just why do the first rock-making silicate minerals to crystallize in a magma become unstable as they are weathered? The explanation lies in their crystal chemistry. They contain the cations Mg^{2+}, Fe^{2+}, and Al^{3+}. For example, olivine contains Mg^{2+} and Fe^{2+} between successive silica tetrahedra; pyroxene contains the same two ions between adjacent single chains of linked silica tetrahedra; hornblende contains Mg^{2+} and Fe^{2+} between adjacent double chains of linked silica tetrahedra and Al^{3+} in some of the Si^{4+} positions within the tetrahedra themselves; feldspars likewise contain Al^{3+} in many of the Si^{4+} positions.

When weathering attacks these minerals the three cations are removed quite easily from their positions in the crystal lattices. Fe^{2+} is removed by oxidation and Mg^{2+} and Al^{3+} by solution. When the cations are removed the lattice structure collapses, decomposing the mineral.

The stubbornness with which quartz yields to decomposition is a part of the reason why quartz is the commonest mineral in sedimentary rocks. Loose quartz grains in the clayey regolith can be picked up by running water, carried far away, spread out in layers, and eventually cemented to form sandstone. If the sandstone is later exposed to erosion and broken up, the quartz grains will slowly grow smaller with abrasion by water and wind but will persist without chemical change for long ages. Much of the sand now on beaches or in desert dunes consists of quartz grains initially set free by the weathering of ancient granitic rocks.

Besides quartz, other minerals, though far less abundant than quartz, are stable in the environment of the Earth's surface and hence are resistant to destruction by weathering. Such minerals as gold, platinum, ilmenite (a titanium ore), cassiterite (a tin ore), and diamond persist in weathered regolith and are therefore included in sediments, in some of which they become concentrated because they are unusually heavy, forming placers (Fig. 23-22) from which they are mined. The widespread occurrence of these minerals in sediments and sedimentary rocks indicates that weathering has been going on throughout all of geologic history.

MECHANICAL WEATHERING: DISINTEGRATION

Before continuing with the general effects of weathering we must examine the ways in which weathering occurs by mechanical means, proceeding at the same time and often in the same places as the more-widespread chemical weathering—indeed hand in hand with it. Although the two groups of processes commonly act together, the action of one promoting the action of the other, a separate look at mechanical weathering clearly reveals that this activity consists of a distinctive group of processes.

Rock strength. In many places the regolith is comprised of rock fragments that contain neither secondary minerals nor altered primary minerals and that are in fact identical with the local bedrock. Evidently such regolith was formed by disintegration rather than by decomposition.

In order to understand how rocks can be broken mechanically, or *ruptured*, by natural forces, we must consider the strengths of both rocks and forces. Laboratory tests show quite consistently that rocks have great compressive strength—that is, they strongly resist great confining pressure—but are weak when subjected to tensile stresses. We can speculate that the comparative ease with which rocks break when subjected to tensional stresses explains the joints and other cracks that cut all rocks, generally into quite small blocks. How does the ease with which rocks break compare with the natural stresses in the outer part of the Earth's crust?

Frost wedging. One conspicuous means of subjecting rock to stress is by freezing water contained within it. Freezing experiments in the laboratory show that a volume increase of 9 per cent occurs when water crystallizes into ice. If this reaction takes place in a confined space the pressure of expansion can reach as high as 2,100 kg/cm^2 provided the temperature is $-22°C$. This pressure is greater by a factor of 40, than that necessary to break granite. We must note that under natural rather than laboratory conditions the extreme pressure mentioned above is probably never reached. However, far smaller pressures would suffice to rupture even a strong rock. When we observe that almost no rocks are free from cracks of various kinds, it seems that freezing should be very effective.

In cold and temperate regions, water freezes in both bedrock and regolith. Thermographs installed in such places show that during the winter, at least, the fluctuating temperature crosses the freezing point frequently—in many places at least twice daily. As ice forms, rock materials are pushed up or pried apart—not only tiny particles but also blocks large and small, some weighing many tons.

Fig. 7-8. Mechanical weathering (chiefly frost wedging along joints) is disintegrating dolerite bedrock in east Greenland, forming large, angular blocks. Curved surface (skyline of foreground) was created by glacial erosion during latest glacial age (compare glacially abraded rock, Fig. 7-15). Radiocarbon dates of samples collected nearby show that glaciers had melted back and that this locality was exposed to weathering about 8,000 years ago. Thus the disintegration shown is the work of about 8,000 years of weathering. Mesters Vig, Greenland. (A. L. Washburn.)

This mechanism is aptly called *frost wedging*. In Fig. 7-8 we can see that movement of the largest block of rock has been lateral, toward the left. This is the direction of easiest relief of pressure exerted in the crack—now a wide gap—the direction in which the least energy would be needed to move the block. This is why frost wedging is particularly effective at the faces of cliffs.

Because most cracks and cavities in rocks are open to the atmosphere, it might be supposed that freezing water would be unconfined and would easily expand along the cracks. But the water in the upper and outer parts of the openings tends to freeze first, creating closed systems in which further freezing can set up bursting pressures.

Probably the wedging apart of minute mineral grains loosens and, in the long run, moves more rock material than the more spectacular pushing of large blocks. The comparison would be very difficult to measure.

Freezing water can affect not only bedrock but also fine-grained regolith, in which it causes *frost heaving*. Early in the winter season water in such material freezes close below ground, as the ground loses heat to the atmosphere. The ice gradually thickens by additions from rainwater and snow melt above and by upward capillary movement from below the frozen zone. Expansion during freezing heaves up the material above the frozen zone and creates bulges in the surface. The bulges, "frost boils," are visible in unpaved roads and other areas at the season of the spring thaw.

Mechanical work of plants and animals. Tree roots gradually extend into crevices in bedrock, and in their growth they wedge apart the adjoining blocks. Shrubs and smaller plants also send their rootlets into tiny openings and slowly enlarge them. There is general agreement that the total amount of rock breaking done in this way must be enormous, but, again, it would be hard to measure. Much of it is obscured by chemical decay of the rock, which takes advantage of the new openings formed by mechanical means.

Burrowing animals large and small bring quantities of partly decayed rock fragments to the surface, where they are exposed more effectively to chemical action. Charles Darwin made close observations in his English garden and calculated that every year worms brought particles to the surface at the rate of more than 10 tons per acre ($22.4g/m^2$). After a study in the Amazon basin, the geologist J. C. Branner wrote that the soil "looks as if it had been literally turned inside out by the burrowing of ants and termites." The mechanical work of burrowing organisms, accumulated throughout hundreds of millions of years of Earth history, must be huge in amount.

The problem of thermal changes. The heat of forest and brush fires breaks large flakes from exposed surfaces of bedrock. Rock is a poor conductor of heat; fire heats only a thin outer shell, which expands and is disrupted. Forest fires set by lightning must have been common during long ages before man came to disturb nature's economy, and doubtless fire has been an agent of considerable importance in the mechanical breaking of rocks at the Earth's surface. The expansion sets up tensional stresses acting at right angles to the surface, and these are relieved with the least expenditure of energy by upward or outward movement, which ruptures the rock parallel with the surface and pulls the expanded rock away.

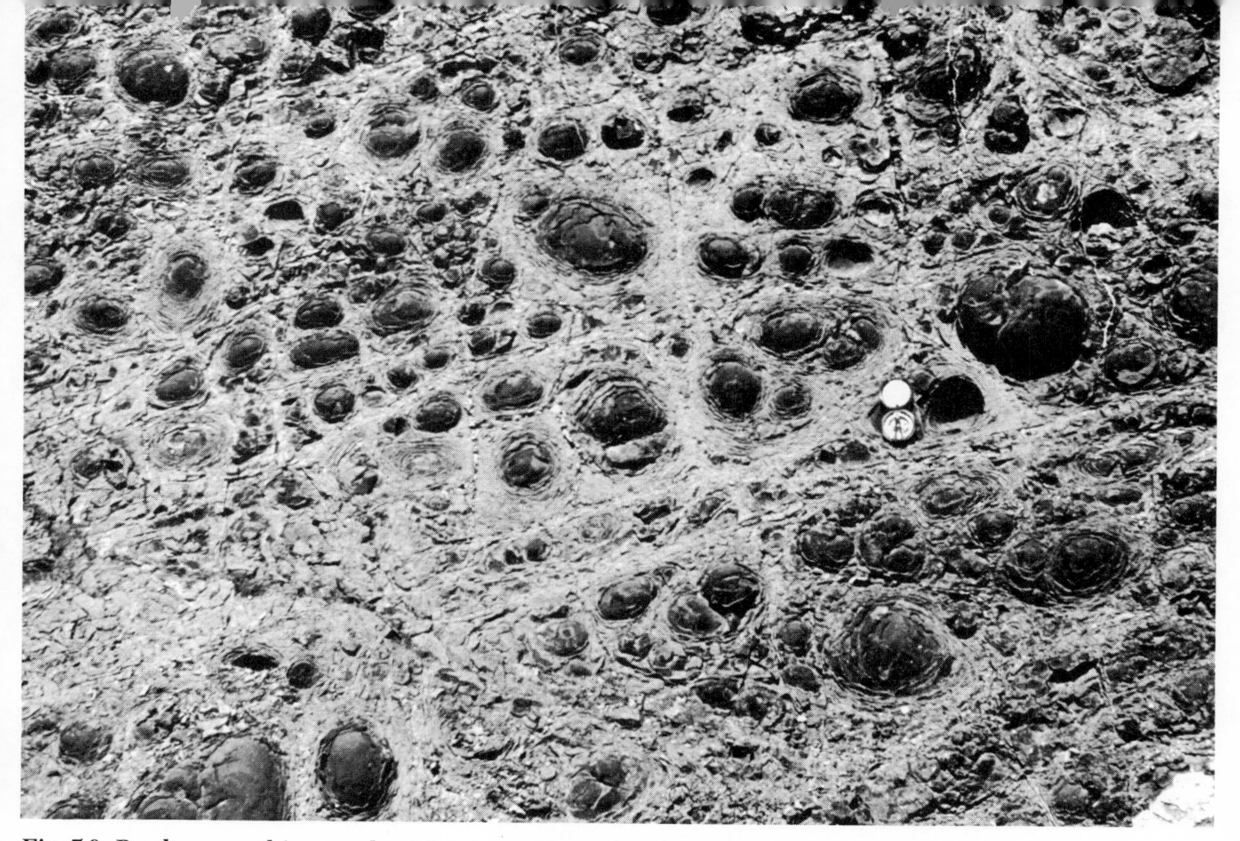

Fig. 7-9. Basalt, exposed in a road cut in western Argentina, is cut by two sets of fractures, one nearly vertical, the other inclined gently from right to left. Solutions penetrating inward from the joints have converted nearly cubic blocks of rock to spheroids, each with several successive shells. Diameter of mirror on the compass (*right center*) is 7.5cm. (R. F. Flint.)

In desert areas having sparse vegetation, rock fragments, including rounded boulders of quartzite that lie shattered on the surface or partly imbedded in the ground, pose a problem that is not yet solved. Fifty years ago and more, it was simply assumed that shattering resulted from mechanical expansion and contraction caused by repeated thermal changes between night and day, the source of energy being the Sun. Ground temperatures in such places are thought to range through as much as 65°C during a 24-hour cycle, although generally the range is much less.

The idea that boulders are shattered by solar heating was tested by heating pieces of rock in a furnace to high temperatures. They reacted by beginning to rupture only at temperatures far higher than those experienced in a desert. Other rock samples were subjected to heating and cooling through 110°C nearly 90,000 times, equivalent to 240 years of change between day and night; yet they showed no significant change. Next holes were bored in bedrock in a desert environment, and temperatures within the rock down to depths of 60cm were read hourly. It was found that within the rock the daily temperature change decreases rapidly inward from the surface and becomes insignificant at around 60cm.

These experiments cast doubt on the long-held idea that solar heat alone can disrupt rocks. Moisture occurs in small amounts between the grains within rocks even in deserts; perhaps its presence during the temperature changes plays a part. Also the experiments, like many others in geology, could not adequately reproduce the time factor; for desert boulders could have been subjected to millions of thermal cycles instead of only tens of thousands. The problem remains unsolved for the present.

Exfoliation. Another process, which commonly occurs in jointed rocks is ***exfoliation,*** *the separation, during weathering, of successive shells from massive rock,* like the "skins" of an onion (Figs. 7-9, 7-10). Most shells range in thickness from a few millimeters to a few centimeters, although some may be thicker. In some cases only a single shell is present, in others ten or more. The outermost shells are likely to have flattish faces, whereas the shapes of the inner ones

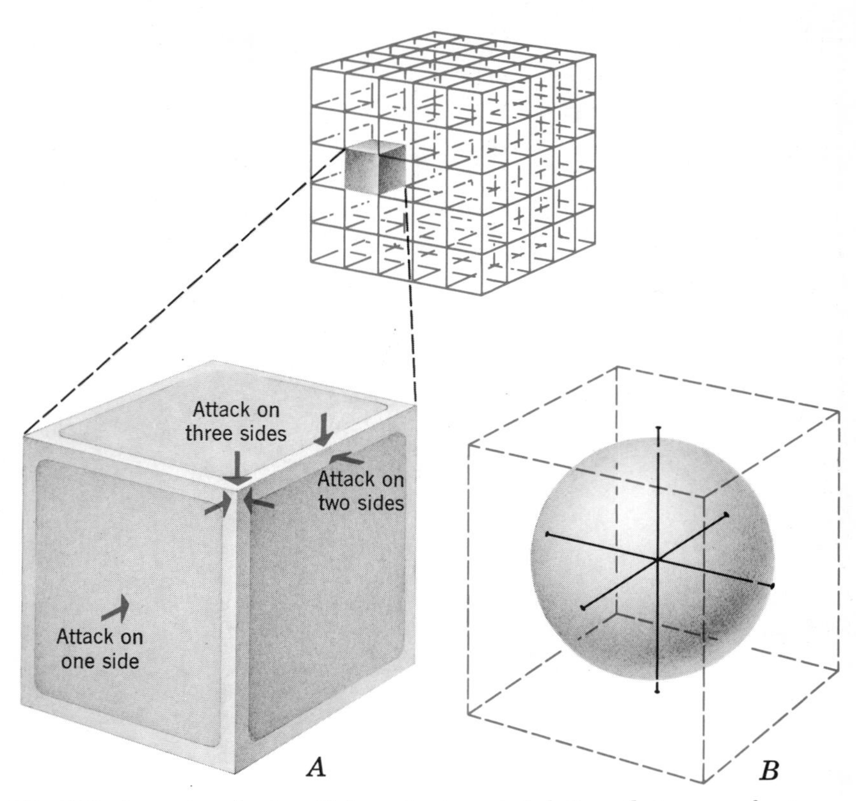

Fig. 7-10. Geometry of spheroidal weathering. *A*. Solutions that occupy fractures separating nearly cubic blocks of rock attack corners, edges, and sides at rates that decline in that order because the numbers of corresponding surfaces are 3, 2, and 1. Corners become rounded; eventually the blocks are reduced to spheres (*B*). Energy of attack has now become distributed uniformly over the whole surface, so that no further change of form can occur.

tend to be spherical. Evidently exfoliation can take place below ground, because its effects are exposed in new road cuts (Fig. 7-9). By contrast, it is seen in progress at the surface, in places where the present surface seems to have been stable and little changed for a long time. It is not confined to a particular type of climate, although it is less commonly seen in very wet than in semiarid climates.

A close look at the exfoliated rock reveals evidence of chemical weathering, such as decomposition of feldspars and ferromagnesian minerals and development of clay minerals and iron oxides, which seems to become less intense from the outermost to the innermost of the series of spheroidal shells. Commonly the exfoliation spheroids are arranged in a rather regular manner; and if the exposure goes deep enough the partings between spheroids are seen to pass downward into joints, like those in Fig. 4-1, which existed in the rock long before weathering began. Evidently the positions of the spheroids are controlled by joints, which have acted as the avenues along which solutions move.

Although the outer surface of exposed rock dries rapidly after wetting, moisture that penetrates between the mineral grains and into minute crevices remains long enough to cause some decomposition. The creation of clay minerals through decomposition of feldspars and other primary minerals in the rock is accompanied by volume increase. Probably the cumulative volume increase sets up small tensional stresses that cause shells to separate from the main body of the rock. Separation is a mechanical process much like the one that is believed to create sheets (described in the following section), but it is set off by chemical changes. Hence it is a mechanical effect of chemical weathering. The reason why shells become more nearly spherical toward the center of a block is demonstrated in Fig. 7-10.

Fig. 7-11. Granular disintegration of granite creates a very rough surface. Near Bulawayo, Rhodesia. Length of pocket knife is 9cm. (R. F. Flint.)

Even where exfoliation does not occur, massive rocks can acquire dull, very rough surfaces owing to the loosening of some of their mineral grains, which fall out and in some cases can be seen covering the ground nearby. This *granular disintegration,* like exfoliation, is a mechanical effect of chemical weathering of rock surfaces (Fig. 7-11). Like exfoliation, also, it can result in spheres or spheroids, creating rounded bosses and even completely detached "boulders," because its geometry is controlled by joints, as shown in Fig. 7-10.

Under arid or semiarid climates, granular disintegration frees feldspar grains from bedrock faster than the feldspar is chemically attacked and converted to clay. This fact helps explain the occurrence of fresh or nearly fresh feldspar in some sedimentary rocks. The products of granular disintegration can be washed into streams and deposited by the streams as sediments. We might think that this explanation would enable us to infer that climates were dry where and when sediments bearing feldspar were deposited. But the matter is more complicated, because grains of feldspar dislodged from bedrock on steep slopes, even in wet climates, can be washed downstream and buried without appreciable decomposition (Chap. 16).

Development of sheets; expansion cracks. Quarries in granite and other massive rocks commonly show groups of continuous partings or cracks that are distinct from the more closely spaced intersecting sets of joints described in Chapter 17. The partings are parallel to the ground surface and are spaced from less than 1m to more than 10m apart. They separate the rock into nearly parallel layers or *sheets* (Figs. 7-12, 7-13) which cut right across rock structures such as foliation. Similar sheets also are naturally exposed in many areas of massive bedrock, regardless of whether the climate is wet or dry or whether any chemical decay has occurred. Therefore, although in a sense they are a kind of exfoliation, sheets are not the result of any process hitherto described, but have originated in some different way.

The cause of the sheets has been much debated. Many geologists believe that cracking results from upward relief of pressure by gradual removal (through erosion) of the load of bedrock that formerly lay above it (Fig. 7-14). Before unloading, the bedrock beneath was compressed by the heavy load of rock above. The atomic structure of its minerals had been determined by the high-pressure environment. With unloading, this structure was distorted; the rock expanded upward; and the tensile stresses cracked it. Where the surface of the ground is curved, the cracks curve with it, because they develop at right angles to the direction of relief of pressure. Seemingly, the curved sheets thereafter help to

Fig. 7-12. Two successive sheets, each 40 to 60 feet thick, have formed on this bare granite hill, 600 feet high, west of Bulawayo, Rhodesia. Sheets curve with slope of the hill. Bare-rock surface is being weathered by granular disintegration. (R. F. Flint.)

control and perpetuate the form of the hill; the processes of erosion and the creation of additional sheets go forward hand in hand.

FACTORS THAT INFLUENCE WEATHERING

At least four factors that control the character and rate of weathering come readily to mind: kind of rock, slope, climate, organisms, and time.

Kind of rock. As Fig. 7-2 shows, the minerals of which a rock is composed influence its decomposition. Quartz is so resistant to decomposition that rocks rich in quartz stoutly resist chemical weathering. For example, in the eastern United States, from Maine to Georgia, granite commonly takes the form of high hills and mountains that stand well above surrounding areas underlain by weaker rocks such as schists. This relationship between mineral composition and topography is mainly the result of differences in rate of weathering.

However, rate of weathering of a rock is influenced not only by minerals but also by structure. If a rock, even though it consists entirely of quartz (sandstone, quartzite), contains closely spaced joints or other partings, it can disintegrate very rapidly, especially when attacked by frost wedging.

Slope. Where slopes are steep, the solid products of weathering readily move downward, continually exposing fresh bedrock to attack. As a result weathering extends only to slight depths below the surface. This is happening to an extreme degree on the bare granite hill shown in Fig. 7-12. As soon as a mineral grain is loosened by decomposition it is washed down the hill by the next rain. On the other hand, gentle slopes favor a much greater depth of weathering—in favorable places 50m or even more.

Climate and organisms. Because abundant moisture and heat promote chemical reactions, weathering generally goes deeper in moist and warm than in dry and cold climates. In high mountains and in cold high latitudes, frost-wedged rock fragments litter large areas of the surface, partly obscuring any effects of decomposition. Rocks such as limestone and marble (Fig. 7-1), which consist almost entirely of calcite, decompose quickly in moist climates owing to the solubility of calcite in carbonic acid. In dry climates, however, such rocks are very resistant. This difference in the behavior of a single kind of rock in two different climatic environments illus-

Fig. 7-13. Large-scale fractures separating a hill of granite into slightly curved sheets parallel with the surface of the hill. Near Shuteye Peak, Sierra Nevada, California. (N. K. Huber, U. S. Geol. Survey.)

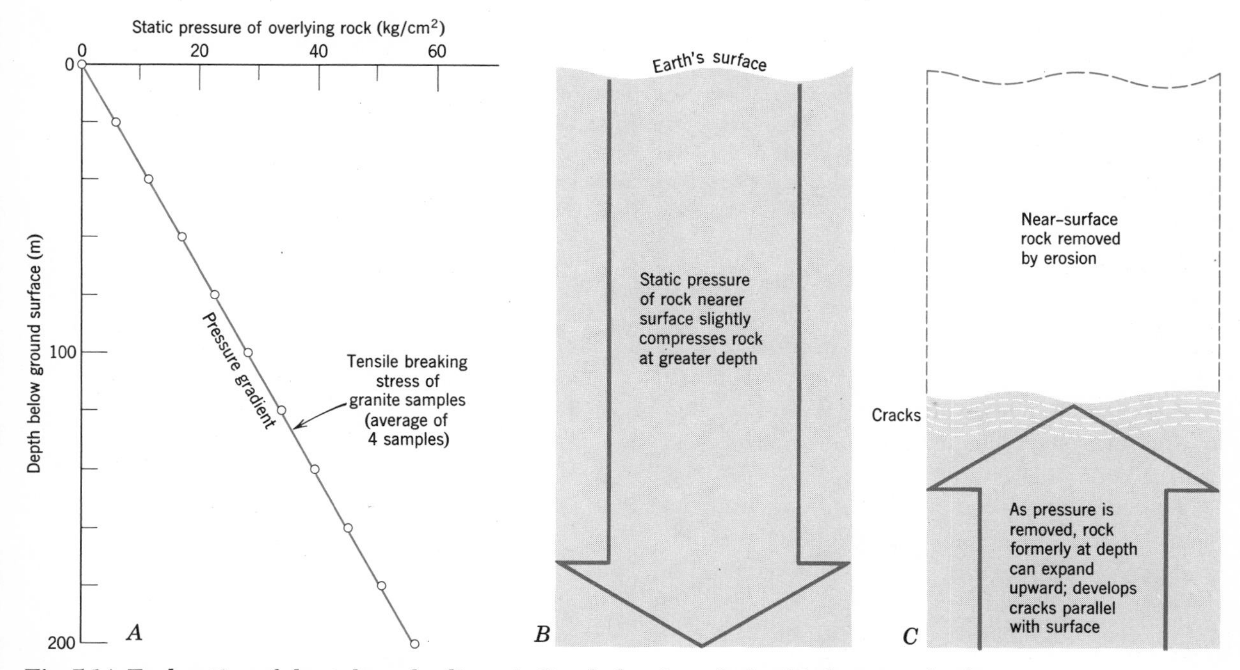

Fig. 7-14. Explanation of sheets by unloading. *A*. Graph showing relationship between depth and static pressure caused by overlying rock of density 2.65. *B, C*. Suggested explanation: A column of the Earth's crust, consisting of granite (density 2.65) (*B*) before and (*C*) after a long period of erosion. From the graph, it appears that in *C* the column of rock removed would have to be nearly 200m long in order to rupture the granite, if we assume that unloading produces tensile stresses equal to the weight of rock removed.

Fig. 7-15. Granitic bedrock high in the Sierra Nevada, California, was smoothed and polished by glacier ice probably more than 15,000 years ago. Although the surface may have been covered for a time with stony debris deposited by the glacier, the bedrock at this locality has been exposed throughout most of the time since then. Yet weathering, by granular disintegration, has destroyed only parts of the polished surface. (Frances Coleberd.)

trates the difficulty experienced in trying to isolate one factor at a time in the analysis of a complex natural system. The kinds and abundance of organisms, especially plants, depend in part on climate. During their lifetimes, plants influence soil by taking up moisture and chemical substances, and when dead, add organic matter to the soil.

Time. Study of decomposition of stones in ancient buildings and monuments readily shows that hundreds or even thousands of years are required for the decomposition of rocks to depths of only a few millimeters. Granite and other kinds of bedrock in the Sierra Nevada, New England, northern Europe, and elsewhere still exhibit polish and fine grooves made by glaciers during the latest glacial age, some 10,000 to 25,000 years ago (Figs. 7-15, 12-8). We can reason that in such rocks and in such cool-temperate climates, it could take many tens of thousands of years (at the very least) to create weathered regolith like that shown in Fig. 7-3. Probably the most promising way to measure the time involved in developing a zone of weathering is by dating of radioactive compounds. At present we have barely made a start at measuring rates of weathering.

SOILS

One of the significant aspects of weathering is that it creates soil, "the great bridge between the inanimate and the living." Soil is the principal natural resource of any country, and the special field of study, ***pedology,*** or *soil science,* is concerned with its origin, use, and protection. In problems of soil origin, pedologists and geologists have worked closely together.

The term *soil* is used in more than one sense.

To an engineer it is a synonym for regolith, the aggregate of all loose rock material above bedrock. Most geologists, however, adopt the definition used in soil science and agriculture: ***soil*** *is that part of the regolith which can support rooted plants.* This definition excludes the lower part of the residual regolith, as can be seen from the comparison in Fig. 7-16. In order to be consistent with currently used terms we subdivide regolith in slightly different ways, depending on whether our interest is mainly in weathering or mainly in soil analysis.

Soil profiles. Where the chief concern is with soil, a subdivision is made into three layers which pedologists call *horizons* (Figs. 7-16, 7-17). *The succession of distinctive horizons in a soil, and the unchanged parent material beneath it, is a* ***soil profile.*** The parent material can be bedrock or transported

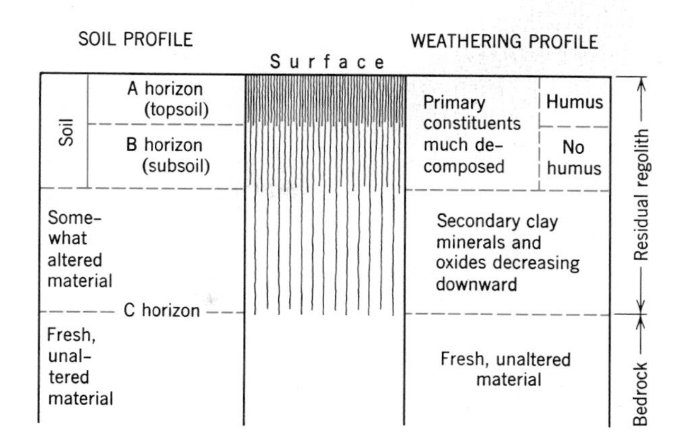

Fig. 7-16. Idealized section illustrating relation between a soil profile and a weathering profile. Because the section is idealized no vertical scale is shown. Soils are generally thin, whereas the thickness of weathered regolith in places amounts to tens of meters.

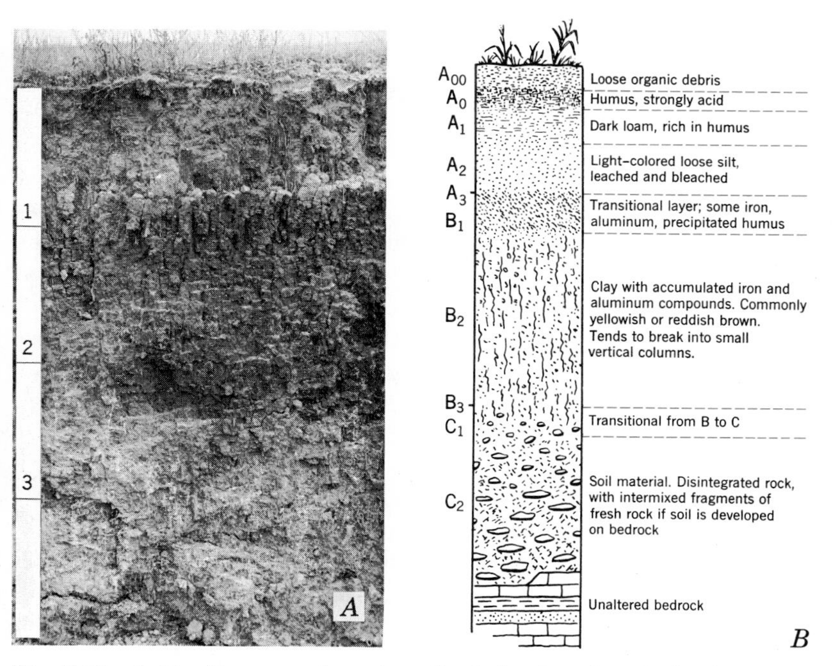

Fig. 7-17. **A. Freshly exposed section of soil developed on bedrock consisting of shale and sandstone. The three horizons of the soil profile are clearly distinguished. The scale shows thickness in feet. (U. S. Soil Conservation Service.)** ***B.*** **Idealized soil profile showing all horizons commonly developed under a moist, temperate climate. Boundaries of adjacent units are gradational. A_{00} and A_0 are present in forests, usually not on grasslands. The transitional unit A_3-B_1 is commonly, though not universally, recognized in well-developed profiles. The B_3-C_1 gradational zone generally is distinct but varies in width. (After T. L. Lyon and H. S. Buckman, 1943, The nature and properties of soils: New York, Macmillan, p. 264.)**

regolith. From top to bottom the horizons are designated A, B, and C. Commonly each horizon is divisible into two or more distinct fractions. Horizons A and B together constitute the soil proper; horizon C, although part of the soil profile, is not part of the soil itself.

The A and B horizons essentially complement each other. The A horizon has lost part of its original substance through leaching of soluble constituents (not only by carbonic acid but also by humic acids) and through mechanical removal (by percolating water) of minute particles of secondary clay. Rather commonly its hue is grayish or blackish because of the addition to it of ***humus,*** *the decomposed residue of plant and animal tissues.*

The B, although poor in organic content, is a site of accumulation, for it has gained part of what the A has lost. Iron oxides derived from the oxidation of pyrite and ferromagnesian minerals, and secondary clay washed down from above commonly accumulate in the B. In a moist climate the dissolved Ca^{2+} and other soluble cations are carried away altogether, but in a dry climate the water in the soil tends to evaporate, precipitating calcium as a carbonate in the B horizon, as noted below.

The C horizon consists of parent material (bedrock or regolith). Accumulation of secondary minerals, slight in the upper part of the C, decreases downward into completely fresh, unaltered material. The lower limit of the C horizon, therefore, is very indistinct.

The several horizons of a soil profile vary considerably with differences in topography, drainage, kinds of vegetation, and other factors. For example, on a treeless slope the soil differs in important details from neighboring soils in forested areas or in swampy lowlands.

The parent material from which a soil is developed can be bedrock or regolith of any kind. In the "bottom lands" of large valleys the soil is exceptionally rich and thick because the parent material itself is in considerable part transported regolith, consisting of topsoil from nearby uplands washed down into the valleys. In much of the northern United States and Canada the parent material of soils is regolith consisting of glacial deposits (Chap. 12), deposited as recently as 10,000 years ago by former ice sheets. Because of their rather short history, the soils of New England and the Great Lakes region are immature, which means that their profiles are only partly developed. They differ radically from soils farther south, in the region untouched by glaciation. Such soils are generally mature. A ***mature soil*** is *a soil having a fully developed profile.*

Climatic soil groups. Parent materials differ widely and strongly influence the character of soils, especially during the earlier part of soil development. The influence of parent bedrock is expressed in such terms as *granite soils* and *limestone soils.* But over a very long time the influence of climate is even stronger than that of bedrock in determining types of soil. Under given climatic conditions the profiles of mature soils developed on widely different kinds of rock become surprisingly alike.

In a general way the soils in the United States are divided into two types on the basis of degree of leaching by circulating water. The most effective leaching is found in areas with average annual rainfall of more than 25 inches. An irregular line extending northward from central Texas to western Minnesota divides the more humid eastern district from the drier western country (Fig. 7-18). East of this line the upper horizons of mature soils have, by leaching, lost large fractions of the calcium and magnesium carbonates that were in the parent material. Moreover, the B horizon in such soils has received much iron and clay carried down from the A horizon. These effects reflect not only a good supply of circulating water but also abundant organic acid furnished by plants. West of the line, where circulating solutions are less abundant and chemically weaker, carbonates accumulate in the upper part of the soil making it strongly alkaline in contrast to the acidic soils of humid regions. An important part of the accumulation results from evaporation of water that rises in capillary channels, bringing dissolved salts from below. In large areas of western Texas and adjoining states the carbonates deposited in this way have built up a solid, almost impervious layer. Such *a whitish accumulation of calcium carbonate developed in a soil profile* is generally known in western North America as ***caliche*** (kă lē′ che); in British literature it is rather commonly referred to as ***calcrete.***

A soil with calcium-rich upper horizons is a ***pedocal*** (pĕd′ ō kăl; Gr. *pedon,* "soil," and the first syllable of *calcium*). *A soil in which much clay and iron have been added to the* B *horizon* is a ***pedalfer*** (pĕd ăl′ fer; symbols for aluminum and iron—Al

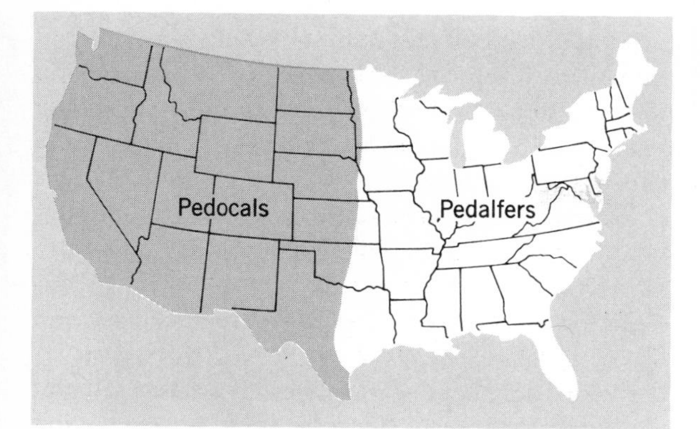

Fig. 7-18. General division of soils in the United States into two major classes determined by climate. West of the critical line, which approximates annual rainfall of 25 inches, many variations in kinds of soil are found in mountainous districts. Compare Fig. 2-12. (A. N. Strahler, after C. F. Marbut.)

and Fe—added to the Greek root). West of the critical line shown in Fig. 7-18 many limited areas exist where rainfall is higher than the average for the region. In some such areas soils belonging to the pedalfer group are present. The soils in many comparatively moist parts of the continent as a whole, especially where conifer forests exist, are of a kind called ***podzols.*** These are *pedalfer soils that have been intensely leached by solutions rich in humic acids.* The acids have removed most of the calcium and iron, changing the hue of the upper part of the soil profile to light gray or nearly white.

In the Arctic region soils are generally immature and stony, because their development has been disturbed by glaciation, and because chemical changes there are slow. In large desert areas, whether warm or cold, real soils are almost absent, and a large share of those that may be present are stony, alkaline, and lacking in organic matter.

Most soils developed at moderate altitudes in humid-tropical lands differ conspicuously from those found in cooler parts of the temperate zones. The basic cause of this difference seems to be organic. In temperate regions organic matter accumulates, but under higher temperatures certain bacteria destroy much of the humus, except in swampy areas. Hence in many tropical soils the acids from decaying vegetation are ineffective in leaching out iron, although living plants remove silica from the soil. As a result iron accumulates steadily as hydrated oxides. Over large areas in southeast Asia, Australia, Brazil, Cuba, and other parts of the tropics having seasonal rainfall, these iron oxides form thick red to blackish regolith called *laterite* (lăt′ ĕr -īte; Lat. *later,* "brick"), because the substance has been cut and used as building bricks. In very mature laterites these oxides are so concentrated and plant food is so deficient that even with abundant rainfall the vegetation grows poorly. In some laterites the concentration and volume of iron oxides make valuable ore deposits.

Laterite, then, is *a reddish residual product of tropical weathering, rich in oxides of iron and aluminum.* It is not a mineral nor is it a complete soil. It can be regarded as part of the B horizon of a peculiar sort of soil profile; it is overlain by a thoroughly leached A horizon, which contains quartz grains if the parent bedrock is granite or some other quartz-bearing rock. Laterite is soft and earthy when formed but becomes extremely hard on exposure. In many places its hard, crustlike layer is 5m or more thick. Manganese and nickel, as well as iron, are metallic concentrates in some laterites, reflecting differences in composition of the parent rocks.

Another kind of deeply weathered regolith, in the humid tropics and subtropics, that contains little iron but large volumes of hydrous aluminum oxide is used widely as an ore of aluminum (Fig. 23-21). This material, known as ***bauxite,*** is not a mineral but *a mixture of hydrated oxides,* usually *expressed by the generalized chemical formula* $Al_2O_3 \cdot nH_2O$. Probably the parent material was a rock consisting largely of feldspar.

Rate of soil formation. Although the making of soil constitutes a part of the complex process of weathering, soil formation and weathering are not synonymous. The time factor in weathering chiefly concerns the decomposition of bedrock, which involves great lengths of time and is essentially a geologic process. The length of time required to form an orderly series of soil horizons in regolith involves far shorter periods, which may be considered primarily in agricultural terms. Yet here, too, little definite evidence exists on which to base any convincing estimate of rate. Soil scientists are sure that acid soils in humid regions are developed as rapidly as any, and, from their studies, they hazard the guess that on loose sandy material (not bedrock) with a forest cover a fair soil might develop in 100 to 200 years.

ROLE OF WEATHERING IN THE GEOLOGIC CYCLE

The group of processes that together constitute weathering are part and parcel of erosion and can not be separated from it. This fact is emphasized in the definition of erosion given in Chapter 1. Indeed weathering occupies a key place in the grand scheme envisaged by James Hutton and within that scheme forms a conspicuous segment of the rock cycle and also a segment of the geochemical cycle.

The place where weathering occurs is like the post office of origin in a worldwide sorting system, where outgoing pieces of mail get their first sorting and are sent off in various directions, only to be sorted again and again as they approach their destinations. In the sorting system begun by weathering, particles are detached or extracted from rocks and are sent on their way down the nearest slope, into a stream, and so onward into a basin, where they are deposited. At every step along the way the particles are sorted according to their size, their weight, their shape, their durability, and other factors. Some are deposited at various points en route; others make it to the end of the particular system through which they are moving—a basin of some kind.

The products of chemical weathering are more stable than are the materials of rocks that have not been weathered. We could state this differently by saying that weathering results in an increased adjustment of rock materials to environments at the Earth's surface. This is because more and more of the unweathered rock, including minerals possessing various degrees of stability, give place to chemically weathered products that are quite stable with respect to the surface environment. Of these products the most abundant are the clay minerals. Broadly speaking, then, chemical weathering is a huge clay-making process. It is also a process in which quartz is unlocked from granites and other quartz-bearing rocks; the quartz, being comparatively stable, persists in an unchanged state. The two most conspicuous products of weathering, new-made clay and old (but now free) quartz, go on into the next phase of the rock cycle, are transported and deposited, and eventually are transformed into claystone and sandstone.

Meanwhile, other products of weathering are the cations Na^+, K^+, Ca^{2+}, Mg^{2+}. Most of the K^+ goes immediately into plants and certain clay minerals. The other cations move slowly through the ground in solution (Chap. 11) and emerge at all points along the line into streams and so eventually reach the sea. There Ca^{2+} and Mg^{2+} are deposited, mostly as carbonates, eventually to form the limestone and dolostone that with claystone and sandstone constitute the common sedimentary rocks. The Na^+ mostly remains in solution.

In the final analysis, therefore, weathering is one of two great links between rocks and the sediments derived from them. The other link consists of the processes that transport weathered detritus from the places of weathering to the places of deposition. Having examined weathering, we shall consider the transporting processes in the chapters that follow.

Summary

Environment

1. The ground surface and a shallow zone beneath it represent an environment of low temperature and pressure, in which water and organic matter are present.

Processes

2. Chemical weathering and mechanical weathering are very different but generally work in close cooperation.

3. Subdivision of large blocks into smaller particles increases surface area and thereby accelerates weathering.

Chemical weathering

4. Carbonic acid is the prime agent of chemical weathering; heat and moisture speed chemical reactions.

5. Minerals are decomposed at rates distinctive for each mineral. While feldspars and some other minerals are converted into clay, their soluble cations are removed in solution. Grains of quartz, however, are resistant and tend to accumulate.

Mechanical weathering

6. The most obvious process of mechanical weathering is frost wedging, in which mineral composition is not altered.

7. Decomposition, however, is one of the causes of mechanical weathering.

Soils

8. A soil profile consists of horizons A, B, and C.

9. In moist climates, profiles of mature soils are similar even though they are developed in bedrock of different kinds.

10. Pedalfers form in areas of high rainfall; pedocals in areas of low rainfall.

11. Laterite and bauxite, rich in iron and alumina, form in tropical climates.

Effect of biosphere

12. In many processes of weathering the biosphere plays a significant part.

The geologic cycle

13. Weathering transforms bedrock into residual regolith, the chief source of sediments, which eventually become sedimentary rocks.

14. Weathering, as a part of the geochemical cycle, is the source of sodium in the sea and of calcium in limestone.

Selected References

Bradley, W. C., 1963, Large-scale exfoliation in massive sandstones of the Colorado Plateau: Geol. Soc. America Bull., v. 74, p. 519–528.

Goldich, S. S., 1938, A study in rock weathering: Jour. Geology, v. 46, p. 17–58.

Jenny, Hans, 1950, Origin of soils, *in* Trask, P. D., Applied sedimentation, a symposium: New York, John Wiley, p. 41–61.

Keller, W. D., 1957, The principles of chemical weathering: Columbia, Mo., Lucas Bros.

Lyon, T. L., Buckman, H. O., and Brady, N. C., 1960, The nature and properties of soils, 6th ed.: New York, Macmillan.

Reiche, Parry, 1950, A survey of weathering processes and products, rev. ed.: Albuquerque, Univ. of New Mexico Publications in Geology.

U. S. Department of Agriculture, 1938, Soils and men: Yearbook for 1938: Washington, U. S. Gov. Printing Office.

U. S. Department of Agriculture, 1957, Soil: Yearbook for 1957: Washington, U. S. Gov. Printing Office.

CHAPTER 8

Mass-Wasting

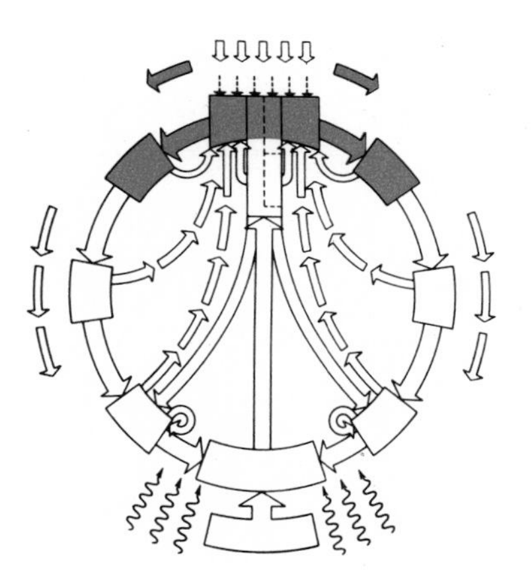

Analysis of movement
Slope angles
Mass-wasting as an open system
Processes
Products
Practical applications

Plate 8

Slump and debris flow,
near Oakland, California, December 9, 1950.
The slopes, although steep, are exaggerated somewhat
by the angle of view. The whole feature is more
than 700 feet long. (Bill Young, from *San Francisco Chronicle.*)

CHAPTER 8 *Mass-Wasting*

DOWNSLOPE MOVEMENT

Universality of mass-wasting. Wherever slopes exist, regolith is displaced downward along them by gravity. In Chapter 7 we read that weathering breaks bedrock down into regolith that can be moved away to lower places by carriers such as streams. However, most regolith is so situated that in order to reach a carrier it must first move downslope (Fig. 8-1). The trip downslope, constituting a phase of the rock cycle distinct from the phase represented by weathering, is caused by a whole series of activities, related to one another as complexly as are those involved in weathering. Although movement ranges from extremely slow to extremely fast, it is controlled primarily by gravity. The activities are known collectively as ***mass-wasting,*** *the gravitative movement of rock debris downslope, without the aid of a flowing medium of transport such as air at ordinary pressure, water, or glacier ice.*

Mass-wasting affects whole bodies of regolith, wide blankets or narrow streams of rock particles large or small, moving down hillsides, cliffs, and other slopes. But at times single particles, small boulders for instance, are detached from a cliff and fall, roll, or bounce downslope. This, too, is mass-wasting, because such movement depends on the force of gravity directly, rather than indirectly through a carrying medium.

Mass-wasting is not confined to the lands. Regolith, in the form of transported sediments, covers vast areas of the sea floor and there in places overlies bedrock. Evidence from sediment cores and other sources shows that such material moves down submarine slopes by gravity today (Fig. 15-15) and has done so in the past. Because such movements are closely analogous to mass-wasting, we think of mass movement as a universal process, submarine as well as subaerial, active wherever slopes exist.

Analysis of movement. Mass-wasting depends on the interplay of the components of gravity g_n and g_t (Fig. 2-8). When g_t exceeds g_n reinforced by cohesion and friction among the particles of rock, movement occurs. Cohesion among rock particles is the result of two forces, electrostatic forces and capillary tension, both of which tend to hold particles together. In gravel and coarse sand, the particles, and the pore spaces between them, are so large that neither of the cohesive forces can act effectively; hence, such materials lack cohesion. This is why dry, coarse sand falls apart when it is dug. But as grain size decreases, pore space increases, surface area increases conspicuously (Fig. 7-5), and cohesion can reach large values. This is why silt and clay, when dug, hold together to form chunks or clods.

When the forces of cohesion and friction that keep particles of rock from moving on a slope are reduced by factors, such as saturation with water, earthquakes, or mechanical weathering, to a value less than the stress (force per unit area) applied by g_t to the surface material on the slope, the particles move downslope. At times we can actually see the movement. But even where movement is slow we can identify, in the mass-wasting debris, particles derived from kinds of bedrock exposed only in the upslope direction; so we know that movement has occurred.

When subjected to the influence of g_t through a period of time, regolith develops small stresses and flows plastically (Chap. 3). In such movement the particles that constitute the framework of the regolith undergo no internal change but shift and rotate slowly with respect to their neighbors. Parts of the regolith flow downslope as their equilibrium is disturbed by various factors, some of which are discussed in the section on *creep.* Rates of motion are slow, partly because capillary tension contributes strongly to cohesion where films of water adhere to the sides of fine particles and air-filled spaces remain between them. But as the spaces

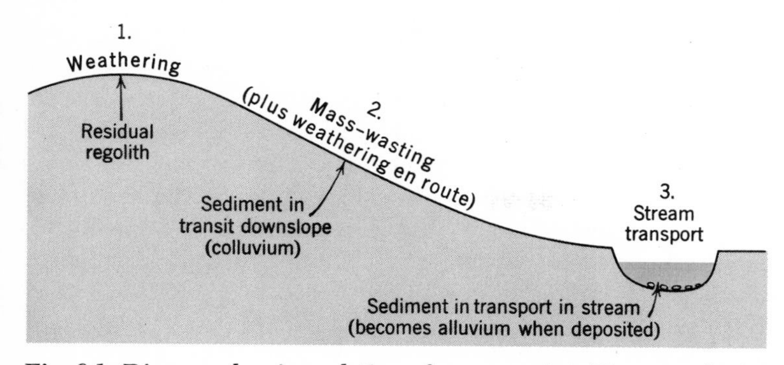

Fig. 8-1. Diagram showing relation of mass-wasting (2) to weathering (1) and to stream transport (3).

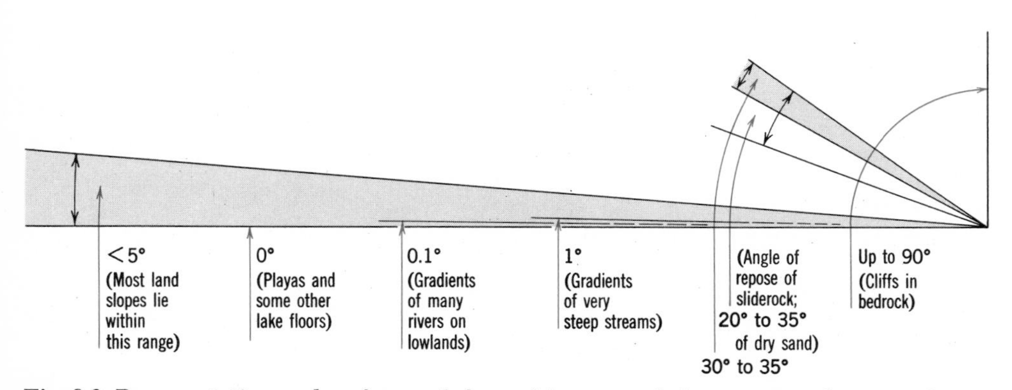

Fig. 8-2. Representative angles of natural slopes. Most natural slopes are gentle; steep slopes are rare.

become filled with water following a heavy rainfall, the pressure of the water destroys capillary cohesion and forces apart the rock particles, even to the extent of lifting them. As a result, flow becomes much more rapid, resembling the flow of a fluid. Thus the material on a slope can move at different rates at different times, depending on changes in its water content. This is the main reason why some mass-wasting activities are especially common after protracted rains. The differences between the two activities called debris flow and mudflow, described hereafter, are related to the amount of water present in the moving mass.

Slope angles. Mass-wasting is well-nigh universal because the configuration of the Earth's land areas (and sea floors too) consists almost entirely of slopes, mostly arranged in systematic groups (Pl. 10). Natural flat surfaces that are truly horizontal are very rare. Furthermore most land slopes are gentle; probably the majority measure less than about 5° from the horizontal (Fig. 8-2). Although steep slopes are conspicuous, they are rare in terms of the amount of land surface they represent.

Most aggregates of loose, coarse particles remain stable at slope angles of 34 to 37°; at steeper angles the forces tending to hold them together are generally exceeded by the force of gravity, because on such slopes the value of g_t is high. *The steepest angle, measured from the horizontal, at which a material remains stable* is its ***angle of repose,*** or *critical slope.* If the slope is steepened, for example by addition of material to its upper part or by removal from its lower part, the material becomes unstable and shifts downslope just enough to restore stability by recreating the angle of repose. This readjustment is seen going on in Fig. 8-6.

In a pile of material consisting of mixed sizes, such as sand and pebbles, momentum may carry pebbles all the way to the bottom of the slope, whereas sand grains moving downward are trapped in the large spaces between the pebbles. This amounts to a very elementary kind of sorting, but as stated earlier, mass-wasting sorts sediment either very little or not at all.

When earth-moving equipment makes a steep-sided excavation in gravel, the material soon rolls down to a gentler slope, whereas silt and clay, being strongly cohesive, tend to retain steep slopes that are cut into them. When attacked by the runoff from rain, however, steep slopes of silt and clay are unstable and in wet climates are soon reduced to smaller angles.

Mass-wasting as an open system. The best way to visualize mass-wasting as a whole is to think of it as one of the natural open systems described in Chapter 3, one that includes many gravity-controlled activities operating on slopes. In this system, the input consists of the solid products of weathering contributed all the way down the slope; the output consists of sediment discharged, at the bottom of the slope, into carriers such as streams. As soon as regolith begins to move downslope, it becomes sediment by definition. Equilibrium—or, better, a steady state—in the system is represented by a balance between input and output, and also by alteration of the slope itself, to an angle that will just permit the quantity of regolith moving down it to maintain that balance.

RESEARCH STUDIES OF MASS-WASTING

The broad process of mass-wasting can be analyzed and subdivided into a varied group of subprocesses. Although many of these differ greatly from one another, they all share one characteristic: they occur on slopes. Many of the subprocesses (which hereafter we will simply call processes) were recognized even before the beginning of the present century, but study of them was accelerated by the soil-conservation movement in North America, which gained impetus in the early 1930s. Scientists began to give close attention to the soil of farmlands and pastures and, hence, to the regolith as a whole. Their first objective was to help reduce rates of soil erosion by running water (Fig. 10-5) and wind. Such rates had been increasing alarmingly, partly through lack of understanding of the various factors that determine the stability of soil on slopes.

An additional impetus to research on mass-wasting was recognition that understanding of mass-wasting could help solve problems in engineering. Some important engineering questions are: How stable are the foundations of proposed buildings, dams, and other structures built on or at the bases of slopes of various angles and in various materials? If unstable, what kind of engineering treatment will cause stability, or should the proposed structure be built elsewhere?

CLASSIFICATION OF PROCESSES

All these studies led to attempts to define and classify the processes of mass-wasting, a task more difficult than might seem at first sight. It is easy to describe a few simple processes like some of those shown in Table 8-1, but extremely difficult to analyze the motion in a rock avalanche, a process that does not appear in the table because our knowledge is inadequate to enable us to draw a meaningful picture of it.

It was recognized early that some processes are so slow that measurements of their velocity must be made over a period of years, whereas others are comparable in speed with the takeoff and landing speeds of airplanes. In Fig. 8-3 measured rates of movement in each of several processes of mass-wasting have been plotted, together with measured rates of flow of streams, glaciers, ground water, and so forth, for comparison. Although based on very few data, the plot shows that, in terms of speed, rock avalanche is in a class by itself, exceeding rivers by an order of magnitude.

It would be satisfying to be able to classify the various processes entirely according to the kind of motion each displays. This is not possible—at least not yet—because some processes involve two or more quite distinct kinds of motion. Again, it would help if we could classify the processes according to their velocities. But one can not, because the velocity of a single process at a single locality can vary with time. Even the gaps between the bars in Fig. 8-3 may be illusory, because some bars are drawn from meager data. Therefore the best we can do in our present inadequate state of knowledge is to describe several of the processes, pointing out—and in some cases only suggesting—the kinds of movement that occur.

Let us take up the more rapid processes first, and then the slower ones. In detail, however, the order of our discussion bears little relation to velocity; it

TABLE 8-1. PROCESSES INVOLVED IN SELECTED KINDS OF MASS-WASTING

Process	Definition and Characteristics	Illustration
Rockfall and debris fall	*The rapid descent of a rock mass, vertically from a cliff or by leaps down a slope.* The chief means by which taluses are maintained.	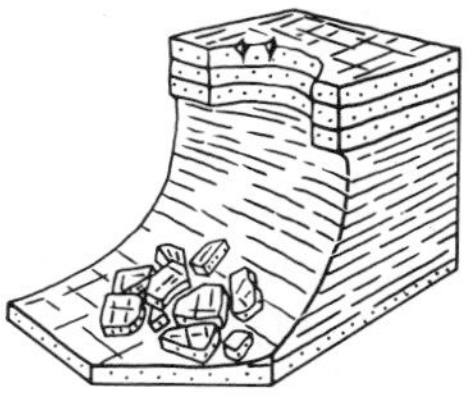
Rockslide and debris slide	*The rapid, sliding descent of a rock mass down a slope.* Commonly forms heaps and confused, irregular masses of rubble.	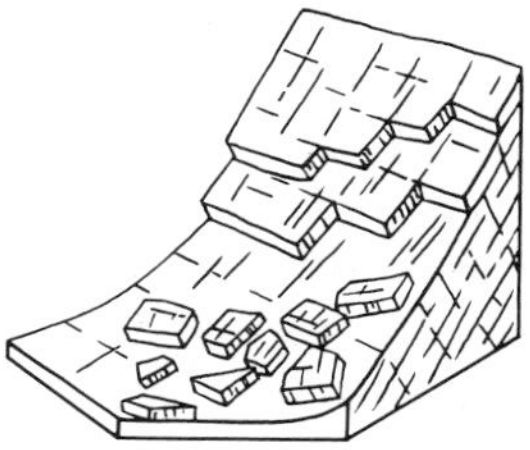
Slump	*The downward slipping of a coherent body of rock or regolith along a curved surface of rupture.* The original surface of the slumped mass, and any flat-lying planes in it, become rotated as they slide downward. The movement creates a scarp facing downslope.	
Debris flow	*The rapid downslope plastic flow of a mass of debris.* Commonly forms an apronlike or tonguelike area, with a very irregular surface. In some cases begins with slump at head, and concentric ridges and transverse furrows in surface of the tonguelike part.	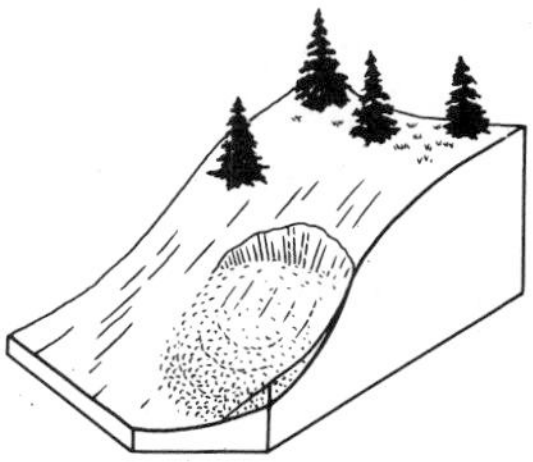
Variety: *Mudflow*	*A debris flow in which the consistency of the substance is that of mud;* generally contains a large proportion of fine particles, and a large amount of water.	

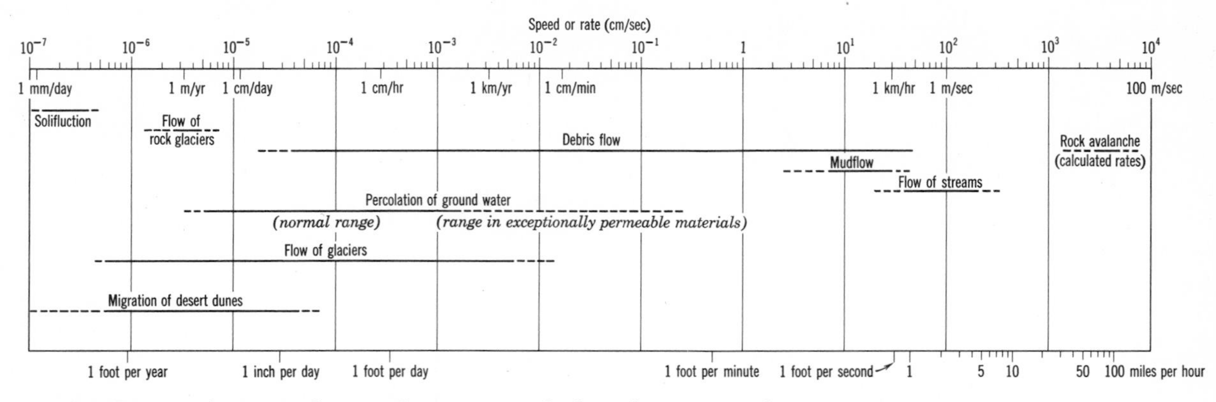

Fig. 8-3. Ranges of measured rates of movement of selected processes of mass-wasting, compared with those of some other external processes. Log scale is identical with the one used and explained in Fig. 4-11. The frame serves as a metric-English conversion scale.

has been adopted only for convenience. Some of the names of the processes we discuss are applied by some writers not only to the processes but also to the *bodies of material* that result from them. However, to avoid ambiguity we shall use these terms only for the processes and shall refer to the resulting geologic products as *deposits.*

DESCRIPTION OF PROCESSES

Falling and sliding. The sketches in Table 8-1 are almost self-explanatory. On cliffs and steep slopes rock particles of various sizes are frequently seen falling or sliding downward, singly or in groups and masses. Sometimes they represent spectacular events. One of these occurred in 1925 on the north slope of Sheep Mountain, south of Yellowstone National Park, Wyoming (Fig. 8-4). At that place clay shale underlies a thick layer of sandstone, both layers inclined toward the valley at about 20°. During heavy spring rains, water penetrated down to the shale, saturating the clay and causing it to swell. This reduced the frictional resistance between shale and sandstone so much that a wide strip of the sandstone broke away. It slid into the valley of the Gros Ventre River at the foot of the mountain and piled up as a mass of debris, damming a stream and creating a lake 4 miles long and 200 feet deep. Some of the debris was moved as much as 2 miles, through a vertical distance of 2,100 feet. Although most of the sliding sandstone broke up and formed a chaotic deposit, one enormous mass slid down without disintegrating, still covered with regolith and a pine forest.

As Fig. 8-4 shows, the strata at this place nearly parallel the steep surface slope, a geometric situation that certainly promoted sliding. Although the process here was rockslide, with probably also some rockfall (Table 8-1), down around the outer margin of the accumulated rubble there may have been some plastic deformation. More important, rock avalanche probably occurred. We shall discuss it presently.

Slump. This process, described in Table 8-1, is likewise simple. Essentially resembling the slipping movement of a fault block, slump is particularly common in places where slopes are kept steep and clifflike by erosion at their bases, as along stream banks (Fig. 8-5) and coastal cliffs (Fig. 8-6).

Rock avalanche. Early in 1903 a remarkable event destroyed much of the coal-mining town of Frank, Alberta and killed 70 people. With a great roar, a mass of rock having a volume of some 40 million cubic yards rushed down the face of Turtle Mountain, 3,000 feet high, at a speed estimated at about 60 miles per hour. Its momentum carried parts of the moving mass across a valley 2 miles wide and 400 feet up the opposite side. Later investigation showed that the configuration of Turtle Mountain almost guaranteed an unstable situation (Fig. 8-7). Joints in the limestone were inclined toward the valley and were being enlarged and extended by solution and frost wedging. When the rock along joints crossing the limestone layers, in the position of the line of section, becomes further weakened, the mass of rock above the line will be in serious danger of sudden movement, in another event similar to that of 1903.

Formerly the Frank event was referred to as a

"landslide," a term we do not use in this discussion because it is ambiguous, implying that only sliding occurred. Rockslide and rockfall certainly were involved, but the dominant process is thought to have been *rock avalanche,* in which the dry, moving debris actually flows. The high velocity attained and the fact that debris was carried uphill in places suggest that mixtures of rock particles and air flowed on a cushion of air trapped and compressed within and beneath the rushing mass. Apparently air pressure forces the moving rock particles apart, exercising a lifting force that overcomes the force g_{n} and the mass may become fluidized. Where the heights involved permit great acceleration, velocities well in excess of 100 miles per hour are thought to be possible.

Possibly the Gros Ventre event also consisted in part of rock avalanche, but the data on it do not enable us to be sure. The Vaiont event, which we shall now examine, is an even more likely candidate for this category.

Vaiont event. A colossal mass-wasting event that occurred in 1963 attained disaster proportions, because in it 2,600 people lost their lives. This occurrence was in the deep Vaiont Canyon, which traverses mountainous country in the Italian Alps

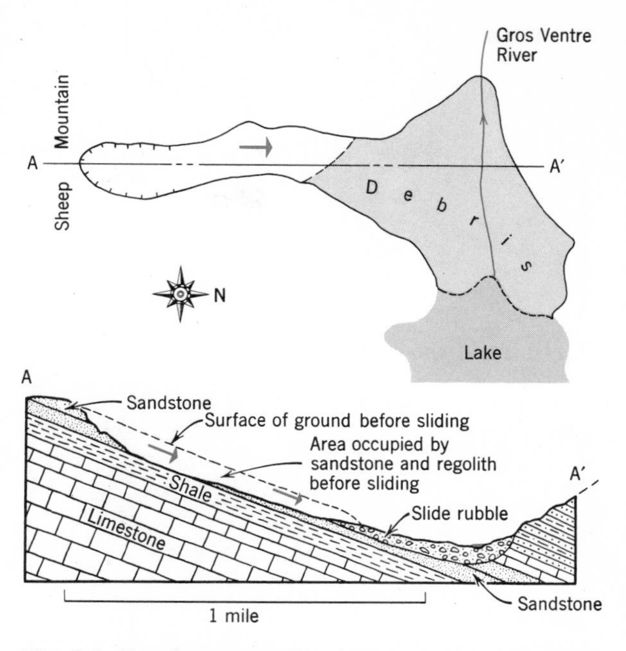

Fig. 8-4. Sketch map, profile, and section of the results of the event at Gros Ventre, Wyoming. Dark blue arrows show direction of sliding. The area of rubble shown in the section does not match the area of former rock up the slope because, as the map shows, much of the rubble was spread both up and down the valley. (After W. C. Alden, 1928.)

Fig. 8-5. Slump in regolith, consisting of unconsolidated silt, clay, and sand, in the bank of a stream. The forward part of the slumped mass has lost its former cohesion and has started to flow. Slump was started by stream erosion, which steepened the bank, creating instability. Spartanburg, South Carolina. (C. F. S. Sharpe, U. S. Soil Conservation Service.)

Fig. 8-6. Slump in coastal cliffs, Point Firmin, California, 1941. Principal surface of slip is curved—cuts cliff and displaces highway at two points. Other surfaces of slip, also curved, are shown by arrows. Erosion of foot of cliff by surf may have been the chief factor in removing support and causing movement. (Spence Air Photos.)

(Fig. 8-8). A great dam, 265m high, had been built in the canyon in 1960, impounding a deep reservoir. On October 9, at 10:41 P.M., a tremendous body of rubble 1.8km long and 1.6km wide, consisting of an estimated $240 \times 10^6 m^3$ of rock debris, thundered down the slopes on the south side of the canyon, setting up seismic waves (Chap. 18) that were recorded throughout much of Europe. Within about 1 minute, debris had filled the reservoir and had piled up to more than 150m higher than the water surface. Air compressed by the rapidly flowing debris moved water and rock material more than 260m up the opposite (northern) side of the canyon.

Had the reservoir not been present, probably the disaster would have been far less serious. But the sudden filling with debris displaced a huge volume of water, which struck the dam in waves so large that they overtopped it by 100m. Although the dam itself withstood the impact, the water swept on as a flood, destroying every structure down the valley through 20km or more, all within a period of 7 minutes.

Thorough investigation showed that two conditions had contributed greatly to the movement. One was an unfavorable geologic situation; the other was the presence of the reservoir. In other words, both natural conditions and human activity were involved. The bedrock consists of layers of limestone and claystone, weakened by much ancient deformation that has bent the layers and created many joints. Back in prehistoric times these conditions had caused earlier mass-wasting on a large scale, as inferred from debris. Then, in 1960, when the reservoir was first filled, water percolated from it into the rock of the valley sides, saturating the rock and changing the environment below ground. Moistening of rock layers previously dry caused the clay to increase in volume and become plastic, weakening the rocks. In 1960 rockslide occurred on a small scale, reflecting the deteriorating conditions.

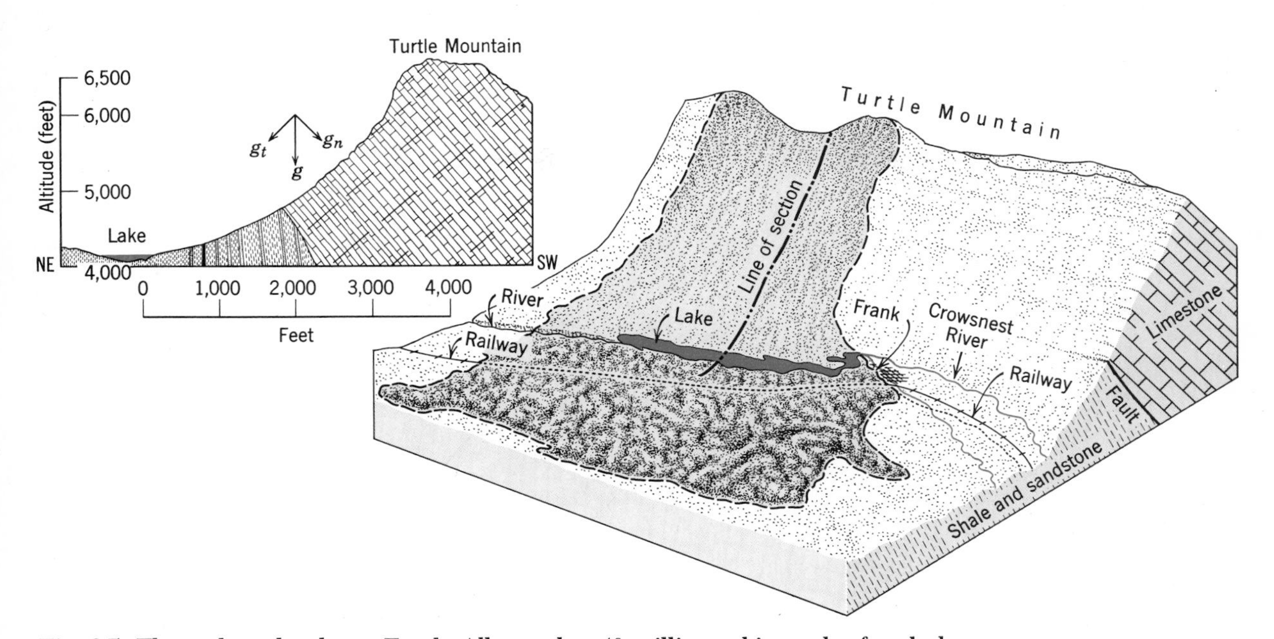

Fig. 8-7. The rock avalanche at Frank, Alberta shot 40 million cubic yards of rock downward 3,000 feet to the base of Turtle Mountain, dammed a river to form a lake, destroyed 7,000 feet of railway line and part of the town of Frank, and spread a great apron of debris up the gentle slope beyond, reaching as high as 400 feet above the lake. Dotted line shows segment of railway that was destroyed. Section at left was made along a line running directly down the mountain slope north of the avalanche area. (Data from Geol. Survey of Canada.)

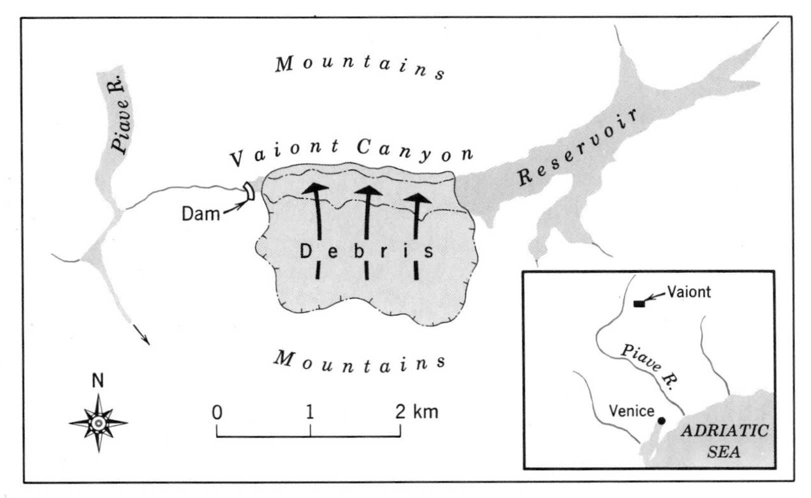

Fig. 8-8. Sketch map showing setting of the Vaiont event in 1963.

During the period between the 1960 and 1963 events, surveys showed that the ground was moving downslope at a rate of around 1cm/week. In September 1963 heavy rains lasting 2 weeks intensified the movement, which increased to about 1cm/day and finally ended in catastrophe.

The activity in the Vaiont event was complex. Probably the slow movement between 1960 and 1963 was principally rockslide accompanied by debris flow, but the 1963 catastrophe was mainly rock avalanche, as indicated by the evidence of the action of compressed air.

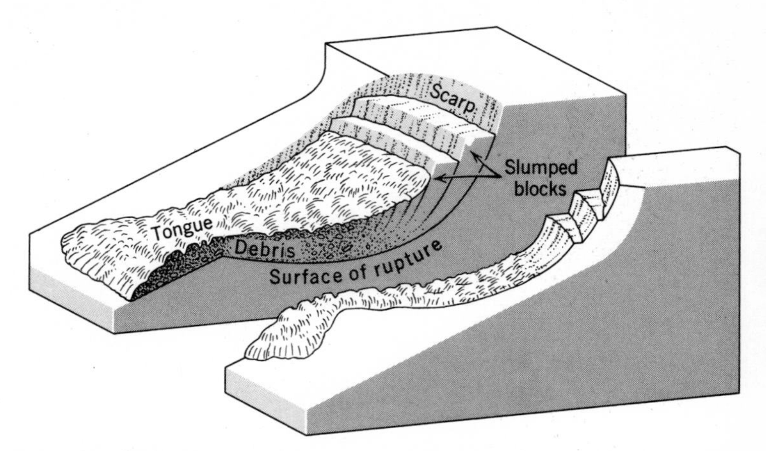

Fig. 8-9. Idealized sketch showing chief characteristics of many debris flows. Compare Plate 8.

Debris Flow. Many if not most of the conspicuous events in mass-wasting involve debris flow (Table 8-1). In many of them movement begins with slump and continues as debris flow. This happens as a result of rapid breakup of the blocklike slump masses and is, of course, promoted by the presence of water in the moving material. Figure 8-9 shows the idealized relationship between slump and debris flow, with well-defined masses, created by slump, breaking up into a single flowing mass that is ordinarily tonguelike in shape. Many such features are 2 to 5km in length, although some are larger and others much smaller. Plate 8 shows the same features: slumped blocks at head, tonguelike flowing mass downslope.

Nicolet debris flow. A clear example of debris flow combined with slump occurred at Nicolet, Quebec, a town built on the flat surface of a body of unconsolidated silt and clay. The Nicolet River had cut a cliff in this material, and late in 1955, in the course of widening a road along its base, the cliff was steepened artificially. Following heavy rains in November of that year, the mass of sediment became unstable, and on November 12, at 11:40 A.M., slump like that shown in Fig. 8-7 occurred back of the cliff. The slump blocks immediately broke up and flowed toward the river as a mass. As rapidly as the flowing away of each mass weakened the support of the slump scarp, further slump occurred, and the blocks in turn were destroyed by flow. Undermined buildings and pavements disintegrated and were incorporated in the flowing mass. The whole event lasted 7 minutes. Yet despite all this activity only three lives were lost.

The movement stopped because the long gentle slope from scarp to river, created by debris flow, had restored stability. The repeated slump destroyed a garage, a school, and several houses but stopped just short of the cathedral, which was left intact. The final result was a bite 500 feet long, 300 feet wide, and 10 to 20 feet deep taken out of the town (Fig. 8-10). The bite had a concave scarp at its head and was floored with a lobate tongue of flowed debris more than 800 feet long, which dammed the river temporarily.

A 7-minute movement that transports material through 500 feet would have a velocity of about 70 feet per minute or more than 1 foot per second. Such a rate falls at the extreme right end of the bar labeled "Debris flow" in Fig. 8-3. It completely overlaps the shorter bar labeled "Mudflow," which, however, originates in a different way.

Mudflow. A variety of debris flow, mudflow is characterized by the presence of predominantly fine particles and a water content that can amount to as much as 30 per cent. It does not originate in marked slump. As a result of its fine grain size and large water content, flowing mud tends to follow valleys as streams do. Of course, rates of flow vary with water content, slope, and other factors, but as Fig. 8-3 shows, measured rates approach and even equal those of rivers.

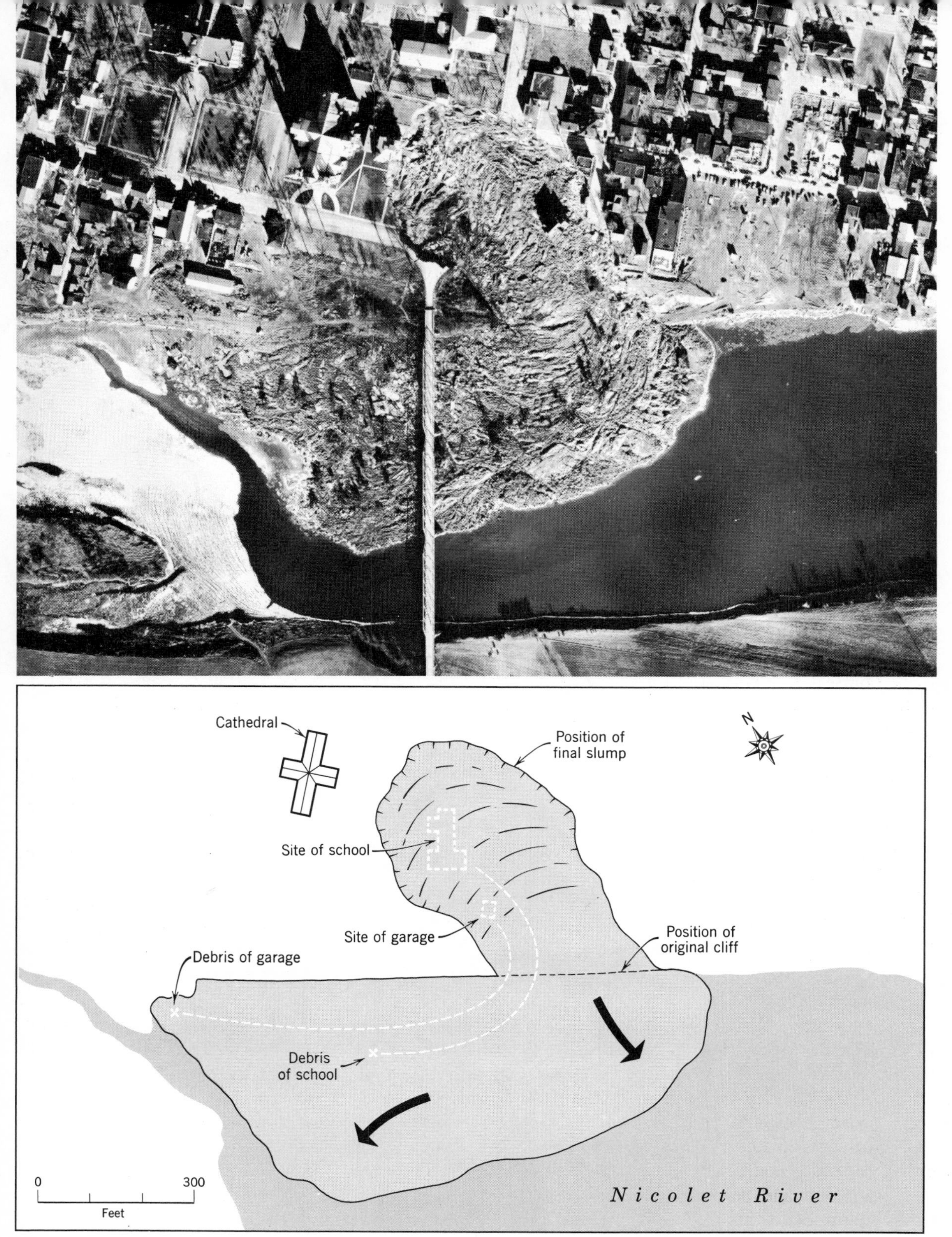

Fig. 8-10. Vertical view from the air, showing result of the Nicolet, Quebec debris flow in November 1955, a few days after the event. The outer edge of the displaced mass is poorly defined because part of it has been washed over by the river. (Spartan Air Services, Ltd.)

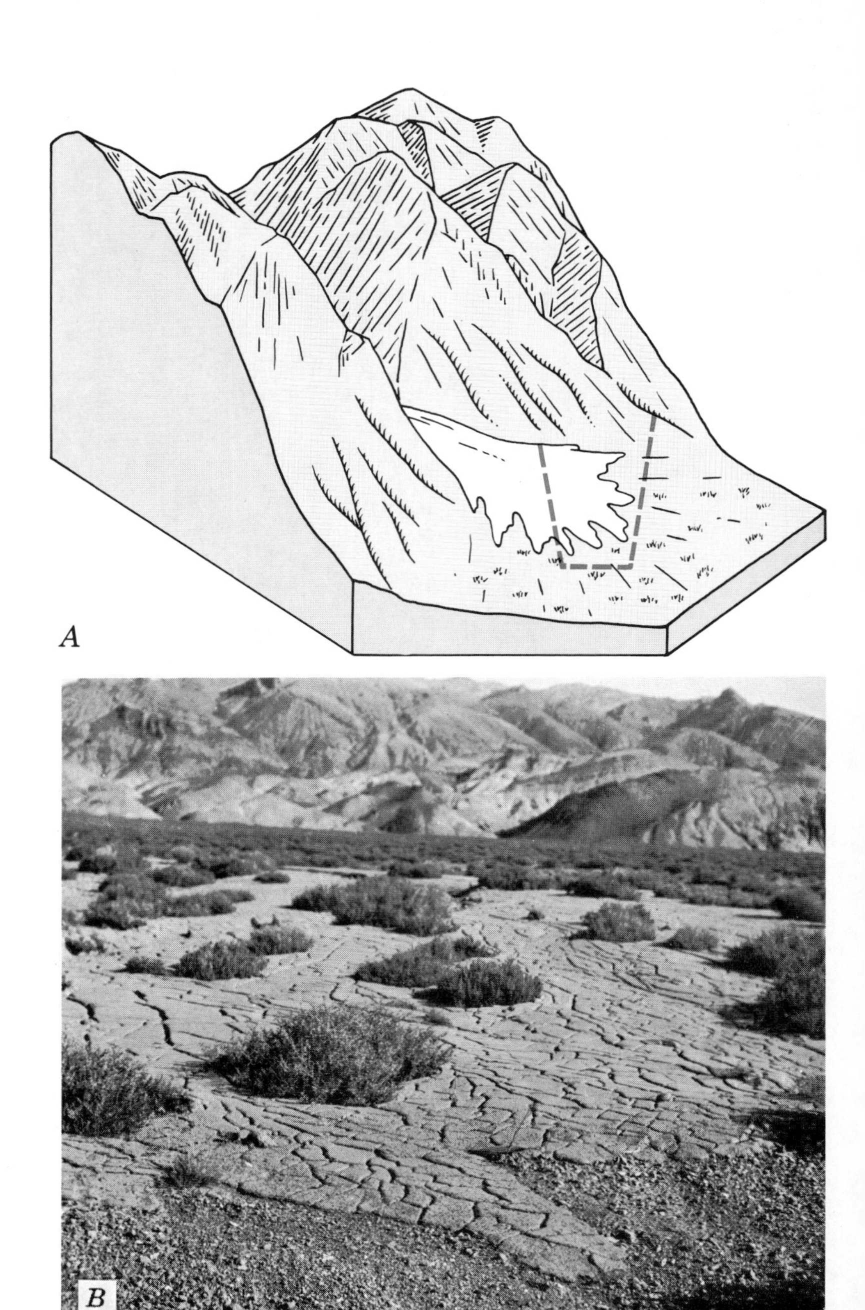

Fig. 8-11. Mudflow. *A.* Typical mudflow setting, at base of mountains in an arid climate. Dashed line shows area like that in *B.* *B.* Thin mudflow sediment at its downslope limit, forming small lobes on a slightly uneven surface of gravel. Mud has shrunk during drying and has been split by cracks. East base of Stillwater Range, Nevada. (Eliot Blackwelder.)

Fig. 8-12. Vertical section of sediment of volcanic mudflow, Mud Mountain, east of Tacoma, Washington. Large stones are concentrated near base, marked by small pickaxe. Mudflow sediment overlies stratified sand deposited by a stream before the mudflow occurred. Logs found in mudflow sediment yielded a C^{14} age of about 4,800 years. (D. R. Crandell, U. S. Geol. Survey.)

Mudflow material grades from mud as stiff as freshly poured concrete to a souplike mixture not much more viscous than very muddy water. In fact, after heavy rains in mountain canyons, mudflow can start as a muddy stream that continues to pick up loose material until its front portion becomes a moving dam of mud and rubble, extending to each steep wall of the canyon and urged along by the pent-up water behind it. On reaching open country at the mountain front, the moving dam collapses, floodwater pours around and over it, and mud mixed with boulders is spread out as a wide, thin sheet (Fig. 8-11, *A*) with destructive effects to farms and towns.

Because of its great density, which enables it to move large, heavy objects, mudflow is destructive. Houses and barns in the paths of some mudflows have been carried from their foundations, and large blocks of rock are pushed along, rolling and sliding in the slimy mixture, some of them finally coming to rest on gentle slopes well beyond the foot of a mountain range. The occurrence of huge, isolated boulders in such positions is not uncommon, and has led to the erroneous inference that the boulders were carried by former glaciers or by huge stream floods. Certainly some and probably many of them are the work of mudflow, the mud having been later removed by erosion.

Such removal is not difficult, because the sediment deposited by mudflow is seldom thick at its lower end—commonly only a few centimeters (Fig. 8-11, *B*) and rarely much more than 1m, although in a narrow valley near the upper end, thickness can be far greater. Thicknesses are easily measured in exposures in the sides of stream channels. In such places thin layers of mudflow sediment sandwiched

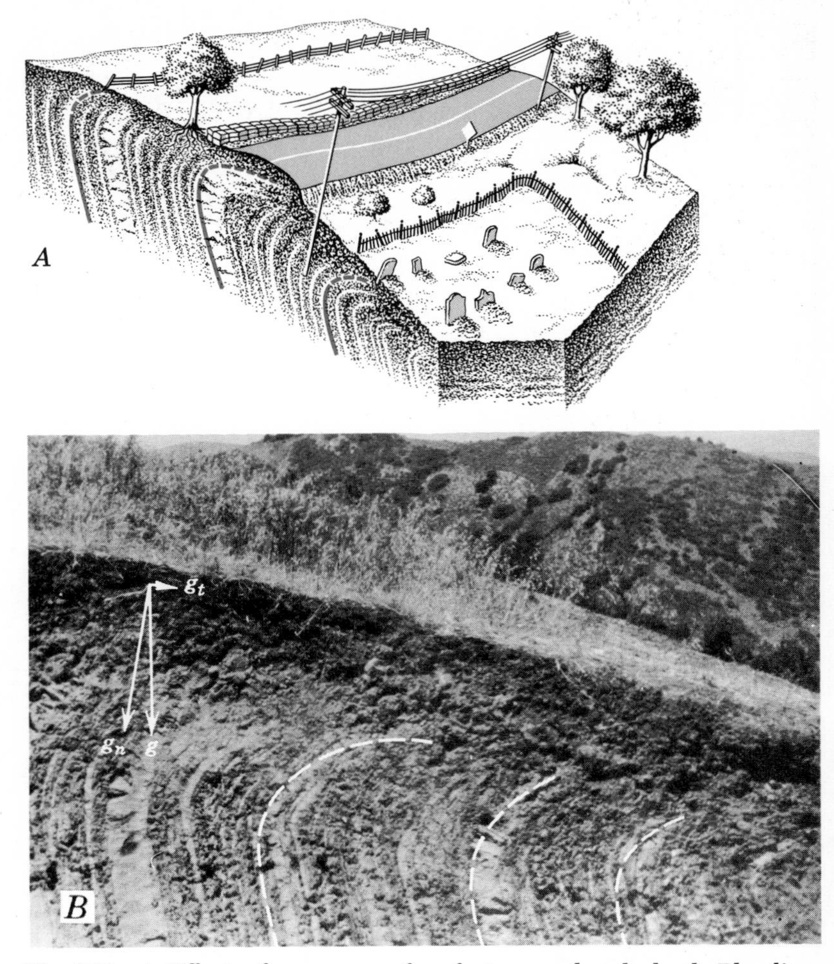

Fig. 8-13. *A.* Effects of creep on surface features and on bedrock. Blue lines emphasize bent strata. (Modified from C. F. S. Sharpe, 1938.) *B.* Regolith created by weathering at the surface of nearly vertical layers of claystone and siltstone, exposed in a road cut, Santa Monica Mountains, California. Drag by creeping regolith has bent all layers strongly downslope. Component g_t is the responsible force. Compare Fig. 2-8. (H. B. Page.)

between layers of alluvium show that mudflow, alternating with stream activity, has contributed to the deposition of sediment along the bases of mountains.

Volcanic mudflow. In regions of explosive volcanic activity, layers of tephra are common. Noncemented fine-grained tephra, mantling slopes, are peculiarly susceptible to mudflow. A sudden heavy rain can bring this loose, easily transportable material into and down valleys in huge volume, as mudflow. Where downvalley slopes are gentle this mushy material, having the consistency of fresh, fluid concrete, can flow long distances and build up thick, flat-topped deposits of sediment (Fig. 8-12) that fill the containing valley from side to side. A valley fill of this kind, on the north side of Mount Rainier in western Washington, is 45 miles long and 20 to 350 feet thick; its estimated volume is 1.5 billion cubic yards. A similar though smaller volcanic mudflow buried and destroyed the Roman city of Herculaneum (Fig. 19-17) during the eruption of Vesuvius in A.D. 79.

Creep. In contrast with those described so far, another group of mass-wasting processes moves at rates that are generally imperceptible. In this group the most widespread process is ***creep,*** *the imper-*

Fig. 8-14. Granite boulders in regolith, softened by chemical decay and deformed into pancake shapes by creep. In the lower part of the exposure (the wall of an artificial cut) boulders retain their original form. Nearer the surface of the ground the section of each deformed boulder appears as a long, thin, white lens. (S. R. Capps, U. S. Geol. Survey.)

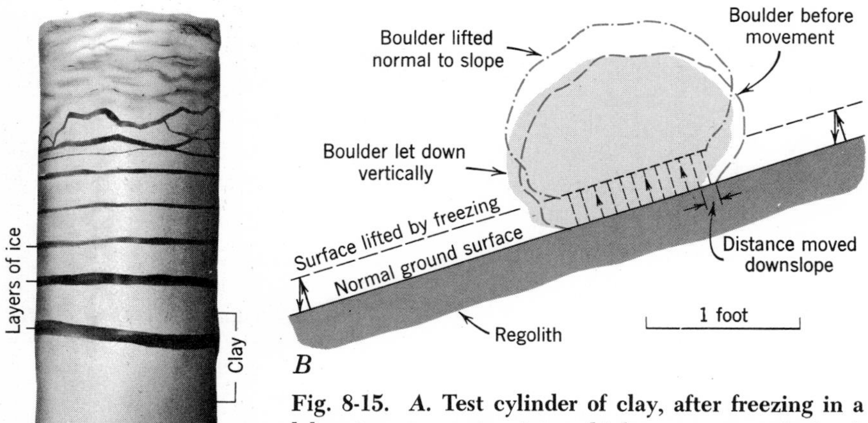

Fig. 8-15. *A.* Test cylinder of clay, after freezing in a laboratory apparatus into which water was admitted. Dark bands are layers of ice, from water drawn in during the freezing. (Stephen Taber.) *B.* A boulder moved downslope by alternate freezing and thawing. Diagram explains mechanism of one step in the movement.

ceptibly slow downslope movement of regolith. Its effects are evident in the consistent leaning of old fences, poles, and gravestones and in the derangement of roads. Cuts made into bedrock show that steeply inclined layers curve strongly downslope in the zone of weathering, and blocks of rock detached from distinctive layers have drifted downhill to considerable distances (Fig. 8-13).

A close cover of grass or other vegetation on sloping ground acts as a protective armor against the making of gullies, and the casual observer may suppose that ground so protected loses nothing to erosion except for a little mineral matter dissolved by percolating water. But closer observation shows clearly that the regolith is creeping downslope, carrying the vegetation with it. In some places boulders of granite have been softened by weathering and drawn out into long thin sheets by movement of the inclosing material (Fig. 8-14). Many factors contribute to the effectiveness of creep. In regions with cold winters, water in the pore spaces of the regolith freezes and increases in volume. Moreover, considerable water becomes segregated and frozen into layers of clear ice that separate layers of fine-grained regolith (Fig. 8-15, *A*). The ice increases the volume and, in the process, pushes up the surface

Fig. 8-16. On this slope, actively moving by solifluction, a line of cone targets, each with a peg extending down into the regolith, was set up two years earlier along a straight line marked by string. The targets have been displaced in the downslope direction, through a maximum of 25cm during the two-year period. Mesters Vig, northeast Greenland. (John Scully, from A. L. Washburn.)

by an amount equal to the combined thickness of the ice layers plus the volume increase in pore spaces. This *lifting of rock waste by expansion during freezing of contained water* is ***frost heaving.*** On a hillside the surface of the ground is lifted essentially at right angles to the slope; but when thawing occurs, each point tends to drop vertically and so moves downhill (Fig. 8-15, *B*). The movement consists of a complex series of zigzags and is very discontinuous.

The persistent activities that combine to make particles in the regolith creep downslope are listed in Table 8-2. Through countless repetitions on every slope the effects of these activities add up to slow persistent transport of regolith downhill. On a worldwide basis, probably all the mass-wasting visibly accomplished by rock avalanche, rockslide, and other rapid processes, no matter how spectacular, is but a small fraction of what is accomplished by creep in the same period of time.

We have spoken of creep as slow, but how slow is it? Its rate depends on steepness of slope. Rates of creep of small rock fragments on shale hillslopes in western Colorado were measured by marking individual fragments and determining their positions, over a period of years, with reference to stakes fixed in bedrock. The slopes have various degrees of steepness, but all have very little vegetation. Rates of creep of individual particles ranged from a few millimeters per year on a slope of 3° to nearly 70mm/year on a slope of 40°—so slow that they lie far to the left, outside the plot shown in Fig. 8-3. All rates proved to be directly proportional to the sines of the slope angles down which they were moving. As we shall see in later chapters, not only creep rate but also rate of flow of streams and glaciers and rate of percolation of water underground are functions of slope.

Solifluction. In cold regions, including some high mountains, another imperceptible activity carries regolith downhill. In such places the ground freezes to a great depth, and when the weather warms enough to thaw the upper part of the regolith the deeper part stays frozen. Surplus water then can not percolate downward, and the thawed layer, a few inches to several feet thick, becomes saturated and will not bear the weight of a man. On slopes the water-soaked material flows like an extremely stiff liquid. This *imperceptibly slow downslope flow of water-saturated regolith* is ***solifluction.*** This process occurs less commonly where the ground is not frozen, but where the regolith is saturated.

TABLE 8-2. CAUSAL FACTORS IN CREEP OF REGOLITH

Frost heaving	Freezing and thawing, without necessarily saturating the regolith, cause lifting and subsidence of particles
Wetting and drying	Causes expansion and contraction of clay minerals; creation and disappearance of films of water on mineral particles causes volume changes
Heating and cooling without freezing	Causes volume changes in mineral particles
Growth and decay of plants	Causes wedging, moving particles downslope; cavities formed when roots decay are filled from upslope
Activities of animals	Worms, insects, and other burrowing animals, also animals trampling the surface, displace particles
Solution	Solution of mineral matter creates voids, which tend to be filled from upslope
Activity of snow	Where a seasonal snow cover is present, it tends to creep downward and drag with it particles from the underlying surface

Fig. 8-17. Rock glaciers at the base of a cliff, Silver Basin, near Silverton, Colorado. Lobate form, steep fronts, and concentric ridges demonstrate that flow has occurred, but the bodies are not moving now. They are believed to have been active late in the last glacial age. The cliff is the headwall of a cirque (Fig. 12-9). (Whitman Cross, U. S. Geol. Survey.)

The essential difference between solifluction and most creep lies in the amount of water the regolith contains. For solifluction the material requires a very high water content; for creep it does not. Indeed creep can occur in regolith that is completely dry.

Water content large enough to eliminate capillary tension lowers the resistance of regolith to the pull of gravity enough to cause solifluction, with rates of movement generally greater than those of creep. Measured rates of solifluction (Fig. 8-16) are as great as 12cm/year.

Flow of rock glaciers. In mountains where temperatures are very low, remarkable tonguelike bodies, some of them exceeding 1km in length and 3km in width, consisting of angular, blocky boulders as well as finer particles, occur at the bases of cliffs and extend outward into valleys (Fig. 8-17). The kind of rock in the boulders is identical with that which forms the cliffs. These bodies are not the taluses described in a following section; they are too extensive and their lobate form, steep fronts, and concentric, wrinklelike ridges suggest that they are flowing or at least have flowed. Because they resemble glaciers in form, they have been given the name ***rock glacier,*** which we can define as *a lobate, steep-fronted mass of coarse, angular regolith, extending out from the feet of cliffs in mountainous regions, whose downslope movement generally is aided by interstitial water and ice.*

In rock glaciers studied in Alaska, ice occupies openings between the rock particles. Measurements made during several successive years show these masses are moving outward at the rates shown in Fig. 8-3. Boulders are fed to the rock glaciers by frost wedging, water accumulates between the boulders and in the finer material freezes, and the whole mass begins to flow like a true glacier. Nevertheless, despite its resemblance to flow as in a true glacier, the movement in a rock glacier is best classed as a process of mass-wasting.

SEDIMENTS DEPOSITED BY MASS-WASTING

Colluvium. Having reviewed the processes of mass-wasting, we are in a position to examine the sediments deposited by these processes and learn the characteristics by which we can distinguish them from those deposited by streams, wind, and other

Fig. 8-18. Aprons of coarse sliderock form coalescing taluses against the base of a basalt cliff about 200 feet high, Grand Coulee, Washington. Basalt is layered, each layer cut by thousands of vertical joints. Sliderock particles have been created by mechanical weathering controlled by the joints.

external processes. *A body of sediment that has been deposited by any process of mass-wasting or by overland flow* (Chap. 9) is ***colluvium.*** Whether the depositing process is rapid or slow, the resulting colluvium is poorly sorted and poorly stratified, or not stratified at all. These characteristics enable us to distinguish colluvium from sediments deposited in streams, lakes, and the sea, and also from windblown sediments. They do not, however, separate colluvium from glacial deposits, and the similarities between these two sediment groups have led to confusion at times. Nevertheless differences do exist, and these are mentioned in Chapter 12.

Particles transported within a fluid medium, such as flowing water or air, become worn and rounded in transit, are sorted according to size and weight, and commonly are deposited in the form of strata. Particles of colluvium are not rounded (unless they were rounded before starting to move) because, in mass-wasting, particles do not collide frequently. However, they are worn by scraping, gouging, and chipping; some large particles show short, gougelike scratches. But such modifications acquired during transport are generally not enough to mask the original shapes, which are likely to be angular if the particles constituted pieces broken from bedrock.

Furthermore colluvium lacks internal structure because its particles moved mostly as falling, sliding, rolling, or flowing masses, not *in* a flowing medium. This distinction explains why sediments resulting from falling, sliding, and rock avalanche tend to be a chaotic jumble. Even in the deposits made by debris flow and mudflow the sediments are not stratified. The nearest approach to orderly arrangement is seen in the deposits of some volcanic mudflows, in which the coarsest fragments, generally large boulders, tend to be concentrated in the lower part and the finer toward the top (Fig. 8-12). Probably this arrangement results from the slow settling of the larger, heavier units while the flowing movement is in progress.

Sliderock; taluses. Having examined the general characteristics of colluvium, we now describe two regionally common varieties, *sliderock* and *solifluction sediments.* In areas where steep cliffs prevail and weathering is dominantly mechanical, accumulations of weathered particles commonly mantle the bases of the cliffs. The particles, commonly angular and ranging in diameter from sand grains to boulders, are loosened from the bedrock of a cliff by mechanical weathering, accumulate, and move downward at various rates. *The apron of rock waste sloping outward from the cliff that supplies it is a* ***talus;*** *the*

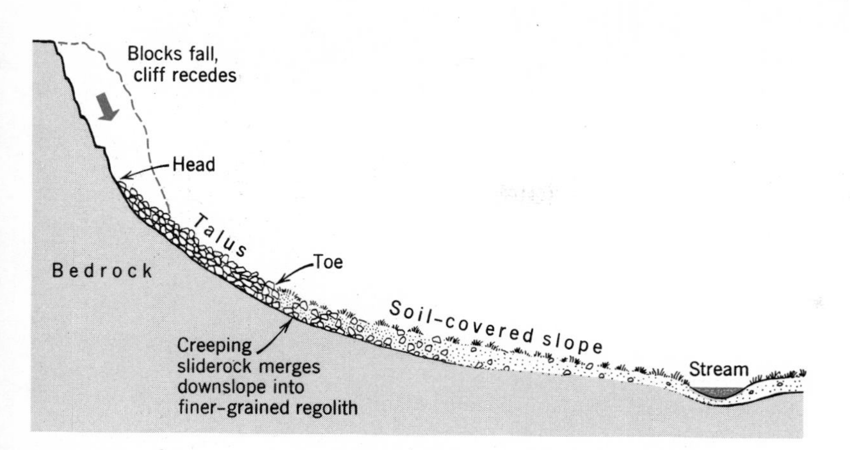

Fig. 8-19. A talus in characteristic relation to bedrock. The sliderock in this talus grades downslope into finer-grained regolith, transformed from the sliderock by mechanical and chemical weathering and added to by weathering of the bedrock beneath. Fine regolith is slowly fed to the stream, which carries it away. (After C. F. S. Sharpe, 1938.)

material composing a talus is ***sliderock*** (Fig. 8-18). From cliff to talus the movement is chiefly falling, sliding, and rolling; within the talus it consists of creep, generally so slow that the rock particles are weathered en route. Water plays little or no part in the movement. Weathering converts the sliderock into fine-grained regolith which, with its pores of extremely small diameter, can hold much more moisture than sliderock and thus acquire both vegetation and soil (Fig. 8-19).

The activity in a talus constitutes a local open system with an input of large fragments at its head, an output of generally finer fragments at its toe, and creep in progress from head to toe. The profile of the talus is concave-up, but with a radius of curvature so long that some profiles seem to be nearly straight. In each short segment down its length, the profile represents the angle of rest for the material in that segment. Because the profile as a whole is the sum of all the short segments, in each of which the sliderock is stable, we call it an *equilibrium profile*, or stability profile. In it the force of gravity and other forces, mainly friction, are balanced so nicely that the material added to each segment from upslope is just equalled by the material that moves out of that segment into the next segment downslope. For a similar reason the soil-covered slope downhill from the talus in Fig. 8-19 also is an equilibrium profile. Such profiles are characteristic not only of hillslopes but also of stream systems, wave activity along coasts, and other dynamic systems.

Solifluction sediments. The sediment that results from solifluction is in some cases crudely stratified but is easily distinguished from the well-sorted and well-stratified sediment deposited by a stream. Solifluction sediments have been found in parts of Europe, North America, and Australia where today's climate is not conducive to solifluction. Such sediments are believed to have been deposited during glacial ages (Chap. 12) when climates in middle-latitude regions were considerably colder and wetter than they are today. This illustrates the fact that if we have the ability to identify the origins of various kinds of sediments, we can reconstruct parts of geologic history.

PRACTICAL APPLICATIONS

Some knowledge of mass-wasting is essential in selecting successful locations for large buildings, bridges, dams, highways, and other engineering works. The St. Francis dam in southern California, which failed in 1928 with an appalling loss of life and property, was built in part on weak, soluble bedrock. This would have been recognized if a competent geologist or an engineer with knowledge of geology had been consulted.

Some engineering works have to be constructed despite troubles with mass-wasting. When work was begun on the deep Culebra Cut for the Panama Canal, slump and debris flow occurred on such a scale that for a time the project appeared to face defeat. It was necessary to reduce the slopes by removing enormous volumes of material beyond the original estimates. Some masses have been stabilized by building drainage tunnels to carry away water that played an important part in the movement. During construction of the Grand Coulee Dam on the Columbia River, debris flow at a critical location gave serious trouble until engineers conceived the ingenious plan of laying pipes through the wet mass and circulating a refrigerant. The ground froze and remained stable until the necessary construction was finished.

Private companies and government departments are put to enormous yearly expense in repairing and rebuilding railroads, highways, aqueducts, and other major structures damaged by rapid mass-wasting, especially in country characterized by long, steep slopes. Some of this cost could be avoided by informed planning; the rest is part of the price we pay for civilized living on an unstable Earth.

Despite the spectacularly rapid movements we have described, the part they play in the overall movement of material down the slopes of the lands is minor. Processes such as creep are slow, but they act continuously and they affect nearly all slopes. Hence they do much more work, in the aggregate, as agents of transport. With these almost universal activities in mind we can turn next to running water, the transporting agent that, sooner or later, receives and handles most of the sediment moved downhill by mass-wasting. In each of the chapters that follow we shall trace, for each external process, an analysis of the process itself, the character of the sediments it creates, and the influence of the process on the form of the Earth's surface and on the activities of man.

Summary

Downslope movement

1. Through mass-wasting, residual regolith created by weathering reaches the carrying agencies.
2. Because of the prevalence of slopes, mass-wasting is nearly universal.
3. Mass-wasting can involve both plastic deformation and fluid flow.
4. Some mass-wasting processes are promoted by increased water content of the regolith.

Processes of mass-wasting

5. Rock avalanche apparently involves fluidization of a mass of rock particles through compression of air; its motion consists partly of flow of mixed rock and air.
6. Factors that cause creep include cycles of freezing and thawing, wetting and drying, and heating and cooling; also solution, and activities of plants and animals.
7. Although rock avalanches and debris flow are spectacular, imperceptible creep and solifluction, because they are widespread, move more material.
8. Mudflow is particularly common in arid climates where rainfall is sporadic and in areas covered with fine-grained tephra.

Sediments

9. *Colluvium,* sediment deposited by mass-wasting, is generally nonsorted and either nonstratified or very poorly so. Its particles, although they may show wear, are not rounded. These characteristics distinguish colluvium from sediments deposited in water.
10. Taluses develop at the bases of cliffs; commonly the material composing taluses grade outward into finer, soil-covered regolith.

Selected References

Crandell, D. R., and Waldron, H. H., 1956, A recent volcanic mudflow of exceptional dimensions from Mt. Rainier, Washington: Am. Jour. Sci., v. 254, p. 349–362.

Daly, R. A., Miller, W. G., and Rice, G. S., 1912, Report of the Committee to Investigate Turtle Mountain, Frank, Alberta: Geol. Survey of Canada, Mem. 27.

Eckel, E. B., ed., 1958, Landslides in engineering practice: Highway Research Board, Special Report 29, National Research Council, Washington, D. C.

Howe, Ernest, 1909, Landslides in the San Juan Mountains, Colorado: U. S. Geol. Survey Prof. Paper 67.

Kiersch, G. A., 1964, Vaiont Reservoir disaster: Civil Eng., v. 34, p. 32–39.

Legget, R. F., 1962, Geology and engineering, 2nd ed.: New York, McGraw-Hill, p. 106–128, 385–443.

Sharp, R. P., and Nobles, L. H., 1953, Mudflow of 1941 at Wrightwood, southern California: Geol. Soc. America Bull., v. 64, p. 547–560.

Sharpe, C. F. S., 1938, Landslides and related phenomena: New York, Columbia Univ. Press. (Reprinted 1960 by Pageant Books, Paterson, N. J.)

Terzaghi, Karl, 1950, Mechanism of landslides: Geol. Soc. America, Berkey Volume, p. 83–123.

Wahrhaftig, Clyde, and Cox, Allan, 1959, Rock glaciers in the Alaska Range: Geol. Soc. America Bull., v. 70, p. 383–436.

Washburn, A. L., 1966, Instrumental observations of mass-wasting in the Mesters Vig district, northeast Greenland: Meddelelser om Grønland, v. 166, no. 4.

CHAPTER 9

Running Water

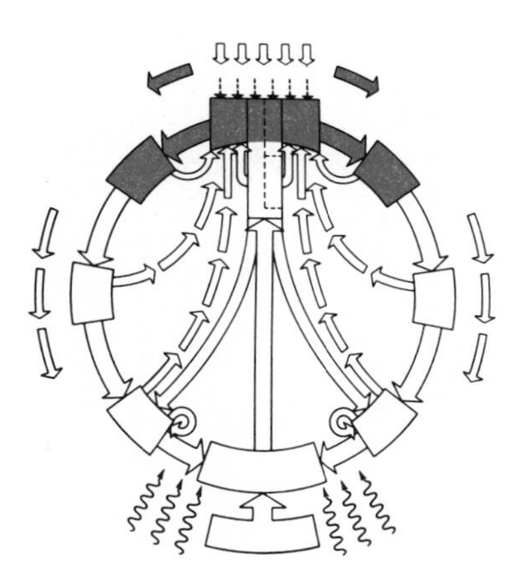

Geologic importance
Mechanics of flowing water
Behavior of rock particles
Economy of a stream
Special features of stream channels
Dynamic equilibrium in streams

Plate 9

Turbulent Potomac River at Great Falls,

upstream from Washington, D. C.

(*New York Times*.)

CHAPTER 9 *Running Water*

GEOLOGIC IMPORTANCE OF RUNNING WATER

A first look at running water. A good way to understand the important part played by streams in moving water and sediment down slopes is to sit beside a moderately rapid small stream a few yards wide and watch it. First you see that the water is flowing at different rates at various points. Out in midstream, flow is more rapid than near the banks, where eddies are usually in sight. It is a period of low water. This particular stream is clear, and its bed consists of sand and gravel. You can see the pebbles rolling and sliding intermittently along the bed, whereas sand grains now and then take low jumps.

Walk along the stream, and you will note the channel winding from side to side in a rather smooth sequence of curves. At a place where the bank is undercut by the current at the outer side of a curve, the bank is slumping into the channel. If the bank is of sand, you can watch the current distributing sand grains along the bed, washed from the slumped mass.

All this activity could have been so interesting that you did not notice the darkening sky, and so you have just time to take refuge under a large tree standing beside a plowed field, as an intense thundershower begins. For a time the tree acts as an umbrella, its leaves and branches holding back the raindrops, which do not reach the ground. But in the field beyond the tree large drops strike the ground and splash up the loose soil, creating tiny craters as they do so. If you were to step out from under the tree, where nothing is happening yet, raindrops would splash soil onto your shoes, a couple of inches above the ground surface.

At first the ground absorbs the water as it falls. But little by little the ground begins to glisten in patches, showing that it is saturated. The excess water starts to flow off over the surface toward the stream. Unless, during plowing, the field was furrowed, the water may be flowing as wide, shallow sheets, which may be slightly turbid with fine particles picked up from the soil. Where the water enters the stream it makes turbid patches that soon coalesce. But from the grassy pasture beside the plowed field the water enters the stream through widely spaced grassy gullies, essentially as little, temporary streams and is not turbid.

The difference in behavior of the water, as it flows across plowed field and pasture, is the difference between overland flow and channel flow, a distinction dealt with in the next section. Meanwhile, observe that the whole stream is swollen, is flowing faster, and has become turbid with silt and clay contributed from plowed fields, roads, and other bare areas unprotected by vegetation.

By now the leaves and branches of the tree can no longer hold back the rain, which is dripping onto the ground, though with less impact than on the unshielded ground outside. The tree being no longer an effective umbrella, you will have to make a dash for better shelter.

It will probably have occurred to you that all these activities are related and that all are responses to a coordinated group of physical principles. Observation and experiment by geologists and engineers indicate that this is indeed the case, and that stream activities are related to each other in a very complex way. These relationships are fundamental to an understanding of geology, because the stream you have been watching is one of the millions of drainageways by which water from rain and snow runs down over the land toward the sea (Fig. 9-1).

Significance of streams to society. Apart from their strictly geologic importance, streams are significant to us because (1) they are an increasing source of water for human and industrial consumption; (2) they are a relatively small but essential source of industrial energy; (3) many rivers are avenues of industrial transportation; (4) they have great recreational value; and (5) the increase of population in river valleys has created the necessity for protection against damage from stream floods and for controlling pollution created by the discharge of wastes into streams. These aspects of streams form part of the science of hydrology, but they will turn up hereafter in connection with geologic problems.

Streams as geologic agents. The significance of streams as geologic agents is threefold.

1. As movers of water from land to sea, streams are an essential part of the hydrologic cycle.
2. By far the greatest importance of streams as geologic agents lies in what they carry. From sampling the loads of many rivers in many countries it has been estimated that, every year, streams transport from the lands to the oceans about 1 billion tons of sediment mechanically, plus about 400 million tons of material in solution.
3. Streams, together with mass-wasting, are the prime sculptors of the land. That is to say, the configuration of an enormous part of the Earth's land area consists of stream valleys with intervening hills and is the work of streams and mass-wasting combined. Other agents, such as glaciers (flowing ice) and wind (flowing air), have shaped only minor parts of the lands, and even in many of those parts the principal configuration is the product of streams. This aspect of streams is the subject of Chapter 10.

RAINDROP IMPACT AND SHEET EROSION

Erosion of the land by water begins even before a distinct stream has been formed. It occurs in two ways: by impact as raindrops hit the ground, and by sheets of water that result from heavy rains.

As raindrops strike bare ground they dislodge small particles, spattering them in all directions. On a slope the result is net displacement downhill.

As Fig. 9-1 shows, rainfall that is not returned to the atmosphere by transpiration either infiltrates (sinks into) the ground or flows off over the surface as runoff. We can subdivide the runoff into ***overland flow****, the movement of runoff in broad sheets or groups of small, interconnecting rills,* and ***stream flow****, the flow of surface water between well-defined banks.* Stream flow is very obvious. Overland flow is not obvious and commonly occurs only through short distances before ending in a stream valley. But

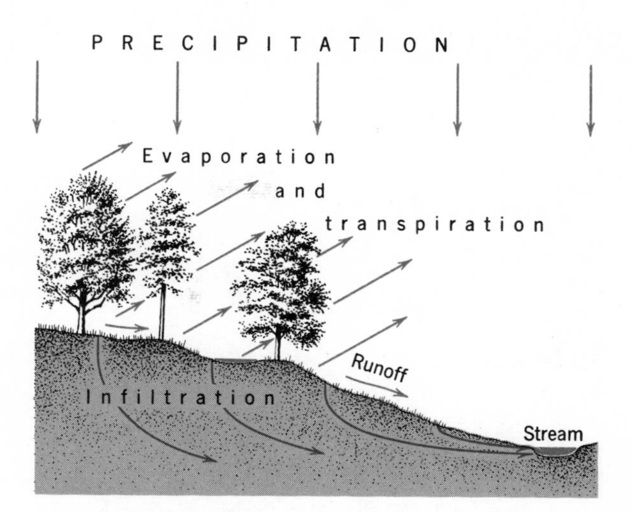

Fig. 9-1 Disposition of water precipitated on land. Water returns to the atmosphere directly by evaporation and transpiration. Other water gets into streams by infiltration into the ground and by runoff over the surface.

it takes place wherever intensity of rainfall exceeds rate of infiltration.

Water flowing downslope as broad sheets or as tiny rills is affected by the opposing forces we examined in mass-wasting. In response to g_t, water flows over rock particles and exerts a drag on them. The drag is resisted by g_n reinforced by cohesion and friction. If drag and g_t together exceed g_n plus cohesion and friction, the particles move downslope. Unlike most processes of mass-wasting, the particles move not as a mass but as individuals. Each particle is entrained in the flowing water and does not move solely as a result of g_t. *The erosion performed by overland flow is* ***sheet erosion.***

The erosive effectiveness of both raindrops and overland flow is reduced by vegetation which, where it forms a truly continuous ground covering, holds erosion to small values. But on bare sloping fields, closely grazed pastures, or areas planted with widely spaced crops, such as corn, rates of erosion can be great. Because sheet erosion creates no obvious valleys, the magnitude of its effect on soil was not fully realized until accurate measurements began to be made. Now many experiment stations (Fig. 9-2) maintained by the U. S. Department of Agriculture measure the erosion of soil (Fig. 9-3). To farmers the measurements are of far more than academic interest, for they show that sheet erosion is a menace

Fig. 9-2. Apparatus for measuring sheet erosion, Marlboro, New Jersey. Runoff, carrying soil particles, flows down a slope of 3 per cent; sediment is trapped in containers and measured by weight. A different kind of plant covers each plot. Results from a similar station are shown in Fig. 9-3. (U. S. Soil Conservation Service.)

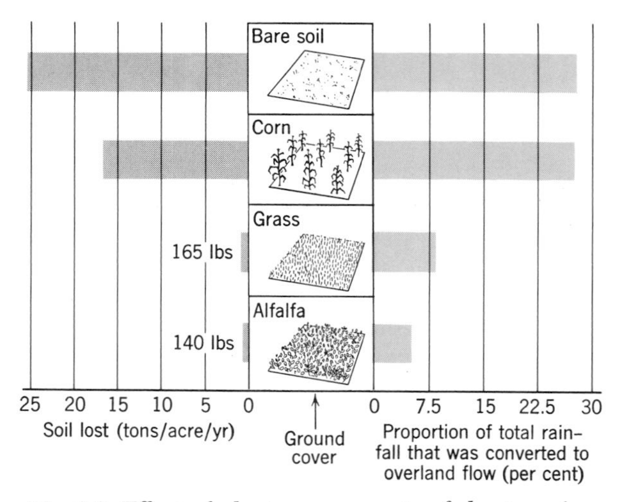

Fig. 9-3. Effect of plant cover on rate of sheet erosion, measured over 4 years at Bethany, Missouri, a station like that shown in Fig. 9-2. Soil is silty, slope is 8 per cent, and annual rainfall is 40 inches. The measurements show that grass and alfalfa, with their continuous network of roots and stems, are nearly 300 times as effective as "row crops" such as corn, in holding soil in place. Erosional loss from bare soil shown here is at a rate of about 1.5 feet/100 years.

to soil left unprotected on slopes. In recognition of this fact, wise farmers reduce areas of bare soil to a minimum and prevent the grass cover on pastures from being weakened by overgrazing. When the protective plant cover is weakened, runoff and erosion are increased (Fig. 9-4). If crops such as corn, tobacco, and cotton must be planted on a slope, strips of such crops are often alternated with strips of grass or similar plants (Fig. 10-4) that resist sheet erosion.

The examples just illustrated should not, however, lead us to suppose that erosion of slopes results wholly from human activities. This is by no means the case. Under some natural conditions, without any agriculture at all, the splash of raindrops and the work of sheet erosion are so effective that they combine to remove large volumes of fine rock particles. For example, in some subtropical grasslands all the rainfall is concentrated within a single rainy season. During the long dry season, evaporation so depletes soil moisture that grass is sparse, covering no more than 40 to 60 per cent of each square meter of ground. Although kept bare by natural causes, the soil is as vulnerable to erosion as soil laid bare

Fig. 9-4. Too many cattle were grazed on one of these two pastures in northern Kentucky, weakening the grass cover. Erosion did the rest. (U. S. Soil Conservation Service.)

by farming. Raindrop impact and overland flow contribute considerable amounts of sediment directly to streams.

Although common on slopes, overland flow takes place only through short distances. The flowing water soon concentrates into channels (as explained in Chapter 10) forming streams whose lengths may reach 3,000km or more. Most sediment transported by running water, therefore, is carried in streams, which likewise deposit sediment. Accordingly streams occupy an important place in the geologic cycle, and we must look at them rather closely.

MECHANICS OF FLOWING WATER AND PARTICLE MOVEMENT

Basic factors. A ***stream*** is *a body of water carrying rock particles and dissolved substances, and flowing down a slope along a definite path.* The path is the stream's channel, and the rock particles are an essential part of the stream itself. We can diagram it as in Fig. 9-5. In any part of its course a stream, then, is a quantity of water flowing down a slope at a certain average velocity.

A good way to begin our analysis is to experiment in the laboratory. In an artificial stream we can manipulate the variables and measure their effects on one another. To begin with we can list five basic factors:

1. ***Discharge*** (*the quantity of water passing a given point in a unit of time*). Discharge is usually expressed in cubic feet per second (cfs) or in m^3/sec.
2. *Average velocity.*
3. *Size and shape of channel.*
4. ***Gradient*** (*the slope measured along a stream, on the water surface or on the bottom*).
5. ***Load*** (*the material the stream carries*). The load consists of rock particles plus matter in solution. Unlike rock particles that constitute the mechanical load, dissolved matter generally makes little difference to the behavior of the stream; so we shall confine our attention mainly to the mechanical load.

Laminar and turbulent flow. The flow of fluids can be studied by means of colored tracer dyes. When water flows over a smooth bed, *the water particles move in straight, parallel paths and slip over one another along parallel plane surfaces.* This is

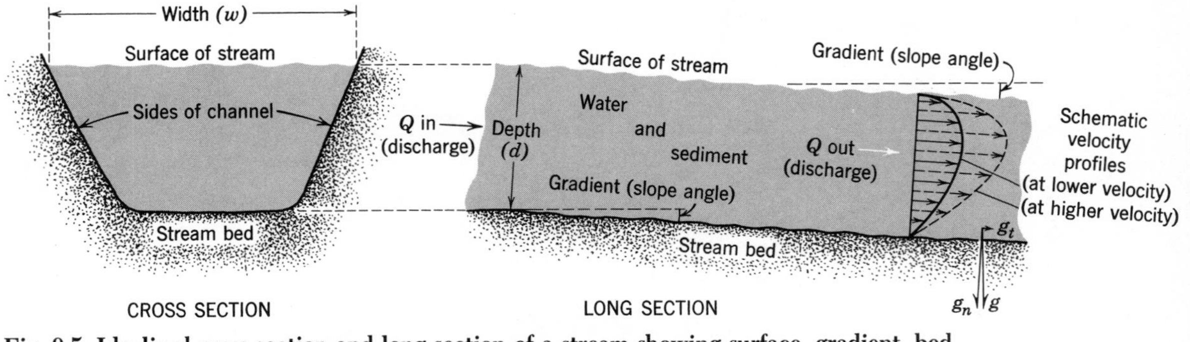

Fig. 9-5. Idealized cross section and long section of a stream showing surface, gradient, bed, depth, width, velocity profiles, discharge (Q) input and output, and gravity forces. The tangential component of gravity, g_t, is responsible for flow of the water; the normal component, g_n, tends to keep rock particles in place on the bed.

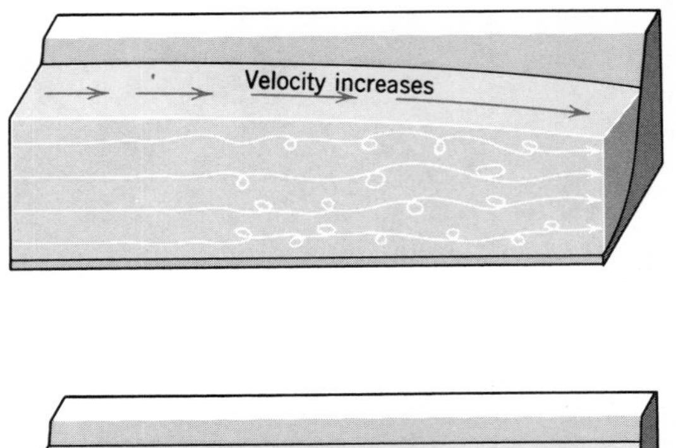

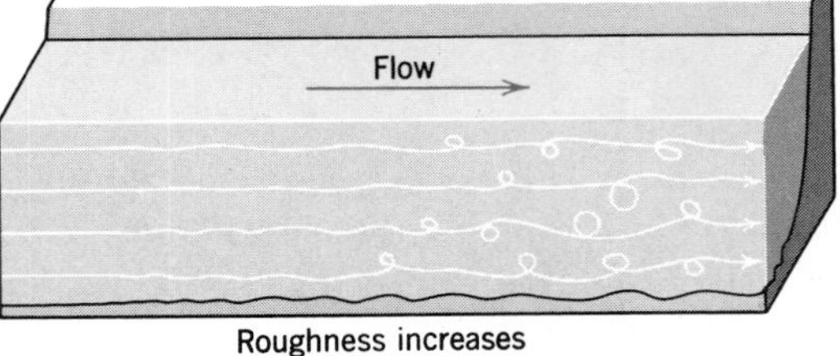

Fig. 9-6. Change from laminar (streamline) flow (*left*) to turbulent flow (*right*) in a stream, with increase of either velocity or channel roughness. The lines represent paths of water particles.

laminar (or *streamline*) ***flow.*** As velocity increases or as the bed becomes rough, the flow paths begin to bend, forming eddies which become more numerous as velocity increases further. This kind of *fluid flow characterized by eddies* is ***turbulent flow*** (Fig. 9-6). In turbulent flow, water particles move in all directions and at many different local velocities. As we shall see presently, this fact is important because it explains why fine sediment is kept in suspension in a stream. Turbulence increases with increasing velocity of flow and also with increasing roughness of the bed. Commonly it is greatest near the sides (Fig. 9-7) and bottom of the channel, where water drags against the bed.

Distribution of velocity in a channel. Measurements with current meters make it possible to plot the distribution of velocity in a stream. In a plan, as sketched in Fig. 9-7, velocity is greatest at the center of the stream, close to the surface, as far as possible from the drag at the sides and bottom of the channel. In cross section, it appears that velocity is greatest where depth is greatest (Fig. 9-8). As depth increases during a flood, the ratio of cross-sectional area to perimeter of channel increases in a spectacular way (Fig. 9-9). This makes possible the great velocities that develop during floods, with consequent erosion and destructive effects.

BEHAVIOR OF ROCK PARTICLES IN A STREAM

The stream's load. Following our discussion of stream flow, we must give special attention to the rock particles in the stream, because our main interest is in the geologic work the stream performs. As Fig. 9-10 shows, the stream's mechanical load ideally consists of *coarse particles moving along or close above the stream bed* (the **bed load**) and *fine particles suspended in the stream* (the ***suspended load***).

Bed load. In our artificial stream channel we can control both discharge and load. Let us set it up with glass sides so that we can watch the water and

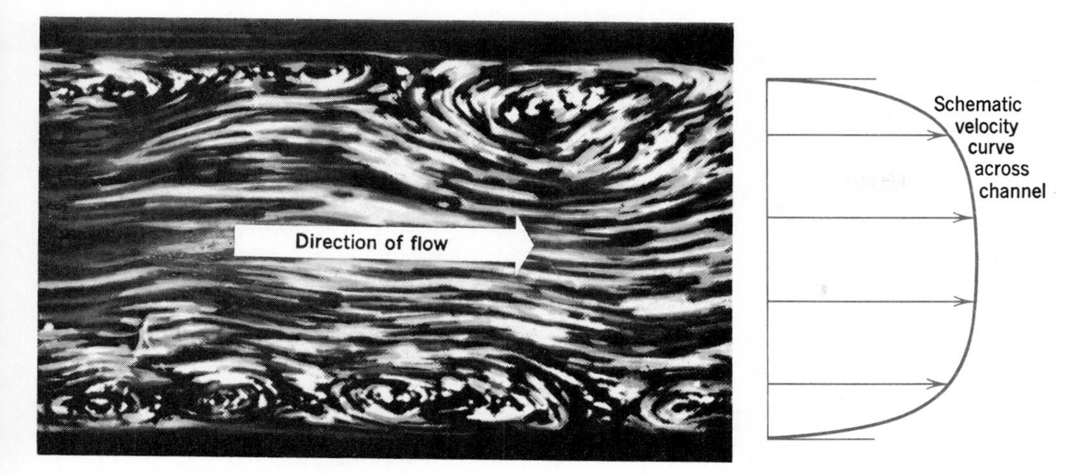

Fig. 9-7. Turbulent flow in an open channel, seen looking down onto surface of stream. Because of drag, turbulence at the surface is greatest near the channel sides. (From a photograph in Prandtl and Tietjens, Applied Hydro- and Aeromechanics; courtesy Engineering Societies Monographs Committee.)

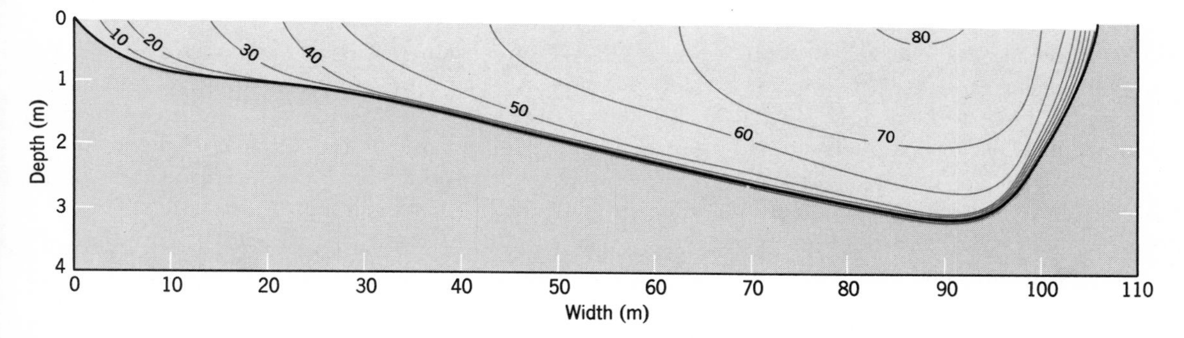

Fig. 9-8. Distribution of velocity through the cross section of a stream, shown by lines of equal velocity (cm/sec). In the River Klarälven in southern Sweden, velocity measurements were made with a current meter at intervals of 10m across the stream and at intervals of 0.5m of depth. (Åke Sundborg, 1956.)

the rock particles as they move through it.

The flowing water exerts a drag force on particles on the bed. The force of drag, a function of velocity, tends to move particles. The inertia of the rock particles and the normal component of gravity (Fig. 9-5) tend to keep the particles in place. When the force of drag exerted on a particle and g_t together exceed the inertia of the particle plus g_n the particle begins to move. At first only a few particles move, by sliding or rolling. Pebbles start to move sooner than sand grains because they project higher into the current and thus into zones of greater velocity (Fig. 9-10).

Now we increase the velocity a little. More particles on the bed start to move and to collide with one another. Some are propelled upward by the collisions and then are pushed forward by the current. Some are sucked upward by the faster current that flows above them, as the wing of an airplane is lifted by the flow of air above it. Gravity pulls the particles down to the bed again, but the result for each particle is a jump forward, downstream. *The jumping movement of rock particles in a current of water or air* is ***saltation.***

As we step up velocity just a bit more, particle movement becomes general. Seen from above, the particles travel parallel with the current.

Ripples. If the bed consists of abundant fine sand, and if we set the velocity of our current at about 25cm/sec, a sudden transformation occurs in the stream bed. The surface of the sand takes on a rhythmic pattern of asymmetrical ripples heading at right angles to the current (Fig. 9-11, *A*). The upstream sides of the ripples slope gently and the downstream (lee) sides slope steeply. The current moves sand grains from the gentle upstream slope to the crest of a ripple. The grains roll down the steep lee slope, and the ripple migrates forward. By the time a ripple has migrated through one wavelength, its internal structure has come to consist entirely of thin, downstream-dipping layers, which represent the successive former positions of the lee slope (Fig. 9-11, *B*). The dipping layers are called *cross-strata*. The direction of dip of cross-strata thus indicates the direction of stream flow. This is important, because cross-strata in ancient sedimentary rocks enable us to reconstruct the directions of flow of ancient rivers.

With further increase in velocity the ripples suddenly disappear, and the sand moves rapidly downstream as a flat-topped mass (Fig. 9-11, *C*). Within it are plane layers parallel with the bed of the stream.

Sand Waves. At this point we increase the scale of our experiment by increasing discharge, stream depth, and quantity of sand. At a velocity of about 50 cm/sec the form of the bed changes again. Large-scale *sand waves* appear. In shape they resemble ripples, but they are much larger. Wavelengths range from a few meters to tens or even hundreds of meters, and wave heights from 50cm to several meters (Fig. 9-12). The upstream slopes of many sand waves are covered with ripples.

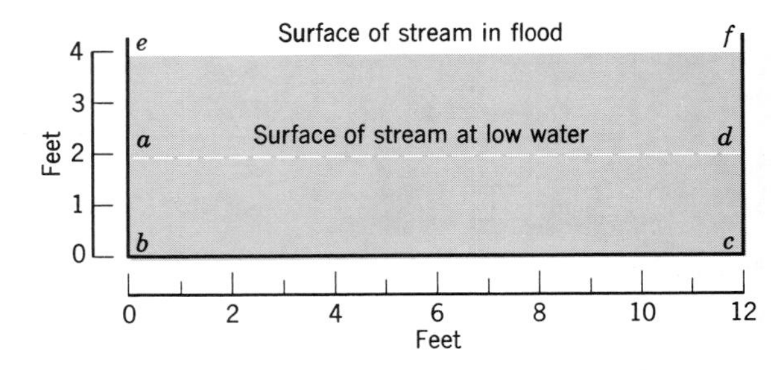

Fig. 9-9. Cross section of an idealized rectangular stream channel having a water surface *ad* during low water and a water surface *ef* when in flood. Cross-sectional area of the stream (width × depth) increases from *abcd* (24 square feet) to *ebcf* (48 square feet), but perimeter of channel (width plus twice the depth) increases only from 2 + 12 + 2 = 16 feet to 4 + 12 + 4 = 20 feet. In this case, therefore, cross-sectional area has increased by 100 per cent, whereas perimeter has increased by only 25 per cent. Ratio of area to perimeter nearly doubles, increasing from 1.5 to 2.9. An irregular channel will show a smaller difference, but the principle is the same.

By greatly increasing the velocity of our stream, we can make it destroy the sand waves and flatten its bed, just as our shallower stream flattened its ripples. The result is the same; sand moves in flat sheets within which the layers are plane and parallel with the bed of the stream. At very high velocities other kinds of sand waves form, but as they are ephemeral and do not form strata that remain in deposited sediments, they are not of interest to us here.

Suspended load. The experiment we have been watching involved clear water flowing over sand. Now if we add silt and clay to the water we can see that these fine particles spread quickly throughout the water, making it turbid. If we look very

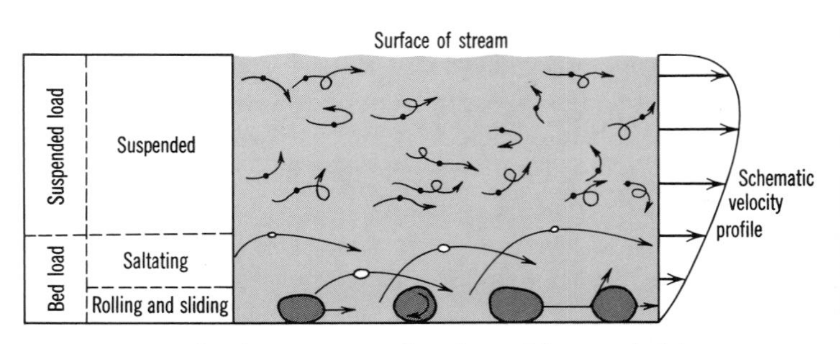

Fig. 9-10. Kinds of movement of rock particles carried in a stream; vertical distribution of bed load and suspended load.

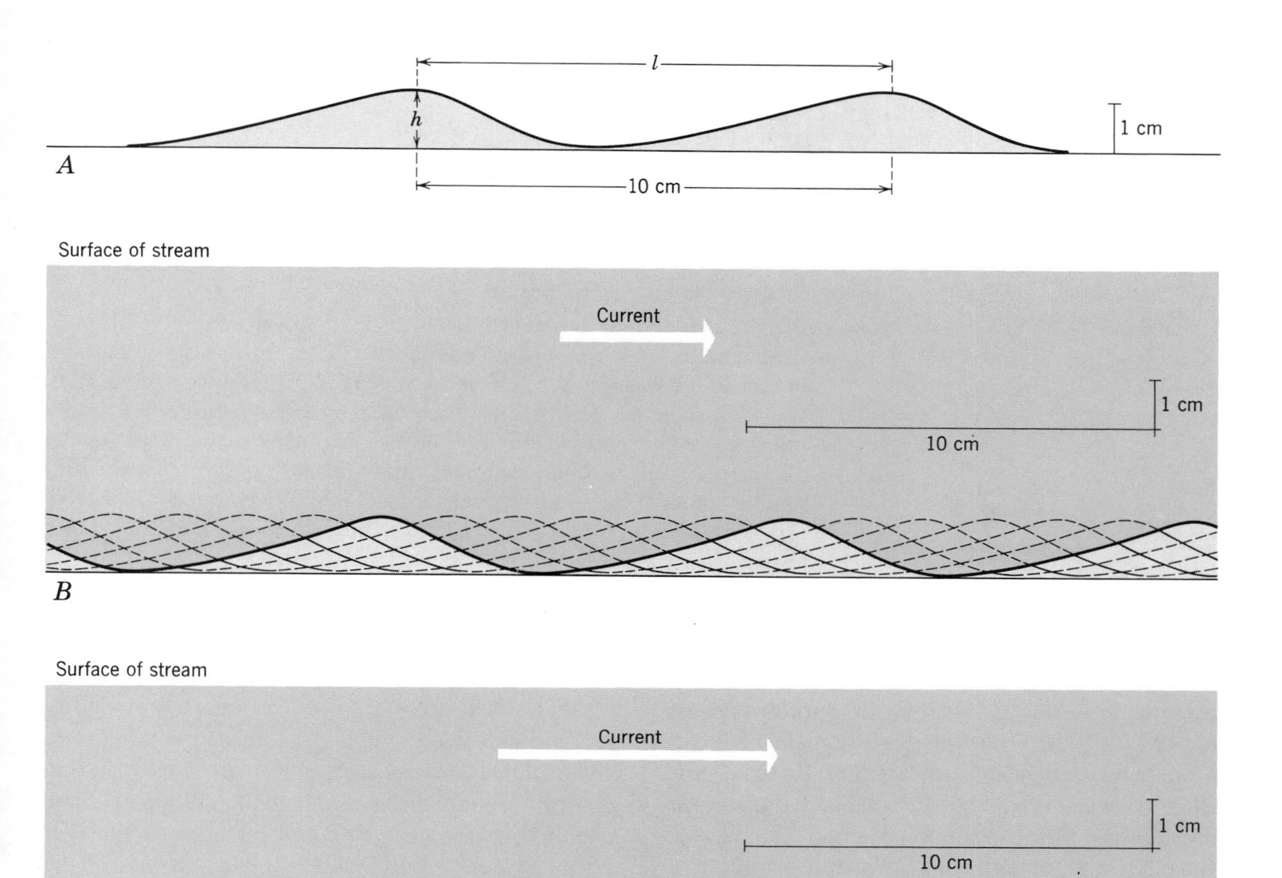

Fig. 9-11. Long profiles through sand on rippled and plane surfaces of a stream bed. *A.* Idealized ripples showing standard dimensions. Ripple length, *l*, is measured from crest to crest of adjoining ripples. Ripple height, *h*, is measured from bottom of trough to crest of ripple, perpendicular to *l*. *B.* Internal cross-strata made by migration of ripples. Dashed lines show successive former positions of ripples that have migrated downstream, left to right. Lines within ripples, sloping downstream, are layers, each of which was the downstream side of a ripple at one moment. Other parts of ripple have been destroyed by migration. *C.* Flat-topped sheet of moving sand, formed at greater velocity than in *B*, which has caused ripples to disappear. Sand moves in plane sheets, the higher sheets moving faster than the lower ones (shown by relative lengths of small black arrows). Movement of flat-topped sheets of sand does not create cross-strata.

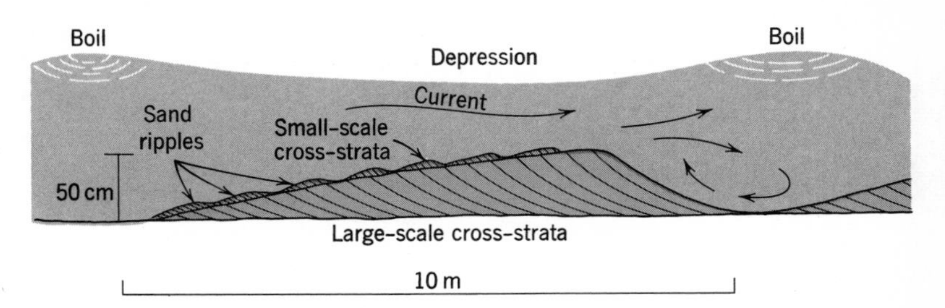

Fig. 9-12. Long profile through a sand wave, showing its internal cross-strata and its influence on water surface. Like the small-scale cross-strata in a ripple, the large-scale inclined layers within a sand wave develop as the wave migrates downstream. These layers represent successive former positions of the downstream side of the migrating sand wave. Ripples on upstream side of sand wave occur only in fine sand at moderate current velocities. Water surface over crests of sand waves is depressed; over troughs it rises and forms boils, eddies within which are shown by white lines. (After U. S. Geol. Survey.)

closely we can see that each particle follows a very irregular path. The particles are *suspended* in the water, because upward-moving threads of current within the general turbulence exceed the *settling velocity,* the constant velocity attained by a particle falling through a still fluid. Settling velocity is determined by particle diameter (Fig. 9-13), shape, and specific gravity, and by density of the fluid. Flaky particles such as mica settle slowly. The specific gravity of most particles found in streams lies between 2.6 and 2.7 (Table C-2). Heavier grains (gold, for instance) settle faster, but only a few such grains are likely to be present.

The vertical component of turbulence which keeps particles in suspension is estimated to be about $\frac{1}{5}$ the average velocity of the stream. In any turbulent stream, therefore, silt and clay always remain in suspension; they move continuously, as fast as the water flows. Such particles can settle and be deposited only where turbulence ceases; for example, on the floodplain shown in Fig. 9-29, *A*, in a lake, or in the sea. Because of these relationships, the transport of particles in the suspended load differs greatly from that of bed-load particles, which tend to move only intermittently.

Most of the suspended load in streams is derived from the erosion of fine-grained regolith in plowed fields and other areas unprotected by vegetation and gets into the streams during rains.

ECONOMY OF A STREAM

We have been studying a laboratory stream in an artificial channel and have been concentrating on the stream itself rather than on the channel, which, because it consists of metal and glass, is virtually unaffected by the stream. However, the channels of most natural streams consist partly or wholly of sediment. Streams move this material from place to place, and so their channels are continually being altered. Because stream and channel are closely related and are ever changing, we have to examine them together as an interrelated system. We can think of the ***economy*** of the stream and its channel as *the input and consumption of energy within a stream or other system and the changes that result.* We noted in Fig. 5-4 that annual rainfall on the area of the United States is equivalent to a layer of water 30 inches thick. Of this layer, 21 inches evaporate and 9 inches form ***runoff,*** defined as *the water that flows over the lands.*

As the average altitude in the United States is about 2,500 feet, our 9-inch layer of water falls through that vertical distance as gravity pulls it down to sea level. The potential energy of the water is tremendous. It is equal to the weight of the water times the height of the land. One cubic foot of water weighs 62.4 pounds, 62.4 times 2,500 feet equals 156,000 foot-pounds. This potential energy is converted into the kinetic energy of stream flow, part

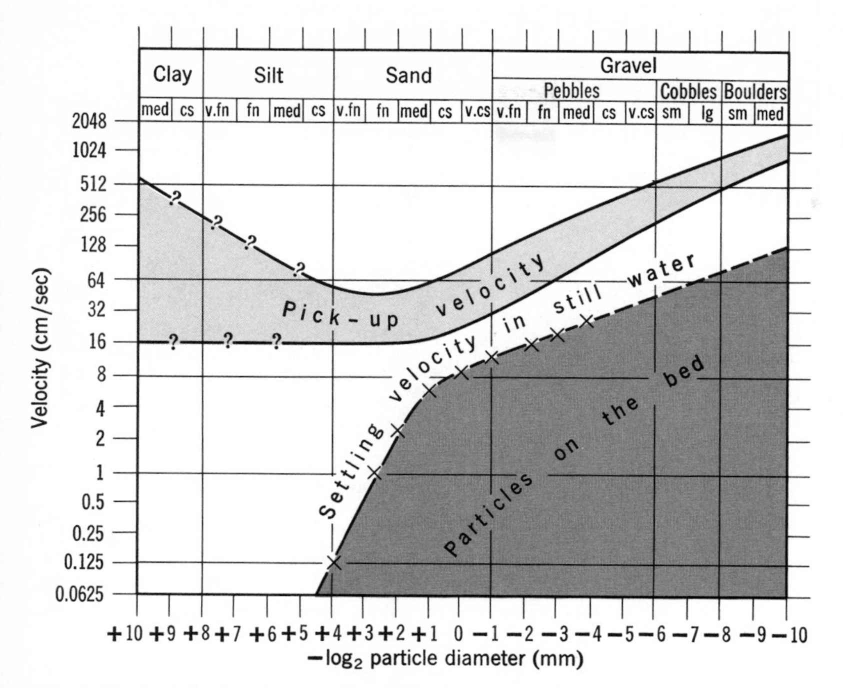

Fig. 9-13. Graph showing relationships among particle diameter, settling velocity, and pick-up velocity. Curve showing settling velocity of particles in still water changes slope within the range of medium sand. In the dark blue area, settling velocity exceeds velocity of upward components of flow, and particles remain on the bed, though they can be moved along it. Pick-up velocity (velocity required to pick up a particle of a given size from the bed) is shown by a band or zone, because it includes the velocities at various heights above the bed. The lower part of the band shows velocities at 1cm above the bed, and the top part shows velocities at 10m above the bed. As long as velocity exceeds pick-up velocity for a given particle size, that size will be continuously moved by the current. Sand, silt, and clay particles will be moved in suspension, but the state of motion of particles coarser than sand is not certain and can not be predicted from the curves. Through the silt and clay sizes the curve of pick-up velocity rises to the left. These small particles present a smooth surface to the current and because of cohesion tend to stick together. Therefore higher velocity is required to lift them from the bed than to keep them suspended once they are off the bed. Silt and clay remain suspended in currents of very low velocity. (Pick-up curve from Sundborg, 1956, p. 177; settling curve after Rubey from Dunbar and Rodgers, 1957, p. 5). Logarithmic scale is explained in Fig. 4-11.

of which is spent in the geologic work of picking up and carrying rock particles. This kinetic energy makes streams the prime movers of rock waste from lands to ocean. This is why streams occupy a key position in the rock cycle, for they are among the chief agencies by which the sedimentary rocks of the Earth's crust were accumulated.

Some of the sediment carried by streams is deposited on the land, at least temporarily. ***Alluvium,*** the general name for *all sediment deposited in land environments by streams,* exists in great quantity, and still larger quantities of it have been converted into sedimentary rocks. Much of the sediment, however, is carried into the sea and deposited there.

Once it reaches the sea, it is no longer alluvium but becomes marine sediment.

Factors in the economy. The economy of the system depends on a continual interplay among the five basic factors mentioned earlier: discharge, velocity, size and shape of channel, gradient, and load. The first factor is calculated from measurements of velocity and channel, made systematically at selected points along large and small streams for purposes of flood control, irrigation, water supply, and the like. The U. S. Geological Survey, for example, maintains nearly 6,500 measurement points, called *gaging stations,* in various parts of the United States.

The measurements clearly show that as discharge changes velocity and channel shape must change also. The relationship is definite and can be expressed by the formula:

$$Q = w \times d \times v$$

Discharge (cubic feet per second) = width (feet) × depth (feet) × velocity (feet per second)

When discharge changes, as it does continually, the product of the other three terms must change accordingly. This relationship can be shown in two ways: by graphs (Fig. 9-14) and by cross sections (Fig. 9-15), both plotted from the measurements. Figure 9-15 does not show velocity, but this can be added numerically from the measurements available.

The main reason why velocity increases when discharge increases is that the geometry of the channel changes in response to changes in the amount of water. With increased discharge, not only velocity but also load increases, as can be inferred from Fig. 9-15; for in that figure the big channel that existed on October 14 could not have been created unless a large load of sediment had been removed. The factor of load, however, is not included in the formula; the relationships involved are so complex that we shall discuss them only in qualitative terms. Another factor, slope, also is not included in the formula, although velocity is closely related to it.

A stream and its channel are related so intimately that we can think of them as a single open system. The channel is so responsive to changes in discharge that the system, at any point along the stream, is continually close to a steady state (Chap. 3).

When discharge increases, the stream erodes and enlarges its channel, instantaneously if it flows on alluvium, more slowly if it flows on bedrock. The

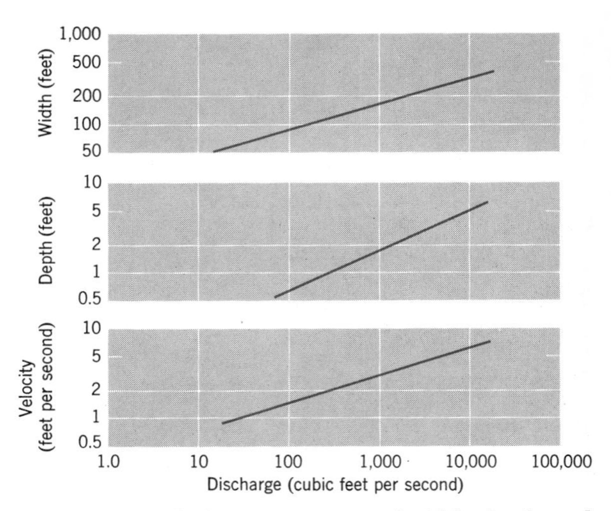

Fig. 9-14. Graph showing increase of width, depth, and velocity at a single point in a stream, when discharge increases during a flood at a single point on a stream. Curves represent values measured at a series of times during a rising flood on Powder River at Locate, Montana gaging station. Both abscissas and ordinates are $\log_{10}$ scales. (After Leopold and Maddock, 1953.)

increased load is carried away. This goes on until the increased discharge can be accommodated. When discharge decreases, some of the load is dropped, making the channel less deep and less wide, and the velocity is reduced by increased friction. In this way width, depth, and velocity are continually readjusted to discharge, and the relationship $Q = wdv$ is maintained.

Other aspects of the economy, especially slope and load, are considered in the following paragraphs.

Geologic activity during floods. Seasonal distribution of rainfall causes many streams to rise in flood seasonally. During a flood, discharge and velocity increase, enabling a stream to transport an increased aggregate load. Increase of velocity results in increase not only of total load but also of maximum diameter of the rock particles a stream can carry, as is evident in Fig. 9-13. In the bed load, particles with larger diameters can be moved because increased drag is applied directly to rock particles on the bed. In the suspended load, larger particles can be carried because of increased turbulence in which part of the force is directed upward. An extreme example resulted from the St. Francis Dam catastrophe mentioned in Chapter 8. As the dam gave way, the water behind it rushed down the valley, moving as bed load blocks of concrete weighing as

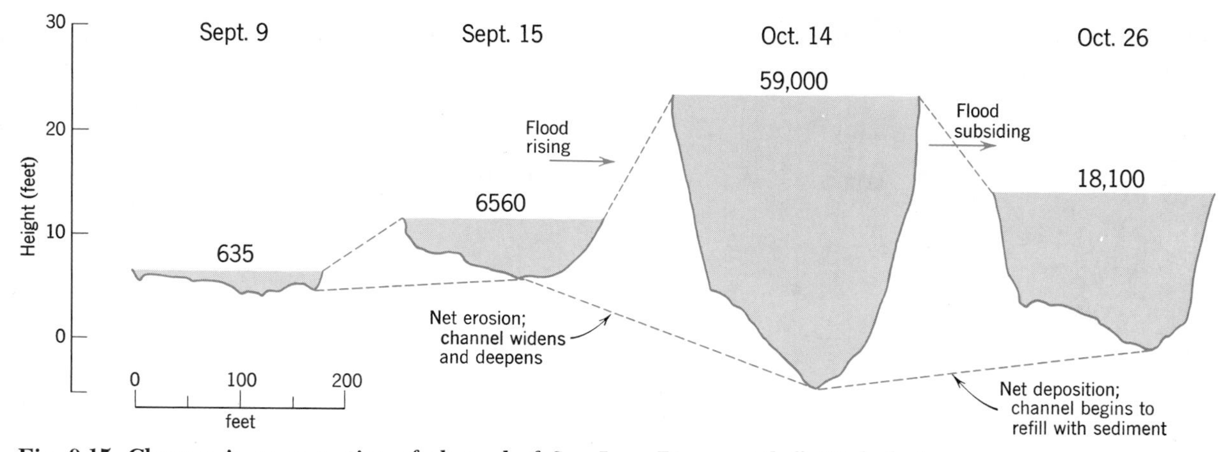

Fig. 9-15. Changes in cross section of channel of San Juan River at Bluff, Utah during a flood in 1944 constructed from detailed measurements. Numbers represent discharge in cubic feet per second. During the flood the water is held in between high banks, not shown above stream surface because their positions were not included in the measurements. (After Leopold and Maddock, 1954.)

much as 10,000 tons through distances of more than 2,500 feet.

In most streams more geologic work is accomplished during regular seasonal floods than during intervals of low water. In addition to seasonal floods, rare, exceptional floods (Fig. 9-16) outside the stream's normal economy occur perhaps only once in several decades or centuries. In these the geologic work done may be prodigious, but such events happen so rarely that their effects probably are less than the aggregate effect of normal activity throughout the very long periods intervening between them. As in the fable, this race, too, is won by the tortoise rather than by the hare.

Changes downstream. When we watched events at a single point only, we saw that a stream adjusts its channel so as to approach a stable condition at all times. Now let us look at the whole length of the stream to see what changes take place. If we go downstream from head to mouth, we see that orderly adjustments take place from point to point. Specifically, four factors change. (1) Discharge increases, (2) width and depth of channel increase, (3) velocity increases slightly, and (4) gradient decreases. A few comments on these changes are in order.

Discharge increases (Fig. 9-17) as entering tributaries contribute additional water. To satisfy the relationship $Q = wdv$, increased discharge must be accompanied by increases in one or more of the factors width, depth, and velocity, as shown in Fig. 9-18. The demonstration that velocity increases downstream seems to contradict the common observation that water rushes down steep mountain slopes and flows smoothly over nearly flat lowlands. But the physical appearance of the water is not a true measure of its velocity, which increases downstream mainly because channels become deeper and wider in that direction, and (as Fig. 9-17 shows) the drag exerted by the channel becomes relatively less.

Long profile. The one factor remaining is gradient, commonly expressed as per cent or as the difference of altitude between two points measured along the stream. The gradients of some mountain streams exceed 300 or even 400 feet per mile, whereas the gradients of downstream parts of the Mississippi River are less than 0.5 feet per mile. Gradients of many streams range between 10 and 20 feet per mile (Fig. 8-2).

Because, as we have noted, gradient decreases downstream, the stream's ***long profile,*** *a line connecting points on the stream surface,* is characteristically concave-up. Figure 9-19 represents the long profile, drawn to scale, of the Platte-South Platte River in Colorado and Nebraska. This stream originates in the Rocky Mountains, enters the Great Plains near Denver, and empties into the Missouri River near Omaha. The profiles of most streams show this general form regardless of length, discharge, or kind of bed material.

Fig. 9-16. Flood of Quinebaug River at Putnam, Connecticut in August 1955, the result of extraordinary rainfall. View looking north, upstream. River channel is at left, just out of view; its edge is seen as darker water in lower-left corner of picture. Stream shown is water that overflowed from main channel and is rejoining it at lower left. Although short lived, the overflow eroded a large area of a bank of sand and gravel about 25 feet thick (remnants are near flooded railroad tracks) and deposited the resulting load as a series of bars and islands, seen in foreground as part of a braided pattern. (Providence *Journal-Bulletin.*)

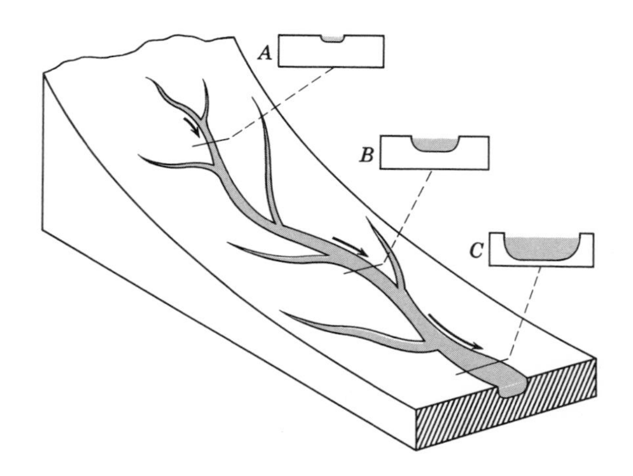

Fig. 9-17. Changes in the downstream direction along a stream system. Discharge is indicated by entrance of successive tributaries; width and depth of channel are shown by cross sections *A, B,* and *C,* and velocity by relative lengths of three arrows. (After Leopold and Maddock, 1953.)

The concave-up profile must reflect an underlying general cause. The matter was pondered in the nineteenth century by G. K. Gilbert, an American geologist, who reasoned that as the channel becomes deeper and wider downstream, friction decreases (Fig. 9-17), and the mechanical efficiency of the stream therefore increases. Hence the stream, with increased discharge, flows with increased velocity and can carry its load on an ever-decreasing slope.

Base level. The vertical position of the mouth of a stream is determined by its ***base level.*** This is *the limiting level below which a stream can not erode the land* (Fig. 9-20). The ***ultimate base level,*** for streams in general, is *sea level,*[1] *projected inland as an imaginary surface underneath the stream.* When a stream cuts down to that surface, its energy quickly approaches zero. For a stream ending in a lake, base level is the level of the lake (Fig. 9-20), for the stream can not erode below it. But, if the lake were destroyed by erosion at its outlet, the base

[1] Although true in principle, this statement is not quite true in detail. A stream confined between channel banks can erode its bed slightly below sea level, as is evident from Fig. 9-30, where depth of base of the topset beds represents depth of stream-channel erosion.

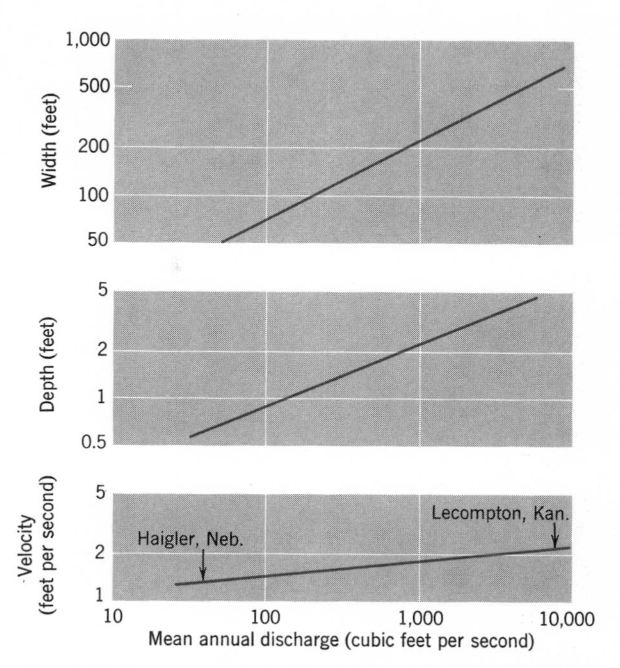

Fig. 9-18. Increase of width, depth, and velocity with discharge, from point to point down a stream system. Curves were constructed from measurements at gaging stations located on the Arikaree, Republican, Smoky Hill, and Kansas rivers, all forming a continuous system. (After Leopold and Maddock, 1953.)

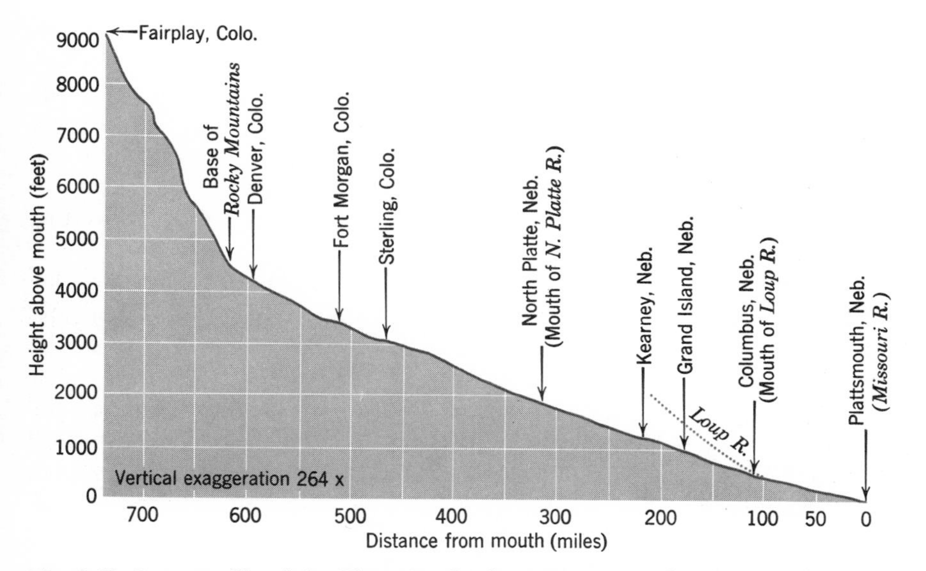

Fig. 9-19. Long profile of the Platte-South Platte River in Colorado and Nebraska, illustrating the common concave-up form. Profile of Loup River, a tributary (dotted line), shows a typical smooth accordant junction with the main stream. Without the large vertical exaggeration the upstream end of the curve, on this horizontal scale, would not be visible above the base line 0-0. (After Henry Gannett, 1901.)

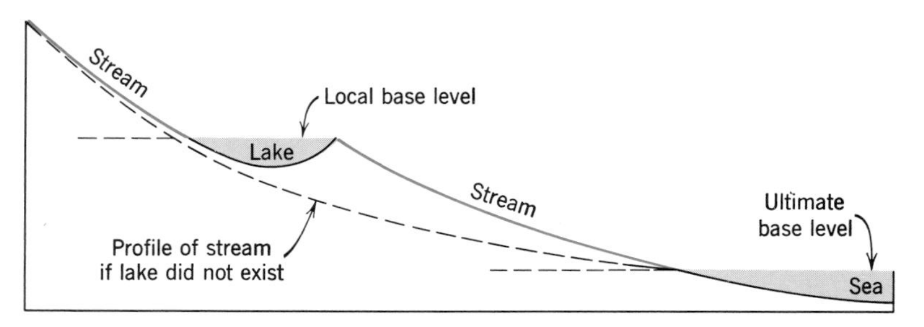

Fig. 9-20. Relation between ultimate base level (the sea) and local base levels such as lakes. Other local base levels are shown in Fig. 9-21.

level represented by the lake surface would disappear, and the stream, having acquired additional potential energy, would deepen its channel. *The levels of lakes and all other base levels that stand above sea level* are ***local base levels.*** A common kind of local base level is the level of a belt of particularly hard rock lying across the stream's path. Even sea level itself changes slowly over long periods (Chap. 14), and this too affects the long profiles of streams.

Lower Colorado River. The adjustments that occur at each point along the length of a stream can be seen clearly when artificial changes act to alter the distribution of the stream's energy and load. A good example is the response of the lower Colorado River to dams built across parts of its course (Fig. 9-21). The Colorado habitually carried a large suspended load and probably a large bed load also. When the Hoover Dam was completed in 1935, this load was trapped in the newly created reservoir behind the dam; hence the water passing the dam was clear. With its potential energy locally increased by an amount determined by the height of the dam and with no load to carry, the stream below the dam increased its velocity of flow and picked up a new load from the materials of its channel. In the process the channel was deepened.

During the first 6 months the river lowered its bed by 2 to 6 feet through the first 13 miles below the dam. The rate of deepening gradually diminished; 10 years later, in 1945, the rate was only a small fraction of its 1935 value. What was happening was this: by lowering its bed in this sector the river reduced its slope. In consequence, its energy was reduced toward a value commensurate with the reduced load. A new condition of stability was developing.[2]

When the Parker Dam was completed in 1938, similar bed erosion occurred through a distance of at least 100 miles downstream. In the same year the Imperial Dam was completed. The sediment picked up by erosion below the Parker Dam, together with sediment washed in from the sides of the valley, was trapped in the reservoir behind the Imperial Dam. By 1950, 12 years after it had been created, this reservoir was almost completely filled with sediment. Of course this had been anticipated when the dams were designed, and a mechanical desilting device was in operation, removing 500 to 5,000 tons of sediment daily from water diverted into a large irrigation canal leading away from the reservoir.

Sedimentation is occurring in the Parker reservoir also. In each of the segments between the two dams, therefore, the Colorado River is adjusting itself to the new conditions by deepening its bed below each dam and by depositing the resulting load farther downstream, where the horizontal water surface of a reservoir has been substituted for a continuous slope. The deposits in the reservoirs are deltas (Fig. 9-30).

The behavior of the Colorado illustrates the general tendency of streams to maintain a steady-state condition. When one of the factors in the stream's economy is altered, the others must adjust themselves so as to approach stability under the altered conditions. Because the tendency toward stability exists at each point along the stream's course, the various factors continually interact to maintain a channel that, with increasing distance downstream, can carry an ever-increasing discharge and load.

[2] This segment of the river no longer exists because, with the closing of the Davis Dam in 1950, it became a lake.

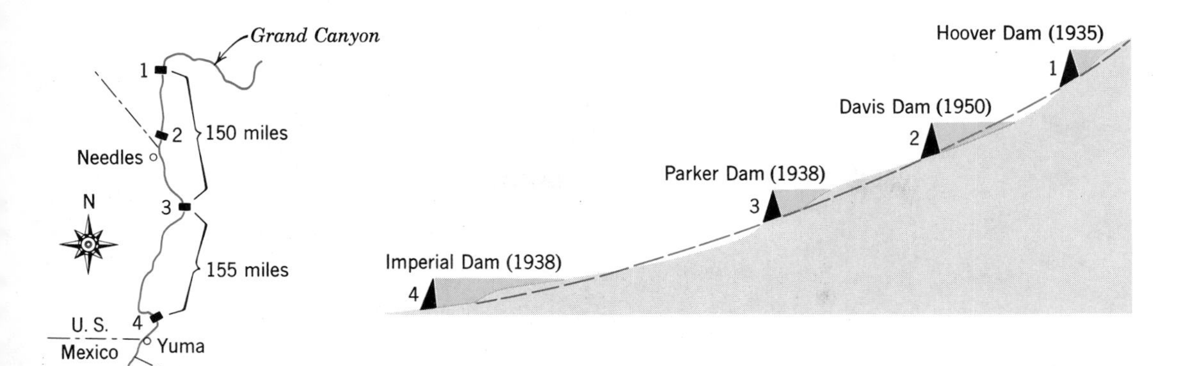

Fig. 9-21. Plan and profile of Colorado River from Hoover Dam to Imperial Dam. Schematic profile, with enormous vertical exaggeration, shows changes in stream bed before Davis Dam was completed in 1950. Broken-line profile shows stream bed before construction of any dams; continuous-line profile after completion of three dams. Channel cross section, not shown, probably changed correspondingly. (Data from J. W. Stanley.)

Fans. Any long profile that is not concave-up is unstable and is altered by the stream until it conforms to that shape. The building of a fan is a common example. When a stream flows down through a steep highland valley and comes out suddenly onto a nearly level valley floor or plain, it encounters an abrupt decrease of slope. It deposits that part of its load which can not be carried on the gentler slope. The material it deposits takes the form of a ***fan***, defined as *a fan-shaped body of alluvium built at the base of a steep slope* (Figs. 13-3, 13-8). The surface of the fan slopes outward in a wide arc from an apex at the mouth of the steep valley. The profile of the fan, from apex to base in any direction, commonly shows the concave-up form characteristic of stream profiles.

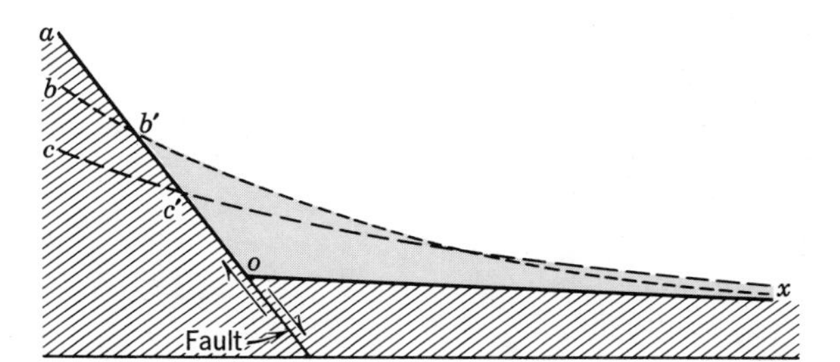

Fig. 9-22. Vertical section showing growth of a fan. Bedrock is shaded; alluvium is blue. Line *aox*, profile of surface before deposition of fan. Line *bb′x*, long profile of stream at an early stage of fan building. Line *cc′x* long profile at a later stage, after stream has cut away apex of fan *bb′x*, increasing fan radius while establishing a continuous, concave profile.

On most fans the stream-channel pattern is braided, and upbuilding is accomplished by the channel filling and cutting described in a following section. When one radius of a fan has been built up in this way, the stream shifts to an adjacent radius and builds up that one. Thus a wide area becomes covered with alluvium, creating a deposit whose form and stratification are remarkably symmetrical.

Upbuilding of the fan steepens the slope of the stream at the fan itself, but upstream from the fan erosion is in progress, reducing the slope there. By the combination of erosion upstream and deposition on the fan, together with changes in channel cross section, the stream and channel are adjusted to a continuous profile, concave up (Fig. 9-22). The form of the profile depends chiefly on discharge and the diameters of particles in the bed load; hence no two fans are exactly alike. A small stream carrying a load of coarse particles builds a shorter, steeper fan than a larger stream carrying a load of finer particles.

Although a fan is originally localized by decrease of slope, as soon as its long profile has become smooth, the chief cause of further deposition on it is the spreading of water through a network of channels, with consequent loss of discharge and velocity in each channel and overall loss of water that percolates down into the underlying sediments.

Unless special circumstances preserve it, the fan will be destroyed piecemeal by continuing erosion downward below the profile *cc′x* (Fig. 9-22). A fan, therefore, is likely to be a temporary deposit, repre-

Fig. 9-23. Meanders of Pecos River near Roswell, New Mexico. View looking south, downstream. The areas covered with rows of bushes are point bars. At right of center is an oxbow lake. (Spence Air Photos.)

senting a stream's quick transition toward a steady-state condition, on a smooth, concave-up profile, at a place where stability on the original slopes is impossible.

SPECIAL FEATURES OF STREAM CHANNELS

Meanders. No stream is straight through more than a short distance. The pattern of most streams is a series of bends, and very commonly the bends are smooth, looplike, and similar in size. Such bends are termed ***meanders*** (Gr. *maiandros*, "a bend"), defined as *looplike bends of a stream channel* (Fig. 9-23). The meanders of the Mississippi and other large streams are under continual study by engineers and geologists concerned with problems such as floods and navigability. These people know that meanders are not accidental, that they occur most commonly in channels with gentle gradients in fine-grained alluvium, and that they occur even in streams having no load at all. The geometry of meanders is distinctive (Fig. 9-24). In a large sample of streams the ratios of radius of curvature of meander to channel width and to wavelength (the distance between two successive meanders), and the ratio of wavelength to channel length, fall within narrow limits and are therefore systematic. From this we can infer that meanders result from the flow of water through a channel, and that they represent the form by means of which, in making a turn, the river experiences least resistance to flow, does the least work, or dissipates energy most nearly uniformly along its course. The meander form is therefore one of stability. Although stable as a form, a meander changes its position almost continually, as shown by year-to-year measurements of river channels and by artificial streams. The shift, or migration, of a meander is accomplished by predominant deposition on one bank and predominant erosion of the opposite bank, as is explained next.

Lateral migration of meanders; point bars. Many meandering streams carry abundant loads and deposit the coarser particles so as to form ***point bars,*** *crescent-shaped bars built out from each convex ("inside") bank of the channel.* The growth of point bars can be seen in experiments with model streams like the one shown in Fig. 9-25. The cross sections in that figure show that both the line of greatest velocity (V) and the location of greatest turbulence (T) are at a maximum along the concave bank. Sand

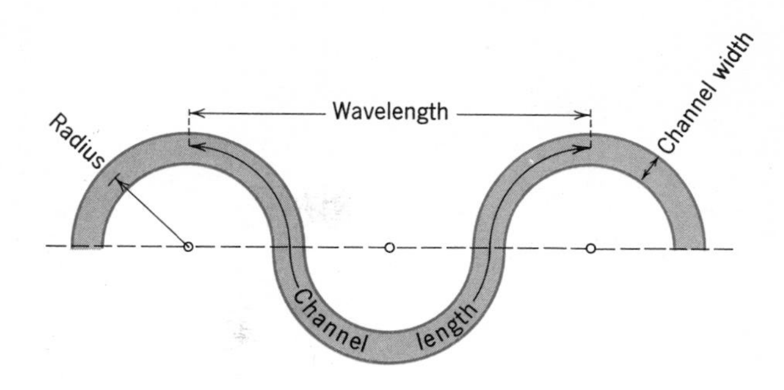

Fig. 9-24. Geometry of an ideal meander, showing wavelength, channel length within one wavelength, channel width, and radius of curvature.

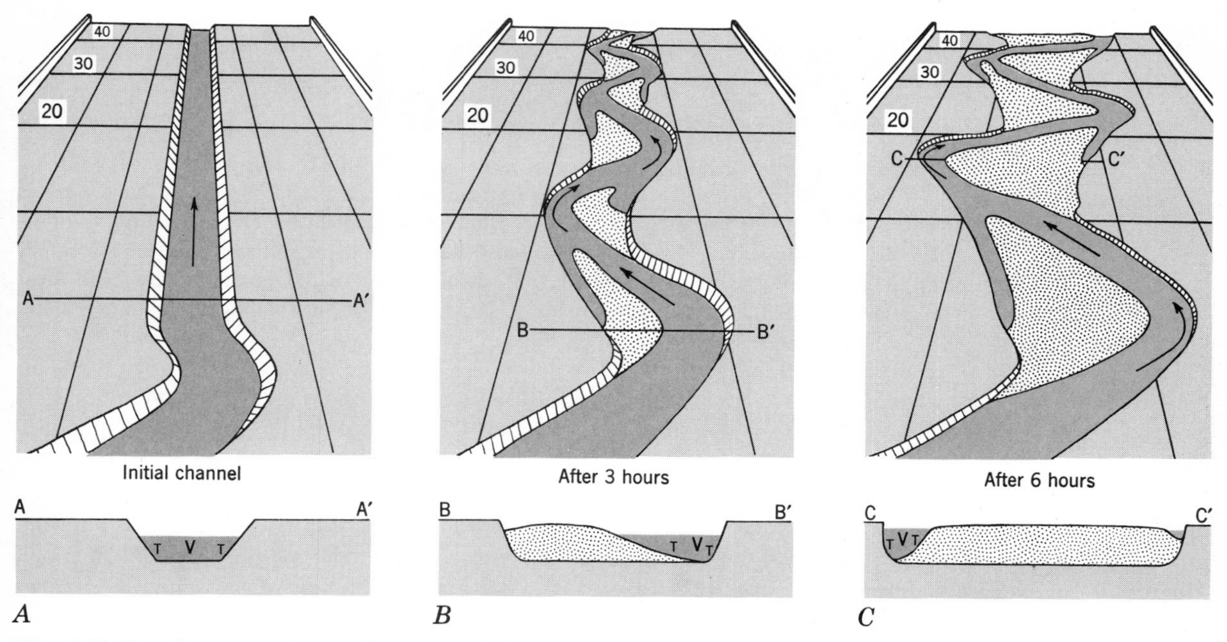

A *B* *C*

Fig. 9-25. Development of point bars in a model stream. *A. Initial channel* in sand, in a wide trough 50 feet long (numbers are distances, in feet, from the observer) sloping in direction of arrow. Channel was made with an initial irregularity (in foreground). Cross section of channel at A-A′ is shown below. Water is introduced into channel head; sizes of T and V suggest relative intensity of turbulence and velocity. Stream starts to meander. *B. After 3 hours* concave banks are eroded and sand is deposited as point bars along the convex banks next downstream. Cross section B-B′ shows asymmetrical channel and point bar. *C. After 6 hours* meanders and point bars have enlarged. Bars indicate that caving of stream banks supplies a greater load than stream can carry away. (After J. F. Friedkin, U. S. Waterways Experiment Station.)

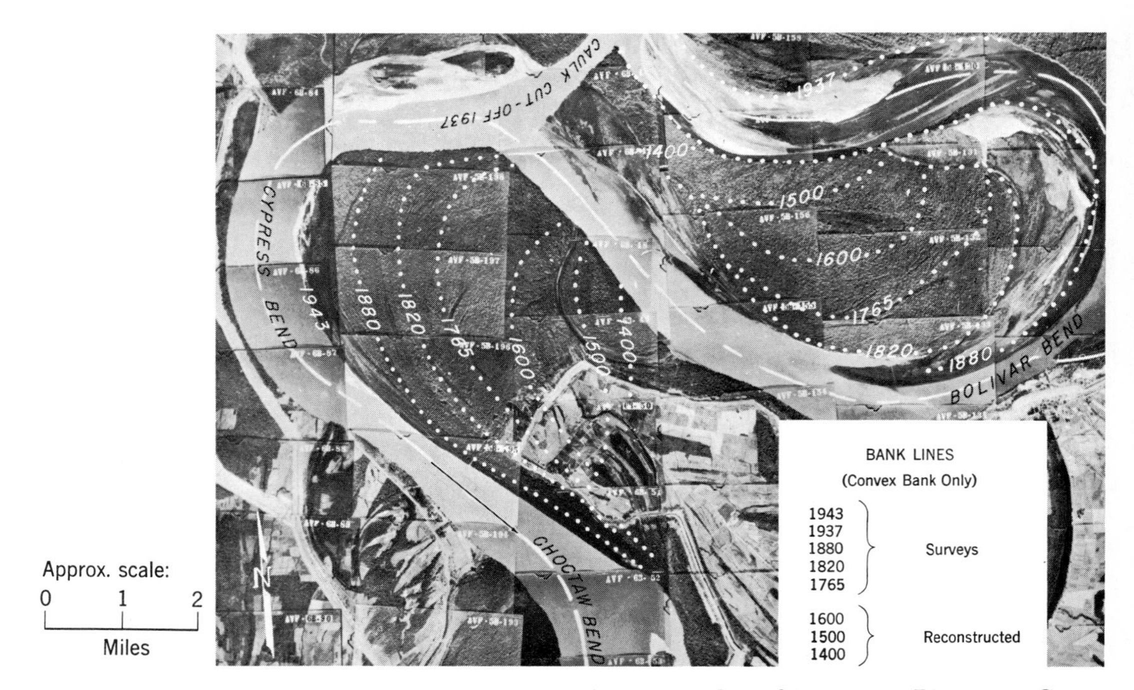

Fig. 9-26. Point bars forming the cores of two meanders of Mississippi River near Greenville, Mississippi in 1943. Dotted lines show successive positions of convex banks. Lines 1765 to 1943 were determined by surveys; lines 1400 to 1600 by reconstruction from archeologic and other data. Channel at Cypress Bend has been shifting nearly 60 feet per year. Caulk cutoff, which occurred in 1937 resulted in gradual abandonment of Bolivar Bend, which by 1952 had become an oxbow lake. (H. N. Fisk, Mississippi River Commission.)

eroded from that bank is deposited as a point bar along the convex bank next downstream, where velocity and turbulence are least. Although the growth of point bars is not necessary to the development of meanders, possibly it speeds the growth of the stable meandering pattern. The rates of growth of point bars, forming the cores of two successive meanders of the Mississippi River channel, have been measured (Fig. 9-26). Such bars are conspicuous features of many broad valley floors.

The migration of point bars is fundamental to the deposition of bed load in a meandering stream. As point bars migrate they leave behind a distinctive succession of layers, which can be seen in borings and trenches. Coarse sediments at the base (Fig. 9-27, *a*) are overlain by sand with large-scale cross-strata and in turn by sand with plane-parallel layers. At the very top is sand cross-stratified on the small scale of ripples.

This arrangement of layers has been found over and over again both in the sediments of modern streams and in the deposits of ancient meandering streams. Remembering (Fig. 9-26) that point bars grow as the channel shifts laterally, let us compare this arrangement of layers with the ways in which sediment is moved through the channel.

The basal coarse layer (Fig. 9-27, *a*) is laterally continuous, on the floor of the adjacent channel, with coarse sediment that the stream is unable to carry away. Bearing in mind that large-scale cross-strata originate in sand waves (Fig. 9-12) migrating downstream along the deep part of the channel, we can visualize the abandonment of these waves and their incorporation into the growing point bar as the channel shifts sideways (Fig. 9-27, *b*). The fact that sand waves form only in the deep part of the channel explains why large-scale cross-strata are found only in the lower layers of point bars.

In the shallow parts of the channel, where sand waves can not form, sand moves either in flat-topped sheets or in ripples (Fig. 9-11). We can therefore visualize how sand with plane-parallel layers and small-scale cross-strata (Fig. 9-27, *c*) becomes stranded as the channel shifts and is incorporated

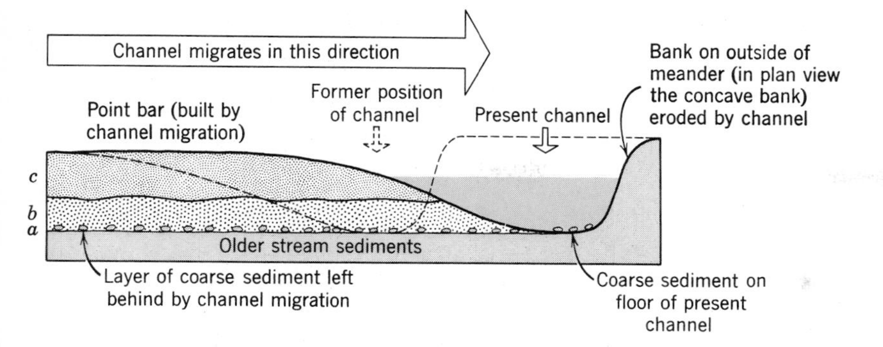

Fig. 9-27. Section, through point bar and stream channel, at right angles to direction of stream flow. Sediments in a point bar consist of (*a*) coarse basal layer, (*b*) sand, with large-scale cross-strata (not shown), and (*c*) sand, with plane-parallel layers having small-scale ripple cross-strata at the top (internal structures not shown). For origin of these layers, see text.

in the upper part of the point bar.

Downvalley migration of meanders: cutoffs. Because of the slope in the downvalley direction, caving is a little more rapid on concave banks that face upvalley than on other banks. Therefore meanders tend to migrate slowly down the valley, subtracting from and adding to various pieces of real estate along the banks, according to location, and causing legal disputes over property lines and even over the boundaries between counties and states.

The behavior of streams in laboratory channels shows that if the bank material is uniform the meanders are symmetrical and migrate downvalley at the same rate. But, because the bank material in a natural stream is not uniform, migration of the downstream limb of a meander can be slowed up by resistant bank material while the upstream limb, migrating more rapidly, intersects it at the "neck" of the meander. This causes a ***neck cutoff,*** defined as *the intersection of a meander bend by the bend next upstream, causing the stream to bypass the loop between the bends* (Figs. 9-26, 9-28, *B*).

The usual result of a neck cutoff is an ***oxbow lake,*** *a curved lake occupying a cutoff meander loop* (Figs. 9-23, 9-28, *B*), as the cutoff part of the channel becomes blocked with alluvium at both ends. Oxbow lakes gradually fill up with clay and silt, which settle out of suspension because the water in them has ceased to flow. The meander pattern in Fig. 9-28, *A* is irregular chiefly because the resistant clay fillings of old oxbow lakes are encountered by migrating meanders.

Another kind of cutoff, a ***chute cutoff,*** is *a new channel cut across a point bar, resulting in abandonment of part of a meander.* The cutoff occurs during floods, when rising water overtops the channel. Instead of following the gentle slope around the meander, the water flows down the steeper slope directly across the point bar and makes a new channel. If the flow continues for long, the new channel replaces (cuts off) the meander, which then fills up with fine sediment. In a large river like the Mississippi the shift is gradual; commonly its completion requires many floods over a period of years (Fig. 9-28, *C*).

Since 1776 the aggregate length of channel abandoned by the Mississippi through cutoffs has amounted to more than 230 miles; yet the river has not been shortened appreciably because the lost mileage has been balanced by the enlargement of other meanders.

Floodplain: natural levees. The Mississippi River habitually floods during the spring season. Before the river began to be restrained by artifical flood-control measures, it frequently overtopped its banks and inundated the lower parts of the valley floor. *That part of any stream valley which is inundated during floods* (Fig. 1-3) is a ***floodplain;*** the area of the natural floodplain of the Mississippi from Cairo to the delta is 30,000 square miles, but more than half of this area is now protected against floods by dikes and other structures.

The channels of the lower Mississippi and many other meandering streams are bordered by ***natural levees,*** *broad, low ridges of fine alluvium built along both sides of a stream channel by water*

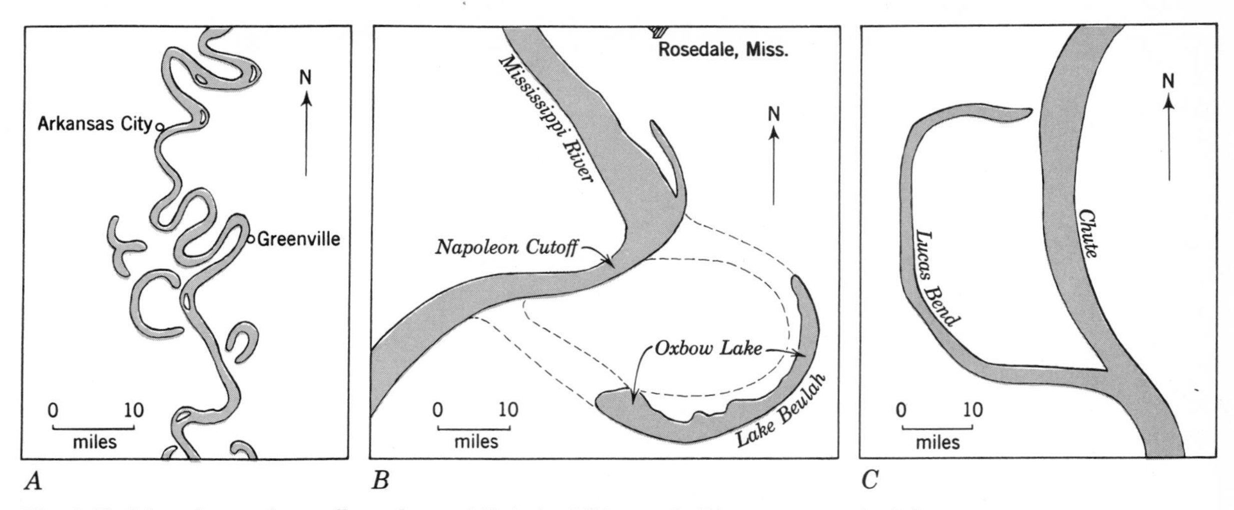

Fig. 9-28. Meanders and cutoffs on lower Mississippi River. *A.* River segment in Arkansas and Mississippi, showing meandering channel and cutoff bend. *B.* Before 1963, when Napoleon Cutoff, a neck cutoff, was made, main channel followed meander at lower right. Ends of cutoff meander were gradually filled with sand, creating an oxbow lake. *C.* Chute cutoff 15 miles south of Cairo, Illinois. Originally, river followed Lucas Bend meander, but, in 1880, flood overflow started the chute. By 1932 almost all the flow was going through the chute, and by 1945 Lucas Bend was abandoned and its upstream end had filled with sand. When its lower end is blocked it will have become an oxbow lake. (After Mississippi River Commission.)

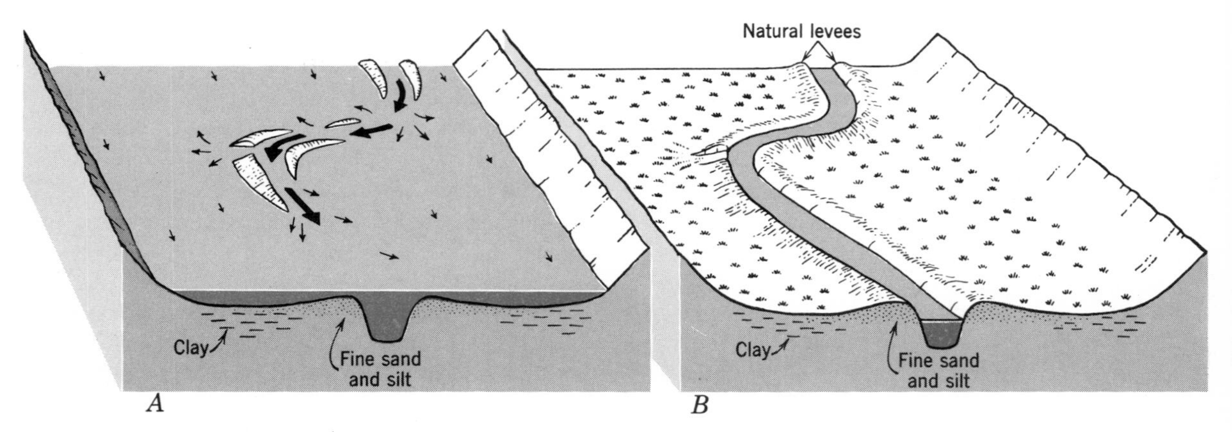

Fig. 9-29. Floodplain with natural levees. *A.* During a big seasonal flood much of the valley floor becomes a lake (Fig. 1-3). Water remaining in the channel flows with high velocity (large arrows). Water escaping from channel flows with diminishing velocity (small arrows) into adjacent broad shallow areas. It deposits silt to form natural levees where it leaves the channel and blankets lower land with clay. Highest parts of levees, added to only during still higher floods, form islands. *B.* At times of low water, levees stand as low ridges along sides of channel; beyond them are swamp lands. Vertical scale much exaggerated.

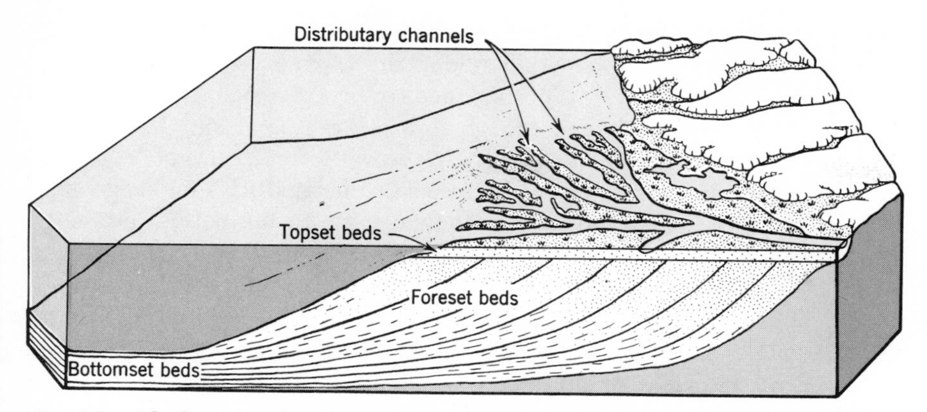

Fig. 9-30. Idealized small delta. Foreset beds consist of sand, which grades outward and downward into silt and clay in bottomset beds. Inclination of foreset beds is identical with slope of delta front, shown in phantom view through the water. The less deep the water offshore, the gentler the slope and the less distinct this stratification. The area of such a delta might be 1 square mile or less.

spreading out of the channel during floods (Fig. 9-29). Along the lower Mississippi natural levees are 15- to 25 feet high. The fine alluvium of which they are chiefly built becomes still finer away from the river and grades into a thin cover of silt and clay over the rest of the floodplain. Natural levees were built and are continually added to only during floods so high that the floodplain is converted essentially into a lake deep enough to submerge the levees. In the water that flows laterally from the submerged channel over the submerged floodplain, depth, velocity, and turbulence decrease abruptly at the channel margins. This results in sudden, rapid deposition of the coarser part of the suspended load (usually fine sand and silt) along the sides of the channel. Farther away from the channel, finer silt and clay settle out in the quiet water. In the vicinity of Kansas City, Missouri, during an exceptional flood in 1952, Missouri River water deposited a layer of silt as much as 6 inches thick over wide areas of the floodplain. In some places fences and other obstacles caused silt and fine sand to accumulate to thicknesses as great as 5 feet. Under ordinary conditions flood-deposited silt is beneficial to agricultural lands, because it contains organic matter, washed from soils on the watershed, which acts as fertilizer.

Flood control. The flood deposits of the Mississippi River, including the natural levees, were built up during a long period, but the process has been interfered with by engineering works (mainly dikes built of earth and concrete) designed to prevent the river from generally inundating its floodplain. The natural levees have been heightened artificially by earth dikes to hold in ordinary floods, and at selected points spillways have been built to allow the water of the highest floods to escape harmlessly into natural channels that parallel the channel of the Mississippi.

Braided streams. A ***braided stream*** is *a stream that flows in two or more interconnected channels around islands of bed-load alluvium* (Fig. 9-16). Braiding, apparently the result of very shallow channel depth combined with very abundant sediment on the bed, is seen commonly in two situations. One is the channel of a large stream at very low-water stages. More and more bed-load particles stop moving, and the water, becoming continually shallower, breaks up into a network of very small shallow channels, all of them confined within the limits of the larger, high-water channel. The other situation is one in which supply of sand or gravel is plentiful, but discharge is insufficient to maintain a single channel wide enough to submerge all the sediment. Such situations exist near glaciers (Fig. 12-19) and on fans in arid country where water is lost by evaporation and by infiltration into the dry alluvium.

The stratification in the sediments deposited by a braided stream is distinctive (Fig. 12-20). Cross-strata are abundant and are cut by channel fills that intersect one another in a complex way.

Deltas. A ***delta*** is *a body of sediment deposited by a stream flowing into standing water.* As the water of the stream diffuses into the standing water of sea or lake, its velocity is checked by friction, it loses energy, and deposits its load as a delta. Although deltas are of several kinds, the type easiest to recognize, and probably most common, is shown in Fig. 9-30. Although in form it somewhat resembles a fan, it differs from a fan because of two factors: (1) stream flow is checked by standing water; (2) the level surface of sea or lake sets an approximate limit to upbuilding of the deposit, the top of which is flatter than the profile of a fan.

The particles in the bed load are deposited first, in order of decreasing weight; beyond this, the suspended sediments drop out. A layer deposited at any one time (as during a single flood) is sorted, grading from coarse at the stream mouth to fine offshore. The deposition of many successive layers creates an embankment that grows outward like a highway fill made by dumping. *The coarse, thick, steeply sloping part of each layer in a delta* is a ***foreset bed.***[3] Traced seaward, the same layer becomes rapidly thinner and finer, covering the bottom over a wide area. This *gently sloping, fine, thin part of each layer in a delta* is a **bottomset bed.**

As successive layers are deposited, the coarse foreset beds one by one overlap the bottomset beds, producing the arrangement seen in Fig. 9-30. The stream gradually extends seaward over the growing delta, erodes the tops of the foreset beds during floods, and at other times deposits part of its bed load in its channel and its suspended load in areas between channels during floods. The channel deposits and interchannel deposits form the ***topset beds*** of the delta. We can define these deposits as *the layers of stream sediment that overlie the foreset beds in a delta.*

During floods the stream spills out of its channel and forms distributary channels, through which the water enters the sea independently, multiplying the topset deposits. Radiating distributary channels give the delta a crudely triangular shape resembling the Greek letter Δ, from which the deposit derives its name.

It may seem surprising that the suspended load, much of which has been carried hundreds of miles through the channel of a large river without being deposited, should drop out so abruptly to form part of a localized delta instead of remaining in suspension long enough to be carried far from land. But the salts dissolved in seawater act to coagulate, or flocculate, the suspended fine particles into aggregates so large that they settle to the bottom promptly.

Some of the world's greatest rivers, among them the Nile, the Hwang Ho, the Mackenzie, the Colorado, and the Mississippi, have built massive deltas at their mouths. Each delta has its own peculiarities, and none is so simple as the small delta shown in Fig. 9-30. The Mississippi delta, with an area of 12,000 square miles, not counting the submarine part, is in reality a complex of several coalescing subdeltas built successively during the last several thousand years. Each subdelta was begun by a flood that created a new distributary. Figure 9-31 shows the sequence of distributaries and resulting subdeltas, identified mainly from abandoned channels flanked by natural levees. In this, as in some other big deltas, the distinctions between topset, foreset, and bottomset beds are more subtle, owing to a number of complicating factors. No steeply dipping foreset beds occur where coarse sediment is not present and where the body of standing water is shallow.

Features of steep, rocky channels. Hitherto we have been discussing streams whose beds consist of sediment. Now let us turn to streams, common in mountain areas, whose channels are steep and consist of bedrock. In them turbulence is extreme, and three processes of channel erosion are evident: *hydraulic plucking, abrasion,* and *cavitation.*

Hydraulic plucking is *the lifting out, by turbulent water, of blocks of bedrock bounded by joints and other surfaces of weakness* (Fig. 9-32, A). The lift is accomplished by suction in strong eddies spiraling up around vertical axes. In some places boulder-sized blocks have been lifted through nearly 100 feet vertically.

Abrasion, cavitation, potholes. In streams, ***abrasion,*** *the mechanical wear of rock on rock,* is caused by friction and impacts between rock particles in the mechanical load and between particles in the load and bedrock in the channel. A stream uses its mechanical load as tools. By rubbing, scraping, bumping, and crushing it erodes bedrock and at the same time smooths and rounds the tools. Even where no bedrock is exposed in the channel, rounding of

[3]Foreset, bottomset, and topset beds should properly be called layers rather than beds because many are thinner than beds in the strict sense. However, because of its wide use, we retain the familiar term bed for the present.

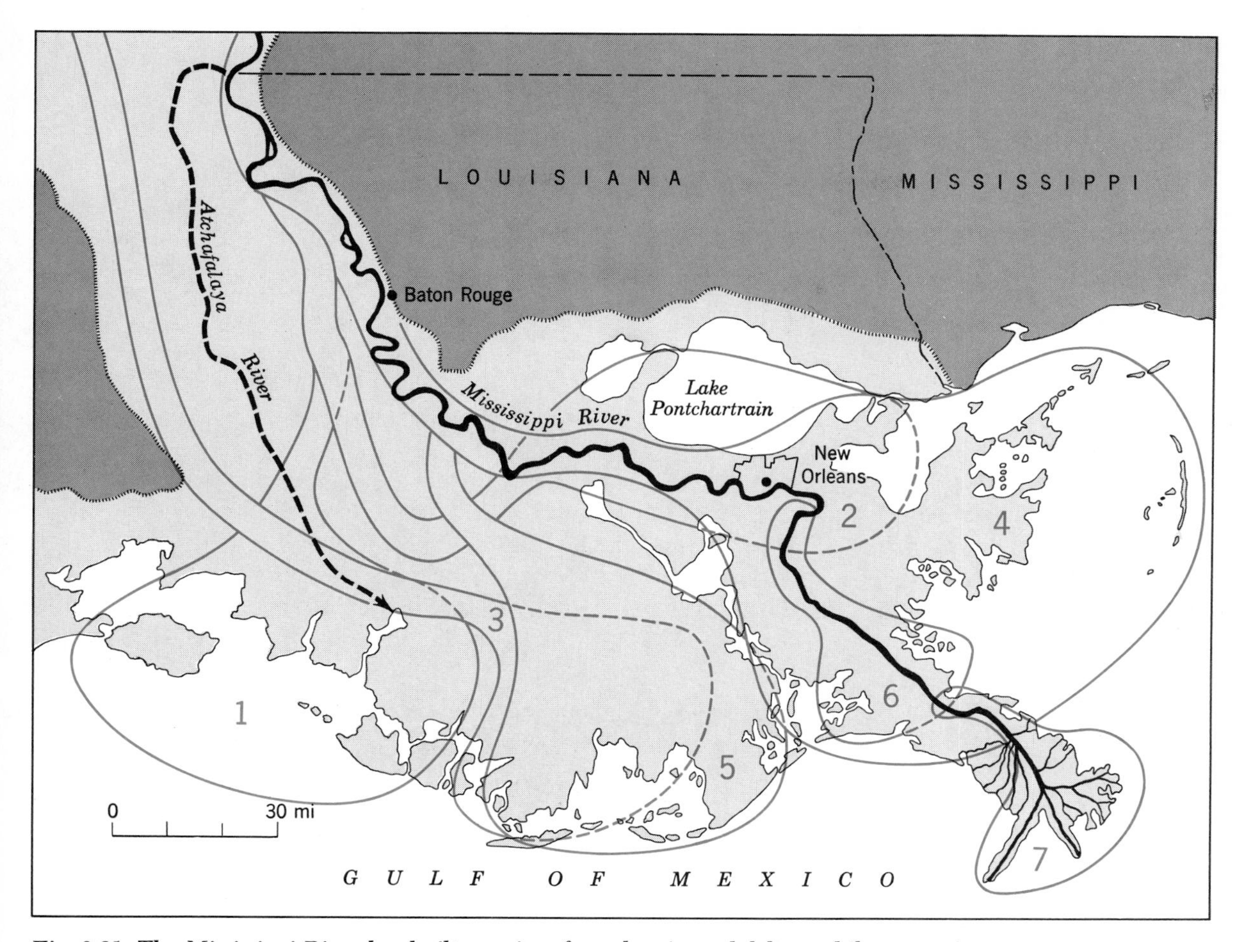

Fig. 9-31. The Mississippi River has built a series of overlapping subdeltas, while occupying successive distributary channels (numbers 1 to 7). Age of subdelta 1 is estimated at 3,000 years, of 3, 1,500 years, and of 5, 1,000 years. Construction of subdelta 7 began more than 100 years ago. Discharge of the Mississippi has been gradually shifting to the Atchafalaya distributary. By 1958, 28 per cent of the discharge was following this new route. Construction of a barrier to stop the diversion was begun in 1955. Had this not been done, the percentage would have increased to 40 by 1975. (After Kolb and Van Lopik, 1958.)

sand and coarser particles goes on, although rounding decreases rapidly as particle diameter decreases (Fig. 4-12).

Cavitation is *the formation and collapse of bubbles in a turbulent liquid.* In a violently turbulent stream, water molecules are pulled apart by momentary shocks to form bubbles that contain no air. When the shock has passed, the bubble collapses, and the energy of the impact of its walls against each other is very great, exerting pressures that can exceed 30,000 atmospheres. Such pressures can knock small chips off bedrock with cumulative results that can be large; but cavitation is difficult to appraise in nature because its effects are not easily separable from those of hydraulic action and impact.

Common features of a steep, rocky stream bed are ***potholes,*** *cylindrical holes drilled in bedrock by a turbulent stream,* where an eddy swirls pebbles and sand grains in spiral paths and abrades the bedrock (Fig. 9-32, *B*, *C*). It is likely that shallow potholes are made by cavitation, and that deep ones are commenced by cavitation and then enlarged and smoothed by abrasion.

Erosion at falls. Under the special conditions in which a stream flows over a rock ledge or cliff, as

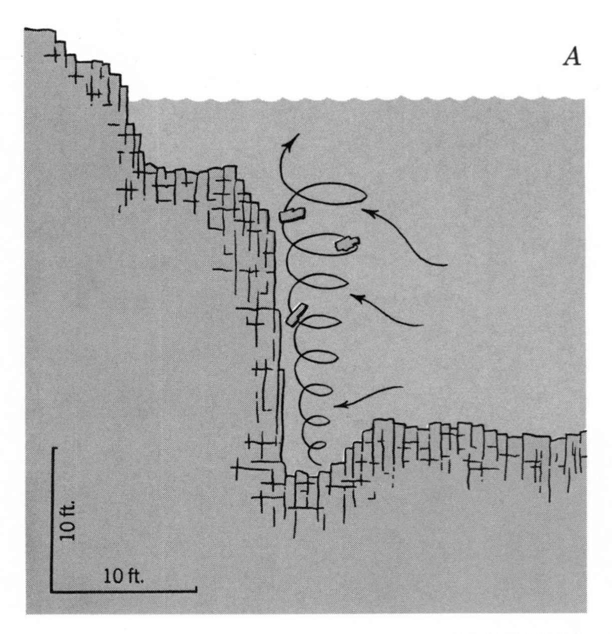

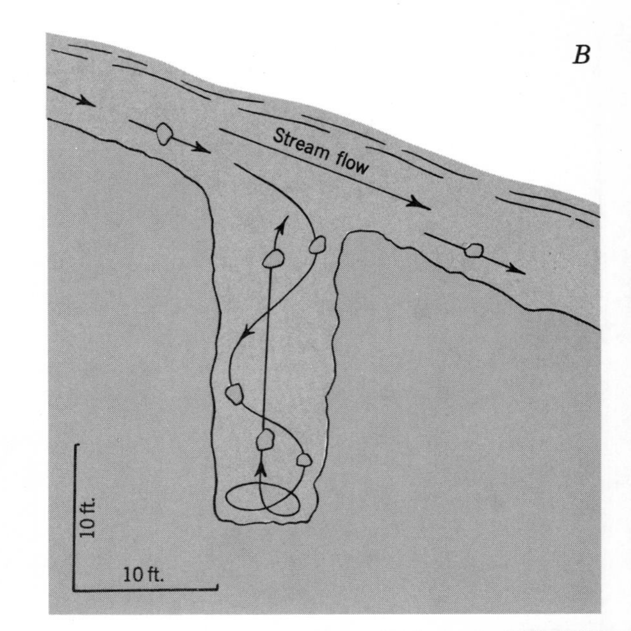

Fig. 9-32. Erosion of rocky stream beds by strong vertical eddies. *A.* Hydraulic plucking of jointed bedrock by upward-moving eddy. Partial cross section of Potomac River at Great Falls, upstream from Washington, D. C. (Pl. 9). Note blocks of rock being lifted. (After Gerard Matthes.) *B.* A deep pothole being drilled by a vertical eddy. Spiral paths of the eddy-transported pebbles are inferred from laboratory observation of model streams. (After Olof Ångeby.) *C.* Potholes in granite, exposed in the bed of a stream after water was diverted. James River, Virginia. (C. K. Wentworth, U. S. Geol. Survey.)

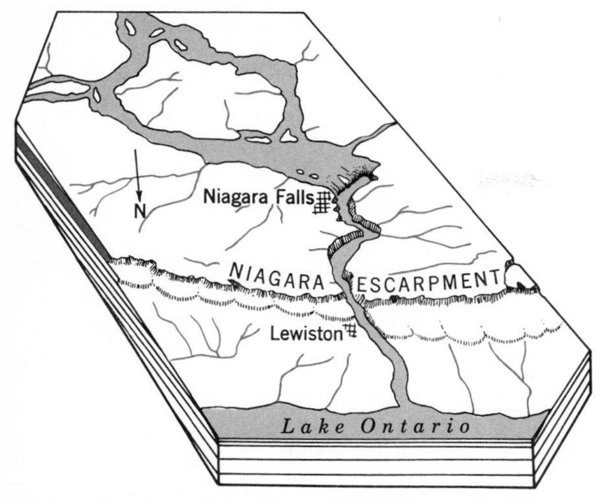

Fig. 9-33. Bird's-eye view of Niagara Falls, looking south, showing south-dipping strata and Niagara Escarpment, a great ledge formed by a resistant stratum, from which the Falls originated. By erosion the Falls has retreated through an aggregate distance of about 7 miles. Greatest length of block, 35 miles. Vertical exaggeration 2×.

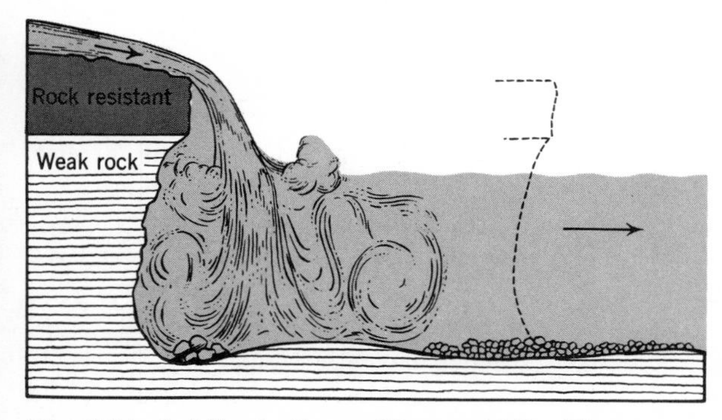

Fig. 9-34. A falls similar to Niagara Falls. The stream bed downstream from the falls was created by successive positions of the retreating cliffs; one former position is shown by the dotted line. Turbulence at base of falls keeps bedrock scoured nearly clean of sediment, but in less turbulent water downstream coarse gravel is deposited.

at Niagara Falls (Fig. 9-33), the increased velocity of the falling water sets up strong turbulence at the base of the falls, where both hydraulic plucking and abrasion deepen the stream bed. The cliff is gradually undermined and the falls retreats upstream. The retreat of the Canadian Falls at Niagara Falls has been rapid. Measurements by survey show that between 1850 and 1950 the rate of retreat averaged around 4 feet per year. This rapid rate is favored by the fact that the lip of the falls consists of strong, resistant dolostone beneath which is weak, easily eroded shale. As the shale cliff is eroded back, the dolostone lip is undermined and caves in piecemeal (Fig. 9-34). The long, deep gorge downstream from the falls was created by retreat of the falls, through successive positions, during periods of many thousands of years.

DYNAMIC EQUILIBRIUM IN STREAMS

We have seen that the characteristics of a stream and its channel are determined by the interplay, at each point along the stream, of factors such as discharge, velocity, channel geometry, and load. These factors continually adjust themselves to one another, and the sum of the adjustments tends toward a steady-state condition at each point. The conditions at each point, added together, combine to determine the concave-up curve that is the stream's long profile. The tendency toward a steady-state condition exists whether the stream's course is steep, narrow, and rocky, or wider and more gently sloping.

Anything that disturbs the steady state sets up a response by the stream, which tends to restore that state. For instance, if velocity is checked by the sea or a lake, the stream responds by building a delta and thereby smoothing the profile at its mouth. If the stream flows over a cliff, it responds by building a fan at the base of the cliff, thereby gradually restoring its smooth profile. Meanders and braided patterns likewise are approaches to equilibrium in response to special conditions.

In Chapter 10 we shall see how this strong tendency toward the steady state in streams is reflected in the sculpture of the lands.

Summary

Importance of running water

1. As part of the hydrologic cycle, streams are the chief means of returning water from land to sea. As geologic agents, stream erosion and mass-wasting are foremost among the processes that erode the land and transport sediments from land to sea.

2. Splash of raindrops and sheet erosion effectively erode regolith on bare, unprotected slopes.

Stream flow and behavior of rock particles

3. Turbulence characterizes the flow of nearly all streams. Turbulence is a prime factor in picking up and transporting sediment.

4. A stream's load consists of bed load, suspended load, and dissolved load.

5. Because cross-strata dip in the downstream direction, study of such strata in sedimentary rocks enables us to reconstruct the patterns of ancient streams.

Economy of a stream

6. Long profiles of streams are concave-up curves. The profiles become gentler with time but are limited downward by base level.

7. Discharge equals width times depth of channel times velocity. Hence stream and channel are intimately related and constantly adjust to each other.

8. Because of increased discharge and velocity, a stream can carry a load both coarser and greater in amount during floods, than it can transport at times of low water. Streams do most of their geologic work during seasonal floods.

Geologic work of streams

9. Features of streams in alluvial valleys include meanders, oxbow lakes, point bars, natural levees, and floodplains.

10. The migration of point bars, a chief factor in sedimentation by a meandering river, creates a distinctive succession of strata.

11. Fans are built at the toes of steep slopes; deltas are built at the mouths of streams. A common kind of delta consists of foreset, bottomset, and topset beds.

12. Streams tend to maintain a condition of dynamic equilibrium. If it is interrupted, the stream will return to it.

Selected References

Bagnold, R. A., 1960, Some aspects of the shape of river meanders: U. S. Geol. Survey Prof. Paper 282, p. 133–144.

Barnes, H. L., 1956, Cavitation as a geological agent: Am. Jour. Sci., v. 254, p. 493–505.

Colby, B. R., 1963, Fluvial sediments—a summary of source, transportation, deposition, and measurement of discharge: U. S. Geol. Survey Bull. 1181, p. A1–A47.

Davis, S. N., and de Wiest, R. J. M., 1966, Hydrogeology: New York, John Wiley.

Fisk, H. N., 1952, Mississippi River valley geology in relation to river regime: Am. Soc. Civil Engrs. Trans., v. 117, p. 667–682.

Hoyt, W. G., and Langbein, W. B., 1955, Floods: Princeton, N. J., Princeton Univ. Press.

Leopold, L. B., and Langbein, W. B., 1966, River meanders: Scientific Am., v. 214, p. 60–70.

Leopold, L. B., and Maddock, T., Jr., 1953, The hydraulic geometry of stream channels and some physiographic implications: U. S. Geol. Survey Prof. Paper 252.

Leopold, L. B., Wolman, M. G., and Miller, J. P., 1964, Fluvial processes in geomorphology: San Francisco, W. H. Freeman.

Livingstone, D. A., 1963, Data of geochemistry, 6th ed., Chap. G., Chemical composition of rivers and lakes: U. S. Geol. Survey Prof. Paper 440, p. G1–G64.

Shirley, M. L., ed., 1966, Deltas in their geologic framework: Houston Geol. Soc., p. 233–251, maps of existing deltas assembled by A. E. Smith, Jr.

Sundborg, Åke, 1956, The River Klarälven. A study of fluvial processes: Geograf. Annaler, v. 38, p. 125–316.

Motion-picture film: *Flow in Alluvial* channels (16 mm color with sound). Shows stream flow and examples of ripples and sand waves formed in a laboratory channel. Available for free loan on application to Map Information Office, U. S. Geological Survey, Washington, D. C.

CHAPTER 10

Sculpture of the Land by Running Water and Mass-Wasting

Relation of valleys to streams
Development of valleys
Sculptural activity as a complex open system
Cycle of erosion
Interruptions in sculptural evolution
Effects of unequal resistance of rocks
Classification and history of streams

Plate 10

Badlands, South Dakota.

(Josef Muench.)

CHAPTER 10 *Sculpture of the Land by Running Water and Mass-Wasting*

DRAINAGE SYSTEMS

Land sculpture as part of the geologic cycle. In Chapter 5 we saw that the "great geological cycle" first visualized by Hutton as the transformation and movement of materials from place to place has several aspects. In Chapters 7, 8, and 9 we followed the creation of rock particles by weathering, their movement downslope by mass-wasting and sheet erosion, and their transport in streams to places where they are deposited as sediments that will form new rocks. These activities are part of the rock cycle.

But the dynamic agents not only create, transport, and deposit sediments, they shape the land surface as well. The progressive removal of rock materials alters the configuration (the form) of the surface, so that several subcycles are in operation at the same time. While the rock cycle is going forward, and in the geochemical cycle, solutes are being carried away from higher to lower places, another cyclic activity is transforming the surface in a systematic way. The present chapter traces the shaping of the land, by weathering, mass-wasting, and running water, into systems of valleys.

Groups or families of valleys are the commonest features of the lands. Throughout wide regions the surface of a continent is little more than a complex of valleys, created by erosion and separated by higher areas that erosion has not yet consumed. Valleys exist in such great numbers that they have never been counted except in sample areas. The enormous number of valleys is commensurate with the huge volume of water runoff over the land.

Relation of valleys to streams. Before the middle of the eighteenth century, valleys were generally thought to be the result of catastrophes that somehow broke the Earth's crust and pulled it apart, creating paths for running water to follow. As far as we know, the idea that streams made their own valleys was first stated by Leonardo da Vinci. In 1802 the idea was more fully affirmed by John Playfair, the Scottish naturalist who did so much to interpret Hutton's "great cycle." Playfair pointed out two important relationships in valleys: (1) the size of a valley is proportional to the size of the stream that flows in it; (2) a stream's tributaries enter it generally at its own water level (as the Loup River enters the Platte; see Fig. 9-19). Playfair reasoned that if streams merely occupy valleys ready-made for them by some other agency these two relationships would be, as he put it, "infinitely improbable." Hence, he concluded, valleys are made by streams.

Today we can add that streams are actually observed to make and enlarge valleys; capacious ones have been created within the last 100 years (Fig. 10-5), and some are reported to be lengthening at rates as great as 1 mile per year. Furthermore, artificial rain creates runoff on model land surfaces set up in laboratories, and the runoff cuts miniature valleys analogous to the natural ones. The origin of valleys is therefore well established.

This does not mean that every depression through which a stream flows was created entirely by the stream. In some places running water does follow depressions already begun by other agencies, such as those which bend and fracture the Earth's crust (Fig. 4-3). In other places a tributary joins a main stream by cascading over a cliff. But special circumstances in the geology of such places explain the exceptions, and the general conclusion still holds true.

Drainage basins and divides. Every stream, or segment of a stream, has its ***drainage basin,*** consisting of *the total area that contributes water to the stream. The line that separates adjacent drainage basins* is a ***divide*** (Fig. 10-1). On the map (Plate B) we can trace approximately the divide that incloses the huge drainage basin of the Mississippi River, an area exceeding 40 per cent of that of the conterminous United States.

Other things being equal, the spacing of the streams in a drainage basin is orderly. When the streams in a large basin are measured accurately on a map, it appears that the distances between the mouths of tributaries are spaced in an orderly way, and that there is a mathematical relationship between the length of a stream and the area of its drainage basin. This orderliness measured on maps is analogous to the orderliness inherent in a stream's long profile, in which gradient decreases systematically from head to mouth, while discharge, velocity, and channel dimensions increase. What all these relationships imply is that in response to a given quantity of runoff, stream systems develop with just the size and spacing required to move the water off each part of the land with maximum efficiency.

Surface slopes converge toward the heads of small valleys, and runoff, moving at first as overland flow, soon concentrates in valleys (Fig. 10-2). We must realize that these slopes and valleys, in fact all the parts of a drainage system and the configuration of the land related to it, were developed under a single group of controls, chief of which are type of underlying rock and climate.

The drainage system of the Mississippi River not only is very large but also is old. Various parts of it came into existence at different times, and the climates under which they developed are not known with certainty. But a system does not necessarily require a long time to develop, as is indicated by the following example. In August, 1959, an earthquake occurred at Hebgen Lake, near West Yellowstone, Montana. The movement tilted the country in such a way that a large area of silt and sand, formerly part of the lake bed, emerged and was subject to runoff. Small-size drainage systems began to develop immediately. Sample areas were surveyed and mapped 1 year and 2 years after the earthquake occurred. The results showed the same basic geometry that characterizes much larger and older systems. The small, newly formed valleys, together with the areas between them, were disposing of the available runoff in a highly systematic way, all within a period of 2 years after the surface had emerged from beneath the lake.

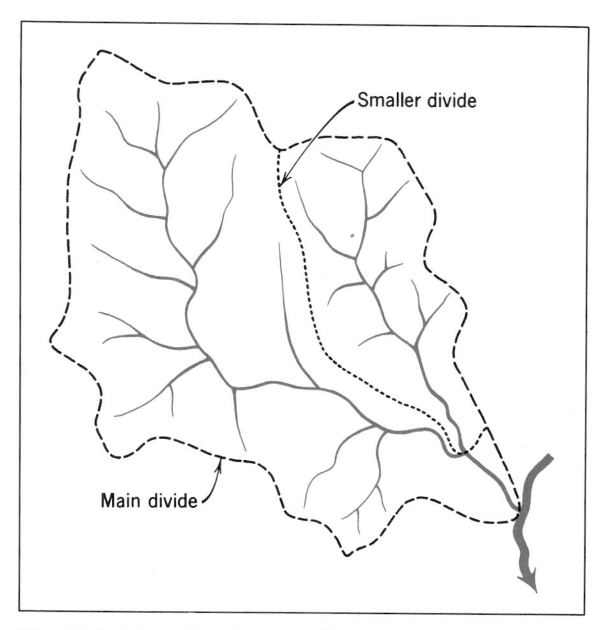

Fig. 10-1. Map of a drainage basin, inclosed by a main divide, ending downstream at a large river. A smaller drainage basin, inclosed on one side by a smaller divide, defines a tributary area. Other, still smaller divides within the basin are not shown. Patterns made by streams are like branching trees.

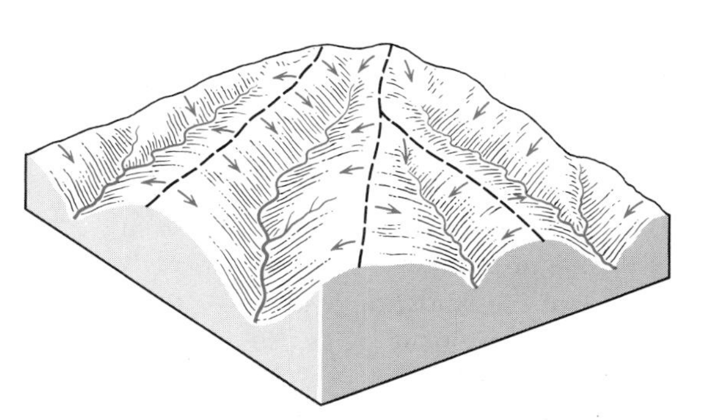

Fig. 10-2. Runoff moves over hillslopes as overland flow (suggested by arrows), soon concentrates in small valleys. Broken lines indicate divides.

Valleys, sheet erosion, and mass-wasting. If the sole agency involved in cutting a valley were the stream that flows through it, the valley should be as narrow and steep-sided as the one shown in Fig. 10-3, resembling a cut made by a saw through a block of wood. The shaping of most of the land surface, including the valley sides themselves, is mainly the work of sheet erosion and mass-wasting of weathered rock material.

Of course the rate of downslope movement of regolith is increased by stream erosion, which, by deepening a valley, increases local differences of

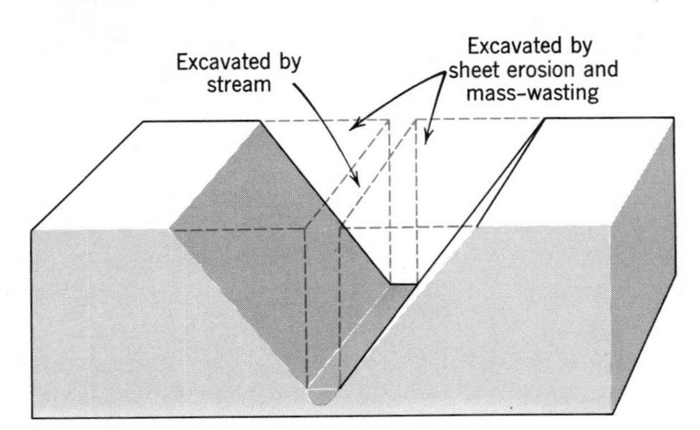

Fig. 10-3. In this idealized valley segment the volume of rock excavated by the stream is compared with the much greater volume excavated by sheet erosion and mass-wasting. However, all the waste from the slopes, while being transported out of the area, had to pass through the stream.

height and slope. Conversely, the load fed to the stream by processes on the slope affects the other variables in the stream. This is the basis of the shaping of valley sideslopes, by continual adjustment between processes acting on slopes and those acting within the stream channel.

SCULPTURAL EVOLUTION OF THE LAND

The sculpture of the lands is a grand destructive process, in which the forces of erosion are pitted against the resisting forces of cohesion within bedrock, regolith, and plant cover. The erosional forces are limited downward by base level, below which they can not work. The raising of any land above base level creates disequilibrium between two parts of the Earth's crust, but transfer of rock particles from the upraised land to the sea works to restore equilibrium. In the process the land is *sculptured* into a series of valleys and hills (Fig. 10-4). The sculptured surface, slowly and continuously changing, is the expression resulting from (1) uplift of the crust modified by (2) destructive erosion of the uplifted area.

Open-system concept. We can think of this whole complex of processes as an open physical-chemical system, which we can assume is operating on a land mass that has been uplifted and thereafter remains stable. Apart from the uplift, the input consists basically of rainfall, occurring on a land surface that varies in height, and new sediment and solutes derived from weathering of bedrock. The output consists of water, sediment, and dissolved substances delivered to the sea. The system therefore consists of a water part, which is continually recycled, and a rock-material part, which is progressively lost. The water is being continually lifted from the sea by solar energy and carried again over the land (Fig. 5-3). When precipitated on the land, the water has potential energy determined by the height of the land above base level. The energy expended in converting the water input to output is potential energy that has been converted into kinetic energy and that is consumed in the transport of rock material down the slopes and down the streams to base level. The progressive loss of rock material gradually lowers the surface and so reduces the potential energy available to the water part. The system as a whole, therefore, undergoes continuous change.

Quickly or slowly, the various segments of the system approach a steady-state condition; that is, one in which average output of water, sediment, and solutes equals average input derived from rainfall and weathering. In the steady-state condition, dynamic equilibrium exists between the land forms and the processes acting on them. On a hillslope underlain by homogeneous rocks, this condition of steady state, or dynamic equilibrium, is expressed in smooth curves without abrupt change in slope. Geologists have long said of such slopes that they are *graded.* This is merely a short way of saying that the slopes are part of an open system that has reached a steady state.

As time passes, more and more segments of the surface become graded, or reach dynamic equilibrium. But, strictly speaking, the whole system, as a single unit, can never reach static equilibrium because it is continually losing rock matter. This fact need not disturb us, however, because we are concerned only with individual parts, or segments, which we can observe and measure and within which input and output can balance.

In our open system, potential energy, measured by height of the country above base level, decreases with time and so do gradients. We can expect that the system will reach a condition of low energy, in which lands are low, slopes are gentle, and loads transported mechanically by streams are smaller than under earlier conditions of higher energy.

Sequence of sculptured forms: cycle of erosion. Having looked at land sculpture in terms of an open physical-chemical system, we can now examine it descriptively in terms of that part of Hutton's "great

Fig. 10-4. In western Pennsylvania, where annual rainfall is about 40 inches, the land surface consists largely of smooth slopes covered with regolith. Some strips of grass parallel to contours on cultivated slopes serve to check erosion of soil. Note strong contrast between this landscape and that shown in Fig. 1-8. (Ewing Galloway.)

geological cycle" which geologists have long called the ***cycle of erosion*** (although "cycle of land sculpture" might be a more precise name for it). We can define this cycle as *the sequence of forms,* essentially valleys and hills, *through which a land mass is thought to evolve from the time it begins to be eroded until it is reduced to near base level.*

In controlled laboratory experiments, in such miniature natural examples as the one we described at Hebgen Lake, and in the man-induced erosion of farm land (Fig. 10-5), it is evident that as rain falls on a new surface, the runoff creates drainage systems whose parts are systematically related, and that the streams *dissect* (cut up) the surface. Aided by weathering, mass-wasting, and sheet erosion, the streams shape, or sculpture, the dissected surface in the familiar pattern of hills and valleys.

We can observe this whole process of sculpture, from beginning to end of the cycle, only in miniature examples, because in a wide region the changes take place much too slowly. However, when we compare a number of land surfaces, each in a different state of dissection, we perceive that their relation to each other is orderly. They can be arranged in a continuous series, each differing only slightly from the one preceding it. It seems probable, therefore, that a single land mass subjected to erosion will progress, in the course of time, through all these states. If this is true, we can predict, at least in a general way, the sculptural evolution that will take place.

The evolution is generally thought of as passing through three stages—*youth, maturity,* and *old age*—in analogy with stages in the life cycle of an individual human being. Although these three terms are useful for general description we can not define them because no real boundaries separate them; they merge into one another. Furthermore the terms, even though we call them *stages,* have nothing to do with the absolute ages we discussed in Chapter 6; they do not refer to specific periods of time. After this caution, we can proceed to characterize the stages in the cycle of erosion.

In the stage of youth, stream gradients tend to

Fig. 10-5. This valley in Stewart County, Georgia is growing headward, swallowing up farm land and threatening a road. Now more than 75 feet deep, it started more than 100 years ago. Many valleys like this one start in a rutted dirt road or in a field with furrows running downslope. (U. S. Soil Conservation Service.)

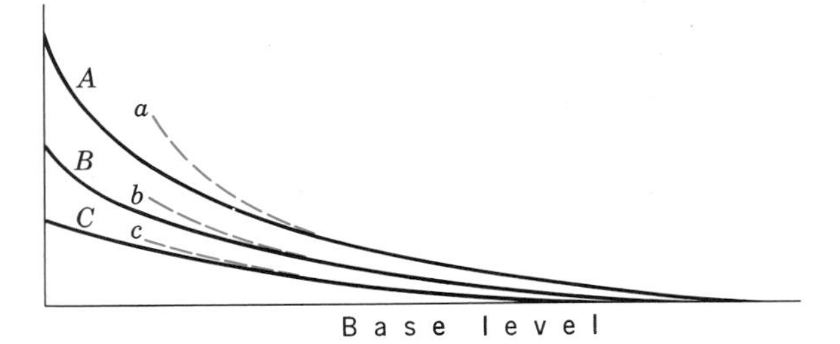

Fig. 10-6. Successive long profiles of a main stream (*A, B, C*) and one of its tributaries (*a, b, c*), changing with the passage of time. Gradients progressively decrease, but at all times the tributaries are nicely adjusted to the main stream, which in turn is adjusted to its base level.

be steep (Fig. 10-6, *A, a*) and erosion rapid. Valleys are actively deepened and make sharp cuts into the land increasing the relief. But they have not reached their full lengths, so that broad areas of the original surface remain uncut.

In the stage of maturity, valleys have lengthened so that the entire surface is dissected, but stream gradients have become gentler and streams are eroding more slowly. Slopes have developed smoothly curved profiles (Fig. 10-4).

The mature surface is lowered toward base level very slowly, with the rate of erosion becoming increasingly slow as stream gradients and hill slopes become ever more gentle. A surface in a late phase of erosion, lying close to base level, is in the stage of old age.

Peneplains. A land surface in a very late phase of old age is a ***peneplain*** ("almost a plain"), defined as *a land surface worn down to very low relief by streams and mass-wasting.* With ever-decreasing energy, erosion of a mature surface becomes so slow that many millions of years are necessary for the creation of an extensive peneplain. Despite its generally low relief, high steep hills can be present on it in places. *Conspicuous residual hills on a peneplain* are termed ***monadnocks*** and, as might be expected, are likely to be underlain by stronger, more resistant rocks than those beneath the area that surrounds them. The name was taken, in 1893, from Mount Monadnock in New Hampshire, which at that time was believed to be a residual hill on an ancient peneplain. Current opinion as to that par-

Fig. 10-7. Result of accelerated erosion. The smooth surfaces of the interstream areas are remnants of a formerly continuous surface in a steady-state condition, essentially a peneplain. Following a sudden and considerable increase in stream energy, streams are deepening their valleys and are extending them headward; the surface is being rejuvenated. Probably the energy increase resulted from faulting, which lowered the extreme foreground relative to the land next behind, thereby increasing the fall from stream heads to stream mouths. View west across south end of Inyo Range to Sierra Nevada, California. (John H. Maxson.)

ticular peneplain is divided, but the name continues in good standing.

We do not know of a single peneplain that exists today in an undisturbed state. Numerous peneplains have been submerged beneath the sea, deeply buried by thick layers of sediment and later partly exposed by erosion of their covers of sedimentary rock so that we can see small parts of them. Others have been subjected to uplift and have been dissected so that again only small parts remain; an example is shown in Fig. 10-7. No peneplain, therefore, is seen today lying unaltered in the position in which it was made. This fact implies considerable instability of the Earth's crust during at least the last few million years of geologic time for, if the crust had remained quiescent, sculptural evolution of the lands should be more advanced than it is.

In conclusion, the sculptural evolution of a land

mass is most accurately thought of in terms of a complex open system. Although the alternative concept of the cycle of erosion can not be described in quantitative terms, it is nevertheless a useful aid in visualizing the kinds of changes to be expected during the slow reduction of a land mass toward base level.

The sculptured forms that constitute the cycle of erosion differ in detail, although not in principle, with differing climates. The forms we have just described characterize regions of moist climate. In Chapter 13 we shall describe the cycle in very arid regions, so that the sequences of forms can be compared.

RATES OF EROSION

The cutting of valleys and the resulting loads of rock particles in streams are known through actual observation. The rate of erosion of the land is known also, more broadly than can be inferred from soil-erosion values like those in Fig. 9-3. The rate is computed from measurements of suspended and dissolved loads in streams over a wide region. From such computation it is calculated that the surface of the conterminous United States is being stripped away at an average rate of about 6cm/1,000 years (2.4 inches/1,000 years). This value is subject to a considerable margin of error because nonmeasurable bed loads are not included and because by no means every stream has gaging stations.

Erosion of 1cm in 166 years, the lifetime of 8 human generations, may seem a slow rate, but it involves the yearly removal to the sea of nearly 1.5 billion tons of rock substance from the area of the United States alone. Since prehistoric, stone-age man hunted big game in the United States 10,000 years ago, rock material equivalent to a layer 60cm thick must have been removed to the sea.

The rate we have just quoted applies only to one large region (the United States) at one time (the geologic present). Rates of erosion in other regions are different, depending on altitudes, kinds of rock present, and climate. Rates in the United States certainly have differed in the past, in accordance with changes in the same factors.

It is not easy to estimate the time needed to accomplish the reduction of a wide region to a peneplain. A recent estimate, based on various assumptions, places the time required at between 15 and 100 million years.

INTERRUPTIONS IN SCULPTURAL EVOLUTION

Unstable systems: rejuvenation and filling. Our description of a land progressing slowly through a cycle of erosion assumed that the progress of erosion was smooth and uninterrupted by any outside influences. However, when we look critically at valleys we can see evidence that the stable system has been interrupted. Because a part of an open system that has attained a steady state does not become unstable unless altered by a disturbing force, we conclude that an interruption, in the form of a pronounced change in the energy of part or all of the system, has occurred. The interruptions fall into two groups, according to whether they have caused conspicuously increased erosion or greatly increased deposition.

Rejuvenation: stream terraces. Figure 10-7 illustrates a marked interruption in the sculptural evolution of an area in southeastern California. The interruption was apparently caused by faulting, which by steepening the slope greatly increased stream energy upstream from the fault. The relations are like those shown in Fig. 9-22. The increased stream energy is applied to erosion (downcutting) in the channels. This steepens the sideslopes of the valleys, so that erosion of them also is increased. The result is general instability. A land mass in this condition is said to have been *rejuvenated* because, after reaching maturity or old age, it has taken on anew the characteristics of youth. In Fig. 10-5, which should be compared closely with Fig. 10-7, rejuvenation is occurring on a small scale. Observe in Figs. 10-7 and 9-22 that the streams are already building fans at the bases of the steepened slopes. In so doing they are beginning to restore the stable condition that prevailed before the interruption occurred. ***Rejuvenation***, then, is *the development of youthful topographic features in a land mass further advanced in the cycle of erosion.*

Stream terraces likewise result from increased erosion. A ***stream terrace*** is *a bench along the side of a valley, the upper surface of which was formerly the alluvial floor of the valley.* In a stream flowing on a broad valley floor sudden increase in rate of erosion results in the cutting of a new valley within the older one. The floor of the older valley is left as a pair of stream terraces (Fig. 10-8), which, of course, will in time be entirely destroyed by erosion.

Stream terraces of another kind common in

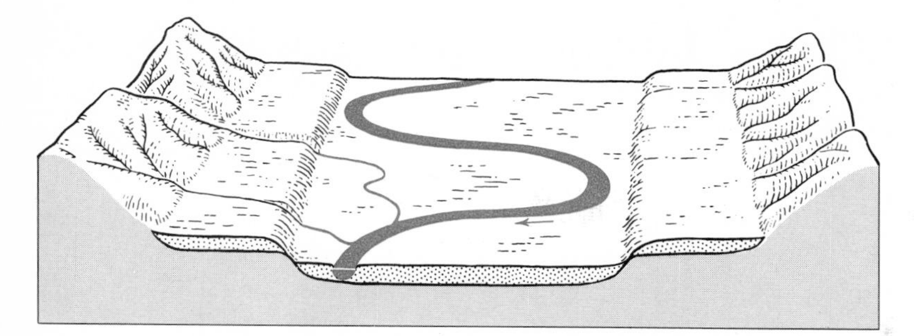

Fig. 10-8. Stream terraces (cut from alluvium-covered bedrock) result from a stream's incision of its valley floor. Such terraces commonly occur in matched pairs (same height on both sides of valley.) Although the terraces shown have been cut from bedrock thinly covered with alluvium, they can be cut from thick alluvial fill as well. Because the stream meanders, alluvium consists mainly of point-bar sediments. Near valley walls it is likely to include much colluvium.

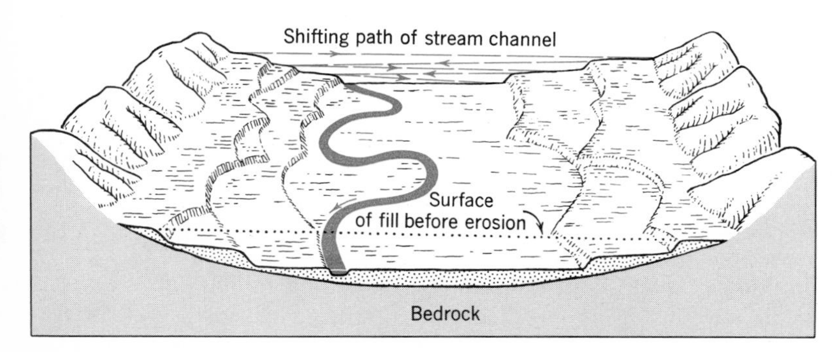

Fig. 10-9. Alluvial fill in a valley, later eroded to form stream terraces (cut from alluvium only). These events are recorded: 1. The stream excavated a broad valley in bedrock. 2. Reduction of energy in relation to load resulted in deposition of alluvium, half filling the valley. 3. Increase of energy in relation to load caused rejuvenation; meandering stream was deflected wherever it encountered bedrock beneath the alluvial fill, leaving terraces at various heights. Such terraces do not match up across the valley. Paired terraces, like those in Fig. 10-8, can occur in fill as well as in bedrock.

valleys that have been filled with alluvium and that do not occur in pairs are shown in Fig. 10-9.

Alluvial fills. The thickness of sediment deposited by a stable stream on its valley floor is slight; at most it can not exceed the depth of the stream when in flood. In small streams it may amount to no more than a few feet. Yet many valleys, even small ones, are filled or partly filled with alluvium scores or hundreds of feet thick. An ***alluvial fill*** (*a body of alluvium, occupying a stream valley, and conspicuously thicker than the depth of the stream*) indicates that some past event must have reduced stream energy below the level necessary to carry the load. As a result, the stream deposited part of its load, building up alluvium, until it had again reached a stable condition. But the new long profile is higher than the former one. Figures 10-9 and 10-10 are examples. Probably the kind of event responsible for most alluvial fills is variation in climate, through which the rate and character of weathering and

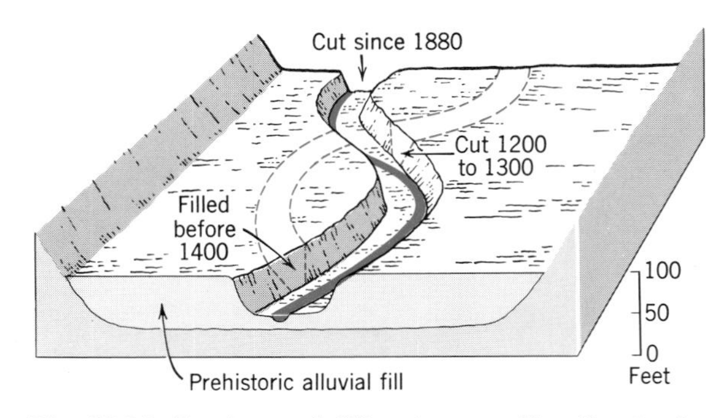

Fig. 10-10. Erosion and filling in a small valley in the southwestern United States, dated as described in the text. Between A.D. 1400 and A.D. 1880 the stream changed its position, so that the older fill is transected by the existing valley. (After Sheldon Judson.)

mass-wasting and the rate of sheet runoff and stream discharge are altered.

Causes of interruptions

1. Movements of the crust. Some interruptions in the stability of streams, valleys, and slopes can be traced directly to movements of the Earth's crust. If the upstream part of a drainage basin is elevated in relation to the downstream part, energy increases because stream gradients increase and rejuvenation results. The streams flowing down the western slope of the Sierra Nevada in eastern California have been rejuvenated repeatedly by successive elevations of the mountain range. Conversely, the western part of the same land mass in central California has been bent down and buried beneath accumulating sediments.

2. Change of base level. Rise and fall of sea level (Chap. 14) change the base level of streams and can cause filling and erosion, respectively, in the segments of valleys that are near the sea. The making of a dam across a valley by a fan, a landslide, a glacier, a lava flow, or even by human construction creates a local base level and can cause alluvial filling in the valley that is dammed. Erosion of the dam then can result in erosion of the fill.

3. Glacial sediments. A melting glacier commonly delivers so large a load of sediment to a stream that the stream can not carry it away and deposits it as a fill. The lower Mississippi valley contains an alluvial fill more than 200 feet thick, believed to have resulted partly from the deposition of a copious load of glacial sediment and partly from rise of sea level.

4. Change of climate: erosion and filling in dry regions. In the dry southwestern part of the United States deepening and headward extension of innumerable small valleys have been going on since about 1880. The valleys are bare, steep-walled canyons cut into soft rock and loose sediment. Some of the canyons have grown headward as rapidly as 1 mile per year and have been eroded to depths as great as 75 feet. Yet Spanish settlers, before the accelerated erosion began, found the land surface stable and covered with vegetation.

Study of the valleys and of weather records suggests that the cause lies mainly in very slight changes of climate. During long periods of drought, with few but heavy rainstorms, the grass cover deteriorates and lays the ground bare to erosion. During long periods with more frequent but lighter rains the grass cover improves and the valleys tend to fill with alluvium.

A period of few though heavy rains during the last half of the nineteenth century is believed to have caused the erosion now in progress, and overgrazing by cattle and sheep is thought to have accelerated the process. Some of the valleys contain clear evidence of repeated erosion and refilling (Fig. 10-10). Fragments of ancient Indian pottery buried in the alluvium, coupled with records of ancient Indian migrations, give approximate dates of an earlier erosion and filling. A still-earlier erosion is probably prehistoric, dating back to several thousand years ago.

Both rejuvenation and alluvial fills are so widely prevalent in drainage systems that land masses evidencing only one period of cutting and no filling are rather rare. From this we conclude that interruptions in the process of sculptural evolution are not the exception but the rule. If this is the case, then the concept of the cycle of erosion represents a general trend, which can be obscured or reversed by the common interruptions.

EFFECTS OF UNEQUAL RESISTANCE OF ROCKS

Effects on profiles of streams and valleys. The extent to which rocks resist erosion by streams and mass-wasting affects streams and valleys in several

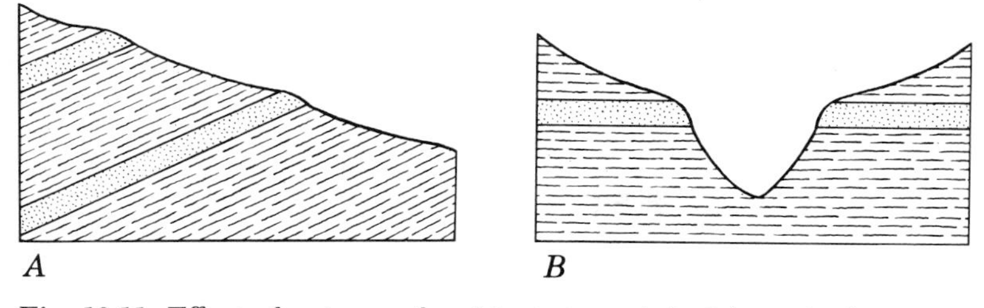

Fig. 10-11. Effect of outcrop of resistant strata (stipple) on the long profile (*A*) of a stream and on the cross profile (*B*) of a valley.

ways. When a stream flows over a layer of resistant rock, its long profile is steepened (Fig. 10-11, *A*). An extreme example is shown in Fig. 9-34. A resistant stratum exposed along the sides of a valley gives the cross profile of the valley a steplike form (Fig. 10-11, *B*). A valley is likely to be narrower where it cuts resistant rock than where it cuts weak rock. A ***water gap*** (*a pass in a ridge or mountain, through which a stream flows*) is a common feature at such a place (Fig. 10-12). The figure illustrates also that mass-wasting, sheet erosion, and stream erosion have lowered the surface underlain by weak rocks so effectively that the narrow belt of resistant rock has been left standing as a ridge above the general surface. The resistant rock is being eroded but at a slower rate. When many such differences in rock resistance are present, a morphology consisting of alternating ridges and lowlands develops.

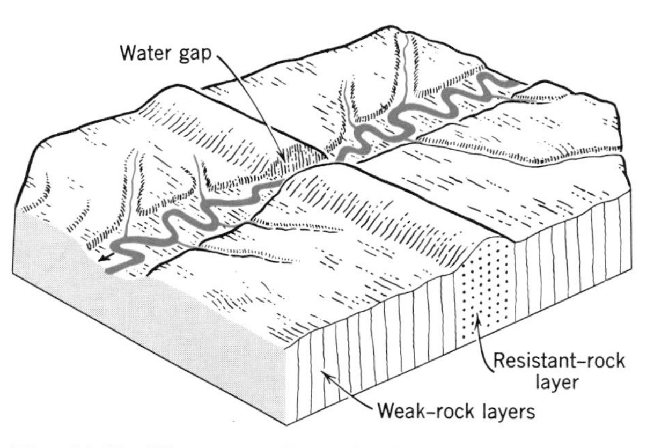

Fig. 10-12. Water gap formed where stream has cut through layer of resistant rock. In such rock, a valley is narrower and has steeper sides and gradient than in weak rock. Water gaps are common in the Appalachian region (Fig. 21-4).

Stream patterns. Not only the profiles of streams and valleys but also their patterns, as seen on a map, are affected by the kinds of rock on which they are developed. Stream patterns, however, are affected also by the history of the areas in which they occur. Three common kinds of stream patterns are shown in Fig. 10-13.

The ***dendritic*** ("treelike") ***pattern*** is *a stream pattern characterized by irregular branching in many directions.* This pattern is common in massive rocks and in flat-lying strata. In such situations, differences in rock resistance are so slight that their control of the directions in which valleys grow headward is negligible.

The ***rectangular pattern*** is *a stream pattern characterized by right-angle bends in the streams.* Generally it results from the presence of joints (Fig. 4-3) and faults (Table 17-1) in massive rocks or from foliation (App. C) in metamorphic rocks; structures such as these, with their geometrical patterns, have guided the directions of headward growth of valleys.

The ***trellis pattern*** is *a rectangular stream pattern in which tributary streams are parallel and very long,* like vines or tree branches trained on a trellis. This pattern is common in areas like the Appalachian region, where the edges of folded sedimentary rocks, both weak and resistant, form long, nearly parallel belts.

CLASSIFICATION AND HISTORY OF STREAMS

Kinds of streams. On the basis of their patterns and other characteristics, streams are classified into four groups, labeled consequent, subsequent, antecedent, and superposed. The streams in each group have distinctive origins and histories.

A ***consequent stream*** is *a stream whose pattern is determined solely by the direction of slope of the land.* Therefore, consequent streams generally occur in massive and flat-lying rocks and commonly have

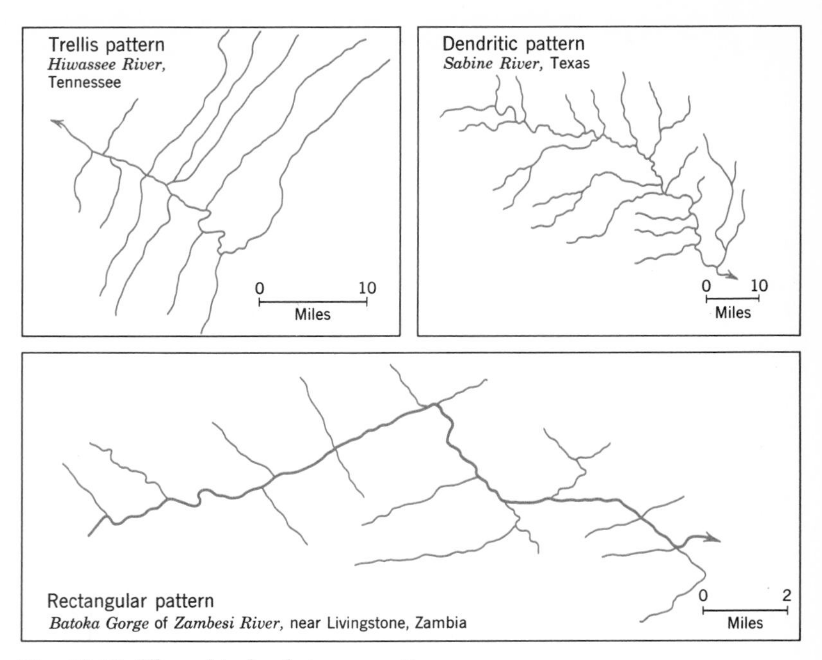

Fig. 10-13. Three kinds of stream patterns.

dendritic patterns (Fig. 10-14).

A ***subsequent stream*** is *a stream whose course has become adjusted so that it occupies belts of weak rock.* When such belts are long and straight, subsequent streams constitute the long straight tributaries characteristic of trellis drainage patterns. Figure 10-14 illustrates the difference between consequent and subsequent streams.

An ***antecedent stream*** is *a stream that has maintained its course across an area of the crust that was raised across its path by folding or faulting* (Fig. 10-15). The name arises from the fact that the stream is antecedent to (older than) the uplifting.

A ***superposed stream*** is *a stream that was let down, or superposed, from overlying strata onto buried bedrock having composition or structure unlike that of the covering strata* (Figs. 10-16, 17-9, *C*). Most superposed streams began as consequents on the surface of the covering rocks. The streams' paths, therefore, were not controlled in any way by the surfaces on which they are now flowing.

Stream capture. When the gradient of one of two streams, flowing in opposite directions from a single divide, is much steeper than that of the other, the steeper stream can extend its valley headward, shifting the divide against the other stream. In this way the steeper stream can capture the other one little by little. Alternatively, it can capture at one stroke a long tributary of the other stream by intersecting the tributary at its mouth. This process of ***stream capture*** (or *piracy*), *the diversion of a stream by the headward growth of another stream,* is illustrated in Fig. 10-17. The Provo River shifted the divide at its head northward and eastward a distance of several miles until the divide intersected and diverted a principal tributary of the Weber River. Evidence of capture is of two kinds: (1) an abandoned segment of the valley of the diverted stream; (2) tributary streams that are barbed with respect to the new (main) stream they have joined. A ***barbed tributary*** is *a tributary that enters a main stream at an angle that is acute in the downstream direction.*

Adjustment of streams. In a long process of sculptural evolution, a stream system tends to adjust itself to the pattern of the rocks it drains, so that more and more stream segments occupy belts of weak rock or follow joints, faults, and other avenues of easy erodibility. A ***well-adjusted stream system*** is *a system in which most of the streams occupy weak-rock positions.* Through the headward growth of subsequent streams and occasionally through capture, the degree of adjustment of a stream system tends to increase with time and with depth of erosion. The result is a surface that reflects the

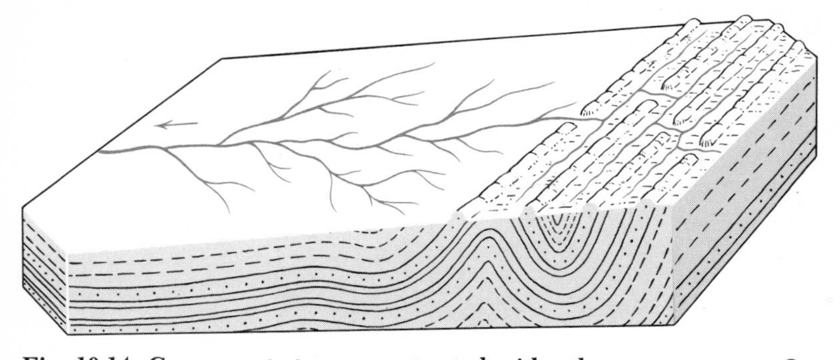

Fig. 10-14. Consequent streams contrasted with subsequent streams. On the left the land surface is underlain by flat-lying strata. Drainage has developed under control of the slope of the land (shown by arrow) and is therefore consequent. To the right the same strata are folded. On them tributaries developed most readily along parallel belts of weak rock which determined stream locations. These tributaries are therefore subsequent streams. The main stream crosses ridges of resistant rock through water gaps.

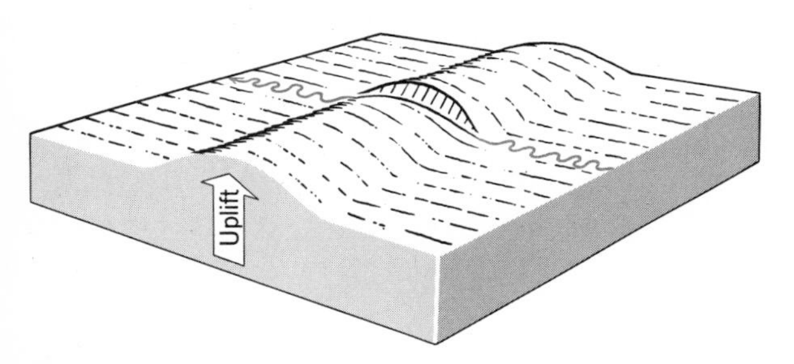

Fig. 10-15. This stream has an antecedent relationship to the present surface because of local uplift across its course. Stream has cut a deep gorge across the uplifted belt.

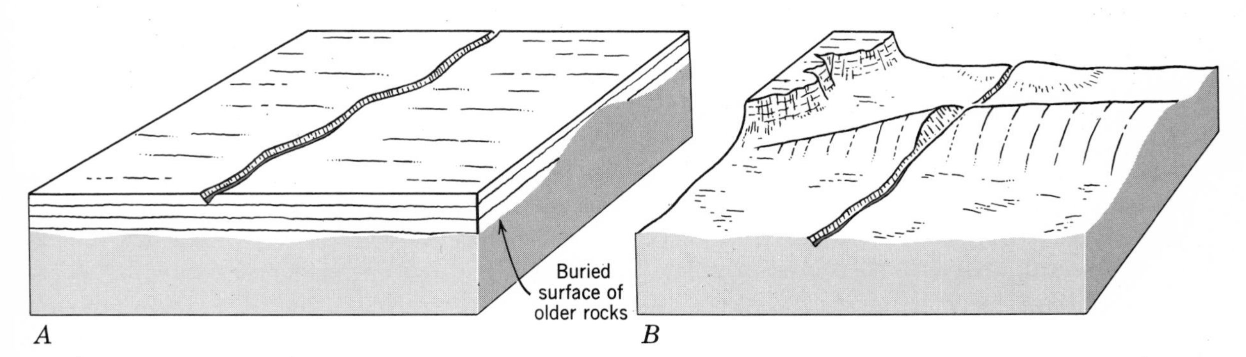

Fig. 10-16. Development of a superposed stream. *A.* Stream consequent on strata that bury a former land surface. *B.* After long-continued erosion the stream has become superposed and has cut a water gap through a hill that formed part of the older surface. Overlying strata have been removed by erosion, except for remnant in upper left. Compare Fig. 10-15.

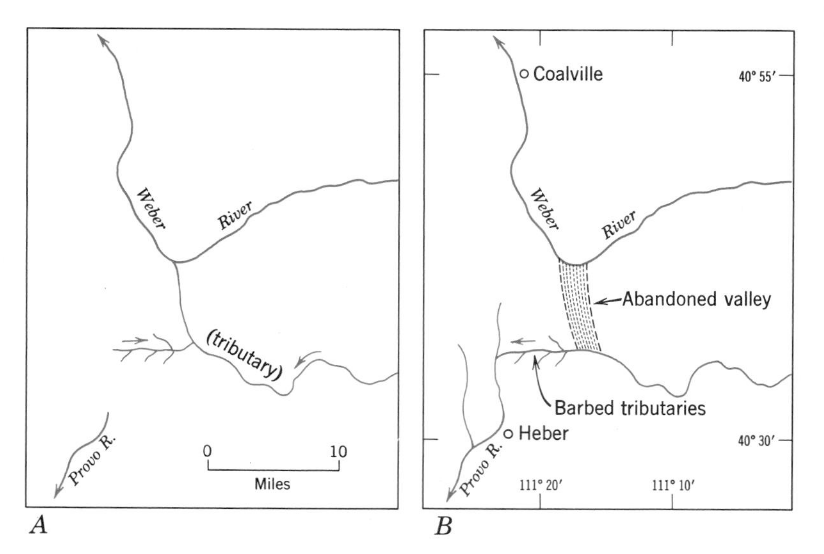

Fig. 10-17. Capture of tributary to Weber River by Provo River, upstream from Coalville, Utah. *A.* Reconstructed drainage pattern of an earlier time. *B.* Present drainage pattern. Provo River has extended its valley headward, capturing several small tributaries to Weber River and also the large tributary now part of Provo River. Abandoned segment of valley of former north-flowing stream is floored with stream gravel derived from the territory of the diverted stream. Small tributaries have barbed pattern, showing former flow toward the east. Probable cause of capture: Provo River, shorter and with a steeper gradient than Weber River, had the greater erosion potential. (After G. E. Anderson.)

pattern of the exposed belts of rock. The resistant rocks form hills, ridges, and other highlands, and the weak rocks underlie valleys and lowlands (Figs. 10-12, 10-14, 21-4).

An example of good adjustment is the segment of the Delaware River drainage shown in Fig. 10-18, in which the principal tributary streams coincide with areas underlain by weak rocks. Even more strikingly, the Delaware River, whose general course lies "across the grain" of the rocks, itself successfully avoids some of the belts of resistant rock and crosses others either at narrow places or at places where the rock is dislocated by faults.

How the tributaries became adjusted is fairly evident. Probably they are the successful competitors among many small streams whose valleys gradually grew headward from the Delaware at an earlier time in history. Those that started on resistant rocks developed very slowly; those on weak rocks lengthened more rapidly and secured most of the potential drainage area. "For he that hath, to him shall be given; and he that hath not, from him shall be taken. . . ." Out of context, this is an apt rule for the competition that occurs in a drainage system.

How the Delaware itself became adjusted is less easy to visualize, and we may never know in detail what led to its pattern. One hypothesis ascribes it to a similar competition between the young Delaware and other parallel streams, as all grew headward (northwestward) up the regional slope. The successful competitor was the one with the least amount of resistant rock to traverse and with a fault to help it at one place—in short, it was lucky. Whether this explanation is the true one is not known. In every area of complex rocks the locations of streams present puzzles, some of which have been solved with a good degree of probability. But others may never be solved, simply for lack of enough surviving evidence. Erosive forces not only sculpture the land but also tend eventually to destroy their own work and with it the evidence of how the work was done.

In contrast to streams such as the Delaware,

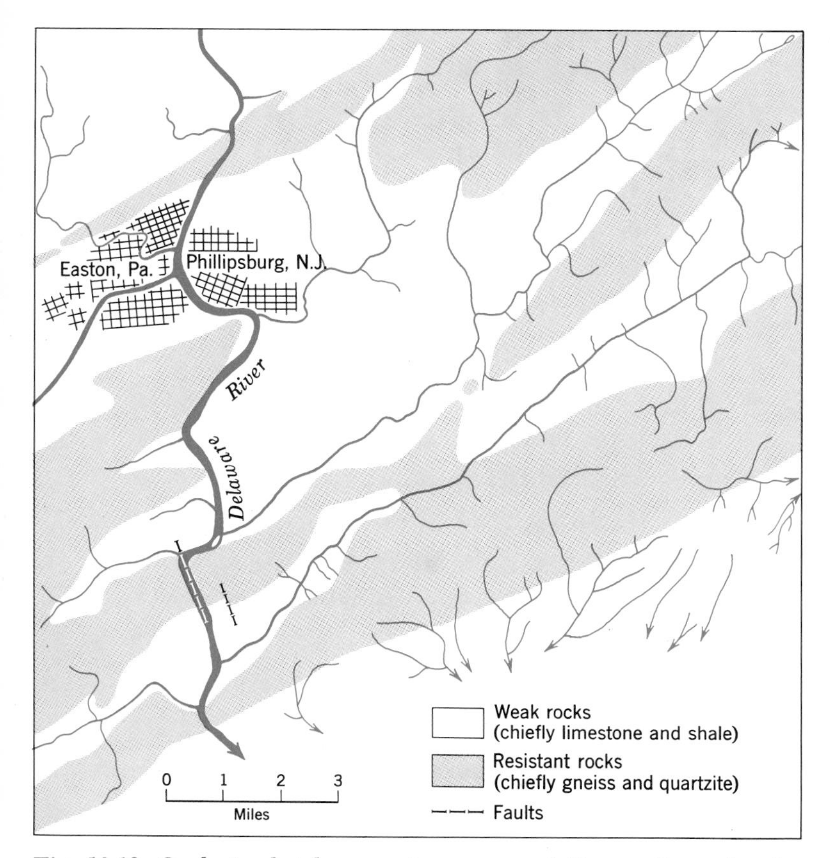

Fig. 10-18. Geologic sketch map of area around Easton, Pennsylvania, showing a stream pattern well adjusted to weak rocks. As the gneiss and quartzite resist not only stream erosion but mass-wasting as well, they are left as ridges about 500 feet higher than the weak-rock areas. (After W. S. Bayley, 1941, Pl. 5.)

antecedent streams and superposed streams are by definition not adjusted. Whatever their earlier degree of adjustment may have been, it was destroyed at the places where the antecedent or superposed relationship developed.

We have been looking at many diverse details in the sculpture of the land. Whatever the details, the cycle of erosion accompanies important parts of the rock cycle. While a land mass passes through a slowly changing sequence of forms and is gradually eroded to form a peneplain, sediments resulting from erosion are transported and spread out to form new strata, on lower land as basin fills or on the floor of a sea. Thus the cycle of erosion and the rock cycle are complementary. Of course the creation of a peneplain destroys most or all of the land forms that preceded it, and of course the peneplain itself can be elevated and destroyed by renewed erosion. But some peneplains sink down and are buried by layers of sediment. A surface entombed in this way is preserved until, at some time, uplift occurs. This movement renews erosion, which cuts away the cover, exposing the buried surface once more. Ancient peneplains created as long ago as Precambrian time are now, after hundreds of millions of years of burial, being exposed to view.

In the sequence represented by Chapters 7 through 10 we have followed the creation of rock particles by weathering, their movement down hillslopes into streams, and their transport and deposition as sediment, thus completing three closely related phases of the rock cycle. We have followed also the sculpture of the lands into characteristic configurations, by the same streams as they move and deposit sediment. The next logical step is to turn to ground water, the immediate source of most of the water of streams, and examine its part in the hydrologic, geochemical, and other cycles.

Summary

Drainage systems

1. Land sculpture is a conspicuous result of the geologic cycle.
2. Most land surfaces consist of complexes of valleys, cut by streams that flow through them.
3. Weathering, mass-wasting, and sheet erosion together erode more rock material than streams do. The main work of streams is to carry away the material fed to them from slopes.

Sculptural evolution of the land

4. The processes operating in a drainage basin form a complex open system, whose parts tend toward a steady-state condition.
5. Uninterrupted sculpture of a land mass follows a broadly predictable cycle of erosion, ending in a peneplain.

Interruptions in sculptural evolution

6. The orderly progress of land sculpture is commonly interrupted. Among the interruptions are movements of the crust and changes in the position of base level. These can cause rejuvenation or deposition.

History of streams

7. Streams tend to follow belts of weak rock. Therefore the pattern of rocks exposed at the Earth's surface influences the pattern of streams.

Selected References

Cotton, C. A., 1952, Geomorphology, an introduction to the study of landforms, 6th ed.: New York, John Wiley.

Leopold, L. B., and Langbein, W. B., 1966, River meanders: Scientific Am., v. 214, p. 60–70.

Leopold, L. B., Wolman, M. G., and Miller, J. P., 1964, Fluvial processes in geomorphology: San Francisco, W. H. Freeman.

Morisawa, M. E., 1964, Development of drainage systems on an upraised lake floor: Am. Jour. Sci., v. 262, p. 340–354.

Schumm, S. A., and Lichty, R. W., 1965, Time, space, and causality in geomorphology: Am. Jour. Sci. v. 263, p. 110–119.

Thornbury, W. D., 1954, Principles of geomorphology: New York, John Wiley.

CHAPTER 11

Ground Water

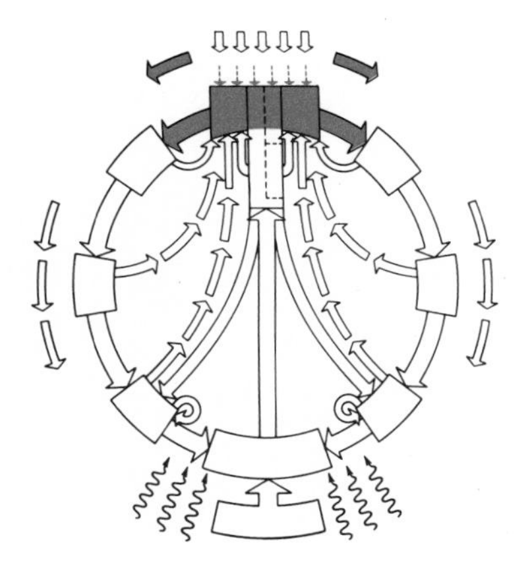

Geologic significance
Distribution and origin
Movement
Economy
Ordinary wells and springs
Artesian water
Economic aspects
Ground water in the geologic cycles

Plate 11

Quarry face about 10 feet high, in Salem Limestone near Bedford, Indiana. Solution along joints and stratification surfaces has created small caverns or sinks, which have been filled with clay-rich regolith washed down from above. Horizontal parallel lines are not stratification; they are marks made by a saw. Irregular, near-horizontal line through center of face is the result of solution along an interface between two strata, involving the removal of an unknown thickness of the limestone. (R. Dee Rarick, Indiana Geol. Survey.)

CHAPTER 11 *Ground Water*

GEOLOGIC SIGNIFICANCE

Below the surface of the lithosphere, pores and larger openings in rocks and regolith contain fluids that are highly important to people. These fluids are oil and gas (which we shall discuss in Chapter 23) and ***ground water,*** defined simply as *the water, beneath the Earth's solid surface, contained in pore spaces within regolith and bedrock.*[1] The geologic significance of ground water is fourfold: (1) it is an essential part of the hydrologic cycle (Fig. 5-3); (2) it performs geologic work by dissolving and depositing substances below ground; (3) it supplies plants and animals (including man) with a sizable fraction of their water requirements; and (4) as a chief factor in controlling the distribution of vegetation, it influences erosion by running water, mass-wasting, and wind.

The questions we must try to answer are these. How is water distributed beneath the ground? How does it get into the ground? How does it move? What geologic work does it do? And last, but not least important, how do we find and develop subsurface water for economic use, and how does our ever-growing demand for water affect the supply? The answers to these questions lie in the realm of *hydrogeology,* the study of ground water, emphasizing its chemistry, its movement, and its relation to the geologic environment. This study is growing rapidly, in large part because of the acute need in industrialized countries for greatly increased supplies of water of good quality.

DISTRIBUTION AND ORIGIN

Amount and depth. One of the reasons why man has been able to establish permanent settlements, not only in well-watered country, but also in desert lands, is that few areas exist in which holes, intelligently located and sunk far enough into the ground, do not find at least some water. In a moist country the depth of an adequate well may have to be only a few meters; in a desert it may have to be hundreds. These facts have been learned by experience. Water is present beneath the land nearly everywhere, but whether it constitutes a usable supply depends on depth of occurrence, kinds of rocks present, kinds and amounts of substances dissolved in the water, and other factors. For this reason some places are much more favorable than others for obtaining ground water.

An up-to-date estimate of the quantity of water in the ground beneath the world's land area, within a depth of 800m, is 4 million km^3. The quantity present below 800m is probably small. The deepest holes drilled for oil have found small amounts of water at depths exceeding 6km, but at depths of around 16km pressure from the weight of overlying rocks causes rock material to flow, closing all open spaces, and thereby excluding water. Recoverable ground water, therefore, is confined to a rather shallow zone with a lower limit that exceeds a depth of 800m only rarely. Because the water in this shallow zone is confined to the open spaces within the rocks, its volume is only a small fraction of the volume of our 800m zone and would be equivalent to a layer of water, spread evenly over the lands, less than 30m thick.

In the hydrologic cycle, ground water acts as a storage reservoir, receiving rainfall at irregular intervals by infiltration from above and transmitting it downward and outward to streams at steadier rates.

Origin of ground water. The ancient Greeks thought that ground water was seawater driven into the rocks by the winds and somehow desalted or that it was created in some manner from rocks and air deep below the surface. Later it was recog-

[1] This term is restricted by many geologists to the water beneath the water table, but the broader definition given here will simplify our discussion.

nized that rivers are fed, at least in part, by springs emerging from the ground and also that the discharge of rivers does not raise the surface of the sea appreciably. The truth that ground water is derived mainly from rain and snow was recognized by Marcus Vitruvius, a Roman architect of the time of Christ, who wrote a treatise on aqueducts and water supply, a matter of great practical importance to the Romans.

Although true, Vitruvius' statement that ground water comes from rain was qualitative only, and not until the seventeenth century was it established on a quantitative basis. Then Pierre Perrault, a French physicist, measured the mean annual rainfall on a part of the drainage basin of the River Seine in eastern France and also the mean annual runoff from it in terms of river discharge. He concluded that the difference was ample enough, over a period of years, to account for the amount of water in the ground. Today we accept rainfall as the source of all ground water, except for a tiny proportion that comes from substances within the Earth's interior. The distribution of tritium (H^3), the radioactive isotope of hydrogen, supports this conclusion. Tritium, like radiocarbon, is created only in the upper atmosphere (Fig. 5-15); it enters into the compound H_2O which falls to the ground as rain and snow. It is present in ground water, which therefore must have been derived from the atmosphere.

Water table. Much of our knowledge of groundwater occurrence has been learned the slow and hard way, by the accumulated experience of many generations of men who have dug or drilled millions of wells. This experience (Fig. 11-1) tells us that a hole penetrating the ground ordinarily passes first through a ***zone of aeration,*** *the zone in which the open spaces in regolith or bedrock are normally filled mainly with air.*

The hole then enters the ***zone of saturation,*** *the subsurface zone in which all openings are filled with water. The upper surface of the zone of saturation* is the ***water table,*** which, at any place, normally slopes toward the nearest stream. Ordinarily the water lies within a few meters of the surface, but it can be at the surface, as for instance, along the shore of a lake or the bank of a river, or, in some mountainous districts, it can lie at a depth of 100m or more. Whatever its depth, the water table is a very significant surface, because it represents the upper limit of most of the ground water that is readily usable. We shall return to it shortly.

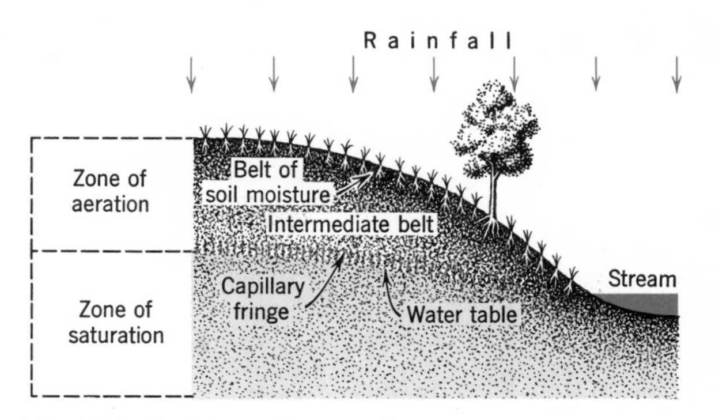

Fig. 11-1. Positions of zone of saturation, water table, and zone of aeration. (After W. C. Ackermann and others, U. S. Dept. of Agriculture.)

MOVEMENT

Most of the ground water within a few hundred meters of the Earth's surface does not lie there inert; it moves. However, its movement is unlike the turbulent flow of rivers measurable in kilometers per hour. It is so slow that velocities are expressed in centimeters per day and in rare cases even in centimeters per year. To understand why this is so, we must understand the porosity and permeability of rocks.

Porosity and permeability. The limiting amount of water that can be contained within a given volume of rock material depends on the ***porosity*** of the material; that is, *the proportion (in per cent) of the total volume of a given body of bedrock or regolith that consists of pore spaces.* A stated volume, V_o, of rock material consists of a quantity of solids, V_s, plus a quantity of voids, V_v. Porosity, n (in per cent) can then be defined:

$$n = 100 \frac{V_v}{V_o}$$

The porosity of some igneous and metamorphic rocks is very low (in some rocks less than 1 per cent) because the crystallization of their minerals at comparatively high pressures and temperatures left few or no open spaces in the rock. In contrast, sediments are quite porous because most of them merely consist of solid particles dropped onto a surface. This is evident in Table 11-1, which also shows that the sedimentary rock is less porous than the sediment of corresponding grain size. Former

TABLE 11-1. COMMON RANGE OF POROSITY IN TYPICAL SEDIMENTS AND SEDIMENTARY ROCKS.

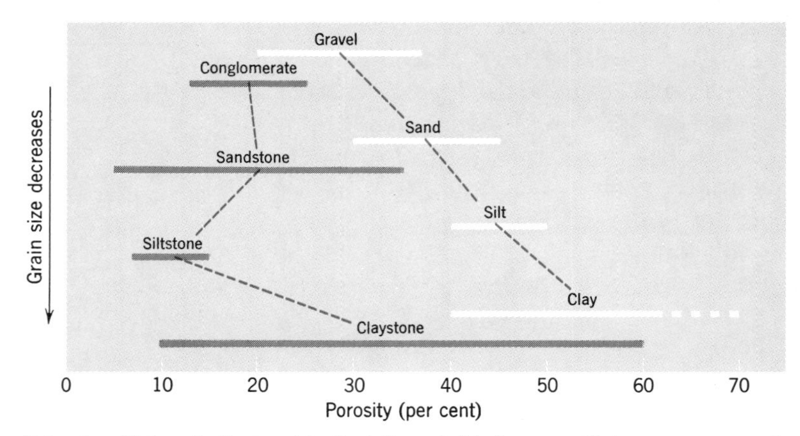

Value for siltstone is displaced to the left, probably because siltstone measurements were based on very few samples. (*Source*: Geol. Soc. America Spec. Paper 36, 1962)

pore space has been lost to compaction and to the cement that converted the sediment into rock. Furthermore, the table shows that in sediments porosity increases regularly with decreasing grain size, whereas in sedimentary rocks it increases overall but with great irregularity. The difference is at least partly the result of irregularities in the deposition of cement from place to place.

Permeability, *the capacity for transmitting fluids,* is expressed in the term *hydraulic conductivity,* the quantity of fluid that passes through material of given cross section, per unit of time, when driven by a given pressure and at a stated temperature. It is measured in units of various kinds. When we drill a well, permeability becomes highly important because we want to tap a zone in which the desired fluid (water, oil, or gas) will be transmitted readily into our well. High porosity values do not necessarily mean high permeability values, because size and continuity of the openings influence permeability in an important way. The relationship between size of openings and the molecular attraction of rock surfaces plays a large part. Molecular attraction is the force that makes a thin coating of water adhere to a rock surface despite the force of gravity; an example is the wet film on a pebble that has been dipped in water. If the open spaces between two adjacent grains is small enough, the films of water adhering to them will come into contact. This means that the force of molecular attraction extends right across the opening, as shown on the left side of Fig. 11-2. At ordinary pressure, therefore, the water is held firmly in place and permeability is very low. This is what happens in clay, whose component grains are less than 0.005mm in diameter (Fig. 4-11).

By contrast, in a sediment with grains at least as large as sand grains (0.06 to 2mm) the open spaces are wider than the films of water adhering to the grains. As the force of molecular attraction does not extend across them effectively, water in the centers of the openings is free to move in response to gravity or other forces, as shown at the right in Fig. 11-2. The sediment is therefore permeable. As the diameters of the openings increase, permeability increases. With its very large openings, gravel is more permeable than sand and yields very large volumes of water to wells.

Movement above water table. Let us return for a moment to Fig. 11-1. Water from a rain shower infiltrates the soil, which usually contains clay resulting from weathering of the bedrock. Because of its fine-grained texture, the soil is generally less permeable than underlying materials. Part of the water, therefore, is held in place by the forces of molecular attraction (Fig. 11-2). This is the belt of soil moisture in Fig. 11-1. Some of it evaporates directly and much is taken up by plants.

The water in the soil that molecular attraction can not hold seeps downward through the intermediate belt shown in Fig. 11-1 until it reaches the water table. In fine-grained material a narrow

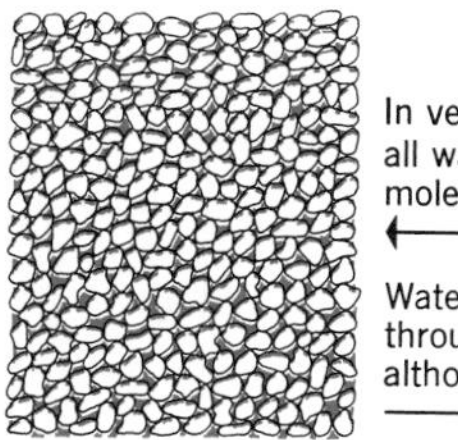
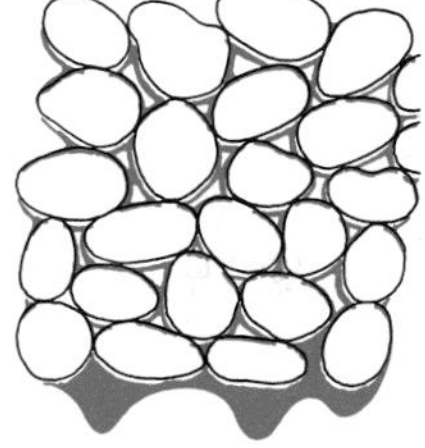

Fig. 11-2. Effect of molecular attraction in the intergranular spaces in fine sediment (*left*) and in coarser sediment (*right*). Scale is much larger than natural size.

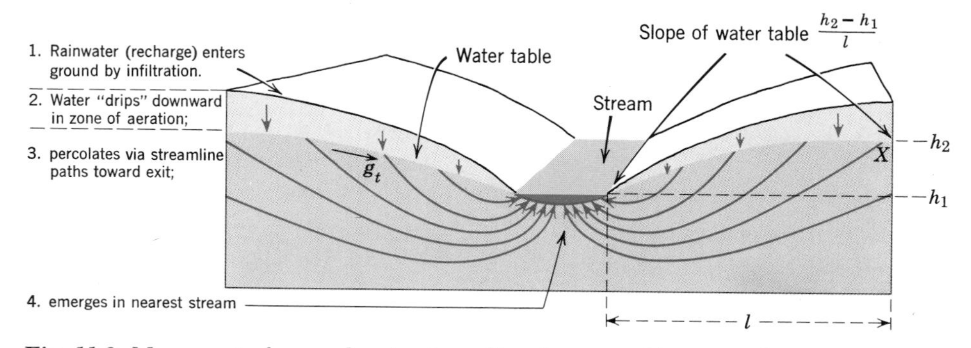

Fig. 11-3. Movement of ground water in uniformly permeable material. Long curved arrows represent only a few of many possible subparallel paths. At any point such as *x*, slope of water table is determined by

$$\frac{h_2 - h_1}{l},$$

where $h_2 - h_1$ is height of *x* above point of emergence in surface stream and *l* is distance from *x* to point of emergence. (After M. K. Hubbert.)

fringe as much as 60cm thick, immediately above the water table, is kept wet by capillary tension wherever open spaces are so narrow that molecular forces can extend across them. Water is drawn upward through these tiny spaces (Fig. 3-2) in the same way that ink is drawn upward through blotting paper.

With every rainfall, more water is supplied from above, but, apart from the belt of soil water and capillary fringe, the zone of aeration, as its name implies, can be nearly dry during the times between rains.

Movement below water table: percolation. In the zone of saturation the flow of ground water is laminar, with the water particles moving along parallel paths. *Laminar flow through interconnected spaces in saturated material* is called ***percolation.*** Responding to the force of gravity the water "tends to seek its own level," percolating from areas where the water table is high toward areas where it is lowest; in other words, toward surface streams (Fig. 11-3). Only part of the water travels by the most direct route, right down the slope of the water table. Other parts follow innumerable long, curving paths that go deeper through the ground. Some of the deeper paths turn upward against the force of gravity and enter the stream from beneath. This behavior is explained by the fact that in the zone of saturation the water along any plane of given height, such as h_1 in Fig. 11-3, is under greater pressure beneath a hill than beneath a stream. The water therefore tends to move toward points where pressure is least. When it does this, we say it is moving down a *pressure gradient.*

Laboratory models have been made in which dye is injected into the percolating ground water at various depths. Paths followed by the dye resemble those in Fig. 11-3 and turn upward beneath a model stream at the base of a hill. The rate at which the water moves along these paths decreases sharply with increasing depth. Most of the ground water that enters a stream, therefore, has traveled to it via shallow paths not far beneath the water table.

Whatever their paths, the water particles pass smoothly from one opening to the next, except for those openings so narrow that the force of molecular attraction extends completely across them, barring the way.

ECONOMY

Velocity of flow: Darcy's law. Having examined the process of percolation and the multitude of curved paths followed by ground water moving in the zone of saturation, we must turn now to the economy of ground water which, like that of surface streams, involves both velocities and quantities.

As Fig. 11-3 shows, the water table is a surface consisting of slopes. Water is fed into the zone of saturation by infiltration, and because of the slope of its surface, it flows down, under the influence of g_t, toward the level of the stream in the nearest valley, where it emerges and empties into the stream. Like the gradient of a stream, the slope of a water table between any point such as X at height h_2 and the point where it emerges at height h_1 is measured by the difference in height $(h_2 - h_1)$ divided by the horizontal distance l.

The relationship between slope and velocity of flow was established in the mid-nineteenth century by Henry Darcy, engineer in charge of public works in the French city of Dijon. In a program of improvement of the city's water supply he experimented with percolation through sand used for filtering the water and, as a result, established the fundamental equation:

$$v = k\frac{h_2 - h_1}{l}$$

where v = velocity of flow, k = a coefficient representing permeability, and $(h_2 - h_1)/l$ of course is slope. The equation, known as Darcy's law, says that in material of given permeability, velocity increases as slope of the water table increases.

We can speak of the *economy* of a ground-water system, therefore, just as we speak of the economy of a stream system. In the ground-water system, however, the terms are simpler. The water is percolating through the openings in a fixed framework of bedrock or regolith with nearly constant porosity; there is no alluvial channel that changes dimensions, as in a stream system. The important variable factor is the slope of the water table, which changes with rainfall as does the discharge of a stream.

Ground-water activity, like stream activity, constitutes an open system, with water input at the top and output discharging into streams, lakes, and the sea. Slope of water table and velocity continually adjust to each other and to varying permeability in the various segments of the system to maintain a state of dynamic equilibrium in each segment, as do the hydraulic factors in a river. Whenever rain adds water from above, the slope of the water table steepens, and if permeability is constant, velocity must increase. In a dry season the slope diminishes (Fig. 11-5) and so does velocity. If rainfall on a certain area were to cease entirely, the water table beneath would flatten out and closely approach the level of the streams in the valleys—assuming, of course, that the streams were still being fed from rainfall outside the area.

Velocity and discharge. Because of the large amount of friction involved in percolation, velocities are slow, commonly ranging between 5 feet per day and 5 feet per year. The largest rate yet measured within the United States, in exceptionally permeable material, is only 770 feet per day (Fig. 8-3).

Velocity of percolation is measured between pairs of wells. In one method two wells with metal casings are connected to form the electric circuit shown in Fig. 11-4. Ammonium chloride, an efficient conductor, is poured into the upslope well and percolates downslope. On its arrival at the downslope well it creates a short circuit between well casing and electrode; this is recorded on an ammeter. Distance between wells divided by elapsed time gives the velocity.

Another means of measuring rate (and, of course, direction) of percolation is to put a strong dye, such as fluorescein, into a well and then time the appearance of dyed water in neighboring wells. Ground-water hydrologists want to determine not only velocity (v) but also discharge (Q)—the quantity of water that flows through a given cross-sectional area (A) in a unit of time—because this

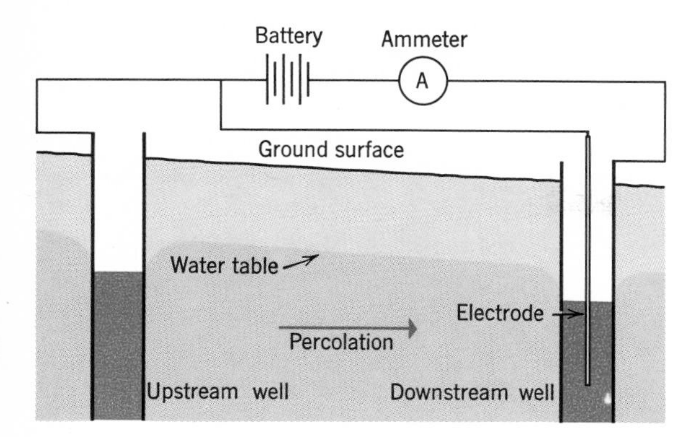

Fig. 11-4. Apparatus for measuring velocity of percolation.

determines the amount available for human use. The discharge equation is analogous to the one representing discharge in rivers but with velocity given by Darcy's law:

$$Q = vA = k\frac{h_2 - h_1}{l}A$$

In this equation Q can be compared with the stream's discharge and A with the cross-sectional area, wd (Fig. 9-9), of the stream channel. The velocity, v, comparison is less direct. In a stream, v is measured directly, but in ground water it is derived through Darcy's law, the elements of which can be compared with the stream's gradient plus the losses to friction on the sides and bottom of the stream channel.

ORDINARY WELLS AND SPRINGS

Having examined the way in which ground water moves, and the reasons for its fluctuating economy, we can learn more about its behavior by studying wells and springs as sources of water supply. We shall find that there are two classes of wells and springs, determined by the geometry of the rock bodies that supply them. In one class, water movement is unconfined and is therefore comparatively simple. In the other class, movement is confined to a particular part of the rock that is present and in consequence resembles the flow of water through the pipes of a system of plumbing. The wells and springs of the first class we call ordinary; the others we call artesian.

Aquifers. We can learn more about the behavior of ground water through the practical examination of wells and springs as sources of water supply. A feature commonly referred to by hydrologists and geologists is an ***aquifer*** (Lat. "water carrier"), *a body of permeable rock or regolith through which ground water moves.* Bodies of gravel and sand are commonly good aquifers and so are many sandstones. But the presence of cement between the grains of sandstone reduces the diameter of the openings and so reduces the effectiveness of these rocks as aquifers.

It might seem that claystones (Table 11-1), igneous rocks, and metamorphic rocks should not be aquifers, because in them the spaces between grains are extremely small, and because samples of them, measured in the laboratory, are impermeable. What is true for laboratory samples, however, does not necessarily apply to large bodies of the same material. Many such bodies contain fissures, spaces between layers, and other openings that are too large for water flow to be controlled entirely by molecular attraction; these bodies can be aquifers. Even so, as aquifers they are less effective than are coarse-grained sediments of incompletely cemented sedimentary rocks. Whatever their effectiveness, it is in aquifers that we find wells and springs.

Ordinary wells. An ordinary well fills with water simply by intersecting the water table (Fig. 11-5). Lifting water from the well lowers the water level and so creates a ***cone of depression,*** *a conical depression in the water table immediately surrounding a well.* In most small domestic wells the cone of depression is hardly appreciable. Wells pumped for irrigation and industrial uses, however, withdraw so much water that the depression can become very wide and steep and can lower the water table in all the wells of a district.

Figure 11-5 shows that a shallow well can become dry at times, whereas a deeper well in the vicinity may yield water throughout the year.

If rocks are not homogeneous, the yields of wells are likely to vary considerably within short distances. Massive igneous and metamorphic rocks, for example (Fig. 11-6, *A*), are not likely to be very permeable except where they are cut by fractures, so that a hole that does not intersect fractures is likely to be dry. Because fractures generally die out downward, the yield of water to a shallow well can be greater than to a deep one. Again, discontinuous bodies of permeable and impermeable material (Fig. 11-6, *B*) result in very different yields to wells. They also create ***perched water bodies*** (*water*

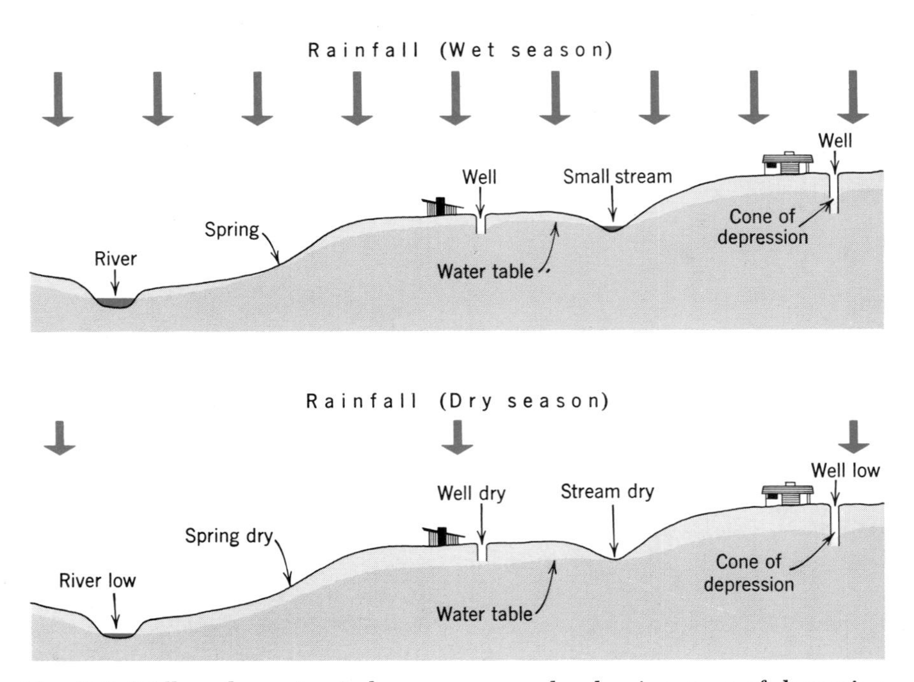

Fig. 11-5. Wells and a spring in homogeneous rocks, showing cones of depression and effect of seasonal fluctuation of water table. Note that the slopes of the water table are steeper in the wet season when input of water into the system is greatest than in the dry season when input is least.

bodies that occupy basins in impermeable material, perched in positions higher than the main water table). The impermeable layer catches and holds the water reaching it from above.

Gravity springs. A ***spring*** is *a flow of ground water emerging naturally onto the Earth's surface.* The simplest spring is an ordinary or *gravity spring,* whose flow results directly from the force of gravity. Three common kinds of gravity springs are illustrated in Fig. 11-7.

ARTESIAN WATER

Confined percolation. In some regions the geometry of inclined rock layers makes possible a special pattern of circulation of ground water. Three essentials of the pattern are shown in Fig. 11-8:

1. A series of inclined strata that include a permeable layer sandwiched between impermeable ones.
2. Rainfall, to feed water into the permeable layer where that layer is cut by the ground surface.
3. A fissure or a well so situated that water from the sandstone can escape upward through the impermeable roof.

When these essentials are present, we have a special kind of open system. The input consists of rain water, which enters the permeable layer (now an aquifer by definition) and percolates through it. The output consists of water forced upward through fissures or wells that perforate the roof.

Observe, in Fig. 11-8, the position of the water table. Except for a thin zone close to the surface, the whole series of strata is saturated with water. In the claystone the water is held immobile by molecular attraction in the tiny spaces between the particles of rock. But in the aquifer it moves, provided only that water can escape through fissures or wells. Percolation in the aquifer, however, is *confined* between the impermeable strata above and below; it moves past the water held immobile in those strata.

If, in the area of *recharge* shown in the illustration, rainfall reaches the ground in greater volume than that of water output through fissures or wells,

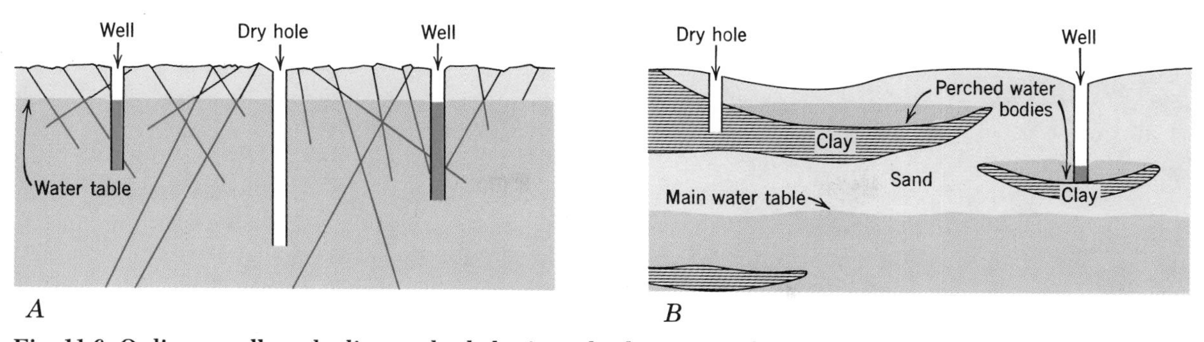

Fig. 11-6. Ordinary wells and adjacent dry holes in rocks that are not homogeneous. *A*. In fractured massive rocks such as granite. *B*. In bodies of permeable sand containing discontinuous bodies of impermeable clay. Two perched water bodies are shown.

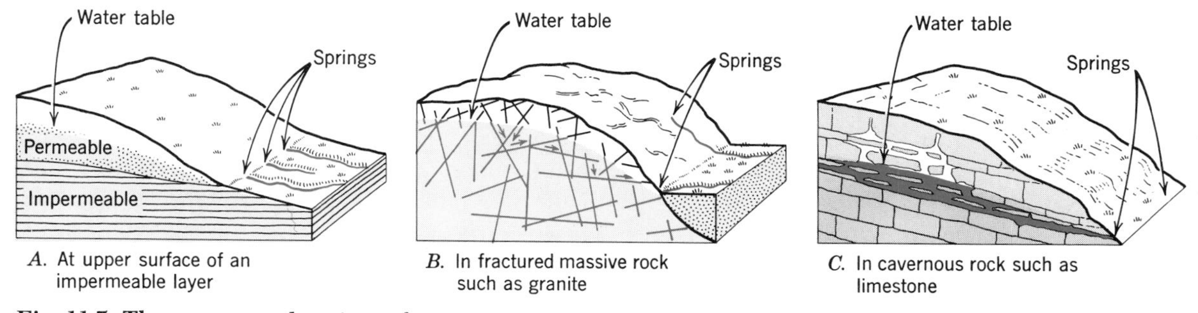

Fig. 11-7. Three common locations of gravity springs.

only enough water to balance output can enter the system; the excess flows away over the surface. On the other hand, if wells draw out of the system more water than can enter it from the available rainfall on the area of recharge, the yield of the wells will diminish to a quantity small enough to be balanced by the recharge.

The aquifer is like a broad, flat, sand-filled pipe or conduit, holding its ground water confined under the pressure of the column of water that extends up to the water table at its upper end. The water rises, in any well, to the level of the recharge area, minus an amount determined by the loss of energy in friction of percolation. Hence the height to which the water will rise in any well depends on distance from recharge area, height of that area above the well head, and permeability of the aquifer.

A well of this kind is an ***artesian well,*** defined

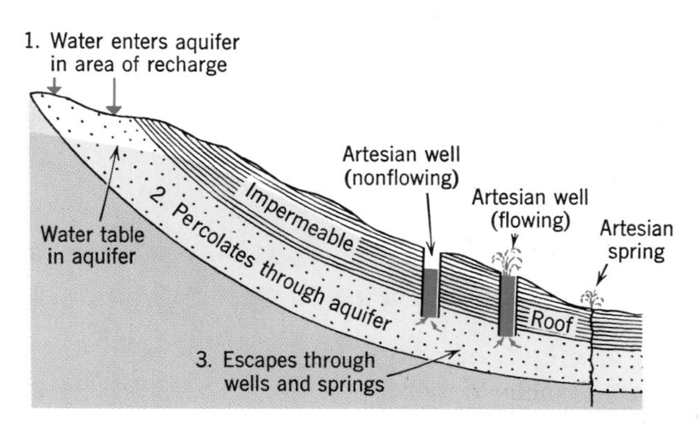

Fig. 11-8. Three essential conditions for artesian wells: an aquifer, an impermeable roof, and water pressure sufficient to make the water in any well rise above the aquifer.

as *a well in which water rises above the aquifer.* The name comes from the French province of Artois, in which, near Calais, the first well of this sort in Europe was bored. When the factors listed above are unusually favorable, pressure can be great enough to lift the water above ground creating fountains as much as 60m high.

Deep wells bored into rock to intersect the water table are popularly called artesian wells, but this is an incorrect use of the term. Such wells are ordinary wells, like those in Figs. 11-5 and 11-6.

Ocala Limestone artesian system. The Ocala Limestone, the principal artesian aquifer in Florida, is an aquifer because it is full of caverns and smaller openings, intricately interconnected, created by solution. It is at the surface in the central and northwestern part of the peninsula and slopes toward both coasts. Impermeable roofs are provided by layers of chert within the limestone.

Velocity of percolation has been determined by measuring the activity of C^{14}, from $(HCO_3)^-$ dissolved in the water, at a series of wells along an 83-mile line generally parallel to the dip of the aquifer. The C^{14} entered the ground in rainfall on the recharge area and moved through the aquifer in the ground water. The activity of C^{14} was found to decrease systematically away from the recharge area. From the sum of the differences in activity between samples from pairs of wells in the series is calculated an average velocity of percolation (through the whole distance) of 23 feet per year. According to this velocity, water in the well farthest from the recharge area has been in the ground for nearly 19,000 C^{14} years.

These C^{14} measurements also support the correctness of Darcy's equation, which was used independently to calculate the velocity between the same two points, with the same result, 23 feet per year.

Tapping an artesian system is a very ancient art. Four thousand years ago many artesian wells as much as 100m deep were in existence. In A.D. 1126 the well near Calais was bored, and it is still flowing today. In that area, at any rate, withdrawal has not seriously exceeded supply.

Artesian systems are not confined to wells. Some are ***artesian springs,*** *natural outflows at the Earth's surface of water from a confined aquifer, usually through a fissure or along a fault,* instead of through a well (Fig. 11-8).

The temperatures of many artesian springs are substantially higher than the local mean annual air temperatures. There are more than a thousand such *thermal springs* in the United States, and even larger numbers in other parts of the world. Ground water can become heated in two ways: (1) By descending so deep that it is warmed by the general internal heat that is measured by the geothermal gradient. (2) By contact with bodies of igneous rock that are slowly cooling within the crust. Because dissolution is more rapid in warm water than in cold, thermal springs are likely to be unusually rich in mineral matter dissolved from rocks with which they have been in contact.

ECONOMIC ASPECTS

The per capita consumption of water, regardless of source, in urban households in the United States is about 160 gallons per day, but when industrial and irrigation consumption are added the daily per capita consumption rises to more than 15,000 gallons. About one-fifth of this water comes from beneath the ground; the rest comes from rivers, lakes, and artificial surface reservoirs all of which, like ground water, are supplied by rain and snow. Among United States cities, Houston, Indianapolis, Memphis, and Spokane derive their water largely or entirely from below ground. With the increase in population and industrialization, the demand for water is growing very rapidly, and the search for new sources of ground water is intensive.

Water finding. Sites for farm and domestic wells are sometimes located by persons who use forked twigs and other kinds of "divining rods" and who claim to possess supernatural powers. The search for water by this means, often called "dowsing," dates back at least to the time of Moses and is still widespread. Although no scientific basis for this kind of claim is known to exist, use of the divining rod persists, partly because in many areas shallow supplies of ground water are so widespread that successful results would be numerous even though sites were located at random. If the diviner were asked to indicate where water is *not* present below the ground and if his predictions were then tested by boring holes, the statistical results would soon reveal the unsoundness of his claims. But, because little money is spent on holes in attempts to avoid water, this has never been done.

Recognizing the need for increasing the supply of ground water, government agencies such as the U. S. Geological Survey, the Geological Survey of Canada, and others in many countries have studied conditions underground in critical areas and have pointed the way to hidden sources of supply. This is one of the many direct ways in which geologic study aids the general economy. The location of a new aquifer or a successful well in a known aquifer is most reliably based on three activities: (1) detailed examination of geologic conditions at the surface, especially kinds and permeabilities of rocks, positions of rock layers, and character of fissures and other large openings; (2) exploration below the surface by seismic, electric, and other techniques used in petroleum exploration, and described in Chapter 23, to find possible buried aquifers; (3) study of the performance records of nearby wells or wells in analogous situations. Kinds of aquifers located by these methods include permeable sandstones, cavernous limestones, jointed and fissured crystalline rocks, extrusive igneous rocks, and coarse alluvium of various kinds.

Recharge. *The addition of water to the zone of saturation* is termed ***recharge,*** whether in a gravity system (Fig. 11-3) or in an artesian system (Fig. 11-8). In a dry region traversed by streams fed from mountains or other areas of substantial rainfall the water table is likely to be far below ground. In such a situation recharge can occur from the streams themselves (Fig. 11-9).

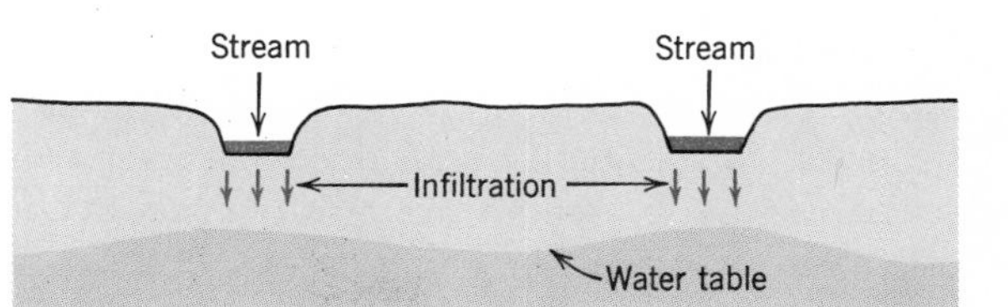

Fig. 11-9. Recharge of ground water in a dry region, by infiltration from streams having their sources in mountains with abundant precipitation. The stream channels are actually leaking. Relation of water table to streams is the reverse of that shown in Fig. 11-1, where leaking is impossible.

The intense demand for water in some areas has led to artificial recharging of the ground. One example is the practice of water spreading in dry parts of the west. A common way to spread water for recharge is to build a low dam across a valley. This holds back water in a surface stream that would otherwise run to waste and allows it to recharge aquifers beneath the stream bed. The water thereby stored underground is withdrawn through wells as needed.

A chemical-industrial plant in the Ohio River valley, near Louisville, Kentucky, was located on a valley fill of sandy alluvium more than 150 feet thick. Water was needed for cooling. The plant bought city water (purified river water) in winter when the water was cold and fed it into wells in the valley fill, thus recharging the sandy aquifer. In summer the cold water stored in the ground was pumped out for industrial use.

In some districts an aquifer is recharged with used water. This practice has increased with the increased use of air conditioning, which requires a large volume of water. Some cities have laws requiring that certain water that has been used for air conditioning be returned to the ground, where it successfully builds up the water table. This illustrates the basic principle of ground-water conservation: that withdrawal of water must, in the long run, be balanced by recharge; if it is not, then either recharge must be increased or withdrawal curtailed.

Subsidence of ground. One of the consequences of continued excess of withdrawal over recharge, from an aquifer that lacks a rigid framework, is compaction of the aquifer. The water itself supports part of the load of the overlying materials and also keeps the grains of the aquifer loosely packed. When the water is removed from the intergranular spaces in such an aquifer, the weight of the overlying rocks packs the grains together more closely, as shown in Fig. 4-14. This can not only reduce permanently the capacity of the aquifer but also cause serious subsidence of the ground overlying the aquifer.

An example is the area of the city of Las Vegas, Nevada, which derives much of its water from pumped wells. Precise instrumental surveys show that during the period 1935 to 1949, the ground within an area 5 miles in diameter sank to form a conical depression that reflected the form of the cone of depression created in the water table underneath. Maximum subsidence was 14 inches.

Pumping of ground water through 9,000 wells resulted in sinking of the area of Mexico City, between 1910 and 1952, by as much as 5m. During 1953 the rate of subsidence reached 50cm/year.

Pollution and sanitation. The most common source of contamination of water in wells and springs is sewage, and the infection most commonly communicated by polluted water is typhoid. Drainage from cesspools, broken sewers, privies, and barnyards contaminates ground water. If the water contaminated with sewage bacteria passes through material with large openings, such as very coarse gravel or the cavernous limestone shown in Fig. 11-10, it can travel for miles without much change. If, on the other hand, it percolates through sand or permeable sandstone, it can become purified within short distances, in some cases less than 100 feet (Fig. 11-11). The difference lies in the aggregate internal surface area of the material through which it percolates (Fig. 7-5). This large aggregate force of molecular attraction holds the water and promotes its purification by (1) mechanical filtering-out of bacteria (water gets through but most of the bacteria do not), (2) destruction of bacteria by chemical oxidation, and (3) destruction of bacteria by other organisms, which consume and oxidize them. Purification goes on both in the zone of aeration and in the zone of saturation. Because particles of clay are much finer than those of sand, we might suppose that clay, with its much larger internal surface area, would be the ideal medium for purification. But it is not, because another result of the small size of its particles is that the rate of percolation through it is extremely slow. Particles of sand are large enough to permit rapid percolation, yet small enough to permit purification within short distances. For this reason treatment plants for purification of municipal water supplies and processing of sewage use artificial layers of sand, through which they percolate these fluids.

A substantial proportion of domestic sewage passes through septic tanks (Fig. 11-11) and then mingles with ground water. It percolates into streams, and, for the reasons we have given, generally reaches them in a pure condition. On the other hand the domestic waste from a very large population, as well as much industrial waste, is dumped unaltered into surface streams. Although purification can be accomplished during stream transport, the distances involved are very much greater than those required for the purification of ground water, and the amounts of sewage in many rivers are far too great to be dealt with by natural processes. In densely populated industrial countries these facts constitute serious public health problems.

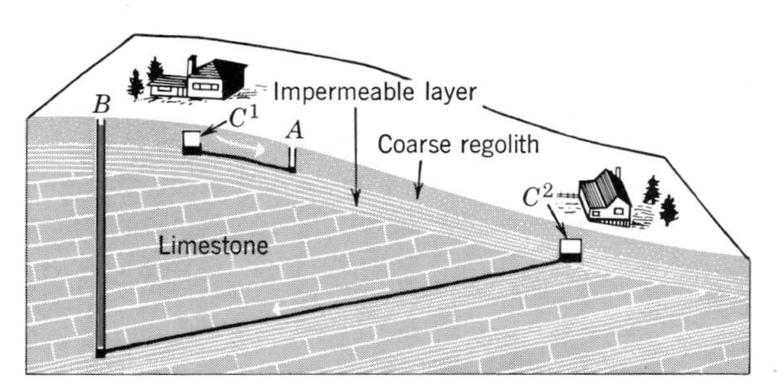

Fig. 11-10. Contamination of wells. The shallow dug well, *A*, was unwisely located downslope from a cesspool, C^1 and received contaminated drainage (black) from it. The owner then drilled a deeper well, *B*. This well tapped layers of cavernous limestone inclined toward it from the lower cesspool C^2. The water flowed through openings in the limestone, and reached the bottom of well *B* unpurified by percolation. The well owner must relocate his cesspool or dig a shallow well upslope from C^1.

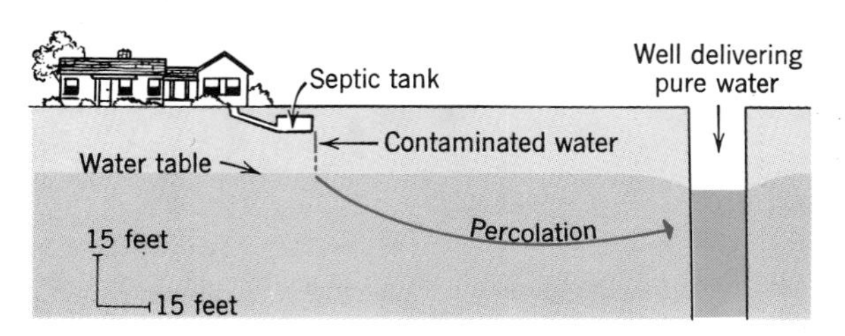

Fig. 11-11. Purification of contaminated ground water in sand and gravel during percolation through a short distance.

A dramatic illustration of the difference between surface and underground conditions is this. In some communities heavily polluted water that has traveled tens of miles through a river is pumped into the ground, where it becomes a part of the local ground-water supply. In one city, percolation through a horizontal distance of 500 feet removed all impurities from the sewage and made the water fit to drink.

GROUND WATER IN THE GEOLOGIC CYCLES

Ground water in the rock cycle. At the beginning of this chapter we viewed ground water as a part of the hydrologic cycle. Now let us look at it as an essential part of the rock cycle, especially the geochemical cycle. Ground water dissolves minerals, transports the solutes, precipitates some of the solutes within rocks and regolith, and pours the rest of them into the sea, where some of them are used to build limestone and other marine sedimentary rocks.

In Chapter 7 we analyzed the decomposition of minerals during weathering. We found that most of this activity is accomplished by ground water, essentially as a weak solution of carbonic acid, and that one of the great results is leaching. The products of leaching, dissolved in the ground water, percolate into surface streams and there constitute the big dissolved load (Fig. 5-13) that the world's rivers carry to the sea.

The process has been quantified in local areas. For example, in an area in the Piedmont region of North Carolina, underlain by diorite, the discharge of small streams and springs was measured and the total quantity of the many mineral solids in solution in this water was determined. From the data it was calculated that in that area ground water is dissolving solids equivalent to a layer of rock 1 centimeter thick every 933 years.

What is going on here is simply chemical weathering. The minerals in the diorite, particularly plagioclase, pyroxene, and amphibole, are being decomposed actively in a mild, moist climate. This is a striking example of the operation of the geochemical cycle, the importance of which is commonly underestimated because its effects are not readily visible.

Chemical composition. Analyses of many wells and springs show that the solutes in ground water consist mainly of chlorides, sulfates, and bicarbonates of calcium, magnesium, sodium, potassium, and iron. We can trace these substances back to the common minerals in the rocks from which they were derived by weathering. As might be expected, the composition of ground water varies from place to place according to the kind of rock in which it occurs. In much of the central United States the water is "hard," that is, rich in calcium and magnesium bicarbonates, because the bedrock includes abundant limestones and dolostones that consist of those carbonates. In some places within arid regions the concentration of dissolved substances, notably sulfates and chlorides, is so great that the ground water is unfit for human consumption. Furthermore, evaporation of water in the zone of aeration precipitates not only calcium carbonate but, in particularly dry regions, sodium sulfate, sodium carbonate, and sodium chloride. Soils containing these precipitates are loosely termed "alkali soils." They are unsuitable for agriculture because crops will not grow in them.

Some ground water is salty. Two of the various causes of saltiness in ground water are these: (1) In some deep-lying sedimentary rocks, salty ground water is ***connate water,*** defined as *seawater that was trapped in a sedimentary deposit when the sediment lay on the sea floor.* It has remained in them ever since. (2) Along some coasts, ground water has become contaminated by the infiltration of seawater. As shown in Fig. 5-4, *A*, fresh ground water, being slightly less dense than seawater, floats on it with an interface or zone of contact that slopes landward, at an angle determined mainly by rainfall. The two bodies of ground water are in equilibrium. If, now, excessive withdrawal of freshwater occurs, through wells situated on the landward side of the interface, the freshwater body thins, the interface flattens, and seawater encroaches farther inland. The position of the encroaching interface can be determined by measuring from year to year the concentration of chloride in wells.

Deposition. One of the great transformations that occur as parts of the rock cycle, the conversion of sediments into sedimentary rocks, is primarily the work of ground water. A body of sediment lying beneath the sea is generally saturated with water, very likely connate water; as is a sediment lying in the zone of saturation beneath the land. Substances in solution in the permeating water are precipitated in the spaces between the rock particles in the sediment as a cement (Fig. 1-10, *B*) that transforms the loose sediment into firm rock. Calcite, silica, and iron compounds (mainly oxides), in that order, are the chief cementing substances.

Less common than the deposition of cement between the grains in a sediment is ***replacement,*** *the process by which a fluid dissolves matter already present and at the same time deposits from solution an equal volume of a different substance.* Evidently replacement takes place on a volume-for-volume basis because the new material preserves the most minute textures of the material replaced. Petrified wood is a common example. But replacement is not confined to wood and other organic matter; it occurs in mineral matter as well. Some ore bodies owe their origin to replacement (Chap. 23).

A distinctive feature resulting from ground-water deposition is a ***geode,*** *a hollow, rounded body having a lining of crystals pointing inward* (Fig. 11-12). The linings are usually quartz or calcite, and the rocks in which geodes occur are generally limestones. Undoubtedly, geodes originated after the sedimentary rocks that inclose them were formed. One explanation of these features is that they are essentially tiny caverns excavated by solution, somewhat as are the caverns described in the next section. Later, owing to changes in physical-chemical conditions, they were partly refilled with mineral substances crystallized from water solution. Other deposits made by ground water include some concretions (Fig. 16-12), some veins (Chap. 23), and the dripstone deposited in caverns, described in the following section.

Caverns. Limestone, dolostone, and marble are carbonate rocks that consist of the minerals calcite and dolomite in various proportions. These rocks underlie millions of square miles of the Earth's surface. Although carbonate minerals are nearly insoluble in pure water, they are readily dissolved by the carbonic acid in ground water, which becomes charged with calcium bicarbonate, as shown by these reactions, the first of which we have already met in our analysis of chemical weathering in Chapter 7:

$$\underset{\text{(Carbon dioxide)}}{CO_2} + \underset{\text{(Water)}}{H_2O} \longrightarrow \underset{\text{(Carbonic acid)}}{H_2CO_3} \text{ which ionizes to } \underset{\text{(Hydrogen ion)}}{H^+} + \underset{\text{(Bicarbonate ion)}}{(HCO_3)^-}$$

The hydrogen ions attack the calcite.

$$\underset{\text{(Calcite)}}{CaCO_3} + \underset{\text{(Hydrogen ion)}}{2H^+} \longrightarrow \underset{\text{(Water)}}{H_2O} + \underset{\text{(Carbon dioxide)}}{CO_2} + \underset{\text{(Calcium ion)}}{Ca^{2+}}$$

The calcium ion released combines with bicarbonate ions. It is worthwhile to compare this reaction with the one in which potassium feldspar is attacked by H^+ ions and water. Again hydrogen ions, H^+, are the active agents. They force an atom of calcium out of the calcite and combine with one oxygen atom to form water, thus leaving a free Ca^{2+} ion, as well as one carbon and two oxygens in the form of CO_2, a gas. The Ca^{2+} ion combines with two bicarbonate ions, $(HCO_3)^-$, to form the calcium bicarbonate, which, together with the CO_2 gas, remains in solution in the ground water.

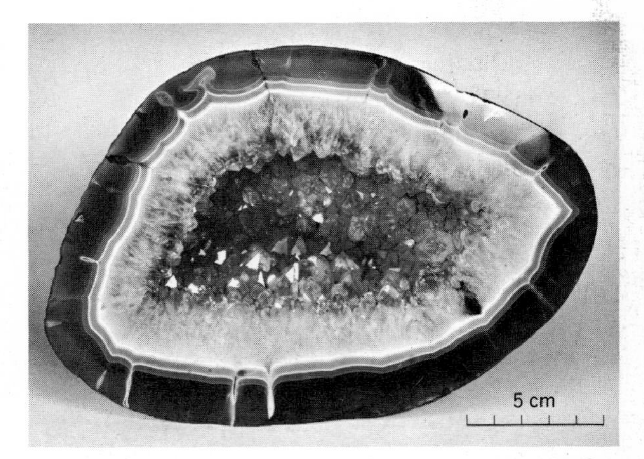

Fig. 11-12. Geode resulting from partial filling of a cavity in rock, first with agate and later with crystals of quartz. Erosion has removed the surrounding rock, but the resistant quartz that lined the cavity remains. (American Museum of Natural History.)

Because calcite and dolomite are very soluble in carbonic acid, ground water dissolves carbonate rocks very effectively. This is a form of chemical weathering just as much as is the decomposition of igneous rocks that contain feldspar and ferromagnesian minerals. In both cases the weathering attack occurs along joints and other partings in the bedrock (Fig. 11-13). But whereas in granite the quartz and other insoluble minerals remain, nearly all the substance of a body of pure limestone can be carried away in solution in the slowly moving ground water. Because impurities in limestone commonly consist of clay, which is insoluble, solution of many limestones leaves accumulations of clay. Because so much of the substance of limestone is soluble, its solution creates cavities of many sizes and shapes. *A large, roofed-over cavity in any kind of rock* is a ***cavern*** (Fig. 11-14).

Although most caverns are small, some are of exceptional size. Carlsbad Caverns in southeastern New Mexico include one chamber 4,000 feet long, 625 feet wide, and 350 feet high. Mammoth Cave, Kentucky consists of interconnected caverns with an aggregate length of at least 30 miles.

Some caverns have been partly filled with insoluble clay and silt, originally present as impurities in the limestone and gradually released by solution. Other caverns contain partial fillings of ***dripstone*** (*material chemically precipitated from dripping water in an air-filled cavity*) and ***flowstone***

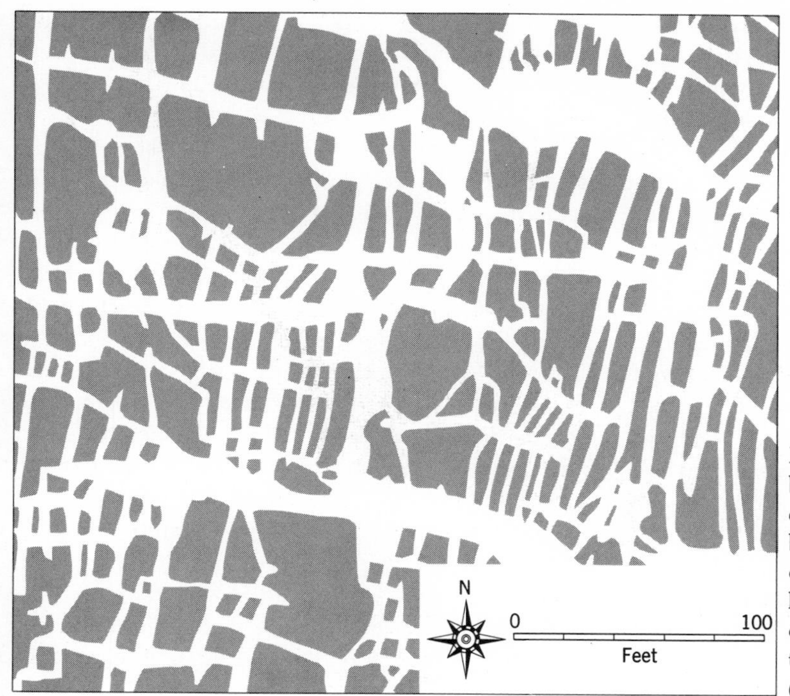

Fig. 11-13. Part of a map made by members of the National Speleological Society, of Anvil Cave, near Decatur, Alabama shows how closely solution was controlled by the pattern of joints in limestone of the Gasper Formation. The cavern extends through an area about ten times greater than the part shown. (W. W. Varnedoe.)

Fig. 11-14. Large cavern in limestone, partly refilled with dripstone and flowstone in the form of stalactites, stalagmites, and columns. Length of view at base of photo about 20 feet. (Luray Caverns, Virginia.)

(*material chemically precipitated from flowing water in the open air or in an air-filled cavity*). The "stones" take on many curious forms, which are among the chief attractions to cavern visitors. The most common shapes are ***stalactites*** (*icicle-like forms of dripstone and flowstone, hanging from ceilings*), ***stalagmites*** (*blunt "icicles" of flowstone projecting upward from floors*), and ***columns*** (*stalactites joined with stalagmites, forming connections between the floor and roof of a cavern*) (Fig. 11-14).

The deposition of dripstone takes place in this way. When ground water containing dissolved calcium bicarbonate and carbon dioxide appears on the roof of a cave, a little carbon dioxide gas escapes from solution because the partial pressure of CO_2 in the cave air is lower than that in the solution. Loss of CO_2 upsets the equilibrium between calcium bicarbonate and carbonic acid. The result is the formation of water and the precipitation of calcium carbonate.

$$\underset{\text{(Carbon dioxide dissolved in water)}}{H_2CO_3} + \underset{\text{(Calcium ion plus bicarbonate ions)}}{Ca^{2+} + 2(HCO_3)^-} \longrightarrow \overset{\text{(Escapes)}}{2CO_2\uparrow} + 2H_2O + \underset{\text{(Precipitates)}}{CaCO_3\downarrow}$$

The calcium carbonate is deposited on roof or floor or both. The reaction involved in its deposition is a simple reversal of the reaction we looked at earlier, by which calcium carbonate is dissolved by carbonic acid.

Because dripstone forms only in air-filled openings, we can be sure that a cavern was *above* the water table while dripstone was being deposited in it. But some caverns, including most of those in the Black Hills of South Dakota, are partly or wholly lined with glittering crystals of calcite or quartz; the caverns are like huge geodes. Now, these crystals form only in water solutions of their substances; so the caverns that contain them must have been *below* the water table at the time the crystals formed. But when the caverns themselves were being formed by solution, were they above or below the water table? We can not determine this either from crystalline linings or from dripstone, both of which are merely refillings of a space that already existed. But the shapes of caverns answer the question. Most caverns narrow and pinch out not only downward and laterally but upward as well. This indicates that solution affected all parts of the cavern circumferences in the same way, and therefore occurred while the caverns were continuously filled with water. This and other features support the belief that most—although not all—caverns were excavated below the water table, in the zone of saturation.

If this is the case, then in at least some districts where caverns are filled with dripstone, the water table must have lowered at some time *after* cavern solution and *before* the deposition of dripstone. What was the cause? We could suggest that rainfall might have diminished, making the water table lower beneath interstream areas (Fig. 11-3). But little evidence of other kinds exists to support this idea. Another possibility is that in such regions the streams deepened their valleys. Deepening could have occurred either with or without rejuvenation of the streams (Fig. 10-7). Without any change in rainfall, this erosion could have resulted in a readjustment of the slopes of the water table, so that beneath hill areas it would be lowered. Some caverns that before rejuvenation were in the zone of saturation would now be located in the zone of aeration (Fig. 11-15).

Sinks. In contrast to a cavern, a ***sink*** is *a large solution cavity open to the sky*. Some sinks are caverns whose roofs have collapsed. Others are formed at the surface, where infiltrating water is freshly charged with carbon dioxide and is at its most effective as a solvent. Many sinks, located at the intersections of joints where movement of water downward is most rapid, have funnel-like shapes. A funnel-shaped sink is shown in Fig. 11-16. This sink is a small one, but some sinks are very large. The area of one, near Mammoth Cave, Kentucky, is 5 square miles.

Karst topography. In some regions of exceptionally soluble rocks, sinks and caverns are so numerous that they combine to form a peculiar topography characterized by many small basins. In this kind of topography the drainage pattern is irregular; streams disappear abruptly into the ground, leaving their valleys dry and then reappear elsewhere as large springs. This has been termed ***karst topography*** (Fig. 11-17) because it is strikingly developed in the Karst region of Yugoslavia, inland from Trieste. It is defined as *an assemblage of topographic forms consisting primarily of closely spaced sinks.* Karst topography is developed through wide areas in Kentucky, Tennessee, southern Indiana, and northern Florida (Fig. 11-18).

Sinks and caverns record the destruction of a

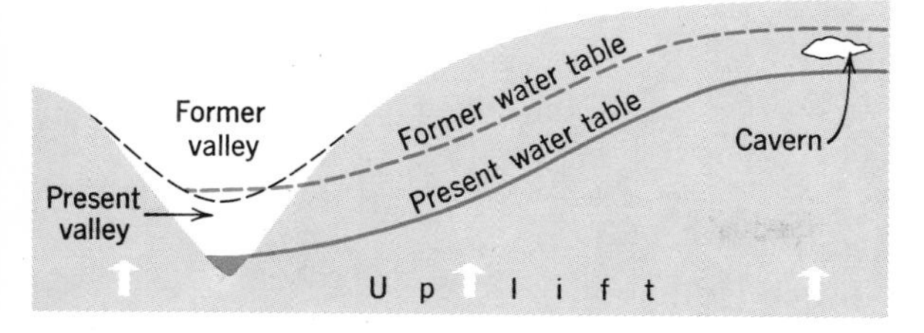

Fig. 11-15. Possible history of a cavern containing dripstone. Cavern was excavated below water table. When streams deepened their valleys, water table was lowered as it adjusted to the deepened valleys. This left cavern above water table.

Fig. 11-16. Sink in limestone near Sunken Lake, Presque Isle County, Michigan. (I. D. Scott.)

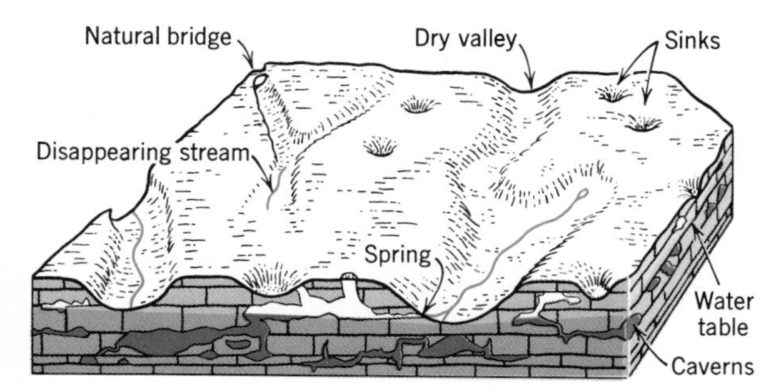

Fig. 11-17. Karst topography. Area of block is between 1 and 2 square miles.

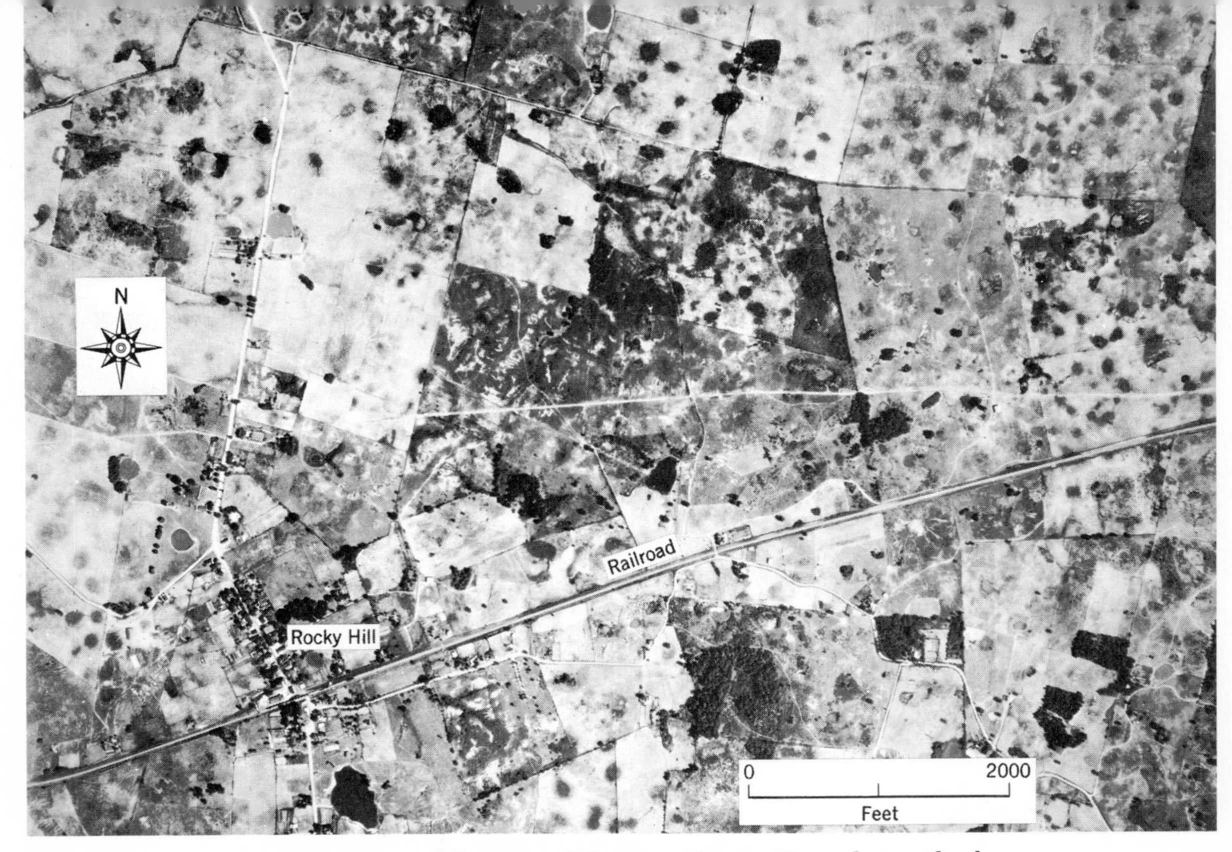

Fig. 11-18. Karst topography in area of limestone, Edmonson County, Kentucky, south of Mammoth Cave National Park, seen in a vertical air photograph. Dark-colored areas with white rims are lakes that occupy some of the sinks. Small gray spots with less definite boundaries are dry sinks. Almost no surface streams are present because most rainfall infiltrates the limestone before it can run off. (Litton Industries, Aero Service Division.)

very large volume of carbonate rock. It is calculated from measured amounts of carbon dioxide in ground water and from the solubilities of limestones that the amount of precipitation on northern Kentucky is capable, as ground water, of dissolving a layer of limestone 1 centimeter thick every 66 years. This potential is far greater than the average erosional reduction of the United States by mass-wasting, sheet erosion, and streams. It depends on the presence of exceptionally soluble rocks.

In our examination of the various external processes we have followed both surface water (streams) and ground water. We have noted some of their interrelationships as well as the sediments they deposit and the sculptured forms for which they are responsible. We shall turn next to frozen water on the lands, in order to compare glaciers with streams in such respects as movement, regimen, sediments, and sculpture.

Summary

Geologic significance

1. In the hydrologic cycle ground water plays a part in the return of water from land to sea.

Distribution and origin

2. Ground water, derived almost entirely from rainfall, occurs almost universally.

3. The water table marks the top of the zone of saturation.

Movement

4. Ground water flows chiefly by percolation, at rates far slower than those of surface streams.

5. Darcy's law states that with constant permeability, velocity of flow of ground water increases as slope of the water table increases.

6. In moist regions ground water percolates away from hills and emerges in valleys. In dry regions it is likely to percolate away from areas beneath surface streams.

Economy

7. A ground-water system is an open system in each segment of which a state of dynamic equilibrium is approached.

Wells and springs

8. Ground water flows into most wells directly by gravity but into artesian wells under hydrostatic pressure.

9. A basic principle of conservation is that withdrawal of ground water must not exceed recharge.

Geologic work

10. In the geochemical cycle ground water dissolves mineral matter from rocks; much of the dissolved product eventually gets into the sea.

11. In the rock cycle, ground water deposits substances as cement between grains and so reduces porosity and converts sediments into rocks.

12. In carbonate rocks ground water not only creates caverns and sinks by solution but also deposits mineral matter in some caverns.

Selected References

Bretz, J H., 1956, Caves of Missouri: Missouri Geol. Survey, v. 39.

Davis, S. N., and DeWiest, R. J. M., 1966, Hydrogeology: New York, John Wiley.

Ellis, A. J., 1917, The divining rod, a history of water witching: U. S. Geol. Survey Water-Supply Paper 416.

Hubbert, M. K., 1940, The theory of ground-water motion: Jour. Geology, v. 48, p. 785–944.

McGuinness, C. L., 1963, The role of ground water in the national water situation: U. S. Geol. Survey Water-Supply Paper 1800.

Meinzer, O. E., 1923, The occurrence of ground water in the United States, with a discussion of principles: U. S. Geol. Survey Water-Supply Paper 489.

—— 1942, Occurrence, origin, and discharge of ground water, *in* Meinzer, O. E., and others, Hydrology: New York, McGraw-Hill, pp. 385–477.

Moore, G. W., and Nicholas, G., 1964, Speleology. The study of caves: New York, D. C. Heath.

Todd, D. K., 1959, Ground-water hydrology: New York, John Wiley.

U. S. Department of Agriculture, 1955, Water: Yearbook for 1955: Washington, U. S. Govt. Printing Office.

CHAPTER 12

Glaciers and Glaciation

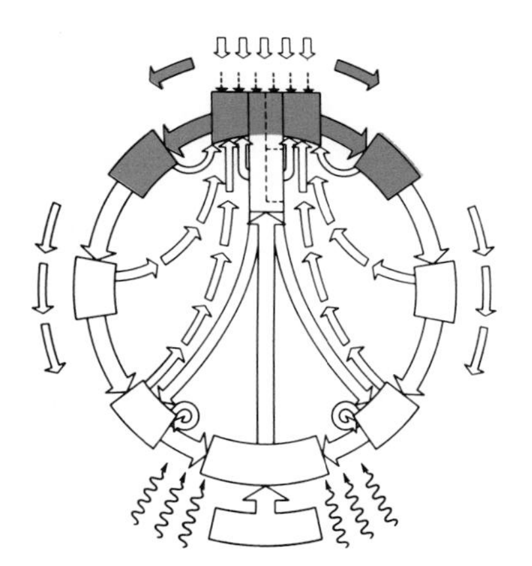

Origin
Movement
Economy
Glaciation
Glacial erosion and transport
Sediments
The glacial ages
Causes of glacial climates

Plate 12

Valley glaciers in Northeast Greenland, flowing down from snowfields on a mountainous highland, spread out on the floor of a major valley to form small piedmont glaciers. Later they began to shrink, but shrinkage was punctuated by short readvances of their margins, during which closely packed, concentric end moraines were built. Streams of meltwater with braided patterns are depositing outwash sediment along moraine margins and down the major valley, toward left. Valley drains into Rhedins Fjord. Three spurs, between the glaciers, have triangular facets made by glacial erosion. They show that the major valley was formerly occupied by a very large glacier. (Ernst Hofer.)

Lateral parts
Lake
End moraine
Terminal part
Outwash sediments

CHAPTER 12 *Glaciers and Glaciation*

GLACIERS

Geologic significance. We can hardly look at a scene like that shown in Plate 12 without realizing that glaciers are among the most spectacular of natural features. Not only are they interesting in themselves as solid crystalline bodies that flow and change form, but the geologic work of erosion and deposition they accomplish is noteworthy. Glaciers are significant further in that during the ice age from which the Earth has been emerging during the last 15,000 to 20,000 years, they spread over many million square miles of territory and then melted, leaving a remarkable geologic record of their activity. Because long-term changes in the dimensions of glaciers are closely related to variations of climate, the study of them has contributed much to what we know of the Earth's recent climatic history. The geologic record raises the problem of the causes of the ice ages, a problem still mainly unsolved.

As in the discussions of streams and ground water, we shall consider first the character and movement of glaciers we can see and study today, then examine the geologic work these glaciers do, and finally compare today's glaciers with those of the ice ages.

Definition and form. Defined as simply as possible, a ***glacier*** is *a body of ice, consisting mainly of recrystallized snow, flowing on a land surface.*

On the basis of their form, we can readily distinguish four kinds of glaciers. Most numerous and possibly most familiar are ***cirque glaciers*** (Fig. 12-9), *very small glaciers that occupy cirques.* In the conterminous United States there are about 950 cirque glaciers, and about 1,500 exist in the Alps.

A ***valley glacier*** (Fig. 12-1) is *a glacier that flows downward through a valley.* Such glaciers vary in size from little things no more than a few acres in extent to mighty tonguelike forms many tens of miles long. In the conterminous United States they number about 50; in the Alps their number is larger.

A ***piedmont glacier*** (Pl. 12; Fig. 21-15, A) is *a glacier on a lowland at the base of a mountain, fed by one or more valley glaciers.* It is shaped like a covered frying pan whose narrow inclined handle represents a feeding valley glacier.

An ***ice sheet*** is *a broad glacier of irregular shape generally blanketing a large land surface. A small ice sheet* is an ***ice cap;*** some ice caps, such as those in Fig. 12-1, are very small. The best known ice sheet is the Greenland Ice Sheet. Its area is about 1,726,000km^2 and its highest part is more than 3km above sea level. Seismic measurement—that is, measurement by a method using artificial earthquake waves (Fig. 18-9)—of the approximate thickness of this glacier was made in 1951 at intervals along certain routes of travel, one of which is shown in Fig. 12-2. The profile and cross section along it reveal that in places the base of the ice sheet is below sea level and that the ice is very thick; at one point its thickness exceeds 3km. Near its margins the ice sheet is held in by mountains, through which it flows along deep valleys to reach the sea. Icebergs floating in the sea are masses broken off from the ends of such glaciers.

The Antarctic Ice Sheet (Fig. 12-3) is far bigger. Its area is 13,100,000km^2 (1.4 times the area of the United States), its highest altitude exceeds 4km, its greatest thickness likewise exceeds 4km in one place, and its volume is estimated at around 26×10^6km^3. With this huge volume the Antarctic Ice Sheet alone contains more than 90 per cent of the world's stock of ice and 75 per cent of all the fresh water in the world.

Valley glaciers, piedmont glaciers, and ice sheets constitute a gradational sequence. Indeed, a great ice sheet like that in Greenland probably began in coastal mountains with the growth of thousands of valley glaciers, which spread out in the interior as piedmont glaciers and gradually thickened to form a body of ice higher in places than the mountain peaks themselves.

Fig. 12-1. South Cascade Glacier, a valley glacier about 2 miles long, in the North Cascade Range, Washington, September 1960. In view are position of snowline and a small ice cap. (A. S. Post, U. S. Geol. Survey.)

Distribution: snowline. Two chief requirements for the existence of glaciers are (1) precipitation and (2) temperatures low enough to permit snow to accumulate. Both requirements depend on climate. They are met with in high latitudes and high altitudes and occur more commonly in wet coastal areas than in the dry interiors of continents. Glaciers are related to each other by the ***snowline,*** *the lower limit of perennial snow* (Fig. 12-1). Above the snowline are ***snowfields,*** *wide covers, banks, and patches of snow,* usually in protected places, *that persist throughout the summer season* (Fig. 12-1). The snowline passes across all active glaciers. It rises from near sea level in polar regions to altitudes of as much as 6,000m in tropical mountains and also rises from coasts toward continental interiors.

Conversion of snow into glacier ice. When temperatures fall below 0°C, some atmospheric vapor changes into the solid state, forming the hexagonal ice crystals we know as snowflakes (Fig. 3-25, *A*). Because newly fallen dry snow is very porous, its specific gravity is low, as little as 0.05, and it is fluffy. High porosity and the intricate shapes of its flakes give a mass of snow an internal surface area much greater than that of ordinary sediment. Also, as the spaces between the flakes are large, snow is easily penetrated by air. Ice evaporates most readily at the points of crystals, where surface area is great-

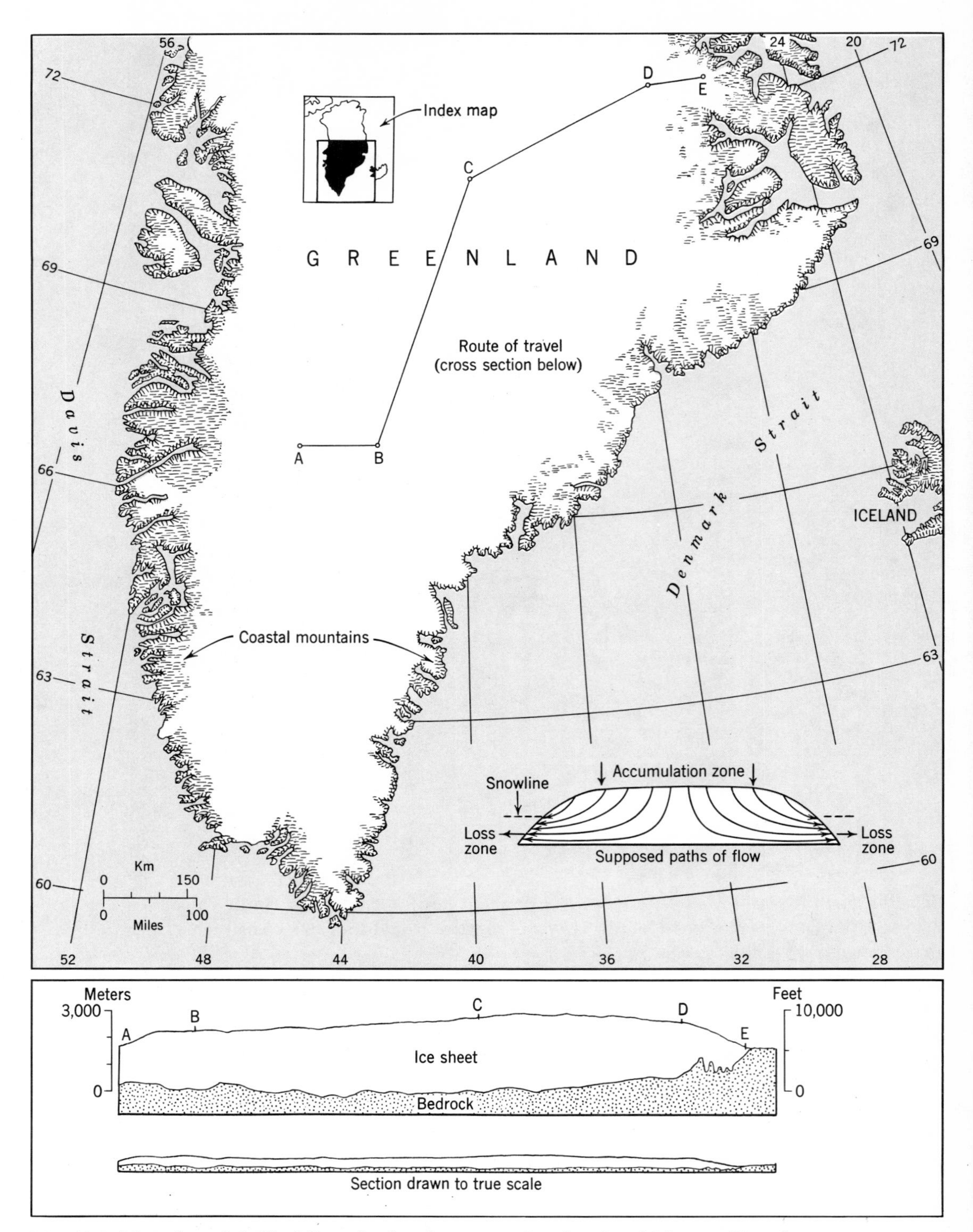

Fig. 12-2. Map of south half of Greenland and cross section showing thickness of ice sheet. (Thickness data from Albert Bauer.)

Fig. 12-3. Part of the Antarctic Ice Sheet, in Marie Byrd Land, with peaks of Edsel Ford Mountains projecting through its surface. In foreground are crevasses, reflecting a change of gradient of the floor beneath the glacier, which in this area is comparatively thin. (P. A. Siple, U. S. Antarctic Service.)

est, and vapor condenses most readily in the reentrants, nearer the crystal centers. The result is transfer of material from the outer edges toward the centers of snowflakes, so that individual particles of snow gradually change from clusters of starlike crystals to nearly spherical grains 3mm or less in diameter (Fig. 3-25, *B*). In this way the whole mass takes on the granular texture we find in old snowdrifts at winter's end. As the snow becomes granular its porosity diminishes, its specific gravity therefore increases, and air is continually forced out from the diminishing spaces between the grains. The material is being changed from a sediment into a sedimentary rock.

When the specific gravity of the body of granular snow has reached 0.8, the snow becomes impermeable to air and is then said to be *ice.* Such ice is a metamorphic rock, for it consists of interlocking crystalline grains of the mineral ice. It is, of course, a rock with a far lower melting point than any other rock. The specific gravity of glacier ice is about 0.9 (Fig. 3-25, *C*).

Mechanics of movement. Ice (including snow) has very low resistance to stress, and so it moves and is deformed quite easily. Although generally spoken of as flow, the movement of ice takes place through the action of gravity on its mass. Two kinds of motion are involved. One is sliding of a body of ice along the ground, like the slow sliding of a blanket of snow down the steeply sloping roof of a building (Fig. 22-9) or like a rapid avalanche. The other is analogous to the creep of metals, in which the crystal grains are slowly deformed. Ice, being weaker than metals, is deformed very easily even under its own weight. In detail, motion resulting from such deformation is complex, but a conspicuous feature of it consists of slipping (on a very small scale) along planes within the lattices of the ice crystals. At this point, a look at the mechanical behavior of solids in laboratory experiments will help in visualizing the motion of ice.

Stresses. Any unconfined solid responds in a characteristic way when subjected to an external force or forces that tend to deform it. The external forces set up various kinds of internal ***stress,*** defined as *force per unit area.* Three kinds of stress (Fig. 12-4, A) may be present:

1. *Tensile stress,* at a maximum in a direction in which the solid tends to be elongated or pulled apart.
2. *Compressive stress,* acting along a direction in which the solid tends to be shortened. The direction of maximum compression is always at right angles to that of maximum tensile stress.
3. *Shearing stresses,* of which there are two, are maximum along two sets of planes whose directions are diagonal to the perpendicular planes of maximum compressive and tensile stresses.

Along the directions of maximum shearing stresses the particles of the mass tend to slip past one another in a translatory motion. The internal-stress network behaves as a unit. Its orientation may vary with respect to the external deforming forces, but the orientation of the three maximum internal stress directions with respect to one another is constant.

Strain. When stresses act on a deformable body, its shape, and/or volume changes. This change is expressed in a ratio called *strain:*

$$\text{Strain} = \frac{a - b}{a}$$

where a = original shape or volume
b = changed shape or volume

In a moving glacier strain takes place continuously, and the rate of strain (amount of strain in per cent per unit of time) governs the velocity of the glacier.

A body of ice that demonstrates by actual measurement, by its outer form, or by the deformation of ice crystals that it has flowed or is flowing slowly under its own weight is a glacier by the definition we have given. The deformation of ice crystals during flow produces in a glacier the effect of foliation like that in metamorphic rocks (Fig. C-11).

Calculation of shearing stress. Stress in a glacier can be calculated by considering the forces acting on a cube of ice, 1cm on a side, at the base of the glacier, caused by the weight of the overlying column of ice (Fig. 12-4, *B*). The weight of the ice column equals the thickness of the glacier times the density of ice times the acceleration of gravity. This weight acts vertically downward. Its component parallel with the slope, w_t, equals the weight times the sine of the angle of slope and is the shearing stress on our cube. Expressed mathematically, this becomes:

$$\tau = \rho g h \sin \alpha$$

where τ is the shearing stress at the base of the glacier (kg/cm^2)
ρ is the density of the ice (g/cm^3) (0.9)
g is the acceleration of gravity (cm/sec/sec) (980)
h is the thickness ("height") of the glacier, measured perpendicularly to the upper surface (cm)
α is the slope of the glacier's upper surface (°)

Let us apply this equation to a glacier whose thickness is 200m (20,000cm) and whose surface slope is 2°, an angle whose sine equals 0.035. Because the values of ρ and g have been given above, we need only substitute numbers for all the terms on the right-hand side of the equation in order to find the value of τ:

$$\tau = \rho \quad g \quad h \quad \sin \alpha$$
(Substituting) $\tau = 0.9 \quad 980 \quad 20{,}000 \quad 0.035$
$= 0.62\text{kg/cm}^2$

This equation tells us that shearing stress increases directly with thickness of the glacier.

Relation of shearing stress to velocity. Laboratory measurements on ice samples show that rate of strain (which, as we have noted, governs velocity resulting from internal motion) is related to shearing stress according to a *flow law,* as follows:

$$\dot{\gamma} = (\tau/B)^n$$

where $\dot{\gamma}$ is the rate of strain and B and n are constants. When the time base for the rate of strain is 1 year, the value of B is about 2 and that of n is about 3, for ice near its melting temperature. In other words, $\dot{\gamma} = (\tau/2)^3$.

Using this relationship, we can now calculate the rate of strain on our cube of ice, subject to a shearing stress, τ, equal to 0.62kg/cm^2:

$$\dot{\gamma} = (0.62/2)^3 \text{ or about } 0.03 \text{ per year}$$

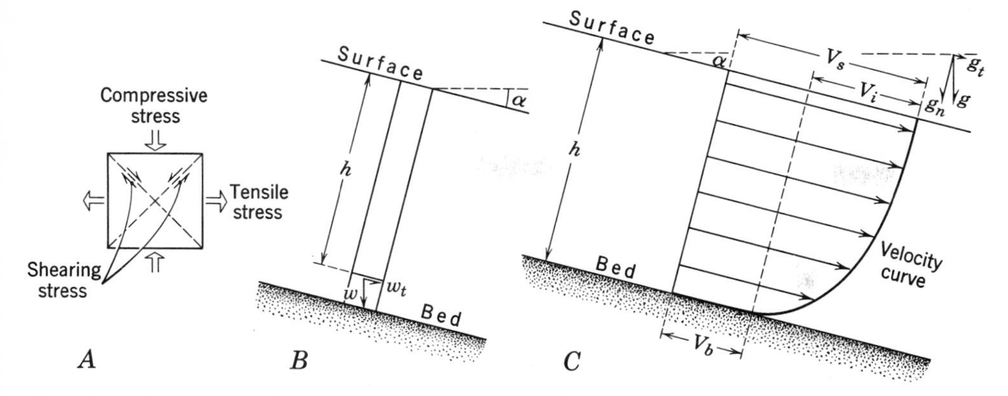

Fig. 12-4. Motion of a glacier shown diagrammatically. *A.* Two-dimensional sketch showing geometric relationships of principal stresses in a solid being deformed by external forces (not shown). *B.* Schematic section through a glacier showing column of ice 1cm wide. For explanation of symbols see text. *C.* Long section of a glacier having thickness *h* and an upper surface that slopes downward at angle α (see Fig. 2-8). Velocity curve for any depth is made up of two components, V_b and V_i. For explanation see text.

The velocity, V_s, of the surface part of a glacier consists of two components (Fig. 12-4, *C*). These are velocity, V_b, of sliding on the bed and velocity, V_i, of internal motions. The flow law allows us to calculate V_i. It is equal to the sum of the strain rates of *all* the 1-cm cubes in the column overlying the original cube. As we go upward from cube to cube, the strain rate becomes progressively less. However, each cube moves forward bodily by an amount equal to the sum of the strain rates of all the cubes below it. Therefore the velocity due to internal motion, V_i, increases upward (Fig. 12-4, *C*). Without entering into the mathematical manipulations involved, we can express this relationship in a statement of surface velocity (V_i in centimeters per second), as follows:

$$V_i = \frac{1}{32}(\rho g)^3 \sin^3 \alpha h^4$$

Because gravity and ice density are essentially constant, we can consider the value $\frac{1}{32}(\rho g)^3$ as a constant, *K*. For most glaciers this value works out to be: $K = 0.68 \times 10^{-18}$. Hence it is evident that the surface velocity of a glacier, caused by its internal motion, is proportional to the third power of the sine of the slope angle and to the fourth power of its thickness.

Using this relationship and assuming $V_b = 0$, we can calculate the surface velocity of the glacier on which we determined basal shearing stress. Substituting the values given above, we get

$$0.68 \times 10^{-18} \times 0.035^3 \times 20{,}000^4$$

This comes out to 1.2×10^{-4}cm/sec, or 38m/year.

Let us consider this relationship between velocity of internal motion and the flow law in the substantial part of a glacier that lies within the area of net accumulation shown in Fig. 12-5. If the thickness of our sample glacier is doubled by the addition of a very large amount of snow to its surface, our velocity equation tells us that V_i would increase sixteen times. A glacier flowing so rapidly would soon be "stretched" in the accumulation area unless a large amount of snow were continually added to it. It would become thinner, V_i would decrease exponentially, and finally *h* and V_i would come into dynamic equilibrium, somewhat as the hydraulic factors in a stream come into a new equilibrium at the end of a flood. The thicknesses of the Greenland and Antarctic Ice Sheets, which are rather comparable, may be explained by the relationship between V_i and *h*. In these glaciers ice spreads outward so rapidly that the existing rate of snowfall can not build them up to greater thickness.

Measurements on glaciers. We can make measurements on glaciers, similar to some of those commonly made on streams. The parameters usually measured are width, slope or gradient, surface velocity, and thickness. Measurement of the first three parameters began more than 100 years ago on glaciers in the Alps and were, and still are, done with surveying

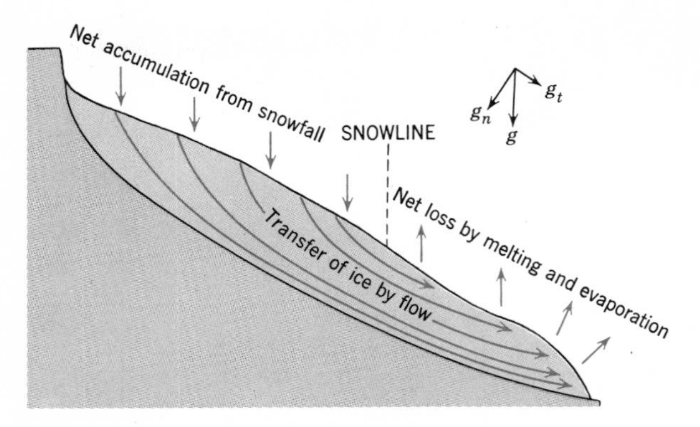

Fig. 12-5. Economy of a glacier. Paths of flowing particles of ice are suggested by long, curved arrows somewhat like those of percolating ground water (Fig. 11-3).

instruments. The surface velocity of a valley glacier is measured by surveying from the sides of the valley, at intervals of time, a line of targets extending from side to side of the glacier. Results show that the central part of the glacier's surface moves faster than the sides, as is true of a stream of water. Velocities normally range from a few millimeters to a few meters per day (Fig. 8-3), the values varying with slope, thickness, and other factors. These low velocities can be compared (though not as to manner of flow) with the rates of percolation of ground water. At such velocities, ice requires a long time to travel the length of a glacier. Probably hundreds or even thousands of years have elapsed since the ice now exposed at the downstream end of a long glacier fell as snow upon the upstream part.

Measurements of thickness began much more recently and are usually made with seismic equipment. It is easier, however, to calculate approximate thickness by using the flow law, once we have measured the slope and velocity at the surface. We are obliged to assume, of course, that $V_b = 0$, so that our calculated thickness is likely to be somewhat too great, but the error from this source is not likely to be large.

Crevasses. Many glaciers are cut by ***crevasses*** (Figs. 12-3, 12-17), which we can define as *deep cracks in the upper surface of a glacier.* The surface part of a glacier, having little weight upon it, cracks as it is subjected to tension when the glacier flows over an abruptly steepened slope. Though fearsome obstacles to a geologist in the field, most crevasses are less than 50m deep; at greater depths, flow of the ice prevents crevasses from forming. A glacier, therefore, like the Earth's crust itself (Chap. 18), can be said to possess an outer shell (a zone of fracture) and an underlying mobile mass (a zone of flow).

Economy. The long profiles of both ice sheets (Fig. 12-2, inset) and valley glaciers (Fig. 12-5) steepen near their lower ends and lateral margins, and become steeply convex-up. The lines of flow of the ice particles in the glacier tend to bend toward the surface, like the paths of percolating ground water beneath a stream (Fig. 11-3). Both ice and ground water are flowing along a pressure gradient.

It is clear in Fig. 12-5 that a glacier consists of two fundamentally different segments and that the boundary between them is the snowline. More snow falls on the "upstream" segment than is removed by melting and evaporation. This tends to thicken the glacier and promote movement. In the "downstream" segment the reverse is true; wastage exceeds snowfall, tending to thin the glacier. The tendency toward thinning in this segment is offset by the supply of ice flowing from the "upstream" segment. The lines of flow bend upward in the "downstream" segment, because the ice is flowing along a pressure gradient toward the surface, where net loss is occurring because of melting.

The combination of wastage and resupply, then, shapes the glacier's profile, and this relationship is the key to the glacier's economy. If resupply is abundant because of ample net accumulations of snow upstream (thicker glacier, larger value of τ), the snowline shifts downstream and the convex profile downstream from it is steepened. If supply decreases owing to deficient snowfall or increased melting upstream (thinner glacier, smaller value of τ), the snowline shifts upstream and the curve near the lower end of the glacier slopes more gently.

These close relationships among thickness, shearing stress, and velocity, as well as snowfall and melting, show us that a glacier, like a stream, is an open system. It has an input of H_2O in the solid state as snow, conversion of snow to ice by elimination of air, downward and outward transport of ice under the influence of g_t, and output of H_2O, mainly as water to streams and as vapor to the atmosphere. All this activity is part of the hydrologic cycle. Besides, a glacier, like a stream, erodes rocks and carries a load, thus performing geologic work as its part of the rock cycle.

Fig. 12-6. Combined Guyot (pronounced Gheé-o) and Yahtse Glaciers, seen from the southeast in 1938. The glaciers flow toward the camera and terminate in Icy Bay, Alaska at West longitude 141° 20′. During the last few decades these glaciers, like their neighbors, have been shrinking, principally by calving. Successive positions of the termini in 4 later years, determined by instrumental surveys, are shown. Terminal recession during the 25-year period 1938 to 1963 totaled 6 miles, a retreat more rapid than would be expected if the glacier had ended on land. By 1963 the two glaciers had become separated by the Guyot Hills. The mountains on the skyline are 120 to 130 miles distant. (Photo by Bradford Washburn; data from M. M. Miller.)

The length and volume of a glacier mirror the dynamic equilibrium (steady-state condition) among three factors: accumulation of snow, loss by melting and evaporation, and transfer by movement (Fig. 12-5). These factors change in response to slow gradual changes in the snowline, in turn caused by changes in the climate. Glaciers, therefore, are a useful yardstick for measuring variations of climate. When the climate in any region becomes cooler and more moist, glaciers expand; when it grows warmer and drier, they shrink. A nearly worldwide shrinkage of glaciers during the first half of the present century implies a general warming of climates during at least that span of time (Fig. 12-6).

During the shrinkage of a glacier in a warming climate, the ice of course does not reverse the direction of its movement. The glacier continues to flow in the same direction, but the terminus melts back faster than ice can be supplied to it by flow (Fig. 12-16). Shifting of the terminus, either backward or forward, is merely the adjustment between supply and melting, tending to reestablish dynamic equilibrium after an interruption caused by a change in the climate.

Meltwater. In the zone below the snowline a good deal of water is usually evident, in the summer season in temperate latitudes. This is ***meltwater,*** defined as *water resulting from the melting of snow and glacier ice.* It is seen as small streams and pools on the surface of a glacier and in channels and tunnels within the glacier, right down to its base. Streams of meltwater flow on the ground, directly along the lateral margins of some valley glaciers. When they reach the glacier terminus, all this meltwater either forms a lake or combines into one stream flowing toward the sea, depending on whether the land slopes toward or away from the glacier.

GLACIATION

The geologic work of glaciers is accomplished by the process of ***glaciation,*** which is *the alteration of a land surface by massive movement over it of glacier ice.* Glaciation includes erosion, transport, and deposition. Because most glaciers have shrunk from former more extended positions, we can see glacially eroded bedrock surfaces being uncovered from year to year by the melting ice and can actually watch the making of glacial deposits at the glacier margins. Similar things were observed early in the nineteenth century by various Europeans, who recognized such features as the work of glaciers. By examining glaciated districts from which glaciers have melted away, we can see that glacial erosion takes place mostly well back beneath the glacier and that deposition occurs mostly at or near the outer margin.

GLACIAL EROSION

Glacial erosion, like stream erosion, includes abrasion and plucking and is generally accompanied by mechanical weathering in the form of frost wedging, just as stream erosion is accompanied by mass-wasting of various kinds.

Frost wedging. Although not a process of glaciation, intense frost wedging accompanies glaciation because both occur in cold climates. When hills or mountains stand higher than the surface of a glacier, wedged-out blocks of rock roll down onto the surface of the glacier, which carries them away. Intense frost wedging is responsible for much of the detail of the sharp, jagged peaks of glaciated mountains (Fig. 12-11).

Plucking and abrasion. It has been said that a glacier is at once a plow, a file, and a sled. As a plow it scrapes up regolith and lifts out blocks of bedrock; as a file it rasps away firm rock; as a sled it carries away the load of sediment acquired by plowing and filing, plus additional rock fragments fallen onto it from adjacent cliffs.

Glacial plucking is *the lifting out and removal of fragments of bedrock by a glacier.* It is broadly equivalent to hydraulic plucking by a stream. The bottom of the glacier breaks off blocks of bedrock and quarries them out, especially at surfaces that are unsupported on their downstream sides (Fig. 12-7). Near the edges of the glacier, at least, water aids plucking by freezing in cracks in the rock and prying out blocks.

Abrasion is the filing process. The under surface of the glacier is studded with rock particles of many sizes, and with these as tools the ice makes long scratches (*glacial striations*) and grooves on bedrock (Fig. 12-8). Fine particles of sand and silt in the base of the glacier, acting like fine sandpaper, polish bedrock to a smooth finish. A striation or groove trending north-south does not as a rule show whether the glacier was flowing south or north. Direction of flow is determined by small features like those in Fig. 12-8 and by rock hills and small knobs that are unsymmetrical. Usually the upstream side of a rock hill slopes gently, because it has been predominantly abraded, whereas the downstream side slopes more steeply, because it has been predominantly plucked (Fig. 12-7).

The great length and straightness of some grooves are made possible by the fact that the flow of ice is not turbulent. In contrast, rock surfaces abraded during the turbulent flow of water show a complex pattern of curves.

Cirques. Among the characteristic features made by glaciation in many mountain areas is the ***cirque*** (pronounced *sirk*). This is *a steep-walled niche, shaped like a half bowl, in a mountain side, excavated mainly by ice plucking and frost action* (Fig. 12-9). A cirque begins to form beneath a snowbank or snowfield just above the snowline. It is at least partly the work of frost wedging. On summer days water from melting snow infiltrates openings in the rock beneath the snowbank. At night the temperature drops and the water freezes and expands, prying out rock fragments. The smaller rock particles are carried away downslope by meltwater during thaws. This activity creates a depression in the rock and enlarges it. If the snowbank grows into a glacier, plucking helps to enlarge the cirque still more, but frost wedging continues as water descends the rock wall of the cirque and freezes there. The floors of many cirques are rock basins, some containing small lakes. In summary, small cirques are excavated beneath snow-

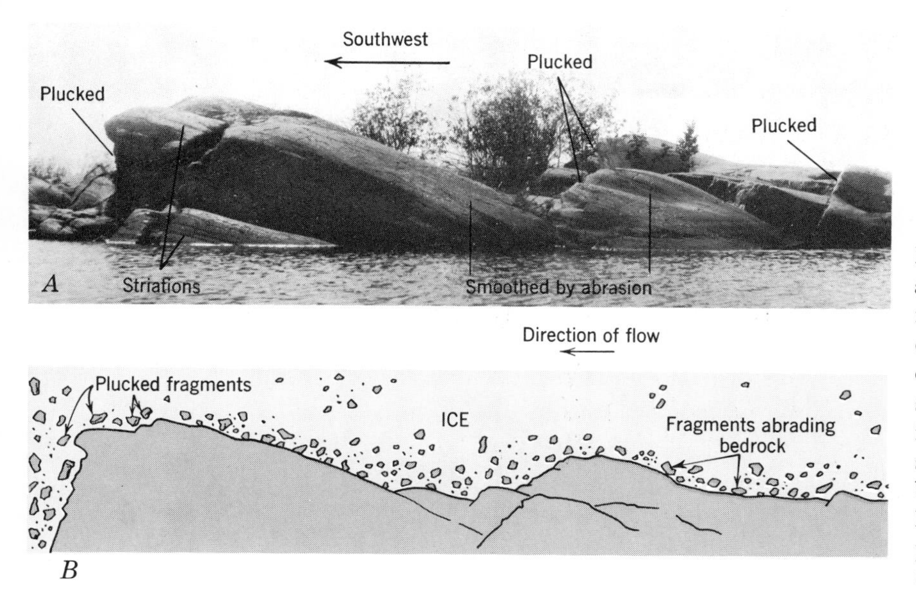

Fig. 12-7. Results of glacial abrasion. A. Knobs of granitic rock abraded and striated on one side and plucked on the other. Asymmetrical form shows former glacier was flowing southwest while erosion occurred. Lake Athabaska, Saskatchewan. B. Plucking and abrasion in progress. (F. J. Alcock, Geol. Survey of Canada.)

Fig. 12-8. Glaciated surface of sandstone bedrock, showing both striations and grooves. The sandstone contains pebbles and cobbles of quartz, which resist erosion better than the sandstone matrix does. The "tails" of sandstone behind them stream away from the camera; hence the former glacier moved in that direction (south). East Haven, Connecticut. (R. F. Flint.)

Fig. 12-9. Six small cirques, containing cirque glaciers (*extreme right*) or the headward ends of small valley glaciers (*left center*). Floors of the cirques lie close to a common altitude, that of the snowline. If all the ice and snow were to melt, the snowline could be reconstructed roughly from cirque-floor altitudes. Mount Lucania district, St. Elias Mountains, Yukon Territory. (Bradford Washburn, 1966.)

banks even where no true glacier exists; large cirques are mostly the work of glaciers that continue the excavation process.

Glaciated valleys differ in several ways from ordinary stream valleys. Their chief characteristics, not all of which are found in every glaciated valley, include (Figs. 12-10, 12-11, *C*): (1) a cross profile that is troughlike (U-shaped), and (2) a floor that lies below floors of tributaries, which therefore "hang" above the main valley. Both (1) and (2) result from erosion by the sides as well as by the base of the glacier, the thickness of which is far greater than the depth of an ordinary stream. In addition (3) the long profile of the floor is marked by steplike irregularities (Fig. 12-17) and shallow basins. Many of these are related to spacing of joints in the rock, which determine ease of plucking, a more effective process in glaciers than in streams. Finally (4) the valley head is likely to be a cirque or a group of cirques.

Some valleys are glaciated from head to mouth. Others are glaciated only in their headward parts, and downstream their form has been shaped by streams, sheet erosion, and mass-wasting. This shows that the valleys were there before the glaciers formed and leads to the conclusion that glaciers, unlike streams, do not cut their own valleys but occupy and remodel valleys already made.

Mountain sculpture. If we examine a mountain area that has been glaciated by a large group of valley glaciers (Fig. 12-11), we find cirques, U-shaped troughs, hanging tributaries, and (in greater detail) striated and polished bedrock. In addition, the forms of the mountain crests are characteristic and are mainly the result of frost wedging in a cold climate. The forms are combinations of three features, for which we use names given to them by Alpine mountaineers. An ***arête*** is *a jagged, knife-edge ridge created where two groups of cirque glaciers have eaten into the ridge from both sides.* A **col** is *a gap or pass in a mountain crest at a place where the headwalls of two cirques intersect each other.* A ***horn*** is *a bare, pyramid-shaped peak left standing where glacial action in cirques has eaten into it from three or more sides.* The Matterhorn is a well-known example.

Fig. 12-10. Glaciated valley. Isterdalen, Norway. (Sawders from Cushing.)

All these features are primarily the work of frost wedging coupled with glacial erosion and transport. All are shown in Fig. 12-11 and some in Fig. 1-15.

Fiords. The deep baylike ***fiords*** (Fig. 1-15) along mountainous coasts such as those of Norway, Alaska, and southern Chile are *glaciated troughs partly submerged by the sea.* The form and depths of many fiords imply depths of glacial erosion of 300m or more.[1] But some areas glaciated by ice sheets still preserve chemically weathered regolith beneath glaciated surfaces. Because ordinarily such regolith is thin, the thickness removed by glacier ice must have been small, possibly only a few feet. So the intensity of glacial erosion, like that of stream erosion, varies according to topography, kind of bedrock or regolith, and thickness and velocity of flow.

[1]The depths of fiord floors below sea level result in part from the rise of sea level since the last glacial age (Fig. 12-27), in part (locally) from subsidence of the Earth's crust, and in part from glacial erosion below the sea's surface. Sea level is not a base level for glaciers as it is for streams, because glacier ice 300m thick, with a specific gravity of 0.9, can continue to erode its bed until it is submerged to a depth of 270m, whereupon it floats and bed erosion ends.

GLACIAL TRANSPORT

In the way in which it carries its load of rock particles, a glacier differs from a stream in two main respects: (1) its load can be carried in its sides and even on its top, and (2) it can carry much larger pieces of rock; also it can transport large and small pieces side by side without segregating them into a bed load and a suspended load (depositing them according to their individual weights). Because of this, deposits made directly from a glacier are neither sorted nor stratified.

The load in a glacier is concentrated in base and sides (Fig. 12-17), because these are the places where glacier and bedrock are in contact and where abrasion and plucking are effective. Much of the rock material on the surface of a valley glacier got there by landsliding from clifflike valley sides.

Much of the load in the base of a glacier consists of small particles such as fine sand and silt. Most of these particles are fresh and unweathered, with angular, jagged surfaces. Such particles are the products of crushing and grinding. *Fine sand and silt produced by crushing and grinding in a glacier* are

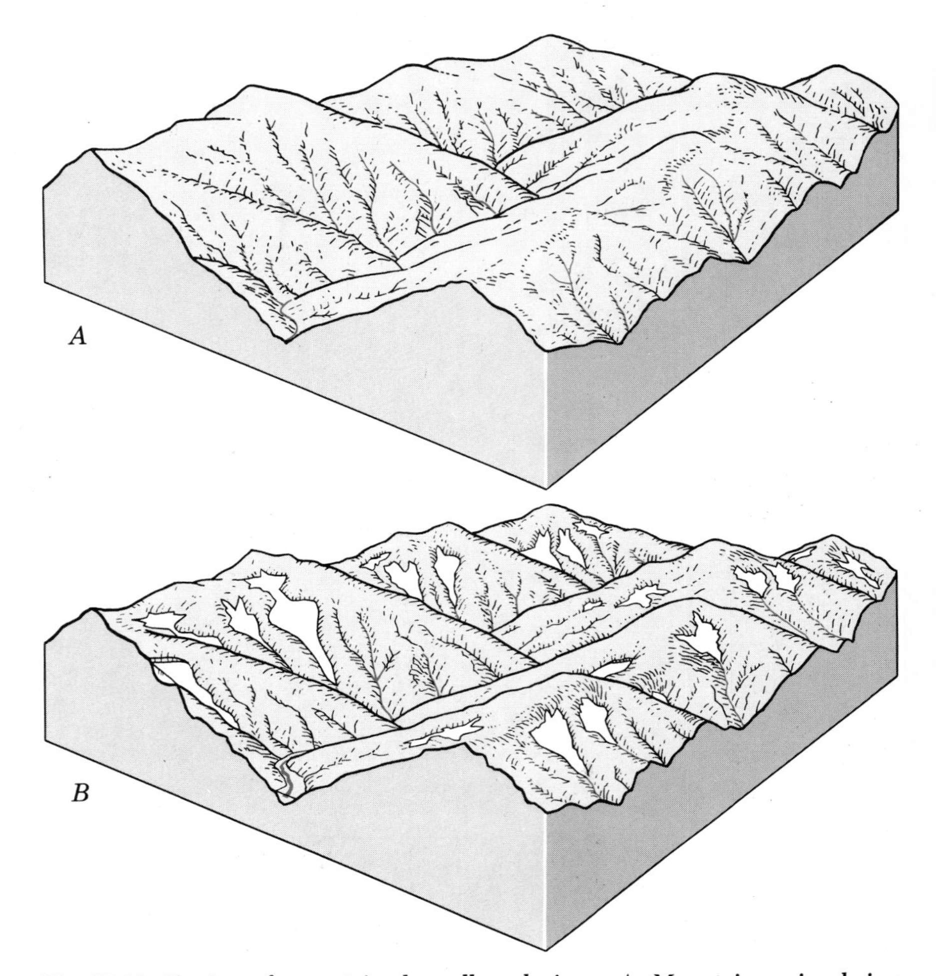

Fig. 12-11. Erosion of mountains by valley glaciers. *A.* Mountain region being eroded by streams. Main valley has many curves. *B.* As climate grows colder, snowfields form, and small cirques are excavated beneath them. Some snowfields become thick enough to form cirque glaciers. *C.* Glaciers merge to form a large valley glacier with tributaries. Frost wedging begins to sharpen the mountain summits. *D.* After climate has warmed again and glaciers have disappeared, geologic work of glaciers is revealed. Valleys have been deepened, widened, and straightened, tributaries have been left hanging above main valley, and empty cirques, some with small lakes, indent the highest areas. Mountain crests have been frost wedged to form knife-edge ridges with pyramid-shaped peaks.

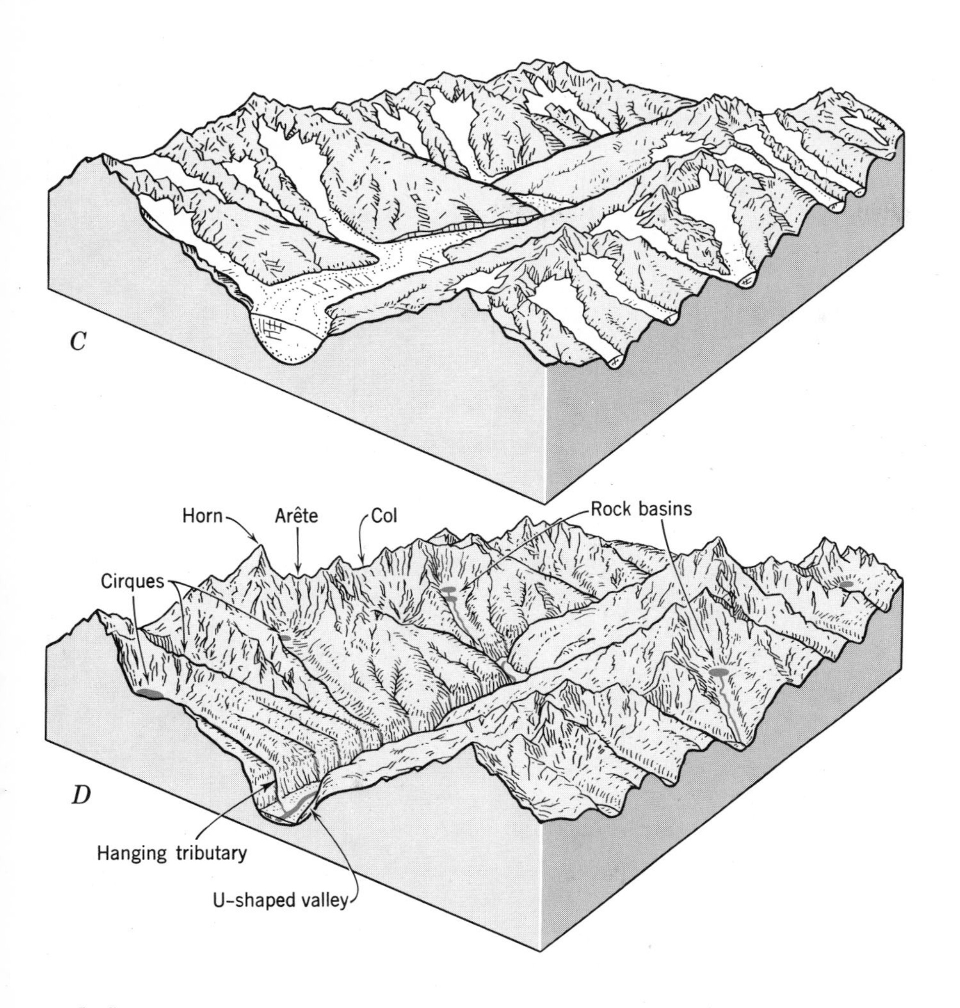

rock flour (Fig. 12-14), a material that differs from the chemically weathered, more rounded particles found in the sediments of nonglaciated areas.

GLACIAL DEPOSITS

Much of the transfer of rock particles from a glacier to the ground occurs by release of particles as the surrounding ice melts. Therefore most glacial deposition takes place in the downstream part of the glacier, below the snowline (Fig. 12-5), where melting is dominant.

Drift, till, and stratified drift. *The sediments deposited directly by glaciers or indirectly in glacial streams, lakes, and the sea* together constitute ***glacial drift***, or simply ***drift***. The name *drift* dates from the early nineteenth century when it was vaguely conjectured that all such deposits had been "drifted" to their resting places by the Flood of Noah or by some other ancient body of water.

Drift consists of two extreme types, *till* and *stratified drift*, which grade into each other. Drift whose constituent rock particles are not sorted according to size and weight but lie just as they were released from the ice (Fig. 12-12) is known as till, a name given it by Scottish farmers long before its origin was understood. **Till** is defined simply as *nonsorted drift* and is deposited directly from ice.

Probably most till is plastered onto the ground, bit by bit, from the base of the flowing ice near the outer margins of glaciers. We say "probably" because no one has yet devised a means of getting underneath a glacier to observe the process. The surfaces of pebbles and larger fragments in till include facets (Fig. 12-13) joining each other along smoothed or rounded edges; some facets are striated. Facets are made by grinding. As pebbles turn in their matrix of ice, new facets are made. The sand and silt particles in till generally consist of rock flour (Fig. 12-14).

Like till, colluvium (Chap. 8) is nonsorted, and some colluvium, on first comparison, looks very

Fig. 12-12. Till exposed in a road cut near Bangor, Pennsylvania. Its nonsorted character is clearly evident. The fine sediment between the stones consists of rock flour. Surfaces of two large boulders have been faceted and polished. (Pennsylvania Geol. Survey.)

Fig. 12-13. Pebbles collected from till, showing characteristic glaciated shapes, with facets. Chittenango Falls, New York. Compare Fig. 13-20. (R. F. Flint.)

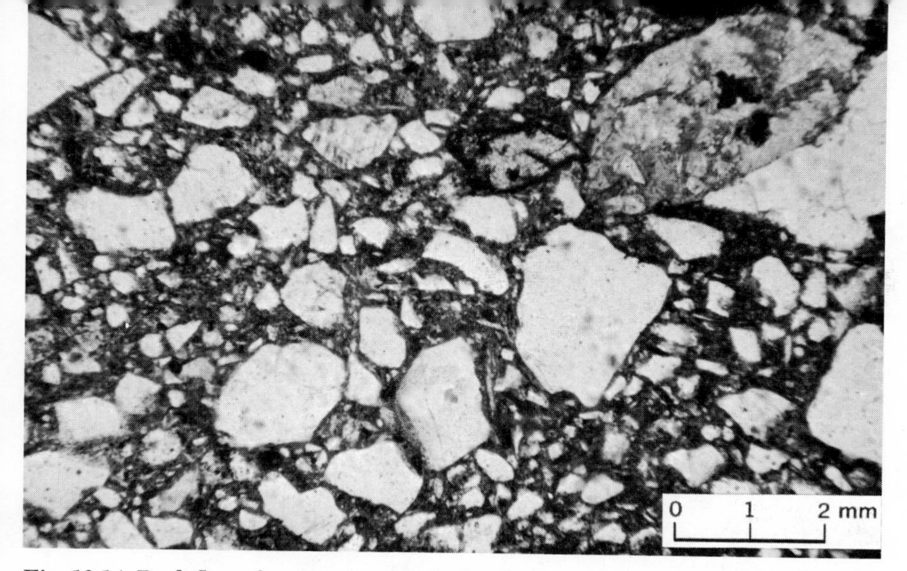

Fig. 12-14. Rock flour, forming part of a body of till near Bethany, Connecticut. The very angular shapes of particles (mostly quartz and feldspar) are the result of crushing. Particles are not sorted. Note similarity to Fig. 12-13. (Photograph by R. W. Powers.)

much like till. However, the presence of faceted, striated stones, rock particles of distant origin, and in some places a grooved or striated pavement underneath help to identify till.

On the other hand, much drift is stratified, indicating that water from melting ice has moved and sorted rock particles carried in the ice and has deposited them in immediate contact with the ice or beyond the glacier itself. ***Stratified drift,*** then, is *sorted and stratified drift;* it is deposited not by a glacier but by glacial meltwater.

Drumlins and other streamline forms. Stratified drift and till are sediments. They occur in various bodies, each having a rather distinctive topographic form. Some of the bodies are described in the following paragraphs.

In many areas drift is molded by streamline flow near the outer edge of an ice sheet into smooth, nearly parallel ridges and troughs that range up to many kilometers in individual length. These forms resemble the streamlined bodies of airplanes and racing cars; they are equilibrium forms, molded to offer minimum frictional resistance to the ice flowing over them. The best-known variety of streamline form is the ***drumlin,*** *a streamline hill consisting of drift, generally till, and elongated parallel with the direction of glacier movement* (Figs. 12-15, 12-25). Not all such forms, however, are built up. Some are shaped by glacial molding of pre-existing drift; these too are drumlins. Others, like the much smaller striations and grooves, are shaped by glacial erosion of bedrock, and even though their form is streamline they are not drumlins. Whether made by building up or cutting out, or both, all these forms reflect streamline molding by flowing ice, and therefore their long axes are reliable indicators of the direction of flow of former glaciers.

Ground moraine and end moraines. *Widespread thin drift with a smooth surface consisting of gently sloping knolls and shallow closed depressions* is ***ground moraine*** (Fig. 12-16). Probably its irregularities result from irregular distribution of rock particles in the base of the glacier.

A ridgelike accumulation of drift, deposited by a glacier along its margin, is an ***end moraine*** (Figs. 12-16, 12-17). It can be built by snowplow- or bulldozer action, by dumping off the glacier margin as the ice melts, by repeated plastering of sticky drift from basal ice onto the ground, or by streams of meltwater depositing stratified drift at the glacier margin. End moraines range in height from a few feet to hundreds of feet. In a valley glacier the end moraine is built not only at the terminus but along the sides of the glacier as well for some distance upstream. The terminal part is a *terminal moraine;* the lateral part is a *lateral moraine;* but both are parts of a single feature (Pl. 12; Figs. 12-1, 12-17, 12-25).

Erratics, boulder trains. Some of the boulders and smaller rock fragments in till are the same kind of rock as the bedrock on which the till was deposited, but many are different rocks, having been brought

Fig. 12-15. Drumlin near Madison, Wisconsin. (C. C. Bradley.)

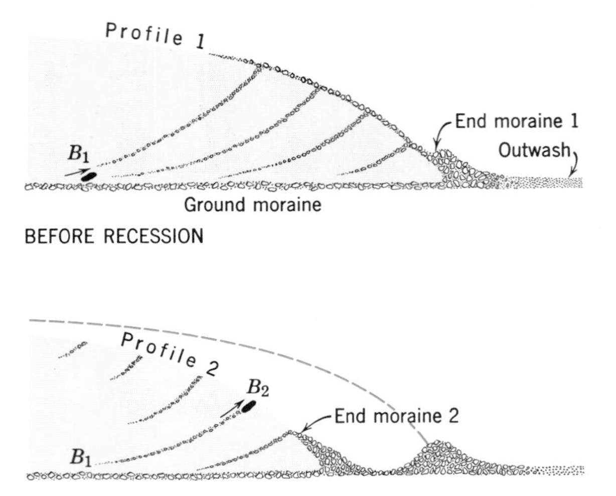

Fig. 12-16. With change to a slightly warmer climate, rate of melting increases and the terminus of a glacier recedes from Profile 1 to Profile 2. Continued flow of glacier during recession moves boulder *B* from B_1 to B_2. We can expect the boulder to be built into End Moraine 2. Repeated fluctuation can pile up a complicated sequence of end moraines and outwash.

from greater distances. *A glacially deposited particle of rock whose composition differs from that of the bedrock beneath it* is an ***erratic*** (Fig. 12-18). The word means simply "foreign," and the presence of foreign boulders was one of the earliest recognized proofs of former glaciation. Some erratics form part of a body of drift; others, such as the one shown in Fig. 12-18, lie free on the ground. Many erratic boulders exceed 3m in diameter, and some are enormous, with weights estimated in the thousands of tons. Such boulders are far larger than those which can be carried by an ordinary stream of water.

In some areas that have been glaciated by ice sheets, erratics derived from some distinctive kind of bedrock are so numerous and easily identified that they can be readily plotted on a map. Generally the plot shows a fanlike shape, spreading out from the area of outcrop of the parent bedrock and reflecting the spreading of the ice sheet. *A group of erratics spread out fanwise* is a ***boulder train,*** so named in the nineteenth century when rock particles of all sizes were called boulders. The boulder train shown in Fig. 12-25 consists of quartzite, very conspicuous in an area in which all the bedrock consists of limestone and sandstone.

Outwash. On the downstream sides of most terminal moraines is *stratified drift deposited by streams of meltwater as they flow away from a glacier.* Sediments of this kind are ***outwash*** ("washed out" beyond

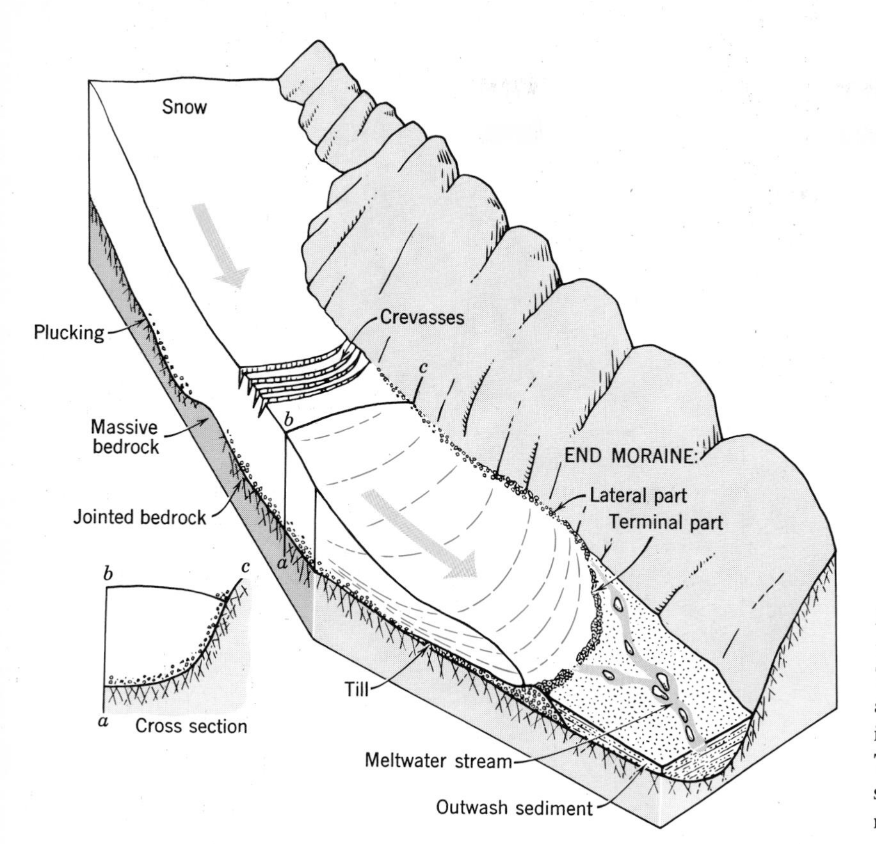

Fig. 12-17. Chief features of a valley glacier and its deposits. The glacier has been cut away along its center line; only half of it is shown. Compare Fig. 12-1. The crevasses shown result from steepened slope of the bed beneath the glacier.

Fig. 12-18. Erratic boulder of granite, perched on top of a high ridge of dolerite, Mount Tom, near Northampton, Massachusetts. The nearest granite bedrock is many miles to the north. (C. R. Longwell.)

Fig. 12-19. Valley train being actively built of outwash sediments by streams of meltwater from glaciers upstream. Tasman Valley, Southern Alps, New Zealand. Tasman Glacier (*right*), dark with stony debris, is the principal source of outwash. Hooker Glacier (*left*) is contributing a fanlike body of outwash that has pushed the streams from Tasman Glacier to one side of their valley. Other, local fans are visible in left foreground. Stream patterns are braided. The valley train has little or no vegetation because a new layer of sediment is deposited on it during each melting season. (V. C. Browne.)

the ice). *A body of outwash that forms a broad plain beyond the moraine is an* ***outwash plain.*** In contrast, a ***valley train*** (Fig. 12-19; Pl. 12) is *a body of outwash that partly fills a valley.* Meltwater generally emerges from the ice as one or more swift streams, turbid with a suspended load of rock flour and with an abundant bed load of pebbles, cobbles, and even boulders. The bed load is invisible, but it is there. The stream is generally braided; its deposits are very thick and have the kind of stratification shown in Fig. 12-20. These characteristics are common in streams with abundant bed loads. The deposition of outwash is analogous to the building of a fan (Fig. 9-22). The stream emerges from an ice-walled valley or tunnel onto a broad smooth surface; part of its load therefore becomes excess; it drops bed load, and its channel takes on a braided pattern. The profile of the entire thick deposit is steep like that of a fan. As the coarse particles are dropped first, the average diameter of particles decreases downstream.

Although the bed load of a meltwater stream is invisible, the suspended load has been measured. The rock flour washed out of the Muir Glacier in coastal Alaska corresponds to a loss of 2cm of bedrock, from the entire area beneath the glacier, every year. That is 332 times faster than the 1cm every 166 years estimated to be lost to the United States by weathering, mass-wasting, and erosion by running water.

Ice-contact stratified drift. When rapid melting and evaporation reduce the thickness of the terminal part of a glacier to 100m or less, movement virtually ceases. Meltwater, flowing over or beside the nearly motionless stagnant ice, deposits stratified drift, which slumps and collapses as the supporting ice slowly melts away. *Stratified drift deposited in contact with supporting ice* is ***ice-contact stratified drift.*** It is recognized by abrupt changes of grain size, distorted, irregular stratification, and extremely uneven surface form (Fig. 12-21). Bodies of ice-contact stratified drift are classified according to their shape: *short, steep-sided knolls and hummocks* are ***kames;*** *terracelike forms along the sides of a valley* are ***kame terraces;*** *long, narrow ridges, commonly sinuous,* are ***eskers*** (Fig. 12-22); and *closed depressions in drift, created by the melting out of masses of underlying ice,* are ***kettles*** and in form are complementary to kames.

THE GLACIAL AGES

History of the concept. As early as 1821 European scientists began to recognize features characteristic of glaciation in places at considerable distances from any existing glaciers. They drew the inference that glacier ice must once have covered wide regions

Fig. 12-20. Outwash sediments consisting of sand and pebbles, exposed in a valley train built during a glacial age, Wallingford, Connecticut. The currents moved generally from left to right. At the time this outwash was deposited the terminus of the glacier stood more than 25 miles north of the locality. (R. F. Flint.)

that are now free of ice. Consciously or unconsciously, they were applying the principle of uniformity of process and materials. The concept of a glacial age with widespread effects was first set forth in 1837 by Louis Agassiz, a Swiss scientist who achieved fame through his hypothesis. Gradually, through the work of many others, information on the character and extent of former glaciation was added to the growing body of knowledge, until today we have formulated a basic picture of former glacial times, although many important questions remain unanswered.

Extent of glaciers. During the second half of the nineteenth century, geologists searched for glacial drift and other characteristic features in order to determine the extent of former glaciers. In most mountain regions the characteristics are those shown in Fig. 12-11, *D*. In most lowland regions, however, they consist generally of rolling ground moraine, end moraines, and related features (Fig. 12-25). By 1900 the general extent of glaciation had been learned, and today our information is much more detailed. Figure 12-23 shows areas in the Northern Hemisphere that were covered by glaciers during the glacial ages. Figure 12-24 shows in more detail the extent of glaciation in the northern United States and southern Canada. These maps were compiled from observations of hundreds of geologists on the distribution of features characteristic of glaciation. On a world scale, the areas formerly glaciated add up to the impressive total of more than $45 \times 10^6 km^2$, more than 30 per cent of the entire land area of

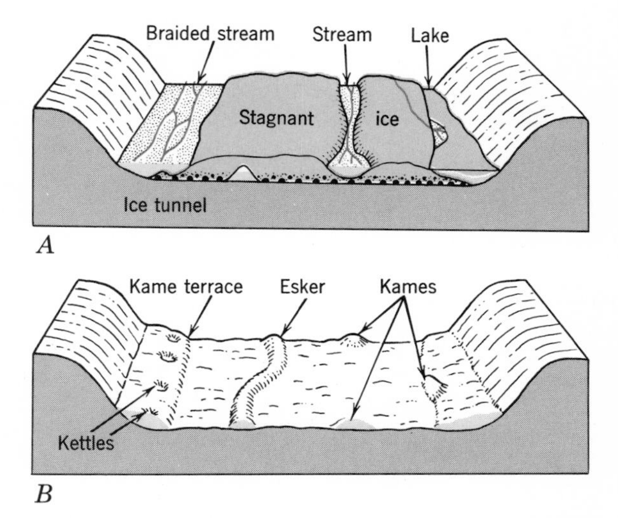

Fig. 12-21. Origin of bodies of ice-contact stratified drift. *A.* Nearly motionless melting ice furnishes temporary retaining walls for bodies of sediment built chiefly by streams of meltwater. *B.* As ice melts, bodies of sediment slump, creating characteristic knolls, ridges, terraces, and closed depressions.

Fig. 12-22. This glacial-age esker, overlying ground moraine in Morrison County, Minnesota, consists of gravel and sand deposited in a winding tunnel (Fig. 12-21) within an ice sheet but near its edge. When supporting ice melted away, the deposit was left as a winding ridge. (W. S. Cooper.)

the world. Today, for comparison, only about 10 per cent of the world's land area is covered with glacier ice, of which nearly 85 per cent (by area) is on the Antarctic Continent. If we neglect that continent and consider only the rest of the world, the glacier-covered area on non-Antarctic lands was more than 13 times larger during the glacial ages than it is on the same lands today.

Directions of flow. In most mountain regions former glaciers simply flowed down existing valleys (Fig. 12-11). In lowland regions streamline forms and end moraines show that ice sheets spread out in a radial manner (Figs. 12-23, 12-25). Flow directions are determined also by tracing erratics of conspicuous kinds to their places of origin in the bedrock. Native copper found as far south as Missouri has been traced to bedrock on the south shore of Lake Superior. In eastern Finland, copper ore traced backward along the line of ice flow led to the discovery of a valuable copper deposit in the bedrock. Several large diamonds of good quality have been found in drift in Wisconsin, Michigan, Ohio, and Indiana. Their source has not yet been found, but the flow directions of former glaciers suggest it is in central Canada.

Amount of erosion. The great central parts of the former ice sheets in north-central Canada and northern Scandinavia and Finland removed underlying regolith and bedrock to a depth averaging perhaps 15 to 25m. Much of the load thus acquired by the ice sheets was deposited as drift beneath their broad outer parts. One such belt of drift reaches from Ohio to Montana; another extends from the British Isles to European Russia. In these belts the average thickness of the drift is believed, from well borings and other measurements, to be as much as 12m. Most of the rest of the load went down the Mississippi River and down various rivers in Europe; some of the fine-grained part was picked up by the wind and blown away.

Where, as in New England and parts of Quebec, the bedrock is resistant and breaks into large chunks, the glacier ice spread boulders liberally over the surface, creating a soil very difficult for agriculture. Where, as in the southern Great Lakes region and on the Plains, the bedrock is weak and crumbles

Fig. 12-23. Areas (white) in the Northern Hemisphere that were glaciated during the glacial ages. Arrows show generalized directions of flow of glacier ice. Shorelines are shown as they were when sea level was 100m lower than it is today. The gray tone with irregular pattern represents floating ice in the Arctic Ocean, extending south into the Atlantic.

easily, the glaciers deposited a thick layer of drift, which now supports a rich soil.

Depression of the Earth's crust. We should note at this point the subsidence of the Earth's crust beneath the weight of the ice sheets. This effect is described in Chapter 21. Probably the surface of the crust beneath the central part of the Greenland Ice Sheet would stand as much as 700 to 900m higher if the ice sheet were not there.

Repeated glaciation. Because most of the drift is fresh and little weathered, it was realized very early that the glacial invasion was recent. But geologists began to find exposures showing a blanket of comparatively fresh drift overlying another layer of drift whose upper part is chemically weathered (Fig. 12-26). This led to the inference that there had been two glaciations separated by a period of time long enough to cause weathering to a depth of several feet. Before the beginning of the twentieth century, accumulated evidence of this kind had established the fact of not merely two but several great glaciations, each covering approximately the same areas.

Recent dating of sediments by K/Ar measurement suggests that the earliest of these glaciations may have occurred between 2 and 3 million years ago. The whole time within which the glaciations occurred is the Pleistocene Epoch. It is often referred to collectively as the *glacial ages,* despite

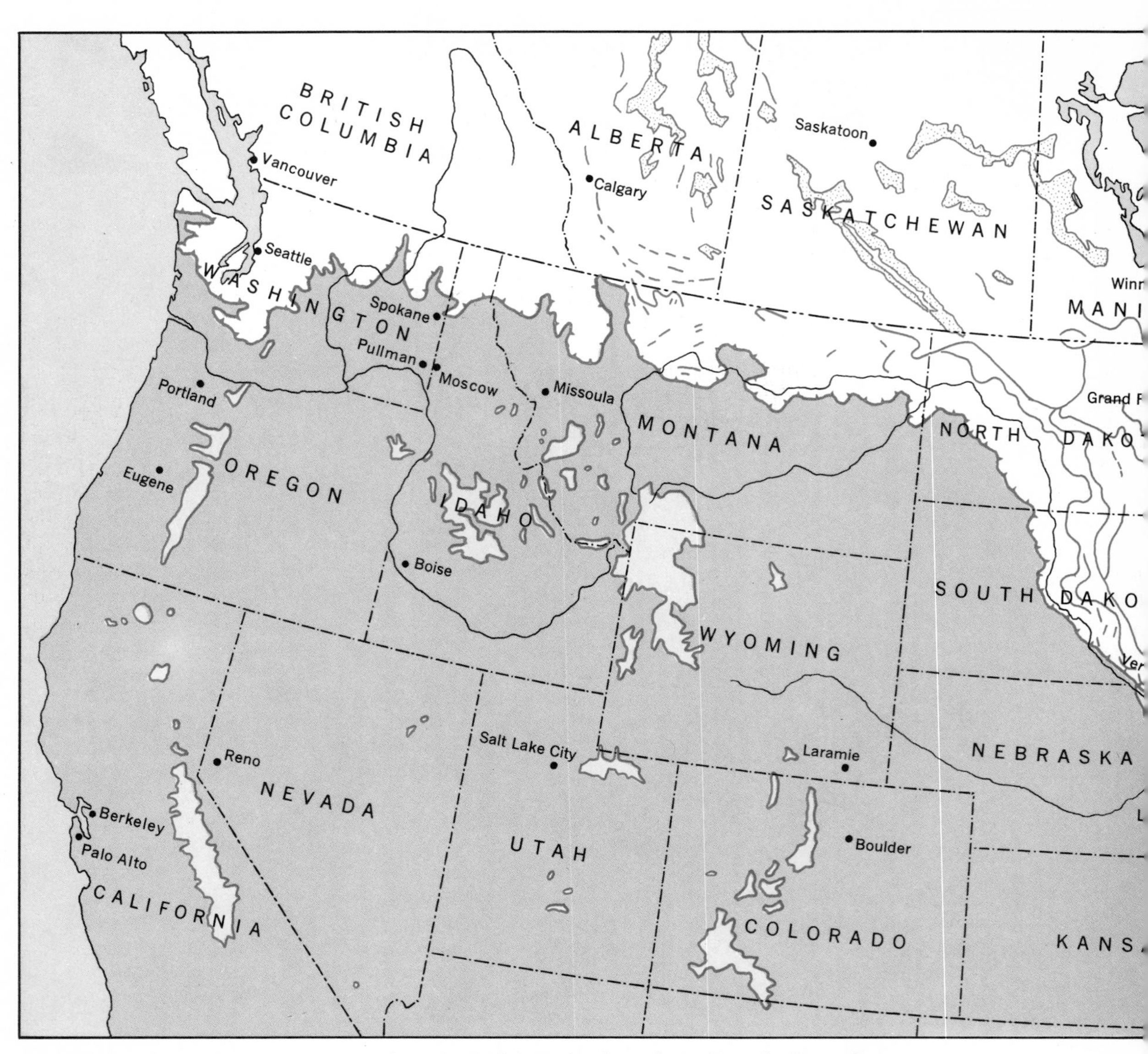

Fig. 12-24. Extent of glaciation in northern United States and southern Canada. Inner line, during latest glacial age; outer line, during earlier glacial ages. (Complied from several sources.)

the fact that between the glaciations were intervals (*interglacial ages*) in which climates were about as warm as today.

Dating by C^{14} indicates that the last time glaciers in North America and Europe reached a great extent was around 18,000 years ago. The subsequent period of shrinkage to their present condition has been marked by repeated, conspicuous expansions. One of these occurred 2,500 to 3,000 years ago and another only 200 to 300 years ago, overwhelming small high-altitude farms and villages in the Alps. Some of the dwellings are today still buried beneath ice.

Effects of glacial-age climates. The formation and spread of glaciers was the most obvious effect of the relatively cold climates of glacial ages. But there were other effects, apparent in the oceans and in lands not covered with ice.

Change of sea level. The first effect was lowering of the oceans. In the hydrologic cycle, snow-

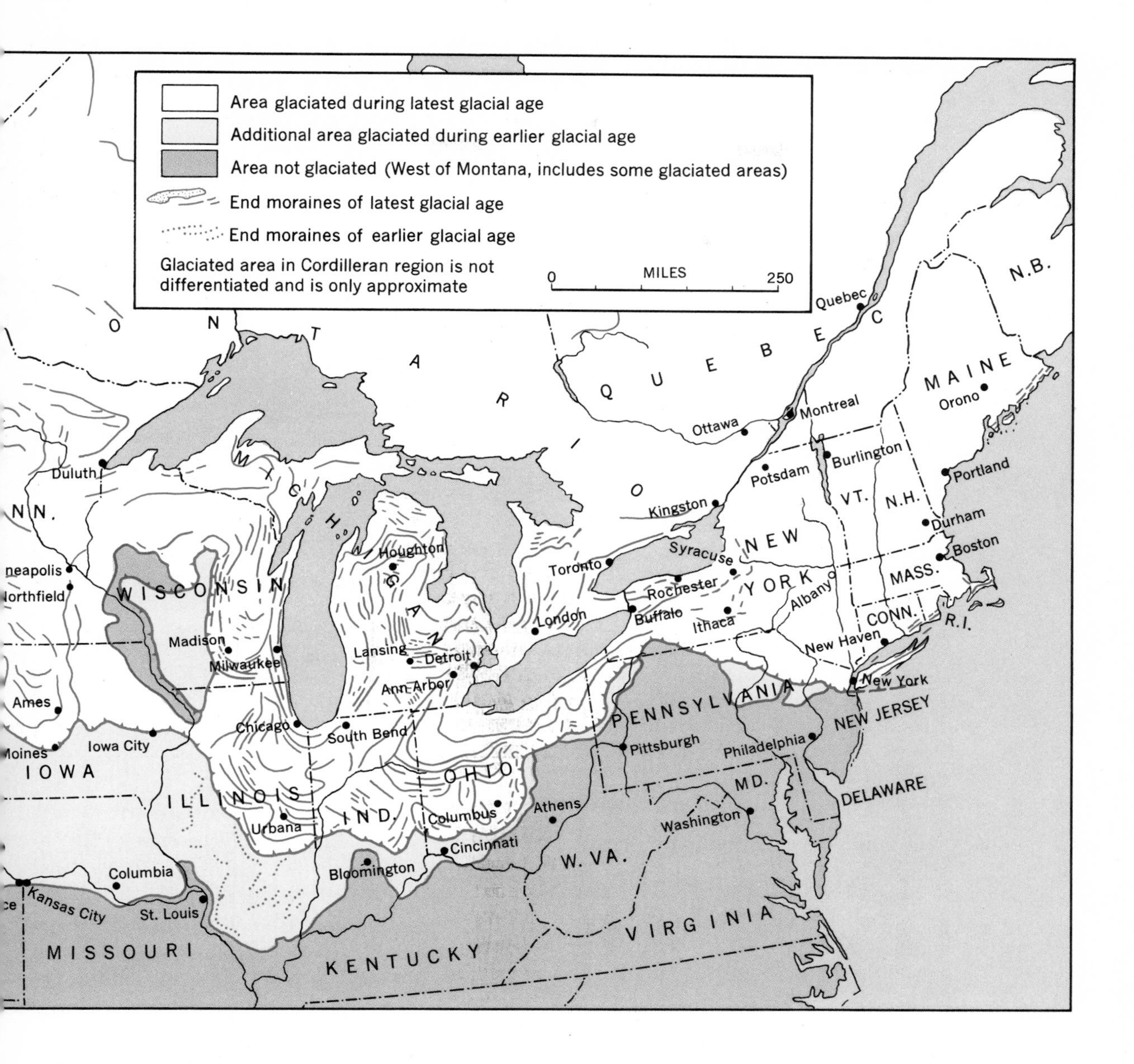

fall to build glaciers comes from the evaporation of water, mainly in the oceans. Although rainfall on the land drains back to the sea rather quickly, snowfall in the form of glaciers remains on the lands for as long as the climate is cold enough to perpetuate the glaciers. So when glaciers increase, sea level must fall; when they decrease, sea level must rise (Fig. 12-27).

Because we do not know precisely the volumes of the Antarctic and Greenland Ice Sheets, we can not calculate their influence on sea level very closely, but it is likely that the complete melting of those huge glaciers would add to the existing oceans a layer of water more than 70m thick.

Along some coasts, old beaches and other shore features, high and dry at altitudes of 35m and more, probably were made during one or more interglacial ages when ice on the lands was less abundant than it is today. Of course, before we fully accept this inference we must make sure the

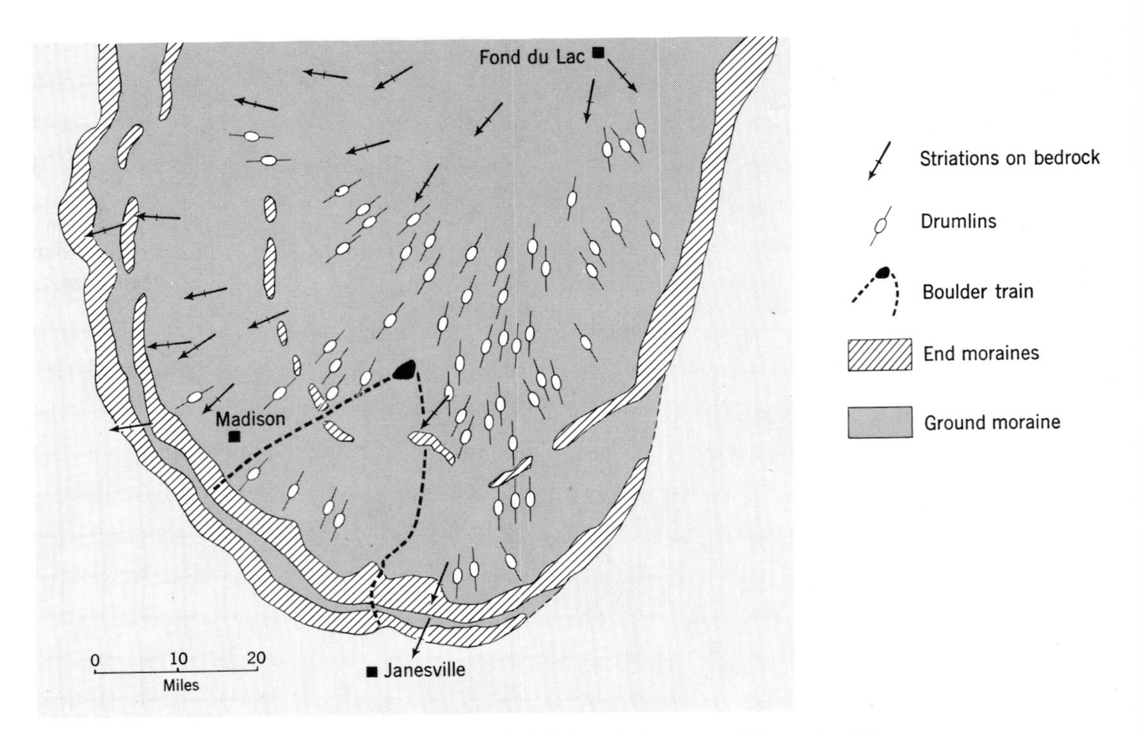

Fig. 12-25. Map of lobe-shaped layer of glacial drift in southeastern Wisconsin. Three successive end moraines (both terminal and lateral parts) mark successive positions of the glacier margin. Directions of drumlins, striations on bedrock, and a boulder train all show spreading of ice toward margin of glacier. (Adapted from Flint and others, 1959.)

beaches have not simply risen during uplift of part of the Earth's surface.

Conversely, during a glacial age when the volume of ice on the lands was far greater than it is today, sea level must have been lower by considerably more than 100m. In their own way these sea-level changes were as dramatic as the overrunning of continents by glaciers.

Quite likely the shallow channel on the continental shelf, connecting the Hudson River with the Hudson Submarine Canyon, is the work of a lengthened Hudson River during a glacial time of lowered sea level (Fig. 15-21).

Pluvial lakes. A second effect of glacial-age climates was the creation of lakes in regions that today are dry or contain saline lakes. Figure 12-28 shows the contrast. The basin of Great Salt Lake, Utah, was occupied formerly by the gigantic Lake Bonneville (Fig. 21-19), more than 1,000 feet deep and with a volume comparable to that of Lake Michigan. Beaches, deltas of tributary streams, and bottom sediments are all there to tell the story. Such lakes are ***pluvial lakes.*** None exist today because, by definition, they are *lakes that existed under a former climate, when rainfall in the region concerned was greater than at present.* Although the name implies a pluvial (rainy) climate, we realize that reduced evaporation caused by reduced temperature was a large factor in the changed water economy, but the name *pluvial* still sticks to the ancient lakes. These lakes were, then, the result of (1) lower temperature, and (2) increased rainfall in the regions where they existed.

Direct geologic evidence and C^{14} dates show that some pluvial lakes were contemporaneous with the most recent of the glacial ages; so it is reasonable to think that this is true of pluvial lakes as a class. It should not be supposed, however, that because pluvial lakes were related in time to glacial ages they were fed by water derived from the melting of glaciers. Some of them were, but only by coincidence. Most of the lakes occupied basins to which glacial waters had no access.

Pluvial lakes were abundant also in other dry

Fig. 12-26. Evidence of repeated glaciation. A layer of "fresh" till, weathered only slightly at its surface, overlying a layer of older till that had been deeply weathered before the overlying till was deposited.

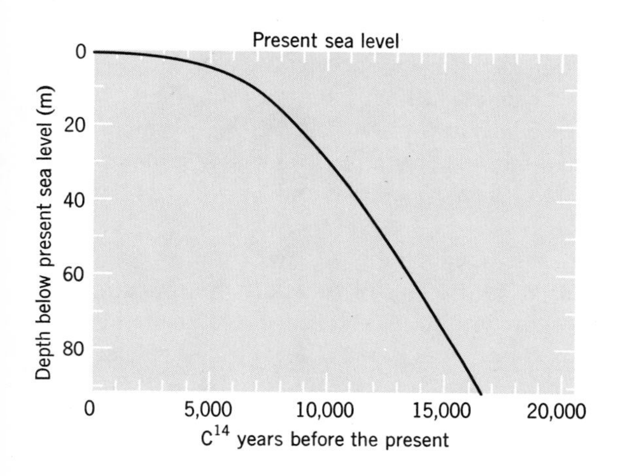

Fig. 12-27. Generalized curve showing position of sea level, relative to the land, through most of the time since the peak of the last glaciation. Constructed mainly from C^{14} dates of wood, peat, and shells deposited at or near sea level but now submerged. (After F. P. Shepard.)

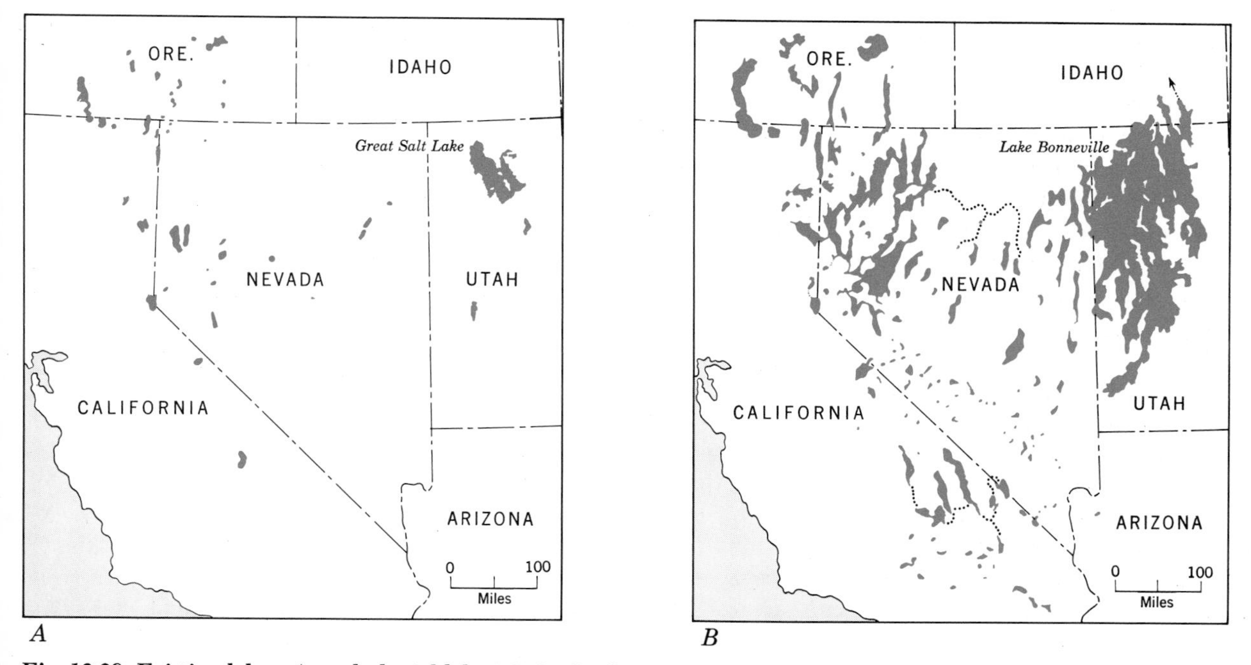

Fig. 12-28. Existing lakes, *A*, and pluvial lakes, *B*, in the dry region of western United States. (Flint, 1957; O. E. Meinzer, 1922.)

regions of the world. The desert area in the Andes of Bolivia, Chile, and Argentina contained many of them. Numerous pluvial lakes were present also in the Saharan region of North Africa and in western and central Asia. The largest in the world was an ancestor of today's Caspian Sea; another that is historically famous was an ancestor of today's Dead Sea.

Spread of permafrost. A third effect of the glacial-age climates was the spread of ***permafrost,*** the prevalent name for *ground that is frozen perennially.* Actually it is ground water that is frozen, to form a firm cement in all openings within soil and rocks. Today the ground is frozen and remains so from year to year, through a wide belt of Arctic country totaling in the Northern Hemisphere about 20 per cent of the land area of the hemisphere. Over this belt, average air temperatures are at or below freezing during most of the year, and loss of ground heat to the atmosphere is the cause of freezing of the ground water. During the short summer melting season only a thin surface zone, usually no more than about a meter thick, thaws out. The thawed layer flows down slopes by solifluction, carrying with it the arctic vegetation that grows at its surface.

Because drilling has shown that in places the thickness of permafrost is 1,000 feet or more (Fig. 2-17), we reason that a long period must have been required for freezing to so great a depth. During glacial ages the southern limit of permafrost stood at a lower latitude than it does now. Today, under a climate that has been growing generally warmer, the area of permafrost is slowly shrinking, just as glaciers are shrinking. Although dwindling, as are glaciers and pluvial lakes, permafrost could be renewed if the climate once more became colder over a period of hundreds of years.

VERY ANCIENT GLACIATIONS

The glacial ages as a group are thought to represent perhaps only the last 2.5 million years of geologic time. However, ***tillite*** (*till converted to solid rock*), with striated surfaces of still-older rock beneath it, occurs in strata hundreds of millions of years old (Figs. 12-29, 16-6). This evidence, fragmentary but unmistakable, implies that glaciation has occurred at least several times in the remote past and is not peculiar to the last few million years. The Earth's climates, evidently, have fluctuated slowly through a range great enough to create large accumulations of ice on the lands from time to time. Another possibility, discussed in Chapter 22, is that lands now in lower latitudes may have lain closer to the poles in former times and may therefore have been colder.

CAUSES OF GLACIAL CLIMATES

No intelligent person can visualize the exchange of so much seawater for land ice and the development of so many lakes and so much permafrost without thinking about what could have caused these changes. Although tens of thousands of pages of reasoning and discussion on this problem have been published, we still do not know the answer; we can only speculate and make and remake hypotheses.

In our speculation we must think of glacial events in two scales of time: (1) the repeated glacial ages within the Pleistocene Epoch, perhaps 2.5 million years long, and (2) the glacial ages recorded in the ancient rocks, separated by hundreds of millions of years of time. Beginning with the smaller-scale events of the Pleistocene, about which we have much more detailed data, we can note three basic things.

1. During glacial ages, mean annual air temperatures, averaged over the Earth, were around 5°C less than are those of today; during interglacial ages temperatures were at least 1° more than today's. The mean range was therefore only about 6°C (Chap. 15). Temperature changes were felt not only in the regions that were glaciated but everywhere, the Equator included, although they varied in amount from place to place.

2. Ice-age glaciers were distributed in the same basic pattern as that of today's glaciers: they originated in the same places but simply were much bigger. The places of origin are highlands so situated that winds can bring moist air to them. True, the big ice sheets spread far outward over low lands, but they are believed to have originated through the coalescence of many glaciers flowing down from high places.

3. Each glaciation was not a single event but a whole group of minor advances and retreats of glacier margins, as shown by the many end moraines

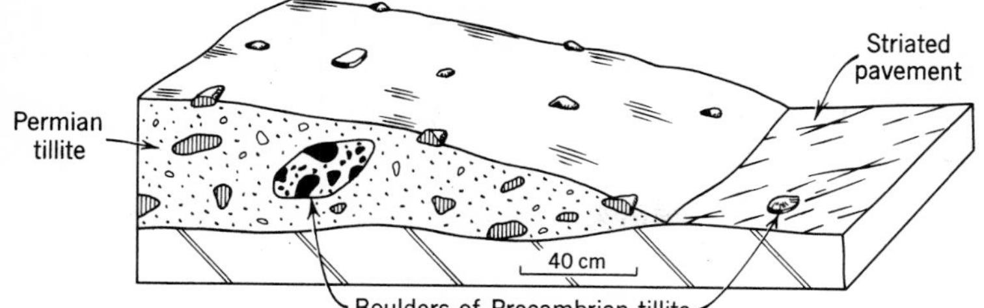

Fig. 12-29. Sketch of exposure near Adelaide, South Australia showing a body of Permian tillite about 250 million years old, containing a boulder broken from a Precambrian tillite more than 600 million years old. This unusual occurrence demonstrates that Australia has experienced at least two very old glacial ages.

and by other data as well. Radiocarbon dates of fluctuations during the later part of the latest glacial age show that these events have been nearly simultaneous in diverse parts of the world. Although we are not yet completely sure, probably each glacial age affected all the continents at the same time.

These three basic considerations enable us to narrow somewhat our field of thinking about the cause of the Pleistocene glacial-age events. The temperature changes affected the whole world, though to different extents in different regions. The glaciers reappeared, at intervals, centered in the same high places, and apparently pulsated nearly in unison. On this basis we can eliminate, as the sole cause, up-and-down movements of the Earth's crust and indeed any other sole cause that is not sensitive and flexible enough to change the behavior of glaciers in a similar manner within periods of only a few hundred years.

An hypothesis, suggested repeatedly, which seems to meet the requirements is that radiant energy, or at least certain wavelengths within it, reaching the Earth's lithosphere from the Sun varies through time. Although measurements thus far do not indicate fluctuation in the Sun itself, changes in the composition of the Earth's atmosphere (H_2O, CO_2, ozone, and very fine volcanic dust have been suggested) might cause changes in the effectiveness of the atmosphere as a filter or insulator. By such changes we might be able to explain the Pleistocene fluctuations of temperature.

This brings us to the longer time scale of the ancient glacial ages. We might expect that if the lithosphere has been receiving solar energy in an irregular way during the last few million years, it would also have been doing so through a much longer time before the Pleistocene Epoch, and that glacial ages should have been frequent throughout geologic history. We do not know whether glaciers in high mountains have been frequent or not, because in high, steep places erosion takes place rapidly and would be likely to destroy glacial sediments. On the other hand the striated surfaces and the sediments made by glaciers that spread out from mountains over lowland areas, as great ice sheets like those of the Pleistocene Epoch, are much more likely to be preserved in the geologic record. Such features do exist in ancient strata, but they are uncommon; so we conclude that important glacial ages have not been frequent.

Looking further, we can suggest that altitude may play a significant part in our hypothesis, because the frequency of glaciers increases with increasing altitude, and times when lands were low would have been times unfavorable for glaciation. During the Pleistocene Epoch lands were (and still are) considerably higher than they were throughout a very long period preceding it. Furthermore, the structure of bedrock tells us that the Earth's surface has been marked by widespread mountains (Chap. 21) at other times much more remote, when extensive glaciers likewise are known to have existed. In short, glacial ages seem to have been times when mountains were conspicuously present.

Building this relationship into our hypothesis, we can speculate that the Earth's lithosphere has been receiving solar energy irregularly throughout a long history, but that only at those times when extensive mountains were present could low temperatures be translated into glaciers so big that they invaded lowlands as great ice sheets.

This two-part idea seems to agree with the facts we have collected up to now, but it is still only an hypothesis. Further research will eventually confirm it or find a better explanation. Many other hypotheses exist, but they do not seem at present to meet the facts quite as well as the one we have outlined.

Summary

Definition and form

1. Glaciers are accumulations of snow and ice, and they flow under their own weight. Their surface parts are brittle; below these parts flow occurs.
2. On a basis of form, glaciers include valley glaciers, piedmont glaciers, and ice sheets.
3. Glaciers require low temperature and adequate snowfall. They are crossed by the snowline, which is low in polar regions and high in tropical regions.

Movement

4. The motion of a glacier includes both internal flow and sliding on its bed.

Economy

5. A glacier can be thought of as an open system, in which thickness, velocity, and other factors approach dynamic equilibrium.

Glacial erosion

6. Glaciers erode rock by plucking and abrasion, transport the product, and deposit it as drift.
7. Valley glaciers convert stream valleys into U-shaped troughs with hanging tributary valleys. Cirques form beneath snowbanks and the heads of valley glaciers. Areas projecting above glaciers are reshaped by frost wedging into arêtes, cols, and horns.

Glacial sediments

8. The load, carried chiefly in the base and sides of a glacier, includes particles of all sizes, from large boulders to rock flour.
9. Till is deposited by glaciers directly. Stratified drift, deposited by meltwater, includes outwash deposited out beyond the ice, and kames, kame terraces, and eskers deposited upon or against the ice itself.
10. Ground moraine is built beneath the glacier, end moraines (both terminal and lateral) at the glacier margins.

The glacial ages

11. During the glacial ages huge ice sheets repeatedly covered northern North America and Europe, eroding bedrock and spreading drift over the outer parts of the glaciated regions.
12. The accumulation of glaciers lowered sea level. The cool, moist climates that prevailed in regions now dry created many pluvial lakes. Permafrost developed in areas free from glaciers but with average temperatures that were below freezing.
13. The cause of the glacial climates is not known. Possibly it is related to the existence of exceptionally high land plus variations in the receipt of solar energy by the Earth.

Selected References

Bentley, C. R., and others, 1964, Physical characteristics of the Antarctic Ice Sheet: Am. Geog. Soc. Antarctic Map Folio 2.

Charlesworth, J. K., 1957, The Quaternary Era: London, Edward Arnold.

Flint, R. F., 1957, Glacial and Pleistocene geology: New York, John Wiley.

Thwaites, F. T., 1946, Outline of glacial geology: Ann Arbor, Edwards Bros.

Wright, H. E., and Frey, D. G., eds., 1965, Quaternary of the United States: Princeton, N. J., Princeton Univ. Press.

Glacial-age maps

Flint, R. F., and others, 1945, Glacial map of North America: Geol. Soc. America Spec. Paper 60. Scale 1:4,555,000.

—— 1959, Glacial map of the United States east of the Rocky Mountains: Geol. Soc. America. Scale 1:750,000.

Wilson, J. T., and others, 1958, Glacial map of Canada: Geol. Assoc. of Canada. Scale 1:3,801,600.

CHAPTER 13

Deserts and Wind Action

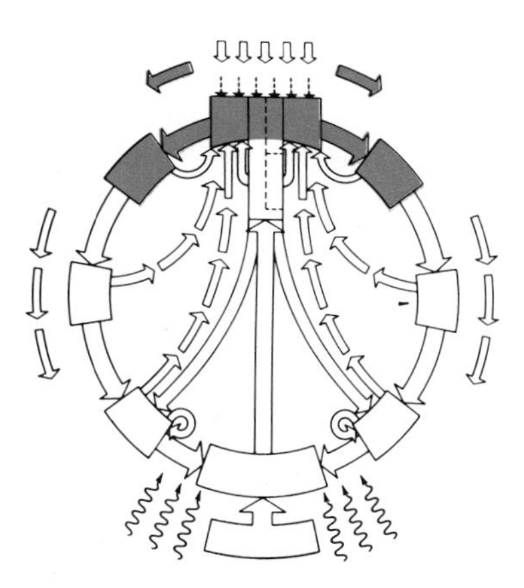

DESERTS
Climate
Geologic processes
Pediments
Cycle of erosion
WIND ACTION
Sediment transport by flowing air
Wind erosion
Wind deposits
Deposits of sand
Deposits of silt
Ancient wind-deposited sediments
Environmental aspects: soil erosion

Plate 13

Field of at least 78 longitudinal dunes, each about 150m high, in southeastern Arabia, seen from Gemini IV spacecraft in 1965. Horizontal distance from top to base of photo is about 280 km. (National Aeronautics and Space Administration.)

Clouds

Exposed bedrock

Stream valleys filled with sediment

Shadows of clouds

Longitudinal dune ridges

CHAPTER 13 *Deserts and Wind Action*

DESERTS

CLIMATE

Importance of climate in geology. The climate of a region plays a fundamental part in determining which geologic process will predominate there. In foregoing chapters we have seen that the relative importance of chemical and mechanical weathering, and the abundance of streams, depend mainly on rainfall. By contrast, the influence of glaciers in the geologic cycle depends mainly on low temperature.

In regions where the climate is dry we find changes in the number and activities of streams and a great increase in the importance of the wind as a carrier of sediment. The present chapter deals with the geologic processes in arid lands and with the activity of the wind in particular, even though winds are effective also along coasts and in some other places that are not arid.

Desert climates. Although the word *desert* means literally a deserted, unoccupied, or uncultivated area, the modern development of artificial water supplies has changed the original meaning of the word by making many dry countries habitable. ***Desert*** has become a synonym for *arid land,* whether "deserted" or not.

A desert, obviously, must be defined in terms of climate. We think of a desert climate as one in which very little rain falls, but rainfall is only part of the picture. As we saw in connection with the former pluvial lakes, evaporation also is important. The higher the temperature, the greater the evaporation, and therefore the more precipitation an area can have and still be arid. For if most of the precipitated water evaporates, little will be left for streams and the growth of vegetation. In parts of the southwestern United States evaporation of exposed water surfaces amounts to as much as 100 inches annually—10 to 20 times more than the annual precipitation. Again, if all the annual rainfall of a region came in a single month, its effects on streams and vegetation would soon be lost, and the region would be classed as arid.

These examples show that a desert or arid climate is defined by three factors: (1) temperature, (2) precipitation, and (3) seasonal distribution of precipitation. Broadly speaking, a *desert* is a land area characterized by low precipitation, high temperature (at least seasonally), and a rather high proportion of evaporation to precipitation. Comparing a map of the world's deserts (Fig. 13-1) with a map of the world's rainfall (Fig. 2-12), we can say that throughout much of the world desert areas coincide fairly closely with those areas that receive no more than 10 inches of precipitation.

Surrounding most desert areas are semiarid or *steppe* areas, in which annual precipitation ranges from about 10 to 20 inches or a little more. Desert grades outward through semiarid country into humid country where most of the human population is concentrated.

The most intensely desert area in the United States is Death Valley in southeastern California. The highest temperature officially recorded there is 42.5°C (the world's record, at el Azizia, Libya, is 43.8°C). Although rainfall averages between 1 and 2 inches annually, Death Valley has experienced years with no rain at all, and at stations in the Atacama Desert in northern Chile periods of more than 10 consecutive years without rain have been recorded.

In addition to low rainfall and high daytime temperatures, most deserts are swept by frequent and rather high winds, commonly the result of convection. During daytime hours air over specially hot places is heated and rises, and this allows surface air to move in and take its place.

Convection likewise is responsible for much of

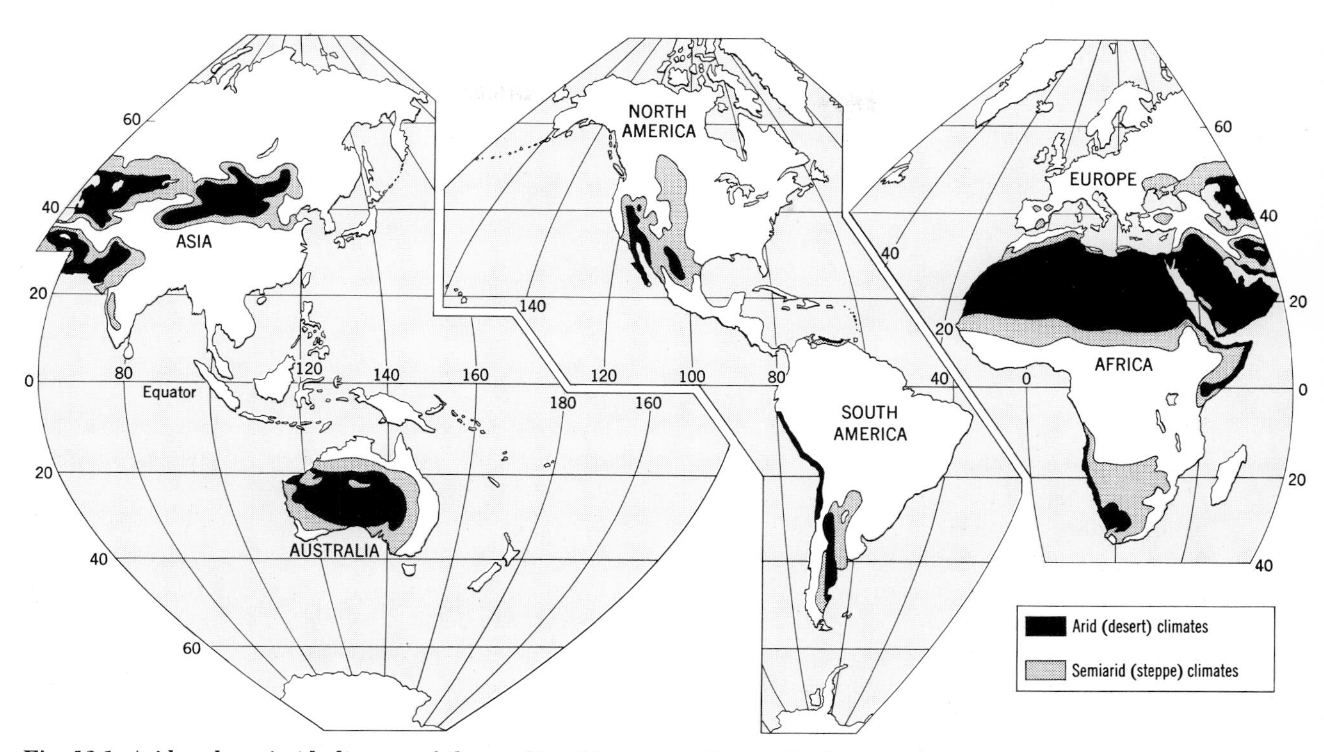

Fig. 13-1. Arid and semiarid climates of the world, plotted according to the Köppen-Geiger system.

the precipitation on deserts. As the columns of air rising over hot places are cooled, the moisture they contain can condense and fall as localized but heavy rains, widely known as cloudbursts. Sooner or later probably every part of a desert is visited by torrential showers of this kind.

The vegetation in deserts is a direct reflection of dry climate. Usually the vegetation is not continuous. Where grass is present, it is likely to be thin and to grow only in patches. More commonly, the plants consist of low bushes, growing rather far apart, with bare areas between them. This pattern of vegetation promotes active movement of sediment by the wind and by running water as well.

WORLD DISTRIBUTION OF DESERTS

Territory classed as arid composes about 25 per cent of the total land area of the world, and a large additional area is semiarid. Deserts are of three principal kinds. (1) Much of the dry territory lies between latitudes 15 and 35°, both north and south of the Equator (Fig. 13-1). In these latitudes (Fig. 2-10) the cause of aridity is the descending air that characterizes them. Heated by compression as it descends, air becomes increasingly able to retain water vapor and so tends to evaporate moisture from the Earth's surface instead of precipitating moisture onto it. Examples of low-latitude deserts are the Sahara and other deserts in northern Africa. (2) Another sort of desert is found in continental interiors, where heating in summer and dry cold continental air in winter prevail. Instances are the deserts of central Asia. (3) Yet another, more local, kind of arid area is one that lies in the lee of a mountain range, which acts as a barrier to rainfall from moist oceanic air ascending over the windward side. Descending and becoming warmer over the territory to leeward, the oceanic air keeps the area dry. Deprived of moisture by the Sierra Nevada and other mountains on the west, the desert country in the Basin-and-Range region in Nevada and adjacent states is of this type.

During former pluvial times, deserts as a group were somewhat less dry and somewhat smaller than they are today, although we still know very little about their actual temperatures and rainfall. If we look still further back into history, we can speculate that for as long as the general circulation of the

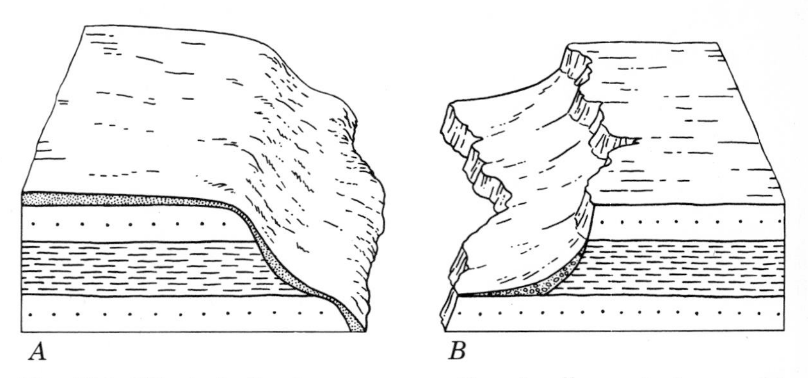

Fig. 13-2. Effect of climate on cross profiles of valleys. *A.* In a moist climate resistant strata are partly masked by a creeping mantle of chemically weathered waste. *B.* In a dry climate resistant rocks stand out as broad platforms and steep cliffs, partly concealed by taluses.

atmosphere has been approximately like that of today there must have existed wide regions with descending, drying air. Probably, therefore, arid climates have persisted wherever those latitudes have coincided with continental masses.

Probably, too, the interiors of large continents have contained at least some deserts throughout much of geologic history. Evidence of former deserts consists mainly of the occurrence among ancient strata of existensive bodies of wind-blown sand converted into sandstone (Fig. 13-29) and extensive layers of salt and gypsum precipitated during the evaporation of water in shallow arms of the sea (Fig. 15-9, *A*).

GEOLOGIC PROCESSES IN DESERTS

No major geologic process is entirely restricted to arid regions. Rather, the same processes operate with different intensities in moist and arid regions. As a result, in a desert the forms of the land, the soils, and the sediments show distinctive differences. Let us look at some of the major processes and note the differences.

Weathering and mass-wasting. In a moist region the regolith is nearly universal, comparatively fine-textured, a product chiefly of chemical weathering, in motion downslope mainly by creep, and covered by almost continuous vegetation. Because of creep, hill profiles as a rule are a series of curves.

In a desert the regolith, much of it a product of mechanical weathering, is thinner, less continuous, and coarser textured. Slope angles developed by downslope creep become adjusted to the average diameter of the particles of regolith; the coarser the particles the steeper the slope required to move them. As the particles created by mechanical weathering tend to be coarse, slopes are generally steeper than in a moist region, and, as mentioned below, the base of a steep slope meets the flatter surface below it at a rather distinct angle instead of in a smooth curve.

Rock fragments tend to break off along joints, leaving steep, rugged cliffs. Hills with clifflike slopes, shaped by erosion of flat-lying rock layers, are the buttes and mesas (Fig. 1-8) common in dry regions. Likewise, as shown in Fig. 13-2, the sides of valleys in dry country are more thinly covered with regolith than are those in moist areas, so that rock structure stands out more distinctly. Finally, mudflow, in the setting illustrated in Fig. 8-11, contributes a distinctive type of sediment to the fans at the bases of desert mountains.

Running water. One of the characteristics of deserts is that most of the streams which originate in them soon disappear by evaporation and by infiltration and never reach the sea. *Drainage that does not persist to the sea is* ***interior drainage.*** Exceptions are long rivers, such as the Nile in Egypt and the Colorado in the southwestern United States, which originate in mountain regions with abundant precipitation. Their discharges are so large that they can keep flowing to the ocean despite great losses where they cross a desert.

In a desert the regolith is generally loose and dry, and where bare it is easily eroded. Because the roots of growing plants offer little obstruction to runoff, the proportion of surface runoff during violent falls of rain is large. Streams are therefore subject to

Fig. 13-3. Fans being built out into Death Valley, California from Black Mountains (*left*). View south from Furnace Creek Ranch (*right foreground*), showing Death Valley, a down-faulted basin with a white salt-incrusted playa. Other fans appear in Figs. 10-7 and 12-19. (Spence Air Photos.)

sudden "flash" floods that move heavy loads of sediment. The loads are deposited as alluvium on fans at the bases of mountain slopes (Fig. 13-3) and on the floors of wide valleys and basins. The deposits of some flash floods are spectacular (Fig. 13-4).

Often, streams in flood effectively undercut the sideslopes of their valleys, causing the slopes to cave. Then, as the flood subsides, the load is deposited rapidly, creating a flat floor of alluvium. The result is a steep-sided, flat-bottomed "box canyon," characteristic of many dry regions (Fig. 13-5).

In a dry region, valleys are generally spaced far apart, because low rainfall does not need many channels to dispose of the runoff. But where impermeable clay and silt are exposed in steep slopes, such as the sideslopes of large valleys, little infiltration occurs and abundant runoff during violent rainstorms soon creates *a system of closely spaced narrow ravines with little or no vegetation.* Such intricately dissected terrain is known as ***badlands*** (Pl. 10), a name given to it by early travelers in the western United States because it was very difficult to cross. On very gentle slopes badlands can not form, and sheet runoff and rills predominate in disposing of the water and in eroding gently sloping surfaces.

Playa lakes. Basins formed by faulting and other movement of the crust play a larger part in determining the general sculpture of the land in an arid region than in a moist one, because only rarely is water abundant enough to fill the basins, overflow them, and establish continuous drainage systems that will reach the sea. Streams flowing down from a highland rarely last until they reach the center of the nearest basin. But after exceptional storms some of them discharge enough water to convert the basin floor into a very shallow lake that may last a few

Fig. 13-4. This deposit of boulder gravel resulted from a single flash flood in a mountain valley. Los Angeles County, California. (U. S. Forest Service.)

Fig. 13-5. Small "box canyon" cut into silty alluvium. Cornfield Wash, Albuquerque district, New Mexico. Compare Fig. 10-10, which shows a valley of a similar kind. (F. W. Kennon and H. V. Peterson, U. S. Geol. Survey.)

days or a few weeks. In the dry region of the western United States *an ephemeral shallow lake in a desert basin* is called a ***playa lake;*** when *dry* (which is most of the time) the *lake bed* is a ***playa*** (Figs. 13-6, 13-7). Many playas are white or grayish because of precipitated salts at their surfaces. But if the lake water can infiltrate the basin floor before evaporation saturates the water with salts, no salts can be precipitated, and the playa sediments consist mainly of clay.

The chain of events in the history of the salts is, in capsule form, solution of rocks by ground water → percolation → stream transport to lake → precipitation on lake floor. Hence the chemical character of the water and of the precipitates depends on the kinds of rocks that underlie the drainage basin. The result can be salt lakes rich in sodium chloride, alkali lakes rich in sodium and potassium carbonates, bitter lakes rich in sodium sulfate and other sulfates, and borax lakes rich in borax and related substances.

Ground water. We noted earlier that in a dry region the water table is likely to be far below ground, but in many places it is highest beneath streams (Fig. 11-9). Under such conditions ground water contributes little or nothing to surface streams as it does in a moist region, where it evens out discharge between rains. Therefore in a desert, the lack of such contribution, and the cloudburst character of the rainfall, produce the flash floods characteristic of desert climates.

Fig. 13-6. Braun's Playa, near Las Vegas, Nevada, nearly 5 miles in greatest diameter. Mountain crest is 15 miles distant. Playa lake on April 10 after an unusually large rainfall. The lake is no more than 2 or 3 feet deep. (C. E. Erdmann.)

Fig. 13-7. Braun's Playa 2 weeks later, after the lake evaporated. Wind is blowing the fine-grained lake sediments (clay and crystals of salts) into dust clouds. Dark spots are desert bushes. (C. E. Erdmann.)

Wind. In dry country, wind is an effective geologic agent. However, contrary to popular belief, deserts are not characterized mainly by sand dunes. Only one-third of Arabia, the sandiest of all dry regions, and only one-ninth of the Sahara are covered with sand. Much of the nonsandy area of deserts is cut by systems of stream valleys or is characterized by fans and alluvial plains. Thus, running water leaves its mark upon a wider territory than does the wind, and we draw the inference that, even in a dry region, more geologic work is done by running water than by wind.

The details of the way in which wind works in deserts and in other regions are set forth in a later section of this chapter.

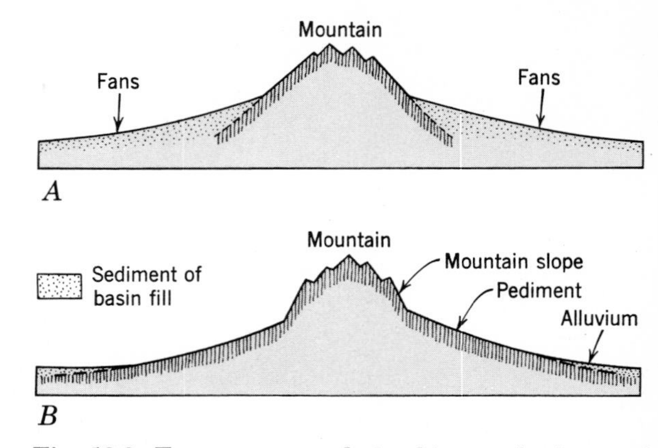

Fig. 13-8. Two common relationships at the bases of mountains in desert country. *A*. Fans built by streams at the foot of a mountain. *B*. Pediment eroded across bedrock at the foot of a mountain.

PEDIMENTS

In deserts, the land forms sculptured by erosion differ from those in country with more rainfall. When the terrain unit consists of a mountain range and an adjacent basin, two kinds of situations are common. One (shown in Figs. 13-3 and 13-8, *A*) consists of a row of fans along the mountain base; the fans merge outward into a general fill of sediment in the basin. The other situation (Fig. 13-8, *B*), more remarkable and less easy to explain fully, consists of a sloping surface at the mountain base that closely resembles a merging row of fans. But the surface is not that of fans. Instead of being constructional (built of alluvium), it is erosional and cuts across bedrock. Scattered over it are rock fragments, some brought by running water from the adjacent mountains and some derived by weathering from the rock immediately beneath. Downslope, the rock fragments become more numerous until they form a continuous cover of alluvium. The bedrock surface, which may be several miles long, has passed beneath a basin fill.

The eroded bedrock surface is called a *pediment* because the surfaces adjacent to the two bases of a mountain mass, seen in profile, together resemble the triangular pediment or gable of a roof (Fig. 13-8, *B*). A ***pediment,*** then, is *a sloping surface, cut across bedrock, adjacent to the base of a highland in an arid climate.* The kinds of bedrock on which pediments are cut are those that yield easily to erosion in such a climate. The profiles of pediments (Fig. 13-9), like those of fans, are concave-up, a form we associate with the work of running water, and pediment surfaces are marked by channels, usually faint and shallow. For these reasons it seems probable that pediments are the work of running water, flowing (according to local conditions) as definite streams, as sheet runoff, or as rills, and in any case flowing mainly during cloudbursts. Between the infrequent times of runoff weathering occurs. A pediment seems to be a form in dynamic equilibrium with the processes of weathering and flow of water that are operating on it. The similarity in form between pediments and fans suggests that both are fashioned under similar controls. Possibly, where rocks upstream are rather erodible and alluvium is abundant, fans are built, and where rocks upstream are less erodible and alluvium is more scanty, the running water, aided by weathering and mass-wasting, creates pediments.

A remarkable thing is that a pediment, regardless of its age, meets the mountain slope not in a curve but at a distinct angle (Fig. 13-8, *B*). Instead of becoming gentler with time, as in a wet region where chemical weathering and creep of regolith are dominant, mountain slopes in the desert seem to adopt an angle determined by the resistance of the rock, and to maintain that angle as they gradually retreat under the attack of weathering and mass-wasting. Retreat of the mountain slope lengthens the pediment at its upslope edge. This growth of the pediment at the expense of the mountain should continue until the entire mountain has been consumed. During the whole time of pediment growth, rock particles are being transferred downslope intermittently from mountain and pediment to the basin fill beyond.

Fig. 13-9. Pediment at south base of Little Ajo Mountains, Arizona. The black butte near center is about 1 mile long. Stripelike rows of bushes are growing along faint, shallow channels. The view represents one-half of what is shown in the section (Fig. 13-10). (James Gilluly, U. S. Geol. Survey.)

This general outline of the way a pediment forms and develops, and its relation to the mountain slope, is not fully established and is partly a matter of opinion. Not until studies of many details have been completed can we expect to establish a satisfactory picture.

CYCLE OF EROSION IN DESERTS

Chapter 10 set forth a theory of sculptural evolution of a land mass in a region of abundant rainfall. The theory is based on the fact that lands in several stages of development could be arranged in sequence to represent a continuous evolution, a cycle of erosion.

The concept of a cycle of erosion applies equally well to desert regions. Here again, because no one district shows all the stages, several districts must be compared. Brought together, the stages fit reasonably into a continuous series; so it is believed that in the course of time a single land mass could evolve through all the stages. Figure 13-10 represents in a very generalized way three stages that can be seen today in parts of the western United States.

The stage represented in *A* commonly occurs in northern Nevada, where mountains are being dissected actively. At the mouths of the mountain canyons streams spread out waste in the form of fans that grade outward into playas. The wind picks up fine waste, sorts it, heaps sand-sized particles into dunes, and lifts some of the finest particles out of the basins altogether. Meanwhile the rocky mountain slops are being worn back.

In *B*, a stage occurring in southern New Mexico in the country north of El Paso, Texas, the mountain slopes have retreated, exposing a belt of bedrock at each mountain base to weathering and running water and so creating pediments. The steep mountain slopes, covered sparsely with coarse weathered rock waste moving slowly down them, maintain their steepness instead of becoming gentler with time, as would result from soil creep in a moist region.

As the area of the mountains diminishes, the sediment contributed to the streams during rainstorms decreases also. This reduces the loads of the streams, which begin to erode the heads of the former fans, planing them down. In the process the streams cut sideways into the bedrock at and near the mountain front, planing it off and adding to the area of pediment.

In *C*, a stage seen in the country northwest of Tucson, Arizona, the mountains have been reduced by gradual retreat of their steep slopes to a series of knobs projecting abruptly above the sloping pediment that surrounds them. Outward beyond the pediment is the surface of the basin fill, beneath which the pediment disappears without any break in the smooth, concave slope. As the basin slopes become gradually gentler, water from the mountains reaches the basin centers more rarely, and increas-

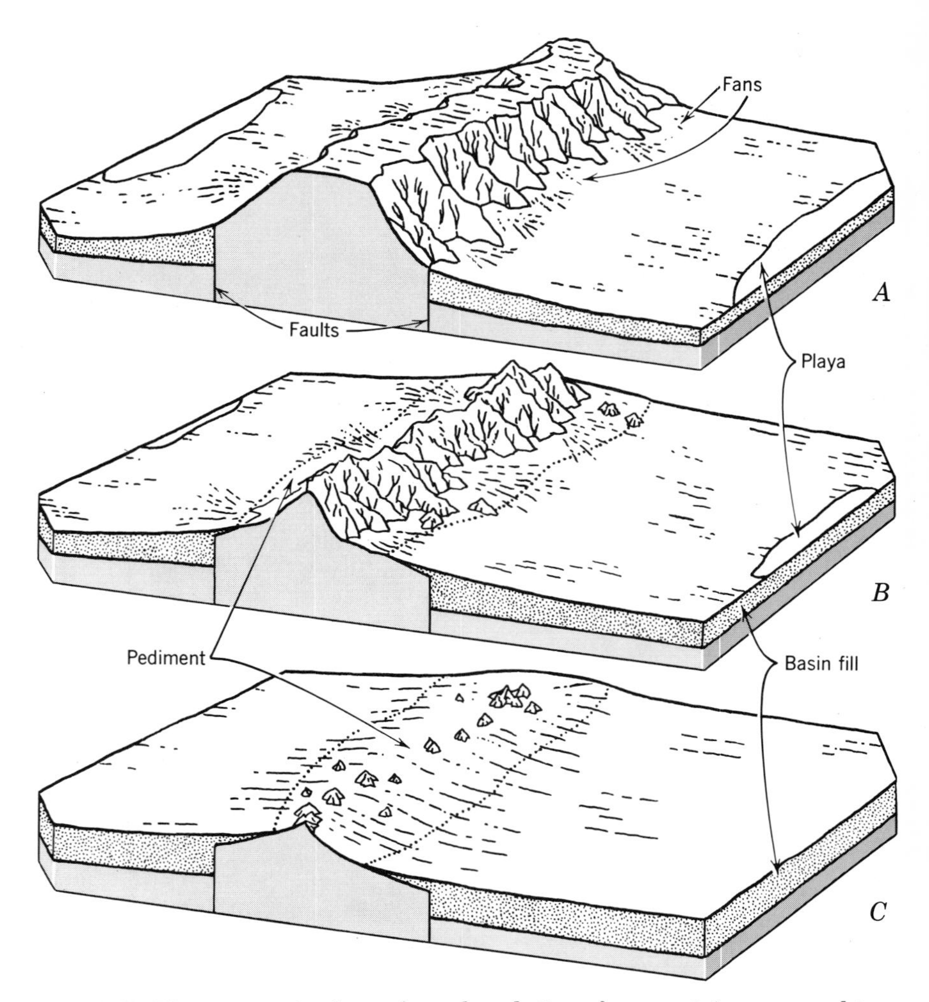

Fig. 13-10. Three stages in the sculptural evolution of a mountain range and two basins (originally created by faulting) in an arid climate.

ingly the wind, sweeping across the basins, picks up fine sediments and carries them away. The stage shown in *C* is essentially the desert equivalent of a peneplain.

Erosion of basins by wind could lower them close to the water table. Here moisture would play a part. By causing rock particles to cohere and by promoting vegetation, moisture would put an end to the picking up of fine sediment by the wind. That would take a very long time, and no example of this condition is known. There is reason to believe that in southern Arizona a cycle of erosion has been in progress with little interruption for millions of years, and it still has a long way to go.

WIND ACTION

SEDIMENT TRANSPORT BY FLOWING AIR

Movement of air and rock particles. Wind is moving air, flowing like water. The most common cause of air movement differs from that of stream flow. Although in some places a thin layer of cold, heavy air flows downward over the surface from highlands to lowlands like a stream of water, the usual cause of movement consists of differences in air pressure created by differences in heating by the Sun. Heated air expands and rises; cooler air descends or flows in along the ground to take its place. Except for high mountain ranges, topographic irregularities influence this flow very little.

The flow of air, like that of water, is turbulent under nearly all natural conditions. Also the average velocity of air movement, like that of a stream (Fig. 9-5) and of a glacier (Fig. 12-4, *C*) increases with increasing height above the ground (Fig. 13-11). But roughness of the surface does more than make the flowing air turbulent; it exerts an influence on velocity close to the ground. Roughness is measured by the heights of obstacles on the ground, whether they are sand grains, grass blades, houses, or trees. Roughness creates a layer of nearly motionless air immediately above the surface. As Fig. 13-11 shows, the thickness of the layer has been found to equal one-thirtieth the height of the obstacles present (for example, the diameters of sand grains on the ground) and remains constant regardless of wind velocity. In bare, open country, with sand covering the surface, the dead-air layer is so thin that it can not influence the movement of sand grains; but we shall return to it later when we follow the behavior of smaller particles.

Observers working in North African deserts have noticed that when strong winds are blowing, two layers of rock particles are present in the air (Fig. 13-12). The lower layer consists of sand grains and extends from only a few inches to a very few feet above the ground. It grades upward into the upper layer, which consists of silt and clay particles and it extends much farther up, often to great heights. The difference between the two layers is the same as that between bed load and suspended load in a stream.

Bed load: movement of sand grains. As we did in analyzing sediment transport in streams, we can profit by observing experiments with sand in glass-sided wind tunnels under controlled conditions. The experiments show that sand grains move by saltation, as they do in streams, in a series of long jumps (Fig. 13-12, *C*). In water, about 800 times heavier than air, saltation involves a hydraulic lift more than a bounce, but in air it is mainly a series of elastic bounces, like those of a ping-pong ball.

A sand grain gets into the air only by bouncing or by being knocked into the air through the impact of another grain. When the wind becomes strong enough, a grain starts to roll along the surface under the pressure of a fast-moving forward eddy. It strikes another grain and knocks it into the air. When the second grain hits the ground, it either splashes up

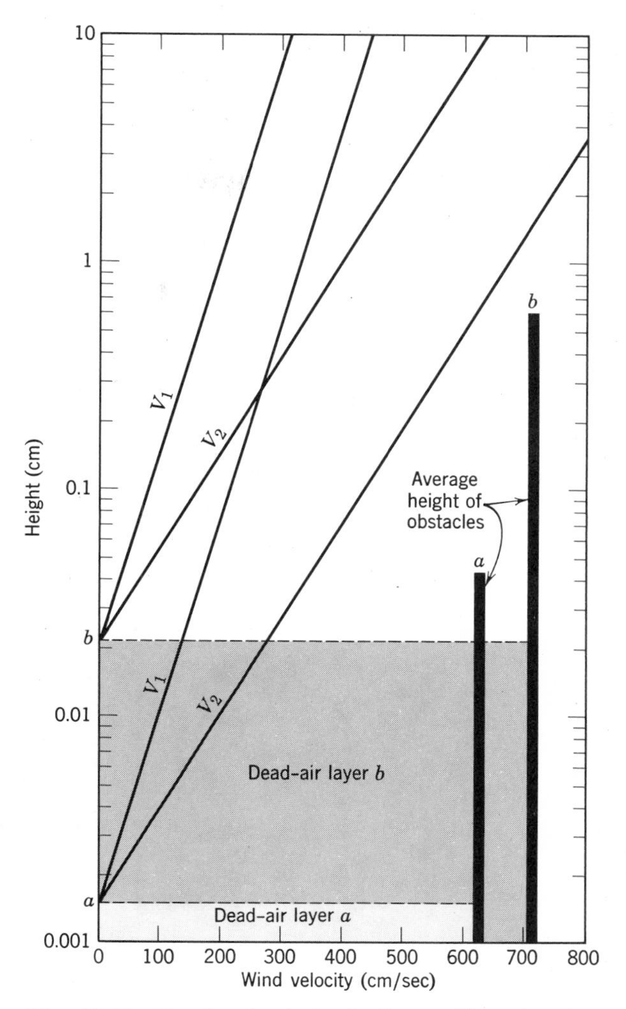

Fig. 13-11. Graph of wind-velocity profiles, showing influence of obstacles of various heights. Because height is plotted on a logarithmic scale, wind profiles show as a straight line. With height plotted on an arithmetic scale, the profiles would be curved, their shape similar to the velocity profile of a stream (Fig. 9-5). Profiles are shown for winds of two velocities, V_1, slower than V_2, blowing across obstacles of two average heights, *a* and *b*. All velocity profiles for winds blowing across obstacles of the same average height intersect at a single point above the ground, marking the top of a layer of dead air. Thickness of the dead-air layer is one-thirtieth the average height of the obstacles. Placement of obstacle-height bars is unrelated to air-velocity scale along abscissa. (After Bagnold, 1941.)

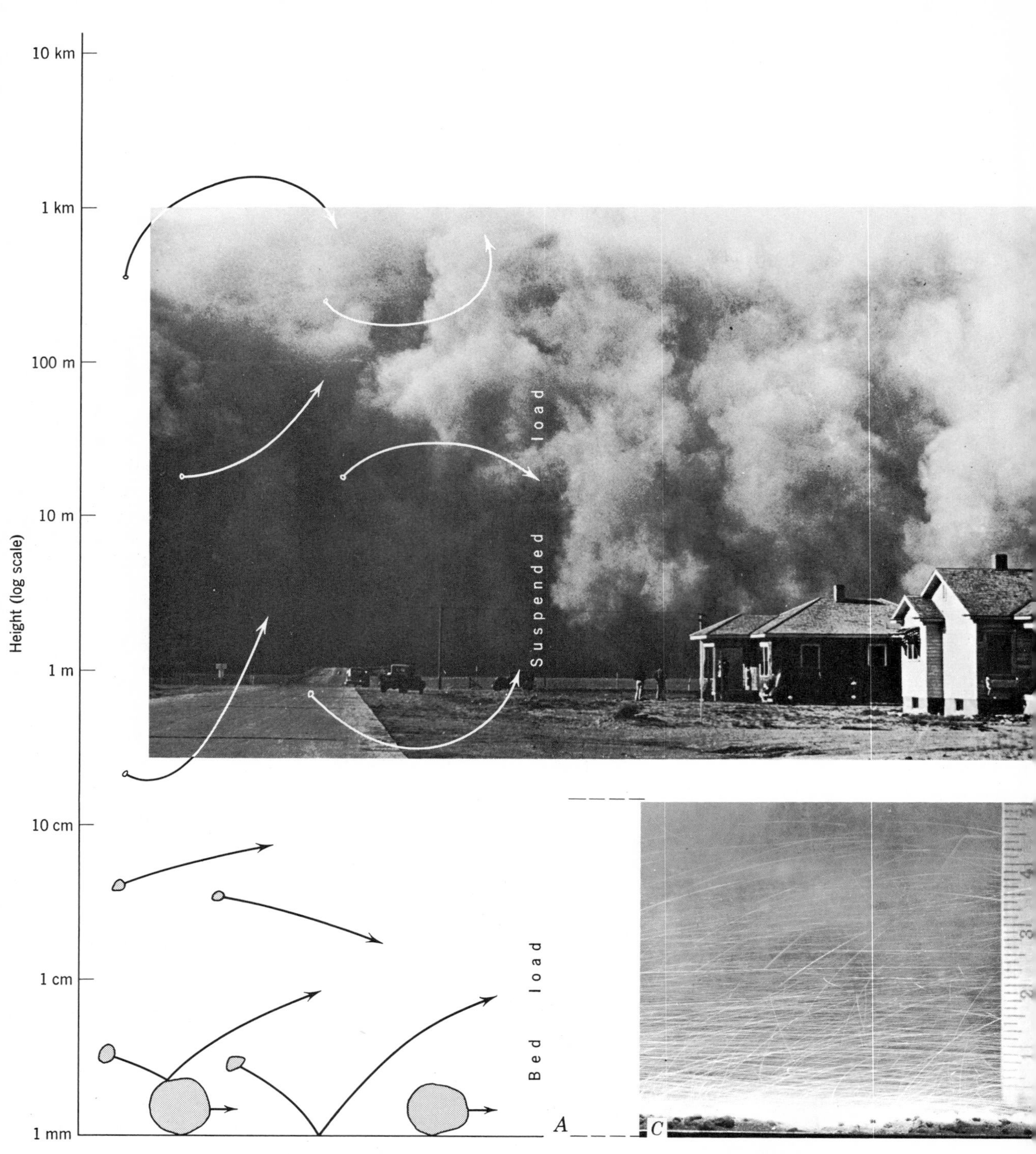

Fig. 13-12. Movement of rock particles carried by wind. (Compare Fig. 9-10). *A.* Graph showing vertical distribution, comparative sizes, and movement of particles in bed load and suspended load. *B.* The "Dust Bowl" during the 1930s. A dust cloud approaching Springfield, Colorado at 4:47 P.M. on May 21, 1937. Total darkness lasted 30 minutes. (U. S. Soil Conservation Service.) *C.* Paths of sand grains being blown through a wind tunnel, photographed in a narrow beam of sunlight. Scale units are inches. Sand grains and small pebbles are visible on tunnel floor. Air current is moving left to right. Traces of both splashing and bouncing impacts are visible (Fig. 13-13). (A. W. Zingg, U. S. Dept. of Agriculture.)

still other grains, making a tiny crater, or bounces into a new jump (Fig. 13-13). In a short time the air close to the ground has become filled with saltating sand grains, which hop and bounce, moving with the wind, as long as wind velocity is great enough to keep them moving.

The jumping sand grains never get far off the ground. They are usually limited to about 10cm, as in the experiment shown in Fig. 13-12, *C*. In desert country they generally jump no higher than about 45cm, as shown by telephone poles, which are sandblasted up to about that height but no higher. Even in the strongest desert winds the height of jump rarely exceeds 1m. This explains why wind-blown sand rarely moves far except on very smooth surfaces. Being always close to the ground, it is easily stopped by obstacles and heaped into dunes. When we measure the diameters of the grains in a deposit of wind-blown sand by passing a sample through a series of sieves of accurately controlled mesh, we find that nearly all the diameters range between 0.3 and 0.15mm; in other words, the sand is of medium size (Fig. 4-11). Observation and experiment show that an air velocity of 5m/sec (about 11 miles per hour) is about the minimum necessary to move sand grains lying on the ground.

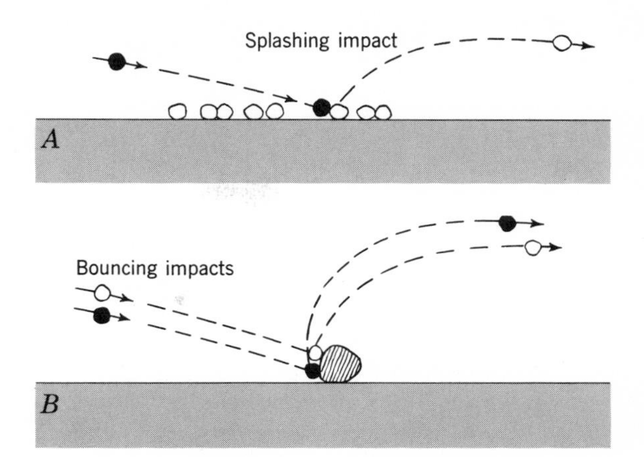

Fig. 13-13. *A*. In splashing impact, saltating sand grain strikes one or more other grains and splashes them into air at slow speeds and to low heights. Splashing happens most commonly when all grains are nearly the same size. *B*. In bouncing impacts, sand grains strike pebble or other wide surface and bounce up at high speed and to greater heights. Angle of rise depends on inclination of surface of impact.

Suspended load: movement of finer particles. Because suspended loads extend to great heights it is not practical to use wind tunnels in analyzing their movement; we must go outdoors and look at silt and clay suspended in the air. We must consider three aspects of the transport: getting the fine particles into suspension, keeping them up in the air, and dropping them out onto the ground.

The dead-air layer immediately above the surface, which we found could be neglected in the movement of medium sand grains, on the contrary plays a controlling part in the movement of all finer particles. The reason why the dead-air layer fails to influence the movement of sand grains is that each individual grain constitutes an obstacle that projects far above the related, microscopically thin dead-air layer. But particles of silt and clay are crowded so closely together that they do not act as individual obstacles but present a very smooth surface to the wind. Whatever roughness is present results from sand grains or other gross irregularities in the surface (Fig. 13-14); the silt particles are down within the dead-air layer related to the gross roughness. Ordinarily the wind can not lift them off the ground directly. It spatters them into the air by the impacts of saltating sand grains.

We can see how this happens by looking at a dusty road in dry country on a windy day. The wind blowing across the road generates little or no dust. But a car driving over the road creates a choking cloud, which is blown a short distance before settling once more. The car wheels have broken up the surface of powdery silt, which was too smooth to be disturbed by the wind.

An additional factor in the relative stability of a surface of fine sediments is the tendency of the particles to cohere because of films of moisture adsorbed onto their surfaces. Such films are what make a silty or clayey soil slightly sticky, whereas in a sandy soil the grains tend to be loose.

Once off the ground, fine particles are subject to two forces: turbulent movements of air, and gravity. Turbulent eddies in air, as in water, move in all directions, upward, downward, sideways, backward, and forward. The velocities of upward eddies, on the average, equal about one-fifth the average forward velocity. Hence in a wind with forward velocity of 5m/sec (the minimum required to move sand grains along the ground) the velocity of upward eddies is 1m/sec. This velocity provides the force available to oppose gravity.

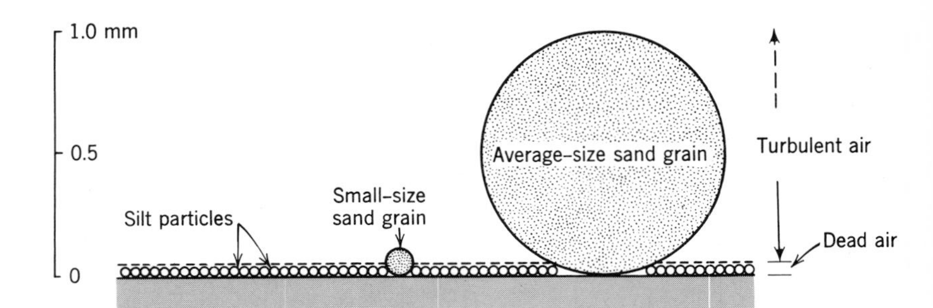

Fig. 13-14. Silt particles form a smooth surface that lies within the dead-air layer caused by sand grains and other large obstacles to flowing air. Thickness of dead-air layer is one-thirtieth the diameter of the average-size sand grain (Fig. 13-11).

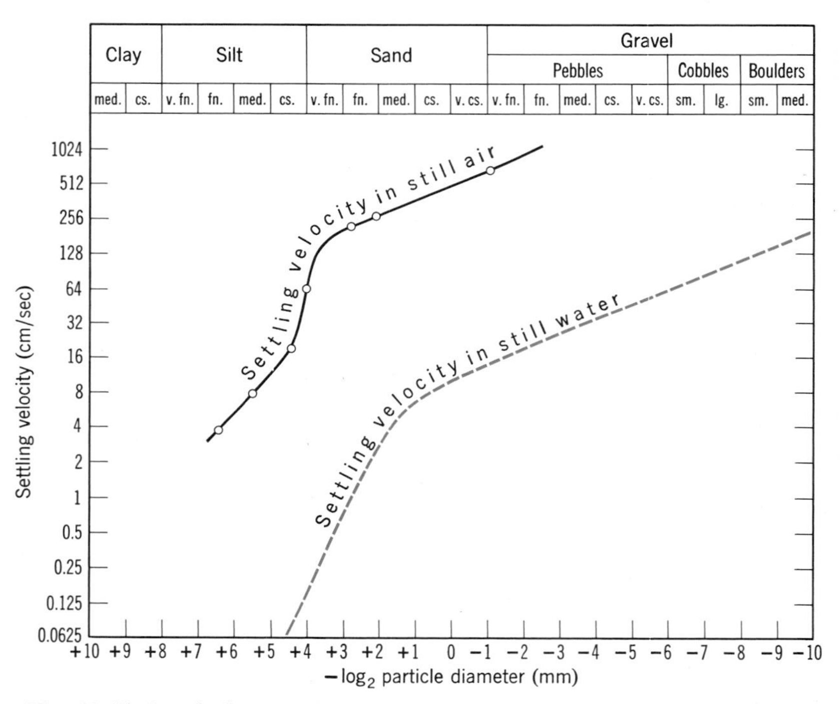

Fig. 13-15. Graph showing diameter versus settling velocity of sand grains in air and in still water. Note abrupt break in slope of the curve for air at the coarse-silt size and the gradual bend in the opposite direction in the very-fine-sand size. For use of $\log_2$ scale see Fig. 4-11. (Data for air from L. Moldvay, 1961, Acta Universitatis Szegedensis (Hungary), Min.-Petrog., v. 14, p. 76, Fig. 3; and Bagnold, 1941; for water from W. W. Rubey, 1933.) (Compare Fig. 9-13.)

Gravity tends to pull a particle toward the ground at a constant settling velocity. As in water, settling velocities are influenced by diameter, shape, and specific gravity of the particles and are faster in air than in water, which is much denser and more viscous. For example, comparison of the two curves in Fig. 13-15 shows that a particle of coarse silt settles at 100cm/sec in air and at 0.125cm/sec in water.

Returning to our upward velocity of 1m/sec (100cm/sec) we can see that it is just sufficient to maintain the particle of coarse silt in suspension. In other words, wind velocity capable of moving sand along the ground can keep silt and finer particles in suspension.

Much of the sediment finer than sand remains at low altitudes and although carried much farther than the bed load of sand, commonly it is deposited in fairly close proximity to its source. An example is the glacial *loess* described in a later section. Sediment deposited in this way ranges in diameter from fine sand to clay size, and both average grain size and thickness of the deposit decrease in the downwind direction.

At times fine particles are lifted by updrafts within the general turbulence, rise into the stratosphere, and can be carried long distances (Fig. 13-16). At such altitudes velocities are large enough to carry even fine sand in suspension. This presents the problem of how suspended particles are sedimented out onto the ground. Although this problem has not been studied fully, it appears that downdrafts of air related to the major air circulation are the cause. Falls of tephra from distant volcanoes are in this category. Such far-traveled sediment is not distinctive in grain size, but unlike that which is carried at lower altitude, average diameter and thickness do not decrease downwind.

In summary, the wind sorts sediment into three groups: bed-load sand, low-altitude suspensions, and high-altitude suspensions.

Amount of load. Of course, a current of air can not hold in suspension nearly so much sediment as a current of water of similar velocity and cross-sectional area, because air is far less dense than water. Nevertheless, it has been estimated that the theoretical sediment-transporting capacity of the winds blowing across the Mississippi River drainage basin is around 1,000 times that of the Mississippi River if both were fully loaded. The apparent anomaly is explained by the frequently greater velocities

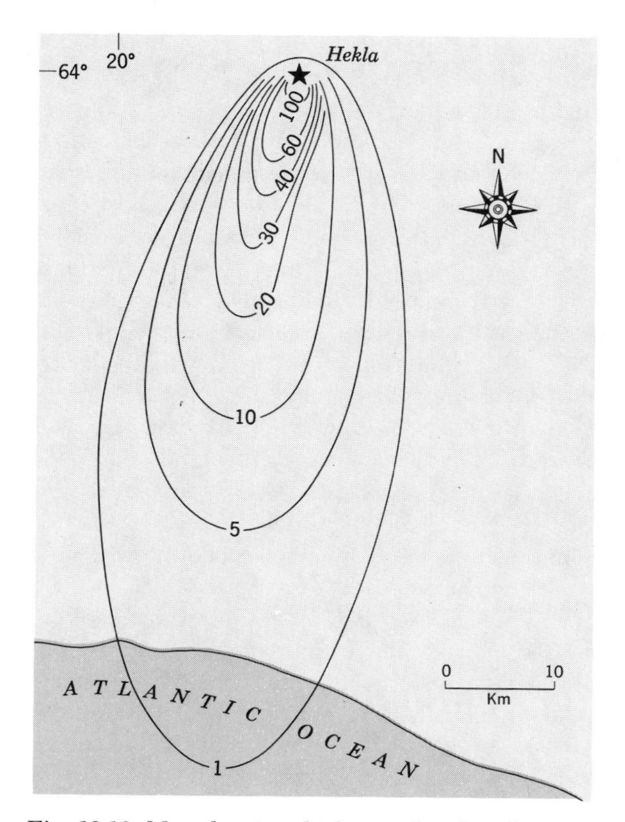

Fig. 13-16. Map showing thickness of tephra deposited downwind from Hekla Volcano, Iceland, during eruption of March 29, 1947. Lines are isopachs in centimeters. The tephra were raised to relatively low altitude and most of them were deposited within an hour after the eruption began. Other tephra were raised 9000m and were transported southward 880km, then northeastward nearly 3,000km, before they were deposited in Finland 51 hours after the eruption. The far-traveled tephra were transported at a mean velocity of 75km/hour (After Thorarinsson, 1954.)

and vastly greater cross-sectional area of the moving air. The amount of sediment actually moved by the atmosphere, year in and year out, is probably only a fraction of 1 per cent of capacity, for the air is rarely if ever fully loaded.

During the great wind storms in the dry years of the 1930s, however, loads became abnormally heavy. In a particularly great storm on March 20, 1935, when the sky looked much as in Fig. 13-12, *B* (compare also Fig. 15-12), the cloud of suspended sediment extended 12,000 feet above the ground, and the load in the area of Wichita, Kansas was estimated at 166,000 tons per cubic mile in the lowermost layer 1 mile thick. But only a little of this huge load was deposited in any one place, for a sample of sediment trapped on the roof of a laboratory building in Lincoln, Nebraska indicated that during that day only about 800 tons of rock particles—around 5 per cent of the load suspended in the lowermost 1-mile layer—were deposited on each square mile. Enough sediment was carried eastward on March 21 to bring temporary twilight over New York and New England, 2,000 miles east of the principal source area in eastern Colorado. The distance and travel time imply wind velocities of about 50 miles per hour.

March 20 was an exceptional day, a "black blizzard," but even if we consider normal conditions, when no sediment is visible in the air, the total amount of rock material moved by the atmosphere in a single year must be enormous. This is shown by the radioactive dust carried to leeward from the localities of nuclear explosions and detected by sensitive instruments right around the Earth. The educated guess has been expressed that every square mile of the Earth's land area probably contains rock particles brought by the wind from each of the remaining square miles. Of course, this statement can not be proved, but possibly it is nearer the truth than is generally realized.

Fine-grained tephra. Not all wind-blown sediments originate by being picked up from the ground. Large quantities are shot into the air during explosive eruptions of volcanoes (Fig. 19-18). Although coarse particles fall out quickly, small particles travel long distances. The bulk of the particles that fall out during an eruption tend to form an elongate body of sediment that trails out downwind from the volcanic source (Fig. 13-16) and thins both laterally and downwind. Smaller amounts can rise into the stratosphere and be carried great distances, even around the world.

If the directions of elongation of a number of these elliptical bodies are plotted on a map of a wide region, they are seen to parallel the directions of the prevailing winds. This fact offers a means of determining, from layers of ancient tephra, whether or not wind belts in the past were oriented relative to the continents as they are today or were different. This would have a bearing on the great problem of continental drifting (Chap. 22).

WIND EROSION

Deflation. The presence of sediment in the atmosphere implies wind erosion, which is of two kinds. The first kind, *the picking up and removal of loose rock particles by wind,* provides most of the wind's load and is known as ***deflation*** (Lat. *deflare,* "to blow away"). The second kind, *abrasion* of rock by wind-driven rock particles, is analogous to abrasion by running water.

Deflation is conspicuous only in the absence of vegetation and only, of course, in material that is capable of being picked up by the wind. The great areas of deflation are the deserts; others are the beaches of seas and large lakes and, of greatest economic significance, bare plowed fields in farming country during times of drought, when no moisture is present to hold the soil particles together.

The effect of deflation on the form of the land is not great. In most areas the results are not easily visible, inasmuch as the whole surface is lowered irregularly. In places, however, measurement is possible. In the dry 1930s deflation in parts of the western United States amounted to several feet within only a few years—a tremendous rate compared with our standard estimate of rate of general erosion. In Fig. 13-17 the tuftlike yucca plants held the soil in place, but elsewhere it was deflated. In the few years before the photograph was taken, 3 feet of soil had been deflated. The silt and clay particles had been exported in suspension, sand grains had been sorted out and had remained behind, and had been added to by further sand moving in from the area upwind.

Probably the most conspicuous evidence of wind work of this kind consists of ***deflation basins,*** which are *depressions excavated by deflation.* These occur by the tens of thousands in semiarid regions, as in the Great Plains region from Canada to Texas. The lengths of most of them are less than a mile and depths are only a few feet. In wet years they are clothed with grass and some even contain shallow lakes; an observer seeing them at such times would hardly guess their origin. But in dry years soil moisture evaporates, grass dies away in patches, and wind deflates the bare soil. At the same time, drifting sand accumulates to leeward, especially along fences and other obstructions.

When sediments are particularly prone to deflation, depths of deflation basins can reach 150 feet,

Fig. 13-17. Part of a large blowout, with a direct record of 3 feet of deflation. The plant roots in the hummock, just above the dark band (the remains of an ancient soil) mark the position of the surface before deflation. Note recent ripples in the sand surface. The date is 1936, in the "Dust Bowl" period. (R. H. Hufnagle from Philip Gendreau.)

as in southern Wyoming, and even more, as in the Libyan Desert in western Egypt, where the floor of the Qattara basin lies about 125m below sea level. Deflation in any basin is limited finally only by the water table, which moistens the surface, encourages vegetation, and inhibits wind erosion.

A natural preventive of deflation is a cover of rock particles too large to be removed by the wind. Deflation of a sediment such as alluvium, which consists of silt, sand, and pebbles, creates its own protective cover for it (Fig. 13-18). The sand and silt are blown away and in places are carried off by sheet erosion also, but the pebbles remain. When the surface has been lowered just enough to create a continuous cover of pebbles, the ground has acquired a ***deflation armor,*** *a surface layer of coarse particles concentrated chiefly by deflation* (Fig. 13-19). Deflation armors are also called *desert pavement* because long-continued removal of the fine particles makes the pebbles settle into such stable positions that they fit together almost like blocks in a cobblestone pavement.

An agency (other than wind) effective in making desert pavement is sheet erosion. By observation of slopes that are nearly free of vegetation we know that sheet runoff washes fine particles away, leaving pebbles behind. The resulting pebble pavements are so similar to those created by deflation that we can be sure deflation has occurred only where distinctive wind-cut stones (Fig. 13-20) are present or where

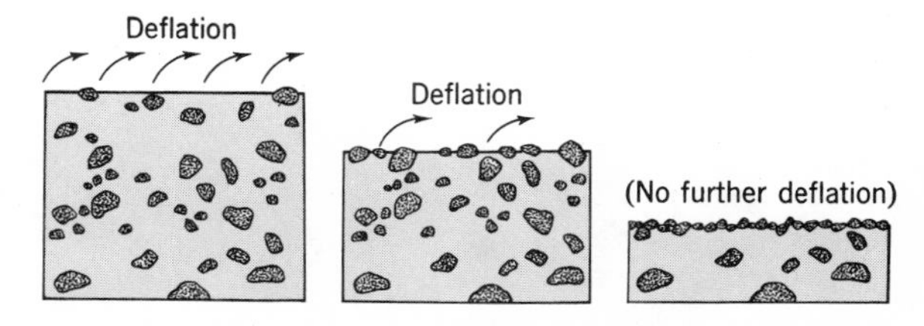

Fig. 13-18. Three stages in the development of a deflation armor.

slopes are too nearly horizontal for water to flow across them.

A more widespread preventive of deflation is a continuous cover of vegetation. This is the principal reason why deflation is evident mainly in desert regions, which lack such cover. However, deflation is very evident, in semiarid country and where coastal dunes have become covered with grass. During seasons of drought or as a result of trampling by animals, the grass is killed off in patches and allows deflation to start, converting the bare patches into irregular, shallow basins known as *blowouts.* A ***blowout*** is merely *a deflation basin excavated in shifting sand.* The view in Fig. 13-17 represents a large blowout.

Abrasion: ventifacts. In desert areas bedrock surfaces, pebbles, and boulders are abraded by wind-driven sand and silt, which can cut and polish them to a high degree. A ***ventifact,*** the name given to *a*

Fig. 13-19. Deflation armor ("desert pavement") on the floor of Death Valley, California. The wind has cut distinct facets on some of the stones. Length of front of photograph is several feet. (Eliot Blackwelder.)

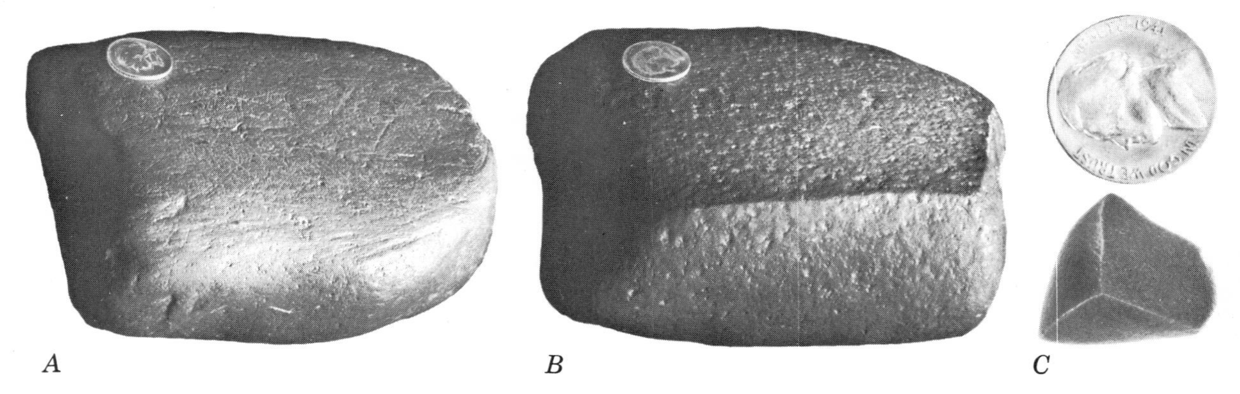

Fig. 13-20. Ventifacts. *A.* Two sides of a basalt cobble, showing a facet and striations made by glacial abrasion. New Haven, Connecticut. *B.* Two more sides of the same cobble, not visible in A. Two facets, smoothed and pitted by wind-driven sand, are visible. Because pits cut the glacial striations, cobble was glaciated first, then wind-cut as it lay on the ground in the lee of a body of outwash. *C.* Ventifact with two facets and a third surface whose curved form suggests it is remnant of original round pebble. From stream gravel on terrace of Bighorn River, Wyoming. (R. F. Flint.)

rock fragment with facets that have been cut by wind action, is recognized by polished, greasy-looking surfaces, which may be pitted or fluted, and by facets separated from each other by sharp edges (Fig. 13-20). Most of the cutting is done by the saltating sand grains of the bed load.

In the laboratory, sand blasting of pieces of plaster of Paris demonstrates that the facets always face the wind (Fig. 13-21). A stone can be worn down flush with the ground by enlargement of a single facet. If the stone is undermined or otherwise rotated or if the wind direction varies from time to time, two, three, or many more facets can be cut on it.

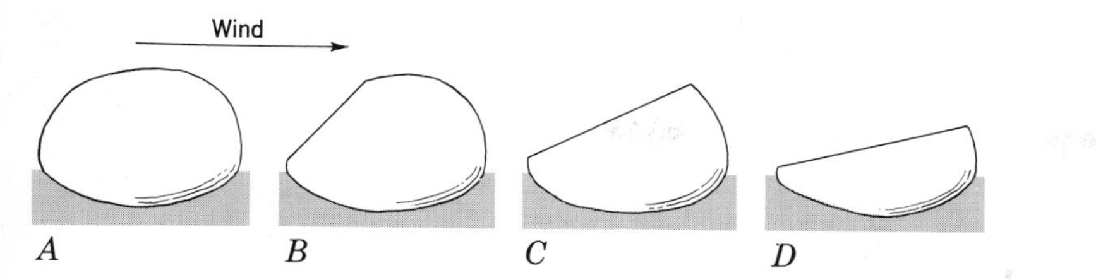

Fig. 13-21. Four stages in the cutting of a ventifact. The pebble becomes a ventifact between stage *A* and stage *B*.

Far from being confined to deserts, ventifacts are found in large numbers at the upper surfaces of layers of glacial drift in the northern United States and Europe. Evidently they were made during glacial ages, just beyond the margins of ice sheets, where surfaces temporarily bare of vegetation were subjected to strong winds.

Although locally striking, the effect of wind abrasion in the aggregate probably is not great. Rocks of various odd shapes in dry regions are often popularly ascribed to wind abrasion, but most of them result from weathering. Wind abrasion, then, is a minor process; the quantitative importance of wind erosion lies not in abrasion but in deflation.

WIND DEPOSITS

Much as in a stream of water, the saltating sand grains close above the ground move relatively slowly and are deposited early, whereas the suspended finer particles travel faster and much farther before dropping to the ground. This is established by systematic sampling of wind deposits derived from a known source. Sampling shows progressive decrease in diameter of particles with increasing distance from the source. Generally also a difference in topography exists between deposits of sand, which occur as dunes near the source, and deposits consisting predominantly of silt and clay, which generally blanket the ground much farther from the source.

DEPOSITS OF SAND

Dunes. A ***dune*** is *a mound or ridge of sand deposited by the wind.* Generally it is localized by an obstacle small or large (Fig. 13-22), which distorts the flow of air. Velocity within a meter or two of the ground varies with the slightest irregularity of the surface. As it encounters an obstacle, the wind sweeps over and around it, but leaves a pocket of slower-moving air immediately behind the obstacle and a similar but smaller pocket in front of it. In these pockets of lower velocity sand grains drop out and form mounds. Once formed, the mounds themselves influence the air flow. As more sand drops out they can coalesce and form a single dune.

Slip face of a dune. A dune is asymmetrical, with a steep, straight lee slope and a gentler windward slope (Fig. 13-23). The wind rolls or pushes sand grains up the windward slope, and at the crest they drop onto the lee slope, building that slope up to the angle of repose. We noted in Chapter 8 that angles of repose for loose sand are close to 34°. When more sand drops onto it the slope becomes unstable, sand slides (or "slips") downward, and the angle is maintained. For this reason *the straight, lee slope of a dune* is known as the ***slip face,*** which, as might be expected, meets the windward slope at a sharp angle (Fig. 13-25).

The angle of slope of the windward side varies with wind velocity and grain size, but it is always less than that of the slip face. The asymmetry of a dune with a slip face, then, indicates the direction of the wind that shaped it.

Height and migration. Many dunes grow to heights of from 30 to 100m, and some desert dunes reach the great height of 200m. Possibly the height to which any dune can grow is determined by upward increase in wind velocity, which at some level will become great enough to whip sand grains off the top of a dune as fast as they arrive there by creeping up the windward slope.

Transfer of sand from the windward to the lee side of a bare dune can cause the whole dune to migrate slowly in the downwind direction (Fig. 13-24). Measurements on desert dunes of the type shown in Fig. 13-25 indicate rates of migration as great as 10 or 20m/year (Fig. 8-3). The migration of dunes, particularly along coasts just inland from sandy beaches, has been known to bury houses and threaten the existence of inhabited places. In such

Fig. 13-22. Wind blowing from left to right deflated the bare dry field, carried away the finer particles in suspension, and dropped sand along the obstacle created by a wire fence with weeds caught in it to form a dune. Hereford, Texas. (A. Devaney, Inc.)

places, sand encroachment is countered most effectively by planting vegetation that can survive in the very dry sandy soil of the dunes. A good plant cover inhibits dune migration for the same reason that it inhibits deflation: if the wind can not move sand grains across it, a dune can not migrate.

Stratification in dunes. The dropping of sand grains over the crest onto the slip face of a dune produces cross-strata much like the foreset beds in a delta (Fig. 9-30) or in a sand wave (Fig. 9-12). Erosion of the windward slope continually erodes the layers already deposited as new ones are added to the slip face (Fig. 13-24). Because of variations in direction and velocity of the wind, no dune, however, shows so uniform an arrangement. As a result, erosion alternates with deposition, and the cross-strata dip in various directions. Vegetation on the windward slope of a dune traps sand and causes deposition there as well; this results in additional irregularity of stratification. Despite the irregularities, however, the general direction of dip of the steep layers in wind-blown sand is the direction toward which the wind was blowing when the layers were built. This enables us to reconstruct ancient wind systems from cross-strata even though the forms of the dunes themselves have been completely destroyed by erosion.

Dune form. The shapes of dunes are diverse. Although the factors that determine them are only partly understood, we can classify dunes roughly into five groups based on their form. These are shown in Table 13-1, and in Figs. 13-25 and 13-26.

In the table we see that some dunes, of which the barchan is the best example, are built by winds blowing from one direction, whereas others are built by winds that shift through an arc. As long as the dune form exists, we can infer wind direction from the position of the slip face. But even after erosion has destroyed the form, and after what is left of the dune sand has been buried, converted into sandstone, subjected to uplift, and reexposed to a new cycle of destruction, we can recover the direction (or directions) of the wind that deposited the sand by measuring the direction of dip of the cross-strata, which are always inclined downwind (Fig. 13-29).

Still other dune shapes exist, many of them partly dependent on the form of the surface on which they lie. For example, long ridges stream downwind from steep narrow buttes, sheets of sand accumulate against cliffs facing the wind, and sand blown across a plateau can drop over a cliff facing to leeward and build up a sloping pile.

It is evident, however, that dunes occur where a supply of sand exists and where vegetation is not continuous enough to prevent deflation. Beaches, floodplains of streams, and exposed bodies of sandstone that disintegrate by weathering are common sources of wind-blown sand.

Composition and shape of sand grains. Virtually all dunes consist of sand-sized particles. Because quartz is the most common mineral in sand-sized sediments, it is not surprising that this mineral predominates in most dunes. Where other minerals are abundant, however, dunes can be built from them. The island of Bermuda mainly consists of wind-blown particles of calcite, derived from the beach, which is underlain by limestone and which has shifted with changes of sea level. The White Sands National Monument, an area of 500 square miles near Alamogordo, New Mexico, is covered

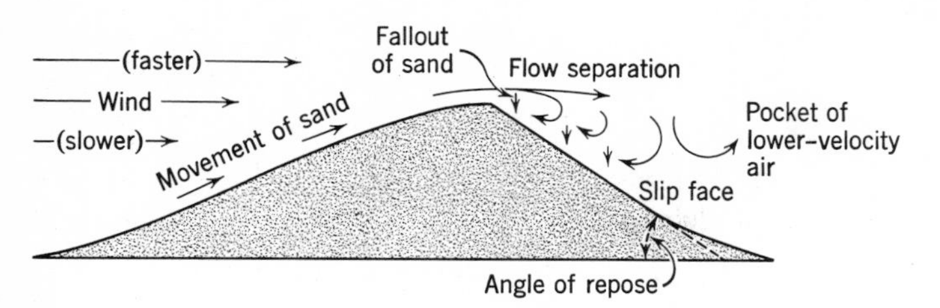

Fig. 13-23. Development of windward and lee slopes of a bare sand dune.

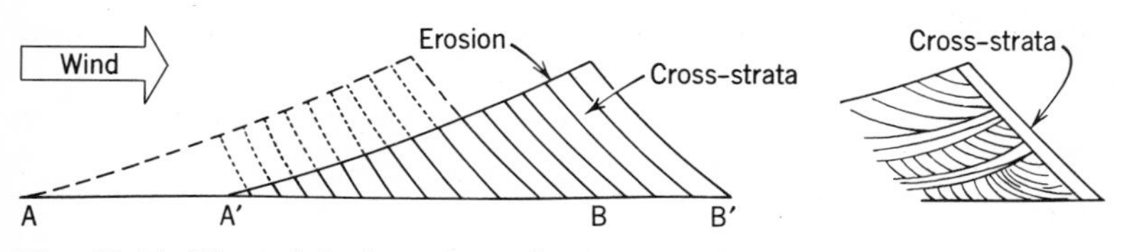

Fig. 13-24. If wind had uniform direction and velocity, it would build (*left*) simple, parallel cross-strata while dune migrates from AB to A′B′. But variations in direction and velocity are common and result in cross-strata like those at right. (Compare Fig. 13-29.)

TABLE 13-1. SOME KINDS OF DUNES BASED ON FORM

Kind	Definition and Occurrence	Illustration (Arrows indicate wind directions)
Beach dunes	Hummocks of various sizes bordering beaches. Inland part is generally covered with vegetation.	Sea Beach
Barchan dune	*A crescent-shaped dune with horns pointing downwind.* Occurs on hard, flat floors in deserts; constant wind, limited sand supply. Height 3 feet to > 100 feet.	
Transverse dune	*A dune forming a wavelike ridge transverse to wind direction.* Occurs in areas with abundant sand and little vegetation. In places grades into barchans.	
U-shaped dune	*A dune of U-shape with the open end of the U facing upwind.* Some form by piling of sand along leeward and lateral margins of a growing blowout in older dunes.	Blowout
Longitudinal dune	*A long, straight, ridge-shaped dune parallel with wind direction.* Occurs in deserts with scanty sand supply and strong winds varying within one general direction. Slip faces vary as wind shifts direction.	1 mile

Fig. 13-25. Barchan dunes between Yuma, Arizona and El Centro, California. (Josef Muench.)

Fig. 13-26. U-shaped dunes near Eltopia, Washington. Centers of some of the U's are indicated by stars. Foreground is about 1,700 feet wide. (Washington National Guard.)

Fig. 13-27. As loess is very cohesive and has vertical joints, it forms cliffs as jointed bedrock does. A typical surface expression of bare loess appears in this cliff, in the east bluff of the Mississippi River valley, Madison County, Illinois. The cliff will stand with little change for many years. (J. C. Frye.)

with dunes built of snow-white gypsum grains deflated from beds of gypsum exposed at the surface. In many areas dunes are built of clay and silt, but these materials were picked up by the wind as aggregates, each the size of a sand grain and each consisting of many particles of clay or silt firmly cohering. The diameters of the aggregates, rather than the diameters of the particles in each, determined the manner of wind transport: the sediment moved as aggregates by saltation, rather than as particles in suspension.

Sand grains become rounded more rapidly in wind than in water because air, being less dense and less viscous, cushions impacts less effectively. The impacts, therefore, wear away edges and corners to produce round shapes.

Ripples. We recall that ripples form on the sandy bed of a stream when current velocity is moderate, and that they migrate in the downstream direction (Fig. 9-11). When winds are moderate to strong, ripples form on the surface of dunes and other areas of bare sand. Coarse sand grains creep up the windward sides, lodge on the lee sides, and become buried by following sand grains. The ripples migrate downwind at rapid rates.

Unlike ripples made by a current of water, ripples made by wind lack cross-strata and generally lack stratification of any kind. Hence, whereas ripples built in currents of water record former current directions even after they have been embodied in sedimentary rocks, wind ripples do not give similar information about former winds. Furthermore wind ripples themselves are so ephemeral that they are not preserved. Their sand grains are continually being reworked into the larger structures of sand dunes.

DEPOSITS OF SILT

Loess. Much of the world's regolith includes small proportions of fine sediment deposited from suspension in the air, but so thoroughly mixed with other materials as not to be distinguishable from them. But over wide areas sediment having this origin is so thick and so pure that it constitutes a distinctive deposit. It is known as ***loess*** (lûs) (Fig. 13-27) and is defined as *wind-deposited silt, usually accompanied by some clay and some fine sand.*

Most loess is not stratified, apparently because

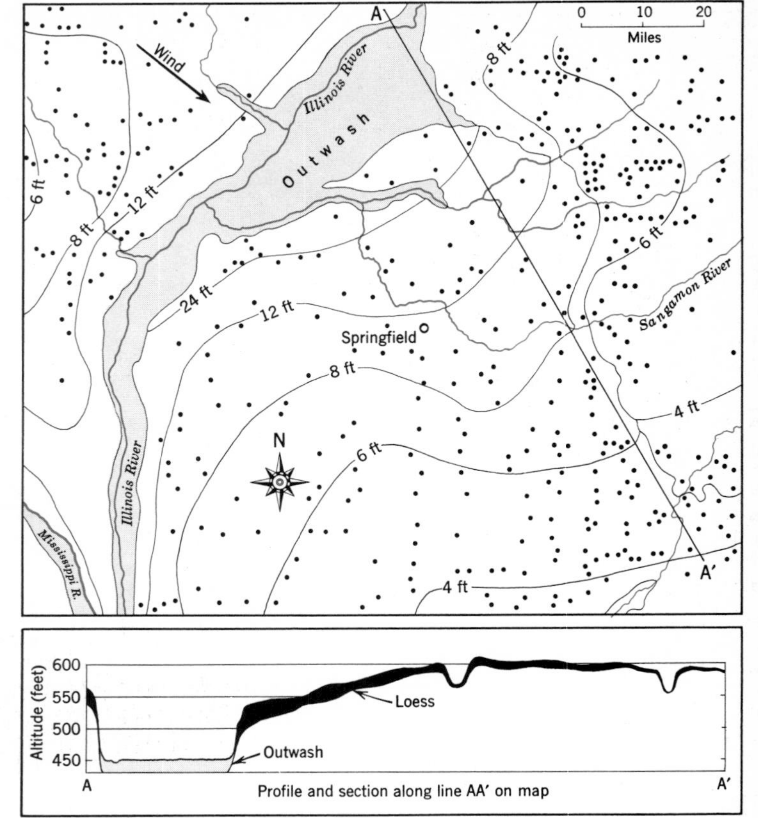

Fig. 13-28. Loess in central Illinois. Thicknesses were determined by borings at places shown by dots. Lines are isopachs. Profile and section were plotted. Thickness decreases away from body of outwash in both directions, but loess is thickest on southeast side, the leeward side for prevailing winds. Grain diameters also decrease in both directions away from outwash, and mineral content of loess and that of outwash are similar. Hence it is inferred that the loess resulted from deflation of the outwash. (Data from G. D. Smith, 1942.)

many grain sizes were deposited together from suspension, and also perhaps because plant roots, worms, and other organisms turned over and churned up the sediment as it was deposited. Where exposed, loess stands at an angle of repose so steep that in many places it is vertical. This results from fine grain size, in which the forces of molecular attraction are strong enough to give great cohesiveness to the particles. Also, because the particles are angular, porosity is extremely high, commonly exceeding 50 per cent. Hence loess accepts and holds water and constitutes a basis for productive soils.

Loess is so widespread and uniform, especially in the region of Nebraska, South Dakota, Iowa, Missouri, and Illinois, that it was once believed to have been deposited by water. But when, after further study, loess was found to mantle hills and valleys alike through an altitude range of 1,500 feet, the belief was abandoned.

Minerals composing loess are chiefly quartz, feldspar, micas, and calcite. The particles are generally fresh, showing little evidence of chemical weathering other than the slight oxidation that has occurred since deposition and that gives a yellowish tinge to the deposit as a whole. The loess in many areas closely resembles the rock flour ground up mechanically by glaciers. The resemblance led geologists to suspect that the loess in those areas consists of glacial drift picked up, sorted, and redeposited by the wind. The suspicion was confirmed by the distribution of loess, much of which occurs immediately to leeward of large areas of glacial outwash (Fig. 13-28).

Glacial loess. Today it is agreed that most of the loess in North America and Europe was derived from glacial drift, chiefly outwash. The loess was deposited during glacial ages when the areas just outside the margins of ice sheets were cold and windy and when glacial outwash—gravel, sand, and silt—was filling up river valleys so fast that plants could not gain a foothold on valley floors (Fig. 12-19). With a windy climate and no vegetation, the wind easily deflated the valley floors. Saltating sand grains splashed silt particles into the air to be

Fig. 13-29. Cross-stratification in wind-blown quartz sand of Jurassic age consolidated to form hard sandstone. Parts of lee slopes of a succession of dunes are exposed. The cavities result from solution at places where calcium-carbonate cement between the grains of quartz is especially soluble. Monument Valley, Arizona. (Tad Nichols.)

carried away in suspension. The silt settled out, forming blankets 15 to 30m thick near the source valleys and thinning downwind to thicknesses of 1 to 2m, spread over thousands of square kilometers.

Why was the silt not picked up again and again and carried even farther by the wind? One reason lies in the stability of a silt surface, which results from its fine grain size (Fig. 13-14). Another, and in the case of loess perhaps a more important reason, is that the loess settled out chiefly in grassland and to some extent even in forested areas. In such environments, once on the ground it would be in no danger of further deflation.

Desert loess. In some regions loess is found through enormous areas that lie to leeward of deserts, an obvious source of mechanically weathered sediments. The loess that covers much of western China, blown from the great desert basins of central Asia, reaches a thickness of more than 70m. Probably the immediate source of such loess is chiefly fine alluvium, washed down by cloudburst streams from the mountains onto desert floors.

ANCIENT WIND-DEPOSITED SEDIMENTS

Sediments deposited by the wind are useful indicators of events in Earth history. Dunes no longer active, because firmly fixed by a continuous cover of vegetation, indicate that when they were built conditions were generally drier than those of today. The grassed-over Sand Hills, extending throughout an area of 20,000 square miles in western Nebraska, record a period or periods when aridity was greater.

From the shapes of dunes and from the directions of their steep cross-strata (always inclined downwind) the directions of former winds can be inferred.

Sheets of loess in the lee of outwash bodies record times when meltwater streams spread out sediments that could be picked up and transported by the wind. They are therefore generally indicative of glacial ages.

In ancient layers of sedimentary rock, in which no traces of dune shape remain, the combination of steep cross-strata (Fig. 13-29) record the deposition of wind-blown sand at times far back in the Earth's history.

Fig. 13-30. Windbreaks on dry sandy farm land in northern Texas. (U. S. Soil Conservation Service.)

ENVIRONMENTAL ASPECTS: SOIL EROSION

With climates that continually fluctuate, regions ordinarily suitable for agriculture have dry periods during which soil erosion by wind reaches tremendous proportions. As we have noted, during the 1930s an enormous volume of soil was blown away from parched, unprotected plowed fields in the "Dust Bowl" region of the Great Plains (Figs. 13-12, 13-17). The sand-sized particles were piled up along fences (Fig. 13-22) and around farm buildings, and finer particles were blown eastward to be deposited over wide areas; a good deal of such material was dropped into the Atlantic Ocean.

What determined the location of the "Dust Bowl"? That area closely approximates the largest area in the United States in which average wind velocity is more than 5m/sec. This is the velocity, as we noted earlier, at which flowing air not only begins to move sand grains but also, as we have seen, can maintain coarse silt in suspension. All that was needed further was a long succession of dry years—the drought in the 1930s. We now realize that with only very slight changes in the Earth's climates deserts expand and contract and can be created or disappear.

In the "Dust Bowl" area good practice includes the planting of windbreaks (Fig. 13-30), consisting of bushes and hardy trees set in strips at right angles to the strongest winds, at intervals of a mile or so. It also involves planting strips of grass alternating with strips of cultivated grain (Fig. 10-4), for the soil must lie bare during alternate years in order to accumulate moisture sufficient to grow grain. The principle on which the rows of plants act to retard deflation is based on the presence, just above the ground, of the dead-air layer we noted earlier, the thickness of the layer being equal to one-thirtieth the height of obstacles on the ground. A group of trees and bushes 10m high, though somewhat permeable to wind, should create a dead-air layer nearly 30cm thick, thus effectively preventing deflation beneath it and through some distance to leeward.

Summary

Deserts

1. Deserts, constituting about a quarter of the world's land area, are areas of slight rainfall, high temperature, great evaporation, relatively strong winds, sparse vegetation, and interior drainage.

2. No major geologic process is confined to deserts, but in them mechanical weathering, impact of rain drops, flash floods, and winds are very effective. The water table is low. Badlands occur locally.

3. Pediments are a conspicuous feature of many deserts. They are shaped by streams, rills, sheet runoff, and weathering.

4. In the arid cycle of erosion, pediments grow headward at the expense of mountain slopes, which appear to retreat without becoming gentler as they do in a moist region. Wind removes fine rock waste.

Wind action

5. Wind carries a bed load of sand grains saltating close to the ground and a suspended load of fine particles higher up. Sorting of sediment results.

6. Wind erodes by deflation and abrasion, chiefly in dry regions and on beaches. It creates deflation basins, blowouts, deflation armor, and ventifacts.

7. With long-continued wind activity, sand grains become rounded.

Dunes

8. Many dunes are localized by obstacles. Bare dunes have steep slip faces and gentler windward slopes. They migrate downwind, forming cross-strata that dip downwind.

9. Dunes are classified by form into beach dunes, barchan dunes, transverse dunes, U-shaped dunes, and longitudinal dunes.

Loess

10. Loess is deposited chiefly in the lee of (a) bodies of glacial outwash, and (b) deserts. Once deposited, it is stable and little affected by further wind action.

Selected References

Bagnold, R. A., 1941, The physics of blown sand and desert dunes, repr. 1954: New York, William Morrow and Co.

Blackwelder, Eliot, 1954, Geomorphic processes in the desert: California Div. Mines Bull. 170, Chap. 5, p. 11–20.

Bryan, Kirk, 1923, Erosion and sedimentation in the Papago Country, Ariz., with a sketch of the geology: U. S. Geol. Survey Bull. 730, p. 19–90.

Cooper, W. S., 1958, Coastal sand dunes of Oregon and Washington: Geol. Soc. America Mem. 72.

Finkel, H. J., 1959, The barchans of southern Peru: Jour. Geology, v. 67, p. 614–647.

Hadley, R. F., 1967, Pediments and pediment-forming processes: Jour. Geol. Education, v. 15, p. 83–89.

Hume, W. F., 1925, Geology of Egypt, v. 1: Cairo, Government Press.

McKee, E. D., 1966, Structures of dunes at White Sands National Monument, New Mexico . . . : Sedimentology, v. 7, p. 1–69.

Neal, J. T., ed., 1965, Geology, mineralogy, and hydrology of U. S. playas: U. S. Air Force, Office of aerospace research, Environmental Res. Paper 96, [Defense Documentation Center, Alexandria, Va.]

Sharp, R. P., 1949, Pleistocene ventifacts east of the Bighorn Mountains, Wyoming: Jour. Geology, v. 57, p. 175–195.

—— 1963, Wind ripples: Jour. Geology, v. 71, p. 617–641.

Thorarinsson, Sigurdur, 1954, The tephra-fall from Hekla on March 29th, 1947: Societas Scientiarum Islandica, v. 2, no. 3.

Map

Thorp, James, and others, 1952, Pleistocene eolian deposits of the United States . . . : Boulder, Colo., Geol. Soc. America.

CHAPTER 14

Coasts and Continental Shelves

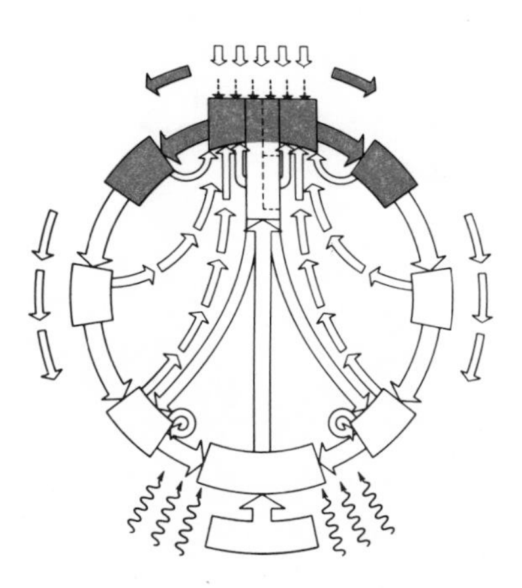

Tides and tidal sediments
Waves and beaches
Strata made when the land builds outward
Sculpture of bedrock coasts
Evolution of coasts
Protection of coasts
Reefs and carbonate sediments
Changes of sea level
Continental shelves

Plate 14

Tomales Bay, 40 miles northwest of San Francisco, California. (Sunderland Aerial Photographs.)

Spit
Waves being
refracted
Wave-cut cliffs

CHAPTER 14 *Coasts and Continental Shelves*

GEOLOGIC VIEW OF THE SEA FROM THE LAND

The sea has been a prime concern of geologists ever since they realized that large areas of the present continents are underlain by marine strata. In 1830 Lyell emphasized the great importance of the geologic processes that take place in the sea, where, he thought, men could never observe them in action. He listed, among hindrances to the progress of geologic study, the fact that man is not an amphibious creature. How delighted Lyell would be if he knew that in the twentieth century men can go down to the sea floor to watch what happens (Fig. 14-1)!

Before we look for ourselves, however, let us examine some inferences drawn by early geologists about the sea and its sediments. Without ever getting wet or seasick these landlubbers learned much about the effects of waves on sediments, of climate on organisms, and of environment on both. Early knowledge of waves, though not quantitative, supported the idea that coarse-grained sediments should be found only in shallow water near shore, and that all sediment deposited offshore should be fine grained. Geologists applied this concept in a search for the positions of shores of ancient seas.

Early in the nineteenth century, biologists had learned that some organisms remain within certain climatic regions. For example, reef-building corals and certain palm trees survive only in the tropics, whereas polar bears cling to the Arctic zone. By 1830 knowledge of the effect of climate on the distribution of organisms was well enough advanced to form a sound basis for Lyell's inference (from fossils) that when the lower Tertiary strata were being deposited, the climate of western Europe was subtropical. Other geologists have drawn many other climatic inferences from their study of fossils elsewhere.

Biologists showed that climate is only one of the aspects of environment that influence the distribution of organisms. Water and air limit the range of some organisms. Certain species of fishes, corals, clams, algae, mangrove, and grass, to name but a few examples, live in salt water. Other organisms, such as insects, birds, horses, trees, and many plants, inhabit the air or the land. No great knowledge of biology was required to support the inference that strata containing fossils of species whose descendants now inhabit the sea had been deposited in an ancient sea. By similar reasoning, geologists decided that strata containing fossils of land-dwelling species had been deposited in nonmarine environments. Even in its grossest form such environmental analysis of strata on the basis of species of fossils proved to be useful. By applying such analysis to strata on the continents, geologists were able to draw maps of the extents of strata of marine and nonmarine origin. From such maps we can see that large areas of the present continents had been submerged beneath the sea at some times and had been emergent at others.

How do the concepts formulated by land-bound geologists compare with the realities of modern oceans? Let us see for ourselves.

TIDES AND TIDAL SEDIMENTS

At almost any point along the east coast of the United States from Long Island to Georgia, and at many places along the Gulf coast, we can see features composed of sediment (Fig. 14-2) deposited by tides and by waves. Let us begin with tides and tidal sediments.

If we watch a ***tidal marsh,*** *a low, flat coastal area thickly grown over with saltwater grasses, in large part submerged at high tide* (Fig. 14-3), for a few hours we see that its surface is alternately inundated and exposed. *The rhythmic rise and fall of the surface of the sea* is the ***tide,*** a phenomenon

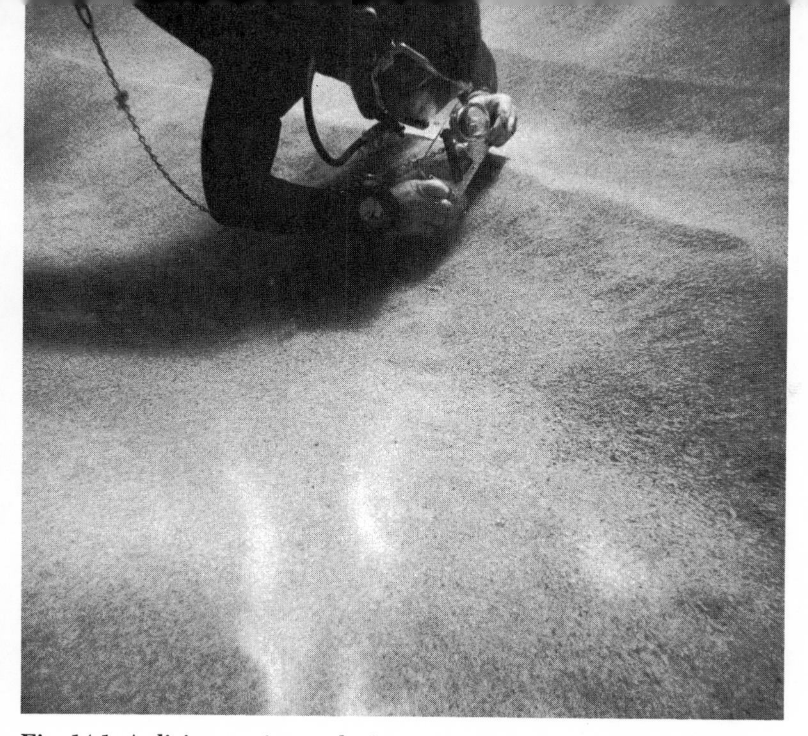

Fig. 14-1. A diving marine geologist examines a ripple made by waves on the sea floor at depth of 10m. (William Bunton, U. S. Navy Electronics Laboratory, courtesy R. F. Dill.)

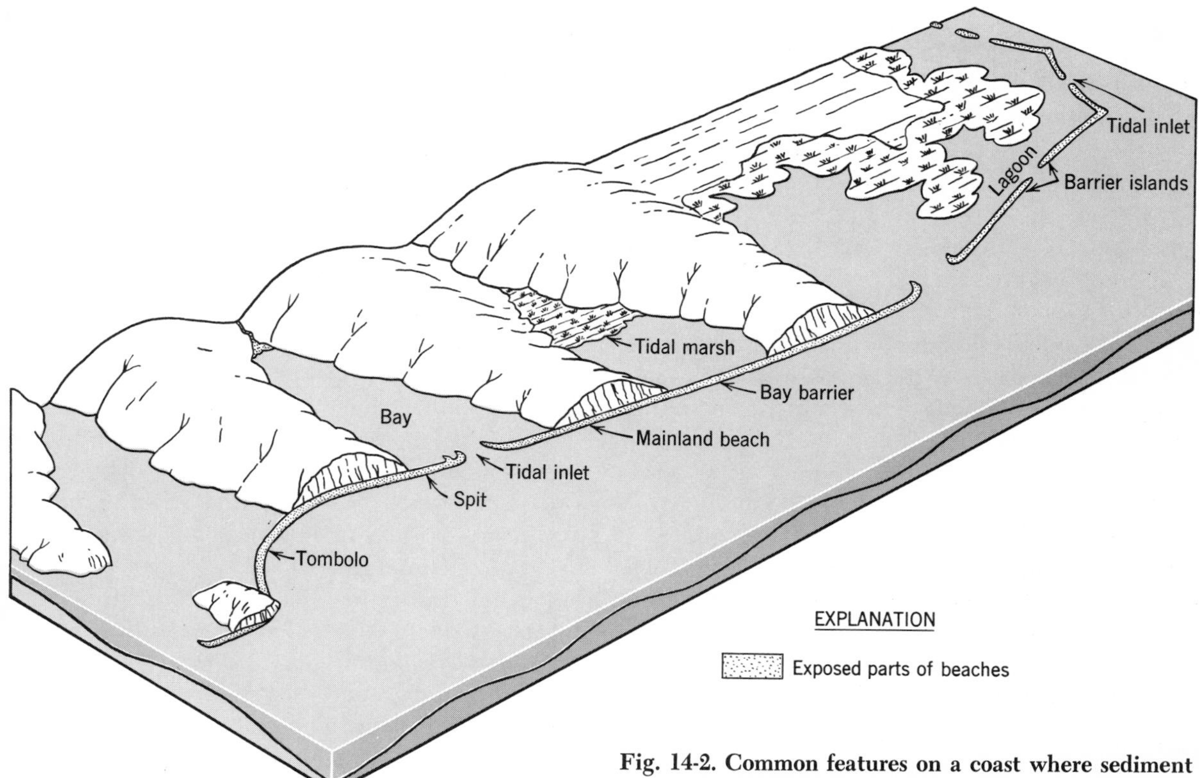

Fig. 14-2. Common features on a coast where sediment is abundant. Features defined in text.

Fig. 14-3. The line of dunes on a barrier spit (*upper left*) separates this tidal marsh and lagoon from the open sea. Cape Cod National Seashore, Eastham, Massachusetts. (J. E. Sanders.)

common to most coasts.

In most localities the water rises and falls twice daily in a cycle, about $12\frac{1}{2}$ hours long, that is displaced by about 50 minutes each day. The range of tidal heights varies with the phases of the Moon (Fig. 14-4) as the distance between Moon and Earth varies (Fig. 2-2).

At each point on Earth the tide-producing force of the Moon is the resultant of the Moon's gravitational forces (black arrows of varying length, Fig. 14-5) and centrifugal forces (white arrows). All the resultants (gray arrows) define the surface of an irregular prolate spheroid (blue areas) which bulges outward beyond the hypothetical oblate spheroid of water that would cover the Earth if continental masses were leveled off.

Wherever the sea rises and falls, it does so because the prolate spheroid always points toward the Moon, whereas the Earth rotates.

The gravitational effect of the Sun creates small tidal effects on Earth. When this effect operates along the same direction as that of the Moon (new and full phases) the tidal effect of the Sun combines with that of the Moon to create tides whose ranges (differences between high tide and low tide) are greater than normal. Operating at right angles (Moon in quarter phases) the tidal effect of the Sun works against that of the Moon and creates tidal ranges less than normal.

Commonly a tidal marsh and the low areas adjacent to it are sheltered from the sea by a ***barrier,*** *an elongate island of sand or gravel parallel with the coast.* Along many coasts, except where interrupted by *tidal inlets,* a line of barriers is continuous. Currents, which flow through an inlet as the water level rises and falls, drag sand along the bottom and

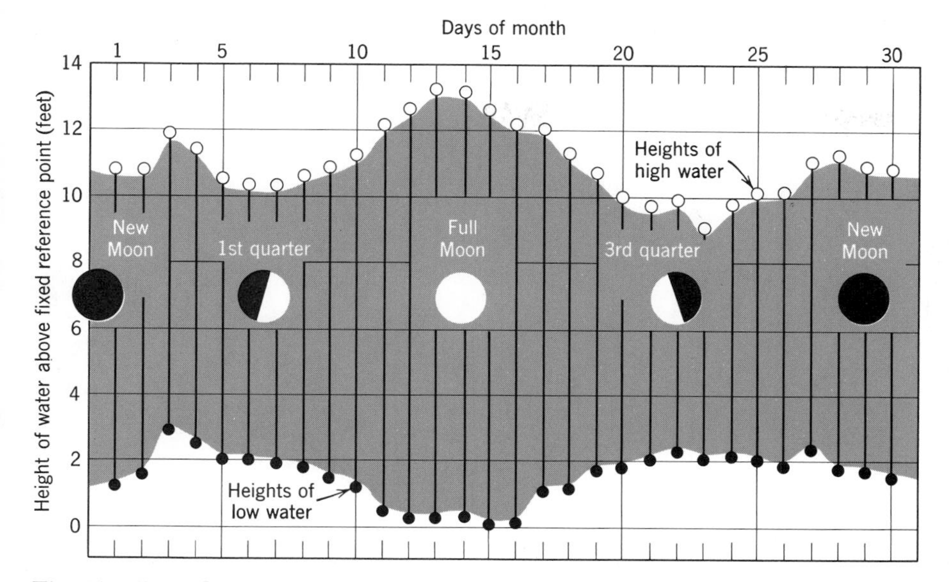

Fig. 14-4. Record of heights of high and low water for 1 month compared with phases of the Moon. (H. A. Marmer, 1929.)

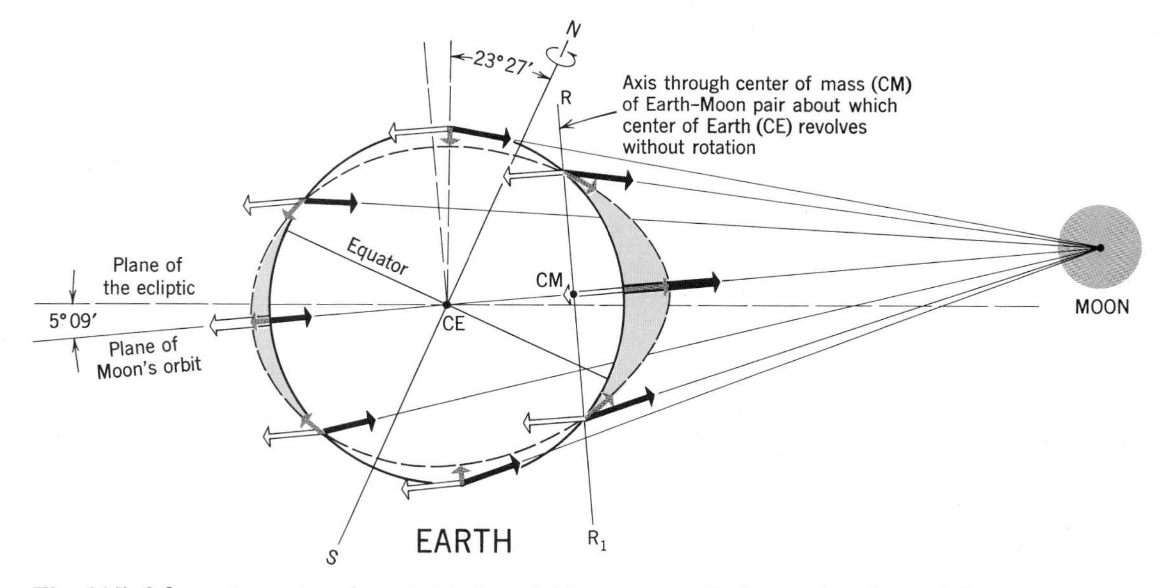

Fig. 14-5. Schematic section through Earth and Moon, perpendicular to the plane of the ecliptic, showing how gravitational attraction and centrifugal force of the Moon's revolution cause tides on Earth. Center of Earth, CE, revolves about center of mass of the Earth-Moon pair, CM, without rotation (Fig. 2-2, *B*). Further explanation in text. (After Darwin, 1898.)

Fig. 14-6. A tidal delta has formed only on the lagoon side of Moriches Inlet, which cuts the barrier along the south side of Long Island, New York. (Litton Industires, Aero Service Corp.)

deposit it opposite each end in *tidal deltas,* bodies of sediment convex away from the inlet (Fig. 14-6). Opposite the seaward end of the inlet, a tidal delta forms only where waves do not remove sand as fast as the tides drag it outward from the inlet. Water flowing through tidal inlets carries not only sand but also silt and clay, thrown into suspension by waves on the seaward side of the barrier. These fine-grained sediments are added to parts of tidal deltas and also build mudflats around the margins of lagoons. Grasses growing on the mudflats trap sediment and continue to grow until they create tidal marshes. By additions of sediment and dead grass the surface of a marsh accretes upward until it is horizontal and lies at or close to the level of high tide. Chemical changes of buried plant material form peat (Fig. 23-3).

Other sediments deposited in the area behind the barrier are not related to tidal action. Sand is blown in by wind and is washed over the barrier during storms. Other sediment drains in from the land. Distinctive organisms inhabit these sediments.

WAVES

Before inquiring into the ways in which waves shape coasts by continuously eroding and depositing sediment, we need to answer some questions about waves. How do waves originate? How far from shore do they influence the bottom? What bottom features are caused by waves? We can find some answers by setting up instruments, by making measurements, and by examining the bottom.

Study of water waves provides us not only with insights into the behavior of coasts but also with a valuable introduction to the principles governing all waves, most of which are not visible. Some important invisible waves include sound waves, waves that travel through the ground during earthquakes, and the waves of the electromagnetic spectrum by which energy is transmitted from Sun to Earth and then is radiated back into space. All waves are expressions of energy being transmitted and their characteristics reflect the sources of their energy.

Knowledge of water waves was improved enormously by studies made during World War II to satisfy requirements for amphibious military operations. These studies were continued after the war to solve problems of destructive beach erosion and to answer questions about the interaction between air and water. Data from laboratory experiments made with small waves in glass-sided tanks, and from field experiments, are now available in great volume. Without entering into all the details we can note some basic facts about wave mechanics.

Wave mechanics. To eliminate all distractions assume that waves of only one size are present. Figure 14-7 shows one of these simple waves in profile. The definitions and standard letter symbols of various attributes, which are fundamental to all kinds of waves, are brought together in Table 14-1.

If we were to take a boat, go out well beyond the breakers, and throw a chip of wood overboard, we could see the chip rise when the crest of a wave passes and then subside as the next trough arrives.

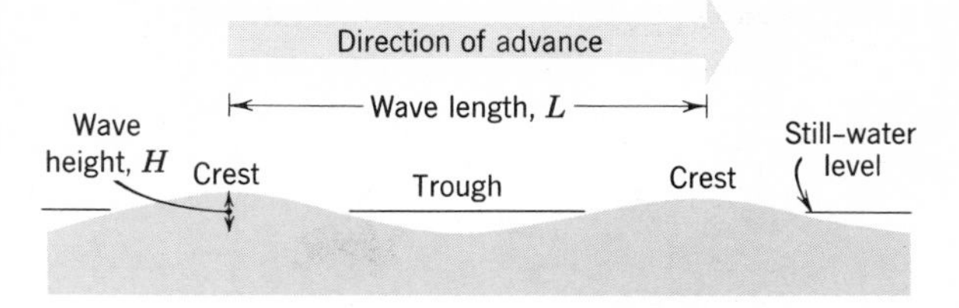

Fig. 14-7. Profile through ideal deep-water waves of oscillation drawn parallel to direction of wave advance. (Terms and symbols defined in Table 14-1.)

If we could measure its position with respect to the bottom, we would find that the chip has not moved, or has moved only a negligible distance. The water particles, therefore, are not moving forward with the wave; only the wave *form* is passing along the water surface. Accordingly, we can classify our waves as ***waves of oscillation***, *waves that cause water particles to oscillate but not to undergo appreciable net displacement.*

As the waves pass, what, if anything, is happening below the water surface? It is hard to determine this in the open sea, but we can watch similar action in a glass-sided laboratory tank. We can create waves by moving a paddle back and forth at one end of the tank. By injecting into the tank droplets of colored material having the same density as the water, we can see how the water particles move and with a movie camera can photograph the movement as waves pass through the tank. The tank experiment shows us that the particles move in circles and that the circles lie in vertical planes parallel to the direction of wave advance.

TABLE 14-1. TERMS AND SYMBOLS USED IN DESCRIPTION OF WAVES

	Symbol	Term	Definition
Applicable to all waves	L	*Wavelength*	*The horizontal distance between successive wave crests or between successive troughs.*
	H	*Wave height*	*The vertical distance from bottom of trough to top of crest.*
	T	*Wave period*	*The time required for wave to advance through distance of one wavelength.*
	c	*Wave velocity*	*The distance traveled by a wave in a unit of time.*
Special terms for water waves	h	**Depth**	The vertical distance from water surface to bottom in the absence of waves.
	H/L	*Wave steepness*	*The ratio of wave height to wavelength.*

Beneath wave crests the water particles curve around the tops of the circles in the direction of wave advance; beneath troughs they curve around the bottoms of the circles in the opposite direction. At the surface the diameters of the circles, or orbits, are equal to wave height, H, but downward they decrease rapidly. If the water is deeper than half the wavelength, $L/2$, the circular motion, which dies out downward no matter what the depth, becomes negligible before reaching bottom. Therefore bottoms covered by water deeper than wave base remain undisturbed by the waves. Waves with lengths greater than twice water depth are *deep-water waves*.

Sizes and origins of waves. Waves ruffling the surface of the sea come in many sizes, travel in various directions, and originate in numerous ways. As in electromagnetic waves (Fig. A-3) the waves on the sea form a continuous spectrum consisting of overlapping groups of waves that share a common origin. In electromagnetic waves the spectrum is based on wavelengths, which are easily measured. In water waves it is difficult to measure wavelengths but easy to measure wave periods. Accordingly the spectrum of waves on the sea is based on their periods.

The energy sources for sea waves include winds (resulting from solar energy and the Earth's rotation), tides (resulting from the gravitational attraction from the Moon and the Sun and from the Earth's rotation), and the Earth's crustal movements and gravitational acceleration (resulting from displacements of the sea floor).

Because of friction and various complex pressure fluctuations not yet fully understood, winds transfer their kinetic energy to the water surface, creating waves ranging in size from tiny ripples to mountainous seas. Three important factors influence the origin of wind-generated waves: (1) wind speed, (2) wind direction, and (3) length of the stretch of open water across which the wind blows, known as *fetch*. The periods of waves generated directly by winds range from fractions of a second to about 10 seconds. Winds not only generate storm waves but, given sufficient time and fetch, also prevent them from growing beyond a maximum size by blowing their tops off.

While a storm is in progress and after the center responsible for its winds has moved away, wind-generated waves radiate outward in all directions but chiefly downwind. In doing so they mutually interfere, with the result that some waves combine and grow larger, whereas others conflict and become smaller or obliterate one another. After such interference during travel beyond the immediate influence of storm winds, sea waves become longer, lower, and more regular than they were while being actively blown about by storm winds. Transformed former sea waves are *swells*, which transfer energy from a storm to distant shores. The periods of most swells are 8 to 12 seconds.

Tides can be visualized as long-period waves. Their periods, which depend chiefly on the Moon and on the Earth's rotation, range from about $6\frac{1}{4}$ or $12\frac{1}{2}$ hours up to 27 days and even longer.

Long, low waves generated by abrupt displacement of the sea floor or by landslides entering the sea are ***tsunami*** (Jap. *tsu*, "harbor," *nami*, "waves"; pronounced tsoo-năh′mē, and spelled the same in both singular and plural). The energy thus transferred to the water from movement of the sea floor is converted into a series of waves having periods of about 12 minutes. Tsunami speed outward from their origin at 650 to 950km/hour, the value depending only on depth of water. Because of their great lengths (130 to 150km) and low heights (less than lm), tsunami are scarcely visible in the open ocean. But on some shallow coasts they can become heaped up into great waves whose effects can be devastating (Fig. 1-2).

Nearly all tsunami originate in the Pacific Ocean in zones of active earthquakes. Because total evacuation of low-lying areas is the only defense against tsunami, an elaborate warning system has been devised for the Pacific Ocean to alert people to move to high ground before the great waves strike the land.

EFFECTS OF WAVES APPROACHING SHORE

Let us pick out a train of large ocean waves with lengths of 225m and heights of 10m (Fig. 14-8) and follow them from deep water to the shore. When, at depth $L/2$, the bottom begins to interfere with the motion of the water particles, we say the deep-water waves become *shallow-water waves*, and we refer to depth $L/2$ as *wave base*. These terms "deep" and "shallow," applied to waves, are not

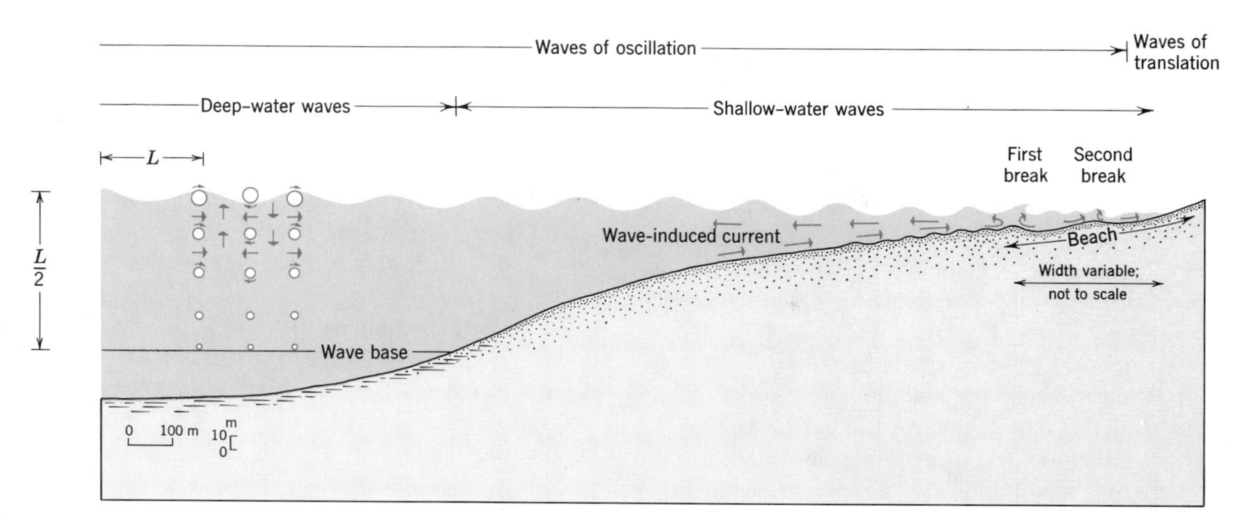

Fig. 14-8. Profile at right angles to coast, showing changes in waves as they travel from deep water through shallow water to shore. Arrows show directions of motion of water in waves (*left*) and wave-induced currents (*right*). Vertical exaggeration is 3×.

absolute depths but are ratios determined by wavelength. The deepest parts of the ocean are "shallow" to extremely long waves and the shallow parts are "deep" to short ones.

The water particles at the bottom can move neither upward away from the bottom nor downward into it. Hence they are unable to describe circular orbits as they did in deep water. Their only possible movement is back and forth, parallel to the bottom. Above the bottom the particles can move in the vertical direction, but instead of circles, they describe ellipses with major axes parallel to the bottom. Near bottom the minor axes of the ellipses are very short, but upward they lengthen. Because the particles make ellipses and not circles, the profile of a shallow-water wave lacks the symmetry shown by a deep-water wave. The crest of a shallow-water wave is narrow and steep, whereas its trough is wide and flat.

When the bottom and the waves begin to interact, two things happen: (1) waves approaching the shore obliquely swing around to become more nearly parallel to the shore, and (2) all waves create a drag on the bottom and thus are capable of transporting sediment. Although the two things happen simultaneously we can best discuss them one at a time.

Refraction of waves. A wave that approaches a coast straight on "feels bottom" (at depth $L/2$) simultaneously along its entire crest. By contrast a wave that approaches a coast obliquely feels bottom piecemeal. Only a small segment of such a wave feels bottom at any one time. Vertically above the point where each segment feels bottom, wavelength decreases and height increases. The wave slows and swings around, part by part, tending to become parallel with the shore (Pl. 14). This process is known as *refraction* of the wave. A deep-water wave that approaches the shore at an angle of 40° or 50° may be refracted to reach shore at only 5°. Rarely is refraction so complete that the wave exactly parallels the shore.

Waves arriving over a submerged ridge off a headland converge on the headland. This convergence, together with increased wave height, concentrates wave energy on the headland. Refraction explains an old sailor's saying that "the points draw the waves." Conversely, refraction of waves approaching a bay or a submerged valley makes them diverge, diffusing their energy at the shore.

In summary, ***wave refraction*** is *the process by which the direction of a series of waves, moving in shallow water at an angle to the shoreline, is changed so that waves become more nearly, but rarely exactly, parallel to the shore.*

Drag on the bottom by waves. The oscillating water particles beneath the shallow-water waves create an oscillating drag on the bottom. Wave drag is much like the drag of a stream on its bed, except that wave drag reverses direction periodically. If drag of any kind exceeds the threshold value of particles of a given size, shape, and density it moves them. If waves move sediment, they drag it toward shore when a crest passes overhead and seaward as the next trough goes by. The oscillating motion can create ripples whose sizes, shapes, and orientations vary, the long axes of ripples oriented at right angles to the direction of the oscillating water. Because of refraction the long axes of many ripples are approximately parallel to shore. Where drag is equal in both directions, the ripples are symmetrical, their crests are narrow and pointed, and their troughs are broadly rounded and concave-up (Fig. 14-1). The shapes of such ripples in an ancient stratum of sandstone tell us which way was up originally, even though the stratum has been deformed (Fig. 17-11).

Laboratory experiments in wave tanks have shown that in addition to the oscillating drag just described shallow-water waves moving across a gently sloping sea floor can create a pulsating current. The current flows along the bottom in the direction of wave advance, to the line where the waves are breaking. There it rises and returns seaward near the surface. Under some conditions not yet understood, the lower current creates enough drag to make large-scale relief on the bottom analogous to the sand waves on the bed of a stream (Fig. 9-12). The relief consists of rounded ridges and intervening troughs. The ridges, one kind of ***bar,*** *an elongate ridge of sediment, built offshore by waves and currents, whose top is always submerged,* migrate slowly landward toward the line of breakers. As in sand waves in a stream channel, sediment eroded from the seaward sides of the bars is deposited on the landward sides as cross-strata, dipping landward (Fig. 14-9).

Breakers, surf, and longshore drift. As waves move into shallower water their height increases. At a depth equal to $2H$ the wave profile becomes very peaked and asymmetrical; its front is steep (Fig. 14-10) and its seaward side slopes more gently. When depth equals $1.28H$ the wave form becomes unstable and is converted into a ***breaker,*** *a wave that is collapsing.* Beneath a breaker, the motion of the water is very turbulent and is predominantly upward. As each wave crest collapses, water from both adjacent troughs flows toward the ***breaker zone,*** *the zone where waves collapse.* The water from many breaking waves flowing toward the lines of breakers builds a bar beneath each. Such breaker bars migrate landward under some wave conditions and seaward under others. Therefore some cross-strata in them dip landward and others dip seaward (Fig. 14-9). During storms breaker bars may disappear.

Shoreward of each breaker bar is a trough where the water deepens. When breakers surge into the deeper water of this trough they are reorganized into smaller waves of oscillation. Where the smaller waves collapse, a line of secondary breakers forms and under it another breaker bar. When the water becomes so shallow that no new waves of oscillation can form shoreward of the breaker line, the breakers are transformed into ***waves of translation,*** *waves that displace the water masses within the moving crests. Waves of translation landward of the breakers and seaward of the backwash* constitute the ***surf.*** This mass of water surges toward the land and flows up the gently sloping shore as *swash,* returning as *backwash.*

The zone between the outermost breaker and the outermost (lower) limit of the backwash is the ***surf zone,*** which can include one or more smaller breaker zones; on steeply inclined shores a surf zone may not be present.

The oblique approach of waves creates a ***longshore current,*** *a current in the surf zone flowing parallel to the shore* (Fig. 14-11). Even the slowest longshore current can transport fine particles thrown into suspension by wave action. By contrast a longshore current can shift sand only when it exceeds the threshold value determined by water depth, and particle size, shape, and density. These relationships are the same as those that governed our experiments with flowing water in the laboratory stream tank. If a longshore current flows fast enough it can create sand waves and drag them along parallel with the shore (Fig. 14-9), as a stream does with the sand on its bed. Cross-strata formed by migration of these sand waves dip parallel to shore.

At intervals along the coast the water of longshore currents returns seaward through the breaker line as narrow *rip currents.* Although these currents cut channels through the breaker bar, they separate from the bottom just seaward of the breakers and

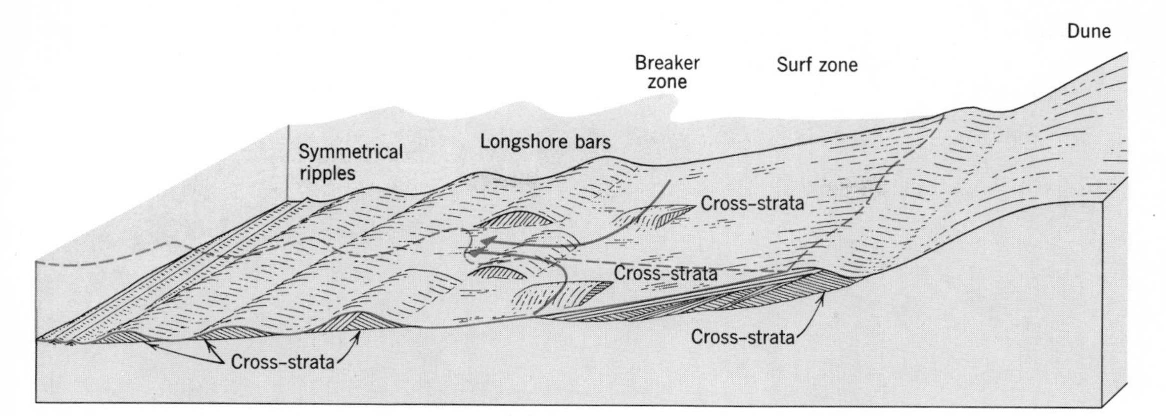

Fig. 14-9. Schematic diagram of beach with water cut away, showing varieties of large-scale cross-strata. Blue arrows show directions of currents that move parallel with shore in surf zone and then seaward, through a rip-current channel, across breaker zone. (Text defines surf zone and breaker zone.)

Fig. 14-10. Wave starting to break (*left*) off Waimea Bay, Oahu, Hawaii tests this surfer's skill. (Don James.)

Fig. 14-11. Directions of transport of sediment along a coast shown by arrows. Finer particles in suspension and, on occasion, coarser particles as bed load are transported parallel to the shore by longshore current. Dotted line shows path of a typical sand grain in the swash zone. Uprush of wave washes grains diagonally up the sloping beach; backwash carries them back at right angles to shoreline. Resulting path is a series of parabolas. (Photo by A. Devaney.)

dissipate in deeper water. Rip currents that flow fast enough can create transverse sand waves just as longshore currents and streams do. These sand waves migrate seaward, depositing cross-strata that dip seaward. Such waves do not exist seaward of the breaker zone because there the rip currents are no longer in contact with the bottom.

BEACHES

The upper limit of the swash defines the landward limit of a ***beach,*** *a body of wave-washed sediment extending along a coast between the landward limit of wave action and the outermost breakers.* Most beaches are parallel with the coast. One kind that does not necessarily parallel the coast is a ***tombolo,*** *a beach that connects an island with the mainland or with another island* (Fig. 14-2).

The movement of particles obliquely up the slope by the swash and nearly directly down it by the backwash shifts particles along the shore; the process is *beach drift* (Fig. 14-11). Measurements of colored tracer particles released at known points show that waves move sediment particles not only on exposed parts of beaches but also in many other zones near the shore. On gently sloping bottoms, waves tend to move coarse particles onshore because the land-

Fig. 14-12. Spit with curved end. Pattern of sand near groins (*left*) results from longshore drift toward upper left. Presque Isle peninsula, Erie, Pennsylvania, June 1959, looking east. (U. S. Army Corps of Engineers.)

ward drag under the crests exceeds the seaward drag under the troughs. On steeper bottoms, the effects of the tangential component of gravity, g_t, combined with the seaward drag on the bottom beneath wave troughs, exceed the landward drag under wave crests and coarser particles move seaward. No matter what the slope of the bottom, waves, and various currents created by waves, tides, and winds, tend to move fine particles offshore. When they approach the coast obliquely, waves transport all particles along the shore. *The net movement of sediment parallel to the shore by waves and wave-induced currents* is ***longshore drift;*** it occurs on the bottom seaward of the breakers, along the breaker bar, in the surf zone, in the zone of swash and backwash, and where backwash collides with surf (Fig. 14-11). When sand in all five zones moves in the same direction, as it generally does, a veritable "river of sand" flows along the shore at everyday rates that range from about 2 to 146m/hour off California. Of the measured rates, 80 per cent lie between 9

Fig. 14-13. Rounded cobbles and boulders on a beach at Montauk Point, Long Island, New York. The coarser particles are wave-washed residue from till, which forms the bluff in the distance. (G. A. Douglas, from Philip Gendreau.)

and 36m/hour. During severe storms the rates doubtless increase, but measurements have not yet been made under such conditions.

Along a coast, sand reacts with waves to create a condition of dynamic equilibrium. Where sand supply and movements by waves are in equilibrium, the long-term position of the shore does not change appreciably. (Of course, without regard to overall sand supply, short-term changes occur almost daily in response to variations in wave characteristics.) Where the supply of sand is abundant the shore shifts outward and where supply of sand is deficient the shore is eroded. We examine both these conditions in later sections of this chapter. No matter how the equilibrium shifts, the profile perpendicular to the shore remains concave-up. Shore profiles thus resemble stream profiles, but this similarity of shape results from contrasting processes. In a stream the water flowing downslope transports sediment along the length of the profile. By contrast, at the shore the bulk of the sediment is transported parallel to shore at *right angles to the profile,* and much is even transported *upslope* along the profile. Rather minor amounts of sediment, much of it including fine particles, are transported in a direction parallel to the profile and toward the downslope end, the predominant direction of transport in a stream.

Longshore drift of sand creates ***spits,*** *elongate ridges of sand or gravel projecting from the mainland and ending in open water* (Pl. 14). Well-known examples are Sandy Hook at the entrance to New York Harbor; the northern tip of Cape Code at Provincetown, Massachusetts; and the Presque Isle

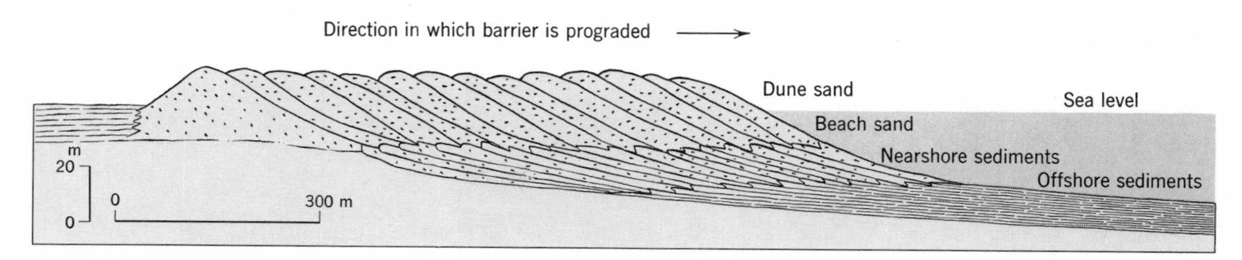

Fig. 14-14. Profile at right angle to coast, showing barrier that has been prograded. Offshore marine sediments are overlain successively by nearshore sediments, beach sand, and dune sand. Growth of a spit parallel to shore creates a similar sequence of strata.

peninsula in Lake Erie at Erie, Pennsylvania (Fig. 14-12). The free ends of all three curve landward in response to currents created by the waves.

A ***bay barrier****, a ridge of sand or gravel that completely blocks the mouth of a bay* (Fig. 14-2), is thought to be a spit greatly lengthened across the mouth of a bay. A bay barrier forms where longshore drift of sediment overpowers currents flowing into and out of the bay. If currents can scour sediment away faster than waves drift it laterally, they keep the bay mouth open.

Beach sediments. Beaches of all kinds consist of the coarsest sediment available locally (Fig. 14-13). Many are composed chiefly of sand-size particles of quartz, which occurs in most rocks as sand-size mineral grains and is the most durable of common rock-forming minerals.

Particles of beach sediment are moved back and forth incessantly, are turned over and over, and may become so well rounded that we are unable to distinguish them easily from rock particles worn by streams. Not only do beach sediments resemble stream sediments, but the cross-strata in the submerged parts of beaches are similar to those found in stream sediments, especially those in point bars. Cross-strata of a kind not made in streams originate on exposed parts of beaches where the surface slopes gently seaward at angles generally less than 10°. These cross-strata, formed only on the exposed parts of beaches, are distinguished by their gentle dip and large lateral extent.

ADVANCE OF THE LAND INTO THE SEA

In the nineteenth century, geologists realized that strata built along ancient coasts were of wide extent. At the shore, however, they saw sediments extending through great distances parallel with the coast but making only narrow belts at right angles to it. Thus they were puzzled. They could not figure out how sediment deposited in coastal environments confined to narrow belts could build widespread strata. They supposed, wrongly, that strata composed of uniform sediments had been deposited nearly simultaneously throughout. Now we are aware of an important process previously overlooked: land can advance seaward, or as we say, be *prograded.* When this happens all coastal environments shift outward, each leaving its sediments behind, just as the sediments of a channel and point bar are left behind when a stream meanders (Fig. 9-27). The strata originate, not all at once, but gradually. Each kind derives from a single environment and is widened gradually as the environments migrate. New parts overlap older parts, somewhat like shingles on a roof.

Progradation of a beach. Where sand is abundant, waves widen the beach by depositing layer upon layer of new sediment on its seaward side. A beach being prograded rather than eroded attracts little attention, but it sheds far more light on the origin of strata than do beaches from which sand is being removed by waves. Only a few prograding beaches are known in the modern world; some occur along the west coast of Mexico. But judging by the extensive strata deposited on some parts of the continents, such beaches must have been much more common in the geologic past. Examples of strata deposited by prograding beaches are numerous among the Cretaceous rocks of the Rocky Mountain region.

A section through a prograded beach (Fig. 14-14) shows how migration of narrow zones of sediment

Fig. 14-15. Small waves dashing against rocks throw spray high into the air. Coast of Ghana, Africa. (Anthony Howarth, from Pix, Inc.)

can build widespread strata. In some localities the bottom, seaward of wave base, consists of mud and in a zone extending from wave base to the line of breakers, of fine sand. The sand is covered completely with active ripples, which, when moved by the currents, deposit small-scale cross-laminae. The seaward boundary of sand moves outward causing cross-laminated sand to cover mud. Near the line of breakers lie various bars. When shifted by waves and currents, they leave large-scale cross-strata. We have already described how the orientations of these cross-strata vary.

Progradation of the shore results in the deposition of a vertical succession of strata possibly as much as 30m thick (vertical distance from everyday wave base to top of the dunes capping a barrier). Fine-grained, offshore marine sediments at the bottom are overlain by nearshore sediments. Next come sediments from submerged and exposed parts of a beach and these are topped by dune sand. Lateral migration of a spit creates a sequence of strata nearly identical to that shown in Fig. 14-14. In our example, sediments with cross-laminae made by ripples lie near the bottom of the sequence. In contrast, rippled sediments occur at the top of a sequence deposited by a migrating point bar.

Progradation of a delta. A large proportion of the terrigenous sediment preserved in the geologic record was delivered to the sea by rivers and much of it accumulated near the shore. Even the sand with which a beach is prograded probably came from a river, yet it is deposited by waves at an unknown distance from its point of entry into the sea. If the river brings in sediment faster than the waves carry it away, a delta forms. Most modern marine deltas consist of fine-grained sediment. Where the shore has been prograded by such sediment, we can be reasonably sure a delta was nearby. When a delta grows forward it leaves behind a vertical sequence consisting of marine strata at the base and nonmarine strata at the top. In between are sediments of varied grain size, among which silt and clay predominate.

EROSION OF BEDROCK BY WAVES

Energy of waves. The energy that waves transmit is both kinetic and potential. Kinetic energy, transferred from moving waves to the water particles, causes them to drag the bottom and set sediment particles in motion. The sediment-laden currents within the surf zone perform most of the geologic work on coasts. Potential energy, represented by the heights of wave crests above the still-water surface, comes into play where waves strike rigid bedrock and man-made structures.

Hydraulic action. When a wave breaks against rocks, the water exerts great force (Fig. 14-15). On the west coast of Scotland, pressures measured by dynamometers averaged as much as 1 ton per square foot and in storms reached 3 tons per square foot. During a great storm at Wick, in northern Scotland, a solid mass of stone, iron, and concrete weighing 1,350 tons was ripped from the end of a breakwater and moved inshore. The damage was repaired with a block weighing 2,600 tons, but 5 years later storm waves ripped it loose and swept it away. Such hydraulic pressures liberate and transport blocks of bedrock much as cavitation in a stream erodes bedrock by hydraulic plucking (Fig. 9-32). Hydraulic pressure also acts indirectly by suddenly compressing the air in fissures and pushing out large blocks of rock. Finally, hydraulic action can be concentrated at one point. A rock fragment or even a large piece of driftwood can be flung violently against a cliff, like an artillery projectile, and actually shatter firm rock.

The vertical distance through which waves can splash water upward against the shore is astonishing. In 1954, near Wick, Scotland, storm waves dashed the bow half of a small steamship against a cliff and left it wedged in a big crevice, 148 feet above sea level.

Abrasion. By creating abrasive sediment-laden water masses and driving them to and fro, waves grind away rock surfaces. Such grinding is vigorous only where the waves create vigorous water motion, namely, within the surf zone; on exposed coasts this occurs at depths of less than about 10m. With its rock tools the surf continuously drills and saws laterally into the land, smoothing and slightly deepening the nearshore bottom. While the land is worn away the tools themselves are smoothed, rounded, and made smaller. Because weak places in coastal bedrock retreat more rapidly than stronger parts, cliffs become scalloped (Pl. 14, *lower right*) and even tunneled. Seaward beyond the surf zone, where waves move only fine particles, wave erosion of bedrock is negligible but considerable erosion of other kinds takes place by submarine mass-wasting (Chap. 15).

The shore profile. Seen in profile, the results of wave attack on a coast are a ***wave-cut cliff,*** *a coastal cliff whose base has been undermined by waves and other marine agencies,* and a ***wave-cut bench,*** *a bench or platform cut across bedrock by waves* (Fig. 14-16). Both cliff and bench are the work of erosion; debris derived from the cliff accumulates as a beach and a ***wave-built terrace,*** *a body of wave-washed sediment lying seaward of a wave-cut bench.*

If waves remove the supporting lower parts of shore cliffs composed of weak materials, (Fig. 8-6), the upper parts slump. As any cliff retreats, the wave-cut bench correspondingly widens landward (Fig. 14-17). Wave erosion has completely destroyed some islands (Fig. 19-22) but the idea, once held, that wave erosion at constant sea level can destroy continents has been abandoned. If sea level does not change, how wide can a wave-cut bench become? Two relevant factors are erodibility of coastal materials and range of tide. As benches widen, the waves break progressively farther from the cliff. Therefore wave-cut benches themselves limit their own width; on the California coast maximum possible bench width is estimated to be about 0.5km. If sea level rises, however, much wider benches can form.

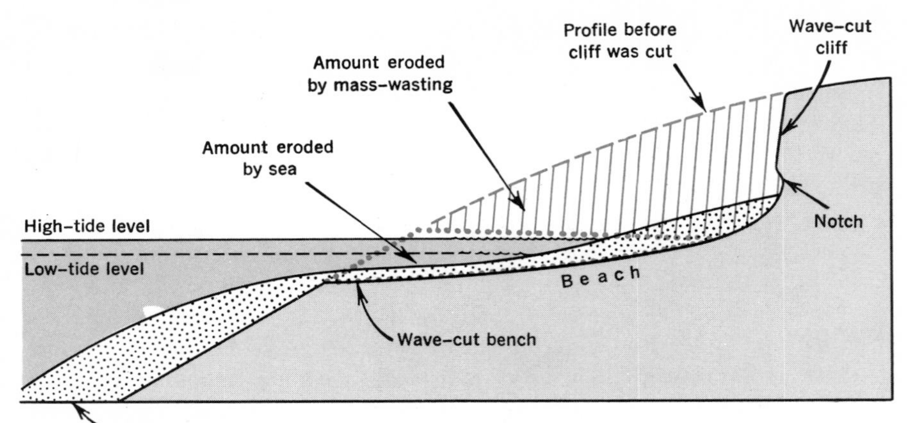

A

B

Fig. 14-16. Erosion by waves and mass-wasting along a coast consisting of bedrock. A. Principal features of the shore profile at right angle to coast, shown in schematic sketch with vertical dimension exaggerated. Terms defined in text. Compare amount of material eroded by the sea with amount removed by mass-wasting. (After Emery, 1960.) *B.* Minor erosional features along a cliffed bedrock shore, seen at low tide. *Sea cave* is drilled out by waves in erodible parts of bedrock. Enlargement of sea cave creates a *sea arch.* Isolated *stack* is left standing after waves erode surrounding rocks to level of wave-cut bench.

Fig. 14-17. Small island surrounded by wave-cut bench eroded across inclined strata and swept clean of sediment by waves. View at low tide, Kyushu, Japan. (U. S. Navy, courtesy J. T. McGill.)

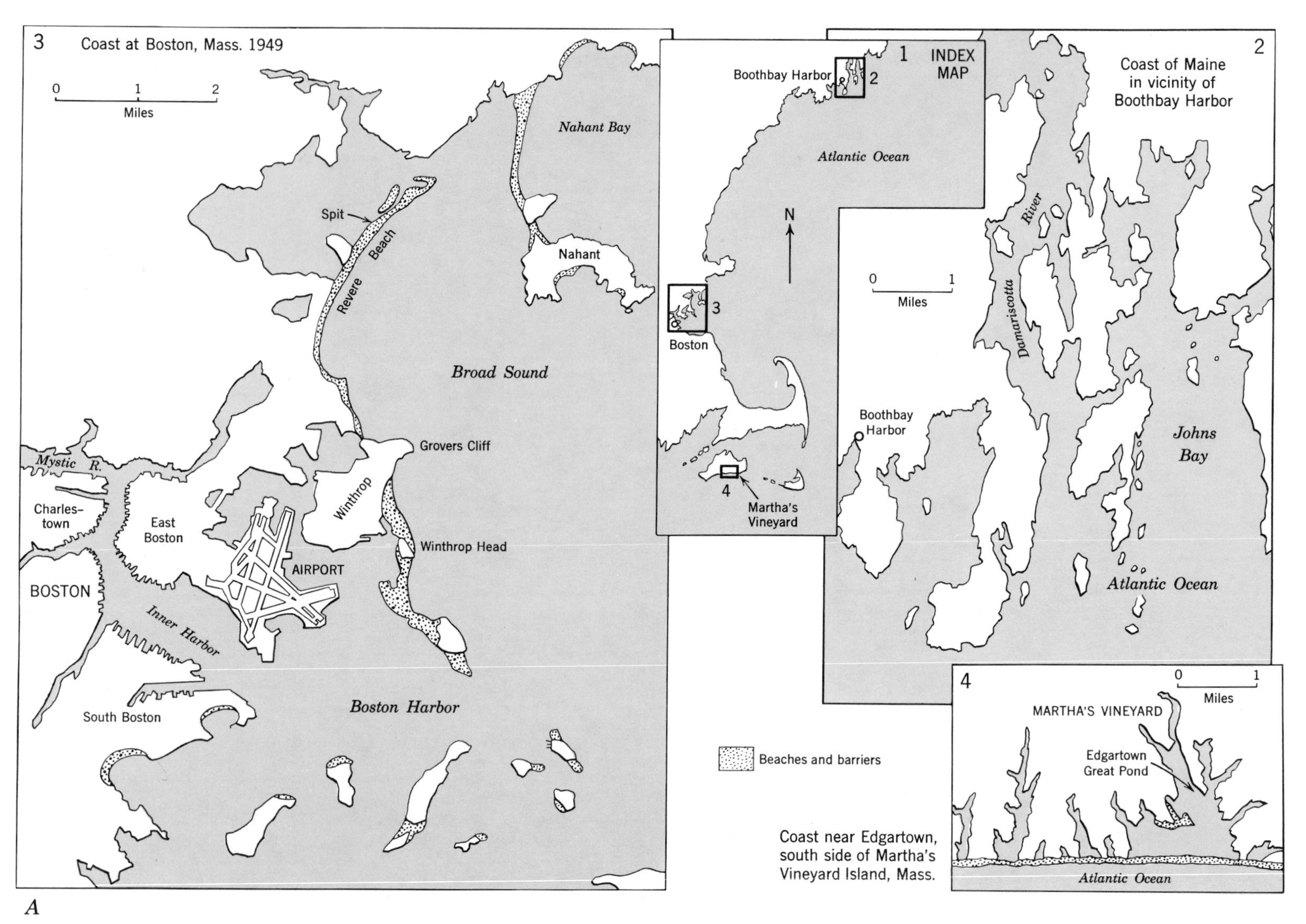

Fig. 14-18. Sculptural alteration of coasts by waves and currents. A. Three areas in New England having shore materials of unequal resistance to waves. 1. Index map showing locations of areas shown in maps 2, 3, and 4. Explanation in text. [Shore features in (3) after Laurence LaForge, 1932; in (4) after J. B. Woodworth and Frank Wigglesworth, 1934.]

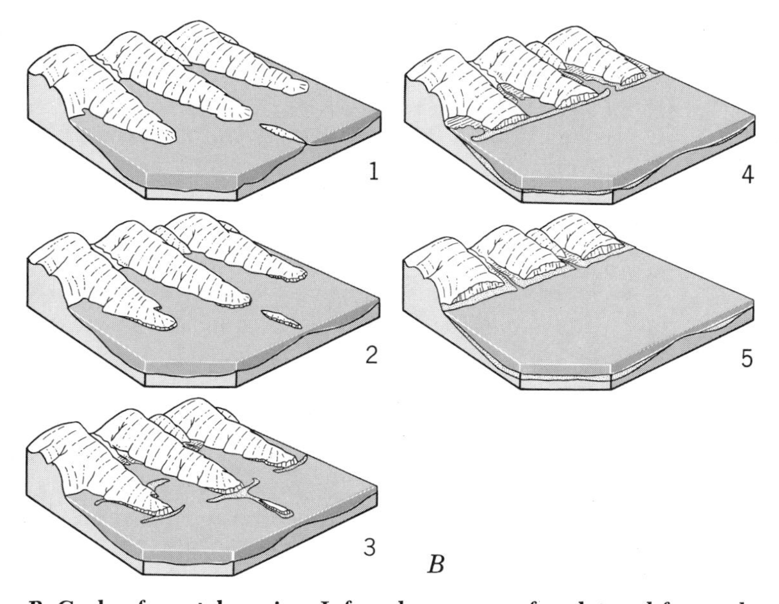

B. Cycle of coastal erosion. Inferred sequence of sculptured forms, developed through time on an embayed coast underlain by homogeneous rocks. 1. Initial condition marked by deep bays, headlands, and islands. 2. Youthful stage; headlands cliffed by surf. 3. Mature stage; various beaches have formed. 4. Old age; truncated headlands are connected by bay barriers to form a much-straightened shoreline. The shortened bays are being filled with alluvium and, where still open to the sea, with tidal sediment. 5. Headlands and bays have been eliminated; shoreline has been simplified to a nearly straight line. Distribution of wave energy along the coast has become as nearly uniform as possible.

EVOLUTION OF COASTS

In their work of smoothing coasts, waves never relent, but their rates of progress vary according to (1) wave size, (2) coastal morphology, and (3) the resistance of coastal materials. We have selected three short segments of the Atlantic coast of the United States (Fig. 14-18, *A*) to illustrate how nonmarine landscapes, composed of various geologic materials, have interacted with varying waves during a short time interval, about 4,000 years. The resistant metamorphic rocks thinly veneered with patches of till at Boothbay Harbor, Maine receive the full force of Atlantic waves. The thick till near Boston, Massachusetts is somewhat protected from large Atlantic waves. The sandy outwash and till along the south shore of Martha's Vineyard, Massachusetts are exposed to the full-force of the Atlantic waves.

At Boothbay Harbor (Fig. 14-18, *A*, 2), rolling bedrock hills with relief of 200 to 300 feet have become islands and former stream valleys (Damariscotta "River," Johns Bay, and the other bays) have become bays. The irregular, indented shoreline affirms that wave work has been negligible. The only beaches (too small to be shown on the map) occur in a few minor coves. Wave-cut cliffs and benches are not present. In 4,000 years the mighty waves have scarcely scratched the rockbound coast of Maine.

At Boston (Fig. 14-18, *A*, 3), an indented coast has resulted from the drowning of a landscape consisting largely of tough, compact, bouldery till with low hills, including many drumlins, having relief of less than 100 feet. As in Maine, hills have become islands and former stream valleys have become bays. Although the irregular coast reflects the nonmarine ancestry of the landscape, the conspicuous smooth segments disclose that wave work has progressed. Seaward shores of drumlin-islands have been cliffed and cut back, leaving beaches consisting of washed boulder residues similar to the one shown in Fig. 14-13. The spit at Revere Beach

Fig. 14-19. Seawall at Winthrop, Massachusetts protects houses from surf associated with storm winds of March 6, 1962. (A. P. Wirephoto, from Wide World Photos.)

has grown so long that it has nearly become a bay barrier. Large tombolos have been built at Nahant and Winthrop Head. Many former islands have been destroyed; their monuments are gravel shoals (not visible on the map). The smaller waves have transformed the erodible coastal materials at Boston to a much greater extent than have the larger waves working on resistant bedrock at Boothbay Harbor.

At Edgartown, on Martha's Vineyard (Fig. 14-18, A, 4), loose sand and gravel underlies an outwash plain standing only 10 to 20 feet above sea level. The irregular shorelines of Edgartown Great Pond and its neighbors reflect the drowning of networks of small, shallow stream valleys. But the smooth ocean shoreline is an unmistakable effect of wave work. Evidently the interstream areas, originally headlands, have been cliffed and cut back through a considerable distance. Figures are available for comparing retreat rates of cliffs eroded in till near Boston and on Martha's Vineyard. During a 48-year period between two surveys, the cliff at Winthrop Head retreated at an average rate of 9 inches per year. During two shorter periods, average rates of retreat at Grovers Cliff were 9 and 12 inches per year, respectively. By contrast, during the 40-year interval between two surveys, the high cliffs west of Edgartown (not shown on the map) retreated at an average rate of 66 inches per year. On Martha's Vineyard the undiminished Atlantic waves have eroded till 5 to 7 times more rapidly than have the smaller waves near Boston. Moreover, along the south shore of Martha's Vineyard, waves have completely transformed a formerly embayed coast into a straight coast; they have deposited bay barriers to create a continuous beach.

By such comparative studies of the effects of waves at many localities where materials of varying resistance are being eroded, geologists have inferred that in the long run a coast at one locality, whatever its materials, undergoes sculptural evolution. This inferred evolution of coasts by waves has been compared to that of a land mass as it is being eroded by mass-wasting, overland flow, and streams. As in cycles of erosion on land, the stages described as youth, maturity, and old age represent continuous transformation of the coast (Fig. 14-18, *B*) and are not clearly marked off from one another.

PROTECTION OF COASTS

Even small waves continually erode noticeable amounts of sediment from beaches and from coastal cliffs consisting of sediments. During storms, erosion increases sharply. On the south shore of Cape Cod the fair-weather rate of retreat of coastal cliffs is not more than 1 foot per year, but a 2-day hurricane in September, 1944, cut back the cliffs in places by as much as 50 feet. When shore property is sufficiently valuable to justify the expense, erosion of a cliff can be stopped, temporarily at least, by building a *seawall* of wood pilings, stone, or concrete. A well-built seawall, with foundations deep enough to prevent undermining by waves, gives complete protection against ordinary storms (Fig. 14-19) and some protection against exceptional storms.

Protection of beaches involves problems of supply as well as erosion of sediment. The stability of the exposed parts of beaches depends on availability of sand in the local longshore-drift system. As long as sand is plentiful the main work of the waves consists of moving it along the shore, and, to a lesser extent, onshore or offshore. If, however, for any reason sand is depleted from the updrift end of the system, serious erosion results all along the shore. Waves continue to feed sand into the downdrift end of the system, but now they take it from the exposed parts of beaches.

Sand can be dredged from offshore to replenish losses, but its proper placement requires full knowledge of the local conditions of longshore drift. Ideally, sand is stockpiled at the updrift end of each drift system. The waves then distribute the sand all along their route, supplying it continuously to exposed parts of the beaches. When the stockpile has been exhausted, further losses occur and dredging must be repeated.

Before longshore drift was understood fully, the common way to combat beach erosion was to build *groins*, low walls on beaches crossing the shoreline at right angles. Sand accumulates on the updrift sides of groins but generally is removed from their downdrift sides (Fig. 14-12). To be effective, therefore, groins must be spaced at short intervals. This requires many groins and skyrockets the expense of protecting the shore.

EFFECTS OF ORGANISMS AND CLIMATE

So far, we have examined features made by waves in sand composed largely of quartz. Climate entered the picture only indirectly; it affected the growth of saltwater grasses and the kinds of animals that burrow into the tidal flats. The effects of climate on coastal features become conspicuous in the tropics, as in southern Florida, the Bahamas, the Caribbean, and the South Pacific. Tidal marshes of temperate coasts and cold coasts give way to mangrove swamps of subtropical and tropical coasts. Branches of the bushlike aquatic mangrove plants that reach down into the water sprout new roots. Continuous growing and sprouting create intricate networks of stems and roots that trap fine sediment; the decaying plant tissue forms mangrove peat.

Reefs. Instead of barrier islands on tropical coasts we commonly find ***reefs,*** *massive wave-resistant structures built by the secretions of organisms.* Typical tropical reef-building organisms include mostly corals and algae that are unable to survive temperatures lower than 18.5°C. The algae living with the colonial corals require light for photosynthesis. This requirement confines them to depths of less than 50m because a thicker layer of water filters out most of the Sun's rays. Although the most vigorous growth of reef-building organisms takes place in water shallower than wave base and the top of the reef itself grows up close to sea level, not all massive organic structures built by corals and other colonial organisms originate in shallow tropical seas. Species of corals not dependent on algae can build massive structures in quiet, deep, cold water in the open ocean.

Carbonate sediments and evaporites. Along many tropical coasts, waves and currents deposit carbonate sediments consisting of particles derived entirely from the sea. Examples include bits of broken corals and shells (Fig. 4-8, *D*) and particles that have accreted by concentric addition of material chemically precipitated (Fig. C-4). These carbonate sediments become limestones. Therefore we can look for nearshore features in ancient limestones and interpret carbonate rocks just as we have interpreted nearshore sandstones consisting of quartz and other silicate minerals.

Within bodies of sediment whose tops lie close above sea level (Fig. 5-3) evaporite minerals are precipitated. Some ancient dolostones formed in this way; they indicate the positions of ancient sea levels.

CHANGES OF SEA LEVEL

Measurement of sea level. Our ability to fix the vertical position of sea level depends on mapping the shapes of islands and continents. Therefore our result is known only *relative to the land.* Unfortunately we are unable to fix an absolute sea level in terms of distance from the center of the Earth.

Factors that control sea level. The level of the sea against the land is determined by three major factors.

(1) *Volume of ocean water.* Variable storage of water on land during the hydrologic cycle creates short-term variations in the volume of water in the sea. The gradual increase of total water by additions from the interior of the Earth creates a long-term variation. Sea level in all oceans responds immediately to all changes in volume of water.

(2) *Volume of ocean basins.* Uplift of the sea floor and accumulation of sediment decreases the volume of an ocean basin. Subsidence of the sea floor increases the volume. Such changes in capacity simultaneously affect sea level in all oceans.

(3) *Subsidence or uplift of the land along a coast.* Local movements of the crust influence the position of the sea against the land but do not change sea level in all oceans.

Because (3) operates independently of (1) and (2), sea level can be rising in all oceans, yet falling relative to a coast being uplifted more rapidly than the rate of rise of the water. Or sea level can be

Fig. 14-20. Emerged and abandoned wave-cut cliff, wave-cut bench, and beach, Portuguese Point, San Pedro, California. Emergence of 150 feet resulted from uplift of this part of the coast since the shoreline now abandoned was active (compare Fig. 17-4). (Spence Air Photos.)

falling in all oceans, yet rising relative to a rapidly subsiding coast. *Sea-level changes that affect all oceans* are ***eustatic changes of sea level.***

Evidence of sea-level changes. The distinctive coastal features described in the previous pages of this chapter furnish evidence by which we can distinguish ***emergence,*** *fall of sea level relative to the land,* from ***submergence,*** *rise of sea level relative to the land.* We can do this even though we do not know whether the world's oceans are rising or falling.

A coast emerges either because the land is elevated or because world sea level drops. Along parts of the California coast emerged former beaches and wave-cut cliffs (Fig. 14-20) have resulted from localized crustal movements. Along lengthy segments of the Atlantic coast four former shorelines, emergent by amounts ranging from 5 to 40 feet, are examples resulting from lowering of world sea level.

A coast becomes submerged because the land subsides or because world sea level rises. Submergence, shown by the drowning of features such as forests that obviously form only on land or of distinctive features that form only near shore, characterizes most coasts because a world-wide rise of sea level followed the last major glaciation. Along a few exceptional coasts, the rate of uplift of the land has exceeded the rate of rise of sea level. Shoreline features provide useful reference levels for measuring crustal movements, as we shall see in Chapters 17 and 21.

During submergence, environments of deposition migrate landward. When this happens sediments of deeper-water belts cover areas of the bottom where only shallow-water sediments accumulated formerly; simultaneously shallow-water sediments spread over the former land surface (Fig. 14-21). We infer by analogy that many ancient strata were deposited during submergence.

CONTINENTAL SHELVES

The lands of the world are surrounded by ***continental shelves,*** *shallow platforms of variable width extending from the shoreline to the first prominent break in slope at a depth of 600m or less.* The shelves have interested geologists for more than a century, but many geologic ideas formulated from the land have been abandoned in the light of modern research. According to one older idea, the shelves represent wave-cut benches. We now know that only small parts of the shelves originated through wave erosion; large parts consist of depositional surfaces.

Some shelves, such as the one off California, are

underlain by deformed strata similar to those exposed on adjacent land. The deformed strata are thinly veneered by undeformed younger sediments. Other shelves, seaward of wide coastal plains, are underlain by wedges of gently inclined sedimentary layers. Borings and geophysical data disclose that the wedge of sedimentary strata underlying coastal North Carolina overlies a sloping floor of ancient metamorphic rocks (Fig. 14-22). The seaward side of the wedge, near the continental slope, is about 3.5km thick. The strata range in age from Lower Cretaceous or possibly Jurassic at the base to Recent at the top, and are in various stages of transformation into rock. The sediments, which were mostly deposited on shallow sea floors but which lie as much as 3.5km below sea level, clearly record subsidence.

Sediments on the continental shelves. In large areas the shelf sediments consist of sands and coarser materials. Coarse sediments are not confined to the vicinity of the modern coast, as many geologists formerly supposed, but occur far offshore as well. Many sediments now covering large parts of the continental shelves were deposited by streams, by glaciers, and by the wind during the last glacial age, when the sea stood lower and the present shelves were mostly land. Other sediments on the shelves were deposited on beaches, between barriers and the mainland, on deltas, and near reefs, at times when the sea halted temporarily during one of its many rises and falls. The rising sea has submerged and reworked all these sediments. Waves have removed at least some of their finer particles and marine organisms have moved in.

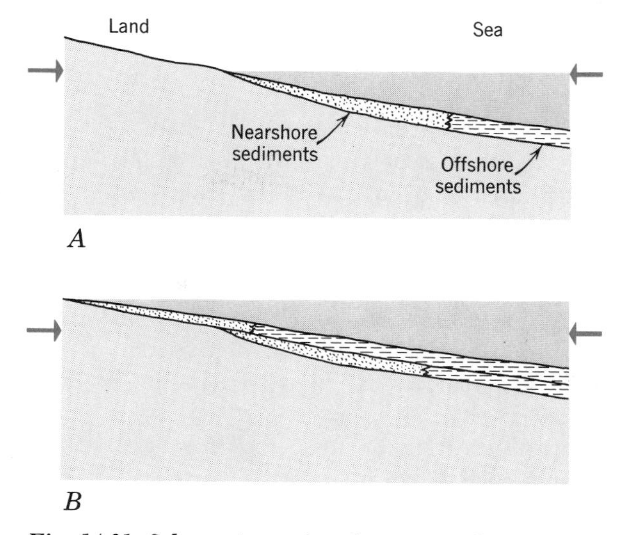

Fig. 14-21. Schematic section drawn at right angles to a coast, showing formation of belts of coastal sediment and their shift during submergence. *A*. Sea remains at one level and deposits sediment nearshore and offshore in belts parallel with coast. *B*. Sea rises to new level and remains there, depositing sediment as in *A*. New sediment overlaps the older.

Fine-grained sediments are abundant only near major deltas and in closed depressions. Nearly everywhere else, waves and currents keep fine sediment in motion and prevent it from accumulating on open shelves. Fine sediment is deposited farther seaward, in deeper water. Let us follow it outward and downward and find out what lies on the deep-sea floor.

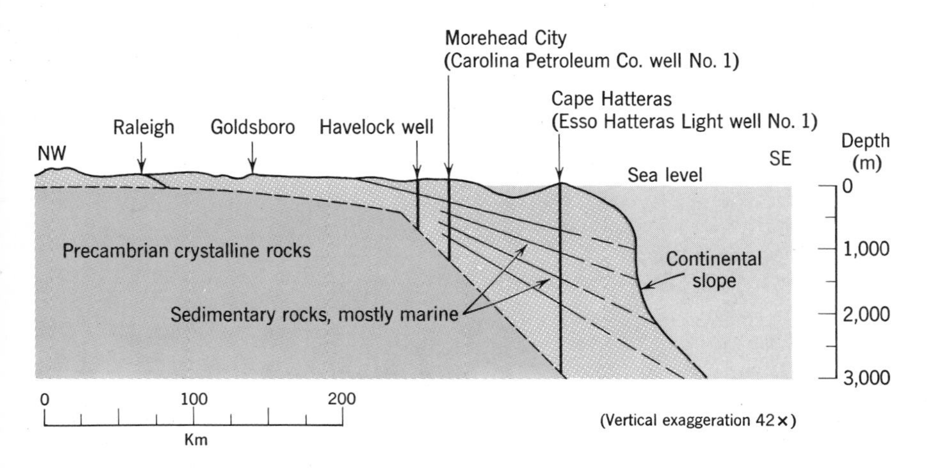

Fig. 14-22. Profile and section through coastal North Carolina and continental shelf, showing underlying strata. (After W. F. Prouty, 1946.)

Summary

Waves

1. Waves are the dominant geologic agents on most coasts. In deep water, waves of oscillation set water particles in motion within circular orbits lying in vertical planes parallel to wave travel. Downward, these orbits become smaller so that water particles exert no drag on the bottom.

2. Shoreward of the point where depth equals half the wavelength, water particles set in motion by waves create an oscillatory drag on the bottom. This drag can move sediment.

3. Waves transport coarse particles toward the shore along the bottom, lift fine particles into suspension, and move much sediment of all sizes parallel with the shore, both on and above the bottom.

4. At a depth of 1.28 times their height, waves become unstable and break; they are transformed into surf and the surf in turn into swash and backwash. Water surging toward shore moves sediment and is capable of eroding bedrock at depths of less than 10m.

Coastal sediments

5. Fine particles lifted into suspension by waves are transported into bays and lagoons by tidal currents and into deeper water offshore by other currents. Continued deposition of fine particles in bays and lagoons builds up tidal mudflats along their shores.

6. Beaches include both submerged and exposed parts; their exposed parts can occur along mainland shores or as spits, bay barriers, barrier islands, and tombolos.

7. When sediment is supplied faster than waves can remove it, the shore is prograded, either as a beach or as a delta. By migrating seaward coastal environments, narrow in a direction at right angles to the shore, can deposit strata of wide extent.

8. Coastal environments that have migrated seaward leave behind a characteristic succession of strata having nearshore marine sediments at the base and nonmarine sediments at the top. Much of the geologic record was built in this way.

Sculpture of bedrock coasts

9. On bedrock coasts waves erode a distinctive profile, consisting of steep wave-cut cliff and gently inclined wave-cut bench. Material swept off the bench builds a terrace in deeper water.

Sea level

10. The position of the sea surface against the land is determined by the relative effects of local crustal movements and by the vertical rise and fall of the water surface.

Continental shelves

11. Continental shelves are zones that border land areas; over them the depth of water is usually less than 600m.

12. The widest shelves are underlain by wedges of little-deformed sediments that accumulated in shallow water as continental margins subsided.

Selected References

Bascom, Willard, 1964, Waves and beaches. The dynamics of the ocean surface: Garden City, N. Y., Anchor Books, Doubleday. (Paperback.)

Darwin, G. H., 1898, The tides; repr., 1962; San Francisco and London, W. H. Freeman. (Paperback.)

Emery, K. O., 1960, The sea off southern California, A modern habitat of petroleum: New York, John Wiley (especially p. 5–37; 116–137; 180–214).

Hill, M. N., ed., 1962-1963, The sea. Ideas and observations on progress in the study of the seas: New York, Interscience; v. 1, 1962, Physical oceanography, Chaps. 15, 16, 17, 18, 19, 23; v. 3, 1963, The Earth beneath the sea. History, Chaps. 21, 22, 24.

Ingle, J. C., 1966, The movement of beach sand: Amsterdam, Elsevier Pub.

Shepard, F. P., 1963, Submarine geology, 2nd ed.: New York, Harper and Row, Chaps. 4, 6, 7, 8, 9, 12.

Shepard, F. P., Phleger, F. B., and van Andel, T. H., eds., 1960, Recent sediments, northwest Gulf of Mexico: Tulsa, Okla., Am. Assoc. Petroleum Geologists.

Steers, J. A., 1946, The coastline of England and Wales: Cambridge Univ. Press.

CHAPTER 15

The Deep-Sea Floor

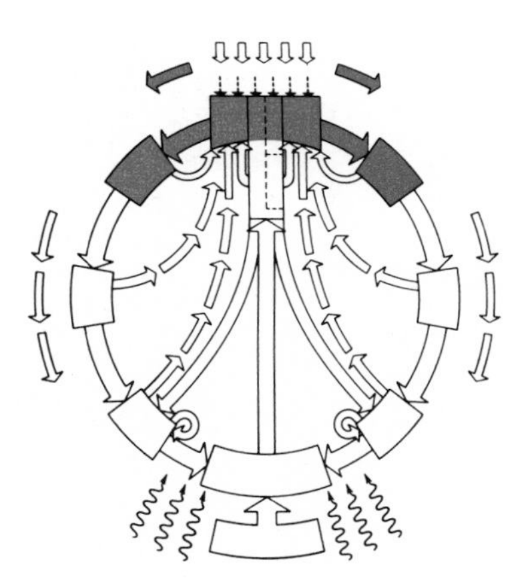

Geologic and practical significance
Soundings made with sound waves
Major features
Water movements at the surface and at depth
Sediments
Varied effects of gravity
Kinds of strata
Thickness and rates of accumulation
Origin of submarine canyons, abyssal fans, and abyssal plains
Are any former deep-sea sediments exposed on land?

Plate 15

The deep-diving research submarine *Alvin* prepares for a descent to the sea floor. (*Woods Hole Oceanographic Institution,* courtesy K. O. Emery.)

WOODS H
CEANOGRA
INST
NOGRAPHIC INSTITUTION
1930
ARCH SUBMAR
OF NAVAL RESE
U.S.A
ALVIN
ALVIN

CHAPTER 15 *The Deep-Sea Floor*

GEOLOGIC SIGNIFICANCE

In the ocean basins, separated from continental masses by the 1,000-m depth contour, typical depths range from 4 to 5km. These deep basins occupy about 60 per cent of the Earth's surface. The remainder consists of land (29 per cent) and shallow seas and continental shelves (11 per cent). How did the ocean basins originate? How old are they? These and other questions have aroused the curiosity of geologists since the nineteenth century when soundings disclosed the great depths.

In the nineteenth century, using only fragmentary data available, geologists formulated various hypotheses about these basins. One hypothesis, widely accepted until the early 1960's, stated that these great depressions were primordial features of the Earth's surface, which had changed little since the time of their origin, except for accumulation of sediment. A corollary of this hypothesis is that deep-sea sediments contain a record of strata extending continuously from the present to a very early time in the history of the Earth. Such a record would be extremely valuable, because the strata exposed on land are separated by numerous gaps.

Other hypotheses, recently formulated on the basis of a large set of data, assert that the ocean basins are neither primordial nor permanent. Rather they have changed with the passage of time. This would imply that the oldest deep-sea sediments were not deposited in very ancient times and that any evidence of changes in the ocean basins would be reflected in the strata. Obviously the sediments can help to evaluate validity of all the hypotheses, and hence great interest attends the oldest of these sediments and the record they preserve.

No matter what we learn from this study of the lower strata of deep-sea sediments, the upper layers present evidence of more recent geologic history, including Pleistocene changes of climate and the growth and decay of glaciers. Deep-sea sediments have been studied not only for their historical information but also to determine their environmental and geochemical significance. Were any ancient sedimentary strata now exposed on land deposited in the deep sea? To what extent has the geochemical cycle been influenced by accumulation of material on the deep-sea floor? These are important scientific questions. Apart from its scientific importance the deep-sea floor has assumed great practical significance in recent years.

PRACTICAL SIGNIFICANCE

The need for knowledge of the deep-sea floor first arose in the mid-nineteenth century, after the telegraph had been invented and a decision was made to lay a submarine telegraph cable from North America to Europe. Expeditions were sent to explore the proposed route. They did so by determining depths, collecting samples of water for chemical analysis, measuring water temperatures at various depths, and dredging sediment and organisms from the bottom. Today many cables cross the sea floor, and corporations and governments concerned with submarine cables have actively supported marine research.

The great value of knowledge of the oceans to naval operations was demonstrated beyond doubt during the Second World War. Subsequently all governments concerned with naval affairs have provided vast support for oceanic research.

Photographs and dredge hauls have shown that nodules containing pure manganese oxides are present through large areas of the floors of all oceans. This discovery has stimulated the development of techniques for direct observation and deep-sea mining.

The deep sea has been looked upon as a possible disposal site for radioactive wastes, a by-product of the atomic age in which we live. Attempts to assess the feasibility of such disposal have stimulated research on water circulation in deep parts of the oceans.

EXPLORATION

It is not easy for persons unacquainted with the sea to visualize the problems involved in exploring the deep-sea floor. Equipment must be lowered to great depths by high-speed winches and must operate reliably when exposed to the tremendous hydrostatic pressures that prevail at depths of 4km or more. Detailed studies are not possible without exact data on ship's position, which are difficult to obtain. Various methods of fixing positions through radio signals transmitted from land have been invented, but the most accurate ones do not extend out beyond 150km. Suitably equipped ships on the open sea can now determine their positions accurately through an important by-product of the space age. Artificial satellites in polar orbits transmit radio signals that can be processed by a shipboard computer to give a ship's position anywhere in the world.

Originally, depth of water was determined by lowering to the bottom a lead weight on a long rope. At the same time tallow smeared on the lead picked up a sample of the bottom sediment. Such soundings were slow, not always accurate, and, of course, they revealed nothing about the depth in the areas between lowerings. Our present understanding of the ocean floor comes from depth determinations made with sound waves.

Sound waves. Water waves on the surface of the sea cause water particles to move in circles lying in vertical planes. In contrast, sound waves make water particles oscillate in the direction of the wave (Fig. 15-1). Therefore sound waves are *longitudinal waves* having frequencies lying within the sensitivity of the human ear, or about 30 to 25,000 cycles per second. By convention physicists now express frequency in *hertz* (abbreviated Hz; 1 hertz = 1 cycle per second).

Sound waves are generated by pressure pulses in the water, as a tuning fork creates pressure pulses in air. The waves radiate outward from the source, following straight paths as long as the physical characteristics of the water remain unchanged. If the waves encounter a solid object, some of them are reflected back toward their source. The reflection time elapsed between the instant the wave goes out and its reflection returns indicates the distance of the object.

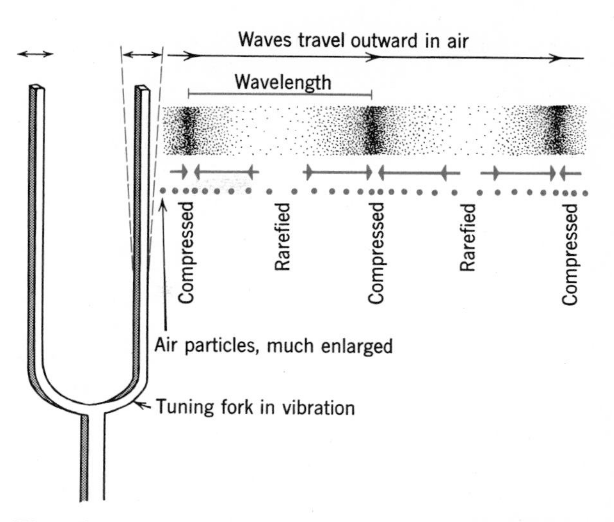

Fig. 15-1. A sound wave moving out through air from one arm of a tuning fork causes particles to oscillate in the direction of wave motion. The air is alternately compressed and rarefied ("stretched") as the sound waves pass. (After O. H. Blackwood, W. C. Kelly, and R. M. Bell, 1963, General Physics, by permission of John Wiley and Sons, Inc.)

Echo sounders. After the disastrous sinking of the liner *Titanic,* which struck an iceberg in the Atlantic in 1912, it was proposed to employ sound waves to scan the sea ahead of a ship. Because sound travels about 1,500m/sec in seawater, only 1 second would be required to locate an iceberg 750m distant. During this time the fastest ocean liner travels only about 15m. Hence, in theory, sound waves can locate icebergs long enough beforehand for the ship to avoid them. Because the properties of seawater are not uniform with depth, sound does not propagate in straight paths in directions parallel to the water surface. Instead sound follows curved paths that are more complex than the inventor of the sonic search method supposed. Nevertheless, in 1914 during sonic iceberg-searching experiments, the inventor aimed his sound-making device downward. He sent out brief pulses of sound and heard regular echoes returning after each. The echo-return times proved that the echoes were coming from the bottom. From this incidental beginning echo sounding was born. Subsequently, sound waves have been applied in many ways for exploring the ocean floor. Results based on such waves

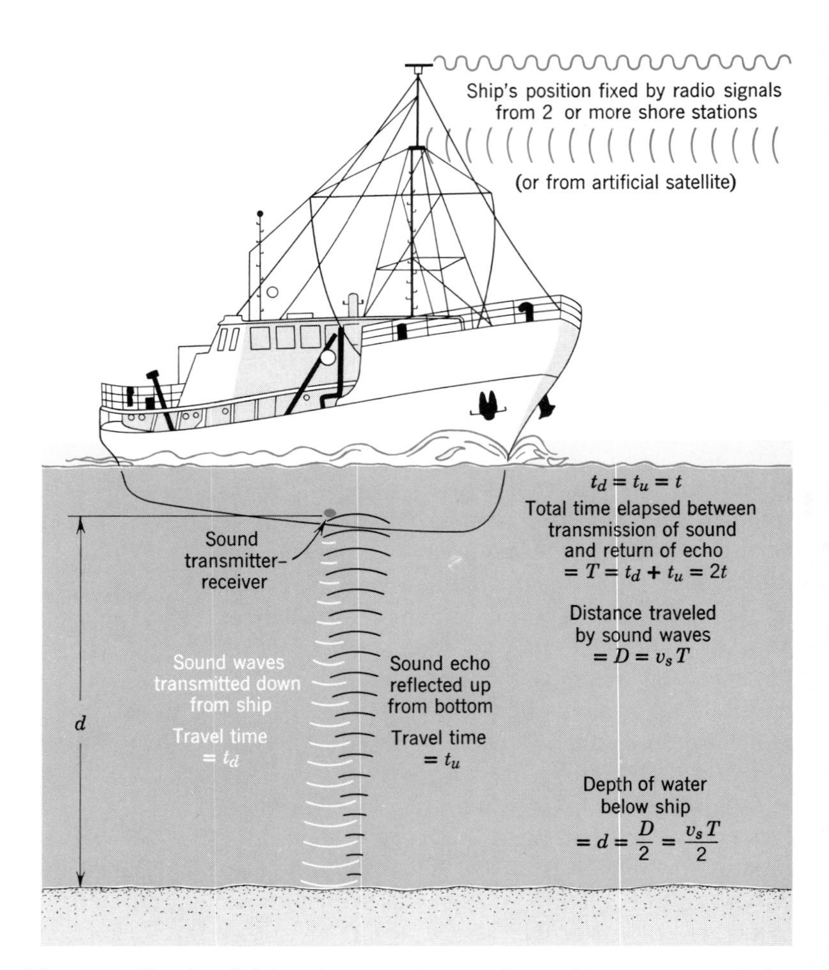

Fig. 15-2. Sketch of ship with an echo sounder making a survey of the bottom illustrates the echo-sounding technique for recording depth continuously.

have revolutionized not only knowledge of the sea floor but also many geologic concepts applicable on land.

Now sound determines depth by means of a special device with a narrow beam pointed at the bottom (Fig. 15-2). An instrument that employs sound energy for determining depth is an *echo sounder.* Various sonic depth sounders became available starting in the 1920's. An important advance came in the 1930's when a special recorder registered continuous profiles of the bottom. An even greater advance occurred in 1953, when the precision recording echo sounder (Fig. 15-3) was invented. With it deep-sea depths can be measured with an accuracy of 2m in 6,000. Modern maps of the deep-sea floor, like those in Plates A and C, are based on precision profiles.

Continuous seismic profilers. By means of sound waves we can determine not only depth of water but also thickness and internal structure of bottom sediments (Fig. 15-4). Many echo sounders generate waves having a frequency of 12 kilohertz (kHz). Such waves penetrate the water but are reflected by even the most water-saturated bottom sediments. By increasing the amount of energy released in the water and lowering the frequencies to the range of 75Hz to 3kHz we can make the waves probe into sediments beneath the bottom. In other words our sound has become ***seismic waves,*** which are *waves traveling within the Earth.* When seismic waves encounter interfaces within or beneath the sediments where physical properties change abruptly, the waves are reflected upward and can be recorded. A device for making profiles of the thickness and internal structure of sediments beneath a body of water is a *continuous seismic pro-*

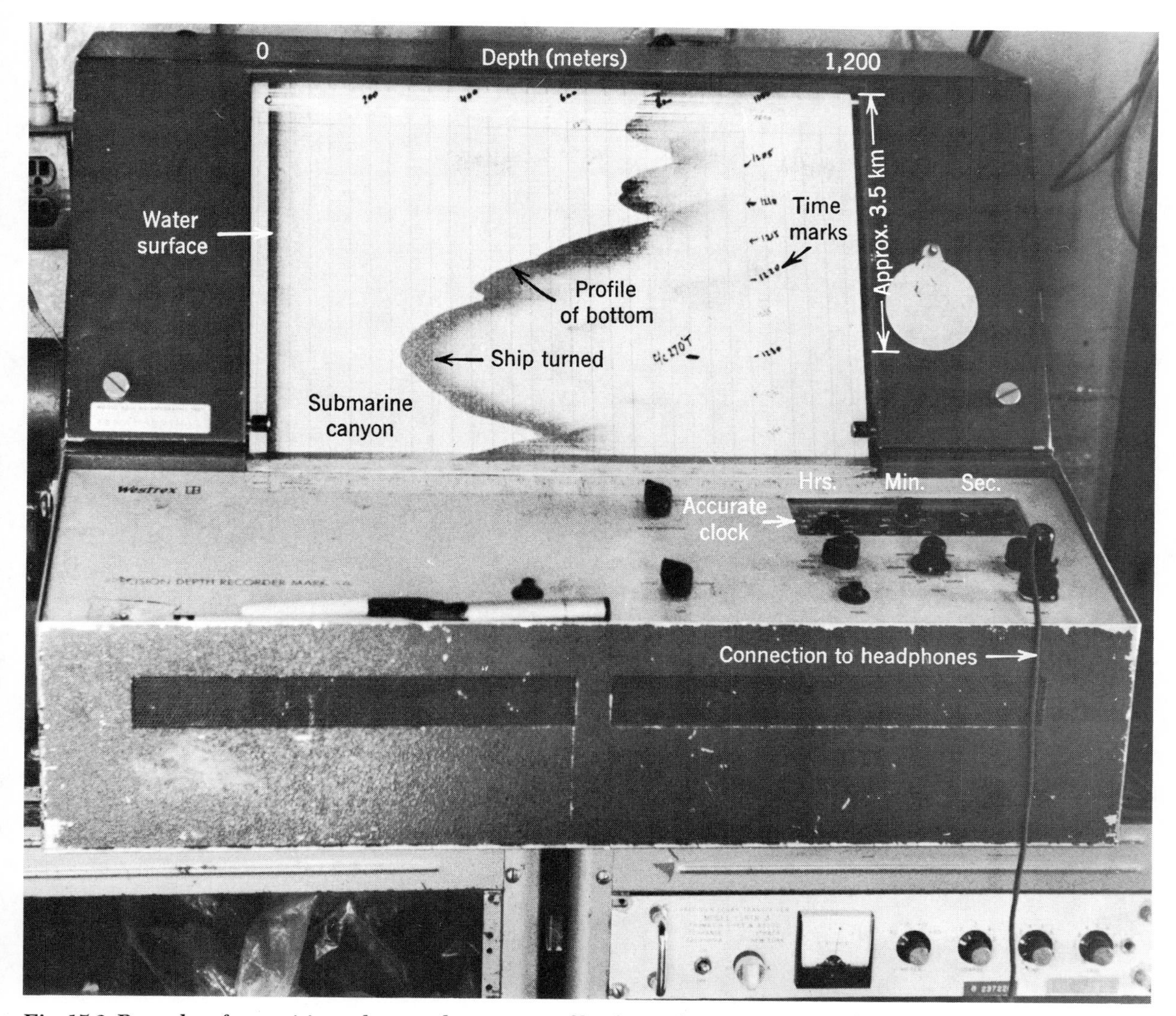

Fig. 15-3. Recorder of a precision echo-sounder traces profile of steep bottom. (**J. E. Sanders.**)

filer. Seismic waves permit us to measure the thickness of ice in glaciers, to search for potential oil-bearing structures, and to determine the structure of the Earth's interior.

Bottom-sediment corers. In the nineteenth century only grab samplers were available; these can collect material at or slightly below the water/sediment interface. Today sediment below this interface can be sampled by forcing in an open-ended tube and drawing up a cylindrical sample known as a *core.* Lowered on a long wire, the coring device (Fig. 15-5) is released near the bottom, so that it falls freely through the last 5m or so. This device can penetrate soft sediment to 20 or 30m. In firmly packed sands penetration is 1 to 3m.

All this is easily accomplished from standard research vessels. We can drill deeper holes, however, by adapting oil-field rotary drilling rigs to sea use on special ships and platforms (Fig. 23-12). This technique was employed to drill the preliminary test hole off Guadalupe Island in the Pacific in March and April, 1961 in the early stages of project MOHOLE, which subsequently has been deactivated. Shipborne rotary drilling equipment successfully penetrated about 170m of bottom sediment and about 1m of underlying basalt in water 3,558m deep.

Deep submersible vehicles. Much important information about the sea and the sea floor has come by direct visual observation. Since 1955 scientists

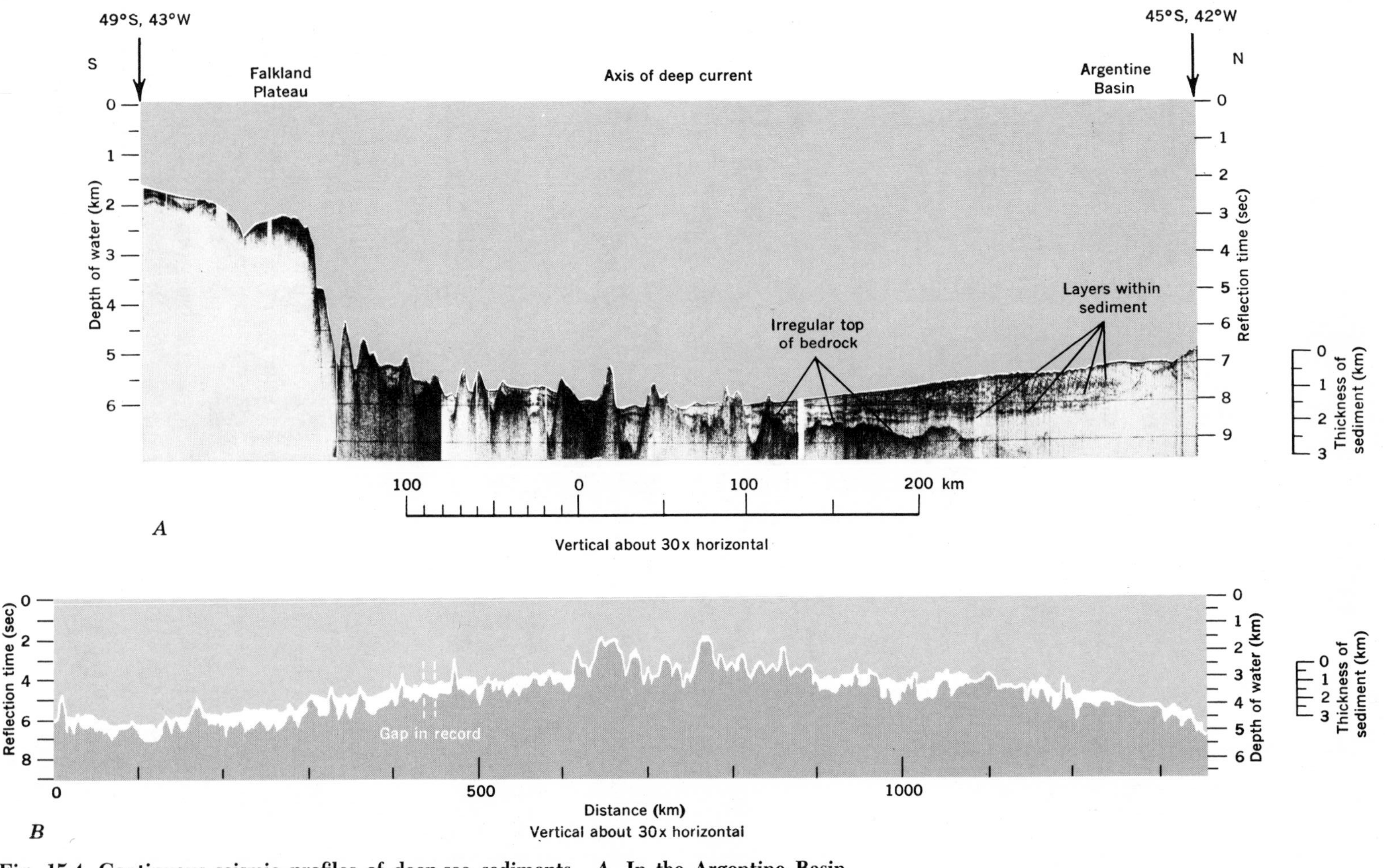

Fig. 15-4. Continuous seismic profiles of deep-sea sediments. *A.* In the Argentine Basin, deep current has heaped sediment into a thick pile at right. (Lamont Geol. Observatory, courtesy Xavier Le Pichon.) *B.* Transect diagonally across Mid-Atlantic Ridge discloses that no sediment (white) is present on central parts of the ridge. (After John Ewing and Maurice Ewing, 1967.)

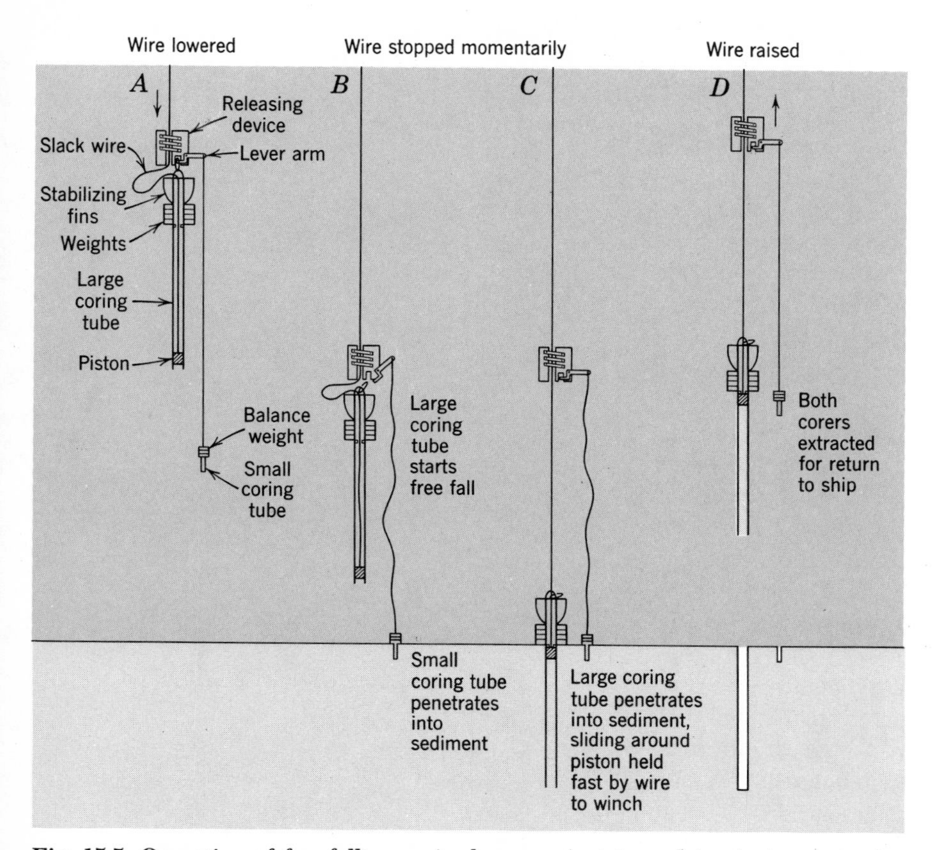

Fig. 15-5. Operation of free-fall corer in deep sea (not to scale). *A.* Apparatus is lowered toward sea floor. Balance weight, acting on lever arm about 2m long, more than counterbalances much heavier, large coring tube, suspended from hook on very short lever arm. *B.* When small coring tube touches bottom, signal is given to stop, then reverse the winch. When cable to balance weight goes slack, the heavy, large coring tube, no longer counterbalanced, rotates the lever arm, releases itself by sliding off the hook, and falls freely to bottom, guided by stabilizing fins. *C.* Large coring tube penetrates into the bottom. *D.* Corers en route back to ship.

have been able to descend to the sea floor in research submarines (Pl. 15). Such deep submersible vehicles will be used more and more in the future.

MAJOR FEATURES

Now that we have considered some of the exploration devices let us examine some of their findings on the deep-sea floor. In particular, we wish to study more closely the features along the margins and on the floors of the ocean basins, especially those that control sedimentation.

Features of ocean-basin margins. Along the irregular margins of the Atlantic and Indian Ocean basins, great bodies of sediment are the rule and deep trenches are the exception. The opposite is true along the margins of the Pacific basin. This arrangement results in part from the distribution of modern rivers, most of which drain into the Atlantic and Indian basins, and in part from geologic history. Recently bodies of sediment that formerly occupied marginal parts of the Pacific floor have been raised onto the continents.

The upper surface of a body of sediment at the margin of an ocean basin slopes seaward at various angles. The landward part is the ***continental slope,*** *the relatively steep (3° to 6°) slope that lies seaward of the continental shelf.* The seaward part is the ***continental rise,*** *the gentle slope with gradient between 1:100 and 1:700 that lies seaward of the continental slope.* The width of a typical continental slope is only a few kilometers; its maximum gradient is 1:40. The width of a typical continental rise is hundreds of kilometers (Fig. 15-6, *A*).

At the seaward side of some continental rises lies an ***outer ridge,*** *a broad, smooth ridge of sediment parallel to the margin of an ocean basin and standing 200 to 2,000m above the adjacent sea floor* (Fig. 15-6, *B*). Outer ridges are important because one of their sides slopes landward, forming an exception to the otherwise uniform seaward slopes on the upper surfaces of bodies of sediment at the margins of ocean basins.

Where bodies of sediment are not present the most common feature at the margin is a ***sea-floor trench,*** *an elongate, narrow, steep-sided depression, generally deeper than the adjacent sea floor by 2,000m or more, extending parallel to the margin of an ocean basin* (Fig. 15-6, *C*). In most trenches, water depths exceed 8,000m. The greatest known depth, 10,850 ±20m, is in the Challenger Deep, Mariana Trench, northwestern Pacific. The presence of trenches where sediment is scarce and their lack where sediment is abundant suggest intriguing questions. Are trenches present beneath the marginal bodies of sediment? At the margins of ocean basins, are trenches universal features that are visible only where they have not been filled with sediment? We must set these questions aside for the present in order to study first other geologic processes.

A common feature, which in many cases indents the outer slopes of all continental masses, is a ***submarine canyon,*** *a sinuous, V-shaped valley, having a variable number of tributaries, that may cross part or all of a continental shelf, and extends down a continental slope.* We shall discuss its origin presently. To complete our list of major features let us turn to those found on the floors of ocean basins.

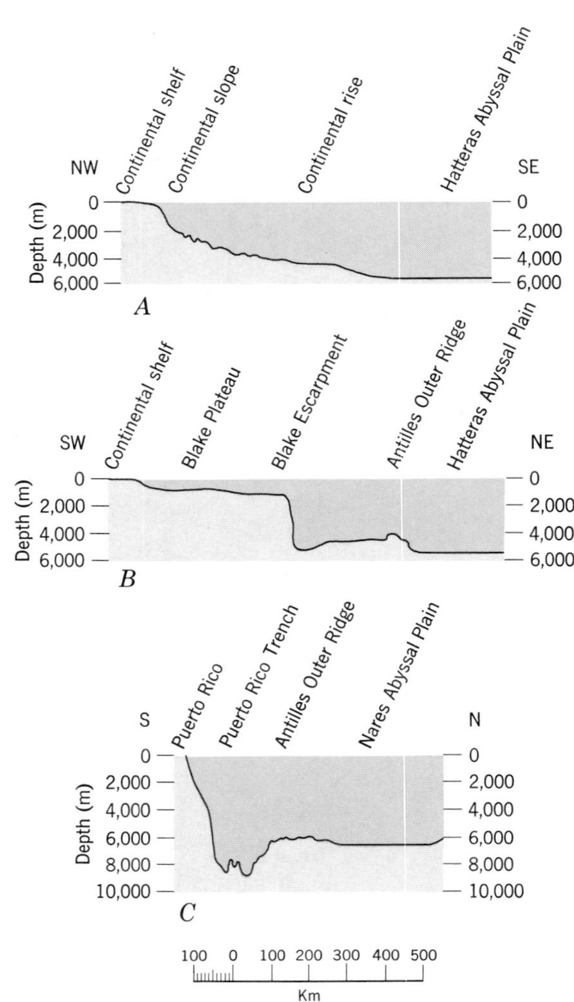

Fig. 15-6. Profiles along western margin of North Atlantic basin, showing three kinds of morphology at margins of ocean basins. *A.* Southeastward from Norfolk, Virginia. *B.* Northeastward from Cape Kennedy, Florida. *C.* Northward from Puerto Rico. (Heezen, Tharp, and Ewing, 1959.)

Features of ocean-basin floors. The floors of the ocean basins consist of low areas and higher standing ridges. Low areas are known collectively as the *abyssal floor.* The higher standing areas with rough topography are classified as distinct features apart from the abyssal floor.

The abyssal floor consists of smooth areas underlain by sediment and rough areas underlain by rock. A smooth area may be an ***abyssal plain,*** *a flat part of the abyssal floor, underlain by sediment, having an imperceptible slope of less than 1:1,000,* or an ***abyssal fan,*** *a fanlike accumulation of sediment at the mouth of a submarine canyon.* The rough parts of the abyssal floor are grouped collectively as ***abyssal-hills provinces,*** *parts of the floor consisting almost completely of irregular rocky hills a few hundred meters to a few kilometers wide and having relief of 50 to 100m.* Abyssal-hills provinces are widespread on the floor of the Pacific basin; their aggregate area exceeds that of any other relief feature on Earth.

The higher parts of the ocean basin floors consist of ***oceanic rises and ridges,*** *continuous rocky*

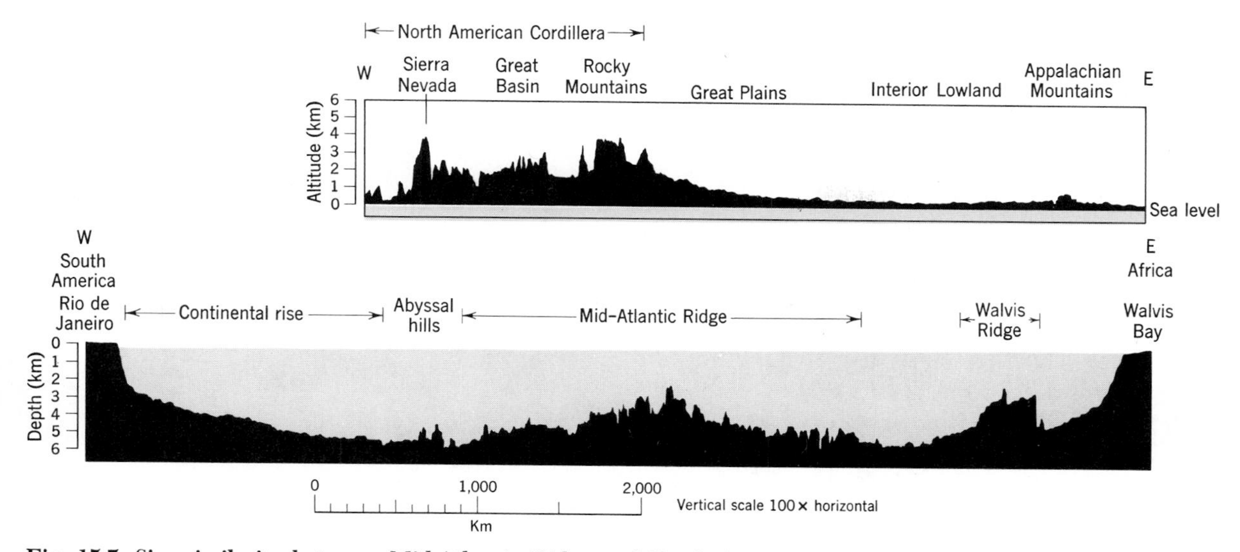

Fig. 15-7. Size similarity between Mid-Atlantic Ridge and North American Cordillera and High Plains, shown by profiles to same scale across South Atlantic near 23° 30′ S (*below*) and across conterminous United States (*above*). Slopes and widths of continental rise off South America and the Great Plains are about the same. (After Shepard, 1963.)

ridges on the ocean floor many hundreds to a few thousands of kilometers wide, whose relief is 600m or more (Fig. 15-7). Continuous seismic profiles have shown that the rough, rocky bottom characteristic of oceanic rises and ridges (on a large scale), and of abyssal-hills provinces (on a smaller scale), extends beneath the sediment underlying abyssal plains. We need only mention these rocky areas here as they are related to movements of seawater and sediment. We discuss their origin in later chapters.

Many conical edifices built by volcanoes occur on the sea floor (Fig. 19-22). Some project above sea level as islands.

MOVEMENTS OF SEAWATER

Seawater moves, not only as waves but also as complex currents. The entire body of ocean water is a dynamic system whose movements have many analogies to those of the atmosphere. Water moves along various paths, at the surface in response to the winds, and below the surface in response to gravity acting on internal differences of density and to the Coriolis effect, as we shall explain farther on.

Surface ocean currents. Prevailing winds blowing over the ocean surface drag the water forward slowly, creating *surface ocean currents* as broad as the currents of air but rarely deeper than 100m. Earlier we remarked on the similarity between the positions of surface ocean currents (Fig. 2-11) and the belts of prevailing winds (Fig. 2-10). In low latitudes, surface seawater moves westward with the trade winds; in higher latitudes, it moves eastward with the westerlies. In each ocean, surface water forms a giant eddy, flowing clockwise in the Northern Hemisphere and counterclockwise in the Southern. The eddy is caused partly by deflection where the currents encounter coasts and partly by the Coriolis effect.

In the Northern Hemisphere the patterns in the Atlantic and Pacific Oceans are similar. In the Atlantic the west-flowing North Equatorial current is deflected northwest by the coast of South America. The warm water enters the Gulf of Mexico, raising the water level there by 19cm. The water curves northeasterly around the Gulf and flows out northward through the Strait of Florida as the Florida current. The Florida current transports $40 \times 10^6 m^3/sec$ of water at average surface velocities ranging from 2 to 3m/sec (7 to 10km/hour). Farther north it is named the Gulf Stream and the North Atlantic current. Part of the North Atlantic current moves into the Arctic Ocean, taking heat with it, and part is deflected south along the European-African coast as the Canary current, which completes the clockwise circulation by flowing back to the Equator. The Canary current is cooler than the surrounding tropical water. In the North Pacific Ocean the corresponding sequence clockwise from the Equator is: North Equatorial current, Kuroshio current, North Pacific current, and, finally, cool California current.

Density currents and deep circulation. At the same time as the great wind-driven currents move slowly through shallow surface waters, the deeper waters circulate in response to gravity acting on water masses differing in density and to the effects of the Earth's rotation. The density of seawater varies inversely with temperature and directly with salinity and concentration of suspended sediment. Dense water sinks, displacing less-dense water. The result is a ***density current***, *a localized current, flowing because it consists of fluid denser than that of the body of fluid through which it moves.* In polar seas cold air chills surface water, causing it to become denser and to sink (Fig. 15-8). If it is denser than all the other water, it reaches the bottom and flows along the sea floor as a density current. Because of the Earth's rotation, however, these currents do not flow as broad sheets along the deepest parts of the ocean basins. Instead, as a result of the Coriolis effect and their velocity, they are banked up along the margins of ocean basins and follow submarine contours as flat, fast-moving tongues. These *contour currents* flow southward along the western margin of the North Atlantic and northward along the western margin of the South Atlantic. Measured velocities of these contour currents range from 5 to 25cm/sec, which as we shall see, is fast enough to transport almost all sizes of deep-ocean sediment particles.

In contrast to polar water, whose density is greater than normal as a result of low temperature, water in the tropics is denser because of high salinity. The Mediterranean Sea, which is nearly isolated from the Atlantic basin by a rocky threshold, exemplifies tropical water of high salinity. A connection between the two only about 400m deep exists at the Strait of Gibraltar. In the warm Mediterranean climate, evaporation exceeds precipitation. As a result the level of the Mediterranean lies 10 to 30cm lower than that of the Atlantic. The temperature of Mediterranean water is warm (13°C) and its salinity is 380,000p.p.m., in contrast to Atlantic water's salinity of 360,000

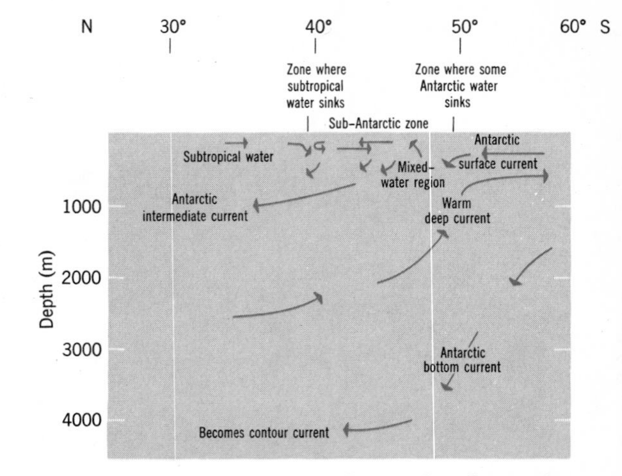

Fig. 15-8. Vertical section in the South Atlantic Ocean, showing movement of density currents. (After J. A. Knauss, 1963.)

p.p.m. Surface water from the Atlantic pours eastward through the Strait into the Mediterranean in response to the lower level. Extending from a depth of about 300m to the bottom, the denser Mediterranean water crosses the Gibraltar threshold, flows westward as a countercurrent, and plunges downward into the Atlantic along the bottom (Fig. 15-9, *A*). At depths ranging from 900 to 1,100m it disperses into the surrounding Atlantic water. Thus differences in density and level between two great bodies of water cause two currents to flow through the Strait of Gibraltar, one above the other. Although the Mediterranean does not deposit evaporites, its circulation is the same as that of basins from which evaporites can be deposited. As in the Strait of Gibraltar, in the open ocean and in many straits, surface currents are generally accompanied by deeper countercurrents.

Life in the sea depends on the availability of oxygen and nutrients in the water. As a result of differences in density, oxygenated surface water flows downward and carries oxygen to ocean depths. Deeper water displaced upward by it brings nutrients to the surface. In a semi-isolated arm of the sea or a bay connected to the open sea by only a narrow, shallow strait, such vertical circulation can be prevented. If the thickness of a surface current of low-density water flowing outward is greater than the depth of the threshold in the strait, the semi-isolated body of water becomes stratified and the denser water below receives no oxygen. In

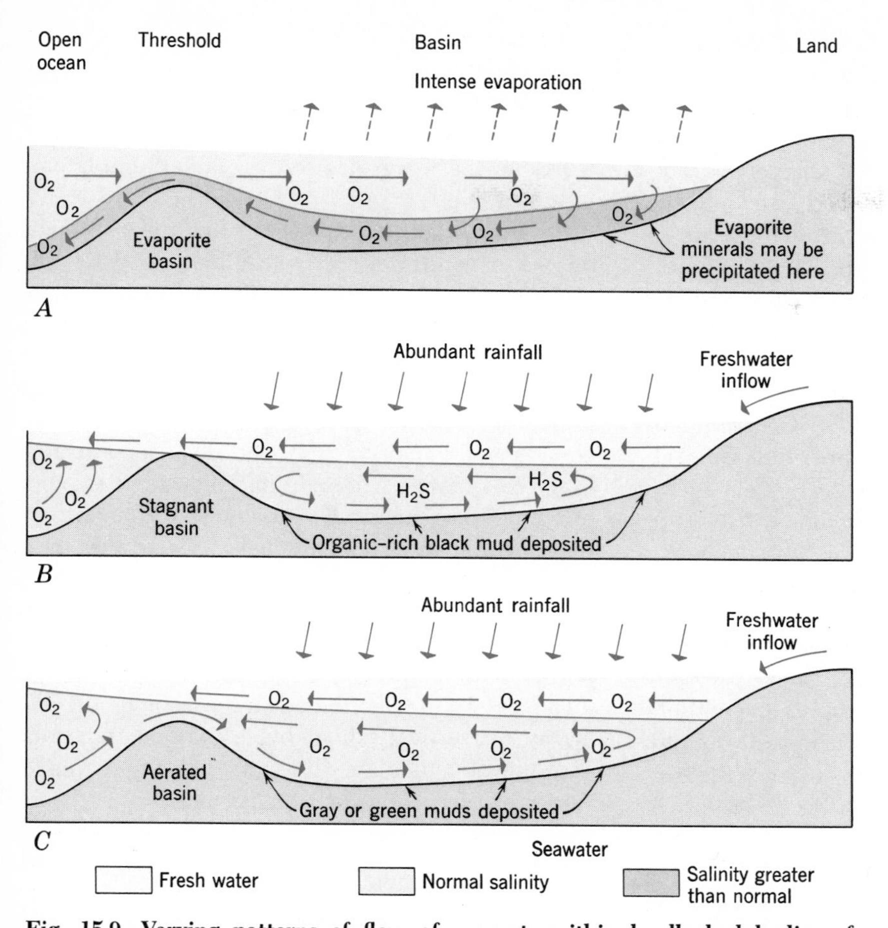

Fig. 15-9. Varying patterns of flow of currents within landlocked bodies of water connected to the sea by a shallow strait, shown by schematic long profiles having vertical scale much exaggerated. *A*. Evaporation exceeds freshwater inflow, as in Mediterranean; basin water is oxygenated throughout. *B* and *C*. Freshwater inflow greatly exceeds precipitation, raising water surface at right and creating surface current out of basin toward left. In *B*, thickness of outflowing current is greater than depth of threshold; in *C*, it is less.

the surface layer, which is in contact with the atmosphere and thus contains oxygen, organisms tolerant of low salinity can survive. But in the lower layer, which is cut off from the atmosphere and oxygen-bearing ocean water and thus receives no oxygen from either one, most organisms perish. Organic matter falls from the surface layer into the lower water; because organic matter has a great affinity for oxygen, it tends to remove all oxygen from the lower water. The lower water, now toxic to organisms requiring oxygen, preserves organic matter which continues to fall from above. *Anaerobic bacteria*, which require no free oxygen, can subsist on the organic matter. They derive oxygen from hydrocarbons or from sulfate radicals in solution. When they reduce sulfates, anaerobic bacteria create hydrogen sulfide. The hydrogen sulfide combines with dissolved iron to form minerals that color bottom sediment black.

In some semi-isolated basins the thickness of the surface layer of outflowing fresh water is less than the depth of the threshold at the strait (Fig. 15-9, *C*). In such basins, inflowing oxygenated seawater ventilates the deep water in the basin.

SEDIMENTS

Kinds of material. Studies made in the nineteenth century disclosed the three principal sources of deep-sea sediment: (1) the lands, (2) the ocean water itself, and, to a much smaller extent, (3) outer space (Table 15-1). The lands contribute particles of terrigenous sediment as well as tephra and material in solution. The larger particles of terrigenous sediment can be seen with the naked eye, but the minute terrigenous particles (Fig. 15-10, *A*) are so small that their study requires special instruments and techniques. The ocean water contributes particles consisting chiefly of the hard parts of tiny organisms, notably the unicellular animals Foraminifera (Fig. 15-10, *B*) and Radiolaria, and the unicellular plants, diatoms and coccoliths. Contributions from outer space consist of tiny spherules of nickel-iron and of glassy materials (*tektites*).

Processes. Nineteenth-century geologists supposed that no currents flowed in the deep sea. There, they thought, an "eternal snowfall" of tiny particles of sediment slowly accumulated at a constant rate to build on the bottom a layer of sediment that was both nonstratified and uniformly thick in all localities. Oceanographic expeditions completed since about 1950 have shown these concepts to be false. We now know that processes of deep-sea sedimentation are complex and that the deep-sea floor is swept by many kinds of currents. What is even more remarkable, we now know that sediment deposited on submarine slopes is subject to displacement by gravity and as a result can be spread widely by submarine mass-wasting. We also know that within the sediments many kinds of strata are present, that the thickness of deep-sea sediments varies from place to place, and that in many places sediments are not present.

Physical processes. Both physical and chemical processes operate in the deep sea. It is convenient, if somewhat arbitrary, to divide the physical processes into the categories vertical fallout and lateral transport.

Vertical fallout. The settling of particles through the water constitutes vertical fallout. Such fallout, formerly thought to be the chief process of sedimentation of all particles in the deep sea, builds layers of uniform thickness throughout the areas where it operates. The rates of fallout depend largely on size, because as particle diameters become smaller their rates of settling decrease. For example, quartz particles of fine silt size (diameters of 10 microns) require 2 years to fall to the bottom through still water 4,000m deep. All particles tend to sink, but as a result of their slow rates of settling fine particles are unable to escape the effects of currents. Fine particles can be suspended, transported, or both, by even the slowest currents. Hence such particles are distributed on the sea floor not primarily by vertical fallout but by lateral transport and they accumulate in bodies having variable thickness. Vertical fallout chiefly affects large particles, such as those dropped from floating ice, which settle so rapidly that they are uninfluenced by currents.

Lateral transport. Particles are transported laterally at all levels in the sea. On the surface, floating ice and other natural rafts, such as tree roots, drift from place to place. Organisms migrate not only by floating on the surface by themselves

TABLE 15-1. KINDS AND SOURCES OF MATERIALS COMPOSING DEEP-SEA SEDIMENTS AND MEANS BY WHICH THEY ARE TRANSPORTED TO THE OCEAN

Sources	Kinds of Material	Means of Transport to Ocean
1. Lands	Rock or mineral particles of older rocks or sediments; tephra	a. By rivers (suspended load) b. By waves eroding coasts c. By wind (most tephra) d. By floating ice e. By submarine mass-wasting
2. Ocean water	Calcareous and siliceous skeletal remains of animals and plants; iron-manganese minerals; phosphates; zeolites; fragmental and other volcanic products	a. By rivers (dissolved load) b. By submarine volcanic activity
3. Outer space	Spherules of magnetic iron and of silicate glass	a. By falling through Earth's atmosphere

Fig. 15-10. Common materials of deep-sea sediment. *A.* Crystals and particles of various minerals, mostly clay minerals, seen in electron micrograph with enlargement about 10,500 times natural size (compare Figs. 3-14 and 7-6). Brown clay from 1.68m beneath floor of Pacific Ocean, at a depth of more than 4,500m, from a station about 1,450km west of Mexico. (U. S. Geol. Survey.) *B.* Shells of calcareous Foraminifera from the floor of the Caribbean Sea, 160km west of Martinique, at a depth of 880m. (Yale Peabody Museum.)

or by attaching to other objects but also within the water by swimming and, along the bottom, by crawling. Within the water, turbulent eddies diffuse fine particles. Some particles transported laterally by these mechanisms reach the bottom by vertical fallout. Of greater concern is lateral transport on or near the bottom. Contour currents can transport sediment parallel to bottom contours, and submarine mass-wasting processes and various currents transport sediment downslope and deposit it, either at the bases of the slopes as abyssal fans or through wide areas as abyssal plains.

Contour currents. Photographs of the deep-sea floor show linear streaks of sediment (Fig. 15-11, *A*) and asymmetrical ripples (Fig. 15-11, *B*). The sediment streaks appear to result from differential deposition or erosion, in which sediment accumulates in the lee of raised areas on the bottom. The asymmetrical ripples originate when a current flows in one direction and exerts drag on a sandy bottom (Fig. 9-11). These sediment streaks and asymmetrical ripples have been attributed to contour currents, which can prevent sediment from being deposited in some areas and can deliver it in great quantities to other areas. Hence contour currents are partly responsible for the irregular thickness of fine-grained deep-sea sediment (Fig. 15-4, *A*).

Symmetrical ripples seen on a few photographs of

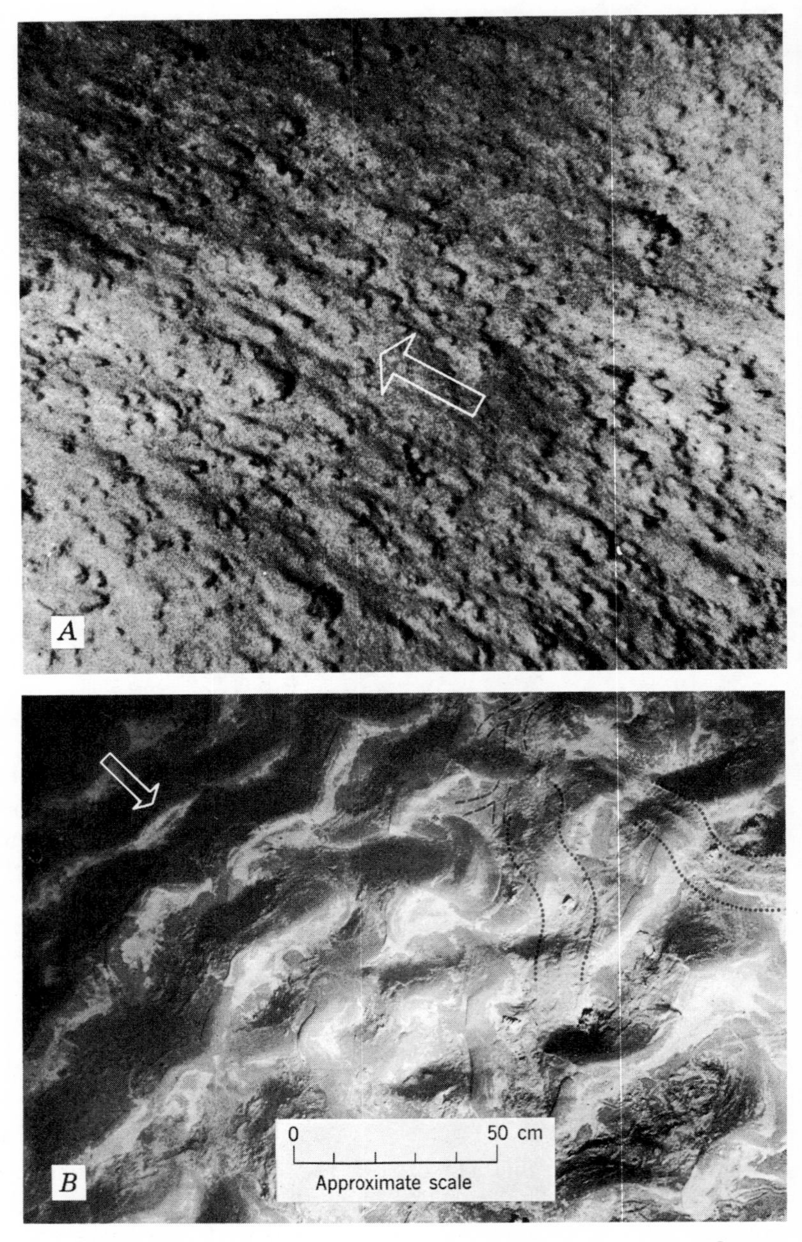

Fig. 15-11. Effects of currents (shown by arrows) on deep-sea sediment. (B. C. Heezen and C. D. Hollister, 1964.) *A.* Current flowing toward upper left created sediment "tails" behind raised objects on sea floor. Indian Ocean, south of Reunion Island; depth 4,909m. *B.* Asymmetrical current ripples in fine sand containing abundant grains of dark-colored minerals. Animal tracks (outlined by dashed and dotted lines) disclose that ripples were not active at time photograph was taken. Floor of Scotian Sea, southeast of Tierra del Fuego; depth 4,010m.

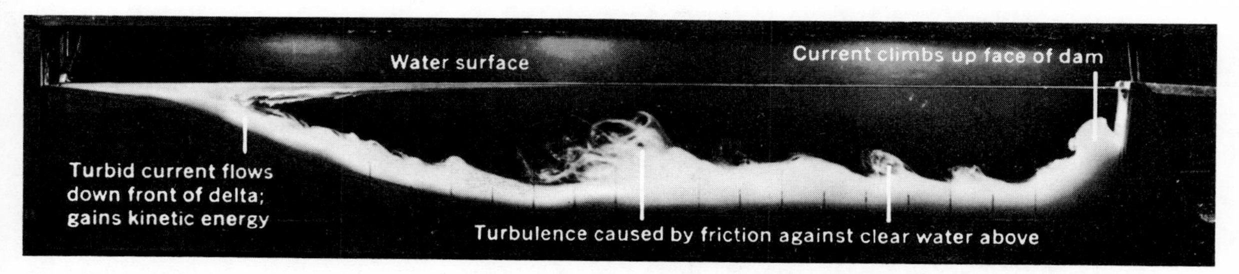

Fig. 15-12. Turbidity current seen in profile through glass wall of water-filled laboratory tank. (U. S. Soil Conservation Service, California Institute of Technology.)

the deep-sea floor are not caused by contour currents. These ripples indicate that the drag oscillated and was equal in both directions, as in wave action. The origin of the oscillatory movements responsible for symmetrical deep-sea ripples is not known.

Turbidity currents. Suspended sediment increases the density of a water mass, forming a body of dense, turbid water that can displace less-dense water surrounding it and flow along the bottom. *A density current whose excess density results from suspended sediment* is a ***turbidity current.*** In a turbidity current gravity propels the sediment and this causes water to flow. The current ceases when the sediment has been deposited. This contrasts with streams in which gravity propels the water, and the water in turn transports sediment. The stream continues to flow even if no sediment is present. Because turbidity currents have proved so fruitful in explaining puzzling features not only of modern sediments but also of ancient strata, we can profit by a closer look at the development of ideas about these currents.

What became known to geologists as turbidity currents were familiar to reservoir engineers by the name density currents. Such currents attracted wide attention in 1935, soon after Hoover Dam had been completed. During short periods the lowest outlet pipes, situated nearly 200 feet above the bottom of the filling reservoir, ceased discharging clear water and poured out muddy water. Even while muddy water flowed through the outlet pipes the surface water of the reservoir remained clear. Observations upstream verified that the muddy water came from the suspended load of the Colorado River. During floods, muddy water poured into the head of the reservoir, sank to the bottom, and flowed all the way to the dam.

An explanation for the muddy discharge from the outlet pipes was provided by laboratory experiments coincidentally in progress at the time. When poured into a tank of clear water, water made dense by dissolving salts or suspending silt and clay in it flowed along the bottom. If released at the head of a short, steep slope, it acquired enough kinetic energy to flow on an extremely small slope along the length of the tank and part way up the far wall, sliding past the clear water above (Fig. 15-12). The boundaries of the current were as distinct as those of a cloud of dust. Someone described it as "a dust storm under water."

In 1948 it was suggested that turbidity currents were one of the mechanisms for depositing graded layers (Fig. 16-5). This idea stimulated new experiments, in which laboratory turbidity currents deposited thin graded layers. Later many long cores exhibiting graded layers (Fig. 15-13) were collected from abyssal fans and some abyssal plains on the deep-sea floor. Graded layers can originate by many processes but one of the commonest of these is thought to be turbidity currents. Accordingly graded layers in the cores are taken as evidence that turbidity currents can operate not only in laboratory tanks and in reservoirs but also on the deep-sea floor.

Submarine mass-wasting. As on the land, gravity moves sediment on the sea floor, not only as coherent masses that preserve stratification but also as incoherent masses in which all internal structure is lost.

Slumping. In the slumping of coherent masses of sediment, preserved strata are folded and broken (Fig. 15-14, *A*). Sediments showing structures formed by slumping were first discovered in ancient marine strata now exposed on land and until recently were unknown in modern deep-sea sediments. Since 1963, specimens displaying effects of small-scale slumping have been brought up from the bottom by large

Fig. 15-13. Graded layer, about 6cm thick, in core of deep-sea sediment from Colombia Abyssal Plain at depth 4,074m. Nonstratified ooze overlies and underlies the graded layer. (Lamont Geol. Obs., courtesy C. D. Hollister.)

rectangular corers, and disturbed strata inferred to have resulted from large-scale slumping have been seen on continuous seismic profiles.

In other kinds of slumping, sediment disaggregates into its component particles, stratification is destroyed, and coarse particles from one layer mix with fine particles from other layers (Fig. 15-14, *B*). Slumped sediments containing mixed coarse and fine particles resemble the deposits made by mudflows or glaciers on land, or by melting icebergs at sea.

Sand flows. Diver-geologists, observing submarine canyons off the Pacific coast of North America, have discovered sand grains flowing along the bottom in fluid-like masses (Fig. 15-15). Acting on steep slopes where sand is abundant, g_t sets the grains in motion. Apparently when sand grains flow down slopes greater than their angle of repose they remain apart from one another because they repeatedly collide and rebound, somewhat like sand grains saltating in the wind. Sand can likewise flow readily if it becomes fluidized.

Evidence of gravity displacement on a continental slope. At the beginning of the chapter we remarked that the desire to lay submarine cables on the deep-sea floor had stimulated exploration of the deep sea. But the matter was not closed when the cables were laid. Now and then cables break and when they do the cable companies try to find out why. Cable breaks, an expensive problem for the companies, have proven a bonanza for scientists. Electric clocks on the teletype machines stop the instant a cable breaks. The distance from points on land to the break can be judged by measuring the electrical resistance, which is proportional to the length of cable from shore to point of break.

The most spectacular example of submarine cable breaks occurred on November 18, 1929, following a severe earthquake on the continental slope southwest of the Grand Banks and southeast of Cabot Strait, between Nova Scotia and Newfoundland (Fig. 15-16). In all, 12 cables broke in 28 places. Seven cables lying near the center of the earthquake broke almost instantly. Accordingly geologists studying the incident supposed that all the cables had broken because the sea floor had subsided in a long, narrow strip lying between two parallel faults extending southeast from the steep margins of the Cabot Strait. This was a reasonable suggestion, because such faults exist in the area and commonly the sea floor is displaced along faults when a submarine earthquake is recorded (Chap. 18). The hypothesis of faulting was accepted until 1952, when the whole problem of cable breaks here and elsewhere was restudied in light of modern knowledge of the oceans.

The following facts were cited as conflicting with the explanation that the cables broke through displacement of the bottom between parallel faults. (1) Although all cables cross the narrow strip that supposedly subsided, only those on the continental slope and deep-sea floor broke; not one of the many cables crossing the strip in the shallow part of the Strait was damaged. (2) All 7 cables that broke simultaneously during the earthquake lay close to the source of the quake. (3) At later times 5 other cables broke sequentially. The order of breaking was from north to south, in the direction of increasing depth and greater distances from the center of the quake. The first of these, cable *H*, 125km distant, broke 59 minutes after the quake; the last, cable L, 480km farther away, broke 12 hours and 18 minutes later (Fig. 15-16, *B*).

The records of the cable-repair operation contain additional information not consistent with the hypothesis. Each cable had broken at 2 or 3

Fig. 15-14. Results of submarine slumping in Miocene marine strata now exposed on land northwest of Los Angeles, California. (J. E. Sanders.) *A.* Thin siltstone layer has been contorted and broken, but has maintained its identity. *B.* Nonsorted sediment resulting from mixing of pebbles and sand that probably were deposited originally in separate layers (compare till, Fig. 12-12, and mudflow sediment, Fig. 8-12). Large chunk of overlying laminated sediment, *left,* has been bent around and downward.

points more than 160km apart. Detached cable segments, between two breaks, had been carried part way down the continental slope or buried beneath sediment on the continental rise and abyssal plain beyond the slope. The area affected measured about 360km wide and at least 720km long.

New information collected on oceanographic and geophysical surveys prompted a new hypothesis, consisting of two parts. (a) The earthquake caused sediment on the continental slope to slump and thus to break 7 cables simultaneously. (b) One or more turbidity currents flowed across the bottom, breaking each of the other 5 cables in turn.

A continuous seismic profile, made in 1963 and showing disturbed strata underlying the continental slope (Fig. 15-16, *B*), has given strong support to part (a) of the hypothesis. Other new evidence strongly supports some of part (b) but leaves important questions unanswered. If one or more turbidity currents formed, where did they originate? Did any other type of gravity displacement also occur?

The hypothesis that turbidity currents alone op-

Fig. 15-15. Sand flowing as a liquid on a steep submarine slope at a depth of about 10m off San Lucas Bay, Baja California. (William Bunton, U. S. Navy Electronics Laboratory, courtesy R. F. Dill.)

erated and broke the cables proposes that currents started at the center of the quake on the continental slope at the time the 7 cables broke. If we take the center of the quake as origin and assume a single current, then from known times of cable breaks and distances between broken cables we can compute the current's speed. From the assumed point of origin to cable H the computed speed is 22.5m/sec. From cable H to those broken still later the calculated speed decreases. Between cables K and L, the 2 most distant from the center of the quake, the calculated speed is 6m/sec. The slope of the bottom decreases regularly from north to south. The slope is 1:50 near the center of the quake and only 1:2,000 between cables K and L.

A velocity of 22.5m/sec for a turbidity current has been greeted with skepticism by some scientists, who place other interpretations on the data. Cable H, which broke 59 minutes after the quake, is the key to the problem. If cable H is ignored or if it was broken by the front of the mass of slumped strata (curved sawtooth line in Fig. 15-16, A) then it does not "time" a turbidity current originating at the center of the quake.

But rejection of the steep continental slope as the source of turbidity current creates a new problem: how can gravity acting on a gentle slope set in motion sediment which presumably moved at least 7m/sec along the bottom to break cables I, J, K, and L?

Still another hypothesis proposes that the quake created a moving wave of unstable conditions which was propagated through the sediments and broke the cables. According to this hypothesis the sedi-

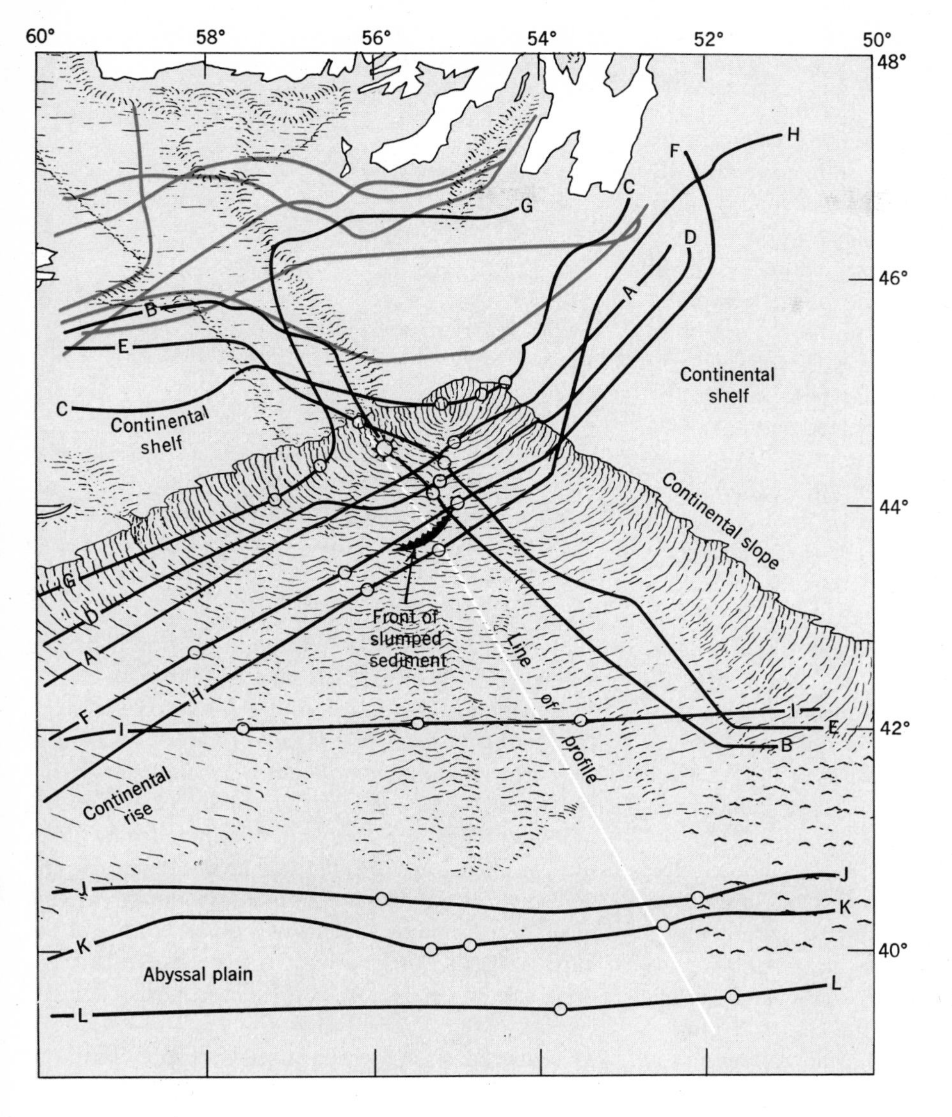

A

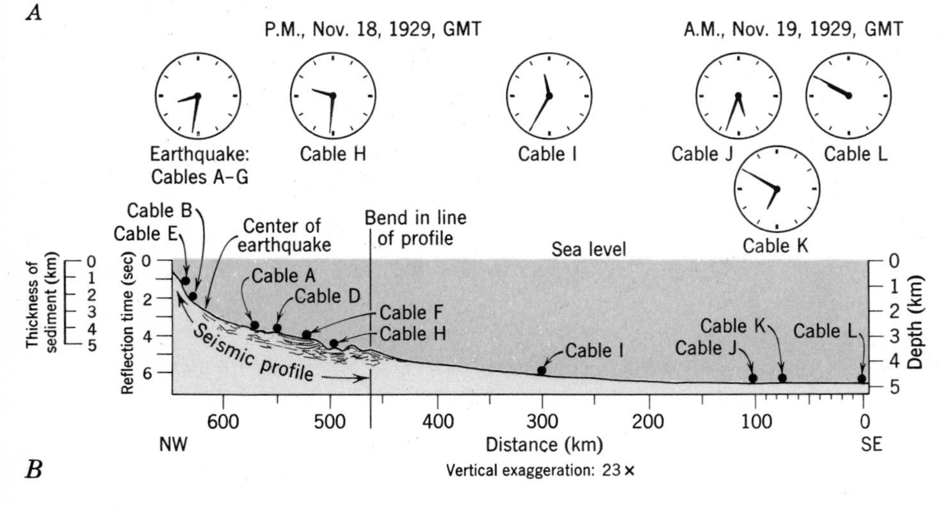

B

Fig. 15-16. Map and profile of sea floor off Nova Scotia and Newfoundland.
A. **Relief sketch of sea floor showing locations of submarine cables (lettered A through L) broken during and just after the Grand Banks earthquake, November 18, 1929. Points of breaks shown by circles; unbroken cables by blue lines; downslope edge of slumped sediment by thick line with sawtooth marks.**
B. **Profile of sea floor along line on map, *A*, showing slumped layers revealed by continuous seismic profile (*left*), submarine cables, and times cables broke (clocks at top). (After Heezen and Drake, 1964.)**

ments themselves moved only locally. This hypothesis does not explain the evidence of long-distance transport of the broken cable segments.

A graded layer of silt and sand 60 to 100cm thick appears near the tops of cores collected from the sea floor in the vicinity of the breaks. The grading has been cited as additional evidence in favor of the hypothesis of turbidity currents. "Sand and small pebbles" buried some of the long pieces of broken cable, according to observers aboard the repair ship. However, no such coarse sediment has yet been collected by corers. Repeated attempts to sample the bottom where the repair ship reported finding coarse sediment have been unsuccessful.

It seems highly probable that the cables were broken by sediment that was set in motion by the earthquake and powered down submarine slopes by gravity. Ambiguity still surrounds the exact processes involved.

Chemical processes. Materials dissolved in seawater reach the bottom by various chemical processes that are not well known. Some substances may be precipitated onto the bottom; others may react with silicate minerals in suspension or be absorbed from the water by various mineral particles lying on the bottom.

All material that reaches the deep-sea floor, either by chemical or physical processes, is lost to the geochemical cycle temporarily, perhaps permanently. We do not know how the cycle may react, for example, to the great losses of calcium and carbon taking place now and within the last 100 million years as calcareous shells of organisms collect on the deep-sea floor.

Classification and distribution. Deep-sea sediments are classified into four major groups: (1) pelagic sediments, (2) terrigenous sediments, (3) tephra, and (4) displaced shallow-water carbonate sediments. Sediments of each group can be subdivided further according to composition (Table 15-2). The relationships among materials, processes, and deep-sea sediments are shown schematically in Fig. 15-17.

Pelagic sediments (Gr. "belonging to the deep sea") are *open-sea deposits containing predominantly skeletal remains of microorganisms and clays or products derived from clays.* Some particles of pelagic sediment settled through the water at a considerable distance from shore or were precipitated out of seawater. Pelagic sediments contain only small percentages of nearshore organisms and less than 20 per cent of terrigenous sediments or of tephra having diameters larger than 10 microns (fine silt).

TABLE 15-2. OUTLINE OF CLASSIFICATION OF DEEP-SEA SEDIMENTS

I. Pelagic sediments
   A. Oozes
      1. Calcareous oozes
      2. Siliceous oozes
   B. Brown clay ("red clay")
   C. Authigenic minerals[a]
II. Terrigenous sediments
   A. Terrigenous muds
   B. Glacial-marine sediments
   C. Shallow-water sediments of varying size[b]
III. Tephra[c]
IV. Shallow-water carbonate sediments[b]

[a] Considered by some to be a distinct group of sediments and not merely a subdivision of pelagic sediments.

[b] Reach deep-sea floor as a result of displacement by gravity.

[c] Reach deep-sea floor by various processes, including displacement by gravity.

Pelagic sediments are subdivided into ***ooze,*** *pelagic sediment consisting of more than 30 per cent of skeletal remains of microorganisms,* and ***brown clay*** (formerly "red clay"), *pelagic sediment containing less than 30 per cent skeletal remains of microorganisms.* Oozes are subdivided into calcareous and siliceous varieties, each named for its dominant kind of organic remains. Calcareous oozes consist of the hard parts of Foraminifera, pteropods, and coccoliths, and siliceous oozes of the hard parts of radiolarians, diatoms, and sponges.

Calcareous ooze is the most widespread kind of sediment in the world today. It occurs on the deep-sea floor (Fig. 15-18) generally at depths less than 4,500 to 5,000m. In deeper water it gives way abruptly to brown clay. Because this change occurs even where calcareous skeletal remains drop down from the surface water, we must infer that calcium carbonate dissolves while the tiny shells are falling through the water or after they reach the bottom. Formerly it was thought that shells made of calcium carbonate dissolved after they reached bottom simply by coming into contact at great depth with water containing much carbon dioxide in solution. Now two other hypotheses have been proposed: (1) the water at and below 4,500m is especially corrosive to calcium carbonate and (2) bottom-dwelling organisms not only churn up the sediment (Fig. 15-19, A,) but also dissolve the calcium carbonate.

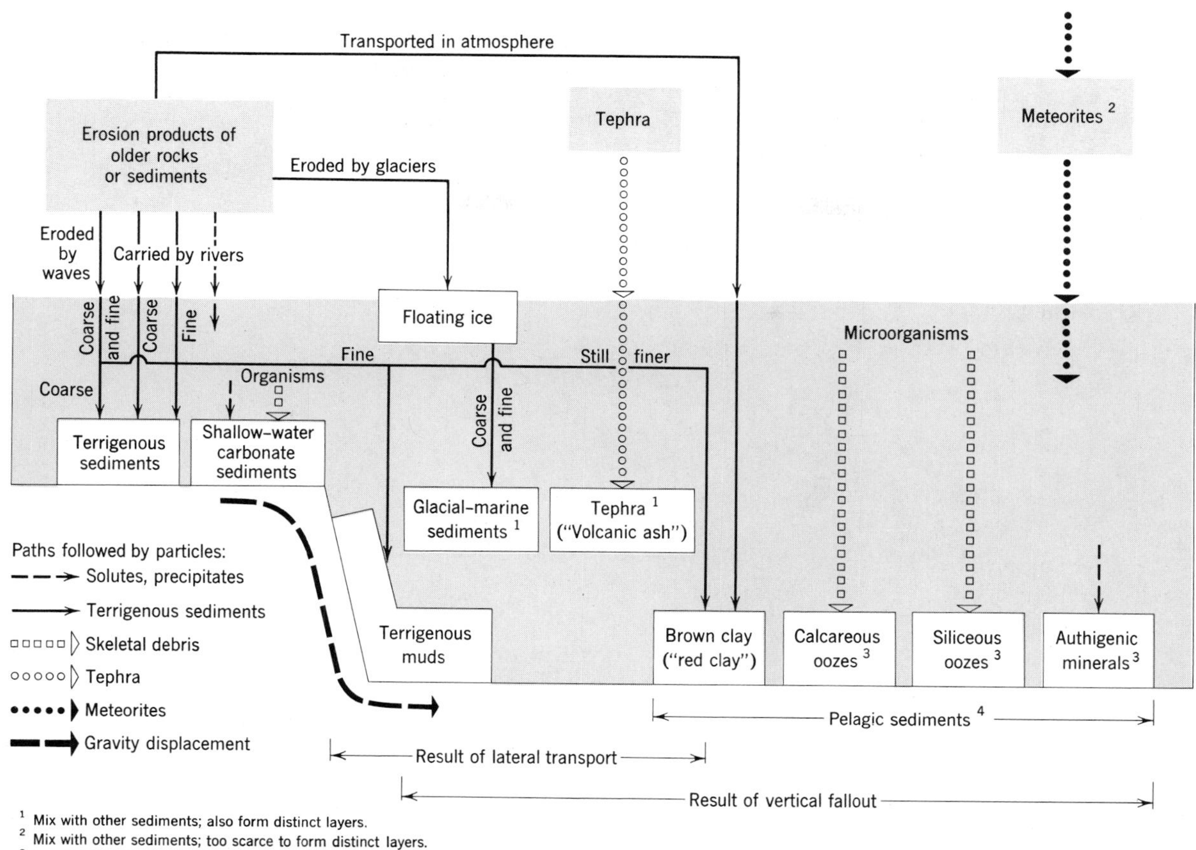

Fig. 15-17. Schematic diagram shows paths followed by particles from their sources to places of deposition as major varieties of deep-sea sediments. Shallow continental shelf at left; deep-sea basin at right. Text and Tables 15-1 and 15-2 define kinds of deep-sea sediments.

Both mechanisms are known to operate but their relative importance is unknown.

A belt of siliceous ooze surrounds the Antarctic Continent. Its dominant organic remains come from diatoms that live in the floating ice marginal to the shores of this continent. Another belt of diatom ooze crosses the North Pacific. Areas of siliceous ooze rich in radiolarian remains occur in the equatorial Pacific.

Authigenic sediments, *sedimentary deposits formed in place, not from physically transported material, and consisting of minerals that crystallized out of seawater* (including zeolites, iron-manganese minerals, shown in Fig. 15-19, *B*, and phosphates) cover broad areas on the Mid-Pacific Ridge and Mid-Indian Ridge. These localities are remote from sources of terrigenous sediment and in the surface water above them concentrations of microorganisms are low. Many manganese nodules are related to areas of strong currents and others to areas of submarine volcanic activity.

Great quantities of terrigenous sediments derive from erosion of the lands. Fine-grained *terrigenous mud* is deposited voluminously in marginal basins and on continental slopes and rises. ***Glacial-marine sediment,*** *terrigenous sediment including nonsorted mixtures of particles of all sizes dropped on the sea floor from floating ice,* can mix with any other kind of sediment. A continuous belt of glacial-marine sediment intervenes between the siliceous ooze and the Antarctic Continent. A smaller area of

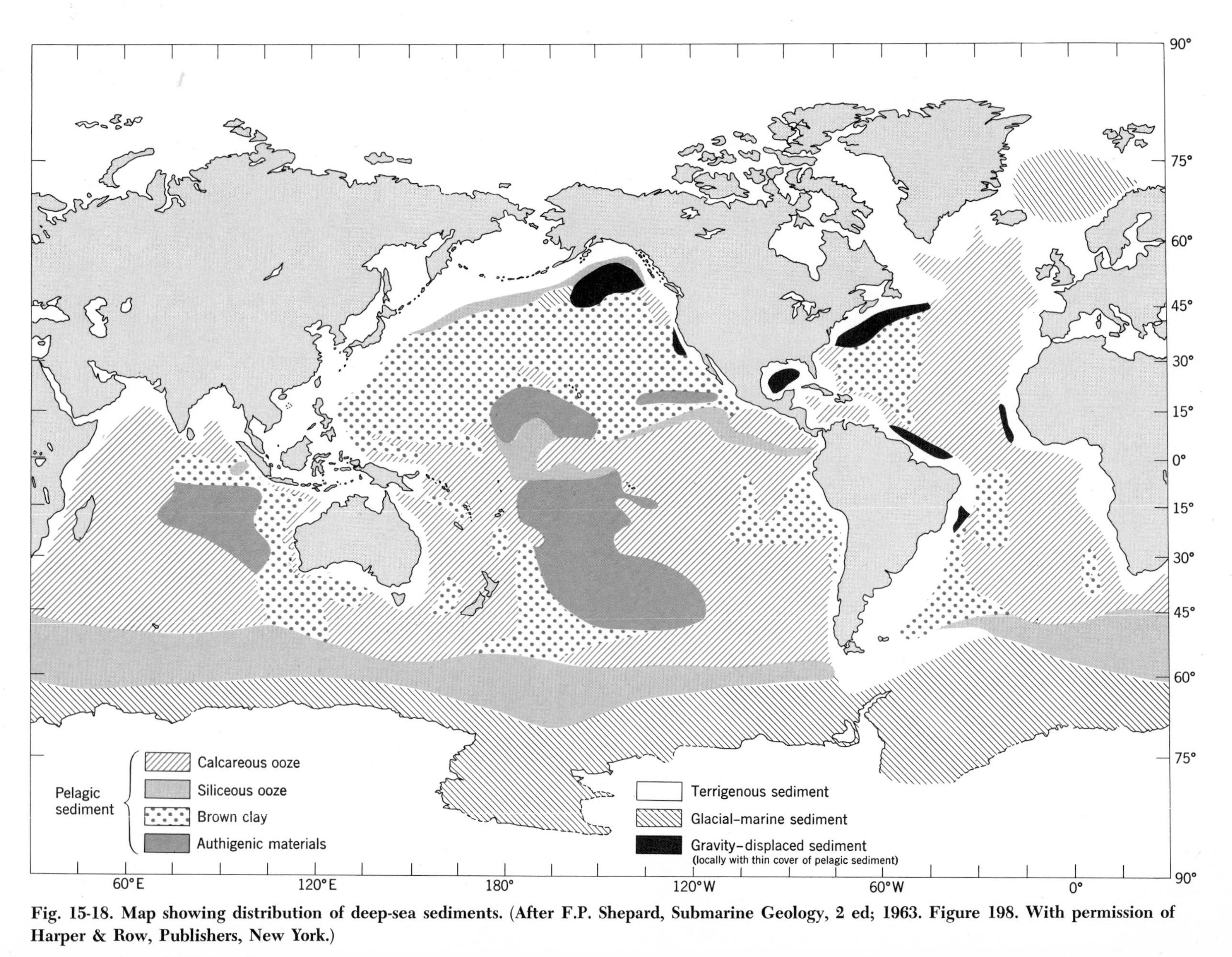

Fig. 15-18. Map showing distribution of deep-sea sediments. (After F.P. Shepard, Submarine Geology, 2 ed; 1963. Figure 198. With permission of Harper & Row, Publishers, New York.)

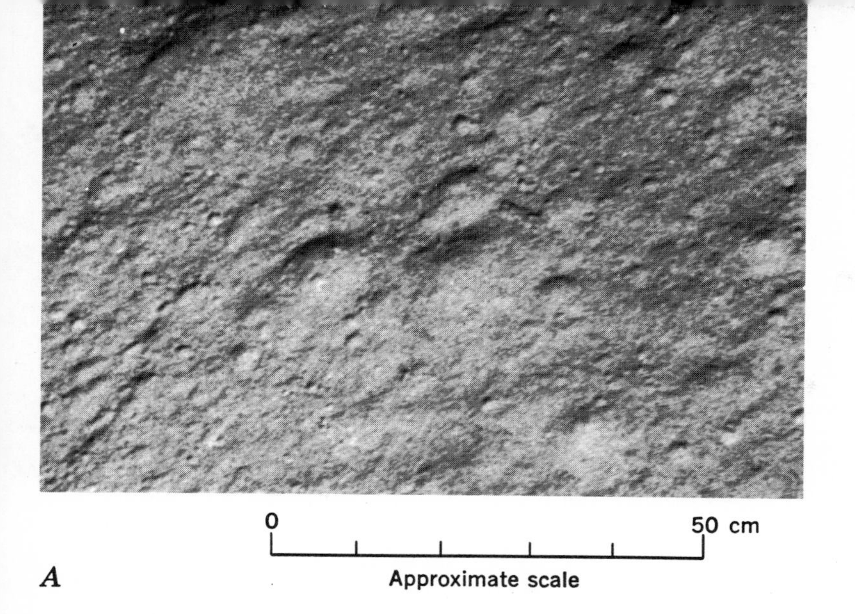

A

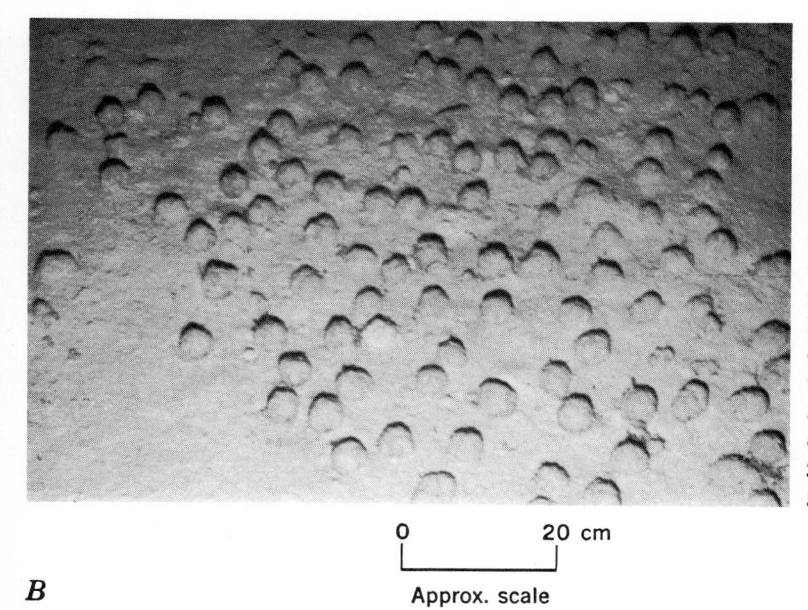

B

Fig. 15-19. Deep-sea floor showing effects of organisms and chemical sedimentation. (B. C. Heezen.) *A.* Mounds of two sizes built by burrowing organisms. Lower east flank of Mid-Atlantic Ridge; depth 5,114m. *B.* Manganese oxides occur not only as spherical nodules ranging in size from a golf ball to a tennis ball but also as irregular lumps, crusts, and coatings. Bermuda Rise, 320km southeast of Bermuda; depth 5,026m.

glacial-marine sediment lies between Greenland and Norway, where glaciers were more abundant in the past than now.

Terrigenous sediments that have been displaced by gravity occur on continental rises opposite the mouths of large rivers. Submarine morphology mainly determines where these sediments go. In some localities where terrigenous sediment derives copiously from the land it is deposited on wide continental shelves. If it flows down submarine canyons it may stop in deep trenches and basins parallel to shore and not reach the deep-sea floor.

Because of their composition tephra qualify for recognition as a distinct group of deep-sea sediments. Those tephra from land volcanoes are neither classified as terrigenous sediments nor distinguished from tephra derived from volcanoes located within ocean basins.

Shallow-water carbonate sediments reach the deep-sea floor in some localities. Their transport has been attributed to displacement by gravity.

Nonstratified deep-sea sediments. Cores of deep-sea sediment from outer ridges, the Argentine basin, and elsewhere consist entirely of nonstratified brown clay or other pelagic sediment. Some cores lacking strata reflect uniform conditions of sedimentation under which no strata ever formed, as nineteenth-century geologists had envisaged for all deep-sea sediments. Evidently deep-flowing density currents have supplied these areas with material of about the same particle size through long periods of time. Other deep-sea sediment, originally stratified, has lost its strata. Pelagic sediment in many cores has been so completely churned by burrowing organisms that any former stratification has vanished. Still other deposits are nonstratified because particles from older layers have slumped into a jumbled mass.

Stratification in deep-sea sediments. We have seen how cores refuted the older idea that deep-sea sediments lack stratification. Now let us examine some of the strata that occur in these sediments.

Alternating layers of various pelagic sediments. Alternating layers of pelagic sediment of two contrasting kinds (1) calcareous ooze and (2) brown clay, underlie a few parts of the floor of the South Atlantic. The calcareous ooze abounds in tiny skeletons of Foraminifera from species that exist today only in warm surface waters. The brown clay contains fossil Foraminifera of species that now live only in cool polar waters. The alternation is one of warm-water calcareous ooze and cool-water brown clay. Such alternation prompted the idea that the layers of ooze were deposited during nonglacial times and the layers of brown clay during glacial ages.

In most parts of the Atlantic floor, only uniform calcareous ooze appears but segments of cores containing species of fossils that live in warm surface waters alternate with segments containing fossils that live in cool surface waters. The topmost sediment in many cores contains warm-water species; ages of the calcareous shells extend back about 11,000 C^{14} years. The underlying sediment contains cool-water species dating from 11,000 to around 40,000 C^{14} years. This arrangement records the last glacial age followed by the postglacial.

From calcareous shells of Foraminifera that lived in surface waters we can learn more about ancient climates. The ratio of two isotopes of oxygen, O^{18} and O^{16}, in shell carbonate varies with (1) the ratio of these two isotopes in the atmosphere, (2) the salinity, and (3) the temperature of the water surrounding the living organisms. If we assume that (1) is constant and can estimate the salinity, then we can use isotope measurements on fossil shells to derive the temperatures of ancient surface waters. The results indicate that temperature changes of 6° to 10°C occurred between the time of deposition of sediment containing "cool" species and that of sediment with "warm" species.

Alternating layers of pelagic and other sediments. Layers of calcareous ooze beneath the floor of the North Atlantic alternate with layers containing terrigenous silt, sand, and small pebbles. Warm-water Foraminifera abound in the layers of calcareous ooze. Radiocarbon dates from the topmost layer of ooze in cores from the North Atlantic are the same as those from the upper sediment containing warm-water species in cores from tropical parts of the Atlantic. The coarse particles, glacial-marine sediment dumped from hosts of floating icebergs, are the same age as the sediment from the tropical Atlantic containing cold-water Foraminifera. Thus alternating layers of pelagic and glacial-marine sediments likewise point to fluctuations in climate.

Thickness. As we remarked at the beginning of this chapter, the aggregate thickness of deep-sea sediments implies important things about the age of the ocean basins. Before the thickness was measured, geologists tried to calculate what it should be by using average rates in the geochemical cycle and by assuming the great age and permanence of the ocean basins. One calculation, widely accepted as the best estimate prior to modern measurements, showed that deep-sea sediments should make a uniform layer, 4 to 5km thick. The MOHOLE test boring off Guadalupe Island in the Pacific encountered only 170m of nonlithified sediment resting on basalt. Seismic measurements indicate that nothing but firm rock lies below the basalt. In the northwestern part of the Atlantic Ocean, continuous seismic profiles indicate that unconsolidated sediments only about 300 to 500m thick overlie a prominent reflector (horizon A) at the top of the Upper Cretaceous deep-sea sediments. The greatest thickness of sediment above horizon A (3.0 to 3.5km) occurs only locally, on the continental rise off southern New England and on the Blake-Bahama Outer Ridges. Through a strip 160 to 240km wide in the crestal area of the Mid-Atlantic Ridge and on other rises and ridges and abyssal hills, unconsolidated sediment is not present at all; in these places the sea floor consists of rock (Fig. 15-4, *B*). Chapter 22 examines some of the implications of the surprising thinness of the carpet of sediment beneath the deep-sea floor.

Rates of accumulation. In order to calculate the rate of accumulation of sediment on the deep-sea floor, we must establish the age of one or more of the strata in a core. If we can prove the date of eruption, we can use a tephra layer to tell us the age. We can also measure certain radioactive isotopes and from the measurements calculate an age. An indirect method is to match strata in undated cores with those in cores where dates have been determined or to match paleomagnetic polarity episodes determined in deep-sea sediments with corresponding episodes discerned from isotopically dated volcanic rocks on land. Use of these methods indicates that rates of sedimentation have varied greatly from place to place (Fig. 15-20). Yet because of the great area of the deep-sea floor, even the slowest measured rate, about 1mm/1,000 years, typical of brown clay on the floor of the Pacific, represents a colossal loss of material from continents. If the deep-sea floor were to be given a uniform paving of steel at the rate of 1mm/1,000 years, it would require 360×10^6 tons/year, more than the world's annual production.

Chronology. Close studies of the uppermost 20m of deep-sea sediments in representative cores have made it possible to construct a composite record of many changes of climate, probably extending far back into the Pleistocene Epoch. The record shows that the Pleistocene climatic changes were worldwide. During glacial ages, cool climates seem to have prevailed simultaneously in both hemispheres. Likewise the warmer climates of the interglacial ages apparently existed everywhere at about the same time.

The earlier expectation of finding in deep-sea sediments an unbroken record of events from early in the history of the Earth down to the present has not been realized. Most of the cores studied thus far seem to represent only parts of the Pleistocene Series. The Guadalupe test boring made in 1961 found sediments recording only a small part of the Cenozoic Era; the basal layer was at most only 45 million years old. The oldest rock fragments dredged so far from the deep-sea floor represent strata of Early Cretaceous age. These fragmentary results from the older parts of the carpet of deep-sea sediment raise more questions than they answer. In Chapter 22 we shall try to evaluate some of these findings.

ORIGIN OF SUBMARINE CANYONS

Most continental slopes are indented by enormous submarine canyons. These huge valleys are hundreds of kilometers long. Their sides slope steeply, and their narrow floors, some lying as much as 1.5km below the valley rims, are inclined seaward as much as 80m/km and extend down to 3.6km below sea level. The walls of many canyons consist of rock. Some canyons have been partly filled with unconsolidated sediments. Some canyons, such as the Hudson (Fig. 15-21) and Congo (Pl. C), appear to be seaward extensions of land rivers; others are not obviously connected with land rivers.

For more than a century submarine canyons have engaged the interest of geologists, who have proposed many hypotheses of origin for these tremendous valleys. Most of their ideas are based on the premise that the canyons originated within a geologically short time interval. One group of hypotheses invokes processes that work on land, another group involves submarine processes.

Hypotheses invoking processes operating on land assert that the canyons were made by streams, either (1) big mountain streams along coasts that later subsided beneath the sea, or (2) rivers flowing over what is now the sea floor when sea level stood thousands of feet lower than today.

Early hypotheses involving submarine processes proposed that the canyons were made by (1) tidal currents, (2) tsunami, or (3) submarine artesian springs. Most of these hypotheses encounter serious objections. Later the hypothesis of turbidity currents become popular. Laboratory experiments show that the erosive power of these currents is great. Proponents argue that turbidity currents eroded the canyons by flowing down them repeatedly, much as streams flow through land valleys.

Direct observations in the upper parts of canyons have disclosed that at least six processes are actively eroding today's canyons: (a) creeping sediment, (b) progressive slumping, (c) flowing sand, (d) turbidity currents (only a few small ones actually have been seen), (e) bottom currents created by tides, and (f) bottom currents created by internal water move-

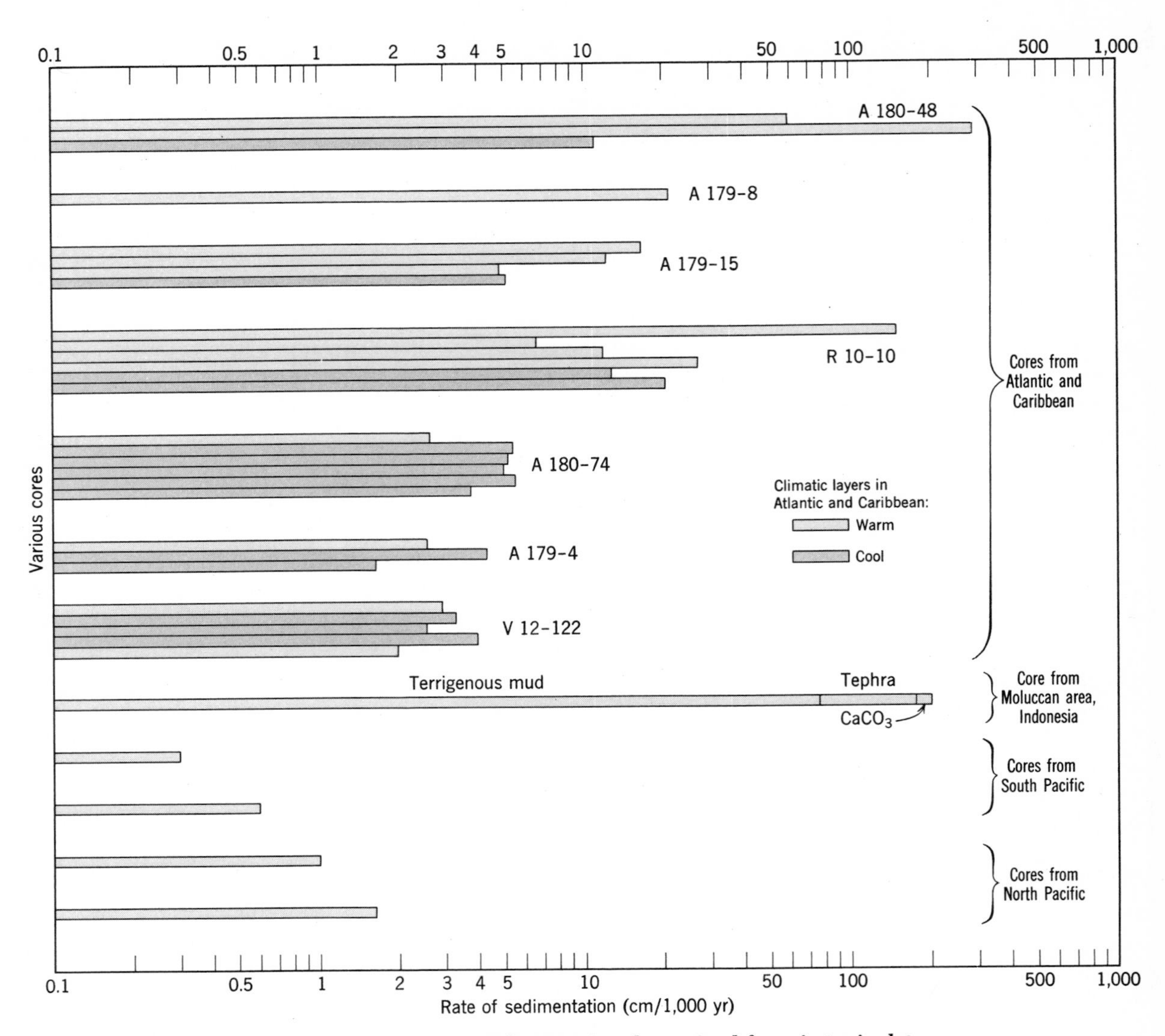

Fig. 15-20. Bar graph shows rates of deep-sea sedimentation determined from isotopic dates and from layers of tephra whose dates of eruption are known (Moluccan area core). Single bars signify that only one average rate of deposition was determined. Stacks of bars denote that many layers from the core have been dated; bars in stack give average rates between pairs of dated layers down the core. Code identifying cores designates name of research vessel (capital letter), cruise number (next to capital letter) and station number (following hyphen). (Data from D. B. Ericson and others, 1964; Teh-Lung Ku and W. S. Broecker, 1966; and Menard, 1964.)

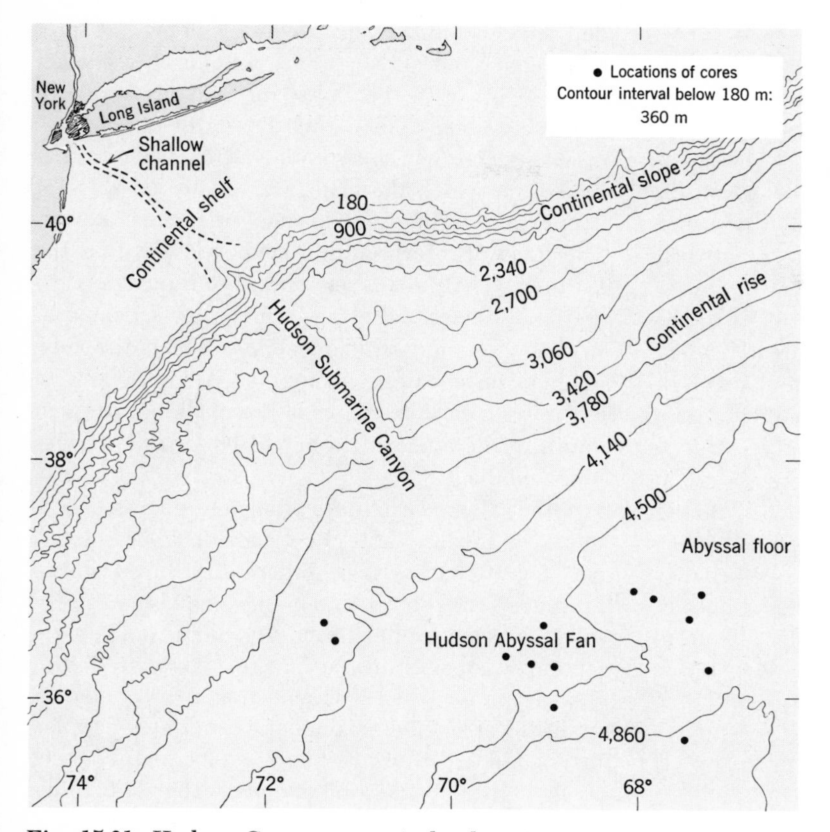

Fig. 15-21. Hudson Canyon, a typical submarine canyon, and Hudson Abyssal Fan portrayed by contours plotted from echo-sounding profiles. (After D. B. Ericson and others, 1952.)

ments. Submarine mass movement of sediment is abrading bedrock walls of some canyons, much as a valley glacier does.

Abyssal fans indicate that great quantities of sediment have been transported through some canyons. Calculations show that the volumes of sediment lying seaward of some canyons are 10,000 times as great as the volumes of the canyons. In other words, even if each 10,000 passing particles had eroded only one particle, the passing sediment could have cut the canyons. Hence submarine processes are capable of great erosion.

Evidently parts of some canyons originated by several processes acting through long periods of time. Their headward parts may have originated as land valleys, but they subsided below sea level and were then modified by submarine processes. Parts of many canyons have never been above sea level.

ORIGIN OF ABYSSAL FANS AND ABYSSAL PLAINS

Abyssal fans, accumulations of sediment with fanlike forms, lie opposite the mouths of many submarine canyons. The Hudson abyssal fan, at a depth of about 4.5km, was first cored in 1951. Since then cores have been raised from other abyssal fans. Coarse gravel displaced from shallow water has been cored at depths of 3.3km at a distance of 220km from the head of Monterey Canyon off California.

Fifteen cores, each 8 to 10m long, show that the Hudson abyssal fan consists of layers of clay appropriate to the depth, alternating with graded layers of sand, silt, and broken shells derived from the continental shelf above. This dual composition suggests that the fan sediments originated through various processes.

The volumes of two abyssal fans off Monterey and Delgada Submarine Canyons, west of California

(Fig. 18-6, *A*) have been calculated to be 450 and 300km^3, respectively. Because the heads of these canyons lie close to shore, they intercept the longshore drift of sand. Each year the waves drift about 1 million cubic meters of sand into the heads of these two canyons. However, the sand does not fill the canyons. Instead it passes through them into deep water, where it comes to rest on the abyssal fans.

Abyssal plains, which are not restricted to the vicinity of submarine canyons but spread widely on the deep-sea floor (Pls. A, C), are the flat tops of bodies of sediment much more extensive than those constituting abyssal fans. Cores show that most abyssal plains on the North Atlantic floor are underlain by alternating layers of graded terrigenous sediment and nongraded pelagic sediment. The graded layers are thought to have been deposited by turbidity currents generated on continental slopes. Continuous seismic profiles show that in most the subbottom sediments contain many widespread parallel layers having contrasting physical properties.

ARE FORMER DEEP-SEA SEDIMENTS EXPOSED ON LAND?

Ever since the first deep-sea sediments were dredged up, geologists have debated whether such sediments occur among the rock strata of the lands. If they do, it follows that great vertical movements of the crust have occurred. Because they doubted that the Earth's crust could move so much, most geologists were skeptical of the idea that former deep-sea sediments could be exposed on land. The few advocates of the idea were unable to convince their more numerous skeptical colleagues, partly because few criteria exist for establishing depth of water at time of deposition. Pelagic sediments, for example, owe their characteristics to remoteness from supply of terrigenous sediments as much as to depth of water. Most skeletal debris in pelagic sediments comes from organisms that live near the surface; such remains are abundant in bottom sediments only where they are not being overshadowed by sediments of other kinds. Today most such sediments happen to lie in deep water, but this need not always have been the case.

Some geologists compared the chalk underlying the white cliffs of Dover and extending widely in northwestern Europe with modern calcareous ooze. They supposed, accordingly, that the chalk had been deposited on an ancient deep-sea floor and later had been lifted above sea level. Among the chief components of the chalk are the remains of coccoliths, tiny unicellular plants. Fossils of other organisms, species that lived on the bottom, suggest that the sea in which the chalk was deposited was not deeper than 1.5km and, in many places, shallower. The modern interpretation acknowledges that the European chalk is a former calcareous pelagic sediment but denies that it is a former deep-sea deposit. In this case, modern knowledge supports the view held by most earlier geologists. We must ascribe the pelagic characteristics of the chalk not to great depth of the ancient sea but to its lack of terrigenous sediment.

Other alleged examples of ancient deep-sea sediments on land, notably the Cenozoic "Oceanic Series" on Barbados, West Indies, and the Mesozoic "red clay" on Timor, Indonesia, likewise were rejected as such by most earlier geologists. Their rejection was based on the fact that associated with the alleged deep-sea sediments are layers of coarse-grained terrigenous sediments containing ripple-marks and debris of fossil land plants, features long thought to be criteria of shallow water. Our fuller knowledge of currents in the deep sea and processes of gravity displacement incline us to be less dogmatic than were earlier geologists about the depth significance of ripple-marks, coarse terrigenous sediment, and debris of fossil land plants. Rippled sediment has been photographed at great depths, and both coarse terrigenous sediment and debris of land plants have been brought up in cores from abyssal plains. Hence, viewed in the light of modern knowledge, these two examples of alleged ancient deep-sea sediments may indeed have been interpreted correctly by the minority of earlier geologists and incorrectly by the majority.

Widespread in many mountain chains are sedimentary strata having characteristics closely similar to those of gravity-displaced sediments cored from modern abyssal fans, continental rises, and abyssal plains. Agreement has not been reached on the depth significance of these strata in mountains. One possibility is that they represent ancient deep-sea deposits raised to form parts of the continents (Chap. 21).

With this chapter we bring to a close our section on external processes. We have studied the mechanisms at work in various processes and the effects of these processes in modern environments as related to the Earth's morphology, Man's activities, and the building of strata. Because strata are so important to geology we devote the entire next chapter to them.

Summary

Morphologic features

1. The margins of ocean basins consist of steep escarpments or sea-floor trenches where sediments from land are not abundant, and of gentler slopes where such sediments are abundant.

2. Great rocky ridges occupy large areas on the deep-sea floor. Elsewhere the floor consists of smooth plains underlain by sediments or of rough areas of varying relief underlain by bedrock.

Movements of seawater

3. Prevailing winds drive surface ocean currents.

4. Differences in water density resulting from variations in temperature and salinity create vertical circulation and deeper currents.

Sediments

5. Deep-sea sediments are derived from the lands, from seawater, and from outer space.

6. The chief physical processes of deep-sea sedimentation are vertical fallout, lateral transport by currents, and submarine mass-wasting.

7. Pelagic sediments, chiefly oozes and brown clays, underlie much of the deep-sea floor.

8. Cores show that many deep-sea sediments are stratified. Layers result from the successive arrival of sediment of varying kinds caused among other processes by changes of climate, passage of turbidity currents, and fallout of tephra.

9. Deep-sea sediments record Pleistocene chronology and changes of climate.

Ancient deep-sea deposits

10. Some alleged examples of ancient deep-sea sediments elevated to form land areas have not withstood modern scrutiny. Other examples are supported by modern knowledge of the sea floor.

Selected References

Emery, K. O., 1960, The sea off southern California. A modern habitat of petroleum: New York, John Wiley (Especially p. 38–61; 214–236; 247–258).

Ericson, D. B., Ewing, Maurice, Wollin, Goesta, and Heezen, B. C., 1961, Atlantic deep-sea sediment cores: Geol. Soc. America Bull., v. 72, p. 193–286.

Heezen, B. C., and Drake, C. L., 1964, Grand Banks slump: Am. Assoc. Petroleum Geologists Bull., v. 48, p. 221–233.

Heezen, B. C., and Ewing, Maurice, 1952, Turbidity currents and submarine slumps, and the 1929 Grand Banks earthquake: Am. Jour. Sci., v. 250, p. 849–873.

Heezen, B. C., Hollister, C. D., and Ruddiman, W. F., 1966, Shaping of the continental rise by deep geostrophic contour currents: Science, v. 152, p. 502–508.

Heezen, B. C., Tharp, Marie, and Ewing, Maurice, 1959, The floors of the ocean. I. North Atlantic: Geol. Soc. America Spec. Paper 65.

Hill, M. N., ed., The sea. Ideas and observations on progress in the study of the seas: New York, Interscience Publishers. v. 1, 1962, Physical oceanography, Chap. 5, Section III; v. 2, 1963, The composition of sea water, comparative and descriptive oceanography, Chaps. 2, 4, 10, 11, 12, 17, 18, and 23; v. 3, 1963, The Earth beneath the sea. History, Chaps. 4, 5, 12, 14, 17, 19, 20, 25, 26, 27, 28, 30, 31, 33, and 34.

Menard, H. W., 1964, Marine geology of the Pacific: New York, McGraw-Hill, Chaps. 1, 2, 7, 8, 9, and 10.

Shepard, F. P., 1963, Submarine geology, 2nd ed.: New York, Harper and Row, Chaps. 1, 2, 11, 12, 13, and 14.

Shepard, F. P., and Dill, R. F., 1966, Submarine canyons and other sea valleys: Chicago, Rand-McNally.

CHAPTER 16

Sedimentary Strata

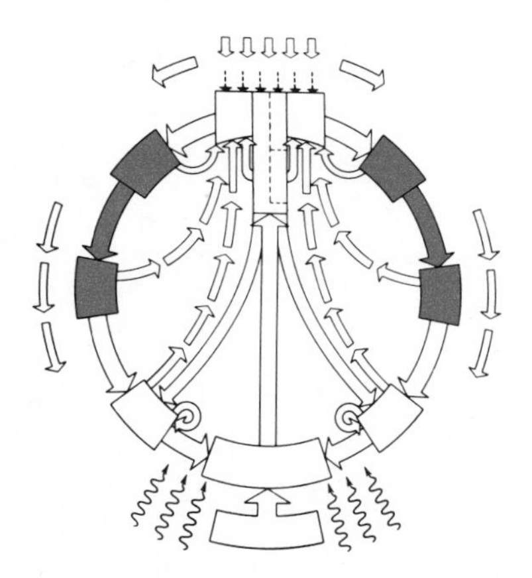

Sedimentary rocks in the rock cycle
Stratification
Sorting
Kinds of layers
Stratigraphy
Environments of deposition
Facies
Formations
Correlation

Plate 16

Students from a geology class examine an exposure of Paleozoic limestone at Hudson, New York. Student at right is using a scale to measure thickness. (*J. E. Sanders.*)

CHAPTER 16 *Sedimentary Strata*

SEDIMENTARY ROCKS IN THE ROCK CYCLE

Strata as a link between external and internal processes. We have postponed any full discussion of sedimentary strata until all the processes of erosion and deposition that lead to the making of strata had been set forth. Now we can gather together the many threads we followed in studying the processes, as we observe in the strata the features characteristic of the sediments deposited by each of the dynamic agents. We can see sedimentary strata as a pivotal phase of the rock cycle, pivotal because it is a normal link between the erosion-transport-deposition complex which precedes it and the folding, faulting, metamorphism, and melting that sooner or later are likely to follow it. Strata, because of their original horizontality, constitute a benchmark from which their later deformation can be measured. This is why the present chapter stands where it does, a link between the external sediment-forming processes we have described up to now and the internal processes that deform and alter strata, processes to be described in chapters still ahead.

Sedimentary strata as former sediments. Before going further we need to recall that all sedimentary rocks are built of the rearranged particles of older rocks. They are the conspicuous product of a vast, complex process of sorting and transport of sediment. The process ends with deposition and is followed by compaction and cementation, which convert the sediment into new, firm rock. We can witness every phase of this segment of the rock cycle: older rocks whose exposed surfaces are being crumbled under the attack of weathering, regolith slowly creeping down hillsides, streams turbid with fine rock particles, sand and silt deposited by a stream and exposed as a flood subsides, sand and mud deposited in shallow water off a sea beach and exposed at low tide, and sediments in all stages of conversion to firm sedimentary rock. Every feature of sediment that geologists have described occurs also in sedimentary rocks. This fact is perhaps the most obvious proof of the principle of uniformity of process.

Principle of original horizontality. We noted at the outset that sediments now being deposited in stream valleys and on the floors of lakes and the sea are spread out in layers that are generally almost horizontal. When the loose particles are immersed in water they are easily moved, and water in motion tends to spread the particles evenly and so to fashion a nearly level surface. In this way, at the end of each stream flood and of each storm offshore, a new sedimentary layer is deposited almost horizontally over the one beneath. Locally some layers of limited extent are deposited in inclined positions (Figs. 9-11, 9-12, 13-24), but these represent special conditions and so are exceptions to the general rule. It is, then, an observed fact that most layers of sediment are nearly level when they are formed.

This fact, first recognized as early as the eighteenth century, embodies the *principle of original horizontality*. We rely on the principle when we analyze sedimentary layers that are no longer horizontal because they have been deformed (Fig. 2-14).

Principle of superposition. We noted in the preceding paragraph that after a new bout of activity a new sedimentary layer is deposited over the one beneath. This fact embodies the *principle of superposition*, defined in Chapter 6, where we characterized it, together with the fossils contained in strata, as fundamental to the building up of the geologic column.

STRATIFICATION

In Chapter 1 we read that the layered arrangement of the constituent particles of a rock is *stratification*. This is an obvious feature of most sedimentary rocks and is seen also in ancient lava flows.

Fig. 16-1. Variation in distinctness of layers. (J. E. Sanders.) A. These strata of limestone (white) and of shale (dark) are distinct because of abrupt changes in grain size, composition, and color. Length of compass is 8cm. (Cambrian; south of Hudson, New York.) *B.* Strata of fine-grained sandstone and siltstone, indistinct because grain-size changes are gradual. (Triassic; Milford, New Jersey.)

For convenience in description, strata are subdivided into ***beds*** (*strata 1cm or more in thickness*) and ***laminae*** (*strata less than 1cm thick;* Fig. C-1, *B*). From these terms are derived the commonly used words *layering, stratification, bedding,* and *lamination.*

If we look closely at rocks that are stratified distinctly, we can see that the strata differ from one another because of differences in some characteristic of the constituent particles or in the way in which the particles are arranged (Fig. 16-1). Very commonly one stratum consists of particles of different diameter from those in another. In a clastic rock such changes of diameter result from fluctuations of energy in a stream, in surf, in wind, in a lake current, or in whatever agency is responsible for the deposit. The energy changes, usually small, are not the exception but the rule.

Sorting. As we noted in Chapter 4, one of the conspicuous results of the transport of particles by flowing water or flowing air is *sorting* in the deposited sediments. Sorting according to specific gravity is evident in mineral placers (Chap. 23). Particles of unusually heavy minerals, such as gold, platinum, and magnetite, are deposited quickly on stream beds and on beaches, whereas lighter particles are carried onward. Most of the particles carried in water and wind, however, consist of quartz and other minerals with a specific gravity rather similar to quartz. Commonly, therefore, such particles are sorted according to size. In a stream, gravel is deposited first, whereas sand and silt are carried farther before deposition. Thin, flat particles are carried farther than spherical particles of similar weight. Long-continued handling of particles by turbulent water and air results in the gradual destruction of the weaker particles. In this way rocks and minerals that are soft or that have pronounced cleavage are eliminated, leaving as residue the particles that can better survive in the turbulent environment. Very commonly the survivor is quartz, because it is hard and lacks cleavage. In this case sorting is based on durability.

Although sorting is the chief cause of stratification, it is not the sole cause. Successive layers that do not differ from each other in grain size, composition, or degree of compaction can still be separated from each other by surfaces of easy splitting representing minor intervals when no deposition occurred. Again two adjacent layers, not otherwise distinct, can differ from each other as to the kind or abundance of cement they contain.

In summary, each stratum nearly always possesses definite characteristics by which it differs from the stratum beneath or above it. With this in mind we can describe two chief kinds of stratification and then examine the particles within a stratum.

Parallel strata. Layers of sediment fall into two classes according to the geometric relation between successive units. One class consists of ***parallel strata,*** *strata whose individual layers are parallel* (Fig. 1-11, *B*). Parallelism indicates that deposition probably occurred in water, and that the activity of waves and (except for the special case of graded layers, described in a following section) currents was minimal. Indeed, the sediments of lakes and the deep-sea floor rather commonly have parallel laminae.

A distinctive variety of parallel strata consists of

Fig. 16-2. Varves in glacial-lake clay and silt, Uppsala, Sweden. In this case each varve is also a graded bed. Thick, pale silt (summer portion) sharply overlies the layer beneath it and grades upward into thinner, darker silty clay (winter portion), making an annual cycle. Pencil (placed vertically at base of exposure) is 6 inches long. (R. F. Flint.)

repeated alternations of laminae of unlike grain size or mineral composition. Such alternation suggests the influence of some naturally occurring rhythm, such as the rise and fall of the tide and a seasonal change from winter to summer. *A pair of laminae deposited during the cycle of the year* with its seasons is a ***varve*** (Swedish: "cycle"). Varves occur in glacial-lake sediments (Fig. 16-2) deposited about 10,000 years ago, near the close of the last glacial age. Radiocarbon dating of related sediments has confirmed that the pairs of laminae are true varves.

Paired laminae deposited in deep glacial lakes are generally very distinct because close to an ice sheet the contrast between summer and winter weather markedly affects the rate of melting of ice. Pairs of laminae very similar to these occur in ancient rocks. Certain shales in South Africa have been interpreted as varves deposited in glacier-dammed lakes during a glaciation more than 200 million years ago.

Varves of different origin characterize rocks of the Green River Formation, which underlies thousands of square miles in Wyoming, Colorado, and Utah. In each varve one lamina consists of calcium carbonate; the other includes dark-colored organic matter. The rhythm is explained as follows: the sediments were deposited in a lake, which warmed in summer, therefore lost carbon dioxide, and precipitated calcium carbonate from solution. During the same warm season floating microscopic organisms reached a peak of abundance. The relatively heavy carbonate sank promptly and formed a summer lamina; the lighter organic matter sank much more slowly to form an overlying winter lamina. Thus the pair of laminae is a varve. It has been estimated from sample counts that between 5 and 8 million varves are present in the Green River Formation; hence we reason that an immense lake occupied the Green River basin for millions of years under remarkably uniform conditions.

Cross-strata. In the other class are *strata inclined with respect to a thicker stratum within which they occur.* These are ***cross-strata.*** We met them first where they were being produced experimentally on the bed of a laboratory stream and again where they showed up in dunes. The term includes both cross-*beds* and cross-*laminae,* depending on thickness. All such strata consist of particles coarser than silt and are the work of turbulent flow of water or air, as in streams, wind, and waves along a shore. As they are driven forward, the particles tend to collect in ridges, mounds, or heaps in the form of ripples and sand waves. These heaps either migrate forward bodily or simply enlarge in the downcurrent direction. In both cases particles continually accumulate on the downcurrent slope of the pile, forming strata with inclinations as great as 30° to 35°.

The geometry of cross-strata is shown in Fig. 16-3. Other examples are seen in deltas (Fig. 9-30), outwash sediments (Fig. 12-20), dunes (Fig. 13-24), and beaches (Fig. 14-9). We must keep in mind that although the foreset layers in deltas and in dunes are commonly parallel with *one another,* nevertheless they are cross-strata because they are not parallel with the larger strata that inclose them. Cross-strata therefore do not violate the principle of original horizontality, which in such occurrences is supported by the larger, inclosing strata.

Under some conditions, however, no larger inclosing strata are present; instead, all the layers deposited at a locality are inclined. For example tephra characteristically accumulate in conical piles surrounding their volcanic source and form layers with steep inclinations (Fig. 16-4). Because *all* such layers are inclined, they are not cross-strata. The same can be said of some ancient lava flows (Fig. 2-15). These are true exceptions to the principle of original horizontality.

The directions of flow of the ancient currents that formed a set of cross-strata can be determined by measuring directions of inclination. In this way aspects of a former environment can be inferred. We must, however, emphasize the obvious point that as not all inclined layers are cross-strata, we

Fig. 16-3. These cross-strata are gently curved; they are truncated at top and become parallel with underlying stratum at base. Current flowed right to left. Silurian sandstone, Hawkins County, Tennessee. (J. E. Sanders.)

Fig. 16-4. Tephra having steep initial dip as a result of deposition on slopes of a volcanic cone near Cerro de Camiro, Mexico. Oldest layers are at lower right, youngest at upper left. Each layer represents an episode of eruption. (Kenneth Segerstrom, U. S. Geol. Survey.)

Fig. 16-5. Graded layer seen in a polished hand specimen. Rock grades from pebbly sandstone at base, where it overlies black shale, through siltstone to claystone at top. Triassic strata, Branford, Connecticut. (J. E. Sanders.)

must look for the origin of the inclination. Layers originally deposited in an inclined position imply no crustal deformation; their direction of inclination is the direction of flow of the currents that deposited them. In contrast, layers deposited in an originally horizontal position but now inclined imply crustal deformation; the direction of their inclination is a function of major geologic structure (Chap. 17) and affords no clue to the direction of flow of the depositing current.

Arrangement of particles within a stratum. In addition to the relationships of layers to each other, several kinds of arrangement of the particles within a single layer are possible. Each kind gives information about the conditions under which the sediment was deposited. Chief among the kinds are *uniform layers, graded layers,* and *nonsorted layers.*

Uniform layers. A layer that consists of particles of about the same diameter is called *uniform.* A uniform layer of clastic rock implies deposition of particles of a single size, with little change in the velocity of the transporting agent. This might occur in the bottomset layers of a delta. A uniform layer of nonclastic rock implies uniform precipitation from solution, which produces crystalline particles of a single size. But, a layer that is laminated because of grain-size differences suggests a transporting agent with fluctuating velocity.

Graded layers. If we put a quantity of small solid particles of different diameters and about the same specific gravity into a glass jar of water, shake the mixture well, and then let it stand, the solid particles will settle out and form a deposit on the bottom of the jar. The heaviest and largest particles settle first, followed by successively smaller particles; the finest ones, if small enough, may stay in suspension for hours or days, keeping the water turbid, before finally settling out. Thus particle size decreases gradually from the bottom upward. This arrangement characterizes a ***graded layer,*** defined as *a layer in which the particles grade upward from coarse to fine* (Figs. 15-13, 16-2, 16-5). The deposit is, as it were, a one-stroke affair. The water receives energy from the shaking of the jar and becomes turbulent; as a result sediment is lifted above the bottom. When the shaking stops, abrupt loss of energy results in continuous deposition.

Although a graded layer produced by simple shaking in a jar is graded only in the vertical dimension, those produced in nature are graded

Fig. 16-6. Tillite, representing a very ancient glaciation, overlying a smooth surface of dolerite, on which many glacial striations are visible. The tillite lacks stratification. The pebbles and cobbles in it are mostly dolerite, further indicating glacial erosion of the underlying dolerite. View looking southwest, the direction of flow of the glacier. Nooitgedacht, Cape Province, South Africa. (R. F. Flint.)

laterally also, because they are made by currents moving from one place to another. As the heaviest particles settle first, grading is not only from bottom to top but laterally in the downcurrent direction.

Graded sediments are widespread beneath abyssal plains of the deep-sea floor and are present in many deposits on the lands. Some of the processes that cause grading are (1) turbidity currents in the sea and in lakes, (2) streams, as they lose energy and deposit bed-load sediment while floods subside, (3) tephra flows (Chap. 19), (4) tephra falls, and (5) dust storms as they die down. All five processes represent rapid, continuous loss of energy, an essential condition for the creation of graded layers.

Nonsorted layers. The particles in some sedimentary rocks are not sorted at all. They consist of mixtures of various sizes arranged chaotically, without any obvious order. Processes that create sediments of this class include avalanching, debris flow, mudflow, solifluction, submarine slump, and transport by glaciers and floating ice. Widely recognized among nonsorted sediments is tillite (Fig. 16-6), of glacial origin.

Rounding, sorting, imbrication. As we noted in Chapters 7 and 8, particles broken from bedrock by mechanical weathering and other processes tend to be angular, because breaking commonly occurs along joints and surfaces of stratification. As they undergo transport by water or air the same particles tend to become smooth and rounded. Figure 16-7 shows what can happen to pebbles on a beach, and Fig. 16-8 indicates how sand grains become shaped during transport. Degree of rounding therefore gives some idea of the distance or time involved in transport by flowing water or flowing air.

Glaciers, however, tend to give irregular shapes to rock particles through crushing and abrasion. The faceted shapes of ideal glaciated coarse particles, and the scratches on them, are illustrated in Fig. 12-13. Flowing air likewise grinds facets on coarse particles, but the facets on ventifacts (Fig. 13-20) commonly are more distinct and meet each other more sharply than do those made by glaciation. Ventifacts are not rounded because they are not transported by wind; they remain immobile while the grains driven by the flowing air abrade them.

We mentioned in this chapter the fact that sorting of sediments is responsible for most stratification. Rock particles become increasingly well sorted while they are becoming increasingly rounded. This change with distance makes it possible to draw general inferences about directions and distances of transport of ancient sediments and therefore about geographic relationships at the times when the sediments were deposited.

The fabric of some sedimentary rocks is inhomogeneous. For instance, streams of high velocity with bed loads of pebbles or cobbles that are platy or

Fig. 16-7. Rounding of pebbles during transport by surf on a beach. These fragments of basalt were collected at random from a talus and an adjacent beach near Clarence, Nova Scotia. They are arranged here to show what can be expected to happen to a plate-shaped piece (*upper left*) and to a spindle-shaped piece (*upper right*) during progressive abrasion. The end product of each could be the same—a spherical pebble. Although this rounding is the work of beach drifting, stream transport produces a similar result. (J. E. Sanders.)

disk shaped (as are some of those shown in Fig. 16-7) tend to deposit them at an angle, with their flat sides sloping steeply downward in the upstream direction. This arrangement of the particles in stream-transported gravel is ***imbrication,*** *the slanting, overlapping arrangement of flat pebbles, like shingles on a roof.* It is visible in Fig. 16-9 and, like cross-stratification, is useful in determining direction of flow of former streams.

Derivation. Through the kinds of minerals or rocks of which they are composed, sediments reflect the kind of parent rock from which they were derived. An uncomplicated example is a boulder train in glacial drift; another is a placer of gold, diamond, or other economic substance. Many boulder trains and placer minerals have been traced backward to point sources; from a boulder train the direction of flow of a glacier can be inferred.

More broadly, we can draw general inferences as to the kind of rock from which a widespread body of sediment or sedimentary rock was derived. Going back to the chemical weathering of granite and basalt (Table 7-1), we recall that granite yields quartz and clay minerals, but that basalt can yield no quartz. If, however, either kind of rock were weathered mechanically, it would yield bits of the rock itself, and the bits would include feldspar. If transported and deposited quickly, the feldspar would not be destroyed by weathering. The presence of fresh feldspar in sedimentary strata would suggest one of two things about the origin of the sediment. Either a dry or very cold climate with a minimum of chemical weathering prevailed, or the cutting of valleys by streams occurred on slopes so steep that rate of chemical weathering could not keep pace with rate of erosion by streams.

Features on surfaces of layers. *Ripple marks.* In foregoing chapters we discussed the making of ripples by streams, by the wind, and by currents and waves in lakes and the sea. Ripples are preserved in some sandstones and siltstones as *ripple marks* (Fig. 16-10).

Mud cracks. Some claystones and siltstones contain layers that are cut by polygonal markings. By comparison with sediments forming today (Fig. 1-11, *A*), such as those in roadside puddles following a rain, we infer that these are ***mud cracks,*** *cracks caused by the shrinkage of wet mud as its surface became dry* (Fig. 16-11). The presence of mud cracks in a rock generally implies at least temporary exposure to air and therefore suggests tidal flats, exposed stream beds, playa lakes, and similar environments. Occurring with some ripple marks and mud cracks, and preserved in a similar manner, are the footprints and trails of animals. Even the impressions of large raindrops made during short, hard showers can be preserved.

OTHER FEATURES

Fossils. The animals and plants (or parts of them) that were buried with sediments, protected against oxidation and erosion, and preserved through the long process of conversion to rock constitute the fossils (Fig. 4-8) that form a very important part of the record of the strata. Not only are fossils an important clue to former environments, but they are the chief basis of correlation and the construction of the geologic column.

Fig. 16-8. Rounding and sorting of mineral grains during transport. *A.* Mineral grains loosened and separated from igneous and metamorphic rocks by mechanical and chemical weathering before transport. The angular shapes of the individual grains, slightly altered by weathering, are the forms assumed by the minerals as they crystallized from a magma. The aggregate of grains is a nonsorted sand. *B.* Sand carried from an area of rocks similar to those which yielded the sand in *A.* Some of the less-durable mineral grains have been broken up and lost, leaving a larger proportion of the durable mineral quartz. Battering in transit has partly rounded the grains. *C.* Sand transported through a long distance. Grains have become well rounded and consist almost entirely of durable quartz. Weaker minerals did not survive transport.

Fig. 16-9. Imbrication of cobbles in middle and upper layers of this conglomerate, of Triassic age, shows deposition by a swift stream that flowed from right to left. Hants County, Nova Scotia. (G. deV. Klein.)

Fig. 16-10. Ripple marks in sandstone of Devonian age, later deformed so that they dip steeply. Visible part of cliff at right is about 6m high. Percé, Quebec. (J. E. Sanders.)

Fig. 16-11. Mud cracks preserved in sandstone of Mississippian age near Pottsville, Pennsylvania. This is the under side of a slab turned up on its edge. The sand was deposited over a layer of river mud in which the cracks had formed and then hardened. The claystone (the former mud) has been destroyed by recent weathering, but chips of it litter the foreground. Most ancient mud cracks, like these, are really fillings of cracks by sand that was deposited in them. (Joseph Barrell.)

Color. The colors of sedimentary rocks vary considerably. Some rocks exposed in cliffs are colored only skin deep by a product of chemical weathering. For example, a sandstone that is pale gray on freshly fractured surfaces may have a surface coating of yellowish-brown limonite, developed during weathering by the oxidation of sparse, iron-rich minerals included with the grains of quartz sand.

The color of fresh rock is the combination of the colors of the minerals that compose it. Iron sulfides and organic matter, buried with the sediment (Fig. 15-9, *B*), are responsible for most of the dark colors in sedimentary rocks. Microscopic examination of red and brown rocks shows that their colors result mainly from the presence of ferric oxides, as powdery coatings on grains of quartz and other minerals or as very fine particles mixed with clay.

These oxides get into strata in at least two different ways. In most red and brown rocks probably the colored oxides are secondary, created in the strata by alteration of ferromagnesian minerals such as hornblende and biotite, during weathering and during the conversion of sediment into rock. Such chemical alteration can take place in any warm climate, dry or moist.

In some strata the colored oxides are believed to be primary, having been deposited as components of the sediments themselves. Such oxides may have been derived from the erosion of red-clay soils like those forming today in the warm climate of the southeastern United States. Washed down from uplands into rivers, the oxides are deposited in places on swampy flood plains, where decaying plant matter creates strong reducing agents that reduce the red ferric oxide to ferrous oxide, which is not red. Rivers carry the rest of the red sediment into the sea, where organic matter on the sea floor likewise reduces it. Probably, therefore, the red color can be preserved only if the sediments escape reduction after deposition. This they could do if deposited in basins where, for one reason or another, organic matter does not accumulate in amounts sufficient to reduce the ferric iron.

Without very detailed research, therefore, all we can tell about the climate where and when an ancient red-colored stratum was deposited is that the climate was warm, either there or in the region from which the sediment was derived.

Concretions. Inclosed in some sedimentary rocks are bodies called *concretions*. They range in diam-

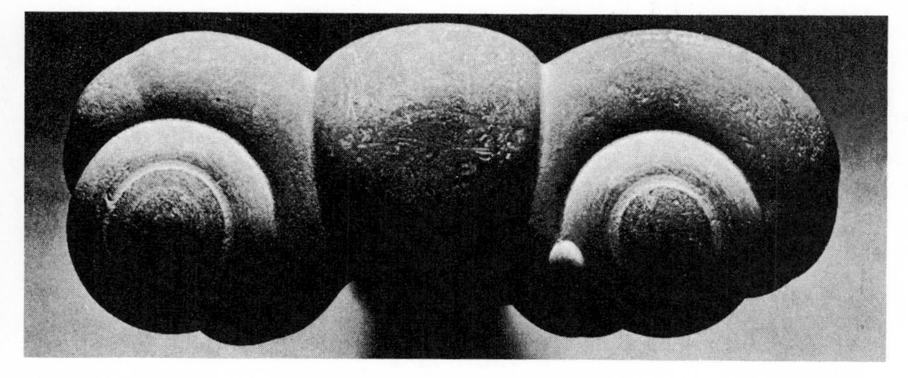

Fig. 16-12. Concretion, 7cm long, consisting of calcium carbonate. It was imbedded in shale. (Andreas Feininger.)

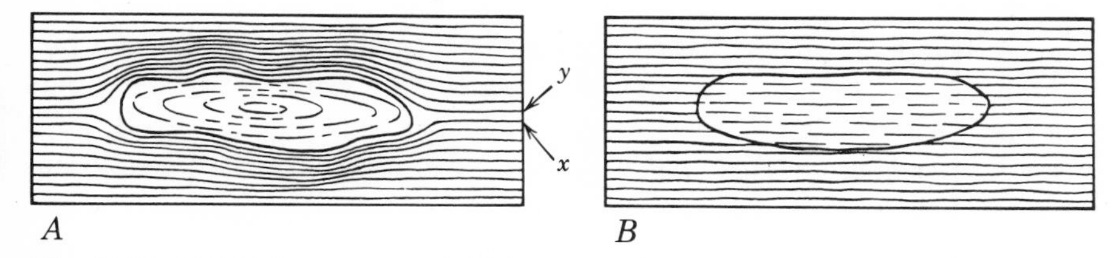

Fig. 16-13. Vertical cross sections of two concretions, each several inches long, in shale, showing different times of origin relative to deposition of the inclosing strata. *A.* Concretion formed after lamina *x* was deposited but before lamina *y* was laid down. Hence the concretion is contemporaneous with deposition of the body of sediment. *B.* Concretion transected by laminae of the inclosing rock. Hence it was formed after all the laminae shown had been deposited.

eter from a fraction of an inch to many feet and in shape from spherical through a variety of odd shapes, many with remarkable symmetry (Fig. 16-12), to elongate lenses that parallel the stratification of the rock. Concretions are composed of many different substances, including calcite, silica, hematite, limonite, siderite (iron carbonate), and pyrite. Small concretions have been dredged up from the sea floor, showing they are forming there today as sediments are being deposited. This origin, contemporaneous with the inclosing sediments, is indicated also by the shapes of some concretions and by their relation to the lamination of the surrounding sedimentary rock. Others formed after the deposition of the sediments, as for example those that retain the stratification of the surrounding rock (Fig. 16-13).

The substances of which concretions are made show that these objects are the result of localized chemical precipitation of dissolved substances from seawater, lake water, or connate water. Once precipitation starts around a fossil or other body that differs from the inclosing rock, the concretion thus formed continues to grow. Indeed, in some rocks perfectly preserved fossils are found at the centers of concretions.

Because we do not yet understand concretions fully, the best definition we can suggest is that a ***concretion*** *is a localized body having distinct boundaries, inclosed in sedimentary rock, and consisting of a substance precipitated from solution, commonly around a nucleus.* Some geologists use the word *nodule* as a synonym for concretion; others restrict it to concretions of small size.

STRATIGRAPHY

Environments of deposition. We noted in Chapter 6 that the study of strata is *stratigraphy.* In the present chapter we have described various aspects of sedimentary rocks. Now we must look at some broader aspects, the first of which concerns our interpretation of the environments in which sediments are deposited.

One of the objectives of research in geology is to reconstruct the *paleogeography* (the ancient geographic relations) of a region, at the time when a particular sequence of strata was deposited as sediments. The reconstruction would show the distribution of land and sea, streams with their directions of flow, mountains and lowlands, deserts, glaciers, and if possible, indications of the climate then prevailing. Information of these kinds is commonly assembled in the form of a *paleogeographic map*, representing a series of inferences drawn from the physical characteristics of the sedimentary rocks and from the fossils they contain. The inferences mainly concern the environment in which each kind of sediment was deposited and are based on an analogy with the environments of today's sediments, by use of the principle of uniformity of process.

Looking at modern sediments and remembering the slopes shown in Fig. 8-2, we conclude that sediments deposited along streams, in lakes, and in the sea are deposited primarily in basins, where the chances of preservation are better than on high places and on steep slopes. To distinguish among the various environments in which a group of strata may have been deposited, we must look closely at exposures of the strata, following the west-east line marked by numbers in Fig. 16-14.

1. In one area we see conglomerate consisting of little-rounded and little-sorted cobbles and pebbles, with very irregular cross-strata dipping east. In places the conglomerate is interbedded with thin layers of mudflow sediment. These characteristics suggest a fan or fans at the foot of the steep slope of a highland.
2. Farther east we find better-sorted sandstone with the cross-strata characteristic of point bars, associated with elongate bodies of siltstone and broad thin layers of claystone. Fossils might include freshwater mollusks and leaves and branches of trees. This assemblage suggests a floodplain on a lowland, with meanders, natural levees, and overbank-flood sediments.
3. Still farther east is well-sorted, well-rounded sandstone, with long, parallel, gently dipping foreset beds overlain disconformably by stream-channel cross-strata. Fossils include plant debris and a few brackish-water mollusks. Evidently we have passed from a former piedmont environment, across a coastal plain drained by sizable rivers, to a low-lying seacoast.
4. Assuming that good exposures of our strata continue eastward, we can see sandy beach sediments (possibly overlain in places by well-sorted sandstone with cross-strata dipping steeply east), suggesting beach dunes of wind-blown sand.
5. Still farther east we find the texture becoming finer, with sandstone grading into siltstone and claystone, containing fossil mollusks of offshore-marine kinds. The increasing fineness of grain reflects the settling velocities of particles of various sizes.

From this information, clearly indicating several environments that differ from one another but that are logically related, we can sketch the crude paleogeographic map shown in Fig. 16-14. In a similar manner we can identify former lakes, estuaries, lagoons, and reefs, as well as sandy deserts and country overrun by glaciers, each feature representing a distinctive environment of deposition.

These environments are no longer present in the district we are studying. But they can be recaptured by virtue of the fact that the various sediments formerly deposited in them were preserved and converted into rocks, as they subsided and were covered by other sediments. In short, they became a part of the geologic record. A common way in which nearshore-marine sediments are preserved in the record is shown in Fig. 14-14.

Sedimentary facies. Implicit in our discussion of environments is the fact that most strata change character from one area to another. In the foregoing discussion we followed a unit or group of strata, all parts of which were deposited in the same interval of time, from the base of a highland across a coastal plain and into a shallow sea that deepened offshore. The changing environments are represented by changing grain size, grain shape, stratification, depositional structures, and fossils in the unit. *A distinctive group of characteristics within a rock unit, that differ as a group from those elsewhere in the same unit,* is a sedimentary ***facies*** (Lat. "aspect"). Two facies merge laterally into each other either gradually or abruptly, depending on the relations between the two former environments of deposition.

If a sedimentary unit were exposed in section

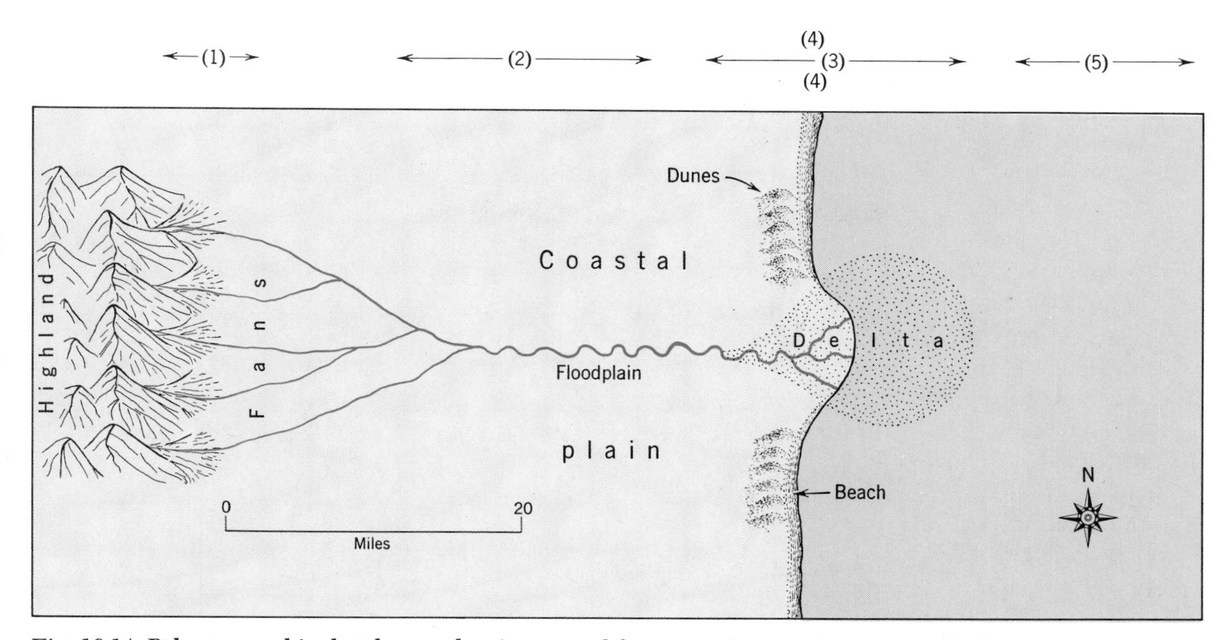

Fig. 16-14. Paleogeographic sketch map showing several former environments, reconstructed from characteristics of strata. Numbers indicate positions along a west-east line, described in text.

from end to end of its extent, it could be identified as a unit despite changes in its facies. But if, as is usual, only widely separated parts are exposed and if each part represents a different facies, its contained fossils would be needed for correlation. A difficulty arises here because the assemblages of fossils in two facies may not be exactly the same, even though the organisms they represent lived at the same time. This happens because the environments in which the organisms lived were different. In the same sea, deep-water shellfish are unlike shallow-water kinds; and on land, animals living in deserts are unlike those living at the same time in moist, forested regions. These variations of fossils with varying facies do not, however, make correlation impossible; they only make it more difficult.

Field study of a sequence of strata. Having followed a single unit laterally through several facies, we can now examine a vertical sequence of several units, such as the sequence shown in Fig. 16-15, exposed in a steep slope. We examine the rocks systematically by climbing up the slope, taking notes as we go. We identify the kinds of rock present and determine the thickness of each of the units, of which the figure shows five, numbered A to E. The surfaces of contact between any two units are not alike. The contacts between D and E and between C and D are sharp and distinct, whereas those between A and B and between B and C are transitional, that is, they represent a gradation between the layer beneath and the layer above. Looking for fossils to help determine the origins of the rocks, we find fossils of marine animals in units A, B, C, and D and fossil bones of land animals in unit E.

Examining sandstone A, we find it is made up of grains of quartz and other minerals. All the minerals in this group occur in igneous rocks; so we infer that the sandstone was built of the sediment resulting from erosion of igneous rocks. The sediment might have been derived from its parent rocks either directly or remotely, for it could have formed part of one or more sandstones older than the one we are examining. In either case the sandstone shows sorting, for the proportion of quartz to other minerals is far greater than in most igneous rocks. Probably this means that the durable quartz grains survived their trip from the region of erosion to the region of deposition in much greater numbers than did their less resistant associates. However, under a hand lens we can see that even the quartz grains have become moderately well rounded. From this

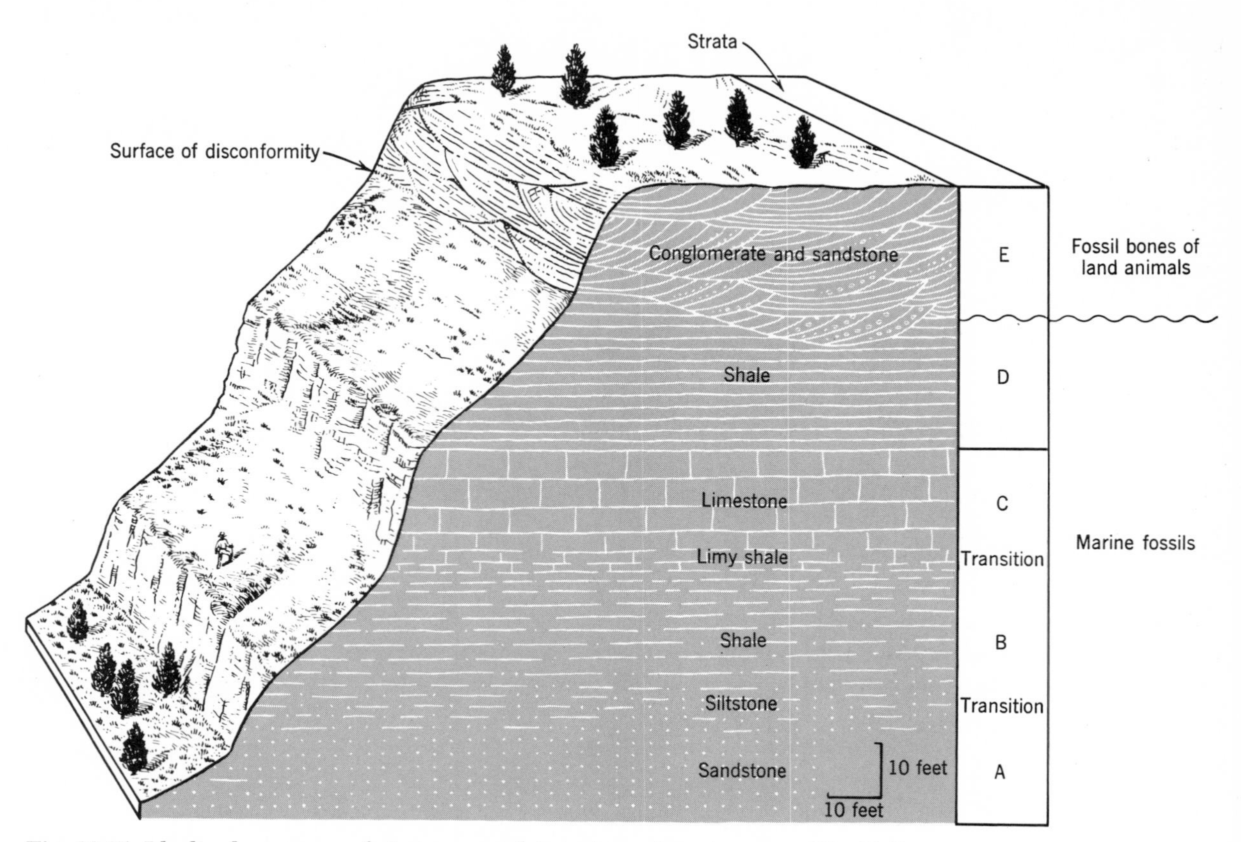

Fig. 16-15. Idealized sequence of strata exposed in a steep slope (compare Fig. D-8).

we judge roughly that the journey involved considerable distances, long periods of time, or both.

Stratification of the sandstone is indistinct. The rock lacks well-marked laminae because the diameters of all the sand grains are nearly the same; hence there is little distinction between one layer and the next. Yet the faint laminae present lie flat and nearly parallel. From this we infer the sand was deposited in water deep enough to be beyond the reach of waves and strong currents. We can suppose the sand was either derived from the wave erosion of cliffs of an older sandstone or brought into the sea by rivers that carried some of the products of weathering of inland rocks. In the latter case the streams must have possessed enough energy to enable them to move sand grains; this in turn suggests the slope was fairly steep and therefore that the tributary region was one of hills or mountains. These are only suggestions, which can hardly be verified from a study of this one exposure. Verification could come only from a wide study of the sandstone and of the tributary region.

We note that upward the sand becomes finer and is accompanied by silt. The silt then increases and is accompanied by clay, so that we pass upward from sandstone through siltstone into shale (unit B). In the shale we find the same near-parallelism of the laminae and much the same kinds of marine fossils. Hence the only properties that have changed, as compared with the sandstone beneath, are sizes and shapes of grains and also kinds of minerals present, because the fine particles that constitute shale consist not of quartz but of clay minerals. We can explain the change in one of two ways. Either the water deepened, shifting the zone of sand deposition farther toward shore and causing deposition of mud here (Fig. 14-21) or the contributing streams lost energy and deposited their sand before reaching the sea. To determine the cause, we would have to study a much wider region.

The upper part of the shale becomes calcareous, that is, it begins to include calcite along with the clay minerals. Also it grades upward into nearly pure limestone (unit C) containing the same general kinds of fossils as those beneath. On close inspection the limestone is seen to be a mechanical deposit of tiny shell fragments. We can conclude with confidence that, in part of the sea, the water cleared and marine organisms became abundant. This could have been brought about by continued deepening.

Examining unit D, we find no transition from C but instead an abrupt change from limestone to shale. Here we find the same kinds of fossils and the same flat-lying, parallel laminae; only the composition of the rock has changed. We infer sudden shoaling of the water or sudden increase in stream energy rather than a gradual change.

A summary of our findings on units A, B, C, and D suggests a former sea, deepening gradually and perhaps somewhat irregularly or fed with sediment by rivers with changing characteristics. Hence we are left with uncertainties. These are something a geologist expects, for the nature of the evidence he must rely on often permits him to do no more than judge the relative probability of alternative explanations of the same feature.

However, from our observation that units A, B, C, and D are parallel and show no sign of intervening breaks, we infer with little doubt that together they represent continuous, unbroken sedimentation on the sea floor. Such strata are said to be *conformable.*

Unit E differs from underlying units. It is part sandstone and part conglomerate and is cross-stratified in wedge-shaped units. The repetition of wedge-shaped beds suggests currents that shifted in position and direction; the occurrence of both sand and gravel indicates changes in energy as well; and finally the common direction of inclination of the cross-strata suggests general flow from right to left.

Also unit E contains the fossil bones of land animals. This fact makes a marine origin unlikely; the presence of conglomerate rules out the wind. Accordingly, we conclude that probably E was deposited by a stream or streams, and the dimensions of the wedge-shaped beds give a clue to the size of the stream channels. Streams today are seen to deposit similar beds.

Evidently E overlies D with a contact that is notably irregular and that cuts through some of the uppermost laminae of D. This relationship could result only from erosion of D before or during the deposition of E. When two parallel strata are separated by an irregular surface of erosion, the relationship between them is said to be *disconformable.* The irregular interface is called a *surface of disconformity.*

From these facts we judge that the sea floor emerged for some reason not shown by the evidence at this place. Unit D thereby became land, and a stream flowed over it, eroding an unknown amount of its upper part. To determine how much of D was removed during this erosion and whether to attribute all of the erosion to the stream or part of it to some agency that antedated the stream, we should have to examine D at many other places. At any rate, the stream possessed rather high energy because it carried a bed load of pebbles as well as sand and because the load was so abundant that some was deposited. Evidently the stream was gradually building up its bed because the wedge-shaped layers are piled one on another.

Like most sand grains, those of D are mainly quartz, but the pebbles are samples of various kinds of rock. Here is a clue to the region from which the stream brought the pebbles. To make use of the clue we should have to identify the kinds of rock in the pebbles and to search (probably in the direction to the right of Fig. 16-15) for areas in which those same kinds of rock occur as bedrock. A further clue to the distance the sediment has been transported is found in the degree of rounding of the particles.

This is the way in which sedimentary rocks are commonly studied in the field. From such a study it is possible to recreate a general sequence of events, but, as we have seen, not all our inferences are firm; we can not always rule out other possibilities; and the study leads to questions that can be answered, if at all, only by extending the inquiry into the surrounding region.

Rock units: formations. We identified unit A (in Fig. 16-15) as sandstone, but a thorough study must go much further. The unit must be distinguished from other sandstones. One respect in which it differs from all others is its position in the vertical sequence of strata. Hence we give it a designation by which its position is fixed and by which it can be catalogued and referred to. The basic rock unit to which such a designation is applied is the *formation* (App. D). A formation must constitute a mappable unit; that is, a unit (1) thick and extensive enough to be shown to scale on a geologic map, and (2) distinguishable from the strata immediately

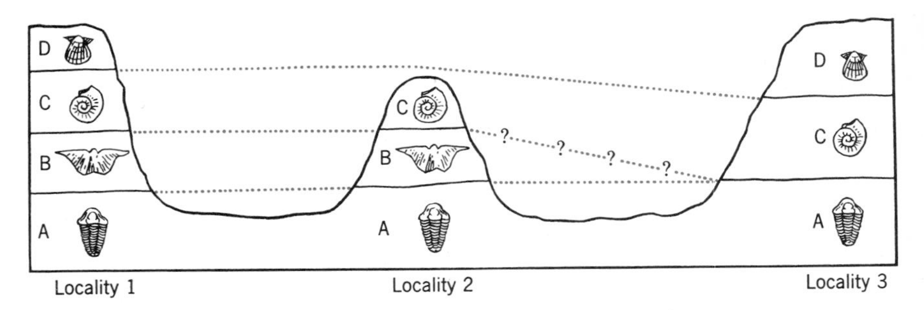

Fig. 16-16. Correlation of strata exposed at three localities, many miles apart, on a basis of similarity of the groups of fossils they contain. The fossil groups show that at Locality 3, stratum B is missing because C directly overlies A. Was B never deposited there, or was it deposited but later removed by erosion, before the deposition of C?

above and below not just at one exposure but generally wherever the unit is exposed. Within these requirements a formation can be thin or thick, to suit the geologist's convenience. Its thickness is likely to depend on the degree of detail of the field study and correspondingly on the scale of the map to be made. In North America each formation is given a name, typically the name of a locality near which it is exposed (Lexington Limestone, Fox Hills Sandstone, Green River Formation). Not only sedimentary rocks but igneous and metamorphic rocks as well are identified as formations (examples: San Juan Tuff, Conway Schist). Figure D-9 is a series of geologic sketch maps showing three formations.

Formations can be subdivided and also grouped into larger rock units. But because we are concerned here only with the way in which strata are identified, the formation alone serves our purpose. For more about this see Appendix D.

Matching of rock units by physical characteristics. Once the layered rock units have been identified in vertical sequence, the extent of each (that is, the area it underlies) must be determined as closely as possible. With few exceptions, layers of sediment are deposited in basins located on land, beneath lakes, or in the sea. A layer of sediment may cover all or only a part of a basin. If deposited in the sea or in a lake, the stratum is likely to extend over the whole basin floor. However, during the time, perhaps very long, since conversion from sediment into sedimentary rock, the stratum may have been eroded so much that only parts or remnants of it may now be preserved. An example is the Pittsburgh Coal Seam, an easily recognized formation, whose original area of perhaps 20,000 square miles has been reduced by long-continued erosion to 5,729 square miles. One of the tasks of the geologist is to study the remnants so thoroughly that he can determine as nearly as possible the original extent of each formation.

This is done by matching the remnants preserved in the hills left by erosion on a basis of physical characteristics such as grain size, grain shape, mineral content, kind of stratification, and color. Matching on this basis is likely to be reliable through short distances but generally becomes less reliable through longer distances, because the physical characteristics tend to change in lateral directions.

Correlation by means of fossils. The usefulness of matching by physical characteristics is virtually limited to the area of the basin in which the strata were originally deposited. In order to determine equivalence in the ages of strata in different basins, and even in different continents, geologists compare fossils, the chief basis of stratigraphic correlation (Fig. 16-16).

Correlation by means of isotopic dates. As we learned in Chapter 6, the ages of fossil-bearing strata are fixed by the isotopic dates of igneous bodies closely related to them. Such dates are useful also in correlation. Where, as is not uncommon, sedimentary layers contain no fossils, we can fix their ages approximately through the isotopic ages of related igneous rocks. In some cases this enables us to correlate dated strata that lack fossils with dated strata that contain them, thus gradually enlarging the known extent of strata whose positions in the geologic column have been fixed by fossils.

We have now examined what strata themselves indicate about geologic history. In Chapter 17 we shall see what strata can reveal about geologic structure.

Summary

1. Sedimentary rocks are the product of complex processes of sorting and transport.

2. Most sedimentary strata were nearly horizontal when their sediments accumulated.

3. The most obvious characteristic of sedimentary rocks is that they are stratified. Strata include parallel strata and cross-strata. Various arrangements of the particles are seen in uniform layers, graded layers, and nonsorted layers.

4. Particles of sediment become rounded and sorted in transport by water and air but not in transport by glaciers and by mass-wasting.

5. Ripple marks, mud cracks, fossils, and, in some cases, red color in sedimentary rocks give evidence of environments of deposition.

6. Stratigraphy is the study of stratified rocks. From it most of our knowledge of the Earth's history has been learned.

7. An extensive unit of strata may possess several facies, each determined by a different environment of deposition.

8. The basic physical rock units are formations, each with a locality name.

9. By means of fossils and in some instances by isotopic dates strata can be correlated through long distances.

10. Correlation takes account of the two principles that strata were horizontal when deposited and that they were formed in sequence from bottom to top.

Selected References

Dunbar, C. O., and Rodgers, John, 1957, Principles of stratigraphy: New York, John Wiley.

Hatch, F. H., and Rastall, R. H., 1965, The petrology of sedimentary rocks, 4th ed., revised by J. T. Greensmith: London, Thomas Murby.

Krumbein, W. C., and Sloss, L. L., 1963, Stratigraphy and sedimentation, 2d ed.: San Francisco, W. H. Freeman.

Lahee, F. H., 1961, Field geology, 6th ed.: New York, McGraw-Hill.

Shrock, R. R., 1948, Sequence in layered rocks: New York, McGraw-Hill.

Walker, T. R., 1967, Formation of red beds in modern and ancient deserts: Geol. Soc. America Bull., v. 78, p. 353–368.

CHAPTER 17

Deformation of Sedimentary Strata

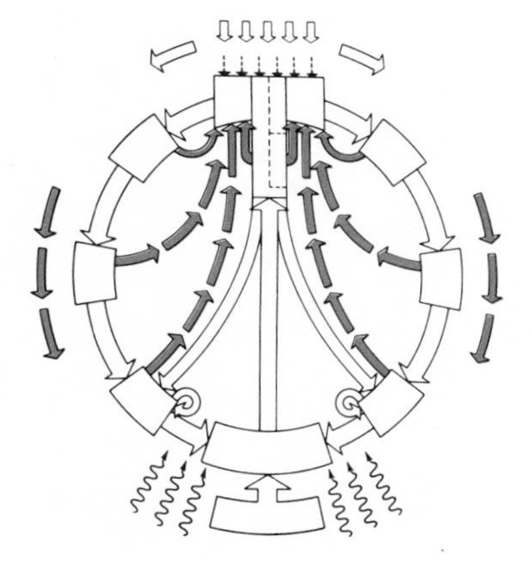

Strata as keys to internal crustal processes
Deformation defined
Strength of rocks
Displacement of man-made structures
Deformation of shorelines and morphologic features
Stratigraphic evidence
Gently warped and closely folded strata
Joints and faults
Relation of joints and folds to faults
Dating structural movements

Plate 17

Students, on class field trip, examining deformed strata (marine sandstones and siltstones) of Ordovician age exposed in roadcut near west end of Rip Van Winkle bridge, near Catskill, New York.

(*J. E. Sanders.*)

CHAPTER 17 *Deformation of Sedimentary Strata*

PROBLEMS IN STUDYING INTERNAL CRUSTAL PROCESSES

Strata serve not only as ties binding together our previous studies of external processes but also as keys by which geologists unlock many secrets of processes inside the Earth's crust. In examining various external processes, we could verify much from our own experience. In the chapters dealing with mass-wasting, running water, wind, and glaciers, for example, we found many familiar landmarks. Ahead of us lie subjects more alien; the way is marked by few, if any, familiar landmarks. We therefore stand at a critical juncture; our further progress must depend on the familiarity with strata we have already gained. Such familiarity in turn depends on knowledge of external processes. To repeat what we said at the beginning, strata are the keys to our further progress.

The subjects we shall examine in ensuing chapters not only are less familiar but also are so closely interrelated that they do not lend themselves readily to the kind of presentation we have used up to now. All aspects of many upcoming subjects do not fit neatly into single chapters. In this chapter, for example, where our study of deformation of the crust begins, it would be desirable to be able to answer the four critical questions: *what? where? when?* and *how?* before moving on to the next chapter. We can cope with the *what* by defining deformation; and, if we confine ourselves largely to the deformation of sedimentary strata, we can encompass some facets of the *where* and most of those of the *when*. This we do by setting forth the evidence that tells us the crust has been deformed, by describing some common structures resulting from deformation, and by showing how the geologic date of deformation is determined. But we will not be able to finish with the *where* nor even to try to understand the *how* of deformation until we have become acquainted with the interior of the Earth, with metamorphism, and with the history of mountains. Therefore we must defer our encounter with the *how* until Chapter 22.

Similarly, the subject of faults is not readily confined to a single chapter, as we already know because we have met faults briefly in several preceding chapters. Here we examine them in much more detail and emphasize their connection with the displacement of strata. Still other aspects of faults lie beyond this chapter. Abrupt movement on faults is a cause of most earthquakes, which create many varied effects both on the surface and within the Earth. Careful instrumental study of earthquakes has revealed much about the interior of the Earth. These subjects are explored in Chapter 18. Some faults have served as conduits through which magma has moved upward from deep within or below the crust to the Earth's surface. Escape of material from within the crust takes place through the many varied activities of volcanism, treated in Chapter 19.

Some great faults have metamorphosed adjacent rocks. We look into this and other kinds of metamorphism in Chapter 20. Within present and former mountain chains are faults, in many places horizontal or gently inclined, along which great displacements have occurred; we shall explore these in Chapter 21. All roads lead to Chapter 22, where, in discussing the evolution of the crust, we synthesize not only internal processes but external ones as well. In that chapter we encounter the tremendous, steeply inclined fracture zones discovered recently on the deep-sea floor—yet another aspect of faults.

To handle such interrelated subjects in many of the following chapters we shall try to pick out and introduce a few principal topics, bring together certain of their aspects, and then show how they relate to subjects treated farther along. We have done this already to some degree with cross-strata, which appeared and reappeared many times in various chapters concerned with external processes.

Hereafter, this treatment will be the rule rather than the exception.

A great difficulty with internal processes is that we are unable to see them in action. Instead we must deal with the results of past activities which become visible to us only after the crust has been elevated and eroded. Accordingly, we shall be using the inferred history of the past in our effort to understand what is going on in the present. This is the reverse of our earlier practice, in which we depended largely on what we could see happening now to guide us through the history of the past. Before we plunge in, let us summarize the importance of deformation.

IMPORTANCE OF DEFORMATION

Deformation is important scientifically because it records many events in the history of the Earth's crust; in fact, some important events are known only by the records of such deformation. Historical analysis of deformation, therefore, is another means by which geologists approach their fundamental goal of working out the history of the Earth.

The results of deformation are of vast economic importance. Layers of valuable materials, such as coal and rock salt, commonly have been displaced by faults. Intelligent exploitation of these layers depends on knowledge of where to find the displaced parts. Petroleum and natural gas have accumulated within the crust in various kinds of traps, many of which are related to geologic structures created by deformation. Fractures within bedrock, resulting from deformation, have been avenues for migration of mineralizing fluids and the sites of deposition of valuable minerals. Finally, knowledge of locations where deformation is active helps us to select the best sites for building man-made structures.

DEFINITION

In Chapter 12 we read how the flow of glaciers can be analyzed in terms of rate of strain of solid ice brought about by shearing stresses created by the weight of the ice itself, and how the upper zone of fracture and lower mobile mass (zone) of flow of glaciers serve as models of the Earth's crust. We can extend this kind of analysis to the deformation of rocks.

To begin with, it will be helpful to mention a few terms. All solids including rocks possess a certain ***strength***, or *the ability to resist stresses created by forces that tend to cause changes of volume, shape or both volume and shape.* **Deformation** of a rock body is *a change of volume, shape, or both volume and shape, or a change of original position within the crust.* Deformation of a rock body, therefore, involves two unrelated subjects (1) its strength, and (2) its original position within the Earth's crust. The geologist's main concern with crustal deformation rests with the second subject, changes of position.

By conducting laboratory experiments with machines designed by structural engineers to test various building materials we can gain insights into the complex subject of strength. As the Earth's crust is far more complex than the uniform specimens tested in the laboratory, the concepts derived from experiments, though useful, are not yet widely applicable to natural examples.

The upper limit of strength is the *elastic limit,* the stress to which a rock can be subjected and still recover its original shape or volume when the forces tending to cause deformation have been removed. If these forces create stress in excess of the elastic limit, a permanent change of shape or volume remains even when the forces have become inactive. Permanent deformation indicates that the strength of a rock has been exceeded or, in other words, that the rock has failed. Failure can take the form of *plastic flow,* or of *rupture.* Plastic substances flow readily, whereas brittle substances rupture easily.

The strength of a rock remains about the same under given conditions but varies if conditions change. Strength must be specified in relation to external conditions, such as (1) magnitude, duration, and rate of application of forces tending to create deformation, (2) confining pressure, and (3) temperature and internal conditions such as interstitial fluids. The same rock can exhibit great strength under some conditions, yet be weak and easily deformed under others. Ordinarily the strength of a rock is great at low temperatures, and if forces tending to produce deformation are applied at slow rates for short periods of time. The strength of a rock is much reduced by raising the temperature, by applying large forces at a great rate for short periods of time, by uniform forces acting through long periods of time, and by adding interstitial fluids. Because the solubility of most substances is increased by increasing the pressure, material is dissolved readily at points of great stress and is precipitated at points of less stress. This internal mobility of the solid mate-

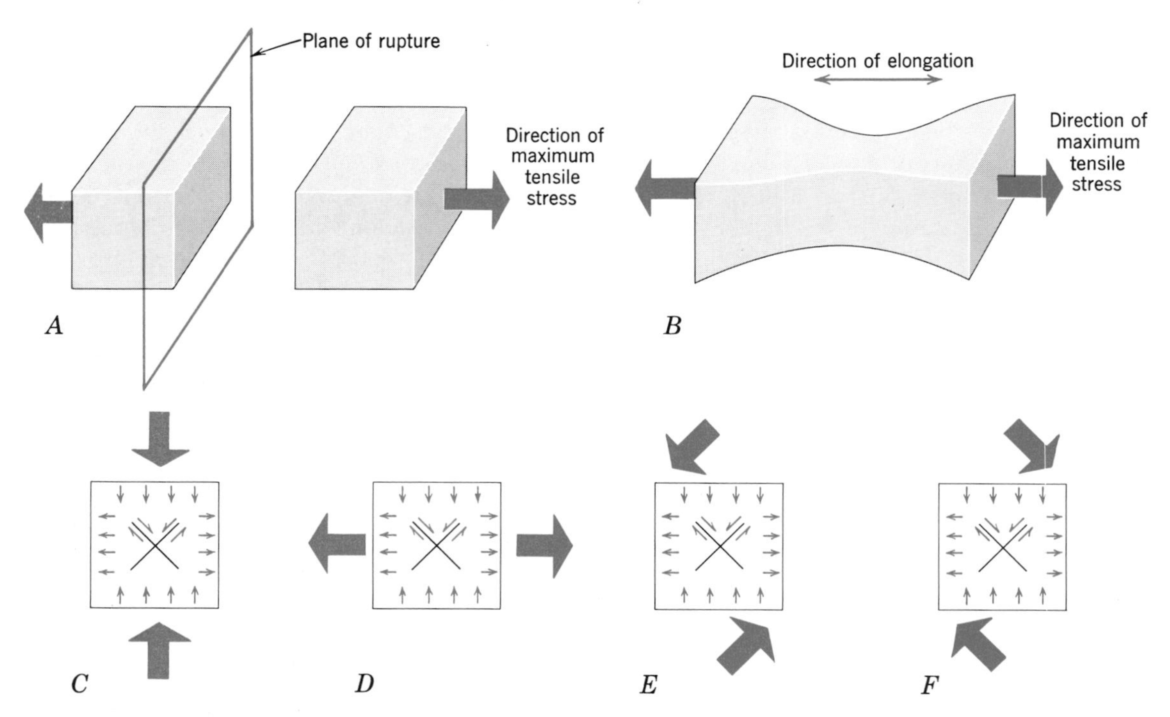

Fig. 17-1. Idealized sketches showing relations of principal stress directions to effects of simple deformation and to various patterns of deforming forces. *A.* Simple tension ruptures a brittle substance. *B.* Simple tension stretches a plastic substance. *C–F.* Identical principal stress directions (small arrows) created by various patterns of forces tending to cause deformation (large blue arrows). Whether or not they actually cause deformation the forces applied create internal stresses as shown. *C.* Simple compressional forces. *D.* Simple tensional forces. *E.* Left-lateral shearing couple. *F.* Right-lateral shearing couple.

rial through solution and reprecipitation reduces the strength of a body of rock. Fluids influence strength in yet another way. Great internal fluid pressures within a rock body can drive the particles of the solid framework apart from one another and thereby reduce strength to very low values. Great pressures on solid particles and high temperatures close all fractures and other openings in a body of rock and can cause the rock to flow, much as glacier ice flows. These factors cause metamorphism to accompany deformation.

Within a rock body that is subjected to forces tending to deform it, the distribution of stresses is complex. A convenient way to analyze such stresses is to concentrate on the principal directions along which stresses reach maximum values (Fig. 12-4, *A*). In simple cases we can relate the products of deformation to principal stress directions. For example, rupture of brittle substances commonly takes place in response to excessive tensile stress; the plane of rupture is perpendicular to the direction of maximum tensile stress (Fig. 17-1, *A*). Most rocks in the outer part of the Earth's crust are brittle. More plastic substances can be stretched and elongated parallel to the direction of maximum tensile stress (Fig. 17-1, *B*). Most rocks deep within the Earth's crust are plastic. Deformation generally is more complex than simple rupturing or stretching, as we shall see presently. Even in the simple cases where we can relate deformation to principal stress directions, we are generally unable to relate these directions to the forces that caused deformation. This unfortunate circumstance results from the fact that identical principal stress directions can be created by more than one pattern of deforming forces (Fig. 17-1, *C–E*). Therefore we must admit our inability to extend complete mechanical analysis to all features created by deforming forces within the Earth's crust. We shall try to relate these features to principal stress directions where possible but shall not attempt to resolve the ambiguities between principal stress directions and patterns of deforming forces.

EVIDENCE

The most common evidence of deformation consists of changes of original position or of level. These can be determined only if we know with certainty the original position or level. Man-made structures serve very well for positive reference; when such structures have been displaced the results are obvious to everyone. Natural reference levels are numerous; their displacement is just as obvious to a geologist as that of man-made structures is to everyone. Displacement can take the form of rupture, of shifting of various parts of the reference object, or of simple raising, lowering, or lateral shifting of the object without other change. Needless to say our conclusions about what has occurred will vary according to the size of the reference object selected. The truth of this statement will become apparent in our discussion of various examples.

Displacement of man-made structures. One of the things we want to know is whether the Earth's crust has been deformed everywhere, or whether deformation is a process that operates in some localities but not in others. We begin our quest for an answer to the *where* of deformation by marshaling the evidence used to recognize displacement of the Earth's crust. Displaced structures made by man, such as streets, buildings (Fig. 17-2), railways, fence lines, rows of trees in orchards, and so forth, permit us to recognize recent crustal deformation. Because many man-made structures are displaced during earthquakes we shall simply mention only a few examples here and save the others for Chapter 18.

Figure 17-2 shows the results of displacement along a plane dipping steeply to the right. Geologists discuss position and geometry in standard terms. These are ***attitude,*** *the orientation* (or position) *of a layer or surface,* measured by *strike* and *dip* (App. D). The street has been cut through and the shop thoroughly bent. The amount of relative displacement is approximately 1.3m.

Some man-made structures have been displaced with respect to sea level. An example, made classic by Lyell's description, consists of the ruins of a Roman marketplace known as the Temple of Serapis, west of Naples. Distinctive borings made by marine clams are found on the three intact columns to a height of about 6m above the floor (Fig. 17-3). Clam shells still line some of the borings. Sediments containing abundant shells of bottom-dwelling clams like those now living in the adjacent bay are exposed along the shore near the ruin, in bluffs rising as much as 7m above present sea level.

Fig. 17-2. Displacement of man-made structures in business district of Anchorage, Alaska resulting from earthquake of March 27, 1964. Overhanging scarp formed along reverse fault. Reverse fault defined in Table 17-1. (Paris Match, from Pictorial Parade.)

We can formulate several hypotheses to explain these observations. One of them states that throughout the world sea level rose and later fell and that the land did not move. The most important deduction from this hypothesis is that a record of such sea-level changes would be present in coastal belts everywhere. This deduction does not stand up to its first test, because the evidence cited of changes in level of land and sea is limited to the vicinity of the old ruin. Accordingly, we reject our first hypothesis and seek another.

Fig. 17-3. Columns in a Roman ruin at Pozzuoli, Italy as they appeared in 1828. Marine-clam borings indicate former submergence. (Drawing copied from Lyell, 1875, Principles of geology, 12th ed.)

Our second hypothesis states that a local part of the Italian coastal belt sank and was reelevated within very recent time. Because this hypothesis accords with all the known evidence we can adopt it with confidence. We infer that the building was constructed with its floor lying some distance above sea level and that it later sank, submerged to a depth at which the water could reach 6m above its former floor. Later, it was elevated but not quite as high as it stood originally, because the floor is still under water. According to historic records part of the uplift occurred after A.D. 1500.

In other parts of the Mediterranean region, whole cities, now submerged as a result of crustal movements, are the subjects of exploration by diving archeologists. In the Tigris and Euphrates valleys numerous cities have been built one above the other in vertical succession as the land subsided and new sediment was deposited on top of the sunken areas.

In these three examples the reference objects were simply moved in a vertical direction but were not otherwise deformed. Likewise the automobile and the lamppost indicating B Street in Fig. 17-2 were displaced bodily with respect to each other but were not damaged in the process.

Displacement of natural features. Various natural features serve as reliable reference objects for determining deformation. The Earth's surface itself, of course, moves when man-made structures are displaced. Small movements of the surface that do not noticeably displace man-made structures can be detected with sensitive instruments. Shoreline features are especially valuable, as are other indications of emergence or submergence that can be shown to have occurred independently of worldwide rises and falls of sea level. Stratigraphic evidence and structure of bedrock complete our list.

Morphologic features. Reliable reference features, originally horizontal, are built along the shores of large lakes and the sea by waves, currents,

Fig. 17-4. Emerged shore profiles eroded by waves, Palos Verdes, southern California. The present altitude of the lowest wave-cut bench (1) is about 150 feet, that of the highest (5) more than 800 feet. Photo, taken in 1933, shows hulk of ship *Gratia*, stranded during a storm in 1932. (Litton Industries, Aero Service Division.)

and organisms. If, in following ancient shoreline features parallel with the former coast, we find any departures from the horizontal, we can use the magnitude of such departures to measure crustal deformation.

Both erosional and depositional morphologic features made along shores are useful for reference. Along the California coast well-developed wave-cut benches standing one above another like stairsteps extend to heights of more than 1,400 feet above the sea (Fig. 17-4). Some of the lowest steps, terminating inland against typical wave-cut cliffs, are decorated with barnacle shells and project above beach deposits. Along some of the higher steps such critical markings have been removed by erosion or covered with younger sediments. The close resemblance of the general form of the higher steps to that of the lower ones indicates a like origin for all. Because the emerged shore features are not found along the entire coast, we infer that the segment of the Coast Ranges containing them has been elevated in a succession of pulses, and that each pulse was separated by a pause long enough to permit distinct wave-cut cliffs and wave-cut terraces to be eroded. No one ancient shoreline can be followed continuously for a long distance, for two reasons. (1) The waves did not shape the shore profile uniformly in all localities. (2) Later erosion has been more destructive in some places than in others. Because the altitude of a given wave-cut bench varies appreciably from place to place, we infer that uplift was irregular. Numerous islands in the tropical parts of the Pacific (Fig. 17-5) and other oceans show shore features built by organisms, which now stand above the sea at various levels.

We leave deformed shorelines for the moment

Fig. 17-5. Emerged shore profiles built by organisms, Aguijan Island, Marianas, southwest Pacific. Prominent steep slopes are seaward sides of reefs; their upper edges approximate former sea levels. Flat areas are floors of ancient lagoons. Trees and graded road (*upper right*) give scale. (U. S. Air Force.)

but shall meet them again in Chapter 21. However, we repeat the caution that studies of emerged and submerged features must take account of worldwide changes of sea level. The postglacial rise of sea level (Fig. 12-27) alone has amounted to more than 100m.

Accurate surveys of the sea floor north of the Aleutian Islands, in water depths greater than 400m, reveal a submarine topography of high ridges and hills, separated by valleys that unite in what appears to be a well-developed drainage system. The only reasonable explanation of this topography is that a wide landscape was eroded by streams and mass-wasting and was later submerged by crustal deformation.

The directions of dip of cross-strata and the directions in which sizes of particles of ancient sediments decrease permit geologists to reconstruct former slopes of the Earth's surface and to locate ancient land areas. A surprising conclusion from studies of this kind is that the inclinations of some ancient slopes were exactly opposite to those of modern slopes. A few of the ancient slopes were inclined directly away from areas now occupied by the sea. Therefore, such areas of the present sea floor formerly must have been lands, and the slopes have been reversed by crustal deformation. Even though we are no longer able to recognize characteristic land topography because it has vanished, we are confident from the geologic evidence that land areas did exist where now only the sea is found.

Stratigraphic evidence. Charles Darwin pioneered the use of fossils to determine depth of water at times when the sediments containing them were deposited. During his exploration of the coasts of South America from H.M.S. *Beagle* in the 1830's, Darwin found fossils of some marine species whose modern representatives live within narrow depth zones. These fossils extend through a thickness of emerged marine strata of Tertiary age that is greater than the depth of water in which their present-day counterparts can survive. Darwin inferred, correctly, that the fossils could not have been distributed in this way if the sediment had been deposited in water as deep as the strata are thick. If this had happened, he reasoned, the water would have been too deep for the organisms (now fossils) to have lived. Furthermore the fossils left behind in such circumstances would have indicated a change in depth of water from deep to shallow as filling by sediment progressed. Darwin's alternative was the far-reaching hypothesis that the depth of water had remained constant and that the thickness of strata resulted from a combination of subsidence of the bottom and deposition of sediment. As the bottom sank, sediment built it up again, thus maintaining a nearly constant depth during the entire time required to deposit the strata. Since Darwin's time this profound conclusion has been widely applied. We have employed it already to interpret the thick strata of the Colorado Plateau and the coastal plain of North Carolina (Fig. 14-22); we shall call upon it again in Chapter 21.

Exposure of rocks formed at depth. Various intrusive igneous and metamorphic rocks originate only deep within the crust. Such rocks become exposed only as a result of great uplift and erosion. Unfortunately we generally are unable to determine how much crustal movement has occurred, but we can be sure that the amount has been large.

Structure of bedrock. The evidence of crustal deformation recorded in displacements of man-made structures and of the Earth's morphologic features, although extremely satisfying, is doomed to destruction by erosion. By contrast stratigraphic evidence persists as long as the strata themselves remain. Other longer-lasting evidence is found in the structure of bedrock. When deposited initially most strata are horizontal and parallel; as James Hutton observed, the slightest deviation from parallelism in sequences of strata can be detected by eye. Slight deviations from the horizontal are more difficult to see in single exposures but can be found by careful measurements at many exposures.

The structure of bedrock is determined quantitatively by making a geologic map (App. D). Lines on the map represent boundaries or *contacts* between rock bodies. In strata that were initially plane and parallel the contact lines represent initially horizontal planes, the most useful natural reference objects.

Structural geology is *the study of rock deformation and the delineation of geologic structural features.* Common geologic structures result from broad bending, close folding, and fracturing of strata.

GEOLOGIC STRUCTURES

Broad warps. *Gentle bending of the crust upward or downward* is ***crustal warping.*** The effects of warping are seen clearly in the plateaus of arid regions (Frontispiece). Here, because of absence of regolith and deep dissection by streams, wide areas of rock layers are exposed. Broad warping has also occurred in the lowlands between the Rocky Mountains and the Appalachian region (Pl. B), but its effects are not readily apparent because regolith and vegetation are abundant and bedrock is visible only in scattered exposures. We know from geologic field study and the records of thousands of borings that strata, horizontal or dipping gently, have been warped upward to form large structural domes (Fig. 17-6) and downward to form basins. Generally these features, not visible in single exposures, become evident when observations at many exposures have been plotted on a geologic map (Fig. 17-7). Inclinations of 1° or 2° are not conspicuous, yet the altitude of a stratum with a persistent average dip of 2° changes by nearly 1,000 feet through a horizontal distance of 5 miles.

Fig. 17-6. A nearly circular dome in Mauretania shown by concentric ridges of low relief made by sedimentary strata having more resistance to erosion than adjoining strata. (U. S. Air Force.)

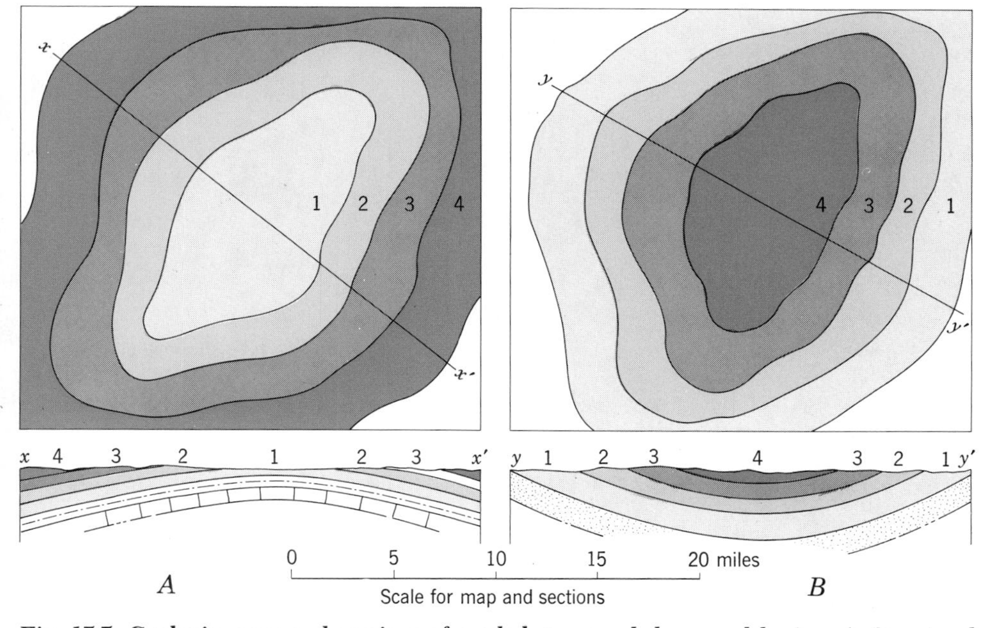

Fig. 17-7. Geologic maps and sections of eroded structural dome and basin. *A.* Structural dome, with oldest formation in center of circular map pattern. *B.* Structural basin, with youngest formation in middle of circular map pattern. The diameters of some large structural domes and basins are several hundred miles. (Appendix D gives further details on geologic maps.)

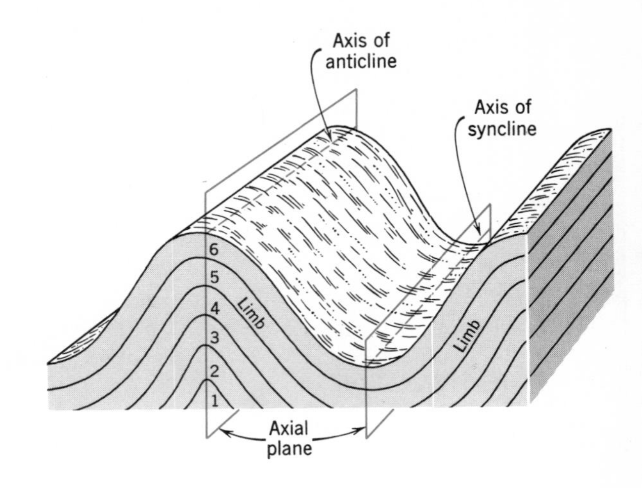

Fig. 17-8. Features of simple upright folds. Upper surface of youngest formation (6) faces *toward* axis of the syncline, *away from* axis of the anticline.

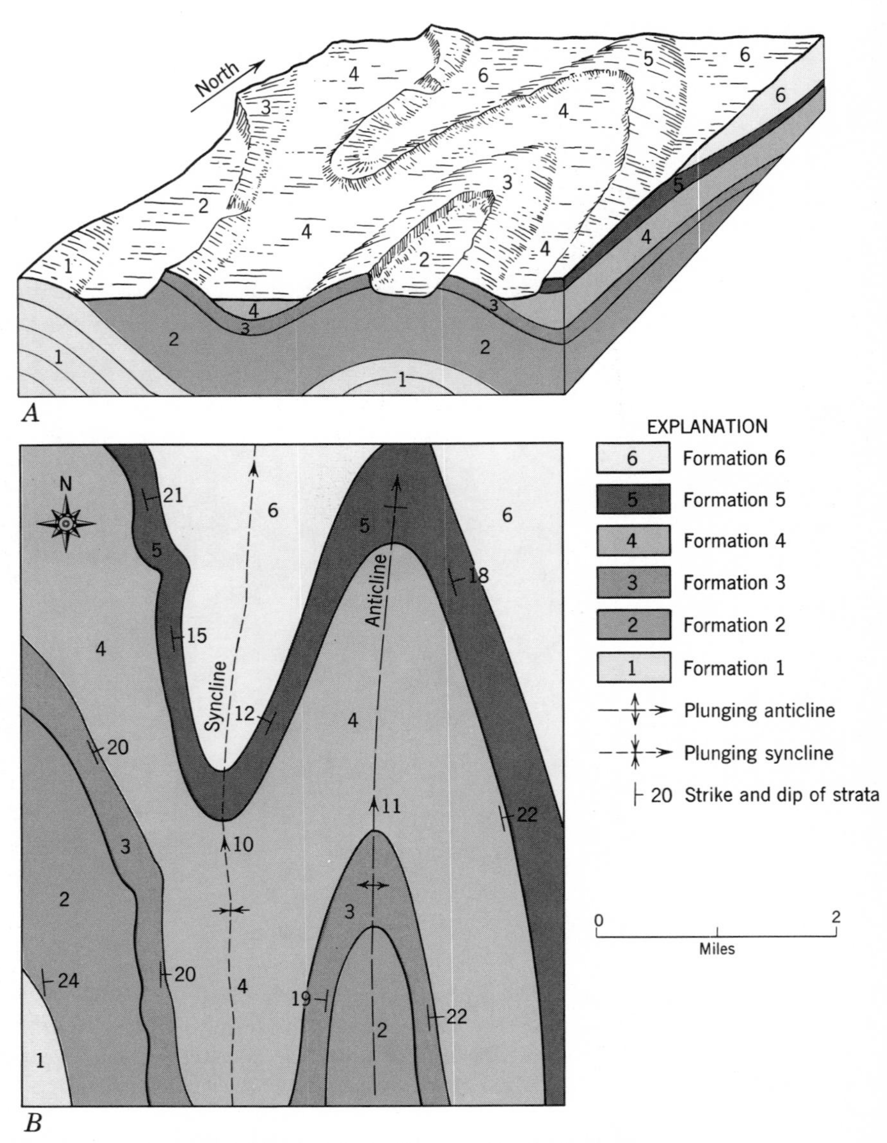

Fig. 17-9. Plunging folds. *A*. Block diagram. *B*. Geologic map.

Folds. *A pronounced bend in layers of rock* is a ***fold*** (Fig. 17-8). *An upfold in the form of an arch and having the oldest strata in the center* is an ***anticline*** (Gr. "inclined oppositely"). *A downfold with troughlike form and having the youngest strata in the center* is a ***syncline*** (Gr. "inclined together").

The sides of a fold are the **limbs**, and *the median line between the limbs, along the apex of an anticline or the lowest part of a syncline,* is the ***fold axis.*** The ***axial plane*** of a fold is *an imaginary plane through the middle of the fold that passes through its axis.*

A fold with inclined axis is a ***plunging fold*** (Fig. 17-9) and *the angle a fold axis makes with the horizontal* is the ***plunge*** of a fold.

A fold with limbs that diverge at an obtuse angle is an ***open fold*** (Fig. 17-9). *A fold with an acute angle between the limbs* is a ***closed fold*** (Pl. 17). *A fold having essentially parallel limbs* is an ***isoclinal fold*** (Gr. "equally inclined").

The dip of strata on the limbs of folds ranges up to vertical (Fig. 17-10, *A*). *A fold having a limb in which the strata have been tilted beyond the vertical* is an ***overturned fold*** (Pl. 17). The limb in which the strata have been tilted beyond the vertical is the *overturned limb.*

Groups of isoclinal folds having vertical limbs present a seemingly endless succession of vertical strata (Fig. 17-10, *B*). Where erosion has cut away the crests of such folds, some skill is required to work out the structure because anticlines and synclines are not easily distinguished in limited exposures. The first procedure is to determine the original top direction of the strata. This can be done by close

C. Air view northwest across Sheep Mountain, a high ridge 15 miles long within the Bighorn basin, Wyoming, through which the Bighorn River has cut a deep canyon. The sharp ridge, made by resistant older strata, lies along the axis of a doubly plunging anticline. Younger strata, composed largely of alternating layers of sandstone and shale, have been eroded from the crest of the fold but make jagged, low ridges along the steep limb. These low ridges curve around each end of the mountain, the curve pointing in the direction of plunge. (J. S. Shelton.)

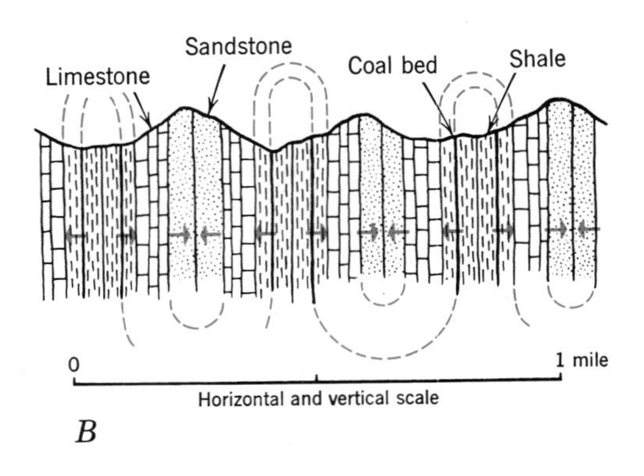

Fig. 17-10. Vertical strata on limbs of folds. *A.* Tops of all strata face toward right (toward syncline) exposed in wave-cut cliff, Percé, Quebec. Two small faults join just left of center to make a wedge-shaped block. Beach sediments in foreground thinly veneer a wave-cut bench (compare Fig. 17-22, *A*). (J. E. Sanders.) *B.* Succession of parallel, vertical strata. Tops of some strata (small arrows) face toward right, tops of others toward left.

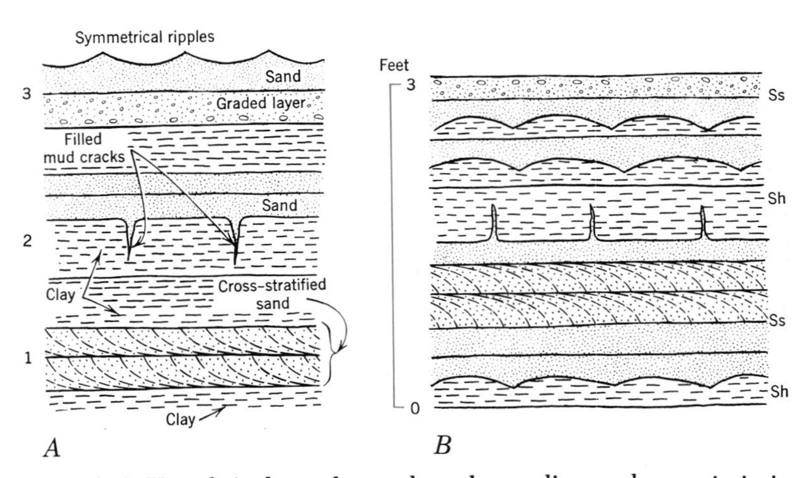

Fig. 17-11. Use of ripple marks, mud cracks, grading, and cross-strata to determine original tops of sedimentary layers. *A.* Layers in place as they were formed; sedimentary structures indicate original tops of layers are toward top of page. *B.* Similar layers, essentially horizontal but upside down on lower limb of recumbent fold; sedimentary structures indicate original tops of layers are toward bottom of page.

Fig. 17-12. View south along monocline lying west of Monument Upwarp (high area at left), Colorado Plateau, southern Utah. Monocline defined in text. (J. S. Shelton.)

study of the primary sedimentary structures (Fig. 17-11). Once the top directions have been determined we can apply the general principle that the tops of the layers face *outward* from the axial plane at an anticline and *inward* toward the axial plane of a syncline.

The axes of some folds have been pushed over until they are horizontal and the strata of one limb are completely upside down. *A fold in which the axial plane is essentially horizontal* is a ***recumbent fold.***

A ***monocline*** (Gr. "one inclination") is *a one-limbed flexure, on either side of which the strata are horizontal or dip uniformly at low angles* (Fig. 17-12).

The folds just described resulted from bending, without appreciable change in length or thickness, of layers that were free to slip past one another, as do the individual sheets composing a thick sheaf of papers when the entire lot is folded. The terms we have defined are applied to all folds, even though many folds originate by processes other than the simple bending of strata (Fig. 20-18, *B*). Folds imply plasticity of rock strata. More brittle behavior is suggested by fractures, to which we turn next.

Fractures. By not separating what is observed from what is inferred geologists encounter difficulty with the names they give to natural features. Examples of such difficulty are the definitions of partings or fractures. As we read in Chapter 4, geologists subdivide fractures into *joints* and *faults*. Objectively, both joints and faults are surfaces of parting, or fractures; inferentially, a joint is a fracture along which no displacement has occurred, and a fault one along which there has been some displacement. Strictly speaking, joints and faults are not objective entities but terms expressing inferred relationships between rock bodies adjacent to fractures. But usage of these terms as if they were objective entities has become so widespread that little would be gained by trying to make any change. In most cases ample evidence exists to establish the inferred relationship beyond reasonable doubt; hence existing usage does not create many serious problems. We raise this seemingly trivial point of nomenclature as a warning because similar difficulties less easily resolved have arisen with other geologic terms.

Joints. The surfaces of many joints are nearly planes. Commonly such planar joints occur in ***joint sets,*** *widespread groups of parallel joints,* or in ***joint systems,*** *combinations of two or more intersecting joint sets.* Joint systems outline rectangular blocks of various sizes (Fig. 17-13). Because systems of joints are widespread, extremely large blocks of bedrock are rare.

The origin of planar joint sets is not well known. Previously geologists supposed, by analogy with small laboratory test specimens, that well-developed joint systems originated through lateral stresses. They accepted this idea even though the widespread distribution of the joints implies that the stresses are transmitted by the rocks through great distances. Recent studies of steeply dipping joint systems in the Colorado Plateau have shown that many joints are continuous in the lower parts of steep canyon walls but do not extend to the top. They must have grown upward from below. Well-developed and widespread vertical joint systems have been found in clastic sediments and in peat. Such joints must have grown upward from below because these materials are so weak that they cannot transmit lateral stresses at all.

In Chapter 7 we read about concentrically curved joints caused by tensional stresses set up during the expansion of rock bodies. Later we shall contrast these with joints formed during the contraction of rock bodies. Where the opposite walls of a joint have been pulled apart they form an open joint or *fissure.* Open and other kinds of joints are related to faults.

Faults. At the outset of the chapter we mentioned many aspects of faults. To these we now add some features to examine when a fault has been identified. We want to determine its attitude, in what direction and how far its adjacent blocks have moved, and what effects, if any, have been created on adjacent rocks as the result of movement.

Attitude. Although faults are inclined at all angles, the dips of many exceed 45°. Because many veins of metallic ore lie along faults, we have inherited some terms used by miners working in tunnels whose *walls* coincide with the boundaries of inclined veins. A miner sees one wall overhanging him and the other beneath his feet. The ***hanging wall*** is *the wall on the block above an inclined fault.* The ***footwall*** is *the wall on the block below an inclined fault* (Fig. 17-14, *A*). These terms, of course, do not apply to vertical faults.

Fig. 17-13. System of nearly vertical joints at right angles (1 and 2) cutting well-bedded, nearly horizontal siltstone (*below*) and fine sandstone (*at top*). East shore of Cayuga Lake, near Ithaca, New York. (B. M. Shaub.)

Movement. Generally we are unable to determine the actual movement of each block adjacent to a fault. Even if the surface of the ground has been displaced (Fig. 17-14, *B* and *C*) or a crystal or pebble in the rock has been cut through by the fracture and the halves carried apart a measurable distance, it is not possible to learn whether one block stood still while the other block moved or whether both blocks shared in the movement (Fig. 17-14, *A*). Conceivably by careful survey we might locate exactly the positions of critical points on the ground on both sides of a known fault. Later, after further movement of the fault, we could resurvey the points. But most faults with which we have had to deal are old, inactive features whose former expression at the Earth's surface was destroyed by erosion long ago, and we have to accept determinations of relative movement as the best we can do with the available evidence.

Effects of fault movement on rocks. Faults cut cleanly through sediments (Fig. 17-15) and through some rock masses without affecting the adjacent

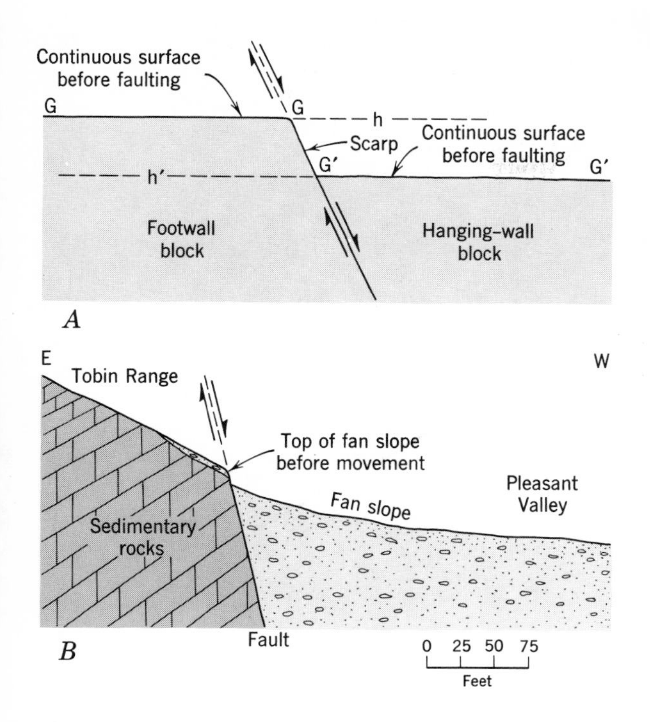

Fig. 17-14. Fault that displaces surface of ground. *A.* Scarp made by displacement of a formerly level surface G-G′ may mean that hanging-wall block dropped from position h, that footwall block moved up from position h′, or that both blocks moved in some measure to create the net displacement shown. Terms defined in text. *B.* Vertical section from lower part of Tobin Range to Pleasant Valley. Half-arrows indicate relative movement of crustal blocks. *C.* Scarp at west base of Tobin Range, Nevada, formed in 1915 by abrupt movement of a fault that generated an earthquake. The whitish band marks a displacement at the top of the fan slope, several miles from the camera. (Eliot Blackwelder.)

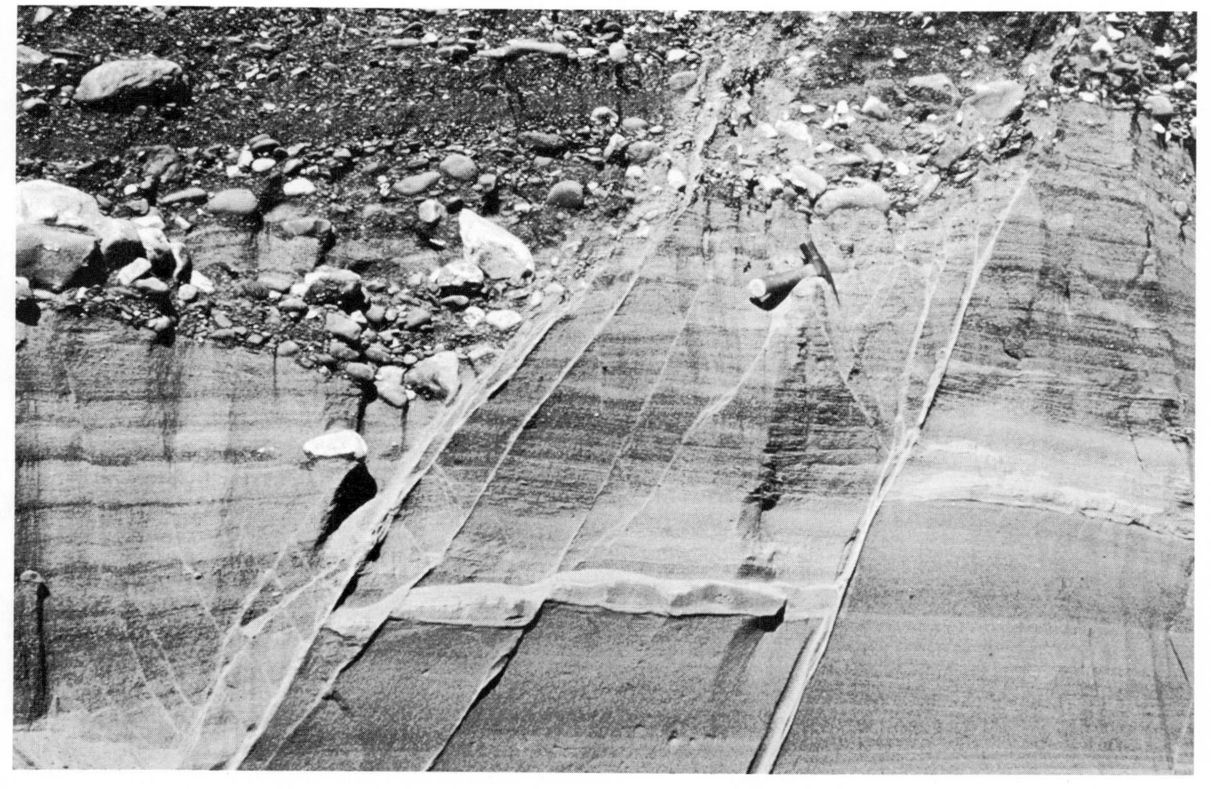

Fig. 17-15. Small faults in Pleistocene sediments near Rensselaer, New York. (G. M. Friedman.)

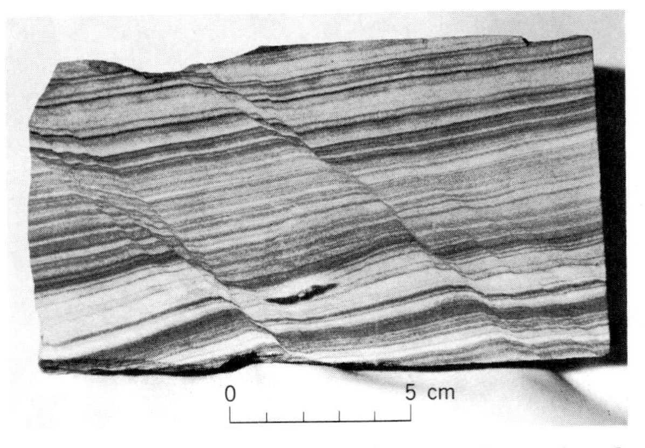

Fig. 17-16. Layers near small faults in specimen of sandstone have been bent during movement along the faults. (B. M. Shaub.)

Fig. 17-17. Polished and grooved surface of an old fault, Spotted Range, southern Nevada, exposed by erosion. (C. R. Longwell.)

Fig. 17-18. Slickensides in dolerite, Mount Tom Range, Westfield, Massachusetts. The block containing grooves and steps moved along the grooves in the direction shown by blue arrow, *away* from the edges formed by tops of steps and grooved surfaces. (B. M. Shaub.)

Fig. 17-19. Scarp formed across 9th Avenue, Anchorage, Alaska during earthquake of March 27, 1964, by movement of oblique-slip fault having normal vertical component of approximately 2m and right-lateral strike-slip horizontal component of approximately 1m. Block on far side of smaller fault in foreground moved about 0.5m toward the right; its component of vertical displacement increases toward the left, making it both a right-lateral strike-slip fault and a hinge fault. Fault displacement is more recent than the snow. (Alaska Pictorial Service.)

material. Adjacent to some faults the layers have been bent, creating large and small structures referred to collectively as *fault drag* (Fig. 17-16). The surfaces of some faults have been smoothed, scratched, and grooved (Fig. 17-17). *Striated or highly polished surfaces on hard rocks abraded by movement along a fault* are ***slickensides.*** Parallel scratches and grooves on such surfaces record the direction of latest movement (Fig. 17-18). Rocks adjacent to a fault can be crushed into irregular pieces, forming *fault breccia.* More intense grinding breaks the pieces into tiny bits, and forms a cataclastic metamorphic rock (Fig. 20-16, *A*).

Classification and names. Faults are classified and named according to inclination of the fault surface and to direction of relative movement of the blocks (Table 17-1). Our standard planes of reference are the vertical and the horizontal. Along many faults movement is confined to the reference planes, but along other faults it occurs in directions oblique to these planes. Commonly during fault movement blocks are rotated. In classifying and naming faults we ignore rotation if the axis of rotation parallels the fault. If, however, the axis does not parallel the fault we classify the fault as a *hinge fault* (Fig. 17-19).

TABLE 17-1. NAMES AND DEFINITIONS OF PRINCIPAL KINDS OF FAULTS

Block diagram	Name of fault	Definition
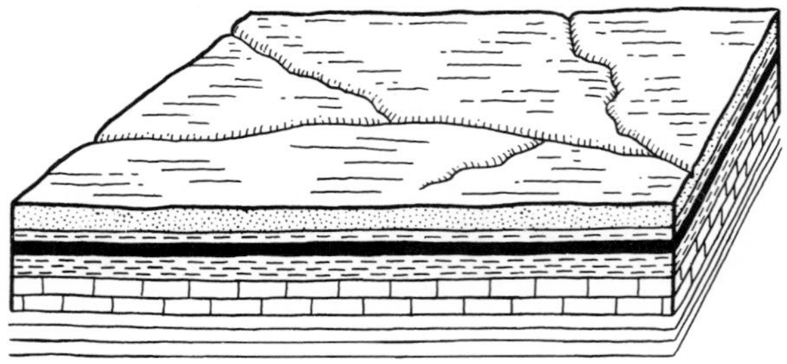		Reference block before faulting. Drainage is from left to right.
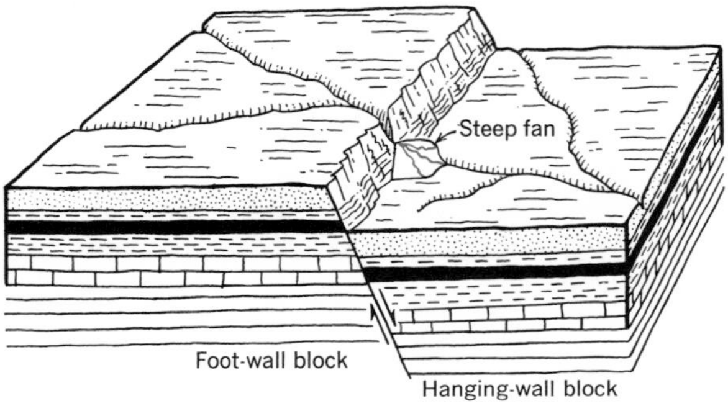 Steep fan Foot-wall block Hanging-wall block	***Normal fault***	*A fault, generally steeply inclined, along which the hanging-wall block has moved relatively downward.*
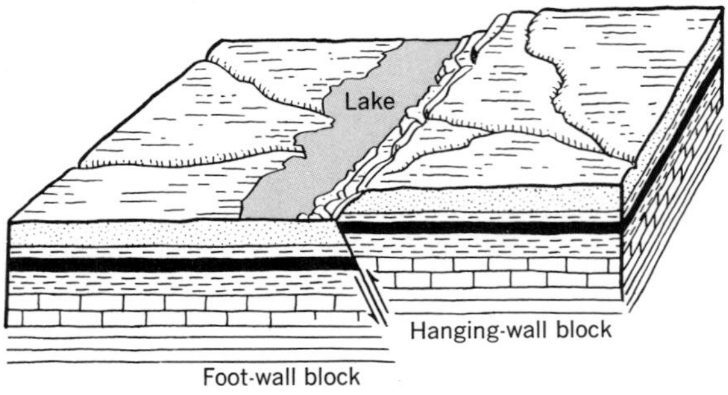 Lake Hanging-wall block Foot-wall block	***Reverse fault***	*A fault, generally steeply inclined, along which the hanging-wall block has moved relatively upward.* See also Fig. 17-2. A normal or reverse fault on which the only component of movement lies in a vertical plane normal to the strike of the fault surface is a *dip-slip fault.*

TABLE 17-1 (Continued)

Block diagram	Name of fault	Definition
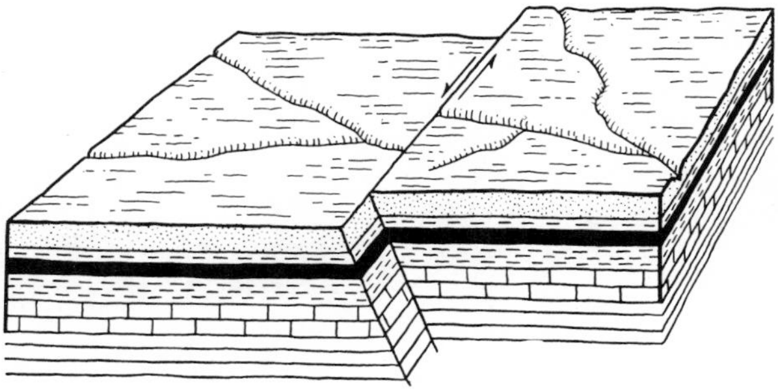	***Strike-slip fault***	A *fault on which displacement has been horizontal.* Movement of a strike-slip fault is described by looking directly across the fault and by noting which way the block on the opposite side has moved. The example shown is a *left-lateral fault* because the opposite block has moved to the left. If the opposite block has moved to the right it is a *right-lateral fault* (Fig. 18-2, *A*). Notice that horizontal strata show no vertical displacement.
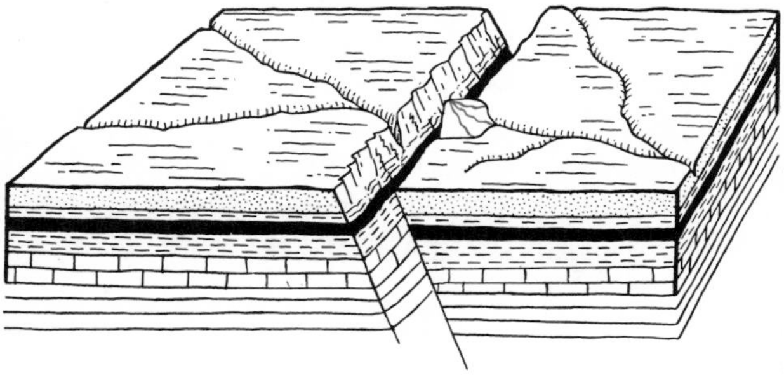	***Oblique-slip fault***	A *fault on which movement includes both horizontal and vertical components.* See also Fig. 17-19.
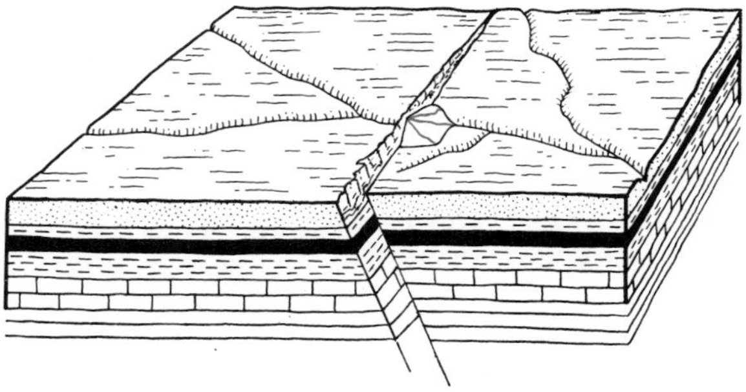	***Hinge fault***	A *fault on which displacement dies out (perceptibly) along strike and ends at a definite point.* Figure 17-19 shows a small example; see also Fig. 17-22.

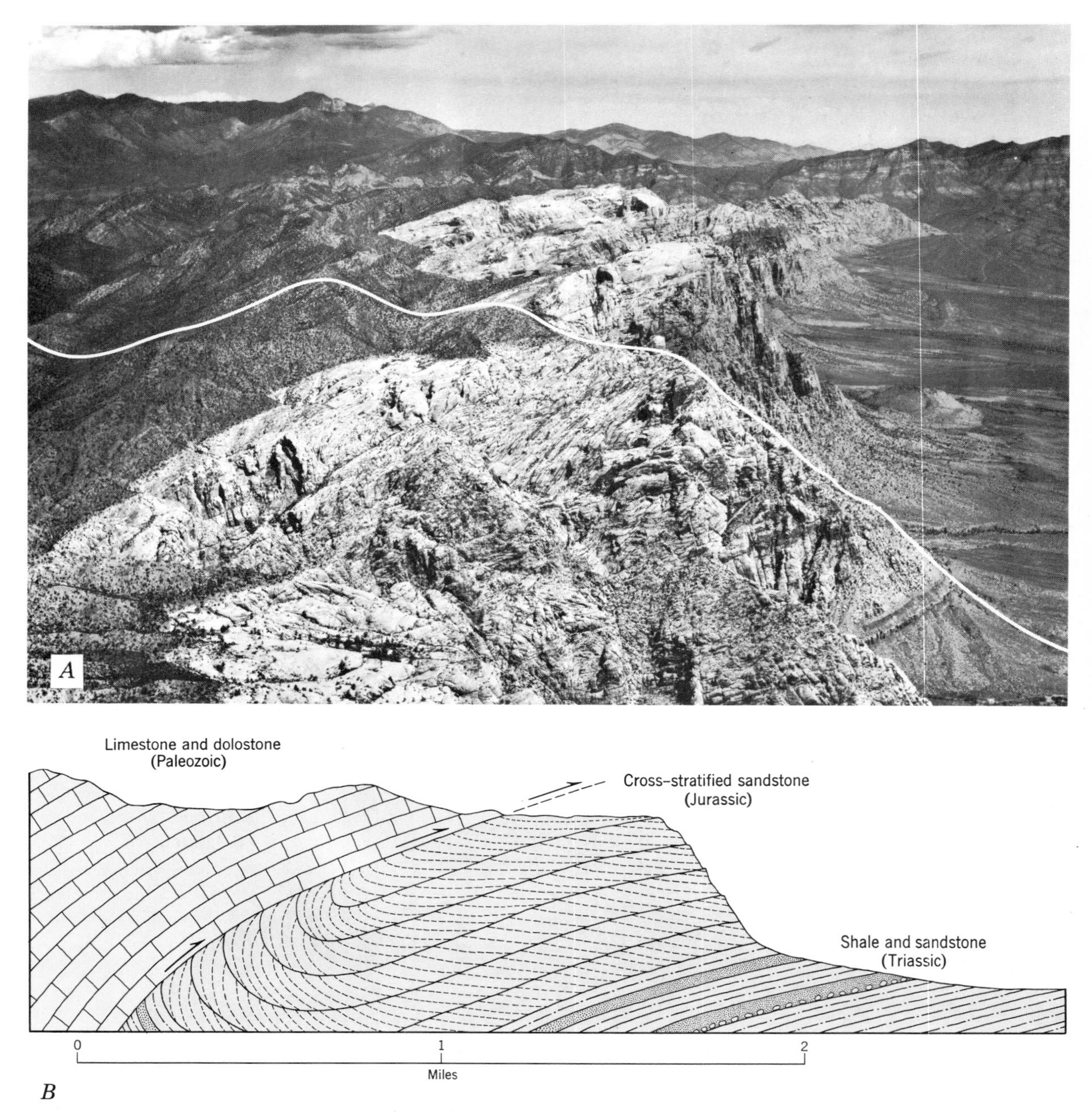

Fig. 17-20. Keystone thrust, west of Las Vegas, Nevada. *A.* Air view northward shows thrust clearly defined by contrast in color of rocks adjacent to it. Light-colored Jurassic sandstone, forming cliff nearly 2,000 feet high (*right*), lies below thrust; dark-colored Paleozoic limestones and dolostones (*left*) lie above thrust. (J. S. Shelton.) *B.* Section drawn along white line in photograph and extending somewhat farther east and west. Canyons crossing the thrust reveal that it steepens downward toward the west and crosses overturned layers of the sandstone. Farther east the thrust becomes essentially parallel with the sedimentary layers, both below and above.

Not shown in Table 17-1 are ***thrust faults,*** generally known as ***thrusts,*** *low-angled reverse faults, with dips generally much less than 45°*. Such faults, common in great mountain chains, are noteworthy because along some of them the hanging-wall block has moved many kilometers over the footwall block, thus shortening and thickening parts of the crust. In most cases the hanging-wall block, thousands of meters thick, consists of rocks much older than those adjacent to the thrust on the footwall block (Fig. 17-20). The strata above some thrusts lie nearly parallel to those beneath, giving a deceptive appearance of an unbroken sequence. Generally, however, if movement was opposed by great frictional resistance, then the beds above and below the thrust surface became folded and broken.

Relation of joints to faults. Some joints are intimately associated with faults. Near and parallel to many faults are joints that are closely spaced; with distance from the faults these joints become spaced more widely. Many strike-slip faults are accompanied by sets of parallel fissures, having *en echelon* arrangement, that are diagonal to the faults (Fig. 17-21). Sets of *en echelon* crevasses, similarly oriented with respect to the valley walls, occur at the margins of many valley glaciers (Fig. 12-1, *right side*). By means of these fissures we can determine its sense of movement if we know the strike of the fault. The basis for this determination is the inference that the strike-slip fault coincides with one of the two planes of maximum shearing stress and that the fissures are situated at right angles to the plane of maximum tensile stresses set up by the shearing movement. Recall that in simple cases the direction of maximum tensile and compressive stresses are disposed at right angles, and that two planes of maximum shearing stresses occupy the intermediate diagonal positions (Fig. 12-4, *A;* 17-1). On one set of shear planes the tendency for slip is *right-lateral,* and on the other set, *left-lateral.*

Relation of folds to faults. Folds and faults can be related in various ways. As we know, near active faults strata may be folded by the effects of frictional drag along the planes of movement (Fig. 17-16). A steeply inclined fault may pass into a fold upward or laterally, as the hinge fault in Fig. 17-22 passes into a monocline. Typically the projected continuation of the fault plane coincides with the axial plane of the related fold.

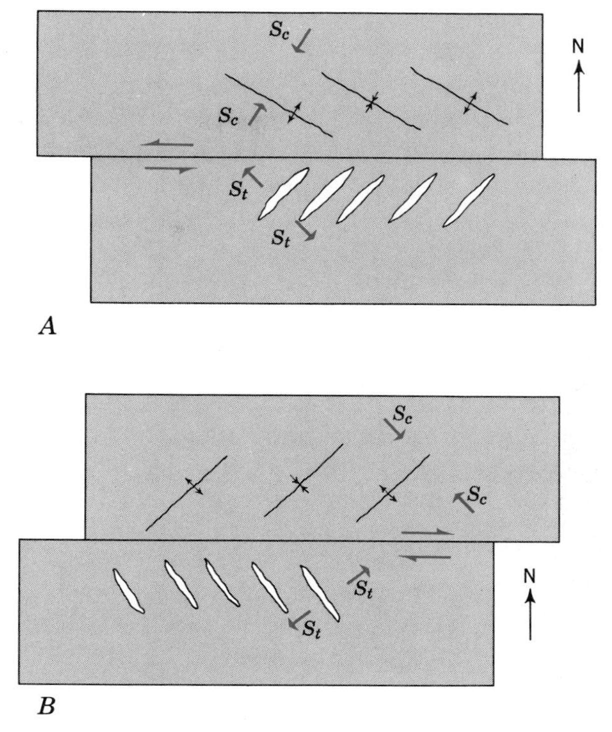

Fig. 17-21. Schematic maps of two strike-slip faults and related *en echelon* parallel fissures (white, shown on south blocks only) and folds (shown by standard map symbols on north blocks only). Small arrows show direction of maximum compressional stress (S_c) and of maximum tensile stress (S_t). Further explanation in text; standard map symbols of folds defined in Appendix D. *A.* Sense of movement is left lateral. *B.* Sense of movement is right lateral.

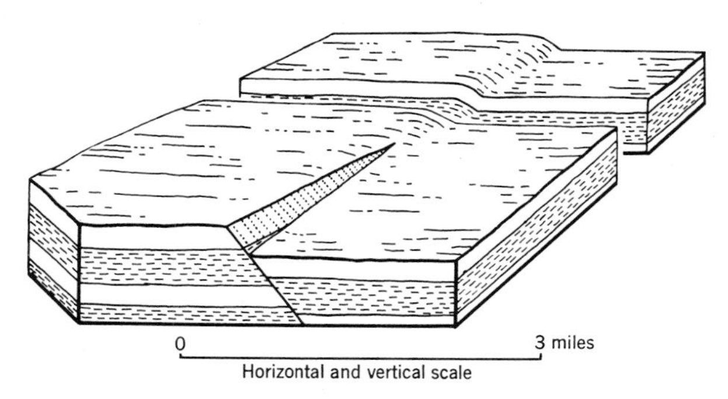

Fig. 17-22. Hinge fault (front block) passes laterally into monocline (rear block) (compare Fig. 17-12).

Many thrusts in sedimentary strata originate after the strata have been compressed into overturned folds and the folds broken by continued compression (Pl. 17; Fig. 17-20, *B*). Commonly the folds break near their axial planes, and one of their limbs is displaced. The thrust shown by the dashed blue line in Pl. 17 cuts across an overturned anticline near its axial plane and has displaced its normal limb to some unknown point outside the field of view. We have added dotted blue lines to the photograph to indicate the original bottom surfaces of some strata. Once we know the bottoms, we can see that the tops of the strata on the normal limb of the contiguous overturned syncline (left) lie toward the top of the photograph. At the axis of this syncline the strata bend sharply upward and become overturned, with their original tops toward the left. At the axis of the overturned anticline the strata right themselves and are cut off by the thrust. Unlike the thrust shown in Pl. 17, which slices through near the axis of an overturned anticline, some thrusts break across overturned folds near the axes of synclines.

Because faults serve as zones of adjustment, strata within blocks bounded by faults can be folded independently of those within other blocks. For example, strata on the hanging-wall block of a thrust can be folded independently of those on the footwall block.

Not only have strata been deformed as the result of active forces of compression but also some thrusts themselves have been folded as if they were strata (Fig. 21-6). In some cases the thrusts have been overturned.

Like diagonal fissures some folds originate from deforming forces that accompany movement on strike-slip faults. The axes of such folds are oriented at right angles to the direction of maximum compressive stress. Therefore from the relationships among principal directions of stress we can infer the sense of movement on the related strike-slip fault, just as we did with the diagonal fissures.

Movement on many faults is not related to the growth of folds. Younger faults cut across and can displace folds in various ways.

GEOLOGIC AGE OF DEFORMATION

We can determine the *when* of deformation from geologic evidence that can be analyzed independently of our conclusions about the *what, where,* and *how*. Ordinarily a geologist faced with the task of determining the *when* of a given deformation does not have suitable specimens for measuring an isotopic age but must be satisfied with relating the age of deformation to that of local geologic features. In some cases he may be able to measure an isotopic age of minerals in a body of rock that has been involved in the deformation and thereby establish a point of known age in the local sequence. In other cases minerals may have recrystallized during deformation and have begun to accumulate products of radioactivity all over again. Where this has happened an isotopic age on the minerals serves as a direct measurement of the age of the deformation itself.

The reasoning geologists employ in determining relative ages of deformation is familiar to us. In Chapter 4 we used it on a smaller scale to analyze textures of mineral particles in hand specimens. In applying it to rock bodies we simply enlarge the scale and add in the features found along geologic contacts. Because geologic contacts contain much valuable information for determining the geologic age of deformation, let us pause to find out more about them and then apply what we have learned to the dating of deformation.

Contacts resulting from burial. The most common type of geologic contact originates when rock bodies, exposed at the Earth's surface by erosion, later disappear from view because they have been covered by sediments or lava. In this manner former parts of the Earth's surface become incorporated into the geologic record; the process is fundamental to the geologic cycle. We shall limit the present discussion to sediments and continue later with lava.

The buried topography, having great or small relief, may be exhumed at a later time so that it forms part of the Earth's surface once again, or it can be cut across by erosion so that we see it later only in sectional view along a valley wall or cliff.

The relative age of the two rock bodies separated by a contact is important. Along a contact resulting from burial the covering sediments clearly are younger than the buried rock bodies. Commonly pieces of the buried rock bodies become incorporated into the covering material.

Stratification surfaces are contacts of burial. Generally where strata have been deposited one above another the stratification surfaces are fairly smooth. This results from the fact that surfaces created by deposition tend to be much smoother than those created by erosion. As we have explained, the relationship between successive strata separated by contacts of burial made during an episode of essentially continuous deposition is one of *conformity*.

Unconformity. Crustal movements can interrupt the *continuity* of depositional episodes in the various ways discussed on preceding pages: blocks of the sea floor can be raised above sea level; parts of the land can be lowered below sea level; and strata can be folded and faulted, both on the sea floor and on land.

Look again at the lower part of Fig. 17-10, *A*. Most of the vertical strata exposed in the wave-cut cliff disappear beneath beach sediments that thinly cover the wave-cut bench in the foreground; the level of high tide is marked by the line of dark-colored debris scattered along the base of the cliff. A few bedrock ledges project through the beach sediments. Nondeformed beach strata, of Recent age, inclined slightly toward the observer, bury older vertical strata of Devonian age. The relationship of the beach sediments to the older rocks is one of ***unconformity,*** *a lack of continuity between units of rock in contact, corresponding to a gap in the geologic record.* We have drawn slightly more area than that shown in Fig. 17-10, *A* into a block diagram (Fig. 17-23, *A*) to give a sectional view of beach sediments lying on the vertical strata. The block diagram illustrates ***angular unconformity,*** which is *unconformity marked by angular divergence between older and younger sedimentary strata.* Figure 17-23, *B* shows angular unconformity between ancient strata exposed on the face of a cliff in Colorado. In 1788, James Hutton was profoundly impressed by finding strata in angular unconformity

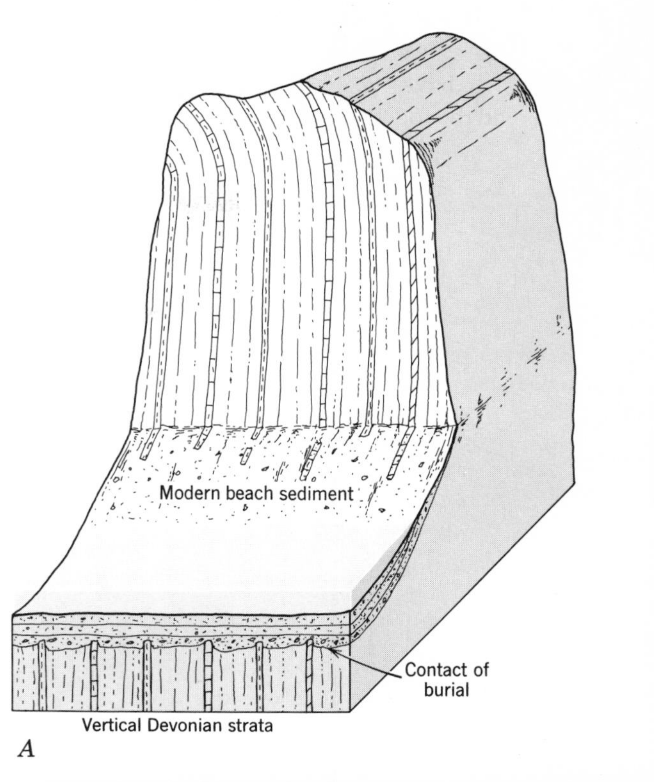

Fig. 17-23. Relationship of angular unconformity. A. Simplified diagram of area in Fig. 17-10, A, showing contact of burial and angular relationship between modern beach sediments and vertical Devonian strata. Explanation in text. *B*. Angular relationship between nearly horizontal strata of Devonian age and vertical strata of Precambrian age, Ouray, Colorado. (Rathbone from Monkmeyer.)

Fig. 17-24. Relationship of angular unconformity between nearly horizontal gravelly stream deposits containing fossil bones of Pliocene age and tilted and eroded carbonate rocks containing marine fossils of Cambrian age resembles large-scale cross-strata (Fig. 16-3). At this exposure deformation can be dated only as being post-Cambrian and pre-Pliocene. Meadow Valley Wash, Lincoln County, Nevada. (C. R. Longwell.)

at Siccar Point, Berwickshire, Scotland; he wrote of it as follows:

"The ruins of an older world are visible in the present structure of our planet, and the strata which now compose the continents have been once beneath the sea, and were formed out of the waste of pre-existing continents."[1]

The angular relationships of strata along some surfaces of angular unconformity (Fig. 17-24) resemble those between truncated cross-strata and their inclosing horizontal strata (Fig. 16-3). Because no major lack of continuity has occurred during deposition of the cross-strata, the angular relationships are not considered to be relationships of angular unconformity.

As we have defined it and as we like to use the term, *unconformity* designates a *relationship* whose existence must be inferred; it does not refer to an objective entity such as a surface. Because of the ambiguities involved with the relationship we think the problem of nomenclature is more serious in this case than it is with joints and faults. Two important corollaries for word use are implied in the definition we prefer. (1) The inferred relationship between two groups of strata is designated by the adjective, *unconformable,* or by the word, *unconformity.* Put another way, we shall employ *unconformable* and its opposite *conformable* but not "*an* unconformity" nor "*a* conformity." (2) The contact of burial between two groups of unconformable strata is a *surface of unconformity,* not "an unconformity." The plural form is *surfaces of unconformity,* not "unconformities." Although we shall adopt this usage, many geologists prefer to define unconformity as an objective entity and to use it as a noun for the surface of contact between unconformable strata, both in the singular and in the plural forms shown in quotation marks.

Variable relationships along surfaces of unconformity. As can be readily visualized, endless variety characterizes contacts resulting from burial; we can treat the present surface of the Earth as a representative sample of what to expect. Likewise, strata having varied structural configurations can be buried. But not all strata that have been buried were crumpled or broken before burial. In some cases strata deposited on the sea floor have been elevated to become land, eroded subaerially, and resubmerged without losing their initial horizontal position. After further deposition has occurred they become covered by younger marine strata that parallel the older, eroded strata. The unconformable relationship between such parallel groups of older, eroded marine strata and the cover of younger marine strata is one of *disconformity.* In other cases the sea floor is not elevated at all, but deposition is interrupted and later resumes for other reasons. An interruption in the geologic record originating without uplift likewise causes unconformity.

Faults compared with contacts resulting from burial. Among groups of strata separated by contacts resulting from burial many of the geometric arrangements are nearly duplicated by the effects of faulting. In the attempt to distinguish these features, field study of actual contacts is the best teacher. In Table 17-2 we have listed a few critical questions that should be asked and have suggested some answers to them. With this summary of geologic contacts in mind let us turn again to our objective of determining the geologic age of deformation.

Dating deformation. We can date crustal movements with our eyes, using a clock and a calendar; with geologic evidence, using stratigraphic and structural relationships, fossils, and radioactive isotopes; and with geophysical instruments.

We can see modern crustal movements; we "catch them in the act," so to speak. All we need do is to note the place, time, and date when the action took place.

[1]Quotation from Hutton, cited in Lyell, Sir Charles, 1830, Principles of geology, v. 1: London, Thomas Murby, p. 61.

TABLE 17-2. SOME QUESTIONS AND ANSWERS CONCERNING CONTACTS RESULTING FROM BURIAL AND FAULTS

Question	Answer	
	Contact of Burial	Fault
1. What is the geometry of the contact surface?	Variable; can be smooth (along surfaces of nondeposition or of disconformity) or very irregular (along surfaces of disconformity or of angular unconformity).	Generally smooth, but can be irregular because some faults (large thrusts) can be folded.
2. What are the relative ages of materials on opposite sides of the contact?	The material on the original upper side is invariably younger than that on the originally lower side.	Variable; material of any age can be faulted against other materials. Along large thrusts older strata are generally pushed above younger strata, but some thrusts follow contacts of burial with the result that younger strata lie above the fault.
3. Does evidence of movement exist along the contact?	No.	Yes. Evidence of movement (such as slickensides and fault breccia) is a criterion of faults.
4. Did the rock bodies adjacent to the contact ever extend beyond the contact to the opposite side?	Yes, for the originally lower material, but all previous extensions were removed by erosion so that after burial they can not be found. No, for the covering material.	Yes, in all cases except faults that occur along contacts, such as surfaces of stratification. Finding the displaced part of a rock body on the opposite side proves that the contact is a fault. Displacement can be difficult to prove where a vertical strike-slip fault cuts horizontal strata (Table 17-1).

Although manifestly we can never witness movement on ancient faults long since immobile, we may be able to reconstruct their history of movement by reference to the preserved geologic record. Closest geologic dating is possible where ancient crustal movements created relief on the sea floor or in other places where sediments were being deposited. Uplifted fault blocks and anticlines create ridges, downdropped fault blocks and synclines create depressions. Relief on the sea floor may influence marine sediments in several ways. Over high-standing parts of the bottom currents flow swifter than they do over low parts. The controlling principle is the same as that regulating the maximum heights of wind-generated waves on the water surface and of dunes. Where sediments of varied particle sizes are introduced into an area of irregular relief, therefore, both thickness and sorting may vary systematically. Small thicknesses of sediment composed predominantly of coarse particles (possibly sand) are the only materials capable of remaining on high areas. By contrast large thicknesses of sediment consisting chiefly of fine particles (possibly sandy silt) may collect in low areas (Fig. 17-25). In some cases sea-floor relief influences the distribution of organisms. Reef-building organisms, which tend to prefer hard, current-swept parts of the sea floor, may grow preferentially on high areas and systematically avoid low areas, where the water is quiet and the bottom likely to be soft. The ages of strata thus influenced by structural relief fixes the dates of the crustal movements.

Along some large faults, movement has continued throughout long ages. Commonly the blocks have moved in the same direction repeatedly, with the result that their aggregate horizontal or vertical displacement can become very large. Where one block is lifted repeatedly and an adjacent block dropped persistently, thick coarse-grained sediments can accumulate. The strata become thick because

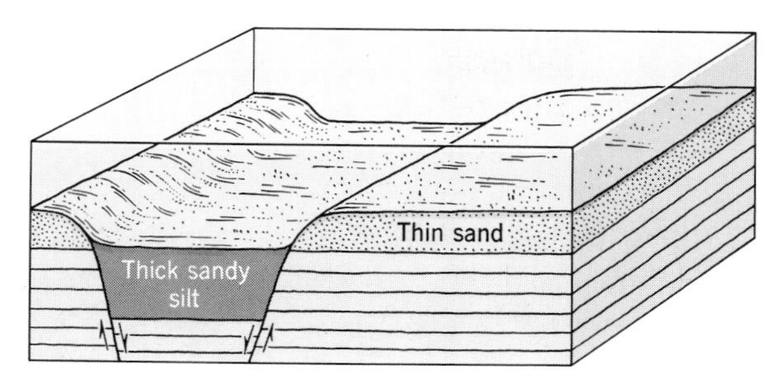

Fig. 17-25. Dating crustal movements by sediments (I). Faulting of the sea floor influences grain size and thickness of sediments. Details in text.

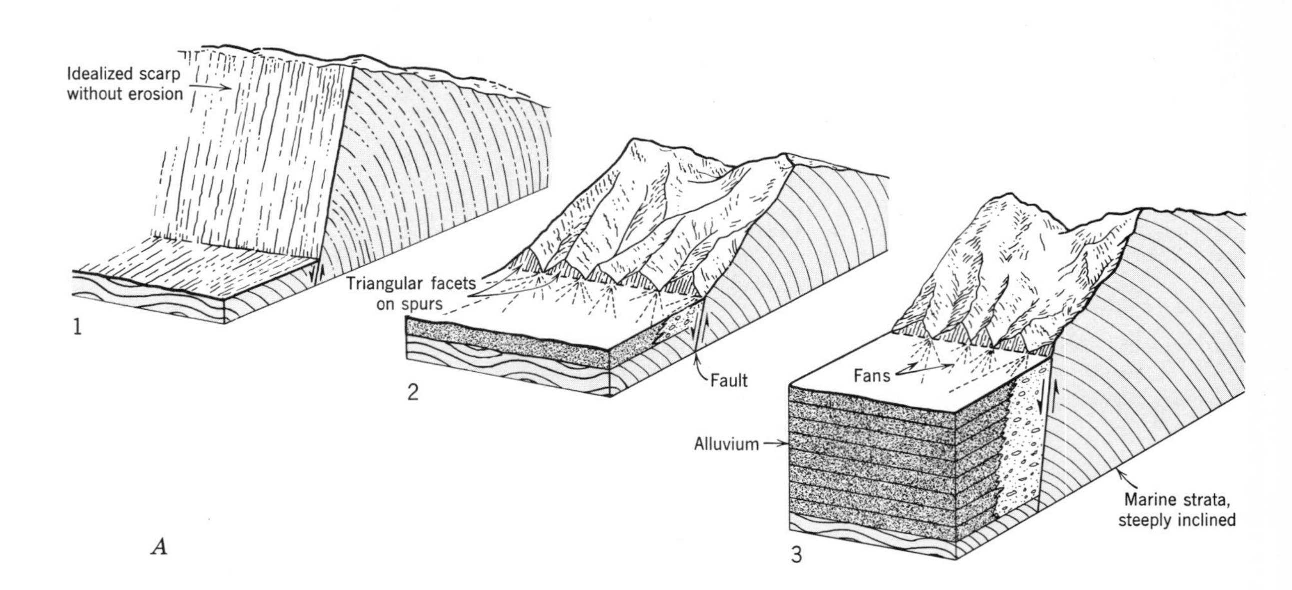

Fig. 17-26. Dating crustal movements by sediments (II). Thick coarse sediments on subsiding fault block. *A.* Stages in development of thick body of coarse sediments deposited by streams on alluvial fans located on downthrown block. 1. Initial movement makes large scarp on upthrown block. 2. Erosion dissects upthrown block; downthrown block receives coarse sediments deposited on fans close to fault and fine sediments deposited on horizontal surfaces farther away. 3. Result of further faulting and sedimentation. Persistence of triangular facets on spurs is evidence that erosion is lagging behind uplift on fault block. The geologic age of the sediments dates some of the fault movement. (After W. M. Davis.) *B.* Collings Ranch Conglomerate, consisting of debris shed from rising Arbuckle Mountains, in late Pennsylvanian time. Average diameter of cobbles is about 150mm. Near Ardmore, Oklahoma. (J. F. Williams and Allen Graffham.)

the dropped block subsides through large distances. Near the fault sediments are coarse because topographic relief on the adjacent block continues to be large as a consequence of repeated uplift. Where such faults occur on land areas vast quantities of extremely coarse-grained sediments deposited on alluvial fans become incorporated into the geologic record (Fig. 17-26). The geologic age of the coarse-grained sediments tells the age of some fault movement but not necessarily the age of all.

Where deformed strata have been buried by undeformed younger strata, we can bracket the time of deformation within a known upper limit. Deformation responsible for the older structures ceased in the interval defined by the youngest strata within these structures and the oldest strata that cover them (Fig. 17-24). In some cases the best we can do is determine the age of the youngest feature deformed. We must be content with an open-ended upper limit for the date of deformation. For example, the evidence shown in Fig. 17-19 permits us to date the faults as being younger than the snow.

Other crustal movements create a lot of noise within the Earth and we can eavesdrop on them with special instruments. To use a popular idiom, scientists have thoroughly "bugged" the Earth and thus can locate closely *where* and tell to the minute *when* noisy crustal movements occur. The next chapter explains how to do this and what the results disclose about the Earth's interior.

Summary

1. The strength of a rock is its ability to resist stresses caused by forces that tend to cause permanent change of volume, shape, or both.

2. Strength is related to many factors; it can be reduced by raising the temperature, by increasing the rate of application of deforming forces, by continuing to apply these forces through long periods of time, and by introducing interstitial fluids.

3. Variable patterns of forces tending to deform a rock body create internal stresses that reach maximum values along directions that are oriented systematically in simple cases. Some features resulting from deformation can be related to the directions along which stresses reach maximum values.

4. Ephemeral evidence of crustal deformation includes displacement of man-made structures and of various natural surface features such as emerged and deformed shorelines.

5. Longer-lasting evidence of crustal deformation is found in strata, in occurrence at the surface of rocks formed deep within the Earth, and in the geologic structure of bedrock.

6. Geologic structures in bedrock resulting from deformation includes domes and basins, close folds, joints, and faults.

7. Some joints and folds originated during movement on faults. Other faults originated from broken folds. Still other joints and folds are not related to faults.

8. By interrupting episodes of sedimentation structural movements create unconformable relationships among groups of rock strata.

9. Modern structural movements can be dated by visual observation and by geophysical measurements. Older movements are dated by fixing the age of younger, undeformed strata lying with angular unconformity on older, deformed strata; by analyzing the effects these movements created in sediments that accumulated during movement; and by measuring radioactive isotopes.

Selected References

Anderson, E. M., 1951, The dynamics of faulting and dyke formation: Edinburgh, Oliver and Boyd.

Billings, M. P., 1954, Structural geology, 2nd ed.: Englewood Cliffs, N. J., Prentice-Hall.

Compton, R. R., 1962, Manual of field geology: New York, John Wiley.

Hills, E. S., 1953, Outlines of structural geology, 3rd ed.: London, Methuen; New York, John Wiley.

Lahee, F. H., 1961, Field geology, 6th ed.: New York, McGraw-Hill.

Longwell, C. R., 1937, Sedimentation in relation to faulting: Geol. Soc. America Bull., v. 48, p. 433–442.

Sitter, L. U. de, 1964, Structural geology, 2nd ed.: New York, McGraw-Hill.

CHAPTER 18

Earthquakes: The Earth's Interior

How can we know what lies inside the Earth?
Earthquakes
Kinds of seismic waves
Causes and distribution of earthquakes
Probing inside the Earth with seismic waves
Architecture of the Earth
Earth's core, source of magnetic field?
Mantle, crust, and lithosphere
Locating sources of earthquakes
Energy released during earthquakes

Plate 18

Cracks and small faults formed in street of Niigata, Japan during earthquake of June 16, 1964. (Paris Match, from Pictorial Parade.)

CHAPTER 18 *Earthquakes: The Earth's Interior*

HOW CAN WE KNOW WHAT LIES INSIDE THE EARTH?

What lies inside the Earth? This question has prompted many answers. Unable as we are to visit the interior in person to see for ourselves, we cannot work out an answer by the methods we found so useful in studying the Earth's exterior. What then can we do? One recourse, actively practiced by men, is speculation.

Early scholars speculated that the Earth's interior was hollow. They did nothing decisive to test this idea, but, by repeating it numerous times to one another, established their faith in it. The concept of a hollow interior became incorporated into one of the theories of Earth history popular in Germany and elsewhere during the eighteenth century within the lifetime of James Hutton. One of the tenets of this theory stated that at a period early in its history the Earth's crust had been completely submerged by a Universal Ocean. The level of this vast body of water had been lowered, alleged proponents of the theory, because cracks had opened in the crust and the water had trickled into the Earth's interior. Vestiges of this idea still persist in the modern world. Even in the 1960's, after serious preparations were underway in the now-defunct project MOHOLE to drill a hole through the Earth's crust underlying the ocean floor, government officials received numerous letters expressing the fear that the proposed hole would empty the oceans like the draining of a bathtub.

If we apply the scientific method we must not only speculate about the composition of the Earth's interior but also test our speculations by observation and measurement. Certain observations that we can make at the surface contain important clues about what lies beneath. For example, vast quantities of molten rock material have been discharged from volcanoes. By studying volcanoes early geologists conceived the idea that the interior of the Earth consisted of molten rock material. As we mentioned earlier, the word *crust* derives from the concept of a solid rocky rind floating on a liquid interior. Although, as we shall see presently, this simple idea is incorrect, the principle of basing ideas about the interior on material discharged at the surface is a sound one and we still use it today.

Gravity offers us another avenue of approach to the problems of the Earth's interior. After Newton formulated the law of gravitation it became possible to calculate from simple experiments the mass and specific gravity of the Earth. Various experimenters have achieved about the same result, approximately 5.5 (Fig. A-4). But we know the specific gravities of rocks at the Earth's surface range from only 2.7 to 3.0 (Table C-2). Therefore the specific gravity of the interior must be greater than 5.5, to counterbalance the light surface rocks.

Most present ideas about the Earth's interior evolve from the study of earthquakes. In this chapter we examine earthquakes, their effects on the Earth's surface and their use in analyzing the interior.

EARTHQUAKES

Earthquakes and their related events are among the most dramatic and disastrous of natural phenomena. During earthquakes eyewitnesses have reported that the ground moves up and down "like waves on the sea;" in regolith small fissures open and close. Unless they are specially designed, buildings may collapse (Fig. 18-1). Gas pipelines break and fires get out of hand because water systems have been disrupted. The ground becomes permanently displaced along faults (Pl. 18); areas are raised, lowered, and shifted laterally (Fig. 18-2). The shaking permits gravity to displace great bodies of rock material down slopes. In mountainous areas large landslides can block valleys and create natural lakes. Commonly the backed-up water suddenly breaches its debris dam and starts a disastrous flood. Displacement and slumping of sea-floor sediments set off dreaded tsunami, whose effects can devastate shores thousands of miles away.

In populous areas, earthquakes are reported and photographed by news-gathering organizations, which detail damage and stories of human tragedy. Teams of scientists then enter the scene, conduct surveys, and make recommendations. We need not repeat familiar stories but shall concentrate on the geologic aspects of the subject.

CAUSES OF EARTHQUAKES

Early Greek and Roman scholars attributed earthquakes to air escaping from underground or to roof collapse of vast subterranean caverns. Scholars debated this and other fanciful views without bothering to gather evidence systematically on the subject until 1755, when the disastrous Lisbon earthquake occurred. Scientific inquiry stimulated by this catastrophe led to the idea, widely accepted by the middle of the nineteenth century, that earthquakes were manifestations of the motions of seismic waves, also known as elastic waves because their speeds are greatly influenced by the elastic properties of materials (App. A).

Fig. 18-1. School building in Long Beach, California wrecked by the earthquake of 1933. Fortunately school was not in session when the earthquake occurred. (Wide World Photos.)

Seismic waves. It is one thing to assert that the Earth quakes because seismic waves pass by and another to determine the characteristics and origins of these waves. *The study of seismic waves and the science of earthquakes* constitute ***seismology*** (sīs-mŏl′ō-jĭ; Gr. "earthquake study"). We shall begin our study of seismology with the recorders that monitor wave motion.

Fig. 18-2. Air view of small valleys, cut into weak materials, offset laterally by strike-slip movement along San Andreas fault, F–F′, probably during the earthquake of 1857. View westward across Carrizo Plain, approximately 125 miles northwest of Los Angeles (map, Fig. 18-6, *A*). (J. S. Shelton.)

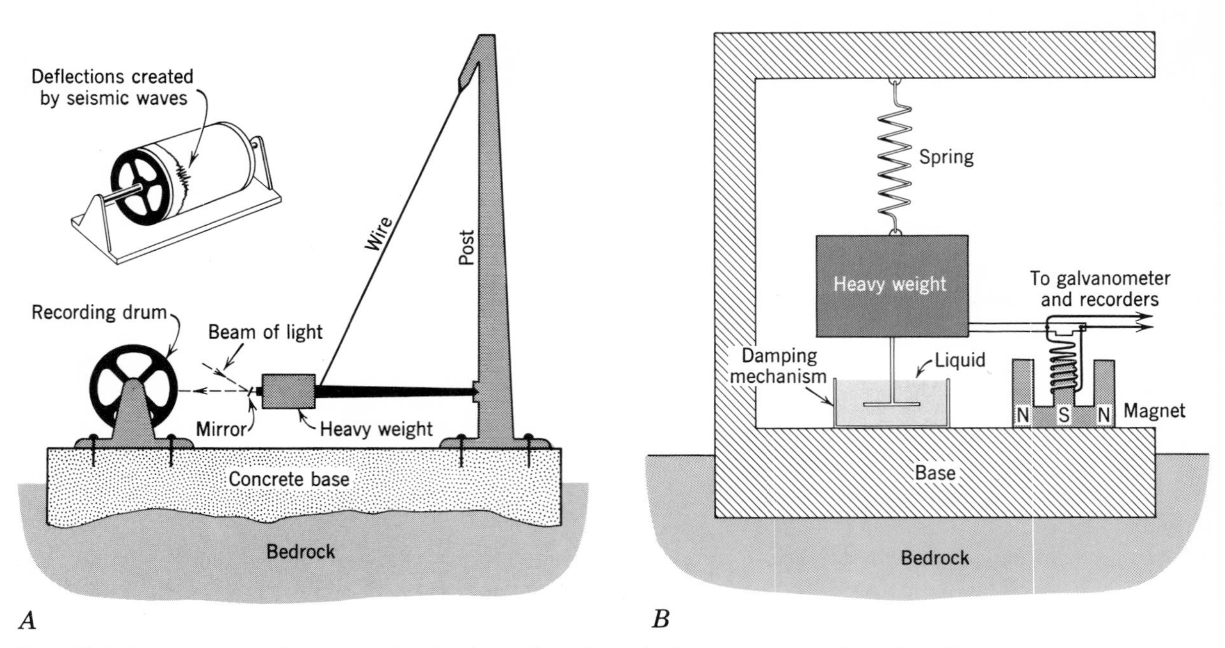

Fig. 18-3. Seismographs for measuring horizontal and vertical components of earthquake waves. A. Horizontal seismograph with recorder based on beam of reflected light. *Upper left.* Trace of reflected light on photographic paper wrapped around drum becomes visible only after photographic development. (After R. M. Garrels, 1951.) *B.* Vertical seismograph for recording long-period seismic waves. Further explanation in text. (After Jack Oliver, "Long earthquake waves." Copyright © March 1959 by Scientific American, Inc. All rights reserved.)

Recording and measurement. Nearly all ***seismographs,*** *instruments for the accurate recording of seismic waves,* depend on a single principle: a heavy mass can be suspended so that its inertia keeps it nearly at rest while the Earth beneath it vibrates. We can understand such seismographs by familiarizing ourselves with two kinds: horizontal and vertical.

To detect horizontal components of motion parallel to one direction a *horizontal seismograph* employs a heavy weight mounted on a horizontal bar. The bar is flexibly connected to a vertical supporting post which is firmly anchored in a concrete base set on bedrock (Fig. 18-3, *A*). Because it tends not to move even if the Earth beneath it does move, the weight serves as a point of reference for registering horizontal motions at right angles to the horizontal bar. To complete our seismograph we need a device to record any differences in motion between the stationary weight and the moving Earth. One such device is a drum mounted on a frame fastened to the concrete base. By rotating the drum at a known and fixed rate we can establish time control. By shifting the drum sideways as it rotates we can record on the drum through many rotations. Records of motion can be transferred to the drum by one of three methods: (1) direct contact using a stylus attached to the weight, (2) indirect contact using a narrow beam of light reflected from the weight onto photographic paper wrapped around the drum, and (3) converting the motion into an electric signal and recording it electronically. Method 2 requires us to develop the paper to see the record. *A record made by a seismograph* is a ***seismogram.*** The drum oscillates when the Earth moves. As a result the trace on the seismogram becomes deflected into a zigzag line. Because each horizontal seismograph records motion parallel to only one direction, we require two such seismographs to record both north-south and east-west components of seismic waves.

To detect the vertical portion of seismic waves a *vertical seismograph* employs a heavy mass hanging from a coiled spring (Fig. 18-3, *B*). A simple vertical seismograph can record waves whose periods are less than the natural period of vibration of the spring, the time required for the spring to

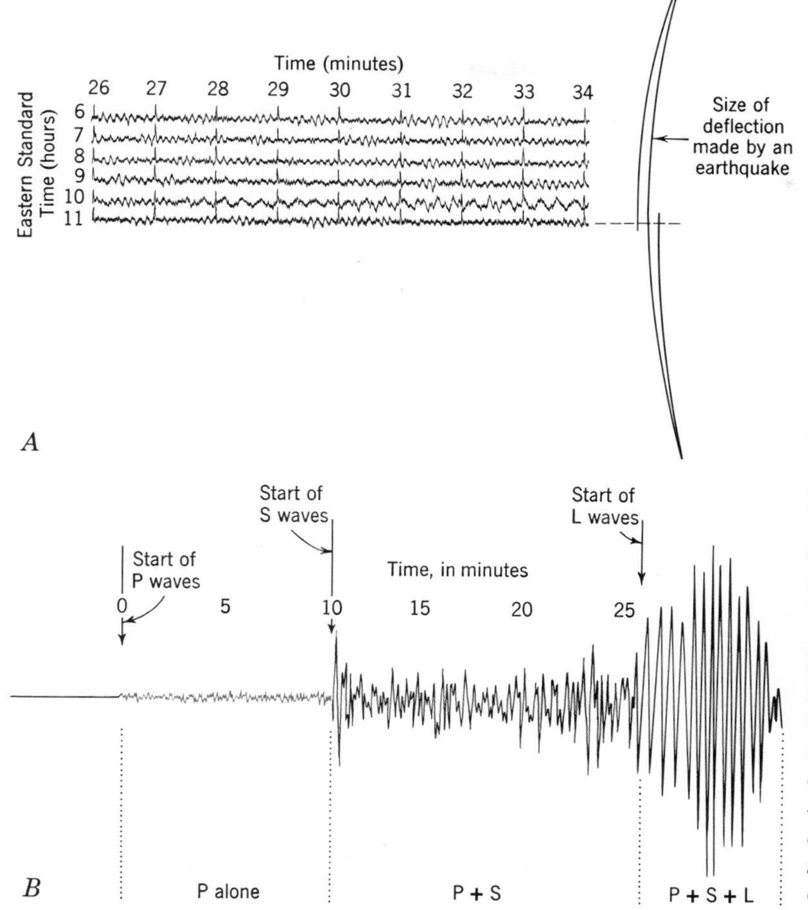

Fig. 18-4. Examples of seismograms. A. Traces of microseisms recorded at Troy, N. Y., January 1, 1967, compared with deflection made by an earthquake on the same day. (Rensselaer Polytechnic Institute.) *B.* Seismogram of earthquake recorded at Cambridge, Mass., December 26, 1939, with epicenter at Erzincan, Turkey, shows small deflections made by P waves alone (blue) and large deflections made by P + S and P + S + L (black). Time elapsed between start of P and that of S, 10 minutes and 45 seconds, registers distance from epicenter to station as 88°30′, or about 9,700km (Fig. 18-12). (After *Geophysics at Harvard.*)

complete one pulse of extension and retraction. The natural periods of most springs are typically less than 15 seconds. Likewise the periods of many seismic waves are less than 15 seconds. This coincidence makes it possible for simple spring seismographs to record many seismic waves. Previously seismologists recorded earthquakes on simple, short-period spring seismographs. Today, however, because they have found that long-period waves, not legible on former seismograms, disclose significant information about the shallower parts of the Earth's interior, seismologists are interested in recording more of the varied spectrum of seismic waves than short-period seismographs register. The natural periods of vertical seismographs can be increased by two methods: (1) damping the spring with a dashpot mechanism, and (2) converting mechanical motion into electrical impulses by means of a coil of wire free to move in a permanent magnetic field (Fig. 18-3, *B*). The electrical impulses can be fed into a graphic recorder and at the same time stored on magnetic tape for analysis by a computer.

On a seismogram (Fig. 18-4) the trace of the first big longitudinal wave may create in the trace a deflection that is either upward or downward from the horizontal line representing an absence of big waves. Upward deflections are created by *positive waves,* downward deflections by *negative waves.* As we explain presently, positive and negative waves originate in contrasting ways.

Kinds of seismic waves. By analyzing seismograms recorded on one vertical and two horizontal seismographs we can reconstruct the kinds of motions created in the ground by passage of seismic waves. From the motions of seismic waves we recognize four varieties: longitudinal waves, shear waves, Rayleigh waves, and Love waves (Table 18-1).

TABLE 18-1. NAMES AND PROPERTIES OF SEISMIC WAVES

Particle Motion Created by Passage of Wave	Synonymous Names of Waves	Standard Letter Designation in Seismology	Names Based on Travel Paths in Earth
Oscillation along lines in direction of wave travel	Longitudinal waves Compressional waves Push-pull waves Sound waves	P (for *Primary*)	Body waves
Oscillation along lines at right angles to direction of wave travel	Shear waves Transverse waves Shake waves	S (for *Secondary*)	Body waves
Around in circles lying in vertical planes; in same direction as wave advance under troughs, in opposite direction under crests	Rayleigh waves	L (for *Large*)	Surface waves
Oscillation at right angles to direction of wave travel along lines lying in horizontal planes.	Love waves	L (for *Large*)	Surface waves

Longitudinal waves consist of alternating pulses of compression and rarefaction acting along the directions in which the waves travel. The commonest kind of longitudinal waves are sound waves (Fig. 15-1); they can propagate through not only water but also solids. The fact that such waves are transmitted by solids through great distances can be of vital importance in perilous situations. On many occasions, miners trapped by cave-ins have sent out messages by rapping with pebbles on the steel rails of tracks, on pipes, or even on bedrock composing tunnel walls.

Even without recourse to a seismograph we can infer from natural noises that longitudinal waves accompany earthquakes. In the early stages of some earthquakes we can actually hear deep rumblings or even loud reports.

The speeds of longitudinal waves are influenced by the properties of the materials through which the waves pass and depend directly on the resistance of a material to changes of both volume and shape, and inversely on its density (App. A).

In contrast to longitudinal waves, which create oscillations in the direction of wave travel, *shear waves* create oscillations at right angles to that direction. We can illustrate the motions of shear waves by the familiar experiment of fastening a rope at one end and shaking the other end vigorously (Fig. 18-5). The analogy of the rope, however, is inadequate to visualize the three-dimensional aspects of seismic waves, which radiate outward in all directions from a point source. The speeds of shear waves are related directly to resistance of a material to change of shape and inversely to density (App. A). Because fluids (liquids and gases) lack ability to maintain distinct shapes, they do not transmit shear waves. We shall make much use of this fact presently when we extend our study of seismology to interpretations of the Earth's interior.

When *Rayleigh waves* pass by, particles are set in motion in circles lying in vertical planes, as they are in water waves, except that in Rayleigh waves the particles move in the opposite direction, backward under crests and forward under troughs.

Love waves make particles oscillate along lines in the horizontal plane at right angles to the direction of wave travel.

Origin. In Chapter 2 we read that electromagnetic waves are expressions of energy being transmitted and that the kinds of waves generated depend on the size of the vibrating object originating the waves. Likewise seismic waves are expressions of energy being transmitted through the Earth. However, the kinds of waves that reach a distant point depend not only on the source mechanism but also, as we have seen, on the properties of the materials through which the waves have passed. Let us first examine the source mechanisms.

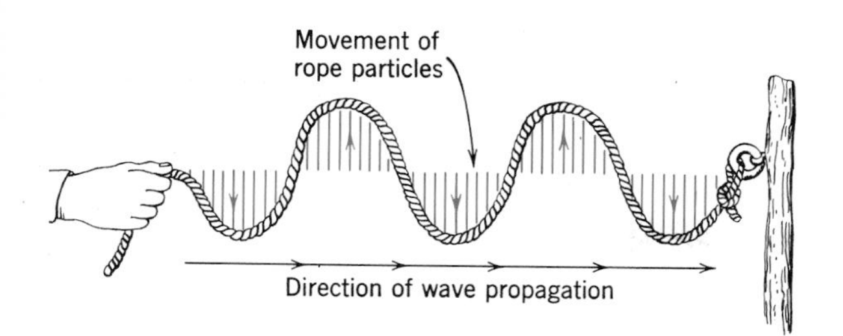

Fig. 18-5. Shear waves displace the particles in a rope (short blue lines and arrows) at right angles to direction of wave propagation. (After R. M. Garrels, A textbook of geology © 1951. By permission of Harper and Row Publishers, Inc., New York, N. Y.)

If energy is transmitted it must have a source. How can enough energy to create seismic waves be released within the Earth? One favored hypothesis states that earthquake energy results from deformation of the lithosphere. According to this hypothesis, elastic energy can be stored in bodies of rock being deformed just as in a spring that is compressed. Eventually deforming forces rupture the rock, creating a fault. If the opposite walls of the fault move suddenly, the stored energy is released and becomes converted into seismic waves. When these waves pass through the Earth they cause it to quake. This is the elastic-rebound concept; because it is now widely held we can profit by examining the evidence from California where the idea was spawned.

Movement on strike-slip faults. Years before 1906, geologists recognized a locus of recent strike-slip movement and named it the San Andreas fault zone; it extends northwest-southeast across California for more than 600 miles (Fig. 18-6, *A*). During careful mapping in central California, beginning in 1874, scientists from the U. S. Coast and Geodetic Survey determined the precise positions of many points adjacent to the fault zone with relation to points remote from this zone miles to the east. On April 18, 1906, the two sides shifted abruptly. The stored energy was transformed into seismic waves, which dissipated radially outward creating the earthquake. Repetition of the surveys soon after the earthquake of 1906 demonstrated that points along the northeast side of the fault had moved southeastward; in other words, the sense of displacement had been right-lateral (Table 17-1). Measurements made after 1906 indicate that the relative movement in the zone of deformation may have occurred at a rate as large as 5cm/year.

We can illustrate the relationships between the gradual elastic deformation and sudden movement on the San Andreas fault which created the earthquake of 1906, by comparing a succession of detailed maps of the fault zone made prior to 1906 with one made after the great quake. On the first map let us suppose that seven points were located along a straight line across the fault (Fig. 18-6, *B*, A to G). Later, gradual elastic deformation occurred with a right-lateral sense of movement. On our second map, points in the former line had shifted to new positions (A′ to G′). Movement of these points reflects the accumulating elastic strain (Chap. 12). As elastic strain increased, enormous elastic energy was stored in the zone of deformation.

A map made after the quake would show that the points had resumed their previous straight alignment (blue dashed line) but that where it crosses the fault the line had been offset (OP). Near San Francisco the total offset was 21 feet.

The abrupt movement on the San Andreas fault in 1906 created a fresh fracture, clearly visible on the ground, extending from San Juan Bautista 210 miles northwestward to Point Arena and another 90 miles under the ocean (Fig. 18-6, *A*). Segments of the San Andreas fault farther south have been displaced at other times (Fig. 18-2). In the Imperial Valley earthquake of 1940, rows of fruit trees and fences were offset horizontally by as much as 17 feet.

Along a strike-slip fault the motion is one of shearing. As we read in Chapter 17, shearing stresses do not exist alone; they are always accompanied by tensile and compressive stresses whose principal directions are at right angles to each other and diagonal to the shearing stresses. When a strike-slip fault shifts abruptly, it releases shear waves that oscillate in a direction parallel to the fault, and two sets of longitudinal waves. The paths of negative longitudinal waves parallel the direction of maximum compressive stress and the paths of positive longitudinal waves parallel the direction of maximum tensile stress (Fig. 18-7, *A*). By careful analysis of the distribution of positive and negative wave traces on seismograms it is possible to reconstruct the direction of first motion along a strike-slip fault.

The locus of first release of the elastic energy, located at various depths beneath the Earth's surface, is the ***earthquake focus.*** *The part of the Earth's surface vertically above the focus* is the ***epicenter***

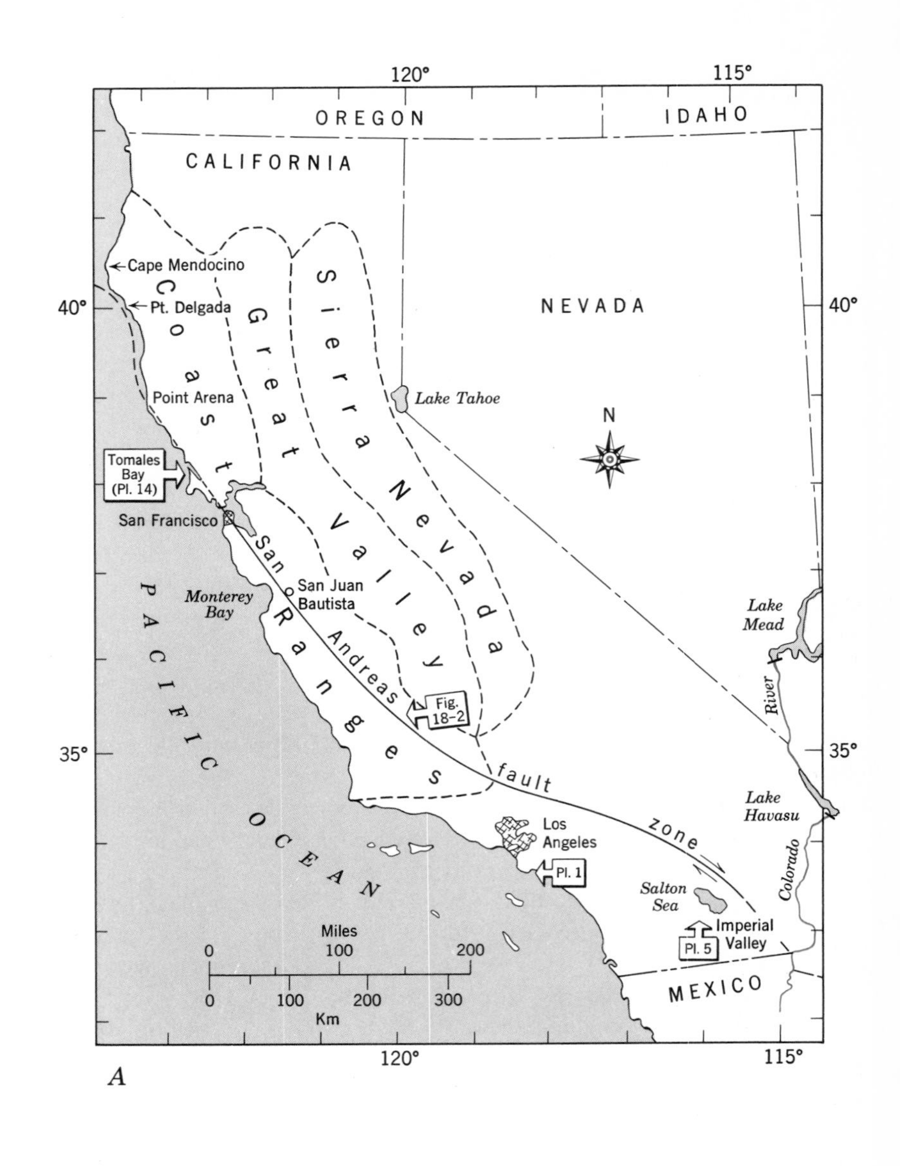

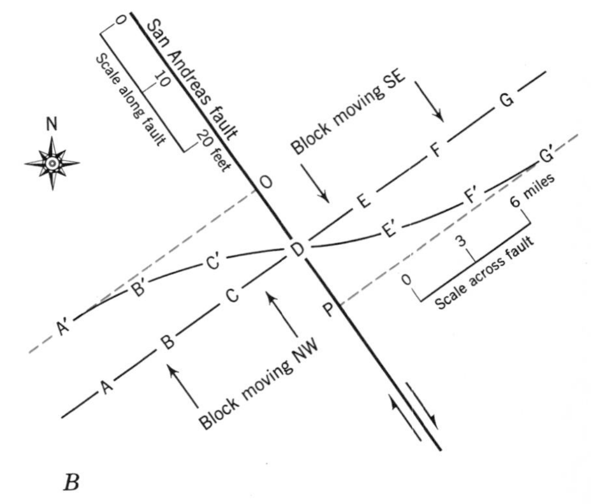

Fig. 18-6. San Andreas fault zone. *A.* Map showing trace of San Andreas fault zone across California from a point near Mexican border, where it is hidden by alluvium, northwestward to Point Arena, where it strikes out to sea. Small squares and arrows locate Plates 1, 5, and 14, and Fig. 18-2. *B.* Sketch, based on detailed surveys conducted before and after abrupt movement, in 1906, along the northwestern segment of the fault. After the two crustal blocks had been bent, movement occurred suddenly and the segments straightened. Further explanation in text.

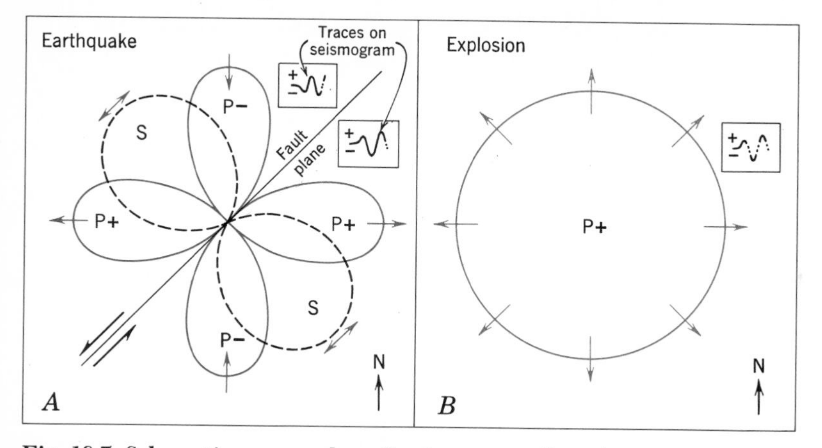

Fig. 18-7. Schematic maps and small seismograms (insets) compare seismic body waves generated by strike-slip fault and by explosion. *A.* Wave pattern generated by abrupt displacement along left-lateral strike-slip fault oriented northeast-southwest. Direction of oscillation of shear waves (S) parallels fault plane. Seismograms display traces of positive (E-W direction) and negative (N-S direction) longitudinal waves (P) generated parallel with directions of maximum compressive and tensile stresses, respectively. Curved lines show complex positions of wave fronts. *B.* An explosion generates only positive longitudinal waves (P+), which radiate outward in all directions. (After Sir E. B. Bullard, "The detection of underground explosions." Copyright © July 1966 by Scientific American, Inc. All rights reserved.)

(Gr. "above the center"). Depending on the characteristics of the focus, the epicenter can be an isolated small circle or a line.

Both longitudinal waves and shear waves radiate outward in all directions from the focus. As long as their paths remain within the Earth these two kinds of waves are known collectively as *body waves*. When body waves reach the Earth's surface, some of their energy is converted into various *surface waves*. The relationship of surface waves and body waves can be illustrated by dropping a pebble into still water. When the pebble strikes the water two things happen. (1) Small surface waves originate and radiate outward from the point of impact in ever-enlarging circles. (2) Sound waves (longitudinal waves) from the impact pass downward into the body of water ahead of the sinking pebble. If we replace the pebble with the body waves from an earthquake focus and trace the sequence upward to the surface from below, we can visualize the origin of surface waves generated by an earthquake. At the water/air interface physical properties change abruptly. These changes can influence the behavior of seismic waves so markedly that the interface represents a seismic discontinuity (Chap. 15). When seismic waves encounter a seismic discontinuity some are refracted and travel along it, whereas others are reflected by it. In this case the refracted waves become surface waves. The reflected waves, meanwhile, are turned downward again, and continue to travel through the Earth as body waves. In the solid Earth the body waves would include not only longitudinal waves but also shear waves, and the surface waves would include both Rayleigh waves and Love waves. In our example of the pebble dropped into water only longitudinal waves (sound waves) are present within the liquid because shear waves do not exist in fluids, and only one kind of surface wave is present.

In an earthquake, then, many kinds of seismic waves are traveling simultaneously. Longitudinal waves and shear waves move through the body of the Earth, and Rayleigh waves and Love waves move along its surface.

Vertical movement on faults. On some faults abrupt movements in the vertical plane create scarps and cause severe earthquakes, as at Pleasant Valley, Nevada, in 1915 (Fig. 17-14) and Prince William Sound, Alaska, on March 27, 1964 (Figs. 17-2, 17-19). Although we lack evidence from the

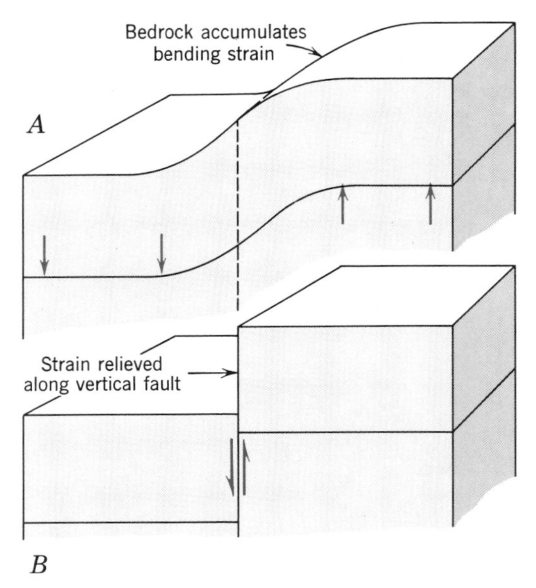

Fig. 18-8. Vertical movement of crustal blocks by bending and faulting. *A.* Arrows show direction of movement by slow bending. *B.* Blocks have slipped along vertical fault, relieving bending strain and creating steep fault scarp. Half-arrows show direction of relative fault displacement. (Vertical scale exaggerated.)

precise surveying points near such faults, by analogy with the San Andreas fault we infer that the great energy released during an abrupt movement had been stored there as a result of previous elastic deformation (Fig. 18-8).

Before discussing other origins of seismic waves, we must add that not all movements on faults are abrupt. A notable exception, which has claimed much attention because of its interference with production in an oil field, occurred near Bakersfield, California. The steel casing pipes lining various oil wells were gradually bent, and finally some of them were sheared off, by movement along a fault having a rather small angle of dip. During a period of 17 years the rate of movement averaged approximately 3cm/year. From our point of view this rate is slow, but if movement should continue at this pace for 20,000 years, only a moment in geologic time, the aggregate displacement would amount to 0.6km.

Movement of bodies of fluid underground. Fluid pressures underground are capable of creating movements within the solid lithosphere. Fluids can raise and lower superincumbent rocks abruptly and the abrupt displacement of adjacent blocks creates seismic waves. When magma moves upward toward the Earth's surface prior to a volcanic eruption, it creates seismic waves intense enough to be perceived as local earthquakes, even without recourse to seismograms. Volcanoes and earthquake zones are closely related geographically, as we shall see farther along.

Near Old Faithful Geyser, Yellowstone Park (Fig. 19-20) distinctive patterns of seismic waves have been recorded during different parts of the cycle of activity. Such seismic waves evidently result from the effects of bodies of fluid moving underground.

In 1966, attention was drawn to the coincidence of small earthquakes near Denver, Colorado with times of disposal underground of lethal liquid wastes from the Rocky Mountain Arsenal. Each time that wastes were pumped down the well, 12,000 feet deep, which had been drilled especially to receive them, the surrounding region, otherwise nearly free of local earthquakes, experienced small tremors almost daily. The mechanism responsible for the small quakes is unknown but is thought to be related to the increased fluid pressures underground.

Explosions. Energy released into the ground by explosions causes seismic waves. We have already described how sound waves can be used to make continuous seismic profiles of the internal structure of sea-floor sediments and of the shape of the surface upon which the sediments lie (Fig. 15-4). By somewhat different techniques we can obtain the same information about the structure of strata underlying land areas.

In the seismic method of exploring beneath the land surface, charges exploded in holes drilled at selected points generate seismic waves. The waves created by the explosion are received by extremely sensitive geophones. The geophones connect to electronic recorders mounted in trucks (Fig. 18-9).

Seismic waves generated by explosions consist of pulses of positive longitudinal waves that radiate outward in all directions (Fig. 18-7, *B*). As in water the waves travel along straight paths as long as they encounter material having uniform physical properties. Likewise when seismic waves encounter materials having contrasting properties, they cease to travel in straight lines. Instead, as we have mentioned previously, when seismic waves encounter an interface where properties change, some are reflected upward by it and others are refracted along it. At some interfaces changes are abrupt, as across a distinct separation between a layer of shale and

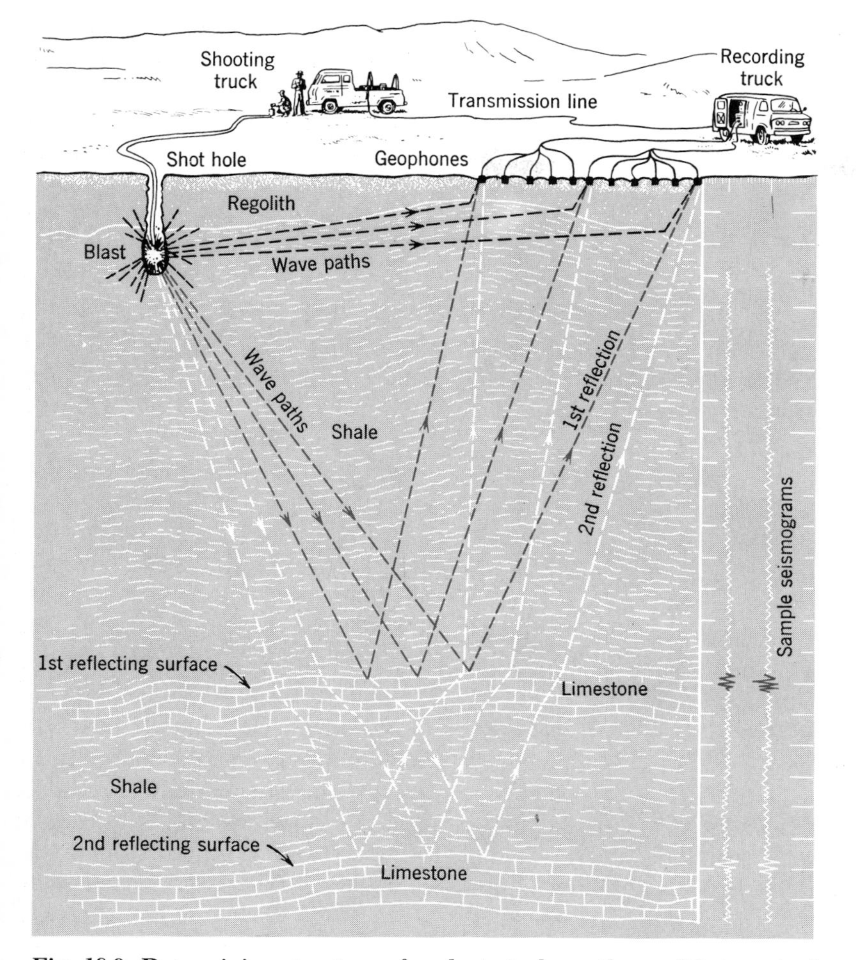

Fig. 18-9. Determining structure of rock strata beneath regolith by seismic exploration with portable equipment. Abrupt contacts with shale at tops of limestone strata are surfaces that reflect seismic waves. Analysis of reflections from such surfaces reveals attitude of strata. (After C. A. Heiland, Geophysical exploration © 1940. By permission of Prentice-Hall, Inc., Englewood Cliffs, N. J.)

a layer of limestone. The pattern of waves returning to the surface from a seismic discontinuity at depth discloses the attitude of the discontinuity, which in stratified rocks is an expression of the subsurface structure. Therefore, the subsurface structure can be determined seismically even where thick regolith hides the bedrock and no hint of the structure is present at the surface. Noises smaller than explosions can be used to measure the thickness of thin regolith; large explosions make it possible to study layers deeper down in the crust.

Miscellaneous causes. At the Earth's surface many disturbances create small seismic waves. Some of these disturbances are the wind, heavy highway traffic, railroad trains, and vibrating industrial machinery. Modern sensitive seismographs register these small seismic waves as *microseisms* (Fig. 18-4, *A*).

PROBING INSIDE THE EARTH WITH SEISMIC WAVES

With a suitable array of seismographs we can "tune in" on the Earth and record natural vibrations. If our instruments are sensitive to very small motions the record will consist of many small wave traces or microseisms, a visual record of the constant creaking and groaning that goes on inside the Earth. In all probability our seismogram will show us nothing more than microseisms for hour after hour. Now and then, however, our patient vigil will be rewarded with something more dramatic. The Earth's louder seismic messages begin abruptly and are transmitted in a code having a patterned format. Seismologists have managed to decipher the format and to decode the messages by which the Earth broadcasts so much about itself.

The format of natural seismic messages. Now that techniques for processing electronic signals and handling data have advanced conspicuously, visual study of seismograms has practically ceased and the electric impulses from the seismographs are processed by computers. Even so, with our eyes we can crack the first parts of the seismic code, just as pioneer seismologists did.

Louder messages begin with three characteristic parts: the first or primary (P) deflections, the second or secondary (S) deflections, and the third or large (L) deflections (Fig. 18-4, *B*). To relieve the boredom we might hope for something different now and then, but in this we seem to be disappointed, for message after message begins in this characteristic way. But are the beginnings of all messages really the same? To find out we must look carefully.

Seismic messages deciphered. After seismologists discovered they could eavesdrop on the Earth, they installed listening devices at various localities so as not to miss any transmissions, organized a central message center where they could report all seismic transmissions, and acquired very accurate clocks, which they check continually by reference to standard time signals broadcast on special radio frequencies. Seismologists from all countries agreed to refer all times to that of the Greenwich Observatory, at the zero meridian, and adopted a standardized procedure for measuring distances between points on the Earth's surface, using angles. We can illustrate this procedure by considering the angle between the hands on a clock. If we fix the minute hand in its position on the even hour and let it represent a line from the center of the Earth to an earthquake epicenter, we can let the hour hand represent a line from the center of the Earth to a seismic station. The angle between the hands measures distance along the circumference of the clock dial, which on the Earth represents the distance from an epicenter to a seismic station. We use the right half of the clock face in a normal way, in which 12 o'clock coincides with 0° at the top and 6 o'clock

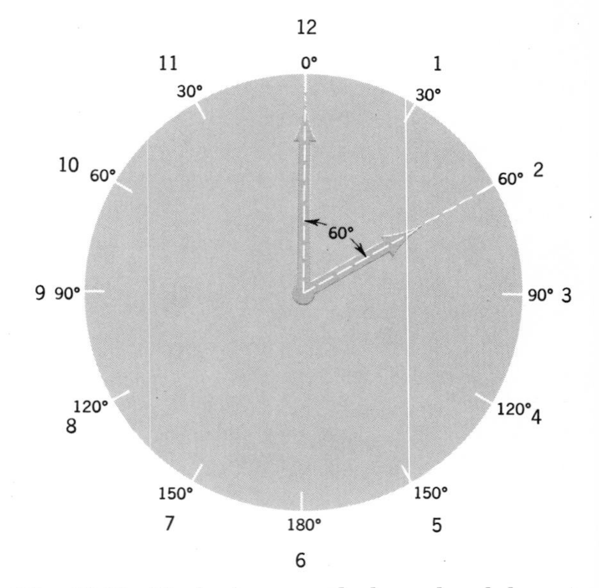

Fig. 18-10. Clock shows method employed by seismologists for expressing distances on Earth's surface from epicenter to stations in terms of angles. Further explanation in text.

with 180° at the bottom (Fig. 18-10). The positions 1-, 2-, 3-, 4-, and 5 o'clock mark off successive 30-degree increments. On the left side of the circle each of the 30-degree positions from 0° to 180° is indicated by the hours 11, 10, 9, 8, 7 and 6, respectively.

After many earthquakes had been recorded, one seismologist noticed that the parts of the same old message, consisting almost invariably of the deflections P, S, and L, were spaced out in a systematic way. Near the focus of an earthquake they came in close succession, like three quick shots. With greater distance from the focus the spacings between deflections P and S and between those of S and L increased.

What do the deflections P, S, and L represent that they should be spaced so systematically? They are the records of motions of seismic waves. The P-deflections are made by longitudinal waves and the S-deflections by shear waves, both influencing our sesimograph at the surface after moving through the body of the Earth. The L-deflections are made by Rayleigh waves that have moved along the surface of the Earth.

In order to understand how seismologists analyze the Earth's interior, let us examine the fanciful hypothesis that the Earth is a uniform sphere having

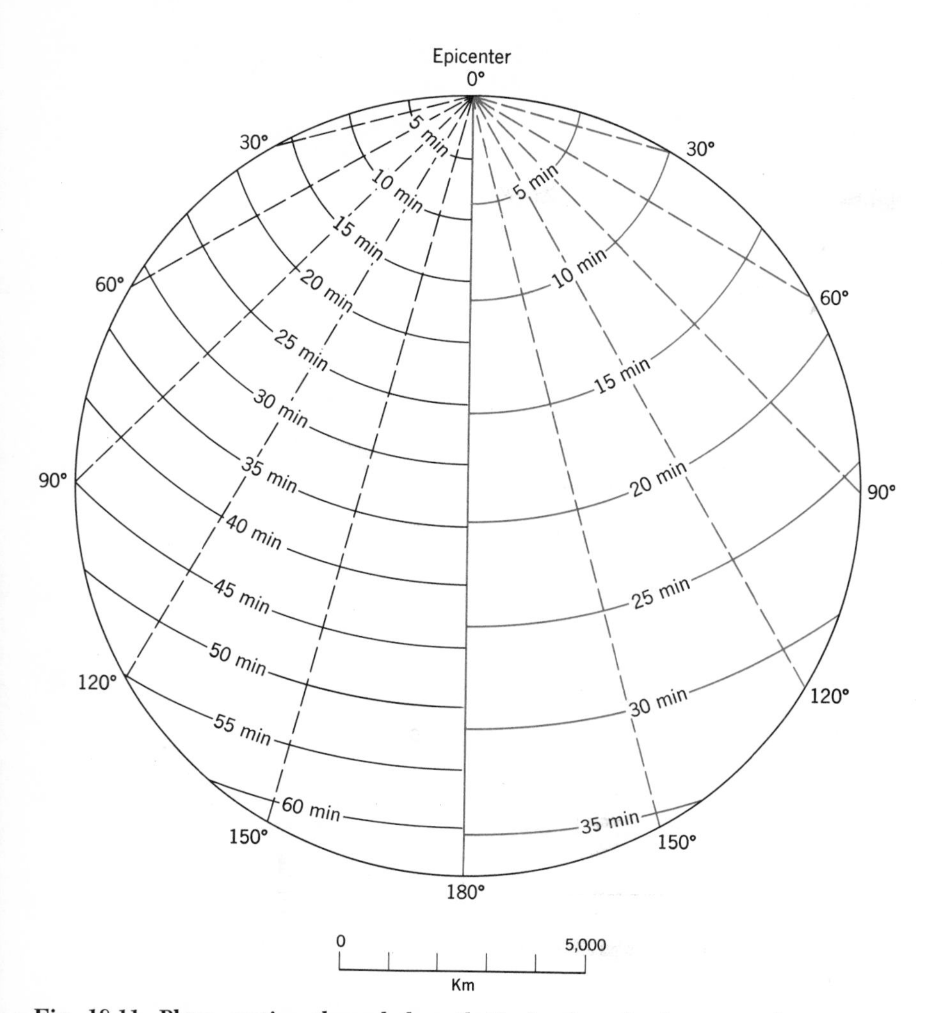

Fig. 18-11. Plane section through hypothetical sphere having same diameter as Earth but composed of uniform material throughout shows travel paths of seismic body waves (dashed lines) to stations at distances from the epicenter 30° apart, and positions of spherical wave fronts (solid curves) at 5-minute time intervals. Speed of P waves (blue) 5.7km/sec; of S waves (black) 3.3km/sec. Travel paths and wave fronts, which in reality extend throughout the circle, have each been drawn on one half only. Paths of surface waves and of body waves reflected from surface of sphere omitted (compare Fig. 18-13.)

elastic properties throughout that are identical with those of rocks exposed at the surface of the real Earth. From comparing the densities of rocks at the surface with the average density of the Earth we know that this hypothesis is incorrect. Nevertheless, let us ignore this evidence of density and evaluate the hypothesis by seismology. From our hypothesis we can draw four corollaries that can be tested by studying earthquake waves. (1) Seismic body waves should travel through our uniform sphere as shown in Fig. 18-11. (2) Longitudinal waves should pass from one side to the other through the center of our hypothetical Earth in slightly less than 37 minutes (Fig. 18-12, dashed blue curve). (3) Shear waves should complete the same trip in slightly more than 63 minutes (dashed black curve). (4) Both kinds of waves should be recorded at each station, whatever its distance from the epicenter.

Assuming that we possessed no prior knowledge of the speeds of seismic waves within the Earth,

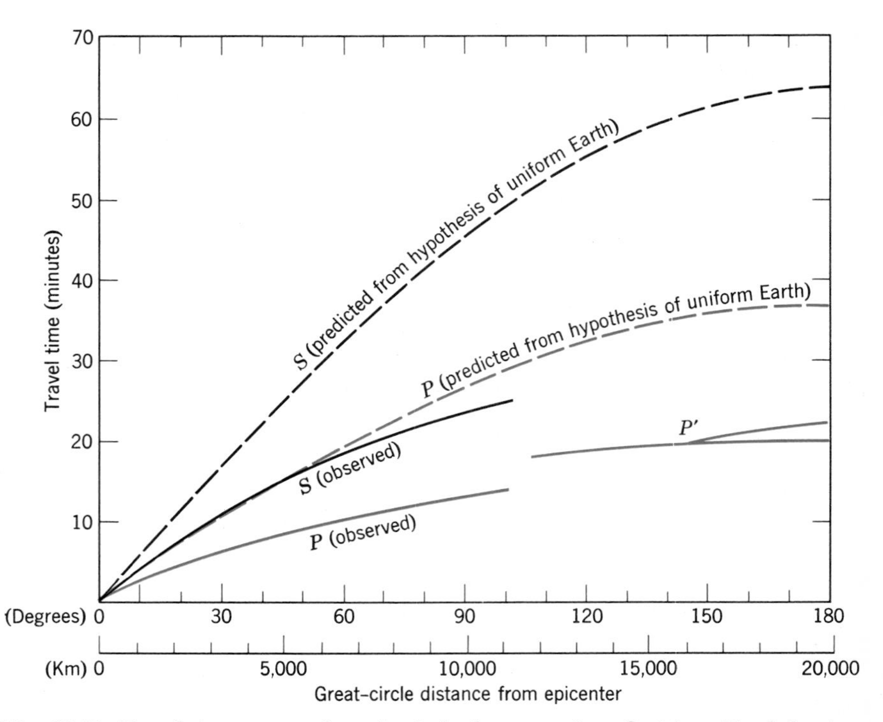

Fig. 18-12. Travel-time curves for seismic body waves in a fictitious Earth having uniform density (dashed curves) and in real Earth (continuous curves). P waves blue, S waves black. Dashed curves are based on linear travel paths and speeds shown in Fig. 18-11. (Data for observed curves from Hodgson, 1964.)

we would have to wait for an earthquake to occur near one of our stations before we could test our corollaries. Such an earthquake would be required so we could fix its time and the location of its epicenter visually and thus know the distance to all stations. From seismograms of this earthquake at many stations we could plot travel-time curves for the seismic waves that make the P- and S-deflections on our records (solid lines in Fig. 18-12; blue for P and black for S).

The seismograms disclose that the arrival times of both P and S waves are earlier than predicted. The divergences between predicted and actual times of arrival increase progressively with distance from the epicenter. Beyond a distance represented by 103° predictions from our hypothesis and observations do not resemble each other even remotely; no traces of the S waves are recorded and a gap appears in the travel-time curve for the P waves. The discrepancies between corollaries based on our hypothesis and observations are so serious that we must abandon our hypothesis and formulate another.

The divergences between the dashed-line curves and solid-line curves of Fig. 18-12 indicate that the velocities of the two kinds of waves are greater in the real Earth than in our fanciful hypothetical Earth in which we assumed that elastic properties were uniform with depth. From the travel time curves we derive a second hypothesis which states that the Earth's elastic properties vary with depth. Recall the evidence that density increases downward within the Earth. If nothing else but a change of density occurred downward, our observed travel-time curves should lie *above* the dashed lines in the graph. The reason is that velocities of seismic waves are related inversely to density (App. A). We have found by measurement that despite the increase of density downward, the velocities of waves *do* increase downward through part of the Earth. Therefore, resistance to deformation, related directly to the velocities of elastic waves, must be increasing downward at a rate greater than that at which density increases.

If elastic properties change with depth in the Earth then our analysis of seismic waves must take account of another phenomenon, wave refraction.

In uniform materials waves travel in straight-line paths, but in materials having elastic properties that vary with depth different parts of the wave fronts travel at different rates. Therefore the wave paths become curves. Deeper within the Earth, seismic waves travel progressively faster. This causes their paths to be refracted, just as rays of light are bent in passing from air into glass or water. Refraction causes seismic waves to curve back toward the Earth's surface (Fig. 18-13, *A*). Because of refraction, the distances waves travel in the real Earth, from focus to each station, become progressively greater than in our hypothetical Earth as distance from epicenter to station increases. Despite the longer travel paths to more distant stations we recall that the waves arrived earlier than our travel-time graph indicated they should. This relationship strengthens our second hypothesis that in at least part of the real Earth elastic properties vary progressively downward.

The observed disappearance of the S waves at stations beyond 103° from the epicenter is exactly the opposite of our prediction from corollary 4 of our first hypothesis. When seismologists became aware of the disappearance of the S waves they suddenly realized they were no longer simply recording earthquakes but that they also could use seismic waves to probe inside the Earth. They noticed that the change within the real Earth which blots out the S waves does not stop the P waves, but permits them to pass. The only condition satisfying this requirement is that the waves encounter fluid material. Fluids do not transmit shear waves but do transmit longitudinal waves.

Seismologists have calculated that the P and S waves coming through to stations just inside the 103-degree limit have penetrated the Earth to a maximum depth of 2,900km (Fig. 18-13, *B*). Stations between 103° and 143° lie in a *shadow zone;* they receive no direct arrivals of waves passing through the body of the Earth, but only waves that have been reflected from the Earth's surface or from some boundary within the Earth. Between 143° and 180° direct arrivals of P waves reappear on the records, but their times of arrival indicate that in the zone between depths of 2,900km and 5,000km, they have slowed down. From a depth of 5,000km to the center of the Earth P waves have speeded up again. Depths of 2,900km and 5,000km mark important seismic discontinuities within the Earth. At 2,900km the S waves disappear and P waves slow down; at 5,000km, P waves speed up again.

The Earth's core. From the speed of the Earth's rotation and from the amounts of its equatorial bulging and polar flattening, physicists have calculated that most of the Earth's mass must lie close to its center. Sun and Moon exert a pull on the Earth's equatorial bulge, causing a slow change in the tilt of the axis of rotation. The amount of this effect, which has been measured precisely, can be explained only by supposing that a large part of the Earth's mass is concentrated near its center. In other words, material of low or moderate density extends far down to a zone having much higher density. The only hypothesis consistent with the seismic observations and these effects of rotation and bulging is that *an inner spherical mass below a depth of 2,900km* constitutes the ***core of the Earth.***

The steady increase of velocity of seismic waves down to a depth of 2,900km suggests that the outer part of the Earth consists of material which exhibits a general similarity of composition but which is compressed more and more with increasing depth. Because this material transmits shear waves, it must be a solid. The abrupt stopping of shear waves at 2,900km indicates a sharp boundary where solid material overlies liquid material.

We know that pressure and temperature increase downward and that increasing the pressure raises the melting points of materials (Chap. 5). At the boundary of the Earth's core, in a zone of great pressure and temperature, liquid material underlies solid material. This can be caused only by a change of chemical composition. The melting point of the material composing the liquid core must be lower than that of the solid material just outside the core.

The P waves continue through the Earth's core, but at the core boundary their velocity drops abruptly and their paths are refracted sharply toward the center of the core. On emerging at the far side of the core, the P waves are refracted again, in a direction that increases the width of the shadow zone (Fig. 18-13, *B*).

The speeding up of P waves and the directions they are refracted at a depth of 5,000km are evidence that at this depth liquid material is underlain by solid material. Presumably increased pressure is responsible. At the depth of the inner core pressure may be great enough to prevent melting even at temperatures higher than those on the outer part

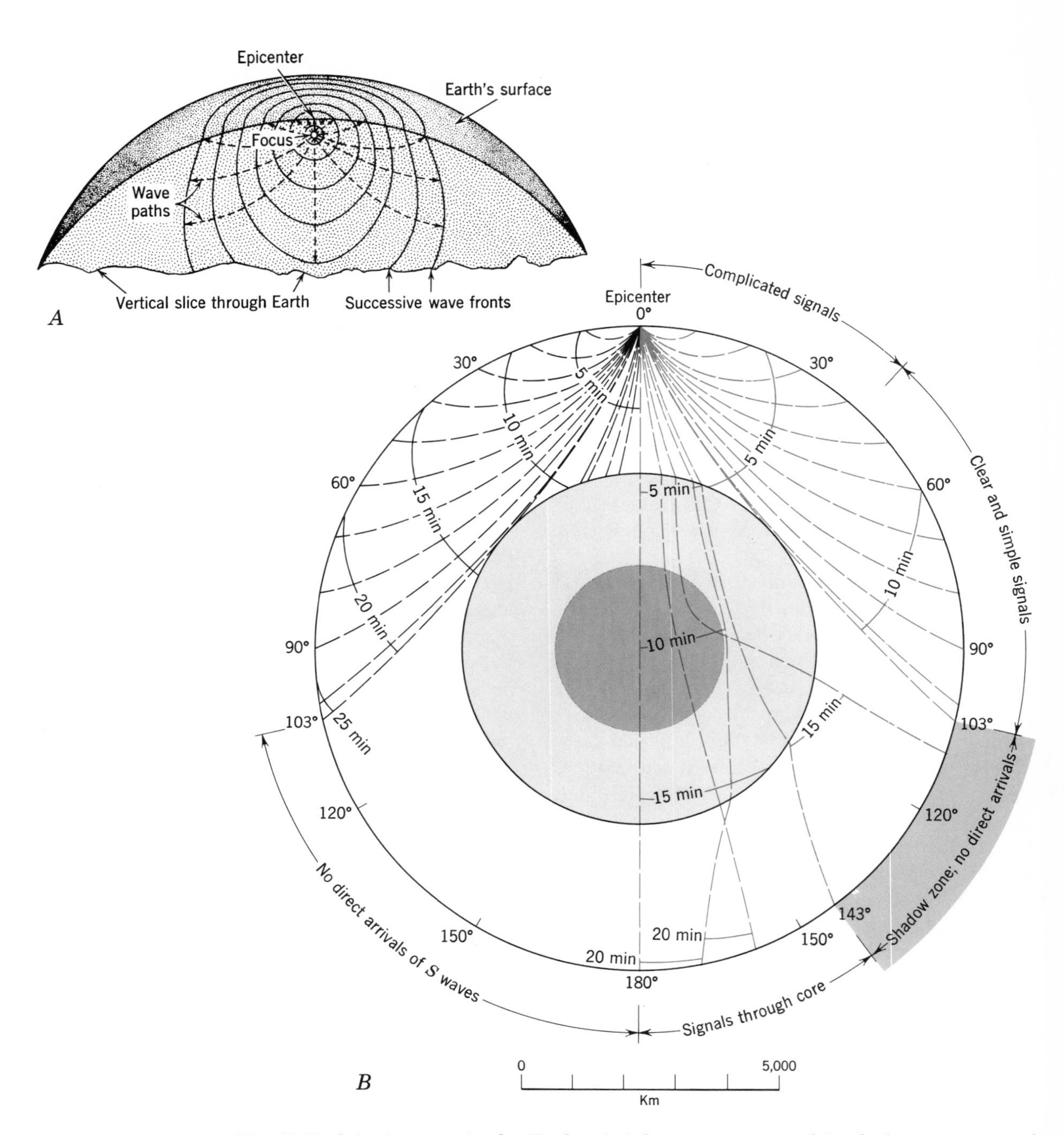

Fig. 18-13. Seismic waves in the Earth. *A.* Schematic segment of Earth shows positions of wave fronts of body waves from an earthquake (continuous curved lines) and travel paths (dashed curved arrows). From focus, waves radiate outward in all directions. Because velocity increases with depth, travel paths, which form right angles to wave fronts, are curves, and wave fronts are ellipsoids. *B.* Plane section through Earth showing paths of seismic body waves (dashed lines; P waves blue, S waves black) from earthquake focus with epicenter at 0° (top). Positions of wave fronts are shown at 5-minute intervals, P waves on right half only; S waves on left half only. Surface waves and paths of body waves reflected from Earth's surface omitted. Further explanation in text. (After Beno Gutenberg, Internal constitution of the Earth © 1950. By permission of Dover Publications, Inc., New York, N. Y.)

of the core. Accordingly we infer that the core consists of two parts: an inner solid sphere, 2,720km in diameter, and an outer fluid part about 2,100km thick.

The outer fluid part of the core may be the source of the Earth's magnetic field. We know from the geothermal gradient and the Curie effect that materials not in motion lying below shallow depths in the Earth must lack magnetic fields. How, then, can a nonmagnetic fluid core generate a magnetic field? According to one hypothesis now much in favor, the fluid outer core is continually stirred by convection, like boiling water in a kettle but, of course, moving much slower than water. If the core consists of metal, then it is an electrical conductor; the motion of an electrical conductor generates electric currents, as in a dynamo. Electric currents, in turn, give rise to a magnetic field. According to this hypothesis the slow shifting of the magnetic poles, which we shall examine in Chapter 22, is caused by slow turbulence in the outer core.

The Earth's crust. The reality of the Earth's crust was demonstrated early in the twentieth century by a seismologist named Mohorovičić (Mō-hō-rō-vĭtch′-ĭck), living in what is now Yugoslavia. He noticed that two distinct sets of deflections from P and S waves appeared on seismograms recorded within 800km of the epicenter of an earthquake having a focus located within 40km of the surface. From these observations he drew two conclusions. First, one pair of deflections resulted from waves that had traveled from the focus to the station by a direct path. Second, the other pair represented waves that had arrived slightly later because they had been refracted. Evidently these had penetrated a deeper zone of higher velocity, had traveled within that zone, and had then been refracted upward to the surface (Fig. 18-14). From his conclusions he inferred that a distinct seismic discontinuity separates the crust from an underlying deeper zone of higher velocity. Later this discontinuity was named in his honor, the ***Mohorovičić discontinuity;*** it marks *the base of the crust.* We shall refer to this feature as the ***M-discontinuity*** for short.[1]

By seismic methods we can determine the thickness of the crust and estimate its probable composition. Under ocean basins the crust is thin, averaging 5km thick in most localities. The elastic properties of oceanic crustal rock are characteristic of those of mafic rocks.

[1]The term "Moho" is also applied in the vernacular as an abbreviation and gave rise to the name of the drilling project MOHOLE.

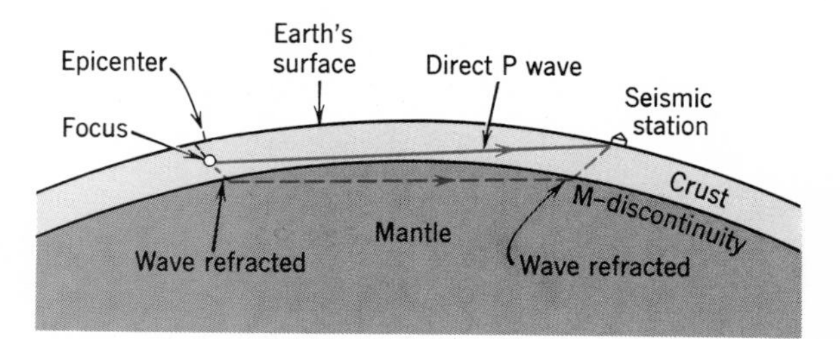

Fig. 18-14. Travel paths of direct and refracted P waves from shallow-focus earthquake to nearby seismic station.

Under continents the thickness of the crust is not uniform but varies from 20 to 60km (Fig. 18-15). The elastic properties of the outer part of continental-crustal material are those characteristic of intermediate and granitic rocks. Typical continental rocks contain minerals with high content of silicon and aluminum; they are sometimes called ***sial*** (Si = silicon; Al = aluminum), *the collective term for silica-rich rocks of continental masses.*

As we shall see in later chapters, many geologic phenomena depend on this systematic distribution of sial and mafic rocks in the Earth's crust. Here we continue with our analysis of the larger features of the Earth's interior.

The Earth's mantle. *The zone between the crust and the central core,* about 2,900km thick, is the ***Earth's mantle.*** The mantle, which constitutes more than 80 per cent of the volume of the Earth, must consist of solid material because it transmits both P and S waves.

Low-velocity zone in upper mantle. Recent studies have confirmed an idea, proposed earlier but not generally accepted at first, that within the mantle at depths ranging between 60 and 300km a zone of lower wave velocities exists. Presumably this zone within which seismic waves travel slower represents a layer having greater plasticity than the material above and below. The top of the low-velocity zone, which lies deeper beneath oceans than beneath continents, is considered to mark the base of the lithosphere. This low-velocity zone in the mantle is of fundamental importance in explaining many geologic processes, as we shall see presently. The relationships among velocities of elastic waves and major zones within the Earth, as they have been most recently judged by seismologists, are shown in Fig. 18-16

OTHER USES OF SEISMIC WAVES

So far we have concentrated on the use of seismic waves to probe the Earth's interior and have not mentioned other information derivable from them. Seismic waves permit us to locate epicenters of earthquakes, determine depth of earthquake focus, measure energy released by an earthquake, and monitor underground nuclear explosions.

Locating earthquake epicenters. In evaluating our hypothesis about the Earth's interior by seismology, we dealt with an earthquake whose time of origin and epicenter were known independently of seismograms. For most earthquakes this is not the case. In fact, seismologists repeatedly employ seismograms to locate earthquake epicenters. This aspect of seismology has been compared with the feat of pulling oneself up by one's bootstraps. Seismograms are the basis for locating an epicenter; thereafter the travel times of waves can be calculated, despite the fact that these times are needed to locate the epicenter. Seismologists have achieved reasonable results because occasionally the time when an earthquake originates and the location of its epicenter can be fixed by direct observation. Even more accurate travel times have been calculated from seismograms made of waves from those nuclear explosions whose times and places of origin are known exactly.

We have seen how the spacing of the deflections resulting from the first arrivals of P, S, and L waves vary with distance from an epicenter out to points represented by 103°, where S waves disappear. The time elapsed between the first arrivals of P, S, and L waves is a function of distance from epicenter to station (Fig. 18-4, *B*). Perhaps without realizing it, many of us may already be familiar with the principle of determining distance by measuring the time gap in the arrival of waves emanating simultaneously from a single source but traveling outward at different speeds. We take advantage of this method, so basic to seismology, when we estimate our distance from a point where lightning strikes, by counting the seconds between the time we see the flash and hear the thunder. We see the flash almost at the instant the bolt strikes because in air light waves travel about 299,800km/sec. But the sound of the thunderclap reaches us approximately 5 seconds later for each mile separating us from the point struck because the air sound travels only at 330m/sec. Thus from a seismogram we can establish the radius of a circle having the station at its center and the epicenter lying somewhere on its circumference. We can locate the epicenter by triangulation; we determine radii from three widely spaced stations and draw circles around each. The epicenter is the point where the three circles intersect (Fig. 18-17).

Many earthquakes occur in uninhabited country or under the sea, so that their epicenters cannot be located without seismic instruments and techniques. In the 1950 Assam shock, which centered in remote mountains lacking rapid communication, announcement of the location of the epicenter came from seismologists before any direct news from the epicentral area itself.

No part of the Earth's surface is exempt from earthquakes, but their intensity is not distributed randomly. In most areas, only occasional shocks of small or moderate intensity have been recorded. By contrast, several large *tracts* are *subject to frequent shocks,* both strong and weak. These are ***seismic belts*** (Fig. 18-18). The most prominent such belt, in which about 80 per cent of all earthquakes originate, is aptly called the *Circum-Pacific belt.* It follows the western highlands of the Americas from Cape Horn to Alaska, crosses to Asia, extends southward down the coast and island arcs related to it, and loops far southeastward to New Zealand and south beyond it. Next in prominence, giving rise to about 15 per cent of all earthquakes, is the *Mediterranean and Trans-Asiatic belt,* a broad east-west zone extending from Gibraltar across the Mediterranean region and through the high mountains of southern Asia. The *Mid-Atlantic belt* follows the Mid-Atlantic Ridge from Arctic to Antarctic waters, and the *Mid-Indian belt* the Mid-Indian Ridge to unite with a belt in eastern Africa. Smaller seismic areas include island groups in the Pacific and Atlantic.

The seismic belts coincide closely with rows of volcanoes and zones of crustal disturbance: active strike-slip faults; deep-sea trenches along the continental borders, as in Chile, the Alaska Peninsula, and Japan; young mountain chains; and mid-oceanic ridges. We explore these areas in subsequent chapters.

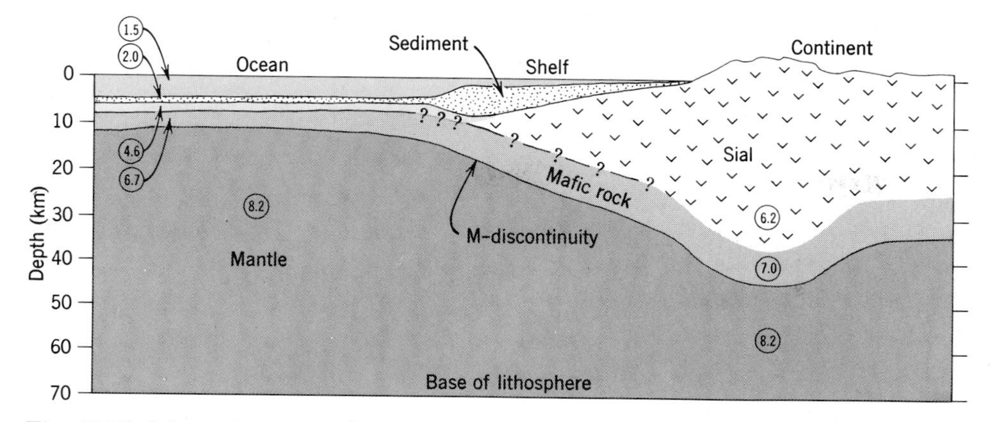

Fig. 18-15. Schematic section through lithosphere compares striking variations in thickness and composition of continental and oceanic crust. Circled numbers are velocities of P waves (km/sec). Further explanation in text.

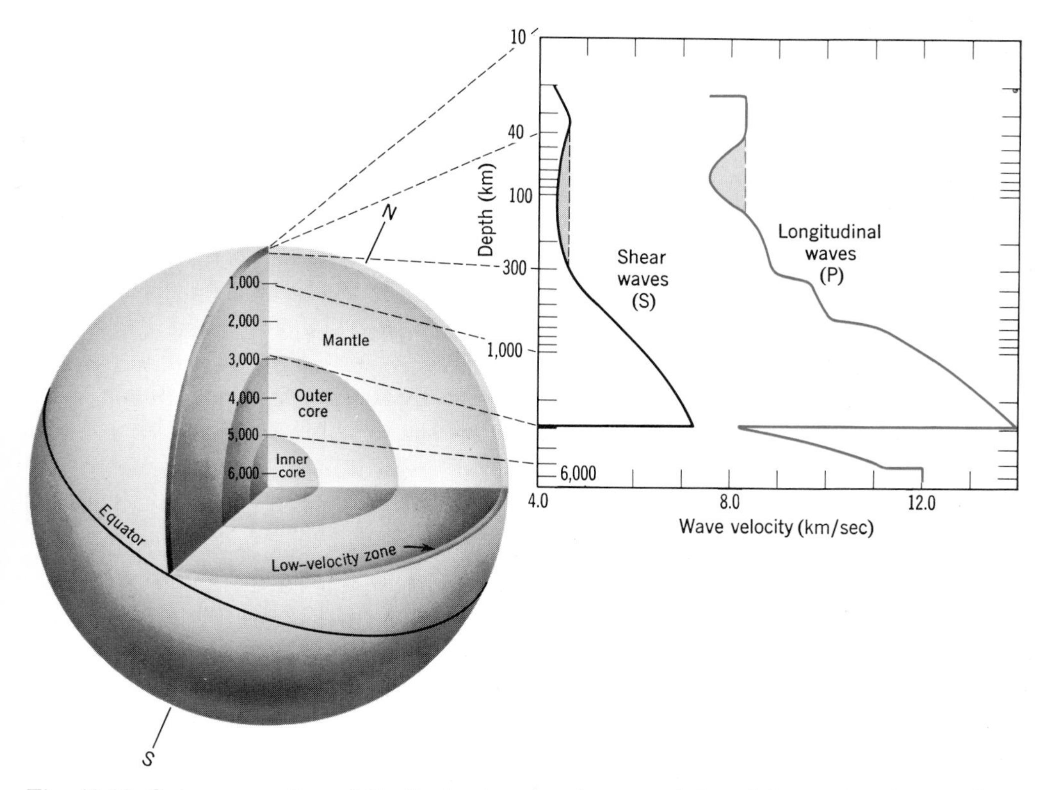

Fig. 18-16. Cut-away section of Earth showing interior parts inferred from seismology and graphs of velocity of seismic waves versus depth. Logarithmic scale of depth expands outer 1,000km to bring out details of reversal of velocity in upper mantle. (Data for velocity curves from Anderson, 1962; Hodgson, 1964.)

Determining depth of focus. If a focus lies only a few kilometers beneath the surface, the P and S waves rise to the epicenter sooner than they reach points 80 to 160km distant. But if the focus lies 150 to 300km deep, the waves arrive almost simultaneously at all points inside a large circle around the epicenter (Fig. 18-13, *A*). In effect, then, the size of the area at the surface within which the shock is felt simultaneously increases rapidly as depth of focus increases. This fact, and accurate determinations of the time intervals between arrivals of P, S, and L waves, enable us to calculate depth of focus for every earthquake.

No earthquake having a focus deeper than about 700km has ever been recorded. All *earthquakes with foci deeper than 300km* are ***deep-focus earthquakes;*** they constitute only about 3 per cent of all recorded quakes and all but one have been noted in the Circum-Pacific belt. ***Intermediate-focus earthquakes*** are *those with depths of focus between 70 and 300km.* They occur in the Circum-Pacific belt but their epicenters are closer to the continental margins than are those of deep-focus earthquakes. A few intermediate-focus earthquakes have been recorded in southwestern Asia and in Mediterranean lands. *Earthquakes with foci at depths of less than 70km* are ***shallow-focus earthquakes.*** Most earthquakes belong in the shallow-focus category. For example, earthquakes generated by abrupt movement along the San Andreas fault originate at depths of 12 to 15km. In the Circum-Pacific belt the epicenters of shallow-focus earthquakes lie nearer to the Pacific basin than do those of intermediate-focus earthquakes (Fig. 22-6).

This classification of earthquakes by depth of focus was made before the low-velocity zone in the upper mantle was acknowledged by most seismologists. Now we think that many shallow-focus earthquakes originate above the low-velocity zone, intermediate-focus earthquakes within it, and deep-focus earthquakes below it.

Because the P wave patterns of deep-focus earthquakes registered on short-period seismographs are similar to those registered by shallow-focus earthquakes, the mechanism of elastic rebound, proposed after the San Francisco earthquake of 1906, was extended to deep-focus earthquakes. However, reconsideration of the strength of rocks and pressures at great depth indicates that beneath the lithosphere faulting may not be possible; if so, the elastic-rebound concept does not apply. At a depth of 600km the pressure from overlying material is equal to about 2×10^{11}kg/cm^2. If two blocks of solid rock occurred along a fault surface they would be forced together by such great pressure that it would be practically impossible to move them. In fact, shearing stresses equal to about 10^{11}kg/cm^2, about two orders of magnitude greater than the shearing strength of known rocks, would be required to shift the blocks. Presumably, then, the rocks themselves would shatter before they could accumulate sufficient stress to cause a fault.

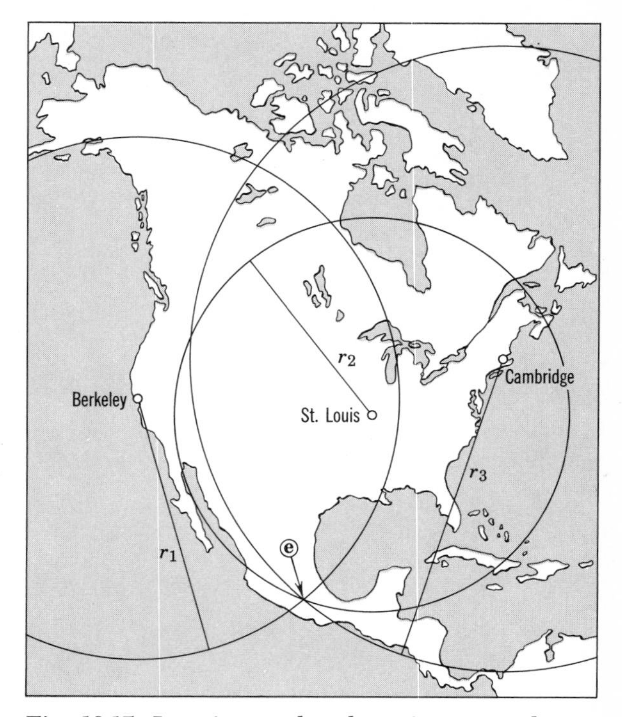

Fig. 18-17. Locating earthquake epicenter, e, by triangulation from three seismic stations. Time elapsed between first records of P and S waves (Fig. 18-4, *B*) determines lengths of radii (r_1, r_2, r_3).

Seismograms from two deep-focus earthquakes in South America, with foci at 600km, made on a new instrument that monitors seismic waves having ultralong periods, show important wave shapes not found on seismograms from earthquakes triggered by abrupt movement on a fault. Instead the seismograms from these deep-focus earthquakes are consistent with an interpretation that a single vertical force was applied suddenly at great depth. It is as if a small volume of rock at the focus had suddenly collapsed and the overlying rock had immediately moved

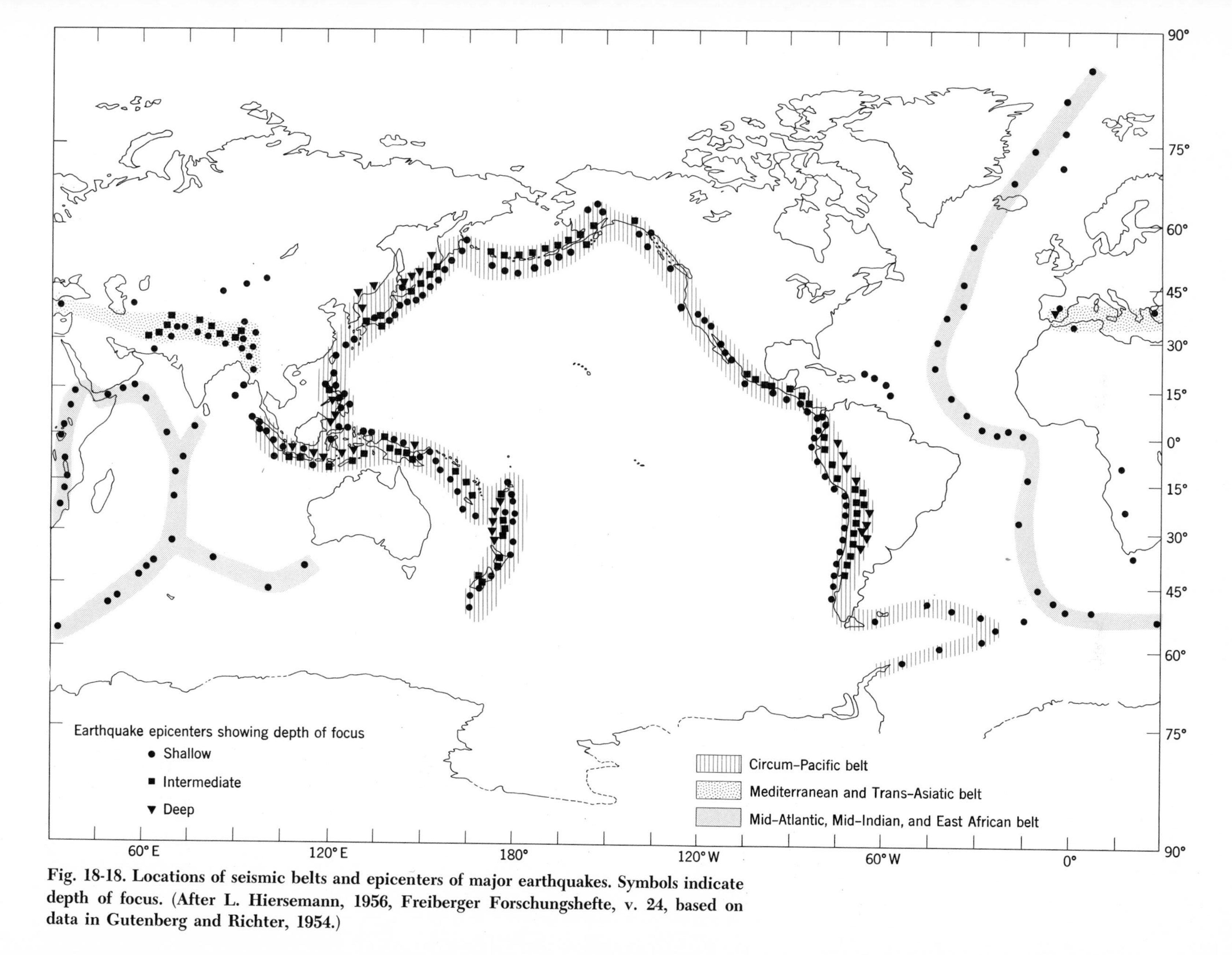

Fig. 18-18. Locations of seismic belts and epicenters of major earthquakes. Symbols indicate depth of focus. (After L. Hiersemann, 1956, Freiberger Forschungshefte, v. 24, based on data in Gutenberg and Richter, 1954.)

downward. Such a change in volume could be caused by an abrupt transformation from one solid to a denser phase. Calculations suggest that collapse of a sphere having a diameter of 0.3km would release the same amount of energy that was liberated by the two South American shocks.

Measuring energy released. Seismologists estimate that in a year as many as a million earthquakes occur, most of them feeble but some of them intense. The average for the world is about two earthquakes per minute. This almost constant trembling confirms other evidence that the Earth is not a stagnant mass bearing only the marks of past activities; it is full of energy which is being released to cause continual change. How much energy is released? How is the energy distributed? As a general principle we can use the relationship between energy and seismic waves; the larger the amount of energy, the larger are the seismic waves. We can get some idea of the size of seismic waves by studying the effects of earthquakes on man-made structures and by making measurements on seismograms.

Generally the effects of an earthquake are most intense near the epicenter and gradually become less intense outward. For comparing ***earthquake intensity,*** *a relative measure of the strength of an earthquake, based on observed destruction or disturbance and on human sensations,* between one area and another during any one shock and among epicentral areas of different shocks, we require some kind of scale. Several scales of intensity have been proposed, and although they use numbers, most are qualitative rather than exact. The Mercalli scale recognizes degrees ranging from I, the weakest, up to XII. Intensity I is felt only by a few people, close to the epicenter; III is felt noticeably indoors, especially on upper floors of buildings; V, felt by nearly everyone, breaks glassware and windows; VII makes everybody run outdoors and breaks some chimneys; IX damages well-designed buildings, cracks the ground, and breaks underground pipes; XI destroys nearly all masonry structures and bends railroad tracks; and XII throws objects into the air and wrecks everything.

After an intense shock the area affected is studied carefully and an intensity map is made (Fig. 18-19). *A line on a map through points of equal earthquake intensity* is called an ***isoseismal line.*** Some isoseismal lines are irregular because effects at the surface reflect not only the intensity of vibrations in bedrock

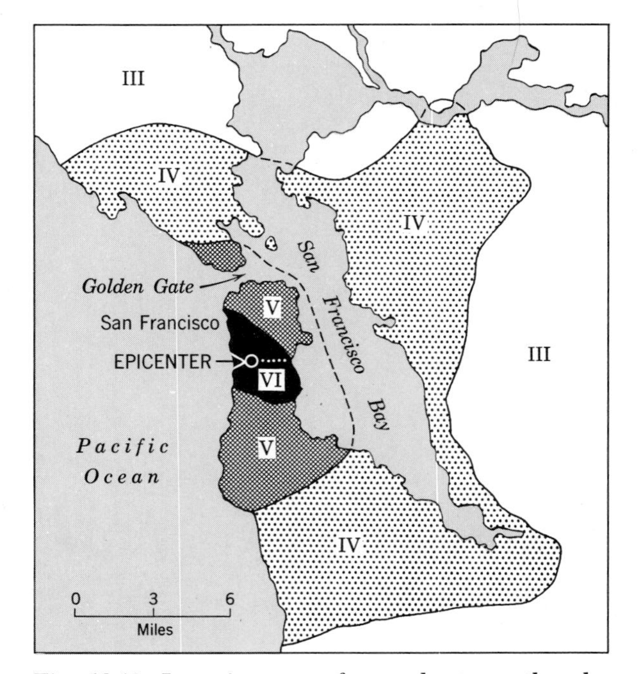

Fig. 18-19. Intensity map of a moderate earthquake with epicenter south of San Francisco, California, October 2, 1934. Isoseismal lines delineate areas of intensity based on Mercalli scale (Roman numerals). Thick alluvium underlies a wide extension of zone IV east of San Francisco Bay. (After Perry Byerly, Seismology © 1942. By permission of Prentice-Hall, Inc., Englewood Cliffs, N. J.)

but also differences in the characteristics of the regolith above the bedrock. Thick, water-soaked alluvium shakes like jelly and causes exceptional damage even at considerable distances from an epicenter. Therefore damage is not related entirely to energy released and we need some other means of measuring the energy.

Precision seismographs permit seismologists to make quantitative comparisons of the amplitude of motion of the bedrock during an earthquake. This is useful because unlike that of regolith amplitude of motion of bedrock depends only on the quantity of energy released at the focus and distance from the focus. In order to compare amplitudes of ground motion in various localities we need a network of stations having standard seismographs accurately calibrated. We can determine amplitude by measuring the heights of deflections on seismograms. To determine the decrease of amplitude outward from the source, we gather the data from many stations and plot amplitude against distance of station from epicenter. Because amplitude varies through a large range, it is more convenient to plot the logarithm of amplitude (Fig. 18-20). The amplitude-distance points for each earthquake define a single curve on the graph; the vertical distance between two curves is an indication of the difference in absolute magnitude of earthquakes plotted. We arbitrarily select for reference a "zero" earthquake, the smallest one we are likely to record. The amplitude of this earthquake gives us a "baseline" curve from which we can measure to determine the magnitudes of other earthquakes. Such curves form the basis for the *Richter scale* of earthquake magnitudes, in which the *magnitude, M,* is defined by:

$$M = \log A - \log A_0$$

where A is the amplitude of any earthquake and A_0 is the amplitude of our "zero" earthquake. Magnitudes on the Richter scale are now determined at many stations and are reported routinely for earthquakes.

In an earthquake the amount of energy released and magnitude on the Richter scale are related empirically by the following equation:

$$\log E = 11.4 + 1.5M$$

where E is the energy released in ergs and M is the magnitude on the Richter scale. The values 11.4 and 1.5 are subject to change as studies continue.

A graph of energy versus magnitude (Fig. 18-21) compares the energy released in various earthquakes. The graph indicates that the greatest known earthquake released about 700,000 times as much energy as the smallest one shown. Each integer of the Richter scale represents a variation by a factor of 100 in the amount of energy released during earthquakes.

Monitoring underground nuclear explosions. As we read on a previous page, waves generated by explosions differ from those formed by abrupt displacement along a fault (Fig. 18-7). Theoretically, then, from study of seismograms we can distinguish an underground nuclear blast from a natural earthquake. Great progress has been made in seismology by the invention and installation of new instruments that permit this distinction to be made. A few underground nuclear blasts have been monitored by seismologists as controlled experiments. Greater precision of interpretation than in natural earthquakes is possible because in such nuclear blasts, the factors of time, place, and amount of energy released

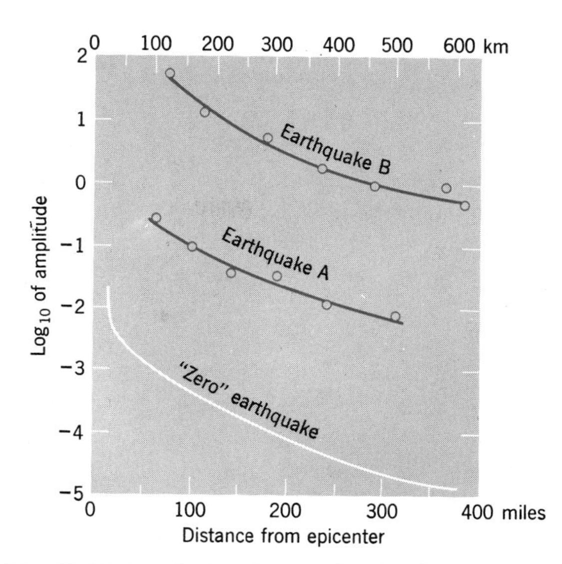

Fig. 18-20. Graph showing amplitude of seismic waves versus distance from epicenter for three earthquakes. (After J. H. Hodgson, Earthquakes and Earth structure © 1964. By permission of Prentice-Hall, Inc., Englewood Cliffs, N. J.)

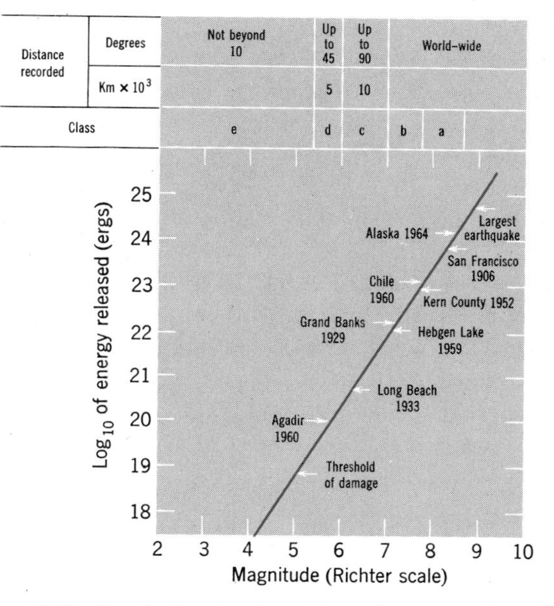

Fig. 18-21. Graph showing logarithm of energy released versus magnitude for selected earthquakes. (After J. H. Hodgson, Earthquakes and Earth structure © 1964. By permission of Prentice-Hall, Inc., Englewood Cliffs, N. J.)

are known exactly, whereas in most natural earthquakes these factors are not known.

As our study of the Earth's interior continues we shall have many occasions to draw upon the lessons we have learned in this chapter. We turn next to the processes and products connected with volcanoes, where material from inside the Earth escapes to the surface.

Summary

1. Earthquakes devastate cities, trigger landslides, and set off submarine slumps that generate destructive tsunami.
2. Abrupt movement on faults is responsible for most earthquakes originating within the Earth's crust. Why fault movements occur is not known.
3. Ninety-five per cent of all earthquakes originate in the Circum-Pacific belt and the Mediterranean and Trans-Asiatic belt. Those in the remaining 5 per cent are distributed widely.
4. Energy released at an earthquake focus radiates outward as body waves of two kinds: longitudinal waves and shear waves. Some energy transmitted to the Earth's surface by body waves creates surface waves; the remainder is reflected downward as additional body waves.
5. From study of seismograms, seismologists infer the internal architecture of the Earth, locate earthquake epicenters, determine depth of focus, measure energy released, and detect most underground nuclear explosions.
6. The thickness and composition of the Earth's crust vary systematically with major relief features. Continental crust, 20 to 60km thick, consists chiefly of granitic rocks. Oceanic crust, only about 5km thick, consists of material having physical properties resembling those of mafic rocks.
7. Beneath the crust and separated from it by a pronounced seismic interface, the M-discontinuity, is the Earth's mantle, composed of dense rocks rich in iron and magnesium. The base of the lithosphere lies within the upper mantle, at a depth of 60 to 100km.
8. The lithosphere rests on a plastic zone whose base lies about 300km beneath the surface. Deeper than this the mantle is rigid and its properties vary gradually downward.
9. Shallow-focus earthquakes originate from within the lithosphere, intermediate-focus earthquakes from within the plastic zone in the upper mantle, and deep-focus earthquakes from the mantle lying below the plastic zone. No earthquake has been recorded having a focus deeper than about 700km.
10. The base of the mantle coincides with a profound seismic interface at a depth of 2,900km below which S waves do not penetrate and P waves are slowed markedly. These changes in behavior of seismic waves signify that the mantle is underlain by fluid.
11. The Earth's core consists of very dense materials, probably metals. The outer part of the core is liquid but its inner part, lying below depths of 5,000km, is solid.
12. Slow motions of the fluid part of the core, driven by convection, may generate electric currents capable of creating the Earth's magnetic field.

Selected References

Anderson, D. L., 1962, The plastic layer of the Earth's mantle: Sci. Am., v. 207, no. 1, p. 52–59. (San Francisco, W. H. Freeman and Company, Repr. 855, 9 p.)

Andrews, Allen, 1963, Earthquake: London, Angus and Robertson, Ltd.

Benioff, Hugo, 1964, Earthquake source mechanisms: Science, v. 143, p. 1399–1406.

Bullard, Sir Edward, 1966, The detection of underground explosions; Sci. Am., v. 215, No. 1, p. 19–29.

Davidson, Charles, 1931, The Japanese Earthquake of 1923: London, Thomas Murby and Co.

Elsasser, W. M., 1958, The Earth as a dynamo: Sci. Am., v. 198, no. 5, p. 44–48. (San Francisco, W. H. Freeman and Company, Repr. 825, 6 p.)

Gutenberg, Beno, and Richter, C. F., 1954, Seismicity of the Earth and associated phenomena, 2nd ed.: Princeton Univ. Press. (Facsimile repr., 1963: New York and London, Hafner Publishing Co.)

Hansen, W. R., and others, 1966, The Alaska earthquake March 27, 1964: Field investigations and reconstruction effort: U. S. Geol. Survey Prof. Paper 541.

Hodgson, J. H., 1964, Earthquakes and Earth structure: Englewood Cliffs, N.J., Prentice-Hall, Inc.

Iacopi, Robert, 1964, Earthquake country: Menlo Park, Calif., Lane Book Co.

Kendrick, T. D., 1956, The Lisbon earthquake: London, Metheun and Co., Ltd.

Lawson, A. C., and others, 1908, Report on the California earthquake of April 18, 1906: Washington, D. C., Carnegie Institution Publication 87, v. 1 and Atlas.

Oliver, Jack, 1959, Long earthquake waves: Sci. Am., v. 200, no. 3., p. 131–143. (San Francisco, W. H. Freeman and Company, Repr. 827, 10 p.)

CHAPTER 19

Volcanoes

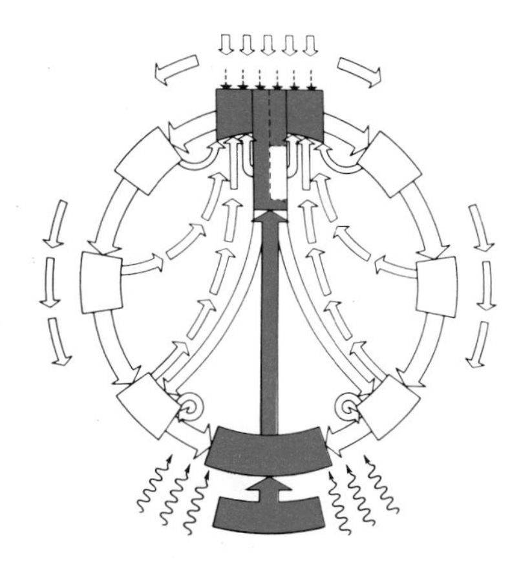

Nature's fireworks
Volcanic products
Volcanic edifices
Activities of selected volcanoes
Hot springs and geysers
Submarine volcanic activity
Can volcanoes influence climate?
Recognition of ancient volcanic activity
Stratigraphy of volcanic rocks
Volcanoes and man

Plate 19

Surtsey, a new volcanic island in the North Atlantic, immediately south of Iceland, broke surface during the night of November 14/15, 1963. This photograph, taken on November 16 when the island had become 40m high and 600m long, shows clouds of steam and tephra emerging from two vents. (Kristinn Helgason.)

CHAPTER 19 *Volcanoes*

NATURE'S FIREWORKS

Nature displays her fireworks in volcanoes (Fig. 19-1). Most new material arrives at the Earth's surface in the midst of proceedings both spectacular and dramatic. Apart from the pyrotechnics, volcanoes build conical mounds and mountains, pour forth lava in tongues and sheets, spread layers of solid particles through varying distances, propel fine particles high above the Earth, and contribute gases to the atmosphere and hydrosphere. Volcanic activity is the clearest manifestation at the Earth's surface of thermal energy derived from the interior. The effects of local hot spots are most apparent at a volcano that pours forth incandescent lava from a subterranean magma chamber. But the motions of the magma and lava cease long before all their heat has been dissipated. Hot gaseous exhalations and hot springs, which may persist for months, years, and even hundreds of years after the magma has been emplaced close to the surface and lava has ceased to flow, reflect the continued upward movement of heat. The hot rock material must adjust itself to the conditions prevailing in environments at or near the Earth's surface. Typically this means that by cooling and solidifying, magma and lava form igneous rocks. This conversion is an important step in the geologic cycle.

By studying modern volcanoes we acquire knowledge that we can apply to ancient volcanic and intrusive rocks in general. However, although they help us to understand the past and provide valuable clues about the Earth's interior, volcanoes confront us with many enigmas.

The foremost geologic riddle posed by volcanoes is how their molten material originates. Seismology discloses that the Earth's crust and mantle transmit shear waves, hence behave as solids, at least for stresses briefly applied. Why does solid material melt to form magma? Once melted, how does it rise to the surface and what determines where it will break out? Do volcanoes on land extrude the same materials as those on the sea floor or different materials? How important are volcanoes on the sea floor? Can volcanoes influence climate? Was the material released from ancient volcanoes similar or dissimilar to that released from modern volcanoes? Can we predict when a volcanic eruption will occur and foretell the train of events that will follow? Can we tap volcanoes as sources of heat energy? We could continue asking questions but our list is long enough; it is time to search for answers.

VOLCANOES AND VOLCANISM

A ***volcano*** is *a vent or a fissure through which molten and solid materials and hot gases pass upward to the Earth's surface.* With the materials they emit, many volcanoes construct conical mounds or mountains around their vents, but these edifices are incidental to the definition. Whether or not the products accumulate to form a cone, a volcano exists from the moment it dispenses the very first gram of material from the Earth's interior. Some volcanoes discharge only gases; others discharge liquid and solid materials that spread out in horizontal layers. The place where such materials come out upon the Earth's surface is as much a volcano as is the orifice whence come the materials heaped into the largest volcanic mountain.

Volcanism designates the aggregate of processes associated with the transfer of materials from the Earth's interior to its surface. Volcanism embraces materials and processes whose interplay builds volcanic edifices and creates varied volcanic activities. Both edifices and activities result from the arrival of a body of magma in the outer part of the crust, the separation of its liquid and gaseous phases (the fundamental volcanic process), and the delivery of these phases to the Earth's surface. As we shall see, the locale of separation and the manner in which it occurs greatly influence volcanic edifices, eruptive activities, and structures in extrusive igneous rocks.

VOLCANIC PRODUCTS

The products of volcanoes are gases, liquids (lava and some tephra), and solids (tephra). The liquid and solid materials tend to attract the greater attention, but in reality gases are a more typical product because they are emitted from all volcanoes, even from those that extrude no lava.

Volcanic gases. Nearly every volcano commences its active life by jetting gases that ream a hole to the surface. During the volcano's most vigorous phases gases continue to stream from the vent. They herald every eruption by clearing the throat of the volcano and accompany the emission of solids, liquids, or both. Moreover, long after other products cease to discharge, gases continue to escape at the surface from the vent, from cracks in its vicinity, and from the volcanic materials dispersed from the vent.

The jets of gases and intense heat create updrafts that produce vertical mushroom clouds and torrential rains. Suspended tephra and general air turbulence give rise to bolts of lightning.

The major constituent of volcanic gases is superheated steam, accounting for 50 to 95 per cent of their volume. Some of the steam derives from ground water heated by the magma and driven upward to the surface. Other steam originates when hydrogen gas from the magma unites with oxygen on contact with the air. Still other steam comes from *water, formerly dissolved in magma, which rises from deep within the Earth to become part of the hydrosphere for the first time;* this is ***juvenile water.*** Ordinary methods do not suffice to distinguish the sources of volcanic steam, but some progress has been made by analyzing the oxygen isotopes in various waters.

Other volcanic gases include carbon dioxide, nitrogen, sulfur dioxide, sulfur trioxide, carbon monoxide, hydrogen, sublimed sulfur, argon, hydrogen sulfide, methane, hydrochloric acid, hydrofluoric acid, and metallic chlorides. Analyses of fourteen samples from Kilauea, Hawaii (Fig. 19-2) substantiate the great variability among volcanic gases. Although they emerge from the same orifice, the contents of no two bubbles are identical.

Sulfur is deposited along the walls of vents or

Fig. 19-1. Incandescent tephra being ejected from Parícutin Volcano, Mexico, in 1943 when the cone was only a few months old. (W. F. Foshag.)

combines with oxygen from the atmosphere to form deadly sulfur dioxide. Some volcanoes emit fluorine and others yield boron, the essential element of commercial borax. Iron, copper, lead, and other metals, transported as volatile chlorides, have been deposited on the walls of fractures. At Vesuvius, in Italy, hematite filled a fissure 1m wide in a few days. After seeing such deposits nineteenth-century geologists conceived the idea, still valid today, that metallic mineral deposits originated primarily by igneous processes.

How much gas is dissolved in magma? At the height of its activity in May, 1945, Parícutin, in Mexico, released an estimated 116,000 tons of material per day, of which 16,000 tons, or 14 per cent, were water vapor and 100,000 tons were lava. The maximum value of volatiles dissolved in laboratory silicate melts is about 11 per cent. If this maximum is a true upper limit, then we must conclude that the estimate of gases released from Parícutin does not represent only the amount of gas dissolved in the magma but includes in addition an unknown quantity of ground water. Chemical analyses of glassy volcanic rocks suggest that the water content of typical volcanic glass is only 1 or 2 per cent, hence does not approach the laboratory maximum. Because glassy volcanic glasses solidify so rapidly, their gas content of 1 or 2 per cent has been

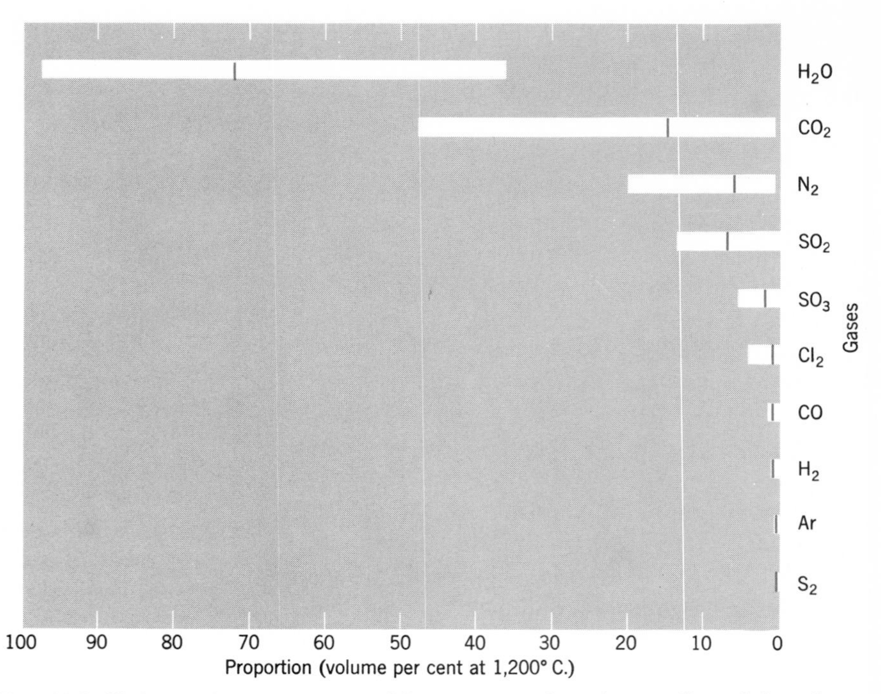

Fig. 19-2. Variation in composition of fourteen samples of gas collected from lava lake at Kilauea, Hawaii, shown by bar graph. Ends of bars mark extreme values and lines inside bars average values. (Data from T. A. Jaggar, 1940.)

treated as an average value for the amount of water dissolved in magma.

Volcanic gases are very hot; temperatures as high as 1,200°C have been reported in volcanoes discharging lava and as much as 700°C from those not discharging lava. Although temperatures can be measured near the vents and samples of volcanic gases can be collected at the same time, doubt prevails that these measurements or samples indicate the true state of gaseous affairs in the magma. Volcanic gases interact with ground water on their way to the surface and react immediately with the atmosphere upon contact with it. Oxidation of hydrogen and methane raises the temperature of the gases and also creates the only true flames connected with volcanic activity.

Lava. The composition of lava extruded from volcanoes determines not only the kind of extrusive igneous rock that forms but also many primary structures in the rock. Some volcanoes, such as those in Hawaii, deal in only one liquid product, lava that solidifies to basalt. Stromboli, in the Lipari Islands in the central Mediterranean north of Sicily, manufactures basalt, whereas Vulcano, only 40km distant in the same island group, has in the recent past generated rhyolite, which stands at the opposite end of the spectrum of igneous rocks. Lava from some volcanoes may change with time. Extrusive rocks derived from lavas recently erupted by Vesuvius differ greatly from rocks created earlier by its ancestor, Somma. The bulk of the Hawaiian islands consists of volcanic mounds composed of silica-saturated basalt lacking olivine. These mounds are capped by small quantities of undersaturated olivine basalt.

Where volcanic zones of Pleistocene and Recent age in the Circum-Pacific belt are more than 50km wide, composition of the volcanic rocks varies systematically from the oceanic to the continental sides. One or both of two contrasting suites of igneous rocks may be involved. On the oceanic side only basalts are represented. The degree of saturation of such basalts varies but all contain more than 1.75 per cent TiO_2. The suite on the continental side includes andesites and basalts containing less than 1.75 per cent TiO_2 and generally more than 15 per cent Al_2O_3. The suite characterized by basalts high in TiO_2 is not restricted to ocean basins but is wide-

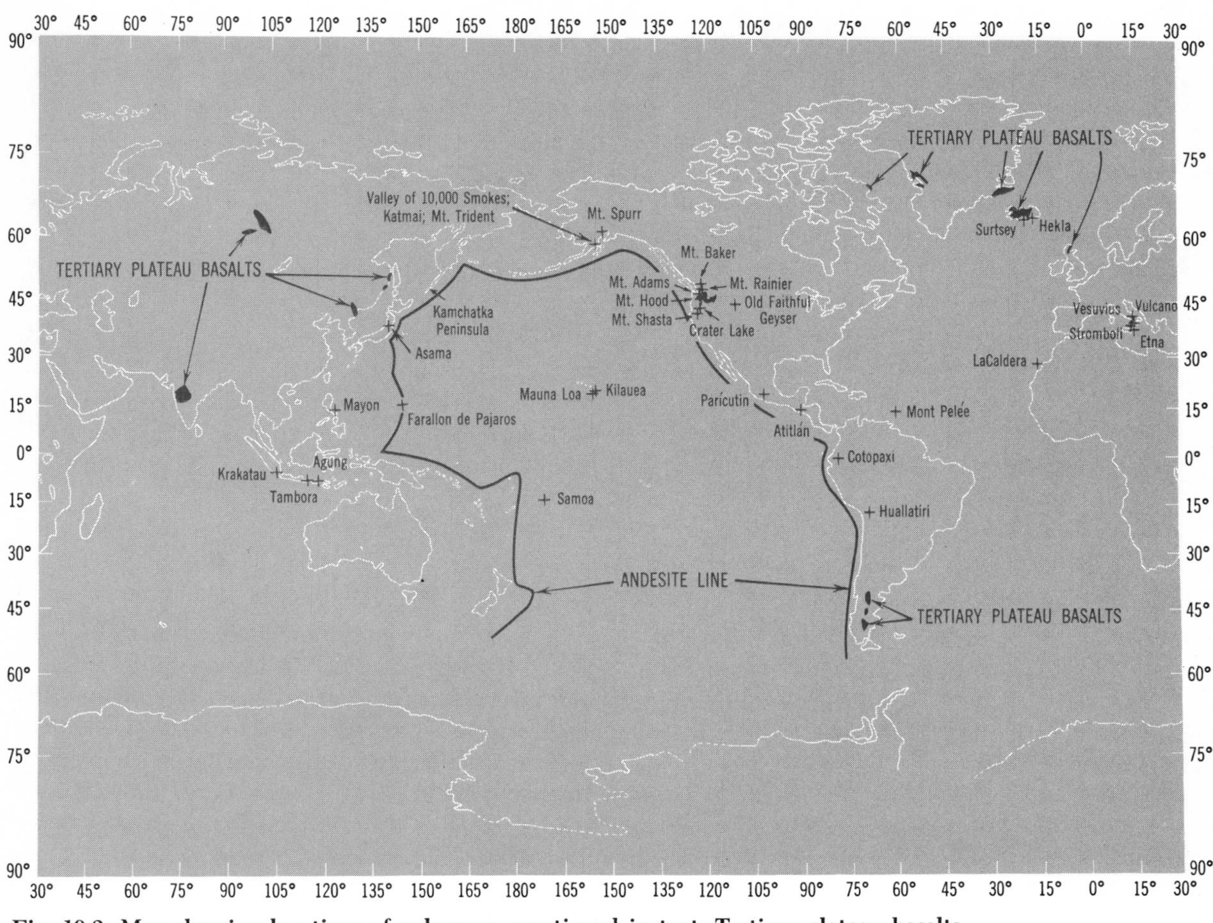

Fig. 19-3. Map showing locations of volcanoes mentioned in text, Tertiary plateau basalts, and andesite line.

spread on continents. By contrast the andesite suite with basalts low in TiO_2 does not occur on islands within the ocean basins. In the Pacific the boundary separating volcanoes creating these contrasting suites of igneous rocks is called the "andesite line" (Fig. 19-3).

The diversity of lavas from a single volcano must stem from changes within the magma chamber beneath. The larger pattern of distribution must be caused by variations within the Earth's upper mantle. Both aspects of variation relate to internal processes examined in the next chapter.

Of more direct concern to volcanic activity is the influence of chemical composition on fluidity of the lava. Fluidity partly determines how fast and how far the lava will flow and to a greater extent the characteristics of both the surface of the flow and the primary structures in the resulting igneous rocks. Fluidity, which is the inverse of viscosity, likewise depends on the temperature and gas content of the lava.

Lavas of great fluidity, such as those that create basalt, flow like rivulets of water without distinctive surface features (Fig. 19-4). Fluid lava traveled with an average speed of 10 miles per hour down the steep slopes of Mauna Loa, Island of Hawaii, in 1850. The speeds of other fluid flows, however, rarely have exceeded 5 miles per hour. As fluidity decreases, either because an originally fluid lava has cooled or because the initial composition made it thus, the surface of the flow assumes curved, ropy, and billowy forms.

Some lavas crust over rapidly while the interior of the flow remains fluid. Such flows are sluggish; their behavior has been compared to that of glaciers. A surge of fluid lava beneath breaks

Fig. 19-4. Fluid lava flows as streams. Mauna Loa Volcano, Island of Hawaii. (Sawders from Cushing.)

the crust into innumerable slabs and blocks that are jostled about as they float along the surface. They are borne to the front of the flow and there tumble to the ground, paving a rough road over which the lava must travel (Fig. 19-5). This pavement is somewhat analogous to ground moraine dumped from the upper surface down the front of a slowly advancing glacier. Where the crust of igneous rock has thickened it may be able to stand intact in the absence of the upward buoyant force exerted by the lava. If the fluid lava beneath such a thick rocky crust drains away, a *lava tunnel* is created through which other lava or ground water may flow later. Some lavas, though red hot, are so viscous that the point of a pick will not indent them.

Tephra. The material ejected explosively from volcanoes consists of clots of lava and particles of solids. Lava clots sprayed into the air create *lava fountains* (Fig. 19-6). Most clots cool rapidly during flight and are solid by the time they reach the ground; in some, solidification is so rapid that glass forms (Fig. 4-9). Sizable clots of lava blown into the air whirl rapidly like pinwheels. This produces ***volcanic bombs,*** *rounded, spindle-shaped masses of volcanic rock* (Fig. 19-7).

The solid rock material ejected from volcanoes consists of particles ranging in size from dust to chunks weighing several tons. Particles are classified and named on the basis of their diameters (Table C-3). Included are crystals, particles of extrusive igneous rocks solidified from earlier lava flows, and particles of rocks, volcanic and nonvolcanic, through which the magma passes on its way to the surface. Rocks underlying volcanic cones are *foundation rocks*. Large angular pieces of foundation rocks are identified easily, but fine particles become hopelessly mixed with the new material. *Large fragments of volcanic rock that were solid when blasted from the vent* are ***volcanic blocks.***

VOLCANIC EDIFICES

The gases, lava, and tephra discharged from volcanoes construct diverse edifices, including cones and conical mountains, and explosion pits. Edifices of much larger scale are enormous plains and plateaus built of horizontal layers of extruded products.

Explosion pits. Most volcanoes begin their careers by drilling a vent through to the surface. A jet of gas armed with particles broken loose from the walls of the vent does the drilling; the particles generally accumulate as a small cone around the orifice. In some cases nothing further happens; no lava appears and all the rocks thrown out may be nonvolcanic. The resulting edifice is an *explosion pit,* which superficially resembles an impact crater created by a large meteorite. An explosion pit lacks the marginal zone of rocks metamorphosed by impact, a feature characterizing a meteor crater.

Volcanic cones. Tephra and lava accumulate in various proportions around a volcanic orifice to form volcanic cones. Clots of lava, which are still liquid after being sprayed upward in lava fountains

Fig. 19-5. Sluggish lava crusts over and advances with a steep front. Parícutin Volcano, Mexico. (F. M. Bullard.)

and after falling back to solid ground, build small *spatter cones.* Other conical edifices consist entirely of tephra that were solid when they fell. Such *small-to-moderate-size cones composed of tephra* are ***tephra cones*** ("cinder cones"; Frontispiece). The steep cone of Parícutin (Fig. 19-8) consists wholly of tephra of assorted sizes lying at their angles of repose. Thousands of such cones are known; in them the initial dip of tephra layers is steep (Fig. 16-4).

Composite cones are *cones consisting partly of tephra and partly of igneous rock both extrusive and intrusive.* Slopes of a typical composite cone are steepest near the summit and decrease outward to the base. The sizes of the tephra largely determine the angles of the slopes. Most of the coarse tephra, which fall near the vent, can form stable slopes as steep as 40°. Some of the smaller tephra are propelled somewhat farther out, whereas still smaller tephra settle within a large radius but in amounts decreasing outward (Fig. 13-16). Lava emerging from the central orifice flows down the steep upper part of the cone.

Viscous lava congeals in tongues on the steep upper slopes (Fig. 2-15) but highly fluid lava continues to move until a large part of it spreads out on the lower slopes. Much of the lava does not spill out from the apex of the cone but breaks out lower on the sides (Fig. 19-9, *A*). When solidified this material strengthens the cone but adds nothing to its steep upper part. Tephra cones are subject to more rapid erosion than composite cones that have been

Fig. 19-6. Great lava fountain, 15m high, at Kilauea Iki, Island of Hawaii, November 1959. (Ward's Natural Science Establishment.)

strengthened internally by sheets of igneous rock.

Steep-sided cones built by active and recently inactive volcanoes are included among the world's loftiest and most majestic peaks. Splendid examples stretch along the South American Cordillera, the Andes, towering 3 to 4km above a platform eroded across much older rocks. The peak of Hualatiri, in Chile, which erupted in 1959, now stands at 6,064m. It has replaced Cotopaxi, in Ecuador, (5,900m) as the highest active volcano in the world.

The North American Cordillera, crossing the western United States, includes in the Cascade Range conical peaks comparable to those in the Andes. Volcanoes not presently active have constructed Mounts Baker, Rainier, and Adams in Washington, Hood in Oregon, Shasta in California, and many more.

Craters. At or near the top of nearly every well-developed volcanic cone is a ***volcanic crater,*** *a funnel-shaped depression from which gases, tephra, and some lava are ejected* (Fig. 19-8). Not all lava spills from the crater; as we mentioned, much lava emerges from fissures located in the sides of the cone or beyond its base. Craters are explosion pits extended upward by the growing cone. The diameter of a crater is that of the vent itself or is slightly larger as a result of slumping of the material forming its sides.

Calderas. If greatly enlarged, the crater becomes a ***caldera*** (Span. "caldron"), *a volcanic crater enlarged to a diameter of several miles.* The name comes from La Caldera, a great circular pit, 5 to 6.5km wide and ringed around by cliffs 500 to 800m high, that is located south of Las Palmas, Gran Canaria, Canary Islands (Fig. 19-3). The diameter of a caldera is conspicuously greater than that of an active vent.

Lying atop a blunt, conical mountain within an enormous circular basin having a diameter of 5 miles and rimmed around by cliffs 600 to 2,000 feet high, Crater Lake, Oregon would have been given the name Caldera Lake by modern geologists, for its diameter far exceeds that of any known vent. The summit of Wizard Island, a small cone in the lake, contains a typical crater.

Calderas are not restricted to volcanic cones; the name is applied to any circular volcanic depression several miles in diameter. Calderas originate through explosion, through collapse by withdrawal of magma from below, or through a combination of both explosion and collapse. Studies of Crater Lake illustrate the way in which geologists analyze the problem of the origin of a caldera.

Fig. 19-7. Pile of eight volcanic bombs of varying sizes collected within a radius of 100 feet on the slope of a cone in Hawaii. Third bomb from base best displays twisted, spindle shape. (C. K. Wentworth, U. S. Geol. Survey.)

The steep cliffs expose layers of tephra and extrusive igneous rock dipping steeply and radially outward. Small U-shaped valleys ending abruptly at the cliffs (Fig. 19-10) prove that valley glaciers flowed from heights vertically above the present lake. Layers of tephra blanket the surrounding countryside. From this evidence we infer that a great volcanic mountain, of about the same size as Mount Adams in Washington, formerly towered above what is now Crater Lake, reaching to an altitude of 12,000 feet. The former mountain has been named Mount Mazama (Fig. 19-11, *A* and *B*). The upper 2,000m of this former peak, representing material aggregating 70km^3 in volume, has vanished. Although layers of tephra are conspicuous in the surrounding area and include some blocks up to 4m in diameter that have moved 32km, the calculated volume of material composing them is only 26.5km^3. Of this, only 6.1km^3 consists of old rocks identified as possibly having come from the vanished peak. We must still account for 63.9km^3 of material com-

Fig. 19-8. Vertical view of tephra cone and volcanic crater. Former lava flows (dark areas with lobate margins) issued not from crater but from base of cone. Base of cone is approximately 1km in diameter. Parícutin Volcano, Mexico. (Servicio Aerotecnico, Mexico, courtesy Wild Heerbrugg Instruments, Inc.)

Fig. 19-9. Conical volcanic edifices. *A.* Farallon de Pajaros (330m) in the Marianas Islands. Its summit consists of tephra, but its lower slopes are extrusive igneous rock. It is therefore a composite cone. Large former lava flow from flank detached a chunk of layered tuff (*left*) from near the summit and the chunk slid or was carried down to the base of the cone. (U. S. Navy, courtesy D. B. Doan.) *B.* A segment of Mauna Loa volcanic shield, 30 miles wide, viewed obliquely from the air. (U. S. Air Force.)

posing the former peak. We must suppose that by explosions the unaccounted for material was propelled high into the air and scattered around the world without leaving a noticeable trace, that collapse occurred, or that both explosions and collapse took place.

The caldera of Mount Mazama is thought to have been the combined work of explosions and collapse. The tephra substantiate that phenomenal explosions and ejections of solids occurred. Yet even though Mazama "blew its top," the mighty explosions alone did not create the caldera. Withdrawal of magma from the underground chamber is thought to have occurred, causing the central part of the mountain to collapse along curved faults. At a later time gases streamed upward through the floor of the caldera to create the small tephra cone that is now Wizard Island (Fig. 19-11, *C*). From charred logs found in the layers of tephra we know that Mount Mazama was decapitated 6,600 C^{14} years ago.

Volcanic shields. Instead of steep-sided cones, some volcanoes build broad gently convexed mounds, in a form suggesting an ancient warrior's circular shield. Examples are found in Hawaii (Fig. 19-9, *B*), Iceland, and Samoa. These mounds are sometimes called "shield volcanoes," but as we have defined volcano, this turnabout is not appropriate. The term *shield* by itself can be confusing; it is likewise used for large stable areas of the continents—for example, *continental shields*. We shall avoid this difficulty

Fig. 19-10. Two U-shaped valleys truncated by inner slope of the caldera of Crater Lake, Oregon. (J. O. Sumner, Monkmeyer Press Photo Service.)

by including the appropriate qualifying adjective *volcanic* or *continental* when we refer to a *shield*. A ***volcanic shield*** is, then, *a broad convex mound of extrusive igneous rock, having surface slopes of only a few degrees.*

All the foregoing features originate when volcanic materials issue from nearly cylindrical vents. Some volcanic materials reach the surface along fractures, which become fissures when the walls are spread apart. *Extrusion of volcanic materials along an extensive fracture* is a ***fissure eruption.***

Volcanic plains and plateaus. We have seen how volcanic products spewed from a pipelike vent accumulate nearly equally in all directions to form a conical edifice. Volcanic products that emerge through fissures tend to spread more widely and to build up plains, plateaus, or broad volcanic shields. Lava and tephra that spread beyond the cone surrounding the orifice from which they were ejected likewise can contribute to plains.

The fissure eruption of Laki, Iceland, in 1783, took place along a fracture about 32km long. Lava flowed as far as 64km outward from one side of the fracture and nearly 48km outward from the other side; altogether it covered an area of $558km^2$. The volume of lava extruded has been estimated at $12km^3$, making this the largest single eruption recorded by man.

Other continental tracts of vast extent have been flooded with lavas that solidified to basalt. One such tract covers 200,000 square miles in Idaho, Nevada, California, Oregon, and Washington (Fig. 19-12, *A*). There, in the deep canyons

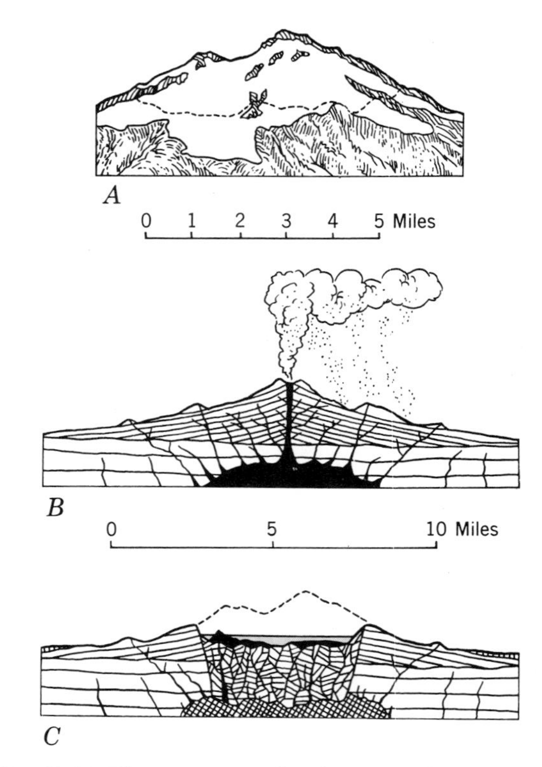

Fig. 19-11. Three stages in development of Crater Lake. A. Sketch of Mount Mazama, prior to its paroxysmal eruption, with summit restored to show glaciers. Dashed line marks present profile. *B.* Idealized section through Mount Mazama during its paroxysmal eruption, showing gases and tephra. *C.* Idealized section through Crater Lake caldera. Volcanic rocks younger than date of collapse black, igneous rocks in former magma chamber crosshatched. (*A* after W. W. Atwood, 1935; *B* and *C*, Howel Williams, 1942.)

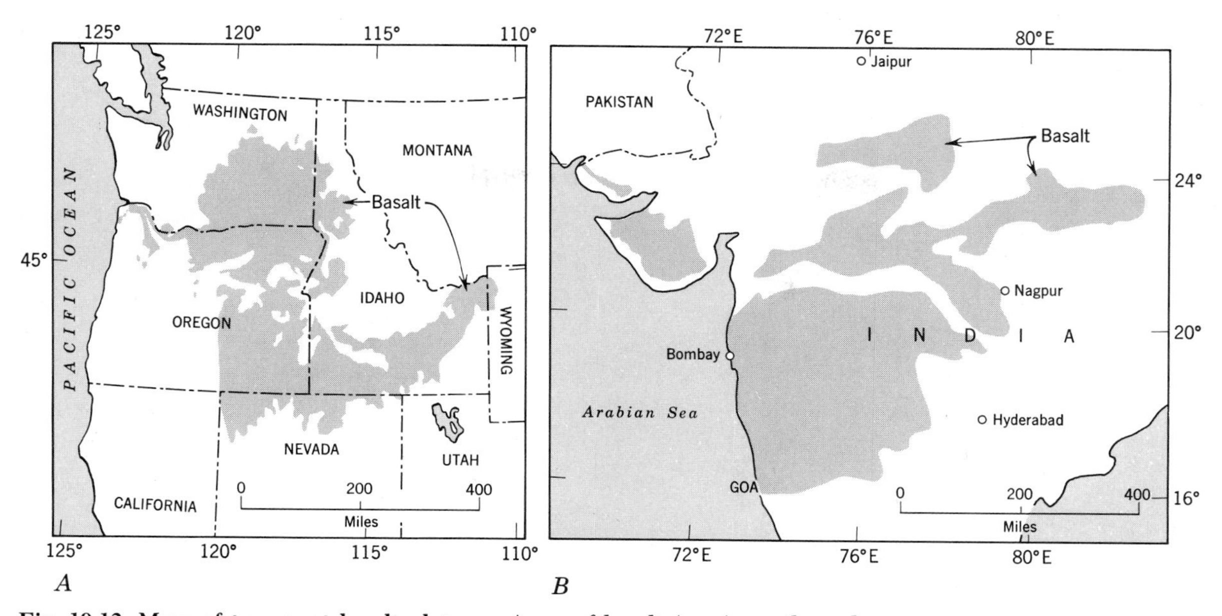

Fig. 19-12. Maps of two great basalt plateaus. Areas of basalt (gray) are the only remnants of the original deposits, reduced by erosion and covered by younger sediments. Rocks older and younger than extrusive basalt shown in white. *A.* Columbia Plateau, northwestern United States. (After A. C. Waters, 1955, and Geologic Map of the United States, U. S. Geol. Survey.) *B.* Deccan Plateau, India. (H. D. Sankalia, 1964.)

cut by the Snake and Columbia Rivers, former lava flows with an aggregate thickness of at least 4,000 feet are exposed. The average thickness of individual former lava flows is about 100 feet. One huge sheet of extrusive basalt ranges in thickness from 360 to 480 feet and extends through an area that measures 120 miles north-south and 50 miles east-west. The molten lava, which must have been very fluid when it originally spread out, filled a broad depression. Around its margins the lava filled valleys and buried a rugged landscape, covering many ridges and peaks. The upper surface of the extrusive sheets forms a vast plateau, which is now being dissected by streams. *Extrusive basalts forming widespread nearly horizontal layers on continental masses* are **plateau basalts.**

Another vast basalt plateau extends nearly continuously through 250,000 square miles in India (Fig. 19-12, *B*). Remnants of basalt sheets scattered through wide surrounding areas suggest that extensive erosion has occurred and that the basaltic rocks originally occupied a land area of about 500,000 square miles. An additional area of unknown but possibly equal size lies beneath the sea.

Other plateau basalts of Tertiary age are shown in Fig. 19-3. If the scattered remnants of plateau basalts in Northern Ireland, Scotland, the Faroes, Iceland, Greenland, and Spitsbergen all belong to a single basalt plateau, as evidence suggests, then they formed from what was probably the world's most extensive flood of basalt. The name Thulean Province designates this collection of basalt remnants and their inferred submarine extensions.

Not only lava but also tephra can spread widely. Some tephra remain aloft long enough to be cool when they land. Other tephra flow so rapidly along the ground that they are still hot when they stop moving. The individual particles, still soft, weld themselves together, forming a rock called *welded tuff.* Hot or cold, they can build horizontal layers (Fig. 19-13). As with the former lava flows underlying basalt plateaus, piles of tephra consist of individual layers that probably accumulated almost instantaneously. However, the layers present evidence that long times elapsed between the successive arrivals of new tephra (Fig. 19-14).

Most volcanic edifices include only one of the features described, but a few consist of more than one. In some volcanoes the lower, gentle slopes of a volcanic shield are supplanted upward by the

Fig. 19-13. Nearly flat-lying layers of welded tuff formed by successive tephra flows. Bandelier Tuff, near Los Alamos, New Mexico. (R. L. Smith, U. S. Geol. Survey.)

steep slopes of a composite cone. Such combinations record changes in the kind of materials ejected: fluid lava built the shield but more viscous lava and tephra constructed the composite cone. Examples include Mount Etna, Sicily and many of the peaks in the Cascades. Likewise, composite cones can be built atop volcanic plateaus.

Lava domes and spines. At the opposite extreme from the horizontal sheets spread by fluid lavas are the upright narrow, cylindrical features constructed by the upward squeezing of sluggish, pasty, silicic lavas, generally those solidifying to rhyolite. Many such features rose into the air surrounded only by a carapace of their own igneous rock. When the edifice grows the carapace cracks, creating blocks that accumulate around its base.

ERUPTIVE ACTIVITIES

The activities accompanying volcanic eruptions result from the sites and conditions whereby gaseous and liquid phases of magma separate. Varied activities result from the ways in which the gaseous phase behaves; gas may escape tranquilly or may cause mighty explosions. Escape of gases depends partly on the fluidity of the lava, which is a function of chemical composition, temperature, and content of volatiles. We can best illustrate eruptive activities by describing the behavior of some well-known volcanoes.

Hawaiian volcanoes. Quiet extrusion of fluid lava, sometimes called the Hawaiian phase of volcanic activity, is admirably exemplified by the active volcanoes Kilauea and Mauna Loa on the Island of Hawaii. The Hawaiian archipelago spreads along a straight line, nearly 1,600 miles long, trending northwest-southeast across the deep Pacific basin (Fig. 19-15, *A*). The Hawaiian Islands, at the southeast end, consist of eight large and dozens of small islands, built almost entirely of extrusive, silica-saturated basalt; undersaturated olivine basalt and fragmental material are present in only minor amounts. These rocks are well exposed in high wave-cut cliffs and on the flanks of lofty volcanic shields. No large Hawaiian volcano, active or inactive, has built a steep-sided cone.

Although many small islands in the northwestern part of the archipelago are coral atolls, probably their superficial reef caps cover piles of volcanic rock. Geologic study and K/Ar isotopic dates indicate that the islands of the archipelago become progressively younger toward the southeast (Fig. 19-15, *A*). Each of the older volcanic shields ceased to be active before the next younger one appeared above sea level. The exposed parts of most were constructed in less than 500,000 years. In Chapter 22 we shall examine some of the ideas concerning the significance of this linearity and systematic decrease in age.

The Island of Hawaii marks the crest of an enormous volcanic shield (Fig. 19-9, *B*) having a maximum diameter of approximately 100 miles and resting on the floor of the ocean in water 20,000 feet deep. The highest points reach to altitudes of nearly 14,000 feet. Therefore the relief of the shield is nearly 34,000 feet; it rises above the sea floor to a greater height than Mount Everest stands above sea level. The total volume of volcanic rock has been estimated at 10,000 cubic miles; the part below sea level is enormously greater than that above (Fig. 19-15, *B*). The oldest extrusive rocks

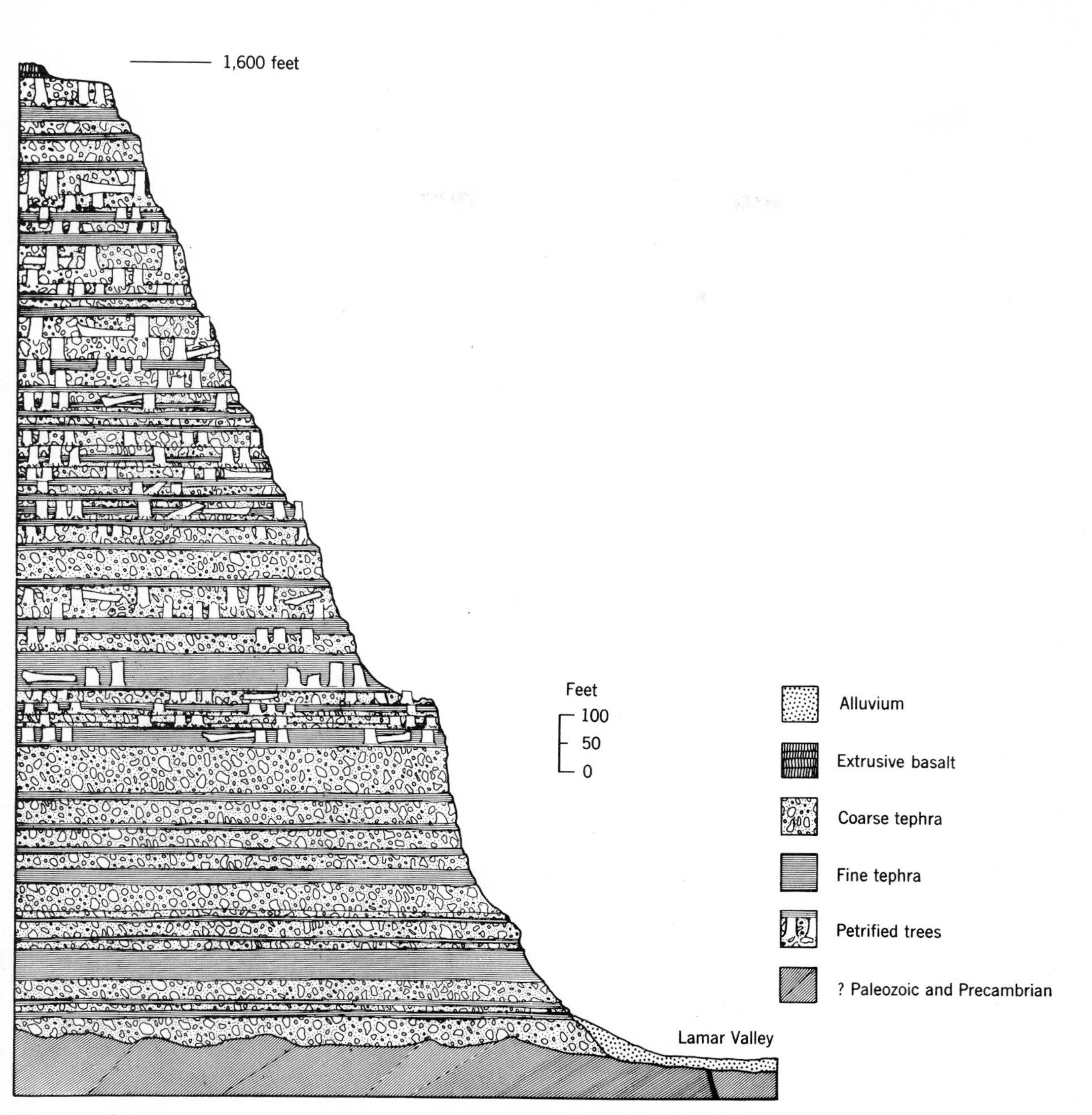

Fig. 19-14. Section in steep wall of Lamar Valley, Yellowstone National Park, Wyoming reveals 27 successive forests buried by tephra. Annual growth rings show that when buried some trees were 500 years old. (After Erling Dorf.)

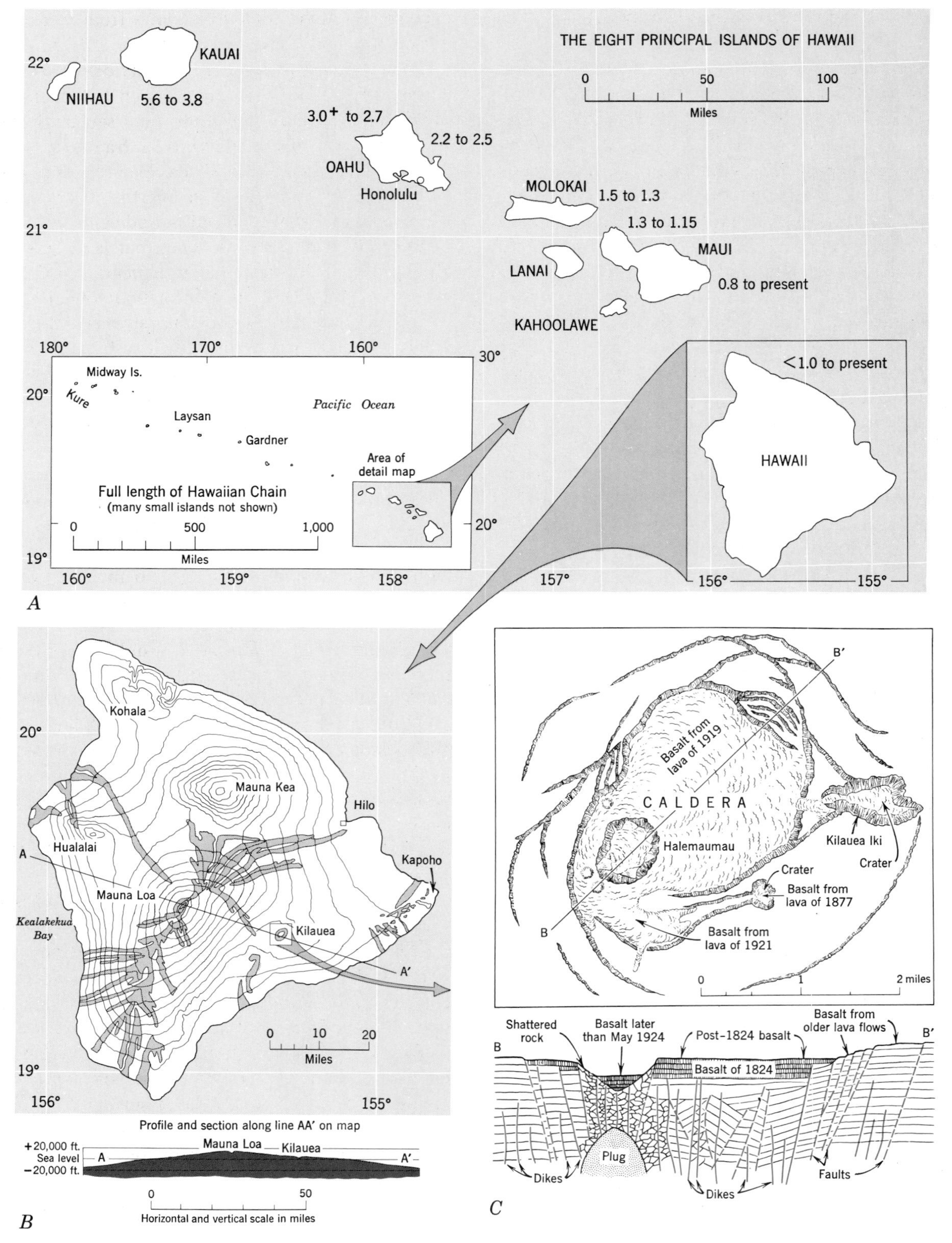

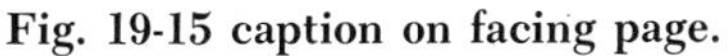

Fig. 19-15 caption on facing page.

on the island formed approximately 800,000 years ago.

Large calderas with almost vertical walls are present at both Kilauea and Mauna Loa. Kilauea has been studied continuously since an observatory was established there in 1912. Lying on the floor of the oval-shaped main caldera of Kilauea, 2 to 3 miles wide, is a much deeper vent, Halemaumau, about $\frac{1}{2}$ mile wide, from which nearly all recent lavas have issued. In 1919 lava welled to the brim of Halemaumau and poured across the floor of the caldera (Fig. 19-15, *C*). The lava then subsided, and for some years was confined to a boiling lake of liquid rock in this vent, which afforded many opportunities to measure temperatures and collect gases. Suddenly in 1924, the level of the lava lake was lowered by several hundred feet. Water from the surrounding rocks drained into the vent, was there converted to steam, and caused a tremendous explosion. The blast tore loose great quantities of rock from the walls and enlarged the vent. This explosive eruption was exceptional in the known history of Kilauea; it illustrates the principle that the phases of volcanic activity can change in response to changes in the conditions that control volcanic gases.

Two other vents connected with Kilauea have been active since 1868. One, Kilauea Iki (Fig. 19-6), extruded lava in 1959 to 1960. Because the vents and the caldera itself are generally circular or elliptical in plan, we reason that a cylindrical column of magma lies beneath each vent.

Mauna Loa lies at the apex of the Hawaiian volcanic shield. Numerous vents lie within its main caldera and along its flanks. Flows from vents far out on the flanks have broadened the shield.

Each of the five volcanic centers on the island consists of a linear belt along which lie dozens of old calderas and smaller vents, now seemingly inactive. These belts mark vertical fractures, each tens of miles long; magma has been pushing up along them for a long time, but the locations of the active vents have shifted periodically.

Stromboli. In most volcanoes, bursts of activity alternate with longer periods of nearly complete inactivity. An exception is Stromboli (Fig. 19-16), which has been active continuously since the beginning of recorded history. Such sustained operation is remarkable; it requires a continuous upward transfer of heat, doubtless accomplished by the gases. Moderate explosions, occurring at intervals of 10 to 15 minutes, hurl clots of red-hot lava upward into an incandescent cloud above the crater. These fall back into the crater and mingle with the pool of molten lava, which does not crust between explosions. The luminous masses, appearing with regularity and visible afar to men on ships, have earned for Stromboli the friendly title "Lighthouse of the Mediterranean."

Vulcano. South of Stromboli stands Vulcano, the supposed chimney of the forge tended by the Roman god Vulcan. Vulcano erupts intermittently; between eruptions the viscous lava in its vent solidifies, and the volcano appears to be inactive. Eruptions begin suddenly and explosively; the crust of igneous rock is broken to pieces and hurled into the air along with masses of new lava, which congeal in flight and descend as bombs. Unlike that of Stromboli, the eruption cloud of Vulcano is not luminous.

Vesuvius. On the Bay of Naples, about 225km north of Stromboli, lies Vesuvius, which has been studied longer than any other volcano. The ancient Romans began the practice of observing and recording its behavior; they established a tradition that has been perpetuated for twenty centuries. A modern observatory high on the slopes of Vesuvius has been manned by trained observers for many decades. The bedrock of a large surrounding area has been mapped in detail. Thick sedimentary strata, including much limestone, dip steeply westward from the crest of the Apennine Mountains several kilometers east but become nearly horizontal beneath the volcanic rocks near Naples.

Like Wizard Island in Crater Lake, Vesuvius is

◀ **Fig. 19-15. Maps and sections of Hawaiian Islands and volcanoes on Island of Hawaii. A. Index map of entire archipelago and locations of eight principal islands of Hawaii. Numbers beside islands represent K/Ar ages (10^6 years) of basalts underlying them. (Data from Ian McDougall, 1963.) *B.* Topographic map and profile, Island of Hawaii. Extrusive basalts congealed from lava flows erupted in the interval 1750 to 1950 are shown in light gray. Compare profile with Fig. 19-9, *B. C.* Diagrammatic map and section, Kilauea caldera. Section shows inferred structure and suggests location of cylindrical mass of intrusive igneous rock beneath Halemaumau. (*B* and *C* after H. T. Stearns and G. A. Macdonald, 1946.)**

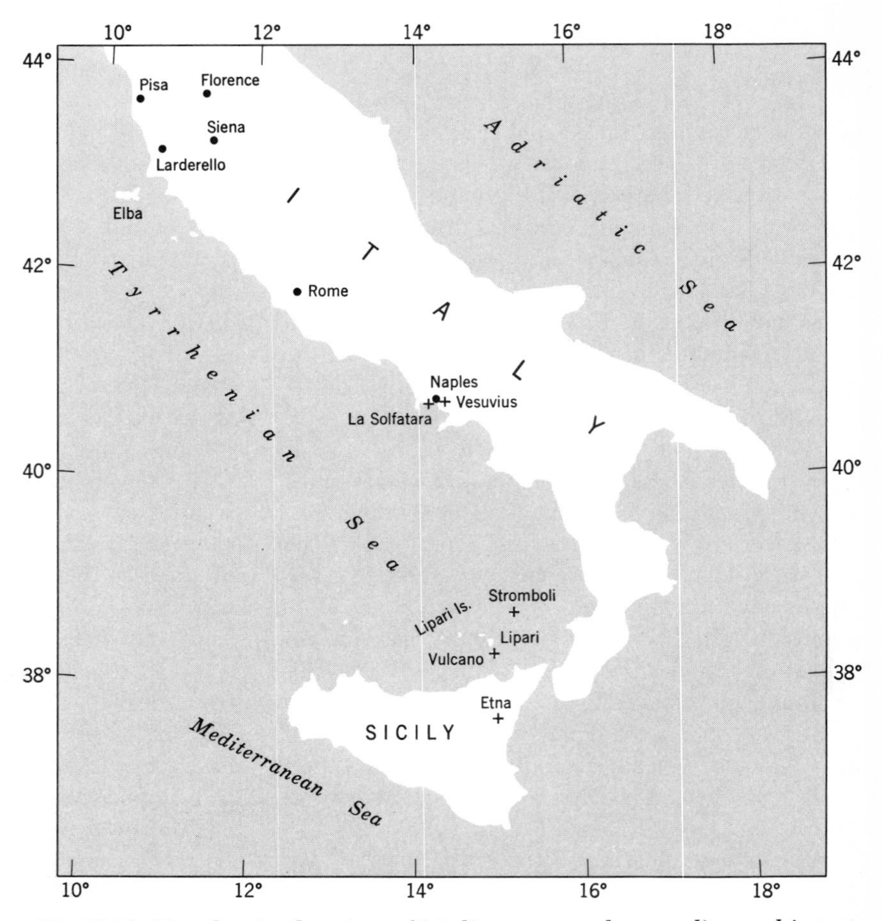

Fig. 19-16. Map showing locations of Mediterranean volcanoes discussed in text.

a cone standing within the caldera at the summit of an older peak. The ancestor of Vesuvius was Monte Somma, whose sides disclose a history of at least 14 eruptions and two long periods of quietude. Somma originated possibly 100,000 years ago; it had been inactive for so long that the ancient Romans, though recognizing it as a volcano, thought its tempestuous days had ended. In the first century A.D. they learned otherwise. Starting in the year 63, numerous earthquakes shook the surrounding country, and in 79 mighty explosions propelled vast quantities of tephra into the air. During this episode the caldera formed. Its southwestern wall has vanished, possibly as a result of explosions in the later stages of the eruption or possibly because of engulfment and subsidence into the magma. The towns Herculaneum on the west and Pompeii on the south (Fig. 19-17, *A*) were destroyed by materials ejected from Monte Somma. Herculaneum was blotted out and buried by three volcanic mudflows originating from the showers of tephra and torrential rains accompanying the eruptions. Pompeii succumbed to voluminous showers of tephra.

Vesuvius grew from the floor of the caldera. During its first millennium Vesuvius built a tephra cone with the products of ten eruptions. During long times between eruptions it was quiet. Lava did not appear until 1139 but has been a consistent product subsequently. A period of repose lasting nearly 500 years was broken by a violent outbreak in 1631. Thereafter the crater commenced its modern regimen of nearly continuous gentle activity, interrupted at intervals ranging from 10 to 40 years by spectacular outbursts, the latest in 1944.

Many features connected with Vesuvius suggest

that the top of the magma chamber lies in limestone at a depth of about 4.8km beneath the base of the volcanic cone (Fig. 19-17, *C*). Blocks of limestone and other sedimentary rocks have been found both among the tephra from Vesuvius and within extrusive igneous rocks formed from its lavas. The exceptionally high content of carbon dioxide (CO_2) in the gases emitted by Vesuvius and the undersaturated lavas can be explained by the reaction between the hot magma and the limestone ($CaCO_3$). Accumulation of carbon dioxide in the magma chamber doubtless contributes to explosivity. The upper part of the magma chamber can be compared to a gigantic bottle of a carbonated beverage; if the gas comes out of solution it escapes vigorously, carrying liquid with it.

Mont Pelée. The explosive violence of Vulcano and Vesuvius has been exceeded by that of several other volcanoes. One, Mont Pelée, in the West Indies, after more than 50 years of quiescence, came to life in 1902 with a succession of major explosions that continued for several months. Two mechanisms distributed the products of each eruption: (1) incandescent clouds, which carried superheated vapor and fine tephra high into the air (Fig. 19-18); and (2) hot avalanches, in which coarse and fine tephra sped silently and swiftly along the ground down the slope of the volcanic cone, some reaching rates as high as 60m/sec. The clouds inspired Alfred Lacroix, a noted French geologist, to give the name *nuée ardente* (Fr. "fiery cloud") for eruptions of this kind. The name is dramatic but deceptive; the eye-catching fiery clouds masked an equally hot, if not hotter, but less visible sheet of tephra, which constituted the bulk of each discharge and swept along the ground beneath the clouds. On the morning of May 8, one of these shrouded sheets became a grim reaper at St. Pierre, capital of Martinique, situated on the coast 8km from the volcano (Fig. 1-5, *B*). It destroyed all buildings and all but two of the 28,000 people in the town; one of the survivors was a prisoner in a dungeon. In the harbor, ships were turned over in a churning sea that literally boiled. Great numbers of fishes and other marine animals were scalded to death.

Witnesses described the forward part of each advancing cloud and tephra sheet as a billowing mass, its base hugging the ground and the clouds swirling upward, with intensely turbulent motions, to heights of more than 4km. The tremendous energy of the May 8 tephra sheet was manifested both by its effects and by the load it carried. The moving sheet sheared off large trees, flattened walls of buildings perpendicular to its path, and in the manner of artillery shells drove particles through the sides of an iron tank and planks through tree stumps. An astonishing total load was spread along its path; and some individual blocks 1 to 8m long were moved as far as 16km.

The crater of Pelée had been blocked by a lava dome, so that the masses of tephra shot out either obliquely upward or in a horizontal direction. Part of their high speeds along the ground has been ascribed to the lateral component of the original explosions.

Katmai. Ten years after the great outbreak at Pelée, Mount Katmai, on the Alaskan Peninsula, erupted in similar fashion and on a much larger scale. As at Pelée, tephra included particles of pumice, indicating the large gas content of the magma. The tephra flows from Katmai partly filled a valley near the crater; fortunately, no large settled community lay in their path. Columns of vapors continued to rise from the hot tephra; these prompted the name Valley of Ten Thousand Smokes. This area is now the largest of the national monuments in the United States. Fine-grained tephra from Katmai were carried high into the air and distributed to all parts of the world.

The hot tephra flows of Pelée and Katmai are two examples of a volcanic phenomenon whose importance has been increasingly recognized in recent years. Vast tracts underlain by welded tuffs and some tuffs not welded, both derived from hot tephra flows which rivaled plateau basalts in areal extent and volume, have been discovered in Utah, Arizona, Nevada, Wyoming, Idaho, Maine, Ontario, Italy, Sumatra, New Zealand, and Armenia. Most sheets of rhyolite have proved to be the products not of solidified lava flows as formerly supposed but of tephra flows. Near their origins many welded tuffs have been intruded by granitic rocks, suggesting a close connection between extrusive and intrusive processes.

The mechanism of a tephra flow is not yet fully understood. When they leave the vent the tephra are *fluidized.* Although initially fluidized as they leave the vent the tephra must maintain this condition in order to flow rapidly. How they do it is the mystery.

Fig. 19-17. Views and section of Vesuvius. A. Sketch showing view north toward Vesuvius and its surroundings. B. Cone and crater of Vesuvius (*lower right*) and northwest rim of caldera of Monte Somma (*cliff, upper left*). Lobate, dark-colored areas at left base of cone are former lava flows. (C. B. Jacobson, U. S. Bureau of Reclamation.) *See facing page for C.*

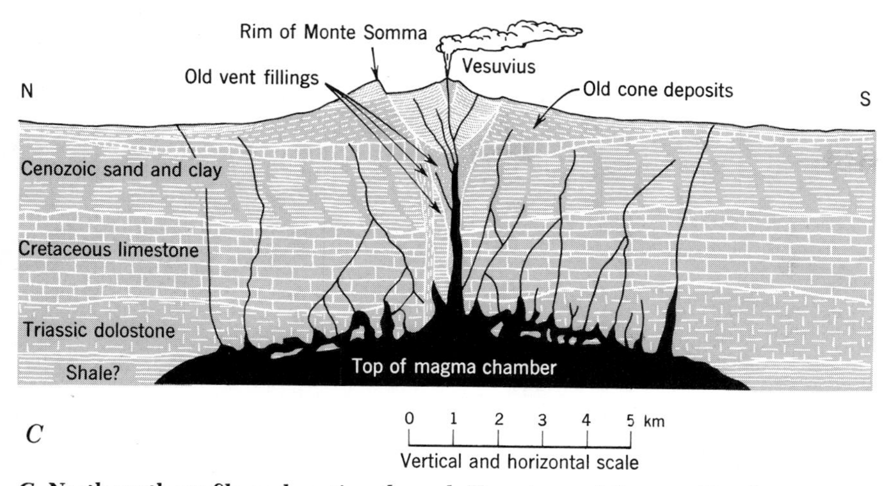

C. North-south profile and section through Vesuvius and Somma. Depth to magma chamber is based on seismic studies. (After Alfred Rittmann, 1960.)

One suggestion supported by small-scale experiments is that the intense heat draws in cooler air from ahead of the flow, forming a cushion of air over which the tephra slide effortlessly. Temperature doubtless enters into the mechanism of tephra flow but evidently it is not the sole factor, because comparably fast-moving but cold, dry avalanches have been recorded (Fig. 8-3). Compressed air generated by the advancing mass of solids has been invoked as the cushioning device for dry avalanches.

By whatever means they do it, many sheets of tephra travel swiftly and reach their destinations while the particles are still hot. The destructive tephra flow that overwhelmed St. Pierre did not melt copper (melting point 1,058°C) but did melt glass (melting temperature 650° to 700°C); hence its temperature is thought to have been about 800°C. Welding of particles requires that the temperature be in the range of 500° to 1,000°C. Evidence of high temperature other than the effects on objects having known melting points is contained in the charred wood commonly occurring with tephra.

Fig. 19-18. ***Nuée ardente*, 4km high, and tephra flow (hidden by base of cloud) reaching the sea on Island of Martinique. One of the many minor eruptions of Mont Pelée during 1902. (Alfred Lacroix.)**

Fumaroles and solfataras. A volcano discharging gas nonexplosively is said to be in the *fumarolic* stage of activity. The name comes from Italy, where smoking cracks were termed *fumaroles.* If the fumes are sulfurous the phase is designated *solfataric* (from La Solfatara, Italy, whose name is derived from the Italian word for sulfur).

Summary. With a circular diagram (Fig. 19-19) we can illustrate the relationships among products of volcanoes, volcanic processes, volcanic edifices,

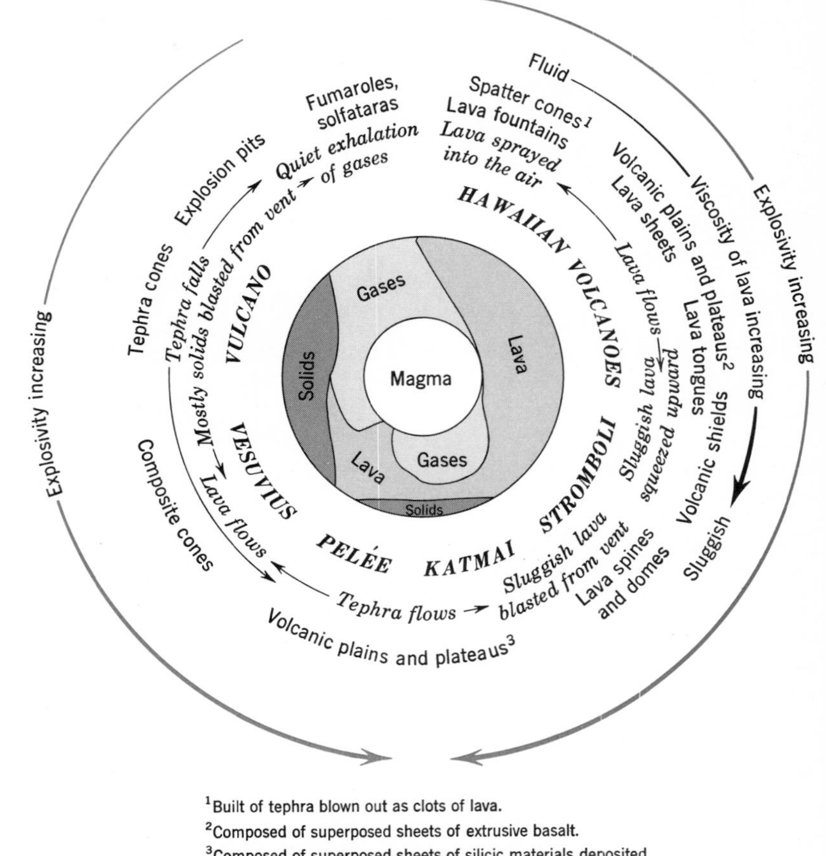

[1]Built of tephra blown out as clots of lava.
[2]Composed of superposed sheets of extrusive basalt.
[3]Composed of superposed sheets of silicic materials deposited by fluidized tephra flows.

Fig. 19-19. Relationships of magma (inner circle) and volcanic products (outer circle) to volcanic eruptive activities (*italic type*), volcanic edifices (roman typeface), and typical volcanoes (BOLDFACE TYPE). Further explanation in text.

and the volcanoes discussed. The center of the diagram represents a magma chamber in the outer part of the lithosphere and the circumference of the larger circle the surface of the Earth. Magma is shown in the small circle; it consists chiefly of molten silicate materials, which become lava at the surface, and gases. Between the magma circle and the outer circle representing the Earth's surface are colored fields for the three major groups of volcanic products: solids, gases, and lava. The fields for solids are not shown in contact with the magma circle because most solid products of volcanoes derive not directly from the magma itself but from masses of rock lying between the magma chamber and the Earth's surface, from igneous rocks that congeal from magma, and from those that solidify from former lava. This arrangement is not strictly correct because magma at times does include solids as crystals. However, the fields for solids have been separated entirely from the magma circle to emphasize the point that most solids are ejected forcibly from volcanoes by the explosivity of gases. Names of volcanoes, kinds of volcanic edifices, and volcanic processes surround the surface circle. Quiet extrusions and exhalations lie at the top, more explosive eruptions toward the bottom along either side. The best way to read the diagram is to imagine that the surface circle is the face of a clock having only an hour hand. As the imaginary hour hand moves around, it points to various volcanoes, volcanic processes, and volcanic edifices. In each position the hand extends from the magma circle through some part of the outer circle showing volcanic products. By following along the length of the imaginary hand of the "clock" we can tell what volcanic products are discharged from a given volcano or were involved in the building of a particular volcanic edifice. For example, at about eleven o'clock the hand would point to fumaroles and solfataras, and in doing so it would pass through a field in the outer circle that includes only gases. At about two o'clock the hand would point to Hawaiian volcanoes, which are characterized by quiet extrusion of fluid lava having small gas content. Typical volcanic edifices from Hawaiian volcanoes are lava sheets if the flow stems from vents, and extensive volcanic plains and plateaus consisting of superimposed sheets of extrusive basalt if the flow comes from fissures. More explosive activities, involving gases plus viscous lava, gases plus lava and solids, or gases plus solids only, occur at the volcanoes shown in the sector between about five and ten o'clock. Vesuvius and similar volcanoes, found in the eight o'clock sector, eject solids explosively and lavas less violently to build composite cones.

Fig. 19-20. Old Faithful Geyser, Yellowstone Park, Wyoming at peak of eruption. (Don Knight, Eastern Photo Service.)

HOT SPRINGS AND GEYSERS

Rising magmatic gases heat underground water. Water near its boiling temperature comes to the surface as a steady flow, a *hot spring*, or as a series of eruptions from a ***geyser*** (Icelandic, "to gush"), *an orifice that erupts steam and boiling water intermittently.* Strictly speaking hot springs and geysers are not volcanic, because much of their water is not newly added from the Earth's interior but consists largely of ground water warmed by magmatic heat. Nor are all hot springs related to magmatic heat; some result from oxidation of large bodies of sulfide minerals, other exothermic chemical reactions, or other sources of local heat.

Nearly all the world's true geysers are located in Iceland, New Zealand, and the United States. Hot springs and geysers are numerous in Yellowstone National Park, Wyoming (Fig. 19-20). No volcanic activity has been recorded by man in the Yellowstone region, but some volcanic rocks within the Park look very recent. Presumably both the geysers and recent-looking volcanic rocks are indications that a body of magma is solidifying beneath the surface of the Earth.

The intermittent action of geysers depends on two factors: (1) an underground reservoir of water distributed through irregularly interconnected passageways, and (2) the increase of the boiling point of water with increasing pressure (Fig. 19-21). From the observed gushing action we can infer that the water does not occupy tiny pores within an aquifer such as a body of sand, but instead must fill interconnected underground passageways, like those in limestone caverns. In both volcanic rocks and limestone large openings occur, but in volcanic rocks the spaces not only result from the effects of removal of rock material but also are primary voids.

If the shape of the reservoir were that of a vertical cylinder, such as a standpipe buried in the ground, convection would equalize the temperature throughout. The lower water, heated first, would circulate upward as in a kettle on a stove and a boiling spring would result. A geyser requires that such large-scale convection be prevented; it can take place only locally in a series of irregular, tortuous passageways.

Within the reservoir hydrostatic pressure, which increases downward at the rate of 1 atmosphere for each 10m no matter how small the passageways, will raise the boiling temperature of water. At the base of a column of water about 150m deep, the boiling temperature is 200°C. Continued heating eventually brings the water throughout the reservoir to its boiling point, which of course varies according to depth. When all the water in the reservoir is at its appropriate boiling temperature steam begins to accumulate against the roof of the uppermost water chamber. The buildup of steam pressure forces some water out of the orifice of the geyser. At Old Faithful in Yellowstone Park this initial jet rises 3 to 8m into the air. Removal of water from the top of the reservoir decreases the hydrostatic pressure at all points beneath. At its boiling temperature under the hydrostatic pressure of a full reservoir, the water lies above its boiling temperature at the lowered pressure within a partially full reservoir. Accordingly all the boiling water that feels the reduced pressure flashes into steam instantaneously. The sudden generation of steam clears the passageways and shoots steam and boiling water high into the air. Old Faithful erupts regularly at about 66-minute intervals, discharging an estimated 10,000 to 12,000 gallons of water, some of which rises to heights of 35 to 50m.

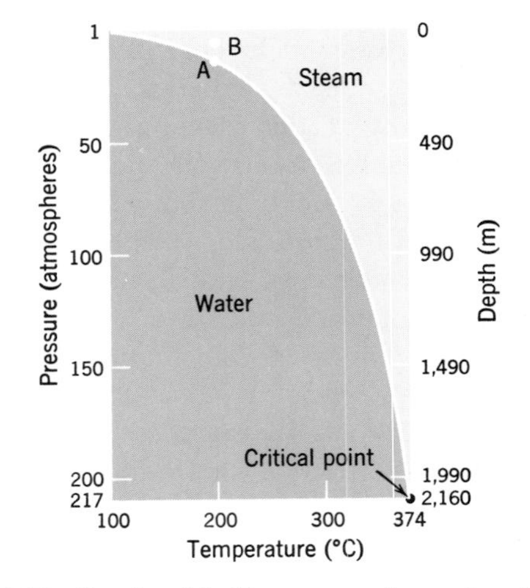

Fig. 19-21. Graph of boiling curve for water showing how pressure affects boiling temperature. Scale at right shows depths of fresh water required to create hydrostatic pressures corresponding to values on scale at left. Further explanation in text. (Compare Fig. 3-3.)

SUBMARINE VOLCANIC ACTIVITY

Most of our ideas about volcanoes are conditioned by observations of subaerial eruptions on continents or oceanic islands. Modern exploration of the sea floor discloses that beneath the surface of the sea many more volcanoes exist than we can see above it (Pls. A and C). Volcanoes on the sea floor build conical edifices of extrusive igneous rock and tephra, and spread lava in sheets, just as do volcanoes on land. But submarine eruptions differ from land volcanoes in at least two respects: (1) The composition of lava from volcanoes lying within the deep-sea basins generally is not the same as that from volcanoes along the margins of these basins; (2) the water exerts great hydrostatic pressure at depth and can react with lava and tephra to create products not found in volcanic rocks on land. We can gain many insights into submarine volcanism by studying new islands, by examining the results of deep-sea expeditions, and by watching lava flow into the sea.

New islands. Submarine volcanism is brought forcibly to our attention when new islands grow out of the sea or when volcanic exhalations discolor seawater or create violent motions within it. Many new islands have appeared within recorded history.

The newcomer among the world's volcanic islands is Surtsey, named after the Icelandic god of volcanoes, Surter. Surtsey broke surface on November 14, 1963; it began by blowing out gases and tephra, some in columns as high as 2km. It built a tephra cone 10m high in one day; three days later the island was 600m long and reached an altitude of 60m (Pl. 19).

During the first $4\frac{1}{2}$ months of its life, Surtsey dispensed only gases and tephra, the estimated amount of solids being 400 to $500 \times 10^6 m^3$ or about $40m^3/sec$. Lava appeared on April 4, 1964. It filled the vent, creating a lake 120m across and then overflowed the rim and cooled to form a protective coating of undersaturated olivine basalt around the tephra island through about three quarters of its perimeter. By April 1965 the island had become circular with a diameter of 1.7km, and included a tephra cone 170m high plus a composite cone 122m high that housed the lava lake.

Volcanic edifices on the sea floor. One of the dramatic early results obtained by continuously recording echo sounders was the discovery that numerous isolated conical mounds dot the deep-sea floor (Fig. 19-22). The name *seamounts* has been given to these cones, of which thousands are known. Almost certainly they are volcanic edifices. A seamount with a conspicuously flat top is called a *guyot*. Some guyots evidently built upward to form volcanic islands. After being planned off by the waves, as was one of the tephra cones near Surtsey, the guyots subsided so that their tops now lie at variable depths below sea level. Possibly some flat tops are initial shapes of the cones, as has been found for some tephra cones built in lakes.

Dredge hauls from seamounts have found extrusive igneous rock to be the predominant building material. Hydrostatic pressure probably inhibits the formation of tephra beneath the sea, but we do not know at what depth the inhibition would occur. Gas pressures of 600 atmospheres have been reported from explosive subaerial volcanoes. Such a pressure would be equalized by a column of water 6km thick. If tephra flows do not occur in the sea, the products of tephra flows should be reliable indicators of a subaerial environment.

Volcanic edifices of another type are suspected on the floor of the central Pacific Ocean. Surrounding some island groups are large areas where the bottom is smooth, slopes gently and radially outward, and is thought to be underlain by extrusive igneous rock. Such areas are *archipelagic aprons;* they may be submarine volcanic shields.

Eruptions at depth. Lava extruded onto the deep-sea floor is influenced by the hydrostatic pressure. In specimens of basalt dredged from the sea floor off Kilauea in depths ranging down to 4,000m, densities and both abundances and sizes of vesicles varied with depth. In specimens from water shallower than 800m the densities were less than 2.8g/cm^3, and more than 10 per cent of the volume of the rocks consisted of vesicles whose average diameters exceeded 0.5mm. At a depth of 800m these properties changed abruptly. At greater depths densities ranged up to 3.0g/cm^3, vesicles were extremely rare, and their diameters averaged less than 0.1mm. No indications were present to suggest that chemical reactions had occurred between lava and seawater.

Despite this lack of evidence of chemical reactions between seawater and modern lava from the sea floor off Hawaii, examples from the geologic record suggest that water from the sea can react with lava. If pyroxene and plagioclase, the two common minerals of basalt, react with seawater, they change dramatically. Pyroxene hydrates to chlorite, and plagioclase transforms into albite and kaolinite. Both reactions release materials into the water: pyroxene supplies calcium bicarbonate that can precipitate as calcite, and both minerals furnish silica that can solidify as chert. Interbedded with unaltered pelagic sediments are examples of ancient submarine volcanic extrusives showing the effects of these changes while preserving original textures. Mineralogic changes similar to these result from metamorphism, but metamorphic processes generally create new textures.

Some deep eruptions do not involve lava at all. In basins along the floor of the Red Sea, three ponded bodies of hot, saline water have been ascribed to submarine volcanic exhalations. Normal Red Sea water having a temperature of 22°C and salinity of 41,000p.p.m extends down to a depth of 1,980m. In the largest of the closed basins, 12km long and 5km wide, water temperature below a depth of 2,040m is about 56°C and salinity is 255,000p.p.m. Oxygen content of the bottom water is only 0.1 milliliters/liter (ml/l), compared to about 1 or 2ml/l in normally aerated water above 1,980m. The bottom sediments beneath the hot brines consist largely of amorphous iron oxides and goethite (hydrated iron oxide).

Flow of lava into water. A few lava flows from volcanic islands have been observed entering the sea. Some proceeded from land to sea without notable activity and without creating much steam. One such flow in the Samoan Islands was visible moving along the sea floor. The lava assumed smooth, curved shapes and its surface cooled quickly. Where the surface cracked, new bulbous masses swelled out and budded off into independent ellipsoids known as *pillows*. In flowing under water some lava from Surtsey has formed pillows. In many ancient extrusive igneous rocks pillows are common (Fig. 19-23); they reveal that the lava flowed under water.

On contact with seawater other lava from Surtsey quickly chills to a glassy crust. This crust fragments nearly as fast as it forms, with the result that the lava is converted into vast quantities of solid particles of volcanic glass.

Some lava flows, mostly of the blocky type on land, react violently with seawater. When such lava and water meet, great quantities of steam are

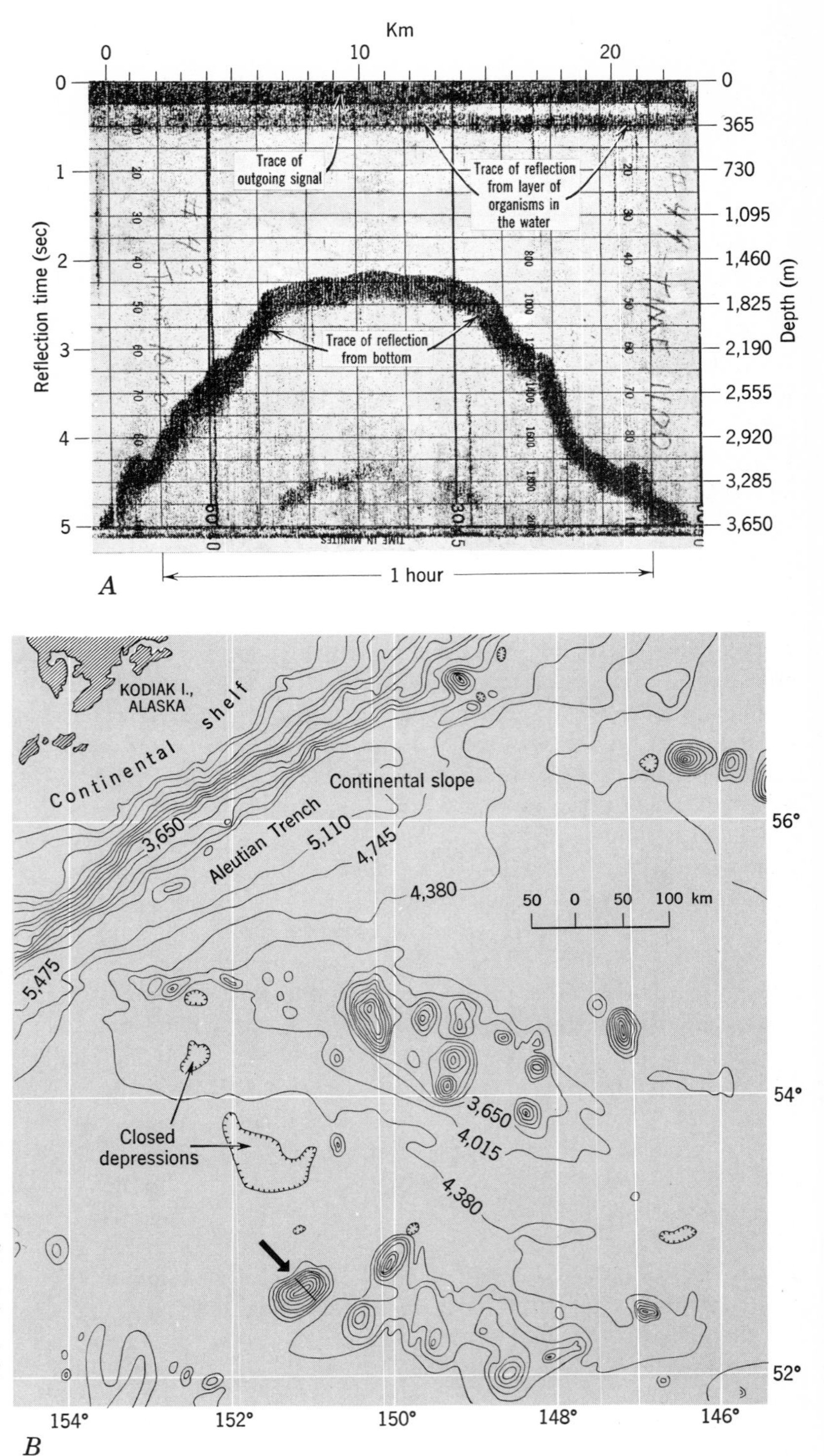

Fig. 19-22. Volcanic cones on the sea floor. A. Echogram showing seamount in Gulf of Alaska made from R/V *Pathfinder*. Vertical lines mark 6-minute intervals; horizontal lines show depth. Location of seamount and survey track are shown in *B*. *B*. Bathymetric chart of part of Gulf of Alaska, made from echograms, shows many volcanic cones. Contour interval 365m. (After H. W. Menard and R. S. Dietz, 1951.) *See facing page for C.*

C. Artist's conception of portion of sea floor in the mid-Pacific with water removed, showing two guyots in the distance. Canyon in foreground is cut into a seamount, most of which does not show. (Painting by Chesley Bonestell, in E. L. Hamilton, 1959.)

generated and clots of lava shoot into the air. The clots solidify before they fall back into the water, where they accumulate as one of the many kinds of volcanic breccia (App. C).

CAN VOLCANOES INFLUENCE CLIMATE?

In 1784 Benjamin Franklin inferred that volcanoes can influence the weather. He attributed the extremely cold winter of 1783/1784 to the effects of explosive eruptions in 1783 from Asama in Japan and Laki in Iceland.

Other great eruptions have sent tephra winging into the stratosphere, where fast-moving air currents can keep them aloft and spread them around the world. Powerful explosions occurred in 1814 and 1815 at Tambora Volcano in Indonesia, and at Mayon in the Philippines. Within a radius of 500km of Tambora, tephra caused total darkness for three days. Smaller amounts spread widely in the stratosphere and caused remarkable sunsets and twilight phenomena in many parts of the world. In 1816, as a possible consequence of the explosions, frosts occurred in Maine in every month of the year and there was a substantial snowstorm in June. The dust cloud from the explosion of August 27, 1883, at Krakatau spread into both hemispheres. It reached Europe in November, 1883, and lowered the values of solar radiation measured at the Montpelier Observatory in southern France by about 10 per cent for a period of 3 years. In the Northern Hemisphere no widespread cool weather followed the Krakatau explosion. Clouds of fine tephra from the 1912 explosion of Katmai, Alaska reduced by 20 per cent the value of solar radiation measured in Algeria.

The Northern Hemisphere experienced a gradual warming trend from 1912 to 1952, during an interval when no mighty tephra ejections occurred. Since 1952 cooling has occurred. Coincidentally three explosions having widespread influence have taken place in the Northern Hemisphere: (1) On June 9, 1953, Mount Spurr, near Anchorage, Alaska, exploded tephra into both lower and upper atmosphere. The low-altitude material fell out as a blanket throughout parts of Alaska. The high-altitude tephra zoomed upward to 22km above the cone

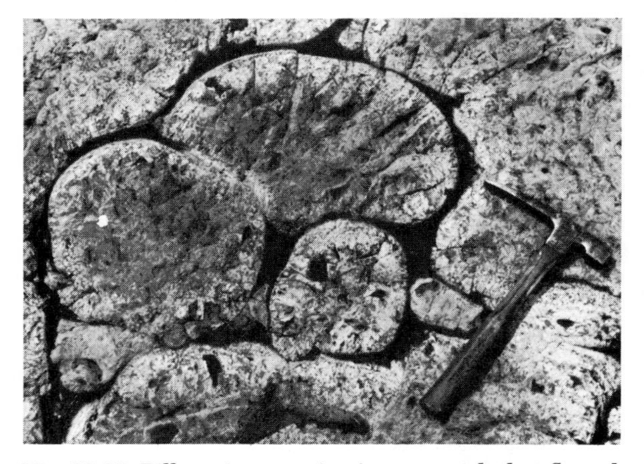

Fig. 19-23. Pillows in extrusive igneous rock that flowed across an ancient sea floor. Ordovician, Eagle Lake, Maine. (Bradford Hall.)

and created a dust cloud that later crossed England at an altitude of 14km. (2) In 1956 a similar dust cloud passed over western England at 16km; it came from a volcano on Kamchatka Peninsula. (3) On April 1, 1963, Mount Trident, at Katmai, Alaska blew out an eruption cloud that could be seen for 100 miles. Also in the spring of 1963 Agung Volcano, on the island of Bali, Indonesia, sent tephra high into the atmosphere; on September 25 the effects of these tephra were recorded as twilight clouds over Colorado at an estimated altitude of 20 to 25km.

Some geologists have attributed Pleistocene changes of climate to the effects of tephra in the atmosphere. This idea, which has not been widely accepted, stands in need of careful reexamination. Most objections to it have been based on the supposition that volcanic activity on the scale required to change the world's climate would necessarily leave behind significant layers of tephra. The absence of abundant tephra in most Pleistocene sediments has been held fatal to the hypothesis. But modern knowledge suggests that tephra layers result almost exclusively from low-altitude suspensions, which are deposited in the vicinity of their source and whose effect on world climate is negligible. Low-altitude tephra suspensions remain aloft for only a few days at most; only high-altitude suspensions of tephra can block the Sun's rays. Such tephra remain aloft for months or even years and seldom build widespread layers. Although their distribution as suspensions can be worldwide, they tend to fall out at only a few localities, where persistent downdrafts occur, and would be so easily mixed with and masked by other sediments that they could escape detection. High-altitude tephra offer little opportunity to settle the relationship of volcanism to climate. We need to map and date low-altitude tephra near volcanoes to see if the abundance of tephra correlates with climatic changes.

RECOGNITION OF ANCIENT VOLCANIC ACTIVITY

Modern volcanoes yield abundant clues by which we can confirm ancient volcanic activity. Conical edifices in varying stages of erosion are an obvious entree into the past from the present. But as cones are soon destroyed, we cannot expect to find very old ones. More permanent evidence is contained in volcanic strata.

Erosion of volcanic cones. From the actively growing conical edifices built by some modern volcanoes we can trace every gradation to the roots of ancient cones. We have seen how some volcanoes, evidently inactive, have sprung to life again and have renewed their youthful vigor. A volcano is classed as *inactive* if it has not erupted within historic time, *dormant* if it has erupted within historic time but not within the previous 50 years, and *active* if it has erupted within the previous 50 years. Active and dormant cones repair themselves by rebuilding eroded parts. Once a volcano has become inactive, however, its cone yields to erosion and is eventually destroyed.

Erosion of inactive volcanic cones discloses their anatomy, which we are otherwise privileged to observe only in the walls of a few calderas, as at Crater Lake, Monte Somma, and elsewhere. From what we can see in dissected inactive cones we are able to envision what must be going on inside an active cone.

Even when all other traces of the cone have been eroded, the ***volcanic neck,*** *a cylindrical filling of an ancient volcano,* generally persists because the material composing it resists erosion. Shiprock, in New Mexico, exemplifies such a neck. The top of the column of igneous rock, which is nearly circular in plan, stands 1,300 feet above the sedimentary strata of the surrounding plateau (Fig. 19-24). Igneous rock fills fissures that radiate outward from Shiprock. These fissures presumably fed

Fig. 19-24. Shiprock, New Mexico, a volcanic neck, 1,300 feet high, from which radiate three prominent ridges consisting of igneous rock that filled fractures. The rest of the volcanic cone has been destroyed by erosion. (J. S. Shelton.)

lava flows that escaped along the flanks of the former cone, all other traces having vanished.

Volcanic strata. Volcanic products that were spread out as sheets can be buried and incorporated into the geologic record. Once they have become part of the bedrock they are less vulnerable to erosion than are cones. Features made by flowing lava can be recognized in ancient bedrock, as we shall see when we compare extrusive and intrusive sheets.

Stratigraphy of volcanic rocks. A resurgence of interest in the study of extrusive igneous rocks has occurred in the wake of modern developments in geophysics and geochemistry, which have permitted us to unlock the full stratigraphic potential of these rocks. The value of tephra layers for local correlation has long been recognized, but widespread piles composed of sheets of tephra and former lava flows previously were ignored or examined only cursorily. Such piles of volcanic rocks are now being studied in detail and with even more zeal than if they were fossiliferous marine strata. The reason stems from at least four advantages offered by volcanic materials: (1) The composition of each layer of volcanic rocks tends to be uniform. (2) The layers spread out rapidly. (3) Basalts are common and from their iron minerals we can derive paleomagnetic information. (4) Many volcanic rocks contain minerals having radioactive isotopes, such as K^{40} that disintegrates to Ar^{40}, from which we can calculate isotopic ages. Although on a global scale the extent of even the most widespread piles of volcanic rock must be considered as local, their paleomagnetic data and radioactive isotopes give us a basis for worldwide comparison with other local deposits. Geophysical and geochemical study of volcanic strata has provided the chronologic basis for the magnetic events that occurred in the later part of the Cenozoic Era. The abundances of these strata make them a fruitful field for research.

VOLCANOES AND MAN

The threats posed by volcanoes to human welfare need no elaboration. Lavas and sheets of hot tephra destroy everything in their paths. In Iceland falls of tephra thicker than 10cm have snuffed out plant life and have interrupted agriculture for a decade or longer. In some areas mudflows cause great damage. Despite the latent danger in Japan and Indonesia, two of the most densely populated countries in the world, as well as elsewhere, many people live "under the gun" of volcanoes. In Japan and Indonesia volcanoes have been studied minutely as they relate to human affairs.

Forecasting volcanic eruptions. By triggering shallow-focus earthquakes magma moving within its subterranean chamber advertises an imminent volcanic eruption. In Hawaii and elsewhere the magma later heaves up the ground just before it erupts. Hence sensitive tiltmeters, recording changes in ground level and in slope, can herald volcanic eruptions. Because it is hotter than its Curie point, magma is nonmagnetic. Accordingly, the near approach of magma to the surface reduces the intensity of the local magnetic lines of force. This change can be detected by sensitive magnetic instruments. The temperatures and chloride contents of some fumaroles increase immediately before an eruption of lava or tephra.

Predicting events. Forecasting a volcanic eruption is easy compared with predicting what will happen once the eruption begins. Close study of some volcanoes suggests that their eruptions follow a predictably cyclic pattern. The outstanding example of a volcanologist's successfully predicting future events occurred during the eruption of Mont Pelée in 1929 to 1932. With the catastrophe of 1902 still fresh in mind, the residents of Martinique could not be blamed for expressing anxiety when many incandescent clouds were projected out of the peak during a period of 5 months. The local volcanologist reassured them by stoutly affirming that no destructive explosion would recur and that the intensity of the eruption would gradually decline and eventually stop. He was right because he was intimately acquainted with the past history of Mont Pelée. Unfortunately one leaky aquifer can change the course of an eruption without warning, and can cause a volcano to abandon what may have been a previously well-established pattern of behavior. Therefore even the best-grounded predictions are always subject to change.

Harnessing volcanic steam. The superheated steam from fumaroles has been looked upon as a source of electric power. Since early in the present century, volcanic steam has turned dynamos in the Larderello district, Italy (Fig. 19-16). The steam comes from wells having a maximum depth of 200m that penetrate only sedimentary strata. Steam is not the only gas used; other gases are recovered and employed for industrial purposes.

Other power developments have been carried out from the fumaroles in Indonesia, Iceland, Japan, New Zealand, and California. In Iceland volcanic heat is piped into residences and into greenhouses, where tropical fruits flourish although the climate outside is bleak. Various areas within the continental United States have been designated as potential sources of geothermal energy.

We have examined many aspects of volcanoes but still have not come to grips with the fundamental question of whence comes the magma. Our answer lies within the Earth, and we shall try to find it by studying intrusive igneous rocks and metamorphic rocks, both of which originate by processes deep within or beneath the crust.

Summary

1. Volcanoes discharge gases, lava, and tephra.
2. Varied volcanic edifices are constructed, depending on size and shape of the volcano, kind of material extruded, viscosity of the lava, and explosivity of the eruptions.
3. Volcanoes that dispense sluggish lava, generally rich in silica, tend to be explosive; those that extrude fluid lava, generally low in silica, erupt less violently.
4. Widespread sheets of basaltic rocks have resulted from fissure eruptions of fluid lava. Sheets of comparable dimensions but of variable composition consist of tephra, transported by a mechanism not yet understood.
5. Geysers result from the combination of volcanic steam, interconnected underground passageways for the water, and the rise in boiling temperature created by increasing pressure. When pressure on boiling water in the geyser's reservoir is lowered because steam forces some water out through the orifice, most of the remaining water in the underground reservoir flashes into steam and erupts at the surface.
6. Submarine volcanic activity has created thousands of volcanic cones and other edifices on the sea floor. Some volcanic cones have become new islands; others became guyots when their tops were planed off by wave erosion and they sank beneath the sea.
7. The composition of lavas extruded from volcanoes within ocean basins is not the same as that of lavas extruded from volcanoes lying at the margins of ocean basins.
8. Some lavas flowing under water create pillows; others disintegrate into breccia.
9. Tephra propelled into the air with great

violence become both low-altitude suspensions and high-altitude suspensions. Dust clouds created by high-altitude suspensions of tephra in the stratosphere can block the Sun's rays from the Earth. Such dust can affect the weather, but its effect on climate is unknown.

10. Ancient volcanic activity is recognized by eroded cones, which permit us to study the anatomy of ancient edifices, and by strata of volcanic rocks. Studies of volcanic rocks combining stratigraphy, geophysics, and geochemistry are contributing much new quantitative information on geologic history.

11. Movement of magma underground, signaling an imminent volcanic eruption, generates shallow-focus earthquakes, reduces the intensity of the local magnetic field, and elevates and tilts the overlying parts of the Earth's surface. Warnings that an eruption is close at hand can be made with more confidence than predictions of the course of events during an eruption.

Selected References

Bullard, F. M., 1962, Volcanoes in history, in theory, and in eruption: Austin, Texas, Univ. Texas Press.

Cotton, C. A., 1952, Volcanoes as landscape forms, rev. ed.: New York, John Wiley.

Hunt, J. M., Hays, E. E., Degens, E. T., and Ross, D. A., 1967, Red Sea: detailed survey of hot brine area: Science, v. 156, p. 514–516.

McTaggart, K. C., 1960, The mobility of nuées ardentes: Am. Jour. Sci., v. 258, p. 369–382 (discussion p. 467–476).

Menard, H. W., 1964, Marine geology of the Pacific: New York, McGraw-Hill Book Co., Chap. 4, p. 55–95.

Richter, D. H., Ault, W. V., Eaton, J. D., and Rime, J. G., 1964, The 1961 eruption of Kilauea Volcano: U. S. Geol. Survey Prof. Paper 474-D.

Smith, R. L., 1960, Ash flows: Geol. Soc. America Bull., v. 71, p. 795–841.

Stearns, H. T., 1946, Geology of the Hawaiian Islands: Honolulu, Hawaii Div. of Hydrography.

Thorarinsson, Sigurdur, 1967, Surtsey: The new island in the North Atlantic: New York, The Viking Press.

Wexler, Harry, 1952, Volcanoes and world climate: Sci. Am. v. 186, no. 4, p. 74–76. San Francisco, W. H. Freeman & Co., Repr. 853, 5 p.)

Williams, Howel, 1951, Volcanoes: Sci. Am., v. 185, no. 5, p. 45–53. (San Francisco, W. H. Freeman & Co., Repr. 822, 11 p.)

CHAPTER 20

Plutonism and Metamorphism

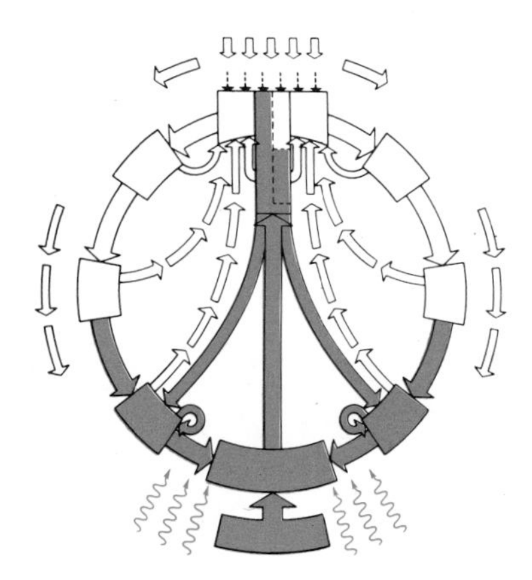

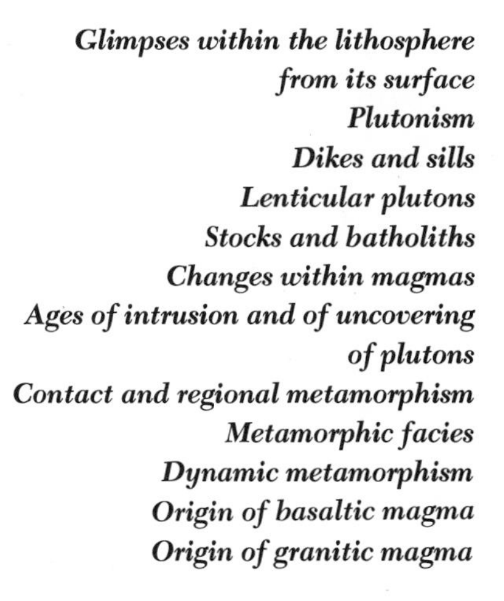

Glimpses within the lithosphere from its surface
Plutonism
Dikes and sills
Lenticular plutons
Stocks and batholiths
Changes within magmas
Ages of intrusion and of uncovering of plutons
Contact and regional metamorphism
Metamorphic facies
Dynamic metamorphism
Origin of basaltic magma
Origin of granitic magma

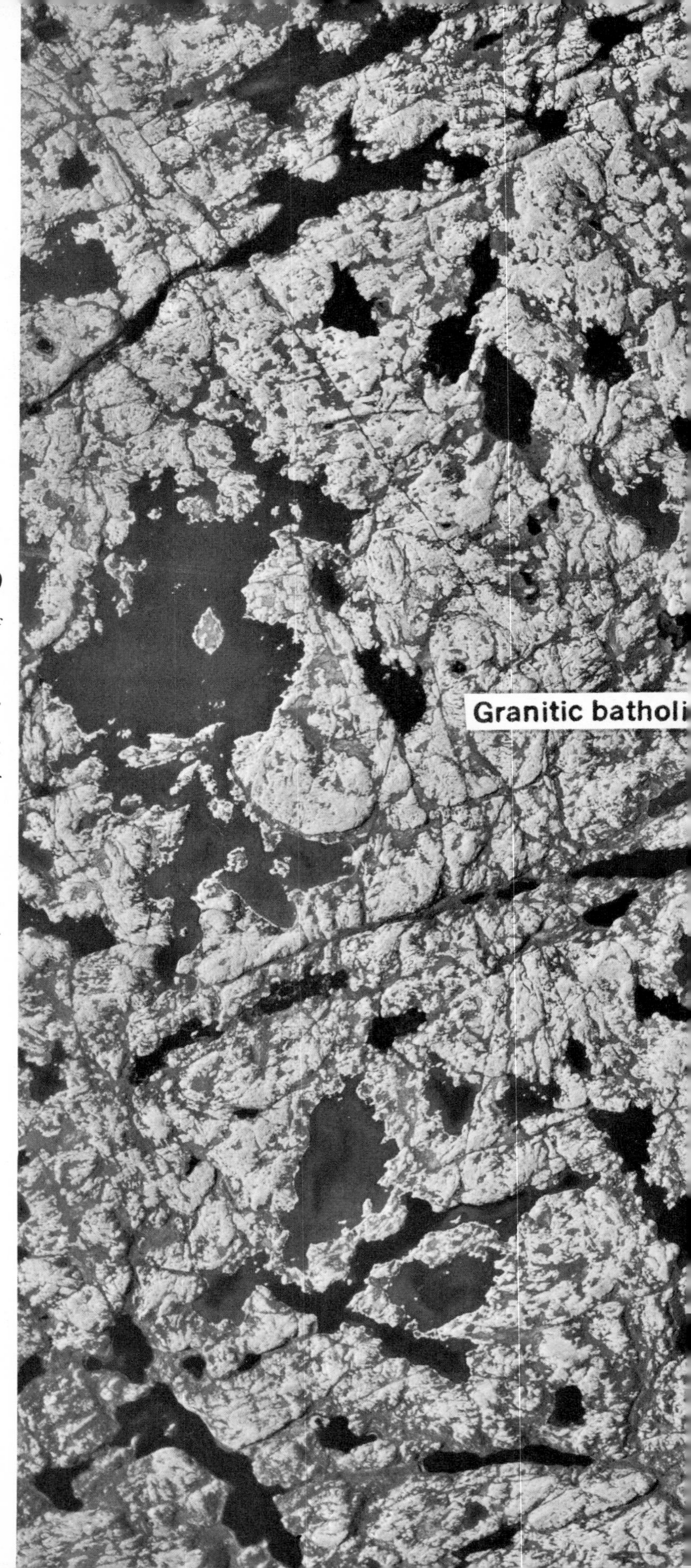

Plate 20

View vertically from the air of a part of the Canadian shield near Yellowknife, Northwest Territories, Canada. Glacier has planed off bedrock surface revealing marginal part of a batholith composed of granitic rock (left) and surrounding ancient metamorphic rocks (right). Small bodies of granitic rock, isolated in this view, probably coalesce with main part of batholith at depth. (Canadian Dept. Mines and Technical Surveys.)

vamp
Small bodies of granitic rocks
Lakes
Metamorphic rocks
Fractures
0
1
Mile

CHAPTER 20 *Plutonism and Metamorphism*

GLIMPSES WITHIN THE LITHOSPHERE FROM ITS SURFACE

Materials at the Earth's surface, varying from glacier ice to lava, reflect environments where temperature ranges from below 0°C to about 1,200°C. So far we have concentrated on the surface effects created near the extreme ends of this temperature range. Now we want to expand the discussion to include what happens not only down inside the lithosphere but also at temperatures lying between these extremes. To do this we must come to grips with the inside of the lithosphere.

The geologist's problems in dealing with inside of the lithosphere resemble those of a pediatrician trying to diagnose an infant's illness. External symptoms manifest the patient's internal condition, but the patient is unable to answer even simple but important questions such as: "Where does it hurt?" We need not pursue this analogy further to realize the problems and limitations confronting a geologist who attempts to analyze the lithosphere's inaccessible parts. Previously we discussed what could be done from the surface in order to comprehend what lies beneath it. We have already described indirect methods made possible by seismology and the direct method of observing what the Earth discharges at volcanoes. Here we continue with the direct method based on the observation of rocks.

We know that despite subsidence and burial by as much as 10km some strata have not been metamorphosed. Such thermally uninfluenced strata extend through wide areas. Many of them still remain essentially horizontal although others have been folded and faulted. Rocks showing the effects of high temperatures tend to be confined to elongate belts. Those manifesting the highest temperatures are localized bodies of intrusive igneous rock. Such bodies proclaim the former existence of magma, confirming what volcanoes lead us to expect.

In between these extremes lie metamorphic rocks. Some metamorphic rocks are confined to the margins of igneous bodies. These are easily explained; they obviously resulted from the effects of magma, which was hotter than its surroundings. The vast tracts where strata have been metamorphosed on a regional extent are the great puzzlers.

Strata that have been regionally metamorphosed pose questions for which there are no certain answers. Were such strata subjected only to average geothermal gradients but buried extremely deep within the lithosphere? Or were they not buried much deeper than nonmetamorphosed strata but subjected to unusually high temperatures accompanying great thermal events which affected the upper parts of the lithosphere? If from independent geologic evidence we could ascertain former depths of burial we could answer these questions. Unfortunately erosion which uncovers such tracts generally removes the geologic evidence for determining their former depths of burial.

In view of these generalizations about the effects of heat on rocks we would do well to commence our detailed study with a consideration of the results created by bodies of magma within the crust. Accordingly we first take up magma and its local thermal effects, and afterward treat thermal effects that are more widespread.

PLUTONISM

Plutonism is a general term referring to the behavior of magmas. This venerable word entered the geologic vocabulary about the beginning of the nineteenth century during a protracted controversy over the origins of granite and basalt. One group of geologists argued that granite and basalt crystallized from seawater; this group acquired the name *Neptunists*, in reference to the god Neptune, ruler of the seas. A second group, led by James Hutton and Sir James Hall (Chap. 4), attributed to granite an igneous origin within the depths of the Earth and to basalt a volcanic origin. Their emphasis on depth earned them the sobriquet *Plutonists,* in reference to Pluto, god of the nether regions. In the modern sense plutonism includes the movements, the internal and external reactions, and the problems of emplacement of magmas.

Shapes and names of plutons. The collective term ***pluton*** designates *any body of intrusive igneous rock,* regardless of shape, size, composition, or number of episodes of intrusion represented. Plutons are classified and named on the dual basis of shape and relationships of their surfaces of contact to the layers of adjacent rocks. The kind of igneous rock composing them and kind of rocks into which they intrude are not involved. Shapes vary from tabular through lenticular to cylindrical and irregular. The surfaces of contact of an intrusive body may be parallel to the layers or other boundaries within the intruded rocks or may cut across these layers at various angles. *Contact surfaces of plutons that are parallel with the layers or other boundaries within the intruded rocks* are **concordant contacts;** *those not parallel with the layers or other boundaries within the intruded rocks* are ***discordant contacts*** (Fig. 20-1).

Fig. 20-1. Irregular discordant contact of a granitic stock (*left*) and layered gneisses dipping gently left. Gneisses resulted from alteration of sedimentary strata. Irregular dike, an offshoot from the stock, crosses layers of gneiss at low angle (*lower right, just above circular road sign*). Telemark area, near Kristiansand, Norway. (R. V. Dietrich.)

Tabular plutons. The common shape of plutons is tabular; the igneous rock occupies a space between walls that are essentially plane and parallel to each other. The magma forces its way into position by spreading apart the containing walls. Thickness of tabular plutons ranges from less than 1cm to 100m or more; their longest dimension can be hundreds of kilometers. According to their contact relationships tabular intrusives are subdivided into *dikes* and *sills.*

Dikes. *A tabular pluton having discordant surfaces of contact* is a ***dike.*** A dike resembles a *vein* (Fig. 23-15) but differs in composition and texture. Minerals of dike rock have crystallized from magma so viscous that crystals could grow throughout, irrespective of their positions with respect to the walls. By contrast, the minerals composing a vein have crystallized from fluids with viscosity so low that the crystals could grow only from the walls inward toward the center.

The pattern of dikes reflects the geometry of the fractures in bedrock. Radiating patterns typically surround volcanic necks, as at Shiprock (Fig. 19-24) and Spanish Peaks (Fig. 20-2). Such patterns suggest that the fractures originated from stresses created by upward deforming forces concentrated near a single point. A large body of magma that domes the rocks above it can create radial fractures.

Groups of associated dikes constitute a *dike swarm.* Some swarms occupy parallel fractures many of which are related to large faults. Parallel fractures are especially abundant in the vicinity of large-scale strike-slip dislocations (Fig. 17-21).

At Spanish Peaks, several sets of nearly parallel dike swarms intersect a group of radial dikes. The dike swarms and radial dikes are younger than the two plutons which form the peaks. Dike swarms, conspicuous in the western islands of Scotland and in the Appalachians (Fig. 20-3), are commonly related to plateau basalts. Dikes have fed the fissure eruptions in the Columbia Plateau and elsewhere. Moreover dikes have supplied many eruptions from central vents, as in Hawaii.

The world's largest known dike is in Rhodesia. Its title, Great Dike, is amply justified by its length of exposure, nearly 300 miles, and by its average width, 5 miles.

Some dikes are *composite;* they consist of more than one generation of igneous rock. Evidently before the first magma had solidified completely, a second magma reopened the fissure, forced apart the walls, and occupied the central space.

Sills. *A tabular pluton having concordant contact surfaces* is a ***sill*** (Fig. 20-4). The name *sill* derives from northern England, where an analogy was made between the foot of a doorframe and the horizontal Whin sill. This tabular intrusive, nearly horizontal and about 90 feet thick, extends for 80 miles. The modern definition of a sill is based on concordant contacts rather than on horizontal position. A concordant pluton inclosed by vertical strata is considered to be a sill as much as one inclosed by horizontal strata. Like dikes, sills can be composite.

The remarkably uniform thicknesses of many sills, particularly those with basaltic compositions, imply that the magma was extremely fluid.

Most plutons are readily identified by their shapes and compositions, and by the relationships of their contacts to the layers of surrounding rocks. Sills are an exception because they resemble buried former lava flows. Superficially both appear the same: a concordant sheet of igneous rock lies within a sequence of other layered rocks. A sill is younger to some unknown extent than the strata it has intruded, whereas a buried extrusive sheet belongs within the stratigraphic succession where it occurs. Like sedimentary strata, extrusive sheets are younger than the layers below and older than those above. Correct interpretation of geologic history obviously depends on accurate diagnosis of the origin of the concordant sheets of igneous rock.

From what we have read, we can compare and contrast the attributes of sills and buried extrusive sheets (Figs. 20-5, 20-6; Table 20-1). The upper contact of the sheets yields diagnostic evidence. But even without seeing the overlying strata, we can recognize some former lava flows by their distinctive structures (Fig. 19-23). Generally only extrusive rocks are conspicuously vesicular (Fig. 20-7); intrusive rocks characteristically lack vesicles.

Lenticular plutons. From what we have learned of lavas we can infer that magmas with intermediate silica content flow less readily than those having low silica content. As the viscosity of the magma increases, lateral flow between the containing walls becomes progressively more difficult. With increasing viscosity the magma must exert more effort to flow between the walls than to bend the strata and enter the space thus formed.

Various shapes of plutons result from the ways in which the strata bend. Of particular interest here is a ***laccolith*** (Gr. "cistern rock"), *a concordant*

Fig. 20-2. Dikes and irregular plutons, Spanish Peaks, Colorado.

A. View toward West Spanish Peak. Dikes, being more resistant to erosion than are the strata they cut, resemble great artificial walls. (G. W. Stose, U. S. Geol. Survey.)

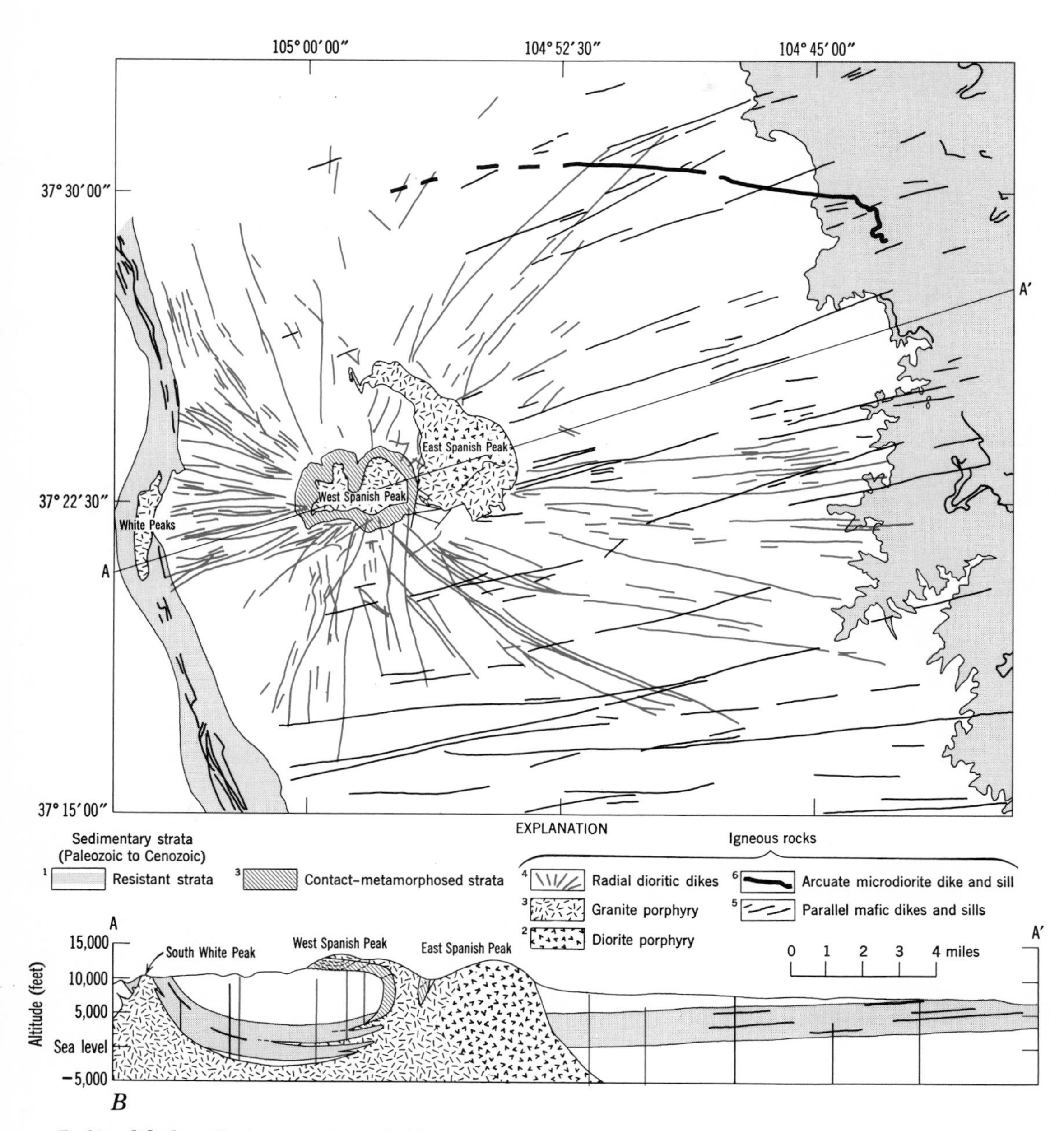

B. **Simplified geologic map. Several dike swarms intersect dikes radiating outward from the larger plutons. Five episodes of intrusion are represented. Rock units are numbered in order of decreasing age. Granite porphyry and contact-metamorphosed strata (both numbered 3) are the same age. The diorite porphyry (No. 2) created insignificant contact metamorphism. (After R. B. Johnson, 1961.)**

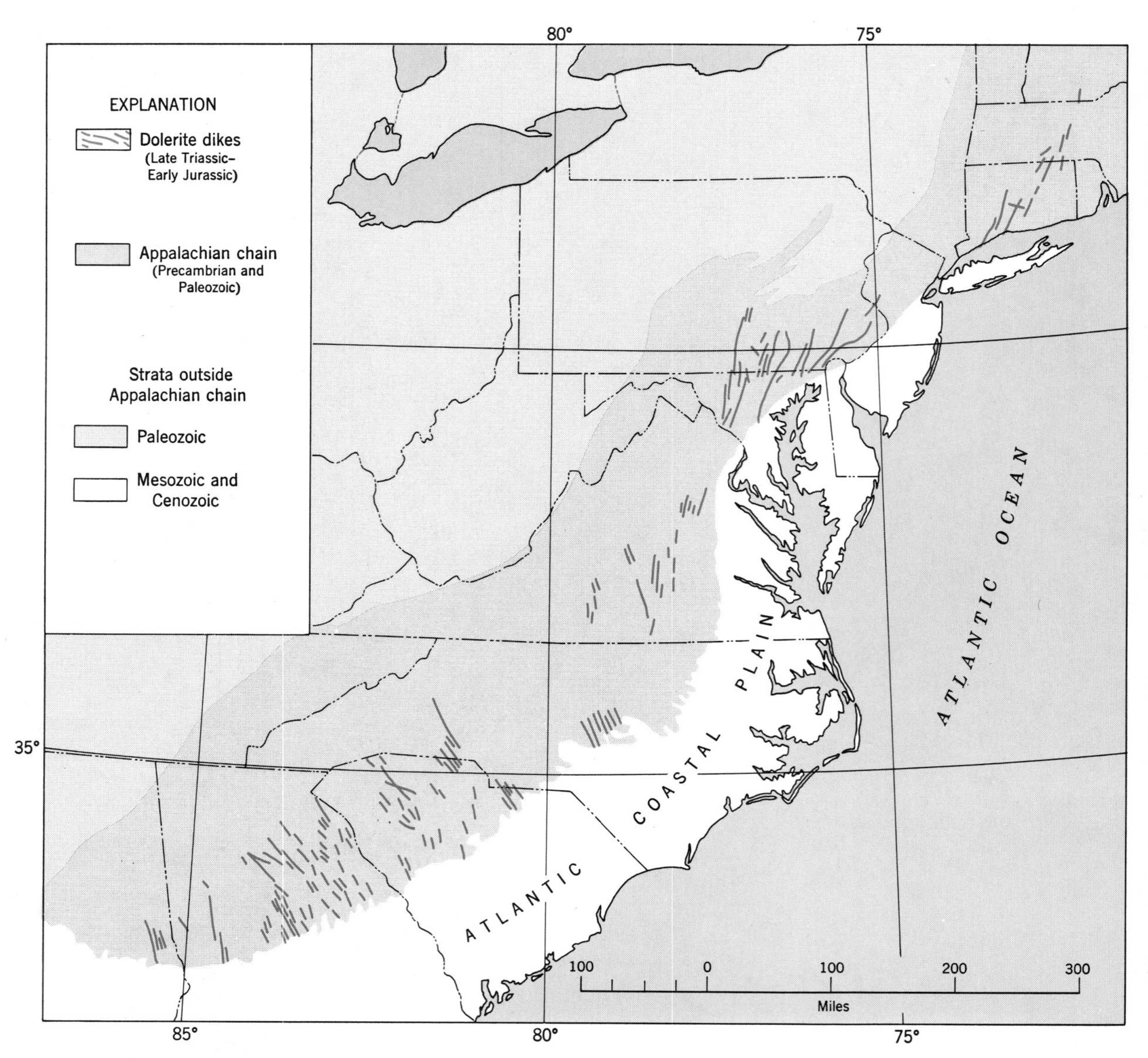

Fig. 20-3. Map showing locations of post-Triassic, pre-Cretaceous dikes in eastern United States. (P. B. King, U. S. Geol. Survey.)

lenticular pluton, circular or elliptical in plan, having an essentially plane floor and a distinctly domed roof. The diameters of large laccoliths can be as much as several miles and the thickest part of the lens can be thousands of feet. In the La Sal Mountains, Utah and Highwood Mountains, Montana, groups of laccoliths make sizable mountain masses.

In the Henry Mountains, southern Utah, where laccoliths were first recognized and studied, groups of them are arranged around larger intrusive bodies like chicks around hens. Because the floors of laccoliths are almost never fully exposed, the relationships of each lenticular body to its supply conduit have remained conjectural. Early students of laccoliths supposed that the magma reamed out a narrow, vertical cylindrical hole and migrated upward to a critical level, where an equilibrium existed among ability to extend the hole, pressure to bend the roof rocks, and pressure to spread laterally. No such inferred narrow vertical conduit (Fig. 20-8) was actually observed. In a modern restudy of the laccoliths in the Henry Mountains, lateral connections were found between laccoliths and the larger intrusive bodies. This suggests that the magma spreading laterally from some of the large parent bodies wedged apart the nearly horizontal sedimentary

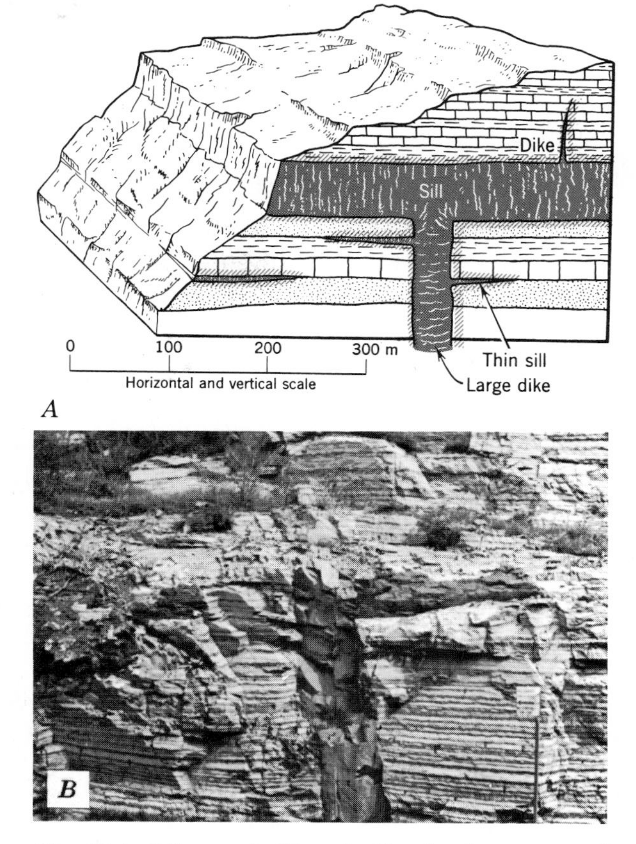

Fig. 20-4. Dikes and sills. A. Sketch of exposure of dike that feeds a large sill (*above*) and several small sills (*below*). B. Dike and sill cutting Ordovician limestones, west side of Mount Royal, Montreal, Canada. (J. E. Sanders.)

Fig. 20-5. Buried extrusive sheet and sill. A. Basal contact beneath thick basaltic extrusive sheet. Purplish white zone, about 2m thick, resulted from effects of heat and hot solutions on Triassic siltstones (*below*), whose normal color is maroon. Strata covering basalt sheet have been eroded from this locality. Gerrish Mountain, Nova Scotia. (Stan Frank.) B. When this thick sill intruded the nearly horizontal strata of fine-grained, dark-colored limestone, it bleached the limestone white within contact zones both above and below. Mount Gould, Glacier National Park, Montana. (Spence Air Photos.)

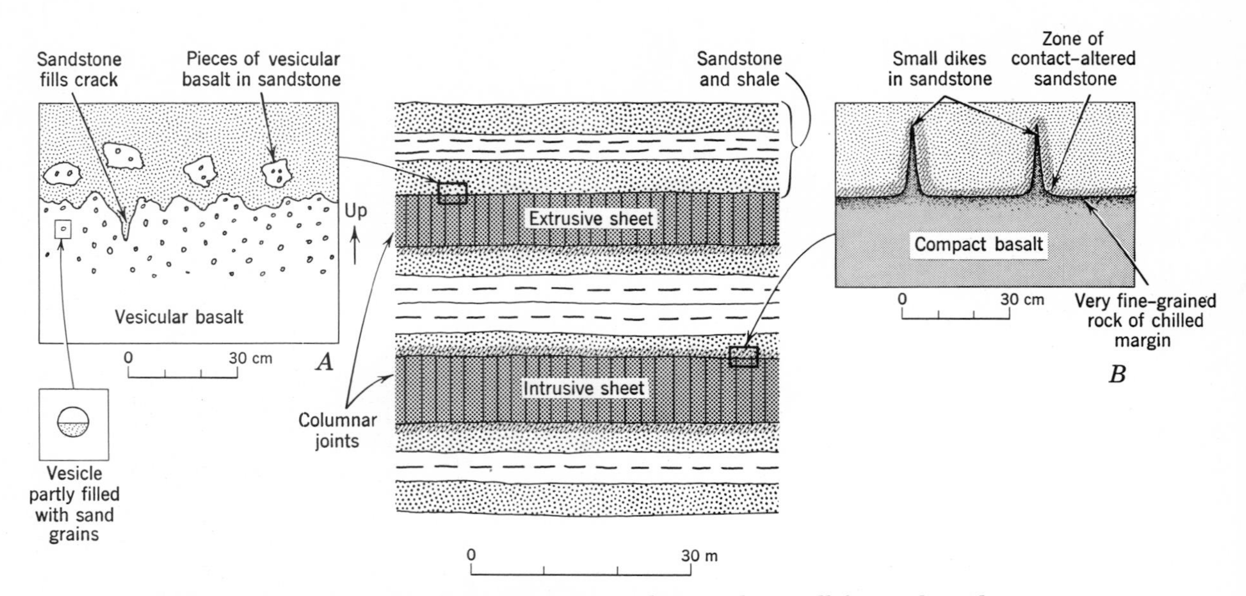

Fig. 20-6. Critical evidence, found at the upper contacts, distinguishes a sill from a buried extrusive sheet. Further explanation in text and in Table 20-1.

TABLE 20-1. SIMILARITIES AND DIFFERENCES BETWEEN A SILL AND A BURIED SHEET OF EXTRUSIVE IGNEOUS ROCK

Similarities	Differences
1. A tabular body of igneous rock having concordant contacts occurs within a sequence of layered rocks.	1. The extrusive igneous rock may show ropy, vesicular, blocky, or pillow structures. These characterize lava flows but are not found in intrusive igneous rocks. (A possible exception: a few vesicles occur locally in some intrusive igneous rocks.)
2. Columnar joints perpendicular to the surface of contact may be present.	2. The basal contact of an extrusive sheet may be irregular if any relief existed on the land surface buried by the lava. The basal contact of a sill generally is smoother, but exceptions exist; likewise lava can flow over a smooth depositional surface.
3. The igneous rock is fine grained at the basal margin and becomes coarser grained away from it.	3. Near the upper contact of both sills and buried extrusive sheets the igneous rock may be aphanitic or glassy; it becomes coarser grained away from the marginal zone. The rocks overlying the intrusive sheet may show bleaching and other thermal effects created by the fluid magma. The intensity of these effects may exceed those of the basal contact. Rocks overlying an extrusive sheet generally do not show thermal effects from the lava that cooled to create the extrusive rock.
4. The rock underlying the sheet of igneous rock may show effects of thermal alteration and bleaching.	4. The material of the covering layer may fill in cracks, crevices, or vesicles in the extrusive igneous rock. In contrast the intrusive rock may extend into cracks within the overlying rocks.
5. If the material below is sand, then sandstone dikes may extend upward into the igneous rock.	

strata. Thus magma from a common source is thought to have supplied not only each of the large bodies but also each of its satellite laccoliths. As the magma became sluggish from loss of heat, it domed the overlying layers and abruptly stopped advancing.

Ordinarily the feeder conduit of a laccolith is not exposed (Fig. 20-8, *C*). Laccoliths isolated from other igneous bodies may have been fed from below, but the existence of a narrow, vertical supply conduit has not been proved despite its generally assumed existence. The deforming force of the magma caused the roofs of some laccoliths not only to bend but also to fracture. Magma entered the fractures, thus creating an irregular surface of contact that is locally discordant.

Other concordant lenticular plutons have forced apart the strata in the axial parts of folds. Some other lenticular plutons are sill-like and locally concordant on a small scale, but overall are discordant. Such a pluton is shaped like a saucer, being planar in the center and turned up around its circumference. The intrusive body that forms the impressive Palisades along the west side of the Hudson River from New York City northward for 35 miles is a saucerlike pluton. The Palisades sheet is nearly 1,000 feet thick and its length originally exceeded 100 miles. Near New York City the sheet resembles a sill, but both north and south it becomes discordant.

Cylindrical and irregular plutons. Nontabular discordant plutons (Fig. 20-1) vary in diameter from cylindrical volcanic necks a few tens or hundreds of feet to irregular bodies measuring tens or hundreds of miles across. According to their size and general form, cylindrical and irregular plutons are classified into *stocks* and *batholiths*.

Stocks. *A pluton, roughly circular or elliptical in plan, with an exposed area of less than 40 square miles (about 100km²)* is a ***stock.*** Characteristically, the surfaces of contact of a stock are curved, dip steeply, and are discordant, though the top may be concordant locally. The name was suggested by the approximately cylindrical form, which resembles that of a tree trunk. Mining operations have shown that the diameters of many stocks increase steadily downward and that others connect with and are merely dome-shaped protuberances of larger intrusive masses lying still deeper (Fig. 20-9). Some stocks may have connected upward with volcanic necks

Fig. 20-7. Upper surfaces of some extrusive sheets are rough and vesicular. Hawaiians call such rock *aa* (aȟ aȟ). Island of Hawaii (compare Fig. 19-5). (G. A. MacDonald.)

serving as vents of volcanoes. The term *neck* is applied only when evidence proves that the igneous body solidified from magma that formerly fed a volcano.

Batholiths. The giant among plutons is the ***batholith*** (Gr. "deep rock"), *a large pluton having an exposed area of more than 40 square miles (about 100km²)*. The distinction between stocks and batholiths based on known area of rock exposed is an arbitrary one, reflecting not only intrinsic differences in size but also accidents of erosion. The exposed area of a pluton might be doubled if erosion had cut another mile or so deeper into it (Pl. 20).

The Boulder batholith in western Montana is exposed through an area approximately 60 by 20 miles (Fig. 20-9, *A*). The Idaho batholith in central Idaho spreads through an exposed area of 16,000 square miles, nearly twice the size of the State of Massachusetts.

Although the Idaho batholith is the largest known in the conterminous United States, it is dwarfed by the Coast Range batholith in western Canada and southern Alaska, which is more than 1,000 miles long and 25 to more than 100 miles wide. Batholiths of comparable size are visible along the western part of South America. Because of their size and complexity, batholiths raise special questions that are best answered in conjunction with our later discussion of the origin of magmas.

Inferred reactions within plutons. In the simplest

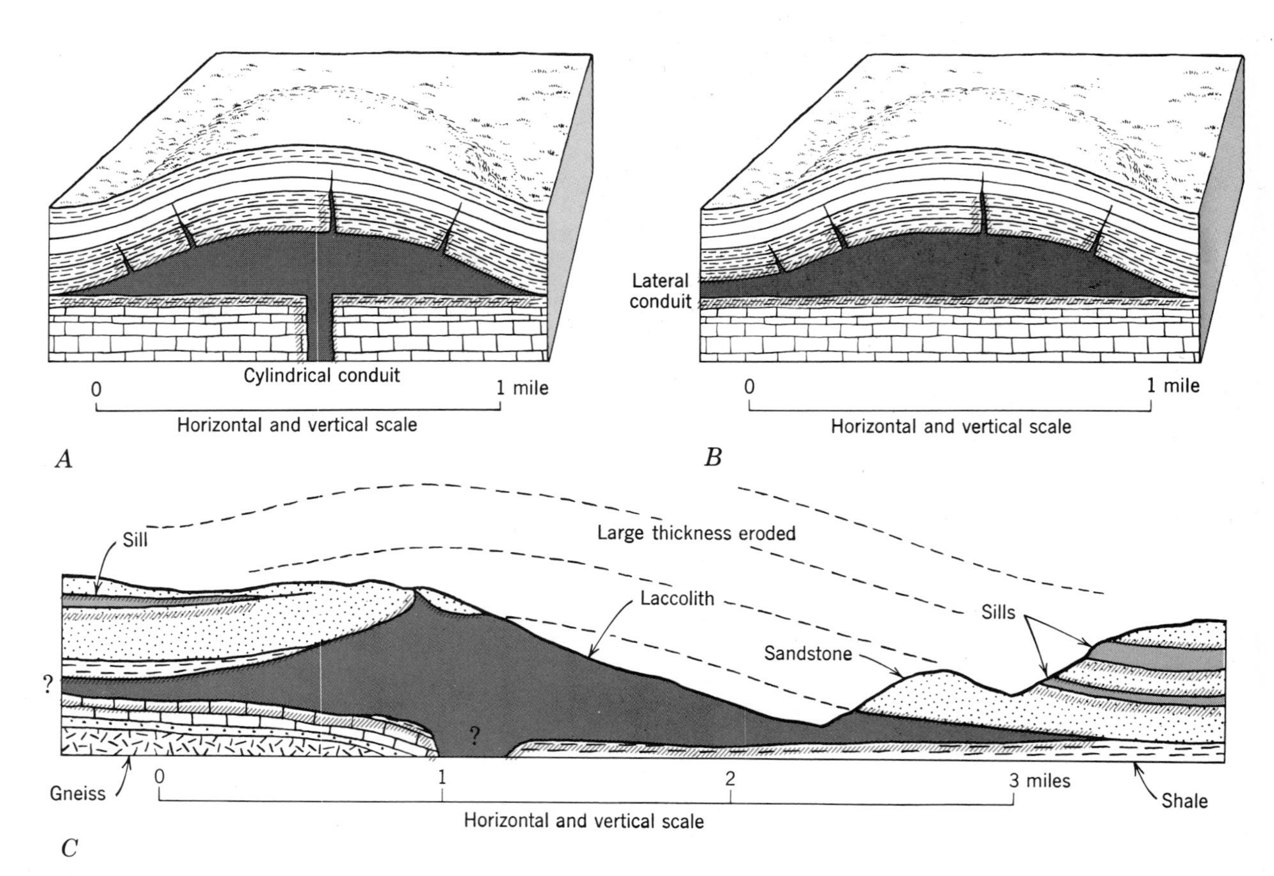

Fig. 20-8. Laccoliths. *A.* Classical laccolith with central cylindrical conduit. *B.* Laccolith with lateral feeder. *C.* Section through large laccolith and thin sills, near Breckenridge, Colorado. Uncertainties concerning location of feeder conduit are shown by including both a sill and a central cylindrical conduit, with question marks alongside each. (Folio 48, U. S. Geol. Survey.)

case, one charge of magma arrives in a magma chamber. The magma cools and solidifies to form a pluton consisting of only a single kind of igneous rock. The final product is nearly homogeneous throughout, though it may be finer grained at the margins if the magma cools against rocks having a temperature lower than its own. In more complicated cases, such as composite dikes, several charges of magma arrive and several kinds of igneous rock are created. Nearly all large plutons are composite.

Magmatic differentiation. How can we explain the multiciplicity of igneous rocks? Does a distinct magma type come from within the Earth to form each of the hundreds of known varieties of igneous rocks? Or are a few magma types capable of reacting and changing so that they generate many kinds of rocks? *The processes within magma by which one magma generates more than one variety of igneous rock* are collectively named ***magmatic differentiation.*** Some kinds of differentiation leave convincing evidence; other kinds must be inferred from indirect evidence.

An example of differentiation for which we have definite evidence is the segregation of early formed mineral crystals according to the relationships of their densities to that of the magma. Minerals denser than the magma sink; those less dense than the magma float. Given enough time and, unless restrained by upward-moving convection currents, crystals denser than the magma will sink and accumulate on the floor of the magma chamber. Crystals that have sunk not only form layers near the bottom of the chamber but also change the composition of the remaining magma by removing from the liquid

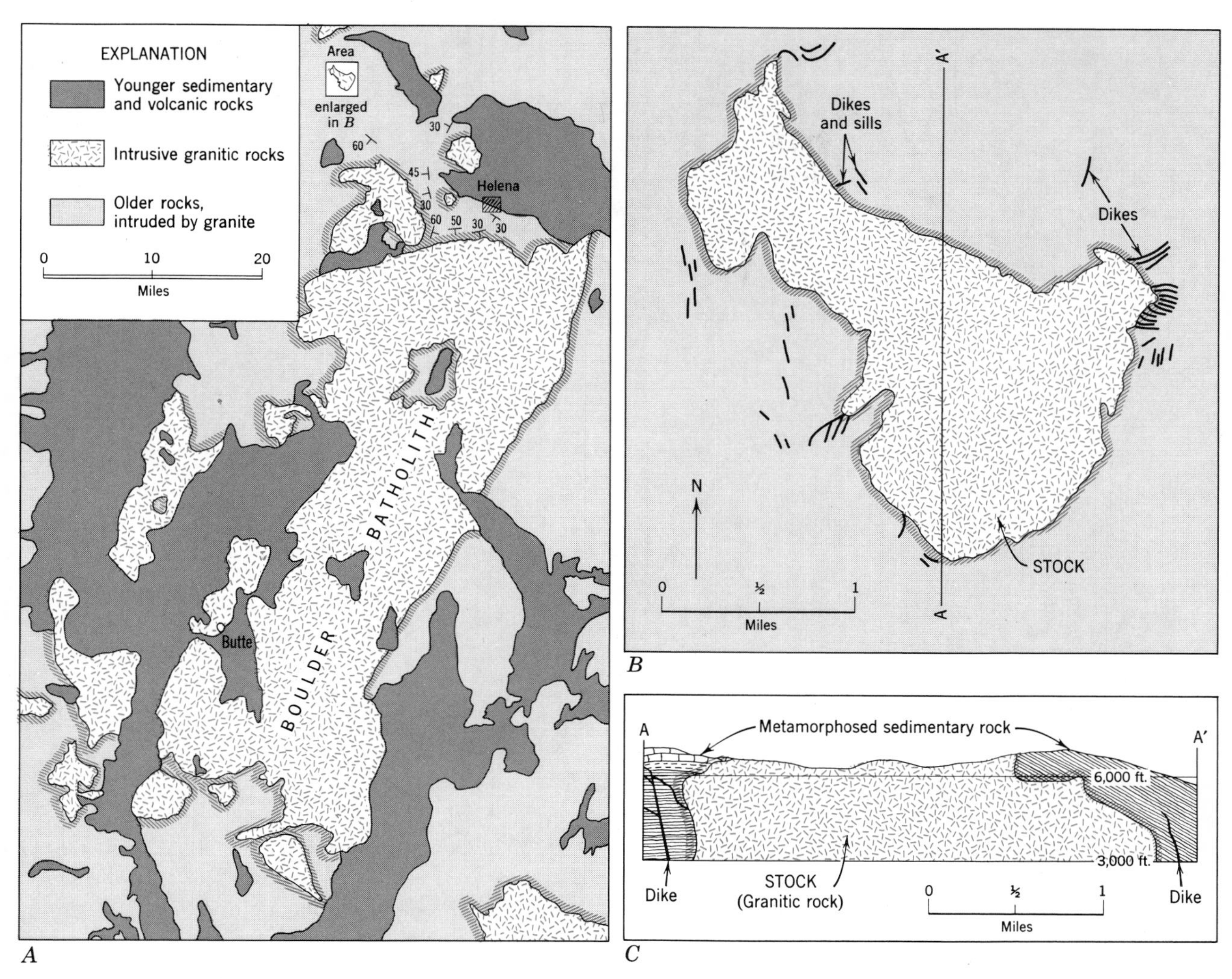

Fig. 20-9. Maps and sections of discordant granitic plutons, northern Rocky Mountains, Montana. *A.* Partly unroofed Boulder batholith and small stocks that probably connect with the batholith at depth. (After Geologic Map of Montana, U. S. Geol. Survey, 1955.) *B.* Marysville stock, a pluton that has been closely studied because valuable mineral deposits surround it. *C.* Section along line A-A′ of map in *B.* Underground relationships based on data from mine shafts and tunnels. (*B* and *C* after Joseph Barrell, 1907.)

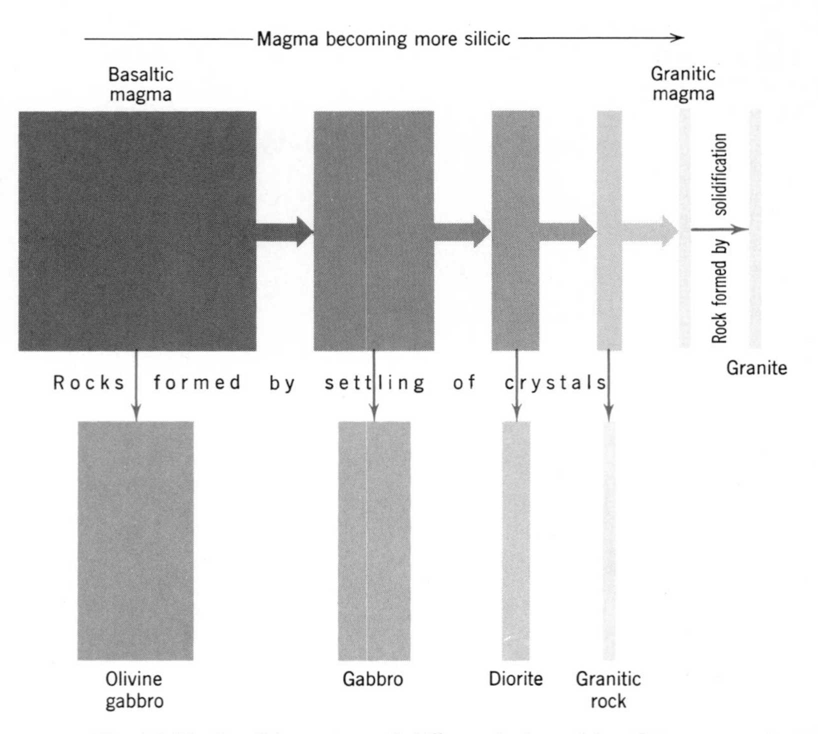

Fig. 20-10. Possible course of differentiation of basaltic magma into granitic magma by settling of crystals, based on scheme of N. L. Bowen. Further explanation in text; rock names in App. C.

the elements of which they are composed. By the settling of its crystals a basaltic magma could manufacture five parts olivine gabbro, three parts gabbro, one part diorite, and one part granitic rock (Fig. 20-10). The layer of olivine and augite crystals, 8 feet thick, that occurs locally within the Palisades sheet about 50 feet above its base has been attributed to crystal settling. In other examples crystals not only of these minerals but also of calcic plagioclase seem to have settled to the bottoms of plutons.

Indirect evidence suggests that differentiation of other kinds also takes place. One possibility is the separation of magma into two or more liquid fractions, as cream rises through standing milk. Each fraction could solidify to form a distinctive igneous rock or could supply a volcano with a variety of lavas. Other changes result from reaction between a magma and its wall rocks.

Joints resulting from tensile stresses created during cooling. Observations of lavas indicate that the molten mass shrinks as it cools just as a mass of water-saturated mud shrinks as the water evaporates. In both cases the results are quite similar. Throughout a cooling tabular mass of intrusive or extrusive igneous rock many centers of contraction develop. Along lines that join any pair of centers of contraction tensile stresses reach a maximum. These stresses create joints perpendicular to the lines (Fig. 17-1). Cooling joints typically form regular sets to outline perfect polygons (Fig. 20-11). Because maximum shrinkage can occur in a plane parallel with the contacts of the sheet, both the joints and the polygons they outline are perpendicular to the contacts. Subsidiary joints develop parallel with the contact surfaces. The intersecting joints split the rock into numerous blocks of various sizes. *Joints that split rocks into long prisms, or columns* are ***columnar joints.***

Geologic dates of plutons. Plutons can be dated both isotopically and through study of their contact relationships. The latter method resembles the one we have already used to determine the geologic date of deformation. Significant dates are (1) date of emplacement, (2) date of unroofing by erosion, and (3) date of covering.

Date of emplacement. The date of emplacement

of a pluton can be determined directly from measurements of radioactive isotopes. If such measurement is not possible, we can fix the lower limit of its age within the local geologic sequence by finding the youngest rock unit penetrated by the pluton. The upper limit is defined by date of unroofing or date of covering.

Date of unroofing. When a pluton is unroofed its particles become available for distribution in surficial sediments. The oldest strata containing particles from the pluton place an upper limit on the pluton's age. For example, the Dartmoor granite, now uncovered at the surface in southwestern England, cuts strata of Devonian and Lower Carboniferous age. Distinctive mineral particles from the granite first appear in strata of Cretaceous age. If such mineral evidence were not present, the age bracket for the Dartmoor granite would be post-lower Carboniferous and pre-Recent. With the mineral evidence we can narrow the bracket to post-Lower Carboniferous and pre-Cretaceous.

Date of covering. Where mineral evidence is lacking, the upper limit of a pluton's emplacement can be bracketed by finding the age of the covering strata. But can we distinguish covering strata, which are younger than the pluton, from the magma chamber's roof rocks, which are older than the pluton? If the critical evidence is present, we certainly can. The procedure involves application of some of the same principles employed to distinguish sills from buried extrusive sheets.

From the geologic evidence we know the granitic batholith near Eastport, Maine was emplaced between Late Silurian and Late Devonian time. The granite cuts Silurian formations and upon its eroded surface the Upper Devonian Perry Formation was deposited.

The remaining facets of plutonism require more knowledge of metamorphism. Hence, before we can complete our discussion of plutonism we must turn to metamorphism.

METAMORPHISM

The products of metamorphism dramatically illustrate the principle that mineral assemblages are governed by environment. This principle is best illustrated by studying the effects accompanying the emplacement of magmas.

Contact metamorphism. A body of magma inserted into cooler strata provides us with an ideal

Fig. 20-11. Structures resulting from stresses created by the shrinking of mud, compared with structures in igneous rocks. *A.* Large columns outlined by cracks when clay shrank after rainwater evaporated from a large puddle. A much smaller set of shrinkage cracks has broken a thin surface layer into a series of tiny polygonal plates. Dome Valley, Arizona. (Fred Taylor from Philip Gendreau.) *B.* Enormous columns outlined by polygonal joints. Devil's Tower, Wyoming. (Philip Gendreau.)

situation for studying metamorphism. Nearly all mineralogic changes result from the sharing of heat and hot fluids between magma and invaded rocks. Not many effects depend on great load pressure or shearing stresses. Ordinarily the changes are limited to a small area; hence we need not travel far to observe them.

Because the zone of altered rocks surrounds the pluton, it merits the designation *contact-metamorphic aureole.* Changes within the zone can be considerable near the pluton, but invariably they fade away in all directions and become imperceptible with distance from the contact. Some plutons lack contact-metamorphic aureoles.

Dimensions of contact-metamorphic aureoles. Contact-metamorphic aureoles are of many sizes. Thickness perpendicular to the contact varies from fractions of a centimeter to a kilometer or more. Thick aureoles surround granitic stocks and batholiths; thin ones abut basalt dikes and sills. Thickness depends on the supply of heat and hot fluids, the ease with which these can spread through the invaded rocks, and sensitivity to metamorphism in the rocks invaded. Width of the contact zone on the ground surface depends on its thickness and attitude. Wide zones need not be thick; width can result from the intersection of the ground surface with a thin zone having gentle dip.

Varied susceptibility of materials. Where a sequence of strata of many kinds has been intruded by a pluton, say a stock, an opportunity arises to study how various materials react to the same set of conditions. Favorable exposures enable us to trace layers from localities where they have not been metamorphosed to other localities where the rocks display metamorphic changes of varying intensities. Bituminous coal becomes anthracite, pure limestones are transformed to marbles, sandstones and cherts to quartzites, shales to hornfelses or mica schists, and so forth.

The change of limestone to marble is particularly striking where ordinary dark gray or black, fine-grained limestone has been converted into glistening white, coarse-grained marble (Fig. C-12, *B*). Because pure limestones contain only $CaCO_3$, the metamorphism involves only recrystallization. But pure limestones represent only a small fraction of the total compositional spectrum of carbonate rocks. In addition to pure limestones some carbonate rocks are dolostones, containing large proportions of the mineral dolomite, $CaMg(CO_3)_2$. Other carbonate rocks contain not only one or both of these carbonate minerals but also mixtures of noncarbonate minerals including quartz, SiO_2, and clay minerals, chiefly aluminum silicates. Figure 3-23 points out the reaction between calcite and quartz that yields wollastonite, $Ca_2\ (Si_2O_6)$. The presence of dolomite changes the end product from wollastonite to the pyroxene *diopside,* $CaMg\ (Si_2O_6)$.

If diopside forms near the pluton it characteristically grades outward into a rock containing the white amphibole *tremolite,* $Ca_2Mg_5\ (Si_4O_{11})_2\ (OH)_2$. Still farther out the rocks contain *talc,* $Mg_3(Si_4O_{10})(OH)_2$. Hydrogen aside, diopside and tremolite contain the same elements. However, where the temperature is higher, diopside is the product; where it is lower, tremolite forms.

Garnets, complex aluminum silicates, appear in contact zones where limestones containing both clay minerals and quartz are subjected to high temperatures (Fig. 20-12). Some garnets are Ca-Al silicates and others are Ca-Fe-Al silicates. The iron-bearing garnets can derive some iron from carbonate minerals and clays, but these garnets may be so numerous that we must conclude they received iron from the magma. Heat from magma can recrystallize soft shales and convert them into *hornfels* (App. C). In some contact zones, aluminous clay minerals have been transformed into *andalusite,* $Al_2(SiO_4)O$.

During simple thermal metamorphism in contact zones, pure quartz sandstones and cherts undergo at most only textural changes. The particles of SiO_2 enlarge and interlock to form quartzite (Fig. C-12, *A*).

Not only sedimentary strata but also igneous rocks, metamorphic rocks, and pyroclastic rocks occur in contact-metamorphic aureoles. Granitic rocks change little. Evidently this stability results from similarity of the contact environment with that in which granites originate. Granites do not change when reexposed to an environment comparable to that in which they originally crystallized.

Near some plutons, basalts alter conspicuously. Pyroxenes in basalt, which are stable at high temperatures, can persist at ordinary temperatures. This persistence results from the slow reaction rates of silicate minerals. Where subjected to magmatic heat, however, the minerals are stimulated to adjust to new conditions. In zones where adjustments occur, pyroxenes alter to hornblende, epidote, and chlorite,

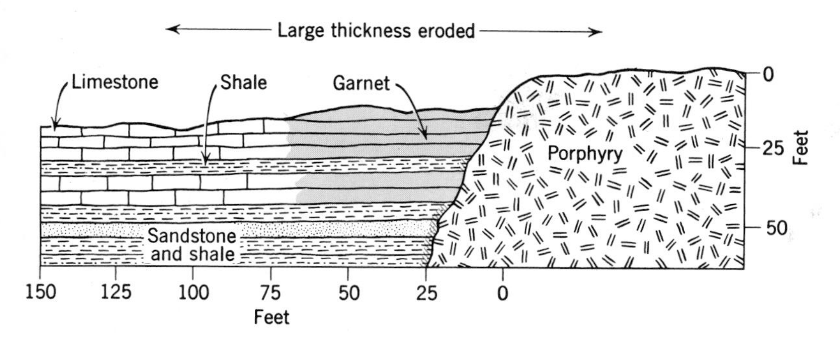

Fig. 20-12. Contact-metamorphic aureole developed in sedimentary strata surrounding a body of porphyry. Near Breckenridge, Colorado. Near the contact, sandstones and shales were merely hardened, but limestone strata as far as 75 feet away from the contact contain metamorphic garnets.

respectively, with decreasing temperature. On the Island of Skye in Scotland, gabbro plutons invade extrusive basalts. Adjacent to the contacts, the basalt has recrystallized to hornfels of pyroxene and plagioclase. Notice that these are the chief minerals of basalt; the only effect of the heat has been recrystallization. With distance from the contact, however, new minerals appear, creating a succession of rocks outward from amphibolite to epidote amphibolite and to chlorite schist. Beyond the chlorite schist stands unaltered basalt.

The foregoing paragraphs describe contact metamorphism of various rocks invaded by individual plutons whose igneous rocks derived from only one kind of magma (but not necessarily the same kind in all cases). In a few situations a single rock is so widespread that it has been intruded by more than one kind of magma and we are enabled to compare the effects of various magmas on the same materials. In northern Minnesota parts of a thick succession of slates have been transformed into biotite schists in aureoles hundreds of yards to several miles wide surrounding granitic plutons. Where the same slates abut a gabbroic pluton the contact-metamorphic aureoles are less than 500 feet thick. This pluton's most intense metamorphism has generated hornfelses; but such rocks are restricted to zones not more than 5 feet thick.

Similarly in the Vermilion district, Minnesota the Keewatin Greenstone, a metamorphosed basalt containing abundant chlorite and other green minerals, acts as host to both granitic and gabbroic plutons. Near the granite the most intense alteration of the greenstone has created a black rock rich in hornblende. Around the gabbroic pluton, hornblende likewise appears but not adjacent to the contact. Between the gabbro and the rock containing hornblende lies a zone containing pyroxene. Because pyroxene is adjacent to the gabbro but absent near the granite we infer that the gabbroic magma was hotter than the granitic magma.

Both greenstone and slate are metamorphic rocks that originate at rather low temperatures. Gneisses and schists, which reflect higher temperatures, are resistant to contact metamorphism. Presumably the reason lies in the similarity of their environment of origin to that of the contact zones, as we concluded was the case with granite.

Fine-grained pyroclastic rocks become coarser grained in contact-metamorphic aureoles. In these rocks the mineral changes vary greatly because the composition of pyroclastic material can range from silicic to mafic.

Concept of metamorphic facies. Various assemblages of minerals occur in metamorphic rocks whose bulk chemical compositions are practically identical. Contemplation of this remarkable fact gave birth to the idea that identical original materials could reach equilibrium during metamorphism under various environmental conditions. A corollary was that each set of environmental conditions would manufacture characteristic assemblages of minerals. Later studies disclosed that despite variations in original materials, say from rocks having the composition of basalt to those having the composition of slate, certain groups of mineral assemblages are recurrently associated. Therefore, students of metamorphic rocks have formulated the concept of ***metamorphic facies,***

TABLE 20-2. METAMORPHIC FACIES AND CHIEF MINERALS IN ROCKS HAVING THE COMPOSITION OF BASALT.[a]

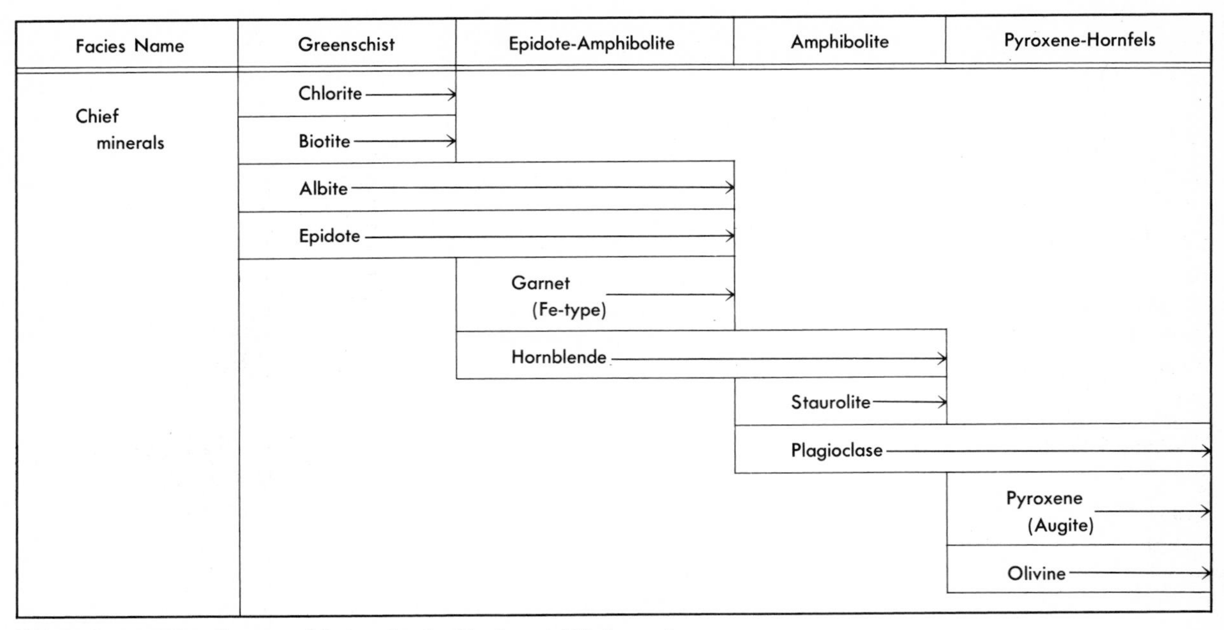

Facies Name	Greenschist	Epidote-Amphibolite	Amphibolite	Pyroxene-Hornfels
Chief minerals	Chlorite →			
	Biotite →			
	Albite →	→		
	Epidote →	→		
		Garnet (Fe-type) →		
		Hornblende →	→	
			Staurolite →	
			Plagioclase →	→
				Pyroxene (Augite) →
				Olivine →

[a] Eskola, Pentti, 1921, The mineral facies of rocks: Norsk geol. Tidsskr., v. 6.

defined as *rocks that reached equilibrium during metamorphism within a single range of environmental conditions.*

The original scheme of metamorphic facies based on basaltic rocks as it appeared in 1921 is shown in Table 20-2. Notice the progression from silicate minerals with sheet structures (Table 3-1) in the greenschist facies to those with chain structures in the pyroxene-hornfels facies.

The localities where a given mineral first appears, such as garnet or staurolite, in metamorphosed basaltic rocks, mark a boundary that can be mapped in the field. All points along a line connecting such occurrences are thought to represent some kind of environmental interface along which conditions are equal. Lines based on first occurrences of minerals in metamorphic rocks are *isograds.*

In theory, tables comparable to Table 20-2 could be prepared for all varieties of rock materials. Experience has shown that most metamorphic rocks fit into four such tables, one for basalt and the other three including compositions represented by aluminum-rich shales, impure carbonate rocks, and granitic rocks and arkoses. The results become complicated because a mineral such as chlorite, which is restricted to the greenschist facies of materials having the composition of basalt, can extend into facies of higher grade in materials having the composition of shale.

We need not delve into all the details of these schemes because the facies classification and names of metamorphic rocks not only are complex but also are in a state of flux. In the original facies analysis, composition was assumed to have been constant and temperature to have been the only variable. Laboratory results have verified the importance of temperature but have also shown that volatiles, particularly water, and total pressure rival the significance of temperature. Moreover, some experiments have raised the suspicion that the constancy of composition assumed in the original analysis may not be valid.

Role of water in metamorphism. To understand the role of water in metamorphism we must consider how water is distributed deep in the crust. At depths greater than 5km the only water remaining in most sediments is that combined within the crystal lattices of hydrous minerals. Although the amounts of such water vary, they typically are never greater than 14 per cent of the mineral by weight. At these depths porosity presumably has vanished and water can move only through crystal lattices at a slow rate.

The kinds of reactions that take place depend on what happens to the combined water. For example, without releasing combined water, muscovite, aluminous serpentine, and quartz can simply recrystallize into a chlorite-muscovite-quartz rock. Other mixtures can create the same end product, but in the process they must eject combined water. Certain clays can combine with quartz to form a chlorite-muscovite-quartz rock, but the temperatures and pressures at which they do so vary according to the water situation. If water is unable to drain away through the surrounding material, its pressure will increase, stopping the reaction unless the temperature is raised.

These two examples illustrate the ambiguities of interpreting the end products of metamorphism. A chlorite-muscovite-quartz rock can arise from various combinations of initial materials, water pressures, and temperatures. Although a geologist might be able to map the chlorite isograd consistently, he would be unable to interpret its temperature significance unless he knew that all the chlorite had resulted from identical reactions.

Similarly anhydrous metamorphic minerals are subject to several interpretations. They need not always be products of temperatures higher than those at which hydrous minerals originate. In the absence of water, anhydrous minerals can form at temperatures lower than certain hydrous minerals.

Addition of new material. Purely thermal effects tell only part of the story of contact metamorphism. Another major part is represented by new material added from the magma. Hot solutions from the magma can introduce into the contact zone calcium, iron, magnesium, alkali elements, and a host of other elements, as well as silica and carbon dioxide. Many elements so enrich the contact zones (Fig. 23-12) that rocks from them can be mined with profit.

Regional metamorphism. Many of the mineral assemblages in contact-metamorphic aureoles surrounding individual plutons exist on a grand scale in nearly all of the world's mountain chains. As mentioned earlier, in regional metamorphism the source of heat is not always evident. Even without knowing this source we can subject vast tracts of metamorphic rocks to systematic analysis. We can illustrate the principle with coals.

Regional changes in coal. Within the nearly horizontal strata underlying the area around Pittsburgh, Pennsylvania are many layers of high-volatile bituminous coal. Within the closely folded strata of the Appalachians near Scranton, 225 miles east-northeast of Pittsburgh (Fig. 20-13, *A*), are numerous layers of *anthracite*. Both the anthracite and the bituminous coal of Pennsylvania are interbedded with shale and sandstone deposited in the same geologic period, the Pennsylvanian. No plutons have been found where the anthracite occurs. Why, then, do the coals of the two districts differ so conspicuously? The most obvious geologic difference is that of structure (Fig. 20-13, *B* and *C*). Was the coal everywhere the same initially but converted by deformation into anthracite in northeastern Pennsylvania? A survey of coal fields around the world favors an affirmative answer. Anthracite typically coincides with mountain belts where strata have been much deformed, whereas bituminous coal occurs where strata have been less disturbed. We infer that anthracite resulted from the compression and heating of bituminous coal. Possibly some of the rise in temperature should be attributed to friction as the layers of rock were bent and crushed. The consequences of the deformation were to expel the gases slowly and to change both the composition and physical properties of the coal.

In northeastern Pennsylvania the shale and sandstone within which the anthracite occurs have not changed appreciably. Even though anthracite is classed as a metamorphic rock because of its great changes from bituminous coal, and even though the sandstone and shale were folded with the anthracite, these two rocks still remain sedimentary. In southern Rhode Island (Fig. 20-14), a granitic batholith has transformed coal interbedded with deformed strata into superanthracite; it has converted some carbon to graphite, which will not burn. People who tried burning Rhode Island "coal" bitterly expressed their opinion of it in the terse statement: "In the final conflagration of the world the Rhode Island coal will be the last thing to catch fire." The strata accompanying the Rhode Island graphite include phyllite, fine-grained schist, and deformed conglomerate layers. The granitic magma in Rhode Island not only changed the coal drastically but also metamorphosed the former sedimentary strata.

If we were to collect and subject to standard laboratory tests many samples of coal from widely scattered localities, we would be able to draw a map

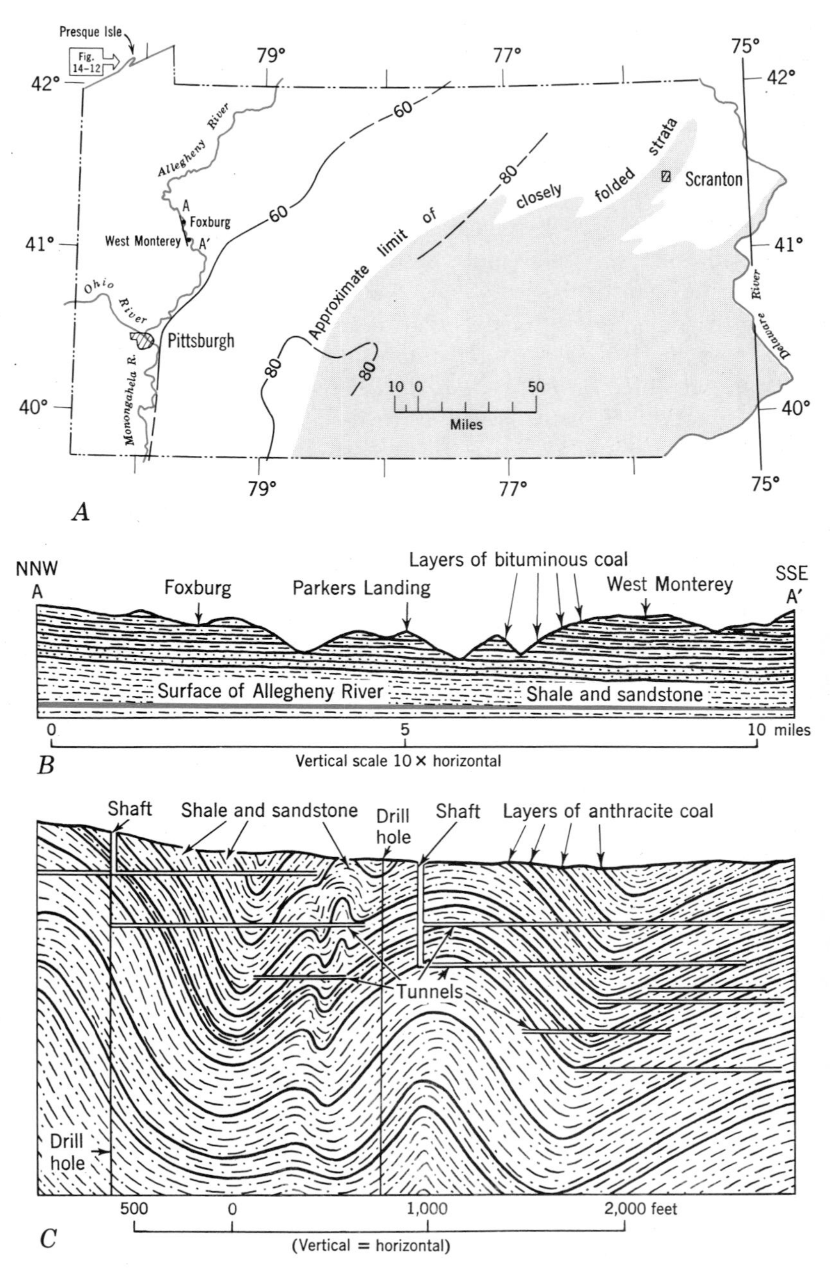

Fig. 20-13. Contrasting occurrences of coal in Pennsylvania. *A.* **Location map. Per cent carbon residue in coal shown by lines, solid where located exactly, dashed where inferred. Further explanation in text. (Lines showing carbon residue from O. C. Postley, 1935.)** *B.* **Section along Allegheny River (line A-A′ on map). Layers of soft coal having less than 60 per cent carbon residue are interbedded with essentially horizontal sandstone and shale. (Folio 178, U. S. Geol. Survey.)** *C.* **Section near Scranton, based on borings, mine shafts, and tunnels. Layers of anthracite are interbedded with shale and sandstone that have been intensely folded. (Coal Investigation Map C3, U. S. Geol. Survey, 1950.)**

showing the locations of various grades of coal. The lines connecting points of equal values of weight of carbon residue are analogous to isograds. Within the areas bounded by two such lines, say 60 and 80 per cent carbon residue, marking the lower and upper limits of medium-volatile bituminous coal, the conditions to which the coal was subjected presumably were nearly the same (assuming constant percentage of carbon in the coaly material to begin with). Of these conditions we think that temperature probably was the most important.

Regional metamorphism in space and time. Regional metamorphic rocks range in age from the oldest rocks yet discovered to those in young mountain chains that crystallized only a few tens of millions of years ago. Recrystallization of minerals permits us to date metamorphic events by isotopic methods. During metamorphism the gaseous products of radioactive decay are driven out of crystal lattices. These products accumulate anew once the rocks cool. Thus we can determine the date of metamorphism of the rock even though we are unable to ascertain the time when the rock was in its original state.

Recrystallization and other evidence demonstrate that regional metamorphism is an expression of great thermal events in the Earth's history. Metamorphism may be accompanied by deformation and may have been a consequence of burial at great depths. But deformation and deep burial have occurred in many places without causing regional metamorphism; thus we conclude that both can be independent of such metamorphism.

Isotopic dates disclose that the duration of thermal activity varies greatly from one regional metamorphic tract to the next. The shortest metamorphic event yet recorded, which ran its course within 30 million years, occurred in the Alps. In other belts, metamorphic events have been longer lived and have included many episodes strung out through spans as long as 800 or 900 million years.

Dynamic metamorphism. If deformation can cause metamorphism then logically we should find the most intense metamorphic effects where deformation has been greatest. In the Appalachians and other mountain chains we find evidence supporting this contention. The evidence consists of slaty cleavage, cataclastic rocks, and structures made by flow of solids.

Slaty cleavage. In western Pennsylvania the strata

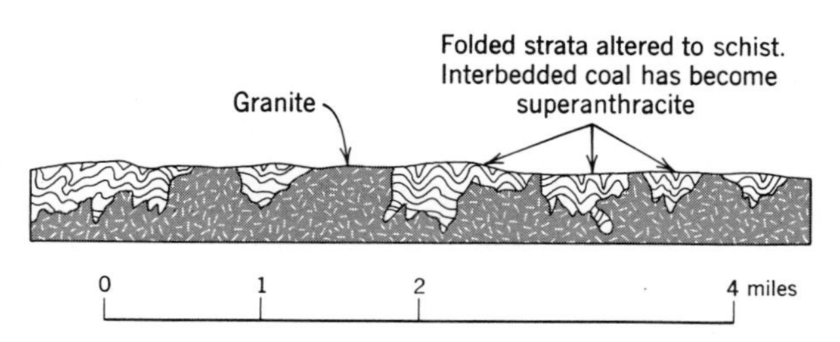

Fig. 20-14. Section of metamorphosed strata intruded by granite. Near Narragansett Pier, Rhode Island. Former coal has been transformed into superanthracite. (After G. F. Loughlin, 1910.)

of shale, sandstone, and bituminous coal are nearly horizontal. Near Scranton, where close folds are present, the coal has become anthracite but the shale looks about like that in western Pennsylvania. Along the Delaware River in eastern Pennsylvania not only have the strata been deformed into closed, overturned folds, but also within the fine-grained rocks stratification has been much obscured by ***slaty cleavage,*** *a closely spaced, plane foliation that divides rock into thin plates* (Fig. 20-15, *A*), and some former siltstone has been converted into slate. Chemically the slate resembles its parent siltstone, but microscopic study of thin sections shows that slate consists chiefly of silt-size quartz particles and slightly larger mica flakes that are remarkably parallel to one another, to the slaty cleavage, and to the axial planes of the folds. Why this orientation? Fold axes generally lie at right angles to the direction of greatest compressive stress. When the mica flakes grew the rocks were being greatly squeezed. The lattices aligned themselves at right angles to the direction of maximum compressive stress. Some rise in temperature doubtless assisted the conversion of clay minerals to mica. Other slaty cleavage resulted from internal shearing movements and is complexly related to mineral particles.

Slaty cleavage is found not only in shales but also in clayey limestones and pyroclastic tuffs. Such cleavage occurs throughout parts of the Appalachians from eastern Canada to Alabama and indeed in all great mountain belts of the world. Much slaty cleavage is related to overturned folds but some originates from the effects of large thrusts (Fig. 20-15, *B*).

Cataclastic rocks. Deformation of coarse-grained massive rocks such as granite may have been purely mechanical, with the result that particles are broken

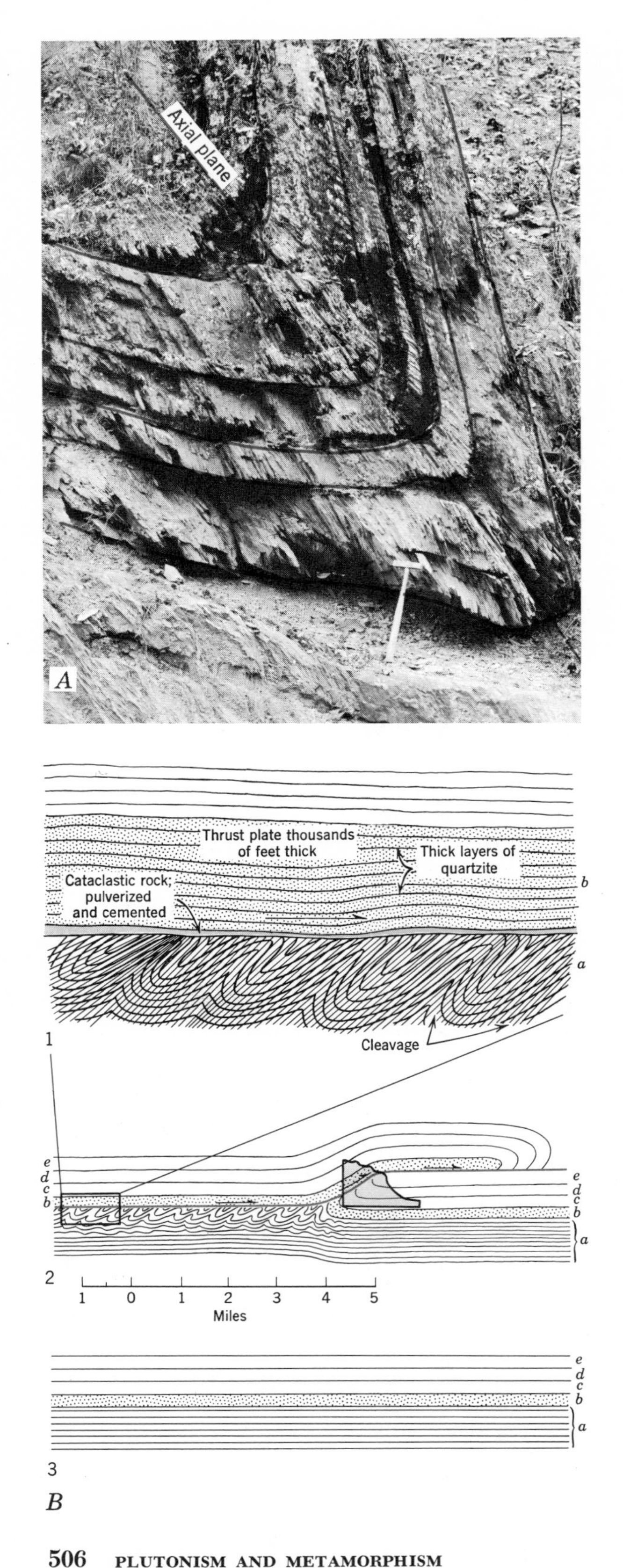

(Fig. 20-16). In some cases, mechanical effects completely penetrated the rock, pulverizing the grains. Intense mechanical grinding can cause initially coarse-grained rocks to become unrecognizable; some such rocks have been milled so fine that they resemble chert.

Structures formed by the flow of solids. Chapter 12 recounted the physical characteristics of glaciers: brittle above but plastic below because of the weight of overlying material. Similarly, stretched and flattened pebbles and cobbles of massive rocks (Fig. 20-17) prove that under some conditions solids flow. Other evidence of changes in the shape of natural materials is found in stretched and flattened fossils and pillows.

Recumbent folds result where the planes of movement of a glacier intersect planes defined by sediment particles of a glacier's load. In the section of the glacier shown in Fig. 20-18, *A*, the zone of maximum flow crosses the axial plane of the fold at a small angle. As the overlying ice moves toward the right it transports the upper limb of the fold both forward in the same direction and slightly downward. As the limbs are stretched the nose of the fold advances slowly downglacier. The dark layers of the upper limb eventually become part of the nose and are then turned over to become part of the lower limb.

No proof exists that folds in rocks have originated by such stretching and rolling out, but the similarity in form between folds in glaciers and those in many gneisses (Fig. 20-18, *B*) is rather remarkable. Extremely complex patterns can result from the flow of rocks along axial planes of folds. In the folds described in Chapter 17 most axial planes were true planes, but for the folds in many metamorphic rocks the median parts are not planes but curve in various directions. Hence the term axial surface is more appropriate. Complex folds nearly always occur in

Fig. 20-15. Field occurrences of slaty cleavage. A. Axial part of closed, overturned syncline. Slaty cleavage, parting paralleling the axial plane of the fold, is nearly as prominent as the partings between successive strata. *B.* Slaty cleavage parallels axes of small folds created by frictional drag at base of thick thrust sheet, near Johnnie, southern Nevada. Nearby zone of cataclastic rock along thrust shale has been transformed into slate and phyllite. Gray Area in 2 locates setting of Fig. 17-20, *B.*

Fig. 20-16. Metamorphic rocks with cataclastic textures. *A*. View, through polarizing microscope, of slice of a rock from vicinity of a major fault, Branford, Connecticut. Particles of quartz (gray and white areas) have been intensely sheared parallel with top and bottom of photograph; dark, shadowlike areas within some are indications of the effects of deformation. (B. M. Shaub.) *B*. Large feldspar particles, possibly phenocrysts of a former granite porphyry (Fig. C-9, *A*), have been compressed into lenticular shapes parallel with the irregular foliation that crosses specimen diagonally from upper left to lower right. (C. R. Longwell.)

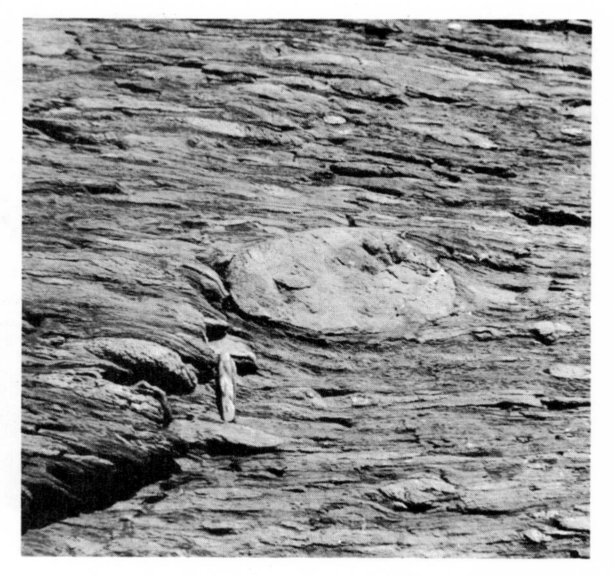

Fig. 20-17. Horizontal surface eroded across edges of vertical layers of metamorphosed conglomerate. During deformation, pebbles, chiefly of light-colored igneous rock, were elongated and flattened. (J. W. Ambrose, Geol. Survey of Canada.)

Fig. 20-18. Folds in glaciers and in gneisses. *A.* Section of recumbent fold outlined by layers of dark silt. (Further explanation in text.) Taylor Glacier, South Victoria Land, Antarctica. (Warren Hamilton and P. T. Hayes, U. S. Geol. Survey.) *B.* Complex folds in metamorphic rock. Quarry along Skeena River, British Columbia. (B. M. Shaub.)

recrystallized rocks. This association suggests that high temperature accompanies movement and is an important factor in softening the rocks and making them plastic.

ORIGIN AND RISE OF MAGMA

The origin of magma and its means of rising through the lithosphere are fundamental problems not yet thoroughly understood. As found in laboratory experiments, the effects on solid rock materials of varying temperatures and pressures furnish fundamental guidelines in our attempts to understand these problems.

Effects on solids of varying temperatures and pressures. From seismology we conclude that no permanent reservoir of magma lurks within the crust or upper mantle. Therefore the average geothermal gradient is not sufficient to melt solids. From our experience with solids at ordinary surface temperatures and pressures we know that energy is required to melt a solid. The energy consists of that required not only to raise the temperature to the melting point but also to exceed the latent heat of fusion. Clearly, then, one of our first concerns must be potential sources of heat within the Earth in excess of the average geothermal gradient. Concentrations of radioactive materials in excess of the average, and friction during deformation are two possible mechanisms for creating local spots hot enough to melt rocks.

Analysis of the problem based solely on temperature, however, is inadequate; we dare not neglect the other factor, pressure. As we have read previously, the effects of pressure on the melting behavior of silicates is complex. Increasing total pressure alone raises the melting temperatures of silicates just as it does the boiling point of water in a geyser (Fig. 19-21). A solid subjected to both great heat and great pressure can melt if nothing more than the pressure is reduced, just as water under pressure in a geyser flashes into steam when the pressure is decreased.

Opposite to the effect of total pressure is that of water pressure. Increasing the pressure of water vapor and other volatiles lowers the melting temperatures of silicates. At great pressures water and other potential volatile materials presumably are bound within the crystal lattices of silicates. If pressure is reduced we can visualize that these substances would escape from the lattices and begin to influence the melting behavior of surrounding materials.

Reduction of pressure, therefore, would bring about a double-barreled effect: (1) some solids would melt and (2) volatiles would be released, and they would tend to lower the melting temperatures of remaining solids. Both magma and volatiles thus generated would tend to move in the direction of least pressure, presumably upward. If fractures exist, the magma and volatiles take the easy way and travel upward along the openings to the surface. If no fractures are available the magma and volatiles *may* be able to melt their way upward. The advance guard of volatiles lowers the melting temperatures of the materials above. When the volatiles have passed, crystallization occurs below in the rear. When material in the liquid state passes into the solid state, heat energy equal to the latent heat of fusion is released. This heat rises and tends to perpetuate the upward movement of the zone of melting. Eventually the melting zone would work its way to the surface and create a volcano.

The third important effect of pressure is that it can determine whether a solid melts congruently or incongruently. Because it results in fractionation of materials incongruent melting introduces another complexity into the problem of the origin of magmas. Where fractionation by partial melting has occurred, the magma reaching the surface or upper levels of the Earth's crust represents a sample not of the kind of material melted but only of its easily melted ingredients. Left behind at depth are the less fusible constituents.

Origin of basaltic magma. Two important facts about basalts limit our ideas on the origin of basaltic magma. (1) Among extrusive rocks basalt predominates by a wide margin. (2) The compositions of oceanic basalts and many basalts from continental interiors differ from the compositions of basalts from continental margins. Plateau basalts on a grand scale exist on all continents and range in age from Precambrian to Cenozoic. Vast bodies of basalt have originated on the floors of the ocean basins. Such copious discharge of basaltic lava has been interpreted to mean that a single uniform basaltic parent magma exists and that it derives from a uniform layer within the Earth's mantle.

From what depths could such apparently uniform material come? Seismic data from Hawaii, the Aleutians, and elsewhere show that the magma chambers beneath some volcanoes lie at depths of about 60km, well within the Earth's upper mantle. In Japan the circumoceanic basalts come from deeper sources, whose depths are 100 to 120km. These seismic data and recent laboratory experiments suggest that one parent substance, mantle rock, partially melts at various depths to create the two major kinds of basaltic magmas. Partial melting at shallower depths is thought to generate magmas that create oceanic basalts; partial melting at greater depths is thought to create circumoceanic basalts.

Origin of granitic magma. In both chemical composition and geologic mode of occurrence, granites contrast with basalts. Granites contain abundant silica and alkali elements and low amounts of mafic materials. Basalts and their coarse-grained equivalents contain low amounts of silica and alkali elements and abundant mafic materials. The predominance of basalts among extrusive rocks is matched by the prevalence of granites among plutonic rocks. Basalts exist both on continental masses and in ocean basins whereas granites are restricted to continental masses. Intrusive basalts and their coarser-grained relatives are largely confined to tabular plutons and to a few stocks, whereas only a minor proportion of granite occurs in tabular plutons and the vast majority of it in stocks and batholiths. These contrasts are so profound that they have fostered the concept that the origin of basalt differs fundamentally from that of granite. According to some geologists the differences mean that granites are not igneous rocks. Let us examine the evidence.

The existence of granitic magma is affirmed by silicic extrusive rocks ranging from rhyolites to andesites and by batholiths that are surrounded by contact-metamorphic aureoles, in which are found some of the world's premier deposits of metallic minerals. Although their contacts can be very irregular the granitic rock distinctly cuts across the intruded rock (Fig. 20-19). The igneous rock may wrench loose and completely engulf blocks of the invaded rock, forming ***xenoliths*** (Gr. "stranger rocks"), *blocks, formerly part of the wall rock of a pluton, that have been broken loose and have become completely surrounded by igneous rock.*

The evidence that the Boulder batholith and others locally have intruded the welded tuffs forming the roof rocks suggests that some granite magmas solidified at high levels in the crust. The coarse grain size can be attributed to slow cooling caused by the insulating effects of the hot extrusive materials.

Fig. 20-19. Irregular contact of coarse, light-colored granitic rock with dark-colored gneiss. Near Guilford, Connecticut. (J. E. Sanders.)

The large size of batholiths raises a "room problem" of how the granite was able to occupy its present space, which was occupied by other rocks prior to the origin of the granite. Some granites were forcibly intruded and shouldered aside the rocks. Others "made room" by *stoping,* the repeated spalling off and removal of xenoliths. In still other granites the mechanism of emplacement is enigmatic.

Other batholiths are surrounded by regionally metamorphosed strata, lack metallic mineral deposits, and grade into a wide, irregular zone of ***migmatite*** ("mixed rock"), *a rock in which thin stringers and threads of granitic material are intertwined with dark schistose layers.* Stringers of schist fade imperceptibly into the granite as if they had been dissolved. In schistose rocks, far outside the borders of some plutons, pockets and clots of feldspar and other minerals like those in the granite are completely isolated. Apparently the rock that occupied these pockets has been replaced bodily by material identical with granite in the pluton. Migmatites, which form wide zones around granitic plutons, have prompted the idea that granite can originate by extreme metamorphism, that is, ***granitization,*** *the transformation, without fusion, of older rocks into granitic rock.*

These processes need not be mutually exclusive.

Recall the reaction between a gabbro pluton and basalt on the Island of Skye. Within the pluton a pyroxene-plagioclase rock (gabbro) resulted from the cooling of a magma. In the contact aureole a pyroxene-plagioclase rock (hornfels) resulted from intense metamorphism of basalt. Such occurrences form the basis for the broader term, ***mineral facies,*** designating *mineral assemblages of any origin or composition that have reached chemical equilibrium under similar environmental conditions.* It is immaterial whether the minerals crystallized from a melt or grew in the solid state during metamorphism. Presumably within any pair of identical mineral assemblages having contrasting origins, texture will reveal which process operated. Granitic rocks, then, would belong to the feldspar-quartz mineral facies, whether the rock-forming process had been solidification of a magma or intense metamorphism.

One of the features of the concept of granitization that rendered it attractive to many geologists was its apparent resolution of the "room problem" represented by supposedly bottomless batholiths. Recent geophysical results have indicated that batholiths are rarely thicker than 10km. This finding dulls the edge of the argument based on the magnitude of the "room problem."

Geochemical data disclose that some of the elements composing granites must have originated in the mantle. Experiments show that artificial mixtures of granitic materials melt and cool in the proportions found in natural granites. This evidence and the geophysical information on the thickness of batholiths strongly support the hypothesis that most granites are magmatic.

The rise of granitic magma through the lithosphere can be explained by the process of zone melting previously mentioned. Presumably, then, the rise of a body of granitic magma represents a great thermal climax. After the magma has been emplaced thermal equilibrium can be restored.

We know that some granitic magmas have caused intense contact metamorphism. We do not know if such magmas are the cause or the effect of regional metamorphism. Whatever their origin, the final rise of granite coincides closely in time with the cessation of regional metamorphism; clearly the two are intimately related.

Obviously our knowledge of granite is still incomplete. We leave the subject in that condition and offer it as a worthy challenge for the best minds of geologists, present and future. Despite these uncertainties we know from the observations of several generations of geologists that granites and associated regionally metamorphosed rocks are the hallmarks of mountain chains. In our study of mountains in the next chapter we shall meet these rocks again.

Summary

Plutonism

1. Bodies of intrusive igneous rock occur as plutons having tabular, lenticular, cylindrical, or irregular shapes.

2. Magma within a pluton can solidify to a single uniform igneous rock, or differentiate and generate several kinds of igneous rock. Settling of crystals and separation of liquids are two mechanisms of differentiation.

3. Three important geologic milestones in the history of a pluton are the date of its intrusion, the date of its unroofing, and the date of its covering.

4. Study of relationships along the upper contact of a body of igneous rock can permit us to distinguish sills from buried extrusive sheets. With favorable exposures we can distinguish between the top of a pluton's magma chamber and younger covering strata deposited on the eroded surface of the pluton.

Metamorphism

5. Heat and hot fluids from many plutons create contact-metamorphic aureoles. Results vary according to the properties of the plutons and the characteristics of the rocks intruded.

6. Where individual plutons cut strata of many kinds we can compare the response of varied materials to the same conditions. Where plutons of more than one kind intrude a single rock type we can compare the response of one material to varied conditions.

7. In regionally metamorphosed tracts where individual plutons are absent, mineral assemblages similar to those in contact aureoles occur. Assemblages formed under the same range of conditions constitute a metamorphic facies.

8. Slaty cleavage, cataclastic rocks, and structures resulting from plastic deformation affirm dynamic metamorphism.

9. Basaltic magmas are thought to originate by partial melting of upper-mantle rock. Basalts from

central parts of ocean basins and interiors of continents are thought to derive from a depth of about 60km; circumoceanic basalts possibly originate deeper down.

10. Some granite plutons resulted from the intrusion of new magma into cooler surroundings; other bodies of granite originated in hot surroundings and locally may have originated by partial melting, replacement, and recrystallization of older rocks.

Selected References

Boyd, F. R., 1964, Geological aspects of high-pressure research: Science, v. 145, p. 13–20.

Carnegie Institution Year Books, Annual Reports of the Director, Geophysical Laboratory.

Clark, S. P., Jr., and Ringwood, A. E., 1964, Density distribution and composition of the mantle: Rev. Geophysics, v. 2, p. 35–88.

Fyfe, W. S., Turner, F. J., and Verhoogen, Jean, 1958, Metamorphic reactions and metamorphic facies: Geol. Soc. America Mem. 73.

Hamilton, Warren, and Myers, W. B., 1967, The nature of batholiths: U. S. Geol. Survey Prof. Paper 554-C, p. Cl–C30.

Hunt, C. B., 1953, Geology and geography of the Henry Mountains region, Utah: U. S. Geol. Survey Prof. Paper 228.

Knopf, Adolph, 1936, Igneous geology of the Spanish Peaks region, Colorado: Geol. Soc. America. Bull., v. 47, p. 1727–1784.

Read, H. H., 1957, The granite controversy: London-New York-Sydney: Interscience Publishers.

Tuttle, O. F., 1955, The origin of granite: Sci. Am., v. 192, no. 4, p. 77–82. (San Francisco, W. H. Freeman and Co., Repr. 819, 6 p.)

Tyrrell, G. W., 1929, The principles of petrology, 2nd ed.: New York, E. P. Dutton and Co., Inc.

Uffen, R. J., 1959, On the origin of rock magmas: Jour. Geophys. Research, v. 64, p. 117–122.

Yoder, H. S., 1955, Role of water in metamorphism, *in* Poldervaart, Arie, Crust of the Earth. A symposium: Geol. Soc. America Spec. Paper 55, p. 505–524.

CHAPTER 21

Mountains

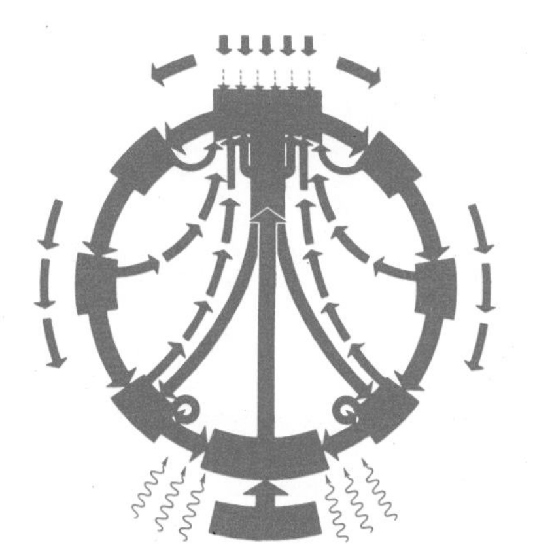

The challenges of mountains
Characteristics of mountain chains
Traverses across three mountain chains
Appalachians
North American Cordillera
Alps
Relief of mountains
Strength of the lithosphere
Principle of isostasy
Variation in Earth's gravitational acceleration
Events in a tectonic cycle

Plate 21

View eastward along the main Alpine frontal thrust in central Switzerland. Rigi is underlain by nonmarine conglomerates of Oligocene age deposited as ancient fans, Pilatus, Bürgenstock, and peaks in upper right corner, by marine strata of Cretaceous and Cenozoic ages. (*Swiss National Tourist Office.*)

Lucerne
Bürgenstock

CHAPTER 21 *Mountains*

THE CHALLENGES OF MOUNTAINS

Mountain peaks never cease to challenge men. One way that men have responded to this challenge is by organizing expeditions to climb the highest peaks. The obstacles to successful ascent are so numerous that each attempt begins with careful preparations. Men, equipment, and supplies are assembled and maps are studied to select the route of march. We need not delve further into the heroic adventures of the great expeditions in order to emphasize the point that an unconquered peak stimulates men to great effort.

Mountains challenge not only our physiques by daring us to climb them but also our intellects by confronting us with the question of how mountains originated. Many generations of geologists working in all parts of the world have assembled an impressive array of data on mountains. Yet some of the questions associated with the origin of mountains remain unanswered. Perhaps we have not been asking the right questions. Nor have we evolved a standardized procedure for presenting and analyzing our evidence. In short, the choice of procedure is up to us; in this chapter we propose to begin by examining the characteristics of mountain chains as exemplified by the Appalachians, the North American Cordillera, and the Alps.

The preceding chapters should have prepared us for the geological analysis of mountains. We shall draw upon our knowledge of strata and of their use in analyzing geologic structure. We shall use our concepts of the Earth's interior, our ability to recognize ancient volcanic rocks, and our physical and chemical interpretations of the thermal effects created during plutonism and metamorphism. If in previous chapters we have not sufficiently exercised our geologic material, in discussing mountains now we shall give all aspects of it a full workout.

CHARACTERISTICS OF MOUNTAIN CHAINS

What characterizes mountain chains? Is it topographic relief, geologic structure, or both? When the relief reaches 1,000m nearly everybody acknowledges that the terrain is mountainous. Yet relief is not the sole criterion used in defining mountain chains. Many plateaus, for instance the Catskill and Pocono "Mountains" in northeastern United States and the Colorado Plateau of southwestern United States, show relief of 800 to 1,000m or more, yet fail to qualify geologically as mountains because their strata are uncrumpled.

Many of the Earth's great linear belts exhibit not only mountainous relief but also three geologic characteristics: (1) their strata have been folded and faulted, (2) some of their rocks have been regionally metamorphosed, and (3) granitic batholiths are present. Rocks displaying these characteristics are found in many linear belts where mountainous relief is not present, as along the south shore of Lake Superior. To a geologist the presence of these three characteristics in the rocks implies "mountains" even though there is no mountainous relief.

In modern mountain chains we see great relief associated with the three geologic characteristics; in ancient mountain chains the relief has disappeared but the deformed and metamorphosed strata and granitic batholiths remain. Accordingly we conclude that relief is ephemeral but that the diagnostic geologic characteristics remain indelibly imprinted upon the rocks of all mountain chains.

In analyzing mountain chains we need to consider their topographic and geologic characteristics separately. Implicit in the geologic term ***orogeny*** (ŏr-ŏj′ĕn-ĭ; Gr. "mountain genesis"), *the deformation of the crust in the development of mountains*, is the connotation that geologic and topographic features are created simultaneously. *Orogenic belts* designate mountain chains and regions where moun-

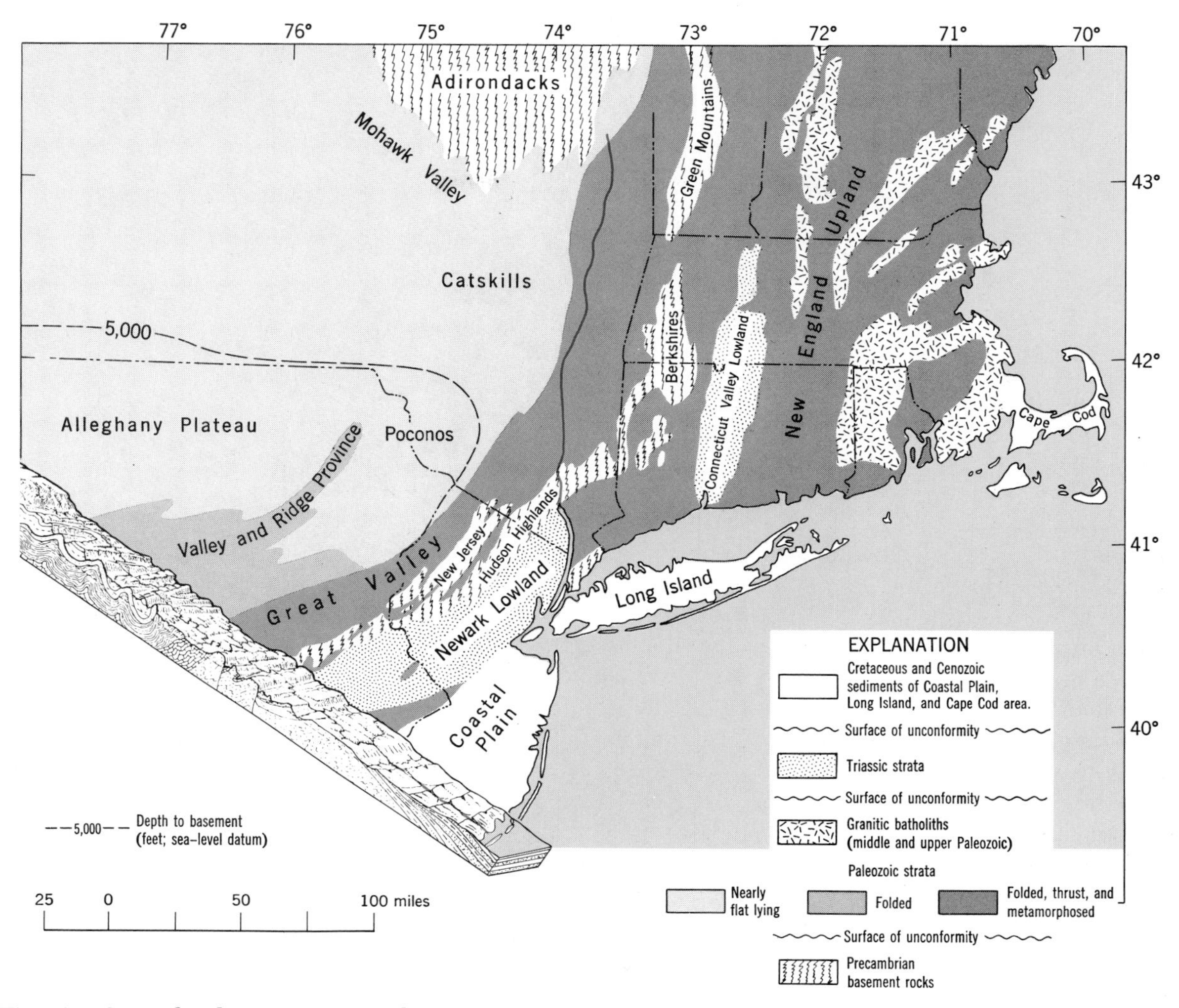

Fig. 21-1. Generalized tectonic map and cross-section through a segment of the Appalachians and adjacent regions. (Map after Tectonic Map of United States; New England after Richard Goldsmith, 1964; cross section from Erwin Raisz, *in* Douglas Johnson, 1932.)

tains are in the process of formation. To distinguish clearly between the origins of the geologic characteristics of mountain rocks and the topographic relief of mountains some geologists employ the term ***diastrophism*** for *the processes of large-scale deformation, metamorphism, and intrusion that occur in orogenic belts.* Lateral squeezing is thought to accompany diastrophism. *Vertical displacement or tilting of the Earth's crust without large-scale crumpling of strata and generally affecting broad areas of a continental mass* is ***epeirogeny.*** Plateaus display the results of epeirogenic movements. Likewise during the development of many mountain chains epeirogeny has been active.

Let us begin our analysis of mountains by making traverses across mountain chains. In this way we can become acquainted with mountain rocks; later we shall take up the problem of relief.

TRAVERSES ACROSS MOUNTAIN CHAINS

In our traverses we shall approach the Appalachians from the west, the North American Cordillera from the east, and the Alps from the north. Along all three, before reaching the mountains we cross terrains of low relief underlain by strata that are essentially horizontal. The land surface is char-

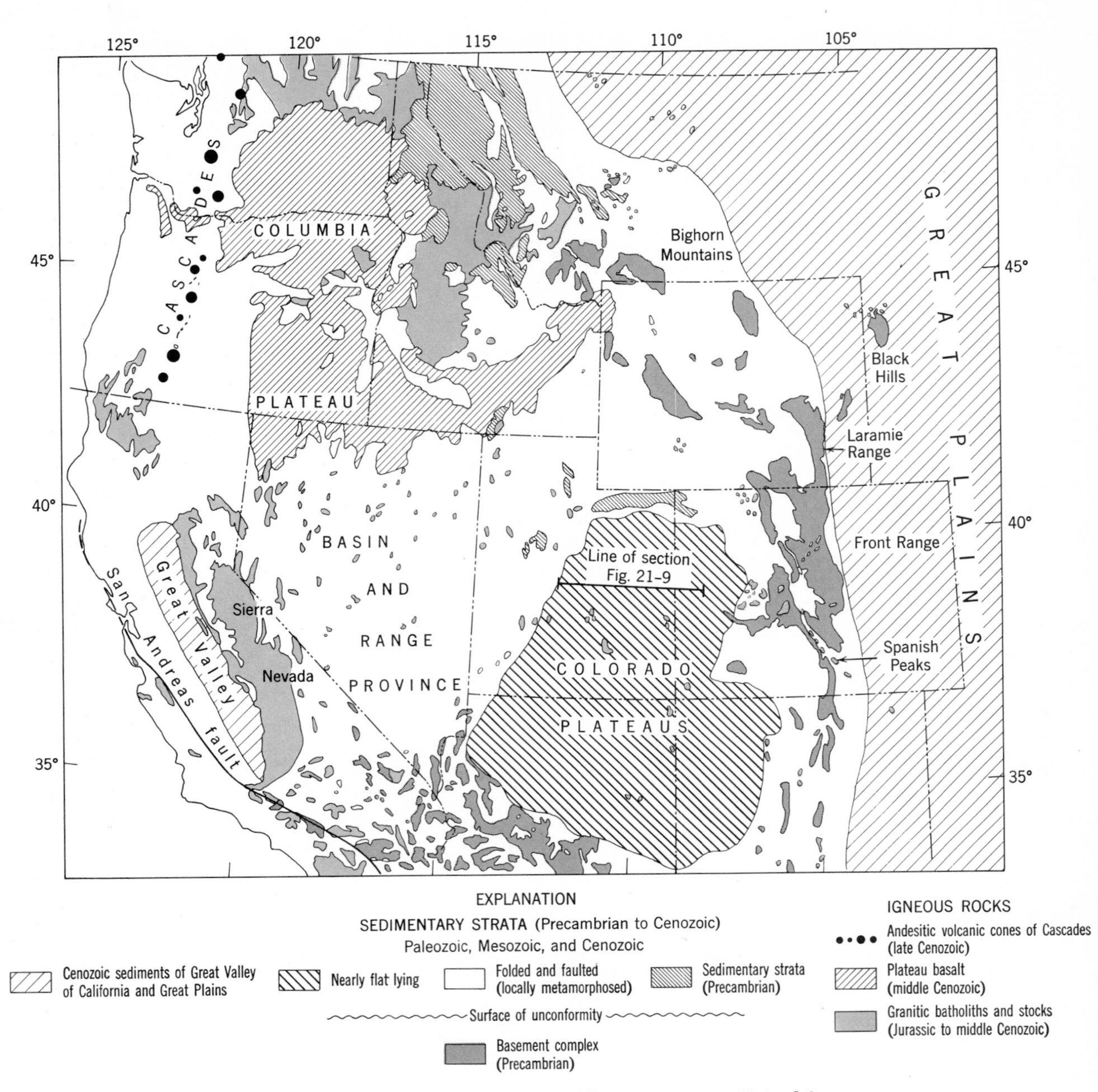

Fig. 21-2. Generalized tectonic map of North American Cordillera in western United States. Compare locations of Precambrian basement rocks with those in the Appalachian segment (Fig. 21-1). (After C. R. Longwell, 1950.)

acterized by dendritic stream patterns (Fig. 10-13). These horizontal strata overlie sialic "basement" complexes, which bear the characteristics of ancient mountain rocks.

Marginal zones. The mountains themselves differ from the lowlands underlain by horizontal strata; their form becomes linear or irregular (Pl. 21), and trellis drainage is common. Within the marginal zones of the mountains we may find great circular or oval areas where basement rocks have been elevated so high that they tower above their surroundings. Examples include the Adirondacks in the Appalachians (Fig. 21-1); the Black Hills, Front Range, Laramie Range, Bighorn Mountains, and others in the Cordillera (Fig. 21-2); and the massifs of the Vosges-Black Forest and French Central Plateau near the Alpine chain (Fig. 21-3).

If the basement does not stand high in the marginal parts of mountain chains it may do just the opposite and lie well below deformed sedimentary strata. Such strata, fully exposed in coastal cliffs, were known to Hutton and to other early geologists. But in the first part of the nineteenth century geologists did not imagine that the deformed strata of mountain chains composed parts of regular geologic structures. In 1847 two brothers, William B. and Henry D. Rogers, demonstrated that the linear ridges and valleys in the western parts of the Appalachians, in Virginia to Pennsylvania (Fig. 21-4), resulted from the erosion of great anticlines and synclines (Fig. 21-5). This region has been aptly named the Valley and Ridge Province. Folds similar to those in that province have been found in nearly every mountain chain, but nowhere are they more magnificently displayed than in this part of the Appalachians. The term *Appalachian structure* refers to parallel anticlines and synclines of great extent.

Beginning in the second half of the nineteenth century, the structural analysis of mountains progressed rapidly. From the Scottish Highlands came the concept of great overthrusts. Such structures, found in nearly every mountain chain, reach their grandest proportions in the Alps. Just as the term Appalachian structure connotes well-developed parallel folds, so the term *Alpine structure* refers to great overthrusts (Fig. 21-6). In many chains the overthrusts are associated with folds.

As we move along the length of a mountain chain in its marginal parts, we may encounter areas displaying any of these features: (1) high-standing basement, (2) folded strata only, and (3) folded strata plus overthrusts. The northwestern parts of the Appalachians exemplify the lateral change from one major feature to another. In northern New York, the marginal part of the Appalachians is formed by the Adirondacks, a circular area of high-standing basement rocks about 130 miles in diameter. The basement plunges toward the southwest and is not exposed south of the Mohawk Valley. Between central Pennsylvania and central-western Virginia the deformed strata display parallel folds; thrusts are not visible. But in Tennessee the marginal part of the Appalachians is characterized by both folds and overthrusts (Fig. 21-7). Metamorphic rocks and granitic batholiths younger than those found in the basement complexes are not present in the marginal zones of most mountain chains.

In many chains the sedimentary strata have been folded independently of the underlying basement rocks. The implication is that the strata were crumpled together as they slipped along one or more planes of easy motion. The Jura Mountains in France and Switzerland are classic examples of such deformation of strata in which the basement rocks are not intimately involved (Fig. 21-8). Some geologists contend that the folds in the Appalachian Valley and Ridge Province and in the adjacent plateau originated because the strata moved independently of their underlying basement.

Interior zones. Before we continue our traverses we turn our attention to the historical information preserved in the strata. The youngest strata found in the approaches to and in the marginal zones of our three mountain chains contrast widely in age but are remarkably similar to one another in appearance. Evidently they were deposited in nearly identical environmental settings. In each case sequences of conglomerates, sandstones, shales, and coals become thinner and include a greater proportion of marine strata in a direction away from the mountain chain. In the Appalachians such strata range from Pennsylvanian to Early Permian age; in the Cordillera they are of Cretaceous to Paleocene age; and in the Alps, of Miocene age.

From these predominantly clastic strata we infer that areas within what is now the mountain belt were being vigorously uplifted while an adjacent marginal area subsided. Streams drained directly away from highland blocks, creating fans and coastal

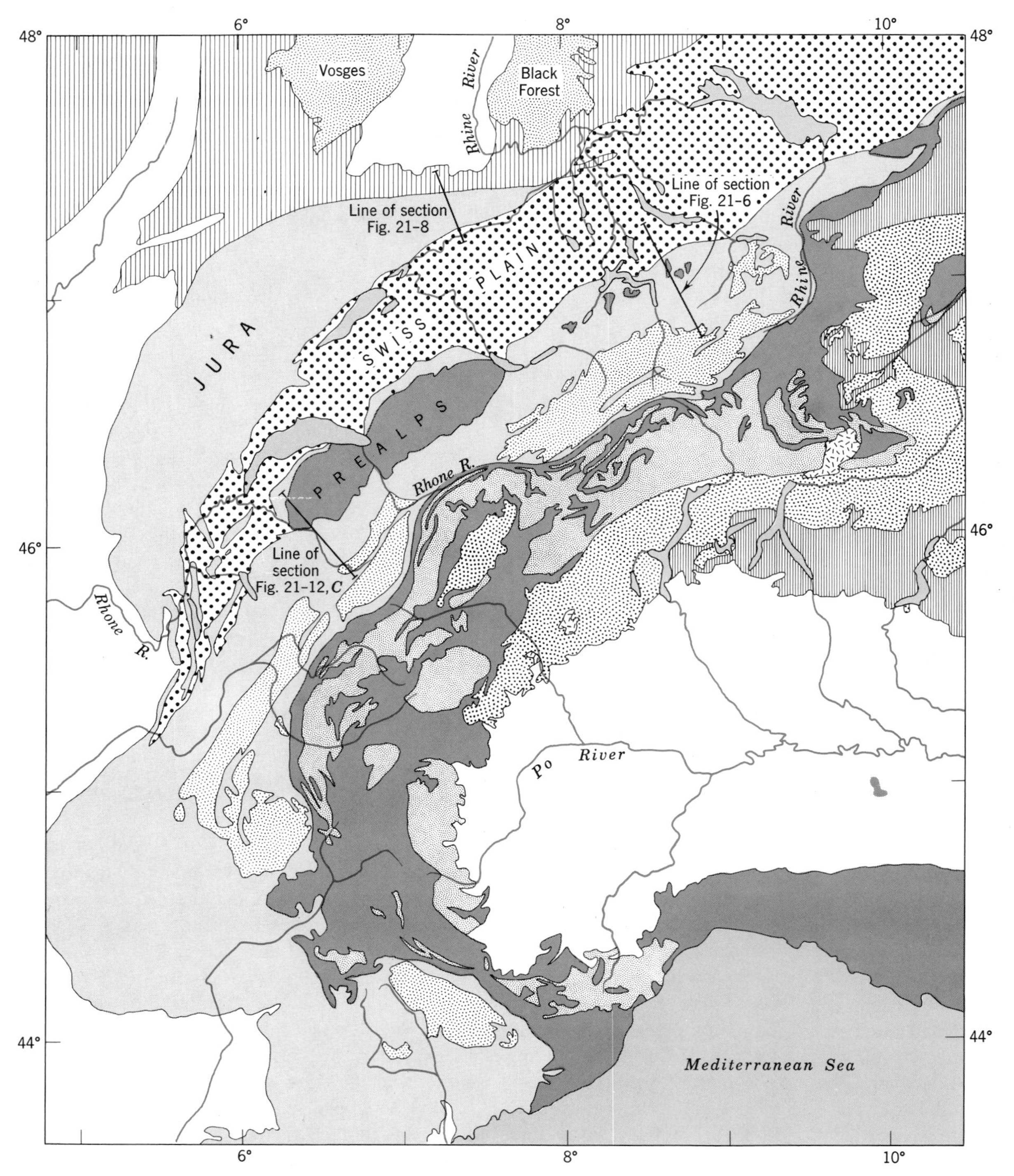

Fig. 21-3. Generalized tectonic map of the region of the western Alps. (After Trümpy, 1960.) Explanation is on facing page.

plains along the shore of a shallow sea (Fig. 16-14) and transporting sediment from the rising to the subsiding areas. In western Colorado and central Utah exceptionally well-exposed strata display the relationships between eroded areas of uplift and sediments deposited in an adjoining area of subsidence (Fig. 21-9). The evidence from this part of the Cordillera is so clear-cut that we are confident the similar sheets of clastic strata of other ages and in other localities likewise resulted from pulses of uplift. Moreover, we can infer that such clastic strata record uplifts, even if we lack direct structural evidence of uplift.

In both marginal and interior zones of all three mountain chains one or more sheets of clastic strata lie above carbonate rocks; locally clastic strata and carbonate rocks are interbedded. In the Appalachians the carbonate rocks crop out through nearly the length of the chain. As a result of the humid climate these rocks form a valley, 10 to 15 miles wide, named the Great Valley of the Appalachians (Fig. 21-4).

The story of the carbonate rocks is unlike that of the sheets of clastic strata. A great sheet of carbonate rocks of Early Paleozoic age, extending from the Appalachians to the Cordillera, was deposited in a shallow sea that covered much of North America (Fig. 21-10). In the marginal zones of both chains, thickness of carbonate rocks is greater than it is beneath the lowlands lying between the chains. We infer that the carbonate rocks were deposited in warm, shallow seas in which terrigenous sediment was sparse.

In the interior zones of both chains the carbonate rocks pass laterally into dark shales and siltstones having persistent interbeds of graded sandstones (Fig. 21-11; Pl. 17). In some localities near great thrusts, blocks of carbonate rocks lie scattered within a matrix of noncalcareous dark shale. Typically these terrigenous strata of the interiors of mountain chains are entirely of marine origin; possibly they were deposited at abyssal depths. The pieces of carbonate rocks evidently slumped down steep submarine escarpments that formerly were inclined *away* from the interior of the continent and toward the mountain chain. Yet at a later time, when thrusting occurred, the rocks of the overthrust block moved *toward* the interior of the continent. In some cases, notably in the Alps, material has been transported up the former slope and well beyond, through a

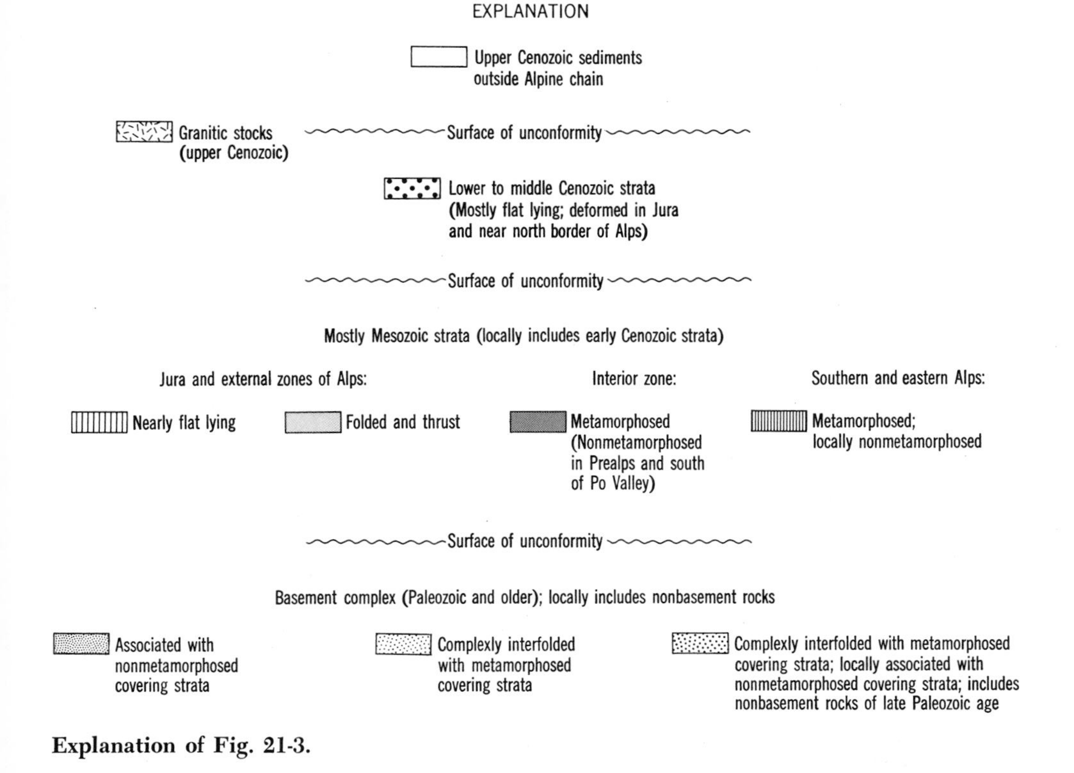

Explanation of Fig. 21-3.

Fig. 21-4. View southeastward across Appalachian Valley and Ridge Province, eastern Pennsylvania. Resistant strata underlie wooded ridges; more erodible strata underlie cultivated lowlands. Distance between Tuscarora Mountain and Blue Mountain is 23 miles. (J. S. Shelton.)

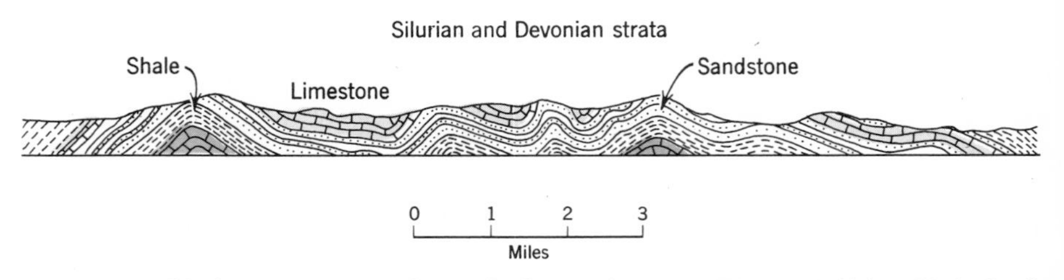

Fig. 21-5. Folded strata in central Appalachians. Compare Fig. 21-7. (After U. S. Geol. Survey Folio 59.)

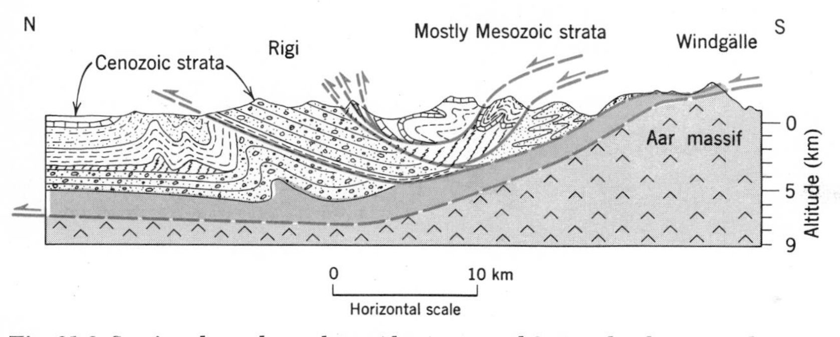

Fig. 21-6. Section through northern Alps in central Switzerland, in area shown on Pl. 21. Strata have moved northward along great overthrusts, which later were themselves folded. Major thrust separates overlying strata from basement rocks. (After Albert Heim, 1922, Geologie der Schweiz, v. 2, pl. 27.)

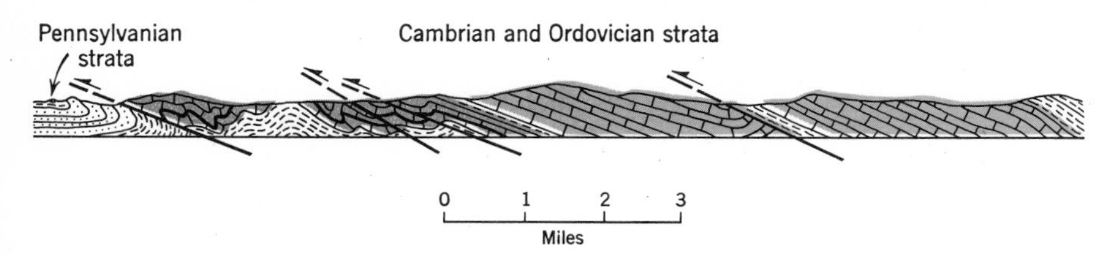

Fig. 21-7. Folds and faults in the southern Appalachians. Many of the thrusts originated from the stretching and breaking of limbs of folds. (After U. S. Geol. Survey Folio 61.)

distance greater than that through which the particles had moved down the ancient slope during sedimentation (Fig. 21-12).

Along the parts of the chain where carbonate rocks give way to marine terrigenous strata, commonly great overthrusts are present. Moreover, in this general zone other elongate patches of uplifted basement rocks are numerous. Evidence in the strata indicates that during deposition some basement blocks moved upward actively. The structural configuration discloses that upward movement also occurred after deposition.

In the Appalachians, an extensive area of basement rocks extends nearly throughout the length of the chain. It forms the Blue Ridge in the southern Appalachians and the New Jersey-Hudson Highlands-Green Mountains in the northern Appalachians (Fig. 21-1). In many localities the basement rocks have been thrust over their younger covering strata (Fig. 21-13). In the Cordillera, exposures of basement within the interior zones are confined chiefly to southern Nevada and southern California (Fig. 21-2). In the Alps uplifted basement rocks within the interior zones underlie Mont Blanc and many of its famous neighboring peaks.

Slaty cleavage (Fig. 21-14) typically occurs in the interior zones of mountain chains. As metamorphic intensity increases we pass into the central zones.

Central zones. In the central zones of mountain chains, plutonic and metamorphic rocks prevail. In the Appalachians these zones coincide with the Piedmont and the New England Upland. Many great batholiths of granitic rocks occur in them. Although common in central zones, batholiths also occur elsewhere. For example, the Boulder batholith in Montana (Fig. 20-9) lies in the outer part of the interior zone of the North American Cordillera. Bodies of ultramafic rock and voluminous quantities of volcanic rock typify the central zones of many chains. Many volcanic rocks display pillows and are associated with bedded cherts containing Radiolaria.

Within the central zones basement rocks may be exposed, but they are difficult to distinguish from

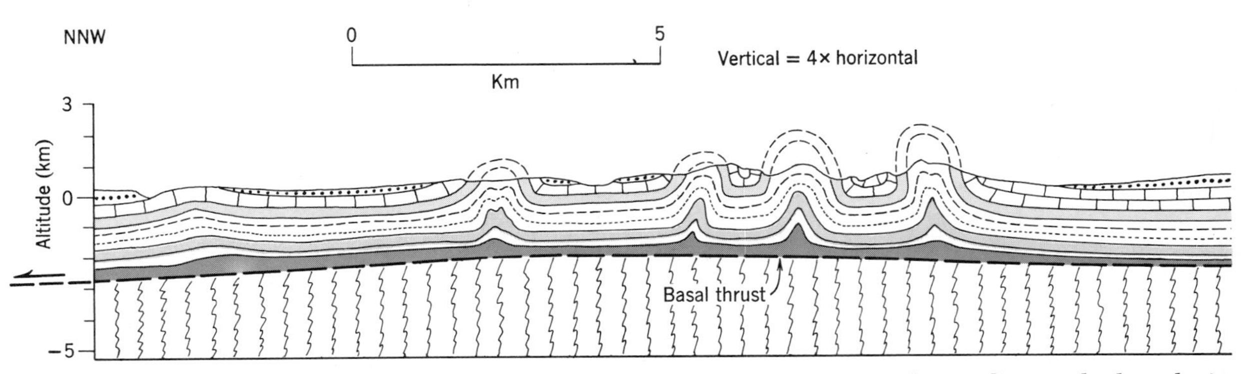

Fig. 21-8. Section through the Jura, Switzerland, interpreted according to the hypothesis that the strata were deformed independently of the underlying basement rocks. If the hy-

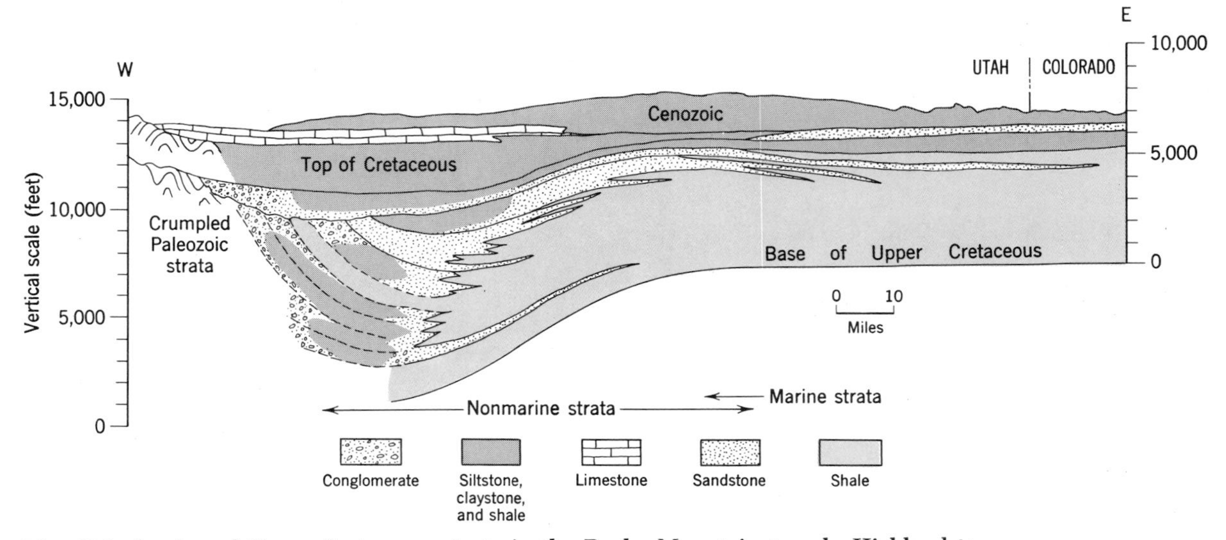

Fig. 21-9. Section of Upper Cretaceous strata in the Rocky Mountain trough. Highland to west, composed of crumpled Paleozoic strata, supplied most of the sediment that fills the trough. (After P. B. King, 1955 and E. M. Spieker, 1946.)

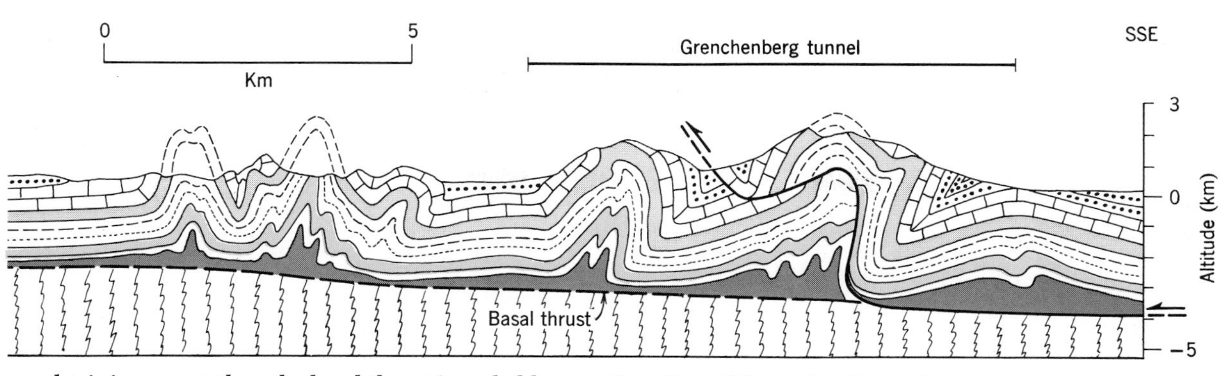

pothesis is correct, then the basal thrust is probably a continuation of the major thrust shown in Fig. 21-6.

their intensely metamorphosed covering strata. In fact, it might be appropriate to visualize the central zones of mountain belts as places where strata can be converted into basement rocks.

Structures in the central zones result from plastic flow of the rocks. Steep fold axes are common. Folds have resulted from flow along planes parallel with each other but not parallel with the strata. The structures in the metamorphic rocks resemble features seen in glaciers (Fig. 21-15; 20-18).

The rocks in the central zones of most mountain chains overwhelmingly convey the results of high temperatures on rock materials. Great thermal activity must have been concentrated in these zones, but we do not yet know why. However, the thermal events can be dated isotopically.

After the great thermal and diastrophic events had run their courses, the rocks in the central zones of mountain chains may have been not only metamorphosed and deformed but also elevated above sea level and not again resubmerged. Yet despite this change, still other strata may have accumulated within the mountain belt. During great differential vertical movement some crustal blocks were elevated into ranges whereas others were depressed as basins. Sediment eroded from rising ranges filled subsiding basins. Deposition occurred on fans, in stream channels and on flood plains, and in lakes. The Rocky Mountain parts of the North American Cordillera display nonmarine basin sediments on a grand scale. During the Cenozoic, vast lakes existed in parts of Colorado, Utah, Wyoming, and Nevada. The Green River Formation (Fig. C-1, *B*) is an example of strata deposited within one of the lakes. In the northern Appalachians, nonmarine basins comparable to those of the Rocky Mountain regions were filled with sediments deposited on fans, on valley floors, and in lakes. The Pennsylvanian strata of Nova Scotia resulted from such activity.

Fault blocks in central zones. Rocks of the central zones of many mountain chains have been broken into numerous blocks bounded by steep faults. Sediment eroded from some elevated blocks continued to accumulate on subsiding blocks. Faulting may have been accompanied by eruptions of vast quantities of basaltic lava from fissures. A splendid example is the Columbia Plateau (Fig. 19-12) created in the central zones of the Cordillera during the Miocene Epoch.

During Late Triassic time elongate fractures cut the rocks of the central zones of the Appalachians. Along some fractures dikes were injected (Fig. 20-3). Nonmarine sandstones, siltstones, and conglomerates, whose aggregate thickness may reach 20,000 feet or more, were deposited in elongate, subsiding fault troughs across whose floors several sheets of basaltic lava spread. Eroded edges of the now-inclined resistant sheets of intrusive rock thus formed constitute the Watchung Mountains of New Jersey and the Hanging Hills of central Connecticut (Fig. 21-16).

The Basin and Range Province of the North American Cordillera displays a spectacular network of faults, some of which are still active, causing earthquakes. In that province, extending south and

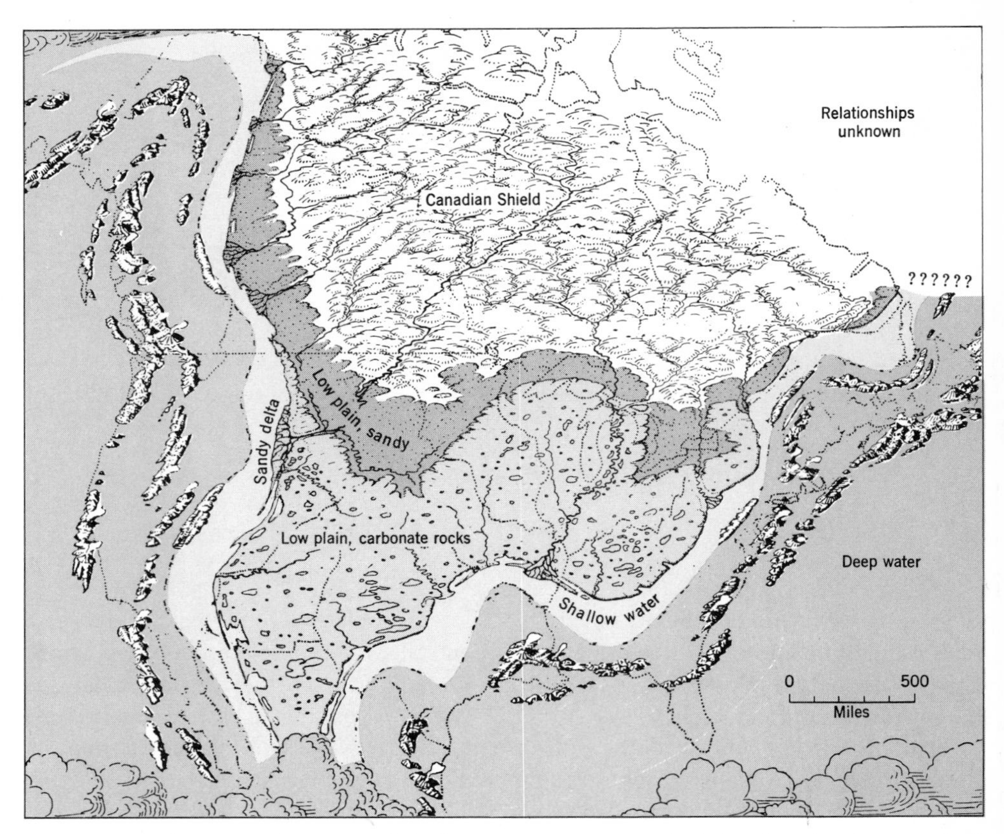

Fig. 21-10. Paleogeographic map of North America during Ordovician Period, after marginal lands had been created from former sea floors and after the sea had withdrawn from much of the interior of the continent. (After G. M. Kay, 1951.)

southeast from southern Oregon through several states into Mexico, movement along thousands of faults has cut the crust into innumerable blocks. The upheaved blocks stand as hundreds of ridges, dozens of them forming mountain ranges of imposing height. Sediments eroded from the ranges are accumulating in the intervening basins (Pl. 5).

Although the high, steep scarps of the Wasatch Range in Utah and of ranges overlooking Death Valley, California suggest that movements were rapid and catastrophic, probably this is not so. Studies of recent faulting imply that high scarps grow in a suc-

Fig. 21-11. Steeply inclined siltstone and sandstone strata of Ordovician age, south shore of St. Lawrence River, Quebec. Light-colored layers are fine-grained limestones 1 to 2cm thick. Strata resemble sediments cored from modern abyssal plains. (B. M. Shaub.)

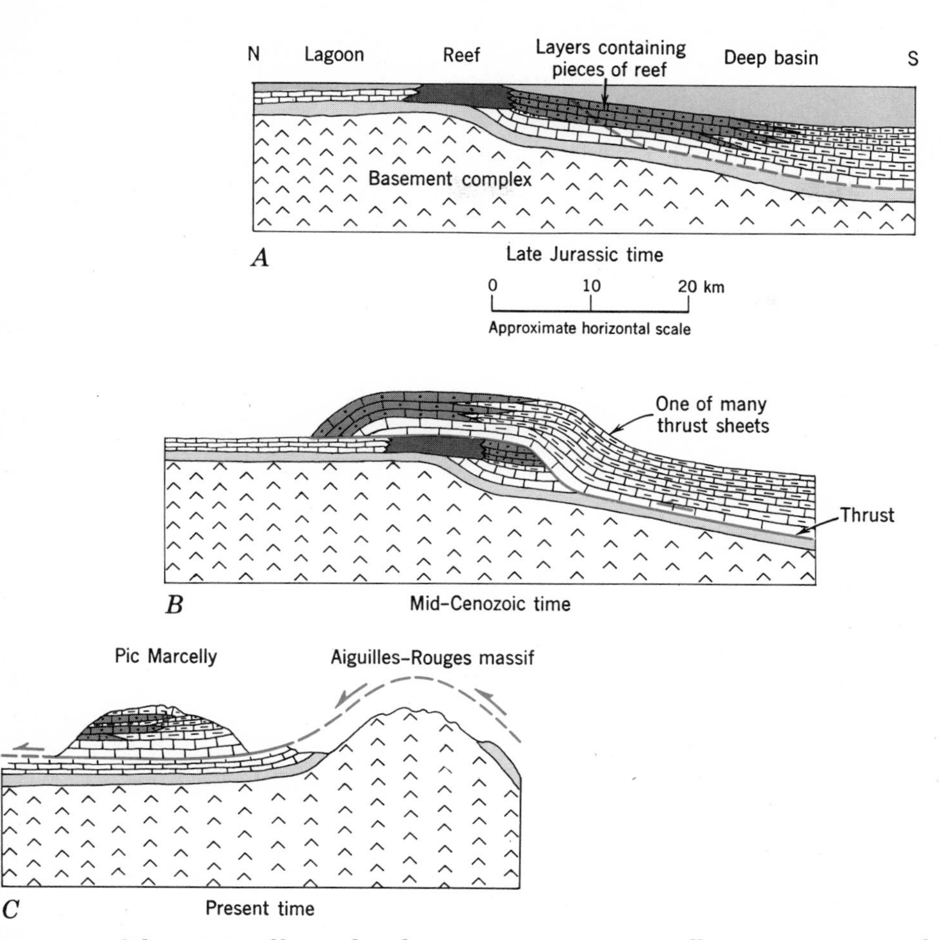

Fig. 21-12. Schematic profiles in the Alps, Haute-Savoie, France, illustrate the concept that thrusts traveled *up* former submarine slopes. *A*. Inferred conditions of sedimentation. Reef, along northern margin of deep marine basin, sheds sediment southward down steep submarine slope. Dashed blue line marks future thrust surface. *B*. During great deformation many thrusts transported strata northward from the deep marine basin past site of former reef. *C*. Distal ends of strata from north slope of former deep marine basin are about 20km north of the inferred site of the reef, which originally stood about 20km north of where the sediments were deposited in the basin. Deformation after thrusting has elevated the basement complex of the Aiguilles-Rouges massif, and extensive erosion has removed its covering strata, including the reef. (After Ph. H. Kuenen and A. V. Carozzi, 1953.)

cession of spasmodic jumps, each adding no more than a few feet, and that long pauses intervene between movements. The steep faults trend north or northwest, in part paralleling the earlier structural grain established by folding and thrusting.

Some range blocks have been raised uniformly between parallel faults; more have been rotated about their long axes. The largest range of all, the Sierra Nevada, 400 miles long and more than 50 miles wide, stands at the western limit of the province. Its eastern side is a high fault scarp but its western side has been thickly covered by sediments (Fig. 21-17, *A*).

The kinds of rocks and types of structures present in the ranges vary greatly. Many ranges consist of steeply tilted sequences of marine strata (Fig. 21-17, *B*). The complex bedrock of the Sierra Nevada block consists partly of closely folded and metamorphosed strata and partly of a large granitic batholith. The folding and intrusion occurred during the Mesozoic

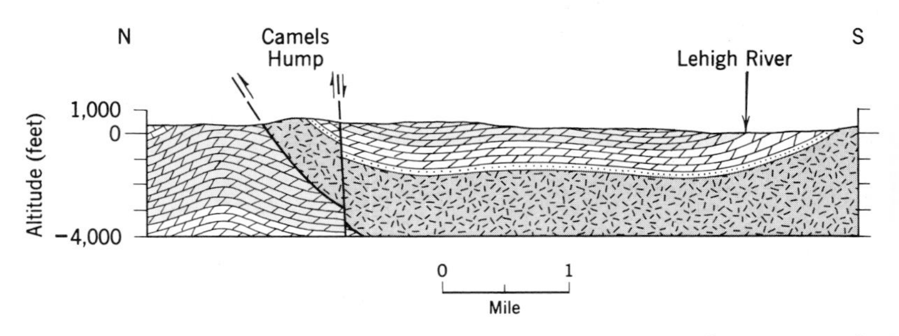

Fig. 21-13. Section in eastern Pennsylvania shows Precambrian basement rocks thrust over Paleozoic carbonate rocks. (After B. L. Miller, 1948.)

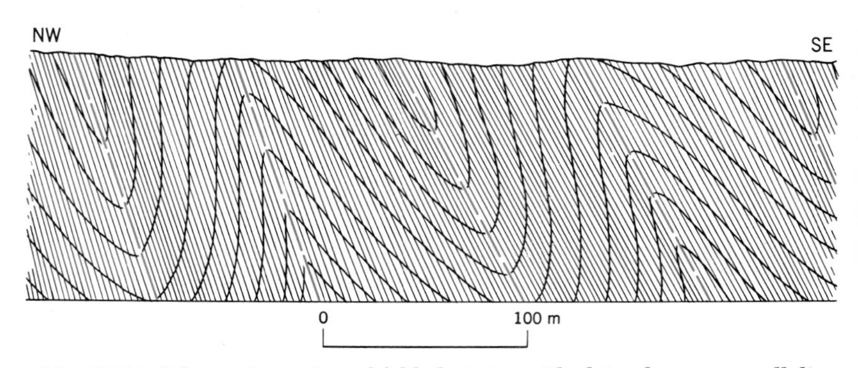

Fig. 21-14. Schematic section of folded strata, with slaty cleavage paralleling axial planes of folds. Axial-plane cleavage is common in folds of many areas in the central zones of mountain chains.

Era. The resulting high mountains were eroded to low altitude. The height of the present range results from uplift and differential rotation that started late in Cenozoic time.

Once the crust has been broken into fault blocks, the processes that create the geologic characteristics of mountains are nearly finished. The mountain belt may not yet have become dormant, but its further activities typically involve large-scale domal uplift and delivery of the sediment eroded from within to places outside the chain. To understand these activities we must examine the reasons why relief appears and disappears.

RELIEF OF MOUNTAINS

The relief of mountains results from continuous vertical uplift of the deformed tract. The Appalachians, for example, began to be domed on a regional scale beginning in Late Cretaceous time. Their present relief has resulted largely from differential erosion of strong and weak rocks during such doming.

From the results of our traverses we can reconstruct for all mountain chains a general progression of events: (a) A region subsides below sea level. (b) Marine troughs deeply subside while adjacent land areas are greatly uplifted. (c) During general uplift above sea level some nonmarine troughs subside. (d) The entire region is elevated above sea level. To understand this progression of events we must study the strength of the lithosphere, even though to do so we must go outside mountains.

Strength of the lithosphere. Is the lithosphere strong enough to sustain the weight of the volume of rock in mountain masses that tower above the surrounding lowlands and also that of the continental masses that loom above the deep-sea floor? How can we attack the problem of the strength of the lithosphere? Useful information comes from two sources: (1) the shape of the Earth and (2) the response of

Fig. 21-15. Complex folds with steep axes in a glacier and in metamorphic rocks. *A.* Steep folds outlined by debris in ice of Malaspina Glacier, Alaska, where it spread out after emerging from narrow valleys in the St. Elias Range. (Bradford Washburn.) *B.* Steep folds in schist (black) and marble (white), southern Sweden. (P. J. Holmquist.)

the lithosphere to natural loads placed upon it.

The shape of the Earth results from the interaction of gravity and the centrifugal force of rotation (Fig. 2-9). If the Earth were initially a cube and its strength infinite, a cube it would remain despite forces resulting from gravity and rotation. If the initial shape were a cube, or any other geometric figure for that matter, and the strength not infinite, then gravity and rotation would mold the Earth into a form governed by the strength of its material. Rapid rotation of weak material would produce a disk, slower rotation of slightly stronger material a much flattened oblate spheroid, and so forth. Calculations based on these considerations suggest that the average strength of the Earth is about the same as steel.

The Earth's response to natural loads provides more specific data about the strength of the lithosphere. Natural loads include gravitational attraction of the Moon, weight of the atmosphere, variable thicknesses of seawater during a tidal cycle, water of lakes, ice of glaciers, and masses of volcanic rock.

Sensitive instruments reveal that the lithosphere responds to all these loads. Small Earth tides have been recorded, and the Earth's surface moves up and down by minute increments in response to changes of barometric pressure and to the height of the tide. Because the lithosphere responds to such

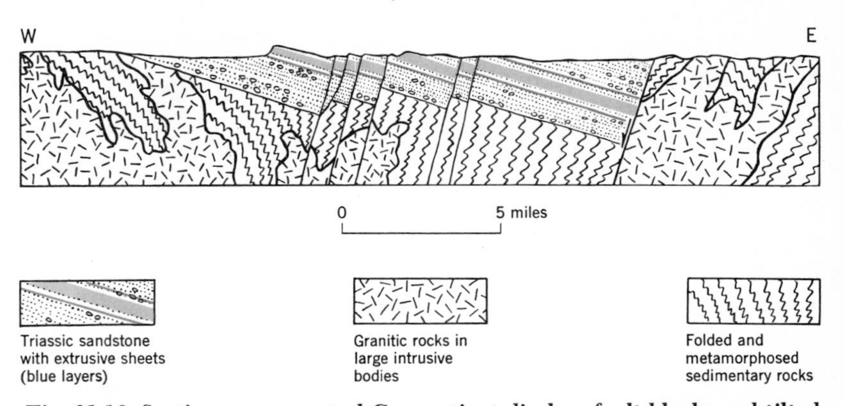

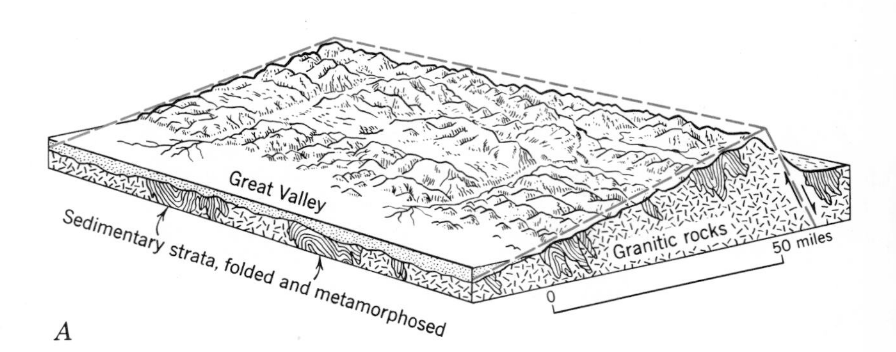

Fig. 21-16. Section across central Connecticut displays fault blocks and tilted nonmarine strata that originated during a late stage of development of the Appalachians.

Fig. 21-17. Contrasting compositions of tilted fault-block ranges in the North American Cordillera. A. Sierra Nevada complex block. As east side has been elevated along a large fault, west side has been depressed. Sediment eroded from the rising part has been deposited in the Great Valley of California. (After F. E. Matthes, 1930.) *B.* Steeply inclined marine strata of Paleozoic age, more than 5,000 feet thick, in Frenchman Mountain east of Las Vegas, Nevada. Block is outlined by large faults that trend north-south; strata have been offset by a smaller east-west fault. (William Belknap, Jr.)

loads we conclude that its strength is small. Subsidence around volcanic cones has been observed, but we are unable to distinguish subsidence caused by load from that caused by changes of volume in the magma chamber beneath. Lakes and glaciers are better sources of data on the lithosphere. The water of a deep lake can create enough load to depress the Earth's surface: the area around Hoover Dam subsided 18cm within 15 years from the weight of the water in the reservoir, and geologic evidence of deformation by weight of water of an ancient lake has been found in Utah.

Fig. 21-18. Five shorelines of Ancient Lake Bonneville along the Wasatch Range near Salt Lake City, Utah. (A. E. Granger, U. S. Geol. Survey.)

Deformed shorelines of Ancient Lake Bonneville. Imprinted on the mountainsides surrounding Great Salt Lake, Utah is unmistakable evidence that water formerly stood as much as 1,000 feet higher than at present (Fig. 21-18). Waves have formed cliffs, beaches, and bars on steep slopes facing the lake basin; streams have built large deltas where their valleys enter the lake basin. The calculated volume of the huge former lake (Ancient Lake Bonneville) is 8,900km^3, about the same as Lake Michigan. The highest shoreline leads to a long stream-cut outlet in southern Idaho, through which overflow water discharged northward into the Snake River (Fig. 21-19).

At 90 places the altitudes of shore features fashioned by the lake at about 5,100 feet above the present sea level have been measured precisely. These altitudes are not all alike; where the lake was deepest they exceed 5,100 feet by amounts up to 200 feet. Radiocarbon dates indicate that the deformation occurred during the last 25,000 years. We conclude from this evidence that the weight of the water that filled the Bonneville basin caused almost immediate local subsidence. After the surface of the lithosphere had subsided, the shore features were created. When the water vanished the lithosphere returned to its former level, doming the shore features.

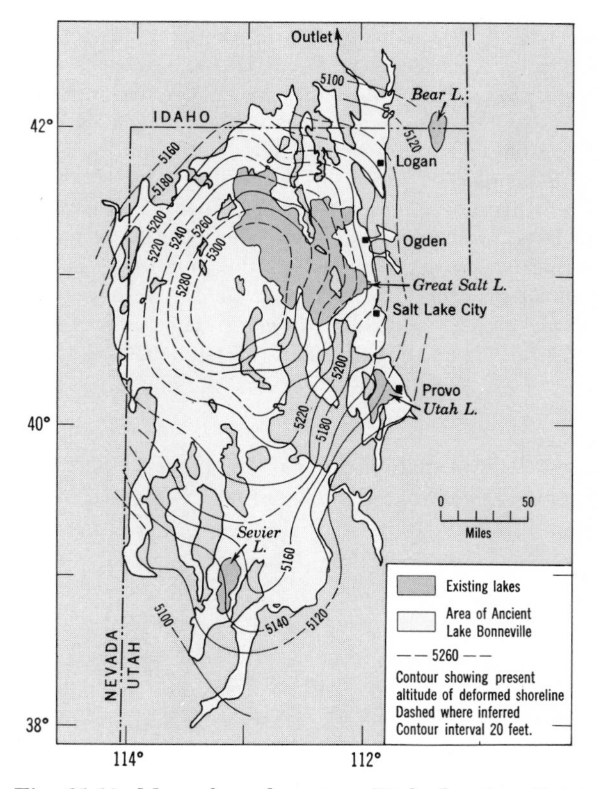

Fig. 21-19. Map of northwestern Utah showing Great Salt Lake and the area covered by its predecessor, Ancient Lake Bonneville. Points on former shoreline, originally at altitude 5,100 feet, have been domed to the present altitudes indicated by the numbered lines. (After Max Crittenden, 1963.)

If the lithosphere can sink under the weight of water from Ancient Lake Bonneville and rise again when the load of water has been removed, it should behave similarly in areas occupied by former glaciers. Deformed shorelines of Scandinavia and in the Great Lakes region of the United States provide striking testimony to such activity.

Deformed shorelines in the Baltic region. Land has been and still is rising rapidly enough out of the Baltic and North Seas to leave clear evidence of

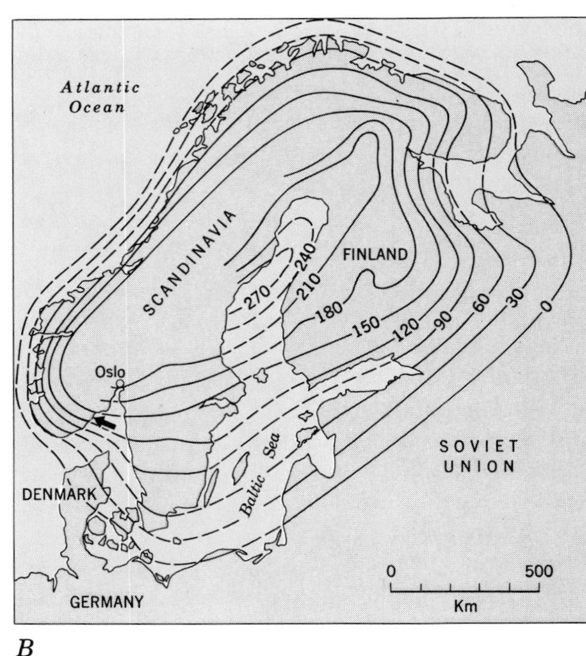

B

Fig. 21-20. Deformed postglacial shorelines in Scandinavia. A. A horizontal line, cut into granitic rock near Oslo, Norway marking position of mean tide in 1839, was 30cm above mean tide in 1939, although sea level had risen measurably during the same period. Location is indicated by arrow on map, *B*. (Olaf Holtedahl.) *B*. Emergence around Baltic Sea since disappearance of latest Pleistocene ice sheet. Lines connect points on former shorelines, elevated by the number of meters indicated. Lines are projected across water-covered areas, and are omitted where evidence of emerged shorelines is lacking. (After R. A. Daly, 1936.)

rise within a span of 50 years or so. In harbors, old seamen have pointed out low rocky islets that originated from what were shoals a few generations previously, and in coastal fields farmers have plowed up numerous marine shells. During the summer of 1834 Sir Charles Lyell journeyed to Sweden to study the evidence, by then well known, that the land was rising out of the sea, and to determine the rate of rise. King Charles XIV of Sweden, perhaps in his great wisdom or in great jest, inquired about the results of the study. He wanted to know how long it would be before Sweden should plan to construct a railway to Great Britain.

Lyell and other students of the Baltic area were convinced that the rate of rise was as much as 1m/100 years. Subsequently this rate has been confirmed by observations on reference marks (Fig. 21-20, *A*) and by systematic recording of the readings from modern tide gages. Rate of uplift reaches a maximum in northeastern Sweden and decreases southward to zero in northern Denmark. Past uplift is greatest where the greatest modern rate has been observed. All shorelines have been domed upward; some have been raised more than 300m above present sea level (Fig. 21-20, *B*).

These deformed shorelines are thought to have originated as follows: The Pleistocene ice sheet that covered Scandinavia weighted down the lithosphere. As the glacier uncovered the land by melting, about 10,000 years ago, the sea inundated the region and created shoreline features. Intermittently the lithosphere has been returning to its preglacial level; each time it moves upward it deforms all active and abandoned shorelines. Because the uplift progresses by steps, new shorelines develop during quiescent intervals. After another pulse of uplift a new shore-

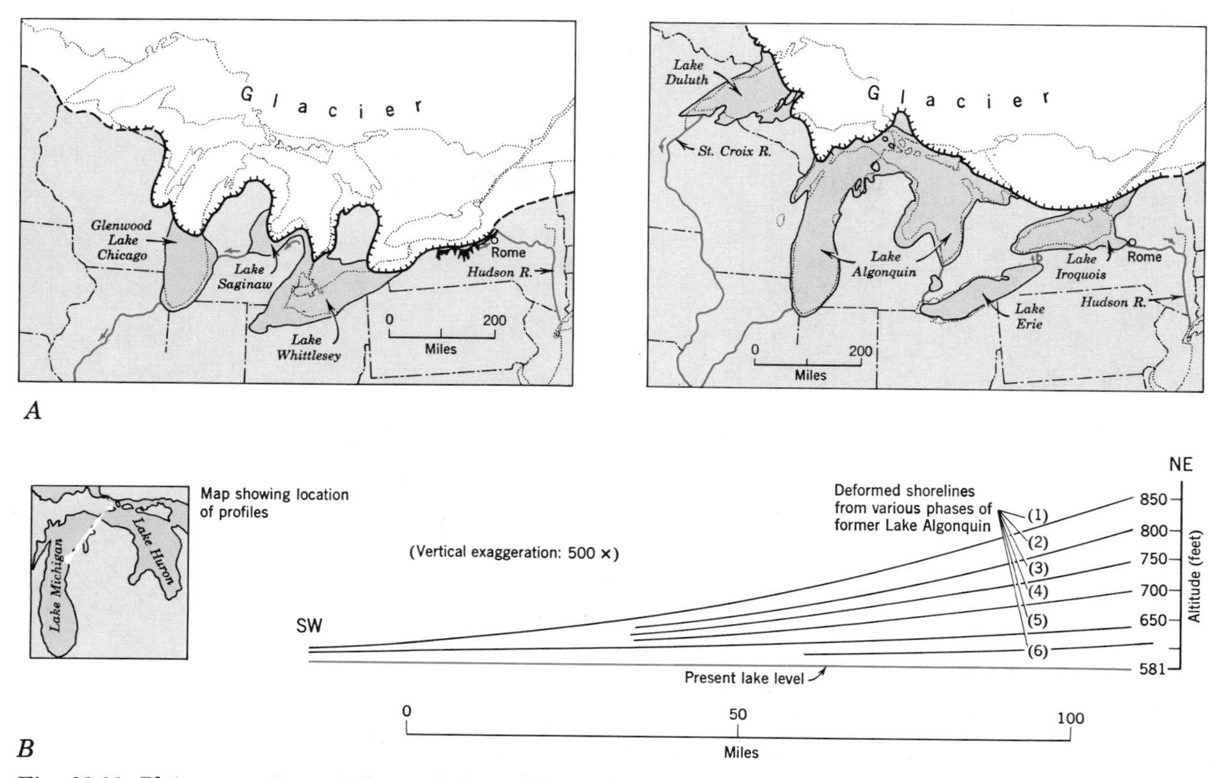

Fig. 21-21. Pleistocene Great Lakes and their deformed shorelines. ***A.*** **Two of the many stages in the development of the glacial Great Lakes during the last 15,000 years. The glacier (white) gradually melted, increasing the sizes of the lakes, which overflowed at various places at different times. The names of former lakes, defined by high-level beaches, differ from those of present Great Lakes. (R. F. Flint, 1957.)** ***B.*** **Profiles of deformed shorelines of several phases of former Lake Algonquin. (After J. W. Goldthwait, 1908.)**

line is added at the base of the uplifted series and those older are raised still higher.

Deformed shorelines in the Great Lakes region. The elevated marine shorelines of Scandinavia are matched by elevated lake shorelines in the area bordering Lakes Superior, Michigan, and Huron. During the retreat of the last ice sheet, various lakes ancestral to the present Great Lakes existed in the area as drainage outlets shifted (Fig. 21-21, *A*). Along the south shore of Lake Michigan are older, higher shorelines. Not only do the altitudes of the former shores increase toward the north but also each higher (and older) feature is inclined more steeply than the next one below. The old shores are not continuous, because in some places they were poorly developed and in others they have been destroyed by erosion. But by careful regional study it has been possible to reconstruct with confidence the relationships shown diagrammatically in Fig. 21-21, *B*. Precise surveys indicate that movement is still in progress.

Principle of isostasy. Evidence of the kind associated with deformed shorelines leads us to conclude that the lithosphere can bob up and down like an iceberg, although more slowly—that it "floats" on an easily deformed substratum. *The ideal condition of flotational balance among segments of the lithosphere* is known as ***isostasy*** (ī-sŏs′tah-sĭ; Gr. "equal standing"). An oversimplified experiment illustrating the principles involved in isostasy can be performed by pouring mercury into a strong vessel and floating in it a number of metal blocks, equal in cross section and weight but differing in length and composition (Fig. 21-22, *A*). Each block is composite, consisting of two or more metals of various densities, the least dense at the top and the most dense

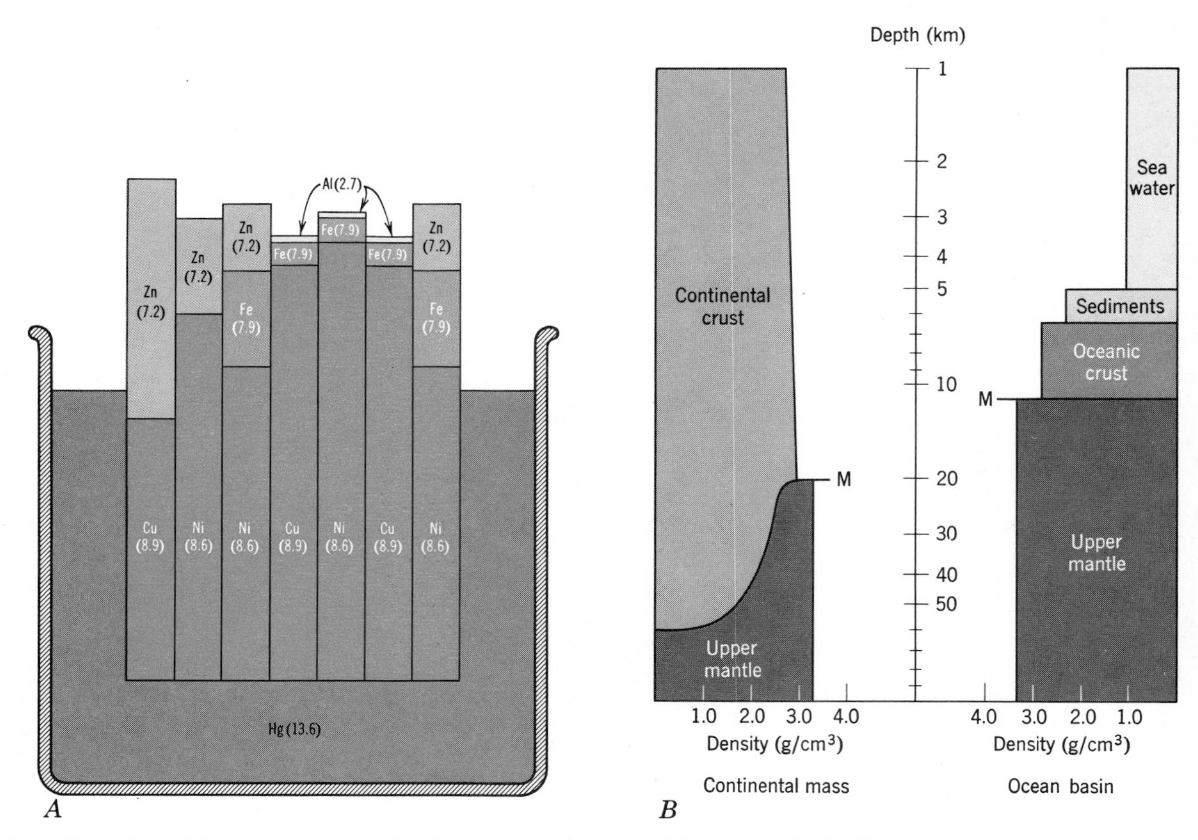

Fig. 21-22. Principle of isostasy. A. Perfect isostasy is attained by metal blocks floating in mercury. Dense metals (copper and nickel) in lower parts of blocks suggest Earth's upper mantle; zinc suggests sialic continental crust; iron, mafic crust; and aluminum, deep-sea sediments. Interface on each block between copper or nickel and zinc or iron suggests M-discontinuity, which lies well below tops of high-standing blocks and near tops of low-standing blocks. *B.* Thickness and density of materials beneath a continental mass compared with those beneath an ocean basin. Logarithmic depth scale greatly exaggerates proportions of upper 10km.

at the bottom. Each block obeys Archimedes' principle just as icebergs do. Because the weights of all blocks are the same, their bases are aligned in the mercury. But because of the various thicknesses and densities, the heights of the block tops and the levels of the interfaces between metals within each block are not the same from block to block. If we were to remove a slice from the top of one block and add it to another, the blocks would adjust to the change of mass. The loaded block would sink and the unloaded block would rise.

In the lithosphere things are not so simple. Nevertheless it is probable that at the base of the lithosphere, say at a depth of 100km, pressure from overlying material is everywhere the same. Therefore a section beneath an ocean basin "weighs" the same as one beneath a continental mass despite their contrasting compositions. A typical section beneath an ocean basin above the M-discontinuity includes about 5km of seawater having a mean density of 1.03g/cm^3, 1.3km of sediments of density 2.3g/cm^3, 4.7km of oceanic crustal rock with density of 2.8g/cm^3. Beneath this discontinuity is an 89km thickness of upper-mantle rock having a density of approximately 3.3g/cm^3 (Fig. 21-22, *B*). Beneath a continent the corresponding section of the lithosphere consists of approximately 34km of continental crustal rock having a density of 2.7g/cm^3 at the surface but increasing downward to 3.0g/cm^3 at the M-discontinuity. Only 66km of upper-mantle rock is present; its average density is $3.3 g/cm^3$. At a depth of 100km the pressures created by these contrasting columns are the same: $31.5 \times 10^3 kg/cm^2$.

The M-discontinuity within the Earth is analogous to the interface at the top of the copper or nickel, and the zone of low velocity within the upper mantle (Fig. 18-17) is analogous to the mercury. In mercury, a liquid, the adjustment to changes in weight is immediate and perfect because the liquid flows easily. Within the Earth we suppose that adjustment requires a longer time and is imperfect because it takes place by plastic flow of solid rock material, as ice flows in a glacier. Another potential natural complication, not shown by the experiment, is that above the base of the lithosphere materials can undergo phase changes. Slight changes of pressure, in response to conditions above the base of the lithosphere, may cause some minerals to undergo changes of phase. For example, at 30 kilobars and 1,530°C plagioclase changes to garnet plus kyanite plus quartz; and at pressures of 30 to 60 kilobars iron-rich olivine assumes a spinel structure.

Despite our incomplete knowledge of the lithosphere and the mantle, evidence justifies our conclusion that the interaction among gravity, flotation, and the variable densities of rock bodies can explain both the Earth's surface relief and vertical movements of the lithosphere. We can learn more about the densities of rock bodies by making precise measurements of the Earth's gravitational acceleration, g.

Variation in Earth's gravitational acceleration. Measurement of g in many localities has shown that small variations exist. Some of these variations are predictable. For example, the value of g at any locality is a function of the distance from the locality to the center of the Earth and of the densities of rocks in the lithosphere. The distance to the center of the Earth varies with latitude and local morphology, and the values used depend on our approximations of the Earth's shape. Calculations take into account the factors that influence distance to the center of the Earth. Rock bodies at the Earth's surface are sampled and their densities determined in the laboratory. From distance and density a prediction of the expected value of g is made. Measured values, which can be made to an accuracy of 1 milligal (1/1,000 of a gal), are then compared with predicted values. Any difference is termed a *gravity anomaly,* an unpredictable value of g.

A gravity anomaly results from variations in the densities of rocks. A positive anomaly denotes a measured value higher than predicted and a negative anomaly a measured value that is lower. In most localities gravity anomalies are small, but positive anomalies as great as 280 milligals and negative anomalies as large as 338 milligals have been detected. Positive anomalies are caused by the presence of rock bodies that are denser than their surroundings. Negative anomalies result from rock bodies that are less dense than their surroundings. Let us examine how gravity anomalies are related to mountains.

Buoying up mountains. Several lines of evidence suggest that mountains stand high because beneath them the crust is thicker than in surrounding areas. The first indication that variations in density occur beneath mountains came in the middle of the nineteenth century from attempts to survey precise points in India near the Himalayas.

An important yet simple tool used in this and other surveys is a *plumb bob,* a heavy conical weight at the end of a string suspended from the tripod that holds sighting instruments. Barring complications, the string is vertical. The Indian surveying party made astronomical observations on the presumption that its plumb lines were truly vertical. Checking its findings with those measured along the ground by standard methods, the group noted that the results disagreed. These inconsistencies could not be attributed to surveying errors.

Finally the explanation adopted was that the plumb lines had not been truly vertical after all. The lines had departed from the vertical because of the proximity of the Himalaya Mountains (Fig. 21-23). The mountain mass towering above the plains created a lateral attraction which deflected the plumb bobs toward the mountains. Although somewhat surprising, the attraction was not nearly so surprising as the results of a calculation made to check the deviation. Assuming equal densities of rock beneath plains and mountains and ascertaining the volume of the mountains from maps, the party computed the expected deflection, which was much larger than that observed. Manifestly, in the calculations too much mass had been assigned to

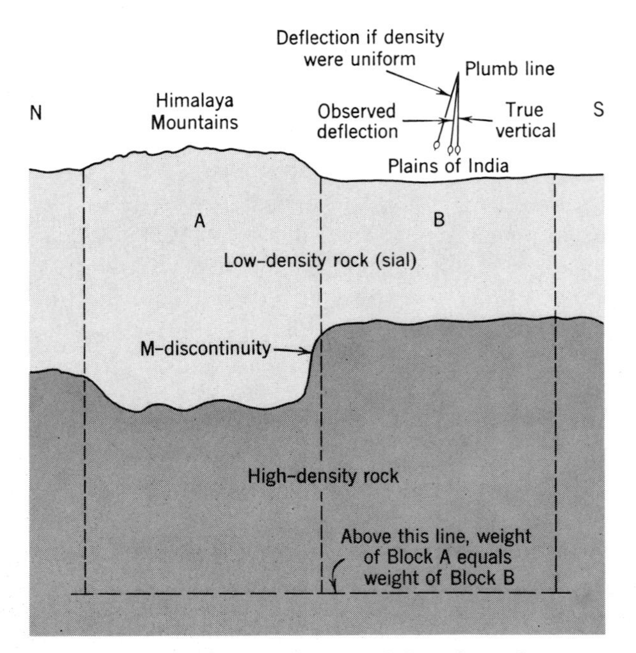

Fig. 21-23. Deviation of this plumb line from the vertical results from lateral gravitational attraction of high mountains. Because mountains are underlain by a thick mass of low-density rock, the observed deflection is less than that calculated by assuming uniform density beneath mountains and plains.

the mountains. The only way to reconcile the differences between the calculated and observed deflections is to suppose that the base of the mountains not only is thick but also extends deeper into heavier rocks, and displaces them. By making this assumption we reduce the computed mass of the mountains relative to the plains.

Seismology provides another line of evidence indicating that a hidden mass lies beneath mountains. Beneath many mountain chains the depth of the M-discontinuity is greater than elsewhere in the crust (Fig. 21–24).

Mountain chains having thickened bases are thought to be eroded in proportion to the density contrasts between mountain rocks and substratum rocks (Fig. 21-24). Let us assume the ratio of the densities is 2.7/3.3, or 9 to 11. Then loss of weight resulting from removal of 11m of rock by erosion would be balanced by deeper plastic flow of 9m of heavier material. If we assume perfect isostasy, for every 11m of rock removed the net lowering of the mountain surface will be only 2m. Eventually the thick part of the crust will be removed, but the constant replenishment of material from below helps prevent erosion from consuming all lands.

Where the crust is thin under mountains (only 20km in the Coast Ranges of California) or thick under plains (50km in the Great Plains east of the Rocky Mountains) we infer that buoyancy and isostasy operate on rocks having densities unlike those shown in Fig. 21-24, or that locally buoyancy and isostasy do not operate at all because of stronger preventing forces within the lithosphere.

Since 1960 geophysical results have indicated that the subcrustal relationships beneath continental masses, particularly beneath mountains, are more complex than was formerly believed. The older interpretations emphasized variations in thickness of the crust above the M-discontinuity and supposed that the density and composition of the upper mantle were everywhere uniform. New gravity and seismic data affirm that the properties of the upper mantle vary from place to place. Therefore explanations based solely on thickness of the crust are inadequate. Instead it is necessary to collect data on the characteristics of the entire lithosphere, including both crust and underlying upper mantle.

EVENTS IN A TECTONIC CYCLE

In Chapter 5 we mentioned the highlights of the tectonic cycle. Now that we are familiar with the three mountain chains we have traversed we can elaborate the tectonic cycle and fill in the gaps left in our earlier discussion. We can organize our list of events around six headings: (1) crustal subsidence and early accumulation of sediment; (2) deformation, further sedimentation, predominantly marine, and possibly large-scale submarine volcanism within elongate narrow belts; (3) great thermal events and large-scale compressive deformation; (4) deformation and further sedimentation, predominantly nonmarine, within elongate belts; (5) extensive faulting along steep fractures, creating elongate crustal blocks, some of which are elevated and eroded and others of which subside and collect nonmarine sediment; and (6) regional doming and transport of sediment outside the mountain belt.

1. As a result of subsidence the following events take place at the beginning of a tectonic cycle.

(a) Strata accumulate at the margin of a continent and become thicker with continued subsidence. The width of the zone of sediment

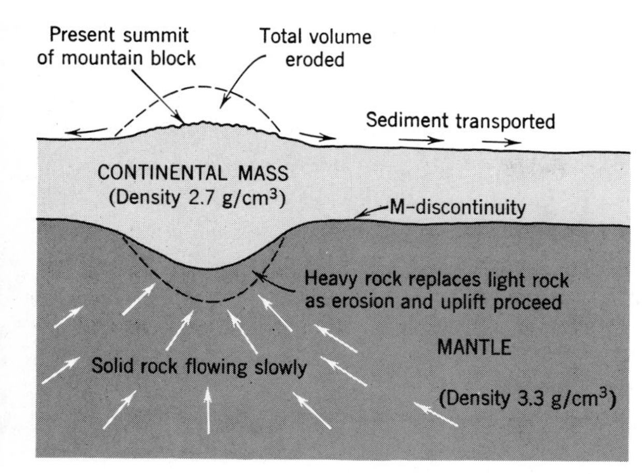

Fig. 21-24. Flow at depth and phase change at M-discontinuity are possible mechanisms for maintaining equilibrium in the lithosphere during erosion of a mountain mass and transport of sediment to distant areas.

accumulation is 600 to 800km. In this and other episodes of sedimentation, net subsidence of the crust largely controls the thicknesses and shapes of bodies of sediment. The relative rates of subsidence and sediment supply control depths of marine waters.

(b) Sediment composing the strata is derived initially from the interior of the continent.

(c) Many of the strata are marine; others are nonmarine.

(d) Among the marine strata several depth zones are represented; where shallow and deep zones occur together the deeper zone lies farther away from the continent.

(e) Sediment from shallow zones may slide or slump down submarine slopes to mix with sediments of deeper zones.

2. Accompanying deformation of various kinds, both during and after accumulation of the strata, the following events may take place:

(a) Parts of the sea floor marginal to the continent may be uplifted, thus locally reversing submarine slopes and creating large islands. The islands supply sediment to be transported toward the continent on one side and toward deeper water on the opposite side.

(b) Submarine extrusives and intrusives may accompany strata of deep-water marine origin.

(c) Subaerial eruptions can occur from upraised islands or from volcanic islands built up above sea level.

3. During the great thermal events, plutonism and metamorphism are active in parts of the areas where strata have been accumulating. The following events typically occur.

(a) Strata become regionally metamorphosed.

(b) Plutons of various sizes originate in intimate association with regionally metamorphosed strata.

(c) Strata are crumpled into folds and are stacked up as great thrust sheets. Along some thrusts strata have moved many kilometers.

(d) Direction of movement on the blocks above most thrusts is toward the continent; basement rocks can be thrust over sedimentary strata.

(e) Where strata from more than one former depth zone have been involved in thrusting, those from former offshore zones have been transported over strata from former shallower zones.

(f) Along a narrow zone near the transition from shallow-water strata to deep-water strata old basement rock underlying the sediments may be domed or thrust upward.

(g) After regional metamorphism the rocks may be upheaved, but the metamorphism need not deform strata in adjacent areas.

(h) Regional metamorphism does not operate after the last major episode of compressive deformation.

4. Deformation and sedimentation that continue after the great thermal and compressional climax involve closely coupled zones of uplift and subsidence. These are active within a general setting of elevation of the mountain belt and expulsion of the sea.

5. During extensive faulting the crust behaves as a brittle solid. Previously it exhibited great plasticity but now rigidity prevails. Crustal breakup typically is the last major event to be recorded in the structural fabric of rocks within mountain chains. Drainage remains predominantly interior; hence sediments accumulate within the mountain belt.

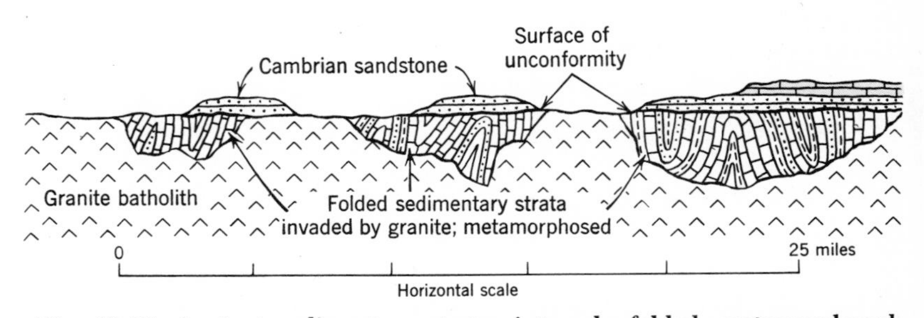

Fig. 21-25. Ancient sedimentary strata, intensely folded, metamorphosed, and intruded by granitic batholiths, lie unconformably beneath horizontal strata of Cambrian age. Such relationships, which reveal the existence of former mountain chains, are common in southern Canada and northern New York. Vertical scale greatly exaggerated.

6. During regional doming the drainage patterns are reorganized. Because external drainage becomes established, sediments eroded from the rising chain are carried outside the mountain belt.

From these complex geologic activities we realize that a tectonic cycle is an event of major importance. Many complications take place in addition to the process we have mentioned. Nevertheless, we can visualize a tectonic cycle as beginning when an area of the crust subsides beneath the sea and receives sediments from elsewhere. The source presumably is an uplifted mountain chain in its last stages of evolution.

Mountains are born of the sea. Once a tectonic cycle is inaugurated, the crust within the orogenic zone reworks itself vigorously. An active orogenic zone feeds upon itself—it has been termed "cannabalistic." Throughout the main phase of the orogenic period, active uplift, subsidence, and sedimentation are confined chiefly to the orogenic zone. The debris derived from the uplifted parts of that zone remains within it.

The final stage commences when the active zone begins to shed its debris elsewhere. Eventually erosion reduces the mountain zone. When this has happened the former mountains may become part of the basement (Fig. 21-25). After subsidence these newly made basement rocks may pass through additional tectonic cycles.

The implications of tectonic cycles have fascinated geologists for more than a century. In Chapter 22 we shall describe attempts to use mountains as a basis for a comprehensive theory of the behavior of the lithosphere.

Summary

1. Mountain chains are elongate belts within which sedimentary strata, many of marine origin, are unusually thick. The strata have been folded, thrust over one another, and in some places regionally metamorphosed.

2. Extrusive igneous rocks, some derived from submarine eruptions and others from subaerial eruptions, are abundant in many mountain chains.

3. Granitic rocks are present in most mountain chains. Such rocks may be (a) greatly elevated parts of the older basement on which the strata were deposited, or (b) younger batholiths intrusive into the strata.

4. The geologic histories of the Appalachians, North American Cordillera, and Alps display many similarities. The sequence of events reconstructed from these chains typifies that from nearly all other mountain chains.

5. Regional subsidence dominates the early stages. Parts of a continent sink beneath the sea; sediment accumulates.

6. Narrow belts of the sea floor are deformed, uplifted, and eroded; adjacent belts subside and receive the products of erosion. The crust behaves plastically.

7. Eventually a climactic deformation occurs; it generates great folds, thrusts, regionally metamorphosed tracts, and granitic batholiths. The entire region may emerge from beneath the sea.

8. Further activity entails closely coupled zones of uplift and erosion and of subsidence and deposition. After the climactic deformation, however, non-

marine strata predominate and the crust displays brittle behavior.

9. The final stage is marked by regional uplift of an elongate dome whose axis parallels that of the mountain chain.

10. Ancient mountain chains become parts of the basement rocks of continental masses. Areas of basement rocks may remain rather stable as parts of continental shields, or may subside to form parts of the floor upon which are deposited the new strata for a new mountain chain.

Selected References

Bailey, Sir E. B., 1935, Tectonic essays, mainly Alpine: Oxford, England, Clarendon Press.

Billings, M. P., 1960, Diastrophism and mountain building: Geol. Soc. America Bull., v. 71, p. 363–398.

Bucher, W. H., 1956, Role of gravity in orogenesis: Geol. Soc. America Bull., v. 67, p. 1295–1318.

Gilluly, James, 1967, Chronology of tectonic movements in western United States: Am. Jour. Sci., v. 265, p. 306–331.

Hsu, K. J., 1965, Isostasy, crustal thinning, mantle changes, and the disappearance of ancient land masses: Am. Jour. Sci., v. 263, p. 97–109.

King, P. B., 1959, The evolution of North America: Princeton, N. J., Princeton Univ. Press.

Pierce, W. G., 1966, Jura tectonics as a décollement: Geol. Soc. America Bull., v. 77, p. 1265–1276.

Thomas, Lowell, 1964, Book of the high mountains: New York, Julian Messner, Inc.

Thornbury, W. D., 1965, Regional geomorphology of the United States: New York, John Wiley.

Trümpy, Rudolf, 1960, Paleotectonic evolution of the central and western Alps: Geol. Soc. America Bull., v. 71, p. 843–908.

Zim, H. S., 1964, The Rocky Mountains: New York, Golden Press.

CHAPTER 22

Evolution of the Lithosphere

Search for a unifying theory
Contraction theory explains all
Decline and fall of the contraction theory
Mountains as keys to the lithosphere
Geosynclinal theory
Studies of remanent magnetism
Pole wandering and continental drift
Surprises from the deep-sea floor
Hypothesis of sea-floor spreading

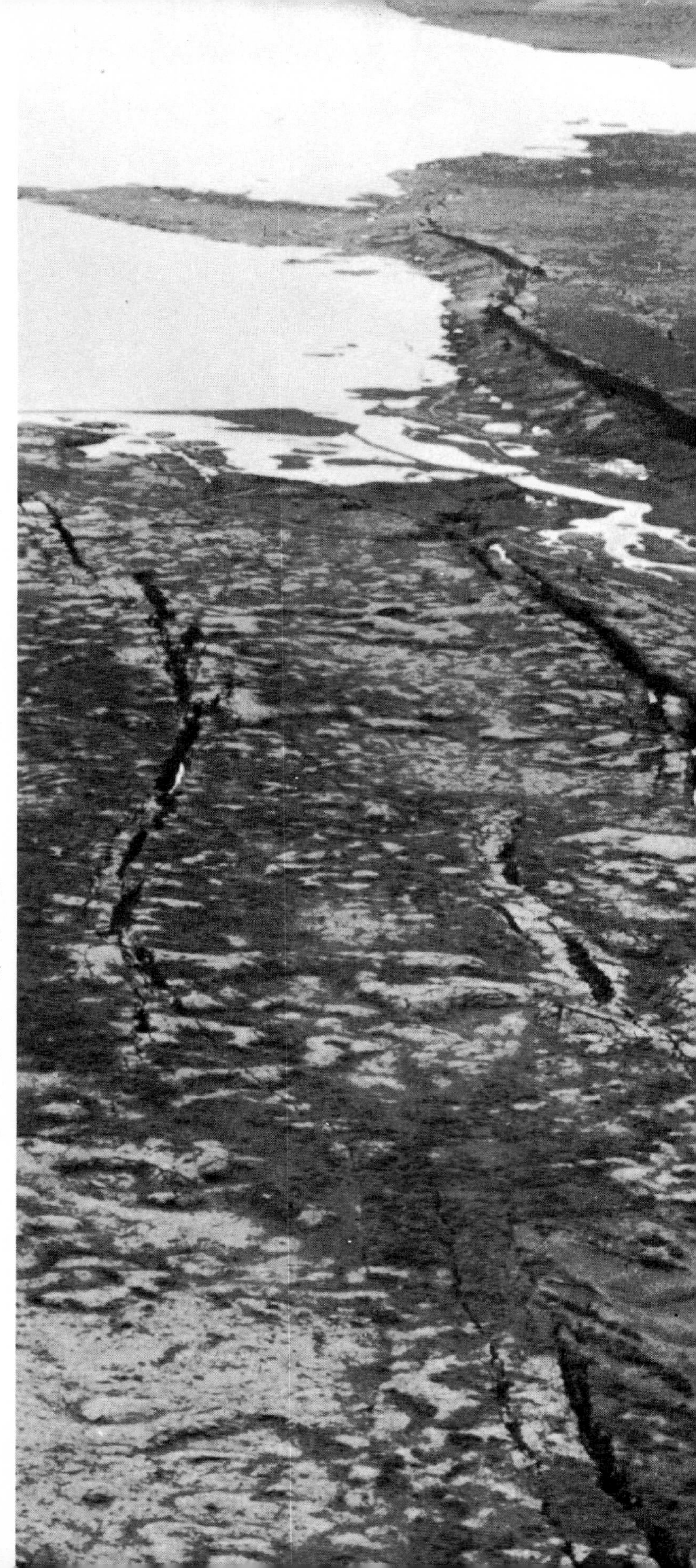

Plate 22

Scarp created by recent movement along fault at margin of Thingvellir rift valley, Iceland, a portion of the Mid-Atlantic Ridge that stands above sea level. (*Sigurdur Thorarinsson.*)

CHAPTER 22 *Evolution of the Lithosphere*

GEOLOGY'S SEARCH FOR A UNIFYING THEORY

Within the last few years new geophysical and geologic discoveries about the deep-sea floor have given rise to a number of important ideas concerning the Earth's behavior. This information offers the prospect of a new unifying theory of the lithosphere's evolution to geologists who, since the late nineteenth century, have operated without the benefit of a generally accepted framework uniting many scattered phenomena. Early in the twentieth century geologists turned to mountains as keys to the lithosphere in a search for a substitute for the older contraction theory which had been thoroughly shattered. Let us review the contraction theory, trace its demise, examine geologists' attempts to replace it through analysis of mountain chains, and finally dwell on the makings of a new theory based on the latest geophysical intelligence.

The nineteenth century's great unifying theory. In the nineteenth century geologists operated within an all-embracing view of Earth history known as the contraction theory. The concept was based on the observed geothermal gradient and the supposition that all heat flowing outward derived from a formerly molten Earth. The inevitable contraction accompanying such cooling of the globe neatly explained the great compression suggested by the folds and thrusts in mountain chains. Because contraction seemed to explain not only the origin of mountains but also many other geologic phenomena it was widely believed.

The eminent physicist Lord Kelvin went a step further, calculating an age for the Earth by mathematical analysis of heat flow (Chap. 6). Few geologists dared dispute Kelvin's result, based as it was on the "incontrovertible" laws of mathematics and physics. Thus, with their account in Kelvin's bank of time "frozen" at about 25 million years, geologists were receptive to corollaries to the hypothesis of cooling and contraction. One such corollary, about the Earth's interior, seemed logical on the basis of surface observations. It stated that a worldwide layer of molten material existed beneath a thin rocky crust. Such a molten layer offered a ready source of lava.

A second corollary, of great significance to geologists working in Precambrian rocks, stipulated that certain rocks in the Precambrian shields had originated under environmental conditions differing radically from those of the present day. Some of these rocks were thought to have formed while the Earth's surface was hotter than 374°C, the critical temperature of water. Above this temperature the liquid state does not exist; hence no ordinary sedimentary rocks could form. All ancient Precambrian volcanic rocks were correlated and assigned to the "great igneous interval" named the *Archean.* By accepting this corollary geologists rid themselves of many field problems. Given an area of Precambrian volcanic rocks and one of metamorphosed sedimentary strata, they did not hesitate to assign an Archean age to the volcanic rocks and a later age to the former sedimentary rocks. After all, the theory stated that no sedimentary rocks could form in Archean time; that sediments did not originate until after cooling and precipitation had occurred on a colossal scale. The great early rainfall visualized by the theory finally created lakes and oceans so that the geologic cycle as we know it could commence.

Decline and fall of the contraction theory. During the late nineteenth and early twentieth centuries the contraction theory fell victim to the often-reenacted scientific tragedy in which a beautiful theory is killed by an ugly fact. In this particular tragedy many facts did the killing. Every prediction derived from the nineteenth-century version of the contraction theory has proven wrong. In the late nineteenth century, the discovery of radioactivity, and the realization that radioactivity is accompanied by the emission of great quantities of heat, revolutionized all of science. In most sciences new vistas appeared immediately and dramatic progress was made.

Over the long run the same can be said of geology, but the immediate impact of radioactivity on geology was devastating. Two props of its unifying theory crumbled: (1) A source of heat, hitherto unsuspected even by the most eminent mathematical physicist, was demonstrated. As the geothermal gradient no longer required original heat to explain its existence, the inevitability of contraction vanished. (2) Precambrian rocks proved to be vastly older than had been supposed. No longer could all "old-looking" volcanic rocks be axiomatically assigned to the Archean. Radioactive age determinations ultimately made possible the correlation of Precambrian rocks. But until isotopic age determinations had been made, geologists were at a loss to classify these rocks and confusion ruled. When ages were finally sorted out Archean sedimentary rocks were demonstrated.

In addition to all these changes, seismology deprived geologists of their layer of molten material close beneath the Earth's surface. Analysis of earthquakes disclosed no evidence of large bodies of molten material outside the Earth's core.

Picking up the pieces. During the first half of the twentieth century some geologists began to analyze mountain chains intensively with the idea that an understanding of orogeny would provide the missing basis for a new theory of crustal behavior. The emphasis on understanding mountains was soundly based because mountain chains are large-scale features whose origin is inseparably connected with the history and behavior of the lithosphere, indeed of the Earth itself.

MOUNTAINS AS KEYS TO THE LITHOSPHERE

If geologists could solve two outstanding problems connected with mountain chains they would have the keys to unlock many mysteries of the lithosphere. These mountain problems are: (1) the relationship between the sites of deposition of the strata of ancient mountains and the environments where modern sediments are accumulating, and (2) the source of the motive power for deformation of the strata composing mountains. Although these two problems can be considered separately they have many interlocking implications. Suppose we begin with the deposition of the strata.

Concept of a geosyncline. In the middle of the nineteenth century, before the demise of the old contraction theory, geologists were well aware that somehow mountains grew out of the sea. In his analysis of the history of the Appalachians, the famous American geologist J. D. Dana recognized that subsidence of the crust and sedimentation had preceded final deformation of the strata and uplift of the chain. Dana proposed the term *geosyncline* for the elongate trough which had subsided in the Appalachian region during the Paleozoic Era, and *geanticline* for the adjacent area which had been uplifted to supply large quantities of the sediments that were deposited in the geosyncline.

Subsequently geologists have expended great effort in attempting to understand orogeny by analyzing geosynclines. In the process they have employed the term in so many ways that it has become nearly impossible to define what a geosyncline is, much less to recognize one in the modern world. To some geologists a geosyncline is any surface that subsides. To others it is the sediment filling a trough that has subsided. To still others it is both the subsided trough and the sediment filling the trough. Most, though not all, geologists restrict the term to mountain chains and agree that a ***geosyncline*** is *a huge linear or arcuate segment of the Earth's crust, which by subsidence below sea level and sediment accumulation incubates a mountain chain.*

In the history of a mountain chain the geosynclinal stage is a time of predominantly marine sediment accumulation and accompanying deformation. Once subsidence below sea level ends and the mountains appear, the geosyncline vanishes. According to some, the only way by which we can recognize a geosyncline with certainty is to wait until it has been obliterated, providing proof that it gave rise to a mountain chain. If we compare the mountain chain to a butterfly the geosyncline becomes analogous to a caterpillar. The concept of a geosyncline began with the butterfly and geologists have been perplexed in their efforts to recognize the characteristics of the caterpillar. Before we attempt to ascertain if any modern geosynclines exist let us summarize some features of ancient geosynclines.

Ancient geosynclines. We have no difficulty in recognizing ancient geosynclines because we can see the mountain chains born of them. Therefore the patterns of these chains disclose the patterns of geosynclines. From a glance at a map of the moun-

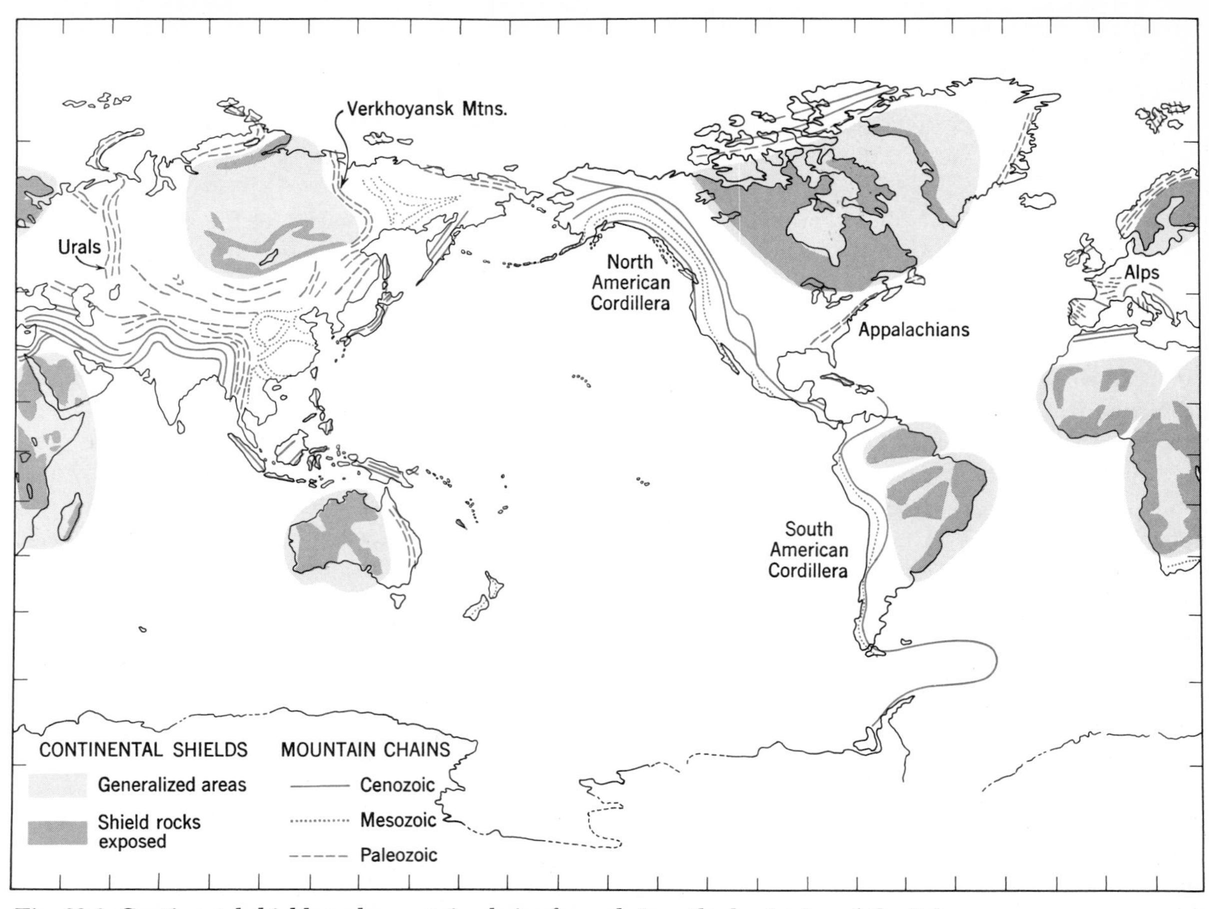

Fig. 22-1. Continental shields and mountain chains formed since the beginning of the Paleozoic Era. Each continental shield itself consists of mountain chains of pre-Paleozoic age. (Shield areas after C. O. Dunbar, 1966.)

tain chains formed since the beginning of the Paleozoic (Fig. 22-1) we can judge three important characteristics of ancient geosynclines. First, their lengths are thousands of kilometers and their widths hundreds of kilometers. Second, they are situated at the margins of continental shields which they therefore frame. Third, where several are parallel, the younger ones tend to lie farther away from the continental shields.

The great lengths of geosynclines, like those of mountain chains, are one of their striking features. Some circular basins have subsided through great vertical distances and have acquired thick sediments. Although such basins contain thick sediments they lack great length; hence nearly all geologists would exclude them from geosynclines.

With respect to continental shields the marginal positions of the geosynclines of North and South America, Europe, and Australia are an obvious feature requiring no further emphasis. In those regions the geosynclines are marginal not only to the continental shields but also to continental masses themselves. By contrast, some geosynclines of Asia, namely those antecedent to the Urals and Verkhoyansk Mountains, though framing two shield areas, are far from being marginal to a continental mass. To the contrary, each marks out about a third of the great Eurasian continent by crossing it at right angles to its length.

The decrease in the age of mountain chains with distance from a continental shield is best demonstrated in eastern Asia. On the mainland of China a pair of fold chains occurs, the chain folded during the Paleozoic lying inland of and parallel with the one folded during the Mesozoic. Parallel with and east of this pair is the main fold belt of Japan, of Cenozoic age.

The same general pattern is evident in the belts of ancient folds within the Canadian Shield. Isotopic-age determinations reveal that the rocks of the North American continent become younger outward from a central area in which the rocks are more than 2,500 million years old (Fig. 22-2). The pattern displayed by the isotopic ages is supported by a systematic change, with age, in the strontium isotopes in granitic rocks associated with the fold chains. The ratio of radiogenic- to nonradiogenic strontium (Sr^{87}/Sr^{86}) in the mantle changes with time through radioactive decay. The only logical explanation for the observed changes in the ratio within crustal rocks is that the materials for these rocks have been derived from the upper mantle at different times. Therefore igneous rocks of various ages are distinctively "tagged" by their Sr^{87}/Sr^{86} ratios.

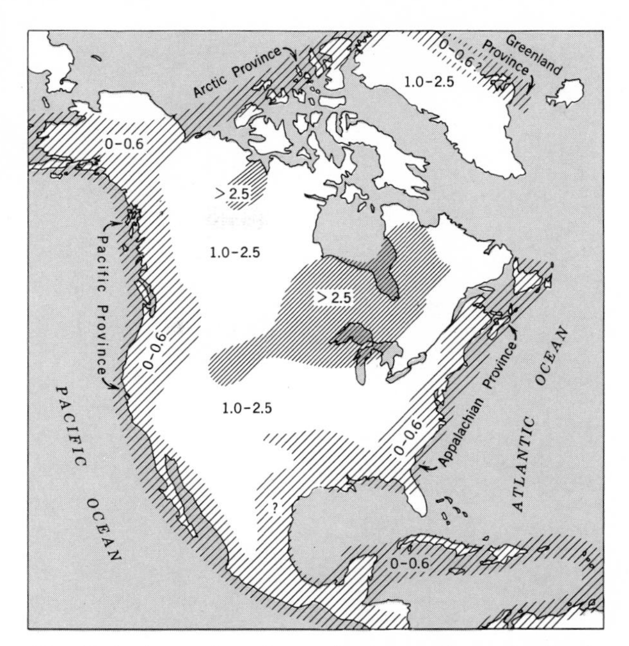

Fig. 22-2. Ages of major geologic provinces in North America, based on isotopic dates of large-scale granite-forming events. Numbers represent isotopic dates of rocks in billions of years. (After Engel, 1963.)

Based on their positions and ages, one important concept of geosynclines is that they develop at continental margins and serve as catalysts by which continents enlarge at the expense of ocean basins. This is an important concept to bear in mind as we proceed.

Modern geosynclines? One way by which we can attempt to identify modern geosynclines is by interpreting the environments in which the strata of ancient geosynclines were deposited and then by looking for comparable environments in the modern world. Most American geologists familiar with the Appalachian Valley and Ridge Province and the Rocky Mountains part of the North American Cordillera have concluded that nearly all the thick pile of marine strata in these two regions had been deposited in shallow water. Accordingly, in their search for modern geosynclines they excluded all deep parts of the present seas and limited themselves to shallow parts. An important corollary of their conclusion is that the great subsidence indicated by the thick strata had proceeded at a rate in equilibrium with sedimentation.

By contrast, most European geologists familiar with the Alps decided that many of the Alpine marine strata had been deposited in deep marine waters. Therefore they searched the modern deeps for examples of geosynclines.

As a result of modern explorations of the deep seas, mounting doubts have gathered that the strata of the Valley and Ridge Province convey the whole Appalachian or geosynclinal story. What is more, the Appalachian geologists may have been too willing to exclude from their search the areas in the modern world where the water is deep. Without becoming involved in the complicated subject of interpreting the depths of water in which ancient strata were deposited let us try another approach and inspect the modern world for features with geosynclinal dimensions. Only two such exist: (1) oceanic rises and ridges, and (2) marginal trenches.

Oceanic rises and ridges, impressive features about which we shall have more to say presently, resemble continental mountain chains, and hence geosynclines, in length and width. Moreover the relief of the ridges is as great as that of typical mountains (Fig. 15-7). The ridges consist of enormous bodies of basalt; on them sediments are rare or absent. Hence the rocks of ridges do not resemble those of any mountain chain on the continents and we conclude that ridges are not examples of modern geosynclines.

Sea-floor trenches merit our consideration on the basis of their dimensions and patterns (Fig. 22-3).

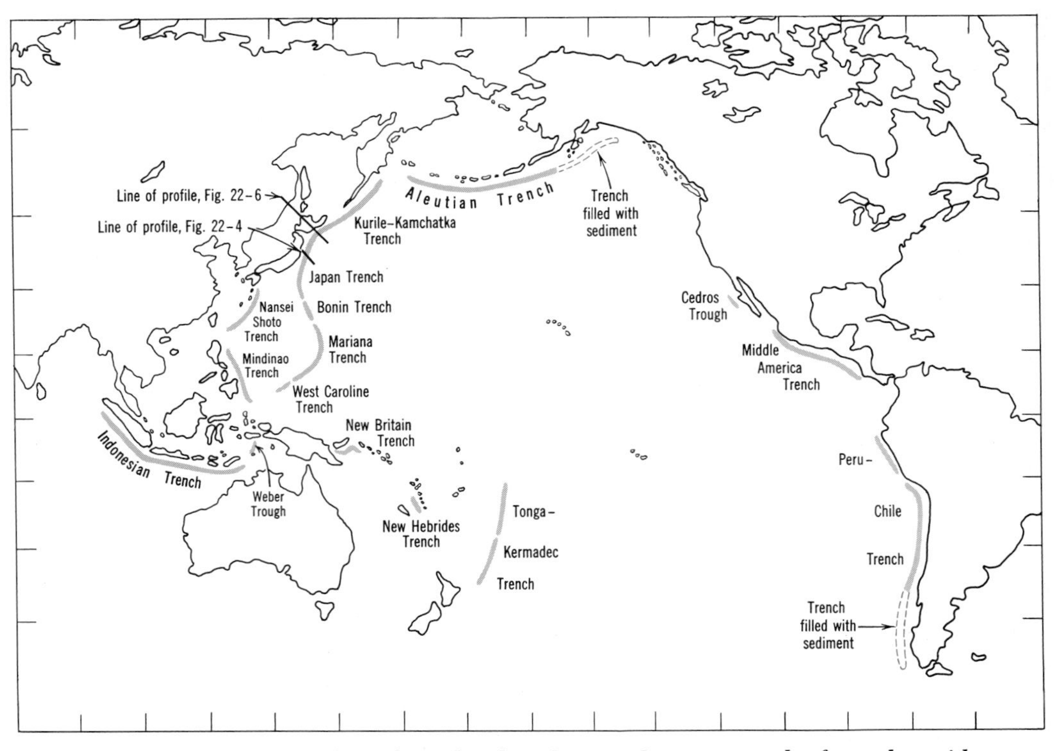

Fig. 22-3. Deep-sea trenches of Pacific and Indian Oceans. Compare trends of trenches with trends of mountain chains (Fig. 22-1).

Sea-floor trenches are thousands of kilometers long and hundreds of kilometers wide. Most modern trenches are situated in positions marginal to continental masses; nearly all of them are along the border of the Pacific basin. Trenches are linear, gently curved, and sharply arcuate, exactly like mountain chains. How do trenches score on our other tests for recognizing geosynclines? Many Pacific trenches lack sediments. This alone would almost seem a sufficient reason for rejecting them, because geosynclines and thick sediments go together hand in hand.

Manifestly a rocky trench must have subsided rapidly. Even if a large supply of sediment is not forthcoming from the nearest continent, sediments tend to fill trenches. In the tropics the sea itself can contribute carbonate skeletal material so fast that any shallow part of the tropical sea floor that subsided slowly would be kept shallow by sediments and could not become a trench.

But suppose a trench had become filled with sediment? Could we then call it a geosyncline and predict that it may become a future mountain chain? The sediments certainly would be thick; average water depth in the Pacific trenches is nearly 10,000m. If such trenches became filled with sediment the thickness would be considered to be of "geosynclinal" proportions by every geologist. If a trench were to become filled with sediment the strata would consist chiefly of deep-sea varieties; shallow-water types would be restricted to the top of the pile.

To settle the status of sea-floor trenches as potential modern geosynclines we need to know whether: (a) any filled trenches are known, and (b) any ancient deep-sea sediments appear in mountain chains. Each subject contains numerous complexities.

The amounts of sediment in trenches vary. Parts of the trenches in the western Pacific, mentioned above, contain almost no sediment. Others, such as the Middle America and Puerto Rico Trenches, contain small prisms of sediment whose upper surfaces form narrow abyssal plains. In the Aleutian and Peru-Chile Trenches the amounts of sediment vary laterally from one end of the trench to the other. The Aleutian Trench disappears eastward

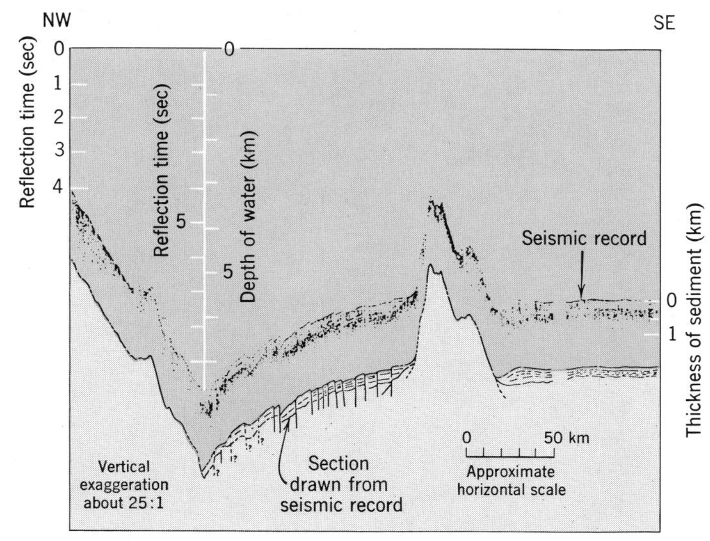

Fig. 22-4. Vertical normal faults on east slope of Japan Trench east of Honshu are shown on a seismic record. For position of profile see Fig. 22-3. (After W. J. Ludwig and others, 1966.)

beneath about 5km of sediment. Beyond the southern end of the Peru-Chile Trench lies a prism of sediment about 3km thick.

Topographically a completely filled trench would not strike us as being a trench. If we used only an echo-sounder we should be unable to identify it as a trench. But seismic exploration shows that, locally at least, buried trenches exist. Despite the difficulties in compiling convincing proof, conviction is growing that bodies of sediment lying beneath continental rises (Chap. 15) may be fillings of former trenches. If so, it may be more than mere coincidence that continental rises rim the Atlantic basin, where trenches are nearly absent, and that trenches surround the Pacific basin, where almost no continental rises are present.

The answer to the question concerning the occurrence of ancient deep-sea sediments in mountain chains likewise is not altogether conclusive. There are many parallels between the sediments of certain parts of mountain chains and recent deep-sea sediments. As we remarked in Chapter 15, the subject of depth of water at time of deposition of sediments is a difficult one. Too often in the past the interpretations have been biased toward shallow water. A reappraisal of depth interpretations should be made in the light of modern knowledge of marine sediments, both shallow and deep. Some such reappraisals suggest that ancient deep-sea sediments are present among the strata of mountain chains.

Continuous seismic profiles in the Kurile-Japan Trench demonstrate that high-angle faults are present and that along them crustal blocks underlying the central part of the trench have subsided relative to the blocks forming the walls of the trench (Fig. 22-4). This finding is contrary to hypotheses stating that trenches have resulted from large-scale compression and downbuckling of the crust. According to the advocates of compression the direction of maximum shortening is at right angles to the axis of the trench. We do not yet know how to relate the evidence of crustal stretching shown in Fig. 22-4 to the subsurface geophysical setting of trenches.

Geophysical measurements indicate that the axes of trenches are marked by great negative gravity anomalies (Fig. 22-5). Extending from the trenches toward the adjacent continental masses and becoming systematically deeper in the direction of the continent are narrow zones in which earthquake foci are concentrated (Fig. 22-6). The depths of focus become deeper away from the trenches, finally reaching a maximum of about 700km at points about 800km away from the axes of the trenches. This arrangement has prompted the interpretation that great landward-dipping shear zones extend from the trenches diagonally downward to the depth below which no earthquakes can occur. Studies of the sense of first motion on earthquakes generated within the inclined zones suggest that all such zones in the Pacific are strike-slip faults (Fig. 18-7, A). Ordinarily such faults are vertical, as is the San An-

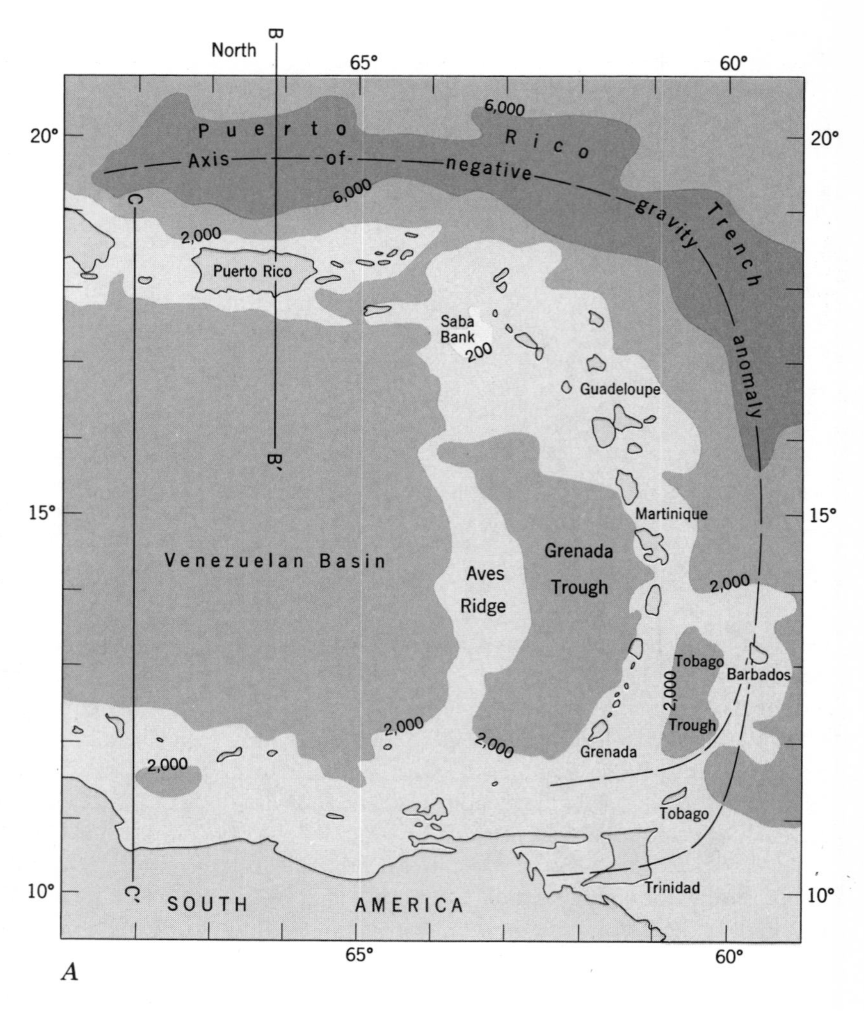

Fig. 22-5. Topography and geophysical sections in the Caribbean region. *A.* Generalized map showing depths of water and locations of geophysical sections. (After R. J. Hurley, 1966.) *B* and *C.* Topography, inferred structure of the lithosphere, and curves of gravity anomalies along north-south transects across Puerto Rico Trench (*B*) and Venezuelan Basin (*C*). (Seismic profile in *B* after Maurice Ewing and John Ewing, 1962; geophysical section and curves of gravity anomalies after Manik Talwani, G. H. Sutton, and J. L. Worzel, 1959; *C* after J. L. Worzel, 1966.)

dreas fault. Strike-slip motion parallel with the axes of trenches could explain the evidence of crustal extension and normal faults (Fig. 17-21).

Rows of volcanic islands paralleling sea-floor trenches presumably derive their molten rock from the inclined zone of earthquake foci. Nonvolcanic islands lying between the rows of active volcanoes and some trenches have subsided and then been uplifted, but many of them have not been deformed otherwise.

On Taiwan, which is not located near a marginal trench nor an inclined zone of concentrated earthquake foci, large-scale folding and thrusting toward the Asiatic mainland has occurred during the later part of the Cenozoic Era. The deformed strata appear to be the filling of a former sea-floor trench. If such a trench formerly existed there it must have been accompanied by an inclined zone of earthquake foci. Yet neither trench nor inclined zone of earthquake foci are present today. Was their disappearance connected with the deformation of the strata?

Causes of compression. Whatever we decide about the results of a comparison between ancient geosynclines and the modern world, we must still come to grips with the problem of the motive power of orogeny. Five hypotheses have been proposed as the causal agent of the folds and thrusts seen in mountain chains: (1) contraction of the mantle, (2) lateral displacement of continents (continental drift), (3) convection currents within the mantle, (4) gravity sliding, and (5) rise of granitic batholiths.

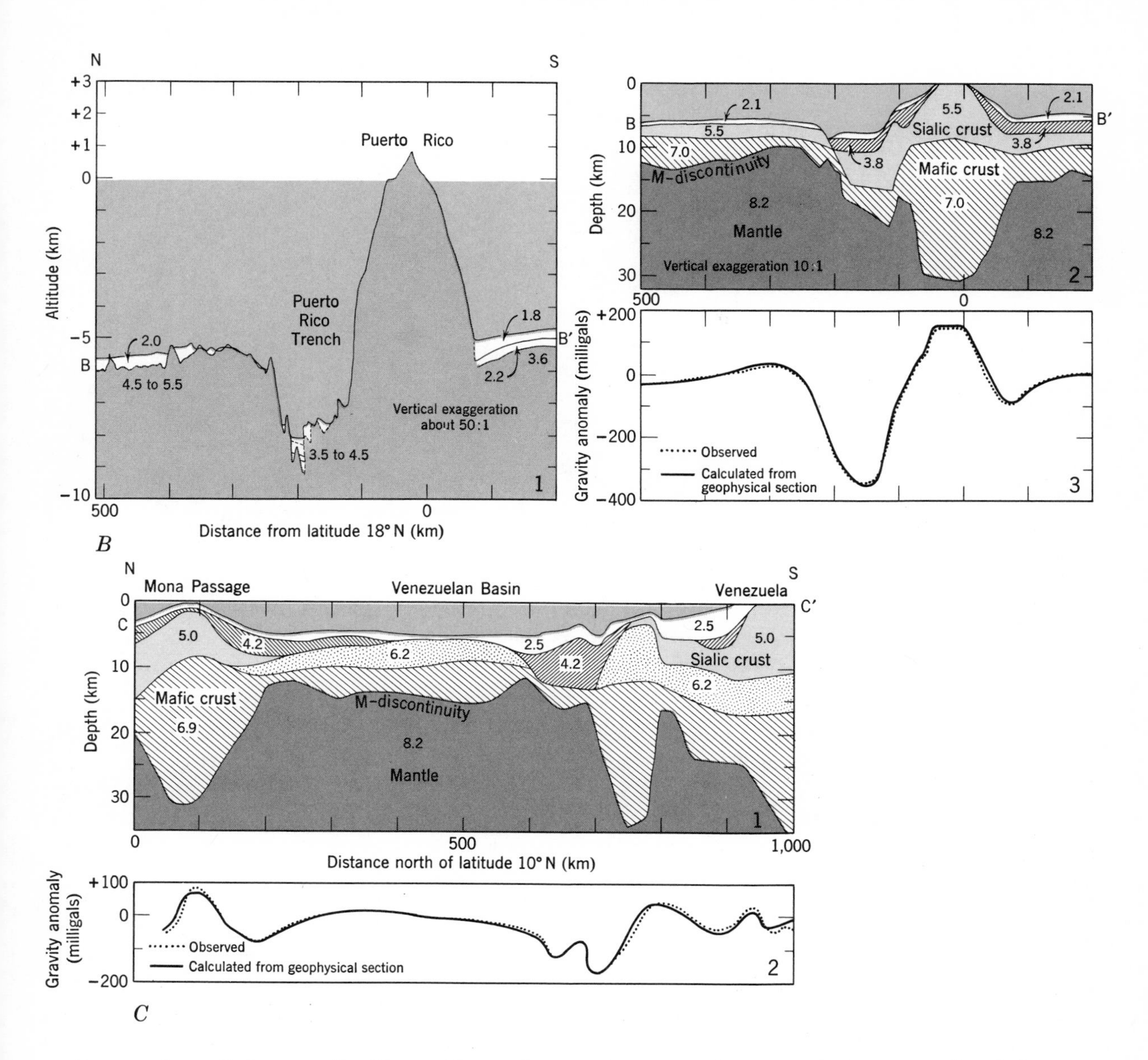

Contraction of the mantle. The evidence for compression within mountain chains has been so impressive to some geologists that they have tried to maintain the contraction hypothesis by amending it to be consistent with modern geophysical information. Evidence exists that the mantle, though solid, has lost and is still losing material upward to the lithosphere and to the Earth's surface. Because of these losses the mantle must be contracting. Here is the evidence.

From seismology we infer that dense metallic material of the core lies beneath less dense silicate material of the mantle. This suggests that at some time the Earth must have passed through a stage when it was hot enough for iron and nickel, originally distributed uniformly, to be fractionated from the silicates and to collect at the center of the Earth. We do not know if the now-solid mantle has ever been entirely molten. But even if the mantle once had been entirely molten we infer that solidification occurred so long ago that contraction resulting from it could not have been a factor in the origin of mountain chains, as visualized by the nineteenth-century version of the contraction hypothesis. Every mountain chain originated after the geologic cycle began, and that could not have occurred until after the mantle had solidified.

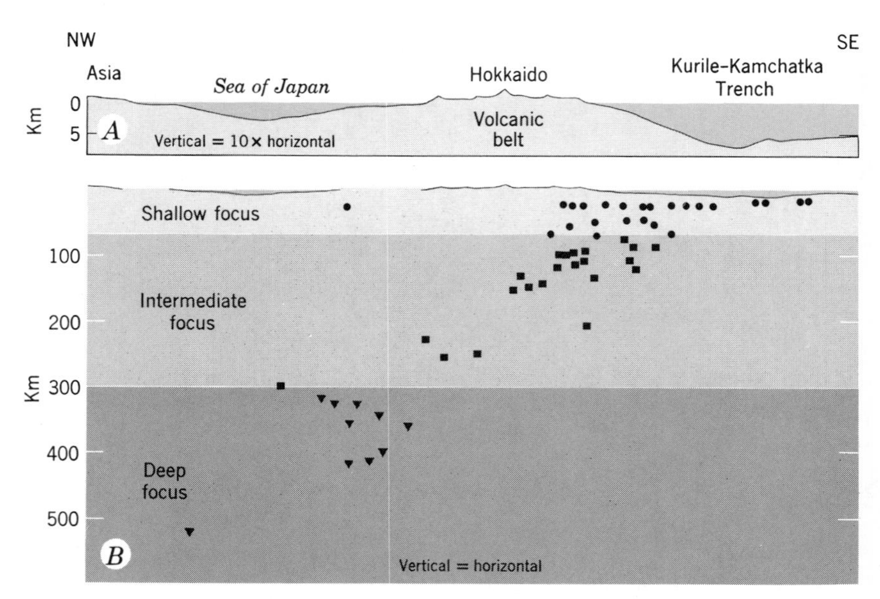

Fig. 22-6. Profile and section from Asian mainland to sea floor southeast of Kurile-Kamchatka Trench. (Fig. 22-3 shows location of profile.) *A.* Profile with vertical scale exaggerated. *B.* Profile and section without vertical exaggeration. Earthquake foci are shown by same symbols as in Fig. 18-18. (After Gutenberg and Richter, 1954; reference at end of Chap. 18.)

More convincing arguments lie closer at hand. Modern geochemical theory favors the concept that the water of the hydrosphere originated from volcanic emanations. If the theory is correct, the mantle has lost and is still losing water and other volatiles to the Earth's surface, because volcanic materials come from the mantle. Many igneous rocks, including nearly all basalt and much granite, are thought to have originated by partial melting of the upper mantle. The unmelted residue must occupy a smaller volume than does mantle material which has not been subjected to partial melting. The mantle is losing not only material but also heat. Radioactive elements are unequally distributed in rocks. Such elements in an average granite generate 10 to 1,000 times more heat than they do in average mafic rocks. Yet despite this unequal distribution of radioactive elements the average geothermal gradient beneath ocean basins, where mafic rocks prevail, has been found equal to that beneath continental masses where granitic rocks predominate. The natural radioactivity of an average granite creates enough heat to account for the average geothermal gradient beneath continents. But the natural radioactivity of average mafic rocks is far too small to account for the average geothermal gradient beneath ocean basins. Therefore we must conclude that the heat flow beneath ocean basins consists largely of contributions from the mantle. By losses of volatiles, silicates, and heat the mantle must be contracting.

Contraction of the mantle would create compressive stresses within the overlying lithosphere. If such contraction were uniform throughout the mantle, we might be right back where we started—the results would be the same as those postulated by the outmoded contraction hypothesis. In other words mountain chains should be scattered at random on the Earth's surface like wrinkles in the skin of a drying apple. The pattern of the world's mountain chains (Fig. 22-1) does not remotely resemble such a pattern; so we conclude that the lithosphere is not underlain by a mantle that is uniformly shrinking. What is more, evidence of crustal extension and large-scale pulling apart, which exists in such places as the Basin and Range Province, speaks against worldwide compression in the lithosphere. We know likewise that such compression could not generate widely spaced zones of folds because the strength of the lithosphere is too small to transmit the stresses through the distances between them. Although we reject universal and continuous compression within the lithosphere as a

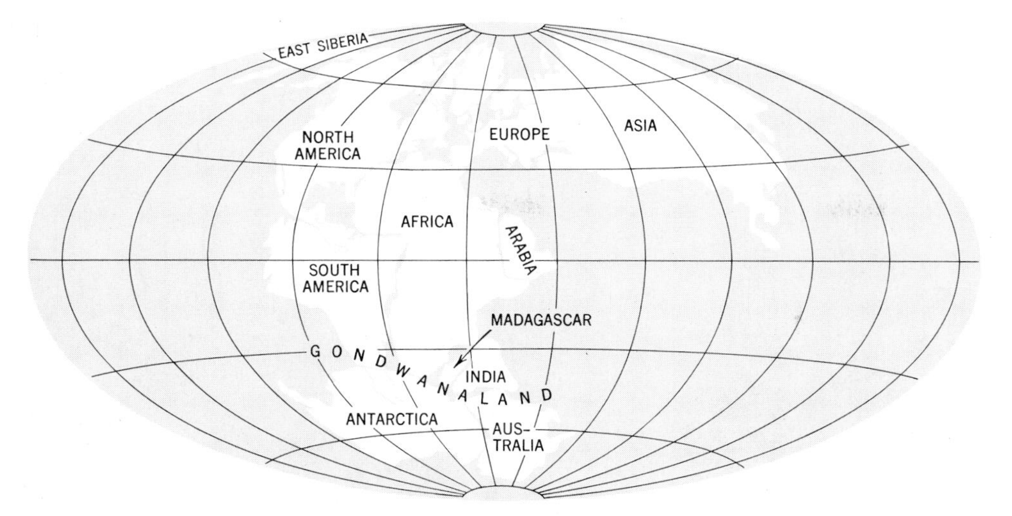

Fig. 22-7. One reconstruction of the hypothetical landmass believed by some to have existed before the continents drifted apart. The southern part of this landmass is called Gondwanaland. (After J. T. Wilson, "Continental drift." Copyright © April 1963 by Scientific American, Inc. All rights reserved.)

result of uniform contraction of the mantle, we have every reason to expect that local contraction may occur within the mantle and be responsible for local compression within the lithosphere. Such compression could contribute to orogeny.

Continental drift. The concept that continents can shift laterally followed close upon the proof that under the weight of continental glaciers the surface of the lithosphere had moved up and down by flowing. After all, ran the argument, if the lithosphere can flow up and down, why can it not also flow laterally? Early arguments in favor of drifting continents derived mainly from the similarity of the shape of the coasts on the opposite sides of the Atlantic Ocean (Fig. 22-7). Further evidence came from parallels in the geology among rocks of Late Paleozoic age in India, Australia, South Africa, and South America, and also among rocks of both Early and Late Paleozoic age in the northern Appalachians and in northwestern Europe. Advocates of drift supposed that sialic continents formed a single mass until late in the Paleozoic but thereafter began to move apart like "rafts" floating in a "sea" of denser mafic rock. Moreover, these sialic rafts were thought to have skimmed off the deep-sea sediments, heaping them up into mountain chains on their forward sides. By this means drifting continents were thought to supply the motive power for folding. Forces of the Earth's rotation were cited as the driving mechanism of the drifting continents. According to enthusiasts some drifting is still in progress.

Formidable objections were raised against the specific contentions of nearly every early scheme of continental drift. But as we shall see, the hypothesis of continental drift has assumed new stature as a result of paleomagnetic studies. During the first rounds, although partisans favoring drift may have been right, they based most of their case on the wrong reasons and were unable to visualize a mechanism consistent with other evidence.

Convection currents within the mantle. The hypothesis of convection currents within the mantle proposes that hot material rises, flows laterally, and after cooling, sinks again, like the air in a hot-air heating system. What evidence is available concerning this hypothesis? We have seen already that heat from the mantle must be rising upward beneath the ocean floors. What becomes of it is suggested by considering another aspect of the geothermal gradient. In mines and borings far from volcanoes and hot springs the temperature gradient, a function of the geothermal heat flow and the thermal conductivity of the rocks, ranges from 10° to 50°C/km; the average value is 30°C/km (Chap. 2). Presumably we can extrapolate the measured gradients downward as far as the thermal conductivity of the rocks remains constant, and as far as any large source of heat. Even though we are uncertain of the exact temperatures at depth, simple extrapolation illustrates that thermal conditions beneath continents

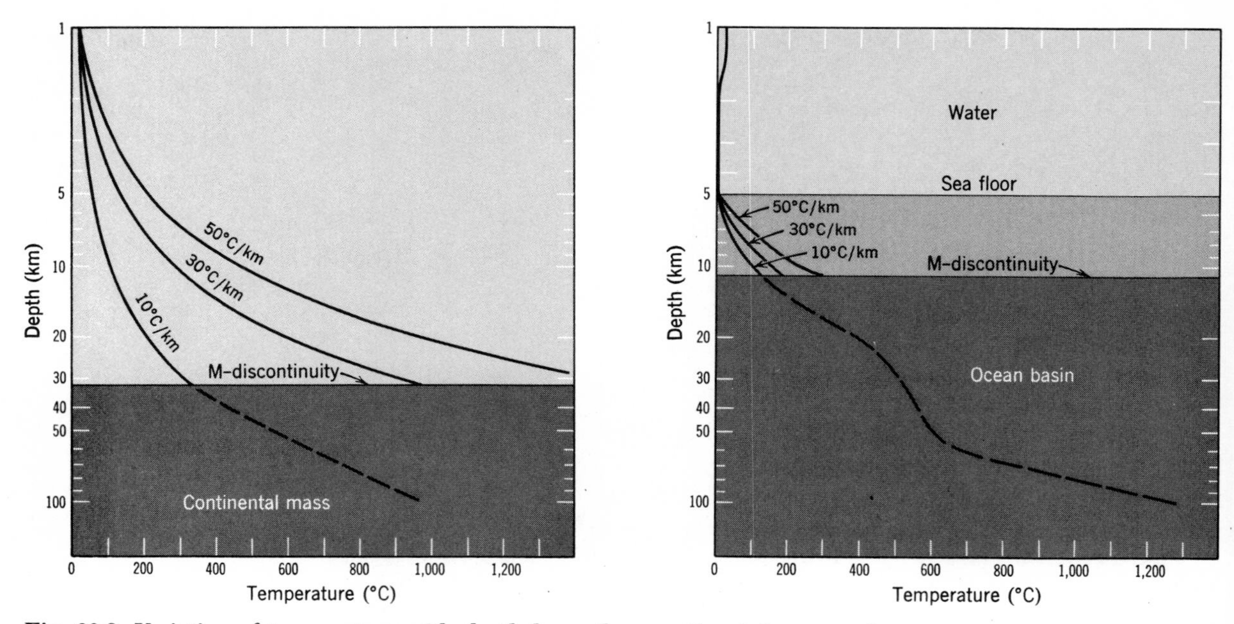

Fig. 22-8. Variation of temperature with depth beneath a continental mass and an ocean basin, based on linear extrapolation of geothermal gradients measured in uppermost 6km beneath continents. Temperature lines curve because depth scale is logarithmic. Curves do not show local variations resulting from changes in thermal conductivity of rocks and from deep-seated sources of heat. (Dashed curve based on F. R. Boyd, 1964.)

contrast with those beneath ocean basins.

The projected temperature at the M-discontinuity beneath a continent ranges from a minimum of 320°C to a maximum of 1,550°C; the average is 950°C. We lack comparable information about the geothermal gradient beneath the deep-sea floor, but by extrapolating from measurements beneath continents we can estimate temperatures within the lithosphere beneath oceans. An important difference arises from the fact that the temperature at the sea floor, about 5km lower than the surface of continents, is close to 4°C, whereas the temperature at the surface of many parts of continents is about 20°C (Fig. 22-8). Therefore the 4°C point on the sea floor stands opposite the level where the average extrapolated temperature beneath continents is about 140°C. Unless drastic changes of thermal conductivity exist within these segments of the lithosphere, we are confident that a lateral temperature gradient exists. Within the outer part of the lithosphere, heat must be flowing laterally from beneath continental masses to beneath ocean basins. Probably but less certainly the temperature at the M-discontinuity beneath the oceans is less than that at the same interface beneath continents.

In theory, convection currents within the mantle can explain the results of both the tensile stresses and the compressive stresses in various parts of the lithosphere. Tension would develop above places where the material from a rising current parted to flow laterally, and compressive stresses would appear where two laterally flowing currents turned downward. The downward-flowing currents could explain both folding and subsidence of narrow belts. Although early schemes emphasizing convection have been abandoned, the idea has been modified in the light of new geophysical information. The revised version forms the basis of many modern theories of crustal behavior. The convection-current hypothesis supposes that the mantle lacks effective strength for long-term stresses and is capable of moving 1 to 2cm/year in response to thermal gradients.

Gravity sliding. Another way to explain large-scale folds in mountain chains is to suppose that under the influence of g_t, large masses of material slide along sloping parts of the Earth's surface. The hypothesis proposes that strata are folded and even stacked up in great sheets as they slide along an inclined surface of displacement, much as a layer of snow is folded when it moves down a sloping roof

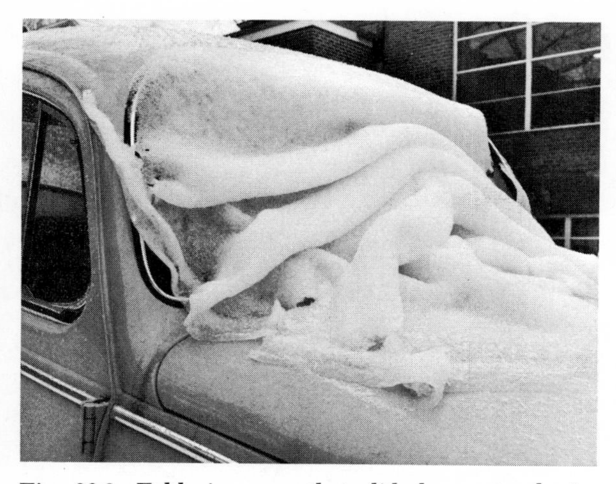

Fig. 22-9. Folds in snow that slid down steeply inclined windshield. (Yonkers *Herald-Statesman*, courtesy Westchester-Rockland Newspaper Group.)

or down the windshield of an automobile (Fig. 22-9). Amplitudes of folds made by slump of sediments on the sea floor range from a few centimeters to hundreds of meters. No line divides large from small. Therefore we are certain that some folded strata have resulted from gravity sliding. The question that is unresolved is whether folds on the scale of those in mountain chains resulted from this mechanism.

The Prealps in Switzerland (Fig. 21-6) consist of numerous sheets of strata moved many miles northward from the places where they were deposited. The higher sheets are thought to have come from the greatest distances. Parts of the sheets dip northward, the direction of their transport. The northward dip has been cited as an argument in support of gravity sliding.

The importance of gravity is not to be denied. Where masses of plastic rock are squeezed upward they become subject to the tangential component of gravity. But an appeal to gravity to explain the folds and thrusts does not solve all the problems of orogeny; the questions of subsidence, sedimentation, deformation during sedimentation, and later uplift require other answers.

Rise of granitic batholiths. The rise of granitic batholiths from deep within the lithosphere or from the mantle to higher levels has been thought to take place by the shouldering aside of strata. When a batholith rises it pushes strata out of the way, crowds them together, and creates folds and thrusts. Although such a mechanism undoubtedly operates, we do not yet know if its scale is large enough to account for the deformation of mountain chains. Many chains display folds and thrusts far removed from batholiths. Therefore it is doubtful that the rise of batholiths constitutes the only mechanism for originating mountain structures.

These contending hypotheses have been much debated and compared with the modern version of the contraction hypothesis. When weighed against geologic evidence from the land, they led to no decisive results. Despite the great progress made in analyzing the puzzle of orogeny, numerous pieces of the puzzle were still missing. Therefore nothing resembling a great unifying theory emerged from the efforts to use mountains as the keys to the lithosphere. But new possibilities have been opened up by recent geophysical intelligence.

RECENT GEOPHYSICAL INTELLIGENCE

The development of new instruments has permitted geophysicists to measure quantities with a precision undreamed of a few decades ago. Great progress has resulted in many fields; of most importance to geology are the magnetic measurements.

Remanent magnetism in rocks. In Chapter 2 we described the principles of remanent ("fossil") magnetism: each magnetic particle in a rock sample retains a tiny field whose orientation and polarity coincide with those of the Earth's magnetic field at the time the rock originated. Paleomagnetic reconstructions based on measurements of remanent magnetism depend upon assumptions about the orientation and behavior of the Earth's present magnetic field. Before we can appreciate paleomagnetic results we need to look closely at the Earth's present field and the assumptions made about it.

Earth's magnetic field. The Earth's magnetic field (Fig. 22-10) resembles that of a huge dipole magnet. Because it extends through the Earth this dipole magnet is said to be an axial dipole.

We describe the orientation of a magnetic line of force by its ***declination,*** *D, the clockwise angle from true north and lying in the horizontal plane,* and its ***inclination,*** *I, the angle with the horizontal measured in the vertical plane passing through the line of force* (Fig. 22-11). If we measure inclination at many points at various latitudes, we find that it is 90° at the North Magnetic Pole, becoming successively smaller southward and 0° at the Magnetic Equator. In the Southern Hemisphere inclination

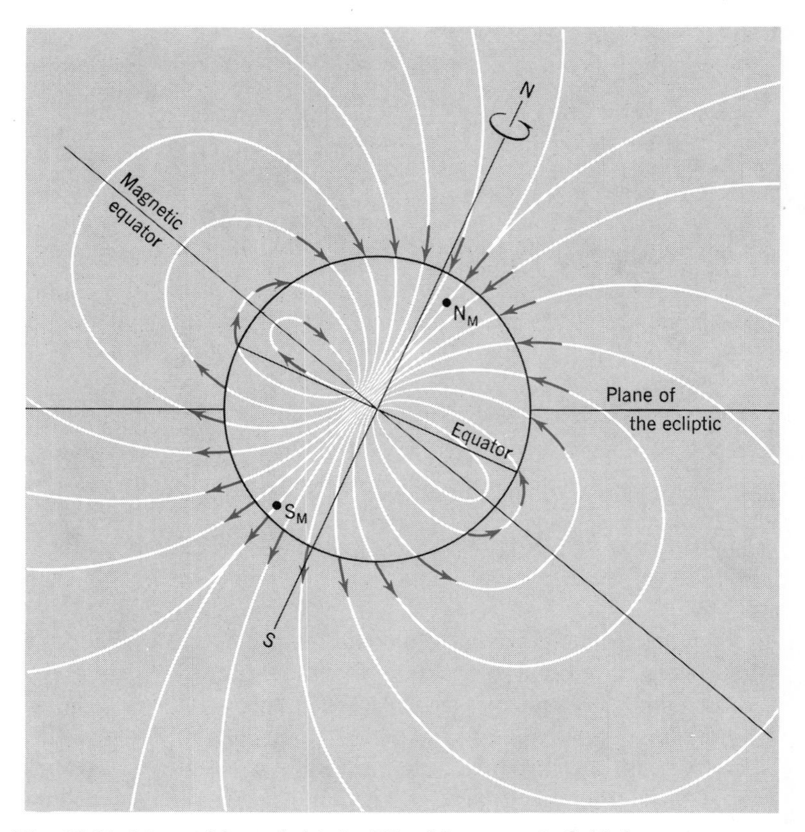

Fig. 22-10. Lines of force (white) of Earth's magnetic field shown on great-circle section along meridian 15°–165°. Line connecting North and South Poles of rotation is the 105° meridian. In 1960 the North Magnetic Pole (N_M) lay at Lat. 73°N and Long. 100°W, 1,920km south of the North Pole of rotation, and the South Magnetic Pole (S) at Lat. 68°S and Long. 143°E. A line connecting the magnetic poles does not pass through the center of the Earth. North-seeking ends of magnetized needles point in directions indicated by blue arrows.

increases from 0° at the Magnetic Equator to −90° at the South Magnetic Pole. Plots of *inclination* versus *declination* define the position of the magnetic pole.

Most interpretations of paleomagnetic measurements depend on two fundamental assumptions: (1) the shape of the Earth's magnetic field has always remained that of an axial dipole, and (2) the axis of the dipole has always been nearly coincident with the Earth's pole of rotation.

(1) The ancient shape of the Earth's magnetic field is difficult to prove. Presumably we could determine its shape at any given time in the past by measuring remanent magnetism in many samples of rocks of that particular age. But as we go farther back, the geologic record becomes more and more fragmentary; hence we are unable to collect all the samples required. Measurements of rocks of Cenozoic age have been made in the necessary quantity to support the inference that the dipole field has persisted for more than 50 million years. Despite uncertainties and lack of proof about the shape of the field at earlier times, most geophysicists accept the assumption that the Earth's magnetic field has always been an axial dipole.

(2) The concept that the magnetic pole always coincides with the rotational axis, likewise, is uncertain. Reconstruction of paleolatitudes from magnetic data on pole positions depends upon the supposition

that the magnetic pole and the axis of rotation lie close together. At present these two poles are as far apart as are Washington, D. C., and Denver, Colorado. Furthermore, we know from careful measurements made since 1540 in European observatories that the magnetic pole has wandered (Fig. 22-12). Because the magnetic pole has demonstrably shifted independently of the pole of rotation, are we justified in concluding that paleolatitudes can be reconstructed from paleomagnetic data? Again the answer, although hedged by uncertainties, is accepted by most geophysicists as affirmative.

Polar wandering supports continental drift. The orientations of remanent magnetism in rocks of varying ages have been determined. Although samples of rocks of Oligocene and younger age show no striking departures between the magnetic poles and the present poles of rotation, samples older than Oligocene do show departures. The apparent

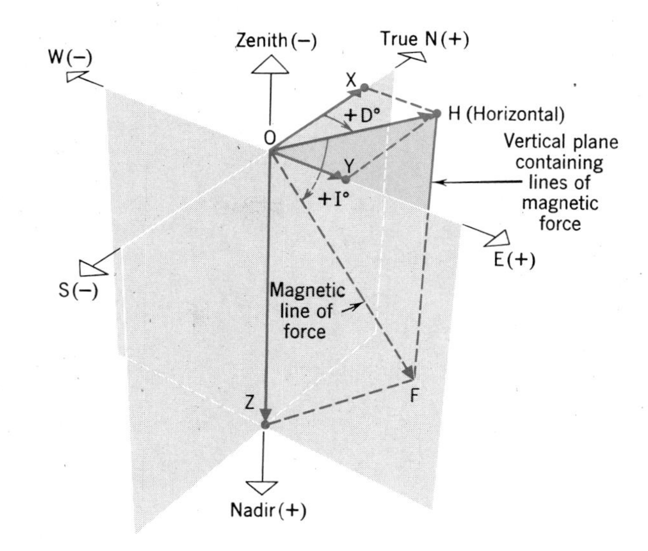

Fig. 22-11. Standard reference planes and directions for lines of force of Earth's magnetic field.

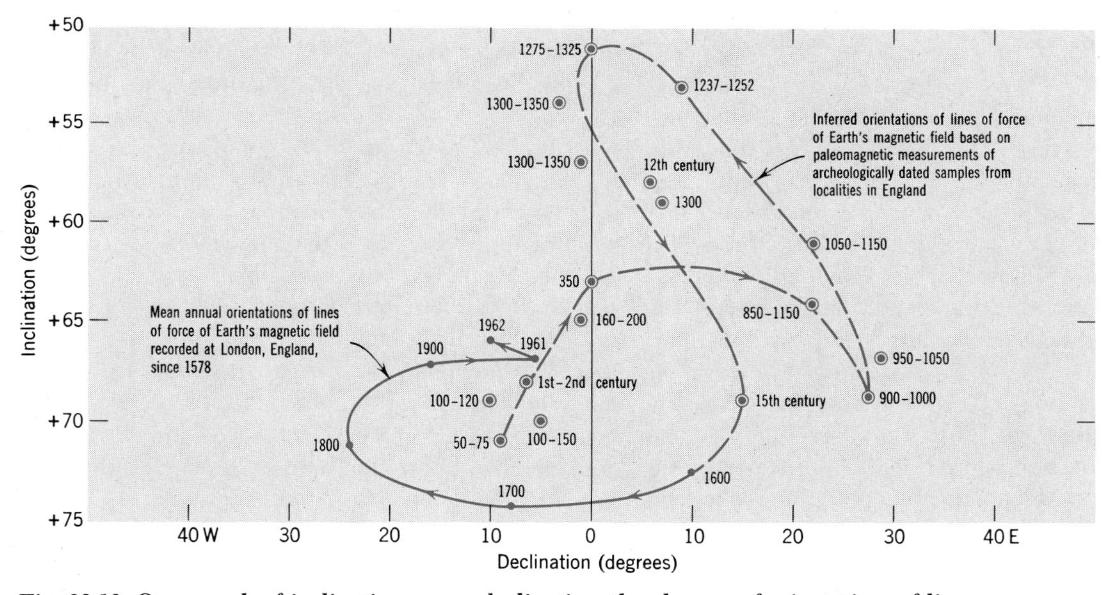

Fig. 22-12. On a graph of inclination versus declination, the changes of orientations of lines of Earth's magnetic field describe an irregular curve. The changes result from shifts in positions of magnetic poles. Data were recorded at London (solid blue curve) or were inferred from paleomagnetic measurements on samples dated historically or archeologically (dashed blue curve). (After Keith Runcorn, 1964.)

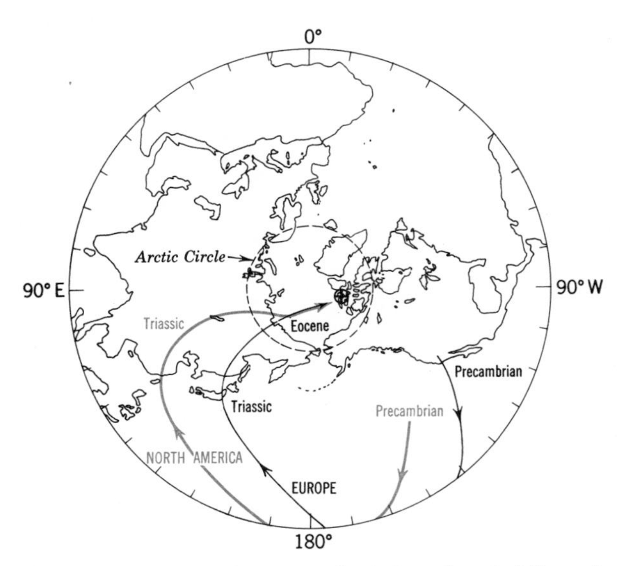

Fig. 22-13. Curves that trace the inferred path followed by the North Magnetic Pole through time, based on paleomagnetic measurements in North America (blue curve) and Europe (black curve). After Eocene time the two curves merge. (After Northrop and Meyerhoff, 1963 based on Allan Cox and R. R. Doell, 1960.)

path of the magnetic pole can be followed in reverse, from successively older samples, across the Arctic Ocean, across Siberia, and through a wide circle across the Pacific Ocean, ending off the coast of California in Precambrian time (Fig. 22-13). The pole positions determined from samples of the same geologic age from Europe and from North America are not identical. How can we explain this divergence?

The most logical explanation is based on the interpretation that Earth's magnetic field has remained an axial dipole and that only one north magnetic pole existed. The divergence in reconstructed positions of this pole could have resulted from progressive drift of the two continents away from each other. Ancient pole positions reconstructed from the two continents can be made to coincide closely if drift is assumed and if the continents are moved back to the positions they are thought to have occupied before the drift began.

Such reconstructions have forced geologists to reexamine their position to the concept of continental drift. The magnetic data strongly support the notion that drift has occurred, but they do not agree with the older idea that all continents were joined together until late in the Paleozoic and thereafter drifted apart, each following a single direction. Instead the magnetic data suggest that although net drift in one direction may have occurred, from time to time the actual directions of drift may have changed and even reversed. Although they lend strong support to the concept of drifting continents, paleomagnetic data shed no light on the mechanism of such drifting.

Paleoclimatic variations. The geologic record contains abundant evidence that climate changed while strata were accumulating. Extreme climates are indicated by glacial deposits and by marine evaporites. In between are other sediments, such as limestones and dolostones, formed in warm seas; laterities and bauxites, resulting from tropical weathering; coal, produced in subtropical and temperate swamps; and dune sands created in low-latitude deserts. In each case it is necessary to establish not only that the feature in question originated in the climatic zone indicated, but also that features similar to those listed have not been confused with them. Organisms provide still other clues. The distribution of many species is restricted by climatic zones bounded by parallels of latitude. Reef-building corals of the kind associated with certain algae indicate tropical zones. The composition of populations of modern organisms yields clues of latitude that are applicable to the fossil record. Cold-climate populations are characterized by few species and enormous numbers of individuals, whereas warm-climate populations are comprised of many species as well as many individuals.

A challenging task for geologists is to reassess the geologic record in order to reconstruct ancient latitudes. Such reconstruction could provide an independent check on paleolatitudes reconstructed from paleomagnetic measurements.

Although paleomagnetic results on accessible rocks have shed much new light on many geologic problems, these results do not answer some important questions about the origin of geosynclines and mountains. Great new developments, including other significant magnetic studies, have come from the sea floor.

Geophysical data from the oceans. The oceans have proved to be an extraordinarily fruitful area for geophysical research. New data of great importance have come from previously known oceanic rises and ridges as well as from newly discovered major features such as fracture zones.

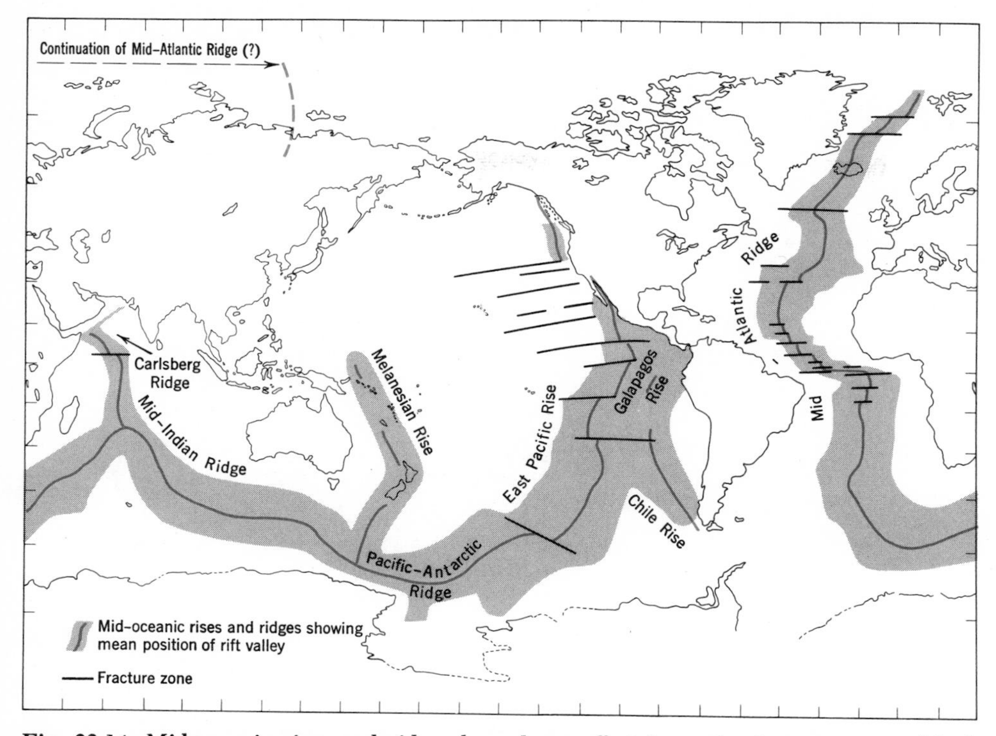

Fig. 22-14. Mid-oceanic rises and ridges have been offset by major fracture zones (black lines). (After B. C. Heezen, 1962.)

Oceanic rises and ridges. Waves of excitement swept through the geologic world when results of modern exploration of the oceanic rise-and-ridge system first appeared in 1959. The ridge system was traced continuously through more than 64,000km (Fig. 22-14). Along many parts of its length the crest of the ridge is characterized by a linear depression with a flat floor, where a narrow block has subsided between two steep faults (*rift valley*) (Pl. 22). Shallow-focus earthquakes are clustered along the rise-and-ridge system; in some parts higher-than-average values of heat flow have been found. Seismic surveys have detected evidence of crustal structure unlike that of a typical deep-sea basin. Despite all these remarkable disclosures, an even greater surprise remained. This came to light from magnetic studies in areas adjacent to the fracture zones.

Fracture zones on the deep-sea floor. When the floor of the eastern Pacific Ocean was explored with the precision echo-sounder, enormous linear scarps were found (Fig. 22-15). Their relief and shapes suggested an origin by faulting; the suggestion was strikingly supported by the results obtained with a sensitive magnetometer. On opposite sides of the scarps a distinctive pattern of linear magnetic anomalies was found, trending at nearly right angles to them. The anomaly pattern was interrupted at each scarp. But by shifting adjoining blocks through varying distances ranging from several hundred to more than a thousand kilometers, the interruptions could be eliminated.

Despite clear evidence that the blocks along the fractures have shifted through great distances, most parts of the fracture zones appear to be inactive at present. Except where the zones offset active mid-ocean ridges, as in the equatorial Atlantic, no earthquake epicenters from them have been recorded.

Ridges and magnetic anomalies. The remarkable linear patterns of magnetic anomalies shown to have been offset by the great fracture zones occur undisturbed along some oceanic rises and ridges. The linearity of the anomalies parallels the trends of the ridges. Magnetic-anomaly curves made transverse to the ridges display other remarkable features. Anomalies having large amplitudes characterize the crests of some ridges (Fig. 22-16). Such central anomalies are thought by some geophysicists to

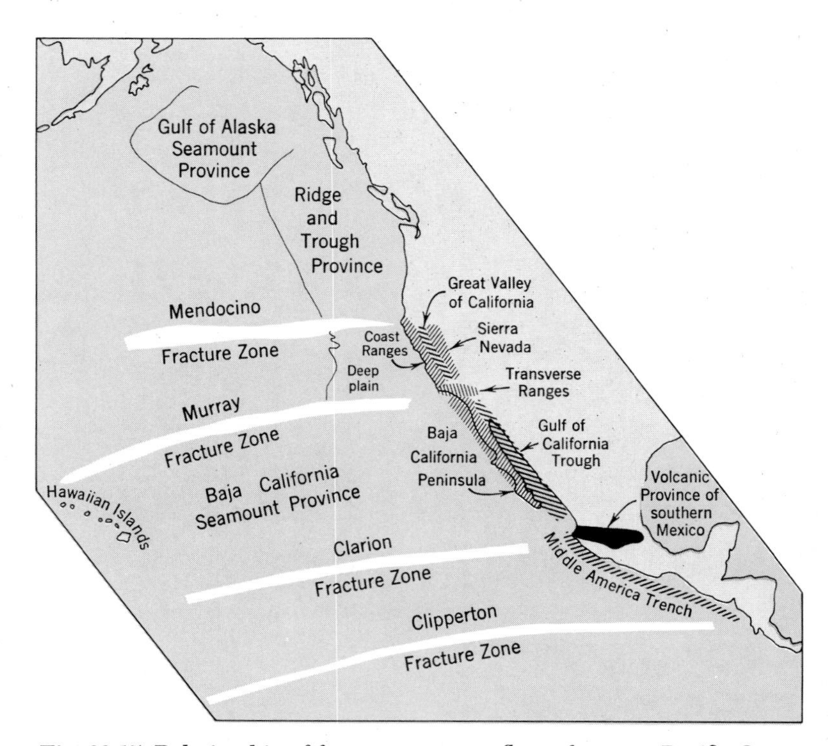

Fig. 22-15. Relationship of fracture zones on floor of eastern Pacific Ocean to major geologic features in western North America. (After Menard, 1955.)

reflect a single vertical body of highly magnetized rock, 10 to 15km wide, extending down to depths of 8 to 10km below the ridge crest. The geothermal gradient sets a lower limit to all anomalies, because at the Curie point magnetic fields vanish. Another explanation is that a body of rock only about 2km thick in the central block is more strongly magnetized because it is the only block composed exclusively of young material having uniform polarization. In either case the anomaly indicates material at a temperature below its Curie point and *not* a large body of magma or lava.

Each flank displays an anomaly curve having many anomalies of nearly constant wavelength but with amplitudes that decrease outward from the crest. The pattern of anomaly curves on the flanks is so symmetrical that the mirror image of the curve for one flank practically serves as the curve for the other.

Of the various rock configurations capable of explaining magnetic-anomaly curves on the flanks of ridges, geophysicists prefer the one based on parallel belts of highly magnetic basalt in a layer about 1.7km thick extending from 3.3 to 5.0km or 2.3 to 4.0km below sea level, depending on depth of water, but with the direction of polarity in each belt being opposite to that in adjacent belts.

If this hypothesis is correct, then it leads to a remarkable chain of further hypotheses. The most logical way to explain the polarities is to suppose that nearly continuous intrusion of magma and extrusion of lava take place along the center of the ridge. As the magma and lava solidify they would assume the orientation of the Earth's magnetic field at the time. Further upwelling of molten material along the centerline of the ridge would drive apart the older rocks and create new ones. When the polarity of the Earth's magnetic field reverses, as it is known to have done, the newly cooling igneous rock would become reversely polarized. When this central belt splits apart, two belts of reversely polarized rock would begin to migrate across the sea floor, each belt spreading away from the ridge down its opposite sides.

The hypothesis implies that an active oceanic ridge is like an open, oozing cut in the skin of the

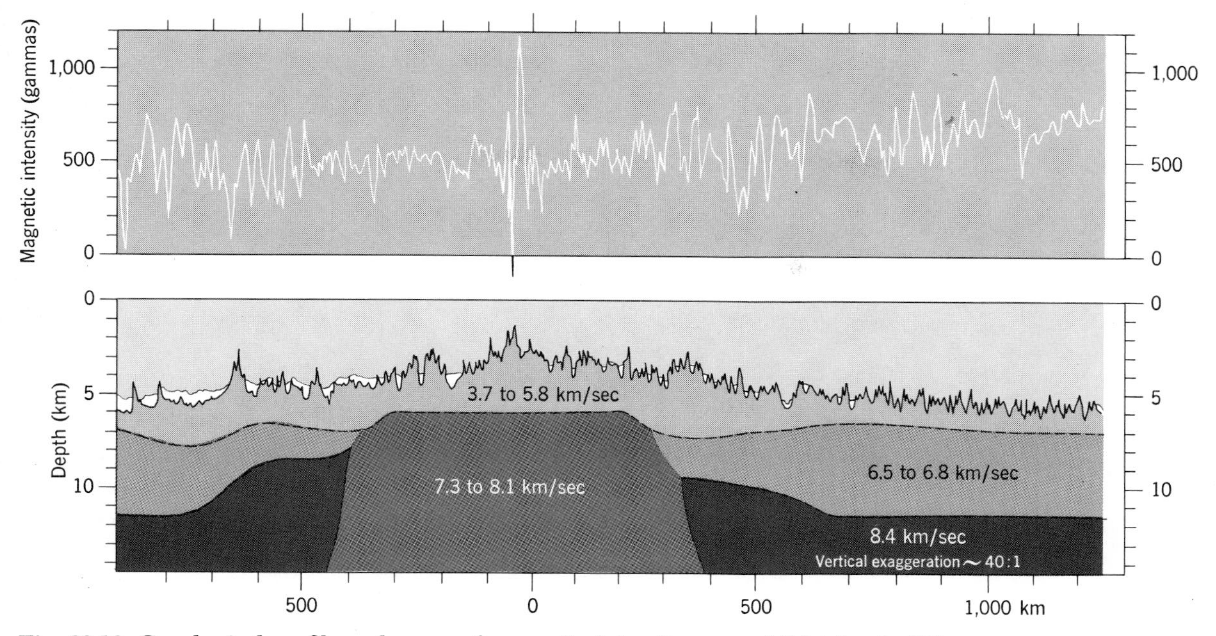

Fig. 22-16. Geophysical profile and curve of magnetic intensity across Mid-Atlantic Ridge. Numbers indicate velocities of longitudinal seismic waves. (J. R. Heirtzler and Xavier LePichon, 1966.)

Earth, and that the sea floor is like a gigantic composite dike. So far no explanation has been offered to reconcile the observed high magnetic anomalies at the crests with nonmagnetic molten rock material presumably upwelling there.

Let us ignore this problem of cool rocks *versus* molten material at the crest of the ridge, and test the hypothesis in another way. We can compare the polarity record of segments of the ocean floor at right angles to a ridge, with that inferred from piles of isotopically dated extrusive sheets in various parts of the continents and from deep-sea sediments. Unfortunately, at present the polarity record is well known only for the later parts of the Cenozoic. The fit of the late-Cenozoic polarity record with the magnetic-anomaly curves is so remarkable that it has been assumed to be the only possible solution. But we have absolutely no other evidence at present to determine whether this remarkable coincidence is the only possible match that could be made. Until a few rock specimens have been drilled out of the sea floor and their ages and polarities determined, we are not justified in supposing that the presently preferred hypothesis is the *only* possible solution. For all we know, the magnetic-anomaly curves might better match the polarity record from some earlier time, say that beginning with the Miocene and working backward, or with the Eocene, or with some other time. Because neither the distribution of magnetic materials in rocks beneath the deep-sea floor nor the rate of creation of new igneous rock is known, various combinations can be invented at will to fit the anomaly curves to polarity-reversal curves. We can be sure that if the first rock sample from the sea floor in an area thought to be of Pleistocene age is determined to be Miocene or any other older age, the present group of hypotheses will have to be abandoned.

POSSIBLE NEW SYNTHESIS

The new oceanic geophysical results have prompted a number of fruitful new hypotheses to explain some large unsolved geologic problems. Although these new hypotheses are in great measure still speculative, their effects have been electrifying and they have stimulated research in new directions. Of these new ideas the concept of sea-floor spreading has been the most influential.

Hypothesis of sea-floor spreading. The hypothesis that the sea floor is spreading laterally from active ridges grew out of the discovery that a rift valley follows the crest of the mid-oceanic ridge system.

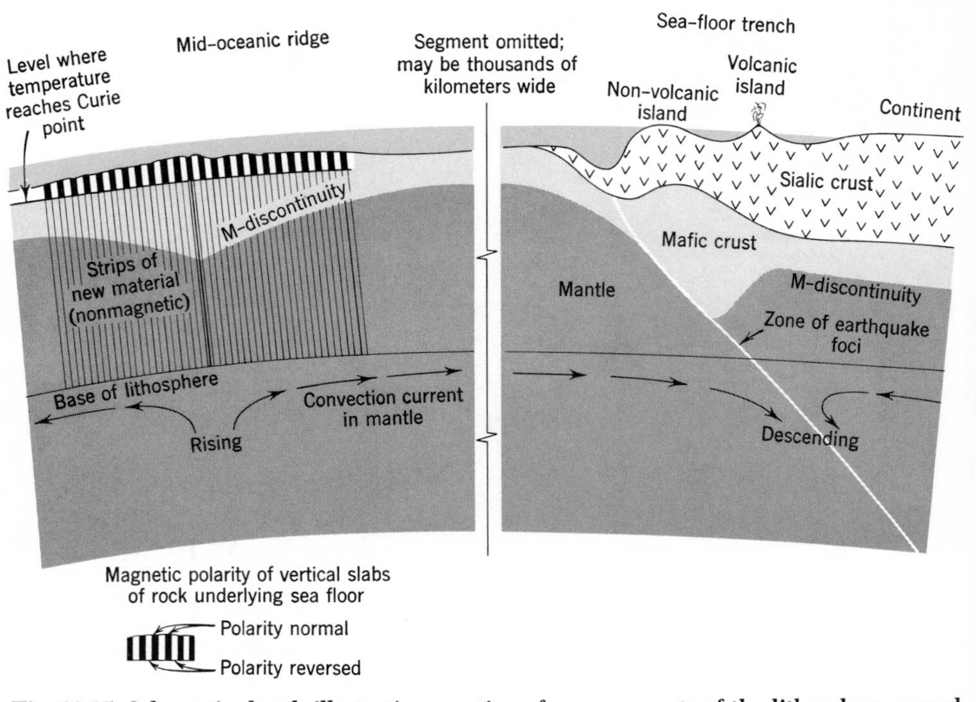

Fig. 22-17. Schematic sketch illustrating creating of new segments of the lithosphere according to the hypothesis of sea-floor spreading. Convection currents in the upper mantle are thought to provide motive power for spreading.

The latest version of the hypothesis combines convection currents, continental drift, and the origin of geosynclines and mountain chains. It states that new oceanic crust keeps appearing above a hot, upward convection current within the mantle at the crest of an active ridge. Then it spreads laterally across the sea floor, and disappears where a cool, descending convection current beneath a sea-floor trench drags it downward (Fig. 22-17). A key corollary of the hypothesis is that rather than being permanent, the oceanic crust is disposable, is in a continual state of flux, and can be renewed completely every 200 to 300 million years. While all this activity takes place on the sea floor the continental masses drift passively along, not on the surface of the oceanic crust as was formerly supposed but on the top of the low-velocity zone in the upper mantle at a depth of 100km or so.

The relationships of Greenland and Rockall Bank to the Reykjanes Ridge southwest of Iceland illustrate the simplest example of spreading and drift (Fig. 22-18, *A*). If we assume that Greenland and Rockall Bank were joined together about 60 million years ago when the basalts of the Thulean Province were extruded, and that they began to drift apart soon afterward, then their average rate of drift has been 2cm/year. A continuous rate of spreading of 1cm/year yields a close fit between anomaly curves and polarity episodes (Fig. 22-18, *B*).

How does this history of the Reykjanes Ridge compare with the rest of the Mid-Atlantic Ridge? Does evidence elsewhere agree with the idea that 1,200km of new sea floor have been created since about the end of Cretaceous time? Continuous seismic profiles in the North American basin show that horizon A, the top of an abyssal plain of Cretaceous age, stops about 700km west of the crest of the Mid-Atlantic Ridge. If this point marks the true eastern limit of the ancient abyssal plain against an older ridge, it places an upper value on the amount of permissible post-Cretaceous spreading. This value coincides closely with that suggested by the Reykjanes Ridge.

A continuous seismic profile across the Mid-Atlantic Ridge (Fig. 15-4, *B*) shows that sediment is present on the flanks but absent near the crest. This

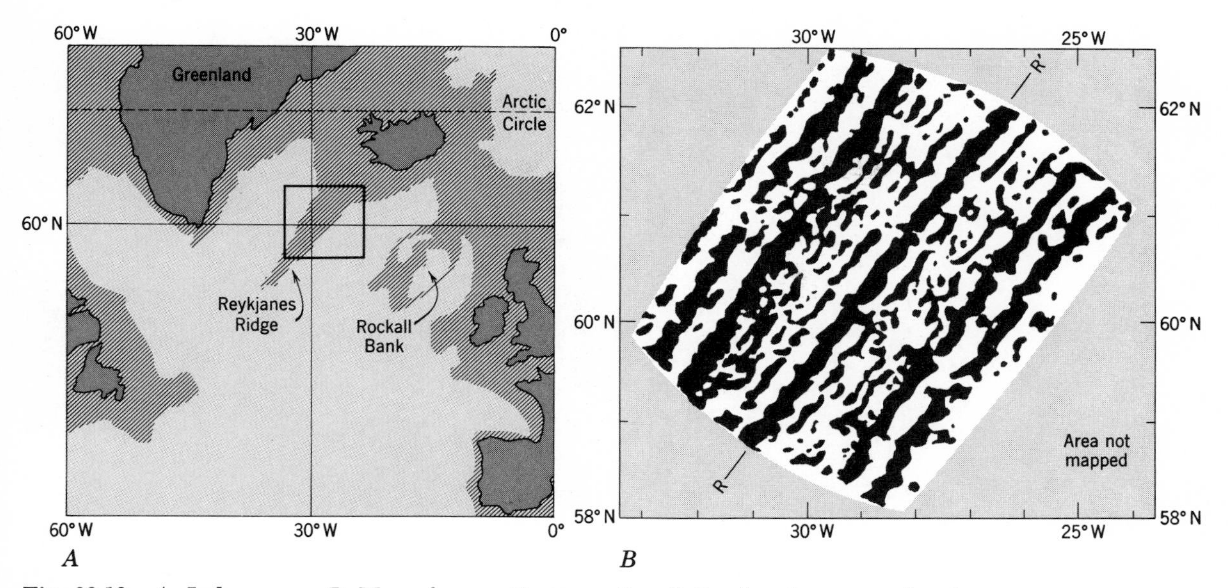

Fig. 22-18. *A.* Index map. *B.* Map of magnetic properties. Belts of rock presumed to be normally polarized (black) alternate with belts presumed to be reversely polarized (white). Line RR′ is centerline of Reykjanes Ridge. (After J. R. Heirtzler, Xavier LePichon, and J. G. Baron, 1966.)

distribution implies that if spreading has occurred, it has done so only since the sediments on the flanks were deposited.

The hypothesis of sea-floor spreading purports to explain rows of volcanic islands varying in age systematically along their length. The explanation states that each row of islands was derived from a single volcanic source that remained fixed. Directly above the source, volcanic cones are intermittently built up and after they become dormant are steadily transported away. The islands most distant from the vent are the oldest (Fig. 22-19). Most island rows fit into a simple pattern of spreading away from modern active ridges, although two in the Pacific are anomalous.

How does the hypothesis of sea-floor spreading relate to the origin of mountains? The hypothesis connects an active Atlantic ridge with active Pacific trenches, and suggests that the trenches are the sites of future mountain chains. If this is correct, we might expect to find an intermediate stage such as an old, inactive Pacific ridge and inactive Atlantic trenches. Observations in the Pacific confirm the presence there of an inactive ridge, 3,000km wide and 14,000km long, the *Darwin Rise* (Fig. 22-19). Numerous atolls (Fig. 22-20) and guyots crown this rise; they indicate subsidence of 1 to 2km during the last 70 million years or so. If the continental rises surrounding the Atlantic basin represent the tops of sediment fillings of inactive trenches, the prediction is fulfilled. If such an inactive system of ancient trenches exists, we can make a further prediction. We can suggest that beneath the eastern side of the North American continent and beneath the western side of the Eurasian continent are the scars of ancient shear zones that slope downward through 700km. These would be the ancient counterparts of the shear zones that extend downward from modern trenches and give rise to deep-focus earthquakes (Fig. 22-6).

A critical and possibly the last major missing piece in the puzzle of the origin of mountains is an explanation of the transition from a filled, dormant trench and adjacent continental shelf (Fig. 22-21) into a mountain chain. Before the sediments that fill a trench can be deformed and uplifted, oceanic crust must be converted into continental crust. Renewed large-scale activity of some kind following the dormant period would appear to be indicated. Presumably such activity results in elevation of the contents of the trench, creation of a land mass where the sea now stands, and shedding of sediments back toward the continent. The sediments will be thrust landward,

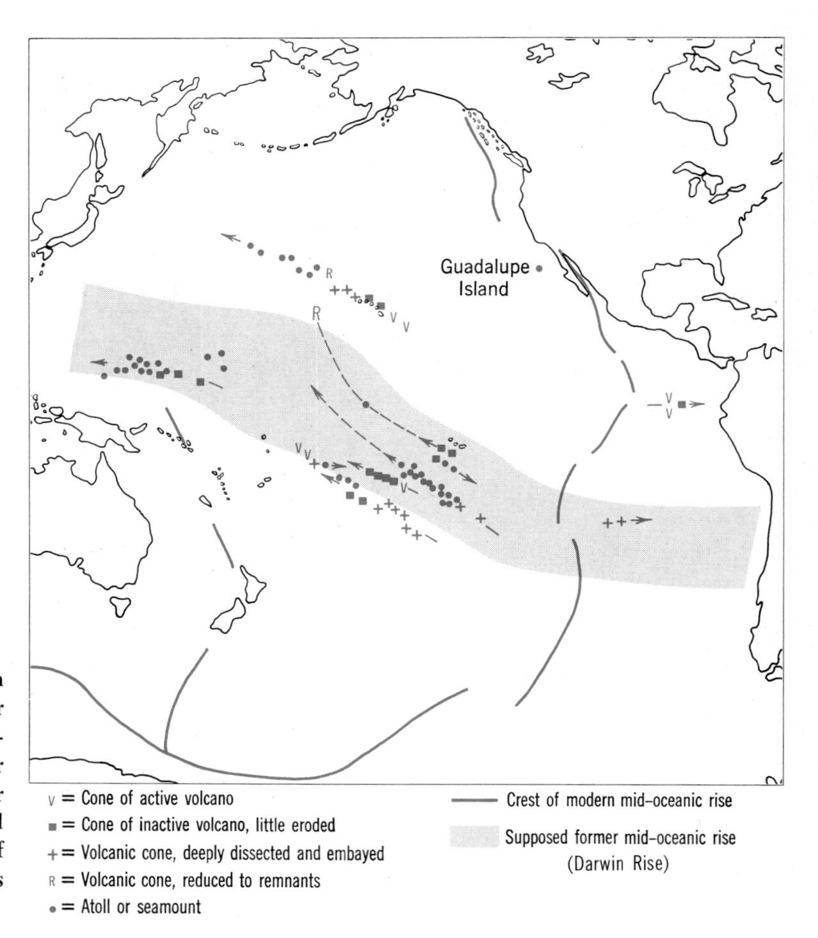

Fig. 22-19. Rows of volcanic islands in central Pacific are related to a linear belt believed to mark a former mid-oceanic rise, now inactive. (Rise after H. H. Hess, 1962; island rows after Arthur Holmes, Principles of physical geology, 2 ed., © 1965, by permission of Ronald Press Co., New York and Thomas Nelson and Sons Ltd., London.

metamorphosed, and intruded by granitic rocks, and a new mountain chain will be born.

Evolution of the lithosphere. Although many details of the hypothesis of sea-floor spreading remain to be worked out, the concept gives us new insights not only into the behavior of the oceanic crust today but also into the process of continental growth through time. Both large-scale activities can be visualized as products of transformations that accompanied rise of material from the upper mantle. Volatiles accumulate as the atmosphere and the hydrosphere. Alkali elements contribute to the making of granitic rocks, whereas partially melted mantle material ascends as basalt. When the geologic cycle began, at least 2,500 million years ago, only small parts of the Earth's surface consisted of continental rocks. Through mechanical and chemical weathering, transport and deposition, crustal material has been reworked over and over again as it passed repeatedly through the rock cycle. New material was continually added from the upper mantle, creating both granite and basalt.

The oldest rocks now exposed have passed through the cycle, which is still in operation. We can predict that it will continue to operate into the indefinite future. Although we already know a great deal about it, fascinating new insights into its workings await future research into all its activities.

Summary

1. In the nineteenth century the concept of contraction by cooling of a molten Earth served as a unifying basis of geology.

2. According to the contraction theory mountains resulted from crustal wrinkling in response to shrinkage of the Earth's interior consequent upon loss of original heat.

3. The discovery of radioactivity and progress in seismology destroyed the basis of the contraction

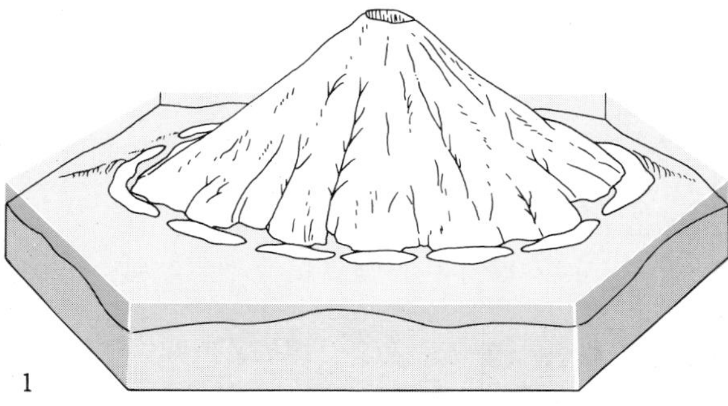

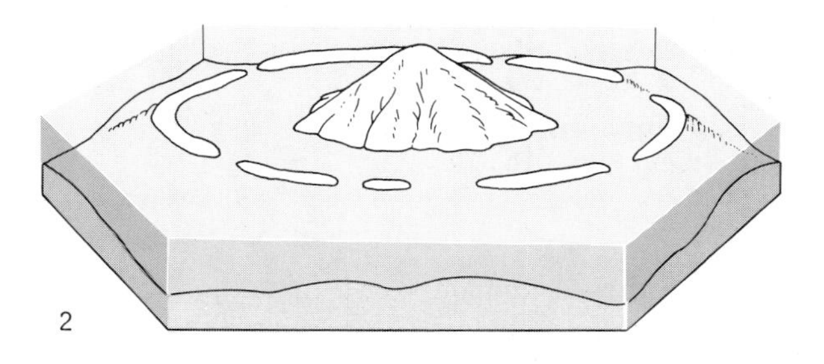

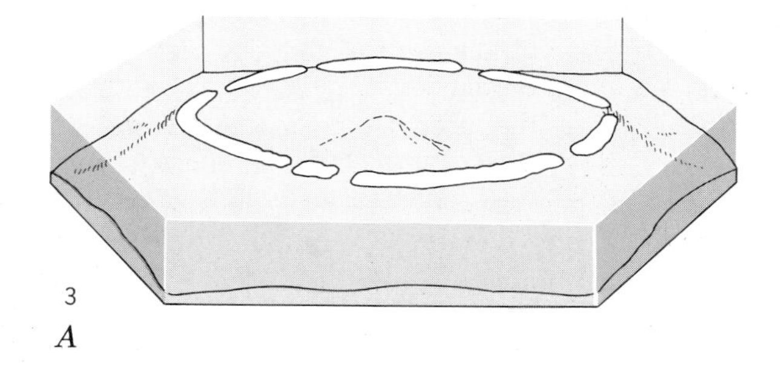

Fig. 22-20. A. Stages in development of an atoll. (1) ***Fringing reef,*** *a reef that is attached to the shore of a landmass* **grows around perimeter of volcanic cone. (2) Reef grows upward as cone subsides, forming a** ***barrier reef,*** *an elongate reef not connected with a landmass.* **(3) Eventually the cone becomes completely submerged; by growing upward the reef has created an** ***atoll,*** *a reef that forms a nearly closed figure within which there is no landmass.* ***B.*** **Lae Atoll, Marshall Islands. Dark areas are tree-covered islands; light areas shoal water above living reef. (D. B. Doan.)**

theory and deprived geology of its unifying framework.

4. From their analysis of mountains, geologists elaborated the geosynclinal concept as a replacement for the older contraction theory.

5. Studies of remanent magnetism in rocks have defined ancient positions of the Magnetic Pole. During late Cenozoic time the North Magnetic Pole lay near the present pole of rotation. At earlier times that pole lay progressively farther from such a position occupying, during Precambrian time, a location off what is now the California coast.

6. Divergence of ancient pole positions reconstructed from Europe and from North America supports the concept of continental drift. The divergent pole positions can be reconciled if these two continents are moved back to their supposed pre-drift positions.

7. Relationships of epicenter location to depth of focus within the Circum-Pacific belt suggest that on their continental sides great inclined shear zones extend from the sea-floor trenches diagonally downward to depths of slightly more than 700km.

8. Although trenches marginal to the Pacific basin contain little or no sediment, their great lengths and

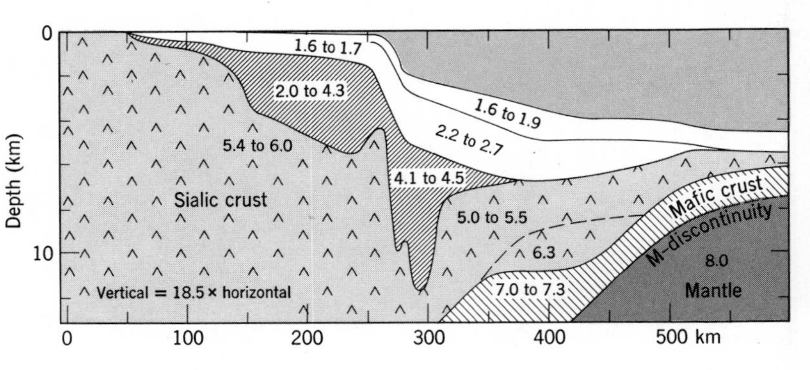

Fig. 22-21. Geophysical section through continental shelf and slope east of New Jersey suggests the presence of a buried marginal trench. Numbers indicate velocities (in km/sec) of longitudinal seismic waves. (C. L. Drake, 1966.)

arcuate patterns suggest that they may be sites of future mountain chains.

9. An oceanic ridge system 64,000km in length encircles the globe almost completely. Through large parts of its length the crest is characterized by a median rift valley.

10. Distinctive patterns of linear magnetic anomalies parallel axes of many mid-oceanic ridges. Anomaly patterns, oriented north-south, have been offset by enormous fracture zones that trend east-west in the eastern Pacific.

11. According to the hypothesis of sea-floor spreading new oceanic crust is continually being manufactured at the crests of mid-oceanic ridges, is shouldered aside, and spreads laterally as new material wells up from behind. Oceanic crust disappears where it is forced downward, possibly into an active sea-floor trench, tens of thousands of kilometers from the ridge.

12. Upper mantle material and erosion products from continents and oceanic islands are transformed into new continental crust during mountain-building along continental margins.

Selected References

Dietz, R. S., 1966, Passive continents, spreading sea floors, and collapsing continental rises: Am. Jour. Sci., v. 264, p. 177–193.

Engel, A. E. J., 1963, Geological evolution of North America: Science, v. 140, p. 143–152.

Ewing, John, Engel, J. L., Ewing, Maurice, and Windisch, Charles, 1966, Ages of horizon A and the oldest Atlantic sediments: Science, v. 154, p. 1125–1132.

Hess, H. H., 1962, History of ocean basins, *in* Engel, A. E. J., James, H. L., and Leonard, B. F., Petrologic studies: a volume in honor of H. F. Buddington: New York, Geol. Soc. America, p. 599–620.

Hurley, P. M., Hughes, H., Faure, G., Fairbairn, H. W., and Pinson, W. H., 1962, Radiogenic strontium-87 model of continental formation: Jour. Geophys. Research, v. 67, p. 5315–5334.

Irvine, T. N., ed., 1966, The world rift system: Geol. Survey of Canada, Paper 66–14.

MacDonald, G. J. F., 1964, The deep structure of continents: Science, v. 143, p. 921–929.

Menard, H. W., 1955, Deformation of the northeastern Pacific basin and the west coast of North America: Geol. Soc. America Bull., v. 66, p. 1149–1198.

Northrop, J. W., III, and Meyerhoff, A. A., 1963, Validity of polar and continental movement hypotheses based on paleomagnetic studies: Am. Assoc. Petroleum Geologists Bull., v. 47, p. 575–585.

Orowan, Egon, 1964, Continental drift and the origin of mountains: Science, v. 146, p. 1003–1010.

Poole, W. H., ed., 1966, Continental margins and island arcs: Geol. Survey of Canada, Paper 66-15.

Vine, F. J., 1966, Spreading of the ocean floor: new evidence: Science, v. 154, p. 1405–1415.

Wilson, J. T., 1963, Continental drift: Sci. Am., v. 208, no. 4, p. 86–100. (San Francisco, W. H. Freeman, Repr. 868, 16 p.)

CHAPTER 23

Geology in Industry

Our high-energy industry
Sources of energy:
fossil fuels, nuclear energy
Sources of materials:
minerals used in industry
Natural concentrations
Origin of minerals deposits
Future of mineral use and exploration

Plate 23

Drilling ore in a New Mexico mine.

(*American Smelting and Refining Co.*)

CHAPTER 23 *Geology in Industry*

OUR HIGH-ENERGY INDUSTRY

Muscles and machines. When people speak of the present epoch in western civilization as "the industrial age," they really mean the age of suddenly and enormously increased use of fuels and metals, the age of high-energy industry. Stone-age man, 30,000 or more years ago, had an industry: he chipped and flaked mineral substances (mostly pieces of quartz) to make tools and weapons. But his was a low-energy industry, with the energy supplied by human muscle. Although he was limited by this fact and by a very narrow choice of materials to work with, we must not underestimate him. He was as intelligent as we, lacking only our accumulated experience and the know-how that comes from experience.

This lack was gradually, though not steadily, overcome by him and by his descendants. Here are a few of their accomplishments, discovered and dated by archeologists:

4000 B.C.: Chaldeans had become skilled workers in metals such as gold, silver, copper, lead, tin, and iron.

3000 B.C.: Eastern Mediterranean peoples were making glass, glazed pottery, and porcelain.

2500 B.C.: Babylonians were using petroleum instead of wood for fuel.

1100 B.C.: Chinese were mining coal and drilling wells hundreds of feet deep for natural gas.

These were arts, learned by experience, and they implied the substitution of metals, glass, and other substances for stone. But most of the energy with which these arts were pursued came from wood fuel and from the muscles of men and animals. This was still true, even in Europe, as late as the seventeenth century, and of the majority of the world's people it is still true today.

The eighteenth century brought the substitution of coal for wood as an energy source and of iron for wood as a construction material, and with the nineteenth century came the steam engine. Thereupon, at an almost incredible speed, a minority of the Earth's peoples substituted machines for muscles, with the result that between 1800 and 1960 the industrial consumption of energy *per person* in the United States increased about five times. Today each person in the United States has working for him continuously more than 20 horsepower of energy, equivalent to the labor of more than two hundred slaves. This explosive industrial revolution has conferred upon its many beneficiaries a leisure and an opportunity for cultural development that could be enjoyed only by the richest people in slave-owning ancient Rome or Greece, Babylon, or Egypt.

Rapid emancipation from heavy labor has its bad aspects as well as its good, but for better or for worse the most highly industrialized nations enjoy its fruits. In the effort to maintain and increase the high consumption of energy, machines, and structures on which our comparative industrial freedom is based, the geologist plays an important part. For we depend on him to find beneath the surface the oil and gas, the coal, and the metals necessary to build machines and to power them.

The supply of materials and energy. Metals and other materials to build machines and energy to power them are complementary. The increasing availability of machines makes it possible to do more things with energy. Furthermore high-energy industry demands also a host of nonmetallic mineral products, such as shale and limestone for making cement, gypsum for making plaster, and asbestos for insulation. All these demands, taken together, have bent up the curve of overall industrial production very steeply (Fig. 23-1).

The majority of industrialized nations possess a

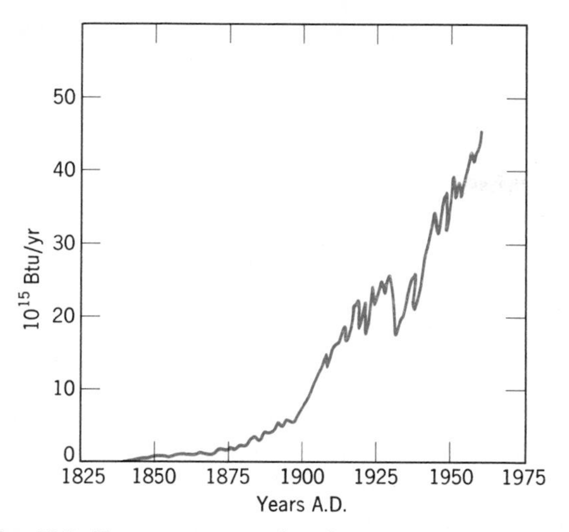

Fig. 23-1. Energy consumed industrially in the United States, 1825 to 1960. A Btu (British thermal unit) is the amount of heat required to raise the temperature of 1 pound of water by 1°F. (Source: U. S. Department of Commerce.)

strong mineral base; that is, they are rich in mineral deposits and are exploiting these vigorously. Mineral resources, however, have certain peculiar aspects that are different from the resources inherent in agriculture, grazing, forestry, and fisheries. First, occurrences of usable minerals are both limited in abundance and distinctly localized at places within the Earth's crust; hence they have to be searched out. Second, the extent of the reserves available in any one country is rarely known with accuracy, and the likelihood that new occurrences will be discovered is difficult to assess. As a result, production over a period of years may be difficult to predict. Third, unlike plants and animals, which are cropped yearly or seasonally, occurrences of minerals are depleted by mining and eventually become exhausted. Minerals therefore have "one-crop" availability per occurrence; this disadvantage can be offset only by finding new occurrences or by making use of scrap—that is, reusing the material repeatedly.

These peculiarities of the mineral industry place a premium on the skill of geologists, who play an essential part in finding and bringing to the surface mineral substances for use in industry. This task is being accomplished by the application of the basic principles that have been set forth in this book and by the use of a great body of specialized knowledge that has no place here. Much ingenuity has been expended by prospectors, geologists, and engineers in bringing the production of minerals to its present state, and more ingenuity will be required in the future.

Let us take a look at what every educated citizen of any industrial nation ought to know about the geology of industrial mineral deposits. We can begin with sources of energy and then continue with mineral sources of materials for machines and structures.

Sources of energy. The chief sources of the energy consumed in highly industrialized nations are few: the fossil fuels (coal, oil, natural gas), hydroelectric power, wood, wind, and a little muscle. Nuclear energy will become an important source, but at present it is just beginning to develop. Excluding wood (used mainly for space heating in some dwellings) and muscle (a resource now getting very little exercise in industrial countries), the energy consumed in the United States in 1960 came from these sources:

Oil	42%
Natural gas	31%
Coal	23%
Water power	4%
	100%

For western Europe the breakdown is different; coal accounts for more than half the energy consumed. For the world as a whole, with its hundreds of millions of agricultural workers, the breakdown would be different again; wood and muscle would be a substantial source of energy, and fossil fuels, especially oil and gas, would bulk less.

COAL

Coal as energy. The black, combustible sedimentary rock we call coal is the most abundant of the fossil fuels. These fuels are so called because they contain solar energy, locked up securely in chemical compounds by the plants or animals of former ages. Coal is also the world's greatest single present source of industrial energy, because most of the coal mined is burned under boilers to make steam, and because coke, a product of the burning of coal, is essential in making steel. Varying amounts of liquid fuel can be derived from coal, and because internal-combustion engines are generally more efficient than steam

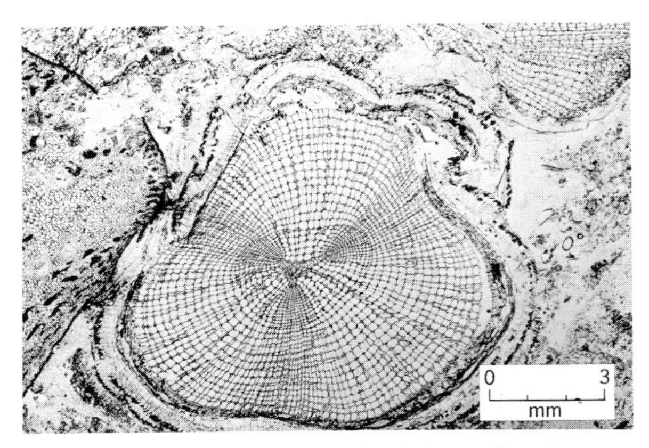

Fig. 23-2. Photograph of coal of Pennsylvanian age from Illinois, showing cellular structure. (General Biological Supply House; Julius Weber.)

engines, probably more and more coal will be used in this way. In addition to its uses as energy, coal is the chief raw material from which nylon, some plastics, and a multitude of chemicals are made.

Origin of coal. Coal occurs in beds or layers (miners call them *seams*) along with other sedimentary rocks, mostly shale and sandstone. A look at a thin slice of coal under the microscope shows that it consists of bits of fossil wood, bark, leaves, roots, and other parts of land plants, chemically altered but still identifiable. This composition leads at once to the inference that coal is fossil plant matter (Fig. 23-2), a characteristic that has been incorporated into the definition of coal. Unlike most other sediments, coal was not eroded, transported, and deposited, but accumulated right where the plants grew, for coal beds include fossil tree stumps rooted in place in underlying shale, evidently a former clay soil. Furthermore, although coal is interbedded with inorganic sediments, it is not usually mixed with them, as inevitably it would be if it had been transported.

It was recognized as long ago as 1778 that the places of accumulation were swamps because (1) a complete physical and chemical gradation exists from coal to peat, which today accumulates only in swamps, and (2) only under swamp conditions is the conversion of plant matter to coal chemically probable. On dry land, dead plant matter (composed of carbon, hydrogen, and oxygen) combines with atmospheric oxygen to form carbon dioxide and water; it rots away. Under water, however, oxygen is excluded from dead plant matter and oxidation is prevented. Instead, the plant substance is attacked by anaerobic bacteria, which partly decompose it, splitting off oxygen and hydrogen. These two elements escape, combined in various gases, and the carbon gradually becomes concentrated in the residue. Although they work to destroy the vegetal matter, the bacteria themselves are destroyed first because the acid compounds they liberate poison them. This could not happen in a stream, whose flowing water would dilute the poisons and permit the bacteria to complete their destructive process.

With destruction of the bacteria, the biochemical phase of coal making comes to an end, the plant matter has been converted to peat, and a geochemical phase, so called because it involves geologic activity, begins. The peat is gradually buried and compressed beneath accumulating sand, silt, or clay (Fig. 23-3). Volatile matter continues to escape, leaving an ever-increasing proportion of carbon. The peat is converted successively into lignite, subbituminous coal, and bituminous coal. These coals are sedimentary rocks, but a still later phase, anthracite, is a metamorphic rock. As it generally occurs in folded strata, anthracite has undergone a further loss of volatiles and concentration of carbon caused by the pressure and heat that accompany folding. Because of its low content of volatiles, anthracite is hard to ignite but burns with almost no smoke. In contrast, lignite, rich in volatiles, ignites so easily that it is dangerously subject to spontaneous ignition (in chemical terms, rapid oxidation) and burns smokily. In certain regions where folding has been intense, coal has been metamorphosed so thoroughly that it has been converted to graphite in which all the volatiles have been lost, leaving nothing but carbon. Graphite therefore will not burn.

Occurrence. We have said that coal occurs in layers or seams, which are merely strata (Fig. 20-14). Each seam is a flat, lens-shaped body corresponding in area to the area of the swamp in which it accumulated originally. Most coal seams are 2 to 10 feet thick, although some reach more than 100 feet. They tend to occur in groups; in western Pennsylvania, for example, there are about sixty beds of bituminous coal. Their occurrence in the rocks of every period later than Devonian indicates that during the last 300 million years or so swamps rich in vegetal growth have been recurrent features of the land. Peat is accumulating today, at an average rate of

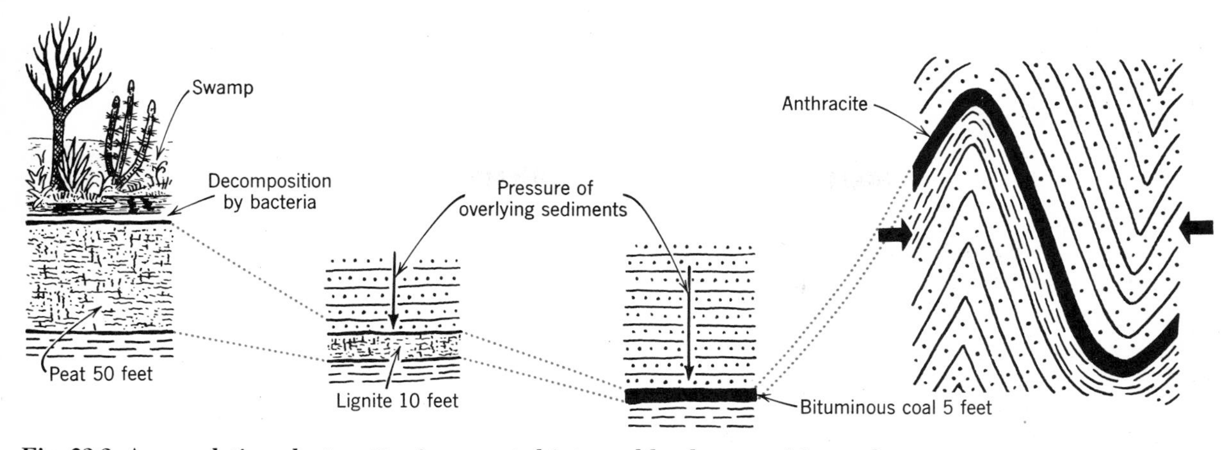

Fig. 23-3. Accumulating plant matter is converted into coal by decomposition and pressure.

about 1 foot in 30 years, in swamps on the Atlantic-Gulf coastal plain of the United States. The swamps now represented by coal beds must have been much the same as these.

Distribution of coal. Coal is not only abundant but also rather widely distributed; it is present even in Antarctica. Although probably most of the world's coal fields have been discovered, world reserves have never been even closely estimated, but experts guess they may amount to 6 trillion tons, of which almost half is in the United States (Fig. 23-4). However, some of this coal would be impossible to mine, and mining much of the rest of it would be expensive, for some beds are thin, some are impure, and some lie far beneath the surface.

Coal mining. Thin beds and deep-lying beds are avoided wherever possible; the average thickness of all beds now mined in the United States is about 5 feet. Increasing costs have stimulated the invention of increasingly efficient methods. In underground mining the automatic coal-cutting and loading machine, which eliminates drilling and blasting, has increased production per man-hour in thick beds 10 to 20 times. Coal within 100 feet of the surface is being mined, without ever going underground, by power shovels that take 10 to 50 cubic yards at a single bite (Fig. 23-5) and by huge augers, 6 feet or more in diameter, that bore horizontally into a seam and move the coal outward between their spiral blades to loading belts in a single operation (Fig. 23-6).

Experiments are now being made with the burning of coal, while it is still in the ground, to produce gas for industrial use. This technique would eliminate mining altogether.

PETROLEUM: OIL AND NATURAL GAS

The completion of the first oil well of modern times, at Titusville, Pennsylvania, in 1859, was an epoch-making event. The well was 69 feet deep and produced 25 barrels per day of a substance that spelled the doom of candles and whale-oil lamps. More than a hundred years later, in 1966, the United States was producing 8.3 million barrels of oil per day from nearly half a million wells, some of which were more than 3 miles deep and at least one of which was 4 miles deep. And the rest of the world was producing three times this amount.[1] Of the huge production, 90 per cent is actually used as fuel; the remainder goes into lubricants, without which the fuel would turn very few wheels, and into a thousand manufactured products of the petrochemical industry. New discoveries are continually being made, but many of these are in areas long recognized as being favorable for oil. Therefore, the search yearly becomes more difficult.

Oil and gas are the two chief kinds of petroleum. We can define ***petroleum*** as *gaseous, liquid, or solid substances, occurring naturally and consisting chiefly of chemical compounds of carbon and hydrogen.* Because both oil and gas occur together and are searched for in the same way, we can follow general practice and talk about oil pools, oil explora-

[1]World Petroleum Report, 1967.

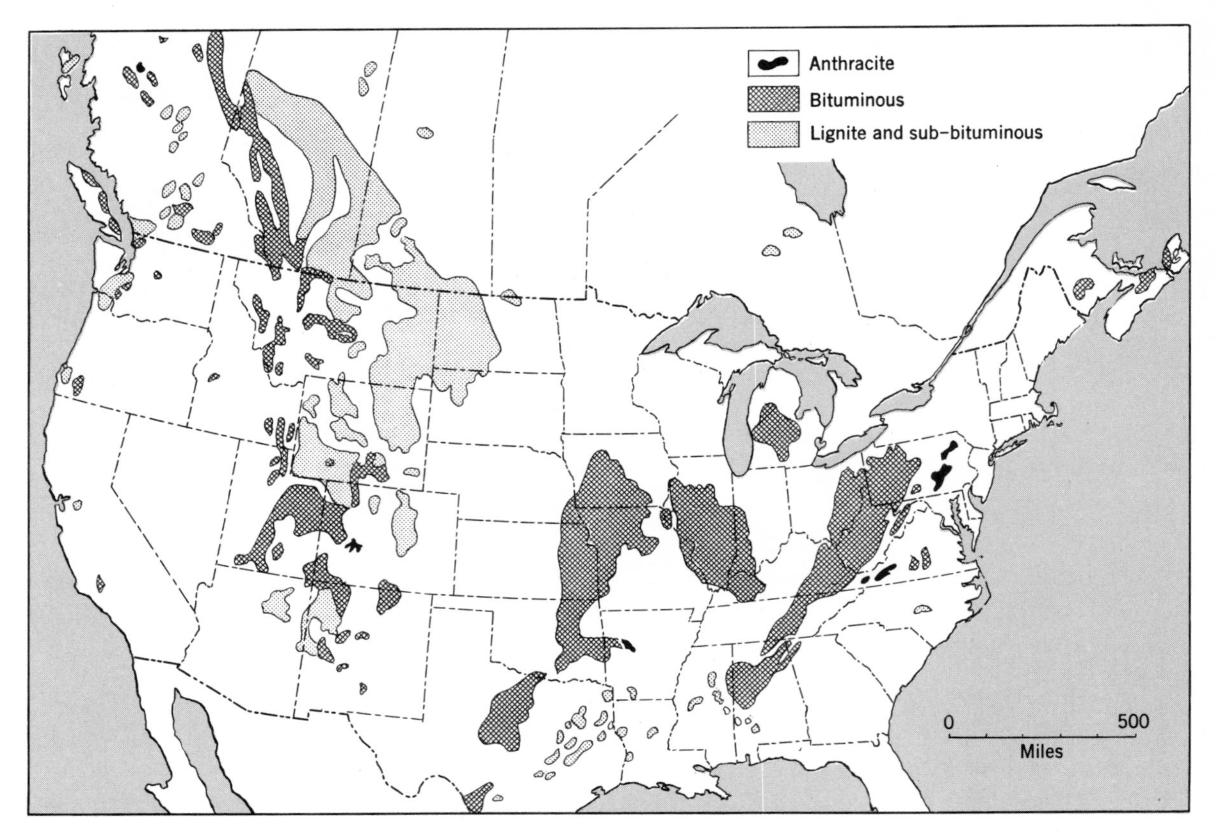

Fig. 23-4. Coal fields of United States and southern Canada. (Adapted from maps by U. S. Geol. Survey and Geol. Survey of Canada.)

tion, and the origin of oil with the understanding that we mean not only oil but gas as well.

Oil is a mixture of various compounds of carbon and hydrogen. It is in a highly reduced state, containing no oxygen; hence like coal it is readily oxidizable and so is a source of energy. In discussing it, we shall deal with its occurrence, its origin, the means used to find new occurrences, and oil production, in that order.

Occurrence: oil pools. The accumulated experience of a century of exploration, drilling, and producing has taught us much about where and how oil occurs, though not very much about how it originates. Oil possesses two important properties that affect its occurrence. It is fluid, and it is generally lighter than water. It is produced from pools. A **pool** is *an underground accumulation of oil or gas in a reservoir limited by geologic barriers.* The word gives a wrong impression because an oil pool is not a lake of oil. It is a body of rock in which oil occupies the pore spaces.

For oil or gas to accumulate in a pool, four essential requirements must be met. (1) There must be a *reservoir rock* to hold the oil, and this rock must be permeable so that the oil can percolate through it under pressure into wells as ground water does. (2) The reservoir rock must be overlain by an impermeable *roof rock,* such as a claystone, to prevent upward escape of the oil, which is floating on ground water. (3) The reservoir rock and roof rock form a *trap* that holds the oil and prevents it from moving any farther under pressure of the water beneath it (Fig. 23-7). These requirements are much like those of an artesian-water system (Fig. 11-8) but with the essential difference that the artesian aquifer connects with the surface, whereas the oil pool does not. Although these three features—reservoir, roof, and trap—are essential, they do not guarantee a pool. In many places where they occur together, drilling has shown that no pool exists, generally because of lack of a source from which oil could enter the trap. So we must add to the three require-

Fig. 23-5. This coal strip-mine at Wyodak, Wyoming produces 300,000 tons a year from a seam 60 to 90 feet thick. The coal is intermediate in rank between lignite and bituminous coal. (U. S. Bureau of Mines.)

ments for a pool: (4) There must be *source rocks* to provide oil.

A group of pools, usually of similar type, or a single pool in an isolated position constitutes an ***oil field.*** The pools can be side by side or one on top of the other.

Origin of oil. Oil is believed to be a product of the decomposition of organic matter, both plant and animal, not because it consists of hydrocarbons but because of two other characteristics: (1) It possesses optical properties known only in organic substances. (2) It contains nitrogen and certain compounds (porphyrins) that chemists believe could come only from organic sources.

Furthermore, oil is nearly always found in marine sedimentary rocks. Indeed, in places on the sea floors of the continental shelves, sampling has shown that fine-grained sediments now accumulating contain up to 7 per cent organic matter, chemically good potential oil substance. Thus, on the basis of the principle of uniformity of process, we can make the general suggestion that the oil we now find in rocks originated as organic matter deposited with marine sediments.

However, the suggestion is only a general one. Analyzing the organic matter in modern sediments, chemists have found significant differences between its composition and that of petroleum, as well as differences among petroleums themselves. The processes of conversion are not yet understood and may be very complex. Some bodies of petroleum may be products of repeated distillation.

Although the steps in the creation of oil are still very poorly known, the following simplified theory is rather widely held and is supported by enough facts to be at least somewhere near the truth.

1. The raw material consists mainly of simple marine organisms, mostly plants, living in multitudes at and near the sea surface. Such material is certainly not lacking. Measurements show that the sea grows at least 350 pounds of protein matter per acre per year, and the most productive inshore waters grow as much as 1 ton per acre. The latter value represents more than could be harvested in a year from the richest farmland.

2. The organic matter accumulates on the bottom, mostly in basins where the water is stagnant and deficient in oxygen (Fig. 15-9, *B*), and where therefore the substance is neither devoured by scav-

Fig. 23-6. A huge multiple auger in eastern Kentucky operates from an artificially excavated bench, cutting, moving out, and loading coal in a single operation. (*Courier-Journal* and *Louisville Times*.)

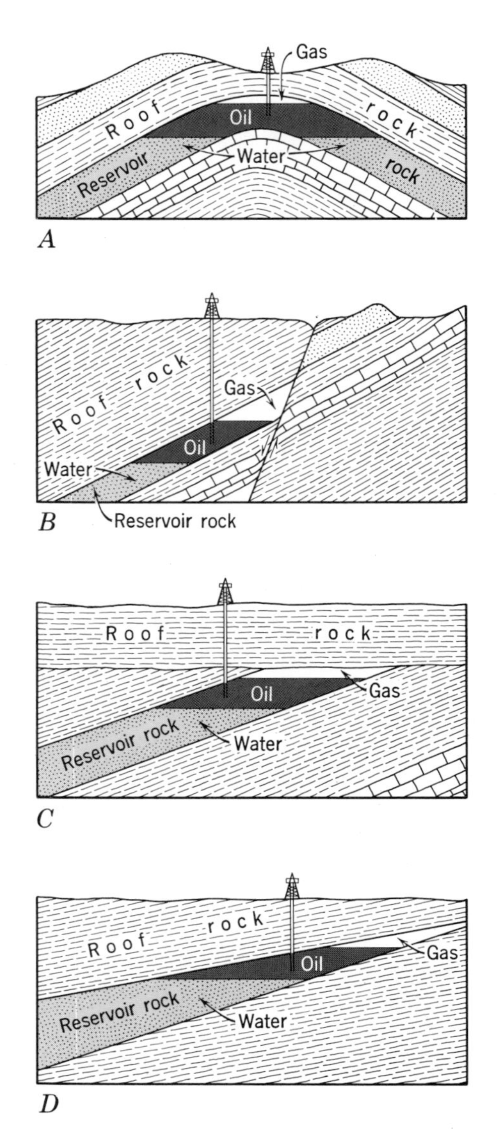

Fig. 23-7. Four of the many kinds of oil traps. *A*, *B*, *C* are structural traps; *D* is a stratigraphic trap. Gas (white) overlies oil (black), which floats on ground water (blue), saturates reservoir rock, and is held down by roof of claystone. Oil fills only the pore spaces in rock.

engers nor destroyed by oxidation. Instead it is attacked and decomposed by bacteria, which split off and remove oxygen, nitrogen, and other elements, leaving residual carbon and hydrogen. Accumulating sediments that are rich in organic matter teem with bacteria.

3. Deep burial beneath further fine sediment destroys the bacteria and provides pressure, heat, and time for further chemical changes that convert the substance into droplets of liquid oil and minute bubbles of gas.

4. Gradual compaction of the inclosing sediments under the pressure of their own increasing weight reduces the space between the rock particles and squeezes out oil and gas substance into nearby layers of sand or sandstone, where open spaces are larger.

5. Aided by their buoyancy and perhaps by artesian water circulation, oil and gas migrate generally upward through the sand until they reach the surface and are lost, or until they are caught in a trap and form a pool.

We repeat that this statement is greatly oversimplified. A long and complex chemistry is involved in conversion of the original organic constituents to crude petroleum. Also chemical changes may occur in oil and gas even after they have migrated into their reservoirs. This may help explain the chemical differences that exist between the oil in one pool and that in another.

The migration of oil needs more explanation. The sediments in which oil substance is accumulating today are rich in clay minerals, whereas many of the rocks that constitute oil pools are mostly sandstones consisting of quartz grains. It seems obvious, therefore, that oil forms in one kind of material and at some later time migrates to another. The migration process involves essentially the principles of movement of ground water. When, as we mentioned above, oil is squeezed out of the clay-rich sediments in which it originated and enters a body of sandstone somewhere above, it can migrate more easily than before, for at least two reasons. One is that sandstone, as explained in Chapter 11, is far more permeable than are clay-rich rocks. The other is that the force of molecular attraction between oil and quartz is less strong than that between water and quartz. Hence, because oil and water do not mix, water remains fastened to the quartz grains while oil occupies the central parts of the larger openings (like those shown in Fig. 11-2). Because it is lighter than water, the oil tends to glide upward past the quartz-held water. In this way it can segregate itself from the water; and when it encounters a trap it can form a pool.

Exploration. The oil industry in the United States dates from about 1860, when commercial production in western Pennsylvania began. The first pools were discovered at places where oil and gas advertised their presence by seeping out at the surface. It soon became apparent that at all the producing wells the rocks were sedimentary and that at most wells the structure consisted of anticlines (Fig. 23-7, *A*). In 1861 the theory was advanced that oil had migrated upward into anticlines and lay trapped beneath their crests. In the 1880's this theory was tested and confirmed by drilling holes in certain anticlines where no oil had been found previously; the holes were successful wells.

Geologic exploration. Proof of the anticline theory led to a period (1890 to 1925) in which geologists searched for and mapped anticlines from surface exposures. During this period traps other than anticlines were discovered (Fig. 23-7, *B–D*), and the search for pools rapidly widened.

The occurrence of buried traps, such as the one in Fig. 23-7, *C*, showed that mapping of exposures would have to be supplemented by methods that would reveal structural and stratigraphic traps not apparent at the surface (Fig. 23-7, *D*). This was done first, and is still being done, by core drilling in a pattern beneath a suspected area. Study of the cores makes it possible to construct a graphic log from each core and to correlate the strata by physical character and contained fossils (Fig. 16-16) almost as well as though the rocks were exposed at the surface. The depths at which a recognizable bed is encountered by a series of drill holes give a strong clue to structure (Fig. 23-8).

Core drilling, however, is expensive. This stimulated the development of electric logging, which can be done using holes drilled by cheaper methods.

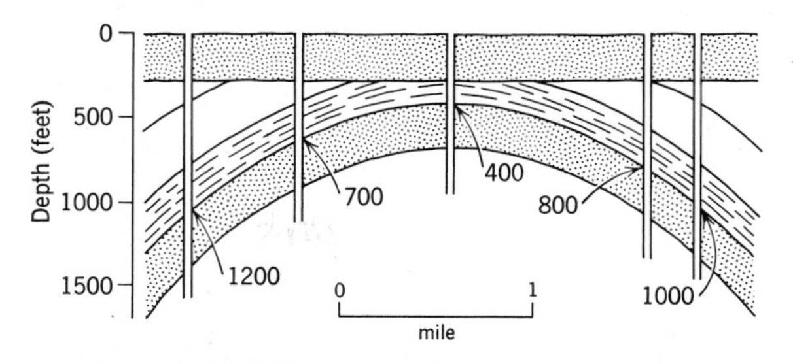

Fig. 23-8. Core-drill holes at five points show that flat-lying layers at surface conceal an anticline lying unconformably beneath it. Numbers show depths in feet at which top of sandstone in the anticline was encountered.

Common salt and other dissolved substances are present in the ground water that saturates all deep-lying sedimentary rocks. As salt solution is a good conductor, the conductivity of a sedimentary layer at depth varies with water content. The conductivity or the resistivity at any desired depth is measured by devices lowered into the hole by a cable. The resulting electric logs are widely used in correlation.

Geophysical exploration. In the 1920's geophysical exploration began to be widely used. Twenty years later this kind of exploration was uncovering most of the new pools simply because the majority of those discoverable by geologic methods alone had been found already. Geophysical exploration is designed to detect variations in distribution within the rocks of some physical property, such as specific gravity, magnetism, or ability to transmit or reflect seismic waves (Fig. 18-9). As most of these variations are related to structures, they help to indicate the presence of possible pools. If, for instance, a stratum having a high specific gravity has been raised during the making of an anticline, the value of gravity will be increased very slightly compared with that in the surrounding area. As the increase is not likely to be more than about one ten-millionth, the weighing instrument used (a gravity meter or gravimeter) has to be extremely sensitive.

Again, distortions of the Earth's magnetic field by rocks of varying magnetic susceptibility are measured with magnetometers (Fig. 23-9). Because igneous rocks, containing magnetite, are much more magnetic than sedimentary rocks, magnetometer surveys are most successful in areas in which igneous rocks project into overlying sedimentary rocks by intrusion, folding, or faulting or as erosional hills beneath surfaces of unconformity (Fig. 23-10). But

Fig. 23-9. Magnetometer, mounted on a helicopter, starting out on a subsurface-exploration traverse in northwestern Canada. Helicopter also carries a scintillation counter (not visible). A magnetometer can also be mounted on a truck and is often so used. (Canadian Aero Service Ltd.)

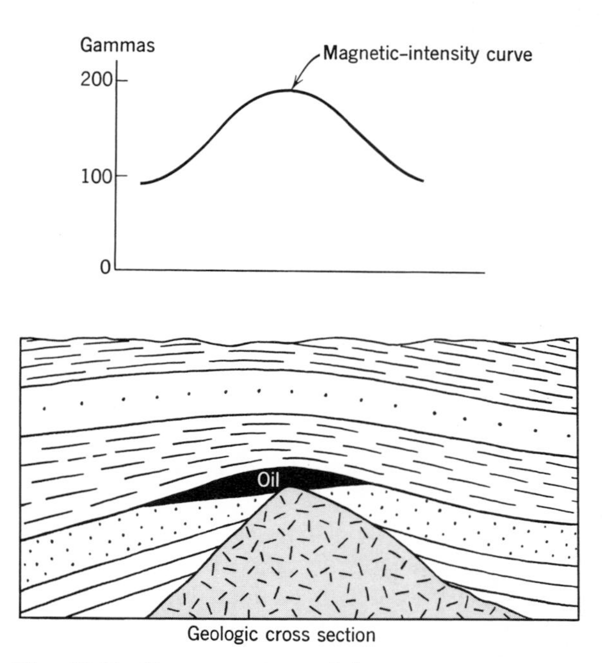

Fig. 23-10. Curve constructed from a magnetometer survey showed an abrupt local increase in magnetic intensity. Later test drilling disclosed a buried hill of relatively magnetic granite, with an oil pool just over its summit. (Sharpe Instruments of Canada.)

the magnetometer also shows differences among sedimentary rocks.

In the search for oil the most useful geophysical exploration method is seismic surveying. An artificial earthquake is created by exploding buried dynamite, and in the common method the waves reflected from the upper surfaces of rock layers with high wave-reflection potential are picked up at ground level by portable seismographs located half a mile or so from the explosion. Velocity of wave transmission being known, depths of the reflecting layers can be calculated from travel times (Fig. 18-9). The principle involved is similar to that used in echo sounding of sea depth (Fig. 15-2) and in continuous seismic profiling (Fig. 15-4).

Areal pattern of exploration. Exploration proceeds by elimination. Experience shows that, essentially, oil occurs only in sedimentary rocks and that the thinner the pile of such rocks the less likely they are to yield oil. Maps showing possibly productive areas are continually being made and revised by exploration companies. On them, areas of igneous and metamorphic rocks are labeled improbable; areas of sedimentary rocks that are very thin or that have unfavorable structure are likely to be put into the same class. With these, too, go areas that, possessing favorable rocks and structure, have been drilled and yet have not yielded oil. Figure 23-11 is a map of this kind for North America. People who risk a million dollars or more in drilling a hole do not rely solely on such general data; they prepare large-scale maps of small areas before making decisions.

The map shows in outline the continental shelf, which adds almost one-third to the potentially productive area of the United States. Both geophysical exploration and drilling have shown that in at least some places the shelf is very favorable territory. The hole at Hatteras Light, shown in Fig. 14-22, was drilled primarily for exploring the strata of the shelf. In the Gulf Coast area, where considerable production has already developed, wells are being drilled as far as 80 miles offshore (Fig. 23-12), and beneath water more than 4,000 feet deep. Continental shelves have large possibilities for future production, as has been shown by the development of two large gas fields beneath the North Sea, one north of The Netherlands and one east of northern England.

The production of oil and the distribution of

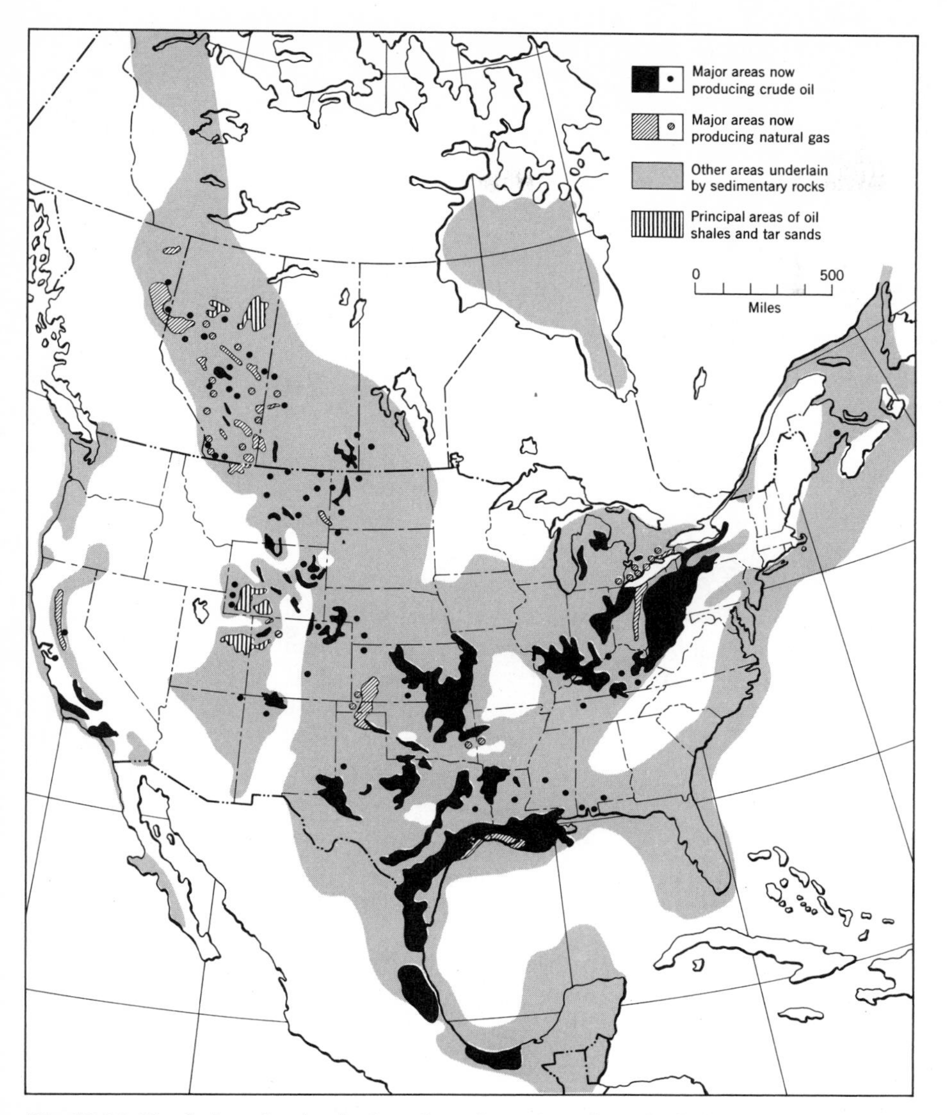

Fig. 23-11. North America (exclusive of northern Canada and Alaska), showing areas of petroleum production, areas of tar sands and oil shales, and areas of sedimentary rocks (possible future production). (Data, as of 1965, from Am. Petroleum Inst., Geol. Survey of Canada, and other sources.)

Fig. 23-12. Drilling rig on a platform in water 100 feet deep, 32 miles offshore near Galveston, Texas, 1965. Crew lives on platform, which has its own heliport (*upper left*). (Shell Oil Company.)

known oil reserves throughout the world are shown in Fig. 23-13.

Tar sands and oil shales. Another source of petroleum, not yet exploited, consists of nonfluid hydrocarbons in the solid state that fill the pores in sedimentary rocks. By far the largest occurrences in North America are the Athabaska tar sands in northern Alberta and the Green River oil shales in Colorado, Utah, and Wyoming. Together they have been estimated to represent potential petroleum reserves four times as great as the continent's presently known reserves of liquid petroleum.

The "sands" in Alberta are sandstone containing asphalt ("tar") that occupies the pore spaces, having migrated into them long ago. The asphalt is too cold to flow. It must be mined, then heated to separate asphalt from quartz grains. A large mining and heating plant is now beginning to operate. Other, much smaller bodies of tar sand occur in parts of the western United States.

The Green River oil shales (Fig. C-1, *B*) contain a much larger volume of hydrocarbon derivatives than the tar sands. They consist of bodies of sediment deposited in very ancient lakes. The hydrocarbon material, consisting of kerogen, is unusual in being of freshwater origin and also in occurring not in a reservoir but in the clay-rich sediments in which it was formed originally. It becomes transformed at high temperature and yields oil. Methods of mining and heat treatment have been devised, and in the future this shaly rock should become a highly important producer of petroleum.

NUCLEAR ENERGY

Although most of the energy we use today comes from fossil fuels, the use of nuclear energy for industrial purposes has already begun and in time will play an important part in the world's economy. So far, such energy has been obtained only from the atomic nuclei of a few unstable, radioactive elements, chiefly uranium.

Like many other economic deposits described later in this chapter, some deposits containing uranium are basically related to magmas, whereas others were deposited by ground water. The most common uranium mineral is uraninite or pitchblende, so called because it is "black as pitch." In some regions pitchblende has been altered, by weathering and redeposition by ground water, to secondary minerals, mostly vanadates and phosphates.

Because uranium is radioactive, prospecting for uranium-bearing deposits employs scintillometers, which respond sensitively to high-energy particles released by radioactive decay.

Deposits from which important North American production is now coming are located in the Colorado Plateau region and Wyoming, and in the Blind River district in Ontario. Important foreign sources include the U.S.S.R., the Witwatersrand district in the Union of South Africa, and France.

SOURCES OF MATERIALS: MINERALS USED IN INDUSTRY

Having reviewed the principal mineral sources of energy, we can turn now to mineral sources of materials. Some of the principal mineral substances used in industry are listed, according to use, in Table 23-1.

The right-hand column lists minerals (example: fluorite), or derivatives (if they are used industrially as such), and metallic elements derived from minerals (if only the metal is used; as for example, iron). The list, much generalized, shows only a few of the hundred or more minerals and metals that enter into commerce.

In the table, subdivision according to use sep-

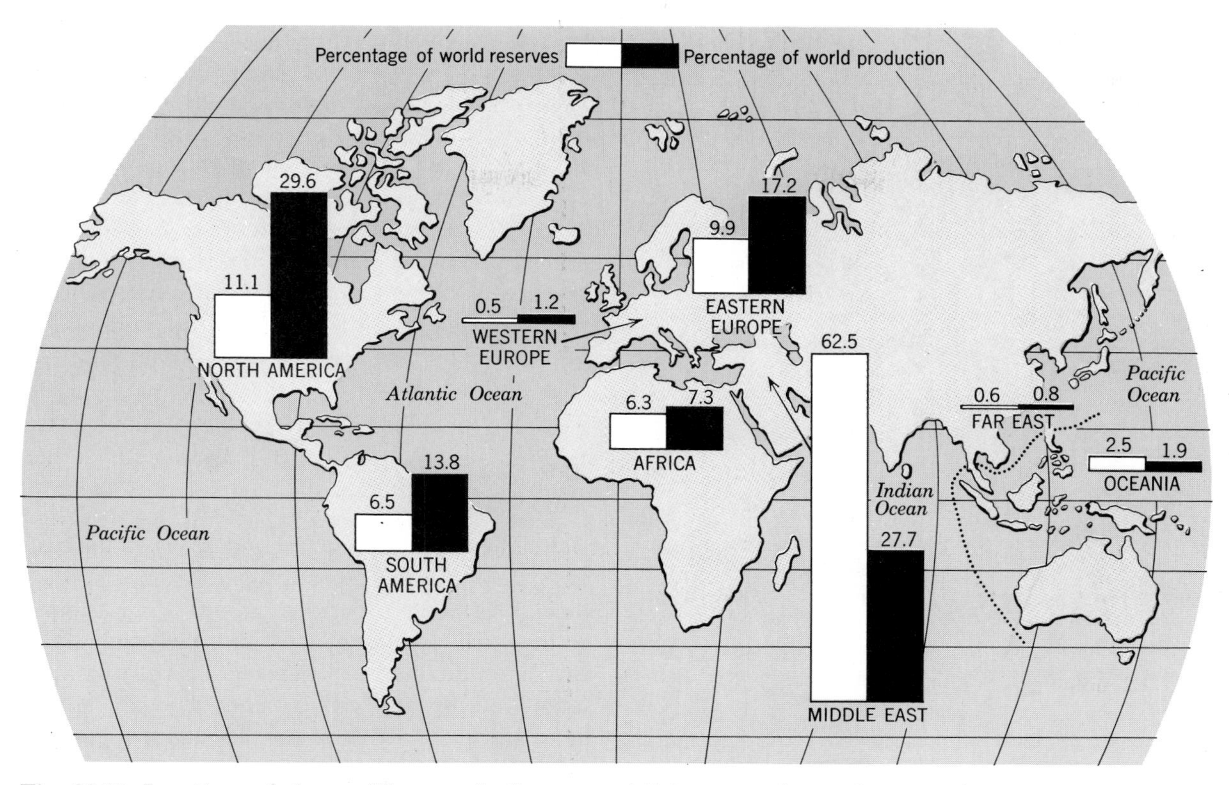

Fig. 23-13. Locations of the world's proved oil reserves (oil known to be in the ground) compared with amount of oil produced, by regions, in 1965. In that year North America possessed 11.1 per cent of the world's known oil but contributed 29.6 per cent of the world's production. (Source: World Oil, 1966.)

arates metals from nonmetals rather distinctly; but we are concerned here mainly with the *origin* of mineral deposits, and in this respect metals are not easily separated from nonmetals, because in many deposits minerals of both groups were formed side by side under much the same conditions. Therefore, we shall discuss the two groups together according to their origin.

NATURAL CONCENTRATIONS

Important though some nonmetals are, to most of us metals are inherently the more interesting group, partly because both the implements (including machines) and the structures on which our material civilization is based are built mainly of metal. Manufacture from metals has been possible—in fact, the transition from the Stone Age to the Age of Metals was possible—only because here and there metals have been *concentrated* in the rocks by natural processes. If the metals present in the Earth's crust were scattered uniformly through the rocks their amounts everywhere would have been so small that it would have been impossible to use them. Whoever the ancient people were, thousands of years before the Christian era, who first used metals, we can be sure they found them in places where the geochemical cycle had brought together enough particles to be easily visible and easily extracted from the rocks. In many places metals like gold and copper occur and are mined as elements in the native form uncombined with other elements; metals such as iron and aluminum, on the other hand, occur in combination with other elements, usually sulfur and oxygen.

What we mean, quantitatively, by concentrations depends on whether our point of view is pure scientific curiosity (in which case we are interested only in percentages) or economic (in which case we are interested also in costs and prices).

The curiosity of scientists has led to the chemical analyses of thousands of samples of igneous rocks of all kinds, and from the analyses the probable abundance of each of the chemical elements present in the crust has been estimated. Of all the useful metals,

TABLE 23-1. PRINCIPAL MINERAL SUBSTANCES USED IN INDUSTRY[a]

Group (according to use)	Partial List of Mineral Substances
METALS	
Iron and steel	Iron ore
Metals used as alloys with iron	Manganese, chromium, nickel, tungsten, molybdenum, vanadium, cobalt
Other important metals	Copper, zinc, lead, tin, aluminum
Minor metals	Mercury, silver, platinum, gold
Fissionable isotopes	Uranium
NONMETALS	
Minerals used in metallurgy	Fluorite, bauxite, graphite
Minerals used in chemical industry	Salt, sulfur, borax, coal
Minerals used in building	Stone, cement minerals, gypsum, petroleum solids
Minerals used in fertilizer	Phosphates, potash, nitrates, calcite
Minerals used in ceramics	Clay, feldspar, quartz
Minerals used in abrasives	Diamond, garnet, quartz, corundum
Fossil fuels	Coal, oil, gas

[a] Modified after A. M. Bateman, 1951.

aluminum (8 per cent), iron (5 per cent), and magnesium (2 per cent) are the only ones that make up more than 1 per cent by weight of the total rock matter. In fact, all the other useful metals lumped together constitute less than 1 per cent of the total. These amounts are so very small that only when they have been greatly increased by natural concentration can metals be mined and used. The degree of concentration required varies with value, accessibility, and other factors. Minable aluminum ore ordinarily contains around 20 per cent pure aluminum; the iron ore now mined contains 25 to 70 per cent pure iron; because of its high value, minable gold ore can be less than 0.001 per cent pure gold. Each represents a concentration equal to many times the average abundance of the metal in the rocks.

Ores. We have just mentioned *ore* and *value.* The two terms go together because the definition of ore is a quantitative one. An ***ore*** is *an aggregate of minerals from which one or more materials can be extracted profitably.* As both costs and market prices fluctuate, a particular aggregate of minerals may be an ore at one time but not at another. Associated with the ore minerals, from which the desired substances are extracted, are other minerals collectively termed the gangue (pronounced *gang*). We define ***gangue*** as *the nonvaluable minerals of an ore.* Familiar minerals common in gangue are quartz, limonite, calcite, dolomite, feldspar, and garnet, although no two ores are exactly alike as to gangue.

The ore problem has always been twofold: first, to find the ores (which altogether underlie an infinitesimally small proportion of the Earth's land area) and, second, to get rid of the gangue as economically as possible. Getting rid of the gangue is merely the completion by man of the process of concentration begun by nature.

ORIGIN AND CLASSIFICATION OF MINERAL DEPOSITS

Relation of most ores to magmas. One of the basic facts about ores is their obvious relation to igneous rocks. Ores tend to occur in groups and clusters in or close to intrusive igneous bodies, especially stocks and batholiths. This relationship, coupled with observed occurrences of metallic deposits such as hematite and galena at active fumaroles and of deposits of gold, silver, copper, lead, and zinc minerals at active hot springs, has led to the opinion that most ores are closely related to magmas, the parent materials of igneous rocks. This idea was dimly suspected as early as the sixteenth century, was clearly set forth in 1788 by James Hutton, and, with increasing support, is held by almost all geologists today. Magma, therefore, is the parent not only of igneous rocks but of most ores as well.

The degree of kinship between ores and magmas varies. Some ores were formed as actual parts of the magmas themselves. Others have been deposited by fluids that moved upward and outward from the magma far into the surrounding rock. Finally a minority of ore deposits is obviously unrelated to magmas, having been concentrated in altogether different ways.

Classification by origin. The search for new deposits is based to a considerable extent on the use of analogy to features characteristic of ores already being mined. Accurate analogy demands a thorough knowledge of known deposits, down to the minutest details visible only under a microscope. Knowledge of these details has made possible a systematic classification of ore deposits based on their conditions of origin. The list below is simplified, but it includes the more important ways in which ores are formed:

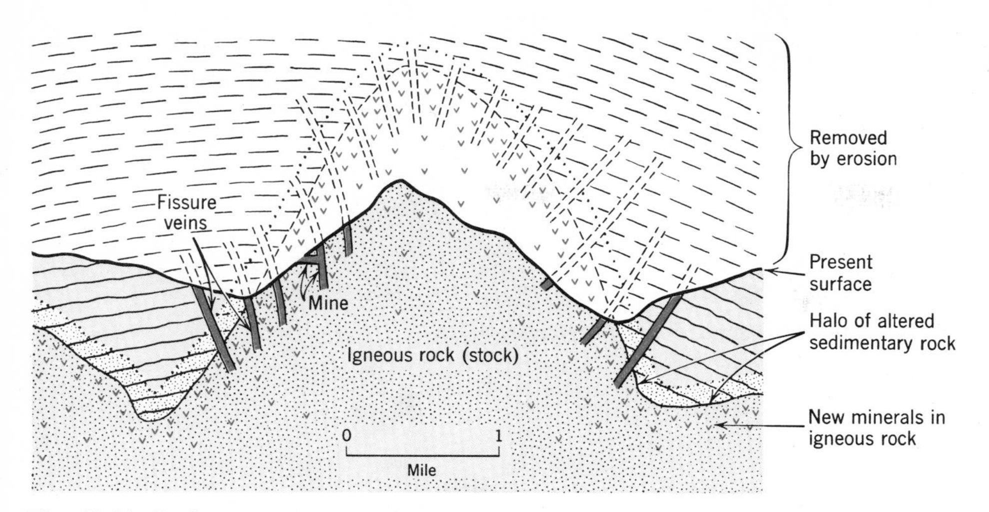

Fig. 23-14. Geologic examination at the surface and in mines shows that hydrothermal solutions coming from a body of cooling magma altered both igneous and sedimentary rocks and converted fissures into veins of ore. Relations before erosion, shown above present surface, are reconstructed by analogy with similar but less-eroded masses elsewhere.

1. Concentration within magma.
2. Alteration of rock in contact with magma.
3. Deposition from hydrothermal solutions.
4. Deposition from solution in seas and lakes.
5. Concentration by weathering and ground water.
   A. Secondary enrichment.
   B. Residual concentration.
6. Mechanical concentration.

1. ***Concentration within magma.*** When some kinds of metallic ore are melted in a smelter, the molten metallic compounds, being heavy, sink to the bottom, whereas the molten silicates, being light, rise to the top as a sort of scum. This simple concentration of metallic compounds occurs in nature as well. The separation and concentration of substances within magma is believed to explain the origin of some iron, chromium, nickel, and copper deposits. The huge deposit of magnetite iron ore at Kiruna, in northern Sweden, the site of one of the most important iron mines in Europe, seems to have been concentrated in a magma; this is true also of some of the much smaller magnetite ore bodies in the Adirondack region of New York. Among nonmetals, corundum, in North Carolina, used for industrial abrasives is believed to have originated through concentration within an ultramafic magma.

2. ***Alteration of rock in contact with magma.*** In the Cananea mining district in Mexico, just across the border from Arizona, part of the copper ore is in limestone at the contact of a body of intrusive igneous rock. The alteration of rock around igneous bodies, in a sort of halo (Fig. 23-14) surrounding the intrusive mass, is common, but when conditions are favorable the halo includes ore. The deposit at Cananea illustrates the two important favorable conditions: first, the minerals composing the rock to be altered must be ready to react chemically when heated; this is especially true of calcite, the constituent mineral of limestone. Second, new elements must be introduced from the magma.

The copper ore consists of chalcopyrite and other sulfides; evidently the copper, sulfur, iron, and other constituents foreign to the unaltered limestone were brought in by hot, gaseous fluids emanating from the magma. The fluids were not magma, for the ore shows their composition was unlike that of any igneous rock; they consisted of relatively few elements. Temperatures were high—400° to 800°C. We know this because the temperatures at which many minerals form have been learned through laboratory experiments with melted igneous rocks that are allowed to cool under controlled conditions; hence, the presence of certain minerals in the gangue is a geologic thermometer. Such thermometers sug-

Fig. 23-15. Quartz vein cutting across a fine-grained metamorphic rock. Ruler is scaled in inches. West Cummington, Massachusetts. (B. M. Shaub.)

gest that in some deposits temperatures may have reached 1,100°C. Attempts to create some of these "magma-contact" minerals synthetically in the laboratory suggest that gases are required for their formation and that therefore the new elements introduced into the rocks were brought from the magmas in a gaseous state.

The chief way in which alteration is brought about is by replacement. In this process the substance of the mineral grains in the rock is replaced, volume for volume, by ore (and other) substances brought in by the gaseous fluids. Commonly this preserves the finest details of the original texture (Fig. C-5), just as the texture of wood is preserved when replacement, under much cooler conditions, converts woody tissues to petrified wood.

Some ores of iron, zinc, and lead, in addition to copper, have been created by alteration of rocks at igneous contacts (Fig. 20-12). Among nonmetals, garnet (used in abrasives) and rubies and sapphires (varieties of corundum) are recovered.

3. ***Deposition from hydrothermal solutions.*** By far the most numerous and best-known ore bodies are those precipitated within pore spaces of rocks by hot-water solutions, perhaps including dissolved gases. Containing dissolved metals, the waters percolate upward. A large proportion of the world's gold (from such famous mining districts as the Mother Lode in California, Kirkland Lake in Ontario, and Kalgoorlie in Australia), silver, copper (as from Cerro de Pasco in Peru), lead, and zinc is mined from deposits originating in this way.

The most obvious of these deposits are ***fissure veins*** (defined simply as *fissures in rock, filled with mineral matter*) (Figs. 23-14, 23-15, 23-16). The original fissures may have been joints or faults. Most fissure veins are no more than a few feet in width, and they are generally lighter in color than the surrounding rock because their common gangue mineral is quartz. But the ores precipitated from hot waters fill openings of other kinds too: solution cavities in limestone, thin spaces between beds in sedimentary rocks, vesicles in extrusive igneous rocks, and angular openings in breccias. The textures of such ores are distinctive, in that they include crystals that have grown inward from the walls toward the centers of the openings, as in a geode (Fig. 11-12). Hydrothermal solutions likewise can form ores by replacement; such ores differ from those deposited in contact with magma in having their own characteristic set of geologic-thermometer minerals, which indicate a watery rather than a gaseous origin and also lower temperatures at the time of deposition.

Why do we believe these hot fluids were watery solutions rather than magma itself? Here is some of the evidence: (1) Many minerals in deposits of this kind have been synthesized in the laboratory by precipitation from *water* solutions. (2) Hot-spring waters are observed to deposit minerals of gold, silver, copper, lead, and other metals, similar to the minerals found in fissure veins. (3) Some of the crystals in fissure veins contain tiny spherical cavities now partially filled with water. These cavities show that when the crystal formed, the fissure was filled with a water solution.

We know the solutions were hot because of the presence of geologic-thermometer minerals. These indicate that deposition begins at around 500°C and that some minerals form only when temperatures have been reduced to 50°C or less.

Many fissure veins, replacements, and related deposits occur in wide zones extending outward away from intrusive igneous bodies. Others pene-

Fig. 23-16. The native gold in this ore, from a mine near Allegheny, California, is worth several thousand dollars per ton of ore. The principal gangue mineral (white) is quartz. Several other minerals are present in small amounts. (A. M. Bateman.)

trate far down into the igneous bodies themselves (Fig. 23-14). At Butte, Montana, an important copper district, veins extend more than 4,000 feet into a batholith; this fact indicates that the batholith had solidified through at least its outermost 4,000 feet before the thick shell began to fracture. Presumably the fractures permitted hot, watery solutions to escape from the still-fluid magma within.

Until recently it was rather widely believed that all the high-temperature water solutions that form ores originated in cooling bodies of magma. According to this theory, as the magma slowly cools and solidifies from the outside inward, the liquids and gases are confined within an ever-decreasing volume of magma, in which also the mineral substances of potential ore deposits become concentrated. Fissures develop in the solidified outer part of the magma body and tap the gas-charged residual fluids within. Under great pressures the fluids escape outward and upward through the fissures into the surrounding rocks. Depending on the pressures and temperatures prevailing, the escaping solutions of metals could be either hot gases or hot liquids. But as they move up through fissures they lose heat, encounter ever-decreasing pressures, react chemically with the rocks through which they pass, and become increasingly mixed with ground water. Accordingly various constituents successively become insoluble and are deposited.

This theory still applies to many ores deposited from hot-water solutions, particularly ores that occur within intrusive igneous bodies. But a different theory is supported strongly by geochemical analysis of the saline, hot-water solutions encountered today in deep steam wells beneath the Salton Sea in southern California. There, although the source of the heat probably is magmatic, the water shows by its chemical character that it is ground water. Furthermore it is *connate water,* seawater in which sediments were deposited, both sediments and water having been then deeply buried beneath younger strata. The solutions from the steam wells contain as much as 25 per cent salt by weight (chiefly sodium chloride, calcium chloride, and potassium chloride) and remarkably high concentrations of copper, lead, zinc, and silver. The metals originally were present in the mineral grains of the sediments, from which they were picked up by the percolating saline ground water. The metals could be collected because they were made very soluble by the formation of metal-chloride complexes in the chloride-rich solution.

We learn from this that magmas need not be the only source of metals precipitated by hot water. Metals may have been present in the constituent minerals of sedimentary rocks and have been concentrated by heated ground water percolating through them. As a result, it appears that metallic ores deposited from hot-water solutions can have come from at least two sources, magma and deep-lying sedimentary rocks. Which was the source for any particular ore body? This problem must be attacked on the basis of the geologic and geochemical relations of the ore body.

4. ***Deposition from solution in seas and lakes.*** Sodium chloride and other nonmetallic compounds precipitated in saline lakes, such as Great Salt Lake, are derived through tributary streams from the products of chemical weathering of the rocks within the lake's drainage area. This process of weathering →stream transport→precipitation in a water body, a prime example of natural concentration, has been responsible for a wide variety of mineral deposits. Many salts of sodium (including common salt), potassium, and boron are recovered from saline lakes and playas. Enormous quantities of common salt, and of the calcium sulfates gypsum and anhydrite used in making plaster, are recovered from sedimentary layers precipitated in shallow, cut-off arms of ancient seas (Fig. 15-9, *A*). Much of the salt production in the United States comes from such layers in rocks of Silurian age, more than 400 million years old.

These and other nonmetallic substances have

Fig. 23-17. Susquehanna location, a deep open-pit iron mine, Hibbing, Minnesota. (Sawders-Cushing.)

been concentrated by evaporation and are therefore evaporites. Metals, too, have been concentrated from solution in the sea, but their abundance in average seawater is so slight that thick beds of iron ore, for example, could not have been concentrated solely by evaporation of ordinary seawater; some other process must have been responsible.

A good example is the Clinton iron ore, occurring as lens-like bodies in several sedimentary units, one of them locally more than 30 feet thick, extending from New York State southward for 700 miles into Alabama. The large steel industry of Birmingham, Alabama is based on the convenient proximity of this ore, coal, and limestone. Fossils in the ore beds show that the strata were laid down in a shallow sea during the Silurian Period. Ripple marks and mud cracks further indicate shallow water. Obviously these beds were parts of the thick pile of sediments accumulated in the Appalachian geosyncline. The iron mineral is red hematite. The proportion of iron is 35 to 40 per cent, except near the surface where greater concentration has resulted from weathering.

It is believed that the iron was dissolved from iron-bearing minerals in mafic igneous rocks, carried by streams to the shallow sea (perhaps as a bicarbonate), and there precipitated as oxides. The Clinton ore concentrated in this way amounts to billions of tons. The enormous iron-ore beds in the province of Lorraine, in eastern France, originated in a similar way. The Lake Superior iron ores (Fig. 23-17), hitherto the mainstay of the North American steel industry, and the similar, newly developed ores in Labrador, on which Canada and the United States will depend increasingly, possibly originated in the same way, but both have been concentrated further by weathering.

The world's biggest deposits of manganese, in the U.S.S.R., originated in a similar manner by precipitation, probably organic, in ancient seas.

5. ***Concentration by weathering and ground water:*** A. *Secondary enrichment.* The related processes of sorting and concentration by ground water are chemical. They depend on the principle (Fig. 7-7) that change in the environment of a mineral may make it vulnerable to chemical attack. Many fissure-vein minerals, though stable at the high temperatures at which they crystallized from solution, become unstable in the zone of aeration at the Earth's surface. In many areas erosion has worn the surface down so that even the deep-lying minerals are brought within the shallow zone in which atmo-

spheric oxygen and ground water can attack them and cause decay in the new environment of low temperature and low pressure. Ground water dissolves pyrite, FeS_2, creating sulfuric acid, H_2SO_4, and residual iron oxides. The acid dissolves copper minerals, forming copper sulfate. Entering the zone of saturation, the solution loses oxygen and reacts with minerals there, depositing copper as sulfides. In a fissure vein the added sulfides enrich the ore already present or convert low-grade deposits into workable ore (Fig. 23-18). This process of addition is known to mining geologists as ***secondary enrichment,*** defined as *natural enrichment of an ore body by addition of mineral matter, generally from percolating solutions.*

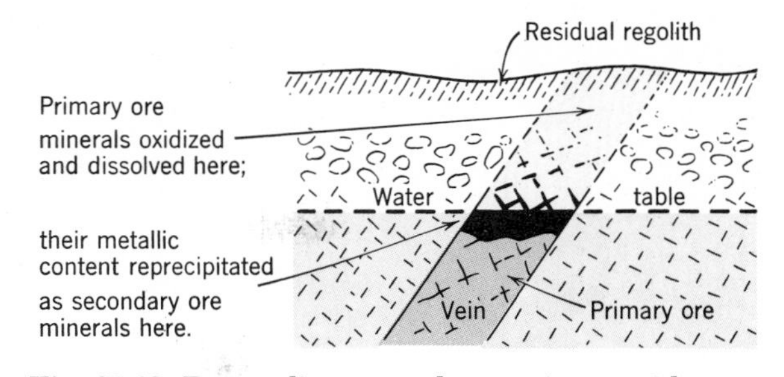

Fig. 23-18. Descending ground water impoverishes ore above water table by removing soluble ore minerals; produces secondary enrichment below.

Secondary enrichment has been important at the Utah Copper Mine (Fig. 23-19) at Bingham Canyon, Utah, the world's second largest copper producer. The ore lies in the upper part of a body of intrusive igneous rock, the silicate minerals of which were replaced by copper, gold, silver and molybdenum carried up by hot solutions from the magma underneath. The upper part of the primary ore has been secondarily enriched through a zone more than 100 feet thick (Fig. 23-20). This enormous low-grade deposit is an ore only because it does not demand underground mining and can be worked by low-cost

Fig. 23-19. Utah Copper Mine at Bingham Canyon, Utah, in 1950. Terraced slope in background is 1,900 feet high. (Kennecott Copper Corporation.)

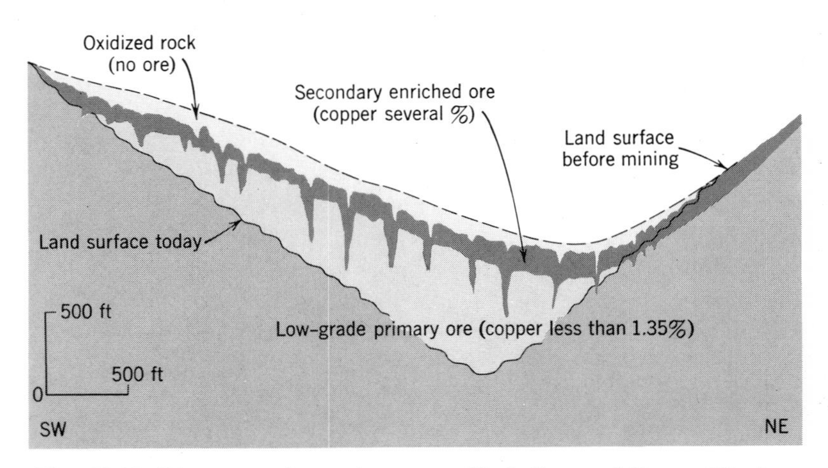

Fig. 23-20. Diagrammatic section across Utah Copper Mine at Bingham Canyon, Utah, shows how weathering has concentrated copper in an enriched zone related to a land surface.

surface methods. The sides of the canyon have been cut into giant steps, 50 to 70 feet high, making a huge pit resembling a Roman amphitheater. The ore is blasted out and is loaded by power shovels into railroad cars. Waste is dumped into side canyons. The ore already mined exceeds 1.25 billion tons, and some ore containing as little as 0.4 per cent copper is being worked profitably.

B. *Residual concentration.* In secondary enrichment ground water adds new material to an existing body. Some ores, however, are developed by the subtraction of old material. This is ***residual concentration,*** defined as *the natural concentration of a mineral substance by removal of a different substance with which it was associated.* An example is bauxite, a hydrous aluminum oxide and the common ore of aluminum (Fig. 23-21).

Fig. 23-21. Bauxite, the ore of aluminum, concentrated by deep, long-continued weathering of syenite. The rounded bodies are not pebbles but concretions, developed during weathering. Bauxite, Arkansas.

Bauxite is a product of chemical weathering. Aluminum was present as a constituent of original, primary silicate minerals in syenite, schist, or clay. During weathering, silica was carried away in solution, gradually increasing the concentration of aluminum. We know this has happened because we can see bauxite grading downward into the underlying rock just as does any weathered regolith. From the locations of bauxite ores we know that the weathering has taken place beneath peneplains and other surfaces of low relief, with the water table close to the surface, with slow ground-water circulation, and always in tropical or subtropical climates. But the chemistry of the process is complex, and we still have much to learn about just what happens. Apparently ground water, charged with substances derived from the soil, decomposes the aluminum silicates in the fresh rock, carries away silica in solution, and leaves aluminum (and iron as well) behind as oxides. Whatever the chemical details, the resulting bauxite may reach a concentration of 20 per cent pure aluminum.

Although some difference of opinion about the matter exists, it is likely that the richer iron ores of the Lake Superior district, and those of the newly developed district in Labrador, originated in a somewhat similar way. The ore bodies are sedimentary beds much like the Clinton iron ores, but at depth they are siliceous, with only about 30 per cent iron, grading upward toward the surface into nonsiliceous hematite with up to 62 per cent iron. Apparently long-continued weathering has dissolved and removed the silica, leaving a residual concentrate of almost pure hematite.

6. ***Mechanical concentration: placers.*** The famous gold rush to California in 1849 resulted from the discovery that the sand and gravel in the bed of a small stream contained bits of gold. Indeed, most mining districts have been discovered by following trails of gold and other minerals upstream to their sources in bedrock. Because its specific gravity is about 19, pure gold is quickly deposited by a stream in preference to quartz, the specific gravity of which is only 2.65. Particles of gold, therefore, are mechanically sorted out and concentrated in the lowermost parts of stream deposits, particularly in point bars (Fig. 9-27) where stream velocities are low. *A deposit of heavy minerals concentrated mechanically* is a ***placer*** (*plăs′-er*). Besides native gold, other heavy, durable metallic minerals form placers. These include metallic minerals such as platinum, tinstone (cassiterite), and native copper, and the nonmetallic diamond and other gemstones as well.

Every phase of the conversion of gold in a fissure vein to placer gold has been traced. Chemical weathering of the exposed vein releases the gold, which then moves slowly downslope by mass-wasting (Fig. 23-22). In some places mass-wasting alone has concentrated gold or tinstone sufficiently to justify mining these metals.

More commonly, however, the mineral particles get into a stream, which concentrates them more effectively than mass-wasting can. Most placer gold occurs in grains the size of silt particles, the "gold dust" of placer miners. Some of it is coarser; pebble-sized fragments are *nuggets* (Fig. 23-23), of which the largest ever recorded weighed 2,280 ounces and at the current price of gold would be worth $79,800. In following placers upstream, prospectors have learned that rounding and flattening (by pounding) increase downstream, just as does rounding of ordinary pebbles; when they find angular nuggets they know they are close to the bedrock source. Mining was done first by hand, simply by swashing stream sediment around in a small pan of water. Later it was done by jetting water under high pressure against gravel banks and washing the sediment through troughs that caught the heavy gold behind cleats. Nowadays the more efficient method of dredging is used (Fig. 23-24). The platinum placers

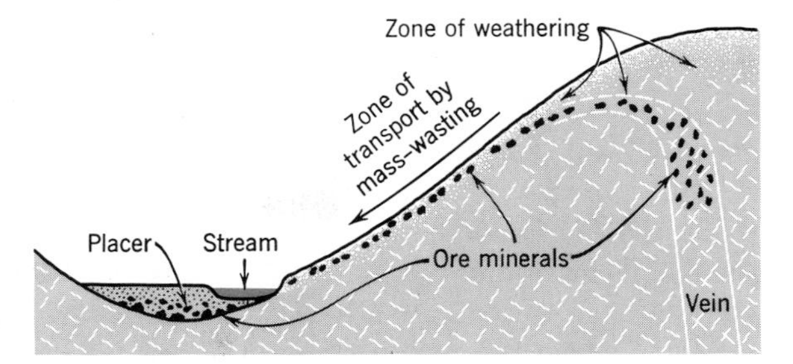

Fig. 23-22. Ore minerals from weathered fissure veins creep down slopes and are redeposited by streams as placers. The cross section shows essentially the same relations as those in Fig. 8-1 but in an economic context.

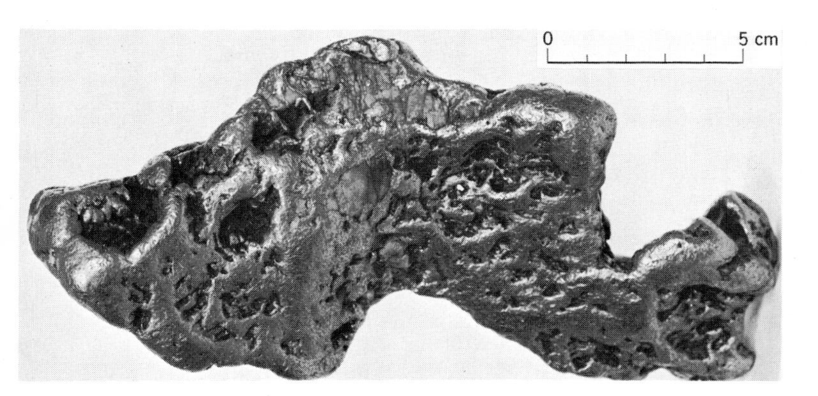

Fig. 23-23. Gold nugget, found near Greenville, California, weighs 82 ounces. Its surface shows clearly the pounding it received during stream transport. (George Switzer; collection U. S. National Museum.)

Fig. 23-24. A gold dredge like this one in the Yukon Valley, Alaska, may cost more than half a million dollars, but it can profitably work gravel containing only 10 to 20 cents in gold per cubic yard. (Bradford Washburn.)

of the Ural Mountains in Russia, the rich diamond placers of the Kasai district in the Congo, and the hundreds of tin placers in Malaya are examples of other mechanical concentrations by streams.

Gold, diamonds, and several other minerals have been concentrated in beaches by surf.

Diamonds are being obtained in large quantities from gravelly beach placers, both above and below present sea level, along a 200-mile strip of the coast of South-West Africa. Weathered from deposits in the interior, the diamonds were transported by the Orange River to the coast and were spread southward by longshore drift. Later some beaches were uplifted, and others were submerged by rise of sea level.

Finally, gold is recovered in Mexico and in Australia from placers concentrated by wind. Quartz and other light minerals have been deflated, leaving the heavier gold particles behind. Therefore, because mechanical concentration results from a variety of natural processes, the principles of erosion and deposition must be thoroughly understood by those who would discover new placers.

FUTURE OF MINERAL USE AND EXPLORATION

The pattern of the world's sources of energy and materials is changing and will continue to change. When the industrial revolution began, mankind had existed for thousands of years on an energy economy of wood fuel and muscle power, both of which are replaceable. Now, in the industrial age, which began only yesterday in man's long history and less than a minute ago in the vastly longer span of geologic history, man is rapidly using up the mineral riches stored in the Earth.

The steep curve of energy consumption shown in Fig. 23-1, and the very similar curve of consumption of materials, can not long continue to steepen if we employ only the materials and energy sources we are using today, because they are "one-crop" products and are not renewable. By persistent search man has found a large proportion of the mineral deposits that are not well hidden. Probably, however, great quantities of ore lie concealed and undiscovered at depths no greater than a few hundred feet. The search for them will involve not old-fashioned surface prospecting but geologic, geochemical, and geophysical research and will require more highly skilled and specialized geologists than have been needed in the easier past. The future, then, should see an increase in the supply of the kinds of materials and energy we are using now, the development of improved ways to apply them, and the addition of entirely new and different kinds.

In the field of energy, water power can never become a major source, for probably little more than 1 per cent of the world's runoff can be made to generate power economically, and about one-tenth of this potential is harnessed and working for us already. Many new oil fields will be found, but their average depth will be greater than at present, and costs will therefore be higher. Eventually the production of petroleum will decline. The amount in the ground is limited, and once used it can not be replaced within the probable lifetime of the human race. In terms of energy, world reserves of uranium probably are greater than reserves of fossil fuels, and the cost of energy from this source may prove to be less than the average cost of energy from fossil fuels. Nuclear energy is likely to come not only from uranium but also from thorium, another unstable, radioactive element. Possibly it will come from still other elements as well.

For the longer future the most promising energy source is the Sun—modern energy, that is, not fossil energy—collected directly and concentrated at the Earth's surface.

When we consider the future of the western world's sources of materials for industry, we again find increasing use and increasing production. As we have seen, if iron, copper, and the other metals essential to our ecomony were distributed evenly throughout the Earth's crust, they would be irrecoverable, unminable. They can be recovered and used only from those few areas, trivial when compared with the area of the Earth, in which igneous activity, erosion and sedimentation, or other geologic processes have concentrated them preferentially.

At some future time the existing phase of our industrial activity, marked by the consumption of stored-up concentrations of minerals, will begin to decline. Then mankind will have to look to materials of other kinds to supplant at least some of the metals now used to build machines and structures. The comparatively recent plastics industry, whose chief raw materials are petroleum, coal, and agricultural products, is a step in that direction. However, the decline of industrial minerals may be far in the future, because we have not yet explored fully the rocks exposed at the surface, let alone the igneous and metamorphic rocks which are widely covered by overlying strata in the continents or are submerged beneath seawater. These unknown or little-known rocks are a challenge to the geologist of today and will test his skill keenly. Only when he and his successors cease to find new deposits to replace those now being used up will the search for substitutes become really intense. When that time comes, chemists are likely to play an important part in the development. Whether their chief co-workers will be geologists or agriculturists, it is too soon to say.

Summary

High-energy industry

1. Ancient industry based largely on wood and muscle has given way to high-energy industry based mainly on fossil fuels and metals.

Fossil fuels

2. Coal originated as plant matter in ancient swamps and is both abundant and widely distributed.

3. Oil and gas probably originated as organic matter sedimented on sea floors and decomposed chemically. Later it moved through reservoir rocks and was caught in geologic traps to form pools.

4. Oil pools are found by the use of both geologic and geophysical methods. Of the latter, seismic surveying is the most commonly used.

Nuclear energy

5. Nuclear energy is derived from atomic nuclei of unstable elements, chiefly uranium.

Mineral deposits: ores

6. The mineral industry is based mainly on local concentrations of useful minerals, concentrated by natural processes to proportions far exceeding their

average abundance in the Earth's crust.

7. Most but not all ores were derived directly or indirectly from magmas. The most numerous were deposited by hot water and gases moving outward from cooling magmas, whether the metals originated in the magmas themselves or were dissolved by ground water from deep-lying rocks.

8. Many important iron ores were originally precipitated in ancient seas and probably later concentrated by ground-water removal of silica.

9. Secondary enrichment by weathering and ground water has been important in further concentrating many copper deposits.

10. Gold, platinum, tinstone, diamonds, and copper have been mechanically concentrated to form placers.

11. Geologic research, as distinct from conventional prospecting, probably will discover many new ore deposits.

Selected References

Bateman, A. M., 1951, The formation of mineral deposits: New York, John Wiley.

Emery, K. O., 1960, The sea off southern California: a modern habitat of petroleum: New York, John Wiley.

Francis, Wilfrid, 1961, Coal: London, E. Arnold.

Hubbert, M. K., 1962, Energy resources: Washington, U. S. National Academy of Sciences Publ. 1000-D.

Lamey, C. A., 1966, Metallic and industrial mineral deposits: New York, McGraw-Hill.

Levorsen, A. I., 1964, Big geology for big needs: Am. Assoc. Petroleum Geologists Bull., v. 48, p. 141–156.

Levorsen, A. I., 1967, Geology of petroleum, 2nd ed.: San Francisco, W. H. Freeman.

McDivitt, J. F., 1965, Minerals and men: Baltimore, Johns Hopkins Press.

Mero, J. L., 1965, The mineral resources of the sea: New York, Elsevier Publ.

Moore, E. S., 1940, Coal: New York, John Wiley.

Russell, W. L., 1960, Principles of petroleum geology: New York, McGraw-Hill.

Van Royen, William, and Bowles, Oliver, 1952, Atlas of the world's resources, II—The mineral resources of the world: Englewood Cliffs, N. J., Prentice-Hall.

Appendices

APPENDIX A

Some Physical and Chemical Concepts

Many phenomena of importance to geology are governed by the fundamental principles of chemistry and physics. Because the text presupposes an acquaintance with many of these principles, a summary of them is included here for convenient reference.

COMPOSITION OF MATTER

Chapter 3 describes some of the physical properties of matter. These and other attributes can be understood in terms of the many particles discovered by chemists and physicists.

The fundamental chemical entity is an ***element.*** Individual elements occupy a middle ground in the field of particles. Elements consist of still smaller particles, yet can join together to form even larger particles. Let us review first the smaller particles.

Atoms and isotopes. *The smallest electrically balanced particle displaying the properties of an element* is an ***atom.*** Each atom consists of a nucleus, only about $1/_{10,000}$ the size of the whole atom, and concentric shell(s) of ***electrons,*** *unit particles of negative electrical charge.*

Atoms themselves are such infinitesimally small particles that it is difficult to conceive how the sizes and masses of still smaller particles can be ascertained. Nevertheless physicists assure us that the mass of an electron is 9×10^{-28}g. The nucleus of most atoms contains two kinds of particles much larger than electrons: (1) ***protons,*** *particles of unit positive electrical charge having a mass of* 1.6730×10^{-24}g and ***neutrons,*** *electrically neutral particles having a mass of* 1.6752×10^{-24}g. Because the particles of the nucleus are densely packed the bulk of the weight of the atom derives from the protons and neutrons, notwithstanding their small volume. The bulk of the space occupied by an atom is taken up by the spherical shells of electrons outside the nucleus.

In standard notation the ***atomic number,*** *the number of protons in a nucleus of an atom,* is represented by the letter Z, and the number of neutrons by *N*. The ***mass number*** of an atom, *A, equals Z plus N.*

Each element is characterized by its value of Z, which also determines the total positive charge. Because an atom is electrically neutral, Z likewise designates the number of electrons.

Variations arise because each number of protons is not always accompanied by the same number of neutrons. In other words a given value of Z can be associated with variable values of *N*. Such variations create ***isotopes,*** *configurations of the same element, all having the same number of protons and generally similar chemical behavior but different mass numbers, A, resulting from different numbers of neutrons in the nucleus* (Fig. A-1).

Isotopes are designated by writing the letter symbol for the chemical element (Table A-1) and the mass number as a raised postscript. (Raised mass numbers should not be confused with the mathematical notation of exponents.) Isotopes of several elements can have the same mass number. Thus the mass number of the carbon isotope, C^{14}, is the same as that of the nitrogen isotope, N^{14}. The difference is that the nucleus of C^{14} contains 6 protons and 8 neutrons, whereas the nucleus of N^{14} contains 7 protons and 7 neutrons. The atomic nuclei of *all* carbon isotopes contain 6 protons, and those of *all* nitrogen isotopes 7 protons.

We can illustrate the parts of an atom by comparing the simplest element, hydrogen, with the next simplest, helium (Fig. A-2). In a hydrogen atom the single proton of the nucleus is orbited by a single electron. Because the attraction of the proton is spherically symmetrical the single electron moves constantly along a surface that describes a sphere. A single neutron is unable to form a nucleus because its lack of electrical charge prevents it from keeping an electron. Deuterium (H^2), a hydrogen isotope with a mass number of 2, originates when a neutron is added to the nucleus of ordinary hydrogen, H^1. Still another isotope of hydrogen, ***tritium,*** originates when a second neutron joins the deuterium nucleus, forming H^3. In a helium atom the nucleus consists of 2 protons and 2 neutrons; it is orbited by 2 electrons.

Both nucleus and surrounding electrons are thought to be arranged in shells. The numbers of neutrons or pro-

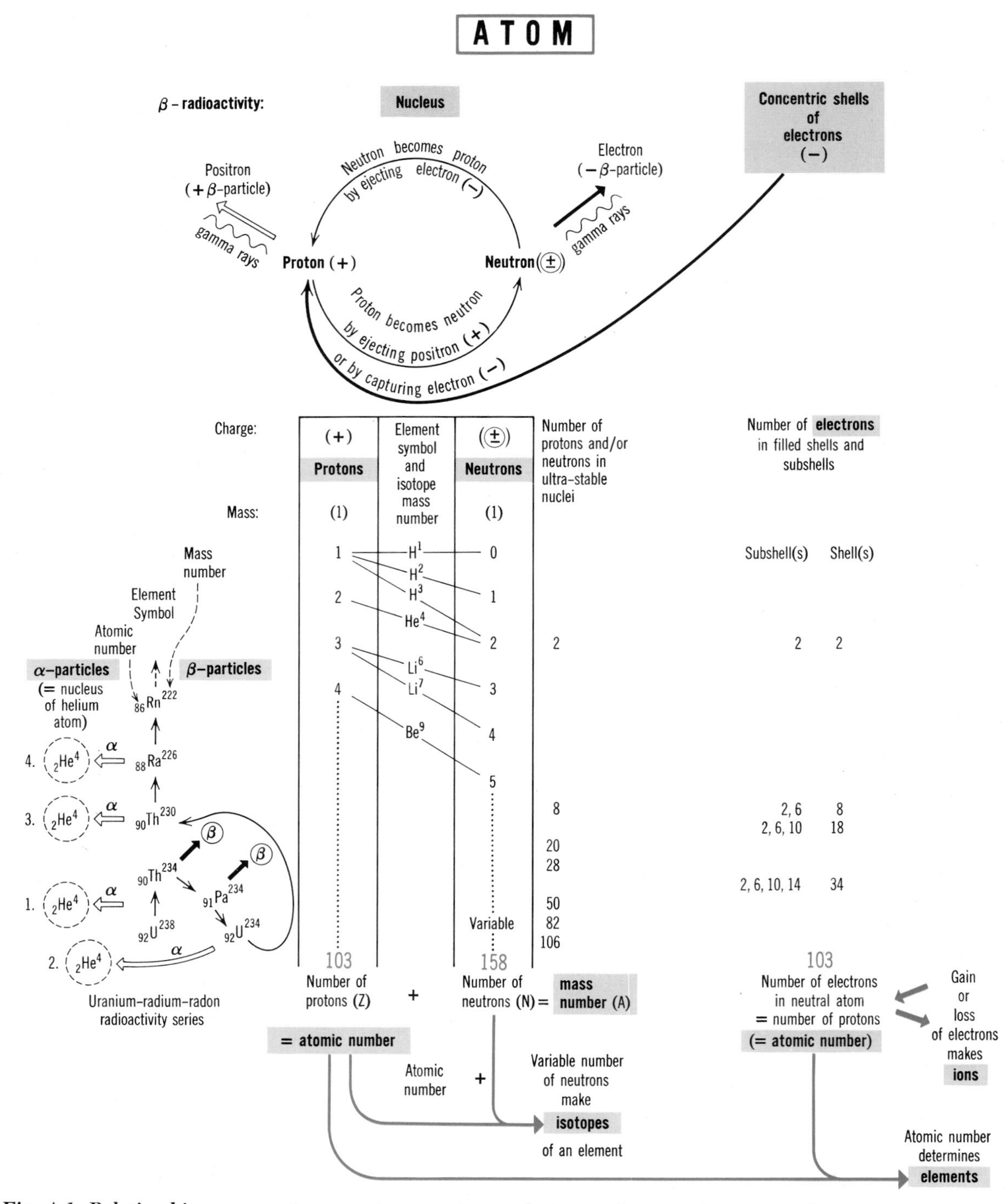

Fig. A-1. Relationships among atoms, protons, neutrons, electrons, elements, isotopes, and ions. At upper left are shown the various types of β-radioactivity. Part of the uranium-radium-radon series, illustrated at lower left, shows changes in both mass number and atomic number when an α-particle (nucleus of helium atom) is emitted from a nucleus. In this series β-radioactivity also takes place, causing changes in atomic number but not in mass number. (Further details in text and in Fig. 6-2.)

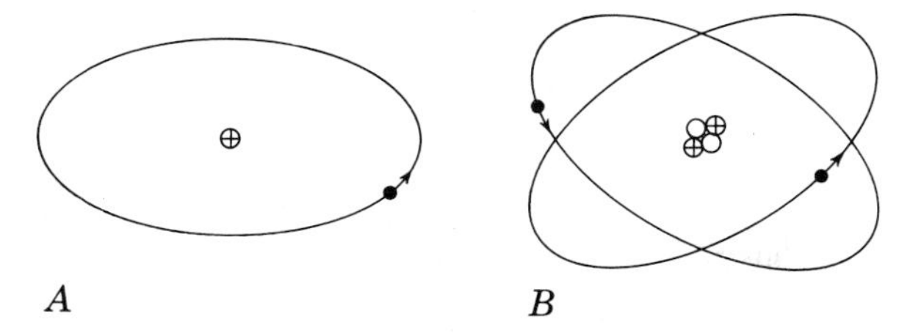

Fig. A-2. Schematic sketches compare simplest atom, hydrogen (*A*) with next simplest atom, helium (*B*). Proton, ⊕ ; neutron, ○ ; electron, • .

tons that fill complete shells in the nucleus are 2, 8, 20, 28, 50, 82, and 126. A nucleus whose particles fill a shell completely is very stable. For example, He^4 illustrates a filled 2-shell, O^{16} a filled 8-shell, and Ca^{40} a filled 20-shell.

The electrons likewise are organized in many shells and subshells, within which the electrons can be oriented in various numbers of ways. In the first shell outside the nucleus, 2 orientations are possible. In the second shell the number of possible orientations becomes 8, with subshells of 2 and 6. In the third shell 18 orientations are possible, with subshells of 2, 6, and 10; in the fourth shell, 32 orientations include subshells of 2, 6, 10, and 14. In the fifth, sixth, and seventh shells, the number of orientations is the same as that in the fourth shell, but no element is known in which these outer shells are completely filled. As each new electron shell appears, the size of the atom increases.

By gaining or losing electrons an atom becomes an ***ion****, a charged particle formed from an atom by adding or subtracting one or more electrons.*

Before we complete our discussion of matter we must consider energy. Despite our separation of them, matter and energy are varied expressions of the same fundamental phenomena.

ENERGY

Energy is defined loosely as *the capacity to do work,* and is usually measured by its mechanical, thermal, and electrical effects. The basic unit of mechanical energy is the ***erg****, the mechanical energy required for a force of 1 dyne to act through 1cm.* A ***dyne*** *is the force required to accelerate a mass of 1g by 1cm/sec*^{2}. An erg is such a small unit that its use in some calculations is cumbersome. Instead a convenient substitute is the ***joule****, 10^7 ergs.*

The basic unit of heat energy is the ***calorie****, the amount of heat energy required to raise the temperature of 1g of water by 1°C at 15°C.* One calorie equals 4.18 joules.

A more precise definition states that energy is *the equivalent of mass according to the Einstein equation,* $E = mc^2$, where E is the energy in ergs, m the mass in grams, and c the velocity of light in centimeters per second. From the measured velocity of light, 2.998×10^{10}cm/sec, we can calculate that the amount of energy in 1g of matter is approximately 9×10^{20} ergs.

A basic unit of electrical energy is the ***electron volt****, the energy possessed by 1 electron which has fallen through a potential difference of 1 volt.* One electron volt equals 1.6×10^{-6} ergs. By analyzing the relationships between matter and energy on a subatomic level, we can understand the basis for expressing energy in various units.

Relationships between mass and energy. One of the precepts of the theory of relativity is that the mass of a particle is related to its velocity. Accordingly, the proper designation of the mass of a particle at rest or moving at low velocity is its *rest-mass.* This term will be understood to apply in the following discussions unless otherwise indicated. If m_0 represents the mass of a particle at low speeds, and m the mass at speed v, then according to the Einstein relationship:

$$m = m_0 \sqrt{1 - \frac{v^2}{c^2}}$$

where c equals the velocity of light. From this Einstein derived his famous equation.

MOLECULES

Atoms of some elements join with atoms of other elements (or in gases with other atoms of elements like themselves) to form ***molecules****, the fundamental entities of chemical compounds.* When atoms join together to create molecules the only changes that take place are in the configuration of the electrons; the nucleus remains unaffected. Because changes in electron configuration involve only small amounts of energy they take place rather readily. Generally these changes are in response to some disturbing influence. Once the influence has been removed, the electron configuration reverts to its neutral condition.

ENERGY AND WAVE MOTION

Electromagnetic waves. Another great unifying principle of physics related to energy states that no ultimate distinction exists between small particles and certain kinds of wave motion. Energy can be analyzed in terms of particles called *photons* as well as of electromagnetic waves. It is possible, therefore, to express energy not

TABLE A-1. THE ELEMENTS

In alphabetic order

Element	Symbol	Atomic Number	Element	Symbol	Atomic Number
Actinium	Ac	89	Molybdenum	Mo	42
Aluminum	Al	13	Neodymium	Nd	60
Americium	Am	95	Neon	Ne	10
Antimony	Sb	51	Neptunium	Np	93
Argon	Ar	18	Nickel	Ni	28
Arsenic	As	33	Niobium (Columbium)	Nb (Cb)	41
Astatine	At	85	Nitrogen	N	7
Barium	Ba	56	Nobelium	No	102
Berkelium	Bk	97	Osmium	Os	76
Beryllium	Be	4	Oxygen	O	8
Bismuth	Bi	83	Palladium	Pd	46
Boron	B	5	Phosphorus	P	15
Bromine	Br	35	Platinum	Pt	78
Cadmium	Cd	48	Plutonium	Pu	94
Calcium	Ca	20	Polonium	Po	84
Californium	Cf	98	Potassium	K	19
Carbon	C	6	Praseodymium	Pr	59
Cerium	Ce	58	Promethium	Pm	61
Cesium	Cs	55	Protactinium	Pa	91
Chlorine	Cl	17	Radium	Ra	88
Chromium	Cr	24	Radon	Rn	86
Cobalt	Co	27	Rhenium	Re	75
Copper	Cu	29	Rhodium	Rh	45
Curium	Cm	96	Rubidium	Rb	37
Dysprosium	Dy	66	Ruthenium	Ru	44
Einsteinium	Es	99	Samarium	Sm	62
Erbium	Er	68	Scandium	Sc	21
Europium	Eu	63	Selenium	Se	34
Fermium	Fm	100	Silicon	Si	14
Fluorine	F	9	Silver	Ag	47
Francium	Fr	87	Sodium	Na	11
Gadolinium	Gd	64	Strontium	Sr	38
Gallium	Ga	31	Sulfur	S	16
Germanium	Ge	32	Tantalum	Ta	73
Gold	Au	79	Technetium	Tc	43
Hafnium	Hf	72	Tellurium	Te	52
Helium	He	2	Terbium	Tb	65
Holmium	Ho	67	Thallium	Tl	81
Hydrogen	H	1	Thorium	Th	90
Indium	In	49	Thulium	Tm	69
Iodine	I	53	Tin	Sn	50
Iridium	Ir	77	Titanium	Ti	22
Iron	Fe	26	Tungsten (Wolfram)	W	74
Krypton	Kr	36	Uranium	U	92
Lanthanum	La	57	Vanadium	V	23
Lawrencium	Lw	103	Xenon	Xe	54
Lead	Pb	82	Ytterbium	Yb	70
Lithium	Li	3	Yttrium	Y	39
Lutetium	Lu	71	Zinc	Zn	30
Magnesium	Mg	12	Zirconium	Zr	40
Manganese	Mn	25			
Mendelevium	Md	101			
Mercury	Hg	80			

In order of atomic numbers

Atomic number	Element	Symbol	Atomic number	Element	Symbol
1	Hydrogen	H	52	Tellurium	Te
2	Helium	He	53	Iodine	I
3	Lithium	Li	54	Xenon	Xe
4	Beryllium	Be	55	Cesium	Cs
5	Boron	B	56	Barium	Ba
6	Carbon	C	57	Lanthanum	La
7	Nitrogen	N	58	Cerium	Ce
8	Oxygen	O	59	Praseodymium	Pr
9	Fluorine	F	60	Neodymium	Nd
10	Neon	Ne	61	Promethium	Pm
11	Sodium	Na	62	Samarium	Sm
12	Magnesium	Mg	63	Europium	Eu
13	Aluminum	Al	64	Gadolinium	Gd
14	Silicon	Si	65	Terbium	Tb
15	Phosphorus	P	66	Dysprosium	Dy
16	Sulfur	S	67	Holmium	Ho
17	Chlorine	Cl	68	Erbium	Er
18	Argon	Ar	69	Thulium	Tm
19	Potassium	K	70	Ytterbium	Yb
20	Calcium	Ca	71	Lutetium	Lu
21	Scandium	Sc	72	Hafnium	Hf
22	Titanium	Ti	73	Tantalum	Ta
23	Vanadium	V	74	Tungsten (Wolfram)	W
24	Chromium	Cr	75	Rhenium	Re
25	Manganese	Mn	76	Osmium	Os
26	Iron	Fe	77	Iridium	Ir
27	Cobalt	Co	78	Platinum	Pt
28	Nickel	Ni	79	Gold	Au
29	Copper	Cu	80	Mercury	Hg
30	Zinc	Zn	81	Thallium	Tl
31	Gallium	Ga	82	Lead	Pb
32	Germanium	Ge	83	Bismuth	Bi
33	Arsenic	As	84	Polonium	Po
34	Selenium	Se	85	Astatine	At
35	Bromine	Br	86	Radon	Rn
36	Krypton	Kr	87	Francium	Fr
37	Rubidium	Rb	88	Radium	Ra
38	Strontium	Sr	89	Actinium	Ac
39	Yttrium	Y	90	Thorium	Th
40	Zirconium	Zr	91	Protactinium	Pa
41	Niobium (Columbium)	Nb (Cb)	92	Uranium	U
42	Molybdenum	Mo	93	Neptunium	Np
43	Technetium	Tc	94	Plutonium	Pu
44	Ruthenium	Ru	95	Americium	Am
45	Rhodium	Rh	96	Curium	Cm
46	Palladium	Pd	97	Berkelium	Bk
47	Silver	Ag	98	Californium	Cf
48	Cadmium	Cd	99	Einsteinium	Es
49	Indium	In	100	Fermium	Fm
50	Tin	Sn	101	Mendelevium	Md
51	Antimony	Sb	102	Nobelium	No
			103	Lawrencium	Lw

only in ergs and volts but also in frequency and wavelength. Frequency (μ) and wavelength (λ) are related to volts (V) as follows:

$$\lambda \text{ (cm)} = \frac{1.2378 \times 10^{-4}}{V} \quad \text{and} \quad \mu = 242.18 \times 10^{12} V$$

A further fundamental relationship between matter and electromagnetic radiation, formulated in 1901 by Max Planck, states that the radiation is emitted or absorbed in discrete bundles (photons or *quanta*), whose energy (E) is proportional to the frequency (μ) of the radiation. By introducing Planck's constant (h), this may be written as an equation:

$$E = h\mu$$
$$h = (6.624 \pm 0.002) \times 10^{27} \text{ erg} \times \text{sec}$$

Light waves are a familiar example of electromagnetic waves. A continuous spectrum exists among many kinds of electromagnetic waves having wavelengths both shorter and longer than light waves. Although discrete cutoff points exist for waves from a particular source, considerable overlap of wavelengths exists between successive sources (Fig. A-3).

Waves are extremely important analytical tools because wavelengths are unique functions of the object whose vibration creates them. For example, long waves, called *Hertzian* waves, are generated by the vibrations of various metal plates, crystals, or other oscillators. The size and geometry of the oscillator determine the wavelengths; the atomic structure of the atoms within the oscillator is not directly involved in the generation of the waves. *Infrared waves* are produced chiefly by motions of molecules and to a lesser extent by some atomic movements within the molecules. Irregular discrete changes of frequency called *quantum jumps* appear in the infrared range, signaling the participation of the electron shells of atoms. Waves of *visible light* and those in the *ultraviolet* range are generated by changes in the electron configuration in the outer shells of atoms.

A remarkable feature of electromagnetic waves shorter than Hertzian waves is that they serve as absolutely reliable guides to the element which produced them. By means of a spectroscope it is possible to record and measure these diagnostic waves and thus to carry out a positive chemical identification of the source of the radiation even if the element responsible is present only in minute quantities and its radiation has come from far away. The importance of this relationship can scarcely be overemphasized. It underlies the science of astrophysics, which has unlocked many of the secrets of the universe. Radiation in the *X-ray* range results from displacements of electrons within the inner shells of atoms. The still shorter *γ-rays* originate from excitations of the atomic nucleus. The origin of *cosmic rays* remains an enigma. One suggestion, not yet proved, is that they result from annihilation of matter.

Because of the continuity of the spectrum, changes of frequency are significant. The ***Doppler shift***, caused by *changes in frequency of electromagnetic waves resulting from relative motion between source and receiver*, has been used to infer the rotation of the Sun, movement of the Sun with respect to other stars, and movements among stars. According to the Doppler principle, frequency increases (hence wavelength decreases) when source and receiver approach each other, and frequency decreases (wavelength increases) when source and receiver move apart. The amount of change of frequency is determined by the relative velocity of movement between source and receiver. A shift of position of characteristic spectral lines of hydrogen, for example, has been observed to take place from the range of visible light into that of the infrared. This represents an increase in wavelength (decrease in frequency), which indicates that the source is receding from the Earth. This phenomenon is the *red shift* of astrophysicists.

Elastic waves. In contrast to electromagnetic waves, which propagate through great distances at a constant speed (the speed of light), certain long Hertzian waves propagate at speeds determined by the elastic properties of the materials through which they travel. Therefore such waves are known as elastic waves. They are important in the exploration of the oceans (Chap. 15) and in study of earthquakes and the interior of the Earth (Chap. 18). To understand elastic waves we must examine the elastic properties of materials.

The elastic properties of a material are expressions of its resistance to forces that tend to change its shape or its volume. As we explained in Chapter 3, solids differ from liquids in having a definite shape; liquids do not possess an independent shape, but assume the shape of the container into which they are poured. The *elasticity of shape* of a solid is expressed by a **modulus** (coefficient) ***of rigidity;*** it is *the stress required to cause a unit change of shape*. The standard Greek letter symbol for the modulus of rigidity is μ.

Both liquids and solids possess elasticity of volume, which is expressed as the ***bulk modulus,*** *the stress required to cause a unit change of volume.* The standard symbol for the bulk modulus is the Greek letter κ.

Certain waves are characterized by the kinds of mo-

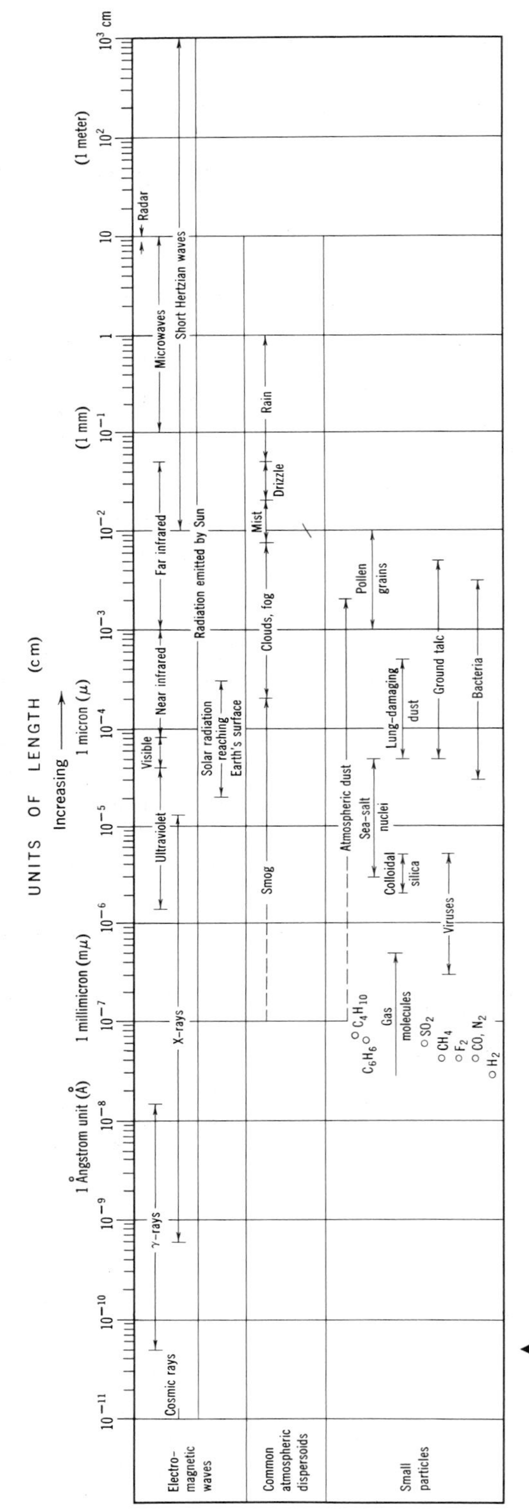

◄ **Fig. A-3. Wavelengths of electromagnetic spectrum and dimensions of various other features compared on a $\log_{10}$ scale in centimeters (*top*). Each long vertical line represents a length 10 times that indicated by the next line to the left. Alternatively the vertical lines could be designated by their logarithms, in which case the divisions would range from −11 to 3. On such a scale the line at 1 cm would be 0 and the 10-cm line would be 1.**

tions they create as they pass by (Table 18-1). These motions involve elasticity of shape, elasticity of volume, or both. The other factor that determines wave velocity is density.

The velocity of longitudinal waves, V_l, in homogeneous elastic solids is governed by the modulus of rigidity, μ, bulk modulus, κ, and the density, d, as follows:

$$V_l = \sqrt{\frac{(4/3)\mu + \kappa}{d}}$$

The velocity of transverse waves, V_t, is governed only by the modulus of rigidity and the density:

$$V_t = \sqrt{\frac{\mu}{d}}$$

In liquids, $\mu = 0$, so $V_l = \sqrt{\kappa/d}$ and $V_t = 0$. The speeds of both longitudinal waves and transverse waves are inversely related to density and directly related to increases in the ability of a material to resist deformation, as expressed by the bulk modulus and modulus of rigidity. The ratio $\sqrt{\mu/d}$ is a common factor in both velocity equations; accordingly, we can make a few algebraic manipulations, eliminate $\sqrt{\mu/d}$, and express κ/d in terms of the velocity of longitudinal and transverse waves:

$$\left(\frac{4}{3}\right) V_t^2 = \left(\frac{4}{3}\right)\frac{\mu}{d}$$

$$V_l^2 - \frac{\kappa}{d} = \left(\frac{4}{3}\right)\frac{\mu}{d}$$

$$\frac{\kappa}{d} = V_l^2 - \left(\frac{4}{3}\right)V_t^2$$

Table A-2 gives the elastic velocities of common geologic materials. It shows that the velocities and densities increase together in apparent contradiction to the position of d in the denominator of the mathematical relationships. The explanation of this apparent contradiction is that the values of the bulk modulus and modulus of rigidity increase at an even greater rate than does den-

TABLE A-2. VELOCITIES OF SEISMIC WAVES IN SELECTED MATERIALS[a]

Material	Speeds of Waves (km/sec)	
	Longitudinal Waves	Shear Waves
Laboratory determinations at atmospheric pressure:		
Granite (Westerly, R. I.)	5.76	3.23
Basalt (Germany)	6.4	3.2
Limestone (Solenhofen, Germany)	5.92	2.88
Steel	5.94	1.84
Glacier ice (0°C)	3.6 to 3.8	1.6 to 1.7
Freshwater (5°C)	1.439	0
Seawater (varies with location and depth; average at surface)	1.5	0
Field measurements:		
Upper continental crust	5.6±	3.3±
Lower continental crust	6.5±	3.75±
Ordinary mantle beneath M-discontinuity	8.0 to 8.2	4.4 to 4.6
"Low-velocity" mantle beneath M-discontinuity	7.4 to 7.8	(Not reported)

[a] Laboratory determinations from Frank Press, 1966; field determinations from Beno Gutenberg, 1955 and K. L. Cook, 1967.

sity. Hence although density does increase, the values of the elastic properties increase even more by comparison, obeying the mathematical relationships, but creating the illusion that speeds of elastic waves are related directly to changes of density.

RADIOACTIVITY

The relationship between changes of mass and energy are continuously on display in the phenomenon of *radioactivity,* which is of great importance to geology in connection with determinations of time (Chap. 6). Radioactivity involves spontaneous changes in the atomic nucleus, which disintegrates by emitting particles and electromagnetic radiations. Before their nature was fully understood two of the kinds of particles were named by Greek letters α and β. An α-particle is identical with the nucleus of an atom of helium, He^4; it consists of 2 protons and 2 neutrons. Close study discloses that what were termed β-particles include not only electrons but also *particles having the same mass as electrons but with a unit positive charge,* named ***positrons.*** Most radioactive disintegrations that involve β-particles and a few of those that involve α-particles are accompanied by the emission of γ-rays, which are electromagnetic waves having very short wavelengths.

Each time an α-particle leaves a nucleus the mass number decreases by 4 and the atomic number by 2 (Fig. 6-2). For example, uranium, U^{238}, loses an α-particle and becomes thorium, Th^{234}.

The subject of emission of β-particles is more complex than that of α-particles and contains an apparent anomaly in that it involves emission of particles that do not exist in the nucleus, namely, electrons or positrons. Physicists resolve this apparent anomaly by postulating that such particles are not created until the instant of emission.

Still another problem arises from the lack of agreement between measured energy and calculated energy associated with various β-particles. The difference appears to indicate a deficit of energy. According to the law of conservation of energy such a deficit is not possible. The energy-deficit problem is resolved by inferring the existence of yet another kind of particle, the ***neutrino,*** *a tiny, electrically neutral bundle of energy lacking intrinsic mass.*

The loss or gain of a β-particle does not change the mass number of the atom, but because the number of protons changes, so does the atomic species. Some examples of the transformations brought about by β-radioactivity are as follows. The nitrogen isotope, N^{13}, emits a positron and becomes carbon, C^{13}. Carbon isotope, C^{14}, gives off an electron and becomes nitrogen, N^{14}. Potassium isotope, K^{40}, can capture an electron and become argon, Ar^{40}, or emit an electron and become calcium, Ca^{40}. What happens in the nucleus, apparently, is that by loss of a negative charge a neutron can change to a proton, or that by capture of an orbital electron a proton can neutralize its positive charge and become a neutron (Fig. A-1).

GRAVITY, G, g, MASS AND MEAN DENSITY OF THE EARTH

Gravity involves two quantities that must not be confused. One is the *universal gravitational constant, G,* the attractive force between two masses of 1g each exactly 1cm apart. The other is the value of the acceleration, designated by g, created by the Earth's gravitational attraction on other bodies. Let us examine how these two gravitational quantities are related.

Consider the relationship between the Earth and another body. According to Newton's law of gravitational attraction, the force of attraction, *F,* is equal to:

$$F = \frac{G M_1 M_2}{d^2}$$

with symbols as explained in Chapter 2. If we replace M_1 by the mass of the Earth, M_E, and d by the radius of the Earth, r, the relation becomes:

$$\frac{F}{M_2} = G \frac{M_E}{r^2}$$

The left side of the equation is now equal to the force per unit mass acting on the body. According to Newton's second law of motion the force, F, on a body equals its mass, M, times acceleration, a: $F = Ma$. Acceleration can be written as F/M. Therefore we can restate the equation of the effect of the Earth on a body as

$$\frac{F}{M_2} = a = G \frac{M_E}{r^2}$$

This acceleration created by the Earth's gravitational attraction is represented by g.

By careful measurements with a pendulum or other device we can ascertain the magnitude of g. Its average value is 980dynes/gram, but it varies from place to place as explained in Chapter 21. From the measured value of g it is possible to calculate the mass and mean density of the Earth as well as the value of G. In the nineteenth century one classical experiment for doing this was performed by a German scientist von Jolly. The method consisted of weighing a spherical flask of mercury by itself and comparing the result with the weight when a heavy lead ball was placed beneath the mercury (Fig. A-4). When the lead ball was in position a tiny weight of 0.589milligram was required to restore balance. This weight became the F of the lead sphere in Newton's gravity equation. It enabled Jolly to solve for G, with a result of 6.47×10^{-8}. A more exact determination of the value of G performed at the U. S. Bureau of Standards in 1942 is $G = 6.673 \pm 0.003 \times 10^{-8}$ (g^3 cm^{-1} sec^{-2}).

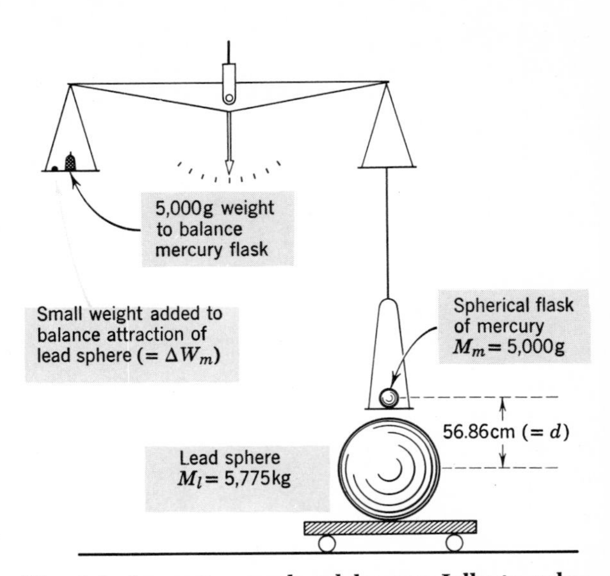

Fig. A-4. Apparatus employed by von Jolly to calculate the mass of the Earth. Details of calculation in text. (After O. H. Blackwood, W. C. Kelly, and R. M. Bell, 1963, General College Physics, Fig. 9, p. 145, by permission of John Wiley and Sons, Inc.)

We know the radius and volume of the Earth and the mean value of g (980dynes/gram). With these values and that of G the mass of the Earth is computed to be 5.976×10^{27}g. The mean density is 5.516g/cm^3.

APPENDIX B

Minerals

Minerals that are abundant in rocks or common as ores number only a few dozens, and many can be identified without special equipment if sizable pieces are available. The techniques to be described consist of direct observations and simple tests. These suffice to recognize groups of rock-making silicate minerals. More exact determinations can be made with laboratory instruments outside the scope of our introductory study.

PHYSICAL PROPERTIES

The simple physical properties useful in mineral identification are: appearance (expressed in luster, color, and form), the ways minerals break, the color of a mineral's powder, its hardness, specific gravity, and magnetic susceptibility.

Luster. *The quality and intensity of light reflected from a mineral* produce an effect known as ***luster.*** Two minerals with almost the same color can have totally different lusters. The more important are described as *metallic,* like that on a polished metal surface; *vitreous,* like that on glass; *resinous,* like that of yellow resin; *pearly,* like that of pearl; *greasy,* as if the surface were covered with a film of oil; and *adamantine,* having the brilliance of a diamond.

Mineral identification will soon become familiar procedure to those who approach it systematically. Minerals with metallic luster should be set aside for tests of streak and magnetic susceptibility. Minerals with vitreous luster should be checked for hardness, inspected for kind of breakage surfaces, and given chemical tests.

Color. The color of a mineral is one of its striking properties, but unfortunately is not a very reliable means of identification. Color commonly results from impurities, which are present in only small amounts. Some minerals display various colors. Quartz, for example, can be clear and colorless (glassy), milky white, rose-colored, violet, and dark gray to black (smoky). Calcite, likewise, can be clear, milky white, pink, green, and gray. Among feldspars flesh-colored, cream-colored, pink, and light green characterize orthoclase and its relatives, whereas dead white and light blue typify plagioclases.

Fig. B-1. Mirrorlike surface on cleavage face of large cube of *galena* (lead sulfide) reflects images of smaller cleavage cubes. (Adolph Knopf.)

Form. As Chapter 3 explains, the form of a mineral depends partly on the crystal lattice (a property unique to the mineral) and partly on conditions of growth (unrelated to the mineral).

We can quickly recognize the cubes of halite (Fig. 3-11), pyrite (Fig. 3-12, *A*), and galena (Fig. B-1), the 12-sided figures of garnet (Fig. 3-12, *B*), and the 6-sided crystals of quartz (Fig. 3-13; Pl. 3).

More commonly the particles are crowded together to form a mineral aggregate. Gypsum and halite are examples.

Fig. B-2. Asbestos, a variety of serpentine, separates into fine, cottonlike threads that are woven during the manufacture of fireproof fabric. (B. M. Shaub.)

Fig. B-3. Botryoidal form, common in various iron oxides, illustrated by a specimen of hydrated manganese oxide.

The form of asbestos and related minerals is *fibrous* (Fig. B-2). Many iron oxides are *earthy,* like crumbly soil. Sheet silicate minerals occur in thin sheets, a form termed *micaceous* because it is well developed in micas. Some ore minerals have grown in *forms resembling grapes closely bunched,* described as ***botryoidal*** (bŏt-rĭ-ŏy′-dăl; Fig. B-3).

Ways in which minerals break. Chapter 3 explains the relationship between the structure of a crystal lattice and ways in which minerals break. Minerals with well-developed cleavage include micas (Fig. 3-15), calcite (Fig. 3-16), amphibole and pyroxene (Fig. 3-17), halite (Fig. 3-11), and feldspars. Minerals lacking perceptible cleavage include garnet, which breaks along irregular fractures (Fig. 3-18, *B*), and quartz and olivine, which fracture irregularly or display ***conchoidal*** (kŏn-koi′-dăl) ***fracture,*** *breakage resulting in smooth curved surfaces* (Fig. 3-18, *A*). Some minerals break along splintery surfaces resembling those of wood.

Because of a distinctive habit of growth within the crystals the cleavage surfaces of plagioclase nearly always appear to be *striated.* These striations are a reliable means of distinguishing plagioclase from orthoclase. Striations are seen to best advantage with a hand lens when the cleavage surface reflects a bright light.

Streak. The ***streak*** is *a thin layer of powdered mineral made by rubbing a specimen on a nonglazed porcelain plate.* The powder diffuses light and gives a reliable color effect that is independent of the form and luster of the mineral specimen. Red streak characterizes hematite whether the specimen itself is red and earthy, like the streak, or black and metallic, like magnetite. Limonite streaks brown, and pyrite and magnetite streak black. Most other minerals streak an undiagnostic white.

Hardness. Comparative *resistance to scratching,* or ***hardness,*** is one of the most noticeable ways in which minerals differ. A series of 10 minerals has been chosen as a standard scale, and any unknown specimen can be classified, 1 to 10, by trying it against the known specimens in the scale, which are arranged in order of decreasing hardness in Table B-1. Fractional values of hardness are common. If a specimen scratches calcite but is distinctly scratched by fluorite, the approximate value of its hardness is 3.5.

In tests of hardness several precautions are necessary. A mineral softer than another may leave a mark that looks like a scratch, just as a soft pencil leaves its mark. A real scratch does not rub off. The physical structure of some minerals may make the hardness test difficult; if a specimen is powdery or in fine grains or if it breaks

TABLE B-1. SCALE OF HARDNESS

	Relative Number in the Scale	Mineral	Hardness of Reference Objects
Decreasing ↓	10.	Diamond	
	9.	Corundum	
	8.	Topaz	
	7.	Quartz	
	6.	Orthoclase	Nail; pocket knife; glass
	5.	Apatite	
	4.	Fluorite	
	3.	Calcite	Copper penny
	2.	Gypsum	Fingernail
	1.	Talc	

easily into splinters, an apparent scratch may be deceptive.

Specific gravity. In comparing unit weights, we use water as a standard. The ***specific gravity*** of any substance is expressed as *a number stating the ratio of the weight of the substance to the weight of an equal volume of pure water at 4°C.* Specific gravity can be approximated by comparing different minerals held in the hand. Metallic minerals such as galena feel "heavy," whereas nearly all others feel "light."

Specific gravity is ascertained by determining the ***density,*** *weight per unit volume.* With an air pycnometer or liquid pycnometer we can determine the volumes of minerals and rocks and with a balance measure their weights. The weight (grams) divided by the volume (cm^3) gives the density (g/cm^3). The specific gravities of selected metals, minerals, and rocks are shown in Table C-2.

Magnetic susceptibility. Of the common minerals only magnetite is strongly magnetic. Specimens of magnetite can be singled out at once by their strong attraction to a small magnet.

CHEMICAL PROPERTIES

Only two chemical tests are commonly used in the beginning study of minerals: (1) taste test for halite and (2) acid test for calcite and dolomite. The salty taste of halite is distinctive. Carbonate minerals effervesce (make bubbles) in dilute hydrochloric acid. Calcite effervesces freely no matter what the size of the particles. Dolomite may not effervesce at all unless the specimen is powdered or the acid is heated. Dolomite powder effervesces slowly in cool dilute acid.

Caution. Hydrochloric acid is hard on teeth and has an unpleasant taste. Where many students share mineral specimens caution in use of acid is necessary. Acid should be applied in small drops and when the test is finished the specimen should be blotted dry. The next user of the specimen may decide to try the taste test.

IDENTIFICATION OF MINERALS

For convenience of reference the most important minerals likely to be encountered in an introductory study of geology are arranged alphabetically in Table B-2. The chemical formulas in the second column are for reference only, not as an aid to making identifications. In complex silicates, only generalized formulas are included; in these formulas various ion groups are designated by the capital letters A, B, and C, as follows:

A = elements with large ionic radius, such as K^+, Ca^{2+}, Na^+
B = elements with intermediate radius, such as Mg^{2+}, Fe^{2+}, Fe^{3+}, Al^{3+}
C = elements with small radius, mainly Si^{4+}, Al^{3+}

Valences greater than +1 are shown by the standard symbol. In this notation 2, 3, and 4 with a plus sign are written as raised postscripts. Particularly diagnostic physical properties have been emphasized by adding blue color to the appropriate box in the table. If the name of a laboratory specimen is known, locate the mineral in the table and be sure that you become familiar with all its physical properties. If the name of a specimen is not known, determine its physical properties and scan the appropriate columns in the table to locate the correct name by elimination.

Selected References

Dietrich, R. V., 1966, Mineral tables—hand specimen properties of 1500 minerals: Blacksburg, Va., Virginia Polytechnic Inst. Bull. 160.

Pearl, R. M., 1962, Successful mineral collecting and prospecting: New York, New American Library.

Vanders, Iris, and Kerr, P. F., 1967, Mineral recognition: New York, John Wiley.

TABLE B-2. PROPERTIES OF SOME IMPORTANT MINERALS
(For explanation of A, B, C in chemical formulas, see text.)

Mineral	Chemical Composition	Luster	Form	Cleavage	Hardness	Specific Gravity	Other Properties; Remarks
Albite (see Plagioclase feldspars)							
Amphibole (complex group of minerals, *hornblende* most common)	Silicates of Ca, Mg, Fe, Al, Na. General formula: $A_{2-3}B_5C_8O_{22}(OH)_2$	Vitreous on cleavage surfaces	In long, 6-sided crystals; also in fibers and irregular grains	Two, intersecting at 56 and 124° (Fig. 3-17, *B*)	5 to 6	2.9 to 3.8	Commonly black, dark and light green, rarely white; some varieties used as asbestos
Apatite	$Ca_5(PO_4)_3(F,Cl,OH)$	Vitreous	In granular masses or as large crystals	Poor, in one direction	5	3.1 to 3.2	Green, brown, blue, violet; accessory mineral in many kinds of rocks, especially in calcite marbles
Aragonite	$CaCO_3$	Vitreous	In slender, needlelike crystals; granular masses	Poor, in two directions	3.5 to 4	2.94	Colorless, white, pale tints; harder and slightly heavier than calcite; effervesces in dilute HCl
Augite (see Pyroxene)							
Azurite	$Cu_3(CO_3)_2(OH)_2$	Vitreous	In crystals, in form of stalactites, in formless masses, or earthy	One perfect, another imperfect	3.5 to 4	3.7 to 3.8	Blue color and streak are diagnostic; occurs with malachite and limonite in copper deposits
Barite	$BaSO_4$	Vitreous to pearly	Crystals tabular or prismatic	Two perfect, one imperfect	2.5 to 3.5	4.3 to 4.6	Translucent to opaque; very heavy for a mineral with nonmetallic luster
Bauxite	Mixture of hydrous aluminum oxides	Dull, earthy	In earthy, claylike masses; also in small spherical forms	None; uneven fracture	1 to 3	2.0 to 2.55	Emits strong clayish odor
Biotite (black mica)	Complex silicate of K, Mg, Fe, Al. General formula: $AB_3C_4O_{10}(OH)_2$	Pearly to nearly vitreous	In perfect thin flakes; 6-sided crystals	One cleavage direction; uniform flakes or sheets	2.5 to 3	2.7 to 3.3	Black, dark brown, or green; nearly or quite opaque; flakes are both flexible and *elastic*
Calcite	$CaCO_3$	Vitreous to dull	In tapering crystals, or granular aggregates	Three perfect, at oblique angles (Fig. 3-16)	3	2.71	Colorless or white; effervesces in dilute HCl
Carnotite	$K_2(UO_2)(VO_4)_2 \cdot 3H_2O$	Earthy	Earthy powder	Not visible	Very soft	4 to 4.7	Powdery form, with brilliant canary-yellow color; ore of vanadium and uranium
Cassiterite (tinstone)	SnO_2	Adamantine to dull	Granular masses; well-formed crystals common; rounded pebbles in stream gravels	Two, indistinct; conchoidal fracture	6 to 7	6.8 to 7	Yellow to red-brown; principal ore of tin; very heavy

TABLE B-2 PROPERTIES OF SOME IMPORTANT MINERALS (Continued)

Mineral	Chemical Composition	Luster	Form	Cleavage	Hardness	Specific Gravity	Other Properties; Remarks
Chalcedony (crypto-crystalline quartz)	SiO_2	Dull	No visible crystals; commonly banded or in formless masses	None; conchoidal fracture	6 to 6.5	2.57 to 2.64	White if pure; variously colored by impurities
Chalcocite	Cu_2S	Metallic	Usually massive, fine-grained; crystals rare	None; conchoidal fracture	2.5 to 3	5.5 to 5.8	Steel-gray to black; dark gray streak; an ore of copper
Chalcopyrite	$CuFeS_2$	Metallic	Massive or granular	None; uneven fracture	3.5 to 4	4.1 to 4.3	Golden yellow to brassy color; streak dark green to black; an ore of copper
Chlorite	Variable silicate of Mg, Fe, Al. General formula: $B_3C_4O_{10}(OH)_2 \cdot B_3(OH)_6$	Greasy to vitreous	In flaky masses or 6-sided crystals	One perfect cleavage	2 to 2.5	2.6 to 2.9	Light to dark green; flakes are weak, *inelastic,* easily separated
Cinnabar	HgS	Adamantine to dull	In veins; also in disseminated grains	One perfect cleavage	2 to 2.5	8.1	Red to red-brown; scarlet streak; chief ore mineral of mercury
Copper (native)	Cu	Dull-metallic	Forms of twisted leaves and wires; also irregular nodules	None; hackly fracture	2.5 to 3	8.9	Copper color, but commonly stained green; ductile and malleable; not now common as an ore
Corundum (ruby, sapphire)	Al_2O_3	Adamantine to vitreous	In separate crystals or in granular masses	Two good cleavages, with striations on planes	9	4.0 to 4.1	Blue, red, yellow-brown, green-violet; valuable as an abrasive (emery) and as gems
Diamond	C	Adamantine to greasy	In octahedral or cubic crystals; faces commonly curved	Good, in four directions	10	3.5	Used as abrasive; now a synthetic as well as natural product; high-quality stones used as gems. High heat conductivity may cause crystals to feel cold momentarily
Dolomite	$CaMg(CO_3)_2$	Vitreous to pearly	In crystals with rhomb-shaped faces; also in granular masses	Perfect in three directions, as in calcite	3.5 to 4	2.85	White, gray, or flesh-colored; some crystals have curved faces; must be scratched or powdered to effervesce in cold dilute HCl
Epidote	Variable silicate of Ca, Fe, Al. General formula: $Ca_2B_3{}^{3+}(SiO_4)_3(OH)$	Vitreous	In small prismatic crystals; also fibrous	One perfect cleavage, another imperfect	6 to 7	About 3.4	Yellow-green to blackish-green; commonly associated with chlorite; distinguished from olivine by cleavage and by form

TABLE B-2 PROPERTIES OF SOME IMPORTANT MINERALS (Continued)

Mineral	Chemical Composition	Luster	Form	Cleavage	Hardness	Specific Gravity	Other Properties; Remarks
Feldspars (see Plagioclase feldspars and Potassium feldspars)							
Fluorite	CaF_2	Vitreous	In well-formed crystals and in granular masses	Good cleavage in four directions	4	3.2	Colorless, green, blue, or nearly black; commonly in veins with lead and silver ores; also in cavities in limestone and dolostone; does not effervesce with dilute HCl
Galena	PbS	Bright metallic	In cubic crystals and granular masses, coarse- or fine-grained	Perfect in three directions at right angles (Fig. B-1)	2.5	7.6	Lead-gray color; streak gray to gray-black; common ore of lead; in many deposits it contains silver
Garnet (complex group)	Isomorphous silicates of Ca, Mg, Fe, Mn, Al. General formula: $(A,B^{2+})_3B_2{}^{3+}(SiO_4)_3$	Vitreous to resinous	Commonly in perfect crystals with 12 or 24 sides (Fig. 3-12; *B*; *C*-6); also granular masses	None; fracture conchoidal or uneven (Fig. 3-18, *B*)	6.5 to 7.5	3.5 to 4.3	Color varies with composition; red, brown, yellow, green to almost black
Gold (native)	Au	Metallic	Massive or in thin irregular scales	None; hackly fracture	2.5 to 3	19.3	Yellow "golden" color; quite malleable; commonly scattered in quartz veins; nuggets occur in stream gravels
Graphite	C	Metallic to dull	In scaly masses	Perfect, in flakes	1 to 2	2.02 to 2.23	Gray to nearly black; black streak; greasy feel; high melting point
Gypsum	$CaSO_4 \cdot 2\,H_2O$	Vitreous to pearly	In tabular, diamond-shaped crystals; also granular, fibrous, or earthly	One perfect cleavage, two imperfect	2	2.3	Usually white or colorless, transparent or translucent; cleavage plates flexible, *inelastic*
Halite (rock salt, common salt)	NaCl	Vitreous	In cubic crystals (Fig. 3-11, *C*) or granular masses	Perfect in three directions at right angles (Fig. 3-11, *D*)	2.5	2.16	Colorless or white when pure, transparent to translucent; strong salty taste
Hematite	Fe_2O_3	Metallic or earthy	Varied: massive, granular, micaceous, earthy	None, uneven fracture	5 to 6	4.9 to 5.3	Red-brown, gray to black; red-brown streak; most important ore of iron
Hornblende (see Amphibole)							
Hypersthene (see Pyroxene)							
Kaolinite (common clay mineral)	$Al_2Si_2O_5(OH)_4$	Dull	Soft, earthy masses. Crystals photographed with electron microscope (Fig. 7-6)	One perfect cleavage, submicroscopic	2 to 2.5	About 2.6	White if pure, usually stained yellow or other colors; plastic; emits clay odor

TABLE B-2 PROPERTIES OF SOME IMPORTANT MINERALS (Continued)

Mineral	Chemical Composition	Luster	Form	Cleavage	Hardness	Specific Gravity	Other Properties; Remarks
Kyanite	Al_2SiO_5	Vitreous to pearly	In groups of blade-like crystals (Fig. C-8, C)	One good, another imperfect	4 to 5 parallel to crystal, 7 across crystal	3.53 to 3.65	White, pale blue, or green; occurs in metamorphic rocks; compare with sillimanite
Limonite (impure oxide)	Mixture of several hydrous oxides of iron	Dull to vitreous	Compact to earthy masses; irregular nodules	None, irregular fracture	1 to 5.5	3.5 to 4	Yellow, brown, black; streak yellow-brown; an ore of iron
Magnetite	Fe_3O_4	Metallic	Varied: massive, granular	None, uneven fracture	5.5 to 6.5	5 to 5.2	Black and opaque; black streak; strongly attracted to a magnet; important ore of iron
Malachite	$Cu_2(CO_3)(OH)_2$	Silky to dull	Rarely in crystals; massive, with mammillary forms on surface	One perfect, another fair	3.5 to 4	3.6 to 4	Green color and streak; an ore mineral of copper that occurs in oxidized parts of copper deposits, commonly with limonite and generally with azurite

Mica (see Muscovite, "white mica," and Biotite, "black mica")
Microcline (see Potassium feldspar)

Mineral	Chemical Composition	Luster	Form	Cleavage	Hardness	Specific Gravity	Other Properties; Remarks
Muscovite (white mica)	Variable silicate of K, Al. General formula: $AB_2C_4O_{10}(OH)_2$	Vitreous to pearly	In uniform thin flakes; rarely in 6-sided crystals	One cleavage direction; perfect flakes or sheets (Fig. 3-15)	2 to 2.5	2.77 to 2.88	Colorless and transparent when pure but commonly greenish and mottled; flakes are flexible *and* elastic
Nepheline	$Na_3KAl_4(Si_2O_8)_2$	Crystals vitreous; cleavage surfaces greasy	Varied; massive or as scattered grains in silica-deficient rocks	Present but poorly developed	5.5 to 6	2.55 to 2.65	Colorless and translucent to white, gray, yellowish, brownish; not so hard as quartz; poorer cleavage than feldspars; gelatinizes in dilute acids
Olivine	Varied proportions of Fe, Mg. General formula: $B_2^{2+}SiO_4$	Vitreous	In small grains or granular masses	None; conchoidal fracture	6.5 to 7	3.2 to 4.3	Olive green to yellow-green; transparent to translucent
Opal (hydrous silica)	$SiO_2 \cdot nH_2O$	Waxy to vitreous	Amorphous; in irregular masses	None; conchoidal fracture	5 to 6.5	2.0 to 2.2	Various colors; translucent to opaque

Note: Other common varieties of silica, most of them containing impurities, are agate, flint, chert, and jasper.

Orthoclase (see Potassium feldspars)

Mineral	Chemical Composition	Luster	Form	Cleavage	Hardness	Specific Gravity	Other Properties; Remarks
Plagioclase feldspars (Na-Ca feldspars)	$NaAlSi_3O_8$(albite) to $CaAl_2Si_2O_8$ (anorthite) General formula: AC_4O_8	Vitreous to pearly	Commonly as irregular grains or cleavable masses; some varieties in thin plates	Two good cleavages, not quite at right angles	6 to 6.5	2.62 to 2.76	White to dark gray and also other colors; cleavage planes may show fine parallel striations; play of colors in some varieties

TABLE B-2 PROPERTIES OF SOME IMPORTANT MINERALS (Continued)

Mineral	Chemical Composition	Luster	Form	Cleavage	Hardness	Specific Gravity	Other Properties; Remarks
Potassium feldspars (orthoclase, microcline, and sanidine)	$KAlSi_3O_8$	Vitreous	In prismatic crystals or grains with cleavage	Two good cleavages, at right angles	6	2.56 to 2.59	Commonly flesh-colored, pink, or gray; one variety green
Pyrite ("fool's gold")	FeS_2	Metallic	In cubic crystals with striated faces (Fig. 3-12, *A*); commonly massive	None, uneven fracture	6 to 6.5	4.9 to 5.2	Pale brass-yellow, darker if tarnished; streak greenish-black; widely distributed; used in manufacture of sulfuric acid
Pyrolusite	MnO_2	Metallic to dull	Rarely in crystals; in coatings on fracture surfaces; commonly in concretions	Crystals have one perfect cleavage	2 to 6.5	4.5 to 5	Dark gray, black or bluish; black streak; important ore mineral of manganese
Pyroxene (complex group, *augite* and *hypersthene* most common)	Silicates of Ca, Fe, Mg, Na, Al. General formula: ABC_2O_6	Vitreous	In 8-sided, stubby crystals; also in granular masses	Two cleavages, nearly at right angles (Fig. 3-17, *A*)	5 to 6	3.2 to 3.9	Light to dark green, or black; alternate crystal faces at right angles (fit into corner of a box)
Quartz	SiO_2	Vitreous to greasy	Six-sided crystals, pyramids at ends (Pl.3; Fig. 3-13); also in irregular grains and masses	None; conchoidal fracture (Fig. 3-18, *A*)	7	2.65	Varies from colorless and transparent to opaque with wide range of colors
Rutile	TiO_2	Adamantine to metallic	In slender prismatic crystals or granular masses	Good in one direction; has conchoidal fracture	6 to 6.5	4.2	Red-brown to black; streak brownish to gray-black; abundant in some beach sands; ore of titanium
Sanidine (see Potassium feldspars)							
Serpentine (fibrous variety is *asbestos*)	Variable silicate of Mg, with (OH). General formula: $B_6C_4O_{10}(OH)_8$	Greasy or resinous	Massive or fibrous (Fig. B-2)	Usually breaks irregularly, except in the fibrous variety	2.5 to 5	2.2 to 2.6	Light to dark green; smooth, greasy feel; translucent to opaque
Sillimanite	Al_2SiO_5	Vitreous	In long crystals or in fibers	Perfect in one direction	6 to 7	3.2	White to greenish-gray; found in high-grade metamorphic rocks; compare with kyanite
Silver (native)	Ag	Bright metallic to dull	In flakes and irregular grains	None; hackly fracture	2.5 to 3	10 to 11	Generally tarnished to dark gray or black; cleans to silvery-white; ductile and malleable
Sphalerite (zinc blende)	ZnS	Resinous to adamantine	Fine- to coarse-granular masses; crystals common	Six directions	3.5 to 4	3.9 to 4.1	Color yellow-brown to black; streak white to yellow or brown; principal ore of zinc

TABLE B-2 PROPERTIES OF SOME IMPORTANT MINERALS (Continued)

Mineral	Chemical Composition	Luster	Form	Cleavage	Hardness	Specific Gravity	Other Properties; Remarks
Staurolite	$Al_4Fe^{2+}(SiO_4)_2(OH)_2O_2$	Vitreous to resinous	Stubby crystals, commonly twinned in form of cross	Distinct in one direction	7 to 7.5	3.7 to 3.8	Red-brown to nearly black; associated with sillimanite, kyanite, garnet
Talc	$Mg_3(OH)_2Si_4O_{10}$	Greasy, pearly	In small scales and compact masses	One perfect cleavage	1	2.58 to 2.83	White to greenish; has greasy feel
Topaz	$Al_2SiO_4(OH,F)_2$	Vitreous	In prismatic crystals; also in granular masses	One perfect cleavage	8	3.49 to 3.57	Colorless to shades of blue, yellow, or brown; found in some pegmatites and quartz veins
Uraninite (pitchblende)	UO_2 to U_3O_8	Submetallic to dull	Massive, with botryoidal forms; crystals cubic or 8-sided	None, uneven fracture	5 to 6	6.5 to 10	Black to dark brown; streak has similar color; a source of uranium and radium
Wollastonite	$CaSiO_3$	Vitreous to silky; pearly on cleavage surfaces	Fibrous or bladed aggregates of elongated crystals	Perfect in two directions	4.5 to 5	2.8 to 2.9	Colorless, white, gray, pinkish, yellowish translucent; soluble in HCl; distinguished from fibrous amphiboles by cleavage and by solubility in HCl

APPENDIX C

Rocks

Laboratory study of rock specimens leads to the ability to identify the most important kinds of materials in the geologic record. The familiarity with rocks thus acquired makes possible the interpretation of exposures encountered in the field.

Chapter 4 explains the basis for the classification of rocks into three major groups and emphasizes the things to look for in studying rocks. Chapter 5 explores some of the chemical aspects of rock composition. Here we include supplementary descriptive information, detail the ways in which rocks are classified and named, and provide hints on how to recognize the common varieties of rocks.

No tools more elaborate than a hand lens and a knife are necessary for identification of the rocks discussed in this appendix. The only important prerequisite is familiarity with the rock-forming silicate minerals and carbonate minerals. In rocks the constituent mineral particles are not only small but also crowded closely together or intergrown with one another. In rocks, therefore, the most important aids to identification of minerals are the ways in which they break, their luster, and their hardness.

As with minerals, a systematic approach forms a basis for rapid progress in the ability to recognize rocks. We suggest that you, the reader, first study the larger features such as layers, partings, texture, fabric, and minor structures; then take a closer look with a hand lens or binocular microscope having 10× magnification. Identify the minerals and estimate their proportions. With this information in hand, turn to the appropriate table of rock identification and establish the name of the rock. This routine is more than an exercise in namesmanship; it is standard practice among geologists. Accurate use of rock names afford geologists a rapid means of communicating significant scientific and practical information.

LAYERS

Begin your examination of a rock specimen by inspecting it for layers. As mentioned in Chapter 4, layering is one of the most conspicuous features of many exposures (Fig. 4-4). Layers characterize sedimentary rocks and their metamorphosed equivalents, but layers are found likewise in some igneous rocks. On the scale of hand specimens, however, layers may or may not be present. Because they convey valuable information layers should be sought and observed closely. Small-scaled layers can occur in igneous rocks (Fig. C-1, *A*), in sedimentary rocks (Fig. C-1, *B*), and in metamorphic rocks (Fig. C-1, *C*). In attempting to classify rocks do not give much weight to absence of layers in a hand specimen. Because the kinds of layers serve as a basis for classifying and naming metamorphic rocks, we defer detailed discussion of layers in metamorphic rocks to a following section of this appendix.

PARTINGS

Partings are uncommon in hand specimens but, when present, form the basis for distinguishing among three varieties of fine-grained rocks. *Siltstone* and *claystone* lack closely spaced partings (Fig. C-2, *A*). *Shale* is characterized by closely spaced partings parallel with the stratification (Fig. C-2, *B*). *Slate* is distinguished by partings that are not parallel with stratification (Fig. C-2, *C*).

TEXTURE

As explained in Chapter 4, texture results from rock-forming processes. Crystalline textures characterize igneous and metamorphic rocks, but likewise occur in some sedimentary rocks. Fragmental textures are typical of sediments and sedimentary rocks.

Textures of igneous rocks. The sizes and uniformity of sizes of particles form the basis for a series of terms applied in the systematic description of igneous rocks and for dividing igneous rocks into several major kinds. (1) *A texture of igneous rocks with all particles about the same size* is ***equigranular texture.*** (2) *A texture of igneous rocks with some particles conspicuously larger than the rest of the particles* is ***porphyritic texture.*** *The particles of por-*

Fig. C-1. Layers in rocks as seen in hand specimens. *A.* Layers in an igneous rock resulting from flow of lava. Obsidian, Guadalajara, Mexico. Pits were caused by removal of soluble parts of the rock. (B. M. Shaub.) *B.* Laminae in a sedimentary rock resulting from variation in rate of deposition. Green River Formation, Garfield County, Colorado. (B. M. Shaub.) *C.* Foliation in metamorphic rock resulting from plastic deformation. Biotite-quartz-feldspar gneiss, Uxbridge, Massachusetts. (Am. Mus. Nat. History.)

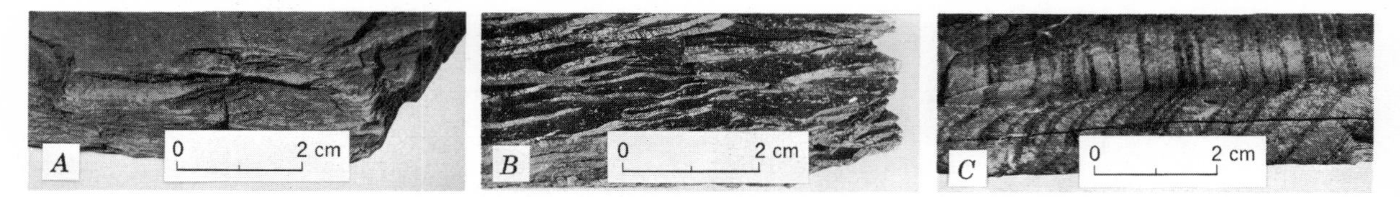

Fig. C-2. Fine-grained rocks without partings compared with those having partings. (B. M. Shaub.) *A.* Claystone that lacks partings. Steplike surfaces reflect parallel arrangement of clay-mineral particles. *B.* Black shale with prominent, closely spaced partings parallel with stratification. *C.* Slate, with slaty cleavage paralleling top and bottom of specimen (emphasized by prominent parting). Stratification shown by light and dark layers inclined steeply to left. Dark line is a crack in the specimen.

phyritic igneous rocks that are conspicuously larger than the remaining particles are ***phenocrysts*** (fēn′-ō-krĭsts; Gr. "visible crystals"). *The particles surrounding the phenocrysts of a porphyritic igneous rock* constitute the ***groundmass.***

In equigranular igneous rocks all gradations exist among particles that are invisible to the unaided eye (smaller than 0.05mm) and those whose dimensions are measured in centimeters or many tens of centimeters. *A texture of equigranular igneous rocks with individual particles not visible to an unaided eye* is an ***aphanitic texture*** (ăf-ăn-ĭt′-ĭk; Gr. "invisible"). At the opposite end of the spectrum of particle sizes are rocks with ***pegmatitic texture,*** *a texture of equigranular igneous rocks with particles larger than 10mm.* In between lie rocks having ***granular texture,*** *a texture of equigranular igneous rocks with particles ranging in size from 0.05 to 10mm.* Granular igneous rocks are further subdivided into *fine grained* (0.05 to 1mm), *medium grained* (1 to 5mm), and *coarse grained* (5 to 10mm) (Fig. C-3).

Igneous rocks that cooled with great rapidity consist not of crystalline particles but of glass. Such rocks display ***glassy texture,*** *a texture of igneous rocks containing only glass and in which crystalline particles are not present.*

Texture influences the naming of igneous rocks in two ways. (1) Aphanitic and glassy rocks carry names unlike granular rocks having identical compositions. For example, the composition of rhyolite is the same as that of granite. The practice of applying one name to aphanitic and glassy rocks and a different name to granular rocks began when geologists believed that texture served as a reliable basis for distinguishing extrusive igneous rocks from intrusive igneous rocks. To a first approximation this principle is valid; intrusive rocks generally are coarser grained than extrusive rocks. So many exceptions are known, however, that the original reason for subdividing igneous rocks into two groups on the basis of particle size has been abandoned. We continue to make a subdivision based on particle size, but no longer suppose that names applied on the basis of texture distinguish extrusive from intrusive igneous rocks. Such a distinction is important, but it must be made on the basis of the available

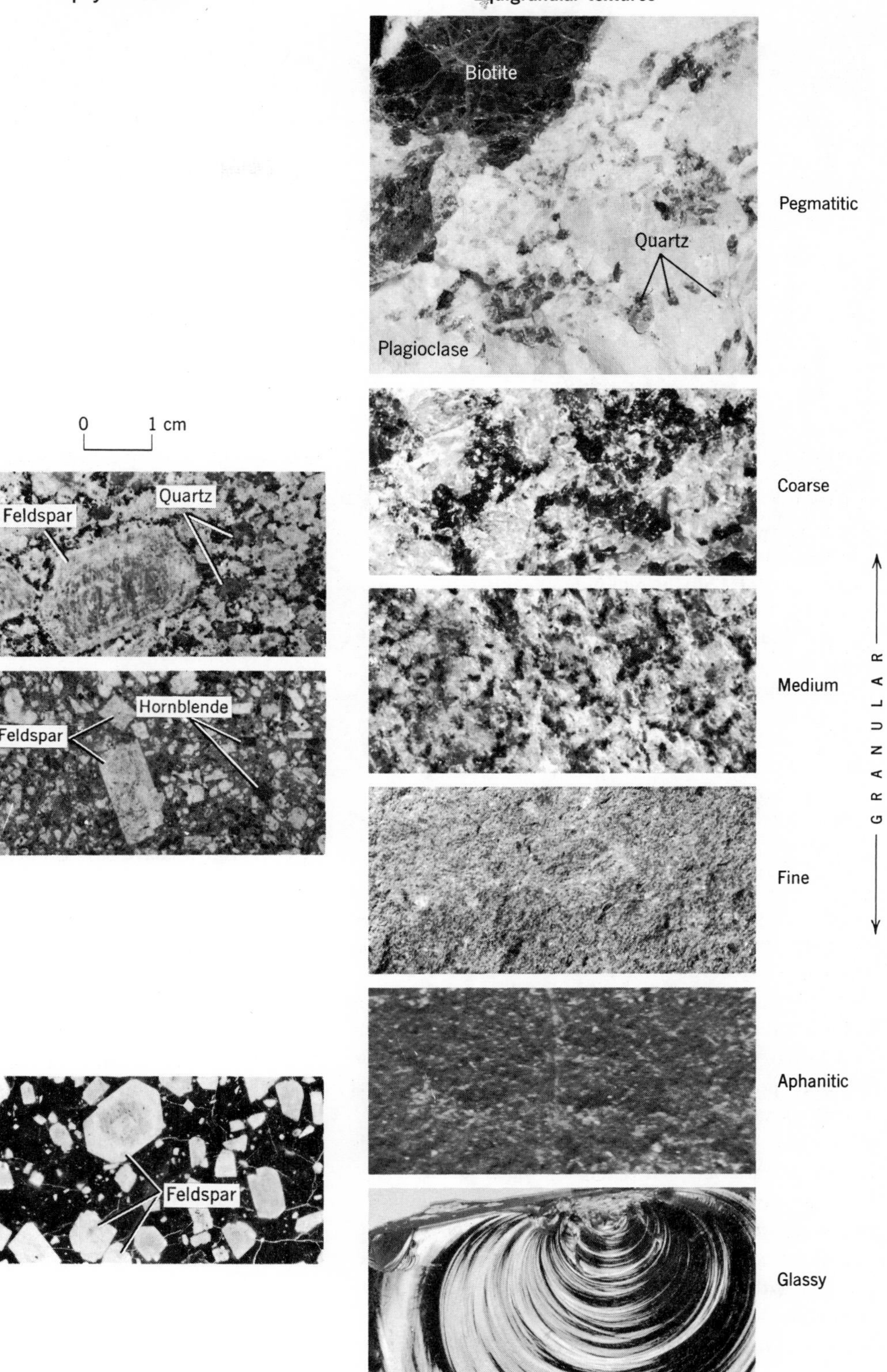

Fig. C-3. Textures of igneous rock. Text defines terms. (Porphyritic rocks and pegmatite, B. M. Shaub; others, C. R. Longwell.)

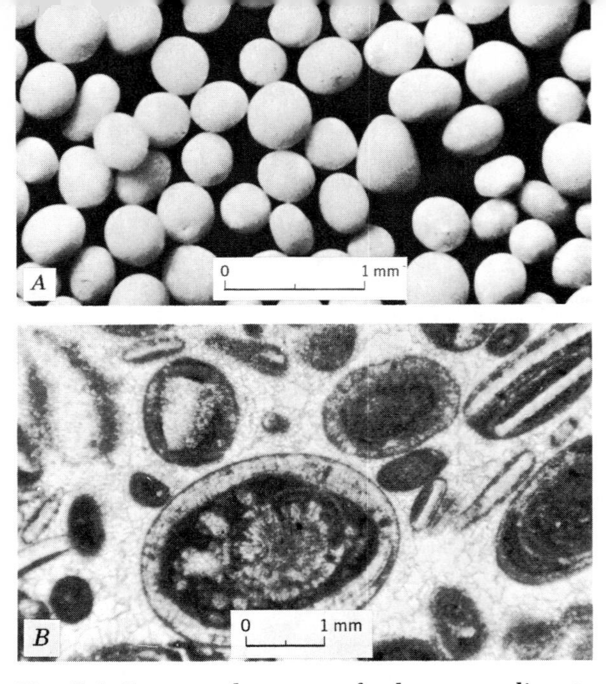

Fig. C-4. Fragmental textures of calcareous sediments and limestones. *A.* Calcareous spheres that have enlarged concentrically by addition of coatings of calcium carbonate to their exteriors. South Cat Cay, Bahamas. (G. M. Friedman.) *B.* View, through microscope, of paper-thin slice cut from limestone with fragmental texture. Particles have been enlarged by addition of fine-grained lime mud (black) and radially oriented clear crystals (on largest particle). Thin, light-colored rectangular particles are shell fragments. Oval particle at upper right includes two shell fragments. Cement is clear calcite (white, between particles). Plattin Formation, Harrell Creek, Missouri. (J. E. Sanders.)

field evidence and not merely upon particle size of hand specimens. (2) Porphyritic texture is expressed in the name of a rock according to the proportion of the total volume of the rock occupied by the phenocrysts. Where phenocrysts are abundant, add to the name of the rock the word ***porphyry,*** *an igneous rock with phenocrysts constituting more than 25 per cent of the total volume.* Where phenocrysts are not so abundant add the word *porphyritic* as a modifier before the name of the rock. Thus ***porphyritic*** describes *an igneous rock in which phenocrysts constitute less than 25 per cent of the total volume.* Accordingly, *granite porphyry* describes a rock containing the minerals of a granite with phenocrysts occupying more than 25 per cent of the total volume, whereas a *porphyritic granite* designates a rock of granitic composition containing phenocrysts that occupy less than 25 per cent of the total volume.

Texture of groundmass in rocks with porphyritic texture varies through a limited range. The groundmass may be granular, aphanitic, or glassy (Fig. C-3).

Textures of sediments and sedimentary rocks. Among sediments and sedimentary rocks the most common texture, indeed the one which is unique to this great group, is ***fragmental texture,*** *a texture of sediments and sedimentary rocks, resulting from physical transport and deposition of particles.* The particles may be of any size or origin, and they may have been broken or, just the opposite, they may have been growing larger at the site of deposition (Fig. C-4). Probably the most common variety is ***clastic texture*** (Fig. 1-10, *B;* 4-6, *B;* 4-12), *a texture of sediments and sedimentary rocks resulting from physical transport and deposition of broken particles of older rocks, of older sediments, and of organic skeletal remains.* Particles of a clastic sediment or sedimentary rock were broken under conditions of low temperature prevailing at the Earth's surface. By contrast, tephra broken by volcanic action display ***pyroclastic texture,*** *a texture of sediments and sedimentary rocks resulting from physical transport and deposition of particles broken by volcanic activity.*

In sedimentary rocks a general distinction should be made among particles and minerals composing the *framework,* and *matrix,* and the *cement.* Otherwise the textures of sedimentary rocks are analyzed along the same general lines followed in igneous rocks. However, the significance of variations in size differs in sedimentary rocks, and no formal textural names comparable to those of igneous rocks have been adopted. Chief interest in the analysis of fragmental textures attaches to the relationships between *sorting* of particles and processes of accumulation. Where particles from several size classes on the standard scale (Fig. 4-11) are present, their proportions should be estimated. Each can be included as a modifier of the sediment name or rock name, which is selected from the dominant size class. Modifiers can be listed in decreasing order of abundance of the sizes present, e.g. pebbly-sandy-siltstone. If desired, the estimated percentage of each can be written as subscripts: pebbly_{10}-sandy_{30}-siltstone_{60} for a rock containing particles of which 10 per cent are of pebble size, 30 per cent of sand size, and 60 per cent of silt size.

A few sedimentary rocks contain crystalline textures identical with textures occurring in igneous and metamorphic rocks. In these rocks, therefore, texture is not diagnostic of sedimentary origin as it generally is with rocks displaying fragmental texture. Sedimentary rocks with crystalline textures include some limestones and dolostones, chert, and evaporites.

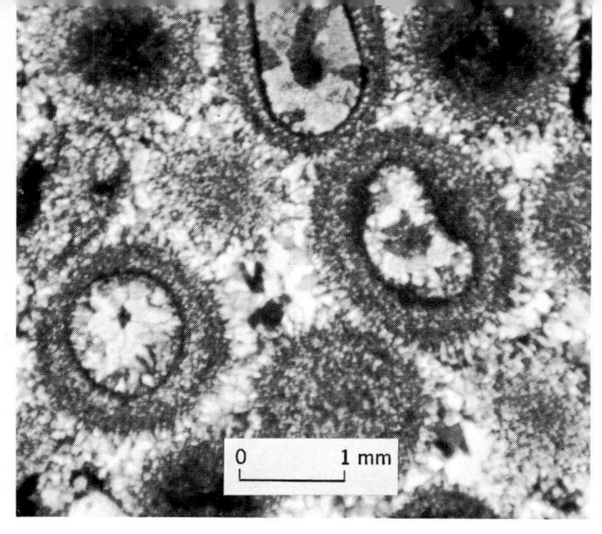

Fig. C-5. View, through microscope, of slice cut from chert formed by replacement of limestone having fragmental texture (compare Fig. C-4, *B*). Even though original calcium carbonate has been completely replaced by quartz, initial texture is still clearly visible. Round, shadowlike areas are sections through what probably were spherical particles originally composed of fine-grained calcium carbonate. Three particles in center with shadowlike halos probably represent former particles of calcium carbonate that were coated with additional fine-grained calcium carbonate before both were replaced by silica. Copper Ridge Formation (Upper Cambrian), near Thorn Hill, Tennessee. (J. E. Sanders.)

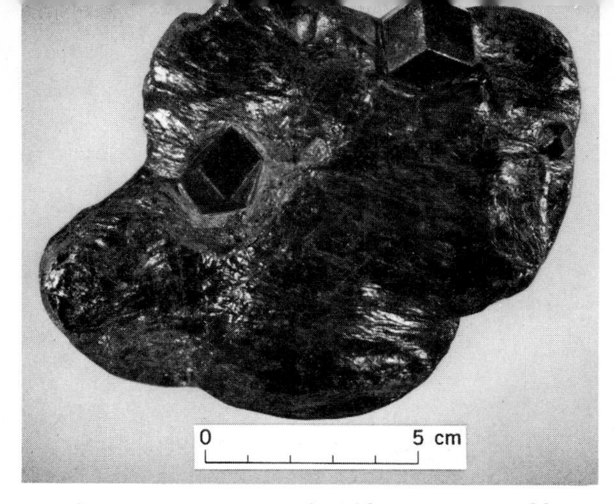

Fig. C-6. Large garnet porphyroblasts in garnet-chlorite schist. (G. M. Friedman.)

Although some limestones display crystalline textures, this variety of sedimentary rocks can include nearly all other kinds of textures as well. Among limestones are those with (1) clastic texture, which consist of broken pieces of older limestones or of broken bits of calcareous skeletal debris (Fig. 4-8, *C* and *D*), (2) fragmental texture in which by addition of concentric coatings of calcium carbonate particles were growing larger at the site of deposition (Fig. C-4, *B*), and (3) ***organic texture,*** *a texture of sedimentary rocks resulting from secretion of skeletal material or from other activity by organisms.* Organic textures are found only in sedimentary rocks (Figs. 4-10; 23-2).

The distinctive original textures of some limestones persist even after the calcium carbonate has been replaced by silica (Fig. C-5). In fact, such persistence is conclusive evidence for the replacement origin of some cherts.

Textures of metamorphic rocks. Many metamorphic rocks exhibit crystalline texture which by itself does not distinguish metamorphic rocks from igneous and some sedimentary rocks. The particles of some metamorphic rocks are conspicuously larger than their neighbors (Fig. C-6). These large mineral particles resemble the phenocrysts in porphyritic igneous rocks. Some of them may indeed have been phenocrysts of a porphyritic igneous rock that survived the processes of metamorphism with little change, but others may have originated by enlargement during metamorphism. Because their origin may not be apparent, they are called *porphyroblasts*

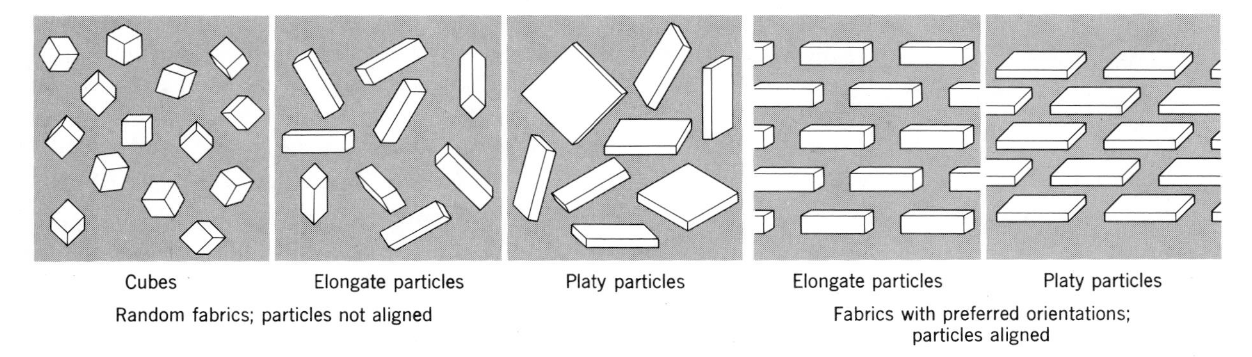

Fig. C-7. Schematic representation of some fabrics of rocks.

rather than phenocrysts. Rocks that have been subjected to vigorous mechanical deformation display ***cataclastic texture*** (Gr. *kata*, "down," and *klastos*, "broken") *a texture of rocks resulting from breakage and pulverization of mineral particles* (Fig. 20-17).

FABRIC

The *fabric* of a rock, the orientation of the particles, reflects the conditions of growth, of deposition, and of deformation. A few kinds of fabrics formed by various orientations of particles are sketched in Fig. C-7. In igneous rocks random fabric is typical, but in some porphyries the phenocrysts have been aligned (Fig. C-8, *A*). In metamorphic rocks, fabrics with preferred orientations are common (Fig. C-8, *B*), but random fabrics can occur (Fig. C-8, *C*).

MINOR STRUCTURES

Minor structures of many kinds occur in rocks. Such structures convey significant information about the history of the rock. The minor structures of many varieties that occur in sedimentary rocks are rare in hand specimens; they are best seen in field exposures.

One kind of minor structure, visible in hand specimens and valuable in classification because it is almost exclusively confined to extrusive igneous rocks, is exemplified by ***vesicles,*** *openings, generally ellipsoidal or cylindrical, formed in rock material by the expansion of gas escaping from solution.* Rocks containing vesicles are *vesicular.* Vesicles are most abundant near the tops of lava flows (Fig. 20-7). Where vesicles are extremely abundant and many of them are open at the top, the structure is said to be *scoriaceous* and the rock is *scoria* (Fig. C-9, *A*). ***Pumice*** is *extremely vesicular natural glass with a high content of silica.* Pumice is so light that it floats on water.

Amygdales (ă-mĭg′ - dāles) are *vesicles that have been filled by minerals.* Rocks containing amygdales are said to be *amygdaloidal* (Fig. C-9, *B*). Amygdales resemble phenocrysts at first sight, but closer study discloses important differences of shape and composition. The shapes of phenocrysts are those of well-formed crystals of rock-forming silicate minerals; these same minerals are also

Fig. C-8. Examples of fabrics in igneous and metamorphic rocks. *A.* Aligned feldspar phenocrysts in granite porphyry. (B. M. Shaub.) *B.* Aligned amphibole porphyroblasts in gneiss. (G. M. Friedman.) *C.* Random orientation of kyanite porphyroblasts lying in plane of foliation of coarse-grained schist. (B. M. Shaub.)

present in the groundmass. Phenocrysts grew *outward* from their centers. By contrast the shapes of amygdales, which reflect the shapes of the vesicles, are circular. Amygdales consist of one or more minerals that typically do not occur in the rest of the rock. The minerals of amygdales grew *inward* from the walls of the spaces they occupy. Any specimen may be vesicular in some parts and amygdaloidal in other parts.

PRINCIPAL KINDS OF IGNEOUS ROCKS

Only a few principal kinds of igneous rocks need be presented in an introductory study based on hand specimens. In Table C-1 we have grouped the chief kinds of igneous rocks according to texture (rows) and composition (columns). Because texture and composition vary gradationally through a wide range the lines in the table are somewhat arbitrary and have been dashed.

We can subdivide minerals of igneous rocks into two great groups: (1) light-colored minerals (including the feldspars, quartz, and muscovite) and (2) dark-colored minerals (including biotite, pyroxene, hornblende, and olivine). Rocks in which light-colored minerals, chiefly feldspars, predominate are collectively designated *felsic rocks,* and those in which dark-colored minerals, chiefly ferromagnesian silicates, predominate are *mafic rocks.* The felsic rocks appear on the left side of the table and the mafic rocks on the right side. The minerals themselves appear within the rectangular field at the bottom of the table. Light-colored minerals are in the upper-left half of the field and dark-colored minerals in the lower-right half.

Among felsic rocks the presence or absence of quartz forms a logical basis for distinguishing between granitic rocks and diorite on the one hand, and between granitic rocks and nepheline syenite on the other. The kinds of feldspars furnish a basis for recognizing still other varieties of felsic rocks.

Use the table by identifying the minerals present and estimating what per cent of the volume of the rock each occupies. These percentages determine a vertical line in the composition field in the lower part of the diagram. Extend this line upward to the appropriate textural row based on particle size and read off the name of the rock. On the table only names of equigranular rocks appear. Modify these as necessary according to the usage explained in the previous section dealing with porphyritic texture.

Granular rocks. Among granular igneous rocks there is continuous gradation in proportion of light- and dark-colored minerals from felsic rocks (granite, granodiorite, and diorite) to the mafic rocks (gabbro and dolerite). Within felsites, abundance of silica varies. Granite is oversaturated with silica whereas nephelene syenite is undersaturated (Chap. 5). These felsic and mafic combinations likewise occur as aphanitic and glassy rocks, both intrusive and extrusive. A few ultramafic rocks, including pyroxenite, peridotite, dunite, and anorthosite, are known only in granular form. The absence of aphanitic and glassy equilavents suggests that ultramafic rocks and anorthosite are not strictly magmatic but originated by the segregation of crystals from a magma or by accumulation of less fusible constituents during partial melting (Chap. 5).

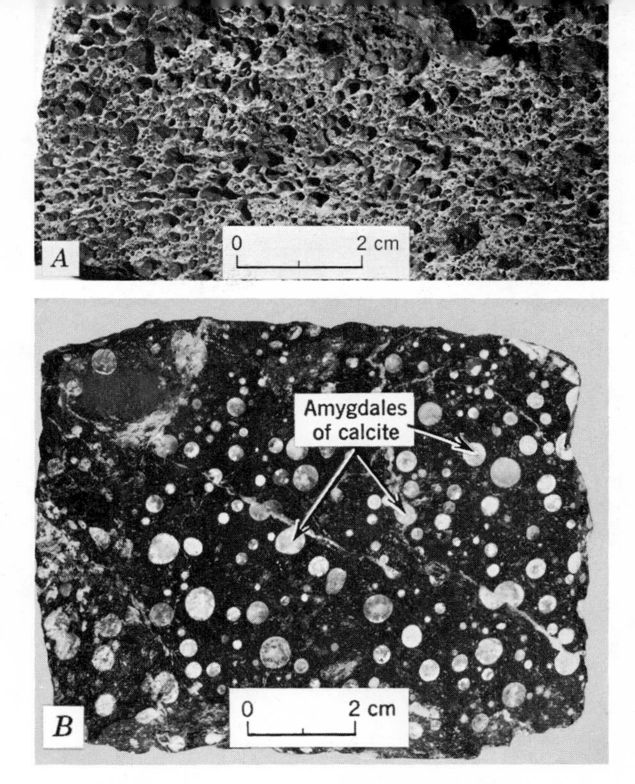

Fig. C-9. Minor structures in igneous rocks. *A.* Extremely vesicular basalt (*scoria*) with numerous irregular vesicles. Near Klamath Falls, Oregon. (B. M. Shaub.) *B.* Amygdaloidal basalt. Compare circular shape of sections of amygdales with crystal shapes of feldspar phenocrysts in porphyry having an aphanitic groundmass, Fig. C-3. (C. R. Longwell.)

Granite. Among felsic rocks the most abundant is ***granite,*** *a coarse-grained igneous rock consisting in large part of potassium feldspar, some sodic plagioclase (albite), and quartz, with minor amounts of ferromagnesian minerals,* chiefly biotite, but including hornblende. The dark minerals commonly occur as nearly perfect crystals, suggesting that they formed first, while surrounded by magma. The feldspars formed next, but

TABLE C-1. PRINCIPAL IGNEOUS ROCKS CLASSIFIED ACCORDING TO MINERAL COMPOSITION AND TEXTURE*

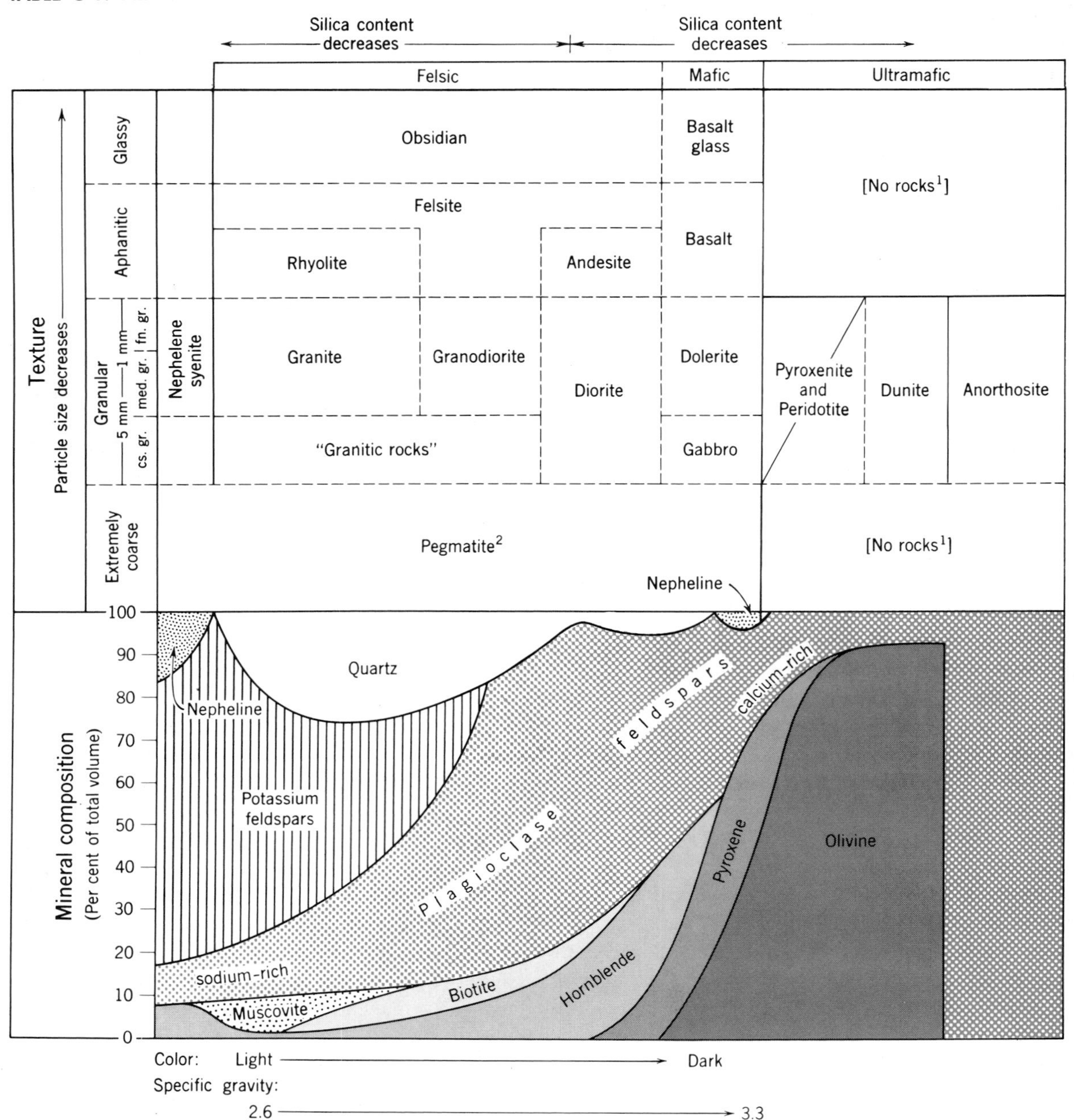

* Gradational boundaries are shown by dashed lines. Only four kinds of aphanitic rocks and two kinds of glassy rocks are shown, because it is not possible to identify others without a microscope or chemical analysis.

[1] No rocks having this composition and texture are known.

[2] Pegmatite designates extremely coarse-grained igneous rocks of any composition. When used alone, *pegmatite* implies granitic composition. Pegmatites having other compositions are named by including the appropriate rock term, as *diorite pegmatite* or *gabbro pegmatite*.

generally are so numerous that the grains crowded together and hindered one another's growth. The surplus silica crystallized as quartz, filling in the spaces among particles already solidified.

A coarse-grained igneous rock, containing quartz, in which plagioclase is the chief feldspar, is not strictly granite; it is ***granodiorite.*** The proportions of feldspars are not readily determined by beginners. Where granodiorite is not recognized separately, the term *granitic rocks* can be extended to include all coarse-grained felsic igneous rocks that contain quartz.

Granite pegmatite, or "giant granite," contains particles larger than 10mm. *Pegmatitic* is a general textural term describing all igneous rocks that are extraordinarily coarse grained, but when used alone as a rock name, *pegmatite* implies granitic composition.

Nephelene syenite. *A light-colored, coarse-grained, felsic igneous rock lacking quartz, and containing nephelene* is ***nepheline syenite.*** Nepheline syenite contains several kinds of ferromagnesian minerals in amounts ranging up to about 25 per cent.

Diorite. As plagioclase and ferromagnesian minerals become more abundant, quartz tends to disappear, and a rock transitional to the mafic group is the result. An example is ***diorite,*** *a coarse-grained felsic igneous rock that lacks quartz, and in which the chief feldspar is plagioclase and ferromagnesian minerals* (chiefly hornblende and biotite) *constitute less than 50 per cent.*

Gabbro and dolerite. *Igneous rocks in which ferromagnesian minerals exceed 50 per cent* are ***mafic rocks.*** We need recognize only one kind of mafic granular rock, with two varieties based on grain size. *Coarse-grained mafic rock* is ***gabbro;*** *medium-grained mafic rock* is ***dolerite.*** In both, the chief ferromagnesian mineral is pyroxene but olivine is likely to be present. Although the silica content of mafic rocks varies, visible quartz is absent. Olivine, nepheline, and other silica-deficient minerals occur in undersaturated varieties. Because ferromagnesian minerals are denser than feldspar, gabbro and dolerite are distinctly denser than granite and average diorite (Table C-2).

Anorthosite and ultramafic rocks. Variation in the proportions of the major constituents of gabbro, plagioclase and ferromagnesian silicates permits us to recognize two additional kinds of igneous rocks. (1) As ferromagnesian minerals disappear plagioclase predominates and the result is ***anorthosite,*** *an igneous rock composed almost entirely of plagioclase.* Although generally rare it occurs abundantly in the Adirondacks. (2) By contrast, as plagioclase disappears ferromagnesian minerals predominate, resulting in ***ultramafic rocks,*** *granular rocks that consist almost entirely of ferromagnesian minerals;* plagioclase may be present, but in negligible amounts.

Ultramafic rocks can be subdivided on the basis of the kinds and proportions of ferromagnesian minerals they contain. ***Pyroxenite*** (pī-rōx′-ĕn-īte) is *an ultramafic rock composed almost entirely of pyroxene;* ***peridotite*** (pĕr′-ĭ-dō-tīte) is *an ultramafic rock consisting of pyroxene and considerable olivine. An ultramafic rock consisting almost wholly of olivine* is ***dunite*** (dŭn′-īte; named for Mount Dun, New Zealand).

Aphanitic rocks. On the basis of hand-specimen study it is virtually impossible to recognize more than two varieties of aphanitic rocks unless phenocrysts are present. Lacking phenocrysts we can distinguish *felsite* from *basalt* by holding freshly broken thin edges of the specimens before a bright light. *An aphanitic igneous rock of granitic composition, whose thin edges transmit some light,* is ***felsite.*** *An aphanitic igneous rock having the composition of gabbro and dolerite, and whose thin edges are opaque* is ***basalt*** (bă-sält′). Because basalt is a common extrusive rock it commonly displays vesicular or amygdaloidal structure (Fig. C-9). The presence of vesicles or amygdales is a criterion of igneous origin. When these structures are not present and the basalt is homogeneous, its appearance is similar to fine-grained, dark-colored limestone and dolostone. Basalt is conspicuously harder than either limestone or dolostone and, if unaltered, does not effervesce with dilute hydrochloric acid. Although the silica content of basalt is generally low, the proportion of silica in basalt can range from oversaturated to undersaturated.

If aphanitic rocks contain phenocrysts it is possible to recognize two varieties of felsites, depending on the presence or absence of quartz. *A fine-grained igneous rock having the composition of granite* is ***rhyolite;*** if prophyritic contains quartz and an aphanitic groundmass, that ranges in color from nearly white through gray, yellow, red, or purple. Many specimens of rhyolite contain small-scale layers caused by flow in the lava shortly before it solidified. Many rhyolites are the products of hot tephra flows (Chap. 19).

A porphyritic rock with aphanitic groundmass, generally similar to rhyolite but lacking quartz phenocrysts, is ***andesite*** (named for the Andes Mountains). Phenocrysts consist of feldspars and ferromagnesian minerals. Common colors range from shades of gray and green to dark gray, even black.

Glassy rocks. Both felsic and mafic rocks have glassy equivalents. Two of these are recognized by the appear-

TABLE C-2. SPECIFIC GRAVITIES, AT ATMOSPHERIC PRESSURE, OF SELECTED METALS, MINERALS, AND ROCKS[a]

		Minerals					
Metals	Sp. Gr.	Metallic Luster	Sp. Gr.	Nonmetallic Luster	Sp. Gr.	Rocks	Sp. Gr.
Gold	19.3						
Mercury	13.6						
Copper	8.9						
Nickel	8.6						
Iron	7.9	Galena	7.57				
Tin	7.3						
Zinc	7.2						
		Hematite	5.26				
		Magnetite	5.2				
		Pyrite	5.01				
				Barite	4.3 to 4.6		
				Corundum	4.0 to 4.1		
				Sphalerite	3.9 to 4.1		
				Garnet	3.58 to 4.31		
				Olivine	3.2 to 4.3		
				Staurolite	3.7 to 3.8		
				Kyanite	3.53 to 3.65		
				Topaz	3.49 to 3.57	Dunite	3.3
				Augite	3.2 to 3.4	Peridotite	3.21
				Fluorite	3.18	Olivine	
						gabbro	3.0
						Gabbro	2.97
				Biotite	2.7 to 3.3		
				Hornblende	2.9 to 3.4	Olivine	
				Aragonite	2.94	basalt	2.95
				Muscovite	2.77 to 2.88	Basalt	2.87
Aluminum	2.7			Dolomite	2.85	Diorite	2.87
				Wollastonite	2.8 to 2.9	Anorthosite	2.75
				Calcite	2.71		
				Talc	2.58 to 2.83		
				Chlorite	2.6 to 2.9		
				Feldspars	2.56 to 2.76		
				Plagioclase	2.62 to 2.76	Granite	2.66
				Albite	2.62	Andesite	2.65
				Orthoclase	2.56 to 2.59	Rhyolite	
				Quartz	2.65	porphyry	2.62
				Kaolinite	2.6		
				Chalcedony	2.57 to 2.64		
				Gypsum	2.3	Rhyolite	2.49
				Serpentine	2.2 to 2.6		
				Graphite	2.02 to 2.23	Obsidian	2.29
				Bauxite	2.0 to 2.55		
				Opal	2.0 to 2.2		
				Halite	2.16		

[a] Metals from Handbook of Chemistry and Physics; rocks from F. F. Grout, 1940; minerals from Vanders and Kerr, 1967.

ance of thin edges in strong light. *Felsic glass that transmits light along thin edges* is ***obsidian.*** Most obsidian is highly lustrous and is dark colored, even black. The dark color results from scattered dark impurities. *Felsic glass having a dull, greasy luster* is ***pitchstone.*** Both obsidian and pitchstone break with conchoidal fracture (Fig. C-3).

Sheets of obsidian commonly are capped with nearly white ***pumice*** (pŭm′ĭs), an extremely vesicular, frothy glass having a high content of silica. Gases escaping through the rapidly cooling viscous lava created the cavities.

Mafic glass opaque along thin edges is ***basalt glass.***

Special varieties of volcanic rocks. The many and varied volcanic activities create special features in rocks that merit names. These names, independent of those based on mineral composition and texture, do not appear in Table C-1. The most important variety for our study is recognized by virtue of its content of broken, angular particles. *A general term for a rock of any origin containing angular particles* is ***breccia*** (brĕtch′-ĭ-ăh). Rocks can be broken into angular fragments by many volcanic processes, by ordinary erosion, by faulting, and by various kinds of collapse caused by removal of support from below. The term breccia should always be used with a modifier to indicate whether it refers to rock of volcanic, sedimentary, or other origin. For example *fault breccia* results from the breaking of rock during movement on a fault.

Many volcanic breccias originate because rapid solidification creates a rocky crust at the tops and sides of lava flows; later surges of the liquid lava break the crust into blocks (Fig. 19-5). The result is ***flow breccia,*** *volcanic breccia created by the breaking up of a hardened crust of extrusive rock as a result of further flow of liquid lava.* When some basaltic lava flows enter water, countless small pieces continuously solidify and spall off as the lava tries to advance.

Other volcanic breccias originate as a result of explosions. Tephra ejected explosively commonly coalesce to form a hot tephra flow, which travels so swiftly that the particles are still hot when they stop moving. Tephra that travel as individual particles behave as sediment, and notwithstanding their igneous origins and high temperatures we treat them as sedimentary rocks.

PRINCIPAL KINDS OF SEDIMENTARY ROCKS

Various combinations of mineral composition, texture, mode of origin, and shape of particles form the basis of classifying and naming sedimentary rocks (Table C-3). We name rocks derived from detrital sediments primarily on the basis of particle size and secondarily on particle shape and composition. Rocks composed of carbonate minerals are classified as *limestones* and *dolostones* regardless of their varied textures and modes of origin. Two large groups of sedimentary rocks are distinguished by their modes of origin, which differ notably from the origins of fragmental materials. One group originates from biologic secretion and the other by precipitation resulting from evaporation of water.

Clastic rocks. The "mainstream" of sedimentary rocks consists of the many varieties that trace their origin to cementation of clastic sediments. Major varieties are based on sizes of particles and minor varieties on shapes and compositions of particles. The four major subdivisions of sediments (gravel, sand, silt, and clay) correspond to the rocks conglomerate (and sedimentary breccia), sandstone (also arkose and graywacke), siltstone, and claystone, respectively.

Conglomerate and sedimentary breccia. When particles of gravel size (coarser than 2mm) predominate, the rock is conglomerate or sedimentary breccia. Although finer sizes are present the dominant particle size can range from fine pebbles up to huge boulders. In conglomerate the gravel particles, coarse or fine, have been at least somewhat rounded by mechanical abrasion during transport. Therefore we define ***conglomerate*** as *a clastic sedimentary rock containing a preponderance of rounded pebbles or larger particles* (Fig. 4-12). In contrast some gravel particles are angular (Fig. C-10). The lack of rounding forms the basis for distinguishing ***sedimentary breccia,*** *a clastic sedimentary rock containing numerous angular particles of pebble size or larger.* As we would expect, there is complete gradation in degree of rounding of particles; hence no sharp boundary separates conglomerate from sedimentary breccia. Some specimens with partly worn fragments may be called by one name or the other, according to individual judgment.

Sandstone, arkose, and graywacke. Sandstone is the medium-grained clastic rock. It is defined as *a clastic rock consisting predominantly of sand-size particles.* In most sandstones the constituent sand grains are quartz. The cementing material is usually calcium carbonate, but in some sandstones it is iron oxide, silica, or some other substance. Color in sandstone, produced partly by the grains and partly by the cementing material, varies within a wide range.

With progressive change in size of grain, coarse sandstone grades into conglomerate, fine sandstone into silt-

TABLE C-3. PRINCIPAL SEDIMENTARY ROCKS CLASSIFIED ACCORDING TO COMPOSITION OF PARTICLES AND TEXTURE

Major classes of sedimentary rocks		Partly igneous, partly sedimentary	Rocks formed from sediments	Precipitates	Accumulations of plant debris and hydrocarbons
Major classes of sediments		Pyroclastic	Clastic	Nonclastic	Nonclastic
Particles	Kinds of particles	Tephra (particles blown out of the Earth's crust)	Detritus (particles of, or mineral grains derived directly from or as a result of weathering of preexisting rocks or sediments)	Whole shells, crystals, and other particles (particles precipitated out of water at site of deposition by biological and/or physical-chemical processes)	Pieces of plants or material derived from plant debris
	Composition of particles	Crystals; crystalline particles; rock particles; glass particles. Minerals are typical of igneous rocks	Particles of silicate rocks; quartz; feldspar; micas; heavy minerals; clay minerals Particles of carbonate rocks; calcite; dolomite	Aragonite Calcite Dolomite Broken shells and shell debris[1] Halite Gypsum Anhydrite	Organic compounds
Texture		Fragmental	Fragmental Clastic	Crystalline	Organic
Method of transport of particles		Physically transported through atmosphere[2]	Physically transported; may be abraded or broken during transport	Material transported in solution; precipitated as particles or as coatings on particles at site of deposition Particles not transported	Some accumulated in place

Name of sediment or rock; size of particles	Rock name	Tephra[4]	Sediment name and limiting diameter of dominant particles	Particle size increases ↑	Sediment name and limiting diameter of dominant particles	Rock name	Rock name	Rock name (Evaporites)	Rock name	Rock name
	Agglomerate	Tephra[4]	Bombs (particles molten when ejected) Blocks (particles solid when ejected)		Gravel	Conglomerate (particles rounded) Sedimentary breccia (particles angular)	[No rocks[3]]	Rock salt (mineral halite) Gypsum rock (mineral gypsum) etc.[4]	Reef rock[4] (mostly limestone and dolostone)	Peat, Coal[4]
			32mm							
	Lapilli tuff		Lapilli		Gravel		Limestone and Dolostone			
			4mm		2mm					
	Tuff		Ash		Sand	Sandstone Arkose Graywacke				
			1/16mm		1/16mm					
	Tuff		Dust		Silt	Siltstone; Shale (if fissile)				
					1/256mm					
					Clay	Claystone; Shale (if fissile)				

[1] Some skeletal material is composed of silica.

[2] Tephra can be transported also by other carrier agents, such as running water.

[3] No rocks having this grain size, composition, and origin are known.

[4] Tephra, salt, gypsum, reef rock, peat, and coal are set vertically to emphasize the fact that they are unrelated to the particle diameters shown in the scale within the diagram.

Fig. C-10. Sedimentary breccia, consisting of angular fragments of numerous kinds of rocks, illustrates deposition without significant abrasion during transport. Upper Triassic, East Haven, Connecticut. (J. E. Sanders.)

stone. In many rocks, grain sizes are mixed, and we speak of conglomeratic sandstones and sandy siltstones and shales.

Two unusual varieties of sandstone deserve special mention. Both are characterized by containing an appreciable proportion of feldspar along with the quartz. As stated in Chapter 7, feldspar is attacked by chemical weathering much more readily than quartz. Therefore feldspar does not ordinarily survive to become part of a sediment of a sedimentary rock. Hence sandstone that contains feldspar implies that it was derived from mechanically weathered source rocks, that the source rocks were eroded mechanically before they could be weathered chemically, or that some other special situation existed. *Sandstone containing at least 25 per cent feldspar as well as quartz* is called ***arkose.*** A composition of feldspar and quartz suggests granite, and some fresh arkose may be mistaken for it. In arkose, the grains do not interlock; they are separated by fine-grained cementing material. *Poorly sorted sandstone, generally dark, containing rock fragments as well as quartz,* is called ***graywacke.*** Feldspar is a common though not essential constituent in graywacke.

Siltstone, claystone, and shale. The clastic rocks with finer grain than sandstone are siltstone, claystone, and shale. Siltstone and claystone are identified by their dominant particle sizes: silt or clay. ***Siltstone,*** then, is *a clastic rock consisting predominantly of silt-size particles.* ***Claystone*** is *a clastic rock consisting predominantly of clay-size particles.* The silt-size particles in siltstone are usually quartz; the clay-size particles in claystone are commonly various clay minerals, but they include quartz and other minerals as well. The particles constituting these rocks are so very small that thorough study of them requires X-ray equipment or an electron microscope.

The basis for recognizing shale is the kind of parting. Some fine-grained rocks break into blocky fragments (Fig. C-2, *A*) whereas others, which split along closely spaced partings (Fig. C-2, *B*), are described as *fissile.* ***Shale*** is *a fine-grained, fissile clastic sedimentary rock.*

Pyroclastic rocks. Tephra, broken and propelled by volcanic explosions, accumulate in various ways and become lithified to form pyroclastic rocks. Particle size is the principal basis for subdividing such rocks (Table C-3). Tephra are classified into size groups that differ from those of clastic sediments. *Bombs* and *blocks* embrace all tephra coarser than 32mm. The term *lapilli* designates tephra between 4 and 32mm; *ash* between $^1/_{16}$mm and 2mm, and *dust,* tephra smaller than $^1/_{16}$mm. The corresponding rock names are ***agglomerate*** for *coarse pyroclastic rocks consisting largely of bombs and blocks;* ***lapilli tuff,*** for *medium-grained pyroclastic rocks consisting of lapilli;* and ***tuff,*** for *fine-grained pyroclastic rocks consisting of ash and dust.*

Some tephra were so hot when deposited that they fused together to create *welded tuff* (Chap. 19). Other tephra, originally propelled by volcanic explosions, were reworked later by running water or the wind. During reworking the tephra may have been mixed in various proportions with nonvolcanic particles.

Carbonate rocks. On the basis of their composition all rocks consisting chiefly of carbonate minerals are grouped together as carbonate rocks. The two chief compositional varieties are limestone and dolostone.

Limestone. The broadest definition of ***limestone*** states merely that it is *a sedimentary rock consisting predominantly of calcite.* Some limestones are clastic, some are chemical precipitates (either inorganic or organic), and still others are mixtures of clastic and precipitated sediments. Clastic limestones (and dolostones) occur in all particle sizes, parallel with the other clastic rocks. Some limestones that are uniformly fine grained probably were formed as chemical precipitates, aided more or less by tiny organisms. Some of the calcareous oozes on present sea floors (Fig. 15-10, *B*) probably represent an early stage in the formation of fine-grained rock. By contrast many limestones are coarse grained, either from crystallization of the calcium carbonate or because they are made largely of shell fragments. All limestone effervesces freely in dilute hydrochloric acid.

A variety of limestone that is mainly clastic is ***coquina*** (Fig. 4-8, *C*), *an aggregation of shells and large shell fragments, cemented with calcium carbonate.* Another variety, ***chalk,*** is *limestone that is weakly cohesive.* ***Fresh-water limestone*** is *limestone that formed in a lake. Marl,* a mixture of calcium carbonate and clay, is also a common deposit in lakes.

Still other limestones form on or beneath the ground by evaporation in an environment of air rather than under water. The dripstone in caverns (Fig. 11-14) is an example. Another is ***calcareous tufa,*** *a light, spongy limestone precipitated by the evaporation of springs and small streams.* ***Travertine*** is used by some as *a name for both dripstone and calcareous tufa.*

Dolostone is *a sedimentary rock consisting chiefly of the mineral dolomite,* the double carbonate of magnesium and calcium. Dolostone and limestone are very intimately related and resemble each other in appearance. Dolostone is slightly harder than limestone and effervesces with acid only on a scratched surface or in powdered form. Not only is there a complete gradation between them, but a single sample of rock can consist of both varieties. Most dolostone originated through the replacement of calcium-carbonate sediments and limestones with dolomite. Because of this origin such dolostone is a secondary rather than a primary rock.

Evaporites. *Nonclastic sedimentary rocks whose constituent minerals were precipitated from water solution as a result of evaporation* are ***evaporites.*** The most common evaporites are *salt* (consisting of the mineral halite, NaCl), *gypsum* (consisting of the mineral gypsum ($CaSO_4 \cdot {}_2H_2O$), and *anhydrite* (consisting of the mineral anhydrite, $CaSO_4$). Typically these rocks contain one mineral only, but even when mixed with impurities salt can be identified by its unmistakeable taste and gypsum by its characteristic hardness. Some limestones and dolostones formed as evaporites.

Rocks with organic textures. The class of sedimentary rocks built largely by the direct effects of animals and plants includes reef rock (built by lime-secreting organisms) and peat and coal (residues of plant tissues). Such rocks record valuable clues to the environment of deposition.

Reef rocks. Massive limestones preserving organic skeletal remains *in situ* typify *reef rock* (Fig. 4-10). In spaces within the framework secreted by the organisms are fragments of skeletal debris of various nonreef organisms, pieces of the reef broken by wave action, and other particles. Reef rock that is calcareous originally can be replaced by dolomite creating massive dolostone.

Peat and coal. *Peat* is a brownish, lightweight mixture of partly decomposed plant tissues in which the parts of plants are easily recognized. *Lignite* (*brown coal*) is much more compact than peat but usually shows some recognizable plant tissues. ***Coal*** is *a black sedimentary rock consisting chiefly of partly decomposed plant matter and containing less than 40 per cent inorganic matter.* Unlike nearly all other sedimentary rocks, most coal consists of substances that were not transported to areas of deposition, but that accumulated in the places where the plants grew. The process is described in Chapter 23.

Two varieties of coal can be distinguished readily. *Bituminous* (soft) *coal* is black and firm; it breaks into blocks and contains alternating layers having dull and bright luster. *Anthracite* (hard coal) breaks with conchoidal fracture and is characterized by its lustrous surfaces.

Chert. Although it does not lend itself readily to classification into a clear-cut category, ***chert,*** *a common micrograined sedimentary rock composed of nonfragmental silica,* is widely associated with carbonate rocks. Much chert occurs in irregular or nodular masses; it originated by replacement of calcium-carbonate sediments, of limestones (Fig. C-5), and of dolostones with silica. Some chert occurs in widespread strata; some of it may have been an initially siliceous deposit.

PRINCIPAL KINDS OF METAMORPHIC ROCKS

Because, as seen in hand specimens, most metamorphic rocks display more or less identical crystalline textures, the chief basis for subdivision of this varied rock group is the presence or absence of layers. The principal layer structure is ***foliation*** (Lat. *folium,* "leaf"), *a parallel or nearly parallel structure in metamorphic rocks along which the rock tends to split into flakes or thin slabs.* Foliation, which results from alignment of platy mineral particles in fine-grained rocks and from alternations of layers having differing mineral composition in coarse-grained rocks, is generally conspicuous. Although some foliation is parallel, other foliation is lenticular (Fig. C-11).

Foliation can originate from stratification by mere recrystallization of the minerals of a stratified parent rock, or it can be a by-product of extreme deformation during metamorphism. On a large scale nearly all metamorphic rocks are foliated. For the purpose of hand-specimen study, however, we can readily distinguish two groups: (1) *metamorphic rocks displaying foliation,* or ***foliates,*** and (2) *metamorphic rocks lacking foliation,* or ***nonfoliates.***

The relationships among the common types of metamorphic rocks possessing crystalline texture are shown in Table C-4. In general appearance that table resembles Table C-1 (igneous rocks) but is not employed in exactly

Fig. C-11. Examples of foliates. (B. M. Shaub.) *A.* Closely spaced foliation in mica schist. North Branford, Connecticut. *B.* Lenticular gneissic foliation in biotite-quartz-feldspar gneiss. Near Bedford, New York.

the same way. With igneous rocks we began by determining mineral composition and entered the lower part of Table C-1 to find a composition line. We extended this line upward to the appropriate texture and read off the name of the rock. With metamorphic rocks we begin at the top of Table C-4. The presence or absence of foliation determines which part of the diagram we follow. Use texture to subdivide the foliates. Finally, determine mineral composition and with it separate foliates into felsic and mafic varieties and nonfoliates into hornfels, quartzite, and marble. The percentages of various minerals shown in the lower part of the table are intended only to suggest some of the many possible variations: such a diagram could not show all the possibilities.

Instead of determining the rock name of foliates as in granular igneous rocks, the minerals present become modifiers of the name of the metamorphic rock. When two or more minerals are abundant, list them in order of increasing abundance and place a hyphen between two adjacent mineral names. For example, *garnet-mica schist* designates a schist in which mica is more abundant than garnet, and *biotite-quartz-plagioclase gneiss,* a gneiss in which plagioclase is the most abundant mineral and quartz is more abundant than biotite. Likewise similar mineral modifiers can be added to the names of nonfoliates. The principal kinds of metamorphic rocks shown in Table C-4 are described briefly in the following paragraphs.

Foliates. The foliates include slate, phyllite, schist, gneiss, and amphibolite.

Slate. A *fine-grained foliate that splits along smooth planes into very thin plates* defines ***slate.*** The luster visible on the surface of the cleavage slabs may be considerable, but individual mineral particles are nearly invisible to the naked eye. "Slate color" means dark bluish gray, but many slates are red, green, or black. Thin slabs of slate ring when they are tapped sharply with a nail. A positive feature identifying slate is angular divergence between the smooth parting surfaces (slaty cleavage) and original stratification (Fig. C-2, *C*).

Phyllite. As particle size increases, slate grades into ***phyllite*** (fĭll′-īte), *an exceptionally lustrous, fine-grained foliate parting along smooth or irregular surfaces.* The luster is caused by parallel micaceous minerals ranging from barely visible to obvious with a hand lens. In addition some phyllites contain visible garnets and other minerals. *Chlorite phyllite* is a mafic variety containing abundant chlorite.

Schist. Because of greater intensity of metamorphism or greater content of coarser-grained quartz and feldspar in the parent rock, phyllite gives way to ***schist,*** *a well-foliated metamorphic rock in which the component flaky minerals are distinctly visible* (Fig. C-11, *A*). Schists are subdivided on the basis of their prominent constituents, which include the micaceous minerals chlorite, muscovite, and biotite, and other minerals, such as hornblende, kyanite (Fig. C-8, *C*), staurolite, garnet (Fig. C-6), and other minerals that grew in the rock during metamorphism. *Chlorite schist* is a mafic variety containing abundant chlorite. Quartz is abundant in all kinds and feldspar may be an important constituent. Both quartz and feldspar tend to be overshadowed by the micaceous minerals. The abundance of quartz and feldspar can be most reliably ascertained by examining the rock along a surface broken across the foliation.

Gneiss. Prominently foliated felsic metamorphic rocks are exemplified by ***gneiss*** (nīce), *a coarse-grained foliate breaking along irregular surfaces and commonly containing prominently alternating layers of light- and dark-colored minerals.* Feldspar, quartz, mica, hornblende, and garnet are common minerals in gneisses. ***Granite gneiss*** is *a distinctly foliated metamorphic rock with the mineral composition of granite.*

Amphibolite. *A coarse-grained mafic metamorphic rock*

TABLE C-4. PRINCIPAL METAMORPHIC ROCKS HAVING CRYSTALLINE TEXTURE, CLASSIFIED ACCORDING TO FOLIATION, TEXTURE, AND MINERAL COMPOSITION.*

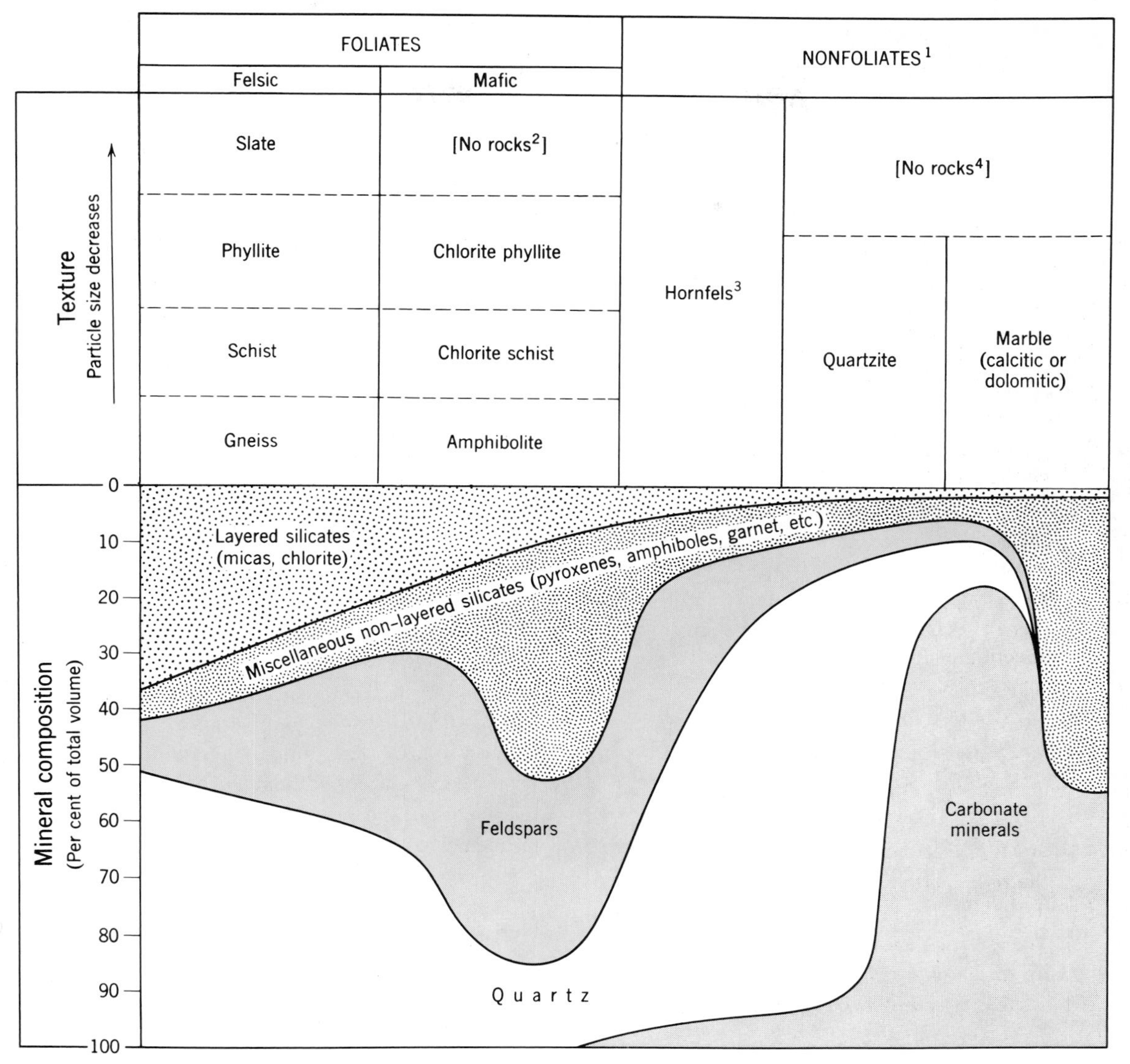

* Horizontal dashed lines show gradational particle-size boundaries.

[1] Most rocks that appear in hand specimens to be nonfoliates actually are foliated on a large scale when seen in large natural exposures.

[2] In a simple classification no rocks of this kind need be considered.

[3] The name hornfels is generally applied only to contact-metamorphosed rocks. Typical hornfels is fine grained, but coarse-grained varieties are known.

[4] Fine-grained quartzites and marbles are rare. Although fine-grained parent rocks exist, recrystallization enlarges their particles.

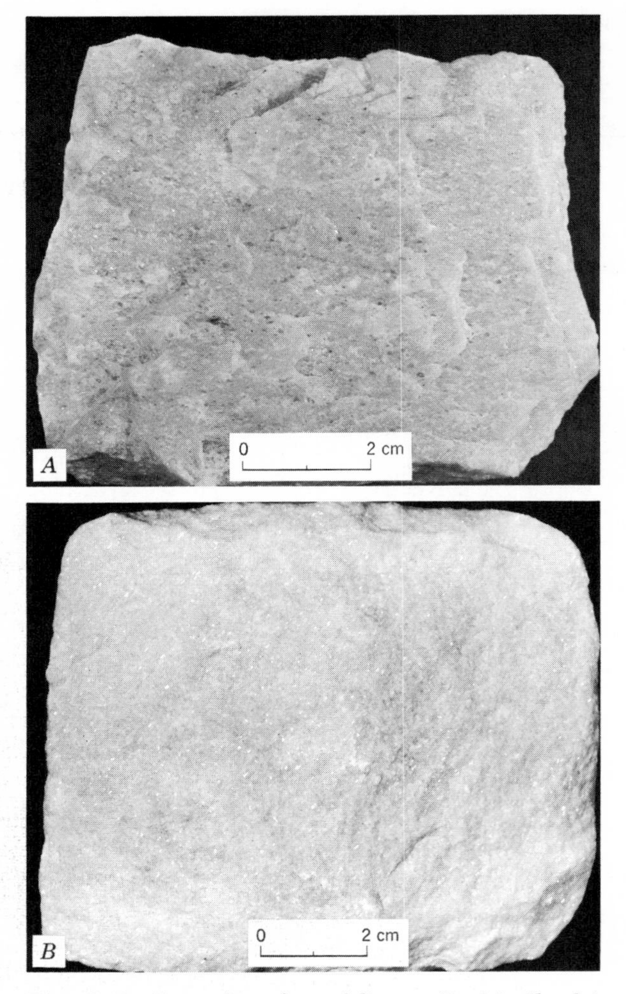

Fig. C-12. Examples of nonfoliates. (B. M. Shaub.) *A.* Quartzite. Southern Vermont. *B.* Dolomitic marble. Lee, Massachusetts.

containing abundant amphibole is an ***amphibolite.*** Many amphibolites are foliates but some are not. Other minerals in amphibolite include feldspars, garnet, micas, and quartz. Amphibolite resembles gneiss but is distinguished from gneiss by its abundance of amphiboles. Because many amphibolites are dark green or black, amphibolite generally is darker colored than gneiss.

Nonfoliates. Included among our list of nonfoliates are hornfels, quartzite, and marble.

Hornfels. In contact-metamorphic aureoles, various sedimentary or igneous parent rocks can be transformed into ***hornfels,*** *a tough, generally massive nonfoliate containing scattered crystals of high-temperature minerals.* Typical hornfels is fine grained and dark colored; it resembles aphanitic basalt, dark chert, and dark-colored, fine-grained carbonate rocks. Hornfels is harder than any carbonate rock and can be distinguished from aphanitic basalt and dark chert by its high-temperature minerals such as garnet and andalusite.

Quartzite. Recrystallization of sandstone produces ***quartzite,*** *a quartz-rich nonfoliate.* Quartzite displays a pronounced vitreous luster and is distinguished by the fact that the rock breaks through its particles rather than around them, as in ordinary sandstone. Accessory minerals in quartzite include micas, feldspars, and garnet.

Some sandstones are composed of nearly 100 per cent quartz and the grains have been cemented by silica. Such silica-cemented sandstones are considered as a variety of quartzite even though they have not been recrystallized and therefore do not strictly qualify as metamorphic rocks. We can be satisfied by applying the name quartzite to any quartz-rich rock that breaks across its particles rather than around them (Fig. C-12, *A*).

Marble. Recrystallization of carbonate rocks during metamorphism creates ***marble,*** *a nonfoliate consisting chiefly of calcite or dolomite.* Because carbonate minerals are very soluble, some ordinary limestones and dolostones have recrystallized without being subjected to high temperatures and pressures. Because any coarsely crystalline carbonate rock capable of being highly polished is known commercially as "marble," some nonmetamorphic rocks are included under that title.

Both marbles and sedimentary carbonate rocks effervesce in dilute HCl, vigorously if calcite is the dominant mineral and only in places where the rock has been scratched or powdered if dolomite predominates. The occurrence of metamorphic minerals, such as serpentine, wollastonite, and diopside, as accessory constituents is a criterion of some marbles. Because recrystallization drives out organic matter and nearly all other pigmenting materials, many marbles are glistening white (Fig. C-12, *B*).

Two other nonfoliates, *anthracite* and *graphite,* result from metamorphism of coal. Anthracite was described in a previous section treating coal; graphite is described in Table B-2.

APPENDIX D

Maps, Cross-sections, and Field Measurements

USES OF MAPS

An important part of the accumulated information about the geology and morphology of the Earth's crust exists in the form of maps. Nearly everyone has used automobile road maps in planning a trip or in following an unmarked road. A road map of a state, province, or county does what all maps have done since their invention at some unknown time more than 5,000 years ago; it reduces the pattern of part of the Earth's surface to a size small enough to be seen as a whole. Maps are especially important for an understanding of geologic relations, because a continent, a mountain chain, and a major river valley are of such large size that they can not be viewed as a whole unless represented on a map.

Furthermore, although most geologic maps are supplemented by text material, they contain information that can not be stated as fully and accurately in words or by any other means. Maps are the core of most geologic reports based on field study and are supplementary to many other kinds of geologic papers. This appendix is designed to outline basic information about maps and about common field measurements used in the construction of geologic maps.

A map can be made to express much information within a small space by the use of various kinds of symbols. Just as some aspects of physics and chemistry use the symbolic language of mathematics to express significant relationships, so many aspects of geology use the simple symbolic language of maps to depict relationships too large to be observed within a single view. Maps made or used by geologists generally depict either one or two of three sorts of things:

1. Hills, valleys, and other surface forms. Most maps of this kind are *geographic maps.*
2. Distribution and attitudes of rock bodies. Maps showing such things are *geologic maps.*
3. Geographic features such as mountains, rivers, and seas, not as they are today but reconstructed as they are inferred to have been at some time in the past (Fig. 16-14). Maps of this kind are *paleogeographic maps.*

The first two express the results of direct observation and measurement and are frequently included on a single map. The third expresses concepts built up from whole groups of observations and necessarily is shown on special maps separate from those representing present-day features.

BASE MAPS

Every map is made for some special purpose. Road maps, charts for sea or air navigation, and geologic maps are examples of three special purposes. But whatever the purpose, all maps have two classes of data: *base* data and *special-purpose* data. As base data most geologic maps show a latitude-longitude grid, streams, and inhabited places; many also show roads and railroads. A geologist may take an existing *base map* containing such data and plot geologic information on it, or he may start with blank paper and plot on it both base and geologic data, a much slower process if the map is made accurately.

Two-dimensional base maps. Many base maps used for plotting geologic data are two-dimensional; that is, they represent length and breadth but not height. A point can be located only in terms of its horizontal distance, in a particular direction, from some other point. Hence a base map always embodies the basic concepts of direction and distance. Two natural reference points on Earth are the North and South Poles. Using these two, the ancient Greeks established a grid by means of which any other point could be located. The grid we use now consists of lines of *longitude* (half circles joining the poles) and *latitude* (parallel circles concentric to the poles) (Fig. D-1). The longitude lines (*meridians*) run exactly north-south, crossing the east-west *parallels* of latitude at right angles. The circumference of the Earth at its Equator and the somewhat smaller circumference through its two poles being known with fair accuracy, it is possible to define any point on the Earth in terms of direction and distance

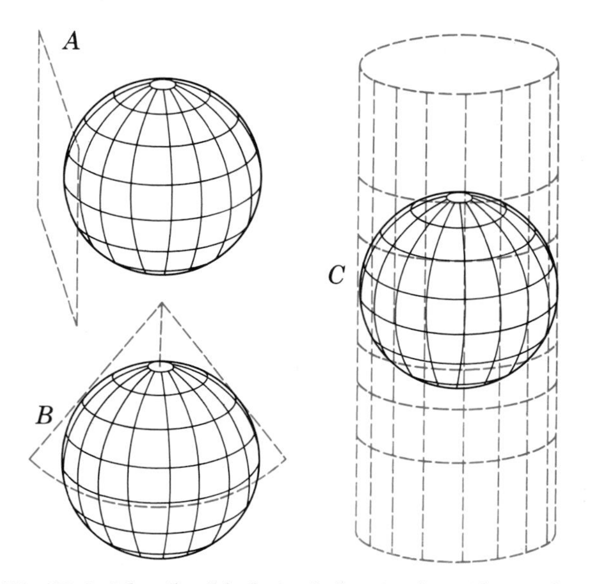

Fig. D-1. The Earth's latitude-longitude grid can be projected onto a plane *A*, a cylinder *C*, or a cone *B* that theoretically can be cut and flattened out.

from either pole or from the point of intersection of any parallel with any meridian.

For convenience in reading, most maps are drawn so that the north direction is at the top or upper edge of the map. This is an arbitrary convention adopted mainly to save time. The north direction could just as well be placed elsewhere, provided its position is clearly indicated.

The accuracy with which distance is represented determines the accuracy of the map. *The proportion between a unit of distance on a map and the unit it represents on the Earth's surface* is the ***scale*** of the map. It is expressed as a simple proportion, such as 1 : 1,000,000. This ratio means that 1 foot, meter, or other unit on the map represents exactly 1,000,000 feet, meters, or other units on the Earth's surface; it works out to 1 inch equals about 16 miles and is approximately the scale of many of the road maps widely used by motorists. Scale is also expressed graphically by means of a numbered bar, as is done on most of the maps in this book. A map with a latitude-longitude grid needs no other indication of scale (except for convenience) because the lengths of a degree of longitude (varying from 69.17 statute miles at the Equator to 0 at the poles) and of latitude (varying from 68.70 statute miles at the Equator to 69.41 at the poles) are known.

Map projections. The Earth's surface is nearly spherical, whereas nearly all maps other than globes are planes, usually sheets of paper. It is geometrically impossible to represent any part of a spherical surface on a plane surface without distortion (Fig. D-2). The latitude-longitude grid has to be *projected* from the curved surface to the flat one. This can be done in various ways, each of which has advantages, but all of which represent a sacrifice of accuracy in that the resulting scale on the flat map will vary from one part of the map to another. Examples of cylindrical projections are Figs. 2-11 and 15-19, and Plates A–C. The most famous of these is the Mercator projection; although it distorts the Polar Regions very greatly, compass directions drawn on it are straight lines. Because this is of enormous value in navigation, the Mercator projection is widely used in navigators' charts.

Figure 19-12, *A* is an example of conic projection. Some commonly used varieties are polyconic, in which not one cone, as in Fig. D-1, *B*, but several cones are employed, each one tangent to the globe at a different latitude. This device reduces distortion.

Figure 23-11 represents a plane projection, and Fig. 2-12 is a projection with certain areas, unimportant for the purpose of this map, eliminated. In a map of a very small area, such as Fig. 10-18, the distortion is of course slight, but it is there nevertheless.

Topographic maps. A more complete kind of base map is three-dimensional; it represents not only length and breadth but also height. Therefore it shows ***relief*** (*the difference in altitude between the high and low parts of a land surface*) and also ***topography,*** defined as *the relief and form of the land. A map that shows topography* is a ***topographic map.*** Topographic maps can give the form of the land in various ways. The maps most commonly used by geologists show it by contour lines.

Contours. A ***contour line*** (often called simply a *contour*) is *a line passing through points having the same altitude above sea level.* If we start at a certain altitude on an irregular surface and walk in such a way as to go neither uphill nor downhill, we will trace out a path that corresponds to a contour line. Such a path will curve around hills, bend upstream in valleys, and swing outward around spurs. Viewed broadly, every contour must be a closed line, just as the shoreline of an island or of a continent returns upon itself, however long it may be. Even on maps of small areas, many contours are closed lines, such as those at or near the tops of hills. Many, however, do not close within a given map area; they extend to the edges of the map and join the contours on adjacent maps.

Imagine an island in the sea crowned by two promi-

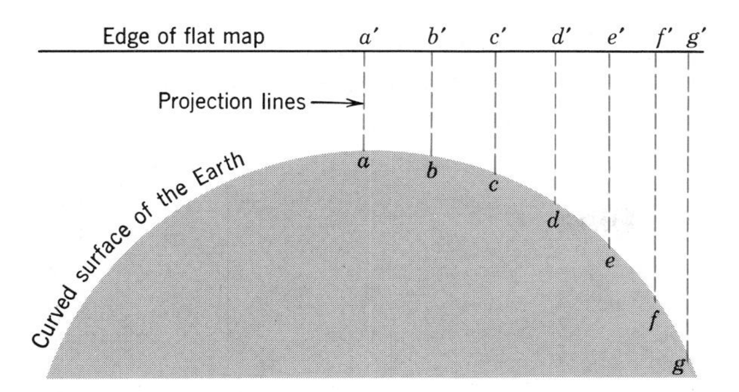

Fig. D-2. Equally spaced points (*a*, *b*, *c*, etc.) along a line in any direction on the Earth's surface become unequally spaced when projected onto a plane. This is why all flat maps are distorted. Compare Fig. 2-5 in which points equally spaced in a plane are distorted by the Earth's curved surface.

nent isolated hills, with much steeper slopes on one side than on the other and with an irregular shoreline. The shoreline is a contour line (the zero contour) because the surface of the water is horizontal. If the island is pictured as submerged until only the two isolated peaks project above the sea, and then raised above the sea 20 feet at a time, the successive new shorelines will form a series of contour lines separated by 20-foot contour intervals. (A ***contour interval*** is *the vertical distance between two successive contour lines,* and is commonly the same throughout any one map.) At first, two small islands will appear, each with its own shoreline, and the contours marking their shorelines will have the form of two closed lines. When the main mass of the island rises above the water, the remaining shorelines or contours will pass completely around the land mass. The final shoreline is represented by the zero contour, which now forms the lowest of a series of contours separated by vertical distances of 20 feet.

As the island is raised, the successive new shorelines are not displaced through so great a horizontal distance where the slope is steep as where it is more gradual. In other words, the water retreats through a shorter horizontal distance in falling from one level to the next along the steep slope than along the gentle slope. Therefore, when these successive shorelines are projected upon the flat surface of a map, they will be crowded where the slope is steep and farther apart where it is moderate. In order to facilitate reading the contours on a map, certain contours (usually every fifth line) are drawn with a wider line. Contours are numbered at convenient intervals for ready identification. The numbers are always multiples of the contour interval and are placed between broken ends of the contour they designate.

Because the contours that represent a depression without an outlet resemble those of an isolated hill, it is necessary to give them a distinctive appearance. Depression contours therefore are *hatched;* that is, they are marked on the downslope side with short transverse lines called *hachures.* An example is shown on one contour in Fig. D-4. The contour interval employed is the same as in other contours on the same map.

Idealized example. Figures D-3 and D-4 show the relation between the surface of the land and the contour map representing it. Figure D-3, a perspective sketch, shows a stream valley between two hills, viewed from the south. In the foreground is the sea, with a bay sheltered by a curving spit. Terraces in which small streams have excavated gullies border the valley. The hill on the east has a rounded summit and sloping spurs. Some of the spurs are truncated at their lower ends by a wave-cut cliff, at the base of which is a beach. The hill on the west stands abruptly above the valley with a steep scarp and slopes gently westward, trenched by a few shallow gullies.

Each of the features on the map (Fig. D-4) is represented by contours directly beneath its position in the sketch.

AIR PHOTOGRAPHS

Plate 14 and Figs. 10-7 and 14-17 are photographs made from airplanes, with cameras pointing obliquely at the Earth's surface. Oblique air photographs enable us to see over a much larger area than could be seen from a single point on the ground and are therefore useful in making clear the broad rather than the detailed relations of various geologic features. The figures referred to above were selected for this very purpose. All oblique photographs are "pictorial" in that they show hills and valleys in perspective.

In contrast, Figs. 11-18 and 19-8 are photographs made with cameras pointing down vertically at the Earth's surface. Unlike oblique views, vertical photographs do not show perspective; they look "flat." But although distortion is always present (Fig. D-2) they show the pattern of the ground with distortion at a minimum. The vertical air photograph, therefore, is one kind of map. It is used widely as a base map on which geologic data are plotted in the field. Generally the scale of such a photograph must be 1:20,000 or larger, for it shows so much detail that on a smaller scale the various features can become blurred.

Used by itself, a vertical air photograph is a two-dimensional map. However, two photographs taken in

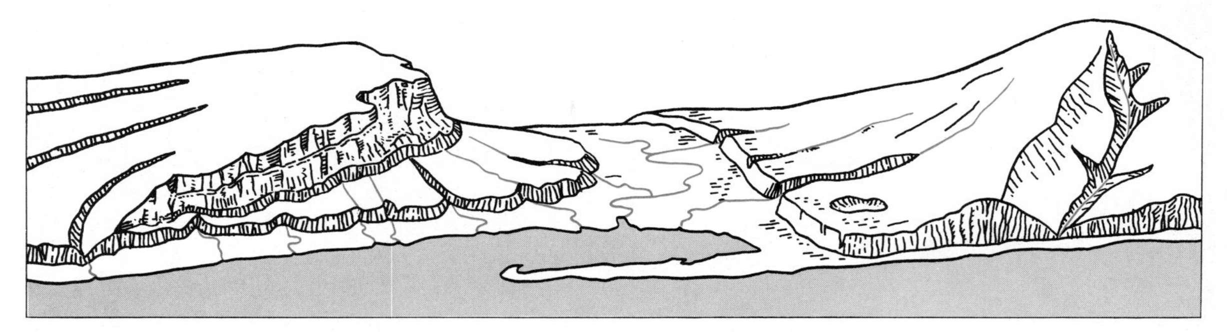

Fig. D-3. Perspective sketch of a landscape. (Modified from U. S. Geol. Survey.)

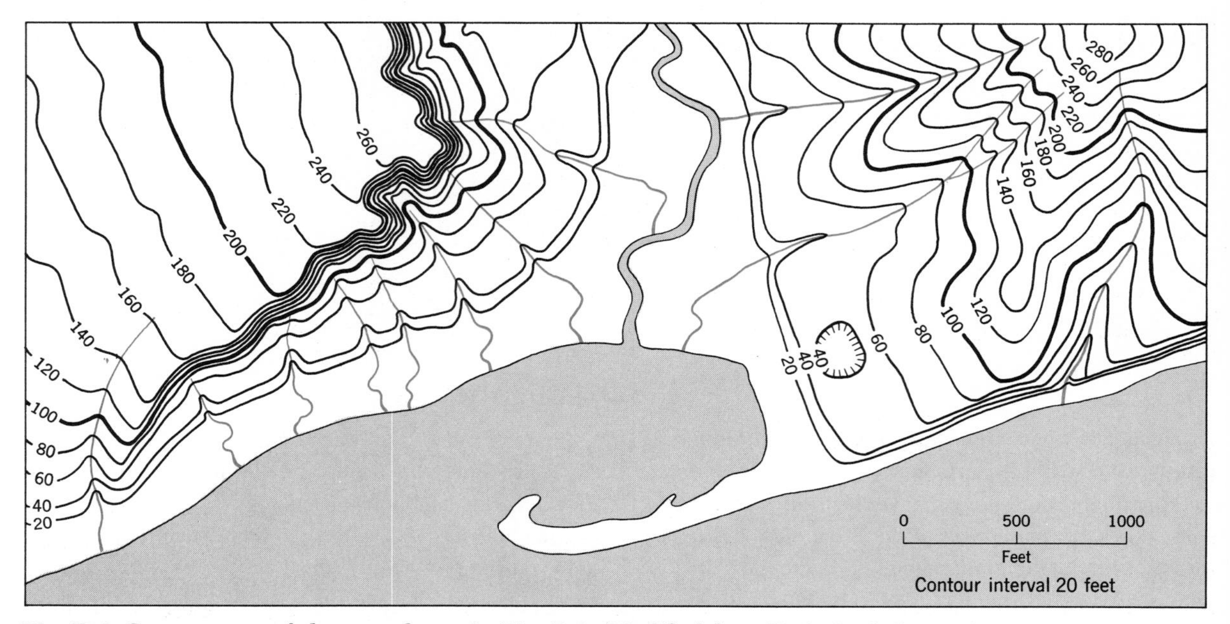

Fig. D-4. Contour map of the area shown in Fig. D-3. (Modified from U. S. Geol. Survey.)

sequence from a flying airplane, so that they overlap by a mile or two, can form a three-dimensional map with the vertical dimension exaggerated. The camera lens acts as a series of "eyes" with an interpupillary separation equal to the distance flown between successive pictures. This separation is so much greater than the distance between the two eyes of a human being that when overlapping photographs are viewed through a stereoscope, the relief of the ground surface becomes startlingly apparent. In fact we can see much more than we could by looking down from above at the same field of view. Not only is this a direct aid in plotting geologic data on the photograph, but also it forms the basis of a method of drawing an accurate contour map of the area photographed without the necessity of making a slow, laborious, and costly ground survey. Topographic contour maps are made largely by methods based on this principle. Finally, an individual air photograph can be used as a map, on which geologic features are then plotted.

FIELD MEASUREMENTS

The most common kind of ***geologic map*** is *a map that shows the distribution, at the surface, of rocks of various kinds or of various ages.* Examples are Fig. 9-31, a sketch map of bodies of alluvium of successive ages, Fig. 17-9, *B*, a sketch of eroded and folded strata, and Fig. 19-12,

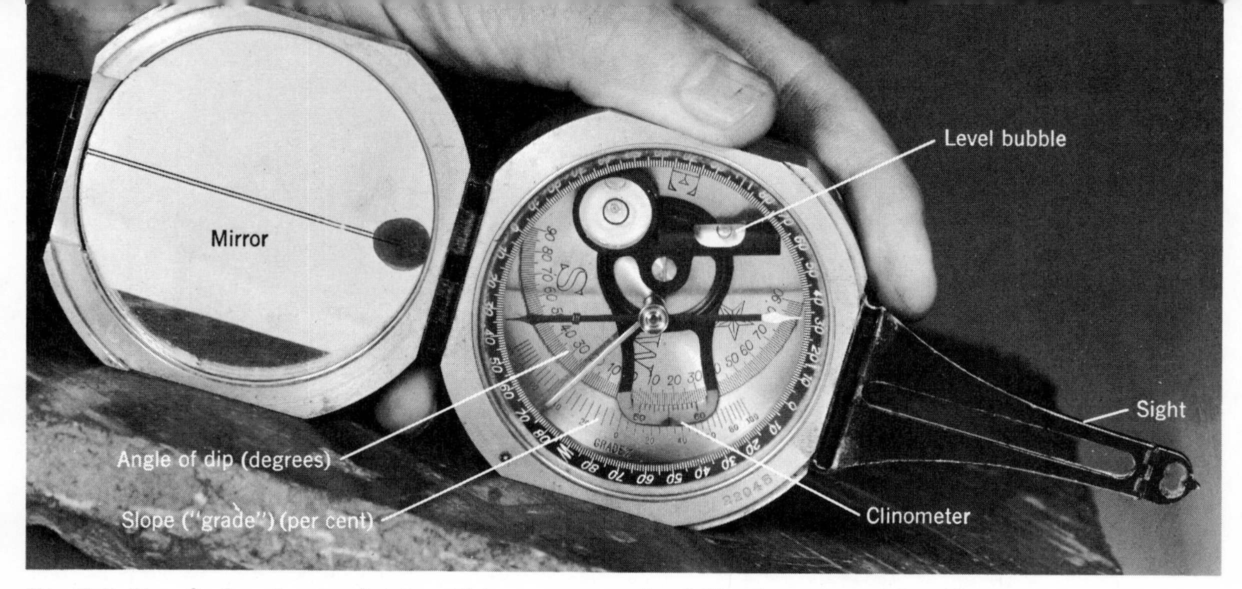

Fig. D-5. Use of a brunton (pocket transit) to measure angle of dip of a stratum. The sides of the brunton are plane, parallel surfaces. With one side placed on a sloping plane surface in the direction of dip, the level bubble of the clinometer is centered by means of a lever on the back of the case (not visible). The angle is read on one of the two arcs below. The inner semicircular arc, calibrated in degrees from 0° to 90°, shows a reading of 17.5°; the outer, shorter scale, calibrated in per cent, reads 31 per cent. Compass directions are read on the dark outer scale when the instrument is held face up and is leveled by use of the circular level bubble. The hinged mirror and sight aid in taking bearings on selected points, and in using the brunton as a hand level (Fig. D-8), with the clinometer set to read zero degrees.

sketch maps showing the distribution of basalt compared with that of other rocks.

Field equipment. A first essential for making a geologic map in the field is a base map, preferably a topographic map with a contour interval no greater than 10 or 20 feet. Other probable necessities are a hammer, a steel tape for measuring thicknesses of strata, and a pocket transit (geologists' compass) known almost universally in the United States and Canada as a Brunton compass, or simply a brunton, after the name of its designer. This compact instrument (Fig. D-5) is a compass, a clinometer, and a sighting device used in reading compass directions and in hand leveling.

Strike and dip. One of the commonest kinds of field measurement is determination of the attitude of a stratum. To represent the orientation of an inclined plane, we need to remember two principles of geometry: (1) the intersection of two planes defines a line; (2) in an inclined plane only one horizontal direction exists. The horizontal line formed by the intersection of the inclined plane with the horizontal plane can be visualized as the water line on a boat-launching ramp or by placing the edge of a carpenter's level in a horizontal position on a sloping plane (Fig. D-6). Instead of the level we can place one edge of a brunton, in the level position, against the inclined surface of a stratum.

The compass direction of the horizontal line in an inclined plane is the ***strike*** of the plane. In the United States and Canada, strike and other compass directions commonly are expressed as angles between 0° and 90° east or west of true north. A strike trending 20° east of north would be written N2OE (Fig. D-7, *B*); one trending 72° west of north would be written N72W. In Europe and elsewhere strikes are measured as angles clockwise from true north (0°), through a full circle of 360°.

Once we know the strike we need only one more measurement to fix the orientation of the plane. That is the ***dip***, *the angle in degrees between a horizontal plane and the inclined plane, measured down from horizontal in a plane perpendicular to the strike.* Dip is measured with a *clinometer,* usually a brunton in the position seen in Fig. D-5. Not only the angle but the direction of dip (always at right angles to the strike) must be noted.

When plotted on a map as symbols (Fig. D-7, *A*), orientation measurements made at each locality graphically convey the significant structural features of an area (Fig. D-7, *C*). In a similar manner dip directions of cross-strata can be plotted on a map to indicate directions

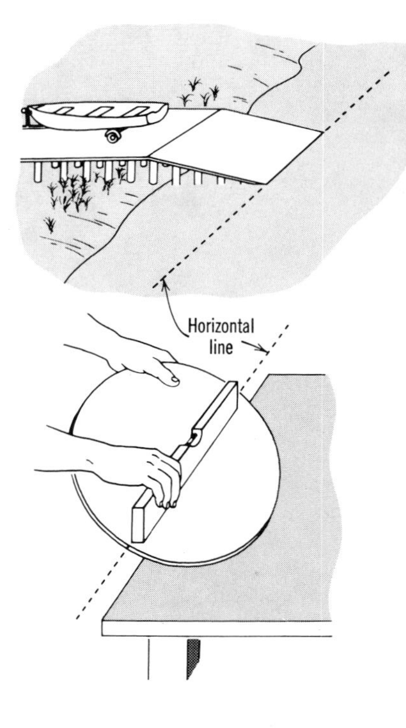

Fig. D-6. **The one horizontal direction in an inclined plane is illustrated by the water line against a boat-launching ramp, and by a carpenter's level held against an inclined board. The two horizontal lines shown are the directions of *strike* of the two inclined planes (see text).**

of flow of ancient currents (Fig. 16-14).

Thicknesses of strata. The perpendicular distance between the upper and lower surfaces of a stratum, the *thickness* of the stratum, can be measured directly with a steel tape or other scale (Pl. 16) or with a hand level (Fig. D-8, *A*) or altimeter. Thickness of an inclined stratum is usually calculated from simple trigonometric data (Fig. D-8, *B*).

GEOLOGIC MAPS

Patterns made by strata. *Horizontal strata.* Figure D-9, *A* is a geologic sketch map showing three units: two shale units with a unit of limestone between them. Each is a *formation* as defined in Chapter 16. The sketch map has been constructed on a topographic base representing two rounded hills with a stream between them. The two black lines represent the traces, along the land surface, of the interface of contact between the base of the limestone and underlying shale and of that between the top of the limestone and overlying shale. Symbols (Fig. D-10, *A*) at five places indicate that the strata exposed at those places were found to be horizontal; that is, their dip is zero.

The geologist who drew the black lines, or "contacts" determined the altitude of the lower one by hand level from a known point farther down the slope and then walked around each hill, following by eye, and then plotting on his map, the change in type of rock from shale below to limestone above. His circuit of each hill brought

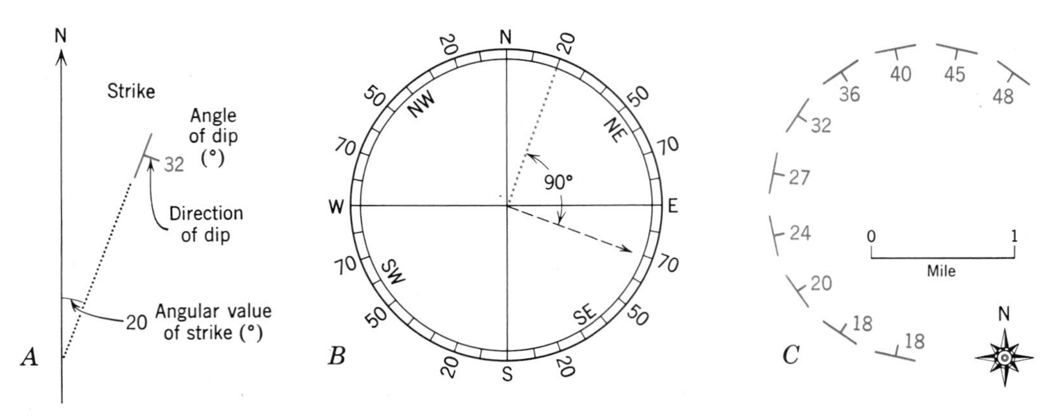

Fig. D-7. **A. Strike and dip are plotted on maps with the T-shaped symbol shown in blue. *B.* Strike and direction of dip shown in A are indicated by blue lines on the face of a compass. Strike is N20E, direction of dip S70E. Angle of dip, being a vertical angle, can not be shown. *C.* Strike and dip symbols, from measurements at several localities on the same stratum, plotted on a geologic map. Evidently the structure is a syncline plunging southeast, the northeast limb dipping more steeply than the southwest (Fig. 17-9).**

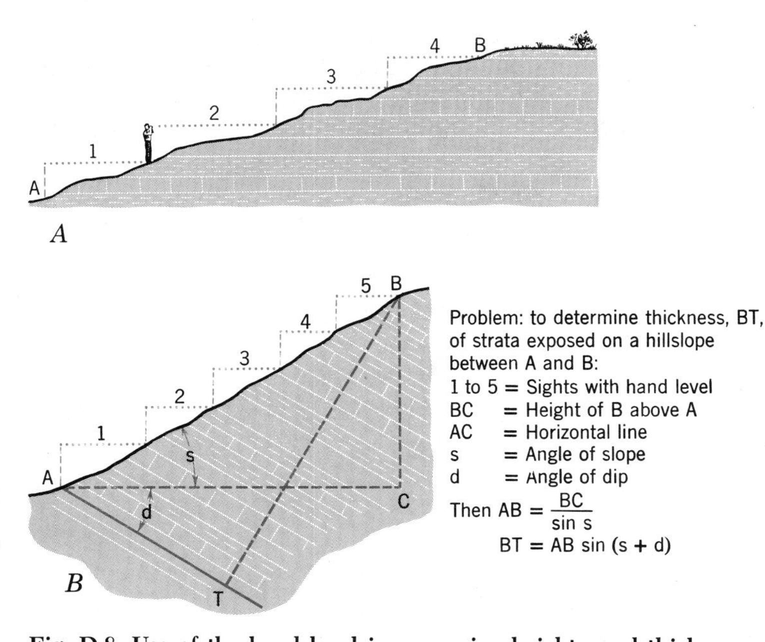

Fig. D-8. Use of the hand level in measuring heights, and thicknesses of strata exposed in slopes. *A.* Thicknesses of horizontal strata are measured directly by hand level. The geologist determines accurately the height of his eyes above the ground, and multiplies this figure by the number of "sights" (1, 2, 3, etc.) he makes. If, instead of a hand level, he uses an altimeter, he computes differences between readings at critical points. *B.* Where strata are inclined, vertical distance between base and top is measured, up the slope, by hand level. Angle of dip is measured with a clinometer. Thickness is determined by the trigonometric computation given above.

him back to his starting point, and the contact he had plotted maintained a constant altitude around both hills. The line representing it therefore extends between the same two contours around both hills and is parallel with the contours.

The geologist then repeated the process with the higher contact, thereby completing the map. By hand leveling (Fig. D-8, *A*) he could measure the vertical distance between base and top of the limestone and thereby find that the stratum is 14 feet thick. For the thicknesses of each of the two layers of shale, however, he could measure only minimum values, because neither the base of one of them nor the top of the other one is exposed within the area of the map.

The area, on a geologic map, shown as occupied by a particular rock unit is the ***outcrop area*** of that unit. An outcrop area, therefore, is the area of the Earth's surface in which some particular rock unit constitutes the highest part of the underlying rock, whether exposed at the surface or covered by regolith derived from it. Generally the aggregate area of actual exposure, free from overlying regolith, is far less than the outcrop area.

Vertical strata. The lines representing contacts between strata whose dip is vertical appear as straight lines on a geologic map (Fig. D-9, *B*). They change direction only where the strike of an interface of contact changes, or where the contact is offset by a fault.

Inclined strata. On a geologic map, the lines of contact between strata that are inclined cross the contours in directions that vary with angle of dip of the strata and with slope of the surface as represented by trend and spacing of contours. In Fig. D-9, *C* we note that the lines representing the top and base of the limestone layer swing west through broad arcs in each of the two hills and bend east in a chevronlike pattern as they cross the valley. Measurements of strike and dip at five localities show that all three strata are dipping east. If the dip were west instead of east, the pattern would be reversed, with arcs swinging east and the chevron pointing west.

Faults. On a geologic map faults (which are, of course,

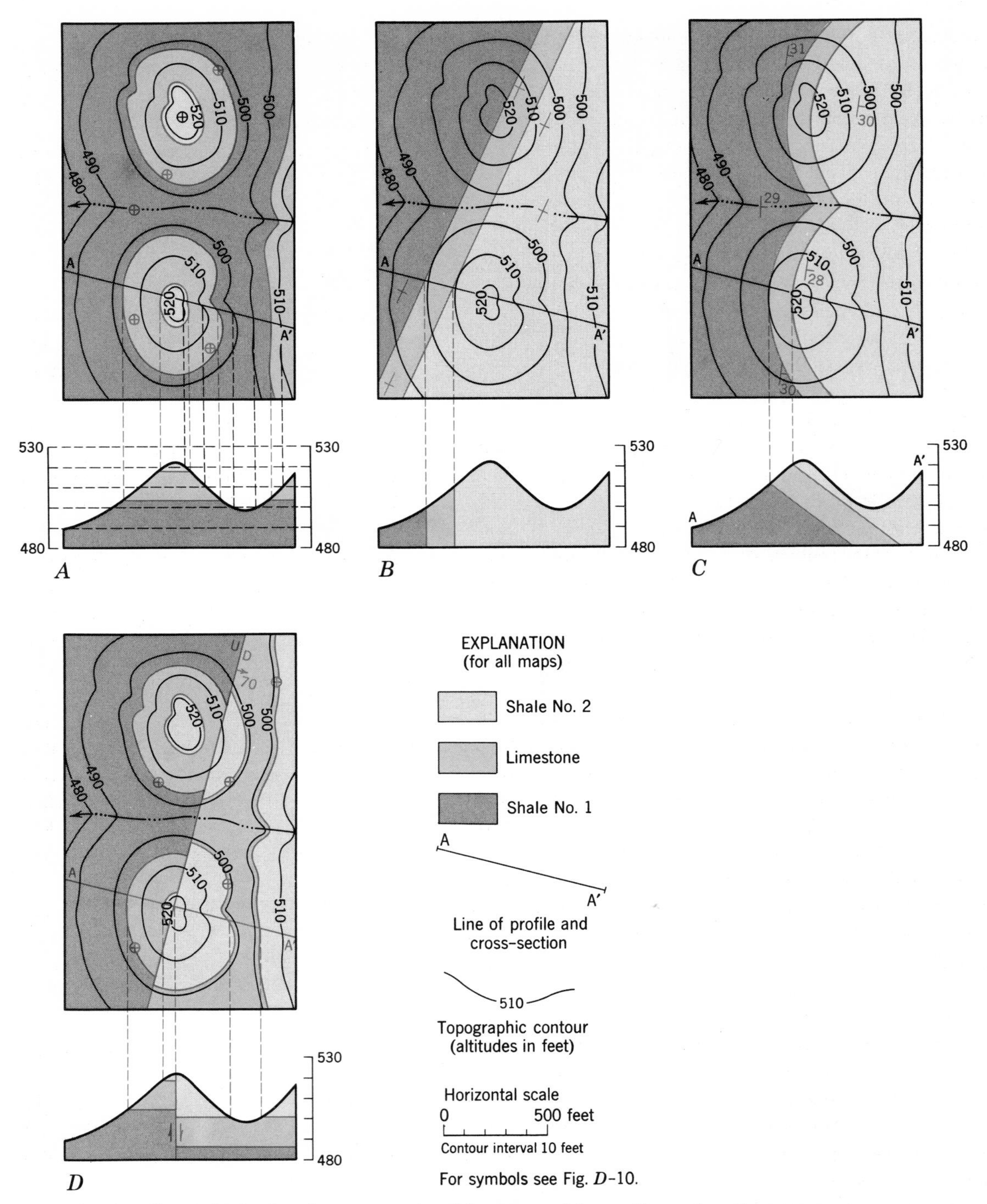

Fig. D-9. Simple geologic sketch maps, each with topographic profile and geologic cross-section below it, showing relation between pattern on map and geometry of section in four situations: *A.* Strata horizontal. *B.* Strata vertical. *C.* Strata dipping east at 30° angle. *D.* Horizontal strata cut by a fault.

one kind of surface of contact) are represented by lines thicker than those used for all other kinds of contact. Where faults displace rock units they also displace other contacts between the units; these appear on a map as offsets (Fig. D-9, *D*). The details of faults and other structures are brought out on geologic maps by means of special symbols (Fig. D-10, *A*).

Geologic cross-sections. Once a geologic map has been constructed, it can be used as the basis for constructing cross-sections along planes that extend downward below the surface, at right angles to the plane of the map. A ***geologic cross-section*** is *a diagram showing the arrangement of rocks in a vertical plane.* It represents what would be revealed, as outcrops, in the vertical wall of a deep trench. In a natural trench such as the Grand Canyon we could construct a geologic cross-section by measuring the exposed rocks directly, with an accuracy as great as would be possible in making a geologic map of a comparable area. Most sections, however, are made by geometric projection of the contacts already drawn on a map. The accuracy of a section constructed in this way varies with the kinds of rock bodies present. Where strata of originally wide extent have been deformed, surfaces of contact are generally parallel and can be projected downward with fair accuracy. But where rock bodies having irregular surfaces of contact are present and where exposures of the rocks are few, sections are less accurate and some are little more than speculative. Sections of this kind can be checked and corrected only by comparing them with subsurface data obtained from the records of drilling or geophysical exploration.

The method of constructing geologic cross-sections is illustrated in Fig. D-9. A distinctive unit of limestone, shown in four geologic sketch maps, is horizontal in *A*, vertical in *B*, dipping at 30° in *C*, and cut by a fault in *D*. What would be the geometry of this unit as seen in a vertical plane?

Below the map we construct a grid of horizontal lines, spaced to represent the contours. Dashed vertical lines are drawn from the intersection of line AA' with each contour on the map to corresponding horizontal lines on the grid. By connecting the points of intersection, we construct an east-west profile of the hill, a topographic profile. Next we draw lines from borders of the outcropping unit to intersect the profile on the grid. By connecting points on the profile that represent, respectively, the top and bottom of the unit, the structure section of the layer is completed.

Because the topography on all four base maps is the same, construction lines for making the topographic profile are shown for *A* only. On the identical profiles for *B*, *C*, and *D*, the lines dropped from lower and upper boundaries of the outcropping limestone unit give the points required for completing the sections.

Patterns commonly used in geologic cross-sections to represent rocks of various kinds are illustrated in Fig. D-10, *B*.

MAPS OF IGNEOUS AND METAMORPHIC ROCKS

Because sedimentary rocks are commonly stratified, with rather distinct upper and lower surfaces of contact, and because many of them are widely extensive, they present fewer mapping problems than do most igneous and metamorphic rocks. An exception is volcanic rocks that occur in widespread sheets like sedimentary strata (Fig. 19-13) and that may be interbedded with layers of tephra. In some districts it is possible to map these units separately just as sedimentary formations are mapped. But in many volcanic fields, bodies of extrusive igneous rock and bodies of tephra, erupted from a number of centers at irregular intervals, are mixed in great confusion. Such complexes can be mapped only as general assemblages, without regard to the various kinds of volcanic rock that compose them.

Intrusive igneous bodies exposed at the surface are highly varied in form and in their relation to associated rocks. Commonly they cut across older sedimentary units and their outcrop areas range widely in shape and in size. Those large enough to be shown clearly to the scale of the map are outlined and marked with distinctive colors or patterns.

Most metamorphic rocks have complex structure and are not easily divisible into distinctive units. Many such rocks therefore are represented on maps by a single color or pattern. In many mountain zones, metamorphic rocks intermixed with bodies of granite or other intrusive rocks are treated as a unit and are identified on many maps as "basement complex."

In some areas, on the other hand, metamorphic rocks are mapped in great detail. Various kinds of schists and gneisses are mapped individually, as formations; attitudes of foliation are recorded at many points by symbols (Fig. D-10, *A*), and faults are mapped, many of them through long distances. In such areas the various intensities of metamorphism that have affected the rocks are identified through critical minerals (Table 20-2). With such identification at a large number of points, it is possible to draw ***isograds,*** in this case *lines on a map, connecting points of first occurrence of a given mineral in metamorphic*

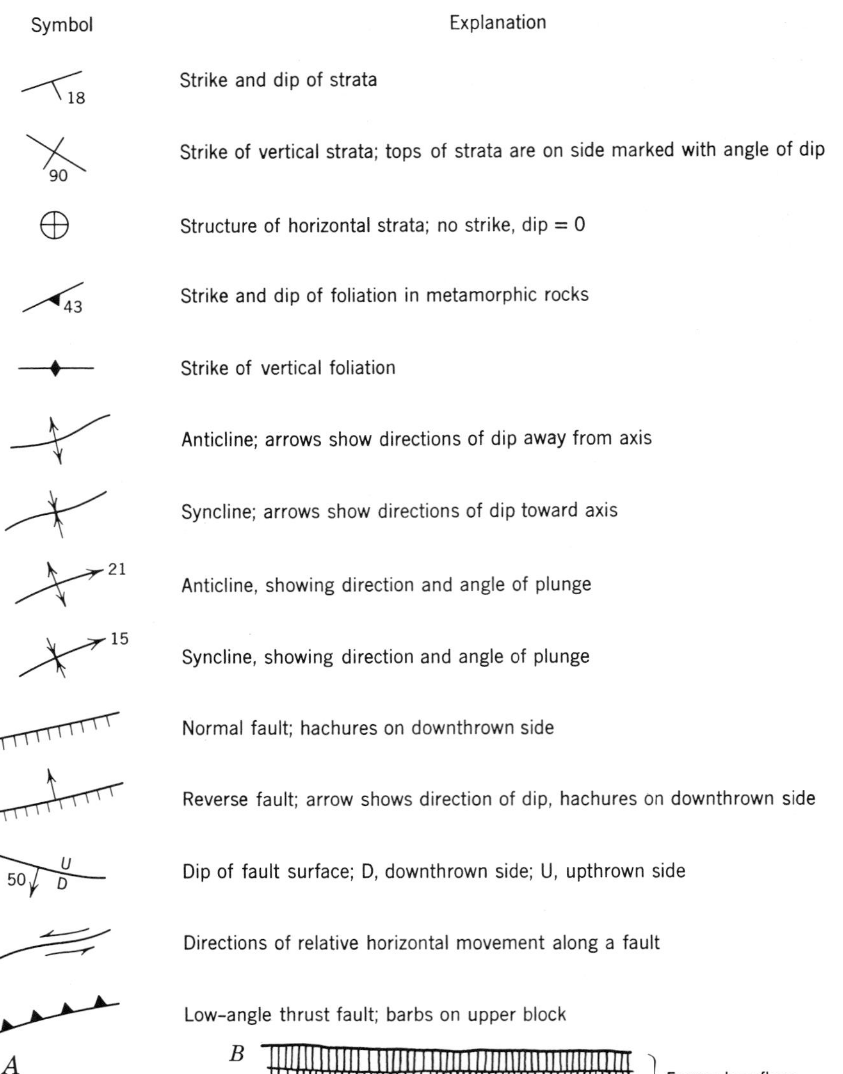

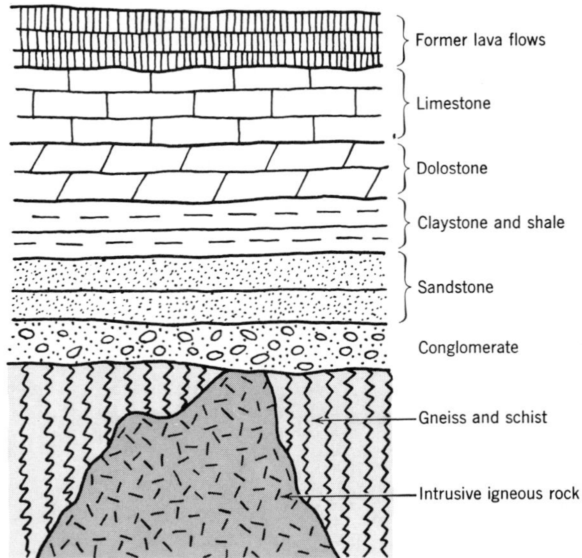

Fig. D-10. *A.* **Symbols commonly used to show structure on geologic maps.** *B.* **Representative patterns commonly, but not universally, used to show kinds of rock in geologic cross-sections.**

rocks. The trends and patterns of isograds give clues to the directions in which the former pressures were applied, and also suggest former temperature gradients.

ISOPACH MAPS

Somewhat analogous to isograds are ***isopachs,*** *lines on a map that connect points of equal thickness of a rock unit.* Isopachs are usually shown on special maps of limited areas where thickness is particularly significant. For example, an isopach map might show the thickness of a surface cover of loess (Fig. 13-28), tephra (Fig. 13-16), alluvium, or glacial drift. Again it might represent the thinning, in some direction, of a buried unit similar to the one appearing in cross-section in Fig. 23-7, *D.* An isopach map of that unit probably would be useful in a local search for petroleum.

Selected References

Base maps

Raisz, Erwin, 1948, General cartography, 2nd ed.: New York, McGraw-Hill.
Robinson, A. H., 1960, Elements of cartography, 2nd ed.: New York, John Wiley.
(See also Lobeck and Tellington, below, p. 1–198.)

Air photographs

American Society of Photogrammetry, 1952, Manual of photogrammetry, 2nd ed.: Washington, American Society of Photogrammetry.
American Society of Photogrammetry, 1960, Manual of photographic interpretation: Washington, American Society of Photogrammetry.
Lobeck, A. K., and Tellington, W. J., 1944, Military maps and air photographs: New York, McGraw-Hill, p. 199–250.
Miller, V. C., and Miller, C. F., 1961, Photogeology: New York, McGraw-Hill.

Geological maps and cross-sections

Blyth, F. G. H., 1965, Geological maps and their interpretation: London, Edward Arnold.

Field techniques

Compton, R. R., 1962, Manual of field geology: New York, John Wiley.

APPENDIX E

Metric-English Conversion Tables

UNITS OF LENGTH

Metric System			U.S.-British Equivalent (Approx.)			
Centimeters (cm)	Meters (m)	Kilometers (km)	Inches	Feet	Yards	Miles
1	10^{-2}	10^{-5}	0.3937	—	—	—
10^2	1	10^{-3}	39.37	3.2808	1.0936	—
10^5	10^3	1	—	—	1,093.60	0.6214

U.S.-British System				Metric Equivalent (Approx.)		
Inches	Feet	Yards	Miles	Centimeters (cm)	Meters (m)	Kilometers (km)
1	1/12	1/36	—	2.54	—	—
12	1	1/3	—	30.48	0.3048	—
36	3	1	—	91.44	0.9144	—
—	5,280	1,760	1	—	1,609.40	1.6094

Micrometric Scale (Fig. A-3)

1 micron $= 10^{-3}$ mm $= 10^{-4}$ cm $= 10^{-6}$ m $= 4 \times 10^{-5}$ inches

1 angstrom $= 10^{-8}$ cm $= 4 \times 10^{-9}$ inches

UNITS OF AREA

Metric System			U.S.-British Equivalent (Approx.)			
Square centimeters (cm^2)	Square meters (m^2)	Square kilometers (km^2)	Square inches	Square feet	Square yards	Square miles
1	10^{-4}	10^{-10}	0.1550	—	—	—
10^4	1	10^{-6}	1,550.0	10.7639	1.1960	—
10^{10}	10^6	1	—	—	—	0.3861

UNITS OF VOLUME

Metric System			U.S.-British Equivalent (Approx.)			
Cubic centimeters (cm^3)	Cubic meters (m^3)	Cubic kilometers (km^3)	Cubic inches	Cubic feet	Cubic yards	Cubic miles
1	10^{-6}	10^{-15}	0.0610	—	—	—
10^6	1	10^{-9}	—	35.3150	1.3080	—
10^{15}	10^9	1	—	—	—	0.2388

UNITS OF WEIGHT

Metric System			U.S.-British Equivalent (Approx.)	
Grams (g)	Kilograms (kg)	Metric tons (M.T.)	Ounces avoirdupois	Pounds avoirdupois
1	10^{-3}	10^{-6}	0.0353	—
10^3	1	10^{-3}	—	2.2046
10^6	10^3	1	—	2,204.62
U.S.-British System			**Metric Equivalent (Approx.)**	
Ounces	Pounds	Tons	Grams (g)	Kilograms (kg)
1	1/16	—	28.3495	—
16	1	—	453.5924	—
—	2,000	1	—	907.1849

Glossary[1]

Abrasion. The mechanical wear of rock on rock.

Abyssal fan. A fanlike accumulation of sediment at the mouth of a submarine canyon.

Abyssal floor. The low areas of ocean basins.

Abyssal-hills province. A part of the abyssal ocean floor consisting almost completely of irregular rocky hills a few hundred meters to a few kilometers wide and having relief of 50 to 100m.

Abyssal plain. A flat part of the abyssal floor, underlain by sediment, having an imperceptible slope of less than 1:1,000.

Accessory mineral. A minor constituent of a rock that does not enter into classifying and naming of the rock.

Active volcano. See ***Volcano: active.***

Aeration, zone of. See ***Zone of aeration.***

Agglomerate. A coarse-grained pyroclastic rock consisting largely of bombs and blocks.

Alluvial fan. See ***Fan.***

Alluvial fill. A body of alluvium, occupying a stream valley, and conspicuously thicker than the depth of the stream.

Alluvium. The general name for all sediment deposited in land environments by streams.

Alpha-particle. One of the kinds of particles involved in radioactivity; identical with the nucleus of an atom of helium, consisting of 2 protons and 2 neutrons. The mass number of an α-particle is 4 and its atomic number is 2.

American Cordillera. The entire broad mountain belt that extends continuously from Alaska to Cape Horn.

Amorphous solid. A solid in which the molecules, atoms, or ions are randomly arranged in an irregular network.

Amphibolite. A coarse-grained mafic metamorphic rock containing abundant amphibole.

Amygdale. A vesicle that has been filled by one or more minerals.

Amygdaloidal. An adjective describing a rock containing amygdales.

Anaerobic bacteria. Bacteria that require no free oxygen but that derive oxygen from organic matter or from sulfate radicals in solution.

[1] The definitions of rocks are intended only for the use of beginning students in naming hand specimens. Many rock terms are approximations that do not coincide with the more precise definitions of professional geologists based on data obtained with a petrographic microscope. Mineral names are not included; see Table B-2.

Andesite. A fine-grained, felsic igneous rock of intermediate composition. Porphyritic varieties contain phenocrysts of feldspar but none of quartz and have aphanitic groundmass. Andesite is a fine-grained equivalent of diorite.

Andesite line. The boundary, in the Pacific Ocean, between volcanoes on the oceanic side discharging only oceanic basalts (q. v.) and those on the continental side discharging both andesite and circumoceanic basalts (q. v.).

Angle of repose. Also known as critical slope. As used in mass-wasting, it describes the steepest angle, measured from the horizontal, at which a material remains stable. The minimum angle along which coarse particles in an aggregate of particles begin to fall under the influence of gravity.

Angular unconformity. See ***Unconformity: angular.***

Anorthosite. An igneous rock composed almost entirely of plagioclase.

Anticline. An upfold of layered rocks in the form of an arch and having the oldest strata in the center. The reverse of a syncline.

Aphanitic texture. A texture of equigranular igneous rocks with individual particles not visible to the unaided eye.

Aquifer. A body of permeable rock or regolith through which ground water moves.

Arête. A jagged, knife-edge ridge created where two groups of glaciers have eaten into the ridge from both sides.

Arkose. A sandstone containing at least 25 per cent feldspar as well as quartz.

Artesian spring. See ***Spring: artesian.***

Artesian well. A well in which water rises above the aquifer.

Ash (volcanic). Tephra ranging in size between $\frac{1}{16}$ and 4mm.

Atmosphere. The gaseous envelope that surrounds the Earth.

Atoll. A reef that forms a nearly closed figure within which there is no land mass.

Atom. The smallest electrically balanced particle displaying the properties of an element.

Atomic mass. The mass of an atom, nearly all of which is contained in the nucleus. The mass of an atomic nucleus is expressed by the *mass number* (q. v.).

Atomic number. The number of protons in an atomic nucleus, represented by the letter Z. Also the number of electrons in a neutral atom.

Attitude. The orientation, or position, of a layer or of a surface. Measured by determining strike and dip.

Authigenic sediment. A sedimentary deposit formed in place, not from physically transported materials, and consisting of minerals that crystallized out of seawater. An ***authigenic mineral*** can be precipitated from interstitial water, not necessarily seawater, in a sediment.

Axial plane. An imaginary plane through the middle of the fold that passes through the axis of the fold.

Backwash. The return sheet flow, down a sloping shore, of water from the spent swash.

Badlands. A system of closely spaced narrow ravines with little or no vegetation.

Bar. An elongate ridge of sediment, built offshore by waves and currents, whose top is always submerged.

Barbed tributary. See ***Tributary: barbed.***

Barchan. See ***Dune: barchan.***

Barrier. An elongate island of sand or gravel parallel with the coast.

Barrier reef. An elongate reef not connected with a land mass.

Basalt. An aphanitic igneous rock having the composition of gabbro and dolerite. Specimens of basalt are opaque even on thin edges.

Basalt: circumoceanic. Basalt, discharged from volcanoes marginal to an ocean basin, which contains less than 1.75 per cent TiO_2 and generally more than 15 per cent Al_2O_3.

Basalt glass. Mafic glass, opaque along thin edges.

Basalt: oceanic. Basalt, discharged by volcanoes within ocean basins (or within continental masses), containing more than 1.75 per cent TiO_2 and generally less than 15 per cent Al_2O_3.

Base level. The limiting level below which a stream can not erode the land.

Base level: local. The level of a lake or any other base level that stands above sea level.

Base level: ultimate. Sea level, projected inland as an imaginary surface underneath a stream.

Basement complex. An assemblage of igneous and metamorphic rocks lying beneath the oldest stratified rocks of a region. Many, but not all basement rocks are of Precambrian age.

Batholith. A large pluton having an exposed area of more than 40 square miles ($100km^2$).

Bauxite. A mixture of hydrated oxides, containing large volumes of hydrous aluminum oxide and widely used as an ore of metallic aluminum. It is generally expressed by the chemical formula $Al_2O_3 \cdot nH_2O$.

Bay barrier. A ridge of sand or gravel that completely blocks the mouth of a bay.

Beach. A body of wave-washed sediment extending along a coast between the landward limit of wave action and the outermost breakers.

Beach drift. The movement of particles obliquely up the slope of a beach by the swash and directly down this slope by the backwash.

Bed. A stratum (q. v.) 1cm or more thick. Also, the floor of a stream channel.

Bedding plane. The top or bottom surface of a bed, or stratum.

Bedding-plane parting. A surface of separation between adjacent strata.

Bed load. See ***Load: bed.***

Bedrock. Continuous solid rock that everywhere underlies regolith and locally forms the Earth's surface.

Beheaded stream. The remaining part of a stream course that has lost its upper part by stream capture (q. v.).

Beta-particle. One of the kinds of particles involved in radioactivity, consisting of either an electron or a positron (q. v.).

Biologic material. A general term for material created by organisms.

Biosphere. A collective term for the somewhat spherical habitat of the Earth's living organisms.

Bituminous coal. Black and firm soft coal that breaks into blocks and contains alternating layers having dull and bright luster.

Blowout. A deflation basin excavated in shifting sand or other easily eroded regolith.

Body wave. A seismic wave that travels entirely beneath the Earth's surface.

Botryoidal. An adjective describing minerals having rounded forms resembling grapes closely bunched.

Bottomset bed. The gently sloping, fine, thin part of each layer in a delta.

Boulder size. Particles of sediment having diameters greater than 256mm (approximately the size of a volleyball).

Boulder train. A group of erratics spread out fanwise.

Bowen's reaction series. See ***Reaction relationship.***

Braided stream. See ***Stream: braided.***

Breaker. A wave that is collapsing.

Breaker zone. The zone where waves collapse.

Breccia. A general term for a rock of any origin containing angular particles.

Brittle solid. A solid that fractures readily.

Brown clay. Formerly called "red clay." Pelagic sediment containing less than 30 per cent skeletal remains of microorganisms.

Brownian movement. The movement, as a result of impacts from moving water molecules, of tiny solid particles (diameters less than 2 microns) suspended in water.

Bulk modulus. The stress required to cause a unit change of volume.

Calcareous. Containing calcium carbonate.

Calcareous tufa. A light spongy limestone precipitated by algae or bacteria or by the evaporation of water from springs and small streams.

Calcrete. See ***Caliche.***

Caldera. A volcanic crater enlarged to a diameter of several miles.

Caliche. A whitish accumulation of calcium carbonate developed in a soil profile. Also known as a *calcrete.*

Calorie. The amount of heat energy required to raise the temperature of 1g of water from 15° to 16°C.

Capillary. A small opening having a diameter less than that of a human hair.

Carbonate rocks. Sedimentary rocks consisting chiefly of carbonate minerals.

Cataclastic texture. A texture of rocks resulting from breakage and pulverization of mineral particles.

Cavern. A large, roofed-over cavity in any kind of rock.

Cavitation. The formation and collapse of bubbles in a turbulent liquid.

Cement. Materials, precipitated from solution, which bind together the framework particles of a sedimentary rock.

Cementation. The binding together of particles of framework and matrix of a sediment by precipitation of mineral cement in former pore spaces. Calcite, quartz, and iron oxides are common cements.

Chalk. Limestone that is weakly cohesive.

Chert. A sedimentary rock composed of silica which is either an original precipitate or a replacement product of calcium-carbonate minerals. Commonly the original textures of the carbonate minerals are preserved.

Chute cutoff. A new channel cut across a point bar, resulting in abandonment of part of a meander.

Cinder cone. See ***Tephra cone.***

Cinder, volcanic. Obsolete name for medium-grained ***tephra*** (q. v.).

Circumoceanic basalt. See ***Basalt: circumoceanic.***

Cirque. A steep-walled niche, shaped like a half bowl, in a mountain side, excavated mainly by ice plucking and frost action.

Cirque glacier. A very small glacier that occupies a cirque.

Clastic rock. A general term for a sedimentary rock having clastic texture.

Clastic sediment. See ***Detritus.***

Clastic texture. A texture of sediments and sedimentary rocks resulting from physical transport and deposition of broken particles of older rocks, of older sediments, and of organic skeletal remains.

Clay size. Dimension of sedimentary particles having diameters less than $\frac{1}{256}$mm (4 microns).

Claystone. A clastic rock consisting predominantly of clay-size particles.

Cleavage (as applied to minerals). The capacity of a mineral to break in preferred directions along surfaces parallel to lattice planes.

Cleavage, rock. See ***Rock cleavage.***

Clinometer. An instrument for measuring degree of inclination, or *dip* (q. v.).

Closed system. A reaction in which no material escapes from the scene of reaction.

Coal. A black sedimentary rock consisting chiefly of partly decomposed plant matter and containing less than 40 per cent inorganic matter.

Cobble size. Sediment particles having diameters greater than 64mm (about the size of a tennis ball) and less than 256mm (about the size of a volleyball).

Cohesion. An electrostatic force of attraction among fine particles.

Col. A gap or pass in a mountain crest at a place where the headwalls of two cirques intersect.

Colluvium. A body of sediment that has been deposited by any process of mass-wasting or by overland flow.

Column. (1) A stalactite connected with a stalagmite. (2) Sometimes used as an abbreviation for ***geologic column.***

Columnar joints. Joints that split rocks into long prisms, or columns. Columnar joints are a common feature in tabular bodies of igneous rock.

Compaction. The reduction in pore space within a body of fine-grained sediments in response to the weight of overlying material or to pressures within the Earth's crust.

Composite cone. A volcanic cone consisting partly of tephra and partly of igneous rock both extrusive and intrusive.

Composite dike. A dike composed of more than one generation of igneous rock.

Compressive stress. See ***Stress: compressive.***

Conchoidal fracture. Breakage resulting in smooth curved surfaces.

Concordant contact. A contact surface of a pluton that is parallel to the layers of the intruded rocks.

Concordant pluton. A pluton having concordant contacts (q. v.).

Concretion. A localized rock body having distinct boundaries, inclosed in sedimentary rock, and consisting of substances precipitated from solution, commonly around a nucleus.

Cone of depression. A conical depression in the water table immediately surrounding a well.

Confined aquifer. An aquifer in which flow of ground water in all directions, particularly upward, is prevented by impermeable material.

Confined percolation. See ***Percolation: confined.***

Conglomerate. A clastic sedimentary rock containing numerous rounded pebbles or larger particles.

Congruent melting. Melting without change of composition.

Connate water. Seawater that was trapped in a sedimentary deposit when the sediment lay on the sea floor.

Consequent stream. See ***Stream: consequent.***

Contact-metamorphic aureole. A zone of altered rocks, surrounding a pluton, and originating from the effects of the magma.

Contact metamorphism. Metamorphism (q. v.) in the vicinity of a pluton and resulting from the effects of the magma on its surrounding rocks.

Continent. A major land area that stands above sea level. Compare ***continental mass.***

Continental crust. See ***Crust: continental.***

Continental drift. A concept that continental masses, composed largely of sialic rock, have moved widely and differentially over and through the denser mafic rock that underlies continental blocks and ocean floors.

Continental glacier. See ***Glacier: continental.***

Continental mass. A major high-standing part of the lithosphere.

Continental rise. The gentle slope with gradient between 1:100 and 1:700 that lies seaward of the continental slope.

Continental shelf. A submerged marginal zone of a continental mass forming a shallow platform of variable width extending from the shoreline to the first prominent break in slope at a depth of 600m or less. On most shelves the depth of this break in slope is 200m or less.

Continental shield. An extensive area in which the Precambrian foundation rocks of a continental mass are exposed.

Continental slope. A relatively steep (3° to 6°) slope that lies seaward of a continental shelf.

Continuous reaction relationship. The exchange of materials, during cooling, between a silicate melt and a continuously growing crystal of one mineral species. An example is the exchange of Fe^{2+} and Mg^{2+} that takes place during the crystallization of olivine from a mafic magma.

Continuous seismic profiler. An acoustic device for making profiles of the thickness and internal structure of sediments beneath a body of water.

Contour. See ***Contour line.***

Contour current. A subsurface density current flowing parallel to the submarine slopes at the margin of an ocean basin.

Contour interval. The vertical distance between two successive contour lines.

Contour line. A line passing through points having the same altitude above sea level; often simply called a contour.

Convection current. The movement of material, within a closed system, as a result of thermal convection arising from the unequal distribution of heat.

Coquina. An aggregation of shells and large shell fragments, cemented with calcium carbonate.

Cordillera. One of the great mountain belts of the Earth.

Cordillera, American. See ***American Cordillera.***

Cordillera, North American. See ***North American Cordillera.***

Core. (1) A cylindrical sample of sediment of rock recovered from a drilled hole. (2) The central part of the Earth. See ***Earth's core.***

Coriolis force. The inertial force that must be added to motions where the frame of reference rotates. The Coriolis force acts to the right where the frame of reference rotates counterclockwise (as in the Northern Hemisphere) and to the left where the frame of reference rotates clockwise (as in the Southern Hemisphere).

Correlation. In stratigraphy, this term is used to mean determination of equivalence, in geologic age and position in the sequence, of strata in different areas.

Covalent bond. A kind of bond in which atoms are held together by sharing electrons.

Crater. See ***Volcanic crater.***

Creep. The imperceptibly slow downslope movement of regolith.

Crevasse. A deep crack in the upper surface of a glacier.

Critical temperature. The temperature at which the distinction between the liquid and the gaseous states of aggregation disappears.

Cross-strata. See ***Strata: cross-strata.***

Crust (Earth's). The outer part of the lithosphere.

Crust: continental. The Earth's crust beneath continents, 20 to 60km thick, consisting of an upper part having the same elastic properties as sialic rocks and a lower part having the same elastic properties as mafic rocks.

Crust: oceanic. The Earth's crust beneath ocean basins, averaging about 5km thick, and consisting of material having the same elastic properties as mafic rocks.

Crustal warping. Gentle bending of the crust upward or downward.

Cryptocrystalline. A texture of rocks in which the crystalline particles are so small that they can not be resolved with an ordinary microscope.

Crystal. A solid bounded by natural, regular plane surfaces formed by growth of a crystal lattice.

Crystal face. A smooth plane face of a crystal.

Crystal lattice. A systematic, regular, symmetrical network of particles within a crystal.

Crystalline particle. A mineral particle of any size lacking well-developed crystal faces. If crystal faces are present the term *crystal* is used whatever its size.

Crystalline solid. A solid in which the pattern of the particles (molecules, atoms, or ions) is repeated and symmetrical.

Crystalline texture. A texture resulting from simultaneous growth of associated crystalline particles.

Crystallization. The process of development of crystals, by condensation of materials in a gaseous state, by precipitation of materials in a solution, or by solidification of materials in a melt.

Curie point. The temperature above which a substance loses its magnetism.

Cycle of erosion. The sequence of forms, essentially valleys and hills, through which a landmass is thought to evolve from the time it begins to be eroded until it is reduced to near base level.

Cycle of volcanic eruption. A repetition of similar activities during an eruption of a volcano.

Darcy's law. The equation for the velocity of flow of ground water which states that in material of given permeability, velocity of flow increases as the slope of the ground water table increases.

Daughter isotope. In radioactivity, the isotope, created by decay of a parent isotope, that continually increases in amount with time.

Debris flow. The rapid downslope plastic flow of a mass of debris.

Debris flow, variety mudflow. A debris flow in which the consistency of the substance is that of mud.

Decay constant. The proportion of radioactive atoms of each isotope that decay in a unit of time; represented by the Greek letter, λ.

Declination (magnetic). The clockwise angle in the horizontal plane between true north and a magnetic line of force. Also the angle between the direction of true north and magnetic north.

Decomposition. The chemical alteration of rock materials.

Deep-focus earthquake. See ***Earthquake: deep-focus.***

Deep-water wave. A wave, on the surface of a body of water, beneath which particles of water move in circular orbits, within vertical planes, and are not influenced by the bottom.

Deflation. The picking up and removal of loose rock particles by the wind.

Deflation armor. A surface layer of coarse particles concentrated chiefly by deflation.

Deflation basin. A depression excavated by deflation.

Deformation. A change of volume, of shape, or of both volume and shape of a rock body, or a change of its original position within the Earth's crust.

Delta. A body of sediment deposited by a stream flowing into standing water of a lake or the sea. The name comes from the similarity of the plan view to the shape of the Greek letter Δ.

Dendritic stream pattern. See ***Stream pattern: dendritic.***

Density. Weight per unit volume.

Density current. A localized current flowing because it consists of fluid denser than that of the body of the fluid through which it moves. The excess density may result from low temperature, high salinity, sediment held in suspension, or from various combinations of these factors.

Desert. Arid land.

Detrital sedimentary rock. A sedimentary rock composed of lithified detritus (q. v.).

Detritus. A collective term referring to broken pieces of older rocks, of minerals, or of skeletal remains of organisms. Also known as ***clastic sediment.***

Diastrophism. The processes of large-scale deformation, metamorphism, and intrusion that occur in orogenic belts.

Differential weathering. The result of variations in the rate of weathering on different parts of a rock body. As a result of differential weathering, resistant parts stand in relief whereas less-resistant parts form recesses.

Diffusion. The transport of particles in the absence of bulk flow.

Dike. A tabular pluton having discordant surfaces of contact.

Dike, composite. See ***Composite dike.***

Dike swarm. A group of associated dikes.

Diorite. A coarse-grained felsic igneous rock that lacks quartz, and in which the chief feldspar is plagioclase with ferromagnesian minerals (chiefly hornblende and biotite) constituting less than 50 per cent.

Dip. The angle in degrees between a horizontal plane and an inclined plane, measured down from horizontal in a plane perpendicular to the strike. Dip is measured with a *clinometer.*

Dipole. An object that contains opposite charges at two points or a magnetic field consisting of both north-seeking and south-seeking components.

Dip-slip fault. A normal fault or a reverse fault on which the only component of movement lies in the vertical plane normal to the strike of the fault.

Discharge. The quantity of water passing a given point in a given unit of time.

Disconformity. A lack of continuity between two groups of parallel strata in contact but separated by an irregular surface of erosion corresponding to a gap in the geologic record.

Discontinuous reaction relationship. The exchange of materials, during cooling, between a silicate melt and crystals, which results in the dissolution of one mineral species and the simultaneous growth of a different mineral species. An example is the dissolution of olivine and the growth of pyroxene that occurs during the cooling of a mafic magma.

Discordant contact. A contact surface of a pluton that is not parallel with the layers or other boundaries within the intruded rocks.

Discordant pluton. A pluton having discordant contacts (q. v.).

Disintegration. The mechanical breakup of rocks.

Disintegration, granular. See ***Granular disintegration.***

Divide. The line that separates adjacent drainage basins.

Dolerite. Medium-grained mafic igneous rock.

Dolostone. A sedimentary rock consisting chiefly of the mineral dolomite.

Doppler shift. Changes in frequency of electromagnetic waves resulting from relative motion between source and receiver.

Dormant volcano. See ***Volcano: dormant.***

Drainage basin. The total area that contributes water to a stream.

Drainage: interior. Drainage that does not persist to the sea.

Drift, continental. See ***Continental drift.***

Drift, glacial. See ***Glacial drift.***

Drift, stratified. See ***Glacial drift: ice-contact stratified.***

Dripstone. Calcite chemically precipitated from dripping water in an air-filled cavity.

Drumlin. A streamline hill consisting of drift, generally till, and elongated parallel with the direction of glacier movement.

Dune. A mound or ridge of sand deposited by the wind.

Dune: barchan. A crescent-shaped dune with horns pointing downwind.

Dune: longitudinal. A long, straight, ridge-shaped dune parallel with wind direction.

Dune: transverse. A dune forming a wavelike ridge transverse to wind direction.

Dune: U-shaped. A dune of U-shape with the open end of the U facing upwind.

Dunite. An ultramafic igneous rock consisting almost wholly of olivine (pronounced *dŭn'ite*).

Dust (volcanic). Tephra smaller than $\frac{1}{16}$mm.

Dyne. The force required to accelerate a mass of 1g by 1cm/sec^2.

Earth's core. The inner spherical mass below a depth of 2,900km, consisting of a fluid outer part, approximately 6,920km thick, and a solid inner part, about 2,720km in diameter.

Earth's mantle. The zone, about 2,900km thick, between the crust and the central core of the Earth. The mantle occupies about 80 per cent of the total volume of the Earth.

Earthquake: deep-focus. An earthquake with focus deeper than 300km.

Earthquake focus. The locus of first release of the elastic energy of an earthquake.

Earthquake intensity. A relative measure of the strength of an earthquake based on observed destruction or disturbance and on human sensations.

Earthquake: intermediate-focus. An earthquake with depth of focus between 70 and 300km.

Earthquake: shallow-focus. An earthquake with focus at depths less than 70km.

Echo sounder. An instrument that employs sound energy to determine depth of water.

Ecliptic. The apparent path of the Sun through space.

Ecliptic, plane of. See ***Plane of the ecliptic.***

Economy. The input and consumption of energy within a stream or other system and the changes that result.

Eh. The oxidation-reduction potential; measured in volts or millivolts.

Elastic deformation. Nonpermanent deformation of a body in which the stresses do not exceed its elastic limit (q. v.).

Elastic limit. The upper limit of strength of a body; the stress to which it can be subjected and still recover its original shape or volume when the forces tending to cause deformation have been removed.

Elastic rebound. The springing back of a deformed solid to its original shape as a result of rupturing under continued deforming forces or by elastic movement when the deforming forces have been removed. Elastic rebound of a body of deformed rock as a result of rupture under continued deforming forces releases great amounts of energy into the Earth's crust, creating earthquakes.

Electron. Unit particle of negative electrical charge.

Electron volt. The energy possessed by one electron which has fallen through a potential difference of 1 volt.

Emergence (of land). A fall of sea level relative to the land.

Energy. Loosely defined as (1) the capacity to do work; or more precisely as (2) the equivalent of mass according to the Einstein equation, $E = mc^2$.

Epeirogeny. Broad movements of uplift and subsidence affecting large portions of continental areas or of ocean floors.

Epicenter. The part of the Earth's surface vertically above an earthquake focus.

Equigranular texture. A texture of igneous rocks with all particles about the same size.

Erg. The mechanical energy required for a force of 1 dyne to act through 1cm.

Erosion. A general term that describes the physical breaking down, chemical solution, and movement of broken-down and dissolved rock materials from place to place on the Earth's surface.

Erosion: sheet. The erosion performed by overland flow (q. v.).

Erratic. A transported rock fragment different from the bedrock beneath it. The agent of transport was commonly glacier ice or floating ice.

Escarpment. A steep slope or cliff.

Esker. A body of ice-contact stratified drift shaped into a long narrow ridge, commonly sinuous.

Essential mineral. A major constituent of a rock used in classifying and naming the rock.

Eustatic change of sea level. A worldwide change in the level of the sea resulting from a change in the amount of water within or in the capacity of the ocean basins.

Evaporite. A nonclastic sedimentary rock whose constituent minerals were precipitated from water solution as a result of evaporation.

Evaporite mineral. A mineral precipitated as a result of evaporation. A few examples are aragonite, dolomite, gypsum, halite, and anhydrite.

Exfoliation. The separation, during weathering, of successive shells from massive rocks. The resulting sheets of rock resemble the "skins" of an onion.

Explosion pit. A vent, drilled to the surface by volcanic gases, from which no lava issued.

Exposure. A body of bedrock not covered by regolith and forming part of the Earth's surface.

Extrusive igneous rock. A rock that originated from solidification of lava.

Fabric. The orientation of particles in a rock or a sediment.

Fabric: preferred orientation. A fabric in which the particles are aligned.

Fabric: random. A fabric in which the particles are oriented in random directions.

Facies. A distinctive group of characteristics within a rock unit, that differ as a group from those elsewhere in the same unit.

Facies, metamorphic. See ***Metamorphic facies.***

Facies, mineral. See ***Mineral facies.***

Fan. A fan-shaped body of alluvium built at the base of a steep slope. Known also as *alluvial fan.*

Fan, abyssal. See ***Abyssal fan.***

Fault. A fracture along which the opposite sides have been relatively displaced.

Fault-block mountain. A mountain bounded by one or more faults.

Fault breccia. A breccia (q. v.) consisting of irregular pieces of rock broken as a result of faulting.

Fault drag. The bending of layers next to a fault as a result of fault movement.

Fault: footwall. The wall on (or boundary of) the block below an inclined fault.

Fault: hanging wall. The wall on (or boundary of) the block above an inclined fault.

Fault: thrust fault (or simply ***thrust***). A low-angle reverse fault, with dip generally less than 45°.

Faunal succession, law of. This law, discovered by William Smith, states that fossil faunas and floras succeed one another in a definite, recognizable order.

Felsic. An adjective used to describe a rock in which light-colored minerals, chiefly feldspars, predominate.

Felsite. An aphanitic igneous rock of granitic composition, whose thin edges transmit some light.

Ferromagnesian silicate mineral. A silicate mineral containing abundant iron and magnesium.

Fiord. Also *fjord*. A glaciated trough partly submerged by the sea.

Fissile. An adjective describing rocks that split along closely spaced parting planes. In sedimentary rocks such planes parallel the stratification.

Fissure. A fracture in rocks along which the opposite walls have been pulled apart.

Fissure eruption. Extrusion of volcanic materials along an extensive fracture.

Fissure vein. A fissure in rocks, filled with mineral matter.

Flint. A popular name for dark-colored chert (q. v.).

Floodplain. That part of any stream valley which is inundated during floods.

Flow breccia. A volcanic breccia created by the breaking up of a hardened crust of extrusive rock as a result of further flow of liquid lava.

Flow: laminar. Flow in which the fluid particles move in straight, parallel paths and slip over one another along parallel plane surfaces. Also known as streamline flow.

Flow law. A statement of the quantitative relationship between rate of strain of a body of flowing ice and the shearing stress within it.

Flow layering. Layers resulting from flow of magma or lava.

Flow: overland. The movement of runoff in broad sheets or groups of small interconnecting rills.

Flow: plastic. See ***Plastic flow.***

Flowstone. Material chemically precipitated from flowing water in the open air or in an air-filled cavity.

Flow: stream. The flow of surface water between well-defined banks.

Flow: turbulent. Fluid flow characterized by eddies.

Fluid. A state of aggregation of liquid or gaseous matter in which shear waves do not propagate.

Fluidized. An adjective describing a body of solid particles in a dilatant state; that is, the particles are no longer in continuous contact with one another.

Focus, earthquake. See ***Earthquake focus.***

Fold. A pronounced bend in layers of rock.

Fold axis. The median line between the limbs of the fold, along the apex of an anticline or the lowest part of a syncline.

Fold: closed. A fold with an acute angle between the limbs.

Fold: isoclinal. A fold having essentially parallel limbs.

Fold: open. A fold with limbs that diverge at an obtuse angle.

Fold: overturned. A fold having a limb in which the strata have been tilted beyond the vertical.

Fold: plunge. The angle a fold axis makes with the horizontal.

Fold: plunging. A fold with an inclined axis.

Fold: recumbent. A fold in which the axial plane is essentially horizontal.

Foliate. A general term for a metamorphic rock possessing foliation on the scale of hand specimens.

Foliation. A parallel or nearly parallel structure in metamorphic rocks along which the rock tends to split into flakes or thin slabs.

Foliation: gneissic. Foliation in which the layers are thicker and parting takes place along rougher surfaces than in the other two types (schistose and slaty).

Foliation: schistose. Foliation in which the layers are thin and parting occurs along generally smooth surfaces.

Foliation: slaty. Foliation characterized by its extremely smooth and plane parting surfaces.

Footwall. See ***Fault: footwall.***

Foreset bed. The coarse, thick, steeply sloping part of each layer in a delta.

Former lava flow. A body of extrusive igneous rock derived from solidification of a lava flow.

Fossil. The naturally preserved remains or traces of an animal or a plant.

Fossil fuel. A fuel that contains solar energy, locked up securely in chemical compounds by the plants or animals of former ages.

Fractional melting. See ***Incongruent melting.***

Fracture (as applied to minerals). The capacity of a mineral to break along irregular surfaces.

Fracture zone. A great linear system of breaks in the Earth's crust.

Fragment. A mineral or rock particle larger than a grain.

Fragmental texture. A texture of sediments and sedimentary rocks, resulting from physical transport and deposition of particles.

Framework. The rigid arrangement created by the particles of a sediment or a sedimentary rock that support one another at their points of contact.

Freshwater limestone. A limestone that formed in a lake, stream, or cave.

Fringing reef. A reef that is attached to the shore of a land mass.

Frost heaving. The lifting of rock waste by expansion during freezing of contained water.

Frost wedging. The mechanism involving the pushing up or apart of rock particles by the action of ice.

Fumarole. A volcano discharging gas nonexplosively.

Fusion, latent heat of. See ***Latent heat of fusion.***

G. The universal gravitational constant, the precise value of the attractive force between two masses of 1g each, exactly 1cm apart.

g. See ***Gravitational acceleration, Earth's.***

Gabbro. Coarse-grained mafic igneous rock.

Gaging station. A point of measurement of various attributes of a stream, such as level of the water, discharge, velocity of flow, and characteristics of the mechanical and chemical loads.

Gangue. The nonvaluable minerals of an ore.

Geochemical cycle. The cyclic chemical changes that accompany the operations of many natural processes.

Geochemistry. The study of the chemistry of natural reactions.

Geode. A hollow rounded body having a lining of crystals pointing inward.

Geologic column. A composite diagram combining in a single column the succession of all known strata, fitted together on the basis of their fossils or of other evidence of relative age.

Geologic cross section. A diagram showing the arrangement of rocks in a vertical plane.

Geologic cycle. The sum total of all internal and external processes acting on the materials of the Earth's crust.

Geologic map. A map that shows the distribution, at the surface, of rocks of various kinds or of various ages.

Geologic record. The archive of Earth history represented by bedrock, regolith, and the Earth's morphology.

Geologic time scale. The time relationships established for the geologic column (q. v.).

Geology. The science of the Earth.

Geology: physical. The study of the composition and configuration of the Earth's physical features and rock masses, their relationships to one another, and the surficial and subsurface processes that operate on and in the Earth.

Geology: historical. The systematic study of geologic history, based on relationships of rock layers, surficial sediments, morphology, measurements of radioactive minerals, and measurements of magnetic properties of rocks; history of life on Earth, including both evolutionary development and relationships of organisms to their environments as expressed in the sedimentary record; and ancient geographic relationships.

Geophysical exploration (also called ***geophysical prospecting***). Exploration to infer subsurface conditions based on the distribution within rocks of some physical property, such as specific gravity, magnetic susceptibility, electrical conductivity, and elastic or other properties. In geophysical prospecting these methods are applied in the search for economically valuable substances.

Geosyncline. A huge linear or arcuate segment of the Earth's crust, which by subsidence below sea level and sediment accumulation incubates a mountain chain.

Geothermal gradient. The rate of increase of temperature downward in the Earth.

Geyser. An orifice that erupts steam and boiling water intermittently.

Glacial drift. Or, simply, ***drift.*** The sediments deposited directly by glaciers (***till,*** q. v.), or indirectly in glacial streams, lakes, and the sea (***stratified drift,*** q. v.).

Glacial drift: ice-contact stratified. Stratified drift deposited in contact with its supporting ice.

Glacial-marine sediment. See ***Sediment: glacial-marine.***

Glacial plucking. The lifting out and removal of fragments of bedrock by a glacier.

Glacial polish. A smooth surface, on bedrock, abraded by a glacier.

Glacial striation. A linear scratch or groove, on bedrock, created by movement of a glacier.

Glaciated valley. A valley that has been modified by a glacier. A glaciated valley may have a troughlike (U-shaped) cross profile, hanging tributaries, steplike irregularities in its long profile, and a cirque or group of cirques at its head.

Glaciation. The alteration of a land surface by the massive movement over it of glacier ice.

Glacier. A body of ice, consisting mainly of recrystallized snow, flowing on a land surface.

Glassy texture. A texture of igneous rocks containing only glass in which crystalline particles are not present.

Gneiss. A coarse-grained foliate breaking along irregular surfaces and commonly containing prominently alternating layers of light-colored and dark-colored minerals.

Graded layer. A layer of sediment in which the particles grade upward from coarse to fine.

Gradient. Applied to a stream, it is the slope measured along the stream, on the water surface or on the bottom.

Grain. A mineral or rock particle having a diameter of less than a few millimeters and generally lacking well-developed crystal faces.

Granite. A coarse-grained igneous rock consisting in major part of potassium feldspar, some sodic plagioclase (albite), and quartz, with minor amounts of ferromagnesian minerals.

Granite gneiss. A distinctly foliated metamorphic rock with the mineral composition of granite.

Granitic rock. A general term for a coarse-grained felsic igneous rock containing quartz and minor amounts of ferromagnesian minerals.

Granitization. The transformation, without fusion, of older rocks into granitic rocks.

Granodiorite. A coarse-grained felsic igneous rock, with composition intermediate between granite and diorite, containing quartz, and in which plagioclase is the chief feldspar.

Granular disintegration. The mechanical loosening, during chemical weathering, of the individual mineral particles of bedrock.

Granular texture. A texture of equigranular igneous rock with particles ranging in size from 0.05 to 10mm.

Gravitation, the universal law of. Sir Isaac Newton's universal law of gravitation states that every particle in the universe attracts every other particle with a force directly proportional to the product of their masses and inversely as the square of the distance between their centers.

Gravitational acceleration, Earth's (***g***). The inward acting force with which the Earth tends to pull all objects toward its center and which tends to make the Earth a sphere; also called the *Earth's gravity*. The Earth's gravitational acceleration, represented by g, is measured in gals (for Galileo; 1 gal = $1 cm/sec^2$ of acceleration).

Gravity anomaly. A difference between the computed and the measured values of gravity at a given location. The anomaly is positive where measured values exceed computed values, and negative where measured values are less than computed values.

Gravity meter (or ***gravimeter***). A sensitive measuring device for determining the value of g at any locality.

Gravity prospecting. The attempt to use sensitive measurements of g to find valuable substances associated with slight variations in the densities of rocks.

Gravity spring. See ***Spring: gravity.***

Graywacke. A poorly sorted sandstone, generally dark, containing rock fragments as well as quartz.

Groin. A low wall on a beach crossing the shoreline at a right angle.

Groundmass. The particles surrounding the phenocrysts of a porphyritic igneous rock.

Ground water. The water, beneath the Earth's solid surface, contained in pore spaces within regolith and bedrock.

Guyot. A seamount having a conspicuously flat top.

Half-life (of a radioactive isotope). The time required to reduce the number of parent atoms by one-half.

Hanging valley (also ***hanging tributary***). A tributary stream whose valley floor lies ("hangs") above that of the valley of a main stream. Hanging valleys are common where main valleys have been glaciated.

Hanging wall. See ***Fault: hanging wall.***

Hardness. Resistance to scratching.

Hertz. A unit for expressing frequency; 1 Hertz = 1 cycle/sec (abbreviated Hz).

Hinge fault. A fault on which displacement dies out perceptibly along strike and ends at a definite point.

Historical geology. See ***Geology: historical.***

Horn. A bare, pyramidal-shaped peak left standing where glacial action in cirques has eaten into it from three or more sides.

Hornfels. A tough, generally massive nonfoliate containing scattered crystals of high-temperature minerals.

Hot spring. A spring (q. v.) from which hot water flows freely. If the water is at its boiling temperature the spring is a ***boiling spring.***

Humus. The decomposed residue of plant and animal tissues.

Hydraulic conductivity. The quantity of fluid that passes through material of a given cross section per unit of time, when driven by a given pressure and at a stated temperature.

Hydraulic gradient. The slope of a water table, found by determining the difference in height between two points and dividing by the horizontal distance between them.

Hydraulic plucking. The lifting out, by turbulent water, of blocks of bedrock bounded by joints and other partings.

Hydrocarbon. An organic compound consisting of hydrogen and carbon. Petroleum and natural gas are examples of natural hydrocarbons.

Hydrogeology. The study of ground water.

Hydrologic cycle. The system driven by Solar energy of water circulation from oceans to atmosphere and back to oceans either directly or via the lands.

Hydrolysis. The combination of water with other molecules.

Hydrosphere. The Earth's discontinuous water envelope.

Ice cap. A small ice sheet.

Ice-contact stratified drift. See ***Glacial drift: ice contact stratified.***

Ice sheet. A broad glacier of irregular shape, generally blanketing a large land surface.

Igneous rock. Rock formed by solidification of molten silicate materials.

Imbrication. The slanting, overlapping arrangement of flat pebbles, like shingles on a roof. Imbrication is likewise applied to inclined, overlapping thrust sheets.

Inactive volcano. See ***Volcano: inactive.***

Inclination (***magnetic***). The angle with the horizontal measured in the vertical plane passing through the magnetic line of force.

Inclusion. An impurity, solid or liquid, within a crystal, resulting from encirclement by growth of a lattice that is larger than the impurity.

Incongruent melting. Melting accompanied by a change of composition.

Intensity, earthquake. See ***Earthquake intensity.***

Interface. A boundary between bodies of matter having different physical states of aggregation, or between bodies of matter having the same physical state of aggregation but different physical properties.

Interfacial tension. The force, created parallel to an interface between a liquid and an unlike substance, which results from unequal distribution of the particles of the liquid on opposite sides of the interface.

Interior drainage. See ***Drainage: interior.***

Intermediate-focus earthquake. See ***Earthquake: intermediate-focus.***

Interstices. See ***Pore spaces.***

Interstitial spaces. See ***Pore spaces.***

Intrusive igneous rock. Rock that originated from solidification of magma emplaced in older bedrock.

Ion. A charged particle formed from an atom by adding or subtracting one or more electrons.

Ionic bond. A kind of bond in which ions are held together by strong nondirectional electrostatic forces of attraction among oppositely charged particles.

Isograd. A line connecting points of first occurrence of a given mineral in metamorphic rocks.

Isoseismal line. A line on a map through points of equal earthquake intensity.

Isostasy. The ideal condition of flotational balance among segments of the lithosphere.

Isotopes. Configurations of the same element, all having the same number of protons and generally similar chemical behavior but different mass numbers, A, resulting from different numbers of neutrons in the nucleus.

Isotopic date. A determination of the age of a sample

by a calculation based on a measurement of its content of a suitable radioactive isotope.

Joint. A fracture, on which no appreciable movement parallel with the fracture has occurred.
Joint set. A widespread group of parallel joints.
Joint system. A combination of two or more intersecting joint sets.
Joule. A unit of mechanical energy equal to 10^7 ergs.
Juvenile water. Water, formerly dissolved in magma, which rises from deep within the Earth to become part of the hydrosphere for the first time.

Kame. A body of ice-contact stratified drift shaped as a short, steep-sided knoll or hummock.
Kame terrace. A body of ice-contact stratified drift shaped into a terracelike form along the sides of a valley.
Karst topography. An assemblage of topographic forms consisting primarily of closely spaced sinks; strikingly developed in the Karst region of Yugoslavia.
Kettle. A closed depression in drift, created by the melting out of a mass of underlying ice.

Laccolith. A concordant lenticular pluton, circular or elliptical in plan, having an essentially plane floor and a distinctly domed roof.
Lamina (plural, ***laminae***). A stratum less than 1cm thick.
Laminar flow. See ***Flow: laminar.***
Lapilli tuff. A medium-grained pyroclastic rock consisting of lapilli.
Lapilli (***volcanic***). Tephra ranging in size between 4 and 32mm.
Latent heat of fusion. The amount of energy added or released in changing state from a solid to a liquid.
Latent heat of sublimation. The amount of energy involved in moving particles across the solid/gas interface.
Latent heat of vaporization. The amount of energy added or released when particles move across the liquid/gas interface.
Laterite. A reddish residual product of tropical weathering, rich in oxides of iron and aluminum.
Lava. Molten silicate materials reaching the Earth's surface.
Lava dome. A dome created by the upward movement of a lava spine (q. v.).
Lava flow. A hot stream or sheet of molten material that is flowing or has flowed over the ground. Not to be confused with ***Former lava flow.***
Lava fountain. Lava clots sprayed into the air from a volcano.
Lava spine. An upright, cylindrical feature created by the upward squeezing of a mass of sluggish, pasty lava.
Leaching. The continued removal, by water, of soluble matter from bedrock or regolith.
Levee: natural. A broad, low ridge of fine alluvium built along the side of a stream channel by water spreading out of the channel during floods.
Lignite. A brownish-black coal, intermediate in composition between peat and bituminous coal.
Limb. The side of a fold.
Limb: overturned. The limb of a fold in which the strata have been tilted beyond the vertical.
Limestone. A sedimentary rock consisting predominantly of calcium carbonate.
Lithification. Literally, rock making. In practice, it is a general term for the conversion of sediments into sedimentary rocks.
Lithosphere. The outer zone of the solid Earth. The lithosphere includes the crust and the upper part of the mantle lying above the low-velocity zone.
Load. The material carried at a given time, by a stream, by a current of water, by the wind, or by a glacier.
Load: bed. The coarse solid particles, within a body of flowing fluid, moving along or close above the bed.
Load: suspended. The fine solid particles turbulently suspended within a body of flowing fluid.
Local base level. See ***Base level: local.***
Loess. Wind-deposited silt, usually accompanied by some clay and some fine sand.
Longitudinal dune. See ***Dune: longitudinal.***
Longitudinal wave. A seismic body wave that causes particles to oscillate along lines in the direction of wave travel. In seismology designated by the letter P; also called P-wave.
Long profile. A line connecting points on the surface of a stream.
Longshore current. A current in the surf zone flowing parallel to the shore.
Longshore drift. The net movement of sediment, parallel to the shore, by waves and wave-induced currents.
Love wave. A seismic surface wave that causes particles to oscillate at right angles to direction of wave travel along lines lying in horizontal planes.
Luster. The quality and intensity of light reflected from a mineral.

M-discontinuity. See ***Mohorovičić discontinuity.***
Mafic rock. A rock in which ferromagnesian minerals exceed 50 per cent.
Magma. Molten silicate materials beneath the Earth's surface including crystals derived from them and gases dissolved in them.
Magmatic differentiation. A collective name for the processes by which one magma generates more than one variety of igneous rock.
Magnetic declination. See ***Declination, magnetic.***
Magnetic field (***Earth's***). The magnetic lines of force surrounding the Earth.
Magnetic pole. A point on the Earth's surface where the magnetic inclination is 90°. A north-seeking needle mounted on a horizontal axis points directly down at the North Magnetic Pole and directly up at the South Magnetic Pole.
Magnetic reversal. A reversal of the polarity of the Earth's magnetic field. The mechanism of reversal is not known, but during the process of reversal, the

Earth's magnetic field is presumed to vanish for a time, and to reappear with polarity reversed.

Magnetism, remanent. The magnetism remaining in rock particles which causes magnetic fields to persist.

Mantle. See ***Earth's mantle.***

Marble. A nonfoliate consisting chiefly of calcite or dolomite.

Marine sediment. Sediment deposited in the sea.

Marl. A mixture of calcium carbonate and clay; a common lake deposit.

Mass number. Represented by the letter-symbol A; the number of protons (Z) plus the number of neutrons (N) in an atomic nucleus.

Mass-wasting. The gravitative movement of rock debris downslope, without the aid of a flowing medium of transport such as air at ordinary pressure, water, or glacier ice.

Matrix. The small particles of a sediment or a sedimentary rock, which occupy the spaces between the larger particles that form the framework.

Meander. A looplike bend of a stream channel.

Meltwater. Water resulting from the melting of snow and glacier ice.

Mercalli scale of earthquake intensity. A scale of earthquake intensity (q. v.) with divisions, based on human sensations and on damage to manmade structures, ranging from I to XII.

Metallic bond. A kind of bond in metals in which positive ions form a fixed framework on which is superimposed a network of electrons that can move freely.

Metamorphic facies. Rocks that reached equilibrium during metamorphism within a single range of environmental conditions.

Metamorphic rock. A rock formed within the Earth's crust by the transformation of a preexisting rock in the solid state without fusion and with or without addition of new material, as a result of high temperature, high pressure, or both.

Metamorphism. The changes, in mineral composition, arrangement of minerals, or both, that take place in the solid state within the Earth's crust at high temperatures, high pressures, or both.

Metamorphosed rock: contact. A rock that has been altered near a pluton. Two varieties are: (1) rocks metamorphosed by high temperature, and (2) rocks metamorphosed by high temperature and by the addition of new material.

Metamorphosed rocks: regional. Metamorphic rocks that extend through great distances in mountain chains and continental shields.

Metastable mineral. A mineral that persists under environmental conditions in which it is not the most stable form.

Meteorite. A particle of solid matter from outer space that has fallen to the ground through the atmosphere.

Microseism. A small deflection, on a seismogram, created by a seismic wave having a period ranging from 1 to 9 seconds, and not generated by an earthquake. Storms, surf, and various other motions create microseisms.

Migmatite. A rock in which thin stringers and threads of granitic material are intertwined with dark schistose layers.

Mineral. A naturally occurring substance, most but not all of which are crystalline solids and whose exteriors may or may not consist of crystal faces; whose atoms or ions of one or more elements or molecules of compounds are arranged regularly in a definite lattice, and whose chemical compositions, though constant or variable within limits, bear fixed relationships to certain physical properties.

Mineral, accessory. See ***Accessory mineral.***

Mineral aggregate. A body consisting of more than one grain of the same or of different mineral species. Mineral aggregates contain more than one crystal lattice; they can occur as regolith or sediment if loosely bound, or as rock if they are tightly bound.

Mineral, essential. See ***Essential mineral.***

Mineral facies. Mineral assemblages of any origin or composition that have reached chemical equilibrium under similar environmental conditions.

Mineralogy. The study of minerals.

Modulus of rigidity. The stress required to cause a unit change of shape.

Mohorovičić discontinuity. The seismic discontinuity marking the base of the Earth's crust. Abbreviated *M-discontinuity* (or, in the vernacular, "Moho").

Molecular bond. A kind of bond in which molecules having electrically balanced internal atomic charges are held together by forces of attraction between unlike partial charges distributed unevenly over the surfaces of the molecules.

Molecule. The fundamental entity of a chemical compound.

Monadnock. A conspicuous residual hill on a peneplain. The name was taken from Mount Monadnock, in New Hampshire, in 1893.

Monocline. A one-limb flexure, on either side of which the strata are horizontal or dip uniformly at low angles.

Moraine: end. A ridgelike accumulation of drift, deposited by a glacier along its front margin.

Moraine: ground. Widespread thin drift with a smooth surface consisting of gently sloping knolls and shallow closed depressions.

Moraine: lateral. An accumulation of drift along the side of a valley glacier.

Moraine: medial. A strip of drift formed by coalescence of lateral moraines at the junction of two valley glaciers.

Morphology (Earth's). The shape of the Earth's surface.

Mountain. In a general sense, any land mass that stands conspicuously higher than its surroundings. Geologically it refers to parts of the Earth's crust having thick, crumpled strata, regionally metamorphosed rocks, and granitic batholiths.

Mountain chain. An elongate unit consisting of numerous ranges or groups, regardless of similarity in form or of equivalence in age.

Mountain making. The creation of elongate highlands by large-scale deformation of rocks in the Earth's crust.

Mountain range. A single large complex ridge or series of clearly related ridges that constitute a fairly continuous and compact unit.

Mountain, residual. See ***Residual mountain.***

Mountain system. A group of ranges similar in general form, alignment, and structure, which presumably originated from the same general causes.

Mountain, volcanic. See ***Volcanic mountain.***

Mud crack. A crack caused by the shrinkage of wet mud as its surface becomes dry.

Mudflow. See ***Debris flow, variety mudflow.***

Natural gas. Gaseous hydrocarbons, predominantly methane, that occur in rocks underground.

Natural levee. See ***Levee: natural.***

Neck cutoff. The intersection of a meander bend by the bend next upstream causing the stream to bypass the loop between the bends.

Nepheline syenite. A light-colored, coarse-grained felsic igneous rock lacking quartz, and containing nepheline.

Neptunists. A name applied to nineteenth-century geologists who believed that granite and basalt crystallized out of sea water. Compare ***Plutonists.***

Neutron. An electrically neutral particle having a mass of $1.6752 \text{x} 10^{-24}$g.

Nonfoliate. A general term for metamorphic rocks lacking foliation on the scale of hand specimens.

Normal component of gravity (g_n). The component of gravity, acting at right angles to a slope, which tends to hold particles in place. The magnitude of the normal component of gravity is determined by the cosine of the angle of slope.

Normal fault. A fault, generally steeply inclined, along which the hanging-wall block has moved relatively downward.

North American Cordillera. All the mountain units in western North America, from the eastern border of the Rocky Mountains to the Pacific coast.

Nuée ardente. An incandescent cloud consisting of superheated gases and of hot, fine-grained tephra.

Oblique-slip fault. A fault on which movement includes both horizontal and vertical components.

Obsidian. Felsic glass that transmits light along thin edges.

Ocean basin. A low part of the lithosphere that lies between continental masses and is covered by seawater.

Oceanic basalt. See ***Basalt: oceanic.***

Oceanic crust. See ***Crust: oceanic.***

Oceanic rise and ridge. A continuous rocky ridge, on the ocean floor, many hundreds of kilometers wide, whose relief is 600m or more.

Oil field. A group of oil pools, usually of similar type, or a single pool in an isolated position.

Oil pool. An underground accumulation of oil or gas in a reservoir limited by geologic barriers.

Oil shale. A body of fine-grained sediment rich in hydrocarbon derivatives. The chief hydrocarbon derivative in the Green River oil shales of the Rocky Mountain region is kerogen.

Oölite. A limestone consisting predominantly of ***oöids,*** spherical particles of calcium carbonate, having many concentric shells, that enlarged at the site of deposition by radial growth.

Ooze. Pelagic sediment consisting of more than 30 per cent of skeletal remains of microorganisms.

Ooze: calcareous. Ooze consisting of the hard parts of Foraminifera, pteropods, and coccoliths.

Ooze: siliceous. Ooze consisting of the hard parts of radiolarians, diatoms, and sponges.

Open system. A reaction in which some material can escape from the scene of reaction.

Order of magnitude scale. A $\log_{10}$ scale.

Ore. An aggregate of minerals from which one or more materials can be extracted profitably.

Organic material. Material composed of hydrocarbon compounds.

Organic texture. A texture of sedimentary rocks resulting from secretion of skeletal material or from other activity by organisms.

Original horizontality, principle of. The principle which states that most strata are nearly horizontal when originally deposited.

Orogenic belt. A mountain chain or a region where a mountain chain is in the process of formation.

Orogeny. Another name for mountain making. The deformation of the crust in the development of mountains.

Outcrop area. The area, on a geologic map, shown as occupied by a particular rock unit.

Outer ridge. A broad, smooth ridge of sediment, generally parallel to the margin of an ocean basin, and standing 200 to 2,000m above the adjacent sea floor.

Outwash. Stratified drift deposited by streams of meltwater as they flow away from a glacier.

Outwash plain. A body of outwash that forms a broad plain.

Overland flow. See ***Flow: overland.***

Oxbow lake. A curved lake occupying a cutoff meander loop.

Packing. The arrangement of particles of a sediment or sedimentary rock.

Paleomagnetism. The study of the Earth's magnetic field in former times.

Parallel strata. See ***Strata: parallel.***

Parent isotope. In radioactivity, the isotope that decays and continually decreases in amount with time.

Particle. A general term used in this book; it refers to anything from electrons to larger sizes without restriction as to shape, composition, or internal structure.

Parting. A surface of separation within a rock body.

Parting, bedding-plane. See **Bedding-plane parting.**

Peat. A brownish, lightweight mixture of partly decomposed plant tissues in which the parts of plants are easily recognized.

Pebble size. Sediment particles having diameters greater than 2mm (about the size of the head of a small wooden match) and less than 64mm (about the size of a tennis ball).

Pedalfer. A soil in which much clay and iron have been added to the B horizon.

Pediment. A sloping surface, cut across bedrock, adjacent to the base of a highland in an arid climate.

Pedocal. A soil with calcium-rich upper horizons.

Pedology. The science of soils, their origin, use, and protection.

Pegmatite. Exceptionally coarse-grained granite with individual particles and crystals ranging in length from 1cm to many meters.

Pegmatitic texture. A texture of equigranular igneous rocks with particles larger than 10mm.

Pelagic sediment. An open-sea deposit containing predominantly skeletal remains of microorganisms and clays or products derived from clays.

Peneplain. "Almost a plain." A land surface worn down to very low relief by streams and mass-wasting.

Perched water body. A water body that occupies a basin in impermeable material, perched in a position higher than the main water table.

Percolation. Laminar flow through interconnected spaces in saturated material.

Percolation: confined. Percolation in an aquifer, between impermeable strata, wherein ground water moves past other ground water held immobile in those strata.

Peridotite. An ultramafic igneous rock consisting of pyroxene and considerable olivine; plagioclase, if present, is only a minor constituent.

Permafrost. Ground that is frozen perennially. Permafrost occurs generally at high latitudes and locally at high altitudes.

Permeability. The capacity of a material for transmitting fluids; expressed as hydraulic conductivity (q. v.).

Petroleum. Gaseous, liquid, or solid substances, occurring naturally and consisting chiefly of chemical compounds of carbon and hydrogen. Petroleum includes both oil and natural gas.

Petrology. The study of rocks.

pH. A measure of acidity or alkalinity of solutions; the negative reciprocal of the logarithm to base 10 of the hydrogen-ion concentration.

Phenocrysts. The particles in porphyritic igneous rocks that are conspicuously larger than the remaining particles.

Phyllite. An exceptionally lustrous fine-grained foliate parting along smooth or irregular surfaces.

Physical geology. See **Geology: physical.**

Physiographic province. A region having unities of bedrock, morphology, and morphologic history.

Piedmont glacier. A glacier on a lowland at the base of a mountain, fed by one or more valley glaciers.

Pillow. An ellipsoidal mass of lava or of extrusive igneous rock, having a fine-grained margin, and formed by the extrusion of lava under water.

Pitchstone. Felsic glass having a dull, greasy luster.

Placer. A deposit of heavy minerals concentrated mechanically. Common methods of mechanical concentration are sorting resulting from stream flow, wind action, and wave action.

Plane of the ecliptic. The plane in which the planets of the Solar System orbit around the Sun.

Plastic flow. A continuous and permanent change of shape in any direction without breakage.

Plastic solid. A solid that flows readily.

Plateau. An extensive upland, underlain by essentially horizontal strata, and having large areas where the surface is nearly flat.

Plateau basalt. Widespread nearly horizontal layers of extrusive basalt on a continental mass.

Playa. A dry bed of a playa lake (q. v.); a nearly level area on the floor of an intermontane basin. Occasional floods deposit silt and clay that mantle a playa.

Playa lake. An ephemeral shallow lake in a desert basin.

Plunge (of a fold). See **Fold: plunge.**

Pluton. Any body of intrusive igneous rock.

Plutonic igneous rock. A coarse-grained igneous rock that cooled slowly at great depth (generally) or at shallow depth (less commonly) within the Earth's crust.

Plutonism. A general term referring to the behavior of magmas, including their movement, internal and external reactions, and emplacement.

Plutonists. A name applied to a group of nineteenth-century geologists who believed that granite originated by igneous processes within the Earth's crust and that basalt was a volcanic product. Compare ***Neptunists.***

Pluvial lake. A lake that existed under a former climate, when rainfall in the region concerned was greater than at present.

Podzol. A pedalfer soil that has been intensely leached by solutions rich in humic acids.

Point bar. A crescent-shaped bar built out from each convex ("inside") bank of a stream channel.

Polar easterlies. The winds common in the polar regions.

Pool. See **Oil pool.**

Poorly sorted sediment. A sediment consisting of particles of many sizes.

Pore spaces. The spaces in a body of rock or of sediment that are unoccupied by solid materials.

Porosity. The proportion, in per cent, of the total volume of a given body of bedrock or of regolith that consists of pore spaces (q. v.).

Porphyritic. An adjective describing an igneous rock in which phenocrysts constitute less than 25 per cent of the total volume.

Porphyritic texture. A texture of igneous rocks with some particles conspicuously larger than the rest of the particles.

Porphyroblasts. Particles of metamorphic rocks that are conspicuously larger than their neighbors.

Porphyry. An igneous rock with phenocrysts constituting more than 25 per cent of the volume.

Positron. A particle having the same mass as an electron but with a unit positive charge.

Pothole. A cylindrical hole drilled in bedrock by a turbulent stream.

Pre-geologic time. The part of Earth history that antedates the oldest rocks.

Primary wave. The first-arriving body wave from an earthquake; a longitudinal wave (q. v.).

Principle of original horizontality. See ***Original horizontality, principle of.***

Principle of stratigraphic superposition. See ***Stratigraphic superposition, principle of.***

Principle of uniformity. See ***Uniformity, principle of.***

Progradation. The process of outward or of forward building by a body of sediment, such as a beach, a delta, or a fan, resulting from addition of successive layers of new sediment.

Proton. Particle of unit positive electrical charge having a mass of 1.6730×10^{-24}g.

Pumice. Extremely vesicular, frothy natural glass with a high content of silica. Pumice is so light it will float on water.

Pyroclastic material. See ***Tephra.***

Pyroclastic rock. Lithified tephra (q. v.).

Pyroclastic texture. A texture of sediments and sedimentary rocks resulting from physical transport and deposition of particles broken by volcanic activity.

Pyroxenite. An ultramafic igneous rock composed almost entirely of pyroxene.

Quartzite. A quartz-rich nonfoliate.

Radiation belts. Inner and outer zones within the Earth's magnetic field having high concentrations of charged particles.

Radioactivity. The spontaneous decay of the atoms of certain isotopes into new isotopes, which may be stable or undergo further decay until a stable isotope is finally created. Radioactivity is accompanied by the emission of α-particles, β-particles, and γ-rays and by the generation of large quantities of heat.

Radiocarbon. The radioactive isotope of carbon, C^{14}, which is created in the upper atmosphere and circulates throughout the biosphere.

Radiocarbon dating. A determination of the age of a sample by a calculation based on a measurement of its content of natural radiocarbon.

Rayleigh wave. A seismic surface wave that causes particles to oscillate in circles lying in a vertical plane parallel to the direction of wave travel. The particles move in the same direction as the wave advance under troughs but in the opposite direction under crests.

Reaction relationship. The interaction between crystals and a silicate melt in which material is exchanged as cooling occurs. This relationship was first stated by N. L. Bowen and commonly is called "Bowen's reaction series." See also ***Continuous reaction relationship*** and ***Discontinuous reaction relationship.***

Recharge. The addition of water to the zone of saturation (q. v.).

Recumbent fold. See ***Fold: recumbent.***

Reef. A massive wave-resistant structure built by the secretions of marine organisms.

Regolith. The noncemented rock fragments, and mineral grains derived from rocks, which overlie the bedrock in most places. Regolith is of two kinds, residual and transported.

Rejuvenation. The development of youthful morphologic features in a land mass further advanced in the cycle of erosion. Regional uplift is the common cause of rejuvenation.

Relative humidity. The ratio of the amount of water vapor present to the maximum amount that the air mass can contain without condensation or precipitation.

Relief. The difference in altitude between the high and low parts of a land surface.

Remanent magnetism. See ***Magnetism, remanent.***

Replacement. The process by which one mineral takes the place of another mineral. During replacement a fluid dissolves matter already present and at the same time deposits from solution an equal volume of a different substance.

Reservoir rock. A porous and permeable rock containing petroleum, natural gas, or both.

Residual concentration. The natural concentration of a mineral substance by removal of a different substance with which it was associated.

Residual mountain. A resistant remnant standing high as a result of long-continued erosion.

Reverse fault. A fault, generally steeply inclined, along which the hanging-wall block has moved relatively upward.

Rhyolite. A fine-grained igneous rock having the composition of granite; if porphyritic, contains quartz phenocrysts and has aphanitic groundmass.

Richter scale of earthquake magnitude. A scale of earthquake magnitude based on the logarithm (base 10) of the amplitudes of the deflections, on a seismogram, created by seismic waves and recorded on a calibrated seismograph.

Rigidity. Resistance to change of shape.

Rip current. A current, flowing seaward through the line of breakers, which returns to the open sea the water of longshore currents.

Ripple mark. A small-scale wave, in sand, created by the effects of the drag of moving water or wind. Two principal kinds are symmetrical and asymmetrical.

Roche moutonnée (plural, ***roches moutonées***). A knob of bedrock, smoothed and generally striated by an overriding glacier, elongate in the direction of ice movement.

Rock avalanche. The rapid downslope flow of a mass of dry rock particles.

Rock cleavage. Closely spaced partings in rocks controlled by platy particles that have been aligned in response to pressures within the Earth's crust.

Rock cycle. That part of the geologic cycle (q. v.) concerned with the creation, destruction, and alteration of rocks during erosion, transport, deposition, metamorphism, plutonism, and volcanism.

Rockfall (and ***debris fall***). The rapid descent of a rock mass, vertically from a cliff or by leaps down a slope.

Rock flour. Fine sand and silt produced by crushing and grinding in a glacier.

Rock flow. The slow movement of a rock body that is in a plastic condition.

Rock glacier. A lobate, steep-fronted mass of coarse, angular regolith, extending from the front of cliffs in a mountainous area. Downslope movement of the mass is aided by interstitial water and ice.

Rockslide. The rapid descent of a rock mass down a slope.

Roof rock. In petroleum geology, an impermeable stratum that prevents the upward escape of petroleum or of natural gas. Also, the layer of rock overlying a coal bed or a confined aquifer.

Runoff. Water that flows over the lands.

Rupture. The mechanical breaking of rocks.

Salinity. The proportion of dissolved solids in a solution.

Saltation. The jumping movement of rock particles in a current of water or air.

Salt dome. A dome, in sedimentary strata, created by the upward movement of a salt plug (q. v.).

Salt plug. A cylindrical body of rock salt that has risen upward from an underlying source bed.

Sand size. Particles of sediment having diameters larger than $\frac{1}{16}$mm (about the lower limit of visibility of the individual particles with the unaided eye) and smaller than 2mm (about the size of the head of a small wooden match).

Sandstone. A clastic sedimentary rock consisting predominantly of sand-size particles.

Sand wave. A large wave, in sand or gravel, created by the effects of the drag of a thick and swift-moving current of air or water. Sand wave is a general term that includes *dune*. In general appearance, a sand wave resembles some varieties of ripple mark, but a sand wave is many times larger than a ripple mark and can consist of coarse sand and gravel. The upcurrent sides of some sand waves are covered with ripple marks.

Saturation, zone of. See ***Zone of saturation.***

Scale. The proportion between a unit of distance on a map and the unit it represents on the Earth's surface.

Schist. A well-foliated metamorphic rock in which the component flaky minerals are distinctly visible.

Scientific method. The analysis of a problem based on formulation of questions, collection of data and observations, formulation of predictions, and testing of the predictions.

Scoria. An igneous rock containing abundant vesicles many of which are open at the surface.

Scoriaceous. An adjective describing an extremely vesicular igneous rock containing many vesicles open at the surface.

Sea arch. A roofed-over opening through a narrow headland, resulting from attack by waves on both sides.

Sea cave. A chasm in a sea cliff, excavated in weak rock by waves and currents.

Sea-floor trench (also ***deep-sea trench***). An elongate, narrow, steep-sided depression, generally deeper than the adjacent sea floor by 2,000m or more, extending parallel to the margin of an ocean basin.

Sea level. The level continuous with that of the oceans at mean tide. Also, the surface of the sea.

Seamount. An isolated conical mound projecting more than 1,000m. above the deep-sea floor.

Sea wall. A wall paralleling the shore, built to prevent coastal erosion by waves.

Secondary enrichment. Natural enrichment of an ore body by later addition of mineral matter, generally from percolating solutions.

Secondary mineral. A mineral that was not present in the original rock or sediment.

Secondary wave. The second-arriving body wave from an earthquake; a shear wave (q. v.).

Secular equilibrium. In radioactivity, the condition of equilibrium in which the rate of decay of the parent isotope is exactly matched by the rate of decay of every intermediate daughter isotope. When secular equilibrium has been established the concentrations of intermediate daughters remain virtually constant.

Sediment. Regolith that has been transported; mineral or organic matter deposited by water, air, or ice.

Sediment, clastic. See ***Detritus.***

Sediment: glacial-marine. Marine terrigenous sediment including nonsorted mixtures of particles of all sizes dropped from floating ice.

Sediment: poorly sorted. See ***Poorly sorted sediment.***

Sediment, terrigenous. See ***Terrigenous sediment.***

Sediment: well-sorted. See ***Well-sorted sediment.***

Sedimentary breccia. A clastic sedimentary rock containing numerous angular particles of pebble size or larger.

Sedimentary rock. A rock formed by cementation of sediment or by other processes acting at ordinary temperatures at or close beneath the Earth's surface.

Seismic. Pertaining to shock waves, natural or artificial, within the Earth.

Seismic belt. A tract subject to frequent earthquake shocks, both strong and weak. The Earth's principal seismic belts are the Circum-Pacific belt, the Mediterranean and Trans-Asiatic belt, the Mid-Atlantic belt, and the Mid-Indian belt.

Seismic discontinuity. An interface within the Earth where physical properties change. Along the discontinuity some seismic waves are reflected, others are refracted.

Seismic prospecting. The search for valuable substances underground by using seismic waves from artificial explosions to infer the subsurface structure.

Seismic seawave. Another name for ***tsunami*** (q. v.).

Seismic wave. A wave traveling within the Earth.

Seismogram. A record made by a seismograph.

Seismograph. An instrument for accurate recording of seismic waves.

Seismology. The study of seismic waves and the science of earthquakes.
Settling velocity. The constant velocity attained by a particle falling through a still fluid.
Shale. A fine-grained, fissile sedimentary rock composed of clay-size and silt-size particles of unspecified mineral composition.
Shallow-focus earthquake. See ***Earthquake: shallow-focus.***
Shallow-water wave. A wave, on the surface of a body of water, beneath which particles of water move in elliptical orbits, within vertical planes, because the bottom influences their motion.
Shearing stress. See ***Stress: shearing.***
Shear wave. A seismic wave that causes particles to oscillate along lines at right angles to the direction of wave travel. Shear waves propagate only through solids. In seismology designated by the letter S; also called S-wave.
Sheet erosion. See ***Erosion: sheet.***
Shield, continental. See ***Continental shield.***
Shield, volcanic. See ***Volcanic shield.***
Shield volcano. See ***Volcanic shield.***
Sial. The collective term for silica-rich rocks of continental masses. Derived from Si (symbol for silicon) and Al (symbol for aluminum).
Sialic rock. A rock rich in silicon and aluminum; also a felsic rock. See ***Sial.***
Sill. A tabular pluton having concordant surfaces of contact.
Silt size. Scale of sediment particles having diameters larger than 4 microns and smaller than $\frac{1}{16}$mm (about the lower limit of visibility of individual particles with the unaided eye).
Siltstone. A clastic sedimentary rock consisting predominantly of silt-size particles.
Sima. A collective term, coined from *Si* (symbol for silicon) and *ma* (abbreviated form of magnesium), designating the mafic rocks beneath the ocean basins.
Simatic rock. See ***Mafic rock.***
Sink. A large solution cavity open to the sky, generally created by collapse of a cavern roof.
Slate. A fine-grained foliate that splits along smooth planes into very thin plates.
Slaty cleavage. A closely spaced, plane foliation that divides rock into thin plates.
Slickensides. Striated and polished surfaces on rocks abraded by movement along a fault.
Sliderock. The material composing a talus.
Slip face. The straight, lee slope of a dune.
Slope, angle of. The angle, measured in the vertical plane, between an inclined portion of the Earth's surface and the horizontal. Slope is expressed in degrees or in feet per mile, the number of feet of vertical descent in a mile of horizontal distance.
Slump. The downward slipping of a coherent body of rock or regolith along a curved surface of rupture.
Snowfield. A wide cover, bank, or patch of snow, above the snowline, that persists throughout the summer season.
Snowline. The lower limit of perennial snow.
Soil. That part of the regolith which can support rooted plants.
Soil horizon. One of the subdivisions of a layered soil. Pedologists (q. v.) usually recognize three chief subdivisions, designated from top to bottom by the letters A, B, and C.
Soil, mature. A soil having a fully developed profile.
Soil profile. The succession of distinctive horizons in a soil and the unchanged parent material beneath it.
Solar constant. The amount of solar energy arriving on 1cm^2 of the Earth's surface per second.
Solar energy. The energy derived from the Sun.
Solifluction. The imperceptibly slow downslope flow of water-saturated regolith.
Solifluction sediment. A sediment that has resulted from solifluction (q. v.).
Sorting. The selection, by natural processes during transport, of rock particles or other particles according to size, specific gravity, shape, durability, or other characteristics.
Source rock. In petroleum geology, a rock within which petroleum or natural gas originates.
Specific gravity. A number stating the ratio of the weight of the substance to the weight of an equal volume of pure water at 4°C.
Spectrum, wave. A collection of waves of different lengths, usually covering a considerable range.
Spit. An elongate ridge of sand or gravel projecting from the mainland and ending in open water.
Spring. A flow of ground water emerging naturally onto the surface.
Spring: artesian. A natural outflow, at the Earth's surface, of water from a confined aquifer, usually through a fissure or along a fault.
Spring: gravity. Also known as an *ordinary spring.* A spring whose flow results directly from the force of gravity.
Stack. A small, prominent island of bedrock, remnant of a former narrow promontory destroyed by wave erosion.
Stalactite. An iciclelike form of dripstone and flowstone that hangs from the ceiling of a cavern.
Stalagmite. A blunt, iciclelike form of flowstone projecting upward from the floor of a cavern.
Steady state. A condition of dynamic equilibrium in which the rate of arrival of some materials equals the rate of escape of other materials.
Stock. A pluton, roughly circular or elliptical in plan, with an exposed area of less than 40 square miles (100km^2).
Stoping. A mechanism of emplacement of large plutons, involving repeated breaking off and engulfment of xenoliths (q. v.).
Strain. A change of shape or of volume, or of both shape and volume created by deforming forces. Strain is defined by the ratio $\frac{a-b}{a}$
where a is the original shape or volume and b is the changed shape or volume.

Strata: cross-strata. Strata inclined with respect to a thicker stratum within which they occur.

Strata: parallel. Strata whose individual layers are parallel.

Stratification. The layered arrangement of the constituent particles of a rock body.

Stratified drift. Sorted and stratified glacial drift.

Stratigraphic superposition, principle of. The principle which states that in any sequence of strata, not later disturbed, the order in which they were deposited is from bottom to top.

Stratigraphic trap. An oil trap resulting from variations in permeability controlled by stratigraphic relationships.

Stratigraphy. The systematic study of stratified rocks.

Stratum (plural, ***strata***). A definite layer of sedimentary or igneous rock consisting of material that has been spread out upon the Earth's surface.

Streak. A thin layer of powdered mineral made by rubbing a specimen on a nonglazed porcelain plate.

Stream. A body of water carrying suspended and dissolved substances and flowing down a slope along a definite path, the stream's channel.

Stream: antecedent. A stream that has maintained its course across an area of the crust that was raised across its path by folding or faulting.

Stream: braided. A stream that flows in two or more interconnected channels around islands of bed-load alluvium.

Stream capture. (Known also as ***stream piracy.***) The diversion of a stream by the headward growth of another stream.

Stream: consequent. A stream whose pattern is determined solely by the direction of the slope of the land.

Stream flow. The flow of surface water between well-defined banks.

Stream pattern: dendritic. A stream pattern characterized by irregular branching in all directions.

Stream pattern: rectangular. A stream pattern characterized by right-angle bends in the stream.

Stream pattern: trellis. A rectangular stream pattern in which tributary streams are parallel and very long. Known also as grapevine pattern.

Stream piracy. See ***Stream capture.***

Stream: subsequent. A stream whose course has become adjusted so that it occupies belts of weak rock.

Stream: superposed. A stream that was let down, or superposed, from overlying strata onto buried bedrock having composition or structure unlike that of the covering strata.

Stream system: well adjusted. A stream system in which most of the streams occupy weak-rock positions.

Stream terrace. A bench along the side of a valley, the upper surface of which was formerly the alluvial floor of the valley.

Strength. The ability of a rock body to resist stresses created by forces that tend to cause changes of volume or of shape, or of both volume and shape.

Stress. Force per unit area.

Stress: compressive. A stress paralleling the direction in which a body subjected to deforming forces tends to be shortened.

Stress: shearing. A stress causing parts of a solid to slip past one another, like cards in a pack.

Stress: tensile. A stress paralleling the direction in which a body subject to deforming forces tends to be elongated or pulled apart.

Striations. Scratches and grooves on bedrock surfaces, caused by grinding of rock against rock during movement of glacier ice. See also ***Glacial striation.***

Strike. The compass direction of the horizontal line in an inclined plane.

Strike-slip fault. A fault on which displacement has been horizontal.

Structural geology. The study of rock deformation and the delineation of geologic structural features.

Sublimation. The process by which particles of a solid pass directly into the gaseous state.

Sublimation, latent heat of. See ***Latent heat of sublimation.***

Submarine canyon. A sinuous, V-shaped valley, having a variable number of tributaries, that crosses part or all of a continental shelf and extends down a continental slope.

Submergence. Rise of sea level relative to the land.

Subsequent stream. See ***Stream: subsequent.***

Surf. Waves of translation landward of the breakers and seaward of the backwash.

Surface wave. (1) A seismic wave traveling along the surface of the Earth. (2) A wave traveling along the surface of a body of water.

Surf zone. The zone between the outermost breaker and the outhermost (lower) limit of the backwash.

Suspended load. See ***Load: suspended.***

Swash. The surge of water, from a breaking wave, that flows as a thin sheet up the sloping shore.

Syncline. A downfold with troughlike form and having the youngest strata in the center.

Talus. An apron of rock waste sloping outward from the cliff that supplies it.

Tangential component of gravity (g_t). The component of gravity, acting along a slope, which tends to pull particles downslope. The magnitude of the tangential component of gravity is determined by the sine of the angle of slope.

Tar sand. A sand or sandstone whose pores have been filled with hydrocarbons in the solid state, generally known as asphalt.

Tectonic cycle. The cycle which relates the larger structural features of the Earth's crust to the kinds of rocks that form in the various stages of development of these features and to gross crustal movements.

Tensile stress. See ***Stress: tensile.***

Tephra. A collective term designating all particles ejected from volcanoes, irrespective of size, shape, or composition.

Tephra cone. A small- to moderate-size cone composed of tephra.

Tephra flow ("Ash flow"). A fluidized mass of tephra, whose particles may be red hot, that flows like a liquid. A tephra flow may create a welded tuff (q. v.).
Terminal moraine. The end moraine deposited by a glacier along its line of greatest advance.
Terrace. A relatively flat, elongate surface, bounded by a steeper ascending slope on one side and a steep descending slope on the other. See also ***Stream terrace.***
Terrigenous sediment. Solid particles of sediment derived from erosion of the lands.
Tetrahedron (plural, *tetrahedra*). A solid with four sides, each of which is a triangle. The arrangement of the unit clusters of oxygen and silicon atoms in silicate minerals describes a tetrahedron.
Texture. The sizes and shapes of the particles, in a rock or a sediment, and the mutual relationships among them.
Thin section. A paper-thin, transparent slice, about 30 microns thick, of a rock, mounted on a glass microscope slide.
Thrust (or ***thrust fault***). See ***Fault: thrust fault.***
Tidal marsh. A low, flat, coastal area thickly grown over with saltwater grasses, in large part submerged at high tide.
Tidal wave. Popular and erroneous name for ***tsunami*** (q. v.).
Tide. The rhythmic rise and fall of the surface of the sea.
Till. Nonsorted glacial drift.
Tillite. Till converted to solid rock.
Tombolo. A beach that connects an island with the mainland or with another island.
Topographic map. A map that delineates surface forms.
Topography. The relief and form of a land surface.
Topset bed. Stream sediment that overlies the foreset beds in a delta.
Trade winds. The prevailing winds in tropical regions.
Transpiration. The passing of water vapor into the atmosphere from pores of plant tissues.
Transverse dune. See ***Dune: transverse.***
Trap. In petroleum geology, an arrangement of reservoir rock and roof rock that localizes a body of petroleum or natural gas.
Travertine. A collective variety of limestone, including dripstone, flowstone, and calcareous tufa.
Trellis pattern. See ***Stream pattern: trellis.***
Tributary: barbed. A tributary that forms an angle acute in the downstream direction at the point where it enters the main stream.
Triple point. The point on a pressure-temperature graph at which solid, liquid, and gaseous states coexist.
Tritium. A radioactive isotope of hydrogen, H^3, which is created in the upper atmosphere and circulates throughout the hydrosphere.
Tsunami. Long, low wave(s) generated by abrupt displacement of the sea floor or by landslides entering the sea. (From the Japanese *tsu,* "harbor" and *nami* "waves," pronounced tsoo-náh′-mē, and spelled the same in the singular and plural.)
Tuff. A fine-grained pyroclastic rock consisting of ash and dust.
Turbidity current. A density current whose excess density results from suspended sediment.
Turbulent flow. See ***Flow: turbulent.***

Ultramafic rock. Granular igneous rock consisting almost entirely of ferromagnesian minerals.
Unconformity. A lack of continuity between units of rock in contact, corresponding to a gap in the geologic record.
Unconformity: angular. Unconformity marked by angular divergence between older and younger sedimentary strata.
Unconformity: surface of. The contact of burial between two groups of unconformable strata. The plural form is *surfaces of unconformity.*
Uniformity, principle of. The concept that relationships established between processes and materials in the modern world can be applied as a basis for interpreting the geologic record and for reconstructing Earth history.
Unit cell. The fundamental building block of a crystal: the smallest part of a crystal lattice that displays the systematic pattern of the particles.
U-shaped dune. See ***Dune: U-shaped.***

Valley glacier. A glacier that flows downward through a valley.
Valley train. A body of outwash that partly fills a valley.
Vaporization, latent heat of. See ***Latent heat of vaporization.***
Vapor pressure. The pressure at which both the liquid and the gaseous states of aggregation exist.
Varve. A pair of laminae deposited during the cycle of the year. *Varve* is the Swedish word for cycle.
Vein. A tabular deposit of minerals, occupying a fracture, in which the particles grew away from the walls toward the middle.
Ventifact. A rock fragment with facets that have been cut by wind action.
Vesicles. Openings, generally ellipsoidal or cylindrical, formed in molten rock material by the expansion of gas escaping from solution.
Vesicular. An adjective describing rocks containing vesicles.
Viscosity. The tendency within a flowing body to oppose flow in any direction by acting in the opposite direction.
Volcanic ash. See ***Ash*** (volcanic).
Volcanic block. A large fragment of volcanic rock that was solid when blasted from the vent.
Volcanic bomb. A rounded, spindle-shaped mass of volcanic rock that was molten when blasted from the vent.
Volcanic crater. A funnel-shaped depression from which gases, tephra, and some lava are ejected.
Volcanic edifice. A feature built by a volcano.
Volcanic mountain. A conical accumulation of volcanic materials.
Volcanic mudflow. A mudflow of water-saturated, predominantly fine-grained tephra.

Volcanic neck. A cylindrical filling of an ancient volcano.
Volcanic shield. A broad convex mound of extrusive igneous rock, having surface slopes of only a few degrees.
Volcanism. A term designating the aggregate of processes associated with the transfer of materials from the Earth's interior to its surface.
Volcano. A vent or a fissure through which molten and solid materials and hot gases pass upward to the Earth's surface.
Volcano: active. A volcano that is erupting or that has erupted within the previous 50 years.
Volcano: dormant. A volcano that has erupted within historic time but not within the previous 50 years.
Volcano: inactive. A volcano that has not erupted within historic time.

Water balance. Also known as ***water economy.*** A quantitative statement of the amounts of water circulating through various paths of the hydrologic cycle.
Water gap. A pass, in a ridge or mountain, through which a stream flows.
Water table. The upper surface of the zone of saturation.
Wave base. The depth (equal to half the wavelength) at which the bottom begins to interfere conspicuously with the motions of the water particles beneath a shoaling wave. At wave base, deep-water waves become shallow-water waves.
Wave-built terrace. A body of wave-washed sediment that lies seaward of a wave-cut bench.
Wave-cut bench. A bench or platform cut across bedrock by waves.
Wave-cut cliff. A coastal cliff whose base has been undermined by waves and other marine agencies.
Wave height. The vertical distance from the bottom of a wave trough to the top of a wave crest.
Wavelength. The horizontal distance between successive wave crests or between successive troughs.
Wave of oscillation. A wave that causes water particles to oscillate but not to undergo appreciable net displacement.
Wave of translation. A wave that displaces the water particles within the moving crest.
Wave period. The time required for a wave to advance the distance of one wavelength.
Wave refraction. The process by which the direction of a series of waves, moving in shallow water at an angle to the shoreline, is changed so that waves become more nearly, but rarely exactly, parallel to the shore.
Wave steepness. The ratio of wave height to wavelength.
Wave velocity. The distance traveled by a wave in a unit of time.
Weathering. The chemical alteration and mechanical breakdown of rock materials during exposure to air, moisture, and organic matter.
Welded tuff. A fine-grained volcanic rock whose particles were so hot when deposited that they fused together.
Well-adjusted stream system. See ***Stream system: well-adjusted.***
Well-sorted sediment. A sediment consisting of particles all having about the same size.
Westerlies. The prevailing winds in the middle latitudes.
Wind gap. A former water gap through which a stream no longer flows.

Xenolith. A block, formerly part of the wall rock of a pluton, that has been broken loose and completely surrounded by igneous rock.

Zone of aeration. The zone in which the open spaces in regolith or bedrock are normally filled mainly with air.
Zone of saturation. The subsurface zone in which all openings are filled with water.

Index

Numbers of pages on which terms are defined are in ***boldface italics.*** Asterisks indicate illustrations.

Aa, volcanic rock, 495*
Abrasion, ***208, 644***
  by surf, 333
Absolute (Kelvin) temperature scale, 52
Absolute zero, 48
Abyssal fan, ***352, 644***
  Delgada, 371–372
  Hudson, 371*
  Monterey, 371–372
  origin, 371
  sediment volumes, 371, 372
Abyssal floor, 352, ***644***
Abyssal-hills province, ***352, 644***
Abyssal plain, ***352, 644***
  graded layers, 372
  Hatteras, 352*
  Nares, 352*
  North Atlantic, 372
  origin, 372
Accessory minerals, *see* Rocks, accessory minerals
Active volcano, *see* Volcano, active
Adirondacks, 517,* 519, 583, 621
Aeration, zone of, *see* Zone, of aeration
Africa, 25*
  Casablanca, 25*
  mountains, 40
  shields and plateaus, 40
Agate, 246, 609
Age determination, 127
Age of the Earth, 131
Ages, apparent, 129, 130, 132*
Agglomerate, 625, 626, ***644***
Aggregation, states, 48
  changes, 50–53
  gases, 50
  liquids, 49–50
  solids, 48–49
Agulhas current, in Indian Ocean, 37*
Aiguilles-Rouges massif, Haute-Savoie, France, thrusting, 527*
Air photographs, 633
Alaska current, in North Pacific, 37*
Alaska earthquake (1964), 7,* 401,* 413,* 435, 449*
Albite, 59, 108, 502, 606, 622
  crystallization temperature, effect of water-vapor pressure, 109
  melting behavior, effect of pressure, 109
Aleutian Trench, 478,* 548*
Alkali elements, 510
  in contact zones, 503
Alkali feldspars, 108,* 143*
Alluvial fan, *see* Fan(s)
Alluvial fill, ***223,*** 224,* ***644***
Alluvium, ***195,*** 223,* 291, 292,* 295,* 422,* ***644***
Al_2O_3, 105,* 458
Alpha-particle, 126, 128,* 596,* 601, ***644***
Alpine structure, 519, 523*
Alps, 40, 514*–515,* 516, 517, 519, 520*–521,* 546
Aluminum, in rock-making silicate minerals, 56
  ions, 145
  ore, 588*
  substitution for silicon in tetrahedra, plagioclase, 60
ALVIN, DSRV, 344, 345*
American Cordillera, ***40, 644***
Amorphous solid, 48, 49,* ***644***
Amphibole(s), 56, 57, 59, 62, 108, 144, 500, 604, 606, 618,* 629
  cleavages, 62, 65*
Amphibolite, 501, ***628,*** 629, 630,* ***644***
Amygdale(s), ***618,*** 619, ***644***
Anaerobic bacteria, 355, 572, 576, ***644***
  hydrogen sulfide, 355
Anchorage, Alaska, 1964 earthquake damage, 7,* 401,* 413*
Ancient marine strata, *see* Marine strata, ancient
Andalusite, 59, 500, 630
Andes, 40
  volcanic mountains, 39
Andesite, 620,* ***621,*** 622, ***644***
Andesite line, 459,* ***644***
Angle of repose, ***163, 644***
Angular unconformity, *see* Unconformity, angular
Anhydrite, 56, 585, 624
Anorthite, 59, 108
Anorthosite, 619, 620,* ***621,*** 622, ***644***
Antarctica, Taylor Glacier, folds, 508*
Antarctic Ice Sheet, 19, 40, 256, 259,* 276, 279
Antecedent stream, ***226,*** 227,* ***660***
Anthracite, 503, 504,* 630
Anticline, 406,* ***407,*** 576,* 577,* ***644***
Anticline, theory of oil accumulation, 577
Apatite, 605, 606
Apennine Mountains, Italy, 496
Apennine Mountains, Moon, 27*
Aphanitic rocks, 621
Aphanitic texture, ***614,*** 615,* 620,* ***644***
Appalachians, 40, 516, 517, 519, 545, 546*
  folded strata, 522,* 523*
  Great Valley, 517,* 522*
  post-Triassic, pre-Cretaceous dikes, 492*
  profile, 353*
  slaty cleavage, 505
  tectonic map and cross section, 517*
  thrusts, 523,* 528*
  Valley and Ridge Province, 517,* 519, 522*
    analysis of strata, 547
Appalachian structure, 519, 522*
Aquifer, ***239,*** 241,* ***644***
Arabia, northeastern, longitudinal dunes, 286,* 287*
Aragonite, 56, 606, 622, 624
Arbuckle Mountains, Oklahoma, debris of Pennsylvanian age, 422*
Archean, "great igneous interval" in Precambrian, 544
Arctic Ocean, cold-water current, 36
Artic zone, cold air masses, 37
Arête, ***266,*** 269,* ***644***
Argentine Basin, continuous seismic profile, 350*
  nonstratified pelagic sediment, 368
Argon, in atmosphere, 107,* 128
  radioactive decay product, 127,* 128, 129, 130, 131
  volcanic gases, 457, 458*
Arid climates, 289*
Arizona, Colorado Plateau, 12,* 14, 15*
  Dome Valley, mudcracks, 499*
  Grand Canyon, 12,* 14,* 15*
  Monument Valley, 10*
  Mount Emma, 13, 14*
  Mount Trumbull, 13
  Pine Mountains, 13, 14*
  Tuweep Valley, 13, 14*
  Uinkaret Plateau, 13, 14*
  Vulcan's Throne, 13, 14,* 15*
  Whitmore Canyon, 13, 14*

Arkose, 625, ***626, 644***
Artesian spring(s), 241,° ***242, 659***
wells, ***241°-242, 644***
Asama Volcano, Japan, 459,° 479
Ash, volcanic, ***644;*** *see also* Tephra
Asia, desert belts, 37
southeast, mountains, 40
Asphalt, 580
Astronomy, 13
Atitlán, Lake, Guatemala, 41°
Atitlán Volcano, Guatemala, 41,° 459°
Atlantic Coastal Plain, Pl. B, 40, 492,° 517
Atlantic Ocean, 24°-25,° 36, 37,°
Equatorial current, 36, 37°
Guinea current, 37°
Gulf Stream and North Atlantic current, 36, 37°
salinity, 354
Atlantic Ocean basin, 40
arcuate trenches, 40
fracture zones, 41, 559°
marginal bodies of sediment, 40, 351, 563, 566°
Atlantic Ocean floor, calcareous ooze, alternations of fossil species, 368
stratified pelagic sediment, 368
Atmosphere, 6, 7, 26, ***30,*** 31,° 35, 104, 114, 115,° ***644***
behavior of elements, 113–114
changes with height, 113°
chemical composition, 19, 35, 104, 107°
circulation, 15, 31, 35
ozone layer, 113°
reactions with solar radiation, 29, 30, 31,° 113°
Atoll, 466, 562, ***565,° 644***
Atom, ***595,*** 596,° 597,° ***644***
Atomic mass, ***644***
Atomic number, ***595,*** 596,° ***644***
Attitude, ***401, 645***
Augite, 59, 108, 144,° 498, 502, 606, 610, 622
Australia, mountains, 40
plains and plateaus, 40
spheroidal-jointed basalt, 147°
tektites, 32°
Authigenic sediment(s), 364, 365,° ***645***
Avalanche(s), 4
speeds, 166°
Axial plane, 406,° ***407, 645***
Azurite, 606

Backwash, 328,° ***645***
Badlands, 214–215,° ***291, 645***
Bar(s), 326, ***645***
breaker, 329
longshore, 327°
Barbed tributary, ***226,*** 228,° ***661***
Barchan dune, ***307,*** 308,° ***648***
Barite, 606, 622
Barrier, 319,° ***320,*** 331,° ***645***
progradation, 331°
Barrier reef, ***565,° 645***
Basalt, 9, 13, 15,° 620,° ***621,*** 622, 630, ***645***
chemical composition, 104, 105°
high TiO_2, 458
low TiO_2, 458
circumoceanic, 510, ***645***
extrusive, 509
intrusive, 510
metamorphism, 500, 501
oceanic, 509, 510, ***645***
origin, 488–489, 509–510
plateaus, 459,° 464, ***465,°*** 509, ***656***
radioactivity, 552
silica-saturated, lacking olivine, 458, 466
silica-undersaturated, containing olivine, 458, 466
spheroidal-jointed, 147
tongues of, 13, 14,° 15°
vesicular, 494,° 619°
Basalt glass, 620,° ***623, 645***
Base level, 7, ***199,*** 200,° 220,° ***645***
change, 224
local, ***200,° 645***
ultimate, ***199,*** 200,° ***645***
Base maps, 631
Basement complex (or rocks), 39,° 486°-487,° ***645***
exposed in interior parts of mountain chains, 523
Aiguilles-Rouge massif, 527°
Blue Ridge, 523
Mont Blanc, 520,° 523
New Jersey-Hudson Highlands-Green Mountains, 517,° 523
Southern Nevada and California, 518,° 523
exposed in marginal parts of mountain chains, 517,° 518,° 519, 520°-521°
Adirondacks, 517,° 519
Bighorn Mountains, 518,° 519
Black Hills, 518,° 519
French Central Plateau, 519
Front Range, 518,° 519
Laramie Range, 518,° 519
Vosges-Black Forest, 519, 520°
thrust over Paleozoic carbonate rocks, eastern Pennsylvania, 528°
Basin, aerated, 355°
evaporite, 355°
stagnant, 355°
structural, 405°
Basin and Range Province, western United States, P1. B, 518,° 552
fault blocks, 525–528
Batholith(s), 486°-487,°***495,*** 497,° ***645***
Boulder, Montana, 495, 497°
Coast Range, 495
Dartmoor, England, 499
Eastport, Maine, 499
Idaho, 495
in mountains, 517,° 518,° 523
rise of as cause of deformation, 555
"room problem," 510, 511
Batholith(s), Sierra Nevada, 527, 530°
thickness, 511
Bauxite, 140, ***155,*** 588,° 606, ***645***
Bay, 319°
Bay barrier, 319,° ***331, 645***
Beach(es), 279, 325,° ***328,*** 334,° 335,° ***645***
erosion, 323, 338–339
mainland, 319°
progradation, 331°-332
tombolo, *see* Tombolo
Beach drift, 328, ***645***
Beach dunes, ***307***
Beach sediments, 331°
Bed(s), ***379, 645***
Bedding, 379
plane, ***645***
Bedding-plane parting, 78, 376°-377,° 379,° 381,° ***645***
Bed load, ***190,*** 192,° 297, 298,° 299,° ***653***
Bedrock, 2–3,° 6, 7, 8, ***9,°*** 11,° 14, 15,° 16, 17, 19, 76, 96°-97,° ***645***
conversion to regolith, 100, 103
distinctive effects of glaciers on, 19, 264, 265,° 266,° 267°
exposed at Earth's surface 2–3,° 9,° 11,° 15,° 77°
geologic structure, 404, 405,° 406,° 407,° 408,° 409,° 410, 411,° 414°-415,° 416,° 417,° 419,° 420,° 522,° 523,° 524°-525,° 528,° 529,° 538°
Benguela current, in South Atlantic Ocean, 37°
Beryl, 57,° 59
crystal, 57°
X-ray diffraction photograph, 57°
Beta-particles, 126, 128, 596,° 601, ***645***
Bicarbonate ions, in average sea water, 107°
in average river water, 107°
Bighorn Mountains, Wyoming, 518,° 519
Biosphere, 103, 114, 115°
chemical composition, 105
Biotite, 47, 57, 59, 82,° 83,° 143,° 144,° 502, 571, 606, 619, 620,° 622, 628
Biotite schist, 501
Bituminous coal, *see* Coal
Black Hills, South Dakota, 518,° 519
Blake Plateau (sea floor), 352°
Blowout, ***303°, 645***
Body waves, *see* Seismic waves, body waves
Bolivian plateau, 40
Bonding, covalent, 61, 62
influence on breakage of minerals, 60
ionic, 61, 62
metallic, 61, 62
molecular, 61, 62
relationship to physical properties, crystalline solids, 62
Boron, in volcanic gases, 457

Bottomset bed, 207,° ***208, 645***
Boulder(s), 89,° 195,° 292,° 300,° 330,° ***645***
Boulder batholith, Montana, 497°
Boulder train, ***272,*** 280,° ***645***
Bowen, N. L., 108, 498
Bowen's reaction series, *see* Reaction principle (relationship)
"Box canyon," 291, 292°
Braided stream, ***207,*** 254, 255,° 274,° 275,° ***660***
Branford, Connecticut, cataclastic rock, 507°
partings in rocks, 77°
Triassic graded bed, 382°
Breaker(s), ***326,*** 328,° 331, ***645***
Breaker zone, ***326,*** 327,° ***645***
Breccia, ***623, 645***
sedimentary, *see* Sedimentary breccia
volcanic, *see* Volcanic breccia
Breckenridge, Colorado, contact zone, 501°
laccolith, 496°
Brenham, Kansas, stony-iron meteorite, 32°
Brown clay, *see* Deep-sea sediments
Brownian movement, 48, 89,° ***645***
Brunton pocket transit, 635°
Bulawayo, Rhodesia, granular disintegration, 149°
joint sheets, 150°
Bulk modulus, ***599, 645***

Cabot Strait, 360
Calcareous ooze, 626
Calcareous tufa, ***627, 645***
Calcite, 56, 62, 69, 85, 92, 603, 604, 605, 606
cleavage, 64°
Calcium, in evaporite minerals, 112
in rock-making silicate minerals, 56
in silicate melts, 106
Calcium carbonate, 10, 500, 623
mineral forms, 56
skeletons of organisms, 112
Calcium ions, 503
in average rainwater, 107°
in average seawater, 107°
in feldspars, 60
Calcrete, ***154;*** *see also* Caliche
Caldera(s), *see* Volcanic edifices, calderas
Caliche, ***154, 645***
California, Carrizo Plain, San Andreas fault, 429°
Death Valley, 291°
Devil's Post Pile National Monument, 78°
fission tracks in clam shell, 120, 121°
Great Valley, 518°
Imperial Valley, 96°–97°
Long Beach earthquake, 1933, 429,° 449°
Miocene strata northwest of Los Angeles, 361°
Mount Shasta Volcano, 459,° 462
California, Oakland landslide, 160, 161°
Pt. Firmin, slump in coastal cliff, 168°
surf and coastal cliffs, 2°–3°
Salton Sea and surroundings, 96°–97,° 585
San Andreas fault, *see* San Andreas fault
San Bernardino Mountains, 97°
Shuteye Peak, joint sheets, 151°
Sierra Nevada, glacially striated bedrock, 152°
Tomales Bay, coastal morphology, 316°–317°
California current, 37,° 354
Calving, 257°
Canada, British Columbia, Skeena River quarry, metamorphic rocks, 508°
Northwest Territories, deformed metamorphic rocks of Canadian shield north of Yellowknife and Great Slave Lake, 39°
Nova Scotia, Gerrish Mountain, extrusive basalt, 493°
Ontario, Parry Sound, igneous and metamorphic rock, 74°–75°
Quebec, Percé, contact of burial, 419°
vertical strata, 408,° 419°
wave-cut cliff, 408,° 419°
Canary current, 37,° 353, 354
Canyon, 7
CaO, 104, 105°
Cape St. Vincent, Portugal, 24°
Capillaries, rise of water, 50°
Capillary, 50,° ***646***
Carbon, in calcium carbonate, 114
in carbon dioxide, 114
diamond, 56
in dolostones, 114
in evaporite minerals, 112
in fats, 105
in geochemical cycle, 114, 115°
graphite, 56
in hydrocarbons, 105, 114
in limestones, 114
mineral forms, 56
in proteins, 105
radioactive isotope created from nitrogen, 113°
stable isotopes created from nitrogen, 114
Carbonate minerals, 629
Carbonate rocks, 500, 626, 630, ***646***
in Appalachian Great Valley, 521
Early Paleozoic age, North America, 521
solution, 250
Carbonate sediments, 85,° 339, 616°
Carbon dioxide, in atmosphere, 107°
in contact metamorphic zones, 503
in hydrosphere, 35
in rocks, 104, 105°
in volcanic gases, 457, 459°
in wollastonite reaction, 69
Carbonic acid, 141
Carbon monoxide, in volcanic gases, 457, 458°
Caribbean basin, arcuate trench at border with North Atlantic basin, 40
geophysical sections, 551°
Carlsbad Caverns, New Mexico, 246
Carlsberg Ridge, 559°
Carnotite, 606
Casablanca, 24, 25°
Cascade Range, air currents forced upward, 37
volcanic mountains, 39, 462, 518°
Cassiterite, 145, 606
Cataclastic rocks, 505–506, 507°
Cataclastic texture, ***618, 646***
Catastrophic changes, 18
Catastrophism, 18
Catskill, New York, deformed Ordovician strata, 396°–397°
Catskills, New York, residual mountains, 39, 516, 517°
Cavern(s), 232°–233,° 241,° ***246,*** 247,° 248,° ***646***
Cavitation, ***209, 646***
Cedros Trough, 548°
Cement, 616, ***646***
Cementation, *see* Lithification, cementation
Central Lowland, crustal warping, 405
Central Plateau, France, 519
Chalcedony, 607, 622
Chalcocite, 607
Chalcopyrite, 607
Chalk, 627, 646
northwestern Europe, 372
Challenger Deep, 352
Changes of state, 50–51
Channels, distributary, 208
Charles XIV, King of Sweden, 532
Chert, 609, 616, 617,° ***627,*** 630, ***646***
metamorphism, 500
Chichén-Itzá, Yucatán, 136°
Chile Rise, 559°
Chlorine ions, in average river water, 107°
in average seawater, 107°
in evaporite minerals, 112
Chlorite, 57, 59, 500, 501, 502, 503, 607, 622, 628, 629
Chlorite-muscovite-quartz rock, 503
Chlorite phyllite, 628, 629
Chlorite schist, 501, 502, 617,° 628, 629
Chute cutoff, ***205,*** 206,° ***646***
Cinnabar, 607
Circulation, atmosphere, 15, 31
deep ocean, 36
Circumoceanic basalt, *see* Basalt, circumoceanic
Circum-Pacific belt, 444, 446, 447°
Cirque, ***264,*** 266,° 269,° ***646***
Cirque glacier, ***256,*** 266,° ***646***
Clastic rock(s), 623–626, ***646***
Clastic sediment, ***84,*** 622, ***646;*** *see also* Detritus

Clastic texture(s), ***86–87, 616,*** 624, ***646***
Clay, 10, 89,° 195,° 300,° 625, ***646***
  clay minerals, 56, 57, 59, 62, 143, 500, 505, 624
    deep-sea sediments, 357°
    metamorphosed to aluminosilicate minerals, 112
  Norway, chemical composition, 104, 105°
  porosity, 236°
Claystone, 10, 613, ***614°***, 625, ***626, 646***
  porosity, 236°
Cleavage, mineral, ***62,*** 604, ***646***
  rock, *see* Rock cleavage
Cliff, coastal, 2–3,° 7, 168,° 317,° 330,° 334°
Climate(s), 4, 31, 36
  ancient, evidence, 558
    Tertiary, western Europe, subtropical, 318
  change, 224
  organisms, 318
  strata in deep-sea sediments, 368, 370°
Climatic regions, 30, 31
Clinometer, ***646***
Clinopyroxene, 59
Clinton iron ore, 586
Closed system(s), 66, 126, ***646***
Clouds, influence on solar energy, 30
Coal, 4, 519, 571, 572,° 573,° 625, ***627, 646***
  carbon residue, 504,° 505
  distribution, 573
  mining, 573, 575,° 576°
  occurrence, 572, 574°
  origin, 572
  regional metamorphism, 503, 504°
  seams, 572
Coast(s), crustal movement, 339
  emergence, ***340, 649***
  protection, 338–339
  sculptural evolution, 336,° 337°–338
  shifting depositional environments, 341,° 342°
  submergence, ***340, 660***
Coastal environments, outward shifting, 331°
Coastal features, 319°
    emerged wave-cut benches, California, 403°
  reefs, Marianas, southwestern Pacific, 403, 404°
Coastal Plain, Atlantic, 40
  Gulf, 40
Cobbles, 89, 195,° 300,° 330,° ***646***
Coccoliths, 356, 364
Coesite, *see* SiO_2 mineral forms: coesite
Cohesion, 91, 162, ***646***
Cohesive "soils," 89°
Col, ***266,*** 269,° ***646***
Collings Range Conglomerate, Ardmore, Oklahoma, 422°
Colluvium, 177, ***178,*** 269, ***646***
Colorado, Breckenridge, contact zone, 501°
Colorado, Breckenridge, laccolith, 496°
  Johnstown, stony meteorite, 32°
  Ouray, strata in angular unconformity, 419°
  Silver Basin, rock glaciers, 177°
  Spanish Peaks, plutons, 490,° 491°
Colorado Plateau, 12,° 14,° 516, 618°
  joints, 79°
  relief, 516
Colorado River, 97°
  regimen, 200, 201°
Columbia Plateau, basalts, 518,° 525
Column, 247,° ***248, 646***
  geologic, *see* Geologic column
Columnar joints, 498, 499,° ***646***
Compaction, *see* Lithification, compaction
Composite cone, *see* Volcanic edifices, composite cones
Composite dike, 489, 561, ***646***
Compressive stress, 260, 400,° 417,° ***660***
Concentrations, natural, 581
Conchoidal fracture, 66, ***604,*** 615,° ***646***
Concordant contact, *see* Geologic contacts, concordant
Concordant pluton, 490, 496,° ***646***
Concretion(s), 386–***387,° 646***
Cone of depression, ***239,*** 240,° ***646***
Confined aquifer, 241,° ***646***
Confined percolation, ***240,*** 241,° ***656***
Conformable strata, 391
Conglomerate, 10, 422,° 519, ***623,*** 625, ***646***
  deformed, 503
  metamorphosed, 507°
  porosity, 236°
Congruent melting, ***106,*** 109–110, 509, ***646***
Conical hills and mountains, volcanic, *see* Volcanic edifices, volcanic cones
Connate water, ***245,*** 585, ***646***
Connecticut, Branford, cataclastic rock, 507°
    partings in rocks, 77°
    Triassic graded layer, 382°
  central, fault blocks, 530°
  East Haven, sandstone containing feldspar, 82, 83°
  gravestones, weathering, 510°
  Guilford, granitic rock, irregular contact, 510°
  Quinebaug River, flood, 198°
Consequent stream, ***225,*** 227,° ***660***
Contact(s), geologic, *see* Geologic contacts
Contact-metamorphic aureole, 500, ***646***
  dimensions, 500, 501
Contact metamorphism, 499–501, ***646***
  basalt, Skye, Scotland, 501
  greenstone, Vermilion district, Minnesota, 501
  slates, Northern Minnesota, 501
    by granitic pluton, 501
Contact metamorphism, slates, by gabbroic pluton, 501
Continent, ***39, 646***
  annihilation, 16
  average altitude, 39
  changes in relative positions, 19; *see also* Continental drift
  crust, *see* Earth's crust, continental
  former, ruins, 15
Continental drift, 553,° ***646***
Continental margin, 352°
  east of New Jersey, geophysical profile, 566°
Continental mass, ***39,*** 346, 537,° ***647***
  continual elevation, 43
Continental rise, ***351,*** 352,° 353,° 371, ***647***
Continental shelf, 39, ***340,*** 346, 363, 371, 478,° 563, 566,° ***647***
  sediments, 341
  strata, deformed, 341
    gently inclined wedges, 341
    records of subsidence, 341
Continental shield(s), 39, 463, 546,° ***647***
  Africa, 40
  Brazil, 40
  Canadian, 39, 486°–487,° 526,° 547°
Continental slope, ***351,*** 352,° 363,° 371,° 478,° 566, ***647***
Continuous reaction, *see* Reaction principle (relationship), continuous reaction
Continuous seismic profiler, 348–349, ***647***
Continuous seismic profiles, 350,° 361, 363°
Contour(s), ***632,*** 634,° ***647***
  current, 354,° ***647***
  interval, ***633, 647***
  line, ***632,*** 634,° ***647***
Contraction theory, 544–545
Convection current, ***647***
  in Earth's core, 443
  in Earth's mantle, 553–554, 562,° ***647***
Copper, 607, 622
Coquina, 85,° ***626, 647***
Cordillera, ***39, 647***
Core(s), 359, 360,° 367, 368, 370,° ***647***
  Earth's, *see* Earth's core
Corers, 349, 351°
Coriolis, Gaspard Gustave de, 34
Coriolis effect, 353
Coriolis force, 34, 35, ***647***
Correlation, ***124,*** 392, ***647***
Corundum, 583, 605, 607, 622
Cosmic dust, 86
Cosmic rays, 113
Cotopaxi Volcano, Ecuador, 459,° 462
Coupled substitution, plagioclose feldspars, 108
Covalent bond, 61, ***647***
Crater(s), volcanic, *see* Volcanic edifices, craters
Crater Lake Volcano, Oregon, 459,° 462, 464°

Creep, 8, 162, ***174,**** 175,* 176, 179,* ***647***
Crevasse(s), 257,* ***262,*** 273,* ***647***
Cristobalite, *see* SiO_2 mineral forms: cristobalite
Critical point, 52
  of water, 51,* 476*
Critical pressure, 52
  of water, 51,* 52–53
Critical temperature, 50–52, 544, ***647***
  of water, 51,* 52
Cross-strata, 192, 193,* 194,* 204, 307, 326, 327,* 328, ***380,*** 381,* 404, 408,* ***660***
  beach, 331
  eolian, 311
  original tops up, 408*
  overturned, 408*
Crust, Earth's, *see* Earth's crust
Crustal movements, 224, 401–404
  warping, 405, ***647***
Crystal(s), 46*–47,* 57, 61,* 62,* 63,* 64,* 66,* 67,* 624, ***647***
  conditions of growth, 60
Crystal face, ***60, 647***
Crystal form, 60
Crystal lattice(s), 56, 58,* 60,* 61,* 65,* 603, ***647***
  water in, 502, 503
Crystalline particle, ***647***
Crystalline solid, 48, 49,* ***647***
Crystalline textures, ***86,*** 613, 617, 624, ***647***
Curie point, ***43,*** 482, 562,* ***647***
Current(s), convection, *see* Convection currents
  cool, 37*
  equatorial, 36, 37*
  rip, *see* Rip current(s)
  surface, ocean, 36, 37,* 353, 354
  warm, 37*
Cutoff(s), 206*
  chute, ***205,*** 206,* ***646***
  neck, ***205,*** 206,* ***646***
Cycle(s), of causation (Dutton's), 13
  daily, daylight and darkness, 115
  of erosion, ***219,*** 229, ***647***
    in deserts, 295–296*
  geochemical, *see* Geochemical cycle
  geologic, *see* Geologic cycle
  "great geological" (Hutton's), 13, 15, 17
  hydrologic, *see* Hydrologic cycle
  Moon's orbital, 116
  polar wobble, 116
  rock, *see* Rock cycle
  sunspot, *see* Sunspot cycle
  tectonic, *see* Tectonic cycle
  volcanic eruption, *see* Volcanic eruption cycle(s)
  yearly, seasonal, 115, 116*

Dana, J. D., 545
Darcy's Law, 238, ***647***
Dartmoor granite, southwestern England, 499
Darwin, Charles, 17, 124, 125, 404
Darwin Rise, Pacific Ocean basin, 563, 564*
Dating, isotopic, 125
Daughter isotope, *see* Isotope, daughter
Death Valley, California, 291*
Debris fall, ***165****
Debris flow, 161,* ***165,**** 170,* ***647***
  rate, 166*
  variety mudflow, *see* Mudflow
Debris slide, ***165****
Decay constant, 125, ***647***
Declination, magnetic, ***555,*** 557,* ***648***
Decomposition, ***140,*** 141, ***648***
Deep-focus earthquake, *see* Earthquake(s), depth of focus: deep-focus
Deep sea, disposal of radioactive wastes, 346
  exploration, 347–351
Deep-sea floor, abyssal floor, 352
  linear magnetic anomalies, 559, 560, 561,* 563*
  linear sediment streaks, 357, 358*
  magnetic-polarity record, 561
  major features, 351–353
  manganese nodules, 346, 365, 367*
    ripples, asymmetrical, 357, 358*
    symmetrical, 357, 359
Deep-sea sediment(s), 346
  authigenic, *see* Authigenic sediment(s)
  brown clay, 357,* ***364,*** 365,* 366,* 368, ***645***
  changes of climate, records, 346
  chemical processes, 364
  chronology, 369
  classification and distribution, 364
  clay minerals, 357*
  Early Cretaceous age, 369
  exposed on land, 346
  Foraminifera, 357*
  former, *see* Former deep-sea sediments
  glacial-marine, *see* Glacial-marine sediment
  graded layers, 359, 360,* 364
  nonstratified, 367–368
  organisms, 356
  oxygen isotopes, 368
  paleomagnetic polarity episodes, 369
  pelagic sediments, *see* Pelagic sediments
  processes, physical, 356–364
    lateral transport, 356–357
    vertical fallout, 356
  radiocarbon dates, 368
  rates of accumulation, 369, 371*
  sand flows, 360, 362*
  shallow-water sediments displaced by gravity, 365,* 366,* 367
  sources, 355–356
  stratification, 368
    alternating layers, pelagic and other sediments, 368
      fossil species, 369
      pelagic sediment, 368
    climatic significance, 368
Deep-sea sediment(s), tephra, *see* Tephra, deep-sea sediments
  terrigenous mud, *see* Terrigenous mud
  terrigenous sediments displaced by gravity, 367
  thickness, 368–369
  Upper Cretaceous, horizon A, 368
Deep submersible vehicles, 344, 345,* 349, 351
Deep-water waves, *see* Waves, water surface: deep-water waves
Deflation, ***302, 648***
Deflation armor, ***303,**** 304,* ***648***
Deflation basin(s), ***302, 648***
Deformation, crustal, evidence, 401–405
    displacement, man-made structures, 401*
      natural features, 402
    exposure of rocks formed at depth, 404
    structure of bedrock, 404–405
  Geologic age, 418–423
    by unconformity, 423
    effects on sediments, 421,* 422*
    isotopic dates, 418
    relative age, 418
Delaware River, Pennsylvania, adjustment, 228–229*
  slaty cleavage exposed, 505
Delta, ***207,**** ***648***
  tidal, *see* Tidal delta
Dendritic pattern, ***225, 660***
Density, ***648***
  rocks and minerals, 622
Density current, ***354,**** 355,* ***648***
Deposition, 102,* 103
Deposits, mass-wasting, 166
Depth of focus, *see* Earthquake(s), depth of focus
Depth sounding, acoustic, *see* Echo sounding
  lead weight, 347
Descending air, 37
Desert(s), 5* ***288, 648***
  climate, 288–289
  cycle of erosion, 295–296*
  distribution, 289–290
  pediments, 294*
  processes, 290
  winds, 294
Detritus, ***84,*** 624, ***648***
Deuterium, 595
Devil's Post Pile National Monument, California, polygonal (columnar joints), 78
Devil's Tower, Wyoming, polygonal (columnar joints), 499*
Diamond, 56, 66, 605, 607
  temperature-pressure graph, 67*
Diastrophism, ***517, 648***
Diatoms, 112, 364
Dickite, 63*
Diffusion, ***49, 648***
Dike(s), ***489,*** 490,* 491,* 492,* 493,* ***648***
  composite, 489, ***646***
  Great Dike, Rhodesia, 489

Dike(s), mineral growth, 489
  patterns, 489
  swarm, 489, ***648***
    eastern United States, 489, 492°
    parallel, 489, 490,° 491,° 492°
    radiating, 481,° 489, 491°
    Scotland, 489
    Spanish Peaks, Colorado, 491
Diopside, 59, 500, 630
Diorite, 498,° 619, 620,° ***621***, 622, ***648***
Dip, ***635***, 636,° ***648***
  initial, 381°
Dipole, 555, ***556***, ***648***
Dip-slip fault, 414,° ***648***
Directions of movement, dunes, 307°
  glaciers, 276
  point bars, 204,° 205°
  sand waves, 194°
Disasters, natural, 5
Discharge, stream, ***189***, 190,° ***648***
Disconformable strata, 391
Disconformity, 420, ***648***
Discontinuous reaction relationship, *see* Reaction principle (relationship), discontinuous reaction
Discordant contact, *see* Geologic contacts, discordant
Discordant pluton, 495, ***648***
  batholiths, *see* Batholith(s)
  stocks, *see* Stock(s)
Disintegration, ***140***, 145, 146, ***648***
Divide, ***216***, ***648***
Divining rods, 242
Dolerite, 619, 620,° ***621***, ***648***
Dolomite, 56, 114, 500, 605, 607, 622, 624
Dolostone, 339, 500, 616, 621, 623, 625, ***627***, ***648***
Dome, 405°
Dome Valley, Arizona, mud cracks, 499°
Dormant volcano, *see* Volcano, dormant
Dowsing, 242
Drag, sea floor, oscillating, wave, 326, 329
  stream bed, 191, 326
Drainage basin, ***216***, 217,° ***648***
Drift, continental, *see* Continental drift
  glacial, *see* Glacial drift
  ice-contact stratified, *see* Ice-contact stratified drift
Drilling ships, 349
Dripstone, ***246***, 247,° 248, ***648***
Drumlin, ***271***, 272,° 280,° ***648***
Dry valley, 249
Dune(s), 7, ***305***, 306,° 316,° 320,° 327,° ***648***
  form, 307°
    barchan, *see* Barchan dune
    longitudinal, *see* Longitudinal dune
    transverse, *see* Transverse dune
    U-shaped, *see* U-shaped dune
  sand grains, 306
  stratification, 306, 307°
Dunite, 619, 620,° ***621***, 622, ***648***
DuQuoin, Illinois, strip mine, 9°
Dust, 7, 298,° 299
  cosmic, 86
  volcanic, 479, 480, 625, ***648***; *see also* Tephra
"Dust bowl," 298,° 303, 312
Dutton, C. E., 13
Dynamic metamorphism, 505–509
  slaty cleavage, 505, 506,° 528,° ***659***
  stretched and flattened pebbles, 507°

Earth, 4, 26, 27,° 29, 30,° 31,° 33,° 321°
  active hidden forces, 44
  age, *see* Age of the Earth
  atmosphere, *see* Atmosphere
  circumference, 33
  core, *see* Earth's core
  crust, *see* Earth's crust
  curvature, 24°–25°
  diameter, 31
  elastic properties, variation with depth, 440, 441
  environments, 4
  equatorial bulge, 34
  external forces, 6
  former surfaces, 17
  general view of, 23–44
  gravity, *see* Earth's gravity
  history, 6, 16
    relation to Earth's present features, 13
    written in rocks, 4, 15
  hydrosphere, *see* Hydrosphere
  interior, *see* Earth's interior
  internal forces, 6
  internal heat, 28, 41, 488, 509, *see also* Geothermal gradient
  lithosphere, *see* Lithosphere
  mantle, *see* Earth's mantle
  mean density, 602
  photographed from Gemini satellites, 24°–25,° 96°–97,° 286°–287°
  rotation, 33, 35, 354
  size and shape, 31, 33
  specific gravity, 428
  surface, *see also* Earth's morphology
    aggregate area, 33
    surface, former, 17
    layers of material spread out on, 10
    subsidence, 100, 103
  theories of, 14
  thermal history, ideas based on radioactivity, 17, 544–545
Earth-Moon pair, 27,° 321°
  center of mass, 27,° 321°
Earthquake(s), 418, 428–429
  California (1857), 429°
  causes, 429, 432–437
  damage, 7°
  depth of focus, 446
    deep-focus, ***446***, 447,° ***649***
      mechanism, 446
Earthquake(s), depth of focus, intermediate-focus, ***446***, 447,° ***649***
    shallow-focus, ***446***, 447,° ***649***
  energy released, 448
  epicenter, ***433***, 434, 440, ***649***
  Erzincan, Turkey (1939), 431°
  focus, ***433***, ***649***
  intensity, ***448***, ***649***
    map, 448°
    Mercalli scale, 448, ***654***
    Richter scale, 449, ***657***
  Lisbon, Portugal (1744), 429
  Long Beach, California (1933), 429°
  magnitude versus intensity, 449°
  Niigata, Japan (1964), 426°–427°
  Pleasant Valley, Nevada (1915), 435
  Prince William Sound, Alaska (1964), ("Alaska earthquake"), 7,° 407,° 413°
  San Francisco (1906), 446
Earthquake waves, *see* Seismic waves
Earth's core, ***441***, 442,° 445,° 551, ***649***
  composition, 441
  fluid outer part, 443
  solid inner part, 443
  source of magnetic field, 443
Earth's crust, ***37***, 428, 443,° 445,° ***647***
  composition, 443
  continental, 443, 445,° ***647***
  movements, energy source for sea waves, 324
  oceanic, 443, 445,° ***647***
  thickness, 443
Earth's gravity, cause of downslope flow of water, 33
    of downslope flow of rock material, 33
    of spherical form, 33°
  components, 33°
    see also g_n; g_t
  interaction with centrifugal force of rotation, 34°
  normal component, *see* g_n
  shaper of surface of sea, 33
  tangential component, *see* g_t
  variations, 34
Earth's interior, 9, 10, 17, 441–444; *see also* Earth's core; Earth's crust; Earth's mantle; and Lithosphere
Earth's magnetic field, 30,° 42, 556,° ***653***
  ancient shape, 556
  changes in polarity, 19
  lines of force, 30
  protection against charged particles from Sun, 30, 42
  reversals, 43, 131, 133,° 562,° ***653***
    dates, 131
  uses, by geologists, 42
    by mariners, 42
  variations, 42
    atmospheric phenomena, 42
    changes within Earth's interior, 42
    differences in content of magnetic-iron compounds, near-surface

Earth's magnetic field, variations, rocks, 42, 577, 578°
Earth's Magnetic Poles, 43, ***653***
changes in positions, 19
Earth's mantle, ***443,*** 445,° 535, 551, ***649***
convection currents, 553–554
low-velocity zone, 443
partial melting, 510, 552
thickness, 443
Earth's morphology, 6, 9, 15, 17, 19, 76, 103, ***654***
changes, 18
distinctive effects of glaciers, 19
East Pacific Rise, 40, 559°
authigenic sediments, 365
Echogram, 478°
Echo sounder, 348,° 559, ***649***
frequency, 348
precision recorder, 348, 349°
Echo sounding, 347–348°
Ecliptic, plane of, *see* Plane of the ecliptic
Edinburgh, city, 13
Royal Society of Edinburgh, the, 14
Universities of Edinburgh, 13
Edmonson County, Kentucky, karst topography, 250°
Edmonton, Kentucky, nickel-iron meteorite, 32°
Eh, ***68,° 649***
Elastic deformation, 433, ***649***
Elastic limit, 399, ***649***
Elastic rebound, 433, 434,° 436, ***649***
Elastic waves, 429, 599–601
Electromagnetic waves, 28–30, 597, 600°
speeds, 30
spectrum, 599,° 600
Electron, ***595,*** 596,° 597,° ***649***
Element, ***595,*** 596°
Emergence, *see* Coast(s), emergence
End moraine, 255,° 257,° ***271,*** 272,° 273,° 280,° ***654***
Energy, 17, ***649***
coal as, 571–572
geothermal, 482
industrial consumption, 570, 571°
nuclear, 580
relation to matter and electromagnetic spectrum, 28, 597, 599
solar, *see* Solar energy
sources, 571
England, Dartmoor granite, 499
Oxford, weathering of limestone, 138°
Environment(s), 19
coastal, *see* Coastal environments
of deposition, 387
problems, 6
relationships to man, 4
Environmental realm, rock-making, 9, 78–79
Eolian sediments, 305–311
ancient, 311°
Epeirogeny, ***517, 649***
Epicenter, see Earthquake(s), epicenter
Epidote, 59, 500, 502, 607
Equator, 31° 33,° 37,° 321, 354
Magnetic, *see* Magnetic Equator
Equatorial bulge, countercurrents, 37°
Earth's currents, 36, 37°
region, more than average receipt of solar energy, 31°
Equigranular texture, ***613,*** 615,° ***649***
Equilibrium, 66, 68, 69
dynamic, 66, 69
coastal, 330
between environment(s) and mineral(s), 64
static, 66
Equilibrium profile, 179
Erosion, ***7,*** 13, 17, 26, 102,° 103, 488, ***649***
agents, 7
accelerated, ***221***
beach, *see* Beach(es), erosion
cycle, *see* Cycle(s), of erosion
rates, 110, 111,° 222
remnants, thick sandstone layer, Monument Valley, Arizona, 10°
sheet, *see* Sheet erosion
Erratic, ***272,*** 273,° ***649***
Eruption cloud, volcanic, 8,° 454°–455,° 463,° 473°
Esker(s), ***274,*** 275,° 276,° ***649***
Essential mineral(s), *see* Rocks, essential minerals
Eurasia, European sector, average altitude, 40
Eustatic change of sea level, ***340, 649***
Evaporite(s), 616, 625, ***627, 649***
Evaporite minerals, ***100,*** 112, 339, ***649***
Excursion, geologic, 12
Exfoliation, 138,° ***147,°*** 148, ***649***
Existing causes, relation to past changes, 16, 18
Explosion(s), nuclear, underground, 17, 449
Explosion pits, 460, 474,° ***649***
resemblance to impact craters, 460
Exposure, 77, ***649***
Extrusive igneous rock(s), ***79,*** 102, 103, ***649***
silicic, 510
see also Volcanic rocks

Fabric, 617,° 618,° ***649***
preferred orientations, 74°–75,° 617,° 618,° ***649***
Fabric, random, 87, 617,° 618,° ***649***
Faceted spur, 254°
Facets, rock particles, 269, 270,° 383
Facies, metamorphic, *see* Metamorphic facies
mineral, *see* Mineral facies
sedimentary, ***388, 649***
Falkland current, 37°
Falls, 209
Fan(s), ***201,°*** 291,° 294,° ***649***
abyssal, *see* Abyssal fan
Farallón de Pájaros Volcano, Marianas, southwest Pacific, 459,° 463°
Fats, 105
Fault(s), ***78,*** 296, 398, 408,° 409, 410, 411,° 412,° 413,° 414°–415,° 417,° 422,° ***649***
abrupt movements, 433, 435, 436
attitude, 410
bending of strata, 412°
blocks, in central zones of mountain chains, 525
central Connecticut, 530°
Frenchman Mountain, Nevada, 530°
Sierra Nevada, California, 530°
breccia, 413, ***649***
drag, 412,° 413, ***649***
footwall, 410, 414°–415,° ***649***
gradual movement, 436
hanging wall, 410, 414°–415,° ***649***
movement, 410
effect(s) on rocks, 410, 412,° 413
relation to folds, 417–418
to joints, 417°
scarps, 401,° 411,° 413,° 414,° 415,° 422,° 435, 436°
strike-slip, *see* Strike-slip fault
thrust, *see* Thrusts (or thrust faults)
Fault-block mountain, 527, 530,° ***649***
Fault scarp, 411,° 413,° 414,° 415,° 417,° 422,° 435
Fauna, 124
Faunal succession, law of, *see* Law of faunal succession
Feldspar(s), 46, 49, 82,° 83,° 604, 608, 618,° 622, 624, 628, 629, 630
Felsic rock, 619, 620,° ***650***
Felsite, 620,° ***621, 650***
FeO, 104, 105°
Fe_2O_3, 105°
Ferromagnesian silicates, 57, 58–59, ***650***
ions in: iron, 57
magnesium, 57
reaction relationships, 106, 108–110
Ferrous ions, in micas, 143
contact-metamorphic zones, 503
ferromagnesian minerals, 143
magnetite, 143
oxidation, 145
Fetch, 324
Field measurements, 634
Fiord(s), ***267, 650***
Fish, fossil, 85°
Fission, 28
Fission tracks, 120, 121°
Fissure, 410, 417,° ***650***
eruption, 464, ***650***
parallel, 417°
veins, 583,° ***584, 650***
Flint, 609, ***650***
Flood(s), 4, 6,° 18
control, 207
Floodplain, ***205,*** 206,° ***650***
Florida, Cape Kennedy, sea-floor profile, 352°
Florida current, average surface speed, 353
quantity of water flowing, 353

Flow, laminar, *see* Laminar flow
  overland, *see* Overland flow
  plastic, *see* Plastic flow
    of regolith, 8
  stream, *see* Stream flow
  turbulent, *see* Turbulent flow
Flow breccia, ***623, 650***
Flow law, 260, ***650***
Flow layering, 614,° ***650***
Flowstone, ***246,*** 247,° ***248, 650***
Fluid, 50, ***650***
Fluidization, 167
Fluidized, 91, ***650***
Fluorine, in volcanic gases, 457
Fluorite, 604, 605, 608, 622
Fold, 396°–397,° 406,° 407,° 408,° 416,° 504,° 506,° ***650***
  closed, 396°–397,° 408,° ***650***
  complex, 506, 507,° 529°
    in glaciers, 508,° 529°
  isoclinal, ***407,*** 408,° ***650***
  open, ***407, 650***
  overturned, 396°–397,° ***407,*** 506,° ***650***
  plunging, 407,° ***650***
  recumbent, 409, 508,° ***650***
Fold axis, 406,° ***407, 650***
  plunge, ***407, 650***
Foliate, ***627,*** 628,° 629, ***650***
Foliation, ***627, 650***
  gneissic, 628, ***650***
  schistose, 628, ***650***
  slaty, 628, ***650***
Footwall, *see* Fault(s), footwall
Foraminifera, 356, 357,° 364
  deep-sea sediments, 357°
Forces, interatomic, 48
  intermolecular, 48
  lines of (magnetic), 30
Foreset bed, 207,° ***208, 650***
Formation, ***391,*** 636
Former deep-sea sediments, 372
Former lava flow(s), ***80,*** 463,° 465, 472,° ***650***
Former surfaces of the Earth, 17
Fossil fuels, 571, ***650***
Fossils, ***85,°*** 384, 389, 390, 392,° ***650***
  carbonaceous residue, 85°
  environmental analysis, 318
  impressions, 85°
  time significance, 122
Foundation rocks, *see* Basement complex (or rocks)
Fractional melting, *see* Incongruent melting
Fracture, conchoidal, *see* Conchoidal fracture
  mineral, ***63, 650***
  zones, ***41,*** 559,° 560,° ***650***
Fractures, in bedrock, 42, 74, 77, 78, 79, 80, 335, 408, 418
  in granite, 151°
Fragmental texture, ***616,°*** 624, ***650***
Framework, 616, ***650***
Frank, Alberta, rock avalanche, 166–167, 169°
Franklin, Benjamin, 479
Freshwater, derivation from conversion of saltwater, 35
  flow into the sea, 99, 101°
  interception and storage on land, 35
Freshwater limestone, ***627, 650***
Fringing reef, ***565,° 650***
Frost heaving, 146, 175,° ***176, 650***
Frost wedging, 145, ***146,° 650***
Frozen ground, perennially, 42, 43°
Fumarole(s), 473, 474,° ***650***
Fusion, of hydrogen into helium inside Sun, 28

g, ***33,*** 34,° 190,° 261,° 262,° 428, 601, ***651;*** *see also* Earth's gravity
  variations, 535
g_w, 33,° 162, 190,° 261,° 262,° ***655***
g_t, 33,° 102, 162, 190,° 237,° 261,° 262,° 329, ***660***
G, ***28,*** 601, ***650***
Gabbro, 498,° 619, 620,° ***621,*** 622, ***650***
Gaging station(s), 196, ***650***
Galápagos Rise, 559°
Galena, 56, 62, 603,° 605, 608, 622
Gamma-rays, 113,° 600,° 601
Gangue, ***582, 651***
Garnet, 59, 60, 62,° 500, 501,° 502, 604, 617,° 622, 628, 629, 630
  change from congruent to incongruent melting at high pressures, 110
  crystal form, 62°
Gas, natural, 573, ***655***
Gases, 50
  motions of particles, 50
  properties, 50
  spacings of particles, 50
  volcanic, *see* Volcanic gases
Geanticline, 545
Gemini satellites, photographs from, 24,° 25,° 96°–97,° 286°–287°
Geochemical cycle, 98, 99,° 103, 113, 114, ***651***
Geochemistry, ***103, 651***
Geode, ***246,° 651***
Geologic column, 122, 123,° ***124, 651***
  calibration, 124
Geologic contacts, 405
  concordant, ***489,*** 494,° ***646***
  discordant, 489, ***648***
  irregular intrusive, 510°
  resulting from burial, 418, 419
    comparison with faults, 420, 421
    unconformity, 419°; *see also* Unconformity
Geologic cross-section, 406,° 638,° ***639, 651***
Geologic cycle, 17, 26, 28, 44, 76, 98, 99,° ***651***
Geologic excursion, 12
Geologic history, reconstruction from study of geologic record, 12
Geologic map(s), 406,° 631, ***634, 651***
Geologic processes, mechanics, *see* Mechanics of geologic processes
  quantities involved, *see* Quantities involved, geologic processes
Geologic processes, rates and speeds, *see* Rates and speeds, geologic processes
Geologic record, 6, 76, ***651***
  incompleteness, 18
  components, 9,° 18
Geologic structure, *see* Bedrock, geologic structure
Geologic thermometers, 66, 583
Geologic time, 122, 130
  great length, 19
  scale, 132,° ***651***
Geologists, 18, 19
  marine, diving, 319°
Geology, ***6,*** 13, ***651***
  central principles, 6, 19
  goal of, reconstruction of past events, 18
  impact on human affairs, 19
  science of human eyes and brains, 17
  scientific method in, 17
Geophysical exploration (prospecting), 577–578,° ***651***
Georgia, Stone Mountain, residual mountain, 39
Geosyncline(s), ***545***–550, ***651***
  ancient, 545, 546,° 547°
    relation to continental shields, 546
  modern, 547–550
Geothermal gradient, ***41,*** 509, 552, ***651***
Geothermal heat, 102
Geyser, 475, ***651***
  basis of intermittent activity, 475–476
Ghana, rocky coast, 328,° 332
Gibraltar, Strait of, 24–25°
Glacial ages, 274–282
  causes, 282–283
  crustal subsidence, 277
  dating, 278
  erosion, 276
  extent of glaciers, 275, 277,° 278°
  loess, 310
  permafrost, 282
  pluvial lakes, 280, 281°
  sea level, 278–279, 281°
Glacial deposits, 269
Glacial drift, ***269, 651***
Glacial erosion, 152°
Glacial-marine sediment, 365,° 366,° 367, ***658***
Glacial polish, 152,° ***651***
Glacial sculpture, 266, 269°
Glacial striation(s), 152,° 264, 265,° ***6***
Glacial transport, 267
Glaciated valley, 266, 267,° 269,° ***651***
Glaciation, 256, ***264, 651***
Glacier, ***256, 651***
Glaciers, directions of flow, 276
  flow law, 260, ***650***
  economy, 262
  erosion, 268,° 269°
  existing, 19
  movement, 259–261,° 262°
  origin, 257–259
  rates of flow, 166°
  relation to climate, 263, 272°

Glaciers, in the rock cycle, 262
shrinkage, 263°
velocity curve, 261
Glassy texture, 614, 615,° 620,° ***651***
Gneiss, 510,° 628, ***629, 651***
contact metamorphism, 501
Gold, 145, 589,° 608, 622
Gold dredge, 590°
Graded layers, ***382,°*** 408,° ***651***
deep-sea sediments, 359, 360°
origin, 359
original top up, 408°
overturned, 408°
turbidity currents, 359, 364
Graded stream, 218
Gradient, stream, ***189,*** 190,° ***651***
Grand Canyon, Arizona, 14,° 15°
Grand Coulee, Washington, 178,° 180
Granite, 9, 82,° 84, 498,° 614, ***619,*** 620,° 622, ***651***
batholith, 499, 510; *see also* Batholith(s)
chemical composition, 104, 105°
Dartmoor, 499
granitic plutons, contact aureoles, 510
in mountain belts, 39, 516, 523
radioactivity, 552
southern Rhode Island, 503, 505°
stocks, 510
Granite gneiss, ***628, 651***
Granite pegmatite, 621
Granite porphyry, 616
Granitic rock, 486,° 487,° 498,° 510,° 619, 620,° ***651***
Granitization, ***510,*** 511, ***651***
Granodiorite, 619, 620,° ***621, 651***
Granular disintegration, 149,° 150,° 152,° ***651***
Granular texture, 614, 615,° 620,° ***651***
Graphite, 56, 66, 67,° 608, 622, 630
Gravel, 10, 89,° 90,° 195,° 300°
porosity, 236°
Gravitation, law of, 13, ***28,*** 428, ***651***
Gravity anomaly, 535, 551,° ***652***
Gravity meter (also gravimeter), 577, ***652***
Gravity prospecting, 577, ***652***
Gravity sliding, cause of folds, 553, 554°
Gravity spring, 240, ***659***
Graywacke, 625, ***626, 652***
"Great geological cycle" (of Hutton), 13, 15, 17
Great Lakes, deformed Pleistocene shorelines, 533°
Great Plains, 518°
profile, 353°
Great Salt Lake, Utah, 585
Great Slave Lake, Northwest Territories, Canada, ancient fracture, 42°
deformed metamorphic rocks, 39°
Great Valley, of Appalachians, 517,° 521, 522°
of California, 518°
Greeks, ancient, ideas on climate, 30–31
Greenland, Mesters Vig, results of frost wedging, 146°
results of solifluction, 176°
Greenland Ice Sheet, 19, 256, 258°
Green River Formation, Colorado, Utah, Wyoming, and Nevada, 525
Groins, 329,° 339, ***652***
Gros Ventre, Wyoming, "landslide," 166, 167°
Groundmass, ***614,*** 615,° 618,° ***652***
Ground moraine, ***271,*** 272,° 280,° ***654***
Ground water, ***234, 652***
chemical composition, 245
in deserts, 292
discharge, 238
distribution, 234
economic aspects, 242
economy, 238
movement, 236, 237°
origin, 234
pollution, 244°
recharge, 237°
in the rock cycle, 244
velocity, 166,° 238, 239°
Guatemala, Atitlán Volcano, 41°
Lake Atitlán, 41°
Guinea current, 37°
Guilford, Connecticut, granitic rock, irregular contact, 510°
Gulf Coastal Plain, 40
Gulf of Alaska, 478°
of Mexico, 37, 353
Gulf Stream, 36, 37,°
Guyot, 563, ***652***
Gypsum, 56, 585, 603, 605, 608, 622, 624

Hachures, 633
Hail, 36, 70
Halemaumau (Kilauea Volcano), Hawaii, 468,° 469
Half-life, ***125,*** 126, 127,° ***652***
Halite, 56, 60, 61,° 68, 603, 604, 605, 608, 622, 624
Hall, Sir James, 87, 488
Hand level, 637°
Hanging tributaries, 266, 269,° ***652***
Hanging wall, *see* Fault(s), hanging wall
Hardness, 604–605, ***652***
Hawaii, Kilauea Volcano, 457, 459,° 466, 468°
Kuhio wharf, destructive waves, 5°
Mauna Loa Volcano, 459,° 463,° 466, 469
volcanic rock, 495
Waimea Bay, breaking wave, 327°
Hawaiian archipelago, 466, 468°
Islands, 466, 468°
volcanic mountains, 39
volcanic rocks, K/Ar ages, 466, 468°
volcanic shield, 463,° 466
volcanoes, 466, 468,° 469, 474°
Heat, internal, Earth's, 28, 41
compared with solar heat received, 43
Heat, radiated as infrared waves, 28
solar, average amount received on Earth, 42
Heat flow, terrestrial, rate, 41, 42,
Heavy minerals, 58–59, 624
Hebgen Lake, Montana, 217
Hekla Volcano, Iceland, 301, 459°
Helium, 601
in atmosphere, 107°
Hematite, 56, 604, 622
Henry Mountains, Utah, 493
Herculaneum, Italy, 174, 470, 472°
Hertz, 347, ***652***
Hertzian waves, 599, 600°
High-quartz, *see* Quartz, high-quartz
Highwood Mountains, Montana, 493
Hills, conical, volcanic, *see* Volcanic edifices, volcanic cones
Himalayan Mountains, 535
change of trend, 40
detailed survey, 535
lateral attraction of plumb bob, 535–536°
Hinge fault, 413,° ***415,°*** 417,° ***652***
Historical geology, ***651***
Horizons, in a soil, 153°
Horizontality, original, of strata, 378; *see also* Principle of original horizontality
Horn, ***266,*** 269,° ***652***
Hornblende, 59, 82,° 108, 143,° 144,° 145, 501, 502, 608, 619, 620,° 622, 628
cleavage, 65°
crystal structure, 65°
Hornfels, 500, 629, ***630, 652***
Hot spring, 475, ***652;*** *see also* Spring(s), thermal
Huallatiri Volcano, Chile, 459,° 462
Humboldt current, South Pacific Ocean, 37°
Humus, ***154, 652***
Hutton, James, 13, 14, 15, 16, 17, 18, 21, 76, 87, 404, 419, 428, 488, 517
Hydraulic conductivity, 236, ***652***
Hydraulic plucking, ***208,*** 210,° ***652***
Hydrocarbons, 105, ***652***
bacterial reduction, 355
biosphere, 114
Hydrochloric acid, in volcanic gases, 457
Hydrofluoric acid, in volcanic gases, 457
Hydrogen, in atmosphere, 107°
in fats, 105
in hydrocarbons, 105
in proteins, 105
in volcanic gases, 457, 458°
Hydrogen sulfide, in volcanic gases, 457
in sediments, 355
Hydrogeology, 234, ***652***
Hydrologic cycle, ***98,*** 99,° 114, 187,° ***652***
cold climate, 100
hot, dry climate, 100°
humid-temperate climate, 99, 100°

Hydrologic cycle, long, 98, 100°
natural distillation of water, 35
short, 98, 100°
Hydrolysis, 142, ***652***
Hydrosphere, 26, 35, 104, 114, 115,° ***652***
chemical composition, 104, 107°
scale model, 35
total quantity of water, 98, 99°
Hypersthene, 59, 108, 608, 610
Hypothesis, formulation as part of scientific method, 18

Iberian Peninsula, 24°–25°
Ice, 6, 7, ***259***
crystal structure, 69°
dense phases, 51, 52°
melting point, lowering by increasing pressure, 52
Ice age, 256
Icebergs, 257°
sonic location, 347
Ice cap, ***256,*** 257,° ***652***
Ice-contact stratified drift, ***274, 651***
Iceland, Hekla Volcano, 301, 459°
Laki fissure eruption 1783, 464
Thingvellir rift valley, 542°–543°
Ice sheet, ***256, 652***
Idaho batholith, 495
Igneous rocks, 9, 10, ***79,*** 115, 488, 489, 490,° 492,° 494, ***652***
chemical composition, 104
extrusive, *see* Extrusive igneous rocks
intermediate, 104
intrusive, *see* Intrusive igneous rock(s)
layers, 78, 80°
mafic, *see* Mafic rocks
silicic, 104
texture, 87, 613–614, 615,° 616
angular and interlocked particles, 10, 11°
aphanitic, *see* Aphanitic texture
coarse-grained, 9, 10
fine-grained, 9
glassy, *see* Glassy texture
granular, *see* Granular texture
oriented fabric, 74°–75°
special varieties, 623
tonguelike bodies, 13, 14,° 15,° 41°
superposed, 41°
Illinois, DuQuoin, strip mine, 9°
Illustrations of the Huttonian Theory, 14, 15, 21
Ilmenite, 145
Imbrication, ***384,*** 385,° ***652***
Inactive volcano, *see* Volcano, inactive
Inclination, magnetic, ***555,*** 557,° ***652***
Inclusion, ***652***
Incongruent melting, ***106,*** 509, ***652***
Indian Ocean, 37
floor, linear sediment streaks, 358°
Indian Ocean basin, resemblance to Atlantic Ocean basin, 40
margins, bodies of sediment, 351
Indian Ocean basin, northeastern curved trench, 40
rocky ridges, 40
Indonesia, Agung Volcano, 459,° 480
Krakatau Volcano, 459,° 479
mountains with east-west trend, 40
Tambora Volcano, 459,° 479
Indonesian Trench, 548°
Infiltration, 187,° 243°
Infrared waves, 28
Initial dip, 381°
Inlet, tidal, *see* Tidal inlet
Instantaneous linear velocity of rotation, 34
Intensity, earthquake, *see* Earthquake(s), intensity
Interface, ***48, 652***
Interfacial tension, 49, ***652***
Interglacial ages, 278
Interior drainage, ***290, 648***
Interior, Earth's, *see* Earth's interior
Interior Lowland, profile, 353°
Intermediate-focus earthquake, *see* Earthquake(s), depth of focus: intermediate-focus
Internal heat, Earth's, *see* Earth, internal heat; Geothermal gradient
International Hydrologic Decade (1965–1974), 98
Interstices, *see* Pore spaces
Interstitial spaces, *see* Pore spaces
Intrusive igneous bodies, *see* Pluton(s)
Intrusive igneous rocks, ***79,*** 102,° 103, 488–501, ***652***; *see also* Pluton(s)
Ion(s), 596,° 597, ***652***
sizes, 55,° 110
Ionic bond, 61, ***652***
radii, 114,
Ionized particles, streaming away from Sun, 30°
Iron, 622
in contact-metamorphic zones, 503
in proteins, 105
in silicate melts, 106
Iron, ferrous, in rock-making silicate minerals, 56, 57
Iron meteorite, 31, 32°
Iron ore, 586°
Iron oxide, 92, 623
Isograd(s), 502, 503, ***639, 652***
chlorite, 503
Isopach(s), 301
Isoseismal line, ***448,° 652***
Isostasy, ***533,*** 534,° ***652***
Isotope, 125, 127, ***595,*** 596,° ***652***
daughter, 125, ***647***
parent, 125, ***655***
Isotopic dates(s), applications to strata, 131,° 545, ***652-653***
emplacement of plutons, 499
metamorphic events, 505
structural movements, 418
volcanic strata, 481
Isotopic dating, 125–130, 127,° 128°
decay rates, 125°
Italy, Herculaneum, 470, 472°
Larderello district, electric power from volcanic steam, 470,° 482
La Solfatara, 470,° 473
Lipari Islands, 458, 470°
Pompeii, 470, 472°
Temple of Serapis, 401, 402°
submergence and emergence, 401–402
Vaiont Canyon "landslide," 167–169°
Vesuvius Volcano, 458, 459,° 469, 470,° 471, 472,° 473, 474,° 475
Vulcano Volcano, 458, 470,° 474

Japan, Asama Volcano, 459,° 479
Kyushu, wave-cut bench, 335°
Niigata earthquake (1964), 426°–427°
Japan Trench, 548, 549°
Jasper, 609
Johnnie thrust, Nevada, 506°
Johnstown, Colorado, stony meteorite, 32°
Joint(s), ***78,*** 409, 410,° ***653***
in Colorado Plateau, 79,° 410
columnar, *see* Columnar joints
origin, 410
resulting from tensile stresses during cooling, 498
Joint set, ***410, 653***
Joint system, ***410,° 653***
Jura Mountains, France and Switzerland, 519, 520,° 524°–525°
Juvenile water, ***457, 653***

Kamchatka Peninsula, U.S.S.R., Volcano, 480
Kame(s), ***274,*** 275,° ***653***
Kame terrace(s), ***274,*** 275,° ***653***
Kansas, Brenham, stony-iron meteorite, 32°
Kaolinite, 142,° 608, 622
Karst topography, ***248,*** 249,° 250,° ***653***
Kelvin, Lord, 125, 544
Kelvin temperature scale, 52
Kentucky, Edmonson County, karst topography, 250°
Edmonton, nickel-iron meteorite, 32°
Kettle(s), ***274,*** 275,° ***653***
Keystone thrust, west of Las Vegas, Nevada, 416°
Kilauea Volcano, Hawaii, 457, 459,° 466, 468°
Kilauea Iki, 461,° 469
observatory, 469
submarine eruptions, 477
Knyahinya, U.S.S.R., stony meteorite, 32°
K_2O, 105°
Krakatau Volcano, Indonesia, 459,° 479
Krypton, in atmosphere, 107°
Kuhio, Hawaii, waves washing over Pier 1, 5°
Kuhio Wharf, Hawaii, destructive waves, 5°

Kurile, Kamchatka Trench, 548,° 552°
Kuroshio current, North Pacific Ocean, 37,° 354
Kyanite, 59, 609, 618,° 622

Laboratory experiments, at high temperatures and pressures, 18
stream flow, 190–191°
waves, 323
wind action, 297, 298°
Laboratory tests, coal, 503, 505
La Caldera, Canary Islands, 459,° 462
Laccolith(s), ***490, 493,*** 496,° ***653***
conduits, 493, 495
Lacroix, Alfred, 471
Lae Atoll, Marshall Islands, 565°
Lagoon, 319,° 320°
Lake Atitlán, Guatemala, 41°
Lake Bonneville, shorelines, 531°
Lake Superior, ancient mountains on south shore, 516
iron ores, 586
Laminae, ***379, 653***
Laminar flow, ***189–190,°*** 237, ***650***
Lamination, 379
Land, 4
sculpture, 216
Landlocked bodies of water, circulation, 354, 355°
Lands, distribution, changes with time, 19
effects on air currents, 36
erosion, 15, 26
heated and cooled more rapidly than adjoining seas, 36
persistence despite erosion, 26, 43
renovation, 15
Landslide, 7,° 18, 161°
Lapilli (volcanic), 625, 626, ***653***
Lapilli tuff, 625, ***626, 653***
Laramie Range, Wyoming, 518,° 519
Larderello district, Italy, volcanic steam power, 470,° 482
LaSal Mountains, Utah, 493
La Solfatara, Italy, 470,° 473
Latent heat, of fusion, ***52, 653***
of sublimation, ***53, 653***
of vaporization, ***51, 653***
Lateral moraine, ***271, 654***
Laterite, ***155, 653***
Latitude, parallels, 31
Latitude-longitude grid, 632°
Lava, ***79,*** 458–460, 470, 474,° 488, 490, 560, ***653***
changes with time, 458
composition, 458, 459
extrusion, 80
flow into water, 477
fluid, 459, 460°
fluidity, 459
sluggish, 459–460, 461°
viscous, 460
Lava domes, *see* Volcanic edifices, lava domes
Lava flow(s), ***80,*** 459, 460,° 461,° 462, 464, 469, 474,° ***653***
speeds, 459
Lava fountain(s), 460, 461,° 474°
Lava spines, *see* Volcanic edifices, lava spines
Lava tongues, 474°
Lava tunnel, 460
Law of faunal succession, ***124, 650***
Laws, natural, 18
constancy, 19
Layering, 379
Layers (of rocks), 78, 80°–81,° 613, 614°
igneous rocks, 80,° 613, 614°
metamorphic rocks, 81,° 613, 614°
sedimentary rocks, 80,° 613, 614°
Leaching, ***142, 653***
Leiden, University of, 13
Leucite, 59
Levee, natural, *see* Natural levees
Life, history of, 17
Light, speed of, 30
Light, visible, part of electromagnetic spectrum, 28
Light waves, generation by vibrations of molecules, 28
Lignite, 572, ***653***
Limb, 406,° ***407,*** 408,° ***653***
Limestone(s) 10, 339, 616, 621, 623, ***626, 653***
average chemical composition, 104, 105°
clastic texture, 617
coral, 87°
crystalline texture, 617
fragmental texture, 616,° 617
organic texture, ***617***
weathering, 144
Limonite, 142, 604, 609
Liquid (state of aggregation of matter), 49–50
angle of contact of on solid/air interface, 50
molecules, spacing, 49
particles of, movements (Brownian movement), 49
properties, 49–50
Lisbon, Portugal, 24°
Lithification, 80, 102, 103, ***653***
cementation, ***92, 646***
of regolith to convert it back to bedrock, 8, 10, 14, 17
compaction, ***92, 646***
recrystallization, 92
Lithosphere, 26, 35, ***37,*** 114, 115,° 336, 445,° 537,° ***653***
Lithosphere, base, 443
chemical composition, 104
dissolution of by water flowing through, 35
particles of suspended in atmosphere and hydrosphere, 35
section through, 534°
strength, *see* Strength, of lithosphere
vertical displacement under load, 531–533
Little Colorado River, Arizona, 12
Load, bed, ***190***–192, ***653***
dissolved, 110, 111,° 187
Load, stream, ***189, 653***
suspended, 110, 111,° ***190,*** 192, 194, ***653***
Load, glacier, 267
Local base level, *see* Base level, local
Loess, 301, ***309,°*** 310,° ***653***
Logarithmic scale, $\log_{10}$ scale, 88, 89,° 107°
Longitudinal dune, 286°–287,° ***307, 648***
Longitudinal wave(s), *see* Seismic waves, longitudinal waves
Long profile (of a stream), ***197,*** 199,° 200,° 201,° ***653***
Longshore current, *see* Waves, water surface: longshore currents
Longshore drift, see Waves, water surface: longshore drift
Love waves, *see* Seismic wave(s), Love waves
Low-quartz, *see* Quartz, low-quartz
Lunar day, length, 26
Lunar night, length, 26
Luster, 603, ***653***
Lyell, Sir Charles, 16, 18, 318, 401, 420, 532

McDonald Lake, Northwest Territories, Canada, ancient fracture, 42°
M-discontinuity, ***443,°*** 534,° 535, 536,° 537,° 554,° 562,° 566,° 653, ***654***
Mafic rocks, 104, 443, 619, 620,° ***653***
Magma, ***79,*** 102,° 103, 474,° 488, 489 490, 493, 495, 496, 498,° 499, 500, 501, 503, 509, 560, 583,° ***653***
alteration of rock, 583°
basaltic, 498°
origin, 509–510
chemical composition, oversaturated, 106
saturated, 106
undersaturated, 106
granitic, 503
origin, 510–511
relation to regional metamorphism, 511
intrusion, 80
origin and rise, 509–511
separation of liquid and gaseous phases, 456
Magma chamber, 496, 499
depths, 473,° 510
Magmatic differentiation, ***496,*** 498, ***653***
crystal settling, 498°
separation of liquids, 498
Magnesium, in contact zones, 503
in evaporite minerals, 112
in rock-making silicate minerals, 56, 57
in silicate melts, 106
Magnesium ions, 45
in average river water, 107°
in average seawater, 107°
exchange with calcium ions in pyroxenes, 106
exchange with ferrous ions, 106

Magnetic anomalies, oceanic rises and ridges, 559–560, 561,° 563°
Magnetic declination, *see* Declination, magnetic
Magnetic Equator, 555, 556°
Magnetic field, *see* Earth's magnetic field
Magnetic fields, surrounding magnetic particles in rocks, 42; *see also* Remanent magnetism
Magnetic poles, Earth's, *see* Earth's Magnetic Poles
Magnetic surveys, 42, 559, 577–578°
Magnetic reversal(s), *see* Earth's magnetic field, reversals
Magnetism, remanent, *see* Remanent magnetism
Magnetite, 56, 143,° 604, 605, 609, 622
  crystallization from silicate melts, 109
Maine, Boothbay Harbor, coastal morphology, 336,° 337, 338
Malachite, 609
Malay Peninsula, mountains with north-south trends, 40
Man, primitive, 4
  relationship to environments and natural materials, 4
Manganese nodules, *see* Deep-sea floor, manganese nodules
Mangrove swamps, 339
Mantle, *see* Earth's mantle
Mantle, Earth's, *see* Earth's mantle
Map projections, 632
Maps, geologic, 631
Marble, 500, ***630,° 654***
Mariana Trench, 352, 548°
Marine sediment, ***654***
  depth of water during deposition, 549
Marine strata, 318
  ancient, 2°–3,° 372
  deformed and uplifted, 43
  depth criteria, 372
  effects of slumping, 361°
  extent, 318
  flat-lying, 43
  pelagic, 372
Marl, 627, ***654***
Marsh, tidal, *see* Tidal marsh
Marshall Islands, Lae Atoll, 565°
Martinique, St.-Pierre, devasted by 1902 eruption of Mont Pelée Volcano, 8°
  Mont Pelée Volcano, 8, 459,° 471, 473,° 474
Marysville Stock, Montana, 497°
Massachusetts, Boston area, coastal morphology, 336,° 337, 338
  Cape Cod National Seashore, barrier, lagoon, and marsh, 320°
  Martha's Vineyard, coastal morphology, 336,° 337, 339
  Mount Tom, Westfield, slickensides in dolerite, 412°
  Uxbridge, coarse metamorphic rock, 82, 83°
Massachusetts, Winthrop, sea wall, 338°
Mass number, 595, 596,° ***654***
Mass spectrometer, 17°
Mass-wasting, ***162,*** 589,° ***654***
  analysis of movement, 162
  coastal, 334°
  in deserts, 290
  kinds, 165°
  as an open system, 164
  processes, 165°
  rates of movement, 166°
  relation to weathering, 163°
  submarine, 359–364
    gravity displacement, 360–363, 364
    sand flows, *see* Deep-sea sediments, sand flows
    slumping, *see* Slumping
  in valleys, 217, 218°
Materials, natural, 4, 17, 19
Matrix, 84, 616, ***654***
Matter, 17
  conversion to energy, 28
  relation to energy and electromagnetic spectrum, 28
  states of aggregation, 48
Mauna Loa Volcano, Hawaii, 459,° 463,° 466, 469
Mauretania, circular dome, 405°
Meadow Valley Wash, Nevada, strata in angular unconformity, 420°
Meanders, ***202,°*** 203,° 204,° 206,° ***654***
  migration, 205
Mechanics of geologic processes:
  coastal erosion, 333
  cross-laminae, ripples, 192, 193°
  cross-strata, dune, 306, 307°
  cross-strata, sand-wave, in streams, 192, 194°
  density currents, 354–355
  elastic waves, 599–601
  electromagnetic waves, 599, 600°
  flowing air, 297°
    sand, 297, 298,° 299,° 300°
    suspensions, high-altitude, 301
      low-altitude, 299–301
    transport, fine particles, 298,° 299, 300,° 301
  glacier flow, 259–260, 261°
  ground water, percolation, 235–238
  landlocked bodies of water, circulation, 355°
  mass-wasting, submarine, 359–364
  prograding shoreline, 331°–333
  ripples, water, 193°
    wind, 309
  rock deformation, 399–400°
  seismic waves, 431–437
  sound waves, 359°
  stream flow, 189
    sediment, transport, 190–194, 195°
  surface ocean currents, 353
  turbidity currents, 359°
  ventifact formation, 303, 304,° 305°
  waves, water surface, 323–324, 325
    breaking, 326
Mediterranean and Trans-Asiatic belt, 444, 447°
Mediterranean Sea, 24,° 25°
  rocky threshold, 354
  warm water of high salinity, 354
  water level, 354
Melanesian rise, 559°
Melting, congruent, *see* Congruent melting
  incongruent, *see* Incongruent melting
  influence of pressure, 66–67,° 109–110, 509, 535
    water-vapor pressure, 109, 502–503
Meltwater, ***263,*** 274, ***654***
Mercalli scale, *see* Earthquake(s), intensity: Mercalli scale
Mercury, 533, 534,° 622
Mesters Vig, Greenland, results, of frost wedging, 146°
    of solifluction, 176°
Metallic bond, 61, ***654***
Metallic chlorides, volcanic gases, 457
Metallic minerals, contact aureoles, 510
Metallic mineral deposits, origin by igneous action, 457
Metals, 4
Metamorphic events, isotopic dates, 130, 131°
Metamorphic facies, ***501–502, 654***
  amphibolite, 502
  epidote-amphibolite, 502
  greenschist, 502
  pyroxene-hornfels, 502
Metamorphic rock(s), 9, 10, 39, ***80,*** 102,° 103, 115, 486, 487,° 627–630, ***654***
  chemical composition, 104
  fabric with preferred orientation, 74°–75,° 91, ***649***
  layers, 78, 81°
  melting to create igneous rocks, 10, 102,° 103
  in mountain belts, 39, 516, 523
  texture, 91
    cataclastic, 505, 507,° 618
    crystalline, 617
Metamorphism, ***80–81,*** 102,° 103, ***654***
  behavior of elements, 112
  crushing and pulverization, 82, 507°
  dynamic, *see* Dynamic metamorphism
  environment, 501–502
  recombination, 82
  recrystallization, 82, 500
  regional, *see* Regional metamorphism
  reorganization, 82
  replacement, 82, 501,° 503
  role of water, 502–503
Metastable mineral, 64, ***654***
Meteorite, ***31,*** 32,° ***654***
  iron, 31, 32°
  micrometeorites (nickel-ion spheres), 86, 356

Meteorite, stony, 31, 32°
  stony-iron, 31, 32°
  tektites, 31, 32,° 356
Methane, in atmosphere, 107°
  in volcanic gases, 457
Mexico, Paricutín Volcano, 8°
  San Juan Parangaricutiro, engulfed by volcanic rock, 8°
  west coast, prograding beaches, 331
MgO, 104, 105°
Mica, 56, 57, 59, 62, 604, 609, 622, 628, 629, 630; *see also* Muscovite; Biotite
  aligned flakes, 505
Microcline, *see* Potassium feldspar(s)
Microscope, polarizing (petrographic), 17
Microseism, 431,° 438, ***654***
Mid-Atlantic belt, 444, 447°
Mid-Atlantic Ridge, 40, 559°
    continuous seismic profile, 350°
    sea-floor profile, 353°
Middle America Trench, 548°
Mid-Indian belt, 444, 447°
  ridge, 559°
    authigenic sediments, 365
Migmatite, ***510, 654***
Mindinao Trench, 548°
Mineral(s), 53, ***56, 654***
  accessory, *see* Rocks, accessory minerals
  anhydrous, 503
  behavior during melting and cooling, 105–106, 108–109
  chemical composition, 56
  essential, *see* Rocks, essential minerals
  hydrous, 503
  indicators of environment, 63
  metastable, 64
  rock-forming silicate, *see* Rock-forming silicate minerals
  secondary, 142
  stable, 64
  unstable, 64
  used in industry, 580–582
Mineral aggregates, 70, ***654***
Mineral deposits, 582
  origin, 582–590
  relation to magma, 583°
  facies, ***511, 654***
  feldspar-quartz, 511
Mineral industry, 571
Mineral particles, shapes, 60
Mineral-stability series, 144°
Mining, careless, 5
  refuse as pollutant of rivers, 5
Minnesota, St. Cloud, granite, 83°
  Vermillion district, contact metamorphism, 501
Mississippi River, delta, 208, 209°
  junction with Missouri River, 112°
Missouri River, in flood, 6°
  junction with Mississippi River, 112°
Modulus of rigidity, ***599, 654***
Mohawk Valley, New York, 517°
MOHOLE, Guadalupe test boring, Pacific, 329, 368, 369, 428
Mohorovičić discontinuity, *see* M-discontinuity
Molecular attraction, 237
Molecular bond, 61, ***654***
Molecules, ***597, 654***
  vibrations as sources of light waves, 28
Molten material, discharged from volcanoes, 9, 10, 13, 41; *see also* Lava
Monadnock(s), ***220, 654***
Monocline, ***409,°*** 417,° ***654***
Montana, Boulder batholith, 495, 497°
  Heavens Peak, Glacier National Park, deformed strata, 39°
  Highwood Mountains laccoliths, 493
  Mount Gould sill, 493°
  Marysville stock, 497°
Monte Somma Volcano, Italy, 458, 470, 472,° 473°
  destructive eruption, 470
Mont Pelée Volcano, Martinique, 459,° 471, 473,° 474
  hot tephra, 471
    temperature, 473
  prediction of 1929–1932 eruptive events, 482
  ruination of St. Pierre, 8°, 471
Montreal, Quebec, Canada, dike and sill, 493°
Monument Valley, Arizona, isolated remnants, thick sandstone layer, 10°
Moon, 26, 320, 321°
  gravitational attraction, 320, 321°
  lunar day, length, 26
  lunar night, length, 26
  orbit, 27°
    plane of, *see* Plane of Moon's orbit
    phases, 321°
  revolution around the Earth, 26
  rock sphere, 26
  surface, 27, 27°
    Apennine Mountains, 27°
    Archimedes Crater, 27°
    Mare Imbrium, 27°
Moraine, 271, ***654***
Morphology, Earth's, *see* Earth's morphology
Mount Adams Volcano, Washington, 459,° 462
Mountain chain(s), ***39,*** 40, ***654***
Mountain making, ***655;*** *see also* Orogeny
Mountain range, ***39, 655***
Mountains, 4, 19, 36, 37, ***39,° 654***
  central zones, 523, 525–528
    batholiths and stocks, 517,° 518,° 520°–521,° 523
    fault blocks, 525–526, 530°
    metamorphic rocks, 517,° 518,° 520°–521,° 523
    nonmarine strata, 525
    plastic flow of rocks, 525, 529°
    ultramafic rock, 523
  depth to M-discontinuity, 536
  former, 40, 538°
Mountains, former, south shore, Lake Superior, 516
  geologic structure, 516
    folded and faulted strata, 39, 516
    granitic batholiths, 39, 516
    regionally metamorphosed strata, 39, 516
  history, progression of events, 528, 536–538
  interior zones, 519, 521, 523
    strata, historical information, 519
      carbonate rocks, lateral passage into dark marine shales and silt stones, 519
      clastic strata, sheets, records of uplift, 519
    thrusts, 522, 523
  marginal zones, 519
    elevated basement rocks, 519
    folded and faulted strata, 519, 523°
  residual, *see* Residual mountain(s)
  topographic relief, 516, 528–529, 535–536
  volcanic, ***39, 661***
  volcanic rocks, 39, 537
Mountain system, ***39, 655***
Mount Baker Volcano, Washington, 459,° 462
Mount Emma, Arizona, 13, 14°
Mount Everest, Himalayas, 40, 466
Mount Gouldsill, Montana, 493°
Mount Hood Volcano, Oregon, 459°
Mount Katmai Volcano, Alaska, 459,° 471, 474,° 479
Mount Mazama Volcano, Oregon, 462, 464°
  destructive eruption, date, 463
  small U-shaped valleys truncated, 462, 464°
  tephra, volumes, 462
Mount Rainier Volcano, Washington, 459,° 462
Mount Shasta Volcano, California, 459,° 462
Mount Spurr Volcano, Alaska, 459°
  1953 explosion, 479
Mount Tom, Massachusetts, slickensides in dolerite, 412°
Mount Trumbull, Arizona, 13, 14°
Mud crack(s), ***384,*** 386,° 408,° ***655***
  original top up, 408°
  overturned, 408°
Mudflat(s), marginal, lagoon, 322
Mudflow, ***165,°*** 170, 172,° ***647***
  rate of travel, 166°
Muscovite, 59, 64,° 83,° 108, 503, 609, 622
  cleavage, 64°

Nansei-Shoto Trench, 548°
Na_2O, 105°
Natural bridge, 249°
Natural concentrations, 581
Natural gas, 573, ***655***
Natural laws, constancy, 19
  basis for predicting unknown from what is known, 18

Natural levees, ***205,*** 206, ***653***
Natural materials, 4, 17, 19
Natural philosophy, 13
Natural resources, industrial capability, dependence, 19
Nearshore sediments, 331,° 341°
coarse-grained, 318
Neck cutoff, ***205,*** 206,° ***655***
Neon, in atmosphere, 107°
Nepheline, 59, 609, 620°
Nepheline syenite, 619, 620, ***621, 655***
Neptunists, 488, ***655***
Neutron(s), ***595,*** 596,° 597,° ***655***
cosmic rays, 113°
Nevada, Frenchman Mountain fault block, 530°
Johnnie, thrust and metamorphism, 506°
Keystone thrust, Las Vegas, 417°
Meadow Valley Wash, strata in angular unconformity, 420°
Pleasant Valley earthquake 1915, 435
Spotted Range, fault surface, 412°
Stillwater Range, mudflow, 172°
Tobin Range, fault scarp, 411°
New Britain Trench, 548°
New Hebrides Trench, 548°
New Mexico, Carlsbad Caverns, 246
Los Alamos, Bandelier welded tuff, 466°
Shiprock volcanic neck, 480, 481°
New York, Adirondacks, 517,° 519, 583, 621
Catskill, deformed Ordovician strata, 396°–397°
Catskills, residual mountains, 39, 517°
Cayuga Lake, Ithaca, horizontal strata, 410°
vertical joints, 410°
Long Island, Montauk Point, 330°
Mohawk Valley, 517°
Moriches Inlet, 322°
Niagara Falls, 211°
Palisades intrusive sheet, 495
Pound Ridge reservoir, low water, 12°
Rensselaer, Pleistocene sediments, faults, 411°
White Plains, metamorphic rocks, layers, 81°
Newton, Sir Isaac, 28, 428, 601, 602
Newtonian philosophy, 16
Niagara Falls, 211°
Nickel, 622
Nicolet, Quebec, debris flow, 170, 171°
Nitrogen, in atmosphere, 35, 104, 107,° 113,° 114
in fats, 105
in proteins, 105
in volcanic gases, 457, 459°
Nitrous oxide, in atmosphere, 107°
Nonfoliate, ***627,*** 629, 630,° ***655***
Nonmarine basin sediments, Rocky Mountains, 525
Nonsorted layers, 383
Normal component of Earth's gravity, *see* g_n
Normal fault, 411,° ***414, 655***
North America, average altitude, 40
cold climate in northeastern, 36
eastern and western parts, north-south-trending mountains, 40
southward flow of cold Arctic air, 37
North American Cordillera, ***40,*** 516 517, 518,° 519, 546,° ***655***
analysis of strata, 547
profile, 353°
North Atlantic Ocean basin, border with Caribbean basin, arcuate trench, 40
North Atlantic current, 36, 37,° 353
North Carolina, coastal profile and section, 341°
North Equatorial current, 37,° 353
Northern Hemisphere, 35, 36
Coriolis right-deflection rule, 35
North Magnetic Pole, 555, 556,°
relationship to pole of rotation, 556. 557
wandering, 557,° 558°
North Pacific current, 37,° 354
Nuclear energy, 580
Nuclear explosions, underground, 17, 449
Nuclear reactions, inside Sun, source of solar energy, 28, 29
Nuée ardente, 471, 473,° ***655***
Nuggets, 589°

Oakland, California, landslide, 160, 161°
Oblique-slip fault, 413,° ***415,° 655***
Obsidian, 620,° 622, ***623, 655***
Ocala Limestone artesian system, 242
Ocean basin(s), ***39,*** 346, ***655***
aggregate area, 40
greatest depths, near margins, 40
hypotheses of origin, 346
Oceanic basalt, *see* Basalt, oceanic
Oceanic crust, *see* Earth's crust, oceanic
Oceanic research, naval operations, 346
Oceanic rise(s) and ridge(s), ***352,*** 353, 547, 559,° ***655***
linear magnetic anomalies, 559, 561,° 563°
Offshore sediments, 331,° 341°
Oil, 4, 573
Oil field, ***575, 655***
Oil pool, ***574,*** 576,° ***655***
Oil shale(s), 580, ***655***
Oklahoma, Ardmore, Collings Ranch Conglomerate, 422°
Arbuckle Mountains, debris of Pennsylvanian age, 422°
Oldest rocks, 16, 133
Olivine, 56, 57, 59, 108, 144,° 502, 604, 620,° 622
iron-rich, high-pressure, transformation to spinel structure, 535
magnesian, 106
reaction with cooling silicate melt, 106
Olivine basalt, 622
Olivine gabbro, 622
Ontario, Prescott County, bedrock exposed, 11°
Ooze(s), *see* Pelagic sediments
Opal, 609, 622
Open system(s), 68, 69, ***655***
components of geologic record, 69
mass-wasting, 164
stream, 69
Optical spectrometer, 17
Order of magnitude scale, 88, ***655***
Ordinary processes, geologic cycle, components, 26
Ore, ***582,*** 589,° ***655***
origin, 582–590
Oregon, Crater Lake, 459,° 462, 464°
Mount Hood Volcano, 459,° 462
Mount Mazama Volcano, 462–464°
Organic skeletal remains, calcium carbonate, 85,° 114, 115,° 624°
Organic texture, 87,° 624, ***655***
Original horizontality, principle of, *see* Principle of original horizontality
Origin of species, Darwin theory, 17
Orogenic belt(s), 516–517, ***655***
Orogeny, ***516, 655***
causes, 550–555
contraction of mantle, 551–553
continental drift, 553
convection currents within Earth's mantle, 553–554
gravity sliding, 554
rise of granitic batholiths, 555
Orthoclase, 59, 143,° 603, 605, 609, 610, 622
crystallization temperature, lowered by water-vapor pressure, 109
Orthopyroxene, 59
Ouray, Colorado, strata in angular unconformity, 419°
Outcrop area, ***637, 655***
Outer Ridge, ***352,° 655***
Antilles, 352°
Outwash, 255,° ***272,°*** 273,° 274,° 275,° ***655***
Outwash plain, ***274, 655***
Overland flow, ***187,*** 217,° ***650***
Oversaturated igneous rocks, 621
Oxbow lake, 204,° ***205,*** 206,° ***655***
Oxford, England, limestone weathering, 138°
Oxygen, in atmosphere, 35, 104, 107,° 113°
in depths of the oceans, 354
in evaporite minerals, 112
in fats, 105
in hydrocarbons, 105
in proteins, 105
isotopes, ancient climates, 368
volcanic steam, 547
partial pressure, effect on crystallization of iron-oxide minerals, 109
with silicon in rock sphere of Moon, 26

Oxygen, tetrahedral arrangement with silicon in silicate minerals, 56, 57,° 58°

Pacific Ocean, 36, 37°
Pacific Ocean basin, average depth, 40
  fracture zones, 41, 559,° 560°
  oceanic rises and ridges, 40, 559°
  proportion of Earth's aggregate surface area, 40
  trenches, 40, 548°
Pacific Ocean floor, brown clay, 357°
Packing, 92,° ***655***
Paleogeographic map(s), 388, 389,° 631
Paleogeography, 388
Paleomagnetism, 556, 557–558, ***655***
  support for continental drift, 558
Palisades intrusive sheet, New York, 495, 498
Paper-thin slices of rocks, 11,° 17
Parallel strata, ***379,*** 380,° ***660***
Parent isotope, 125, 126, 128,° ***655***
Parent rock, 81,° 82
Paricutín Volcano, Mexico, 8,° 457,° 459,° 463°
  quantity of material discharged, 457
  sluggish lava, 461°
Paris, University of, 13
Parry Sound, Ontario, Canada, igneous and metamorphic rocks, 74°–75°
Particle(s), ***655***
  arrangement, degree of order, 48
  igneous rocks, angular and interlocked, 10, 11°
  ionized, streaming from Sun, 30°
  liquid, Brownian movement, 49
    forces of attraction, 49
      uneven distribution along interfaces, 49
  metamorphic rocks, aligned, elongated, flattened, 10, 11°
  sediment, *see* Sediment(s), particles
  spacing, 48
  regolith, togetherness, 91
Partings, 77,° 78,° 80,° 613, 614,° ***655***
  alignment of streams, 79°
  bedding-plane, *see* Bedding-plane parting
Peat, 322
  origin, 572
Pebble(s), 8, 89,° 90,° 195,° 300,° ***656***
Pedalfer, ***154,*** 155,° ***656***
Pediment, ***294,*** 295,° 296,° ***656***
Pedocal, ***154,*** 155,° ***656***
Pedology, ***152,*** ***656***
Pegmatite, 614, 615,° 619, 620,° ***656***
Pegmatitic texture, 614, 615,° 619, 620,° ***656***
Pelagic sediments, ***364,*** 365,° ***656***
  ooze(s), ***364,*** 365,° ***655***
    calcareous, 364, 365,° 366,° ***655***
      disappearance at great depths, 364–365
    siliceous, 364, 365,° 366,° ***655***
Peneplain, ***220,*** 221,° ***656***
Pennsylvania, Delaware River, slaty cleavage, 505
  Erie, Presque Isle spit, 329°
  Pittsburgh, flat-lying sedimentary strata and coal, 503, 504,° 505
  Scranton, closely folded sedimentary strata and anthracite, 503, 504°
Percé, Quebec, Canada, contact of burial, 419°
  vertical strata, 408,° 419°
  wave-cut cliff, 408,° 419°
Perched water bodies, ***239,***–241,° ***656***
Percolation, ***237,°*** 238, ***656***
Percolation, confined, ***240,*** 241,° ***656***
Perennially frozen ground, 42, 43,° *see also* Permafrost
Peridotite, 619, 620,° ***621,*** 622, ***656***
Permafrost, ***282,*** ***656***
Permeability, ***236,*** ***656***
Peru-Chile Trench, 548,° 549
Petrographic (polarizing) microscope, 17
Petroleum, ***573,*** ***656***
  exploration, 577, 578°
  occurrence, 574, 576,° 579°
  origin, 575
  production, 579,° 580°
  reserves, 581°
Petrology, ***76,*** ***656***
pH, ***68,°*** ***656***
Phenocrysts, ***614,*** 615,° 618,° 619, ***656***
Phosphorus, in fats, 105
  in proteins, 105
Phyllite, 503, ***628,*** 629, ***656***
Physical geology, ***651***
Physiographic province, ***656***
Pick-up velocity, 195°
Piedmont glacier, ***256,*** ***656***
Pillows, volcanic, 477, 480,° 494, 523, ***656***
Pine Mountains, Arizona, 13, 14°
Pitchblende, 580
Pitchstone, ***623,*** ***656***
Pittsburgh, Pennsylvania, flat-lying sedimentary strata and coal, 503, 504°
Placer, 145, ***589,°*** 590,° ***656***
Plagioclase feldspar(s), 59, 60, 108, 143,° 144,° 498, 501, 502, 603, 604, 609
  high-pressure transformation to garnet plus kyanite and quartz, 535
  incongruent melting at high pressures, 109
  reaction principle, 106
  transformation to albite and kaolinite, 477
    release of silica, 477
Plains, underlain by horizontal strata of sediments, 40
Planck, Max, 597
Plane of Moon's orbit, 27,° 321°
Plane of the ecliptic, 26, 27,° 30,° 31,° 33,° 34,° 36,° 321,° ***656***
Planets, 26
Plastic flow, ***49,*** 399, ***656***
Plastic solid, 49, ***656***
Plateau, ***656***
  Bolivian, 40
  Colorado, 12,° 14,° 15,° 79°
  Tibetan, 40
  Uinkaret, 13, 14°
Plateau basalts, *see* Volcanic edifices, volcanic plains and plateaus, plateau basalts
Plateaus, underlain by horizontal strata of bedrock, 40
Platinum, 145
Playa, ***292,*** 293,° 296, ***656***
Playa lake, ***292,*** ***293,°*** ***656***
Playfair, John, 14, 15, 18, 21
Plunge (of a fold), ***407,*** ***656***
Pluton(s), ***489,*** ***656***
  contact metamorphism, 500–501
    by gabbroic, 501
    by granitic magma, 501, 510
  cylindrical and irregular, 495
    batholiths, *see* Batholith(s)
    stocks, *see* Stock(s)
  date, of covering, 499
    of emplacement, 498–499
    of uncovering, 499
  inferred reactions within, 495–496, 498
  isotopic dates, 499
  lenticular, 490, 493, 495
    laccoliths, *see* Laccolith(s)
    others, 495
  shapes and names, 489
  tabular, 489–490
    dikes, *see* Dike(s)
    sills, *see* Sill(s)
  thermal effects, 488, 493,° 494°
Plutonic igneous rock, ***656***
Plutonism, 488–499, ***656***
Plutonists, 489, ***656***
Pluvial lake(s), ***280,*** 281,° ***656***
Poconos, northeastern United States, 516, 517°
  relief, 516
Podzol(s), ***155,*** ***656***
Point bar(s), 202,° ***203,°*** 204,° 205,° 223,° ***656***
Point Firmin, California, coastal cliffs, 2°–3°
  slump, 168
Polar easterlies, 35, 36,° ***656***
Polarity of Earth's magnetic field, changes, 19, 561
Polarizing (petrographic) microscope, 17
Polar regions, 31
Pole, magnetic, Earth's, *see* Earth's Magnetic Poles
Pole of rotation, Earth's, *see* Earth, rotation
Pollen, 86
Pompeii, Italy, 470, 472°
Pool, oil, ***574,*** ***655***
Poorly sorted sediment, 89, ***656***
Pore spaces, 92, ***656***
Porosity, ***235,*** 236,° ***656***

Porphyritic granite, 616
Porphyritic texture, ***613–614***, 615, 618,° ***656***
Porphyroblasts, 617,° 618,° ***656***
Porphyry, 491,° 501,° ***616, 657***
Positron, 596,° 601, ***657***
Potassium, in evaporite minerals, 112
Potassium-argon dating, 128
Potassium feldspars, 108, 610, 619, 620°
Potassium ions, in feldspars, 60
  in alkali feldspars, 143
  in average river water, 107°
  in average seawater, 107°
  in micas, 143
Pothole(s), ***209***, 210,° ***657***
Pound Ridge, New York, reservoir, low water, 12°
Prealps, Switzerland, 520,° 555
Precipitation, of moisture from atmosphere, 36
  world's average annual, 38°
Prediction, scientific method, based on data collected, 18
Pre-geologic time, 131, ***133, 657***
Prescott County, Ontario, Canada, bedrock exposed, 11°
Present, key to past, 18, 19
Pressure, differential, rock particles aligned, 10
  effect on melting of silicates, 109–110, 509, 535
  effect on minerals, 64, 66–67°
  static, metamorphic rocks, 10
Pressure gradient, 237
Pressure-temperature graph, graphite-diamond, 67°
  physical states of water, 51°
  SiO_2 minerals, 67°
Primary ore, 587,° 588°
Primary wave, *see* Seismic wave(s), P-deflections
Principle of original horizontality, 378, 380, ***655***
Principle of stratigraphic superposition, ***122***, 378, ***660***
Principles of Geology, 16, 420
Processes, ordinary, 26
Progradation, beach, 331–332
    Cretaceous strata, Rocky Mountains, 331
    Mexico, west coast, 331
    vertical succession of strata, 331,° 332
  delta, 332–333
Proton, ***595***, 596,° 597,° ***657***
Pteropods, 364
Puerto Rico, sea-floor profile, 352°
Puerto Rico Trench, 352,° 548,° 550°
Pumice, 618, ***623, 657***
Pyrite, 56, 60, 62,° 68,° 603, 604, 610, 622
  crystal form, 62°
Pyroclastic material, *see* Tephra
Pyroclastic rock(s), 501, ***657***
  metamorphism, 500
Pyroclastic texture, ***616***, 624, ***657***
Pyrolusite, 610
Pyroxene, 56, 57, 59, 62, 108, 143,° 144,° 500, 501, 610
  cleavages, 62, 65°
  crystal structure, 65°
  hydration to chlorite, 477
  metamorphism, 500, 620,° 629
  reaction with cooling silicate melt, 106
  release, of calcium bicarbonate, 477
    of silica, 477
Pyroxenite, 619, 620,° ***621, 657***

Quantities involved, geologic processes:
  age of Earth, 131, 133
  Ancient Lake Bonneville, volume, 531
  angles, natural slopes, 163°
  Antarctic Ice Sheet, volume, 256
  deep-sea fans, volumes, 371–372
  deep-sea floor, sediment delivered to, 369
  energy released in earthquakes, 449°
  Florida current, water transport, 353
  Frank, Alberta, avalanche, 169°
  geothermal heat flow, 42, 553
  glacial erosion, 276
  ground water, total volume, 234
  load pressure, base of lithosphere, 535
  permafrost, thickness, 43,° 282
  Pleistocene glaciers, former areal extent, 275
  rock debris, volume, 1963 Vaiont event, 168
  sheet erosion, regolith removed, 188°
  solar energy received on Earth, 29
  stream discharge, 196,° 197,° 199°
  tephra, thickness, 1947 eruption, Hekla, Iceland, 301°
  water in hydrosphere, 99°
  wind-eroded sediment, volume, 301–302
Quartz, 56, 57, 59, 60, 61, 82,° 83,° 92, 108, 143,° 145, 246, 500, 503, 617,° 619, 620,° 622, 623, 624, 626, 628, 629
  crystal form, 46°–47,° 63°
  crystal structure, 60°
  fracture, 66°
  high-quartz, 66, 68,° 69
  low-quartz, 66, 68
Quartzite, 500, 629, ***630,° 657***
Questions, scientific, 18
Quinebaug River, Connecticut, flood, 198°

Radiation, electromagnetic, 28
  solar, 28
Radiation belts, 30,° ***657***
Radioactivity, 28, 125, 544–545, 601, ***657***
  basis for measuring geologic time, 17, 18
  materials, concentrations, in Earth's crust, 509
Radioactivity, materials, concentrations, in rocks, 552
  relation to Earth's thermal history, 17
Radiocarbon, 601, ***657***
Radiocarbon dates: charred logs, Mt. Mazama tephra, 463
  deep-sea sediments, 368
  frost wedging, Mesters Vig, Greenland, 146
  glacial deposits, 278
  ground water, Ocala Limestone, Florida, 242
  logs, glacial deposits, Ohio-Indiana-Illinois region, 129
  sea-level curve, 281°
  shorelines, ancient Lake Bonneville, Utah, 531
  volcanic mudflow, Mud Mountain, Washington, 173°
Radiocarbon dating, 129, ***657***
Radiolaria, 356, 364
  in chert, 523
Raindrops, impact, 187
Rates and speeds, geologic processes:
  continental drift, 562
  contour currents, in sea, 354
  convection currents in Earth's mantle, 554
  creep of regolith, 175
  crustal subsidence, Hoover Dam, 531
  debris flow, 166,° 169
  deep-sea sediment accumulation, 369, 370°
  desert dunes, migration, 166°
  dissolution, limestone, 250
    diorite, 245
  erosion of lands, 110
  erosion velocity, 195°
  fault movement, 433, 436
  Florida current, 353
  glacier flow, 166°
  ground-water percolation, 166,° 235, 238, 242
  high-altitude tephra transport, 301
  longshore drift, sand, 329–330
  mudflow, 166°
  retreat, coastal cliffs, 338
    Niagara Falls, 211
  rock avalanche, 166,° 167
  rock glacier flow, 166°
  sea-floor spreading, 562
  seismic waves, 439, 440,° 442,° 445,° 601
  settling, particles in air, 300°
    in water, 159,° 300°
  soil formation, 155
  solifluction, 166,° 176°
  stream channel, lowering, 200
  stream flow, 166°
  temperature, change with depth in Earth, 554°
  uplift of Scandinavia, 532
  weathering of minerals, 144, 145
Rayleigh wave, *see* Seismic wave(s), Rayleigh waves
Reaction principle (relationship), 106, 108,° 109,° ***657***

Reaction principle (relationship),
continuous reaction, 106, 108,° 109,° ***647***
discontinuous reaction, 106, 108,° 109,° ***648***
exchange of ions, 108, 109°
ferromagnesian silicates, 106
Recharge, 240, 241,° ***243,°*** ***657***
Recrystallization, *see* Lithification, recrystallization; Metamorphism, recrystallization
Rectangular pattern, streams, ***225,*** ***660***
Recumbent fold(s), 409, 506, 508,° ***650***
Red Sea, bodies of hot, saline water, 477
Reef(s), ***339,*** 404,° 565,° ***657***
Reef rock, 87,° 626, 627
Regional metamorphism, 488, 503–505
relation to granitic rocks, 511
Regolith, ***7–8,*** 9,° 14, 76, 140,° 162, 232°–233,° ***657***
conversion to bedrock, 76
creep, 8
flow, 8
lithification, 88
particles, 88
fluidized, 91
residual, 8
slide of, 8
transported, 8
Rejuvenation, 221,° ***222,*** 223,° ***657***
causes, 224
Relative time, 122
Relief, ***632,*** ***657***
of mountains, 528–529, 535–536
of plateaus, 516
Remanent magnetism, ***43,*** 555, ***654***
Rensselaer, New York, Pleistocene sediments, faults, 411°
Reorganization, *see* Metamorphism, reorganization
Replacement, ***245,*** ***657***
by hydrothermal solutions, 584
Reservoir rock, 574, ***657***
Residual concentration, ***588,*** ***657***
Residual mountain(s), ***39,*** ***657***
Reverse fault, ***414,°*** ***657***
Rhode Island, granite batholith, 505°
graphite, 503
metamorphosed sedimentary strata, 503, 505°
superanthracite, 503, 505°
Rhodesia, Bulawayo, granular disintegration, 149°
joint sheets, 150°
Rhyolite, 614, 620,° ***621,*** 622, ***657***
porphyry, 622
Richter scale of earthquake intensity, *see* Earthquake(s), intensity, Richter scale
Rift valley, 542°–543,° 559
Rip current(s), 326, ***657***
Ripple marks, 384, 386,° 408,° ***657***
original tops up, 408°
overturned, 408°
Ripples, 192, 193,° 309
asymmetrical, deep-sea floor, 357, 358°
Ripples, symmetrical, 319,° 326, 327°
axes parallel to shore, 326
cross-laminae, 332
deep-sea floor, 357, 359
Rock avalanche, 166,° 167, ***657***
Rock bodies, passage of fluids through, 77
relative ages, 418
Rock cleavage, ***657;*** *see also* Slaty cleavage
Rock cycle, 98, 99,° 100, 102,° 103, 114, 229, ***658***
Rockfall, ***165,°*** ***658***
Rock flour, ***267–269,*** 274, ***658***
Rock flow, 506, 508,° 525, 535, 537°
Rock-forming silicate minerals, 56–60, 84
arrangement of silica tetahedra, 58°
breakdown, 110
classification, 56–57, 58,° 59,° 60
ions in: aluminum, 56
calcium, 56
ferrous iron, 56, 57
magnesium, 56, 57
melting behavior, effects, of pressure, 509
of water-vapor pressure, 509
reaction with dissolved ions in seawater, 112
Rock glacier, ***177,°*** ***658***
Rock-making environmental realms, 9, 78
Rock-making processes, 9
Rock material, 13
Rocks, 4, 76
accessory minerals, 82, ***644***
destruction of to create sediments, 10, 15
environments of origin, 9, 10, 18
essential minerals, 82, ***649***
fabric, ***87,*** 617,° 618,° ***649***
fragments, 7
broken along partings, 78°
sizes, 77
igneous, *see* Igneous rocks
language of, 4, 76
layers, 76, 78, 80,° 81°
magnetic particles, tiny magnetic fields surrounding, 42
major classification, 76, 78–82
metamorphic, *see* Metamorphic rock(s)
oldest, 16, 133
paper-thin slices, 11,° 17
particles, 82
kind and origin, 84–86
see also Particle(s), igneous rocks
partings, 76, 77
bedding-plane partings, 78
faults, *see* Fault(s)
increased surface area, 78
joints, *see* Joint(s)
sedimentary, *see* Sedimentary rock(s)
solid, *see* Bedrock
stratified, 78
volcanic, *see* Volcanic rocks
Rock salt, 10
Rockslide, ***165,°*** 168, ***658***
Rock strength, 145
Rocky Mountains, 39
deformed strata, Glacier National Park, Montana, 39°
Rocky Mountain trough, Cretaceous strata, 524°
Rogers, Henry D., 519
Rogers, William B., 519
Roof rock, 574, ***658***
Rotation, of Earth, *see* Earth, rotation
Royal Society of Edinburgh, 14
Rubidium, radioactive decay to strontium, isotopic ages, 127, 128
Running water, 186
Runoff, 187,° ***194,*** 217,° ***658***
Rupture of rocks, 145, 151,° ***658***
Rutile, 610

St. Francis dam, California, 179, 196
St. Pierre, Martinique, devastation created by 1902 eruption, Mont Pelée Volcano, 8°
Salinity, ***68,*** ***658***
Salt, 10, 585
Saltation, ***191,*** 192,° 297, ***658***
Salton Sea, California, 96°–97,° 585
Saltwater, conversion to freshwater, 35
organic habitat, 318
San Andreas fault, California, 433, 434,° 518°
elastic rebound theory of earthquakes, 433
rate of movement, 433
shallow-focus earthquakes, 446
Sand, 7, 10, 89,° 90,° 195,° 300,° 625, ***658***
porosity, 236°
Sandstone, 10, 519, ***623,*** 625, ***658***
average, chemical composition, 83°
containing feldspar, 83,° 84; *see also* Arkose
porosity, 236°
Sand waves, 192, 194,° 326, 328, ***658***
Sanidine, 610
San Juan Parangaricutiro, Mexico, engulfed by volcanic rock, 8°
Santa Monica Mountains, California, creep, 174°
Saturation, zone of, *see* Zone, of saturation
Scale, of a map, ***632,*** ***658***
Scandinavian Ice Sheet, 532
Scarp(s), *see* Fault(s), scarps
Schist(s), ***628,°*** 629, ***658***
biotite, 501
chlorite, 501
contact metamorphism, 501
fine-grained, 503, 505°
Scientific method, 18, ***658***
in geology, 17–18
Scotian Sea floor, asymmetrical ripples, 358°
Scotland, Siccar Point, strata in angular unconformity, 420

Scotland, Skye, contact metamorphism, 501, 511
Scranton, Pennsylvania, closely folded sedimentary strata and coal, 503, 504°
Sculptural evolution, 218
Sea, 2°–3°, 7, 18, 318
surface, fundamental physical barrier, 7
liquid/gas interface, 48
shaped by Earth's gravity, 33
Sea arch, 334,° ***658***
cave, 334,° ***658***
Sea-floor spreading hypothesis, 561, 562°
relation to origin of mountains, 562,° 563, 564°
Reykjanes Ridge area, Iceland, 562, 563°
volcanic island rows, 563, 564°
Sea-floor trench(es), ***352,°*** 547, 548,° ***658***
geophysical profile, 551°
high-angle faults, 549,° 551°
inclined zone of earthquake foci, 549, 552°
negative gravity anomalies, 549, 550,° 551°
rows of volcanic islands, 550
sediments, 548–549°
seismic profile, 549,° 551°
Sea level, ***658***
base level for streams, 7
controlling factors, 339
eustatic changes, 340
in glacial ages, 278–279, 281°
measurement, 339
Seamount, 477, 478°–479,° 564,° ***658***
Seashore, 2°–3,° 7
Seasonal freezing and thawing, regolith, 42
Sea wall, 338,° ***658***
Seawater, movements, 36, 353
proportion of Earth's surface covered by, 48
salinity, 110
result of geochemical cycle, 35
surface ocean currents, 36, 353
result of frictional drag created by prevailing winds, 36
Secondary enrichment, ***587,° 658***
Secondary mineral, ***658***
Secondary ore, 587,° 588°
Secondary wave, *see* Seismic wave(s), S-deflections
Secular equilibrium, 126, ***658***
Sediment(s), ***8,*** 13, 15, 19, 84,° 85,° 90,° 96°–97,° 102, 103, 195,° 300,° ***658***
animal footprints, 12°
beach, *see* Beach sediments
carbonate, *see* Carbonate sediments
clastic, *see* Detritus
color, 386
continental shelf, *see* Continental shelf, sediments
deep-sea, *see* Deep-sea sediment(s)
Sediment(s), derivation, 384
distinctive effects of glaciers on, 19
fabric, 383, 385°
glacial-marine, *see* Glacial-marine sediment
marine, *see* Marine sediment
nearshore, *see* Nearshore sediments
nonclastic, 624
offshore, *see* Offshore sediments
particles, 10, 11,° 14
cementation, *see* Lithification, cementation
deposition, 14
fossils, *see* Fossils
framework, 92
friction and gravity, 91°
matrix, 92
packing, 92°
rounding, 383, 384,° 385°
shapes, 89, 90°
sorting, 379, 383, 385°
poorly sorted, *see* Poorly sorted sediment
"well graded," 89
well-sorted, *see* Well-sorted sediment
tephra, *see* Tephra
shrinkage cracks, 12°; *see also* Mud crack(s)
sizes, 88–89,° 90,° 195,° 300°
surficial, 1, 3
terrigenous, *see* Terrigenous sediment
valley-floor, 13
Sedimentary breccia, ***623,*** 625, 626,° ***658***
Sedimentary facies, ***388, 649***
Sedimentary rock(s), 9, 10, 76, ***80,*** 115, ***658***
built *in situ* by organisms, 86, 87°
clastic, *see* Clastic rock(s)
layers, *see* Sedimentary strata
most ancient, implications, 17
nonclastic rock
carbonate, *see* Carbonate rocks
chert, *see* Chert
evaporites, *see* Evaporite(s)
organic, 627
pyroclastic, *see* Pyroclastic rock(s)
Sedimentary strata, 13, 15,° 76
deformed, in mountains, 39°
erosion remnants, 10°
flat-lying, 5,° 9,° 10,° 12,° 13, 14, 15°
indicators, of crustal subsidence, 43
of crustal uplift, 43
gently inclined, 2°–3,° 14
in the rock cycle, 378
Sedimentary structures, 408,° 409
Sedimentation, 13
Seismic belts, ***444,*** 447,° ***658***
Seismic discontinuities, 435, 437, ***658***
deep within Earth, 441
Seismic exploration (prospecting), 436, 437° ***658***
Seismic seawave, *see* Tsunami
Seismic wave(s), 168, ***348,*** 429, ***658***
abrupt movement on strike-slip faults, 433
vertical movement on faults, 435–436
amplitudes, 448–449
amplitude versus distance from epicenter, 449°
arrivals, systematic spacing, 438
body waves, 432, 435, 439,° 442,° ***645***
disappearance of S waves at 103,° 441, 442°
explosions, 435,° 436, 437°
heights of deflections, 448
in Earth, 440,° 442°
in Earth's core, 441, 442°
in Earth's mantle, 443
in hypothetical uniform Earth, 439,° 440°
L-deflections, 438
locating epicenters, 444, 446°
longitudinal waves, 431, 432, 435,° 438, ***653***
Love waves, 431, 432, 435, ***653***
microseisms, 431,° 437, ***654***
miscellaneous causes, 437
motions, 432, 438
movements of bodies of fluid underground, 436
near Denver, Colorado, 436
Old Faithful Geyser, Yellowstone National Park, Wyoming, 436
negative traces, 431, 433, 435°
origin, 432, 433
patterned format, 438
P-deflections, 438
penetration into Earth, 441, 442°
positive traces, 431, 433, 435,° 436
probing inside the Earth, 437–444
Rayleigh waves, 431, 432, 435, 438, ***657***
recording and measurement, 430–431
reflection, 435, 436, 437°
refraction, 435, 436, 437,° 440, 441,° 442,° 443°
relation to density of materials traversed, 440, 599, 600, 601
S-deflections, 438
shadow zone, 441, 442°
shear waves, 431, 432, 433,° 435,° 438, ***659***
surface waves, 432, 435, ***660***
travel-time curves, 440°
underground nuclear explosions, 449
Seismogram, ***430,*** 431,° ***658***
Seismograph, ***430,° 658***
horizontal, 430°
long-period, 431
vertical, 430°
Seismology, ***429,*** 545, 551, ***659***
Semiarid climates, 289°
Serpentine, 604,° 610, 622, 630
Settling velocity, 194, 195,° ***659***
Shale(s), 519, 613, 614,° 625, ***626, 659***
oil, 580, ***655***

Shallow-focus earthquakes, volcanic, 481–482; *see also* Earthquake(s), depth of focus, shallow-focus
Shallow-water wave, *see* Waves, water surface, shallow-water waves
Shearing stress, 260, 400,° ***660***
Shear wave, *see* Seismic waves, shear waves
Sheep Mountain anticline, Wyoming, 407°
Sheet erosion, ***187, 188,°*** 189,° ***649***
Sheets, 149, 150,° 151°
Shields, continental, *see* Continental shield(s)
  volcanic, *see* Volcanic edifices, volcanic shields
Shiprock, New Mexico, volcanic neck, 480, 481°
Shore, 2°–3,° 4
Shorelines, 328–340
  deformed, 340,° 403°–404,° 531–533
    Ancient Lake Bonneville, Utah, 531°
    California coast, 340,° 403°
    Pleistocene Great Lakes, 533°
    Scandinavia, 531, 532,° 533
    southwest Pacific, 404°
Shore profile, 330, 334°
Shuteye Peak, California, joint sheets, 151°
Sial, ***443, 659***
Siccar Point, Berwickshire, Scotland, strata in angular unconformity, 420
Sierra Nevada, California, warm air currents forced upward, 37
  fault block, 530°
  glacially striated bedrock, 152°
Silica, 503, 510, 609, 623; *see also* SiO_2
  in river water, 110
  in seawater, 110
Silica tetrahedron, 56, 57,° 145
    arrangement in rock-forming silicate minerals, 56, 58°
Silicate minerals, rock-forming, *see* Rock-forming silicate minerals
Silicon, tetrahedral arrangement, with oxygen, 56, 57,° 58°
    with oxygen in rock sphere of Moon, 26
Sill(s), ***490,*** 491,° 493,° 497,° ***659***
  comparison with buried extrusive sheets, 490, 493,° 494°
  Whin sill, England, 490
Sillimanite, 59, 610
Silt, 10, 89,° 195,° 292,° 300,° 625, ***659***
  Mississippi River, chemical composition, 104, 105°
  porosity, 236°
  settling rate, 356
Siltstone, 10, 505, 616, 625, ***626, 659***
  porosity, 236°
Silver, 610
Silver Basin, Colorado, rock glaciers, 177°
Sink(s), 232°–233,° ***248,*** 249,° 250,° ***659***
SiO_2, 106°
  in average river water, 107°
  in average seawater, 107°
  in rocks, 104, 105,° 500
  melting point, lowered by water-vapor pressure, 109
SiO_2 mineral forms, 66, 67°
    coesite, 66, 67°
    cristobalite, 66, 67°
    quartz, *see* Quartz
    stishovite, 66, 67°
    temperature-pressure graph, 67°
    tridymite, 66, 67°
Sixmile Canyon, Utah, borrow pit, components of geologic record exposed, 9°
Skeena River, British Columbia, Canada, metamorphic rocks, 508°
Skeletal material, 85°
Skye Island, Scotland, contact metamorphism, 501, 511
Slate, 505, 613, 614,° ***628,*** 629, ***659***
Slaty cleavage, ***505,*** 506,° 528,° ***659***
  in interior zones of mountains, 523, 528°
Slickensides, 412,° ***413, 659***
Sliderock, 178,° ***179, 659***
Slip face, ***305,*** 307,° ***659***
Slope(s), 8, 163°
  angles, 33,° 163°
Slump, 161,° ***165,°*** 166, 167,° 168,° ***659***
Slumping, effects in ancient marine sediments, 361°
  in deep-sea sediments, 359–360
Smith, William, 124
Snow, 36
  granular, 70°
Snowfield(s), ***257,°*** 268,° ***659***
Snowflake, 70°
Snowline, ***257,*** 258,° 262,° 266,° ***659***
Sodalite, 59
Sodium, in evaporite minerals, 112
  ions, in alkali feldspar, 60, 143
    in average seawater, 107°
    in micas, 143
    in plagioclase feldspars, 143
Soil(s), 4, ***153, 659***
  horizons, ***659***
  mature, ***154***
  rate of formation, 155
Soil conservation, 164
Soil erosion, 219°
  by wind, 312°
Soil profile, ***153,° 659***
Solar constant, 29, ***659***
Solar energy, 26, 28, 29, 30, 31, 35, 102, ***659***
  source of sea waves, 324
  transmission, 323
  uneven distribution on Moon's surface, 26
Solar heat, average amount received on Earth, 42
Solar radiation, 28, 29, 30, 31, 35
Solar radiation, reaction with Earth's atmosphere, 29, 30, 31
  uneven distribution over Earth's surface, 31°
Solfatara, 473, 474°
Solid state, 48
  amorphous solid(s), 48, ***644***
  brittle solid(s), 49, 400, ***645***
  changes to, from, and within, 52°
  crystalline solid(s), 48, ***647***
  effects of varying temperatures and pressures, 509
  plastic flow, 506, 508,° 509, 529,° 535, 537,° ***656***
  plastic solid, 49, 400, ***656***
  reactions to passage of waves, 49
  slow rates of diffusion, 49
Solifluction, ***176,°*** 177, ***659***
  rate, 166°
Solifluction sediment(s), 176,° 179, ***659***
Sorting, of elements, 114
  of sediment particles, ***89,*** 379, 616, ***659***
Sound waves, 322
  frequencies, 347
  generation, 347
  longitudinal waves, 347°
  reflection, 347
Source rock, petroleum, ***659***
South American Cordillera, 546°
  volcanoes, 462
South Atlantic Ocean, density currents, 354°
South Atlantic Ocean basin, southeastern margin, arcuate trench, 40
South Carolina, Spartanburg, slump in regolith, 167°
South Cascade Glacier, Washington, 257°
South Equatorial current, 37°
Southern Hemisphere, 35, 36
  Coriolis left-deflection rule, 35
South Magnetic Pole, 556°
Space age, 35
Spanish Peaks, Colorado, plutons, 490,° 491,° 518°
Spartanburg, South Carolina, slump in regolith, 167°
Spatter cones, *see* Volcanic edifices, spatter cones
Species, Darwin's theory of origin, 17
Specific gravity, ***659***
Spectrometer, mass, 17°
Spectrum, *see* Electromagnetic waves, spectrum; Waves, water surface, spectrum
Sphalerite, 610, 622
Spheroidal weathering, 147,° 148°
Spinel, origin by incongruent melting of garnet at high pressures, 110
Spit, 317,° 319,° ***330, 659***
  barrier, 320°
  Presque Isle, Erie, Pennsylvania, 329,° 330
  Provincetown, Massachusetts, 330
  Sandy Hook, New Jersey, 330

Sponges, siliceous, 364
Spores, 86
Spotted Range, Nevada, fault, 412°
Spring(s), ***240,*** 241,° 249,° ***659***
artesian, 241,° ***242, 659***
gravity, 240, 241,° ***659***
thermal, 41, 42, 242
Stack, 334,° ***659***
Stalactite(s), 247,° ***248, 659***
Stalagmite(s), 247,° ***248, 659***
States of aggregation, changes, 50–53
influence of pressure, 52
gases, 50
liquids, 49–50
relationships of three states, 55°
solids, 48–49
Staurolite, 59, 502, 611, 622
Steady state, 69, ***659***
Steppe, 288, 289°
Stillwater Range, Nevada, mudflow, 172°
Stishovite, *see* SiO_2 mineral forms: stishovite
Stock(s), ***495, 659***
in Alps, 520,° 521°
contact metamorphism, 500
Marysville, Montana, 497°
Stone Mountain, Georgia, residual mountain, 39
Stony-iron meteorite, 31, 32°
Stony meteorite, 31, 32°
Stoping, ***510, 659***
Storms, 18
Strain, 260, ***659***
Strait of Gibraltar, 24, 25°
subsurface countercurrent, 354
surface current, 354
Strata, ***10,*** 14, 122, 390,° ***660***
clastic, sheets of, records of uplift, 521
cross-strata, *see* Cross-strata
environmental analysis, 318
horizontal, 414°
marine, *see* Marine strata
original tops, 408°
parallel, 14, 15,° ***379,*** 380,° ***660***
regional metamorphism, 488
sedimentary, *see* Sedimentary strata
terminated by valley fill, Sixmile Canyon, Utah, 9°
thickness measurement, 637°
vertical, 490
volcanic, *see* Volcanic strata
Stratification, 378, 379, ***660***
deep-sea sediments, *see* Deep-sea sediment(s), stratification
Stratified drift, ***271, 660***
ice-contact, ***274, 651***
Stratigraphic superposition, principle of, *see* Principle of stratigraphic superposition
Stratigraphic trap, 576,° ***660***
Stratigraphy, ***122,*** 387, ***660***
Stratum (plural Strata), ***660***
Streak, ***604, 660***
Stream(s), 18, ***189,*** 237, 286°–287,° ***660***
Stream(s), adjustment, 226
agents of dissection of land, 7, 187
antecedent, ***226,*** 227,° ***660***
bed, 8, 190,° ***645***
braided, 198,° ***207, 660***
carriers of rock particles, 7, 187, 195
channel, 189, 190°
consequent, ***225,*** 227,° ***660***
dynamic equilibrium, 211
economy, ***194,*** 196, ***649***
flood(s), 196,° 197,° 198,° 292°
fluctuation, 192°
geologic agents, 187
gradient, 197
kinds, 225
load, ***189,*** 190
bed, ***190,*** 191, 192, 193,° 194,° ***653***
dissolved, 110, 111°
suspended, 110, 111,° ***190,*** 192, 194, ***653***
long profile, ***197,*** 199,° 220,° 225,° ***653***
open system, 218
rejuvenation, 221,° ***222, 657***
significance, 186
spacing, 217
subsequent, ***226,*** 227,° ***660***
superposed, ***226,*** 227,° ***660***
velocity distribution, 191°
profile, 190,° 192°
Stream capture, ***226,*** 228,° ***660***
Stream flow, ***187, 660***
rate, 166°
Streamline forms, 271
Stream patterns, 225, 226°
dendritic, ***225,*** 226,° ***660***
rectangular, ***225,*** 226,° ***660***
trellis, ***225,*** 226,° ***660***
Stream system, 198,° 199°
well-adjusted, ***226,*** 228, 229,° ***660***
Stream terrace, ***222,*** 223,° ***660***
Stream valley, 220°
Strength, ***399***–400, ***660***
of lithosphere, 528–536
deformed shorelines, 531,° 532,° 533°
natural loads, 529
Earth tides, 529
subsidence around volcanic cones, 531
under weight of lakes, 531°
glaciers, 531
shape of Earth, 528–529
Stress, ***260,*** 400, ***660***
compressive, *see* Compressive stress
shearing, *see* Shearing stress
tensile, *see* Tensile stress
Striations, glacial, 152,° 264, 265,° 280,° ***651***
on faults, 412°
Strike, 401, ***635,*** 636,° ***660***
Strike-slip fault, 413, ***415,° 660***
direction of first motion, 433
movement, 433
see also San Andreas fault, California
Stromboli Volcano, Lipari Islands, Italy, 458, 459,° 469, 470,° 474°
eruptive activity, 469
Strontium, isotope ratios in granitic rocks, 547
radioactive isotope, use in dating, *see* Rubidium
Structural geology, ***405, 660***
Structure, of bedrock, *see* Bedrock, geologic structure
Structure symbols, 640°
Sublimation, 53, ***660***
latent heat of, *see* Latent heat of sublimation
Submarine cables, breaks, 360
Grand Banks earthquake (1929), 360–362, 363,° 364
trans-Atlantic, 346
Submarine canyon, ***352,*** 371,° ***660***
Hudson, 371°
origin, 369, 371
hypotheses of subaerial origin, 369
submarine origin, 369
turbidity currents, 369
Submarine volcanic activity, *see* Volcanic activities, submarine
Submerged lands, Aleutian basin sea floor, 404
suggested by directions of cross-strata, 404
Submergence, *see* Coast(s), submergence
Subsequent stream, ***226,*** 227,° ***660***
Subsidence, 103, 115
of ground, 243
Sulfate ions, in average seawater, 107°
bacterial reduction, 355
Sulfur, in evaporite minerals, 112
in fats, 105
in proteins, 105
mineral with composition of an element, 56
oxidation of pyrite, 68
sublimed, in volcanic gases, 457, 458°
Sulfur dioxide, in volcanic gases, 457, 458°
Sulfuric acid, oxidation of pyrite, 68
Sulfur trioxide, in volcanic gases, 457, 458°
Sun, 26, 27,° 28, 29
gravitational attraction, 320
Sun-Earth-Moon system, 26, 27°
nuclear reactions, 28, 29
scale model, 26, 29°
Sunspot cycle, 29, 116
Sunspots, 29
Superposed stream, ***226,*** 227,° ***660***
Superposition of strata, *see* Principle of stratigraphic superposition
Surf, 2–3, ***326, 660***
Surface, Earth's, *see* Earth's morphology
of disconformity, 391
of unconformity, 420
Surface currents, of oceans, 36, 37,° 353, 354
Surface tension, 49, 50°
Surface waves, *see* Seismic wave(s), surface waves; Waves, water surface
Surficial sediment, 6, 13; *see also* Regolith
deposition, 17

Surf zone, 327,° 328,° ***660***
Surtsey Volcano, Iceland, 454°–455,° 476
quantity of materials discharged, 476
Surveys, magnetic, *see* Magnetic surveys
Suspended load, streams, ***190,*** 192,° 208, ***653***
wind, 297, 298,° 299, 301–302
high-altitude, 301
low-altitude, 298,° 299, 301°
Swash, 326, 328,° ***660***
Swells, 324
periods, 324
Swiss Plain, 520°
Switzerland, Alps, 514°–515°
Syncline, 396°–397° 406,° ***407,*** 408,° 636, ***660***

Tacoma, Washington, volcanic mudflow, 173°, 174
Talc, 500, 605, 611, 622
Talus, ***178,°*** 179,° ***660***
Tambora Volcano, Indonesia, 459,° 479
Tangential component of Earth's gravity, *see* g
Tar sand(s), 580, ***660***
Taylor Glacier, Antarctica, folds, 508°
Tectonic cycle, 98, 99,° 114, 115, ***660***
events, 536–538
Tektite (meteorite), 31, 32°
Temperature, critical, effect on minerals, 64
influence on motions of particles, 48
see also Critical temperature
Temperature scale, absolute (Kelvin), 52
Temperature versus pressure graphs, 64
Tensile stress, 260, 400,° 417,° ***660***
Tephra, ***86,*** 301, 302, 356, 460, 623, 624, ***660***
cones, *see* Volcanic edifices, tephra cones
deep-sea sediments, 365,° 367
dated layers, 369
flows, 465,° 623, ***661***
hot, 471, 473°
speeds, 471
fluidized, 471
high-altitude suspensions, 479–480
Agung, Indonesia (1963), 480
climatic effects, 480
Krakatau, Indonesia (1883), 479
Mount Katmai, Alaska (1912), 479
Mount Spurr, Alaska (1953), 479–480
Tambora, Indonesia (1815), 479
Volcano on Kamchatka Peninsula, U.S.S.R. (1956), 480
lava clots, 460, 461°
layers, buried forests, Yellowstone National Park, Wyoming, 467°
low-altitude suspensions, 479–480
Hekla, Iceland (1947), 301,° 302
Mount Spurr, Alaska (1953), 479
tuff, welded, *see* Welded tuff
volcanic blocks, ***460, 661***
Tephra, volcanic bombs, ***460,*** 462, ***661***
Terminal moraine, ***271, 661***
Terrace, stream, *see* Stream terrace
Terrigenous mud, 365°
Terrigenous sediment, ***84,°*** 356, 364, 365,° 366,° 367, ***661***
Tertiary History of the Grand Canyon District, The, 13
Tetrahedron (plural Tetrahedra), 56, 57,° 58,° 60,° 65,° ***661***
Texture, ***86***–88, 613–618, ***661***
igneous rocks, *see* Igneous rocks, texture
metamorphic rocks, *see* Metamorphic rocks, texture
sedimentary rocks, see Clastic texture(s); Crystalline textures; Fragmental texture; and Pyroclastic texture
Theory of the Earth . . . , 14, 21
Thermal conductivity, rock materials, 42
Thermal springs, *see* Spring(s), thermal
Thermometer, geologic, 583
Thingvellir rift valley, Iceland, 542°–543°
Thin section, ***661;*** *see also* Paper-thin slices of rocks
Thorium, 127, 128,° 601
Thrusts (or thrust faults), 397,° 416,° ***417,*** 506,° 514°–515,° 525,° 527,° 528,° ***650***
Thulean Province basalts, continental drift, 562
Tibetan Plateau, 40
Tidal delta, 322°
Tidal inlet, 319, 320, 322°
Tidal marsh, ***318,*** 319,° 322, ***661***
Tidal wave, *see* Tsunami
Tide, ***318, 661***
cycle, 320, 321°
energy for sea waves, 324
period, 324
range, 320, 321°
Tide-producing force, 320, 321°
Till, ***269,*** 270,° 271,° 273,° 330,° ***661***
Tillite, ***282,*** 283,° 383,° ***661***
Tiltmeters, 482
Time, geologic, 122, 130
great length, 19
Tin, 622
TiO_2, 458
TITANIC, sinking, 347
Tobin Range, Nevada, fault scarp, 411°
Tombolo, 319,° ***328, 661***
Tonga-Kermadec Trench, 548°
Topaz, 605, 611, 622
Topographic map, ***632,*** 634,° ***661***
Topography, 19, ***632,*** 634,° ***661***
Topset bed(s), 207,° ***208, 661***
Tornadoes, 36
Trade winds, 35, 36,° ***661***
Transpiration, ***99,*** 187,° ***661***
Transverse dune, ***307, 648***
Trap, oil, 574, ***661***
Travertine, ***627, 661***
Tree rings, growth cycles related to sunspot cycle, 29
Trellis pattern, streams, ***225, 660***
Tremolite, 500
Trenches, *see* Sea-floor trenches
Tridymite, *see* SiO_2, mineral forms: tridymite
Triple point, ***50, 661***
of water, 51, 52
Tritium, 113,° 114, 595, ***661***
Tsunami, ***324, 661***
devastating waves, 5,° 324
heights, 324
lengths, 324
origin, 324
periods, 324
speeds, 324
warning system, 324
Tuff, 625,° ***626, 661***
Turbidity current(s), ***359,° 661***
Grand Banks submarine cable breaks, 361, 362
Hoover Dam, 359
laboratory tank, 359°
origin of graded layers, 359, 364
submarine canyons, 369
Turbulent flow, 185,° ***190,°*** 191,° ***650***
Turnagain-by-the-Sea, Anchorage, Alaska, earthquake damage, 7°
Tuweep Valley, Arizona, 13, 14°

Uinkaret Plateau, Arizona, 13, 14°
Ultramafic rock(s), 619, 620,° ***621, 661***
in mountain chains, 523
Ultraviolet rays, 28, 113°
Unconformable relationships, beneath horizontal Cambrian sandstone, 538°
Unconformity, ***419,°*** 420, ***661***
angular, ***419,°*** 420,° ***661***
comparison with faults, 420
surface of, 420, ***661***
variable relationships, 420
Underground nuclear explosions, 17, 449
Undersaturated igneous rock, 106, 621
Uniformity, principle of, 19, ***661***
Uniform layers, 382
United States Geological Survey, 13
Universal Ocean, 428
University of Edinburgh, 13
University of Leiden, 13
University of Paris, 13
Unloading, erosional, 149
Uplift, crustal, 15, 103
Ural Mountains, U.S.S.R., 546°
Uraninite, 611
Uranium, 601
Uranium dating, 128°
U-shaped dune, ***307,*** 308,° ***648***
U.S.S.R., Kamchatka Peninsula Volcano, 480
Knyahinya, stony meteorite, 32°
Ural Mountains, 546°
Verkhoyansk Mountains, 546°
Utah, Ancient Lake Bonneville, deformed shorelines, 531°
Henry Mountains laccoliths, 493

Utah, LaSal Mountains laccoliths, 493
Sixmile Canyon, borrow-pit exposure, 9°
Utah Copper Mine, 587°

Vaiont Canyon, Italy "landslide," 167–169°
Valley, cross profile, 225°
dry, 249
Valley fill, 9,° 223,° 224,° 287°
Valley floor, 5°
Valley-floor sediments, 13
Valley glacier, 254,° ***256,*** 257, 269,° 273,° ***661***
Valleys, relation to streams, 216
Valley train, ***274,° 661***
Varve, ***380,° 661***
Veins, 583,° 589,° ***661***
Ventifact, ***303,*** 304,° 305,° ***661***
Verkhoyansk Mountains, U.S.S.R., 546°
Vesicles, 490, 495,° ***618,*** 619,° ***661***
submarine volcanic rock, 477
Vesuvius Volcano, Italy, 458, 459,° 469, 470,° 471, 472,° 473,° 474,° 475°
carbon dioxide in volcanic gases, 471
cone and crater, 472°
cross section, 473°
depth of magma chamber, 471, 473°
eruptive activities, 470
Virginia, Norfolk, sea-floor profile, 352°
Volcanic activities, 474°–475
ancient, 480–481
devastation created by, 8°
submarine, 476–479
chemical reactions, 477
discharge of gases, 477
hydrostatic pressure, 477
mineralogic changes, 477
Volcanic ash, *see* Ash, volcanic; Tephra
Volcanic belt, circum-Pacific, 458
Volcanic blocks, *see* Tephra, volcanic blocks
Volcanic bombs, *see* Tephra, volcanic bombs
Volcanic breccia, 479
Volcanic cones, *see* Volcanic edifices, volcanic cones
Volcanic crater, *see* Volcanic edifices, craters
Volcanic edifices, 460–466, 474,° ***661***
calderas, ***462,*** 463, 470, 480, ***645***
Kilauea, 468,° 469
La Caldera, 462
Mauna Loa, 469
Monte Somma, 470, 472°
Mount Mazama (Crater Lake), 462, 464°
composite cones, ***461,*** 463,° 466, 474,° ***646***
craters, ***462,*** 463,° 472,° ***661***
erosion, 480
Shiprock, New Mexico, 480, 481°
explosion pits, *see* Explosion pits
lava domes, 466, 471, 474,° ***653***
lava spines, 466, 474,° ***653***
Volcanic edifices, on sea floor, 476–477, 478,° 479°
archipelagic aprons, 477
guyots, 477, 478,° 479°
seamounts, 477, 478,° 479°
spatter cones, 474°
tephra cones, ***461,*** 474,° ***661***
volcanic cones, 8,° 9, 13, 15,° 41,° 460–462
islands within oceans, 41
volcanic plains and plateaus, 464, 474°
plateau basalts, 459,° 464, ***465,°*** 466,° 509, ***656***
Columbia Plateau, northwestern United States, 464, 465°
Deccan Plateau, India, 465°
Thulean Province, 465
volcanic shields, 463,° ***464,*** 465, 474,° ***662***
Hawaii, 459,° 463
Iceland, 459,° 463
Samoa, 459,° 463
Volcanic eruption, fissure eruption, ***464, 650***
Laki, Iceland (1783), 464
Volcanic eruption cycle(s), 115
Volcanic eruptions: forecasting, 481–482
changes in fumaroles, 482
magnetic effects, 482
shallow-focus earthquakes, 481–482
tiltmeter surveys, 482
Volcanic eruptions: prediction of events, 482
Volcanic eruptive activities, 466, 469–471, 473–475
Volcanic gases, 457–458, 474
composition, 458°
oxidation of hydrogen and methane, 458
superheated steam, 457
temperatures, 458
Volcanic islands, 454°–455,° 459,° 463,° 468,° 470,° 476, 565°
Volcanic materials discharged, quantities: Island of Hawaii, 466
Laki, Iceland, 464
Mount Mazama, 462–463
Parícutin, 457
Surtsey, 476
Volcanic mountain, ***39, 661***
Volcanic neck, ***480, 662***
Volcanic products, 457–460, 474°
gases, *see* Volcanic gases
lava, *see* Lava
tephra, *see* Tephra
Volcanic rocks, in mountain belts, 39, 523
Volcanic shield, *see* Volcanic edifices, volcanic shields
Volcanic steam electric power, 482
Volcanic strata, 466,° 481
composition, 481
isotopic dates, 481
paleomagnetic data, 481
spreading, 481
Volcanism, 456, ***662***
Volcano, 13, ***456, 662***
active, 480, ***662***
dormant, 480, ***662***
inactive, 480, ***662***
Volcanoes and human affairs, 481–482
von Jolly, apparatus for weighing the Earth, 602°
Vosges-Black Forest, 519, 520°
Vulcan's Throne, Arizona, 13, 14,° 15°
Vulcano Volcano, Lipari Islands, Italy, 458, 459,° 469,° 470, 474°
eruptive activity, 469

Washington, Columbia Plateau, Moses Coulee, igneous rock layers, 80°
Creston Bog, tephra, 86°
Grand Coulee, 178,° 180
Mount Adams Volcano, 459,° 462
Mount Baker Volcano, 459,° 462
Mount Rainier Volcano, 174, 459,° 462
Tacoma, volcanic mudflow, 173,° 174
Wastes, industrial, polluting rivers, 5
Water, 4, 6, 7
boiling curve, 476°
changes of state, 53°
critical temperature, 544
crystal structure, 70°
Eh-pH graph, 68°
metamorphic rock, 70°
mineral aggregate, 70°
molecular structure, 69°
molecules, 70
physical states, temperature-pressure graph, 51°
role in metamorphism, 502–503
in crystal lattices, 502, 503, 509
Water balance, 98, ***662***
conterminous United States, 99, 101°
Water gap, ***225, 662***
Water table, ***235,°*** 239,° 240,° 241,° 243,° 249,° 587,° ***662***
slope, 237
Water-vapor pressure, effect on melting of silicates, 109, 502, 503
Wave-built terrace, ***333,*** 334,° ***662***
Wave-cut bench, ***333,*** 334,° ***662***
limiting width, 333
Wave-cut cliff, 317,° ***333,° 662***
Wave height, ***323,° 662***
Wavelength, ***323,° 662***
Wave period, ***323, 662***
Wave refraction, *see* Seismic wave(s), refraction; Waves, water surface, refraction
Waves, analysis by physicists, chemists, and geophysicists, 29
elastic, *see* Elastic waves
electromagnetic, *see* Electromagnetic waves
seismic, *see* Seismic waves
sound, *see* Sound waves
water surface, 322–328
abrasion, 333
amphibious military operations, 323

Waves, water surface:
breaking, 326, 327,° 332°
crests, 323°
deep-water waves, 324, 325,° ***648***
drag on bottom, 324
energy of, 333
energy sources for, 324
erosion, 333, 334,° 337–338
hydraulic action, 333
plucking, 333
interference, 324
longshore currents, ***326,*** 328,° ***653***
longshore drift, ***329,*** 339, ***653***
rates, 329
mechanics, 323
refraction, 317,° ***325,*** 340,° ***662***
sediment transport, 329
shallow-water waves, 324, 325,° ***659***
sizes and origins, 324
spectrum, 324
troughs, 323°
water-particle orbital motion, 324, 325°
oscillatory motion, 325
wave base, 324, 325,° 331, ***662***
wave-induced currents, 325,° 326
waves of oscillation, ***323, 662***
waves of translation, ***326, 662***
Wave spectrum, *see* Electromagnetic waves, spectrum; Waves, water surface, spectrum
Wave velocity, ***323, 662***
Weather, 4
Weathering, 138, ***139, 662***
chemical, 139,° 140, 141, 143°
Weathering, chemical, effect on surface area, 141°
influence of crystal chemistry, 145
of limestone, 144
mineral stability, 144°
products, 143°
environment, 138–139
in deserts, 290
in the geologic cycle, 156
mechanical, 140, 145, 146
by plants and animals, 146
by thermal changes, 146
processes, 140
thermal, 146–147
controlling factors, 150
spheroidal, 147,° 148°
Weber Trough, 548°
Welded tuff, 465, 466,° 626, ***662***
Wells, contamination, 244°
ordinary, 239, 240,° 241°
Well-sorted sediment, 89, ***662***
West Caroline Trench, 548
Westerlies, 35, 36,° ***662***
West-wind drift, Antarctic Ocean, 37°
Whin sill, northern England, 490
Whitmore Canyon, Arizona, 13, 14°
Wind, 7, 18
dead-air layer, 297,° 299, 300°
generation of water-surface waves, 324
movement of sand, 297, 298,° 299°
settling velocities, 300°
suspended load, 298,° 299, 300–302
Wind action, 296
Windbreaks, 312°
Wind deposits, 305
Wind-velocity profiles, 297
Wizard Island, Crater Lake, Oregon, 462, 463, 469
Wollastonite, 59, 69, 500, 611, 622, 630
temperature-pressure graph, 69°
Wyoming, Devil's Tower polygonal (columnar) joints, 499°
Laramie Range, 518,° 519
Sheep Mountain anticline, 407°
Yellowstone National Park, tephra layers and buried forests, 467,°
Old Faithful Geyser, 475°

Xenolith, ***510, 662***
Xenon, in atmosphere, 107°
X-ray diffraction photograph, 57°
X-rays, 113
analysis of minerals by, 18, 56

Yellowknife, Northwest Territories, Canada, deformed metamorphic rocks and granites of Canadian shield, 39,° 486°–487°

Zeolites, 59
Zero, absolute, 48
Zinc, 622
Zone melting, 510, 511
Zone, of aeration, ***235,° 662***
of saturation, ***235,° 662***

80°
60°
40°
20°
0°
20°
60°
60°
40°
40°
20°
20°